# Merrill
# Algebra 2
## *With Trigonometry*

**Applications and Connections**

## *GLENCOE*
### McGraw-Hill

New York, New York
Columbus, Ohio
Mission Hills, California
Peoria, Illinois

Send all inquiries to:
Glencoe/McGraw-Hill
936 Eastwind Drive
Westerville, OH 43081

ISBN: 0-02-824227-0 (Student Edition)
ISBN: 0-02-824228-9 (Teacher Edition)

10    071/043    02 01 00 99

**Alan G. Foster** is chairperson of the mathematics department at Addison Trail High School, Addison, Illinois. He has taught mathematics courses at every level of the high school curriculum. Mr. Foster obtained his B.S. degree from Illinois State University and M.A. in mathematics from the University of Illinois. He is active in professional organizations at the local, state, and national levels. He is a past president of the Illinois Council of Teachers of Mathematics and a recipient of the Illinois Council of Teachers of Mathematics T.E. Rine Award for excellence in the teaching of mathematics. He also was a recipient of the 1987 Presidential Award for Excellence in the Teaching of Mathematics in the state of Illinois. Mr. Foster is a coauthor of *Merrill Geometry, Merrill Algebra Essentials*, and *Merrill Algebra 1*.

**Berchie W. Gordon** is a Mathematics/Computer coordinator for the Northwest Local School District in Cincinnati, Ohio. Dr. Gordon has taught mathematics at every level from junior high school to college. She received her B.S. degree in Mathematics from Emory University in Atlanta, Georgia and M.A.T. in education from Northwestern University, Evanston, Illinois. She has done further study at the University of Illinois and the University of Cincinnati where she received her doctorate in curriculum and instruction. Dr. Gordon has developed and conducted numerous in-service workshops in mathematics and computer applications. She has served as a consultant for IBM. She has traveled nationally to make presentations on the graphing calculator to teacher groups.

**Leslie J. Winters** is the Secondary Mathematics Specialist for the Los Angeles Unified School District. He has thirty years of classroom experience in teaching mathematics at every level from junior high school to college. Mr. Winters received bachelor's degrees in mathematics and secondary education from Pepperdine University and the University of Dayton, and master's degrees from the University of Southern California and Boston College. He is a past president of the California Mathematics Council-Southern Section and was a recipient of the 1983 Presidential Award for Excellence in the Teaching of Mathematics in the state of California. Mr. Winters is a coauthor of *Merrill Algebra Essentials* and *Merrill Algebra 1*.

**James N. Rath** has 30 years of classroom experience in teaching mathematics at every level of the high school curriculum. He is a mathematics teacher and former head of the mathematics department at Darien High School, Darien, Connecticut. Mr. Rath earned his B.A. degree in philosophy from the Catholic University of America and his M.Ed. and M.A. degrees in mathematics from Boston College. He is active in professional organizations at the local, state, and national levels. Mr. Rath is a coauthor of *Merrill Pre-Algebra, Merrill Algebra Essentials*, and *Merrill Algebra 1*.

**Joan M. Gell** is a mathematics teacher and department chairperson at Palos Verdes Peninsula High School in Palos Verdes Estates, California. Ms. Gell has taught mathematics at every level from junior high school to college. She received her B.S. degree in mathematics education from The State University of New York-Cortland, and M.A. degree in mathematics from Bowdoin College in Brunswick, Maine. Ms. Gell has developed and conducted in-service classes in mathematics and computer science and is past president of the California Mathematics Council. She serves as the chairperson of the 1992 MATHCOUNTS Problem Writing Committee. Ms. Gell was a finalist for the 1984 Presidential Award for Excellence in the Teaching of Mathematics in the state of California.

## CONTRIBUTING AUTHOR

**Lee E. Yunker**
Mathematics Department Chairman
West Chicago Community High School
West Chicago, Illinois

## CONSULTANTS

**Donald W. Collins**
Department of Mathematics and Informational
   Sciences
Sam Houston State University
Huntsville, Texas

**Timothy D. Kanold**
Mathematics/Science Division Chairman
Adlai Stevenson High School
Prairie View, Illinois

## REVIEWERS

**Pamela G. Adamson**
Coordinator of Mathematics
Clayton County Schools
Morrow, Georgia

**Trina Price**
Mathematics Teacher
Carmel Junior High School
Carmel, Indiana

**Ann Cumbie**
Mathematics Teacher
Colen Cliff High School
Nashville, Tennessee

**Peggy B. Stuart**
Mathematics Teacher
Emmerich Manual High School
Indianapolis, Indiana

**Susan Heicklen**
Mathematics Department Chairperson
State College Area Senior High School South
State College, Pennsylvania

**Darryl Beissler**
Mathematics Teacher
Johnson High School
St. Paul, Minnesota

**Patricia Stallings**
Mathematics Department Chairperson
Farragut High School
Knoxville, Tennessee

**Edward O'Reilly**
Mathematics Teacher
Westerville North High School
Westerville, Ohio

**Neal E. Anderson**
Mathematics Department Chairperson
Irondale High School
New Brighton, Minnesota

We wish to acknowledge the contributions made to this text during the 1991–1992 school year by Mr. Louis P. La Mastro and his AMP Algebra 2 class at North Bergen High School in North Bergen, New Jersey.

# Table of Contents

# Technology

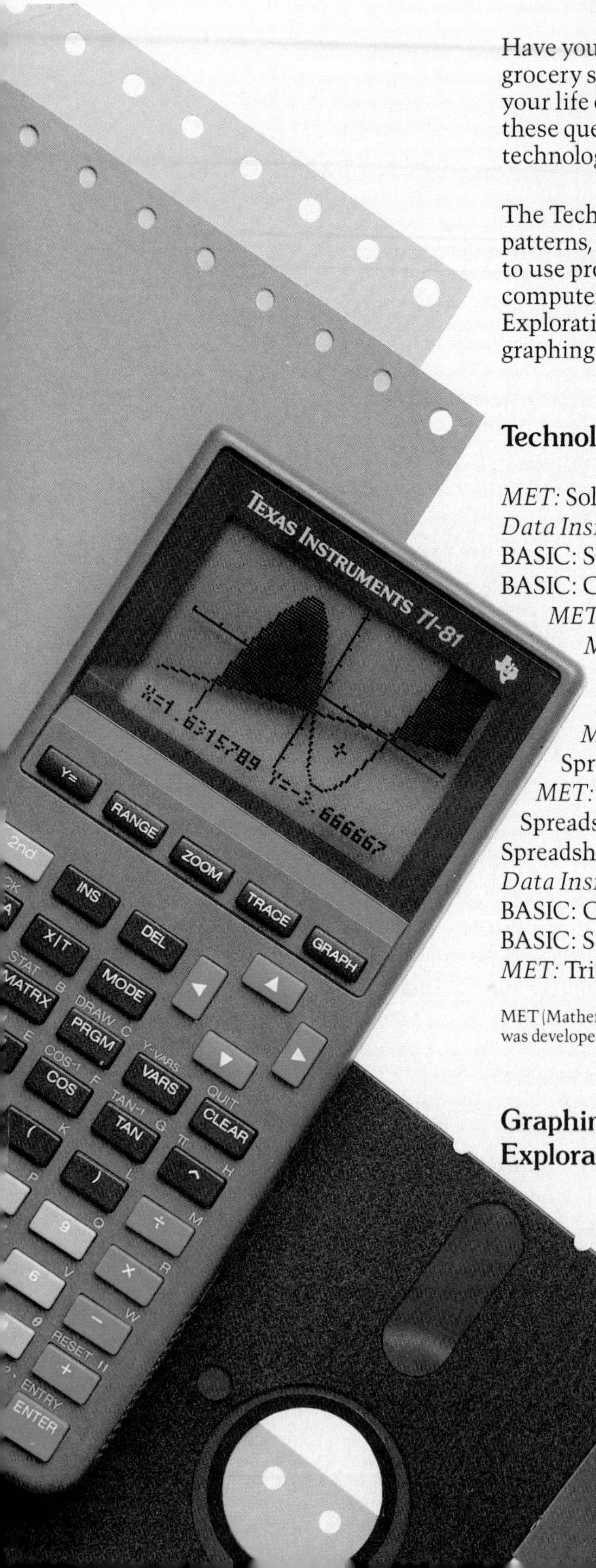

Have you recently purchased something at a department store or grocery store? Do you have a library card? Are credit cards a part of your life or your parents' lives? If you answered yes to even one of these questions, perhaps you're already aware of the growing role technology plays in your everyday life.

The Technology pages in this text let you use technology to explore patterns, make conjectures, and discover mathematics. You will learn to use programs written in the BASIC computer language as well as computer software and spreadsheets. In the Graphing Calculator Explorations, you will investigate mathematical concepts using a graphing calculator.

## Technology

MET (Mathematics Exploration Toolkit) was developed for IBM by Wicat Systems, Inc. Data Insights was developed by Sunburst Communications, Inc.

## Graphing Calculator Explorations

Keystrokes are provided for both Casio and Texas Instruments graphing calculators.

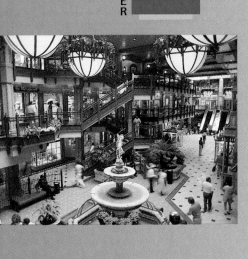

# CHAPTER 3

## Systems of Equations and Inequalities

102

# CHAPTER 4

## Matrices

154

## APPLICATIONS AND CONNECTIONS

### APPLICATIONS

### CONNECTIONS

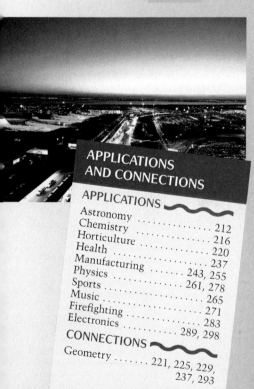

# CHAPTER 5 Polynomials 208

# CHAPTER 6 Irrational and Complex Numbers 250

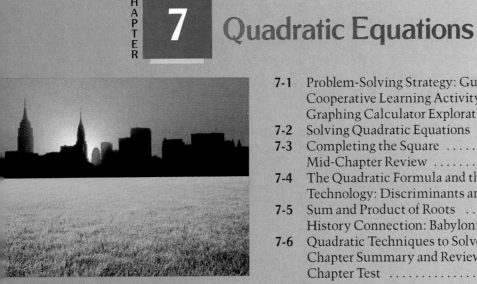
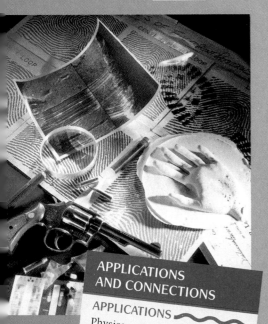

# Special Features

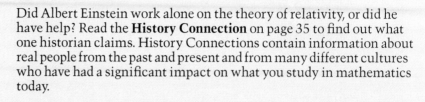

Did Albert Einstein work alone on the theory of relativity, or did he have help? Read the **History Connection** on page 35 to find out what one historian claims. History Connections contain information about real people from the past and present and from many different cultures who have had a significant impact on what you study in mathematics today.

When am I ever going to use this stuff? It may be sooner than you think. You'll find mathematics in most of the subjects you study in school. In the **Biology Connection** on page 578, you'll see how mathematics can be used to estimate wildlife population. The **Applications in Music** feature on page 641 shows how music and trigonometry are related.

## History Connections

## Connections

## Applications

# College Entrance Exam Preview

What should I do when I "grow up?" Will I go to college? If so, which one? What should I major in? These are questions that you must consider more and more as you approach your graduation from high school.

There are many things that you can do to prepare yourself for college. One of the things you will need to do before entering college is take the SAT (Scholastic Aptitude Test) or the ACT (American College Test). To help you practice for the mathematics portion of these tests and other similar tests, you can use the College Entrance Exam Previews given after every other chapter in this text. They are on pages 100-101, 206-207, 306-307, 392-393, 500-501, 590-591, 678-679, and 784-785.

# SYMBOLS

| | | | |
|---|---|---|---|
| $a^n$ | the $n$th power of $a$ | ! | factorial |
| $|a|$ | the absolute value of $a$ | $f \circ g$ | composition function $f$ of $g$ |
| $-a$ | additive inverse of $a$ or the opposite of $a$ | $>$ | is greater than |
| $\text{Cos}^{-1}$ | Arccosine | $<$ | is less than |
| $C(n, r)$ | combinations of $n$ elements taken $r$ at a time | $\geq$ | is greater than or equal to |
| $a + b\mathbf{i}$ | complex number | $\leq$ | is less than or equal to |
| $\circ$ | degrees | $\log_b x$ | the logarithm to the base $b$ of $x$ |
| det | determinant | | |
| $e$ | base of natural logarithms | $P(n, r)$ | permutations of $n$ things taken $r$ at a time |
| $\in$ | is an element of | | |
| $\emptyset$ | empty set | $\pm$ | positive or negative |
| $=$ | equals or is equal to | $\{ \ \}$ | set |
| $\neq$ | does not equal | $\sqrt{\ }$ | the principal square root of |
| $\approx$ | approximately equal to | | |
| $f(x)$ | $f$ of $x$ or the value of $f$ at $x$ | $\sqrt[n]{\ }$ | the $n$th root of |
| $f^{-1}$ | inverse function of $f$ | $\Sigma$ | (sigma) summation symbol |

INSIDE YOUR BOOK

# Understanding the Lesson

*Each chapter is organized into lessons to make learning manageable. The basic plan of the lesson is easy to follow, beginning with a relevant application, followed by the development of the mathematical concept with plenty of examples, and ending with various types of exercises for you to complete.*

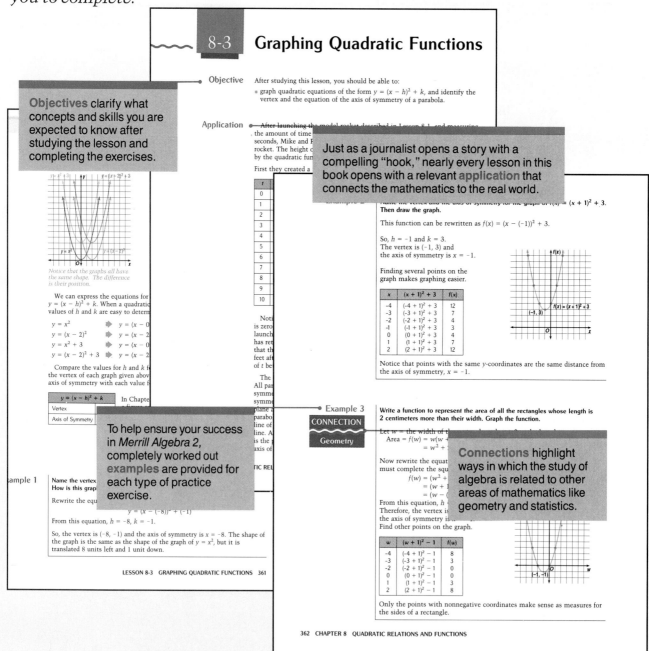

**Objectives** clarify what concepts and skills you are expected to know after studying the lesson and completing the exercises.

Just as a journalist opens a story with a compelling "hook," nearly every lesson in this book opens with a relevant **application** that connects the mathematics to the real world.

To help ensure your success in *Merrill Algebra 2*, completely worked out **examples** are provided for each type of practice exercise.

**Connections** highlight ways in which the study of algebra is related to other areas of mathematics like geometry and statistics.

2

## CHECKING FOR UNDERSTANDING

**Communicating Mathematics**

Read and study the lesson to answer these questions.

1. Did the model rocket described in the beginning of the lesson reach its maximum height 5 seconds after launch? Justify the answer.

2. The graph of a quadratic function is a __?__. It is symmetric about a line called the __?__.

3. Write a quadratic function whose graph has a vertex in the third quadrant.

4. Compare the graphs of the functions $f(x) = (x - 2)^2 + 2$ and $f(x) = (x + 2)^2 + 2$.

5. Write the equation of the function whose graph is shown at the right.

(-2, 1)

**Guided Practice**

Name the vertex and the axis of symmetry for the graph of each equation.

6. $f(x) = x^2$    7. $g(x) = (x - 4)^2$    8. $f(x) = (x + 6)^2$

9. $h(x) = x^2 + 3$    10. $f(x) = x^2 - 9$    11. $y = (x - 6)^2 + 4$

12. $y = (x + 10)^2 - 7$    13. $y = \left(x - \frac{1}{5}\right)^2 + 1$    14. $f(x) = (x + 1.5)^2 - 3.2$

Write the equation of the quadratic function for each graph.

15.

16.

(-2, 0)

(0, 2)

Write each equation in the form $f(x) = (x - h)^2 + k$. Then name the vertex and the axis of symmetry for the graph of each function.

25. $f(x) = x^2 + 8x + 20$      26. $f(x) = x^2 - 3x + 3$

Graph each equation.

27. $y = (x - 5)^2$      28. $y = (x + 7)^2$

29. $f(x) = x^2 + 6$      30. $g(x) = x^2 - 4$

31. $y = (x - 3)^2 + 5$      32. $f(x) = (x - 8)^2 + 3$

33. $g(x) = (x + 2)^2 - 3$      34. $y = (x + 4)^2 + 1$

35. $f(x) = (x - 1)^2 - 4$      36. $y = (x + 11)^2 - 1$

37. $y = x^2 + 6x + 2$      38. $f(x) = x^2 + 10x + 27$

39. $g(x) = x^2 - 2x + 7$      40. $y = x^2 + 3x$

41. $h(x) = x^2 - 5x$      42. $f(x) = x^2 - x - 3$

...ame the vertex and

$f(x) = x^2 + 9$

$f(x) = x^2 - 12x$

...TIC FUNCTIONS  363

**Critical Thinking**

Complete the following table for parabolas with equations of the form $y = (x - h)^2 + k$.

| | Axis of Symmetry | Contains the Point | Vertex | Equation of the Parabola |
|---|---|---|---|---|
| 43. | $x = 0$ | $(2, -1)$ | | |
| 44. | $x = -2$ | $(-5, 9)$ | | |
| 45. | $x = -3$ | $(1, 18)$ | | |
| 46. | $x = 1$ | $(-1, -2)$ | | |

**Applications**

47. **Physics** An arrow is shot upward with an initial velocity of 80 feet per second. The... the time since the arrow...

  a. Draw the graph of th... arrow to the time.

  b. How long after the a... height? What is that...

48. **Sports** Shawn hit a fou... the ball over the level o... $- 16t^2$, where $t$ is the t...

  a. Draw the graph of th... level of the bat to th...

  b. If the catcher is going to attempt to catch the ball, how long does she have to get ready?

  c. Where is the ball when the height found by the function is zero?

**Mixed Review**

49. Identify the quadratic term, the linear term, and the constant term of the function $f(x) = 4x^2 - 8x$...

50. Find a quadratic equation hav...

51. Simplify $\sqrt[3]{2}(3\sqrt[3]{4} + 2\sqrt[3]{32})$.

52. Find the degree of the polynom... (**Lesson 5-4**)

53. Determine the slope of the lin... (-3, 9). (**Lesson 2-4**)

# Getting into the Chapter

---

The list of chapter objectives lets you know what you can expect to learn in the chapter.

Detailed information about each career is given as well as an address where you may write to obtain even more information about the career.

### CHAPTER 12

## Exponential and Logarithmic Functions

**CHAPTER OBJECTIVES**

In this chapter, you will:
- Solve equations involving logarithmic and exponential functions.
- Find common and natural logarithms of numbers.
- Solve problems using estimation.
- Use logarithms to solve problems.

**CAREERS IN SYSTEMS ANALYSIS**

The world depends on computers today, depends on them far more than anyone have believed possible even thirty years ago. Research, business, government, education, sports, entertainment—almost every company every field now has a computer system to he function and grow. Thus it follows that the

If the ant is $\frac{3}{8}$ inch long, estimate the s microchip.

544

You will be introduced to many different careers that are available to you after graduation from college.

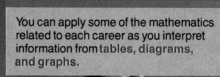

You can apply some of the mathematics related to each career as you interpret information from tables, diagrams, and graphs.

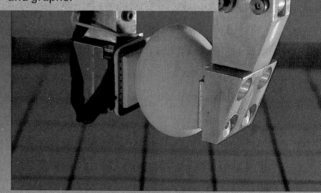

**MORE ABOUT SYSTEMS ANALYSIS**

| Degree Required | Related Math Subjects: | For more information on the various careers available in the field of Systems Analysis, write to: |
|---|---|---|
| ▪ Bachelor's degree in Computer Science | ▪ Advanced Algebra | |
| | ▪ Geometry | |
| | ▪ Statistics/Probability | Association of the Institute for Certification of Computer Professionals |
| **Some systems analysts like:** | **Some systems analysts dislike:** | 2200 East Devon Avenue Suite 268 Des Plaines, IL 60018 |
| ▪ challenge and variety in their work | ▪ the need to update their knowledge with advances in technology | |
| ▪ good employment opportunities | ▪ working overtime to meet deadlines | |
| ▪ good salaries | ▪ frustrating problems in computer programming | |
| ▪ interaction with people in solving problems | | |

545

# Wrapping Up the Chapter

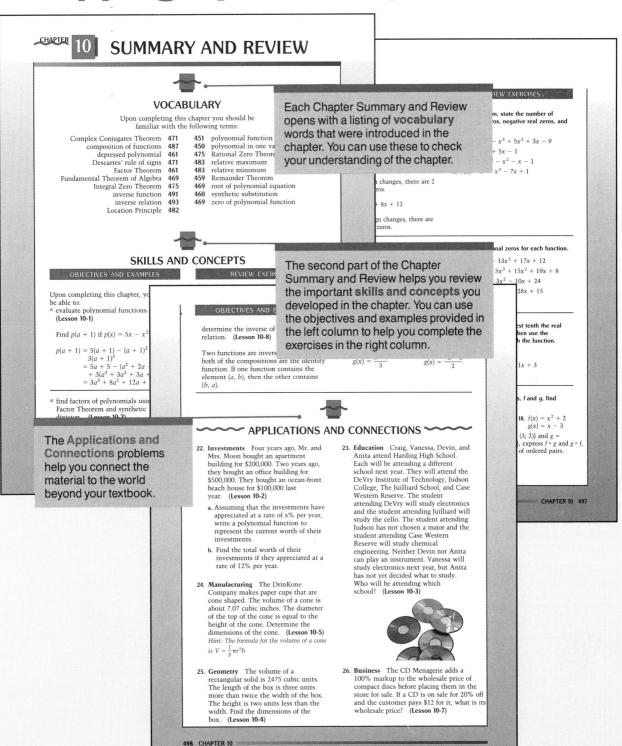

## CHAPTER 10

## SUMMARY AND REVIEW

### VOCABULARY

Upon completing this chapter you should be familiar with the following terms:

| | | | |
|---|---|---|---|
| Complex Conjugates Theorem | 471 | 451 | polynomial function |
| composition of functions | 487 | 450 | polynomial in one va... |
| depressed polynomial | 461 | 475 | Rational Zero Theore... |
| Descartes' rule of signs | 471 | 483 | relative maximum |
| Factor Theorem | 461 | 483 | relative minimum |
| Fundamental Theorem of Algebra | 469 | 459 | Remainder Theorem |
| Integral Zero Theorem | 475 | 469 | root of polynomial equation |
| inverse function | 491 | 460 | synthetic substitution |
| inverse relation | 493 | 469 | zero of polynomial function |
| Location Principle | 482 | | |

> Each Chapter Summary and Review opens with a listing of **vocabulary** words that were introduced in the chapter. You can use these to check your understanding of the chapter.

### SKILLS AND CONCEPTS

> The second part of the Chapter Summary and Review helps you review the important **skills and concepts** you developed in the chapter. You can use the objectives and examples provided in the left column to help you complete the exercises in the right column.

#### OBJECTIVES AND EXAMPLES

Upon completing this chapter, yo... be able to:
- evaluate polynomial functions. (Lesson 10-1)

Find $p(a + 1)$ if $p(x) = 5x - x^2$

$p(a + 1) = 5(a + 1) - (a + 1)^2$
$3(a + 1)^3$
$= 5a + 5 - (a^2 + 2a ...$
$+ 3(a^3 + 3a^2 + 3a ...$
$= 3a^3 + 8a^2 + 12a + ...$

- find factors of polynomials usin... Factor Theorem and synthetic division. (Lesson 10-2)

#### OBJECTIVES AND E...

determine the inverse of ... relation. (Lesson 10-8)

Two functions are invers... both of the compositions are the identity function. If one function contains the element $(a, b)$, then the other contains $(b, a)$.

$g(x) = \frac{}{3} \qquad g(x) = \frac{}{2}$

#### ...IEW EXERCISES

...n, state the number of ...ros, negative real zeros, and

$... - x^3 + 5x^2 + 3x - 9$
$... + 5x - 1$
$... - x^2 - x - 1$
$... x^3 - 7x + 1$

...changes, there are 2 ...ros.

$... + 8x + 12$

...n changes, there are ...zeros.

...onal zeros for each function.

$... - 13x^2 + 17x + 12$
$... 5x^3 + 15x^2 + 19x + 8$
$... 3x^2 - 10x + 24$
$... 28x + 15$

...est tenth the real ...hen use the ...h the function.

$...1x + 3$

..., f and g, find

18. $f(x) = x^2 + 2$
$g(x) = x - 3$
(3, 2)} and $g =$
}, express $f \circ g$ and $g \circ f$, of ordered pairs.

### APPLICATIONS AND CONNECTIONS

> The **Applications and Connections** problems help you connect the material to the world beyond your textbook.

**22. Investments** Four years ago, Mr. and Mrs. Moon bought an apartment building for $200,000. Two years ago, they bought an office building for $500,000. They bought an ocean-front beach house for $100,000 last year. (Lesson 10-2)

a. Assuming that the investments have appreciated at a rate of $x$% per year, write a polynomial function to represent the current worth of their investments.

b. Find the total worth of their investments if they appreciated at a rate of 12% per year.

**24. Manufacturing** The DrinKone Company makes paper cups that are cone shaped. The volume of a cone is about 7.07 cubic inches. The diameter of the top of the cone is equal to the height of the cone. Determine the dimensions of the cone. (Lesson 10-5) *Hint: The formula for the volume of a cone is $V = \frac{1}{3}\pi r^2 h$*

**25. Geometry** The volume of a rectangular solid is 2475 cubic units. The length of the box is three units more than twice the width of the box. The height is two units less than the width. Find the dimensions of the box. (Lesson 10-4)

**23. Education** Craig, Vanessa, Devin, and Anita attend Harding High School. Each will be attending a different school next year. They will attend the DeVry Institute of Technology, Judson College, The Juilliard School, and Case Western Reserve. The student attending DeVry will study electronics and the student attending Juilliard will study the cello. The student attending Judson has not chosen a major and the student attending Case Western Reserve will study chemical engineering. Neither Devin nor Anita can play an instrument. Vanessa will study electronics next year, but Anita has not yet decided what to study. Who will be attending which school? (Lesson 10-3)

**26. Business** The CD Menagerie adds a 100% markup to the wholesale price of compact discs before placing them in the store for sale. If a CD is on sale for 20% off and the customer pays $12 for it, what is its wholesale price? (Lesson 10-7)

# 1

# Equations and Inequalities

## CHAPTER OBJECTIVES

In this chapter, you will:

- Use the properties of real numbers to simplify expressions.
- Solve equations using properties of equality.
- Solve inequalities and graph their solution sets.
- Solve absolute value equations and inequalities.

**Notice the differences between the temperatures of occupied and unoccupied areas? Explain why you think the changes are greater in some areas than in others.**

| BUILDING TEMPERATURE RANGES | 50 55 60 65 70 75 80 85 |
|---|---|
| Classrooms | |
| Cafeteria | |
| Gymnasium | |
| Corridors | |
| Industrial | |
| Home Ec. | |
| Offices | |
| Auditorium | |
| Media Center | |

Unoccupied ▬▬▬     Occupied ▬▬▬

## CAREERS IN ENERGY ENGINEERING

A new generation of intelligent buildings are being "born"—and if you choose the career of energy engineer, you might be the obstetrician for one of them.

These buildings can sense and then adjust to changes in their environment very much like living beings can. Their brains are computer systems, managing the buildings' communications, security, and energy needs. For example, one new building in Boston, Massachusetts, has no fuel-burning furnaces at all, yet it maintains a comfortable indoor temperature all winter long. How? The computer-controlled energy management system circulates the heated air from sunny window sills and the heat generated by people and office machines—including lights and coffee makers.

Now, suppose one of these intelligent buildings develops a maintenance problem—maybe the air-conditioning compressors stop working early one Sunday morning in August. Though no one is in the building to notice the rising temperature and to telephone for repairs, an energy engineer miles away has already noticed the system's overload on his or her monitor. Even though the Service Center for a dozen or more of these intelligent buildings may be hundreds of miles away from each building, the engineers on watch there can easily diagnose and correct the problem. It's a job with responsibility, will it one day be your job?

## MORE ABOUT ENERGY ENGINEERING

**Degree Required:**

- Bachelor's Degree in Electrical Engineering

**Some energy engineers like:**

- solving problems
- working in a new field that gives them opportunities to experiment and be creative
- the challenge in finding ways to conserve energy
- the variety and challenge of their work

**Related Math Subjects:**

- Geometry
- Trigonometry
- Advanced Algebra
- Calculus

**Some energy engineers dislike:**

- having to try to keep up with an ever-changing field
- working long hours to finish a project
- having to meet deadlines
- working against skeptical opinions

For more information on the various careers available in the field of Energy Engineering, write to:

Association of Energy Engineers
4025 Pleasantdale Road
Suite 420
Atlanta, GA 30340

# Expressions and Formulas

**Objectives**

After studying this lesson, you should be able to:
- use the order of operations to evaluate expressions, and
- use formulas.

**Application**

To determine the efficiency of the insulation in a school building, an energy engineer needed to evaluate the expression 4 + 3.7(3). Her calculator gave a result of 23.1. Use your calculator to evaluate the expression. Is your result the same?

ENTER: 4 [ + ] 3.7 [ × ] 3

If you used a scientific calculator, it probably multiplied 3.7 by 3 and then added 4, giving an answer of 15.1. If you used a non-scientific calculator, as the engineer did, it probably added 4 and 3.7 before multiplying by 3. This can be very confusing. Do you add first or multiply first? A numerical expression should have exactly one value. In order to find that value, you must follow the established rules for the **order of operations.**

*Order of Operations*

1. **Simplify the expressions inside grouping symbols, such as parentheses, brackets, and fraction bars.**
2. **Evaluate all powers.**
3. **Do all multiplications and divisions from left to right.**
4. **Do all additions and subtractions from left to right.**

Using these rules, the value of 4 + 3.7(3) is 15.1.

**Example 1**

Find the value of $5 + 8^2 \div 4 \cdot 3$.

$$5 + 8^2 \div 4 \cdot 3 = 5 + 64 \div 4 \cdot 3 \qquad \textit{Evaluate all powers.}$$
$$= 5 + 16 \cdot 3$$
$$= 5 + 48 \qquad \textit{Do all multiplications and divisions left to right.}$$
$$= 53 \qquad \textit{Do all additions and subtractions left to right.}$$

The value is 53.

Grouping symbols can be used to change or clarify the order of operations. Frequently used grouping symbols are parentheses, ( ), and brackets, [ ]. When calculating the value of an expression, you should begin with the operation in the innermost set of grouping symbols.

**Example 2**

Find the value of $[(4 + 8)^2 \div 9] \cdot 5$.

$$
\begin{aligned}
[(4 + 8)^2 \div 9] \cdot 5 &= [(12)^2 \div 9] \cdot 5 && \textit{First add 4 and 8.} \\
&= [144 \div 9] \cdot 5 && \textit{Then find } 12^2. \\
&= [16] \cdot 5 && \textit{Now divide 144 by 9.} \\
&= 80 && \textit{Multiply 16 by 5.}
\end{aligned}
$$

The value is 80.

**Algebraic expressions** contain at least one variable. You can evaluate an algebraic expression by replacing each variable with a value and then applying the rules for the order of operations.

**Example 3**

Evaluate $4x^2 + 3xy$ if $x = -3$ and $y = 5$.

$$
\begin{aligned}
4x^2 + 3xy &= 4(-3)^2 + 3(-3)(5) && \textit{Replace x by } -3 \textit{ and y by 5.} \\
&= 4(9) + 3(-3)(5) && \textit{Find } (-3)^2. \\
&= 36 + (-45) && \textit{Multiply left to right.} \\
&= -9 && \textit{Add.}
\end{aligned}
$$

The value is $-9$.

A **formula** is a mathematical sentence that expresses the relationship between certain quantities. If you know a value for every variable in the formula except one, you can find the value for that remaining variable.

**Example 4**

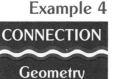

**CONNECTION**

**Geometry**

The volume of a sphere is calculated by using the formula $V = \frac{4}{3}\pi r^3$. In the formula, $r$ represents the measure of the radius. Use a calculator to find the volume of a sphere with a radius of 7.6 centimeters. Use the $\pi$ key on your calculator.

$$
\begin{aligned}
V &= \frac{4}{3}\pi r^3 \\
&= \frac{4}{3}\pi (7.6)^3
\end{aligned}
$$

ENTER:  4 [÷] 3 [x] [π] [x] 7.6 [$y^x$] 3 [=]   8.78
*The display is rounded to the nearest hundredth.*

The volume of the sphere is approximately 1838.78 cubic centimeters.

Example 5

APPLICATION

Banking

You are about to buy a new car. The sales associate offers you a simple interest loan to finance it. Simple interest is calculated using the formula $I = prt$. In the formula, $p$ represents the principal in dollars, $r$ represents the annual interest rate, and $t$ represents the time in years. Find the amount of interest you would pay for a two-year loan if the principal is $6000 and the rate is 12%.

$$I = prt$$
$$= 6000(0.12)(2) \qquad 12\% = 0.12$$
$$= 1440$$

The interest on $6000 at 12% for 2 years is $1440.

Example 6

APPLICATION

Chemistry

The boiling point of the metal zinc is 787.1° on the Fahrenheit scale. Change the temperature to its equivalent on the Celsius scale.

The relationship between Celsius temperature, $C$, and Fahrenheit temperature, $F$, is given by $C = \dfrac{5(F - 32)}{9}$.

Find $C$ if $F = 787.1$.

$$C = \frac{5(F - 32)}{9} \qquad \textit{Write the formula.}$$
$$= \frac{5(787.1 - 32)}{9} \qquad \textit{Replace F by 787.1.}$$
$$= 419.5$$

The Celsius temperature is 419.5°.

# CHECKING FOR UNDERSTANDING

**Communicating Mathematics**

**Read and study the lesson to answer these questions.**

1. When you evaluate $(15 - 4)^3 \div 3$, which operation is performed first?

2. Tell how you would find the perimeter of a rectangle.

3. Where would you insert grouping symbols so that the value of $7 \cdot 4 + 2 \cdot 8$ is 240?

**Guided Practice**

**Find the value of each expression.**

4. $7 - 8 \div 2$
5. $9 - 4(3)$
6. $4(7 + 3)$
7. $9(4 + 2)$
8. $5 - 4 \div 2$
9. $7 - (3 + 2)$

**Find the value of each expression if $a = 2$, $b = -3$, and $c = 4$.**

10. $a + b - c$
11. $a + 2b - c$
12. $a(b + c)$

Find the value of *C* in each formula if the value of *F* is 98.6.

**13.** $C = \dfrac{5F - 160}{9}$

**14.** $F = \dfrac{9}{5}C + 32$

**15.** A rectangle has length $(y + 5)$ cm and width $(y - 5)$ cm. Write a formula to represent its area.

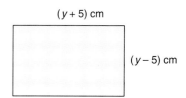

(y + 5) cm

(y − 5) cm

**16.** A triangle has height $(a + 2)$ ft and base $(a + 6)$ ft. Write a formula to represent its area.

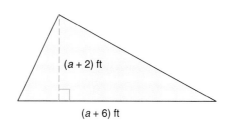

(a + 2) ft

(a + 6) ft

# EXERCISES

**Practice**    **Evaluate each expression.**

**17.** $3(2^2 + 3)$

**18.** $2(3 + 8) - 3$

**19.** $(6 + 5)4 - 3$

**20.** $5 + 3^2 - 16 + 4$

**21.** $(5 + 3) - 16 \div 4$

**22.** $12 + 18 \div 6 + 7$

**23.** $4 + 8(4) \div 2 - 10$

**24.** $[19 - (8 - 1)] \div 3$

**25.** $3 + [8 \div (9 - 2(4))]$

**26.** $[(-8 + 3) \times 4 - 2] \div 6$

**27.** $15 \div 3 \times 5 + 1$

**28.** $3 + (21 \div 7) \times 8 \div 4$

**29.** $0.2(0.5 + 2.2) \div 6$

**30.** $\dfrac{1}{3} - \dfrac{12(77 \div 11)}{9}$

**Evaluate each expression if $a = 3$, $b = 7$, $c = -2$, $d = 0.5$, and $e = 0.3$.**

**31.** $\dfrac{3ab}{cd}$

**32.** $\dfrac{5a + 3c}{3b}$

**33.** $(5a + 3d)^2 - e^2$

**34.** $(3b - 21d)^2$

**35.** $a(b - 7)^3$

**36.** $(a + c)^2 - de$

**37.** $\dfrac{4a - 6d}{2b + c}$

**38.** $c + de^2$

**Find the simple interest, *I*, given each of the values for the principal, rate, and time.**

**39.** $p = \$2500$, $r = 7.37\%$, $t = 4$ years

**40.** $p = \$5280$, $r = 8.2\%$, $t = 30$ months

**41.** $p = \$65,283.21$, $r = 9.32\%$, $t = 78$ months

**42.** $p = \$20,005$, $r = 7.9\%$, $t = 2$ years, 3 months

The formula for the area of a trapezoid is $A = \frac{h}{2}(b_1 + b_2)$. $A$ represents the measure of the area, $h$ represents the measure of the altitude, and $b_1$ and $b_2$ represent the measures of the bases. Find the measure of the area of each trapezoid given the following values.

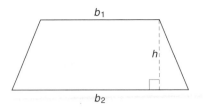

**43.** $h = 8$, $b_1 = 8.6$, and $b_2 = 14.8$

**44.** $b_1 = 4$, $b_2 = 11$, and $h = 7$

**45.** $h = 9$, $b_2 = 7\frac{2}{3}$, and $b_1 = 4\frac{1}{6}$

**46.** $h = 12$, $b_2 = 9.7$, and $b_1 = 6.2$

## Critical Thinking

**47.** Create a formula that will determine the total surface area of this rectangular prism. What is the surface area if $a = 4$? if $a = 6.2$?

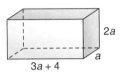

## Applications

**48. Banking**  Mary invests $7500 in a certificate of deposit at the Lombard Bank. The simple interest rate is 7.2%. How much interest will she have earned at the end of three years?

**49. Construction**  A contractor plans to paint the floor and the outside wall of a circular parking garage. The radius of the floor is 300 feet. The height of the wall is 20 feet. If one gallon of paint will cover 425 square feet, how many gallons of paint will it take to complete the job?
*Hint: Draw a picture to help solve this problem.*

## Computer

The BASIC computer language follows the same order of operations as algebra. However, parentheses and brackets do not indicate multiplication in BASIC. The symbol * must be used to multiply. This program finds the areas of three trapezoids. READ and DATA statements are used to assign values for the height and bases.

```
10   PRINT "H","B1","B2","A"
20   READ H, B1, B2
30   IF H = 0 THEN 90
40   DATA 8, 12, 20, 7, 4, 11
50   DATA 4.8, 5.6, 6.4, 0, 0, 0
60   LET A = H/2 *(B1 + B2)
70   PRINT H, B1, B2, A
80   GOTO 20
90   END
```

**50.** Write and run a program to find the area of each triangle. The formula for the area of a triangle is $A = \frac{1}{2}bh$ where $b$ is the measure of the base (in cm) and $h$ is the measure of the height (in cm).

**a.** $b = 6.7$, $h = 13.8$

**b.** $b = 127.2$, $h = 82.6$

**51.** Write and run a program to find the area of each parallelogram. The formula for the area of a parallelogram is $A = bh$ where $b$ is the measure of the base (in ft) and $h$ is the measure of the height (in ft).

**a.** $b = 7.1$, $h = 3.5$

**b.** $b = 97.2$, $h = 26.9$

# Properties of Real Numbers

**Objectives**

After studying this lesson, you should be able to:

- determine the sets of numbers to which a number belongs, and
- use the properties of real numbers to simplify expressions.

All the numbers that we use in everyday life are **real numbers.** Each real number corresponds to exactly one point on the number line, and every point on the number line represents exactly one real number.

### FYI ...

When the Greek society of mathematicians, the Pythagoreans, discovered irrational numbers, they tried to keep their discovery a secret. According to legend, one of their members was drowned for telling the secret to outsiders.

Every real number can be classified as either **rational** or **irrational.** A rational number can be expressed as a ratio $\frac{m}{n}$, where $m$ and $n$ are integers and $n$ is not zero. The decimal form of a rational number is either a terminating or repeating decimal. Some examples of rational numbers are $\frac{2}{3}$, $1.\overline{23}$, $5.8$, $-7$, and $0$. Any real number that is not rational is irrational. $\sqrt{2}$, $\pi$, and $\sqrt{7}$ are irrational numbers.

The sets of natural numbers, $\{1, 2, 3, 4, 5, ...\}$, whole numbers, $\{0, 1, 2, 3, 4, ...\}$, and integers, $\{..., -2, -1, 0, 1, 2, ...\}$ are all subsets of the rational numbers.

The Venn diagram at the right shows the relationships between all of these sets of numbers.

Reals, R

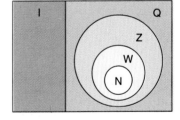

R = reals      Q = rationals
I = irrationals      Z = integers
W = wholes      N = naturals

**Example 1**

**Name the sets of numbers to which each number belongs.**

a. $\sqrt{6}$

irrationals (I)
reals (R)

b. $\frac{3}{8}$

rationals (Q)
reals (R)

c. $-9$

integers (Z)
rationals (Q)
reals (R)

**Example 2** | **Evaluate each expression. Then name the sets of numbers to which each value belongs.**

**a.** $\sqrt{9}$

$\sqrt{9} = 3$

natural numbers (N),
whole numbers (W), integers (Z),
rationals (Q), reals (R)

**b.** $6 \div 10$

$6 \div 10 = 0.6 \text{ or } \dfrac{3}{5}$

rationals (Q),
reals (R)

Operations with real numbers have several important properties. One of the basic properties of addition and multiplication is **commutativity.** The order in which two real numbers are added or multiplied does not change their sum or their product.

$$5 + 9 = 9 + 5 \qquad\qquad 12 \cdot 4 = 4 \cdot 12$$
$$14 = 14 \qquad\qquad\qquad 48 = 48$$

Another basic property of addition and multiplication is **associativity.** The way three or more real numbers are grouped, or associated, does not change their sum or product.

$$(7 + 6) + 9 = 7 + (6 + 9) \qquad (10 \cdot 5) \cdot 2 = 10 \cdot (5 \cdot 2)$$
$$13 + 9 = 7 + 15 \qquad\qquad 50 \cdot 2 = 10 \cdot 10$$
$$22 = 22 \qquad\qquad\qquad 100 = 100$$

The sum of any real number and 0 is the original number. So, for real numbers, the **additive identity** is 0.

$$6.7 + 0 = 6.7 \qquad\qquad\qquad 0 + \sqrt{3} = \sqrt{3}$$

Each real number has a unique **additive inverse** or **opposite.** The sum of a number and its opposite is 0.

$$8 + (-8) = 0 \qquad\qquad\qquad -\frac{1}{3} + \frac{1}{3} = 0$$

The **multiplicative identity** for real numbers is 1, since the product of any real number and 1 is the original number.

$$\frac{5}{8} \cdot 1 = \frac{5}{8} \qquad\qquad\qquad (1)(3.9) = 3.9$$

Each real number, except 0, has a unique **multiplicative inverse** or **reciprocal.** The product of a real number and its reciprocal is 1.

$$\frac{1}{8}(8) = 1 \qquad\qquad\qquad (-0.2)(-5) = 1$$

*Why does zero not have a reciprocal?*

The chart below summarizes the properties of real numbers for addition and multiplication.

| For any real numbers $a$, $b$, and $c$ | | |
|---|---|---|
| | Addition | Multiplication |
| Commutative<br>Associative<br>Identity<br><br>Inverse | $a + b = b + a$<br>$(a + b) + c = a + (b + c)$<br>$a + 0 = a = 0 + a$<br><br>$a + (^-a) = 0 = (^-a) + a$ | $a \cdot b = b \cdot a$<br>$(a \cdot b) \cdot c = a \cdot (b \cdot c)$<br>$a \cdot 1 = a = 1 \cdot a$<br>If $a$ is *not* zero, then<br>$a \cdot \dfrac{1}{a} = 1 = \dfrac{1}{a} \cdot a.$ |

*$^-a$ is read "the opposite of a."*

**Example 3**

CONNECTION

Statistics

**A fast-food restaurant offered a special on their Biggie Burger to entice customers at lunch time. The 99¢ special ran Monday through Friday. The number of Biggie Burgers sold each day are recorded in the table below. What was the average amount received from the sale of Biggie Burgers during each day?**

| M | T | W | T | F |
|---|---|---|---|---|
| 246 | 303 | 182 | 341 | 378 |

An average is calculated by dividing the total amount by the number of items. In this case, it would be the total dollar amount divided by the number of days, 5.

There are two ways to find the total dollar amount.

(1)  $T = 0.99(246 + 303 + 182 + 341 + 378)$
 $T = 243.54 + 299.97 + 180.18 + 337.59$
 $\quad + 374.22$ or $1435.50$

(2)  $T = 0.99(246 + 303 + 182 + 341 + 378)$
 $T = 0.99(1450)$ or $1435.50$

Now find the average by dividing the total by 5.
$1435.50 \div 5 = 287.10$

The average daily revenue received from Biggie Burgers was $287.10.

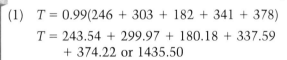

This is an application of a property that involves both addition and multiplication. It is called the distributive property.

*Distributive Property*

**For all real numbers $a$, $b$, and $c$,
$a(b + c) = ab + ac$ and $(b + c)a = ba + ca$.**

**Example 4**

Simplify $5(3m - 7n) + 3(4m + n)$.

$5(3m - 7n) + 3(4m + n)$

$\qquad = 5(3m) - 5(7n) + 3(4m) + 3(n)$ $\qquad$ *Use the distributive property.*

$\qquad = 15m - 35n + 12m + 3n$ $\qquad$ *Multiply.*

$\qquad = 27m - 32n$ $\qquad$ *Combine like terms.*

# CHECKING FOR UNDERSTANDING

**Communicating Mathematics**

Read and study the lesson to answer these questions.

1. What property is used to simplify $3y(4y + 2z)$?

2. Explain why there is no commutative property for subtraction and division.

3. Draw a geometric model to show how $4 \times 3$ is similar to $3 \times 4$.

4. List five irrational numbers.

5. List five rational numbers.

**Guided Practice**

Name the sets of numbers to which each number belongs.

6. $-8$
7. $\frac{4}{3}$
8. $-2.6$
9. $\pi$

10. $-\frac{7}{2}$
11. $10$
12. $\sqrt{10}$
13. $5.11$

14. $0$
15. $-\frac{12}{3}$
16. $-1.0$
17. $\sqrt{16}$

State the property illustrated in each equation.

18. $8 + (6 + 4) = (8 + 6) + 4$

19. $7(5) = 5(7)$

20. $a(3 - 2) = a \cdot 3 - a \cdot 2$

21. $8 + (1 + 6) = 8 + (6 + 1)$

22. $3 + (-3) = 0$

# EXERCISES

**Practice**

Evaluate each expression. Then name the sets of numbers to which each value belongs.

23. $8 - 7$
24. $7 - 8$
25. $-54 \div 6$
26. $68 \div 100$

27. $-2.4 \times 10$
28. $3.9 + 2.6$
29. $6 \div 2^2$
30. $\sqrt{36 + 5}$

Determine whether each statement is *true* or *false*. If *false*, give an example of a number that shows the statement is false.

31. Every whole number is an integer.

32. Every integer is a whole number.

**Determine whether each statement is *true* or *false*. If *false*, give an example of a number that shows the statement is false.**

33. Every rational number is an integer.

34. Every real number is irrational.

35. Every irrational number is a real number.

36. Every integer is a rational number.

37. Every real number is either a rational number or an irrational number.

**State the property illustrated in each equation.**

38. $(4 + 11) \cdot 6 = 4(6) + 11(6)$

39. $(a + b) + [-(a + b)] = 0$

40. $11 + a = a + 11$

41. $(3 + 9) + 14 = 14 + (3 + 9)$

42. $3 + (a + b) = (a + b) + 3$

43. $3\left(\dfrac{1}{3}\right) = 1$

44. $(4 + 9a)2b = 2b(4 + 9a)$

45. $a + b + 0 = a + b$

**Simplify each expression.**

46. $3(5a + 6b) + 8(2a - b)$

47. $3a + 5b + 7a - 3b$

48. $2(7c - 5d) - 3(d + 2c)$

49. $\dfrac{1}{4}(12 + 20a) + \dfrac{3}{4}(12 + 20a)$

50. $\dfrac{1}{2}(17 - 4x) - \dfrac{3}{4}(6 - 16x)$

51. $\dfrac{2}{3}\left(\dfrac{1}{2}a + 3b\right) + \dfrac{1}{2}\left(\dfrac{2}{3}a + b\right)$

52. $\dfrac{3}{4}(2x - 5y) + \dfrac{1}{2}\left(\dfrac{2}{3}x + 4y\right)$

53. $7(0.2m + 0.3n) + 5(0.6m - n)$

**Critical Thinking**

**Use the definitions of the properties of real numbers to answer these questions.**

54. If $a + b = a$, what is the value of $b$?

55. If $ab = 1$, what is the value of $b$? What is $b$ called?

56. If $ab = a$, what is the value of $b$?

**Applications**

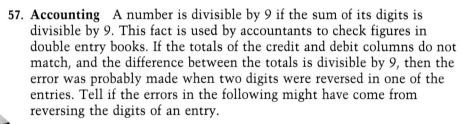

57. **Accounting** A number is divisible by 9 if the sum of its digits is divisible by 9. This fact is used by accountants to check figures in double entry books. If the totals of the credit and debit columns do not match, and the difference between the totals is divisible by 9, then the error was probably made when two digits were reversed in one of the entries. Tell if the errors in the following might have come from reversing the digits of an entry.

   a. credit = \$638, debit = \$577        b. credit = \$1050, debit = \$1095

   c. Why does this check work? *Hint: What is the difference between a number and the number whose digits are reversed?*

**Mixed Review**

**Evaluate each expression.   (Lesson 1-1)**

58. $3 + (3 - 3)^3 - 3$

59. $-8 \div [20 \div (16 - 11)]$

60. Find the value of $12a^2 + bc$ if $a = 3$, $b = 7$, and $c = -2$.   (Lesson 1-1)

61. **Banking** Find the interest earned in 6 years on a savings account containing \$20,000 if the interest rate is 14.5%.   (Lesson 1-1)

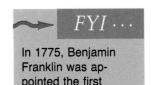

# 1-3 Solving Equations

**Objective**

After studying this lesson, you should be able to:
- solve equations using the properties of equality.

**Application**

*FYI* ...

In 1775, Benjamin Franklin was appointed the first Postmaster General by the Second Continental Congress.

Janet Graves needed to buy some stamps for her graduation announcements. She bought some 25-cent stamps and three times as many 29-cent stamps. She paid a total of $22.40. How many of each type of stamp did she buy?

This problem can be solved by writing and then solving an open sentence. Sentences with variables to be replaced, such as $3x - 7 = 21$ and $3x + 4 > 9$, are called **open sentences.** When you solve an open sentence, you find replacements for the variables that will make the sentence true. Each of these replacements is called a **solution** of the open sentence.

An **equation** states that two mathematical expressions are equal. Solving an equation is like solving an open sentence. You find the values that you can put in place of the variables so that the equation is true.

Real numbers have certain properties that we can use when we solve equations or open sentences. Some of those properties are listed below.

| | |
|---|---|
| *Reflexive Property of Equality* | **For any real number $a$, $a = a$.** |
| *Symmetric Property of Equality* | **For all real numbers $a$ and $b$, if $a = b$, then $b = a$.** |
| *Transitive Property of Equality* | **For all real numbers $a$, $b$, and $c$, if $a = b$ and $b = c$, then $a = c$.** |

**Example 1**

**Name the property of equality illustrated in each statement.**

**a.** If $36 \cdot 2 = 72$, then $72 = 36 \cdot 2$.   **a.** symmetric property

**b.** $21.4 = 21.4$   **b.** reflexive property

**c.** If $8 = 6 + 2$ and $6 + 2 = 5 + 3$, then $8 = 5 + 3$.   **c.** transitive property

Some equations can be solved by making a substitution. The substitution property allows you to replace an expression with another equivalent expression.

**Example 2**

Solve $y = 8(0.3) + 1.2$.

$y = 8(0.3) + 1.2$

$y = 2.4 + 1.2$      *Substitute 2.4 for 8(0.3).*

$y = 3.6$      The solution is 3.6.

Sometimes an equation can be solved by adding or subtracting the same number on each side.

**Example 3**

Solve $x + 28.3 = 56.0$.

$x + 28.3 = 56.0$      **Check:**

$x + 28.3 + (-28.3) = 56.0 + (-28.3)$      $x + 28.3 = 56.0$

$x = 27.7$      $27.7 + 28.3 \stackrel{?}{=} 56.0$

The solution is 27.7.      $56.0 = 56.0$ ✓

*This equation could also be solved by subtracting 28.3 from each side.*

| Addition and Subtraction Properties of Equality | **For any real numbers $a$, $b$, and $c$, if $a = b$,** **then $a + c = b + c$ and $a - c = b - c$.** |
|---|---|

Some equations may be solved by multiplying or dividing each side by the same number.

**Example 4**

Solve $8x = 48$.

$8x = 48$      **Check:** $8x = 48$

$\dfrac{1}{8} \cdot 8x = \dfrac{1}{8} \cdot 48$      *Multiply each side by $\dfrac{1}{8}$,*      $8(6) \stackrel{?}{=} 48$

$x = 6$      *the reciprocal of 8.*      The solution is 6.      $48 = 48$ ✓

*This equation could also be solved by dividing each side by 8.*

**Example 5**

Solve $-\dfrac{2}{3}k = 14$.

$-\dfrac{2}{3}k = 14$      **Check:** $-\dfrac{2}{3}k = 14$

$-\dfrac{3}{2}\left(-\dfrac{2}{3}\right)k = \left(-\dfrac{3}{2}\right)(14)$      *Multiply each side by $-\dfrac{3}{2}$,*      $-\dfrac{2}{3}(-21) \stackrel{?}{=} 14$

$k = -21$      *the reciprocal of $-\dfrac{2}{3}$.*      The solution is $-21$.      $14 = 14$ ✓

| Multiplication and Division Properties of Equality | **For any real numbers $a$, $b$, and $c$, if $a = b$, then $a \cdot c = b \cdot c$ and,** **if $c$ is not zero, $\dfrac{a}{c} = \dfrac{b}{c}$.** |
|---|---|

In order to solve some equations, it may be necessary to apply more than one property. The following examples illustrate the use of several properties in solving equations.

**Example 6**

Solve $0.75(8a + 20) - 2(a - 1) = 3$.

$$0.75(8a + 20) - 2(a - 1) = 3$$
$$6a + 15 - 2a + 2 = 3 \qquad \textit{Distributive and substitution properties}$$
$$4a + 17 = 3 \qquad \textit{Commutative, distributive, and substitution properties}$$
$$4a = {}^-14 \qquad \textit{Subtraction and substitution properties}$$
$$a = {}^-3.5 \qquad \textit{Division and substitution properties}$$

The solution is $^-3.5$. *Check this result.*

**Example 7**

Use a calculator to solve $68x + 373 = 802$.

$$68x + 373 = 802 \qquad \textit{Rewrite the equation to isolate the variable, x.}$$
$$x = \frac{802 - 373}{68}$$

ENTER:  ⌐ 802 ⊟ 373 ⌐ ⊡ 68 ⊟ `6.30882353`

The solution is approximately 6.309.

**Example 8**

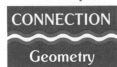

CONNECTION

Geometry

The formula for the volume of a right circular cone is $V = \frac{1}{3}\pi r^2 h$, where $r$ represents the radius of the circular base, and the height is represented by $h$. Solve the formula for $h$.

$$V = \frac{1}{3}\pi r^2 h$$
$$3 \cdot V = 3 \cdot \frac{1}{3}\pi r^2 h \qquad \textit{Multiply each side by 3.}$$
$$\frac{3V}{\pi r^2} = \frac{\pi r^2 h}{\pi r^2} \qquad \textit{Divide each side by } \pi r^2.$$
$$\qquad\qquad \textit{This form of the formula could be used to find the height of a cone if its volume and radius are known.}$$
$$\frac{3V}{\pi r^2} = h$$

# CHECKING FOR UNDERSTANDING

**Communicating Mathematics**

Read and study the lesson to answer these questions.

1. Describe in your own words the meaning of the transitive property of equality.

2. Is the following an example of the transitive property of equality? Juanita has $37.62. Juanita has as much money as Earl. Earl has $37.62.

3. Explain how using a balance scale is similar to adding or subtracting a number from each side of an equation.

**State the property illustrated in each statement.**

4. If $x + 3 = 7$, then $x = 4$.

5. $(4 + 7) + 8 = (4 + 7) + 8$

6. If $6 + 5 = 11$, then $11 = 6 + 5$.

7. If $3x = 10$, then $3x + 6 = 10 + 6$.

8. If $3x = 5y$ and $5y = 10z$, then $3x = 10z$.

9. If $5x = 10$, then $15x = 30$.

**Solve each equation.**

10. $\frac{1}{3}q = 872$

11. $7x + 2 = 23$

12. $2x + 7 = 8x - 11$

# EXERCISES

**Practice**

**Solve each equation.**

13. $\frac{3}{8} - \frac{1}{4}x = \frac{1}{16}$

14. $1.2x + 3.7 = 13.3$

15. $4.5 - 3.9m = 20.1$

16. $1.1x - 0.09 = 2.22$

17. $9 = 16d + 51$

18. $5t + 4 = 2t + 13$

19. $2y - 8 = 14 - 9y$

20. $3x + 5 = 9x + 2$

21. $3x - 4 = 7x - 11$

22. $\frac{3}{4}s - \frac{1}{2} = \frac{1}{4}s + 5$

23. $\frac{2}{3} - \frac{3}{5}x = \frac{2}{5}x + \frac{4}{3}$

24. $8 - x = 5x + 32$

25. $3 = -3(y + 5)$

26. $5(3x + 5) = 4x - 8$

27. $2x - 4(x + 2) = -2x - 8$

28. $285 - 38x = 2033$

29. $2(6 - 7k) = 2k - 4$

30. $8x - 3 = 5(2x + 1)$

**Journal**

Complete this sentence. "Doing my homework is an important part of algebra because ___."

**Critical Thinking**

31. Write an equation that will have no solution. Explain why there is no solution.

32. Write an equation that will have an infinite number of solutions. Explain why.

**Applications**

33. **Banking**   The formula, $A = p + prt$, gives the amount, $A$, in an account when $p$ dollars are invested at a rate of simple interest, $r$, for $t$ years. Solve this formula for $t$.

34. **Transportation**   The director of the Department of Transportation wants to know if the pile of salt for use on the icy roads is too tall to cover with their tarpaulin. The salt is in a pile that is shaped like a right circular cone. The formula for the volume of a right circular cone is $v = \frac{1}{3}\pi r^2 h$. The radius, $r$, of the circular base of the pile is 2.5 meters. The volume of the cone, $V$, is 40 cubic meters. If the tarpaulin can cover a cone up to 7 meters tall, can it cover the pile?

**Mixed Review**

35. Evaluate $\sqrt{9 \div 3}$ and tell the sets of numbers to which the value belongs.   **(Lesson 1-2)**

36. State the property illustrated by $11(3a + 2b) = 11(2b + 3a)$. **(Lesson 1-2)**

37. Evaluate $8a - 3bc$ if $a = 0.3$, $b = 7$, and $c = -2$.   **(Lesson 1-1)**

# Graphing Calculator Exploration
# Evaluating Expressions

The *Graphing Calculator Explorations* in this text will focus on two graphing calculators, the Texas Instruments TI-81® and the Casio fx-7000G®. In these lessons you will be introduced to the keying sequences that allow you to perform mathematical computations and to graph equations. You should always refer to your user's manual for more detail on all of the features of your calculator.

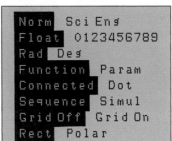

You do not need another scientific calculator since the graphing calculator also performs those computations you previously used a scientific calculator to do.

*Casio's fx-7500G, fx-7700G, fx-8000G, and fx-8500G operate in a similar fashion to the fx-7000G.*

Each calculator has a mode screen. The Casio displays this screen as the start-up screen. The TI-81 requires you to press the MODE key to display these settings. The mode setting shown below are commonly used for scientific calculations.

|  Casio  |  TI-81  |
|---------|---------|

```
**** MODE ****

sys mode  :  RUN
cal mode  :  COMP
   angle  :  Rad
 display  :  Norm

    Step     0
```

```
Norm  Sci Eng
Float    0123456789
Rad  Deg
Function   Param
Connected   Dot
Sequence   Simul
Grid Off   Grid On
Rect   Polar
```

If your mode settings do not match these, you need to change them.

CASIO:  Press [MODE] 1 to select RUN.

Press [MODE] + to select COMP.

Press [MODE] 5 [EXE] to select RAD.

Press [MODE] 9 [EXE] to select Norm.

TI-81:  Press [▶] or [◀] and [ENTER] to change the highlighted mode.

Press [CLEAR] to return to home screen.

These mode settings allow you to use your graphing calculator in the same ways as your scientific calculator. The Norm or Float mode places the decimal point with no set number of decimal places. The table at the top of the next page lists some common calculations and the keys on each calculator that correspond to the function. Just as with a scientific calculator, the graphing calculator observes the assumed order of operations.

| Mathematical Operation | Casio | TI-81 |
|---|---|---|
| Evaluates the expression, acting as an $=$ key. | Press EXE | Press ENTER |
| Clears the screen. | Press AC | Press CLEAR |
| To evaluate $3^7$. | Press 3 $x^y$ 7 EXE | Press 3 ^ 7 ENTER |
| To use a function listed above the regular calculator key. | Press SHIFT to get the function shown in orange. | Press 2nd to get the function shown in blue. |
| Parentheses used as grouping, or in multiplication, such as 3(6) and (3)6. | Press ( and ) to group operations. Will evaluate 3(6), but not (3)6. | Press ( and ) to group operations. Will evaluate both 3(6) and (3)6. |
| Making a number negative. *This not the same as the minus key.* | Press the (-) key and enter the number. | Press the (-) key and enter the number. |

*Note that the procedure for entering a negative number is different from that used with a scientific calculator.*

In each example in the *Graphing Calculator Explorations,* we will show both the Casio and TI-81 keystrokes.

**Example**

Evaluate $[(-5 + 13)^3 \div 6]3$.

*Casio*

ENTER: ( ( (-) 5 + 13 ) $x^y$ 3 ÷ 6 ) × 3 EXE $256$

*TI-81*

ENTER: ( ( (-) 5 + 13 ) ^ 3 ÷ 6 ) 3 ENTER $256$

You may note that there are differences in the ways each calculator displays some operations. For example, the Casio displays $5 \div 2$ as $5 \div 2$, while the TI-81 displays $5 \div 2$ as $5/2$. The Casio also displays the multiplication symbol as it appears on the key, while on the TI-81 it appears as a *.

# EXERCISES

**Use your graphing calculator to evaluate each expression.**

**1.** $2 \left[ \dfrac{5 + \dfrac{4(3 + 7)}{5}}{4} \right] + 13$

**2.** $(42 \times 5)^3 + \dfrac{89}{12}$

**3.** $(6.23 \times 10^{-7})(4.23 \times 10^3)$

**4.** $543.2^4$

# Applications of Equations

Objectives

After studying this lesson, you should be able to:

- translate word expressions into mathematical expressions,
- translate word sentences into equations, and
- use equations to solve problems.

The language of algebra provides a powerful way to translate word expressions into algebraic or mathematical expressions. Variables are used to represent numbers that are not known. Any letter can be used as a variable.

| Verbal Expression | Algebraic Expression |
|---|---|
| a *number* increased by 5 | $x + 5$ |
| twice the cube of a *number* | $2n^3$ |
| the square of a *number* decreased by the cube of the *same number* | $x^2 - x^3$ |
| three times the sum of a *number* and 7 | $3(b + 7)$ |

Equations can be used to represent verbal mathematical sentences.

| Verbal Sentence | Equation |
|---|---|
| Eight is equal to five plus three. | $8 = 5 + 3$ |
| A number decreased by 7 is –3. | $y - 7 = -3$ |
| A number divided by 3 is equal to $\frac{3}{4}$. | $\frac{x}{3} = \frac{3}{4}$ |

Example 1

APPLICATION

Construction

**The Conte family is planning to put a railing around the deck in their backyard. The deck is in two pieces, a square and an equilateral triangle, which share a common side. They will buy 75 feet of railing to go around the outside of the deck. If the lumberyard will custom cut pieces of railing, how long should they have the pieces cut so that there is one piece for each side of the deck?**

EXPLORE    First, explore the problems and choose a variable to represent the unknown number. The problem asks us to find the length of each side.

Let $s$ = the length of each side.

PLAN    Now let's plan the solution. Write an equation that represents the relationship in the problem The diagram at the right may be helpful. The perimeter is five times the length of a side, so $5s = 75$.

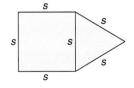

SOLVE     Next, solve the equation.

$$5s = 75$$
$$s = 15 \qquad \text{The length of a side is 15 feet.}$$

EXAMINE     Check your solution against the words of the problem and the diagram we drew. Since there are five equal sides that make up the perimeter of the deck, and $5 \times 15 = 75$, the solution is correct.

The four steps for solving problems that we just used are summarized below.

*Problem-Solving Plan*

1. **Explore the problem.**
2. **Plan the solution.**
3. **Solve the problem.**
4. **Examine the solution.**

**Example 2** | **A number increased by 17 is 41. Find the number.**

EXPLORE     Let $n$ stand for the number.

PLAN     A number increased by 17 is 41.
$$n \qquad\qquad + \qquad 17 = 41$$

SOLVE     $n + 17 = 41$
$$n = 24 \qquad \text{The number is 24.}$$

EXAMINE     Does 24 increased by 17 equal 41?     $24 + 17 \stackrel{?}{=} 41$
$$41 = 41 \quad \checkmark$$

**Example 3**

**Travel**

*FYI ...*

The city of Raleigh, North Carolina, is named for Sir Walter Raleigh. He established the first English settlement in North America on Roanoke Island in the 1580s.

**David Haghiri drove from his home in Raleigh to the beach at Cape Hatteras for his vacation. He drove for an hour and a half at 50 miles per hour. When he realized that soon it would be too dark to set up his tent at the campground, he sped up to 55 miles per hour. He drove the rest of the 174-mile trip at that speed. How long did he drive 55 miles per hour?**

EXPLORE     Let $t$ stand for the length of time he drove at 55 miles per hour.

PLAN     total distance = distance at 50 mph + distance at 55 mph
$$174 \qquad = \qquad 50(1.5) \qquad + \qquad 55 \cdot t$$

SOLVE     $174 = 50(1.5) + 55 \cdot t$
$$174 = 75 + 55t$$
$$99 = 55t$$
$$1.8 = t \qquad \text{David drove 1.8 hours, or 1 hour 48 minutes,}$$
$$\text{at 55 mph.}$$

EXAMINE     If David drives for 1.5 hours     $1.5(50) + 1.8(55) \stackrel{?}{=} 174$
at 50 mph and for 1.8 hours at            $75 + 99 \stackrel{?}{=} 174$
55 mph, will he travel 174 miles?               $174 = 174 \quad \checkmark$

## Example 4

**CONNECTION**

Geometry

**The perimeter of an isosceles triangle is 106 cm. The length of the base is 35 cm. What is the length of one of the equal sides?**

*EXPLORE*   Draw a diagram and let $n$ represent the length of one of the equal sides.

*Sides of equal length are said to be congruent.*

*PLAN*   The perimeter equals the sum of the lengths of the sides. So we can write the following equation.

$$n + n + 35 = 106$$

*SOLVE*
$$n + n + 35 = 106$$
$$2n + 35 = 106$$
$$2n = 71$$
$$n = 35.5$$

The length of one side is 35.5 cm.

*EXAMINE*   If one of the two equal sides has length 35.5 cm, then the perimeter is $35.5 + 35.5 + 35 = 106$ cm. The answer is correct.

# CHECKING FOR UNDERSTANDING

**Communicating Mathematics**

**Read and study the lesson to answer these questions.**

1. Explain the difference between these word expressions: *the sum of twice a number and 2* and *twice the sum of a number and 2.*

2. Write an equation to find the length of a side of a regular octagon if its perimeter is 3000 centimeters.

3. Write an algebraic expression to represent the sum of 3 times a number and twice the sum of the number and 3.

**Guided Practice**

**Choose a variable and write an algebraic expression to represent each verbal expression.**

4. twice the sum of a number and 7

5. five times a number decreased by 4

6. the sum of twice a number and 7

7. three decreased by twice a number

8. twelve decreased by the square of a number

9. the product of the square of a number and six

10. one-fifth the sum of four and a number

11. four times the sum of eight and a number

12. eight times the sum of a number and its square

13. the sum of 8 and four times a number

14. the square of the sum of a number and 11

# EXERCISES

**Practice**

**Write an equation and solve each problem. Be sure to identify the variable.**

15. You have $32 to spend on supplies for your science fair project. If you buy two plants for experiments, you will have $18 left for other supplies. How much is each plant?

16. Ida and Ron make doughnuts for the Homestyle Bakery. The order for Tuesday is two trays of glazed doughnuts and one tray of cinnamon rolls. If each tray contains 3 dozen items, how many pastries do they need to prepare?

17. If you subtract 89 from a number, the result is 29. Find the number.

18. The dealer's asking price for a new car is $9750. Maria Blackmunn offered to pay $7800. Maria's price is what percent of the dealer's price?

19. Darian's dad is 28 years older than Darian. The sum of their ages is 64. How old is Darian?

20. The sum of two consecutive odd integers is 124. What are the integers?

21. Mrs. Gampp was 24 years old when Brittany was born. In three years the sum of their ages will be 68 years. How old is each now?

22. Julie's scores on four English tests were 78%, 98%, 67%, and 90%. What must she score on the fifth test so that her average will be 85%?

23. Don Owens bought a microwave oven for $60 more than half its original price. He paid $274 for the oven. What was the original price?

24. The Wilderness Club's treasurer, Eagan, bought some supplies for the fall rafting trip. He bought some raisins for 99¢ per pound and twice as many pounds of peanuts for $1.29 per pound. If the total bill was $24.99, how many pounds of peanuts did he buy?

25. Tickets to *A Midsummer Night's Dream* cost $10.50 for adults and $7.50 for students. Reba McGowen ordered $192 worth of tickets for a field trip for her English class. She ordered five times as many student tickets as adult tickets. How many of each did she order?

26. The Forest Park High School Drama Club sold 320 adult tickets and 153 student tickets for their last performance. Adult tickets were 75¢ more than student tickets. If total receipts were $949.50, what was the price of each ticket?

27. Michael Werthan is on his way home to Fort Wayne for a family reunion. It is a 360 mile trip. He drives 65 mph for 3 hours. When the speed limit changes, he slows down to 55 mph for the rest of the trip. How long does Michael drive at 55 mph?

**CONNECTION**

**Geometry**

28. The width of a rectangular window frame is 12 inches less than its length. If you add 30 inches to both the length and the width, you double the perimeter. Find the length and the width of the original rectangle.

**Critical Thinking**

29. Write a verbal expression to represent the algebraic expression: $2y(y + 4) + 2(y + 6)$.

**Applications**

**30. Entertainment** The Wheaton Theater sold 379 tickets on Tuesday for the movie *Home Alone 2*. On Wednesday, 532 tickets were sold. The total income for the two days was $5238.25. What was the price of each ticket? **$5.75**

**31. Photography** The perimeter of a rectangular photograph is 24 inches. The photographer wants to enlarge the photograph so that the perimeter is 42 inches. She will add twice as much to the length as to the width. How much should she add to the length and the width?

**Mixed Review**

State the property illustrated in each equation.

**32.** $6 + a = 6 + a$   (**Lesson 1-3**)

**33.** $9 + (2 + 10) = 9 + 12$   (**Lesson 1-3**)

**34.** Simplify the expression $3(2x + 2) - 2(x - 1)$.   (**Lesson 1-2**)

**35.** Find the value of the expression $\dfrac{3a + 4c}{b}$, if $a = -3$, $b = 2$, and $c = 0.5$.   (**Lesson 1-1**).

## MID-CHAPTER REVIEW

**Evaluate each expression.   (Lesson 1-1)**

**1.** $(6 + 3)5 - 6$

**2.** $4 + (3 + 6)9 - 2$

**Evaluate each expression if $a = 6$, $b = -2$, and $c = 0.5$.   (Lesson 1-1)**

**3.** $ab - cb + a$

**4.** $(ab)a + cb + abc$

**5.** Given the formula for simple interest, $I = prt$, find the value of $p$ when $I = \$23.45$, $r = 8\%$, and $t = 3$ years.

**Name the sets of numbers to which each number belongs.   (Lesson 1-2)**

**6.** $13\pi$

**7.** $12 \div 7$

**8.** $67$

**State the property illustrated in each equation.   (Lesson 1-2)**

**9.** $3(4 - 2) = 3(4) - 3(2)$

**10.** $(3 + 4) + 5 = 3 + (4 + 5)$

**Solve each equation.   (Lesson 1-3)**

**11.** $1.2x + 3.7 = 34.6$

**12.** $2x - (3x - 6) + 4x = 3x + 45$

**Write an algebraic expression to represent each verbal expression.   (Lesson 1-4)**

**13.** the sum of three times a number and nine

**14.** the square of the sum of a number and 2

**Write an equation and solve each problem. Be sure to identify the variable.   (Lesson 1-4)**

**15.** A number increased by 45 is 213. What is the number?

**16.** The sum of three times a number and 7 is 46. What is the number?

# Problem Solving Strategy: List the Possibilities

**Objective**

After studying this lesson, you should be able to:

- solve problems by making lists.

**Application**

The telephone area codes in the United States and Canada are three-digit numbers.

1st digit: 2, 3, 4, 5, 6, 7, 8, or 9
2nd digit: 0 or 1
3rd digit: any digit but 0

How many different area codes can start with the digit 3?

One way that we can find the answer is to list the possibilities. Let's make an organized list of possible area codes to answer this question.

The second digit of any code must be a 0 or 1. Assuming that the first digit is 3, list all of the possible codes with a second digit of 0.

| 301 | 302 | 303 | 304 | 305 | *Why is 300 not a possible* |
|-----|-----|-----|-----|-----|------|
| 306 | 307 | 308 | 309 | | *code?* |

List all of the possible codes with 1 as a second digit.

311   312   313   314   315   316   317   318   319

There are 18 possible area codes that begin with the digit 3.

**Example**

**Lee forgot the identification number for his automatic teller bank card. He remembered that he had rearranged the digits from his house number to program the code. If his house number is 1256, what codes should he try in the automatic teller machine?**

List the possible codes.

Possible codes starting with 1:                    Possible codes starting with 2:

  1265    1526                      2156    2165    2516
  1562    1625    1652   *Why isn't*  2561    2615    2651
                          *1256 listed?*

Possible codes starting with 5:                    Possible codes starting with 6:

  5126   5162   5216           6125   6152   6215
  5261   5612   5621           6251   6512   6521

These are the 23 possible codes that Lee should try in the automatic teller machine.

*FYI · · ·*

Most automatic teller machines will keep your card after the third incorrect entry.

# CHECKING FOR UNDERSTANDING

**Communicating Mathematics**

Read and study the lesson to answer these questions.

1. When is the strategy of listing possiblities useful?

2. Why is it helpful to make your list organized when using this strategy?

3. What is another way that we could have organized our list to answer the question in the example?

**Guided Practice**

Use the strategy of listing possibilities to answer these questions.

4. Your Algebra 2 quiz consists of five true-or-false questions. How many different patterns of answers are possible?

5. The Charlotte Hornets have won 4 and lost 2 games this season. List the patterns of records that are possible.

**Exercises**

Solve. Use any strategy.

### Strategies
Look for a pattern.
Solve a simpler problem.
Act it out.
Guess and check.
Draw a diagram.
Make a chart.
Work backwards.

6. Each letter in the addition problem at the right represents a digit. Find the digits that form this sum.

$$\begin{array}{r} TWO \\ + TWO \\ \hline FOUR \end{array}$$

7. Your salary is to be raised 10 percent and then a month later reduced by 10 percent. You may elect to have the cut first and then the raise. Which is the better choice? *Hint: Look at the total income for the year.*

8. The telephone number of a local business is 555-1829. They are trying to make a word from the last three digits of their number so that customers will remember it easily. The digit 8 can be T, U, or V; 2 can be A, B, or C; and 9 can be W, X, or Y. List the possible combinations of letters that their number can represent.

9. The students in a gym class are standing in a circle. When they count off, the students with numbers 5 and 23 are standing exactly opposite one another. Assuming the students are evenly spaced around the circle, how many students are there in the class?

10. Place operation symbols and any necessary grouping symbols in the sentence below to make it correct.

$$7 \quad 7 \quad 7 \quad 7 \quad 7 \quad 7 \quad 7 = 43$$

## COOPERATIVE LEARNING ACTIVITY

**Work in groups. Each person in the group must understand the solution and be able to explain it to any person in class.**

The chemistry teacher at Stevenson High School is ordering equipment for the laboratory. She wants to order sets of five weights totaling 121 grams for each lab station. Students will need to be able to weigh every integral weight from 1 to 121 grams using these weights and a two-pan balance. What weights should the teacher order? *Hint: Recall that weights can be placed on both sides of the balance.*

# 1-6 Solving Absolute Value Equations

**Objective**

After studying this lesson, you should be able to:
- solve equations containing absolute value.

**Application**

Suppose you and your friend, Terry, each live on the same street as your school, but on opposite sides of the school. You each live 5 miles from the school. What can you say about your trips to school each day? Do you each travel the same direction? Do you travel the same distance?

Consider placing both houses and the school on a number line with the school at the origin. Your house is located at 5, and Terry's is located at −5.

Certainly −5 and 5 are quite different, but they do have something in common. They are the same distance from 0 on the number line. This means that you and Terry travel the same distance, but in different directions, when you go to school.

We say that −5 and 5 have the same **absolute value.** The absolute value of a number is the number of units it is from 0 on the number line. We use the symbol $|x|$ to represent the absolute value of a number $x$.

The absolute value of −5 is 5.     The absolute value of 5 is 5.
$$|-5| = 5 \qquad\qquad |5| = 5$$

We can also define absolute value in the following way.

| | |
|---|---|
| *Absolute Value* | **For any real number $a$:**<br>**If $a \geq 0$, then $|a| = a$.**<br>**If $a < 0$, then $|a| = -a$.** |

*The symbol ≥ means "is greater than or equal to."*

**Example 1**

**Find the absolute value of 8 and the absolute value of −12.**

$$|8| = 8 \qquad |-12| = -(-12) \text{ or } 12$$

**Example 2**

**Find the absolute value of $x - 9$.**

Let's make a list of the possible cases.
If $x$ is 9 or greater, then $x - 9 \geq 0$. So, $|x - 9| = x - 9$.
If $x$ is less than 9, then $x - 9 < 0$. So, $|x - 9| = -(x - 9)$ or $9 - x$.

**Example 3**

Evaluate $|2x - 4| + 1.2$ if $x = -3$.

$$
\begin{aligned}
|2x - 4| + 1.2 &= |2(-3) - 4| + 1.2 \\
&= |-6 - 4| + 1.2 \\
&= |-10| + 1.2 \\
&= 10 + 1.2 \\
&= 11.2 \qquad \text{The value is 11.2.}
\end{aligned}
$$

Some equations contain absolute value expressions. The definition of absolute value is used in solving the equations. When more than one solution occurs, they are often written as a set, $\{a, b\}$.

**Example 4**

Solve $|x - 7| = 12$. Check each solution.

$$|x - 7| = 12$$

*If $x - 7$ is positive or zero* $\rightarrow$ $x - 7 = 12$   or   $x - 7 = -12$ $\leftarrow$ *If $x - 7$ is negative*

$$x = 19 \quad \text{or} \quad x = -5$$

**Check:**
$$
\begin{aligned}
|x - 7| &= 12 \\
|19 - 7| &\stackrel{?}{=} 12 \quad \text{or} \quad |-5 - 7| \stackrel{?}{=} 12 \\
|12| &\stackrel{?}{=} 12 \qquad\qquad\; |-12| \stackrel{?}{=} 12 \\
12 &= 12 \;\checkmark \qquad\qquad\; 12 = 12 \;\checkmark
\end{aligned}
$$

The solutions are 19 and $-5$. The solution set is $\{19, -5\}$.

**Example 5**

*FYI* · · ·

Each American eats an average of 62 pounds of sugar in one year.

A machine that fills boxes of sugar is to fill the boxes with 16 ounces of sugar. After the boxes are filled, another machine weighs the boxes. If the box is more than 0.2 ounces above or below the desired weight, the box is rejected. What is the weight of the heaviest and the lightest box that the machine will let pass?

Let $w$ = the weight of the box. $|w - 16| = 0.2$

$$
\begin{aligned}
w - 16 &= 0.2 \quad \text{or} \quad w - 16 = -0.2 \\
w &= 16.2 \quad \text{or} \qquad\quad w = 15.8
\end{aligned}
$$

The heaviest box allowed to pass is 16.2 ounces. The lightest box allowed to pass is 15.8 ounces.

An absolute value equation may have no solution. For example, $|x| = -4$ is never true. Since the absolute value of a number is always positive or zero, there is no replacement for $x$ that will make that sentence true. The solution set has no members. It is called the **empty set** and is symbolized by $\{\}$ or $\varnothing$.

**Example 6**

Solve $|3x + 7| + 4 = 0$.

$$|3x + 7| + 4 = 0$$

$$|3x + 7| = -4 \qquad \text{This sentence is } never \text{ true, so the equation has } no \text{ solution. The solution set is } \varnothing.$$

It is important to check your answers when solving absolute value equations. Even if the correct procedure for solving the equations is used, the answers may not be actual solutions to the original equation.

**Example 7**

Solve $|x - 8| = 3x - 4$.

$$|x - 8| = 3x - 4$$

$$
\begin{array}{lll}
x - 8 = 3x - 4 & \text{or} & x - 8 = -(3x - 4) \\
\phantom{x} -8 = 2x - 4 & & x - 8 = -3x + 4 \\
\phantom{x} -4 = 2x & & 4x = 12 \\
\phantom{x} -2 = x & & x = 3
\end{array}
$$

Check: $\qquad\qquad\qquad |x - 8| = 3x - 4$

$$
\begin{array}{lll}
|(-2) - 8| \stackrel{?}{=} 3(-2) - 4 & \text{or} & |(3) - 8| \stackrel{?}{=} 3(3) - 4 \\
|-2 - 8| \stackrel{?}{=} -6 - 4 & & |3 - 8| \stackrel{?}{=} 9 - 4 \\
10 = -10 \quad no & & 5 = 5 \checkmark
\end{array}
$$

The only solution is 3.

**Example 8**

Use a calculator to evaluate $9|2y - 4.25|$ when $y = 1.5$.

First evaluate $2y - 4.25$.

ENTER:  2 $\boxed{\times}$ 1.5 $\boxed{-}$ 4.25 $\boxed{=}$ $-1.25$

Since an absolute value cannot be negative, change the sign of the number in the display before multiplying by 9.

ENTER:  $\boxed{+/-}$ $\boxed{\times}$ 9 $\boxed{=}$ $11.25$    The value is 11.25.

# CHECKING FOR UNDERSTANDING

**Communicating Mathematics**

Read and study the lesson to answer these questions.

1. Explain why the equation $|2x - 3| + 4 = 2$ has no solution.

2. Explain why there is no solution for the equation in Example 6.

3. The second part of the definition of absolute value is "If $a < 0$, then $|a| = -a$." To some people $-a$ may appear to represent a negative number. How would you explain that it always represents a positive number?

4. Is $-x < x$ always true, sometimes true, or never true? Explain your answer.

## Guided Practice

Evaluate each expression if $x = -4$.

**5.** $|4x|$

**6.** $|-2x|$

**7.** $-|3x - 4|$

**8.** $|-2x - 5|$

**9.** $7 - |3x + 10|$

**10.** $2|x + 4| + |2x|$

Determine which numbers in the set $\{-2, -1, 0, 1, 2\}$ are solutions for each equation.

**11.** $|-x| = 2$

**12.** $|x - 2| = 1$

**13.** $|x| = -x$

**14.** $|x| = x$

**15.** $-x = |x - 2|$

**16.** $|x| = |x - 4|$

# EXERCISES

## Practice

Solve each equation.

**17.** $|x + 6| = 19$

**18.** $|x + 11| = 42$

**19.** $|x - 4| = 11$

**20.** $|x - 3| = 17$

**21.** $3|x + 6| = 36$

**22.** $8|x - 3| = 88$

**23.** $5|x + 4| = 45$

**24.** $11|x - 9| = 121$

**25.** $|2x + 9| = 30$

**26.** $|2x - 37| = 15$

**27.** $|2x + 7| = 0$

**28.** $|4x - 3| = -27$

**29.** $3|3x + 2| = 51$

**30.** $8|4x - 3| = 64$

**31.** $-6|2x - 14| = -42$

**32.** $4|6x - 1| = 29$

**33.** $7|3x + 5| = 35$

**34.** $|2a + 7| = a - 4$

**35.** $|x - 3| = 2x$

**36.** $3|x + 6| = 9x - 6$

**37.** $|7 + 3a| = 11 - a$

**38.** $|3t - 5| = 2t$

**39.** $|x - 3| + 7 = 2$

**40.** $5|3x - 4| = x + 1$

## Critical Thinking

**41.** Solve $|x + 2| = |2x - 4|$ and explain your method of solution.

## Applications

**42. Cartography** Columbus is between Cincinnati and Cleveland, and all three cities are located on Interstate 71. Cleveland is 140 miles from Columbus, and Cincinnati is 108 miles from Columbus. Draw a diagram of the situation and find the distance from Cleveland to Cincinnati.

**43. Manufacturing** Mary Lou's Fudge Factory sells fudge in one pound tins. Each tin is weighed before it is packaged for shipping. If a tin weighs 0.05 pounds more or less than one pound, it is not shipped. What are the largest and smallest amounts of fudge that are allowable?

**44.** This program finds approximate solutions of the equation $|x^2 - 2| = 0$. Two values for $x$ are entered, and two values for $|x^2 - 2|$ are printed as $y$-values. When one $y$-value is negative and the other is positive, a solution exists between the two $x$-values. Next, enter $x$-values between the previous two and check again for $y$-values with different signs.

```
10  DEFFNF(X) = ABS(X^2-2)
15  PRINT "ENTER 2 VALUES."
20  INPUT X1, X2
30  PRINT "X", "Y"
40  PRINT X1, FNF(X1)
50  PRINT X2, FNF(X2)
60  PRINT "TRY AGAIN?(Y/N)";
70  GET A$
80  IF A$ = "Y" THEN 15
90  END
```

**Journal**

Tell how making a drawing might help you write the equation to solve a problem.

Continue this process until a $y$-value is obtained that is as close to zero as desired. The corresponding $x$-value is the approximate solution. For $|x^2 - 2| = 0$, the approximate solutions are 1.4 and $-1.4$ to the nearest tenth. Use the program above to approximate to the nearest tenth the number of solutions indicated for each equation. You will need to change line 10 of the program for each equation. All solutions are between $-10$ and 10.

**a.** $x^2 - 2 - 4 = 0$; 2 solutions

**b.** $x^3 - 3x = 0$; 3 solutions

**c.** $|3x - 2| - 4 = 0$; 2 solutions

**d.** $5x^3 + 3x^2 - 25x - 15 = 0$; 3 solutions

**Mixed Review**

**45.** Lisa forgot the 3-digit address of her aunt's house, but remembered that the digits are 1, 3, and 9. What are all of the possible addresses? **(Lesson 1–5)**

**46.** What are all the possible arrangements of a class schedule containing English, Algebra 2, and American History? **(Lesson 1-5)**

**47.** Write an algebraic expression to represent the sum of a number and its square. **(Lesson 1-4)**

**48.** Evaluate $\sqrt{16} + \sqrt{9}$ and tell the sets of numbers to which the solution belongs. **(Lesson 1-2)**

## HISTORY CONNECTION

**Mileva Einstein**

In 1905, Albert Einstein published three papers that changed the world of science forever. His theory of relativity, quantum theory of light, and proof of the existence of the atom made him the symbol of genius. However, in recent years it has come to light that Einstein may have had some help with these discoveries . . . from his wife. Mileva Maric Einstein was a brilliant physicist, who attended the Swiss Federal Institute of Technology. In a biography of Mrs. Einstein, Abram Joffe claims to have seen the original manuscript for a paper on relativity that was signed Einstein-Maric. Maybe Albert Einstein wasn't a "solitary" genius after all.

## 1-7 Solving Inequalities

**Objectives**

After studying this lesson, you should be able to:
- solve an inequality and graph the solution set, and
- use inequalities to solve problems.

**Application**

José and Kyle are soccer players on the Union High School team. If you compare their scoring for the season, only one of the following statements will be true.

José scored fewer goals than Kyle.
José scored the same number of goals as Kyle.
José scored more goals than Kyle.

Let $j$ represent the number of goals José scored and $k$ represent the number of goals Kyle scored. You can compare the scoring using an inequality or an equation.

$$j < k \qquad\qquad j = k \qquad\qquad j > k$$

This is an illustration of the trichotomy property.

| *Trichotomy Property* | **For any two real numbers, $a$ and $b$, exactly one of the following statements is true.** <br> $\qquad\qquad a < b \qquad\quad a = b \qquad a > b$ |
|---|---|

Adding the same number to each side of an inequality does not change the truth of the inequality.

| *Addition and Subtraction Properties for Inequalities* | **For any real numbers, $a$, $b$, and $c$:** <br> 1. If $a > b$, then $a + c > b + c$ and $a - c > b - c$. <br> 2. If $a < b$, then $a + c < b + c$ and $a - c < b - c$. |
|---|---|

These properties can be used to solve an inequality. The solution set of an inequality can be graphed on a number line.

**Example 1**

**Solve $8x + 5 < 7x - 3$. Graph the solution set.**

$$8x + 5 < 7x - 3$$
$$-7x + 8x + 5 < -7x + 7x - 3 \qquad \textit{Add -7x to each side.}$$
$$x + 5 < -3$$
$$x + 5 + (-5) < -3 + (-5) \qquad \textit{Add -5 to each side.}$$
$$x < -8$$

*A circle means this point is not included.*

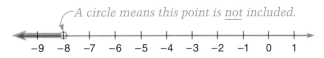

Any real number less than $-8$ is a solution.

To check, substitute $-8$ for $x$ in $8x + 5 < 7x - 3$. The two sides should be equal. Then substitute a number less than $-8$. The inequality should be true.

You know that $15 > -6$ is a true inequality. What happens if you multiply the numbers on each side by a positive number or a negative number? Is it still true?

*Try 7.*

$$15 > -6$$

$$7(15) \overset{?}{>} 7(-6)$$

$$105 > -42 \quad \textit{True}$$

Multiply the inequality by other positive numbers. Do you think that the inequality will always remain true?

*Try $-\dfrac{1}{3}$.*

$$15 > -6$$

$$-\frac{1}{3}(15) \overset{?}{>} -\frac{1}{3}(-6)$$

$$-5 > 2 \quad \textit{False}$$

If you reverse the inequality, the statement is true.

$$-5 < 2 \quad \textit{True}$$

Try other negative numbers as multipliers.

This suggests that when you multiply each side of an inequality by a negative number, the order of the inequality must be reversed.

These examples suggest the following properties.

*Multiplication and Division Properties for Inequalities*

**For any real numbers $a$, $b$, and $c$:**

1. **If $c$ is positive and $a < b$, then $ac < bc$ and $\dfrac{a}{c} < \dfrac{b}{c}$.**

2. **If $c$ is positive and $a > b$, then $ac > bc$ and $\dfrac{a}{c} > \dfrac{b}{c}$.**

3. **If $c$ is negative and $a < b$, then $ac > bc$ and $\dfrac{a}{c} > \dfrac{b}{c}$.**

4. **If $c$ is negative and $a > b$, then $ac < bc$ and $\dfrac{a}{c} < \dfrac{b}{c}$.**

Examples 2 and 3 show how to use these properties when solving inequalities.

**Example 2**

**Solve -0.5y < 6. Graph the solution set.**

$$-0.5y < 6$$
$$(-2)(-0.5y) > (-2)(6)$$ *Reverse the inequality sign because*
$$y > -12$$ *each side is multiplied by a negative.*

Any real number greater than $-12$ is a solution.

The solutions in Example 2 can be written using set-builder notation. This solution set can be written as $\{y|y > -12\}$. This is read as *the set of all numbers y such that y is greater than $-12$.*

**Example 3**

**Solve $-x \geq \dfrac{x + 4}{7}$. Graph the solution set.**

$$-x \geq \frac{x + 4}{7}$$

$$-7x \geq x + 4 \quad \text{\textit{Multiply each side by 7.}}$$

$$-8x \geq 4 \quad \text{\textit{Add -x to each side.}}$$

$$x \leq -\frac{1}{2} \quad \text{\textit{Divide each side by -8, reversing the inequality sign.}}$$

*A dot means this point is included.*

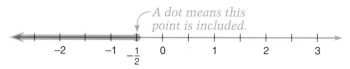

The solution set is $\left\{x|x \leq -\dfrac{1}{2}\right\}$.

Inequalities can be used to solve many verbal problems. You solve problems with inequalities the same way that you solve problems with equations.

**Example 4**

**Judy Kildow received a $10,000 inheritance that she wishes to invest. She wants to earn at least $780 in interest this year so she can buy a stereo system with her earnings. She will invest some of the money in bonds that earn about 6% interest annually and the rest in stock that she expects to earn 9% interest annually. What is the minimum she should invest in the stock?** *The phrase at least 780 means greater than or equal to 780.*

*EXPLORE*   Let $n$ = the amount invested in stocks.
Then $10,000 - n$ = amount invested in bonds.

*PLAN*   (rate)(amount) + (rate)(amount)   $\geq$ minimum desired
$(0.09)(n)$        $(0.06)(10,000 - n) \geq$   780

*SOLVE*   $0.09n + 600 - 0.06n \geq 780$
$0.03n \geq 180$
$n \geq 6000$

Judy must invest at least $6000 in stock.

## CHECKING FOR UNDERSTANDING

**Communicating Mathematics**

**Read and study the lesson to answer these questions.**

1. Explain in your own words what you must do if you are going to multiply each side of an inequality by a negative number.

2. Translate *twice the sum of a number and 3 is less than 6* into algebraic symbols.

3. Translate *eighteen is at least 3 three times the product of 5 and a number* into algebraic symbols.

4. The graph below is a graph of the solution set of an inequality. What is the solution set?

**Guided Practice**

**Graph the solution set of each inequality.**

5. $x \le 0$    6. $x > -3$    7. $x > 4.5$

8. $0.75x < 4$    9. $-3x > 6$    10. $x < -7.5$

## EXERCISES

**Practice**

**Solve each inequality. Graph the solution set.**

11. $6x + 4 \ge 34$    12. $8 - 3x < 44$

13. $6s - 7 < 29$    14. $5r + 8 > 24$

15. $15 - 5t \ge 55$    16. $x - 5 < 0.1$

17. $11 - 5y < -77$    18. $28 - 6y < 23$

**Solve each inequality.**

19. $3(4x + 7) < 21$    20. $5(2x - 7) > 10$

21. $5(3z - 3) \le 60$    22. $-49 > 7(2x + 3)$

23. $7x - 5 > 3x + 4$    24. $40 \le -6(5r - 7)$

25. $7 - 2x \ge 0$    26. $2(r - 4) + 5 \ge 9$

27. $2(3m + 4) - 2 \le 3(1 + 3m)$    28. $2(m - 5) - 3(2m - 5) < 5m + 1$

29. $7 + 3y > 2(y + 3) - 2(-1 - y)$    30. $3b - 2(b - 5) < 2(b + 4)$

31. $0.01x - 4.23 \ge 0$    32. $0.75x - 0.5 < 0$

**Solve each inequality.**

**33.** $2.55x - 4.25 \leq 0$

**34.** $\dfrac{4x + 2}{5} \geq -0.04$

**35.** $\dfrac{2x + 3}{5} \leq 0.03$

**36.** $20\left(\dfrac{1}{5} - \dfrac{w}{4}\right) \geq -2w$

**37.** $\dfrac{3x - 3}{5} < \dfrac{6(x - 1)}{10}$

**38.** $\dfrac{x + 8}{4} - 1 > \dfrac{x}{3}$

**Solve each problem.**

**39.** Rhonda has $110.37 in her checking account. The bank does not charge for checks if $50 or more is in the account. What is the greatest amount for which she can write a check and not be charged?

**40.** The Oklahoma City Municipal Parking Garage charges $1.50 for the first hour and $0.50 for each additional hour or part of an hour. For how many hours can you park your car if you only have $4.50 in cash?

**41.** The Indiana Pacers play 84 games this season. At midseason, they have won 30 games. If they must win at least 60% of all of their games to play in the tournament, how many more games must they win?

**42.** Karl Weekley invested part of $8000 in stock that lost 2%. He invested the rest at 8% annually. If his gain for the year was at least $400, what was the most he could have invested in stock?

**43.** You are enrolled in an algebra course where five tests will be given in the first quarter. You need 450 points to get an A. Your scores on the first four tests were 89, 87, 95, and 98. What is the minimum score that you can earn on the last test and still get an A for the quarter?

**44.** Reliable Rentables rents a car for $12.95 per day plus 15¢ per mile. Your company has limited you to $90 a day for car rental. What is the maximum number of miles that you can drive each day?

**45.** Jerome and Nicola Rugola left an estate that is estimated to be worth at most $300,000. Their will stated that one-fourth of the estate be given to their church and the remainder be divided equally among their four children. What are the maximum amounts to be paid to the church and to each child?

**Critical Thinking**

**46.** Find the set of all numbers that satisfies $3x - 2 \geq 0$ and $5x - 1 \leq 0$.

**Applications**

**47. Manufacturing**   A company manufactures auto parts. The diameter of a piston cannot vary more than 0.001 cm. Write an inequality to represent the diameter of a piston if its diameter is supposed to be 10 cm.

**48. Manufacturing**   A factory can make a table in 30 minutes and a chair in 12 minutes. They produce dining sets with a table and 4 chairs. What is the maximum number of sets that can be produced in an 8-hour shift?

**Mixed Review**

**49.** Evaluate $|7(-3) + 10|$.   **(Lesson 1-6)**

**50.** Solve $3|2x - 5| = -1$.   **(Lesson 1-6)**

**51.** Solve $3 - 2x = 18$.   **(Lesson 1-3)**

**52.** Simplify $7x + 8y + 9y - x(7 - 2)$.   **(Lesson 1-2)**

**53.** Simplify $\sqrt{9} \div \sqrt{4}$.   **(Lesson 1-1)**

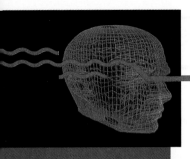

# Technology

## Solving Equations and Inequalities

The *Mathematical Exploration Toolkit (MET)*, produced by IBM, can help you review the steps for solving equations and inequalities. Since the computer will perform all calculations and simplifying steps, you can concentrate on deciding which steps to choose. You will use these CALC commands. The abbreviations in the second column can also be used instead of typing the entire command.

| ADD | add | MULTIPLY | mul |
|---|---|---|---|
| SUBTRACT | sub | DIVIDE | div |
| SIMPLIFY | simp | | |

**Example 1**

Solve $6x - 5 = 3x + 7$.

ENTER:
$6x - 5 = 3x + 7$
sub $3x$
simp
add 5
simp

div 3

simp

$6x - 5 = 3x + 7$
$6x - 5 - 3x = 3x + 7 - 3x$
$3x - 5 = 7$
$5 + (3x - 5) = 7 + 5$
$3x = 12$
$\dfrac{3x}{3} = \dfrac{12}{3}$
$x = 4$

**Example 2**

Solve $x(x - 1) > 3x - (4x - 5)$.

ENTER:
$2(x - 1) > 3x - (4x - 5)$
simp
add $x$
simp
add 2
simp

mult 1/3

simp

$2(x - 1) > 3x - (4x - 5)$
$2x - 2 > -x + 5$
$x + (2x - 2) > x + (-x + 5)$
$3x - 2 > 5$
$2 + (3x - 2) > 2 + 5$
$3x > 7$
$\dfrac{1}{3} \cdot (3x) > \dfrac{1}{3} \cdot (7)$
$x > \dfrac{7}{3}$

# EXERCISES

**Use CALC to solve each equation or inequality. Record your steps and solution.**

1. $2 - 5x = -8x + 4$     2. $\dfrac{x}{4} = \dfrac{x}{6} + \dfrac{1}{3}$      3. $\dfrac{2}{5}x + 1 < -7$

4. $2x - 3(1 - 6x) = 3(2x + 1) - 8$

5. $1 - [2 - (3x - 1)] > 4x - 2(6 + 2x)$

## 1-8 Solving Compound Sentences and Absolute Value Inequalities

Objectives

After studying this lesson, you should be able to:

■ solve compound sentences using *and* and *or*, and
■ solve an inequality involving absolute value and graph its solution.

Application

April calculated her state income tax using the table at the left. Her tax is $29.00. According to the table, this means her taxable income is at least $3925, but less than $3950.

| If Ohio taxable income (Line 5) is: | | |
|---|---|---|
| At least: | But less than: | The tax is: |
| $ 3,800 | $ 3,825 | $ 28 |
| 3,825 | 3,850 | 29 |
| 3,850 | 3,875 | 29 |
| 3,875 | 3,900 | 29 |
| 3,900 | 3,925 | 29 |
| 3,925 | 3,950 | 29 |
| 3,950 | 3,975 | 29 |
| 3,975 | 4,000 | 30 |
| $ 4,000 | | |
| 4,000 | 4,025 | 30 |

Let *I* stand for April's taxable income. The two inequalities, $I \geq 3925$ and $I < 3950$, describe her taxable income. A sentence like this is called a **compound sentence.** A compound sentence containing *and* is true only if both parts of it are true.

Another way of writing $I \geq 3925$ and $I < 3950$ is $3925 \leq I < 3950$. The sentence is read "*I* is greater than or equal to 3925 and is less than 3950." To solve a compound sentence, you must solve each part of the sentence.

Example 1

**Solve $7 < 4x + 3 < 19$.**

Write the compound sentence using the word *and*. Then solve each part.

$$7 < 4x + 3 \quad \text{and} \quad 4x + 3 < 19$$
$$4 < 4x \quad \text{and} \quad 4x < 16 \qquad \text{The solution set is}$$
$$1 < x \quad \text{and} \quad x < 4 \qquad \{x | 1 < x < 4\}.$$

Another way to solve this inequality is to solve both parts at the same time by adding −3 to each part of the sentence and then dividing each part by 4.

$$7 < 4x + 3 < 19$$
$$4 < 4x \quad < 16$$
$$1 < x \quad < 4$$

A compound sentence containing *or* is true if at least one part of it is true.

Example 2

**Solve $3x + 1 < 7$ or $7 < 2x − 9$.**

Solve each part separately.

$$3x + 1 < 7 \quad \text{or} \quad 7 < 2x − 9$$
$$3x < 6 \quad \text{or} \quad 16 < 2x$$
$$x < 2 \quad \text{or} \quad 8 < x \qquad \text{The solution set is } \{x | x < 2 \text{ or } x > 8\}.$$

*There is no short way to write a sentence containing "or."*

The absolute value of a number is its distance from 0 on the number line. We can use this idea to solve inequalities involving absolute value.

**Example 3**

**Solve |x| < 6.**

$|x| < 6$ means that the distance between $x$ and 0 is less than 6 units. To make $|x| < 6$ true, you must substitute values for $x$ that are less than 6 units from 0.

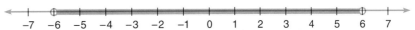

All of the numbers between −6 and 6 are less than 6 units from zero. The solution set is $\{x|-6 < x < 6\}$.

**Example 4**

**Solve |2x − 7| ≥ 11. Graph the solution set.**

This inequality says that $2x - 7$ is more than or equal to 11 units from 0.

$$
\begin{array}{lll}
2x - 7 \geq 11 & \text{or} & 2x - 7 \leq -11 \\
2x \geq 18 & \text{or} & 2x \leq -4 \\
x \geq 9 & \text{or} & x \leq -2
\end{array}
$$

The solution set is $\{x|x \leq -2 \text{ or } x \geq 9\}$.

**Example 5**

**APPLICATION**

**Broadcasting**

**The radio station in Winchester sends out signals for a radius of 100 miles. The Theesville radio station sends out signals for a radius of 60 miles. You are driving from Winchester to Theesville, a distance of 200 miles. Write an inequality that represents your location while you can listen to one of the radio stations on your car radio.**

*EXPLORE*   Let $d$ = the distance you have traveled from Winchester.

*PLAN*   The station from Winchester can be heard up to 100 miles away, so you can listen to that station for the first 100 miles of the trip. The station in Theesville can be heard up to 60 miles away, so you can listen to it when you get to within 60 miles of Theesville. Write a compound inequality that describes the situation.   *Remember that the distance, d, is measured from Winchester.*

*SOLVE*

| Listening to Winchester station | | Listening to Theesville station |
|---|---|---|
| $d \leq 100$ | or | $d \geq (200 - 60)$ |
| | | $d \geq 140$ |

The solution set is $\{d \leq 100 \text{ or } d \geq 140\}$.

*EXAMINE*   You draw a diagram of the situation to check this solution.

Some absolute value inequalities have no solution. For example, $|4x - 3| < -6$ is never true. Since the absolute value of a number is never negative, there is no replacement for $x$ that will make this sentence true. So, the solution set to this inequality is the empty set.

Other absolute value inequalities are always true. One such inequality is $|x + 5| > -10$. The solution set of this inequality is all real numbers. Can you see why? *Think of the definition of absolute value.*

# CHECK FOR UNDERSTANDING

**Communicating Mathematics**

**Read and study the lesson to answer these questions.**

1. Explain why $|x + 2| \geq -4$ has all real numbers as its solution set.
2. Explain why $|x - 5| < 0$ has no solution.
3. Explain the meaning of a compound sentence containing the word *and*.
4. Explain the meaning of a compound sentence containing the word *or*.

**Guided Practice**

**State an absolute value inequality for each of the following. Then graph each inequality.**

5. all numbers less than 7 and greater than $-7$
6. all numbers between $-3$ and 3
7. all numbers greater than 11 or less than $-11$
8. all numbers less than or equal to 5, and greater than or equal to $-5$

**State an absolute value inequality for each of the following graphs.**

9. 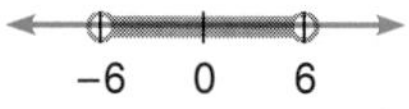
$-6 \quad 0 \quad 6$

10. 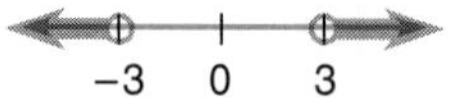
$-3 \quad 0 \quad 3$

11. 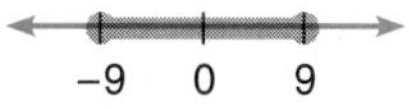
$-9 \quad 0 \quad 9$

12. 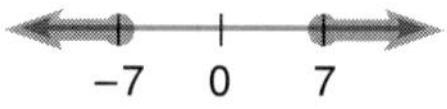
$-7 \quad 0 \quad 7$

13. 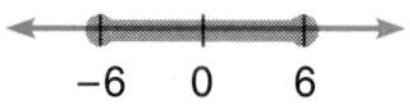
$-6 \quad 0 \quad 6$

14. 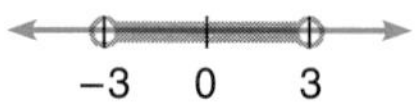
$-3 \quad 0 \quad 3$

# EXERCISES

**Practice**

**Solve each inequality. Graph each solution set.**

15. $|x + 2| > 3$
16. $2 < x + 4 < 11$
17. $|x| < 9$
18. $-2 \leq x - 10 \leq 6$
19. $|2x| < 6$
20. $1 < x - 2 < 7$
21. $|7x| \geq 21$
22. $|x - 9| > 5$
23. $|x| > 5$
24. $x - 4 < 1$ or $x + 2 > 1$
25. $|2x| \geq 28$
26. $|2x| \geq -64$
27. $|5x| < -25$
28. $|x + 3| > 17$
29. $4 < 2x - 2 < 10$
30. $|x - 6| \leq -12$
31. $|x - 15| < 45$
32. $x + 6 \geq -1$ or $x - 2 \leq 4$

33. $|2x - 9| \leq 27$

34. $|3x + 12| > 42$

35. $-1 < 3x + 2 < 14$

36. $5x < 9 + 2x$ or $9 - 2x > 11$

37. $|5x - 7| < 81$

38. $|3x + 11| > 1$

39. $|2x - 5| \leq 9$

40. $-4 \leq 4x + 24 \leq 4$

41. $4 + |2x| > 0$

42. $|3x| + 3 \leq 0$

43. $|x| \leq x$

44. $|x| > x$

45. $2x - 1 < -5$ or $3x + 2 \geq 5$

46. $|x + 2| - x \geq 0$

**Critical Thinking**

47. Solve $|x + 1| + |x - 1| \leq 2$.

**Applications**

48. **Transportation**  On some of the interstate highways, the maximum speed a car may drive is 65 miles per hour. A semi-truck may not drive more than 55 miles per hour. The minimum speed for all vehicles is 45 miles per hour.

 a. Write an inequality to represent the allowable speed for a car on an interstate highway.

 b. Write an inequality to represent the speed at which a semi-truck may travel on an interstate highway.

Portfolio

A portfolio is representative samples of your work, collected over a period of time. Begin your portfolio by selecting an item that shows something new you learned in this chapter.

49. **Manufacturing**  A manufacturer of boat motors has specifications for parts with given tolerance limits. If a part is to be 3.2 inches wide with a tolerance of 0.01 inches, this means that it must be at least 3.19 inches wide or at most 3.21 inches wide. This tolerance limit can be expressed by the absolute value inequality, $|w - 3.2| \leq 0.01$, where $w$ represents the width of the part.

 a. Find the maximum and minimum acceptable dimensions of a part that is supposed to be 7.32 centimeters long with a tolerance of 0.002 centimeter.

 b. Find the tolerance if a part must satisfy the inequality $5.18 \leq w \leq 5.24$.

**Mixed Review**

50. If you were to compare your height to the height of your best friend, you could make one of three comparative statements. What are those statements?  (**Lesson 1-7**)

51. One number is twice another. Twice the lesser number increased by the greater number is at least 85. Find the least possible value for the lesser number.  (**Lesson 1-7**)

52. Use a calculator to evaluate $48|7k - 30|$ if $k = 14$.  (**Lesson 1-6**)

53. **Transportation**  San Francisco and Los Angeles are 470 miles apart by train. An express train leaves Los Angeles for San Francisco at 5:00 P.M., the same time a passenger train leaves San Francisco for Los Angeles. The express train travels 10 miles per hour faster than the passenger train. The two trains pass each other at 7:30 P.M. How fast is each train traveling?  (**Lesson 1-4**)

54. What property is illustrated by $(11a + 3b) + 0 = (11a + 3b)$?  (**Lesson 1-2**)

55. **Banking**  Find the annual simple interest rate on an account containing $5000 if the interest accumulated in 3 years is $1200.  (**Lesson 1-1**)

# SUMMARY AND REVIEW

## VOCABULARY

Upon completing this chapter, you should be
familiar with the following terms:

| | | | |
|---|---|---|---|
| absolute value | **31** | **13** | irrational numbers |
| absolute value inequality | **42** | **14** | multiplicative identity |
| additive identity | **14** | **14** | multiplicative inverse |
| additive inverse | **14** | **18** | open sentence |
| algebraic expressions | **9** | **8** | order of operations |
| associative property | **14** | **13** | rational numbers |
| commutative property | **14** | **13** | real numbers |
| compound sentence | **42** | **14** | reciprocal |
| empty set | **32** | **18** | solution |
| integers | **13** | | |

## SKILLS AND CONCEPTS

### OBJECTIVES AND EXAMPLES

Upon completing this chapter, you should
be able to:

■ use the order of operations to evaluate
expressions  (**Lesson 1-1**)

$$12 \div 3 + 3 \cdot 2^2$$
$$= 12 \div 3 + 3 \cdot 4$$
$$= 4 + 12$$
$$= 16$$

■ use formulas  (**Lesson 1-1**)

The distance, $d$, in feet that an object
falls in $t$ seconds is found by the formula
$d = 16t^2$. How far will an object fall in
10 seconds?

$$d = 16t^2$$
$$d = 16(10)^2 \quad \text{or} \quad 1600$$

The object will fall 1600 feet.

### REVIEW EXERCISES

Use these exercises to review and prepare for
the chapter test.

**Find the value of each expression.**

1. $(3 + 7)^2 - 16 \div 2$    2. $3 + 7^2 - 16 \div 2$

**Evaluate each expression if $a = -0.5$, $b = 4$,
$c = 5$, and $d = -3$.**

3. $\dfrac{4a + 3c}{3b}$         4. $\dfrac{3ab^2 - d^3}{a}$

5. **Physics**  Sandbags are used to weigh
down a hot air balloon. If a sandbag is
dropped from a balloon, use the formula
$d = 16t^2$ to determine how far it will
fall in 20 seconds.

■ classify numbers  (**Lesson 1-2**)

Name the sets of numbers to which $\sqrt{2}$ belongs.

irrational numbers, real numbers

**Name the sets of numbers to which each number belongs. (Use N, W, Z, Q, I, and R.)**

**6.** $-32.4$

**7.** $73 \div 5$

**8.** $-8.0$

**9.** $34\pi$

---

■ use the properties of real numbers to simplify expressions  (**Lesson 1-2**)

$0.2a(3.5 - 3.2b) - 4a(3.6 + 2b)$
$= 0.7a - 0.64ab - 14.4a - 8ab$
$= -13.7a - 8.64ab$

**State the property illustrated in each equation.**

**10.** $4 + (a + r) = (4 + a) + r$

**11.** $(j + k) + 0 = (j + k)$

**Simplify each expression.**

**12.** $(8 + 49)7 + 3$

**13.** $7a + 2b - 5a - 6b$

**14.** $3(a + 4b) - 2(4a + 2b)$

---

■ solve equations using the properties of equality  (**Lesson 1-3**)

$2(a - 1) = 8a - 6$
$2a - 2 = 8a - 6$
$-6a = -4$
$a = \dfrac{2}{3}$

**Solve each equation.**

**15.** $15x + 25 = 2(x - 4)$

**16.** $2(3x - 1) = 3(x + 2)$

**17.** $6 = \dfrac{3x - 6}{3}$

**18.** $\dfrac{3a + 3}{4} = \dfrac{5}{2}$

---

■ use equations to solve problems  (**Lesson 1-4**)

**Problem-Solving Plan**

1. Explore the problem.
2. Plan the solution.
3. Solve the problem.
4. Examine the solution.

**Write an equation and solve each problem. Be sure to identify the variable.**

**19. Geometry**  The width of a rectangular rug is four feet more than one-third of its length. The perimeter is 64 feet. What are the length and the width?

---

■ solve equations containing absolute value  (**Lesson 1-6**)

$|q - 3| = 2$
$q - 3 = 2$  or  $q - 3 = -2$
$q = 5$ $\qquad$ $q = 1$

**Solve each equation.**

**20.** $|2x - 36| = 14$

**21.** $|q - 3| + 7 = 2$

**22.** $8\,|2b - 3| = 64$

■ solve an inequality and graph its solution
set  (**Lesson 1-7**)

$3 - 4x \le 6x - 5$
$-10x \le -8$
$x \ge \frac{4}{5}$

**Solve each inequality. Graph the solution set.**

**23.** $9(x + 2) < 72$

**24.** $3(3x + 2) > 7x - 2$

**25.** $8(2x - 1) > 11x + 31$

---

■ solve compound sentences using *and* or
*or*  (**Lesson 1-8**)

$-2 \le x - 4 < 3$
$-2 \le x - 4$ and $x - 4 < 3$
$2 \le 4$  and  $x < 7$

**Solve each compound sentence.**

**26.** $-1 < 3(y - 2) \le 9$

**27.** $4x - 10 < -10$ or $6x + 4 \ge 10$

---

■ solve absolute value
inequalities  (**Lesson 1-8**)

$|3x + 7| \ge 26$
$3x + 7 \ge 26$ or $3x + 7 \le -26$
$3x \ge 19$ or    $3 \le -33$
$x \ge \frac{19}{3}$ or    $x \le -11$

**Solve each inequality. Graph the solution set.**

**28.** $|2x + 6| \le 4$

**29.** $7 + |9 - 5x| > 1$

**30.** $|4x| + 3 \le 0$

---

# APPLICATIONS AND CONNECTIONS

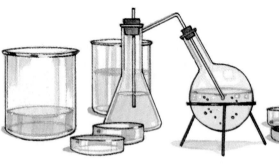

**31.** Your Algebra 2 quiz consists of 6 true/
false questions. How many different
patterns of answers are possible?
(**Lesson 1-5**)

**32. Chemistry**  How much pure ammonia
will a chemist need to add to 10
gallons of a 50% solution to obtain a
solution that is at least 80% ammonia?
(**Lesson 1-7**)

**33. Consumerism**  Where Tertius lives,
gasoline is selling for between $1.40
and $1.60 per gallon. If he has $8.40 to
spend on gas, how many gallons can he
buy?  (**Lesson 1-8**)

# 1 TEST

**State the property illustrated in each equation or statement.**

**1.** $(7 \cdot s) \cdot t = 7 \cdot (s \cdot t)$

**2.** $(7 \cdot s) \cdot t = t \cdot (7 \cdot s)$

**3.** $\left(3 \cdot \dfrac{1}{3}\right) \cdot 7 = \left(3 \cdot \dfrac{1}{3}\right) \cdot 7$

**4.** $(6 - 2)a - 3b = 4a - 3b$

**5.** If $(r + s)t = rt + st,$
then $rt + st = (r + s)t.$

**6.** If $5(3) + 7 = 15 + 7$ and $15 + 7 = 22,$
then $5(3) + 7 = 22.$

**Find the value of each expression.**

**7.** $[2 + 3^3 - 4] \div 2$

**8.** $(2 + 3)^3 - 4 \div 2$

**9.** $(4^5 - 4^2) + 4^3$

**10.** $[5(17 - 2) \div 3] - 2^4$

**Evaluate each expression if $a = -9$, $b = \dfrac{2}{3}$, $c = 8$, and $d = -6$.**

**11.** $\dfrac{db + 4c}{a}$

**12.** $\dfrac{a}{b^2} + c$

**13.** $2b(4a + a^2)$

**14.** $\dfrac{4a + 3c}{3b}$

**Name the sets of numbers to which each number belongs. Use N, W, Z, Q, I, and R.**

**15.** $\sqrt{17}$

**16.** $0.86$

**17.** $-10 \div 2$

**18.** $\sqrt{64}$

**Solve each equation.**

**19.** $2x - 7 - (x - 5) = 0$

**20.** $5t - 3 = -2t + 10$

**21.** $5r + 7 = 5r - 9$

**22.** $5m - (5 + 4m) = (3 + m) - 8$

**23.** $|8w + 2| + 2 = 0$

**24.** $|4y - 5| + 4 = 7y + 8$

**Solve each inequality. Graph each solution set.**

**25.** $4 > b + 1$

**26.** $3q + 7 \geq 13$

**27.** $5(3x - 5) + x < 2(4x - 1) + 1$

**28.** $-12 < 7s - 5 \leq 9$

**29.** $|9y - 4| + 8 > 4$

**30.** $|5 + k| \leq 8$

**31. Statistics** To receive a B in his English class, Dale must earn at least 400 points on five tests. He scored 87, 89, 76, and 77 on his first four tests. What must he score on the last test to receive a B in the class?

**32. Sports** Kathy's softball team has won 3 games and lost 2. How many patterns of records are possible?

**33. Geometry** The formula $A = \dfrac{180(n - 2)}{n}$ relates the measure of an interior angle, $A$, of a regular polygon to the number of sides, $n$. If an interior angle of a regular polygon measures 150 degrees, find the number of sides.

**Bonus**

**Find the set of all numbers $x$ satisfying the given conditions.**

$3x - 2 \geq 0$ and $5x - 1 \leq 0$

# 2 Linear Relations and Functions

## CHAPTER OBJECTIVES

In this chapter, you will:

- Identify different types of relations and functions.
- Graph relations and functions on the coordinate plane.
- Graph inequalities on the coordinate plane.
- Solve applications of equations and inequalities.

**Can you see a pattern in the points plotted on this graph? Explain why you think the graph takes this shape.**

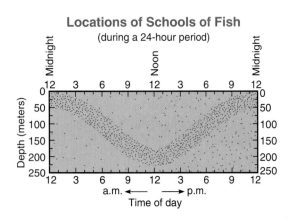

## CAREERS IN OCEANOGRAPHY

Are you constantly being nagged by relatives and friends to choose a career? Just tell them you've chosen oceanography. It will keep them quiet, and only you will know that the "career" you've named includes dozens of different careers from which you will still be able to choose.

Are you good at physics and math? Then physical oceanography might suit you. You'd study tides, currents, temperatures, water densities, sound propagation through water, and water's optical transparency.

Are chemistry and math your strong points? Try chemical oceanography. You'll examine many of the elements that make up seawater, their distribution, and their influence on marine life.

Do biology and math attract your interest? Biological oceanographers study marine animal and plant populations.

Earth's oceans are the biggest unexplored territories we have left. From studying the biology, chemistry, and physics of oceans, we may find new foods, new sources of energy, new clean water to drink, new sources of mineral ores, new medicines, and other unexpected gifts. The ocean is like a new continent, full of vast riches. Will you find some of these riches?

## MORE ABOUT OCEANOGRAPHY

### Degree Required:

- Bachelor's Degree in Oceanography, Marine Biology, or Geology

### Some oceanographers like:

- knowing that their work may benefit others
- working outdoors, research
- variety in their work
- good salaries

### Related Math Subjects:

- Geometry
- Advanced Algebra
- Trigonometry

### Some oceanographers dislike:

- bad weather conditions at sea
- sometimes working long, irregular hours
- being separated from their families
- the physical demands of ocean exploration

For more information on the various careers available in the field of Oceanography, write to:

National Ocean Industries Association
1050 17th Street NW, Suite 700
Washington, D.C. 20036

# Relations and Functions

**Objectives**

After studying this lesson, you should be able to:

■ graph a relation, state its domain and range, and determine if the relation is a function, and

■ find the values of functions for given elements of the domain.

**Application**

The number of pounds of waste recycled in recent years in the United States can be shown using **ordered pairs.** The first number of the ordered pair is the year, and the second number is the number of pounds recycled for every 10 people.

| Year | 1965 | 1970 | 1975 | 1980 | 1985 | 1990 |
|------|------|------|------|------|------|------|
| Pounds Recycled | 17 | 21 | 23 | 32 | 35 | 37 |
| Ordered Pair | (1965, 17) | (1970, 21) | (1975, 23) | (1980, 32) | (1985, 35) | (1990, 37) |

You can **graph** these ordered pairs by creating a **coordinate system** with two axes. The horizontal axis represents the years and the vertical axis represents the pounds of waste. Each point represents the ordered pairs shown in the chart above.

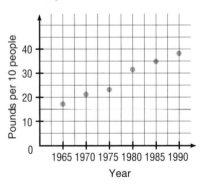

**Recycled Waste in the U.S.**

Remember that each point in the coordinate plane can be named by exactly one ordered pair, and that every ordered pair names exactly one point in the coordinate plane.

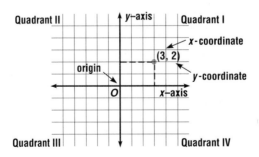

*The points on the two axes do not lie in any quadrant.*

When graphing real-world data, you usually use positive numbers. However, to graph the real numbers you must use the **Cartesian coordinate plane,** which is composed of the **x-axis** and the **y-axis** meeting at the **origin** $(0, 0)$ and dividing the plane into four **quadrants.** The ordered pairs graphed on this plane can be represented by $(x, y)$.

A set of ordered pairs forms a **relation.** The set of first members of the ordered pairs is called the **domain** of the relation. The set of second members of the ordered pairs is called the **range** of the relation. A **mapping** shows how each member of the domain is paired with a member in the range.

{(3, 2), (2, 7), (5, 8)}

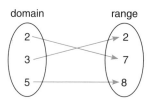

{(8, −4), (−3, 9), (1, 2), (8, 5)}

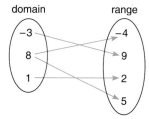

A **function** is a relation in which each element of the domain is paired with *exactly one* element of the range. The first relation shown above is a function. In the second relation, you see that 8 in the domain is paired with both −4 and 5. This relation is not a function.

## Example 1

*Assume that each square on a graph represents 1 unit unless otherwise labeled.*

**Graph the relation {(8, 1), (4, −2), (1, 1), (−3, 2), (−6, 8)}. State the domain and the range of the relation. Is the relation a function?**

Graph the ordered pairs on a coordinate plane.

The domain is {8, 4, 1, −3, −6}.
The range is {−2, 1, 2, 8}.

Each member of the domain is paired with exactly one member of the range, so this relation is a function.

## Example 2

Recycling

**A recycling center received aluminum cans of varying weights which were sent to a smelter to be melted down for use in other aluminum products. The records for one week in June are shown below. Graph this information and determine if it represents a function.**

| Day | Number of Cans | Pounds of Aluminum |
|-----|----------------|--------------------|
| M   | 75,000         | 3000               |
| T   | 80,000         | 3150               |
| W   | 70,000         | 2850               |
| Th  | 75,000         | 3050               |
| F   | 85,000         | 3400               |

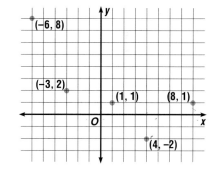

Since 75,000 in the domain is mapped to two members of the range, 3000 and 3050, the relation is not a function.

## Example 3

**Is the set of ordered pairs that satisfy $y^2 = 4x$ a function?**

Prepare a table to find ordered pairs that satisfy the equation. In this case, it is easier to select values for $y$ and then find the corresponding value of $x$.

| x | y |
|---|---|
| 0 |   |
| 2 |   |
| -2 |  |
| 4 |   |
| -4 |  |

→

| x | y |
|---|---|
| 0 | 0 |
| 1 | 2 |
| 1 | -2 |
| 4 | 4 |
| 4 | -4 |

Now graph these ordered pairs.

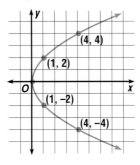

Since $x$ can be any real number, there are an infinite number of ordered pairs that can be graphed. If all of them were graphed, they would take on the shape of a curve, called a *parabola*.

You can see from the graph that there can be two $y$ values for almost all of the $x$ values on the graph. This set of ordered pairs is *not* a function.

---

Look at the graph in Example 3. Suppose you drew a vertical line that intersects the graph. How many times would the line intersect the graph? If you can draw a vertical line anywhere so that it intersects the graph of the relation in more than one point, then the relation is not a function. This is often called the **vertical line test for a function.**

*Letters other than $f$ can be used to represent a function. For example, the equation $y = 4x + 3$ can also be written as $g(x) = 4x + 3$.*

Equations that represent functions are often written in a special way. The equation $y = 2x + 1$ can be written as $f(x) = 2x + 1$. The symbol $f(x)$ is read "$f$ of $x$." Suppose you want to find the value in the range that corresponds to the element 3 in the domain. This is written as $f(3)$ and read "$f$ of 3." The value of $f(3)$ is found by substituting 3 for $x$ in the equation. Therefore, $f(3) = 2(3) + 1$ or 7.

## Example 4

**Find the value of $f(15)$ if $f(x) = 100x - 5x^2$.**

$$f(x) = 100x - 5x^2$$
$$f(15) = 100(15) - 5(15)^2 \quad \textit{Substitute 15 for x.}$$
$$= 1500 - 5(225) \quad \textit{15}^2 = 225$$
$$= 375$$

Therefore, $f(15) = 375$.

**Example 5**

Use your calculator to find $f(1.6)$ if $f(x) = 3x^3 - 4x^2$.

ENTER: 3 $\boxed{\times}$ 1.6 $\boxed{y^x}$ 3 $\boxed{-}$ 4 $\boxed{\times}$ 1.6 $\boxed{x^2}$    $2.048$

Therefore, $f(1.6) = 2.048$.

**Example 6**

Find $g(a + 2)$ if $g(x) = x^2 - 7$.

$$g(x) = x^2 - 7$$
$$g(a + 2) = (a + 2)^2 - 7 \qquad \textit{Substitute } (a + 2) \textit{ for x.}$$
$$= a^2 + 4a + 4 - 7 \qquad \textit{(a + b)}^2 = a^2 + 2ab + b^2$$
$$= a^2 + 4a - 3 \qquad \textit{Combine like terms.}$$

Therefore, $g(a + 2) = a^2 + 4a - 3$.

# CHECKING FOR UNDERSTANDING

**Communicating Mathematics**

Read and study the lesson to answer each question

1. What is the domain and range of the recycling relation shown on page 52?

2. Tell the difference between a relation and a function.

3. A *counterexample* is an example that shows that a given statement is not true. Sam says every straight line is a function. Find a counterexample for this statement.

4. If $f(x) = 2x - 1$, explain how to find $f(7)$.

5. Draw a Cartesian coordinate plane. Name the seven possible locations on the plane where a point might be graphed. Then graph a point in each of these locations.

**Guided Practice**

Tell in which quadrant the graph of $(x, y)$ will lie for each situation.

6. $x$ is positive and $y$ is negative

7. $x$ is negative and $y$ is negative

8. $x$ is negative and $y$ is positive

9. $x$ is 0 and $y$ is positive.

State whether each mapping is a function, or simply a relation.

10.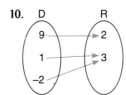

11.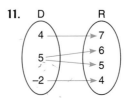

12.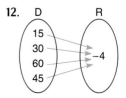

Graph each set of ordered pairs on the same coordinate plane. State the domain and range of each relation. Is the relation a function?

**13.** {(4, 3), (8, −2), (−10, 4), (4, 8)}    **14.** {(−3, −3), (−2, −2), (2, 2), (4, 4)}

**15.** {−3, 3), (−2, 2), (2, −2), (−3, −4)}    **16.** {(−3, 3), (−2, 3), (2, 3), (4, 3)}

**17.** Find $f(4)$ if $f(x) = x^2 − 3x$.

# EXERCISES

**Practice**    Graph each set of ordered pairs on the same coordinate plane. State the domain and range of each relation. Is the relation a function?

**18.** {(3, 9), (2, 4), (1, 1), (0, 0), (−1, 1), (−2, 4), (−3, 9)}

**19.** {(0, −15), (5, −20), (0, −1), (12, 12)}

**20.** $\left\{\left(-2, \frac{1}{2}\right), \left(-1, 1\frac{1}{2}\right), \left(0, \frac{3}{4}\right), \left(1\frac{1}{2}, -2\right)\right\}$

Find each value if $g(x) = \dfrac{7}{x − 2}$.

**21.** $g(12)$             **22.** $g(5.5)$             **23.** $g(−5)$

**24.** $g(0)$              **25.** $g(2)$              **26.** $g(u + 2)$

Use the vertical line test to determine if each relation is a function.

**27.**

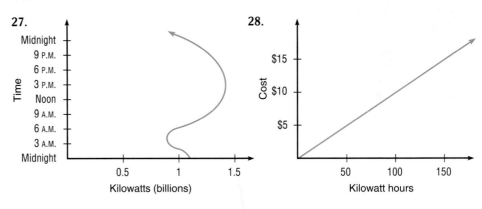

**28.**

**29.**

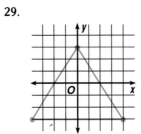

**30.**

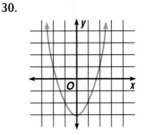

**31.**

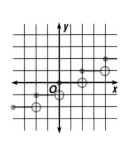

Use your calculator to find each value if $j(x) = x^4 − 3x + 1$.

**32.** $j(5)$        **33.** $j(^{-}7)$        **34.** $j(0.25)$        **35.** $j\left(\dfrac{-12}{5}\right)$

**Three vertices of a rectangle are given. Graph them and find the coordinates of the fourth vertex.**

**36.** $(3, 1)$, $(3, -3)$, $(-5, -3)$

**37.** $(-3, 4)$, $(5, 4)$, $(5, -3)$

**Find each value if $h(x) = \dfrac{x^2 + 5x - 6}{x + 3}$.**

**38.** $h(3)$

**39.** $h(-2)$

**40.** $h\left(\dfrac{1}{2}\right)$

**41.** $h(a - 1)$

**Critical Thinking**

**42.** When a fraction contains a variable in the denominator, there are some values of the variable for which the fraction is undefined. Find the domain of $f(x) = \dfrac{14}{x^2 - 4}$.

**Applications**

**43. Education** Jacob is saving for college. To anticipate the money he may need for tuition, he collected information on the average tuition for public colleges for the years 1986 to 1990. Make a graph of this information. Do the points form a straight line?

| Year | 1986 | 1987 | 1988 | 1989 | 1990 |
|---|---|---|---|---|---|
| Cost per Year | $3963 | $4190 | $4242 | $4736 | $4960 |

**44. Aviation** The air pressure in the cabin of a fighter jet decreases as the plane ascends. Graph the data below. Project what you think the air pressure would be at 60,000 feet.

| Altitude (ft) | 10,000 | 20,000 | 30,000 | 40,000 | 50,000 |
|---|---|---|---|---|---|
| Air Pressure (lb/in²) | 10.2 | 6.4 | 4.3 | 2.7 | 1.6 |

**Mixed Review**

**Solve each inequality. (Lesson 1-8)**

**45.** $4x + 3 < -9$ or $7 < 2x - 11$

**46.** $|y + 1| < 7$

**47.** The number of posters $(x)$ the Art Club can produce in a week can be expressed by the sentence $3x + 7 < 43$. Solve this inequality to find out the maximum number of posters they can produce. **(Lesson 1-7)**

**48.** Name the sets of numbers to which $\sqrt{36}$ belongs. **(Lesson 1-2)**

**Find the value of each expression. (Lesson 1-1)**

**49.** $(-4.8)^2 - 144 + 36$

**50.** $[-2 + (-18)]^3 - (\sqrt{10000})(16 + 2)$

# Graphing Calculator Exploration: Graphing Linear Equations

The graphing calculator is a powerful tool for studying graphs of a wide variety of functions. Let's take a look at some linear functions and their graphs.

The **viewing window** for a graph is that portion of the coordinate grid displayed on the **graphics screen** of the calculator; that is, the portion of the domain and range for the $x$ and $y$ variables shown on the screen. The viewing window can be written as [left, right] by [bottom, top]. So [-10, 10] by [-10, 10] denotes the domain values $-10 \leq x \leq 10$ and the range values $-10 \leq y \leq 10$. This viewing window is called the **standard viewing window.** In order to set the viewing window on both the Casio fx-7000G and the TI-81, press the RANGE key. The following **range screen** display settings are for the standard viewing window. If your calculator is not already set for the standard viewing window, use these values to set it now. You can easily set the TI-81 viewing window to the standard window by pressing ZOOM 6.

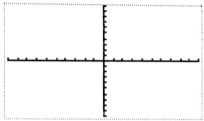

The standard viewing window is [-10, 10] by [-10, 10].

**Casio**

```
Range
Xmin: -10
Xmax: 10
scl: 1
Ymin: -10
Ymax: 10
scl: 1
```

**TI-81**

```
RANGE
Xmin = -10
Xmax = 10
Xscl = 1
Ymin = -10
Ymax = 10
Yscl = 1
Xres = 1
```

The **Xscl, Yscl,** and **scl** values set the frequency of the tick marks along each of the axes. For example, Xscl = 1 means that there will be a tick mark for every one unit along the $x$-axis. The **Xres** for the TI-81 establishes the quality of the resolution on the graphics screen. The Xres value can be set between 1 and 8, with 1 being the best resolution.

The key sequences for graphing with the Casio and the TI-81 are slightly different. The example below demonstrates how to graph a linear equation on each calculator.

**Example 1**

Graph $y = 3x + 5$ using the standard viewing window.

First, make sure that your calculator is set for the standard viewing window. Then enter the appropriate key sequence.

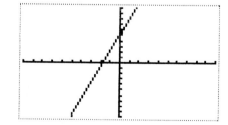

Once you have plotted a graph on the graphics screen of the Casio, it remains on the screen until it is cleared by pressing [SHIFT] [Cls] [EXE] or until the range values are changed. The TI-81 will automatically clear the graphics screen when a new equation is entered.

A graph that appears in the viewing window of the graphics screen is said to be a **complete graph** if all of the important characteristics of the graph are displayed. For a linear equation, this would include the points where the graph crosses the axes.

**Example 2**

Graph $y = -x + 20$.

Use the standard viewing window.

Casio [GRAPH] [(-)] [ALPHA] [X] [+] 20 [EXE]

TI-81 [Y=] [(-)] [X|T] [+] 20 [GRAPH]

None of the graph is shown when you graph this equation in the standard viewing window, so we must change the range values to view a complete graph. There are several viewing windows that will allow us to view the complete graph. Change your viewing window to $[-10, 25]$ by $[-5, 25]$ with scale factors of 1 for each scale. Now graph by pressing [EXE] on the Casio or [GRAPH] on the TI-81.

# EXERCISES

Use your graphing calculator to graph each equation. State the range values that you used to view a complete graph for each equation.

1. $y = 6x + 7$
2. $y = -5x - 6$
3. $y = -3x + 12$
4. $y = -7x + 22$
5. $y = 5x - 35$
6. $y = -8x + 32$
7. $y = 0.1x - 1$
8. $y = -12x$
9. $y = 0.5x + 12$
10. $y = 0.01x$
11. $y = x - 55$
12. $y = 100x - 126$

# Linear Functions

**Objectives**

After studying this lesson, you should be able to:

- identify equations that are linear and graph them, and
- write linear equations in standard form.

**Application**

The employees at the Speedy Repair Shop use a table like the one below to compute their customers' bills.

| Labor Hours | $\frac{1}{2}$ | $\frac{3}{4}$ | 1 | $1\frac{1}{4}$ | $1\frac{1}{2}$ | $1\frac{3}{4}$ | 2 |
|---|---|---|---|---|---|---|---|
| Charge | $29 | $33.50 | $38 | $42.50 | $47 | $51.50 | $56 |

*Usually x is the independent variable and y is the dependent variable. The value of y depends on the value of x.*

The **open sentence** that describes this relationship is $y = 18x + 20$, where $x$ represents the number of hours and $y$ represents the charge to the customer. When this information is graphed, the points lie in a line. This graph is a relation and a function.

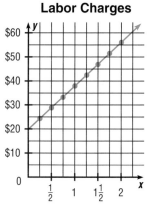

**Labor Charges**

*The independent variable is graphed on the horizontal axis and the dependent variable is graphed on the vertical axis.*

Suppose we connect the points with a line. The line would contain an infinite number of points, all of which satisfy the equation $y = 18x + 20$. An equation whose graph is a line is called a **linear equation.**

*When variables other than x and y are used, assume that the letter coming first in the alphabet represents the domain or horizontal coordinate.*

Linear equations may contain one or two variables with no variable having an exponent other than 1.

Linear equations     $4x + 3y = 7$     $y = 8$     $5m - n = 1$

                    $y = 7 + 2x$     $x = -7$

NOT linear equations     $3x + y^2 = y$     $\frac{1}{x} + y = 4$     *Remember $\frac{1}{x} = x^{-1}$.*

Every equation can be written in many forms. For example, $5m - n = 1$ could also be written as $5m = n + 1$ and $n = 5m - 1$. Any linear equation can be written in **standard form.**

*Standard Form of a Linear Equation*

> **The standard form of a linear equation is**
> $$Ax + By = C,$$
> **where A, B, and C are real numbers and A and B are not both zero.**

*Usually A, B, and C are given as integers whose greatest common factor is 1.*

**Example 1**

> Write the equation $x = \frac{3}{4}y - 1$ in standard form.
>
> $$x = \frac{3}{4}y - 1$$
> $$4x = 3y - 4 \qquad \textit{Multiply each side by 4 to eliminate the fraction.}$$
> $$4x - 3y = -4 \qquad \textit{Add } -3y \textit{ to each side.}$$

To graph a linear equation it is often helpful to make a table of ordered pairs that satisfy the equation. Since any line can be defined by only two points, only two ordered pairs are needed. Finding a third ordered pair is a good idea to check the accuracy of the first two.

**Example 2**

> Graph $3y = -2x - 6$.
>
> First solve the equation for $y$.
>
> $$3y = -2x - 6$$
> $$y = -\frac{2}{3}x - 2 \qquad \textit{Divide each side by 3.}$$
>
> Find three ordered pairs that satisfy the equation. Then graph the ordered pairs and connect the points with a line.
>
>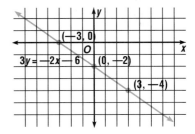
>
> | $x$ | -3 | 0 | 3 |
> |---|---|---|---|
> | $y$ | 0 | -2 | -4 |

**Example 3**

APPLICATION

Geology

> Geothermal energy is generated wherever water comes into contact with heated underground rocks. This heat turns the water into steam that can be used to make electricity. The underground temperature of rocks varies with their depth below the surface. The deeper the rocks are, the hotter they are. The temperature $t$ in degrees Celsius is estimated by the equation $t = 35d + 20$, where $d$ is the depth in kilometers. Complete the table and graph the linear equation. What would be the temperature of the rocks at a depth of 3 km?
>
> | Depth ($d$) in km | Temperature ($t$) in °C |
> |---|---|
> | 0 | 20 |
> | 0.5 | 37.5 |
> | 1 | 55 |
> | 1.5 | 72.5 |
> | 2 | 90 |
>
>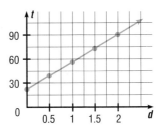
>
> By substituting 3 for $d$ in the equation, $t = 35(3) + 20$ or 125°C.

Any function whose ordered pairs satisfy a linear equation is called a **linear function.**

| *Definition of* *Linear Function* | **A function is linear if it can be defined by $f(x) = mx + b$,** **where $m$ and $b$ are real numbers.** |
|---|---|

In the definition of a linear function, $m$ or $b$ may be zero. If $m = 0$, then $f(x) = b$. The graph is a horizontal line. This function is called a **constant function.** If $f(x) = 0$, the function is called the *zero function.*

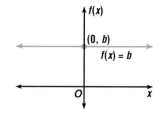

**Example 4**

**Tell whether each function is a linear function.**
**a.** $f(x) = x^4 + 18$

This cannot be written in the form $f(x) = mx + b$, since $x$ has a power of 4. Thus, $f(x)$ is not a linear function.

**b.** $g(x) = 4 - x$

This can be written in the form $f(x) = mx + b$, where $m = -1$ and $b = 4$. Thus, $g(x)$ is a linear function.

# CHECKING FOR UNDERSTANDING

**Communicating Mathematics**

Read and study the lesson to answer each question.

1. An equation whose graph is a line is called a ___?___.
2. An equation in the form $Ax + By = C$ is said to be in ___?___ form.
3. Write the equation $3x + 5y = -2$ in two other forms.
4. How many points are on any line? Choose the best answer.
    a. 2    b. 4    c. 100    d. an infinite number

**Guided Practice**

State whether each equation is a linear equation. If it is not a linear equation, explain why.

5. $x^2 + y^2 = 7$    6. $x + y = 4$    7. $x - 2y = 5$

State whether each function is a linear function. If it is not a linear function, explain why.

8. $f(x) = x^2 + 3$    9. $g(x) = 7$    10. $h(x) = 1.2 - 3.7x$

# EXERCISES

**Practice**

Write each equation in standard form.

11. $y = 3x - 2$    12. $y = -5x + 1$    13. $x = 10$

14. $y = \frac{5}{8}x + 1$    15. $x = \frac{1}{3}y - 4$    16. $y = 14x$

17. Solve $4x + 2y = 9$ for $y$.

18. Solve $5a + 3b = 7$ for $b$.

19. Graph $y = x$.

20. Graph $y = 5x - 4$.

**Solve each equation for the specified variable.**

21. $-5m - 3 = n$, for $m$

22. $x - y = 3$, for $y$

23. $\frac{3}{4}t + \frac{2}{3}r = 12$, for $r$

24. $2a = 3b - 4$, for $b$

**Graph each equation.**

25. $b = 2a - 3$

26. $x - y = 4$

27. $2a + 3b = 6$

28. $5 = 5x$

29. $x + y = 7$

30. $4x + 3y = 12$

31. $f(x) = 2x + 1$

32. $3y + 7 = 12$

33. $f(x) = 3x - 1$

34. $\frac{1}{3}x + \frac{1}{2}y = 1$

35. $\frac{x}{4} - \frac{y}{3} = 2$

36. $\frac{x}{3} + \frac{y}{2} = \frac{15}{2}$

**Critical Thinking**

37. Emilio earns a salary of $150 per week for selling computer game magazines by phone. He receives an extra 30¢ for each magazine over 50 that he sells. Write the equation that represents his total salary for any week.

38. The graph of a vertical line has the form $x = c$, where $c$ is any number. Is this linear equation a linear function? Why or why not?

**Applications**

39. **Finance**  Frontier Auto Shop has a standard $12 shop charge for every job it takes. In addition, the mechanic working on the job charges $20 per hour. The equation that represents the total charge for a job is $c = 12 + 20t$. Find the total charge for a repair requiring 4.5 hours labor.

40. **Consumer Awareness**  Monique's Fine Fashions allows its customers to charge their purchases on a delayed payment plan. When the customer receives a bill from Monique's, there is a $5 charge for using the delayed payment plan plus 2% interest on the purchase. Calculate the bill for a purchase of $110.

**Mixed Review**

41. Graph $\{(3, 4), (4, 5), (-5, -2), (-6, 4), (3, -5)\}$. Is this a function? Explain your answer.  (**Lesson 2-1**)

42. Find the value of $f(-3)$ if $f(x) = x^2 - 3x - 9$.  (**Lesson 2-1**)

43. There are four points on a line that are labeled with the letters M, A, T, and H. How many different ways can these points be labeled? (**Lesson 1-5**)

**Solve.**  (**Lesson 1-4**)

44. **Geometry**  The perimeter of a square is 42 inches. Find the length of one side of the square.

45. **Zoology**  The population of gorillas at the zoo was decreased by moving 5 of them to a neighboring city's zoo. Write an expression to represent the original number of gorillas at the zoo if there are $p$ gorillas there now.

# Problem-Solving Strategy: Look for a Pattern

**Objective**

After studying this lesson you should be able to:

■ Solve a problem by identifying the pattern and using it to find the missing information.

**Connection**

In geometry class, Darlene was daydreaming. She began doodling on paper, drawing various shapes, and connecting the points to draw the diagonals. As she continued to draw, she wondered if there was a way to find how many diagonals there were in any figure. She made a chart to record her findings.

*These figures are convex polygons. Diagonals of convex polygons lie inside the figure.*

| Number of Sides | 3 | 4 | 5 | 6 | 7 | 8 |
|---|---|---|---|---|---|---|
| Number of Diagonals | 0 | 2 | 5 | 9 | ? | 20 |

+2  +3  +4  +?  +?

Notice that a pattern seems to be developing. Darlene didn't want to take time to draw a polygon with 7 sides. She looked at the pattern and said that a 7-sided figure must have 14 diagonals. Was she right?

**Example**

**Look for a pattern to find the next number in (1, 2, 4, 8, 16, 32, □).**

There is more than one way to interpret this pattern.

*Method 1:* Look for an operation from one number to the next.

1  2  4  8  16  32  □

×2  ×2  ×2  ×2  ×2  ×2     The next number is $2 \times 32$ or 64.

*Method 2:* Look for an expression that describes each term.

1  2  4  8  16  32  □

↑  ↑  ↑  ↑  ↑  ↑  ↑     The next number is $2^6$ or 64.

$2^0$  $2^1$  $2^2$  $2^3$  $2^4$  $2^5$  $2^6$

**Strategies**

Look for a pattern.
Solve a simpler problem.
Act it out.
Guess and check.
Draw a diagram.
Make a chart.
Work backwards.

# CHECKING FOR UNDERSTANDING

**Communicating Mathematics**

Read and study the lesson to answer each question.

1. What is a diagonal?

2. How many diagonals does a 12-sided polygon have?

3. What are the next 3 numbers after 64 in the Example?

**Guided Practice**

Find the next number in each group of numbers.

4. 6, 10, 15, 21, 28, __?__

5. 1, 4, 9, 16, 25, __?__

# EXERCISES

Solve. Use any strategy.

6. Find the total number of triangles in Figure A.

7. Find the total number of squares in Figure B.

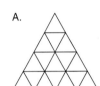

8. Place a piece of paper over the figure at the right. Can you trace it without lifting your pencil or tracing over any line you have already drawn?
   *Intersecting a line is permitted.*

9. If you had one bill from each type of bill from the game of Monopoly®, how much play money would you have?

10. How many aluminum cans does it take to build a pyramid 120 cm tall? Each can is 12 cm tall. The base of the pyramid is a single row of cans and each can in the rows above it rests on two cans below it, and so on.

## COOPERATIVE LEARNING ACTIVITY

**Work in groups. Each person in the group must understand the solution and be able to explain it to any person in class.**

At Dunbar High School, there are 500 students and 500 lockers, numbered 1 through 500. Suppose the first student opens each locker. Then the second student closes every second locker. The third student changes the state of every third locker (that is, closes the ones that were open and opens the ones that were closed). The fourth student changes the state of every fourth locker. This process continues until the 500th student changes the state of the 500th locker. Which lockers are open?

## 2-4 Slopes and Intercepts

**Objectives**

After studying this lesson, you should be able to:

- determine the slope and intercepts of a line,
- use the slope and intercepts to graph a linear equation, and
- determine if two lines are parallel, perpendicular, or neither.

**Application**

Caryn and Brad used a long board and bricks to build a ramp for their radio-controlled model car. At every 24 inches, they placed bricks to create an incline of 4 more inches. The steepness, or **slope,** of the ramp is the ratio of the vertical change to the horizontal change.

$$\text{slope} = \frac{\text{change in the vertical units}}{\text{change in the horizontal units}}$$

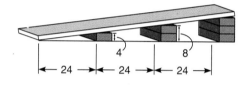

The slope of the car ramp is $\frac{4}{24}$ or $\frac{1}{6}$.

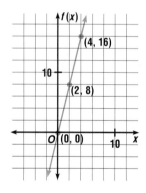

Slope is also defined for the graphs of linear functions. In the graph of $f(x) = 4x$ shown at the left, *look for a pattern* in the relationship of the change in the $y$-coordinates to the change in the $x$-coordinates of the points on the graph.

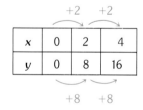

| $x$ | 0 | 2 | 4 |
|---|---|---|---|
| $y$ | 0 | 8 | 16 |

*The pattern for x-coordinates is a difference of 2.*

*The pattern for y-coordinates is a difference of 8.*

The $y$-coordinates increase 8 units for each 2-unit increase in the $x$-coordinates. The slope of the line whose equation is $f(x) = 4x$ is $\frac{8}{2}$ or 4. The vertical change is the difference between the $y$-coordinates of any two points on the graph. The horizontal change is the difference between the corresponding $x$-coordinates. You can use the following formula to find the slope of a line if you know the coordinates of two points on the line.

*Definition of Slope*

**The slope $m$ of a line passing through points $(x_1, y_1)$ and $(x_2, y_2)$ is given by $m = \dfrac{y_2 - y_1}{x_2 - x_1}$.**

**Example 1**

Determine the slope of the line that passes through points (1, −3) and (0, −5). Then graph the line.

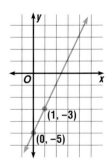

$$m = \frac{y_2 - y_1}{x_2 - x_1}$$

$$= \frac{-5 - (-3)}{0 - 1} \qquad (x_1, y_1) = (1, -3)$$
$$\qquad\qquad\qquad (x_2, y_2) = (0, -5)$$

$$= \frac{-2}{-1} \text{ or } 2 \qquad \textit{The slope of the line is 2.}$$

Graph the two points and draw the line. Use the slope to check your graph by selecting any point on the line and then go up 2 and right 1. This point should also be on the line.

**Example 2**

Graph the line passing through the point (−4, −1) with a slope of $\frac{-2}{3}$. Describe the manner in which the line rises or falls.

First graph the ordered pair (−4, −1). Since the slope of the line is $\frac{-2}{3}$, the vertical change is −2 and the horizontal change is 3. From (−4, −1) move 2 units down and 3 units to the right. This point is (−1, −3).

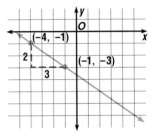

Connect the points to draw the line.

Notice that the line falls to the right.

The slope of a line tells the direction in which it rises or falls. In Example 2, the slope is negative and the line falls to the right. The graphs below show the three other possibilities for a linear graph.

If the line is horizontal, then the slope is zero.

If the line rises to the right, then the slope is positive.

If the line is vertical, then the slope is *undefined*.

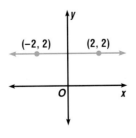

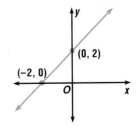

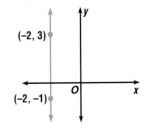

$$m = \frac{2 - 2}{2 - (-2)} = 0$$

$$m = \frac{0 - 2}{-2 - 0} = 1$$

$$m = \frac{3 - (-1)}{-2 - (-2)} = \frac{4}{0}$$

**Example 3**

Graph $f(x) = 3x + 2$, $g(x) = 3x$, and $h(x) = 3x - 5$ on the same coordinate plane. Find the slope of each line and describe what you notice about the graphs.

Find ordered pairs to satisfy each function. Connect the ordered pairs to draw each line.

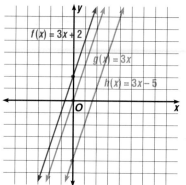

The slope of each line is 3. The lines appear to be parallel.

*Equations whose graphs have similar characteristics are often called families of equations.*

In Example 3, you saw that lines that have the same slope are parallel.

| *Definition of Parallel Lines* | **In a plane, lines with the same slope are parallel. All vertical lines are parallel and all horizontal lines are parallel.** |
|---|---|

If you know the slope of one line, you can use it to graph a line through a given point, parallel to that line.

**Example 4**

Graph the line that goes through the point (0, 3) and is parallel to the line whose equation is $6y - 10x = 30$.

Graph the equation $6y - 10x = 30$. Find the slope of the line. The slope of the line is $\frac{5}{3}$.

Now use the slope $\frac{5}{3}$ and the point (0, 3) to graph the line parallel to the graph of $6y - 10x = 30$.

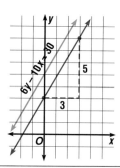

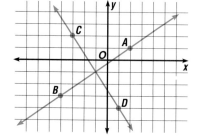

The figure at the left shows the graphs of two lines that are perpendicular. We found that parallel lines have the same slope. Is there a special relationship between the slopes of two perpendicular lines?

slope of line $AB = \dfrac{1 - (-3)}{2 - (-4)} = \dfrac{4}{6} = \dfrac{2}{3}$

slope of line $CD = \dfrac{2 - (-4)}{-3 - 1} = \dfrac{6}{-4} = -\dfrac{3}{2}$

The slopes are negative reciprocals of each other. That is, when you multiply the slopes of two perpendicular lines, the product is always $-1$.

| Definition of Perpendicular Lines | In a plane, two nonvertical lines are perpendicular if and only if the product of their slopes is −1. Any vertical line is perpendicular to any horizontal line. |
|---|---|

*Another way to state the definition is to say that the slopes of perpendicular lines are negative reciprocals of each other.*

**Example 5**

**The consecutive sides of a rectangle are perpendicular. In rectangle WXYZ, the coordinates of point X are (2, 0) and the coordinates of point Y are (5, 1). Find the slope of the line containing side $\overline{YZ}$ of the rectangle.**

In rectangle $WXYZ$, side $\overline{XY}$ is perpendicular to side $\overline{YZ}$. First find the slope of side $\overline{XY}$.

slope of $\overline{XY} = \dfrac{1 - 0}{5 - 2} = \dfrac{1}{3}$

Let $m$ = slope of $\overline{YZ}$. Since the product of the slopes must be −1, you can use this equation.

$m\left(\dfrac{1}{3}\right) = -1$

$m = -1\left(\dfrac{3}{1}\right)$ or −3

The slope of the line containing $\overline{YZ}$ is −3.

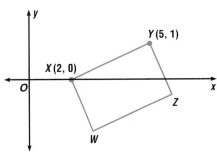

Knowing where a graph crosses each axis often provides a quick way to graph an equation. The point where a graph crosses the *y*-axis is called the **y-intercept**. Likewise, the point where it crosses the *x*-axis is called the **x-intercept**. These two points can be found without graphing the equation.

**Example 6**

**Without graphing, find the y-intercept and the x-intercept of the graph of 3x + 5y = 30.**

Since all points on the *y*-axis have an *x*-coordinate of 0, you can find the *y*-intercept by substituting 0 for *x* in the equation.

$\begin{aligned}
3x + 5y &= 30 \\
3(0) + 5y &= 30 \qquad \text{\textit{Substitute 0 for x.}} \\
5y &= 30 \\
y &= 6 \qquad \text{The \textit{y}-intercept is 6.}
\end{aligned}$

*The ordered pair for the y-intercept is (0, 6).*

Likewise, to find the *x*-intercept, you substitute 0 for *y*, since all points on the *x*-axis have 0 as a *y*-coordinate.

$\begin{aligned}
3x + 5(0) &= 30 \\
3x &= 30 \\
x &= 10 \qquad \text{The \textit{x}-intercept is 10.}
\end{aligned}$

*The ordered pair for the x-intercept is (10, 0).*

*To graph 3x + 5y = 30, place a point at 6 on the y-axis and another point at 10 on the x-axis. Then connect the points to draw the line.*

# CHECKING FOR UNDERSTANDING

**Communicating Mathematics**

Read and study the lesson to answer each question.

1. Explain in your own words how to find the slope of a line.

2. Describe the slopes of vertical lines, horizontal lines, and other nonvertical, nonhorizontal lines.

3. What is the geometric definition of parallel lines?

4. What is the geometric definition of perpendicular lines?

5. Use a dictionary to find other definitions of slope.

6. What is the relationship between the slopes of two perpendicular lines?

7. What is true of the slopes of two parallel lines?

**Guided Practice**

State the *y*-intercept, *x*-intercept, and slope of each line.

8.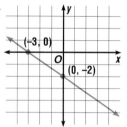

9.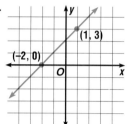

10.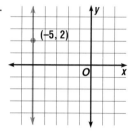

11.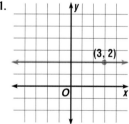

12.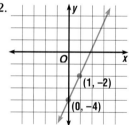

13.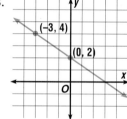

14. Suppose all the graphs shown above were placed on one coordinate plane. Which graphs, if any, would be parallel? Which graphs, if any, would be perpendicular?

# EXERCISES

**Practice**

Determine the slope of the line passing through each pair of points.

15. $(8, -4)$ and $(6, 1)$

16. $(5, 7)$ and $(4, -6)$

17. $(-5, -4)$ and $(5, 2)$

18. $(1, 8)$ and $(7, 8)$

19. $(2.5, 3)$ and $(1, -9)$

20. $(b, 2)$ and $(b, -2)$

Determine whether the graph of each equation rises to the right, falls to the right, is horizontal, or is vertical.

**21.** $x + y = 3$

**22.** $2x + 12 = 0$

**23.** $2x - y = 6$

**24.** $2x + 3y + 32 = 0$

**25.** $2x - 3y = 0$

**26.** $2y + 7 = 3.5(4)$

Find the *y*-intercept and the *x*-intercept of the graph of each equation.

**27.** $y = 6x + 9$

**28.** $y = -3x - 5$

**29.** $y = -2$

**30.** $x = 8$

**31.** $y + 6 = 5x$

**32.** $3x = y$

**33.** $f(x) = x - 2$

**34.** $g(x) = 4x - 1$

**35.** $5x + 3y = 15$

**36.** Graph a line that passes through $(0, 0)$ and has slope 3.

**37.** Graph a line that passes through $(-1, 1)$ and is parallel to a line whose slope is $\frac{1}{4}$.

**38.** Graph a line that passes through $(-4, 1)$ and is perpendicular to a line whose slope is $\frac{-5}{3}$.

**39.** Graph a line that passes through $(2, 2)$ and is perpendicular to the graph of $y = 2$.

**40.** Graph a line that passes through the origin and is parallel to the graph of $x + y = 12$.

**41.** Graph a line that passes through $(-3, -1)$ and has an undefined slope.

**42.** One line has a slope of 0 and another line has an undefined slope, but they both pass through $(1, 1)$. Graph the lines.

**43.** Graph the line that is perpendicular to the graph of $3x - 2y = 24$ at its *x*-intercept.

CONNECTION
Geometry

**44.** The graphs of $2y + x = 6$ and $y = 2x + 3$ contain two sides of a rectangle. If one vertex of the rectangle has coordinates $(8, 4)$, draw the rectangle.

**Critical Thinking**

**45.** Graph this family of equations on the same coordinate plane: $y = x$, $y = 2x$, $y = 5x$, $y = \frac{1}{2}x$, and $y = \frac{1}{5}x$ . Study the graphs and make a conjecture about how the slope of the line affects how steeply it slants.

**Applications**

46. **Demographics** The population of Ashville, North Carolina was 172,000 in 1988. In three years the population grew to 183,000. If $x$ represents the year and $y$ represents the population, find the rate of increase (slope) for the growth of Ashville.

47. **Travel** At 10:00 A.M., Eric traveled 195 miles across the plains states on his way to California. At 2:00 P.M., he had traveled 455 miles. Use slope to calculate his rate of travel.

**Mixed Review**

48. Find the next number in the set $\{100, 95, 85, 70, \underline{\ ?\ }\}$. **(Lesson 2-3)**

49. A board is 12 feet long. It is to be cut into pieces 2 feet long. How many cuts are needed? **(Lesson 2-3)**

50. Solve $|x - 6| = 3x + 4$. **(Lesson 1-6)**

51. What property is illustrated by $3(x - 4) = 3x - 12$? **(Lesson 1-2)**

## ~~~ MID-CHAPTER REVIEW ~~~

1. **Meteorology** When the temperature is 30°F, the speed of the wind makes the temperature feel colder. This is called the windchill factor. The chart below shows how the wind effects your perception of how cold it is when the temperature is 30°. **(Lesson 2-1)**

| Wind Speed (mph) | 0 | 5 | 10 | 15 | 20 | 25 | 30 | 35 | 40 |
|---|---|---|---|---|---|---|---|---|---|
| Windchill Factor (°F) | 30 | 27 | 16 | 9 | 4 | 1 | −2 | −4 | −5 |

   a. State the domain and range of the relation.

   b. Graph the relation. Is it a function?

2. Find $g(7)$, if $g(x) = \dfrac{x^2 - 2x + 3}{x - 5}$. **(Lesson 2-1)**

**Solve for the indicated variable.** **(Lesson 2-2)**

3. $2x + 4y = 7$, for $y$

4. $\dfrac{3}{2}t + \dfrac{4}{7}s = 12$, for $s$

5. $3a + 2b - 4c = 24$, for $c$

6. Determine if $g(x) = x(2 - x)$ is a linear function. **(Lesson 2-2)**

7. Graph $2x - 5y = 10$. **(Lesson 2-2)**

8. **Geometry** In the ordered pairs below, the first coordinate is the number of sides in a regular polygon, and the second coordinate is the degree measure of the angle at each vertex. Find the pattern to complete the last ordered pair. **(Lesson 2-3)**

   $(3, 60), (4, 90), (5, 108), (6, 120), \left(7, 128\frac{4}{7}\right), (8, \square)$

**Graph each equation. Find the slope, the $x$-intercept, and the $y$-intercept.** **(Lesson 2-4)**

9. $2x - 3y = 6$

10. $2x + 3y = 9$

| 2-5 | # Writing Linear Equations |
|---|---|

**Objectives**

After studying this lesson, you should be able to:

- write the slope-intercept form of an equation given the slope and a point, or two points,
- write the standard form of an equation given the slope and a point, or two points, and
- write an equation of a line that is parallel or perpendicular to the graph of a given equation.

In Lesson 2–2 you learned if a function can be written in the form $y = mx + b$, then it is a linear function. But what numbers do $m$ and $b$ represent?

Look at the graph below. The line passes through points $A(0, b)$ and $C(x, y)$. Notice that $b$ is the $y$-intercept of $\overleftrightarrow{AC}$. Suppose you need to find the slope of $\overleftrightarrow{AC}$.

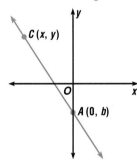

$$m = \frac{y - b}{x - 0}$$

$$m = \frac{y - b}{x}$$

Now solve the equation for $y$.

$$m = \frac{y - b}{x}$$

$mx = y - b$     *Multiply each side by x.*

$mx + b = y$     *Add b to each side.*

or    $y = mx + b$    *Symmetric property of equality*

You may recognize this as the form of a linear function. When an equation is written in the form $y = mx + b$, it is in **slope-intercept form.**

| *Slope-Intercept Form of a Linear Equation* | **The slope-intercept form of the equation of a line is $y = mx + b$, where $m$ is the slope and $b$ is the $y$-intercept.** |
|---|---|

If you are given the slope and the $y$-intercept of a line, you can find an equation of the line by substituting the values of $m$ and $b$ into the slope-intercept form of the equation. Then the equation can be written in standard form. For example, if you know that the slope of a line is $\frac{2}{3}$ and the $y$-intercept is 6, an equation of the line is $y = \frac{2}{3}x + 6$, or, in standard form, $2x - 3y = -18$.

**Example 1**

Find the slope-intercept form of the equation of the line that has a slope of $\frac{3}{4}$ and passes through (8, 2).

You know the slope and the $x$ and $y$ values of one point on the graph. Substitute for $m$, $x$, and $y$ in the slope-intercept form.

$y = mx + b$

$2 = \left(\dfrac{3}{4}\right)(8) + b$

$2 = 6 + b$

$-4 = b$   The $y$-intercept is $-4$.

The equation in slope-intercept form

is $y = \dfrac{3}{4}x - 4$.

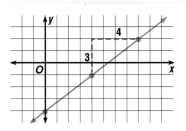

---

Remember that the standard form of the equation of a line is $Ax + By = C$. Suppose we write this general equation in slope-intercept form.

$Ax + By = C$

$\quad By = -Ax + C$     *Subtract Ax from each side.*

$\quad\ y = -\dfrac{A}{B}x + \dfrac{C}{B}$     *Divide each side by B.*

The slope is $-\dfrac{A}{B}$ and the $y$-intercept is $\dfrac{C}{B}$, for $B \neq 0$.

This can be used to write an equation in standard form when you are given the information you usually use to find the slope-intercept form.

**Example 2**

Find the standard form of the equation that passes through (−2, 5) and (3, 1).

First use the two given points to find the slope of the line.

$m = \dfrac{1 - 5}{3 - (-2)}$ or $\dfrac{-4}{5}$

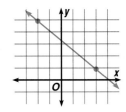

If the slope is $\dfrac{-4}{5}$ or $-\left(\dfrac{4}{5}\right)$, then $-\dfrac{A}{B} = -\left(\dfrac{4}{5}\right)$.
Thus, $A = 4$ and $B = 5$.

Substitute these values into the standard form. The resulting equation is $4x + 5y = C$. Since one of the points on the line is (3, 1), you can substitute these values into the equation to find $C$.

$\quad Ax + By = C$     *Standard form of a linear equation*

$\quad\ 4x + 5y = C$     *Substitute values for A and B.*

$4(3) + 5(1) = C$     *Substitute values for x and y.*

$\quad\quad 12 + 5 = C$

$\quad\quad\quad\ 17 = C$

The standard form of the equation is $4x + 5y = 17$.

## Example 3

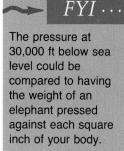

**The atmospheric pressure at sea level is 14.7 pounds per square inch. As divers go deeper into the ocean the pressure increases. Use the chart below to write an equation in slope-intercept form that approximates this relationship. Then find the pressure at 30,000 feet below sea level.**

| Depth (in feet) | Pressure (lb/in²) |
|---|---|
| (Sea level) 0 | 14.7 |
| 600 | 269 |
| 1200 | 536 |
| 3000 | 1338 |
| 7200 | 3208 |
| 18,000 | 8019 |

Let $x$ represent the ocean depth and $y$ represent the pressure.

Use your calculator and a pair of points to find the slope of the line.

$$m = \frac{269 - 536}{600 - 1200} = 0.445$$

You may want to use another pair of points to confirm your slope.

$$m = \frac{3208 - 1338}{7200 - 3000} = 0.445238095$$

The slope of the line is approximately 0.445. The $y$-intercept corresponds to an ocean depth of 0 (at sea level). So the $y$-intercept is 14.7. An equation that approximates the pressure at certain ocean depths is

$$y = 0.445x + 14.7.$$  *The equation that is derived may differ based on the set of points used to determine the slope.*

Use your calculator again to find the approximate pressure at 30,000 feet below sea level.

ENTER: 0.445 [×] 30000 [+] 14.7 [=] $13364.7$

The result is 13,364.7 pounds per square inch.

The slope-intercept form can also be used to find the equations of lines that are parallel or perpendicular.

## Example 4

**Write an equation of the line that passes through (4, 6) and is parallel to the line whose equation is $y = \frac{2}{3}x + 5$.**

Parallel lines have the same slope, so the slope of both lines is $\frac{2}{3}$. Use the slope-intercept form and the point (4, 6) to find the equation.

$y = mx + b$

$6 = \left(\frac{2}{3}\right)(4) + b$   *Substitute $\frac{2}{3}$ for m, 4 for x, and 6 for y.*

$6 = \frac{8}{3} + b$

$\frac{10}{3} = b$   *Subtract $\frac{8}{3}$ from each side.*

The $y$-intercept is $\frac{10}{3}$.

An equation of the line is $y = \frac{2}{3}x + \frac{10}{3}$.   *The standard form is $2x - 3y = -10$.*

**Example 5**

Write an equation in standard form for the line that passes through (4, 6) and is perpendicular to the line whose equation is $y = \frac{2}{3}x + 5$.

The slope of the given line is $\frac{2}{3}$. Since the product of this slope and the slope of the perpendicular line is $-1$, the slope of the perpendicular line is $-\frac{3}{2}$. You can use the slope $-\frac{3}{2}$ and the point (4, 6) to write the equation in standard form.

| | |
|---|---|
| $Ax + By = C$ | *Standard form of linear equation* |
| $3x + 2y = C$ | $m = -\frac{A}{B} = -\left(-\frac{3}{2}\right)$ or $\frac{3}{2}$, $A = 3$ and $B = 2$ |
| $3(4) + 2(6) = C$ | *Substitute 4 for x and 6 for y.* |
| $12 + 12 = C$ | |
| $24 = C$ | |

The standard form of the equation is $3x + 2y = 24$.

*Use a graphing calculator to verify that the equations in Examples 4 and 5 are correct.*

# CHECKING FOR UNDERSTANDING

**Communicating Mathematics**

**Read and study the lesson to answer each question.**

1. What is the slope-intercept form of an equation and what does each variable mean?

2. On which axes are the variables graphed?

3. Explain how to write an equation of a line if you know two points on the line.

4. The slope of a line is $\frac{1}{2}$. What is the slope of a line parallel to this line? What is the slope of a line perpendicular to this line?

5. What is the slope of the line whose equation is $cx - dy = k$?

**Guided Practice**

6. What is the slope and $y$-intercept of the graph of $y = -3x - 4$?

7. Write $y = -3x - 4$ in standard form.

**The slope and $y$-intercept of a line are given. Write the slope-intercept form of the equation for each line described.**

8. $m = 5, b = 6$

9. $m = 2.5, b = 0$

10. $m = -\frac{1}{4}, b = -9$

11. $m = 0, b = 0$

12. Write an equation of a line that passes through (0, 5) and is parallel to the graph of $y = 4x + 12$.

# EXERCISES

**Practice** State the slope and *y*-intercept of the graph of each equation.

**13.** $y = -\frac{3}{4}x - 3$      **14.** $y = \frac{1}{3}x$      **15.** $-y = 0.2x + 6$

**16.** $6y = 3x - 12$      **17.** $-5y = 3x - 30$      **18.** $y = cx + t$

**19.** The slope of $\overleftrightarrow{CD}$ is $\frac{3}{2}$. Line *FG* is perpendicular to $\overleftrightarrow{CD}$ and has a *y*-intercept of 4. Write the equation of $\overleftrightarrow{FG}$.

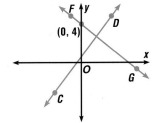

Find the slope-intercept form of each equation.

**20.** $2x - 5y = 10$      **21.** $3x - y = 6$

**22.** $2x - 2y = 4$      **23.** $2x = 11$

Write an equation for the line that satisfies each of the given conditions in slope-intercept form and in standard form.

**24.** slope $= \frac{1}{2}$, passes through $(6, 4)$

**25.** slope $= -\frac{4}{5}$, passes through $(2, -3)$

**26.** slope $= 5$, passes through the origin

**27.** passes through $(6, 1)$ and $(8, -4)$

**28.** passes through $(6, 1)$ and $(6, 7)$

**29.** passes through $(4, 6)$ and $(0, 0)$

**30.** *x*-intercept $= -3$, *y*-intercept $= 6$

**31.** *x*-intercept $= \frac{1}{3}$, *y*-intercept $= -\frac{1}{4}$

**32.** *x*-intercept $= 0$, *y*-intercept $= 2$

**33.** passes through $(-1, -1)$ and $(8, -1)$

**34.** Write an equation of a line that passes through $(4, 2)$ and is parallel to the line whose equation is $y = 2x - 4$.

**35.** Write an equation of a line that passes through $(-2, 0)$ and is perpendicular to the line whose equation is $y = -3x + 7$.

**36.** Write an equation of a line that passes through $(-3, -1)$ and is parallel to the line that passes through $(3, 3)$ and $(0, 6)$.

**37.** Write an equation of a line that passes through $(6, -5)$ and is perpendicular to the line whose equation is $3x - \frac{1}{5}y = 3$.

**Find the value of $k$ in each equation if the given ordered pair is a solution of the equation.**

**38.** $5x + ky = 8$, $(3, -1)$

**39.** $4x - ky = 7$, $(4, 3)$

**40.** $3x + 8y = k$, $(0, 0.5)$

**41.** $kx + 3y = 11$, $(7, 2)$

**Critical Thinking**

**Applications**

**42.** Three vertices of a parallelogram are $(10, 3)$, $(-1, 2)$, and $(1, -1)$. Find the coordinates of the fourth vertex.

**43. Oceanography** The Mariana Trench is the deepest point in any of the oceans. It is located in the western Pacific Ocean north of Australia. The deepest point in the trench is 35,840 feet below sea level. Find the approximate pressure at this point.

**44. Science** Crickets vary the number of chirps they produce with the temperature. If you count the number of chirps in a minute, you can tell the temperature. The chart below records the number of chirps in a minute and the approximate temperature. Write an equation to describe this linear relationship. Then calculate the temperature when the number of chirps is 130.

| Chirps | 50 | 60 | 70 |
|---|---|---|---|
| Temperature | 16°C | 18°C | 20°C |

**45. Nature Studies** A nature preserve worker calculates there are 6000 deer in Sharon Woods Park. She also estimates that 75 more deer die than are born each year. How many deer will be in the park in $x$ years?

**46. Economics** The Serves-You-Best Rental Car Company has two rental offers. The first offer gives the renter a compact car that costs 25¢ a mile to drive plus an initial fee of $20. The second offer gives the renter a luxury car that costs 25¢ a mile plus an initial fee of $35.

a. Write an equation to represent each offer.

b. What is the relationship between the graphs of these offers?

c. If both cars are driven 750 miles, what is the cost difference between renting the compact car and the luxury car?

**Computer**

**47.** The BASIC program at the right finds the slope and $y$-intercept of the line passing through two given points. Use the program to find the slope and $y$-intercept of the line passing through each pair of points. Then write the equation of the line.

**a.** $(-1, 4)$, $(2, -2)$

**b.** $(-2, 3)$, $(1, 7)$

**c.** $(2, 2)$, $(-1, -2)$

**d.** $(-1, -4)$, $(3, -2)$

```
10   PRINT "ENTER THE
       COORDINATES"
20   PRINT "OF TWO POINTS"
30   INPUT X1, Y1, X2, Y2
35   PRINT
40   IF X1 = X2 THEN 100
50   LET M = (Y2 - Y1)/(X2 - X1)
60   PRINT "SLOPE = ";M
70   LET B = Y1 - M * X1
80   PRINT "Y-INTERCEPT = ";B
90   GO TO 120
100  PRINT "UNDEFINED SLOPE"
110  PRINT "NO Y-INTERCEPT"
120  END
```

**Mixed Review**

**48.** Find the slope of the line passing through $(5, 4)$ and $(2, 2)$.   (**Lesson 2-4**)

**49.** Find the $x$-intercept and the $y$-intercept of the graph of $3x - 2y = 12$. (**Lesson 2-4**)

**50.** Find the value of $f(-2)$, if $f(x) = x^2 - 4$.   (**Lesson 2-1**)

**51.** Solve $3x - 5 > -26$.   (**Lesson 1-7**)

**52. Probability**   You have a red button, a green button, a blue button, and a white button. These buttons are to be sewn in a row as decoration on a pocket of a shirt. In how many ways can these four buttons be arranged in that row?   (**Lesson 1-5**)

## HISTORY CONNECTION

**René Descartes** (1596–1650) was a French mathematician and philosopher. He is credited with the invention of a branch of mathematics called analytic geometry. Analytic geometry combines the following from algebra and geometry.

**1.** the coordinate plane

**2.** the correspondence of ordered pairs of numbers to points in the coordinate plane

**3.** graphs of functions like $f(x) = 2x + 1$

Descartes, in 1637, became the first mathematician to put the three ideas together.

*Sometimes, ordered pairs are referred to as Cartesian coordinates. The word Cartesian is taken from the name Descartes.*

## 2-6 Scatter Plots and Prediction Equations

### Objectives

After studying this lesson, you should be able to:
- draw a scatter plot and find a prediction equation, and
- solve problems using prediction equations.

When data is collected in real-life situations, the relation determined by the variables usually does not form a straight line. However, the graph may *approximate* a linear relationship. When this is the case, a **best-fit line** can be drawn and a **prediction equation** can be determined. A prediction equation can be determined by employing a process similar to that used to determine an equation of a line when you know two points.

*The best-fit line does not necessarily contain any points from the data.*

### Application

The Zimco Bottling Company is promoting a continuing education program for its employees. The personnel director, Ms. Dirr, would like to be able to predict an employee's salary if she knows the number of years the employee attended college. From the current personnel files, Ms. Dirr randomly selected the files of ten employees. She recorded each employee's salary and the corresponding number of years of college for the employee.

| Years of College | 3 | 2 | 4 | 6 | 2.5 | 7.5 | 7 | 1 | 5.5 | 4 |
|---|---|---|---|---|---|---|---|---|---|---|
| Salary (in $1000) | 15 | 20 | 22 | 47 | 19 | 18 | 32 | 10 | 30 | 28 |

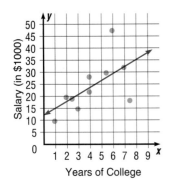

To determine the relationship between the number of years of college and the salary, Ms. Dirr graphed the data points to obtain a **scatter plot.** She found that the points did not lie in a straight line, but clustered in a linear pattern. She drew a line suggested by this pattern of points.

She then selected the points (2.5, 19) and (7, 32) on that line to determine the equation of the line. To find the equation of this line, she first used the slope formula and the two points.

$$m = \frac{32 - 19}{7 - 2.5} = \frac{13}{4.5} \text{ or about } 2.9$$

Let $s$ represent an employee's annual salary. Let $c$ represent the number of years of college education. Use one of the points and the slope to find the prediction equation.

$$s = 2.9c + b \qquad \text{\textit{y = mx + b, where s = y, m = 2.9, and c = x}}$$
$$32 = 2.9(7) + b \qquad \text{\textit{Point (7, 32) is used for the values of c and s.}}$$
$$11.7 = b \qquad \text{\textit{A prediction equation is s = 2.9c + 11.7.}}$$

By using her prediction equation, Ms. Dirr can encourage the employees with little college education to go back to school. For example, she can predict that with five years of college education, their salary might be $26,200.

**Example 1**

APPLICATION

Health

**The table below shows the heights and the corresponding ideal weights of adult women. Find a prediction equation for this relationship.**

| Height (inches) | 60 | 62 | 64 | 66 | 68 | 70 | 72 |
|---|---|---|---|---|---|---|---|
| Weight (pounds) | 105 | 111 | 123 | 130 | 139 | 149 | 158 |

First graph the data. Draw a line that appears most representative of the data. Use two points, (62, 111) and (66, 130), from that line to find the slope.

$$m = \frac{130 - 111}{66 - 62} = \frac{19}{4} \text{ or about } 4.8$$

Now use the slope and one of the points in the slope-intercept form to find the value of $b$.

$$y = mx + b \qquad \textit{x = 62,}$$
$$111 = 4.8(62) + b \qquad \textit{y = 111,}$$
$$-186.6 = b \qquad \textit{m = 4.8}$$

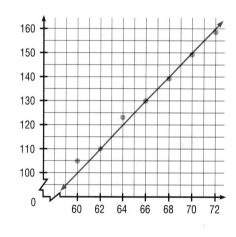

Let $h$ be the independent variable and represent the height. Let $w$ be the dependent variable and represent the corresponding weight. Using the slope and intercept values, we obtain the prediction equation $w = 4.8h - 186.6$.

*The Technology feature on page 91 presents one of these software programs for scatter plots.*

The procedure for determining a prediction equation is dependent upon your judgement. You decide where to draw the best-fit line. You decide which two points on the line are used to find the slope and intercept. Your prediction equation may be different from someone else's. The prediction equation is used when a rough estimate is sufficient. For better analysis of the data, statisticians normally use other, more precise procedures, often relying on computers and high-level programming.

## Example 2

Draw a scatter plot and find two prediction equations to show how typing speed and experience are related. Predict the typing speed of a student who has 11 weeks of experience.

| Experience (weeks) | 4 | 7 | 8 | 1 | 6 | 3 | 5 | 2 | 9 | 6 | 7 | 10 |
|---|---|---|---|---|---|---|---|---|---|---|---|---|
| Typing Speed (wpm) | 33 | 45 | 46 | 20 | 40 | 30 | 38 | 22 | 52 | 44 | 42 | 55 |

*Typing speed is dependent on experience, so the first number in the ordered pair is the number of weeks of experience.*

The pattern of dots suggests a possible line that passes through (5, 36) and (8, 49).

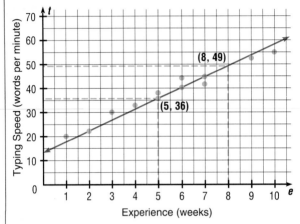

$$\text{slope} = \frac{49 - 36}{8 - 5}$$

$$= \frac{13}{3} \text{ or about } 4.3$$

Let $e$ stand for experience.
Let $t$ stand for typing speed.

$$y = mx + b$$
$$t = 4.3e + b$$
$$36 = 4.3(5) + b$$
$$14.5 = b$$

One prediction equation is $t = 4.3e + 14.5$.

Another line can be suggested by using points (2, 22) and (9, 52). Using these points results in a prediction equation of $t = 4.3e + 13.4$.

If a student had 11 weeks experience, the first equation would predict that the student could type approximately 62 words per minute. For the second equation, the prediction would be 61 words per minute. Thus, either prediction equation produces a good estimate.

# CHECKING FOR UNDERSTANDING

**Communicating Mathematics**

**Read and study the lesson to answer each question.**

1. What is the purpose of a prediction equation?

2. What do you call the graphs of data that form a cluster of dots?

3. In Example 2, why is the weight the dependent variable?

**Guided Practice**

In a study of the relationship between the number of times (*t*) a plant is watered per month and the height (*h*) of the plant in centimeters, the prediction equation is $h = 0.5t + 0.5$. Predict the height for each number of waterings.

**4.** 2                **5.** 5                **6.** 12

**7.** According to a certain prediction equation, if Luther's Soap spends $20,000 on advertising, sales will be $10,000,000. If Luther's Soap spends $50,000 on advertising, sales will be $22,000,000. Let *A* be advertising expenditures and *S* be sales revenues.
   **a.** Find the slope of the prediction equation.
   **b.** Find the *y*-intercept of the prediction equation.
   **c.** Find the prediction equation.
   **d.** Predict the sales if $10,000 is spent on advertising.
   **e.** Predict the amount of money spent on advertising if $16,000,000 of sales revenue was generated.

# EXERCISES

**Practice**

**8.** The distance measured around your head is related to your height. According to the prediction equation, a person 72 in. tall has a head measurement of about 24 in., and a person 60 in. tall has a head measurement of about 20 in. Let *h* be the height of an individual and *m* be that person's head measurement.

   **a.** Find the slope of the prediction equation.
   **b.** Use the value of the slope to complete the following statement: As the person's height (increases/decreases), the head size (increases/decreases).
   **c.** Find the *y*-intercept of the prediction equation. Write a prediction equation.
   **d.** Predict the head size if a person is 66 in. tall.
   **e.** Predict the head size if a person is 76 in. tall.
   **f.** Predict the height of an individual whose head size is 18 in.

**9.** The table below shows the years of experience for eight encyclopedia sales representatives and the amount of sales during a given period of time.

CONNECTION
Statistics

| Amount of Sales | $9000 | $6000 | $4000 | $3000 | $3000 | $5000 | $8000 | $2000 |
|---|---|---|---|---|---|---|---|---|
| Years of Experience | 6 | 5 | 3 | 1 | 4 | 3 | 6 | 2 |

   **a.** Draw a scatter plot to show how the years of experience and the amount of sales are related.
   **b.** Write a prediction equation from this data.
   **c.** Predict the amount of sales for a representative with 8 years of experience.

**d.** Predict the amount of sales for a representative with no experience.

**e.** Predict the year of experience for a representative who sells $7300 of encyclopedias.

10. The table below shows the ideal weight for a man for a given height.

| Height (inches) | 66 | 68 | 70 | 72 | 74 | 76 | 78 |
|---|---|---|---|---|---|---|---|
| Weight (pounds) | 143 | 153 | 164 | 171 | 183 | 198 | 206 |

**a.** Use the information from the chart above to write a prediction equation for the relationship between a man's height and his ideal weight.

**b.** Predict the weight of a man who is 71 inches tall.

**c.** Predict the height of a man who weighs 190 pounds.

**d.** Write a paragraph to compare and contrast the prediction equation for women in Example 1 with your prediction equation for men.

**Critical Thinking**

11. Describe how you could test your prediction equation to see if it is an accurate representation of the data in a scatter plot.

12. Refer to Example 2. Is the prediction equation true for all values of $t$? Why or why not?

**Applications**

13. **Agriculture** Farmers will sometimes hold their crops from market until the price goes up to a level they think is satisfactory. The table below records the price per bushel and how many thousand bushels of wheat were sold at that price during a 10-day selling period in Iowa.

| Price ($ per bushel) | 3.84 | 3.66 | 3.87 | 3.96 | 3.60 | 4.05 | 3.63 | 3.60 | 3.72 | 3.87 |
|---|---|---|---|---|---|---|---|---|---|---|
| Bushels Sold (thousands) | 50 | 47 | 38 | 28 | 49 | 23 | 47 | 46 | 39 | 42 |

**a.** Draw a scatter plot and find a prediction equation for the data.

**b.** If next week, the market price of wheat is $3.90/bushel, how many bushels of wheat can you predict will be sold?

**c.** Estimate what the price of wheat was when 25,500 bushels were sold.

14. **Personal Finance** Sonia works at a clothing store. She earns $8 an hour plus 40¢ for every item over 20 that she sells. She works 30 hours a week. How much money will she make if she sells $c$ items?

**Mixed Review**

15. Write an equation in standard form of a line that passes through (5, 1) and (8, −2). **(Lesson 2-5)**

16. What is the slope of a line perpendicular to the line that passes through (0, 0) and (−4, −2)? **(Lesson 2-5)**

17. Graph $y - 3x = 2$ using the slope and $y$-intercept. **(Lesson 2-4)**

18. Evaluate $|30x - 20| + 72$ if $x = -2$. **(Lesson 1-6)**

19. Simplify $3(x + 2y) - 4(3x - 2y)$. **(Lesson 1-2)**

# Graphing Calculator Exploration: Lines of Regression

You can use your graphing calculator to draw scatter plots and a line that best fits the points in the scatter plot. This line is called a **regression line.** Once you have drawn the regression line, you can use the tracing function on the graphing calculator to make predictions about the data.

**Example 1**

Draw a scatter plot and a regression line for the data about the orbits of ten asteroids that is given in the following table.

| Asteroid | Ceres | Pallas | Juno | Vesta | Astraea | Hebe | Iris | Flora | Metis | Hygeia |
|---|---|---|---|---|---|---|---|---|---|---|
| Mean Distance from Sun (millions of miles) | 257.0 | 257.4 | 247.8 | 219.3 | 239.3 | 225.2 | 221.4 | 204.4 | 221.7 | 222.6 |
| Orbital Period (in years) | 4.60 | 4.61 | 4.36 | 3.63 | 4.14 | 3.78 | 3.68 | 3.27 | 3.69 | 5.59 |

First, set the range parameters. The values of the data suggest that we use the viewing window [200, 260] by [3, 6] with a scale factor of 10 for the *x*-axis and 0.5 for the *y*-axis. *Can you see why this window was chosen?*

Next, set your calculator in the proper statistical mode and clear the statistical memories and the graphics screen.

*Casio*  ENTER:  (SHIFT) (Cls) (EXE) (SHIFT) (MODE) (÷) (SHIFT) (SCL) (EXE)

*TI-81*  ENTER:  (2nd) (STAT) (◄) 2 (ENTER) (2nd) (STAT) (◄) (ENTER)

Now enter your data.

*Casio*  ENTER:  257.0 (SHIFT) (,) 4.60 (DT) . . . 222.6 (SHIFT) , 5.59 (DT)

*TI-81*  ENTER:  257.0 (ENTER) 4.60 (ENTER) . . . 222.6 (ENTER) 5.59 (ENTER)

Now draw the scatter plot and the regression line.

*Casio*  ENTER:  (GRAPH) (SHIFT) (LINE) 1 (EXE)      *Draws the regression line.*

*TI-81*  ENTER:  (2nd) (STAT) (►) 2 (ENTER)      *Draws the scatter plot.*

(2nd) (STAT) 2 (ENTER)      *Calculates the coefficients of the regression line.*

(Y=) (VARS) (►) (►) 4 (ENTER)      *Writes the equation of the regression line.*

(GRAPH)      *Graphs the regression line.*

(2nd) (STAT) (►) 2 (ENTER)      *Restores the scatter plot.*

**Example 2**

Use the tracing function on the regression line found in Example 1 to predict the orbital period of an asteroid that is an average of 236.0 million miles from the sun.

First activate the tracing function.

*Casio* ENTER: [SHIFT] [TRACE]

*TI-81* ENTER: [TRACE]

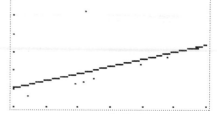

Now use the arrow keys to find the *x*-coordinate value that best approximates 236.0. Then find the corresponding *y*-value to determine the orbital period. The TI-81 will display the *x*- and *y*-coordinates simultaneously. To find the *y*-coordinate on the Casio, find the

*x*-coordinate value closest to 236.0 and then press the [x-y] key to display

the corresponding *y*-coordinate value. Based on this regression line, an asteroid that is an average of 236.0 million miles from the sun would have an orbital period of about 4.22 years.

# EXERCISES

Use your graphing calculator to draw a scatter plot and a regression line for the data in the following tables.

1.

| x | 0.1 | 2 | 3 | 4 | 5 |
|---|-----|---|---|---|---|
| y | 1 | 0.2 | −1 | −1.5 | −2 |

2.

| x | −2 | −1 | 0.5 | 1 | 2.5 |
|---|----|----|-----|---|-----|
| y | −2 | 1 | −1 | 2 | 0.5 |

3. The following table shows the number of thousands of men and women who graduated from college in the years 1981–1988.

| Year | 1981 | 1982 | 1983 | 1984 | 1985 | 1986 | 1987 | 1988 |
|------|------|------|------|------|------|------|------|------|
| Men | 470 | 473 | 479 | 482 | 483 | 486 | 481 | 472 |
| Women | 465 | 480 | 490 | 492 | 497 | 502 | 510 | 517 |

a. Use a graphing calculator to draw a scatter plot and regression line to show how the year is related to the number of thousands of men who graduated from college in the years 1981 to 1988.

b. Use the tracing function to predict the number of men who will graduate from college in 2000.

c. Use a graphing calculator to draw a scatter plot and regression line to show how the year is related to the number of thousands of women who graduated from college in the years 1981 to 1988.

d. Use the tracing function to predict the number of women who will graduate from college in 2000.

## 2-7 Special Functions

**Objective**

After studying this lesson, you should be able to:

- identify and graph special functions (direct variation, constant, identity, absolute value, and greatest integer).

**Application**

During a thunderstorm, Kelly recorded how long it was between seeing the lightning and hearing the thunder. The distance $d$ in kilometers between Kelly and the lightning can be estimated by $d = \frac{1}{3}s$, where $s$ is the number of seconds between seeing the lightning and hearing the thunder.

| Time (sec) | 2 | 4 | 6 | 8 | 12 |
|---|---|---|---|---|---|
| Distance (km) | 0.6 | 1.3 | 2.0 | 2.6 | 4 |

From the equation written in slope-intercept form, you find that the slope is $\frac{1}{3}$ and the $y$-intercept is 0. Whenever a linear function in the form $y = mx + b$, has $b = 0$ and $m \neq 0$, the function is called a **direct variation.** In this situation, the distance varies directly as the number of seconds. In other words, if you hear the thunder soon after you see the lightning, you are fairly close to the lightning. On the other hand, if you don't hear the thunder for a long time after you see the lightning, then the lightning is far away.

There are other special cases of linear functions. Two of these, the **constant function** and the **identity function,** are shown below.

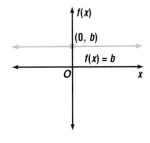

constant function
$m = 0$

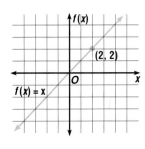

identity function
$m = 1, b = 0$

Several other functions are closely related to linear functions. An **absolute value function** is one example. Consider $f(x) = |x|$ or $y = |x|$. Look for a pattern when studying the values in the chart.

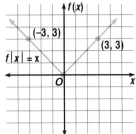

absolute value function

| $x$ | $-3$ | $-2$ | $-1$ | 0 | 1 | 2 | 3 |
|---|---|---|---|---|---|---|---|
| $y$ | 3 | 2 | 1 | 0 | 1 | 2 | 3 |

You see that when $x$ is positive or zero, the absolute value function looks like the graph of $y = x$. When $x$ is negative, the absolute value function looks like the graph of $y = -x$.

**Example 1**

Graph $f(x) = |x| + 3$ and $g(x) = |x + 3|$ on the same coordinate plane. Determine the similarities and differences in the two graphs.

Find several ordered pairs that satisfy each function.

| $x$ | $|x| + 3$ |
|---|---|
| 0 | 3 |
| -1 | 4 |
| 1 | 4 |
| -2 | 5 |
| 2 | 5 |

| $x$ | $|x + 3|$ |
|---|---|
| 0 | 3 |
| 1 | 4 |
| -1 | 2 |
| 2 | 5 |
| -2 | 1 |
| -3 | 0 |

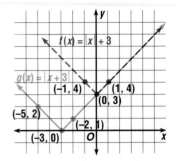

Graph the points and connect them. Both graphs have the same shape and form congruent angles, but have their vertices at different points.

**Step functions** like the ones shown below are also related to linear functions. The open circle means that the point is not included in that part of the graph.

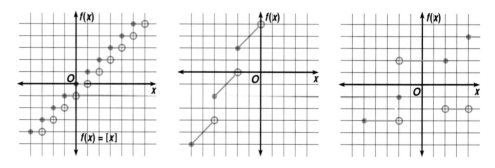

One type of step function is the **greatest integer function.** The symbol $[x]$ means the greatest integer not greater than $x$. For example, $[6.2]$ is 6 and $[-1.8]$ is $-2$, because $-1 > -1.8$. The greatest integer function is given by $f(x) = [x]$. Its graph is the first step function shown above.

**Example 2**

APPLICATION

Commerce

The Speedy-Fast Parcel Service charges for delivering packages by the weight of the package. If the package weighs less than 1 pound, the cost of delivery is $2. If the package weighs at least 1 pound but less than 2 pounds, the cost is $3.50. For each additional pound the cost of delivery increases $1.50. Graph the function that describes this relationship.

This is an example of an application of the greatest integer function. The equation that describes this function is $f(x) = 1.50[x] + 2$.

Make a table of values to help you draw the graph.

| x | [x] | f(x) |
|-----|-----|------|
| 0.1 | 0 | 2.00 |
| 0.5 | 0 | 2.00 |
| 0.7 | 0 | 2.00 |
| 1.0 | 1 | 3.50 |
| 1.4 | 1 | 3.50 |
| 1.9 | 1 | 3.50 |
| 2.4 | 2 | 5.00 |
| 2.7 | 2 | 5.00 |
| 3.1 | 3 | 6.50 |
| 3.7 | 3 | 6.50 |

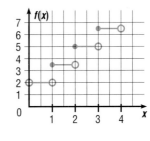

# CHECKING FOR UNDERSTANDING

**Communicating Mathematics**

Read and study the lesson to answer each question.

1. How far away would the lightning be if there was 5 seconds between seeing the lightning and hearing the thunder?

2. Which of the following is an example of a constant function?
   a. $f(x) = 4$    b. $x = 6$    c. $y = {}^-3$

3. Discuss what the difference would be in the graphs of $y = |x|$ and $y = |x - 5|$.

4. What is the greatest integer function?

5. The value of $[4.1]$ is 4, but the value of $[{}^-4.1]$ is $-5$. Why?

**Guided Practice**

Identify each function as C for constant, D for direct variation, A for absolute value, or G for greatest integer function.

6. $f(x) = |3x - 2|$    7. $f(x) = [{}^-x]$    8. $g(x) = 3x$

9. $f(x) = \left| x - \dfrac{2}{3} \right|$    10. $m(x) = \dfrac{2}{3}$    11. $f(x) = -\dfrac{1}{2}x$

12. If $g(x) = [x - 4]$, find $g(3)$.

# EXERCISES

**Practice**

Identify each type of function.

13. $p(x) = x$    14. $h(x) = {}^-7$    15. $g(x) = [2x + 1]$

If $h(x) = [2x + 1]$, find each value.

16. $h(2)$    17. $h({}^-3)$    18. $h(1.4)$    19. $h\left(\dfrac{2}{3}\right)$    20. $h\left(-\dfrac{9}{7}\right)$

Graph each function.

21. $f(x) = x + 2$    22. $f(x) = |x + 2|$    23. $f(x) = [x + 2]$

24. $g(x) = |x| + 2$    25. $g(x) = [x] + 2$    26. $g(x) = 2|x|$

**Journal**

Describe something new you learned in this lesson. Be sure to give examples.

**Graph each pair of equations on the same coordinate plane. Discuss the similarities and differences in the two graphs.**

**27.** $y = |x + 2|$, $y = |x - 2|$

**28.** $y = |x| + 4$, $y = |x| - 4$

**29.** $y = |x + 2|$, $y = |x + 2| - 1$

**30.** $y = 2[x]$, $y = [2x]$

**31.** $y = [x + 5]$, $y = [x] + 5$

**32.** $y = |3x|$, $y = 3|x|$

**33.** $y = -2|4x|$, $y = 4|-2x|$

**34.** $y = -3[x]$, $y = [-3x]$

**35.** Compare and contrast the graphs of $y = |ax|$ and $y = a|x|$.

**36.** Compare and contrast the graphs of $y = |x + b|$ and $y = |x| + b$.

**Graph each equation.**

**37.** $y = [|x|]$

**38.** $y = |[x]|$

**39.** $y = x - [x]$

**40.** $y = x + |x|$

**Critical Thinking**

**41.** Draw a graph that represents how all rational numbers from 0 to 10 are rounded to the nearest whole number. What kind of function is this? Explain your answer.

**Applications**

**42. Sports** In the 1988 Winter Olympics, Bonnie Blair set a world record for women's speed skating by skating approximately 12.79 meters per second in the 500 meter race. If she could maintain that speed, what would be the equation that represents how far she could travel for a given time? What type of function does the equation represent?

**43. Business** The Fix-It Auto Repair Shop has a sign in the Service Department that states that the labor costs are $35 per hour or any fraction thereof. What type of function does this relationship represent?

**Mixed Review**

**44. Statistics** A developer surveyed families in a suburb of Raleigh, North Carolina, to find out each household's monthly income and what percentage of that income was spent on housing. The table below shows the data from eight families. **(Lesson 2-6)**

| Average Monthly Income | $870 | $1430 | $1920 | $2460 | $2850 | $3240 | $3790 | $4510 |
|---|---|---|---|---|---|---|---|---|
| Percentage Spent on Housing | 44 | 39 | 40 | 35 | 43 | 38 | 37 | 33 |

a. Graph a scatter plot and find a prediction equation for this data.

b. Predict the percentage of income spent on housing for a family with an average monthly income of $3000.

c. Predict the income for a family who spends 41% of their money for housing.

**45.** Find the slope and $y$-intercept of the graph of $3x - 4y = -10$. **(Lesson 2-4)**

**46.** Evaluate $\dfrac{3a^2 + 2b}{c^2}$, if $a = 1$, $b = 2$, and $c = 3$. **(Lesson 1-1)**

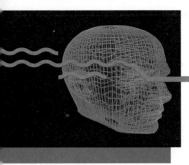

# Technology
## Median-Fit Lines

BASIC
▶ Software
Spreadsheets

In Lesson 2–6, you learned to graph scatter plots and find the prediction equation that can describe the linear pattern for a group of ordered pairs. A software product from Sunburst, Inc., called *Data Insights*, enables you to graph data. Then the program will draw a line suggested by the points, if one exists. *Data Insights* describes this line as the **median-fit line.**

The data below is a list of years and the millions of students in elementary and secondary schools in the United States. The graph done by *Data Insights* shows the graphed data and median-fit line. Notice that *Data Insights* labels your axes and gives you the equation of the median-fit line. It will also title the graph.

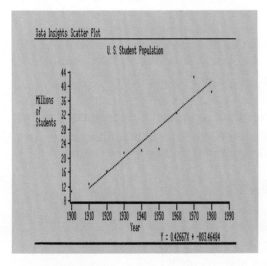

| Year | Millions of Students |
|------|---------------------|
| 1900 | 10.6 |
| 1910 | 12.6 |
| 1920 | 16.2 |
| 1930 | 21.3 |
| 1940 | 22.0 |
| 1950 | 22.3 |
| 1960 | 32.5 |
| 1970 | 42.5 |
| 1980 | 38.2 |
| 1990 | 38.0 |

In the median-fit line equation, or prediction equation, $Y = 0.42333X + -796.93359$, the $X$ is the year and the $Y$ is the number of students (in millions).

# EXERCISES

1. Use the median-fit line given above and your calculator to estimate the number of students in each year.

   a. 1922     b. 1955     c. 1978     d. 1991     e. 2005

2. Use *Data Insights* to enter the data below for the fuel efficiency for city driving of various 1990 cars. What is the median-fit line equation?
   *HP = horsepower, MPG = miles per gallon*

   | Buick Regal | 135 HP | 19 MPH | Chevrolet Camaro | 245 HP | 16 MPG |
   |-------------|--------|--------|------------------|--------|--------|
   | Dodge Colt | 75 HP | 27 MPG | Ford Probe | 145 HP | 21 MPG |
   | Geo Prizm | 102 HP | 25 MPG | Honda Accord | 125 HP | 22 MPG |
   | Mercury Sable | 140 HP | 18 MPG | Pontiac Bonneville | 165 HP | 18 MPG |
   | Toyota Corolla | 102 HP | 25 MPG | | | |

# Graphing Linear Inequalities

**Objectives**

After studying this lesson, you should be able to:

- draw graphs of inequalities in two variables, and
- write an inequality to solve problems.

**Application**

Mr. Harris wants to rent a car for a business trip. Reasonable Car Rental advertises that their daily rental rate is $30 plus $0.25 a mile. Mr. Harris would like to compare this rate with the rates offered by other car rental agencies.

First he determines the equation containing the points that represent the relationship between the number of miles driven ($d$) and the total cost of the rental ($r$).

The initial cost of the car is $30. Since this is the point where no miles are driven, it would be the $y$-intercept of the graph. The slope would be the rate of change in the total cost. In this case, the rate is $0.25 per mile. Since $0.25 = \frac{1}{4}$, the slope is $\frac{1}{4}$. Thus, an equation of the line is $r = \frac{1}{4}d + 30$.

The graph of $r = \frac{1}{4}d + 30$ separates the coordinate plane into two regions. The line is called the *boundary* of the regions. To graph an inequality, first you graph the boundary and then determine which region to shade.

| Miles Driven | Total Cost |
|:---:|:---:|
| 0 | $30 |
| 40 | $40 |
| 100 | $55 |
| 200 | $110 |

The graph of $r > \frac{1}{4}d + 30$ contains points that are located *above* the boundary. In that region, the value of the dependent variable $r$ is greater than the value of $\frac{1}{4}d + 30$. This graph represents car rental costs that are greater than those offered by Reasonable Car Rental. For example, Executive Rental charges $70 for a car rental with 100 miles. The point (100, 70) lies above the boundary.

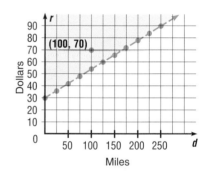

**FYI ···**

About 344 trillion miles are traveled in cars each year by people on business-related trips.

The graph of $r < \frac{1}{4}d + 30$ contains points that are located *below* the boundary. In that region, the value of $r$ is less than the value of $\frac{1}{4}d + 30$. This graph represents car rental costs that are less than those offered by Reasonable Car Rental. For example, Econo-Rental charges $40 for a car rental with 100 miles. The point (100, 40) lies below the boundary.

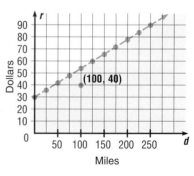

When graphing an inequality, the boundary you draw may be solid or dashed. If the inequality uses the symbol $\leq$ or $\geq$, which include equality, the boundary will be solid. Otherwise, it will be dashed. After graphing the boundary, you must determine which region is to be shaded. Test a point on one side of the line. If the ordered pair satisfies the inequality, that region contains solutions to the inequality. If the ordered pair does not satisfy the inequality, the other region is the solution.

**Example 1**

**Graph $2y - 5x \leq 8$.**

The boundary will be the graph of $2y - 5x = 8$. Let's use intercepts to graph the boundary more easily.

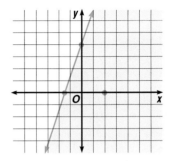

The boundary is included.

| *x-intercept* | *y-intercept* |
|---|---|
| $2(0) - 5x = 8$ | $2y - 5(0) = 8$ |
| $-5x = 8$ | $2y = 8$ |
| $x = -\dfrac{8}{5}$ | $y = 4$ |

Draw a solid line connecting the two intercepts. This is the boundary.

*Now test a point.*

Try $(2, 0)$.
$$2y - 5x \leq 8$$
$$2(0) - 5(2) \leq 8$$
$$-10 \leq 8 \quad \textit{true}$$

The region that contains $(2, 0)$ should be shaded.

**Example 2**

**Graph $y > |x| - 1$.**

The absolute value has two conditions to consider.

$$y > |x| - 1$$

| *when $x < 0$* | | *when $x \geq 0$* |
|---|---|---|
| $y > -x - 1$ | and | $y > x - 1$ |

Graph each inequality for the specified values of $x$. The lines will be dashed.

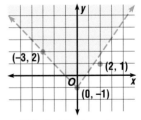

Note that the boundary is *not* included.

Test $(0, 0)$: $\quad y > |x| - 1$
$\qquad\qquad\quad 0 > |0| - 1$
$\qquad\qquad\quad 0 > -1 \quad \textit{true}$

The shaded region should include $(0, 0)$.

Inequalities can sometimes be used to analyze a situation and determine the trends in business and profitability.

Example 3

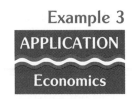

APPLICATION

Economics

The No-Drip Sponge Company must produce a certain number of sponges each day to keep the assembly line staff busy. If production falls, then layoffs may be possible. The equation that describes this relationship is $s > 50e + 25$, where $e$ is the number of employees and $s$ is the number of sponges. Graph this inequality.

Graph the equation $s = 50e + 25$.

Select a point, such as $(2, 50)$, and substitute it into the inequality.

$$s > 50e + 25$$
$$50 > 50(2) + 25$$
$$50 > 125 \qquad false$$

The region that does not contain $(2, 50)$ should be shaded.

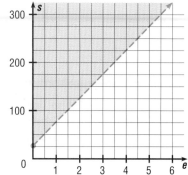

*Any ordered pair that lies in the shaded portion represents an acceptable production rate in relation to the number of employees.*

# CHECKING FOR UNDERSTANDING

**Communicating Mathematics**

Read and study the lesson to answer each question.

1. What is the purpose of the boundary? How do you find it?

2. Suppose Mr. Harris found a rental rate that had points that were the same as those of the boundary. Explain what kind of deal this is.

3. Why is $(0, 0)$ a convenient point to use to test a solution for an inequality?

4. How do you decide if a boundary should be solid or dashed?

5. Explain why $(0, 2)$ satisfies $y \geq -8x + 2$.

**Guided Practice**

State which points, $(0, 0)$, $(2, -3)$, or $(-1, 2)$, satisfy each inequality.

6. $x + 2y < 7$     7. $3x + 2y \leq 0$     8. $4x + 2y \geq 7$     9. $y > 0$

Graph each inequality.

10. $y < 3$       11. $x > -1$       12. $x - y \geq 0$

# EXERCISES

**Practice**

Graph each inequality.

13. $y + 1 < 5$       14. $x - 3 < -5$       15. $y > 5x - 3$

16. $x - 7 \leq y$       17. $y \geq -3x + 1$       18. $y - 3 < 2x$

19. Graph all the points on the coordinate plane to the right of $x = 4$. Write an inequality to describe these points.

**Graph each inequality.**

20. $y > \frac{1}{3}x + 7$

21. $y \geq \frac{1}{2}x - 3$

22. $2 \geq x - 2y$

23. $-2x + 5 \leq 3y$

24. $y \geq |x|$

25. $|x| - 3 \leq y$

26. $|x| + y \geq 3$

27. $y + |x| < 2$

28. $y \geq |3x|$

29. Describe the graph of $|y| < x$ in your own words.

30. Graph all second quadrant points bounded by the lines $x = -2$, $x = -5$, and $y = 3$.

31. Graph all points in the first quadrant bounded by the two axes and the line $x + 2y = 4$.

32. Graph all points in the fourth quadrant bounded by the two axes and the lines $3x - y = 4$ and $x - y = 5$.

**Draw a graph of each inequality.**

33. $|x| \leq |y|$

34. $|x| - |y| = 1$

35. $|x| + |y| \geq 1$

36. $|x + y| > 1$

**Critical Thinking**

37. Draw the graph of a region where $2 \leq x \leq 8$ and $0 \leq y \leq 5$, excluding the region where $4 \leq x \leq 6$ and $2 \leq y \leq 3$.

**Application**

38. **Manufacturing**   The Hoosier Auto Company has a daily production quota of $100,000 worth of cars per day. They produce two types of cars. Their compact model ($C$) is valued at $10,000 and their luxury car ($L$) is valued at $20,000. The equation $10,000C + 20,000L = 100,000$ describes the production quota. Make a graph of the quota equation.

   a. On January 14, the factory produced 5 compacts and 2 luxury cars. Was the company above, below, or on target with their quota? Write equation or inequality that contains this point.

   b. On February 15, the factory produced 6 compacts and 2 luxury cars. Write the equation or inequality that contains this point.

   c. On March 9, the factory produced 9 compacts and 1 luxury car. Write the equation or inequality that contains this point.

**Mixed Review**

39. Graph $y = 2|x| + 7$.   **(Lesson 2-7)**

40. Graph $y = [x] - 4$.   **(Lesson 2-7)**

41. Find the value of $h(a - 3)$ if $h(x) = x^2 + 5$.   **(Lesson 2-1)**

42. Solve $|4x + 2| \geq -10$. Graph the solution.   **(Lesson 1-8)**

43. How many 3-digit numbers are possible to complete the license plate number CNH-☐ ☐ ☐?   **(Lesson 1-5)**

---

**Portfolio**

Select one of the assignments from this chapter that you found especially challenging and place it in your portfolio.

# VOCABULARY

Upon completing this chapter, you should be familiar with the following terms:

| | | | |
|---|---|---|---|
| absolute value function | 87 | 53 | range |
| constant function | 62 | 53 | relation |
| direct variation | 87 | 80 | scatter plot |
| domain | 53 | 66 | slope |
| function | 53 | 73 | slope-intercept form |
| greatest integer function | 88 | 60 | standard form |
| identity function | 87 | 88 | step functions |
| linear equation | 60 | 54 | vertical line test |
| linear function | 62 | 69 | $x$-intercept |
| prediction equation | 80 | 69 | $y$-intercept |

# SKILLS AND CONCEPTS

## OBJECTIVES AND EXAMPLES

Upon completing this chapter, you should be able to:

- graph a relation, state its domain and range, and determine if the relation is a function   (**Lesson 2-1**)

A function is a relation in which each element of the domain is paired with exactly one element of the range.

- find the values of functions for given elements of the domain   (**Lesson 2-1**)

If $f(x) = x^2 - 2$, find $f(2)$.
$f(2) = 2^2 - 2$ or 2

- identify equations that are linear and graph them   (**Lesson 2-2**)

$3x + y^2 = 1$ is not a linear equation.
$3x + y = 1$ is a linear equation.

Find ordered pairs and graph them.

## REVIEW EXERCISES

Use these exercises to review and prepare for the chapter test.

**Graph each relation. Then state its domain and range. Is the relation a function?**

1. $\{(-3, -2), (4, -6), (3, -3), (6, 2)\}$
2. $\{(4.5, 1), (-4.5, 2), (4.5, 3), (-3.5, 4)\}$
3. $\{(1, 4.5), (2, -4.5), (3, 4.5), (-3.5, 4)\}$

**Find the value of each function.**

4. If $f(x) = 2x^3 + 4x^2 + 4x + 1$, find $f(-3)$.
5. If $f(x) = 3x^2 - 2x - 1$, find $f(2a)$.

**Determine if each equation is a linear equation. If it is, graph it.**

6. $x^2 + y^2 = 4$
7. $y = 5$
8. $y = 2x^2 - 1$
9. $4y = x + 8$

- determine the slope of a line
  (**Lesson 2-4**)

  Determine the slope of the line that
  passes through $(-4, 2)$ and $(4, 0)$.
  $$m = \frac{0 - 2}{4 - (-4)}$$
  $$= \frac{-1}{4}$$

**Determine the slope of the line passing through each pair of points.**

**10.** $(5, 1)$ and $(3, 7)$

**11.** $(-3, 2)$ and $(5, -1)$

**12.** $(2, -1)$ and $(-7, -5)$

- use the slope and intercept to graph a
  line and determine if two lines are
  parallel, perpendicular, or
  neither   (**Lesson 2-4**)

  The slope of parallel lines are the same.
  The product of the slope of perpendicular
  lines is $-1$.

**Graph each equation. Then determine if the lines are parallel, perpendicular, or neither.**

**13.** $x + 4y = 8$ and $4x - y = -2$

**14.** $2x - 4y = 8$ and $2x - y = 4$

**15.** $2y = -4x - 5$ and $2x + y = 10$

- Write the slope-intercept form and the
  standard form of an equation given the
  slope and a point, or two points
  (**Lesson 2-5**)

  The slope-intercept form of the line that
  has a slope of $\frac{2}{3}$ and a $y$-intercept of $3$ is
  $y = \frac{2}{3}x + 3$. The standard form of this
  equation is $-2x + 3y = 9$.

**Write the slope-intercept form and the standard form of an equation for each graph described.**

**16.** slope of $5$ and $y$-intercept of $-7$

**17.** slope of $-3$ and passes through $(1, -4)$

**18.** passes through $(-3, 0)$ and $(1, -4)$

**19.** $x$-intercept $= -2$ and $y$-intercept $= 5$

- write the equation of a line that is
  parallel or perpendicular to the graph of a
  given equation   (**Lesson 2-5**)

  The equation of a line parallel to
  $y = 2x - 2$ is $y = 2x + 1$.

  The equation of a line perpendicular to
  $y = 2x - 2$ is $y = -\frac{1}{2}x + 1$.

**Write the slope-intercept form and the standard form of an equation for each graph described.**

**20.** passes through $(2, 4)$ and is parallel to
  the line whose equation is $y = 3x - 5$

**21.** passes through $(-1, -1)$ and is
  perpendicular to the line whose
  equation is $2y + 3x = 10$

- draw a scatter plot and find the
  prediction equation   (**Lesson 2-6**)

  Draw a scatter plot and draw a line
  suggested by the pattern of dots. Then
  select two points on the line and
  determine the equation of the line. Use
  this equation to predict.

On the average, a person 180 centimeters
tall weighs about 76 kilograms. A person
who is 160 centimeters tall weighs about
57 kilograms. Let $h$ represent the height
and $w$ represent the weight.

**22.** Find the prediction equation.

**23.** Predict the weight of a person who is
  174 centimeters tall.

**24.** Predict the height of a person who
  weighs 88 kilograms.

■ identify and graph special functions to include direct variation, constant, identity, absolute value, and greatest integer   (**Lesson 2-7**)

Make a table of values to help you draw the graph.

**Graph each function.**

**25.** $f(x) = |x + 4|$     **26.** $f(x) = 0.5|x|$

**27.** $f(x) = [x] - 1$     **28.** $g(x) = 2[x]$

■ draw graphs of inequalities in two variables   (**Lesson 2-8**)

Replace the inequality symbol with an equals sign to find the equation for the boundary line. Graph the boundary line, using a solid line for ≤ or ≥, or a dashed line for < or >. Test a point to determine which region to shade.

**Graph each inequality.**

**29.** $y - 3 > -2x$     **30.** $2x - 5y \geq 4$

**31.** $3x + 4y < 9$     **32.** $y \leq |x| + 5$

## ~~ APPLICATIONS AND CONNECTIONS ~~

33. **Postal Services**   The cost of mailing a package is determined by its weight. The overnight cost (c) for mailing a package is $10 plus 15¢ an ounce. Write an equation to describe this relationship. Is this a linear function?   (**Lesson 2-2**)

34. Beginning at the letter S and moving only up, down, right, or left, how many different paths can you find to spell the word SLOPE?   (**Lesson 2-3**)

```
              E
            E P E
          E P O P E
        E P O L O P E
      E P O L S L O P E
        E P O L O P E
          E P O P E
            E P E
              E
```

35. **Personal Finance**   As an employee of the Yogurt Delight, Maria receives a salary of $5.50 an hour less $10 a week for cleaning her uniforms. Write an equation to describe her weekly salary and then find her salary for 32 hours of work.   (**Lesson 2-5**)

Graph each relation. State the range and domain. Is the relation a function?

**1.** $\{(-8, 1), (-4, 8), (3, 0), (8, 8), (3, -5)\}$

**2.**

| Year | 1988 | 1989 | 1990 | 1991 | 1992 |
|------|------|------|------|------|------|
| Expenses | 4000 | 4200 | 4000 | 4300 | 4100 |

Use the vertical line test to determine if each relation is a function.

**3.** 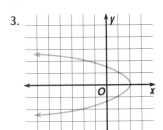  **4.** 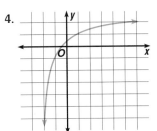  **5.**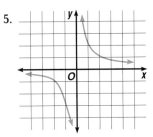

**6.** Given $f(x) = 3x^2 - 5x - 4$, find $f(4)$. Then find $f(c + 2)$.

**Graph each sentence.**

**7.** $y = \dfrac{8}{7}x$

**8.** $y = -7$

**9.** $x = 6$

**10.** $y = -3x + 4$

**11.** $5x - 2y = 12$

**12.** $2x + 3y > 9$

**13.** $2x + 6y \geq 18$

**14.** $f(x) = |x - 2|$

**15.** $y = 2[x] - 1$

**Write the slope-intercept form and standard form of an equation for each graph described.**

**16.** slope $= \dfrac{1}{3}$, $y$-intercept $= -7$

**17.** slope $= 4$, passes through $(-3, 3)$

**18.** $x$-intercept $= -6$, $y$-intercept $= -6$

**19.** passes through $(0, 7)$ and $(5, 2)$

**20.** passes through $(7, 7)$ and is parallel to a line whose equation is $2x + 3y = 6$

**The table below shows the age in years and the systolic blood pressure for a group of ten people tested at a hospital.**

| Age | 50 | 35 | 24 | 34 | 55 | 48 | 26 | 30 | 41 | 37 |
|-----|-----|-----|-----|-----|-----|-----|-----|-----|-----|-----|
| Blood Pressure | 135 | 128 | 108 | 119 | 146 | 140 | 104 | 122 | 132 | 121 |

**21.** Draw a scatter plot to show how age and systolic blood pressure are related.

**22.** Find a prediction equation to show how age and systolic blood pressure are related.

**23.** Predict the systolic blood pressure of a person who is 45 years old.

**24.** Predict the age of a person who has a systolic blood pressure of 120.

**25.** Find the next number in the pattern 2, 5, 9, 14, 20, $\underline{\ \ ?\ \ }$.

**Bonus** Olivia earns $275 per week at a shoe store. She also gets an additional 40¢ for each pair of shoes over 60 that she sells. How much will she earn if she sells $p$ pairs of shoes?

The test questions on these pages deal with a variety of concepts from arithmetic to algebra.

**Directions: Choose the one best answer. Write A, B, C or D.**

1.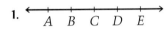
   A  B  C  D  E    *Figure is not drawn to scale.*

   Suppose numbers $\frac{5}{6}$, $1$, $\frac{1}{3}$, $\frac{3}{2}$, and $\frac{6}{5}$ are arranged from least to greatest on the number line above, with each number corresponding to a letter. If the greatest number corresponds to $E$, which corresponds to $\frac{5}{6}$?

   (A) $A$  (B) $B$  (C) $C$  (D) $D$

2. If $8x + 10y$ represents the perimeter of a rectangle, and $x + 3y$ represents it width, the length is

   (A) $3x + 2y$  (B) $7x + 7y$

   (C) $6x + 4y$  (D) $3.5x + 3.5y$

3. In the series 2, 6, 11, 17, 24, ___, ___, ___, the eighth term is

   (A) 41  (B) 45  (C) 51  (D) 62

4. If 9 less than the product of a number and $-4$ is greater than 7, which of the following could be that number?

   (A) $-5$  (B) $-3$  (C) 4  (D) 5

5. The value of $(4 + 3)\, 2^2 - 5$ is

   (A) $-7$  (B) 11  (C) 23  (D) 291

6. Which of the following is not equivalent to $-\frac{6}{8}$?

   (A) $\frac{-3}{4}$  (B) $\frac{3}{-4}$

   (C) $-\frac{-12}{-16}$  (D) $-\frac{6}{-8}$

7. Which of the following is the least?

   (A) $\frac{1}{2}$  (B) $\frac{7}{13}$  (C) $\frac{4}{9}$  (D) $\frac{8}{15}$

8. If the radius of a circle is tripled, then the area is multiplied by

   (A) 3  (B) 6  (C) 9  (D) 27

9. A point on the graph of $x + 3y = 13$ is

   (A) $(4, 4)$  (B) $(-5, 6)$

   (C) $(-2, 3)$  (D) $(4, -3)$

10. Which represents an irrational number?

    (A) $\frac{-2}{3}$  (B) $\sqrt{4}$  (C) $\pi$  (D) 0

11. For what values of $y$ will $2y - 4$ be equal to $2y + 6$?

    (A) all negative values

    (B) 0

    (C) all positive values

    (D) no value

12. The average of 7, 5, 9, 3, and $2x$ is $x$. What is the value of $x$?

(A) 2.4          (B) 4

(C) 6            (D) 8

13. If a number is increased by 5 and the result is multiplied by 8, the product is 168. What is the original number?

(A) 16           (B) 42

(C) 26           (D) 128

14. The radius of a wheel is 4 cm. How many revolutions will it make if it is rolled a distance of $400\pi$ cm?

(A) $25\pi$      (B) 25

(C) 50           (D) $100\pi$

15. Which of the following is the difference of two consecutive prime numbers less than 40?

(A) 1            (B) 5

(C) 8            (D) 11

16. How many integers between 199 and 301 are divisible by 4 or 10?

(A) 26           (B) 31

(C) 35           (D) 37

17. A patient must be given medication every 7 hours starting at 7:00 A.M. Monday. On what day will the patient first receive medication at 6 P.M.?

(A) Monday       (B) Tuesday

(C) Wednesday    (D) Friday

18. Which of the following is not always true?

(A) $a + b = b + a$

(B) $a - b = b - a$

(C) $a - b = a + (-b)$

(D) $a - b = (-b) + a$

19. The value of $|a| > a$ if

(A) $a > 0$      (B) $a < 0$

(C) $a = 0$      (D) $a \neq 0$

20. Which of the following is true if $|x + 3| > 5$?

(A) $-8 < x < 2$     (B) $x > 2$

(C) $x < -8$         (D) none of these

21. Evaluate $\dfrac{\frac{1}{a} + \frac{1}{b}}{ab}$ if $a = 3$ and $b = 5$.

(A) $\dfrac{1}{120}$      (B) $\dfrac{8}{225}$

(C) 8                     (D) $28\frac{1}{8}$

22. When a certain number is divided by 3 there is no remainder. If when the number is divided by 6 there is a remainder, the remainder must be

(A) 1            (B) 2

(C) 3            (D) 4

23. A point off the graph $2x - y < 5$ is

(A) $(-2, 11)$       (B) $(1, 7)$

(C) $(2, 6)$         (D) $(3, -11)$

# Systems of Equations and Inequalities

## CHAPTER OBJECTIVES

In this chapter, you will:

- Solve systems of equations in two or three variables.
- Solve systems of inequalities.
- Use linear programming to find maximum and minimum values of functions.
- Graph linear equations in space.

The breakeven point is the level of sales where the revenue from the sales equals the cost of manufacturing those products. Can you list some of the costs of manufacturing a product or providing a service?

**Cost and Revenue of Sales**

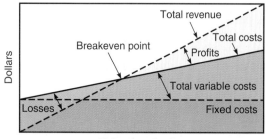

Units of production

## CAREERS IN MARKET RESEARCH

Can you work accurately with detail? Are you patient and persistent? Can you work objectively and systematically to discovery solutions to problems no one has ever solved before? If so, perhaps you could be a successful market research analyst.

These days no company can afford to provide products or services to the public without first making sure exactly what the public wants. So the early steps of any product development rely on market research. A market researcher first designs a survey—that is, decides what questions to ask the public.

Say a fast-food restaurant is being planned for a shopping mall. The researcher will need to ask lots of questions. *What kinds of foods will mall shoppers buy—hot things or cold things, full meals or just snacks? Should the new restaurant have real ice cream or frozen yogurt? Which issues rank highest with mall shoppers: nutrition, taste, or speed?*

The market research analyst heads a team that asks many, many mall shoppers all these questions and more. Then the team analyzes the data. They may need to perform a second or even a third survey, if results indicate that important questions were omitted. Finally, they present their findings both orally and in writing, with visual aids such as charts and graphs. From their work, businesses can know more surely than ever before what the marketplace wants.

## MORE ABOUT MARKET RESEARCH

**Degree Required:**

- Bachelor's Degree in Marketing

**Some market research analysts like:**

- working with people
- lots of travel
- the variety and challenge of their work
- good salaries

**Related Math Subjects:**

- Statistics
- Calculus
- Applied Math

**Some market research analyst dislike:**

- working long hours, including evenings and weekends
- working under high pressure
- the competitive nature of this field

For more information on the various careers available in the field of Market Research, write to:

Marketing Research Association
111 East Wacker Drive
Chicago, Illinois 60601

# Graphing Calculator Exploration:
# Graphing Systems of Equations

You can use a graphing calculator to graph and solve systems of equations, since several equations can be graphed on the screen at the same time. If the system of equations has a solution, it is located where the graphs intersect. The coordinates of this intersection point, $(x, y)$, can be determined by using the trace function.

Graph the system of equations $y = -5.01x + 3.12$ and $y = 3.78x - 2.56$ on the standard viewing window.

*Casio*

ENTER: [GRAPH] [(-)] 5.01 [ALPHA] [X] [+] 3.12 [:] [GRAPH] 3.78 [ALPHA] [X] [−] 2.56 [EXE]

*TI-81*

ENTER: [Y=] [(-)] 5.01 [X|T] [+] 3.12 [ENTER] 3.78 [X|T] [−] 2.56 [GRAPH]

Now use the trace function to determine the coordinates of the intersection point.

*Casio*

ENTER: [SHIFT] [TRACE]

*TI-81*

ENTER: [TRACE]

The trace function on the Casio accesses the last function graphed, while the TI-81 allows you to access any of the functions by pressing the up or down arrow keys. Use the arrow keys to move the cursor along one of the functions to the intersection point and determine the coordinates of the point.

The "zoom-in" feature of the calculator is very useful for determining the coordinates of the intersection point with greater accuracy. Begin by setting the cursor on the intersection point and observing the coordinates of this point. Then zoom-in and place the cursor on the intersection point again. Any digits that are unchanged since the last trace are accurate. Repeat this process of zooming-in and checking digits until you have the number of accurate digits that you desire.

To zoom-in on the Casio, return to the text screen and insert a factor command before the original function. This will reduce each range value by the factor entered.

*Casio*

ENTER: [G↔T] [⇒] [SHIFT] [INS] [SHIFT] [Factor] 10 [:] [EXE]

Once the lines have been replotted, trace to the intersection point again and check the coordinates. If you need to zoom-in again, simply press [G↔T] and [EXE] .

To zoom-in on the TI-81, use the [ZOOM] key and set the factors to 10. This will reduce the values in the range setting by a factor of 10.

*TI-81*

ENTER: [ZOOM] 4 10 [ENTER] 10 [ZOOM] 2 [ENTER]

Now, trace to the point of intersection again and check the coordinates. If you need to zoom-in again, press [ZOOM] 2 and [ENTER] .

This process determines that the $x$- and $y$-coordinates of the intersection point of the system $y = -5.01x + 3.12$ and $y = 3.78x - 2.56$ are (0.646188, -0.117406), accurate to six digits. You will need to zoom-in about six times to obtain this degree of accuracy.

You can also graph systems of inequalities on a TI-81 graphing calculator. Prepare to graph by resetting the range values to the standard viewing window. Then clear any functions from the Y= list. Do this by pressing [Y=] and then using the arrow keys and the CLEAR key to select and clear all functions. Next, return to the home screen by pressing [2nd] and then [QUIT] .

Let's graph the system of inequalities $y \geq x + 3$ and $y \leq -2x - 1$.

We will graph the system of inequalities with the **Shade** function. It graphs functions and shades above the first function entered and below the second function entered. The "greater than or equal to" symbol in $y \geq x + 3$, indicates that values on the line and above the line $y = x + 3$ will satisfy the inequality. Similarly, the "less than or equal to" in $y \leq -2x - 1$ indicates that values on the line and below the line $y = -2x - 1$ will satisfy the inequality. Therefore, the function $y = x + 3$ will be entered first and $y = -2x - 1$ will be entered second.

*TI-81*

ENTER: [2nd] [DRAW] 7 [X|T] [+] 3 [ALPHA] [,] [(-)] 2 [X|T] [−] 1 [ )] [ENTER]

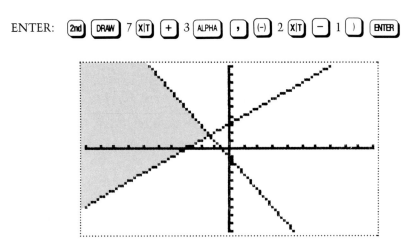

The shaded area indicates points which will satisfy the system of inequalities $y \geq x + 3$ and $y \leq -2x - 1$.

Before you graph another system of inequalities, you must clear the graphics screen.

*TI-81*

ENTER: [2nd] [DRAW] 1 [ENTER]    *Clears the graphics screen.*

**Example**

**Graph the system of inequalities** $\begin{cases} y < 0.5x - 2 \\ y > -4x + 1 \end{cases}$.

Values below the line $y = 0.5x - 2$ will satisfy the inequality $y < 0.5x - 2$.

Values above the line $y = -4x + 1$ will satisfy the inequality $y > -4x + 1$.

So, we will enter the function $y = -4x + 1$ first and the function $y = 0.5x - 2$ second.

ENTER: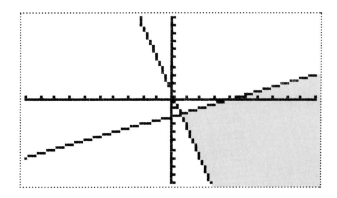

The points in the shaded area satisfy both $y < 0.5x - 2$ and $y > -4x + 1$.

# EXERCISES

Use your graphing calculator to solve the following systems of equations by graphing. Determine the $x$- and $y$-coordinates accurate to six decimal places.

**1.** $y = 1.236x - 1.0825$
$y = -0.7896x + 5.1783$

**2.** $y = 2.5x - 3$
$y = -1.8x + 2$

**3.** $2.1x + 3.2y = 4.3$
$1.4x - 1.8y = 1.6$

**4.** $3.12x + 4.68y = 5$
$-4.38x + 9.21y = 1.6$

**5.** $y = 8x + 1.27$
$y = -5x - 3.61$

**6.** $y = 2.345x + 1$
$y = 0.8765x - 3$

If you have a TI-81 graphing calculator, graph each system of inequalities and sketch the graph.

**7.** $y \geq x$
$y \leq 3$

**8.** $y \geq 5x$
$y \leq 8x$

**9.** $y \leq 4x - 2$
$y \geq 0.5x$

**10.** $y \leq -0.1x - 5$
$y \geq 0.2x - 5$

**11.** $y \geq 5 - x$
$y \leq 0.8x - 7$

**12.** $y \geq 12 - 4x$
$y \leq -3x + 9$

**13.** $y \geq 3x + 0.5$
$y \leq -6x - 2.8$

**14.** $12x + 6y \geq 12$
$y \leq x$

# 3-1 Graphing Systems of Equations

**Objective**    After studying this lesson, you should be able to:
- solve a system of equations by graphing.

**Application**

Reliable Rentables rents moving trucks for $40 a day plus 35¢ per mile driven. The Mover's Helper rents trucks for $36 a day plus 45¢ per mile driven. When is the total cost for a day's rental the same for both companies? When is it better to rent from Reliable Rentables?

Let $d$ = the total cost of a day's rental.
Let $m$ = the miles driven.

We can write the following equations.

$d = 40 + 0.35m$    *Total cost of renting from Reliable Rentables*
$d = 36 + 0.45m$    *Total cost of renting from The Mover's Helper*

By graphing these two equations, we can see how the rental rates compare. Each point on a line has coordinates that satisfy the equation of the line. Since (40, 54) is on both lines, it satisfies both equations. So, if you rent a truck from either company and drive 40 miles, the price will be $54. It is better to rent from Reliable Rentables when you drive more than 50 miles.

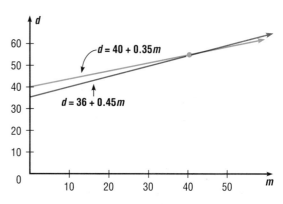

Together the equations $d = 40 + 0.35m$ and $d = 36 + 0.45m$ are called a **system of equations.** The **solution** of this system is (40, 54).

**Example 1**

**Solve this system of equations by graphing:** $\begin{cases} x + y = 4 \\ 2x + 3y = 9 \end{cases}$.

The slope-intercept form of $x + y = 4$ is $y = -x + 4$.

The slope-intercept form of $2x + 3y = 9$ is $y = -\frac{2}{3}x + 3$.

Since the two lines have different slopes, the graphs of the equations are intersecting lines. They intersect at (3, 1).

The solution of the system is (3, 1).

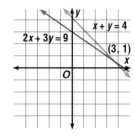

**Example 2**

Solve this system by graphing: $\begin{cases} 2y + 3x = 6 \\ 4y + 6x = 12 \end{cases}$.

$y = -\dfrac{3}{2}x + 3$ is the slope-intercept form of $2y + 3x = 6$.

$y = -\dfrac{3}{2}x + 3$ is the slope-intercept form of $4y + 6x = 12$.

Since the lines have the same slope *and* $y$-intercept, their graphs are the same line. Any ordered pair on that line will satisfy both equations. So, there are *infinitely many* solutions to this system.

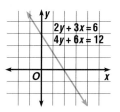

*The solution set is*
$\{(x, y) \mid 2y + 3x = 6\}$.

A system of equations that has at least one solution is called a **consistent** system of equations. The systems in Examples 1 and 2 are consistent. If a system has exactly one solution, it is an **independent** system. The system is **dependent** if it has an infinite number of solutions. So, the system in Example 1 is consistent and independent, and the system in Example 2 is consistent and dependent.

**Example 3**

APPLICATION
Consumerism

Perry's Plumbing charges $35 for any service call plus an additional $40 an hour for labor. A service call from Rapid Repair Plumbing costs $45 plus an additional $40 an hour for labor. When is the total price for a service call the same for both companies? When is it better to use Perry's Plumbing?

Let $h$ represent the hours of labor and $p$ represent the total price of the repair.

Write and graph a system of equations.

$p = 40h + 35$     *Total price of service call from Perry's*

$p = 40h + 45$     *Total price of service call from Rapid Repair*

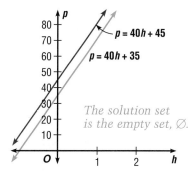

*The solution set is the empty set, ∅.*

These lines have the same slope but different $y$-intercepts. Their graphs are parallel lines. Since they never intersect, there are no solutions to this system. So, the total price of a service call from these two companies is never the same. It is always less expensive to use Perry's Plumbing.

A system with no solutions, like the one in Example 3, is called an **inconsistent** system.

This chart summarizes the possibilities for the graphs of two linear equations in two variables.

| Graphs of Equations | Slopes of Lines | Name of System of Equations | Number of Solutions |
|---|---|---|---|
| lines intersect | different slopes | consistent and independent | one |
| lines coincide | same slope, same intercepts | consistent and dependent | infinite |
| lines parallel | same slope, different intercepts | inconsistent | none |

# CHECKING FOR UNDERSTANDING

**Communicating Mathematics**

Read and study the lesson to answer these questions.

1. Does the graph at the right represent a system of equations that is *consistent and independent, consistent and dependent,* or *inconsistent?*

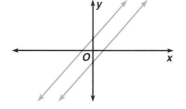

2. Describe the slope and $y$-intercepts of the lines whose graphs are shown in Exercise 1.

3. Explain the difference between an independent and a dependent system of equations.

4. Write a system of equations that is inconsistent.

**Guided Practice**

State the number of solutions to each system of equations graphed below. State whether the system is *consistent and independent, consistent and dependent,* or *inconsistent.* If the system is consistent and independent, estimate the solution.

5. $y = 2x - 3$
   $4x + 5y = 7$

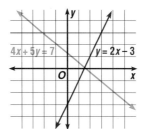

6. $y = -\dfrac{1}{2}x + 1$
   $2x + 4y = 4$

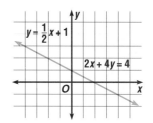

# EXERCISES

**Practice**

Graph each system of equations and state its solution. Also, state whether the system is *consistent and independent, consistent and dependent,* or *inconsistent.*

7. $x + y = 6$
   $3x + 3y = 3$

8. $x + 2y = 5$
   $3x - 15 = -6y$

9. $x + 1 = y$
   $2x - 2y = 8$

10. $2x + 4y = 8$
    $x + 2y = 4$

11. $y = -3x$
    $6y - x = -38$

12. $x + 5y = 10$
    $x + 5y = 15$

13. $x + y = 1$
$3x + 5y = 7$

14. $\frac{1}{2}x + \frac{1}{3}y = 2$
$x - y = -1$

15. $3x - 8y = 4$
$6x - 42 = 16y$

16. $\frac{3}{4}x - y = 0$
$\frac{y}{3} + \frac{x}{2} = 6$

17. $x + y = -6$
$2x - y = 2$

18. $2x + 3y = 5$
$-6x - 9y = -15$

19. $3x + 6 = 7y$
$x + 2y = 11$

20. $y = \frac{x}{2}$
$2y = x + 4$

21. $\frac{2}{3}x = \frac{5}{3}y$
$2x - 5y = 0$

22. $-2x + 5y = -14$
$x - y = 1$

23. $9x + 8y = 8$
$\frac{3}{4}x + \frac{2}{3}y = 8$

24. $9x - 5 = 7y$
$4\frac{1}{2}x - 3\frac{1}{2}y = 2\frac{1}{2}$

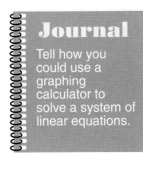

**Journal**

Tell how you could use a graphing calculator to solve a system of linear equations.

**Find a and b so that each of the following are true.**

25. $ax + 5y = b$ and $6x + 10y = 16$ are consistent and dependent equations.

26. The solution of the system $2x + 3y = 7$ and $ax + by = -10$ is $(2, 1)$.

27. There is no solution to the system $5x - 4y = 12$ and $bx + ay = 3$.

**Critical Thinking**

28. Mr. and Mrs. Leshin have more than ten children. The sum of the squares of the number of boys and the number of girls in the family equals 100. How many children do Mr. and Mrs. Leshin have?

**Applications**

29. **Business**   Mr. George bought 7 drums of two different cleaning fluids for his dry cleaning business. One of the fluids cost \$30 a drum and the other was \$20 a drum. The total price of the supplies was \$160. How much of each fluid did Mr. George buy? Write a system of equations and solve by graphing.

30. **Consumer Awareness**   Mrs. Katz is planning a family vacation. She bought 8 rolls of film and 2 camera batteries for \$23.00. The next day, her daughter went back and bought 6 more rolls of film and 2 batteries for her camera. This bill was \$18.00. What is the price of a roll of film and of a camera battery? Write a system of equations and solve by graphing.

**Mixed Review**

31. Name which points, $(0, 0)$, $(-1, -3)$, or $(4, 0)$, satisfy $4x - |y| \leq 12$. **(Lesson 2-8)**

32. Draw a graph of the inequality $y > x + 4$.   **(Lesson 2-8)**

33. Determine the slope of the line that passes through $(2, 0)$ and $(-3, 5)$. **(Lesson 2-4)**

34. State the domain and range of the relation $\{(9, 3), (2, -7), (1, 1)\}$. Is this relation a function?   **(Lesson 2-1)**

35. Solve the inequality $3x + 7 > 43$.   **(Lesson 1-7)**

# 3-2 Solving Systems of Equations Algebraically

**Objectives**

After studying this lesson, you should be able to:

- use the substitution method to solve a system of equations, and
- use the elimination method to solve a system of equations.

**Application**

Liz bought 12 feet of oak framing material to make a frame for the painting she is planning to enter in the Community Art Fair. If the difference between the length and the width of the rectangular frame is to be 18 inches, what will the dimensions of the frame be?

Let $\ell$ represent the length and $w$ represent the width.

$2\ell + 2w = 12$   *The perimeter of the frame will be 12 feet.*
$\ell - w = 1.5$   *18 inches is 1.5 feet.*

Let's try solving by graphing.

It is very difficult to determine an exact solution from the graph. However, we can use the graph to estimate the solution. The width is a little over 2 feet and the length is a little under 4 feet.

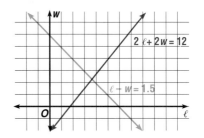

When the solution is a decimal or a fraction, it is usually easier to solve the system by using algebraic methods rather than by graphing. Two algebraic methods are **substitution** and **elimination.** Let's try using substitution to solve this problem.

First, solve one of the equations for one of the variables. Solving for $\ell$ in the second equation is a good choice.   *Why?*

$\ell = 1.5 + w$

Next, find $w$ by substituting $1.5 + w$ for $\ell$ in the first equation.

$2\ell + 2w = 12$
$2(1.5 + w) + 2w = 12$   *Substitute $1.5 + w$ for $\ell$.*
$3 + 2w + 2w = 12$   *The resulting equation now has only one variable, $w$.*
$4w = 9$
$w = 2.25$   *Solve for $w$.*

Now, find $\ell$ by substituting 2.25 for $w$ in $2\ell + 2w = 12$.

$2\ell + 2w = 12$
$2\ell + 2(2.25) = 12$   *Substitute 2.25 for $w$.*
$2\ell + 4.5 = 12$
$2\ell = 7.5$
$\ell = 3.75$   *Solve for $\ell$.*

The length of the frame will be 3.75 feet and the width will be 2.25 feet. Our estimate from the graph was very close to the solution.

The second algebraic method is the elimination method. There are two approaches to using this method.

**Example 1**

Use the elimination method to solve this system: $\begin{cases} x + 2y = -2 \\ 3x - 2y = 10 \end{cases}$.

Add the second equation to the first to eliminate $y$.

$x + 2y = -2$   *The y-coefficients, 2 and -2, are additive inverses.*
$\underline{3x - 2y = 10}$   *Add.*
$4x \quad\quad = 8$   *The variable y is eliminated.*
$\quad\quad x = 2$   *Solve for x.*

Now, find $y$ by substituting 2 for $x$ in either original equation.

*First equation*

$x + 2y = -2$
$2 + 2y = -2$   *Substitute 2 for x.*
$\quad 2y = -4$
$\quad\quad y = -2$   *Solve for y.*

*Second equation*

$3x - 2y = 10$
$3(2) - 2y = 10$
$\quad\quad -2y = 4$
$\quad\quad\quad y = -2$

The solution is $(2, -2)$.    **Check:**    *First equation: $2 + 2(-2) = -2$ ✓*

*Second equation: $3(2) - 2(-2) = 10$ ✓*

**Example 2**

Use the elimination method to solve this system: $\begin{cases} 2x + 3y = 2 \\ 3x - 4y = -14 \end{cases}$.

This time, adding the two equations will not eliminate either of the variables. However, if we multiply the first equation by 4 and the second equation by 3, the variable $y$ can be eliminated by addition.

$2x + 3y = 2$    *Multiply by 4.*    $8x + 12y = 8$

$3x - 4y = -14$    *Multiply by 3.*    $9x - 12y = -42$

*We could also solve the system by eliminating x first. Multiply the first equation by 3 and the second by -2.*

Now, add to eliminate $y$.

$8x + 12y = 8$
$\underline{9x - 12y = -42}$   *Add.*
$17x \quad\quad = -34$   *The variable y is eliminated.*
$\quad\quad x = -2$

*$6x + 9y = 6$*
*$\underline{-6x + 8y = 28}$*
*$17y = 34$*
*$y = 2$*

*Then solve for x.*
*$2x + 3(2) = 2$*
*$2x = -4$*
*$x = -2$*

Find $y$ by substituting $-2$ for $x$ in $2x + 3y = 2$.

$2(-2) + 3y = 2$    *Substitute -2 for x.*
$-4 + 3y = 2$
$\quad 3y = 6$
$\quad\quad y = 2$    *Solve for y.*

The solution is $(-2, 2)$.

**Example 3**

**CONNECTION**

~~~~~~~

**Geometry**

The sides of an angle are parts of two lines whose equations are $y = -\frac{3}{2}x - 6$ and $y = \frac{2}{3}x + 7$. Find the coordinates of the vertex of the angle.

Solve the system of equations.

$$y = -\frac{3}{2}x - 6$$
$$y = \frac{2}{3}x + 7$$

Substitute $\frac{2}{3}x + 7$ for $y$ in the first equation.

$$\frac{2}{3}x + 7 = -\frac{3}{2}x - 6$$
$$4x + 42 = -9x - 36 \qquad \textit{Multiply each side by 6 to eliminate fractions.}$$
$$13x = -78$$
$$x = -6$$

Substitute $-6$ for $x$ in $y = -\frac{3}{2}x - 6$.

$$y = -\frac{3}{2}(-6) - 6$$
$$y = 9 - 6$$
$$y = 3$$

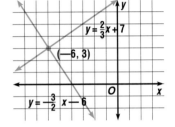

The coordinates of the vertex are $(-6, 3)$.          *The graph verifies the solution.*

# CHECKING FOR UNDERSTANDING

**Communicating Mathematics**

Read and study the lesson to answer these questions.

1. Describe how we could have solved the framing problem using the elimination method.

2. Explain when you would use the substitution method to solve a system of equations.

3. Would you use the substitution or elimination method to solve this system of equations? Why?
$$3x + 5y = 12$$
$$-3x + y = 34$$

**Guided Practice**

For each system, state the multipliers you would use to eliminate each variable by addition. Then solve each system of equations.

4. $x - y = 1$
   $3x - y = 3$

5. $2x + 3y = 7$
   $3x - 4y = 2$

6. $3x - 2y = 10$
   $4x + y = 6$

7. $3x + 4y = 6$
   $2x + 5y = 11$

8. $2x + 4y = 6$
   $5x - 3y = 2$

9. $x + 8y = 12$
   $3x - 7y = 5$

# EXERCISES

**Practice**

**Solve each system of equations using the substitution method.**

**10.** $y = 3x$
$x + 21 = -2y$

**11.** $x + y = 2$
$x - 2y = 0$

**12.** $3x - 2y = -3$
$3x + y = 3$

**13.** $2r + s = 1$
$r - s = 8$

**14.** $5s - 2t = 16$
$s + 3t = 10$

**15.** $2.5x - y = 11$
$3.25x + y = 12$

**Solve each system of equations using the elimination method.**

**16.** $m + n = 6$
$2m - n = 3$

**17.** $3x - 6y = 15$
$-3x + 5y = -8$

**18.** $4s + t = 9$
$3s - 2t = 4$

**19.** $4a + 3b = -2$
$5a + 7b = 17$

**20.** $4x - 6y = 12$
$x - 7y = 14$

**21.** $8x + 3y = 4$
$4x - 9y = -5$

**Solve each system of equations. (Use either algebraic method.)**

**22.** $6x + 4y = 80$
$x - 7y = -2$

**23.** $m + n = 6$
$m - n = 4.5$

**24.** $3x + 2y = 8$
$y - x = 2$

**25.** $9x + y = 30$
$6x - 15 = y$

**26.** $2x - y = 36$
$3x - \frac{1}{2}y = 26$

**27.** $3y - 2x = 4$
$\frac{1}{6}(3y - 4x) = 1$

**28.** $5a + 2b = -8$
$4a + 3b = 2$

**29.** $4x + 4y = -6$
$5x + 3y = 6$

**30.** $a - b = 0$
$3a + 2b = -15$

**31.** $\frac{1}{4}x + y = \frac{7}{2}$
$2x - y = 4$

**32.** $\frac{s + 3t}{7} = 3$
$11s - t = -7$

**33.** $\frac{2x + y}{3} = 15$
$3x - y = 5$

**34.** Find the coordinates of the vertices of the triangle whose sides are contained in the lines whose equations are $x - y = 7$, $3x - 11y = -11$, and $x + y + 1 = 0$.

**35.** Find the coordinates of the vertices of the parallelogram whose sides are contained in the lines whose equations are $2x + y = -12$, $2x - y = -8$, $2x - y - 4 = 0$, and $4x + 2y = 24$.

**Critical Thinking**

**36.** Solve the system $\begin{cases} \dfrac{1}{x} - \dfrac{1}{y} = \dfrac{5}{8} \\ \dfrac{3}{x} + \dfrac{2}{y} = -\dfrac{5}{8}. \end{cases}$   *Hint: Let $m = \dfrac{1}{x}$ and $n = \dfrac{1}{y}$.*

**Applications**

**37. Photography**   The perimeter of a rectangular picture is 86 inches. Twice the width exceeds the length by 2 inches. What are the dimensions of the picture?

**38. Construction**   The steel braces for a new bridge are in the shape of a right triangle. The hypotenuse is 75 meters. The length of one leg is four times one-third of the length of the other leg. Find the lengths of the legs.

**Computer**   **39.** If two people are different ages, was the older ever twice as old as the younger? The answer to any question of this type is yes. This BASIC program will find the year and ages at which one person's age was or will be any factor times a younger person's age. Let $a_1$ be the current age of the older person, $a_2$ be the age of the younger, and $f$ be the factor. Let $x_1$ be the age at which the older person is $f$ times $x_2$, the age of the younger person. The following equations are used to find the ages.

```
 10  PRINT "ENTER CURRENT YEAR"
 20  INPUT Y1
 30  PRINT "ENTER AGES OF YOUNGER
     AND OLDER"
 40  INPUT A1, A2
 50  PRINT "ENTER FACTOR"
 60  INPUT F
 70  LET X2 = (A1-A2)(F-1)
 80  LET X1 = X2 + A1 - A2
 90  LET Y2 = Y1 - A1 + X1
100  PRINT "YEAR", "OLDER",
     "YOUNGER"
110  PRINT Y2, X1, X2
120  END
```

$x_1 = f \cdot x_2$   *The age of the older person is $f$ times the younger person's.*
$x_1 - x_2 = a_1 - a_2$   *The difference in age is the same, then and now.*

Solve the system of equations by substitution.

$f \cdot x_2 - x_2 = a_1 - a_2$
$x_2(f - 1) = a_1 - a_2$   *Distributive property*
$x_2 = \dfrac{a_1 - a_2}{f - 1}$

Once $x_2$ is found, the formula $x_1 = x_2 + a_1 - a_2$ is used to solve for $x_1$.

**a.** Maggie is 12 years old and Dwayne is 16. How many years ago was Dwayne five times as old as Maggie? How old were they at that time?

**b.** Barry is 9 years old and Karen is 27. How old were they when Karen was ten times as old as Barry? How long ago was that?

**c.** Tiffany is 17 years old and her mother is 45. How old will they be when Tiffany's mother is twice as old as Tiffany? How many years from now will that be?

**Mixed Review**   **Draw the graph of each system of equations and state its solution. (Lesson 3-1)**

**40.** $2x + 3y = -16$
    $2y = 4x$

**41.** $3x + 4y = 8$
    $6y - 8x = 12$

**42.** Find the slope-intercept form of the equation of the line that has slope $-2$ and passes through $(3, 1)$.   **(Lesson 2-5)**

**43.** Find a value of $a$ for which the graph of $y = ax + 9$ is perpendicular to the graph of $x + 3y = 14$.   **(Lesson 2-4)**

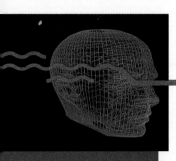

# Technology

▶ **BASIC**
Spreadsheets
Software

## Solving Systems of Equations

You have studied several ways to solve a system of equations. You can also use the BASIC program below to solve a system of two equations. It gives one of three outputs: the ordered pair solution, a message saying there are infinitely many solutions, or a message saying there is no solution. The ordered pair solution is the ordered pair that satisfies both equations. The second response tells you that the two equations are equivalent and there are an infinite number of solutions that satisfy both equations. The third response states that there are no ordered pairs that satisfy both equations.

The BASIC program requires you to first write each equation of the system in standard form. The first equation will be named $Ax + By = C$ and the second equation will be named $Dx + Ey = F$. The program will ask you to enter the values for $A$, $B$, and $C$ and then for $D$, $E$, and $F$.

```
100   PRINT "ENTER A, B, AND C FOR THE EQUATION AX + BY = C."
110   INPUT A,B,C
120   PRINT "ENTER D, E, AND F FOR THE EQUATION DX + EY = F."
130   INPUT D,E,F
140   IF A*E = B*D GOTO 190
150   X = (C*E-F*B)/(A*E-B*D)
160   Y = (A*F-D*C)/(A*E-B*D)
170   PRINT "(";X;", ";Y;")"
180   GOTO 230
190   IF A*F=C*D GOTO 220
200   PRINT "THERE IS NO SOLUTION."
210   GOTO 230
220   PRINT "THERE ARE INFINITELY MANY SOLUTIONS."
230   END
```

To run the program, type RUN and then hit the enter key. Enter each value requested when a question mark appears.

# EXERCISES

**Use the BASIC program to determine what type of solution each system of equations has. If it has a unique solution, state it.**

1. $x - 3y = 6$
   $2x + 6y = 24$

2. $3x - y = 1$
   $-6x + 2y = -2$

3. $x - 5y = 2$
   $-2x + 10y = 4$

4. $x + 4y = 2$
   $-x + y = -7$

5. $3x - 6y = 12$
   $2x + 3y = 1$

6. $9x + 3y = 9$
   $3x + y = 3$

# Cramer's Rule

**Objectives**

After studying this lesson, you should be able to:

- find the value of a second order determinant, and
- solve a system of equations using Cramer's rule.

*FYI* · · ·

The theory of determinants is attributed to a German, Gottfried Wilhelm Leibniz. His work expanded upon the earlier work of Japanese mathematician Seki Kōwa.

Another method for solving systems of equations is to use Cramer's rule. The rule gives us a quick way to find the solution to a system of two equations with two unknowns. Cramer's rule makes use of determinants.

A **determinant** is a square array of numbers or variables. Vertical bars are used to enclose the array and to signify a determinant. The determinant below has two rows and two columns and is called a **second order determinant.**

$$\text{rows} \begin{vmatrix} a & b \\ c & d \end{vmatrix}$$

columns

*The quantities in a determinant are called elements.*

We find the value of a second order determinant as follows.

| *Value of a Second Order Determinant* | $\begin{vmatrix} a & b \\ c & d \end{vmatrix} = ad - bc$ |
|---|---|

Did you notice that the value of a second order determinant is found using products along the diagonals?

$$\begin{vmatrix} a & b \\ c & d \end{vmatrix} \quad \Rightarrow \quad ad - bc$$

*bc*   *ad*

**Example 1**

Find the value of each determinant.

**a.** $\begin{vmatrix} 4 & 2 \\ 6 & 8 \end{vmatrix}$

$\begin{vmatrix} 4 & 2 \\ 6 & 8 \end{vmatrix} = 4 \cdot 8 - 2 \cdot 6$

$= 20$

**b.** $\begin{vmatrix} -2 & 0 \\ 7 & -6 \end{vmatrix}$

$\begin{vmatrix} -2 & 0 \\ 7 & -6 \end{vmatrix} = -2(-6) - 0(7)$

$= 12$

To discover how Cramer's rule uses determinants to solve a system of linear equations, let's consider the following system.

$$ax + by = c$$
$$dx + ey = f$$

*a, b, c, d, e, and f represent constants, not variables.*

Solve for $y$ using the elimination method.

$$adx + bdy = cd$$
$$\underline{-adx - aey = -af}$$
$$bdy - aey = cd - af$$
$$(bd - ae)y = cd - af$$
$$y = \frac{cd - af}{bd - ae}$$
$$y = \frac{af - cd}{ae - bd}$$

*Multiply the first equation by d.*
*Multiply the second equation by −a.*
*Add.*
*Factor.*
*Notice that bd − ae cannot be zero.*
*Multiply numerator and denominator by −1.*

Solving for $x$ the same way, we find the following expression for $x$.

$$x = \frac{ce - bf}{ae - bd}$$

So, the solution to the system $\begin{cases} ax + by = c \\ dx + ey = f \end{cases}$ is $\left[\dfrac{ce - bf}{ae - bd}, \dfrac{af - cd}{ae - bd}\right]$.

Notice that the two fractions have the same denominator. It can be written as a determinant. The numerators can be written as determinants too.

$$ae - bd = \begin{vmatrix} a & b \\ d & e \end{vmatrix} \qquad ce - bf = \begin{vmatrix} c & b \\ f & e \end{vmatrix} \qquad af - cd = \begin{vmatrix} a & c \\ d & f \end{vmatrix}$$

So, we can find the solution to a system of two linear equations in two variables using determinants. This method is **Cramer's rule.**

**Cramer's Rule**

The solution to the system $\begin{cases} ax + by = c \\ dx + ey = f \end{cases}$ is $(x, y)$,

where $x = \dfrac{\begin{vmatrix} c & b \\ f & e \end{vmatrix}}{\begin{vmatrix} a & b \\ d & e \end{vmatrix}}$, $y = \dfrac{\begin{vmatrix} a & c \\ d & f \end{vmatrix}}{\begin{vmatrix} a & b \\ d & e \end{vmatrix}}$, and $\begin{vmatrix} a & b \\ d & e \end{vmatrix} \neq 0$.

**Example 2**

Use Cramer's rule to solve the system $\begin{cases} 3x + 4y = -7 \\ 2x + y = -3 \end{cases}$.

$$x = \frac{\begin{vmatrix} -7 & 4 \\ -3 & 1 \end{vmatrix}}{\begin{vmatrix} 3 & 4 \\ 2 & 1 \end{vmatrix}} \qquad y = \frac{\begin{vmatrix} 3 & -7 \\ 2 & -3 \end{vmatrix}}{\begin{vmatrix} 3 & 4 \\ 2 & 1 \end{vmatrix}}$$

$$= \frac{-7(1) - 4(-3)}{3(1) - 4(2)} \qquad = \frac{3(-3) - (-7)(2)}{3(1) - 4(2)}$$

$$= \frac{5}{-5} \qquad\qquad = \frac{5}{-5}$$

$$= -1 \qquad\qquad\quad = -1$$

The solution is $(-1, -1)$.

Example 3

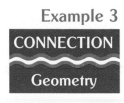

CONNECTION

Geometry

Two sides of a parallelogram are contained in the lines whose equations are $3x + 2y = 8$ and $4x - 2y = 20$. Find the coordinates of a vertex of the parallelogram.

Solve this system of equations using Cramer's rule $\begin{cases} 3x + 2y = 8 \\ 4x - 2y = 20 \end{cases}$.

$$x = \dfrac{\begin{vmatrix} 8 & 2 \\ 20 & -2 \end{vmatrix}}{\begin{vmatrix} 3 & 2 \\ 4 & -2 \end{vmatrix}} \qquad y = \dfrac{\begin{vmatrix} 3 & 8 \\ 4 & 20 \end{vmatrix}}{\begin{vmatrix} 3 & 2 \\ 4 & -2 \end{vmatrix}}$$

$$= \dfrac{8(-2) - 2(20)}{3(-2) - 2(4)} \qquad = \dfrac{3(20) - 8(4)}{3(-2) - 2(4)}$$

$$= \dfrac{-56}{-14} \qquad = \dfrac{28}{-14}$$

$$= 4 \qquad = -2 \qquad \text{The vertex is } (4, -2).$$

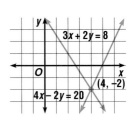

# CHECKING FOR UNDERSTANDING

**Communicating Mathematics**

**Read and study the lesson to answer these questions.**

1. What would be the value of a determinant if both elements in a row or column are zero?

2. Describe some possible situations where a determinant would have a value of zero when none of its elements are zero.

3. Given the system $\begin{cases} 345x + 678y = 0.8765 \\ 234x - 0.459y = 1836 \end{cases}$, which method, elimination or Cramer's rule, would be the best way to solve this system? Why? Would you use your calculator with either method?

**Guided Practice**

**Find the value of each determinant.**

4. $\begin{vmatrix} 6 & 1 \\ 4 & 3 \end{vmatrix}$

5. $\begin{vmatrix} 1 & 0 \\ 1 & 0 \end{vmatrix}$

6. $\begin{vmatrix} 7 & -3 \\ 0 & 1 \end{vmatrix}$

7. $\begin{vmatrix} 1 & 0 \\ 0 & 1 \end{vmatrix}$

8. $\begin{vmatrix} -5 & -2 \\ -3 & 11 \end{vmatrix}$

9. $\begin{vmatrix} -8 & -7 \\ -4 & -6 \end{vmatrix}$

10. $\begin{vmatrix} 4 & 2 \\ 1 & -3 \end{vmatrix}$

11. $\begin{vmatrix} 5 & -3 \\ -1 & -2 \end{vmatrix}$

12. $\begin{vmatrix} 4 & -11 \\ 0 & -9 \end{vmatrix}$

**Write the determinants that you would use to solve each system using Cramer's rule. Then solve.**

13. $4x - y = 3$
    $3x + 2y = 5$

14. $4m + 2n = 8$
    $6m - 3n = 0$

15. $x - 2y = 8$
    $3x - 5y = 21$

16. $a - b = 0$
    $4a + 10b = -6$

17. $s + t = 6$
    $s - t = 2$

18. $3x - 5y = -7$
    $x + 2y = 16$

# EXERCISES

Find the value of each determinant.

19. $\begin{vmatrix} -4 & 6 \\ -13 & 24 \end{vmatrix}$

20. $\begin{vmatrix} -6 & 7 \\ -9 & 10 \end{vmatrix}$

21. $\begin{vmatrix} 2 & -5 \\ -1 & 11 \end{vmatrix}$

22. $\begin{vmatrix} -13 & -11 \\ 17 & -12 \end{vmatrix}$

23. $\begin{vmatrix} 0.9 & 0.12 \\ 89 & -23 \end{vmatrix}$

24. $\begin{vmatrix} 0.007 & 0.873 \\ 0.063 & 7.857 \end{vmatrix}$

Use Cramer's rule to solve each system of equations.

25. $x - 4y = 1$
$2x + 3y = 13$

26. $s + t = 5$
$3s - t = 3$

27. $m - n = 4$
$m + 2n = 1$

28. $3x + 2y = 9$
$2x - 3y = 19$

29. $r + 11 = 8s$
$8(r - s) = 3$

30. $2x - y = 7$
$x + 3y = 7$

31. $3x + 8 = -y$
$4x - 2y = -14$

32. $5a + 4b = -1$
$2a - b = 10$

33. $3x - 7y = 2$
$6x - 13y = 4$

34. $6x + 5y = -7$
$2x - 3y = 7$

35. $0.2a = 0.3b$
$0.4a - 0.2b = 0.2$

36. $3.5x + 4y = -5$
$2(x - y) = 10$

37. $\frac{1}{6}x - \frac{1}{9}y = 0$
$x + y = 15$

38. $\frac{x}{2} - \frac{2y}{3} = 2\frac{1}{3}$
$3x + 4y = -50$

39. $7y + 4x = 22$
$8x - 2y = -5$

40. Explain why Cramer's rule will not work if a system of equations is dependent or inconsistent.

41. **Consumer Awareness**   Handy Hal's Hardware Store sells packages of screws that contain two different sizes of screws. One package contains 5 half-inch screws and 12 quarter-inch screws. It sells for 56¢. A package of 8 half-inch screws and 15 quarter-inch screws sells for 77¢. How much does each type of screw cost?

42. **Banking**   Donna Bowers has a total of $4000 in her savings account and in a certificate of deposit. Her savings account earns 6.5% interest annually. The certificate of deposit pays 8% if the money is invested for one year. How much does she have in each investment if her interest earnings for the year will be $297.50?

Solve each system of equations by using either the elimination or substitution method.   (Lesson 3-2)

43. $2x + y = 0$
$5x + 3y = 1$

44. $4a - 3b = -4$
$3a - 2b = -4$

45. Find the slope of a line that is perpendicular to the line whose equation is $x = 4y + 7$.   (Lesson 3-1)

46. State the domain and range of the relation $\{(9, 0), (3, 1), (12, 7), (1, -4), (12, 8), (-11, -3), (0, -6)\}$. Is this relation a function?   (Lesson 2-1)

47. Solve the equation $|x - 8| = 9$.   (Lesson 1-6)

# 3-4 Graphing Systems of Inequalities

**Objectives**

After studying this lesson, you should be able to:
- graph a system of inequalities, and
- solve a system of inequalities.

**Application**

Ben Yeo intends to major in market research at Oakdale College. He must graduate from high school with at least a 2.5 grade point average and earn a total score of at least 900 points on the Scholastic Aptitude Test (SAT) to be admitted to this program.

This situation can be represented by a **system of inequalities.** To solve a system of inequalities, we must find the ordered pairs that satisfy both inequalities. One way of doing that is to graph both inequalities on the same coordinate plane. The intersection of the two graphs contains the ordered pairs in the solution set. If the graphs do not intersect, then the system has no solution. Let's try graphing the system described above.

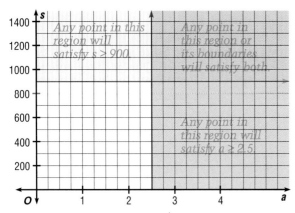

First, we need to write a system of inequalities to represent the situation.

Let $a$ represent Ben's grade point average and $s$ represent his SAT score.

$a > 2.5$     *Ben's average must be 2.5 or greater.*
$s \geq 900$     *His test score must be a 900 or greater.*

Graph the system. Any point in the intersection of the two graphs is a solution to the system. If Ben has a 2.75 grade point average and scores a 900 on the SAT, will he be admitted to Oakdale College?

**Example 1**

Solve this system of equations by graphing: $\begin{cases} y > 2x + 1 \\ y < 2x - 2 \end{cases}$.

Graph each inequality.

The graphs of the two inequalities have no points in common. So, no ordered pair will satisfy both inequalities. The solution set is the empty set, $\varnothing$.

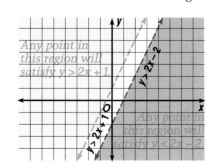

*The broken lines indicate that the boundaries are not part of the graphs.*

**Example 2**

Solve this system of inequalities by graphing $\begin{cases} y \le 3 \\ y > x \\ x \ge -2 \end{cases}$

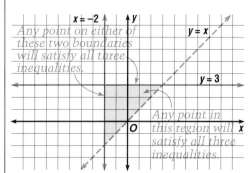

*Any point on either of these two boundaries will satisfy all three inequalities.*

*Any point in this region will satisfy all three inequalities.*

As you recall, an absolute value inequality can be restated as two inequalities using an *and* or an *or*. So, an absolute value inequality can be graphed like a system of inequalities.

**Example 3**

Graph $|y| \le 4$.

First, rewrite the inequality as two inequalities.
$y \ge -4$ and $y \le 4$     *We could also write this as $-4 \le y \le 4$.*
Now graph.

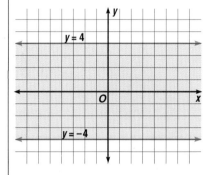

*The "and" tells us that the solution set is the intersection of the two graphs. Both inequalities must be satisfied for an ordered pair to be a solution.*

**Example 4**

Graph $|x| > 3$.

This inequality can be rewritten as $x < -3$ or $x > 3$.

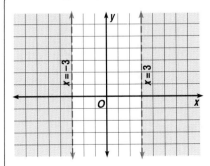

*The "or" tells us that an ordered pair only has to satisfy one inequality or the other to be a solution. So, the solution is the union of the two graphs.*

# CHECKING FOR UNDERSTANDING

**Communicating Mathematics**

Read and study the lesson to answer these questions.

1. How do you determine whether a point is a solution to a system of inequalities?

2. When you draw a graph of a linear inequality, how do you determine which side of the equation to shade?

3. Can you draw graphs of linear inequalities on a graphing calculator?

4. Name which points, $(0, 0)$, $(2, 5)$, $(-1, 3)$, or $(-2, -4)$, are solutions to the inequality graphed at the right.

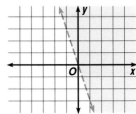

**Guided Practice**

Does the given point satisfy the system of inequalities?

5. $y < x - 2$; $(0, 1)$
   $y > -x$

6. $y < 2x + 4$; $(0, 0)$
   $y > 3x - 2$

7. $y < 3$; $(4, 4)$
   $x \geq -1$

8. $x \leq 1$; $(1, 1)$
   $y > -2$

Determine which, if any, of the ordered pairs $(3, 1)$, $(-3, -1)$, $(2, 1)$, $(1, 2)$, and $(-1, -2)$ satisfy each of the following.

9. $|y| \leq 2$

10. $|x| > 3$

11. $|y| \leq 2x + 2$

12. $y < 2x - 1$
    $y > 3x + 2$

13. $y \geq 3x - 7$
    $y \leq 2x + 4$

14. $|x| \geq 1$
    $|x| \leq 2$

# EXERCISES

**Practice**

Solve each system by graphing.

15. $x + y > 2$
    $y > 3$

16. $x \leq 1$
    $y > 3$

17. $y < -2$
    $y - x > 1$

18. $y \geq x - 3$
    $y \geq -x + 1$

19. $y - x \leq 3$
    $y \geq x + 2$

20. $y \geq 2x - 2$
    $y \leq -x + 2$

21. $y < -x - 3$
    $x > y - 2$

22. $x \geq 3 - y$
    $2x - 3y \leq 6$

23. $x + 2y \geq 7$
    $3x - 4y < 12$

24. $y > x + 2$
    $2y < x - 3$

25. $|x| > 5$
    $x + y < 6$

26. $|x + 2| < 3$
    $x + y \geq 1$

27. $x > 1$
    $y < -1$
    $y < x$

28. $y < 2$
    $y \geq 2x$
    $y \geq x + 1$

29. $x \geq -2$
    $2y \geq 3$
    $x - y \leq -5$

30. $y < 2x + 1$
    $y > 2x - 2$
    $3x + y > 8$

31. $x + y < 9$
    $x - y > 3$
    $y - x > 4$

32. $y > x + 3$
    $y < x - 4$
    $2y + 3x > 4$

**Critical Thinking**

**33.** Write a system of three inequalities that will have no intersection. The system will have no solution. Then, explain how to change one of the inequalities so that a solution exists.

**Applications**

**34. Design** Joe is designing a new dart board. The center of the board is defined by the inequality $|x| + |y| \leq 2$. Draw the graph of this inequality to see what Joe's new dart board will look like.

**35. Sports** In a baseball game, a ball that lands to the right of the right field baseline or to the left of the left field baseline is a foul ball. Think of placing a baseball diamond on a coordinate plane with home plate at the origin. Let first base be on the $x$-axis and third base be on the $y$-axis. Write a system of inequalities that will describe foul territory.

**Mixed Review**

Solve each system of equations using Cramer's rule. **(Lesson 3-3)**

**36.** $2x + 3y - 8 = 0$
  $3x + 2y - 17 = 0$

**37.** $6a + 7b = -10.15$
  $9.2a - 6b = 69.944$

**38. Statistics** The prediction equation in a study of the relationship between minutes spent studying, $s$, and test scores, $t$, is $t = 0.36s + 61.4$. Predict the score a student would receive if she spent 1 hour studying. **(Lesson 2-6)**

**39.** Write the linear equation $y = -\frac{1}{3}x - 18$ in standard form. **(Lesson 2-2)**

**40.** Write a mathematical expression for the verbal expression "the theater can hold no more than 400 people." **(Lesson 1-8)**

## MID-CHAPTER REVIEW

**1.** Solve this system of equations by graphing. **(Lesson 3-1)**
  $3x + 2y = 8$
  $4x - 3y = 5$

Solve each system of equations by substitution or elimination. **(Lesson 3-2)**

**2.** $4x + y = 7$
  $2x - 3y = -7$

**3.** $4r - 2s = 13$
  $2r + 2s = 7$

**4.** $x + y = 6$
  $-2x + y = -3$

**5.** $6x - y = 20$
  $4x + y = 6$

**6.** Find the value of the determinant $\begin{vmatrix} 3 & 4 \\ -7 & 2 \end{vmatrix}$. **(Lesson 3-3)**

Solve each system using Cramer's rule. **(Lesson 3-3)**

**7.** $5x - 2y = 13$
  $2x + 4y = 10$

**8.** $2x - 3y = 7$
  $3x - y = 2$

Solve each system of inequalities by graphing. **(Lesson 3-4)**

**9.** $x + y \geq 3$
  $x - y \leq 1$

**10.** $x \geq -3$
  $2y \geq 4$
  $x - y \leq -5$

# 3-5 Problem-Solving Strategy: Solve a Simpler Problem

**Objective**

After studying this lesson, you should be able to:

■ solve problems using the strategy of solving a simpler problem.

**Application**

Sandy is a marketing research executive for a soda company. The research team has arranged to perform a survey in 16 different local shopping malls. To insure that competing soda companies will not learn of the survey results, Sandy will arrange for telephone lines to be set up so that each of the survey stations has a direct line to each of the other stations. How many telephone lines does Sandy need to have installed?

Sometimes if a method for solution is not obvious, solving a problem directly can be difficult. In these cases, it may be helpful to set aside the original problems for a moment to solve one or more simpler, similar problems. After you have solved the simpler problems, you can use the concepts that were used to solve those on the original. Let's try using this strategy to find the number of telephone lines that Sandy will have to have installed.

Draw diagrams of lines required for different numbers of survey stations.

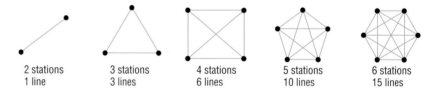

| 2 stations | 3 stations | 4 stations | 5 stations | 6 stations |
| 1 line | 3 lines | 6 lines | 10 lines | 15 lines |

Let's organize our results in a table and look for a pattern.

| Number of stations | 2 | 3 | 4 | 5 | 6 | . . . | $n$ |
|---|---|---|---|---|---|---|---|
| Number of lines | 1 | 3 | 6 | 10 | 15 | . . . | $\dfrac{n(n-1)}{2}$ |

Using the formula that we have discovered by solving simpler problems, we can see that Sandy will need to have $\dfrac{(16)(15)}{2}$ or 120 telephone lines installed.

<table>
<tr><td>Example</td><td>

**Find the sum of the whole numbers 1 to 1000.**

We could add all of those numbers directly, but even with a calculator that would be time consuming and tedious. Let's look at the sum of the whole numbers 1 to 10 to see if we can find a faster way.

$$S = 1 + 2 + 3 + \ldots + 10$$
$$S = 10 + 9 + 8 + \ldots + 1$$
$$2S = 11 + 11 + 11 + \ldots + 11$$
$$2S = 10 \cdot 11$$
$$S = 5 \cdot 11 \text{ or } 55$$

Now, extend this concept to the original problem.

$$S = 1 + 2 + 3 + \ldots + 1000$$
$$S = 1000 + 999 + 998 + \ldots + 1$$
$$2S = 1001 + 1001 + 1001 + \ldots + 1001$$
$$2S = 1000 \cdot 1001$$
$$S = 500 \cdot 1001 \text{ or } 500{,}500$$

The sum of the whole numbers from 1 to 1000 is 500,500.
</td></tr>
</table>

# CHECKING FOR UNDERSTANDING

**Communicating Mathematics**

Read and study the lesson to answer these questions.

1. When is the strategy of solving a simpler problem useful?

2. How many telephone lines would Sandy need to have installed if the number of survey stations was increased to 20?

3. What is the sum of the whole numbers from 1 to 2000?

**Guided Practice**

4. A team is eliminated from the All-City basketball tournament if they lose one game. If there are 30 teams playing in the tournament, how many games will need to be played to determine a champion?

5. A sewer drain pipe is 750 inches long. A spider climbs up 100 inches during the day but falls back 80 inches during the night. If the spider starts at the bottom of the pipe, on what day will it get to the top?

# EXERCISES

Use any strategy.

6. Find the digits $X$, $Y$, and $Z$ that make the following equation true. Assume that $XX$ and $YYZZ$ are two-digit and four-digit base-ten numbers.
$$(XX)^2 = YYZZ$$

7. A total of 3001 digits were used to print the page numbers on the Kettering High School Yearbook. How many pages are in the book?

8. The mailing list for the Discovery Record Club has 50,000 names. Their clerk is trying to find one of the names to update the address. She is trying to locate the name by cutting the alphabetical list in half and keeping the half with the name. If she continues this procedure, how many times will she have to cut the list before she finds the name?

9. Arrange twelve toothpicks as shown at the right. Form five squares by moving three toothpicks.

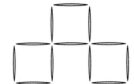

10. The community food pantry had a number of cans of vegetables to distribute. If they put ten cans in a bag, there would be one bag with only nine cans. If they put nine cans in a bag, there would be one bag with only eight. Bags of eight would leave one bag with only seven, and so on, down to bags of two that would leave one bag with only one can. What is the smallest possible number of cans that the food pantry could have?

11. Simplify the following expression.
$(99 - 9)(99 - 19)(99 - 29) \cdot \cdot \cdot (99 - 199)$

12. If you could say one number in a second, about how long would it take you to count to one billion?

13. Triskaidekaphobia is the fear of the number thirteen. Can there be a year with no "Friday the thirteenth?" What is the greatest number of "Friday the thirteenth"'s that can occur in one year?

14. A cube that is 3 inches on each edge is painted green on all six faces. If you were to cut the cube into 27 smaller cubes of 1 inch on each edge, how many of these cubes would have exactly three faces painted green? How many would have two faces painted? 1 face painted? 0 faces painted?

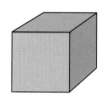

## COOPERATIVE LEARNING ACTIVITY

**Work in groups. Each person in the group must understand the solution and be able to explain it to any person in class.**

How many distinct convex polygons can you find whose vertices are some, or all, of the ten points that are marked on this circle?

*Hint: Distinct polygons can have the same shape, but cannot have the same vertices.*

# 3-6 Linear Programming

Objective

After studying this lesson, you should be able to:

- find the maximum and minimum values of a function over a region using linear programming techniques.

Application

The Northern Wisconsin Paper Mill can convert wood pulp to either notebook paper or newsprint. The mill can produce, at most, 200 units of paper a day. At least 10 units of notebook paper and 80 units of newsprint are required daily by regular customers. Write inequalities to show the possible daily production of the paper mill.

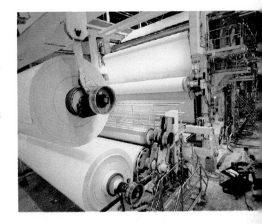

Let x represent the number of units of notebook paper produced.
Let y represent the number of units of newsprint produced.

Since the mill cannot produce negative units of paper, x and y must be nonnegative numbers.

$$x \geq 0 \text{ and } y \geq 0$$

The mill cannot produce more than 200 units of paper in a day.

$$x + y \leq 200$$

The plant must produce at least 10 units of notebook paper and 80 units of newsprint.

$$x \geq 10 \text{ and } y \geq 80$$

If we graph these inequalities, all of the points in their intersection are possible combinations of notebook paper and newsprint that the mill can produce. This area of intersection of the graphs is called the **feasible region** of production for the mill. The inequalities are called the **constraints.**

Let's graph the constraints.

$x \geq 0$
$y \geq 0$
$x \geq 10$
$y \geq 80$
$x + y \leq 200$

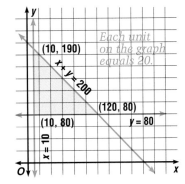

*Each unit on the graph equals 20.*

(10, 190)

x + y = 200

(120, 80)

(10, 80)   y = 80

x = 10

*The first two constraints indicate that the graph is in first quadrant, so we only need to graph the last three constraints.*

The shaded region is the feasible region. It contains all possible solutions.

If we choose any point within the region it should be a solution to each inequality. Let's try (30, 130).

**Check:**
$$30 \geq 0$$
$$130 \geq 0$$
$$30 \geq 10$$
$$130 \geq 80$$
$$30 + 130 \leq 200 \qquad \text{All are true.}$$

Choose a point outside the region and test it in each inequality. Is it a solution to all of the inequalities?

The manager of the mill needs to decide how many units of each kind of paper the plant should produce. There are many options, as we have seen by graphing the constraints. Of course the company would like to make as much profit as possible. If the profit on a unit of notebook paper is $500 and the profit on a unit of newsprint is $350, how much should the manager have the mill produce?

The profit can be defined by the function: $f(x, y) = 500x + 350y$.

Mathematicians have shown that the maximum and minimum values of a function, $f(x, y)$, occur at the vertices of the feasible region. So, the points we need to try are (10, 80), (10, 190), and (120, 80). Let's make a chart to organize our results.

| $(x, y)$ | $500x + 350y$ | profit($) |
|---|---|---|
| (10, 80) | 500(10) + 350(80) | 33,000 |
| (10, 190) | 500(10) + 350(190) | 71,500 |
| (120, 80) | 500(120) + 350(80) | 88,000 |

Check some other values within the feasible region to convince yourself that we have found the maximum profit for the mill. According to our results, the manager should have the mill produce 120 units of notebook paper and 80 units of newsprint in order to maximize profit.

We have just used **linear programming.** This procedure is used to find the maximum or minimum value of a function subject to given conditions on the variables, called constraints. The constraints are usually expressed as linear inequalities.

Example 1

APPLICATION

Business

Jerry works no more than 20 hours a week during the school year. He is paid $10 an hour for tutoring geometry students and $7 an hour for delivering pizzas for Pizza King. He wants to spend at least 3 hours, but no more than 8 hours, a week tutoring, Find Jerry's maximum weekly earnings.

Let $t$ represent the number of hours spent tutoring and $d$ represent the number of hours spent delivering pizzas.

The constraints are:
$t + d \leq 20$
$3 \leq t \leq 8$
$d \geq 0$.

Graph the constraints.

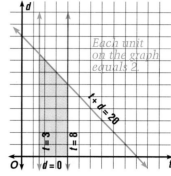

*Each unit on the graph equals 2.*

*Jerry can work no less than zero hours at either job, so the graph is in the first quadrant.*

The vertices of the region are (3, 0), (8, 0), (8, 12), and (3, 17). Now, find Jerry's earnings for each of these combinations to find his maximum weekly earnings. The function $f(t, d) = 10t + 7d$ defines Jerry's weekly earnings.

| $(t, d)$ | $10t + 7d$ | earnings |
|----------|------------|----------|
| (3, 0)   | 10(3) + 7(0) | 30 |
| (8, 0)   | 10(8) + 7(0) | 80 |
| (8, 12)  | 10(8) + 7(12) | 164 |
| (3, 17)  | 10(3) + 7(17) | 149 |

Jerry can earn $164 a week if he tutors for 8 hours and delivers pizzas for 12 hours.

Example 2

CONNECTION

Geometry

Use your calculator to find the maximum and minimum values of $f(x, y) = 3x + 4y$ for the polygonal region determined by these inequalities. Identify the polygon.
$x \leq 6$    $y \leq 3$    $x - 3y \leq 9$    $3x + y \leq 6$

First we must find the vertices of the feasible region. Graph the inequalities.

The coordinates of the vertices are (1, 3), (6, 3), (2.7, −2.1), and (6, −1).

Now use your calculator to evaluate $f(x, y) = 3x + 4y$ for each vertex.

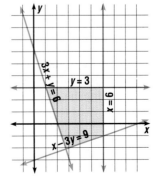

*You may use a graphing calculator to find the vertices if one is available.*

ENTER: 3 ⊠ 1 ⊞ 4 ⊠ 3 ⊟ *15*

*The other values are found in a similar manner.*

The maximum value of $f(x, y) = 3x + 4y$ for this region is 30 at (6, 3) and the minimum value is $-0.3$ at (2.7, $-2.1$). The polygon is a quadrilateral.

# CHECKING FOR UNDERSTANDING

**Communicating Mathematics**

**Read and study the lesson to answer these questions.**

1. In your own words, define linear programming.

2. Why was it important that the shaded region in Example 1 be in the first quadrant?

3. In a business like the one described in the introductory problem, is it always possible or desirable to obtain the maximum profit each day?

4. Name the points where a maximum or minimum value of a function could occur for the constraints graphed at the right.

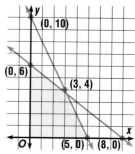

**Guided Practice**

**Given the function $f(x, y) = 3x + 2y$, find each value.**

5. $f(4, 1)$  6. $f(3, 3)$  7. $f(-2, 1)$  8. $f(6, 0)$

**A polygonal region has vertices (0, 0), (4, 0), (5, 5), and (0, 8). Find the maximum and minimum values of each function over this region.**

9. $f(x, y) = x + 3y$  10. $f(x, y) = -x - 3y$  11. $f(x, y) = 0.5x - 1.5y$

# EXERCISES

**Practice**

**Given the function $f(x, y) = 5x - 2y$, use a calculator to find each value.**

12. $f(0, 0)$  13. $f(4, 1)$  14. $f(-2, -6)$

15. $f(5, 4)$  16. $f(3, 1.5)$  17. $f(-0.2, -1)$

**Graph each system of inequalities. Name the coordinates of the vertices of the polygon formed. Find the maximum and minimum values of the given function for this region.**

CONNECTION

Geometry

18. $y \geq 1$
    $x \leq 6$
    $y \leq 2x + 1$
    $f(x, y) = x + y$

19. $y \geq 2$
    $1 \leq x \leq 5$
    $y \leq x + 3$
    $f(x, y) = 3x - 2y$

20. $4y \leq x + 8$
    $x + y \geq 2$
    $y \geq 2x - 5$
    $f(x, y) = 4x + 3y$

**21.** $x + y \geq 2$
$2y \geq 3x - 6$
$4y \leq x + 8$
$f(x, y) = 3y + x$

**22.** $y \leq x + 6$
$y + 2x \geq 6$
$2 \leq x \leq 6$
$f(x, y) = -x + 3y$

**23.** $y \leq 2x + 1$
$1 \leq y \leq 3$
$y \leq -0.5x + 6$
$f(x, y) = 3x + y$

**24.** $y \leq 7$
$y \geq -x + 6$
$y \leq x + 4$
$x \leq 5$
$f(x, y) = 2x - 3y$

**25.** $y \geq x$
$y \leq x + 5$
$x \geq -3$
$y + 2x \leq 5$
$f(x, y) = x - 2y$

**26.** $0 \leq x \leq 5$
$y \geq 0$
$-x + y \leq 2$
$x + y \leq 6$
$f(x, y) = 5x - 3y$

**Use a calculator to find the maximum and minimum values of each function for the polygonal region determined by the given inequalities.**

**27.** $y \leq 1$
$y \geq -2$
$5x \leq -2$
$1.2x - y \geq -2.9$
$f(x, y) = 4x + 2y$

**28.** $x \geq 0$
$y \geq 0$
$x + 2y \leq 6$
$2y - x \leq 2$
$x + y \leq 5$
$f(x, y) = 3x - 5y$

**29.** $x \leq 3$
$y \leq 5$
$x + y \geq 1$
$x \geq 0$
$y \geq 0$
$f(x, y) = 2x + 8y + 10$

**Critical Thinking**

**30.** Create a system of inequalities that determines a feasible region in the shape of a parallelogram. Two of the vertices must be on the $x$-axis and one must be on the $y$-axis.

**Applications**

**31. Agriculture** A farmer has 20 days in which to plant corn and soybeans. The corn can be planted at a rate of 10 acres per day and the soybeans at a rate of 15 acres per day. The farm has 250 acres available for planting. If the profit on corn is $30 per acre, and the profit on soybeans is $25 per acre, how much of each should the farmer plant for maximum profit?

**32. Community Service** A theater where a drug abuse program is being presented seats 150 people. The proceeds will be donated to a local drug information center. Admission is $2.00 for adults and $1.00 for students. Every two adults must bring at least one student. How many adults and students should attend in order to raise the maximum amount of money?

**Mixed Review**

**Solve each system of inequalities by graphing. (Lesson 3-4)**

**33.** $x + y > 5$
$x - y \leq 3$

**34.** $y > x + 1$
$y < x - 3$

**35.** Graph the function $f(x) = |x - 3|$. **(Lesson 2-7)**

**36.** State the $x$- and $y$-intercepts of the graph of the line with equation $3x - 12y = 24$. **(Lesson 2-4)**

**37. Statistics** There will be four tests given in your Algebra 2 course this quarter. On the first three, you scored 87, 92, and 81. If you must have at least 350 points to earn an A, what must you score on the fourth test to earn an A? **(Lesson 1-7)**

## 3-7 Applications of Linear Programming

**Objective**

After studying this lesson, you should be able to:

- solve problems involving maximum and minimum values using linear programming techniques.

Linear programming can be used to solve many types of problems. These problems have certain restrictions placed on the variables, and some function of the variables must be maximized or minimized. Listed below are the steps necessary to solve a problem using linear programming.

*Linear Programming Procedure*

1. **Define the variables.**
2. **Write a system of inequalities.**
3. **Graph the system of inequalities.**
4. **Find the coordinates of the vertices of the polygon formed.**
5. **Write an expression to be maximized or minimized.**
6. **Substitute values from the vertices into the expression.**
7. **Select the greatest or least result. Answer the problem.**

**Example 1**

APPLICATION

Manufacturing

TeeVee Inc. makes console and wide screen televisions. The equipment in the factory allows for making at most 450 console televisions and 200 wide screen televisions in one month. The chart below shows the cost of making each type of television and the profit. During the month of November, the company can spend $360,000 to make these televisions. To maximize profit, how many of each type should they make?

First define the variables.

Let $c$ = the number of consoles.
Let $w$ = the number of wide screens.

| Television | Cost per Unit | Profit per Unit |
|---|---|---|
| Console | $600 | $125 |
| Wide screen | $900 | $200 |

Then write the inequalities.

$$0 \le c \le 450 \qquad \text{\textit{The number of consoles made is at most 450.}}$$
$$0 \le w \le 200 \qquad \text{\textit{The number of wide screens is at most 200.}}$$
$$600c + 900w \le 360{,}000 \qquad \text{\textit{The cost of the consoles plus the cost of the wide screens cannot exceed \$360,000.}}$$

Now graph the system.

Any point in the shaded region or its boundaries will satisfy the conditions of the problem. The vertices of the polygon are (0, 0), (0, 200), (300, 200), (450, 100), and (450, 0).

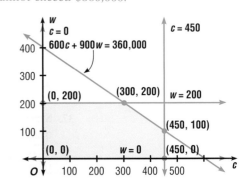

FYI...

The longest pre-scheduled television program was the broadcast by GTV 9 of Melbourne, Australia covering the 1969 Apollo XI moon mission. It was 163 hours 18 minutes long.

We need to maximize the profit. An expression for profit is as follows.

profit = profit on consoles + profit on wide screens
$P(c, w) = 125c + 200w$

Since the maximum or minimum value is always a vertex, substitute the values from the vertices into the expression.

| $(c, w)$ | $(0, 0)$ | $(0, 200)$ | $(300, 200)$ | $(450, 100)$ | $(450, 0)$ |
|---|---|---|---|---|---|
| $125c + 200w$ | $0 | $40,000 | $77,500 | $76,250 | $56,250 |

TeeVee Inc. will make the greatest profit by making 300 console televisions and 200 wide screen televisions. This produces a profit of $77,500.

## Example 2

APPLICATION

Education

The Algebra 2 quiz consists of computation problems and graphing problems. Computation problems are worth 6 points each and graphing problems are worth 10 points each. You can answer a computation problem in 2 minutes and a graphing problem in 4 minutes. You have forty minutes to take the quiz and may choose no more than 12 problems to answer. Assuming you answer all the problems attempted correctly, how many of each type should you answer to get the highest score?

Let $c$ = the number of computation problems.
Let $g$ = the number of graphing problems.

You can write the following inequalities.
$c \geq 0$
$g \geq 0$
$2c + 4g \leq 40$     *You can spend no more than 40 minutes on the quiz.*
$c + g \leq 12$     *You can answer no more than 12 questions.*

Graph the system.

The vertices are $(0, 0)$,
$(0, 10)$, $(4, 8)$, and $(12, 0)$.

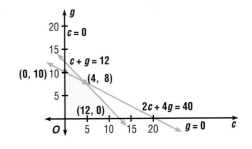

We want to maximize your score.
Write the expression to maximize.

total score = computational score + graphing score
$S(c, g) = 6c + 10g$

Substitute the values of the vertices in the expression.

| $(c, g)$ | $(0, 0)$ | $(0, 10)$ | $(4, 8)$ | $(12, 0)$ |
|---|---|---|---|---|
| $S(c, g)$ | 0 | 100 | 104 | 72 |

You should answer 4 computation and 8 graphing problems. Your score will be 104 points, if you answer them all correctly.

# CHECKING FOR UNDERSTANDING

**Communicating Mathematics**

**Read and study the lesson to answer these questions.**

1. Find the slope of the profit function in Example 1. Think of a line with that slope and move it up and down by changing the $y$-intercept. As this line moves, where does it touch the shaded region first? Where does it touch last? How do these points compare with the maximum and minimum values for the profit function?

2. Is it possible to have more than one value that produces a maximum or a minimum for a given function? Explain.

3. Can a function have no maximum value for a region? Explain.

**Guided Practice**

Two raw materials are needed to make one of the products produced by Dartmouth Inc. The product must contain no more than 9 units of material *A* and at least 18 units of material *B*. The company can spend no more than $300 on materials for each piece produced. Material *A* costs $4 per unit and weighs 10 pounds per unit. Material *B* costs $12 per unit and weighs 20 pounds per unit. How much of each material should be used to maximize the weight of the product?

4. Write an expression to represent the weight.

5. Write an inequality to represent the amounts of materials *A* and *B* allowed.

6. Write an inequality to represent the cost of a product.

7. Graph the system of inequalities.

8. Name the vertices of the polygon.

9. Evaluate the expression from Exercise 4 for each vertex.

10. To maximize weight, how much of each material should be used?

11. What is the maximum weight?

# EXERCISES

**Applications**

12. **Business** The available parking area of a parking lot is 600 square meters. A car requires 6 square meters of space and a bus requires 30 square meters of space. The attendant can handle no more than 60 vehicles.

   a. Let $c$ be the number of cars and let $b$ be the number of buses. Write a system of inequalities to represent the amount of space available and the total number of vehicles allowed.

   b. If a car is charged $2.50 to park and a bus is charged $7.50, how many of each should the attendant accept to maximize income?

   c. The parking lot prices for special events are $4.00 for cars and $8.00 for buses. How many of each vehicle should the attendant accept during a special event?

13. **Manufacturing** The Oklahoma City division of SuperSport Inc. produces footballs and basketballs. It takes 4 hours on machine $A$ and 2 hours on machine $B$ to make a football. Producing a basketball requires 6 hours on machine $A$, 6 hours on machine $B$, and 1 hour on machine $C$. Machine $A$ is available 120 hours a week, machine $B$ is available 72 hours a week, and machine $C$ is available 10 hours per week. If the company makes $3 profit on each football and $2 profit on each basketball, how many of each should they make to maximize their profit?

14. **Veterinary medicine** The table below shows the amounts of nutrient $A$ and nutrient $B$ in two types of dog food: $X$ and $Y$.

| Food Type | Amount of Ingredient A | Amount of Ingredient B |
|---|---|---|
| X | 1 unit per pound | $\frac{1}{2}$ unit per pound |
| Y | $\frac{1}{3}$ unit per pound | 1 unit per pound |

The dogs in Ken's K-9 Kennel must get at least 40 pounds of food per day. The food may be a mixture of foods $X$ and $Y$. The daily diet must include at least 20 units of nutrient $A$ and at least 30 units of nutrient $B$. The dogs must not get more than 100 pounds of food per day.

a. Food $X$ cost $0.80 per pound and food $Y$ costs $0.40 per pound. What is the least possible cost per day for feeding the dogs?

b. If the price of food $X$ is raised to $1.00 per pound, and the price of food $Y$ stays the same, should Ken change the combination of foods he is using?

15. **Retail** The sales associate at a paint store plans to mix as many gallons as possible of colors $A$ and $B$. She has exactly 32 units of blue dye and 54 units of red dye. Each gallon of color $A$ requires 4 units of blue dye and 1 unit of red dye. Each gallon of color $B$ requires 1 unit of blue dye and 6 units of red dye.

a. Let $a$ be the number of gallons of color $A$ and let $b$ be the number of gallons of color $B$. Write the inequalities.

b. Find the maximum number of gallons, $a + b$, possible.

16. **Manufacturing** Oaken Treasures makes two different kinds of chairs, rockers and swivels. Work on machines $A$ and $B$ is required to make both kinds. Machine $A$ can be run no more than 20 hours a day. Machine $B$ is limited to 15 hours a day. The following chart shows the amount of time on each machine that is required to make one chair. The profit made on each chair is also shown.

| Chair | Operation A | Operation B | Profit |
|---|---|---|---|
| Rocker | 2 h | 3 h | $12 |
| Swivel | 4 h | 1 h | $10 |

How many chairs of each kind should Oaken Treasures make each day to maximize their profit?

17. **Manufacturing** Stitches Inc. can make at most 30 jean jackets and 20 leather jackets in a week. It takes a worker 10 hours to make a jean jacket and 20 hours to make a leather jacket. The total number of hours worked by all of the employees can be no more than 500 hours per week.

   a. If the profit on a jean jacket is the same as the profit on a leather jacket, how many of each should be made to maximize profit?

   b. How many of each should be made if the profit on a leather jacket is three times the profit on a jean jacket?

18. **Education** Your semester test in English class consists of short answer and essay questions. Each short answer question is worth 5 points and each essay question is worth 15 points. You may choose up to 20 questions of any type to answer. It takes 2 minutes to answer each short answer question and 12 minutes to answer each essay question.

   a. You have one hour to complete the test. Assuming that you answer all of the questions that you attempt correctly, how many of each type should you answer to earn the highest score?

   b. You have two hours to complete the test. Assuming that you answer all of the questions that you attempt correctly, how many of each type should you answer to earn the highest score?

**Critical Thinking**

19. Create a system of inequalities for which the graph will be a regular hexagon, and its interior will be located in the first quadrant.

**Mixed Review**

Graph each system of inequalities. Name the vertices of the polygon formed. Find the maximum and minimum values of the given function.   (**Lesson 3-6**)

20. $x \geq 0$
   $y \geq 3$
   $y \geq 2x + 1$
   $y \leq -0.5x + 6$
   $f(x, y) = 3x - 2y$

21. $0 \leq x \leq 50$
   $0 \leq y \leq 70$
   $60 \leq x + y \leq 80$
   $f(x, y) = 4x + 3y$

22. State whether $y = x^2 - 4$ is a linear equation.   (**Lesson 2-2**)

23. Solve $|a + 5| + 5 = 3$.   (**Lesson 1-7**)

24. State the property illustrated by $(6 + 4) = (6 + 4)$.   (**Lesson 1-3**)

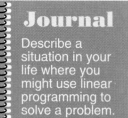

**Journal**

Describe a situation in your life where you might use linear programming to solve a problem.

# 3-8

# Graphing Equations in Three Variables

**Objectives**

After studying this lesson, you should be able to:

■ determine the octant in which a point in space is located, and

■ graph linear equations in space and determine the intercepts and traces.

**Application**

One part of Kim Grant's job as an air traffic controller is to see that airplanes maintain a safe distance from towers, from each other, and from any other obstacles. She can use ordered triples to represent positions of planes and other objects to help her keep track of planes and flight paths.

The equation $3x + 4y + z = 12$ is an open sentence in three variables, $x$, $y$, and $z$. You can write the solutions of open sentences in three variables as sets of **ordered triples**. For example, $(1, 2, 1)$ is a member of the solution set of this open sentence.

To draw the graph of an equation in three variables, it is necessary to add a third dimension to our coordinate system. The graph of an equation of the form $Ax + By + Cz = D$, where $A$, $B$, $C$, and $D$ are real numbers, is a plane.

| *Equation of a Plane* | **If $A$, $B$, $C$, and $D$ are real numbers, such that $A$, $B$ and $C$ are not all zero, then the graph of the equation $Ax + By + Cz = D$ is a plane.** |
|---|---|

When graphing in space, it is necessary to separate space into eight regions, called **octants.** Think of three coordinate planes intersecting at right angles as shown at the right. The octants are numbered as shown at the right. Any point lying in one of the coordinate planes is not in any octant.

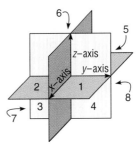

**Example 1**

**Locate the point (3, 6, 1).**

First, locate 3 on the positive $x$-axis, 6 on the positive $y$-axis, and 1 on the positive $z$-axis. Complete a "box" by drawing lines parallel to the axes through each intercept.

This point is in octant 1.

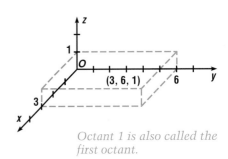

*Octant 1 is also called the first octant.*

It is not necessary to show the entire "box" when you graph an ordered triple. The desired point will always be the corner farthest from the origin.

*If one of the coordinates is zero, the sketch will be a rectangle instead of a box.*

**Example 2**

Locate the point (−3, −5, 2).

First, locate −3 on the x-axis. Then draw a segment 5 units long in the negative direction, parallel to the y-axis. From that point, draw a segment 2 units long in the positive direction, parallel to the z-axis.

*The point is in octant 6.*

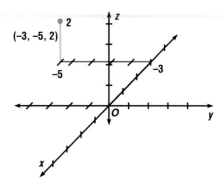

**Example 3**

CONNECTION

Geometry

Find the volume of the "box" formed when locating the point (3, 7, −1).

Let's graph the point.

The lengths of the sides of the "box" are 3 units, 7 units, and 1 unit.

The volume of the "box" is $3 \times 7 \times 1$ or 21 cubic units.

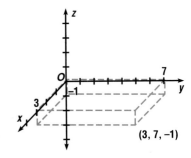

To graph a linear equation in three variables, first find the intercepts of the graph. Connect the intercepts on each axis. This forms a portion of a plane that lies in a single octant. Study the following example.

**Example 4**

Graph $2x + 4y + 3z = 12$.

To find the x-intercept, let $y = 0$ and $z = 0$.
$$2x = 12$$
$$x = 6$$

To find the y-intercept, let $x = 0$ and $z = 0$.
$$4y = 12$$
$$y = 3$$

To find the z-intercept, let $x = 0$ and $y = 0$.
$$3z = 12$$
$$z = 4$$

To indicate the plane, connect the intercepts. Remember, a plane extends indefinitely.

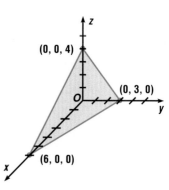

*The portion in octant 1 is shown.*

A **trace** is the intersection of a plane with one of the coordinate planes. The *xy*-trace is the line formed by the intersection of a plane with the *xy*-plane. All points in the *xy*-trace have a *z*-coordinate of zero, so we can find the equation of the *xy*-trace by letting $z = 0$ in the equation of the plane. We can find the equations for the other two traces in a similar manner.

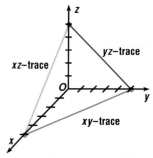

*The portion of the plane in octant 1 is shown.*

**Example 5**

*Each intercept can be found by substituting 0s for two of the variables and solving for the third.*

**Graph $15x - 5y + 6z = 30$ and find the equation of the trace in each coordinate plane.**

The *x*-, *y*-, and *z*-intercepts are 2, −6, and 5. To find the equation of the *xy*-trace, let $z = 0$.
$$15x - 5y = 30$$
$$3x - y = 6 \quad \textit{Simplify.}$$

To find the equation of the *xz*-trace, let $y = 0$.
$$15x + 6z = 30$$
$$5x + 2z = 10 \quad \textit{Simplify.}$$

To find the equation of the *yz*-trace, let $x = 0$.
$$-5y + 6z = 30$$

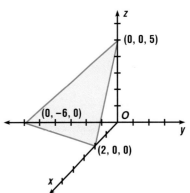

*The portion of the plane in octant 2 is shown.*

If you know the equations of two traces, you can write the equation of the plane containing the two traces.

**Example 6**

**The equations of two traces of a plane are $-2x + 8y = -16$ and $2x + 4z = 16$. Find the equation of the plane.**

First find the intercepts of the two traces.
The *xy*-trace is $-2x + 8y = -16$. So the *x*- and *y*-intercepts are 8 and −2.
The *xz*-trace is $2x + 4z = 16$. So the *x*- and *z*-intercepts are 8 and 4.

Now use the intercepts to write the equation of the plane.
The *x*-, *y*-, and *z*-intercepts of the plane are 8, −2, and 4 respectively. Use 8, the least common multiple of the absolute values of 8, −2, and 4, as the constant in the equation so that the coefficients in the equation will be integers. Divide 8 by each of the intercepts to obtain the coefficients.

$$\frac{8}{8} = 1 \qquad \frac{8}{-2} = -4 \qquad \frac{8}{4} = 2$$

The equation of the plane is $x - 4y + 2z = 8$.

*Check this result by finding the equations of the trace in each coordinate plane.*

# CHECKING FOR UNDERSTANDING

**Communicating Mathematics**

Read and study the lesson to answer these questions.

1. What is the difference between an equation whose graph is a plane and an equation whose graph is a line?

2. Write two ordered triples that satisfy the equation $3x + 4y + z = 12$.

3. Write the equation of the $xy$-trace of $3x + 4y + z = 12$.

**Guided Practice**

In which octant does each point lie?

4. $(5, 2, 3)$      5. $(7, 5, -6)$      6. $(3, 0, 1)$      7. $(3, -7, 2)$

Find the $x$-, $y$-, and $z$-intercepts for each equation.

8. $2x + y - z = 12$             9. $20x - 4y + 10z = 20$
10. $2x + y - 3z = 10$           11. $3x + 5y + 2z = 30$

# EXERCISES

**Practice**

Given the following conditions, name the octant in which the point $(x, y, z)$ lies.

12. $x < 0, y = 4, z > 0$          13. $x > 0, y = 3, z < 0$

14. $x > 2, y > 1, z = 7$          15. $x < 0, y < 0, z < 0$

16. $x > 3, y < -3, z = 3$        17. $x > 2, y < -1, z < -2$

Graph each equation. Find the $x$-, $y$-, and $z$-intercepts and the traces in the coordinate planes.

18. $3x + 6y + 2z = 9$          19. $4x - y + 2z = 10$

20. $3x - y = 3$                  21. $5y + 2z = 20$

22. $3z - 2x = 5$               23. $5x - 8y = -12$

Write an equation of the plane given its $x$-, $y$-, and $z$-intercepts.

24. $2, -2, 5$                    25. $\frac{1}{2}, 3, -2$

Write an equation of the plane given two of its traces in the coordinate planes.

26. $5x + 3z = 15, y + z = 5$      27. $x - 4y = 1, 8y + z = -2$

28. $3y - 4z = 6, y = 2$          29. $x = -2, z = 5$

**Critical Thinking**

30. Describe the equations of two parallel planes. Write an example of two equations whose graphs are parallel planes.

**Application**

31. **Manufacturing** Conner's Corrugated Crates makes cardboard boxes for shipping goods. If a corner of their smallest box is placed at the origin of a coordinate system with 1-inch units on all three axes, the coordinates of the corner farthest from the origin are (12, 10, 10).

   a. Find the volume of the box.

   b. Find the amount of material needed to make the box.

   c. Conner's manager decided that this box would be much more useful if they double all of the dimensions. How does the volume of this new box compare to the original?

**Computer**

32. The BASIC program below finds the distance between two points in space. Line 70 uses the formula for distance.

$$d = \sqrt{(x_2 - x_1)^2 + (y_2 - y_1)^2 + (z_2 - z_1)^2}$$

Enter the coordinates $(x_1, y_1, z_1)$, for the first point. Then enter the coordinates $(x_2, y_2, z_2)$ for the second point.

```
10 PRINT "ENTER THE            50 PRINT "THE SECOND POINT"
   COORDINATES OF"             60 INPUT X2, Y2, Z2
20 PRINT "THE FIRST POINT."    70 LET D = SQR((X2-X1)^2 +
30 INPUT X1, Y1, Z1               (Y2-Y1)^2 + (Z2-Z1)^2)
40 PRINT "ENTER THE            80 PRINT "DISTANCE = ";D
   COORDINATES OF"             90 END
```

**Use the BASIC program to find the distance between each pair of points to the nearest hundredth.**

   a. (3, 0, 0), (0, 4, 0)           b. (0, 0, 0), (1, 2, 3)

   c. (5, 1, 3), (3, -2, 6)          d. (-1, 7, 1), (-2, -1, -3)

   e. Use the program to find a set of four positive whole numbers that satisfy the equation $a^2 + b^2 + c^2 = d^2$. Use (0, 0, 0) as one point and $(a, b, c)$ as the other point.   *Hint: See Exercise 32b.*

**Portfolio**

Select an item from this chapter that you feel shows your best work and place it in your portfolio. Explain why you selected it.

**Mixed Review**

33. **Agriculture** A North Carolina farmer has 25 days to plant cotton and corn. The cotton can be planted at a rate of 9 acres per day and the corn at a rate of 12 acres per day. The farm has 275 acres available. The farmer estimates that the profit on an acre of cotton will be $25 and the profit on an acre of corn will be $18.   **(Lesson 3-7)**

   a. Let $c$ represent the number of acres of cotton and let $r$ represent the number of acres of corn. Write the inequalities.

   b. How many acres of each crop should the farmer plant?   **(Lesson 3-7)**

34. Name which points, (0, 0), (1, -2), or (-3, 1), satisfy the inequality $x + 2y \leq 7$.   **(Lesson 2-8)**

35. Find the next number in the group of numbers $\frac{1}{2}, \frac{2}{3}, 1, \frac{8}{5}, \frac{8}{3}, \underline{\ ?\ }$.
   **(Lesson 2-3)**

36. Solve $|m - 4| + 2 \geq 0$.   **(Lesson 1-8)**

## 3-9 Solving Systems of Equations in Three Variables

**Objective**

After studying this lesson, you should be able to:

■ solve a system of equations in three variables.

**Application**

Courtney has a total of 256 points on three algebra tests. His score on the first test exceeds his score on the second by 6 points. His total score before taking the third test was 164 points. What were Courtney's scores on the three tests?

*Explore*

Problems like this one can be solved by using a **system of equations in three variables.** Solving these systems is very similar to solving systems of equations in two variables. Let's try solving this problem.

Let $f$ = Courtney's score on the first test.
Let $s$ = Courtney's score on the second test.
Let $t$ = Courtney's score on the third test.

*Plan*

Write the system of equations from the information given.

$f + s + t = 256$    *The total of the scores is 256.*
$f - s = 6$    *The difference between the first and second is 6 points.*
$f + s = 164$    *The total before taking the third test is the sum of the first and second tests.*

*Solve*

Now solve. First use elimination on the last two equations to solve for $f$.

$$\begin{array}{r} f - s = 6 \\ \underline{f + s = 164} \\ 2f\phantom{ + s} = 170 \\ f = 85 \end{array}$$

   The first test score is 85.

Then substitute 85 for $f$ in one of the original equations to solve for $s$.

$$\begin{array}{r} f + s = 164 \\ 85 + s = 164 \\ s = 79 \end{array}$$

   The second test score is 79.

Next substitute 85 for $f$ and 79 for $s$ in $f + s + t = 256$.

$$\begin{array}{r} f + s + t = 256 \\ 85 + 79 + t = 256 \\ t = 92 \end{array}$$

   The third test score is 92.

Courtney's test scores were 85, 79, and 92.

*Examine*

Now check your results against the original problem.
Is the total number of points on the three tests 256 points?
   $85 + 79 + 92 = 256 \checkmark$
Is one test score 6 more than another test score?   $79 + 6 = 85 \checkmark$
Do two of the tests total 164 points?   $85 + 79 = 164 \checkmark$
Our answers are correct.

You know that a system of two linear equations in two variables does not always have a solution that is a unique ordered pair. Similarly, a system of three linear equations in three variables does not always have a solution that is a unique ordered triple.

*Recall that a system of two equations in two variables can have a unique solution, an infinite number of solutions, or no solution.*

The graph of each equation in a system of three linear equations in three variables is a plane. Depending on the constraints involved, one of the following possibilities occurs.

The three planes intersect at one point. So the system has a unique solution.

The three planes intersect in a line. There are an infinite number of solutions to the system.

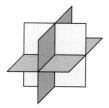 

Each of the diagrams below shows three planes that have no points in common. These systems of equations have no solutions.

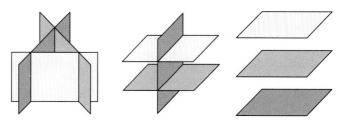

Example 1

**Solve this system of equations:** $\begin{cases} x + 2y + z = 9 \\ 3y - z = -1 \\ 3z = 12 \end{cases}$ .

Solve the third equation, $3z = 12$.

$$3z = 12$$
$$z = 4$$

Substitute 4 for $z$ in the second equation, $3y - z = -1$, to find $y$.

$$3y - (4) = -1$$
$$3y = 3$$
$$y = 1$$

Substitute 4 for $z$ and 1 for $y$ in the first equation, $x + 2y + z = 9$, to find $x$.

$$x + 2(1) + (4) = 9$$
$$x + 6 = 9$$
$$x = 3 \qquad \text{The solution is } (3, 1, 4).$$

**Check:** *First equation:* $\qquad (3) + 2(1) + (4) = 9$ ✓

*Second equation:* $\qquad 3(1) - 4 = -1$ ✓

*Third equation:* $\qquad 3(4) = 12$ ✓

## Example 2

Consumerism

FYI · · ·

French fries did not come from France. They originated in Belgium and were first sold by street vendors in the 1800's.

Carmen, Jack, and Carol went to Bernie's Burgers to get food for their friends at school. Carmen spent $5.95 on two burgers, one order of french fries, and two sodas. Jack's order of 3 burgers, 3 orders of french fries, and 3 sodas totaled $10.41. Carol ordered 1 burger, 2 orders of french fries, and 2 sodas. Her bill was $5.35. How much is a burger, an order of french fries, and a soda?

Let $b$ = the price of a burger, $f$ = the price of an order of french fries, and $s$ = the price of a soda.

Write a system of equations.

$2b + f + 2s = 5.95$     *Carmen's order*
$3b + 3f + 3s = 10.41$     *Jack's order*
$b + 2f + 2s = 5.35$     *Carol's order*

Use elimination to make a system of two equations in two variables.

$2b + f + 2s = 5.95$     Multiply by −1     $-2b - f - 2s = -5.95$

$b + 2f + 2s = 5.35$     Multiply by 2     $\underline{2b + 4f + 4s = 10.70}$

*Add to eliminate b.*     $3f + 2s = 4.75$

$3b + 3f + 3s = 10.41$     Multiply by −1     $-3b - 3f - 3s = -10.41$

$b + 2f + 2s = 5.35$     Multiply by 3     $\underline{3b + 6f + 6s = 16.05}$

*Add to eliminate b.*     $3f + 3s = 5.64$

The result is two equations with the same two variables.

$3f + 2s = 4.75$
$3f + 3s = 5.64$

Use elimination to solve for $s$.

$3f + 2s = 4.75$     Multiply by −1     $-3f - 2s = -4.75$
$3f + 3s = 5.64$     $\underline{3f + 3s = \phantom{-}5.64}$
$s = \phantom{-}0.89$

Now substitute 0.89 for $s$ in one of the equations to find $f$.

$3f + 2(0.89) = 4.75$
$3f + 1.78 = 4.75$
$3f = 2.97$
$f = 0.99$

Now substitute 0.99 for $f$ and 0.89 for $s$ in one of the original equations.

$b + 2f + 2s = 5.35$     *Substituting the values for f and s*
$b + 2(0.99) + 2(0.89) = 5.35$     *in either of the other original*
$b + 3.76 = 5.35$     *equations would give the same*
$b = 1.59$     *result.*

A burger costs $1.59, an order of french fries costs $0.99, and a soda costs $0.89.

# CHECKING FOR UNDERSTANDING

**Communicating Mathematics**

Read and study the lesson to answer these questions.

1. What other methods might we have used to solve the system of two equations in two variables in Example 2?

2. Why is substitution the best method to use to solve the system in Example 1?

3. Assume that a system of three equations in three variables has no solution. What might happen while you are trying to solve the system that would tell you that there is no solution?

4. What might happen while you are trying to solve a system of equations that would tell you that there are infinitely many solutions?

**Guided Practice**

For each system of equations an ordered triple is given. Determine whether it is a solution of the system.

5. $x + 2y + z = 0$
   $x - y + z = 0$
   $2x + 3y - z = 0$; $(0, 0, 0)$

6. $x - 3y + 2z = 1$
   $2x - y + 2z = 0$
   $x + y + z = 6$; $(2, 2, 2)$

7. $4x + y - 2z = 0$
   $x - 2y = 0$
   $2x - y - z = 0$; $(3, 0, 6)$

8. $2x + y - z = 2$
   $3x + 2y + z = 5$
   $x + y + z = 0$; $(8, -11, 3)$

9. $x - z = 1$
   $x + y + z = 3$
   $z - y = 4$; $(3, -2, 2)$

10. $x + y = -6$
    $x + z = -2$
    $y + z = 2$; $(-4, -2, 2)$

# EXERCISES

**Practice**

Solve each system of equations.

11. $x - 2y + z = -9$
    $2y + 3z = 16$
    $2y = 4$

12. $2a + b = 2$
    $5a = 15$
    $a + b + c = -1$

13. $x + y - z = -1$
    $x + y + z = 3$
    $3x - 2y - z = -4$

14. $b + c = 4$
    $2a + 4b - c = -3$
    $3b = -3$

**15.** $r + s + t = 15$
$r + t = 12$
$s + t = 10$

**16.** $2x + y - 2z = 31$
$x - 2y - 3z = 23$
$x - 2y + z = 3$

**17.** $a + b + c = 0$
$2a + 2b + c = 5$
$2a + b - c = 2$

**18.** $x + y + z = 4$
$x - y - z = -2$
$x - y + z = 0$

**19.** $r + s - 2t = 4$
$r - 3s - 4t = -2$
$2r + s + 2t = 0$

**20.** $2x + 3y + z = 28$
$3x + 4y - 2z = 24$
$x + y + z = 16$

**21.** $3x - 2y + 2z = -2$
$x - 3y + z = -2$
$2x - y + 4z = 7$

**22.** $a + 8b + 2c = -24$
$3a + b + 7c = -3$
$4a - 3b + 6c = 9$

**23.** $x + y + z = -1$
$3x - 2y - 4z = 16$
$2x - y + z = 19$

**24.** $4x + 3y + 2z = 34$
$2x + 4y + 3z = 45$
$3x + 2y + 4z = 47$

**25.** $3a + b + 2c = 6$
$6a - 2b = 2$
$3a + b - 2c = 0$

**26.** $x + y + z = 1$
$2x - y = 0$
$-3x + z = 0$

**27.** $2r + 3s + 4t = 3$
$5r - 9s + 6t = 1$
$\dfrac{1}{3}r - \dfrac{1}{2}s + \dfrac{2}{3}t = \dfrac{1}{6}$

**28.** $2x + y + z = 7$
$12x - 2y - 2z = 2$
$\dfrac{2x}{3} - y + \dfrac{z}{3} = -\dfrac{1}{3}$

**29.** Three numbers have a total of 6. The first number is twice the second, and the third is three times the second. What are the three numbers?

**30.** The sum of three numbers is 20. The first number is the sum of the second and the third. The third number is three times the first. What are the three numbers?

**Critical Thinking**

**31.** Since you know how to solve a system of three equations in three variables, use what you know to try to solve this system of four equations in four variables.
$w + x + y + z = 2$
$2w - x - y + 2z = 7$
$2w + 3x + 2y - z = -2$
$3w - 2x - y - 3z = -2$

**Applications**

**32. Surveying**  The perimeter of a triangular lot is 180 meters. The longest side is twice as long as the shortest side. The length of the remaining side is the average of the lengths of the longest and shortest sides. Find the lengths of the three sides.

33. **Consumer Awareness** Melissa works at Angela's Pizza. Her last three orders were 5 slices of pizza, 2 salads, and 2 sodas for $9.75; 3 slices of pizza, 2 salads, and 1 soda for $7.15; and 2 slices of pizza, 1 salad, and 1 soda for a total of $4.35. What are the individual prices for pizza, salad, and soda at Angela's?

34. **Banking** Pat Juarros has $5000 to invest in a certificate of deposit, stocks, or bonds. Since the investment is more risky, the estimated interest rate of the stocks is 1.5% more than that of the certificate of deposit. If she uses all of the money to buy a certificate of deposit, the interest earned in one year will be $400. The interest earned in one year on a $2000 certificate of deposit, $2000 worth of savings bonds, and $1000 in stocks would be $385. What are the interest rates of the three investments?

**Mixed Review**    Find the $x$-, $y$-, and $z$-intercepts for each equation.   (Lesson 3-8)

35. $-3x + 6y - 4z = 24$          36. $9x + 6y - 3z = 36$

37. Solve the following system of equations using the substitution method.   **(Lesson 3-2)**     $2x + 3y = -4$
$-3x + y = -5$

38. The Forest Park Animal Clinic charges $35 for any office visit. In addition, the veterinarian charges $25 an hour after the first one-half hour. Write a linear equation to describe the cost of an office visit of $t$ hours. Assume $t$ is greater than one-half hour.   **(Lesson 2-5)**

39. Find an equation of the line that passes through $(-3, 4)$ and is perpendicular to the line with equation $3y = 2x + 3$.   **(Lesson 2-4)**

## HISTORY CONNECTION

Who developed Cramer's rule? Most of us would answer "Cramer," since his name is attached to the rule. However, Chinese mathematicians had developed a system of determinants long before the Swiss mathematician Cramer published his rule in 1750. Further, it was a Japanese mathematician, **Seki Kowa,** who first applied the Chinese system of determinants to solving systems of equations. According to Japanese traditions, Seki make a pilgrimage to the ancient shrines at Nara to study the Chinese mathematical works preserved in the Buddhist temples there. He is said to have spent three years mastering the contents of those works. Seki's work with the application of determinants laid the foundation for all later work with determinants, including Cramer's.

## CHAPTER 3

# SUMMARY AND REVIEW

## VOCABULARY

Upon completing this chapter, you should be familiar with the following terms:

| | | | |
|---|---|---|---|
| consistent | **109** | **130** | linear programming |
| constraints | **129** | **139** | octants |
| Cramer's rule | **119** | **139** | ordered triple |
| dependent | **109** | **118** | second order determinant |
| determinant | **118** | **108** | solution of a system |
| elimination method | **112** | **112** | substitution method |
| equation of a plane | **139** | **108** | system of equations |
| feasible region | **129** | **122** | system of inequalities |
| inconsistent | **109** | **141** | trace |
| independent | **109** | | |

## SKILLS AND CONCEPTS

| OBJECTIVES AND EXAMPLES | REVIEW EXERCISES |
|---|---|

Upon completing this chapter, you should be able to:

Use these exercises to review and prepare for the chapter test.

■ solve a system of equations by graphing **(Lesson 3-1)**

Graph each equation. The intersection of the graphs is the solution.

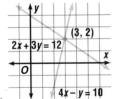

$$4x - y = 10$$
$$2x + 3y = 12$$

The solution is $(3, 2)$.

**Graph each system of equations and state its solution. Also, state whether the system is** *consistent and independent, consistent and dependent,* **or** *inconsistent.*

**1.** $y = 3x - 1$
     $3x - y = 1$

**t, 2.** $x - 2y = 4$
         $y = x - 2$

**3.** $2x - 3y = 7$
     $x - y = 5$

**4.** $x + y = 5$
     $2x - y = 4$

■ solve a system of equations using the substitution or elimination method **(Lesson 3-2)**

$$x + 3y = 2 \qquad 1 + 3y = 2$$
$$\underline{2x - 3y = 1} \qquad \quad 3y = 1$$
$$\frac{3x}{3} = \frac{3}{3} \qquad \quad y = \frac{1}{3}$$
$$\qquad x = 1$$

**Solve each system of equations using the elimination or substitution method.**

**5.** $x + y = 8$
     $x - y = 4.5$

**6.** $3x - 5y = -13$
     $4x + 3y = 2$

**7.** $2x + 3y = 8$
     $x - y = 2$

**8.** $\frac{1}{6}x - \frac{1}{9}y = 0$
     $\frac{1}{3}x + \frac{1}{3}y = 5$

■ find the value of a second order determinant **(Lesson 3-3)**

$$\begin{vmatrix} -2 & 0 \\ 3 & 1 \end{vmatrix} = -2(1) - 0(3) = -2$$

**Find the value of each determinant.**

9. $\begin{vmatrix} 2 & 4 \\ 3 & 5 \end{vmatrix}$

10. $\begin{vmatrix} 8 & -1 \\ 3 & -2 \end{vmatrix}$

---

■ solve a system of equations using Cramer's rule **(Lesson 3-3)**

The solution to the system $\begin{cases} ax+by=c \\ dx+ey=f \end{cases}$ is $(x, y)$, where

$$x = \dfrac{\begin{vmatrix} c & b \\ f & e \end{vmatrix}}{\begin{vmatrix} a & b \\ d & e \end{vmatrix}}, \ y = \dfrac{\begin{vmatrix} a & c \\ d & f \end{vmatrix}}{\begin{vmatrix} a & b \\ d & e \end{vmatrix}}, \ \text{and} \ \begin{vmatrix} a & b \\ d & e \end{vmatrix} \neq 0.$$

**Use Cramer's rule to solve each system of equations.**

11. $4x - 8y = 12$
    $3x + y = 23$

12. $x + 2y = 8$
    $-5x - 3y = -5$

---

■ solve a system of inequalities by graphing **(Lesson 3-4)**

$x + y < 1$

$y \geq x - 1$

**Solve each system by graphing.**

13. $y < -2$
    $y > x + 1$

14. $x + 2y > -3$
    $x + y < 2$

15. $y \geq -x + 1$
    $y \geq x - 3$

16. $y < 4 - x$
    $y \geq -3x + 1$

---

■ find the maximum and minimum values of a function over a region using linear programming techniques **(Lesson 3-6)**

Graph the constraints. Then evaluate the function for each vertex to find the maximum and minimum values.

**Graph each system of inequalities. Name the coordinates of the vertices of the polygon formed. Find the maximum and minimum values of the given function.**

17. $x \geq 0$
    $y \geq 0$
    $y \leq 3 - x$
    $3x + y \leq 6$
    $f(x, y) = 2x + 4y$

18. $0 \leq x \leq 5$
    $0 \leq y \leq 6$
    $x + y \leq 9$
    $f(x, y) = 2x + 3y$

---

■ determine the octant in which a point in space is located **(Lesson 3-8)**

| Octant | Signs of Coordinates |
|--------|----------------------|
| 1 | (+, +, +) |
| 2 | (+, −, +) |
| 3 | (+, −, −) |
| 4 | (+, +, −) |
| 5 | (−, +, +) |
| 6 | (−, −, +) |
| 7 | (−, −, −) |
| 8 | (−, +, −) |

**In which octant does each point lie?**

19. $(9, -3, 1)$

20. $(-5, -2, -11)$

21. $(0, 3, 0)$

22. $(4, 2, -1)$

23. $(2, 7, 10)$

24. $(-9, -1, 9)$

25. $(7, -4, -2)$

26. $(-2, -1, 7)$

| OBJECTIVES AND EXAMPLES | REVIEW EXERCISES |
|---|---|

■ graph linear equations in space and determine the intercepts and traces   (**Lesson 3-8**)

To find the *x*-intercept, let $y = 0$ and $z = 0$. To find the *y*-intercept, let $x = 0$ and $z = 0$. To find the *z*-intercept, let $x = 0$ and $y = 0$. To find the equation of the *xy*-trace, let $z = 0$. To find the equation of the *xz*-trace, let $y = 0$. To find the equation of the *yz*-trace, let $x = 0$.

**Graph each equation. Find the *x*-, *y*-, and *z*-intercepts and the traces in the coordinate planes.**

27. $x + y - 2z = 4$
28. $5x - 2y + 3z = 6$

---

■ solve a system of equations in three variables   (**Lesson 3-9**)

Use elimination to make a system of two equations in two variables. Then use elimination to solve the two equations. Substitute to find the value of the third variable.

**Solve each system of equations.**

29. $x + 2y - 3z = -3$
    $x + y + z = -1$
    $2x + 4y + z = 1$
30. $a + b + 3c = 7$
    $2a - 2b - 3c = 2$
    $3a - b - 2c = 1$

---

# ~~~~~ APPLICATIONS AND CONNECTIONS ~~~~~

31. How many numbers from 10 to 1000 read the same forward or backward?   (**Lesson 3-5**)

32. **Manfacturing**   Denim Duds makes denim jackets and jeans. Each garment must be cut from a pattern and sewn. There are 40 worker-hours per day available for cutting and 52 worker-hours per day for sewing. The chart below shows the number of hours for each operation needed to make both garments, as well as the profit on the garment.

| Garment | Cutting Hours | Sewing Hours | Profit |
|---|---|---|---|
| jacket | 1 | 4 | $14 |
| jeans | 2 | 2 | $8 |

How many of each garment should the company make to maximize profit? (**Lesson 3-7**)

33. **Education**   You may answer up to 30 questions on your final exam in history class. It consists of multiple-choice and essay questions. Two 48-minute class periods have been set aside for taking the test. It will take you 1 minute to answer each multiple-choice question and 12 minutes for each essay question. Correct answers on multiple-choice questions earn 5 points and correct essay questions earn 20 points. If you are confident that you will answer all of the questions you attempt correctly, how many of each type of question should you answer to receive the highest score?   (**Lesson 3-7**)

Graph each system of equations and state its solution.

1. $2x + y = 11$
   $x - y = 1$

2. $x - 2y = -1$
   $2x + 3y = -16$

Solve each system of equations using the elimination or substitution method.

3. $2x + 3y = 5$
   $3x - 6y = -12$

4. $2x + y - 5z = 4$
   $x + 3y + z = 5$
   $3x - y - 4z = -11$

5. Find the value of $\begin{vmatrix} -3 & 1 \\ -2 & -4 \end{vmatrix}$.

6. Find the value of $\begin{vmatrix} 6 & 5 \\ 0 & -9 \end{vmatrix}$.

Use Cramer's Rule to solve each system of equations.

7. $4x + 7y = -1$
   $2x + y = 7$

8. $8x - 3y = -10$
   $4x + y = 5$

9. Solve $\begin{cases} x + y \leq 6 \\ x - y \geq 4 \end{cases}$ by graphing.

10. Solve $\begin{cases} y \leq -3x + 4 \\ y > 2x + 1 \end{cases}$ by graphing.

Graph each system of inequalities. Name the coordinates of the vertices of the polygon formed. Find the maximum and minimum values of the given function.

11. $x \geq 1$
    $y \geq 1$
    $x + y \leq 6$
    $f(x, y) = x + 3y$

12. $-2 \leq x \leq 10$
    $y \geq -2$
    $2x + y \leq 21$
    $y - 2x \leq 9 \quad f(x, y) = 3x - 2y$

A toy manufacturer makes a $3 profit on yo-yo's and a $3 profit on tops. Department *A* requires 3 hours to make parts for 100 yo-yo's and 4 hours to make parts for 100 tops. Department *B* needs 5 hours to make parts for 100 yo-yo's and 2 hours to make parts for 100 tops. Department *A* has 450 hours available and department *B* has 400 hours available.

13. How many yo-yo's and tops should be made?

14. What is the maximum profit the company can make from these two products?

15. In which octant does $(-9, -2, 7)$ lie?

16. In which octant does $(3, -1, 8)$ lie?

Graph each equation. Find the *x*-, *y*-, and *z*-intercepts and the traces in the coordinate planes.

17. $3x - 5y + 6z = 30$

18. $4x - 2y + 3z = 6$

A new housing development has sixty lots available for building houses, where the builder can build colonial or ranch style houses. Sales experience has taughter her that should should plan to build at least three times as many ranch-style houses as colonial.

19. If she will make a profit of $5000 on each colonial and $4500 on each ranch, how many of each kind should she build?

20. Local regulations require that the builder use different materials than she planned, so her profit will be $3500 on each colonial and $4000 on each ranch. How many of each should she build?

**Bonus**   Graph $|x| + |y| \leq 5$.

# 4

# Matrices

## CHAPTER OBJECTIVES

In this chapter, you will:
- Create matrices to represent data and algebraic expressions.
- Perform operations with matrices.
- Use matrices to achieve transformations of geometric figures.
- Use matrices to solve systems of equations.

**Study the graph. For the engineer what seems to be the most important thing to concentrate on when solving an engineering problem? Do you as a consumer have the same priority in a product?**

### How Design Engineers Solve Engineering Problems

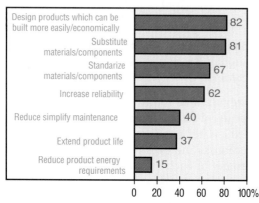

| | |
|---|---|
| Design products which can be built more easily/economically | 82 |
| Substitute materials/components | 81 |
| Standarize materials/components | 67 |
| Increase reliability | 62 |
| Reduce simplify maintenance | 40 |
| Extend product life | 37 |
| Reduce product energy requirements | 15 |

0   20   40   60   80   100%

## CAREERS IN INDUSTRIAL DESIGN

When you buy a tool or product, of course, you are interested in how much it costs. But do you also care whether it works well and looks good? If you do, then you care about good design, and you just might make a good designer.

A designer needs creativity, a strong color sense, a visual imagination, a sense of proportion, a feel for balance, and—even in industrial design—a sensitivity to beauty. When a new car or a monorail transportation system is designed right, it is beautiful. Form fits function.

Most industrial designers, especially those in the aerospace, automative, and electronic industries, work on the computer. Computer-aided design (CAD) lets the operator insert, edit, and replace images on screen, long before a model is manufactured for testing.

When you're designing, you ask questions. *Who will use it? What will it do? What size should it be? What weight? What shape? What color? Of what materials? How can I make it maintenance-easy? How can I make it safe? How can I keep costs down? What are my design's competitors? How will my design be better?* Market research results need to be considered before you can answer many of these questions. Then you'll make sketches or computer images of several possible designs, which you'll offer to your product development team. It may take several tries to satisfy them. Last, you'll make a model, a sample, or detailed plans drawn to scale.

## MORE ABOUT INDUSTRIAL DESIGN

### Degree Required:

- Bachelor's Degree in Industrial Engineering

### Some industrial designers like:

- pleasant working conditions
- good salaries
- opportunity to use their creative and artistic talents
- variety in their work

### Related Math Subjects:

- Geometry
- Advanced Algebra
- Trigonometry
- Calculus

### Some industrial designers dislike:

- having their ideas rejected
- working under pressure
- working overtime to meet deadlines

For more information on the various careers available in Industrial Design, write to:

Industrial Designers Society of America
1142 Walker Road, Suite E
Great Falls, VA 22066-1836

# 4-1 Problem-Solving Strategy: Using Matrix Logic

**Objective**  After studying this lesson, you will be able to:

- solve problems using matrix logic.

Many problems can be solved using a method sometimes referred to as **matrix logic.**

When you use matrix logic, you create a table, or **matrix,** that helps you organize all the information in the problem. By using the matrix, you can eliminate one possibility after another until you eventually arrive at a solution.

**Example**

**Fred, Ted, and Ed are taking Mary, Cari, and Terri to the homecoming dance. Use these clues to find out which couples will be attending the homecoming dance.**

1. Mary is Ed's sister and lives on Fifth Avenue.
2. Ted drives a car to school each day.
3. Ed is taller than Terri's date.
4. Cari and her date ride their bicycles to school every day.
5. Fred's date lives on State Street.

First create a table that allows you to record all that you can learn from each clue. It should look something like the one below. You can then use an **X** to mark any square that is not a valid conclusion and a √ to mark any square that is a valid conclusion.

Look at the first clue. Since Mary is Ed's sister, she's not his date. Put an **X** in Mary's column beside Ed's name. Are there any other clues about Ed?

Look at the third clue. Since Ed is taller than Terri's date, he's not Terri's date. Put an **X** in Terri's column next to Ed's name.

|      | Mary | Cari | Terri |
|------|------|------|-------|
| Fred |      |      |       |
| Ted  |      |      |       |
| Ed   | X    |      | X     |

You can now deduce that Ed's date must be Cari. Put a √ in her column in Ed's row. You can also eliminate the other boys as Cari's date. Put X's in her column for their names.

Continue to evaluate each clue, making decisions, and recording your deductions until you can identify each couple. When you finish you should find that the couples are Ed and Cari, Ted and Mary, and Fred and Terri.

|      | Mary | Cari | Terri |
|------|------|------|-------|
| Fred |      | X    |       |
| Ted  |      | X    |       |
| Ed   | X    | ✓    | X     |

# CHECKING FOR UNDERSTANDING

**Communicating Mathematics**

**Read and study the lesson to answer these questions.**

1. Why is this problem-solving strategy called matrix logic?

2. How do you keep track of the decisions you have made?

**Guided Practice**

**Use matrix logic to solve this problem.**

3. Rae, Carol, and Dena are neighbors. Their hobbies are sculpturing, fixing cars, and gardening. Their occupations are doctor, teacher, and lawyer. Use these clues to find each person's hobby and occupation.
   - The gardener and the teacher both graduated from the same college.
   - Both the lawyer and Rae have poodles, as does the sculptor.
   - The doctor bandaged the sculptor's broken thumb.
   - Carol and the lawyer have lived next door to each other for five years.
   - Dena beat both Carol and the gardener in tennis.

# EXERCISES

**Strategies**

Look for a pattern.
Solve a simpler problem.
Act it out.
Guess and check.
Draw a diagram.
Make a chart.
Work backwards.

**Solve. Use any strategy.**

4. Find a two-digit number such that the sum of its digits is 10 and 3 times the tens digit is twice the ones digit.

5. How many capital letters, when folded, have halves that match exactly? Name them.

6. How many times is the hour display on a digital clock greater than the minute display in a 12-hour period?

**7.** Determine if each pattern can be folded to form a rectangular solid.

a.  b.  c. 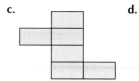 d.

**Each figure can be viewed from two different perspectives. Describe the one you see first. What is the other perspective?**

**8.** 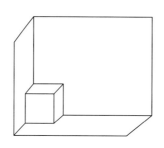 **9.**

**10.** Insert parentheses into the following expression to make the equation valid.

$$2 \cdot 1 + 2 \cdot 3 - 2 \div 2 - 1 = 12$$

**Work in groups. Each person in the group must understand the solution and be able to explain it to any person in class.**

Five rock band members are named Bobbie, Bebe, Bruno, Bart, and Benito. Their last names are Cassady, Casto, Coffman, Crosby, and Cortez. Each band member has a different color trunk for their stage costumes. The colors are red, blue, black, white, and tan. Use the following clues to identify each band member by first and last name and which trunk each owns.

1. Bebe is not a Coffman and does not own a black trunk.
2. Bruno, Bart, and Crosby do not own a red trunk or a white trunk.
3. Of Bobbie and Bart, one is named Cassady and one owns a tan trunk.
4. The owner of the red trunk and Cortez are not named Bebe or Benito.
5. Either Casto or Coffman (who owns a white trunk) is named Bart.

# Graphing Calculator Exploration: Matrices

The TI-81 graphing calculator is designed to handle most of the matrix operations and procedures introduced in this chapter. It will find determinants and inverses of matrices, as well as perform operations with matrices.

The $\boxed{\text{MATRX}}$ key accesses the matrix operations menus. The first menu to appear is the MATRIX menu, which lists the matrix functions available. The EDIT menu allows you to define matrices. When the EDIT menu is accessed, the dimensions of the matrices $A$, $B$, and $C$ are listed. A matrix dimension of $2 \times 3$ indicates a matrix has 2 rows and 3 columns. The TI-81 will accommodate a maximum of 6 rows and 6 columns in a matrix.

To enter a matrix into your calculator, choose the EDIT menu and select matrix [A]. Then enter the dimensions and elements of the matrix. As you enter the elements, notice the small graphics box to the right of the dimension definition that shows the size of the matrix and the current element position.

Define matrix $A = \begin{bmatrix} 3 & -1 \\ 4 & 2 \end{bmatrix}$ in your TI-81 graphing calculator.

ENTER: $\boxed{\text{MATRX}}$ $\boxed{\blacktriangleright}$ $\boxed{\text{ENTER}}$ 2 $\boxed{\text{ENTER}}$ 2 $\boxed{\text{ENTER}}$     *Enter the matrix dimensions.*

3 $\boxed{\text{ENTER}}$ $\boxed{\text{(-)}}$ 1 $\boxed{\text{ENTER}}$     *Enter the matrix elements.*

4 $\boxed{\text{ENTER}}$ 2

You can display the matrix by pressing $\boxed{\text{2nd}}$ $\boxed{\text{QUIT}}$ to return to the home screen, then $\boxed{\text{2nd}}$ $\boxed{[A]}$ $\boxed{\text{ENTER}}$ to display the matrix.

The graphing calculator can find the determinant and the inverse of a matrix.

**Example 1**

**Find the determinant and the inverse of the matrix A.**

First find the determinant.     *The determinant of A is denoted det A.*

ENTER: $\boxed{\text{MATRX}}$ 5 $\boxed{\text{2nd}}$ $\boxed{[A]}$ $\boxed{\text{ENTER}}$   $\mathsf{10}$

Now find the inverse.     *The inverse of A is denoted $A^{-1}$.*

ENTER: $\boxed{\text{2nd}}$ $\boxed{[A]}$ $\boxed{x^{-1}}$ $\boxed{\text{ENTER}}$ $\begin{matrix} [.2 & .1] \\ [-.4 & .3] \end{matrix}$

The determinant of matrix $A$ is 10 and the inverse is $\begin{bmatrix} 0.2 & 0.1 \\ -0.4 & 0.3 \end{bmatrix}$.

You can also use the graphing calculator to perform operations on matrices.

**Example 2**

Enter matrix $B = \begin{bmatrix} 2 & 4 & 8 \\ -4 & -2 & 6 \end{bmatrix}$. Then find $2B$, $AB$, $A^2$, and $B + AB$.

Use the procedure shown on page 159 to enter matrix $B$. *The dimension of the matrix is 2 × 3.*

Find $2B$.

ENTER: 2 (2nd) ([B]) (ENTER)
$$\begin{bmatrix} 4 & 8 & 16 \\ -8 & -4 & 12 \end{bmatrix}$$

Then find $AB$.

ENTER: (2nd) ([A]) (2nd) ([B]) (ENTER)
$$\begin{bmatrix} 10 & 14 & 18 \\ 0 & 12 & 44 \end{bmatrix}$$

Next find $A^2$.

ENTER: (2nd) ([A]) ($x^2$) (ENTER)
$$\begin{bmatrix} 5 & -5 \\ 20 & 0 \end{bmatrix}$$

Now find $B + AB$.

ENTER: (2nd) ([B]) (+) (2nd) ([A]) (2nd) ([B]) (ENTER)
$$\begin{bmatrix} 12 & 18 & 26 \\ -4 & 10 & 50 \end{bmatrix}$$

Therefore, $2B = \begin{bmatrix} 4 & 8 & 16 \\ -8 & -4 & 12 \end{bmatrix}$, $AB = \begin{bmatrix} 10 & 14 & 18 \\ 0 & 12 & 44 \end{bmatrix}$, $A_2 = \begin{bmatrix} 5 & -5 \\ 20 & 0 \end{bmatrix}$,

and $B + AB = \begin{bmatrix} 12 & 18 & 26 \\ -4 & 10 & 50 \end{bmatrix}$.

# EXERCISES

Enter the matrices into your graphing calculator in the appropriate locations. Then find each of the following.

$$A = \begin{bmatrix} 2 & 8 & 3 \\ 1 & 0 & -1 \\ 3 & 4 & 3 \end{bmatrix} \qquad B = \begin{bmatrix} 5 & -2 & 6 \\ 4 & 7 & -1 \end{bmatrix} \qquad C = \begin{bmatrix} 1 & -5 \\ 4 & -3 \\ 7 & -6 \end{bmatrix}$$

| | | |
|---|---|---|
| 1. $-C$ | 2. $5B$ | 3. det $A$ |
| 4. $-2C$ | 5. $A^{-1}$ | 6. $CB$ |
| 7. $BC$ | 8. det $BC$ | 9. $BA$ |
| 10. $CB - A$ | 11. det $CB$ | 12. $A + CB$ |
| 13. $(BC)^{-1}$ | 14. $A^2$ | 15. $(BC)^2$ |
| 16. $B + BA$ | 17. $BAC$ | 18. $CBA$ |

## 4-2 An Introduction to Matrices

**Objectives**

After studying this lesson, you should be able to:
- create a matrix and name it using its dimensions,
- perform scalar multiplication on a matrix,
- add matrices, and
- find unknown values in equal matrices.

In Lesson 4-1, you learned that a **matrix** is a system of rows and columns. A matrix is a problem-solving tool that organizes numbers or data so that each position in the matrix has a purpose.

**Connection**

In algebra, a matrix is not expressed as a table, but as an array of values. Each value is called an **element** of the matrix. Suppose we want to write the coordinates of the vertices of $\triangle ABC$ as a matrix. Let row 1 be the $x$-coordinates and row 2 be the $y$-coordinates. Each column represents the coordinates of vertices $A$, $B$, and $C$.

*3 columns*

$$2 \text{ rows} \left\{ \begin{bmatrix} 3 & -2 & 1 \\ 2 & 1 & -4 \end{bmatrix} \right.$$

*A matrix containing coordinates of a geometric figure is often called a coordinate matrix.*

A matrix is usually named using an uppercase letter. We might call the matrix for $\triangle ABC$ matrix $T$ to stand for triangle. A matrix can also be named by using the matrix **dimensions** with the letter name. The dimensions tell how many rows and columns there are in the matrix. The matrix above would be named $T_{2\times 3}$, since it has two rows and 3 columns.

Certain matrices have special names. A matrix that has only one row is called a *row matrix*, and a matrix that has only one column is called a *column matrix*. A matrix that has the same number of rows and columns is called a *square matrix*.

*The plural of matrix is matrices.*

Two matrices are considered equal if they have the same dimensions and each element of one matrix is equal to the corresponding element of the other matrix. This definition can be used to find values when elements of the matrices are algebraic expressions.

| *Definition of Equal Matrices* | **Two matrices are equal *if and only if* they have the same dimensions and their corresponding elements are equal.** |
|---|---|

## Example 1

Solve for $x$ and $y$.

$$\begin{bmatrix} 6x \\ y \end{bmatrix} = \begin{bmatrix} 62 + 8y \\ 6 - 2x \end{bmatrix}$$

Since the matrices are equal, the corresponding elements are equal. When you write the sentences that show this equality, two linear equations are formed.

$$6x = 62 + 8y$$
$$y = 6 - 2x$$

The second equation gives you a value for $y$ that can be substituted into the first equation. Then you can find a value for $x$.

$$6x = 62 + 8y$$
$$6x = 62 + 8(6 - 2x) \qquad \textit{Substitute } 6 - 2x \textit{ for y.}$$
$$6x = 62 + 48 - 16x \qquad \textit{Use the distributive property.}$$
$$22x = 110 \qquad \textit{Add 16x to each side and combine terms.}$$
$$x = 5 \qquad \textit{Divide by 22.}$$

To find a value for $y$, you can substitute 5 into either equation.

$$y = 6 - 2x$$
$$y = 6 - 2(5) \qquad \textit{Substitute 5 for x.}$$
$$y = -4$$

Check your solutions by substituting the values into the equation you *did not* use to find $y$.

**Check:**
$$6x = 62 + 8y$$
$$6(5) \stackrel{?}{=} 62 + 8(-4) \qquad \textit{Substitute 5 for x and } -4 \textit{ for y.}$$
$$30 = 62 - 32 \text{ or } 30 \quad \checkmark$$

You can multiply any matrix by a constant. This is called **scalar multiplication.** When scalar multiplication is performed each element is multiplied by that constant, and a new matrix is formed. This is summarized by the following rule.

| *Scalar Multiplication of a Matrix* | 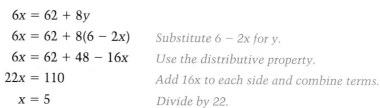 |
|---|---|

*If the perimeter of a figure triples, this does not mean its area triples.*

Scalar multiplication of matrices can be used to find the coordinates of the vertices of a geometric figure when it is enlarged or reduced. This type of change is called a **dilation.** When the size of a figure changes, the measures of its sides change in the same proportion. For example, if a figure triples in perimeter, its sides triple in length.

**Example 2**

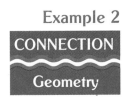

CONNECTION

Geometry

Enlarge △*ABC*, with vertices *A*(3, 2), *B*(−2, 1), and *C*(1, −4), so that its perimeter is twice the perimeter of the original figure.

Graph △*ABC*, then multiply the coordinate matrix by 2.

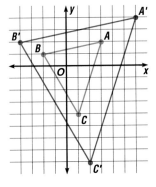

$$2\begin{bmatrix} 3 & -2 & 1 \\ 2 & 1 & -4 \end{bmatrix} = \begin{bmatrix} 6 & -4 & 2 \\ 4 & 2 & -8 \end{bmatrix}$$

The coordinates of the vertices of △*A′B′C′* are (6, 4), (−4, 2), and (2, −8). Graph △*A′B′C′*.

The two triangles are similar. The perimeter of △*A′B′C′* is twice the perimeter of △*ABC*. *You can measure to verify this result.*

Matrices can also be added. In order to add two matrices, they must have the same dimensions.

*Addition of Matrices*

**If *A* and *B* are two *m* × *n* matrices, then *A* + *B* is an *m* × *n* matrix where each element is the sum of the corresponding elements of *A* and *B*.**

$$\begin{bmatrix} a & b & c \\ d & e & f \\ g & h & i \end{bmatrix} + \begin{bmatrix} j & k & l \\ m & n & o \\ p & q & r \end{bmatrix} = \begin{bmatrix} a+j & b+k & c+l \\ d+m & e+n & f+o \\ g+p & h+q & i+r \end{bmatrix}$$

When a figure is moved from one location to another on the coordinate plane without changing its orientation, size, or shape, a **translation** occurs. You can use matrix addition to find the coordinates of the translated figure.

**Example 3**

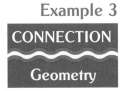

CONNECTION

Geometry

Find the coordinates of the vertices of quadrilateral *QUAD* if the figure is moved 5 units to the right and 1 unit down.

Write the coordinates of quadrilateral *QUAD* in the form of a matrix.

$$\begin{bmatrix} -2 & 1 & 3 & -2 \\ -2 & -2 & 4 & 1 \end{bmatrix}$$

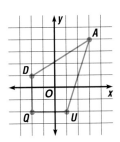

To translate the quadrilateral 5 units to the right means that each *x*-coordinate increases by 5. Translating the figure 1 unit down decreases each *y*-coordinate by 1.

The matrix, called a *translation matrix*, that increases each $x$-value by 5 and decreases each $y$-value by 1 is $\begin{bmatrix} 5 & 5 & 5 & 5 \\ -1 & -1 & -1 & -1 \end{bmatrix}$.

To find the coordinates of the translated quadrilateral $Q'U'A'D'$, add the two matrices.

$$\begin{bmatrix} -2 & 1 & 3 & -2 \\ -2 & -2 & 4 & 1 \end{bmatrix} + \begin{bmatrix} 5 & 5 & 5 & 5 \\ -1 & -1 & -1 & -1 \end{bmatrix} = \begin{bmatrix} 3 & 6 & 8 & 3 \\ -3 & -3 & 3 & 0 \end{bmatrix}$$

Now graph the coordinates of quadrilateral $Q'U'A'D'$ to check the accuracy of your coordinates. The two quadrilaterals have the same size and shape. Quadrilateral $Q'U'A'D'$ is $QUAD$ moved right 5 units and down 1 unit.

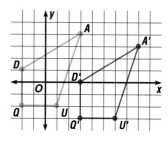

*The two figures are congruent.*

# CHECKING FOR UNDERSTANDING

**Communicating Mathematics**

Read and study the lesson to answer each question.

1. What is a matrix?

2. Describe the matrix $A_{3 \times 2}$.

3. Under what conditions are two matrices equal?

4. Under what conditions can matrices be added?

5. Explain the difference between a dilation and a translation.

**Guided Practice**

Name each matrix using the dimensions. Then multiply the matrix by $-2$.

6. $A = \begin{bmatrix} 7 & -2 & -4 \\ -8 & 9 & 10 \\ 1 & 17 & 10 \\ 21 & -3 & 6 \end{bmatrix}$

7. $V = \begin{bmatrix} -7 & 5 & 26 \end{bmatrix}$

Solve for the variables.

8. $\begin{bmatrix} 2 & x \\ y & 5 \end{bmatrix} = \begin{bmatrix} 2 & 1 \\ 3 & z \end{bmatrix}$

9. $\begin{bmatrix} 2x & 3 & 3z \end{bmatrix} = \begin{bmatrix} 5 & 3y & 9 \end{bmatrix}$

10. Translate $\triangle DEF$ with $D(7, -2)$, $E(4, 5)$, and $F(-3, 4)$ 5 units left and 1 unit up. Graph $\triangle DEF$ and $\triangle D'E'F'$. State the coordinates of the translated triangle.

# EXERCISES

**Practice**

Perform the indicated operations. Name the resulting matrix by its dimensions.

**11.** $4\begin{bmatrix} 2 & -3 \\ 4 & 1 \\ 0 & 3 \end{bmatrix}$

**12.** $\begin{bmatrix} 3 & 7 \\ -2 & 1 \end{bmatrix} - \begin{bmatrix} 2 & -3 \\ 5 & -4 \end{bmatrix}$

**13.** $\begin{bmatrix} 4 & 1 & -3 \end{bmatrix} + \begin{bmatrix} 6 & -5 & 8 \end{bmatrix}$

Solve for the variables.

**14.** $\begin{bmatrix} 2x & 3y \\ 40 & 50 \end{bmatrix} = \begin{bmatrix} -12 & 36 \\ 8z & 2.5w \end{bmatrix}$

**15.** $\begin{bmatrix} 2x \\ y+1 \end{bmatrix} = \begin{bmatrix} y \\ 3 \end{bmatrix}$

**16.** $y\begin{bmatrix} 3 & -4 \\ 2 & x \end{bmatrix} = \begin{bmatrix} 15 & -20 \\ z & 5 \end{bmatrix}$

**17.** $4\begin{bmatrix} x & y-1 \\ 3 & z \end{bmatrix} = \begin{bmatrix} 20 & 8 \\ 6z & x+y \end{bmatrix}$

**18.** $\begin{bmatrix} x^2 & 7 & 9 \\ 5 & 12 & 6 \end{bmatrix} = \begin{bmatrix} 25 & 7 & y \\ 5 & 2z & 6 \end{bmatrix}$

**19.** $\begin{bmatrix} x+3y \\ 3x+y \end{bmatrix} = \begin{bmatrix} -13 \\ 1 \end{bmatrix}$

**20.** Translate $\triangle RST$ with $R(5, -2)$, $S(8, 4)$, and $T(-3, -1)$, so that $R'$ is located at $(3, 4)$. Graph both triangles and state the coordinates of $S'$ and $T'$.

Perform the indicated operations.

**21.** $3\begin{bmatrix} 3 & -2 & 5 \\ 2 & 7 & -5 \end{bmatrix} + 2\begin{bmatrix} -1 & 3 & 4 \\ 2 & -3 & 0 \end{bmatrix}$

**22.** $\frac{1}{2}\begin{bmatrix} 4 & 12 & 9 \\ 3 & 6 & 0 \end{bmatrix} - \frac{2}{3}\begin{bmatrix} 9 & 27 & 6 \\ 0 & 3 & 4 \end{bmatrix}$

**23.** $3\begin{bmatrix} 4 \\ 1 \\ 7 \end{bmatrix} + 2\begin{bmatrix} 3 \\ -2 \\ 6 \end{bmatrix} - 5\begin{bmatrix} -2 \\ 3 \\ 6 \end{bmatrix}$

**24.** $5\begin{bmatrix} -2 & 4 \\ 1 & -1 \\ 3 & 0 \end{bmatrix} - 2\begin{bmatrix} 5 & 3 \\ -3 & 2 \\ 8 & -9 \end{bmatrix} + \begin{bmatrix} 0 & -5 \\ 9 & -3 \\ -2 & 7 \end{bmatrix}$

**25.** Quadrilateral $BURT$ has vertices with coordinates $B(6, 1)$, $U(3, 5)$, $R(-1, 4)$, and $T(-3, -5)$.

    **a.** What translation matrix would you need to translate $BURT$ so that $R'$ has coordinates $(3, 2)$?

    **b.** Use the translation matrix to find the coordinates of $B'$, $T'$, and $U'$.

**26.** The vertex of the right angle of a right triangle is located at the origin with its other vertices at $(0, 12)$ and $(5, 0)$. Find the coordinates of the vertices of a similar triangle whose perimeter is four times that of the original triangle.

**Solve for the variable.**

27. $\begin{bmatrix} x \\ 7z \\ 2y \end{bmatrix} - \begin{bmatrix} 4z \\ -3y \\ 3x \end{bmatrix} + \begin{bmatrix} -2y \\ 2x \\ -5z \end{bmatrix} = \begin{bmatrix} -4 \\ 11 \\ 18 \end{bmatrix}$  28. $\begin{bmatrix} r^2 - 24 & 17 \\ 7 & t^3 \end{bmatrix} = \begin{bmatrix} 1 & 2y + 3 \\ z^2 - 12 & 27 \end{bmatrix}$

29. $\begin{bmatrix} 5x - 7 & 11 \\ 5 & 23 \end{bmatrix} = \begin{bmatrix} 8 & 21 - m \\ r^3 - 3 & 4y + x \end{bmatrix}$

30. $\begin{bmatrix} 13 - 7y & a \\ 1 & 2b - 38 \end{bmatrix} = \begin{bmatrix} 5x & 2 - 6b \\ 2x + 3y & 5a \end{bmatrix}$

**CONNECTION**
**Geometry**

31. Find the coordinates of quadrilateral $MNPQ$ that is congruent to quadrilateral $XYZW$ whose vertices have coordinates $X(5, -3)$, $Y(2, 7)$, $Z(-3, 3)$, and $W(-5, 1)$, if $M$ is located at the origin.

**Critical Thinking**
**CONNECTION**
**Geometry**

32. The coordinate matrix for triangle $XYZ$ is $\begin{bmatrix} -2 & 4 & -1 \\ -1 & 2 & 3 \end{bmatrix}$. Explain what happens to the triangle when the matrix is multiplied by $-\frac{1}{2}$. Make a drawing to justify your answer.

**Applications**

33. **Business**  A local bakery keeps a log of each type of donut sold at three of their branch stores so that they can monitor their purchases of supplies without having extra inventory. Two days of sales are shown below.

| Monday | jelly | glazed | plain | frosted |
|---|---|---|---|---|
| Big Donut | 120 | 97 | 64 | 75 |
| Cal's Donut | 80 | 59 | 36 | 60 |
| Donuts Inc. | 72 | 84 | 29 | 48 |

| Tuesday | jelly | glazed | plain | frosted |
|---|---|---|---|---|
| Big Donut | 112 | 87 | 56 | 74 |
| Cal's Donuts | 84 | 65 | 39 | 70 |
| Donuts Inc. | 88 | 98 | 43 | 60 |

a. Write a matrix for each day's sales. Name each matrix. Then find the sum of the two days' sales expressed as a matrix.

b. Each type of donut takes approximately one-fourth cup of flour. If there are four cups of flour in a pound, how many pounds of flour were needed for these two days of baking?

**Mixed Review**

34. Find an equation of the line that passes through $(-1, 2)$ and is parallel to the graph of $2x - 7y = 11$.  **(Lesson 2-4)**

35. If $f(x) = 2x^2 + 5x - 3$, find $f\left(\frac{2}{3}\right)$.  **(Lesson 2-1)**

36. Solve $5a + 1 \le 8a - 3$.  **(Lesson 1-7)**

37. Solve $\frac{3}{4}(x - 5) = \frac{4}{5}(x + 4)$.  **(Lesson 1-3)**

# 4-3 Matrices and Determinants

**Objectives**

After studying this lesson, you should be able to:
- evaluate the determinant of a 3 × 3 matrix, and
- find the area of a triangle given the coordinates of its vertices.

Every square matrix has a **determinant.** The determinant has the same elements as the matrix, but they are enclosed between vertical bars instead of brackets. In Chapter 3 you learned a method for evaluating a 2 × 2 determinant.

The determinant of $\begin{bmatrix} 3 & -2 \\ 17 & 11 \end{bmatrix}$ is $\begin{vmatrix} 3 & -2 \\ 17 & 11 \end{vmatrix}$. To evaluate the determinant, use the rule for second-order determinants.

$$\begin{vmatrix} 3 & -2 \\ 17 & 11 \end{vmatrix} = 3(11) - (-2)(17)$$

*Remember that* $\begin{vmatrix} a & b \\ c & d \end{vmatrix} = ad - bc.$

$$= 33 - (-34) \text{ or } 67$$

*Determinants of 3 × 3 matrices are called third order determinants.*

A method called **expansion by minors** can be used to evaluate the determinant of a 3 × 3 matrix. The **minor** of an element is the determinant formed when the row and column containing that element are deleted. For the determinant $\begin{vmatrix} 1 & 3 & 7 \\ 4 & 8 & 2 \\ 9 & 5 & 6 \end{vmatrix}$, the minor of 5 is $\begin{vmatrix} 1 & 3 & 7 \\ 4 & 8 & 2 \\ 9 & 5 & 6 \end{vmatrix}$ or $\begin{vmatrix} 1 & 7 \\ 4 & 2 \end{vmatrix}$. The minor of 1 is $\begin{vmatrix} 1 & 3 & 7 \\ 4 & 8 & 2 \\ 9 & 5 & 6 \end{vmatrix}$ or $\begin{vmatrix} 8 & 2 \\ 5 & 6 \end{vmatrix}$.

To use expansion by minors with third-order determinants, each member of one row is multiplied by its minor. The signs of the products alternate, beginning with the second product. The definition below shows an expansion using the elements in the first row of the determinant. However, any row can be used.

*Expansion of a Third-Order Determinant*

$$\begin{vmatrix} a & b & c \\ d & e & f \\ g & h & i \end{vmatrix} = a\begin{vmatrix} e & f \\ h & i \end{vmatrix} - b\begin{vmatrix} d & f \\ g & i \end{vmatrix} + c\begin{vmatrix} d & e \\ g & h \end{vmatrix}$$

## Example 1

Evaluate the determinant of $\begin{bmatrix} 2 & 3 & 4 \\ 6 & 5 & 7 \\ 1 & 2 & 8 \end{bmatrix}$ using expansion by minors.

Decide which row of elements you will use for the expansion.
Let's use the first row.

*After entering this matrix in your TI-81 calculator, try this. Press MATRIX, select #5: det, press 2nd [A], ENTER. What is your answer!*

$$\begin{vmatrix} 2 & 3 & 4 \\ 6 & 5 & 7 \\ 1 & 2 & 8 \end{vmatrix} = 2\begin{vmatrix} 5 & 7 \\ 2 & 8 \end{vmatrix} - 3\begin{vmatrix} 6 & 7 \\ 1 & 8 \end{vmatrix} + 4\begin{vmatrix} 6 & 5 \\ 1 & 2 \end{vmatrix}$$

*The determinant of* $\begin{bmatrix} 2 & 3 & 4 \\ 6 & 5 & 7 \\ 1 & 2 & 8 \end{bmatrix}$ *is* $\begin{vmatrix} 2 & 3 & 4 \\ 6 & 5 & 7 \\ 1 & 2 & 8 \end{vmatrix}$.

$$= 2(40 - 14) - 3(48 - 7) + 4(12 - 5)$$
$$= 52 - 123 + 28 \text{ or } -43$$

You can check your work by evaluating the determinant again using a different row of elements.

Another method for evaluating a third-order determinant is using diagonals. In this method, you begin by repeating the first two columns on the right side of the determinant.

$$\begin{vmatrix} a & b & c \\ d & e & f \\ g & h & i \end{vmatrix} \Rightarrow \begin{vmatrix} a & b & c \\ d & e & f \\ g & h & i \end{vmatrix}\begin{matrix} a & b \\ d & e \\ g & h \end{matrix}$$

Now draw a diagonal from each element in the top row diagonally downward. Find the product of the numbers on each diagonal.

$aei \quad bfg \quad cdh$

Then draw a diagonal from each element in the bottom row diagonally upward. Find the product of the numbers on each diagonal.

$gec \quad hfa \quad idb$

To find the value of the determinant, add the products in the first set of diagonals and then subtract the products from the second set of diagonals.

The value is $aei + bfg + cdh - gec - hfa - idb$.

**Example 2**

Evaluate $\begin{vmatrix} -1 & 4 & 0 \\ 3 & -2 & -5 \\ -3 & -1 & 2 \end{vmatrix}$ using diagonals.

First, rewrite the first two columns along side the determinant.

Next, find the values using the diagonals.

$\begin{vmatrix} -1 & 4 & 0 \\ 3 & -2 & -5 \\ -3 & -1 & 2 \end{vmatrix}\begin{matrix} -1 & 4 \\ 3 & -2 \\ -3 & -1 \end{matrix}$

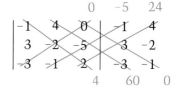

Now add the bottom products and subtract the top products.

$4 + 60 + 0 - 0 - (-5) - 24 = 45$

The value of the determinant is 45.

---

Determinants can be used to find the area of a triangle when you know the coordinates of the three vertices. The area of a triangle whose vertices have coordinates $(a, b)$, $(c, d)$ and $(e, f)$ can be found by using the formula

$A = \dfrac{1}{2}\begin{vmatrix} a & b & 1 \\ c & d & 1 \\ e & f & 1 \end{vmatrix}$, and then finding $|A|$, since the area cannot be negative.

**Example 3**

**CONNECTION**

**Geometry**

Find the area of the triangle whose vertices have coordinates $(-4, -1)$, $(3, 2)$, and $(4, 6)$.

Assign values to $a$, $b$, $c$, $d$, $e$, and $f$ and substitute them into the area formula and evaluate.

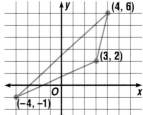

$A = \dfrac{1}{2}\begin{vmatrix} a & b & 1 \\ c & d & 1 \\ e & f & 1 \end{vmatrix}$     $a = -4, b = -1,$
$c = 3, d = 2,$
$e = 4, f = 6$

$= \dfrac{1}{2}\begin{vmatrix} -4 & -1 & 1 \\ 3 & 2 & 1 \\ 4 & 6 & 1 \end{vmatrix} = \dfrac{1}{2}[(-4)2 + (-1)4 + (3)6 - (-4)6 - (-1)3 - (2)4]$

$= \dfrac{1}{2}(-8 - 4 + 18 + 24 + 3 - 8)$

$= \dfrac{1}{2}(25)$ or $12\dfrac{1}{2}$     $\left| 12\dfrac{1}{2} \right| = 12\dfrac{1}{2}$

The area of the triangle is $12\dfrac{1}{2}$ square units.

Sometimes one or more of the elements of a determinant may be unknown, but the value of the determinant is known. You can use expansion to find the values of the variable.

Example 4

Solve for $n$ if $\begin{vmatrix} 4 & 3 & 6 \\ 2 & 2n & 7 \\ -4 & -3n & 3 \end{vmatrix} = -582.$

By expanding the determinant, you get this equation.

$24n - 36n - 84 + 48n - 18 + 84n = -582$

$\qquad\qquad 120n - 102 = -582$   *Combine like terms.*

$\qquad\qquad\qquad 120n = -480$   *Add 102 to each side.*

$\qquad\qquad\qquad\qquad n = -4$   *Divide each side by 120.*

The value of $n$ is $-4$.

# CHECKING FOR UNDERSTANDING

Communicating
Mathematics

**Read and study the lesson to answer each question.**

1. Under what conditions does a matrix have a determinant?

2. Describe the steps in finding the minor of 6 in $\begin{vmatrix} 1 & 2 & 3 \\ 4 & 5 & 6 \\ 7 & 8 & 9 \end{vmatrix}$.

3. Name an application of geometry that may use determinants.

4. In Example 4, explain how you can check the solution.

Guided Practice

**Determine whether each matrix has a determinant. If it does, find the value of the determinant.**

5. $\begin{bmatrix} 2 & 5 \\ -3 & 8 \end{bmatrix}$

6. $\begin{bmatrix} 7 \\ -2 \end{bmatrix}$

7. $\begin{bmatrix} 5 & 4 & -7 \\ 8 & 4 & 1 \end{bmatrix}$

8. $\begin{bmatrix} 6 & 9 \\ 2 & 3 \end{bmatrix}$

**Solve for $x$.**

9. $\begin{vmatrix} 5 & x \\ 2x & 7 \end{vmatrix} = -63$

10. $\begin{vmatrix} 3 & -2 & x \\ x & 1 & -5 \\ 2 & 0 & -1 \end{vmatrix} = 1$

11. Write an expression for finding the area of a triangle whose vertices have coordinates $(4, -5)$, $(3, 8)$, and $(-2, 3)$ using determinants.

# EXERCISES

Practice

Determine whether each matrix has a determinant. If it does, find the value of the determinant.

12. $\begin{bmatrix} 7 \\ 2 \\ 1 \\ -5 \end{bmatrix}$
13. $\begin{bmatrix} 9 & -3 \\ 17 & 4 \end{bmatrix}$
14. $\begin{bmatrix} -2 & 9 \\ 7 & -4 \\ -6 & 1 \end{bmatrix}$
15. $\begin{bmatrix} 4 & -6 & -9 & 3 \\ 0 & -4 & 2 & -8 \end{bmatrix}$

Solve for x.

16. $\begin{vmatrix} 5 & 7 \\ -2 & 2x \end{vmatrix} = 54$

17. $\begin{vmatrix} x & 7 & 5 \\ 0 & 3 & 4 \\ 3 & 2 & -x \end{vmatrix} = 11$

18. Find the area of a triangle whose vertices have coordinates $(0, 0)$, $(5, -6)$, and $(3, 7)$.

Determine the value of the determinant of each matrix.

19. $\begin{bmatrix} 6 & 4 \\ -3 & 2 \end{bmatrix}$
20. $\begin{bmatrix} 2 & -3 & 4 \\ -2 & 1 & 5 \\ 5 & 3 & -2 \end{bmatrix}$
21. $\begin{bmatrix} 6 & 5 & -2 \\ -3 & 0 & 6 \\ 1 & 4 & 2 \end{bmatrix}$

Solve for the variable.

22. $\begin{vmatrix} 5a & 3 \\ a & 5 \end{vmatrix} = 7$

23. $\begin{vmatrix} x^2 & x \\ 3 & 1 \end{vmatrix} = 4$

24. Find the value of $x$ so that the area of a triangle whose vertices have coordinates $(6, 5)$, $(8, 2)$, and $(x, 11)$ is 30.

Solve for x.

25. $\begin{vmatrix} x & 5 & 2 \\ -6 & 4 & 1 \\ 3 & 1 & x \end{vmatrix} = x^2 + 22x - 1$

26. $\begin{vmatrix} 2x & 4 & 1 \\ 2 & 3 & -1 \\ 0 & -2 & x \end{vmatrix} = 6x^2 - 10$

27. How can you convince someone that the formula for the area of a triangle could be used to show that three given points are collinear (lie on the same line)?

Critical Thinking

28. Find a third order matrix in which no element is 0, but the value of the determinant of the matrix is 0.

29. By multiplying only rows or columns of $\begin{vmatrix} 8 & 3 \\ -3 & 5 \end{vmatrix}$ by a constant, create a determinant whose value is six times the value of $\begin{vmatrix} 8 & 3 \\ -3 & 5 \end{vmatrix}$.

**30. Metallurgy** Mary Sitzco has a blueprint of a metal plate in the shape of a quadrilateral having vertices with coordinates $(-5, 2)$, $(4, -1)$, $(3, 8)$, and $(-3, 7)$ that is to be used as part of a sculpture. She needs to know the area of the quadrilateral in order to calculate the amount of ore needed to make the metal plate.

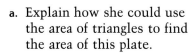

   **a.** Explain how she could use the area of triangles to find the area of this plate.

   **b.** Find the area of the plate.

**31. Horticulture** A rose garden is being planted as a border around two sides of a triangular shaped lawn in a city park. Two of the vertices of the triangular lawn have coordinates $(-2, 4)$ and $(3, -5)$. The gardener wishes to locate the third vertex so that the lawn's area is 25 square feet. Find the value of $f$ if the coordinates of the third vertex are $(3, f)$.

**Mixed Review**

**32.** Find $\begin{bmatrix} -3 & 14 & 12 \\ -2 & -1 & 7 \end{bmatrix} + \begin{bmatrix} 1 & -5 & 10 \\ 22 & 13 & -8 \end{bmatrix}$. **(Lesson 4-2)**

**33.** Find $4\begin{bmatrix} -7 & 5 & -11 \\ 2 & -4 & 9 \end{bmatrix}$. **(Lesson 4-2)**

**34.** Graph $15x + 10y + 6z = 30$. **(Lesson 3-8)**

**35.** Sherri bought 24 cans of soda at the store. She bought $r$ cans at $0.19 per can and $t$ cans at $0.29 per can. Find $r$ and $t$ if she spent $5.46 on soda. **(Lesson 1-4)**

**36.** Find the standard form of the equation whose $x$-intercept is 6 and whose $y$-intercept is $-5$. **(Lesson 2-5)**

---

## ∿∿∿ CHALLENGE ∿∿∿

In this lesson, you learned that one way to find the area of a triangle is to find half the value of the determinant containing the coordinates of the vertices of the triangle.

Use the figure at the right to help you write a convincing argument as to why the formula for the area of a triangle using determinants is valid.

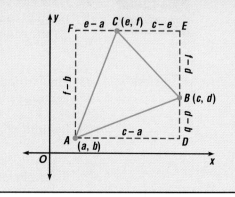

# Multiplication of Matrices

**Objective**

After studying this lesson, you should be able to:

■ multiply two matrices and interpret the results.

**Application**

Mac McDonough owns three fruit farms on which he grows peaches, apricots, plums, and apples. When picked, the fruit is sorted into layered boxes in which they will be sold. The chart below shows the number of boxes for each type of fruit.

| Location | Peaches | Apricots | Plums | Apples |
|----------|---------|----------|-------|--------|
| Farm 1 | 165 | 217 | 430 | 290 |
| Farm 2 | 243 | 190 | 235 | 175 |
| Farm 3 | 74 | 150 | 198 | 0 |

Suppose he sells peaches for $26 a box, apricots for $18 a box, plums for $32 a box, and apples for $19 a box. The total income from this picking of fruit could be found by multiplying matrices. *The solution to this problem is shown in Example 3.*

You can multiply two matrices only if the number of columns in the first matrix is equal to the number of rows in the second matrix.

**Multiplying Matrices**

> The product of an $m \times n$ matrix, $A$, and an $n \times r$ matrix, $B$, is the $m \times r$ matrix $AB$. The element in the $i^{th}$ row and the $j^{th}$ column of $AB$ is the sum of the products of the corresponding elements in the $i^{th}$ row of $A$ and the $j^{th}$ column of $B$. Multiplication of matrices is *not* commutative.

The steps in Example 1 will illustrate how two matrices can be multiplied.

**Example 1**

If $A = \begin{bmatrix} 2 & -1 \\ 3 & 4 \end{bmatrix}$ and $B = \begin{bmatrix} 3 & -9 & 2 \\ 5 & 7 & -6 \end{bmatrix}$, find $AB$.

*Matrix BA is not defined since B has 3 columns and A has 2 rows.*

$$AB = \begin{bmatrix} 2(3) + (-1)(5) & 2(-9) + (-1)(7) & (2)(2) + (-1)(-6) \\ 3(3) + 4(5) & 3(-9) + 4(7) & (3)(2) + 4(-6) \end{bmatrix}$$

$$= \begin{bmatrix} 6 - 5 & -18 - 7 & 4 + 6 \\ 9 + 20 & -27 + 28 & 6 - 24 \end{bmatrix}$$

$$= \begin{bmatrix} 1 & -25 & 10 \\ 29 & 1 & -18 \end{bmatrix}$$

*$A_{2 \times 2} \cdot B_{2 \times 3} = (AB)_{2 \times 3}$*

**Example 2**

If $A = \begin{bmatrix} 6 & 8 & 10 \\ 2 & 3 & -5 \end{bmatrix}$ and $B = \begin{bmatrix} 2 & 5 & -3 \\ 3 & -6 & 2 \end{bmatrix}$, find $AB$.

$A$ has 3 columns and $B$ has 2 rows. In order to find $AB$, $A$ must have the same number of columns as $B$ has rows. Since this is not the case, $AB$ is not defined.

Example 3 shows how matrix multiplication can be used to find the total income for Mr. McDonough's fruit farms.

**Example 3**

APPLICATION

Agriculture

**Find the total income of the three fruit farms owned by Mr. McDonough.**

The first matrix represents the numbers of boxes of each type of fruit for each farm. The second matrix will list the prices per box for each type of fruit.

$$A_{3\times4} \qquad \cdot \qquad B_{4\times1} = \quad (AB)_{3\times1}$$

$$\begin{bmatrix} 165 & 217 & 430 & 290 \\ 243 & 190 & 235 & 175 \\ 74 & 150 & 198 & 0 \end{bmatrix} \cdot \begin{bmatrix} 26 \\ 18 \\ 32 \\ 19 \end{bmatrix} = \begin{bmatrix} 27{,}466 \\ 20{,}583 \\ 10{,}960 \end{bmatrix}$$

Farm 1 earned $27,466, Farm 2 earned $20,583, and Farm 3 earned $10,960. The total income is $59,009. *Verify the resulting totals.*

Another use of matrix multiplication is in transformational geometry. You have already learned to translate a figure and change the size of a figure using matrices. When you wish to move a figure by rotating it, you can use a **rotation matrix.**

The matrix $\begin{bmatrix} 0 & -1 \\ 1 & 0 \end{bmatrix}$ will rotate a figure on a coordinate plane 90° counterclockwise about the origin. In the figure at the right, segment $AB$ is rotated 90° counterclockwise, using the origin as the point of rotation. The result is segment $A'B'$.

*Segments $AO$ and $OA'$ form a 90° angle. Likewise, segments $OB$ and $OB'$ form a 90° angle.*

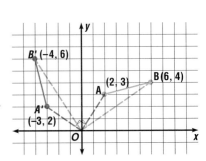

## Example 4

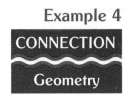

CONNECTION

Geometry

**Triangle RST has vertices with coordinates R(-1, -2), S(2, -4), and T(5, 3). Find the coordinates of the vertices of this triangle after it is rotated counterclockwise 90° about the origin.**

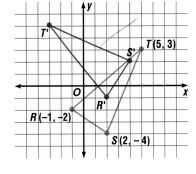

Let each column of a matrix represent an ordered pair of the triangle with the top row containing the x-values. Then multiply the coordinate matrix by the rotation matrix.

$$\begin{bmatrix} 0 & -1 \\ 1 & 0 \end{bmatrix} \cdot \begin{bmatrix} -1 & 2 & 5 \\ -2 & -4 & 3 \end{bmatrix} = \begin{bmatrix} 2 & 4 & -3 \\ -1 & 2 & 5 \end{bmatrix}$$

The coordinates of the vertices of the rotated triangle are $R'(2, -1)$, $S'(4, 2)$, and $T'(-3, 5)$.

# CHECKING FOR UNDERSTANDING

**Communicating Mathematics**

Read and study the lesson to answer each question.

1. Under what conditions can two matrices be multiplied?

2. How do the conditions for adding matrices differ from those for multiplying matrices?

3. Use the definition of matrix multiplication to find the dimension of matrix $M$ if $M = A_{3\times2} \cdot B_{2\times4}$.

4. Describe in your own words the process of multiplying two matrices.

5. What is the rotation matrix for a 90° counterclockwise rotation about the origin?

**Guided Practice**

Find the dimension of each matrix M.

6. $A_{3\times2} \cdot B_{2\times5} = M$

7. $C_{4\times4} \cdot D_{4\times2} = M$

Find each matrix N, if it exists.

8. $N = \begin{bmatrix} 3 & -1 \\ 2 & 4 \end{bmatrix} \cdot \begin{bmatrix} 4 & 0 & -3 \\ 7 & -5 & 9 \end{bmatrix}$

9. $N = \begin{bmatrix} 4 & 0 & -3 \\ 7 & -5 & 9 \end{bmatrix} \cdot \begin{bmatrix} 3 & -1 \\ 2 & 4 \end{bmatrix}$

10. $N = \begin{bmatrix} 2 & 3 & 4 \\ 1 & 0 & -1 \end{bmatrix} \cdot \begin{bmatrix} 1 & 2 \\ 0 & 0 \\ 1 & -1 \end{bmatrix}$

11. $N = \begin{bmatrix} 2 & -1 \\ 0 & 1 \end{bmatrix} \cdot \begin{bmatrix} 3 & 0 \\ 1 & 2 \end{bmatrix}$

# EXERCISES

**Practice**  Determine the dimension of each matrix product.

**12.** $M_{4 \times 2} \cdot N_{2 \times 3}$

**13.** $A_{1 \times 4} \cdot b_{4 \times 2}$

**14.** $A_{3 \times 4} \cdot B_{4 \times 1}$

**15.** $A_{3 \times 2} \cdot B_{3 \times 2}$

Perform the indicated operations.

**16.** $3\begin{bmatrix} 4 & -2 \\ 5 & 7 \end{bmatrix} + 2\begin{bmatrix} -3 & 5 \\ -4 & 3 \end{bmatrix}$

**17.** $\begin{bmatrix} 3 & -1 \\ 2 & 5 \end{bmatrix} \cdot \begin{bmatrix} 4 & -1 & -2 \\ -3 & 5 & 4 \end{bmatrix}$

**18.** $\begin{bmatrix} 4 & 0 & -8 \\ 7 & -2 & 10 \end{bmatrix} \cdot \begin{bmatrix} -1 & 3 \\ 6 & 0 \end{bmatrix}$

**19.** $\begin{bmatrix} 6 & 4 & 1 \end{bmatrix} \cdot \begin{bmatrix} 2 & 5 \\ -3 & 0 \\ -1 & 3 \end{bmatrix}$

**20.** $\begin{bmatrix} -6 & 3 \\ 4 & 7 \end{bmatrix} \cdot \begin{bmatrix} 2 & -5 \\ -3 & 6 \end{bmatrix}$

**21.** $\begin{bmatrix} 2 & 7 \end{bmatrix} \cdot \begin{bmatrix} 5 \\ -4 \end{bmatrix}$

**22.** $\begin{bmatrix} 0 & 8 \\ 3 & 1 \\ -1 & 5 \end{bmatrix} \cdot \begin{bmatrix} 3 & 1 & -2 \\ 0 & 8 & -5 \end{bmatrix}$

**23.** $\begin{bmatrix} 3 & 4 \\ 1 & 0 \\ 2 & -5 \end{bmatrix} \cdot \begin{bmatrix} -2 & 4 & 5 \\ 3 & 0 & -1 \\ 1 & 0 & -1 \end{bmatrix}$

**Find the new coordinates of the vertices of each polygon after the polygon is rotated 90° counterclockwise about the origin.**

**24.** triangle $ABC$ with vertices $A(3, 4)$, $B(6, 5)$, $C(0, 0)$

**25.** rectangle $DEFG$ with vertices $D(-1, -1)$, $E(-4, -1)$, $F(-4, -3)$, and $G(-1, -3)$

Use matrices *A*, *B*, *C*, and *D* to evaluate each expression.

$A = \begin{bmatrix} 3 & -1 \\ 2 & 4 \end{bmatrix} \quad B = \begin{bmatrix} 4 & 0 & -3 \\ 7 & -5 & 9 \end{bmatrix} \quad C = \begin{bmatrix} -6 & 4 \\ -2 & 8 \\ 3 & 0 \end{bmatrix} \quad D = \begin{bmatrix} -1 & 0 \\ 3 & 7 \end{bmatrix}$

**26.** $AB + B$  **27.** $CB + B$  **28.** $AD + CB$  **29.** $AD + BC$

**30.** Find the new coordinates of the vertices of square $MOBY$ with vertices $M(-2, -2)$, $O(2, -2)$, $B(2, 2)$, and $Y(-2, 2)$, when the square is rotated 90° counterclockwise about the origin. Graph the original square and its rotation $M'O'B'Y'$. Describe the result.

31. After a triangle was rotated 90° counterclockwise about the origin its vertices had coordinates (-3, -5), (-2, 7), and (1, 4). What were the coordinates of the triangle in its original position?

**Critical Thinking**

CONNECTION
Geometry

32. Find the new coordinates of quadrilateral *ROSE*, with vertices *R*(-2, -1), *O*(3, 0), *S*(2, 2) and *E*(-1, 2), if it is rotated 90° counterclockwise about the origin *twice*. Compare the new coordinates to the original ones. Make a conjecture about what effect this rotation has on any figure.

33. Find the values of *w*, *x*, *y*, and *z* to make the statement

$$\begin{bmatrix} 1 & 2 \\ 3 & 4 \end{bmatrix} \cdot \begin{bmatrix} w & y \\ x & z \end{bmatrix} = \begin{bmatrix} 1 & 2 \\ 3 & 4 \end{bmatrix}$$ true. If the matrix containing *w*, *x*, *y*, and *z*

were multiplied by any other matrix containing two columns, what do you think the result would be?

**Applications**

34. **Sports** In a three team track meet, the following numbers of first-, second-, and third-place finishes were recorded.

| School | First Place | Second Place | Third Place |
|---|---|---|---|
| Birmingham | 4 | 10 | 6 |
| Chatsworth | 7 | 6 | 9 |
| Monroe | 8 | 3 | 4 |

If 5 points are awarded for first, 3 for second, and 1 for third, use matrices to find the final scores for each school.

35. **Coin Collecting** Jana Wyssman is a coin collector. She specializes in coins of the early 20th century. In her collection, she has 64 nickels, 37 dimes, and 73 quarters. Because these coins are collectors items they are worth more than face value. Each nickel is worth $3.80, each dime is worth $5.40, and each quarter is worth $7.15. Use matrices to find the total value of each group of coins and the value of Miss Wyssman's whole collection.

**Journal**

Is multiplication of matrices commutative? Write a convincing argument to support your answer.

**Computer**

36. The BASIC program at the top of the next page multiplies two matrices. First enter the dimensions of each matrix. Next, enter each row of the first matrix. Continue on to enter each row of the second matrix. The computer will print the product of the two matrices. Note that line 50 makes sure that the multiplication is possible.

Use this program to check your answers to Exercises 17–23.

```
 10 PRINT "ENTER SIZE OF L."        180 FOR Q = 1 TO Z
 20 INPUT X,Y1                      190 FOR P = 1 TO X
 30 PRINT "ENTER SIZE OF R."        200 C(P,Q) = 0
 40 INPUT Y2,Z                      210 FOR I = 1 TO Y1
 50 IF Y1 <> Y2 THEN 500            220 C(P,Q) = C(P,Q) + L(P,I)
 60 PRINT "ENTER MATRIX L."             * R(I,Q)
 70 FOR M = 1 TO X                  230 NEXT I
 80 FOR N = 1 TO Y1                 240 NEXT P : PRINT
 90 INPUT L(M,N)                    250 NEXT Q
100 NEXT N : PRINT                  260 PRINT "L * R" : PRINT
110 NEXT M                          270 FOR M = 1 TO X
120 PRINT "ENTER MATRIX R."         280 FOR N = 1 TO Z
130 FOR M = 1 TO Y2                 290 PRINT C(M,N);" ";
140 FOR N = 1 TO Z                  300 NEXT N : PRINT
150 INPUT R(M,N)                    310 NEXT M
160 NEXT N : PRINT                  320 END
170 NEXT M                          500 PRINT "TRY AGAIN."
                                    510 GOTO 10
```

**Mixed Review**

**37.** Evaluate $\begin{vmatrix} 2 & -3 & 1 \\ 3 & 5 & 2 \\ 1 & 0 & -3 \end{vmatrix}$. **(Lesson 4-3)**

**38.** Solve $\begin{cases} 2x - 3y = -9 \\ x + 7y = -13 \end{cases}$. **(Lesson 3-2)**

**39.** Graph $y < 2x + 6$. **(Lesson 2-8)**

**40.** Graph $f(x) = |x - 2|$. **(Lesson 2-7)**

**41. Consumerism** Leon bought a 10-speed bicycle on sale for 75% of its original price. The sale price was $41 less than the original price. Find the original price and the sale price. **(Lesson 1-4)**

## HISTORY CONNECTION

The most influential mathematics book in China was *Nine Chapters on the Mathematical Art,* written about 250 B.C. This book contained many problems that are solved using matrices. The Chinese matrix was a large counting board that resembled a checkerboard. We read our matrices a row at a time from left to right. The Chinese organized their data differently because they read from top to bottom and from right to left.

# Identity and Inverse Matrices

**Objectives**

After studying this lesson, you should be able to:
- write the identity matrix for any matrix, and
- find the inverse matrix of a $2 \times 2$ matrix.

There are certain properties of real numbers that are related to special matrices. Remember that 1 is the identity for multiplication because $1 \cdot a = a \cdot 1 = a$. The **identity matrix** is a square matrix that, when multiplied by another matrix, equals that same matrix.

With $2 \times 2$ matrices, $\begin{bmatrix} 1 & 0 \\ 0 & 1 \end{bmatrix}$ is the identity matrix because

$$\begin{bmatrix} a & b \\ c & d \end{bmatrix} \cdot \begin{bmatrix} 1 & 0 \\ 0 & 1 \end{bmatrix} = \begin{bmatrix} a & b \\ c & d \end{bmatrix} \text{ and } \begin{bmatrix} 1 & 0 \\ 0 & 1 \end{bmatrix} \cdot \begin{bmatrix} a & b \\ c & d \end{bmatrix} = \begin{bmatrix} a & b \\ c & d \end{bmatrix}. \text{ The identity}$$

matrix is symbolized by $I$. In any identity matrix, the principal diagonal extends from upper left to lower right and consists only of 1's.

| *Identity Matrix for Multiplication* | **The identity matrix $I$ for multiplication is a square matrix with a 1 for every element of the principal diagonal and a 0 in all other positions.** |
|---|---|

**Example 1**

Find $I$ so that $\begin{bmatrix} 3 & 2 & -1 \\ -8 & 4 & 1 \end{bmatrix} I = \begin{bmatrix} 3 & 2 & -1 \\ -8 & 4 & 1 \end{bmatrix}$.

In order for you to multiply the matrices, remember that the number of columns of the first matrix must equal the number of rows in the second one.

The dimensions of the first matrix are $2 \times 3$. So $I$ must have 3 rows. Since all identity matrices are square, it also has 3 columns. The principal diagonal contains 1's. Complete the matrix with 0's.

$$\begin{bmatrix} 1 & & \\ & 1 & \\ & & 1 \end{bmatrix} \Rightarrow \begin{bmatrix} 1 & 0 & 0 \\ 0 & 1 & 0 \\ 0 & 0 & 1 \end{bmatrix} \quad \text{The } 3 \times 3 \text{ identity matrix is } \begin{bmatrix} 1 & 0 & 0 \\ 0 & 1 & 0 \\ 0 & 0 & 1 \end{bmatrix}.$$

Another property of real numbers is that any real number, except 0, has a multiplicative inverse. That is, $\dfrac{1}{a}$ is the multiplicative inverse of $a$ because $a \cdot \dfrac{1}{a} = \dfrac{1}{a} \cdot a = 1$. Likewise, if matrix $A$ has an **inverse** named $A^{-1}$, then $A \cdot A^{-1} = A^{-1} \cdot A = I$. The following example shows how the inverse of a $2 \times 2$ matrix can be found.

**Example 2**

*The TI-81 calculator can be used to find the inverse of a matrix. Enter your matrix as matrix A. Clear the screen. Then press 2nd [A] and $x^{-1}$.*

If $A = \begin{bmatrix} 3 & 2 \\ 5 & 7 \end{bmatrix}$, find $A^{-1}$ and check your result.

Let $A^{-1} = \begin{bmatrix} x & y \\ z & w \end{bmatrix}$. By the definition of an inverse, $A \cdot A^{-1}$ must equal $I$.

So, $\begin{bmatrix} 3 & 2 \\ 5 & 7 \end{bmatrix} \cdot \begin{bmatrix} x & y \\ z & w \end{bmatrix} = \begin{bmatrix} 1 & 0 \\ 0 & 1 \end{bmatrix}$.

Multiply the two matrices.

$$\begin{bmatrix} 3 & 2 \\ 5 & 7 \end{bmatrix} \cdot \begin{bmatrix} x & y \\ z & w \end{bmatrix} = \begin{bmatrix} 3x + 2z & 3y + 2w \\ 5x + 7z & 5y + 7w \end{bmatrix}$$

Thus, $\begin{bmatrix} 3x + 2z & 3y + 2w \\ 5x + 7z & 5y + 7w \end{bmatrix} = \begin{bmatrix} 1 & 0 \\ 0 & 1 \end{bmatrix}$.

When matrices are equal the corresponding elements are equal. So the following equations can be generated from the two equal matrices.

(1) $3x + 2z = 1$    (2) $3y + 2w = 0$    (3) $5x + 7z = 0$    (4) $5y + 7w = 1$

Use equations (1) and (3) to find values for $x$ and $z$.

*First solve for x.   Then substitute the x value into one of the equations to find z.*

$3x + 2z = 1$                   $21x + 14z = 7$                   $3\left(\dfrac{7}{11}\right) + 2z = 1$

$5x + 7z = 0$                   $\dfrac{-10x - 14z = 0}{11x \qquad = 7}$

$\qquad\qquad\qquad\qquad\qquad\qquad x = \dfrac{7}{11}$                   $\dfrac{21}{11} + 2z = 1$

$\qquad\qquad\qquad\qquad\qquad\qquad\qquad\qquad\qquad\qquad z = -\dfrac{5}{11}$

Use equations (2) and (4) to find values for $y$ and $w$.

*First solve for y.   Then substitute the y value into one of the equations to find w.*

$3y + 2w = 0$                   $21y + 14w = 0$                   $3\left(-\dfrac{2}{11}\right) + 2w = 0$

$5y + 7w = 1$                   $\dfrac{-10y - 14w = -2}{11y \qquad = -2}$

$\qquad\qquad\qquad\qquad\qquad\qquad y = -\dfrac{2}{11}$                   $-\dfrac{6}{11} + 2w = 0$

$\qquad\qquad\qquad\qquad\qquad\qquad\qquad\qquad\qquad\qquad w = \dfrac{3}{11}$

Thus $A^{-1} = \begin{bmatrix} \dfrac{7}{11} & -\dfrac{2}{11} \\ -\dfrac{5}{11} & \dfrac{3}{11} \end{bmatrix}$.

**Check:** $\begin{bmatrix} 3 & 2 \\ 5 & 7 \end{bmatrix} \cdot \begin{bmatrix} \dfrac{7}{11} & -\dfrac{2}{11} \\ -\dfrac{5}{11} & \dfrac{3}{11} \end{bmatrix} = \begin{bmatrix} \dfrac{21}{11} - \dfrac{10}{11} & -\dfrac{6}{11} + \dfrac{6}{11} \\ \dfrac{35}{11} - \dfrac{35}{11} & -\dfrac{10}{11} + \dfrac{21}{11} \end{bmatrix} = \begin{bmatrix} 1 & 0 \\ 0 & 1 \end{bmatrix}$ ✓

The same method used in Example 2 can be used to develop the general form of the inverse of a $2 \times 2$ matrix.

The inverse of $\begin{bmatrix} a & b \\ c & d \end{bmatrix}$ is $\begin{bmatrix} \dfrac{d}{ad-bc} & \dfrac{-b}{ad-bc} \\ \dfrac{-c}{ad-bc} & \dfrac{a}{ad-bc} \end{bmatrix}$ or $\dfrac{1}{ad-bc}\begin{bmatrix} d & -b \\ -c & a \end{bmatrix}$.

Notice that $ad - bc$ is the value of the determinant of the matrix. Remember that $\dfrac{1}{ad-bc}$ is not defined when $ad - bc = 0$. Therefore, if the value of the determinant of a matrix is 0, the matrix cannot have an inverse.

| Inverse of a $2 \times 2$ Matrix | Any matrix $M$, $\begin{bmatrix} a & b \\ c & d \end{bmatrix}$, will have an inverse $M^{-1}$ if and only if $\begin{vmatrix} a & b \\ c & d \end{vmatrix} \neq 0$. Then $M^{-1} = \dfrac{1}{ad-bc}\begin{bmatrix} d & -b \\ -c & a \end{bmatrix}$. |
| --- | --- |

**Example 3**

If $A = \begin{bmatrix} -3 & 5 \\ 1 & -4 \end{bmatrix}$, find $A^{-1}$ and check your result.

Compute the value of the determinant.
$$\begin{vmatrix} -3 & 5 \\ 1 & -4 \end{vmatrix} = 12 - 5 = 7$$
Since the determinant does not equal 0, $A^{-1}$ exists.

$$A^{-1} = \frac{1}{7}\begin{bmatrix} -4 & -5 \\ -1 & -3 \end{bmatrix} \qquad \frac{1}{ad-bc}\begin{bmatrix} d & -b \\ -c & a \end{bmatrix}$$

**Check:** $\dfrac{1}{7}\begin{bmatrix} -4 & -5 \\ -1 & -3 \end{bmatrix} \cdot \begin{bmatrix} -3 & 5 \\ 1 & -4 \end{bmatrix} = \dfrac{1}{7}\begin{bmatrix} 12-5 & -20+20 \\ 3-3 & -5+12 \end{bmatrix} = \begin{bmatrix} 1 & 0 \\ 0 & 1 \end{bmatrix}$ ✓

# CHECKING FOR UNDERSTANDING

**Communicating Mathematics**

**Read and study the lesson to answer each question.**

1. What is the result when matrix $A$ is multiplied by the identity matrix?

2. What letter is used to represent the identity matrix?

3. How would you represent the inverse of matrix $Z$?

4. Write the $4 \times 4$ identity matrix.

Determine whether each matrix can have an inverse. If not, explain why.

5. $\begin{bmatrix} 8 & 5 \\ 3 & 3 \end{bmatrix}$    6. $\begin{bmatrix} 3 & 6 \\ 2 & 4 \end{bmatrix}$    7. $\begin{bmatrix} 6 & 6 \\ -6 & -6 \end{bmatrix}$    8. $\begin{bmatrix} -6 & 8 \\ 2 & 1 \end{bmatrix}$

9. Find the inverse of $\begin{bmatrix} 1 & 2 \\ 2 & 1 \end{bmatrix}$, if it exists.

# EXERCISES

## Practice

Find the inverse of each matrix, if it exists.

10. $\begin{bmatrix} 4 & -3 \\ 3 & 8 \end{bmatrix}$    11. $\begin{bmatrix} 2 & -5 \\ 6 & 1 \end{bmatrix}$    12. $\begin{bmatrix} 2 & -4 \\ -1 & 2 \end{bmatrix}$

13. $\begin{bmatrix} 3 & 1 \\ -4 & 1 \end{bmatrix}$    14. $\begin{bmatrix} 4 & 0 \\ 0 & 1 \end{bmatrix}$    15. $\begin{bmatrix} 8 & 4 \\ -4 & -2 \end{bmatrix}$

Determine whether each statement is true or false.

16. $\begin{bmatrix} 0 & 1 \\ 1 & 1 \end{bmatrix} \cdot \begin{bmatrix} -1 & 1 \\ 1 & 0 \end{bmatrix} = I$    17. $\begin{bmatrix} 2 & 1 & -4 \\ -3 & 6 & 5 \end{bmatrix} \cdot \begin{bmatrix} 1 & 0 & 0 \\ 0 & 1 & 0 \\ 0 & 0 & 1 \end{bmatrix} = \begin{bmatrix} 2 & 1 & -4 \\ -3 & 6 & 5 \end{bmatrix}$

18. $\begin{bmatrix} 1 & 5 \\ 1 & -2 \end{bmatrix} \cdot \begin{bmatrix} \frac{2}{7} & \frac{5}{7} \\ \frac{1}{7} & -\frac{1}{7} \end{bmatrix} = I$    19. $\begin{bmatrix} \frac{1}{3} & -\frac{2}{3} \\ \frac{2}{3} & -\frac{1}{3} \end{bmatrix} \cdot \begin{bmatrix} 1 & 2 \\ 2 & 1 \end{bmatrix} = I$

20. Determine if $-\dfrac{1}{64}\begin{bmatrix} -20 & 8 & 4 \\ 16 & 0 & -16 \\ -10 & -12 & 2 \end{bmatrix}$ is the inverse of $\begin{bmatrix} 3 & 1 & 2 \\ -2 & 0 & 4 \\ 3 & 5 & 2 \end{bmatrix}$.

## Critical Thinking

CONNECTION
Geometry

21. Find the inverse of the rotation matrix given in Lesson 4-4. Then make a conjecture about what movement this describes on the coordinate plane. Make a drawing to verify your conclusion.

## Applications

22. **Topology** The figure at the left shows a directed network. The small arrows tell the direction you can travel on each pathway from point to point. These diagrams are used to plan airline routes. A matrix can be designed to represent this network. Each number in the matrix shows how many *direct* paths there are from one number to the next. For example, the 1 in the first row, second column states that there is one path from $A$ to $B$. The last number in the second row states that there are 2 paths from $B$ to $C$. Copy the matrix and complete it using the network diagram.

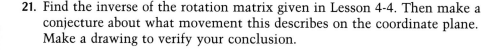

$$\begin{array}{c} \\ A \\ B \\ C \end{array}\begin{array}{ccc} A & B & C \\ \begin{bmatrix} 1 & 1 & ? \\ ? & ? & 2 \\ 1 & ? & 0 \end{bmatrix} \end{array}$$

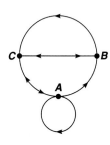

**Mixed Review**   23. Find $\begin{bmatrix} -2 & 1 \\ 3 & -6 \\ 4 & 5 \end{bmatrix} \cdot \begin{bmatrix} 1 & 2 & -3 & 7 \\ -3 & 2 & 9 & -1 \end{bmatrix}$.   (Lesson 4-4)

24. Graph $5x - 3y = 6$.   (Lesson 2-7)

25. Find the slope, $x$-intercept, and $y$-intercept of the line whose equation is $9x - 2y = 4$.   (Lesson 2-4)

26. How many different ways are there to arrange the letters, $A$, $B$, $C$, and $D$ such that $A$ is never the first letter?   (Lesson 1-5)

## MID-CHAPTER REVIEW

1. Three women and their husbands were given a total of $5400. The women received $2400 in all. Sue had $200 more than Jane, and Liz had $200 more than Sue. Lou got half as much as his wife, Bob got the same as his wife, and Matt got twice as much as his wife. Who is married to whom?   (Lesson 4-1)

2. **Geometry**   Use matrix addition to translate quadrilateral $MATH$ 4 units left and 3 units up if the vertices have coordinates $M(3, 5)$, $A(2, -6)$, $T(-3, -1)$, and $H(-7, 5)$. (Lesson 4-2)

**Perform the indicated operation.**   (Lesson 4-2)

3. $\begin{bmatrix} -2 & 1.5 \\ 3 & -0.25 \end{bmatrix} - \begin{bmatrix} -6 & 2 \\ 3 & 1.25 \end{bmatrix}$

4. $-4 \begin{bmatrix} -1 & -\dfrac{1}{4} \\ 0 & 2 \\ \dfrac{1}{2} & 4 \end{bmatrix}$

5. Evaluate the determinant of $\begin{bmatrix} -1 & 3 & 4 \\ 0 & 5 & 1 \\ 6 & -2 & 3 \end{bmatrix}$.   (Lesson 4-3)

6. **Geometry**   Use determinants to find the area of a triangle with vertices having coordinates $(0, 0)$, $(12, 0)$, and $(0, 5)$.   (Lesson 4-3)

**Perform each operation.**   (Lesson 4-4)

7. $\begin{bmatrix} -2 & 3 \\ 1 & 10 \\ 0 & -6 \end{bmatrix} \cdot \begin{bmatrix} 9 & 3 \\ 1 & 4 \end{bmatrix}$

8. $\begin{bmatrix} 1 & 0 & 2 \\ 0 & 4 & 2 \\ 3 & 5 & 0 \end{bmatrix} \cdot \begin{bmatrix} 1 & 0 \\ 0 & 1 \end{bmatrix}$

9. Find the inverse of $\begin{bmatrix} -2 & 5 \\ 3 & 1 \end{bmatrix}$.   (Lesson 4-5)

# Using Inverse Matrices

**Objective**

After studying this lesson, you should be able to:

- write a system of linear equations as a matrix and use the inverse to solve the system.

You have learned to solve systems of equations by several methods already. Matrices can be used to provide several other ways of solving systems of equations. In this lesson you will learn to solve a system of equations by using the inverses of matrices.

The matrices below represent three parts of a system of equations written in standard form. Matrix $A$ is a matrix showing the coefficients of the variables in the two equations. Matrix $B$ shows the variables in the system, and matrix $C$ shows the constant terms.

*Matrix A is called a coefficient matrix.*

$$A = \begin{bmatrix} 5 & 3 \\ 7 & 5 \end{bmatrix} \quad B = \begin{bmatrix} x \\ y \end{bmatrix} \quad C = \begin{bmatrix} -5 \\ -11 \end{bmatrix}$$

When the first two matrices are multiplied and set equal to the third matrix, the result is a system of equations.

$$\begin{bmatrix} 5x + 3y \\ 7x + 5y \end{bmatrix} = \begin{bmatrix} -5 \\ -11 \end{bmatrix} \quad \Rightarrow \quad \begin{array}{l} 5x + 3y = -5 \\ 7x + 5y = -11 \end{array} \qquad A \cdot B = C$$

The equation $\begin{bmatrix} 5 & 3 \\ 7 & 5 \end{bmatrix} \cdot \begin{bmatrix} x \\ y \end{bmatrix} = \begin{bmatrix} -5 \\ -11 \end{bmatrix}$ is called a **matrix equation.**

This equation represents the system $\begin{cases} 5x + 3y = -5 \\ 7x + 5y = -11 \end{cases}$.

**Example 1**

Write each system of equations as a matrix equation.

**a.** $6x + 5y = 8$
$3x - y = 7$

**b.** $2x + y - z = 9$
$x - 3y + 2z = 16$
$3x + 2y - z = 5$

The matrix equation is

$$\begin{bmatrix} 6 & 5 \\ 3 & -1 \end{bmatrix} \cdot \begin{bmatrix} x \\ y \end{bmatrix} = \begin{bmatrix} 8 \\ 7 \end{bmatrix}.$$

The matrix equation is

$$\begin{bmatrix} 2 & 1 & -1 \\ 1 & -3 & 2 \\ 3 & 2 & -1 \end{bmatrix} \cdot \begin{bmatrix} x \\ y \\ z \end{bmatrix} = \begin{bmatrix} 9 \\ 16 \\ 5 \end{bmatrix}.$$

Study the steps in solving this equation.

$$3w = 42$$
$$\left(\frac{1}{3}\right)3w = \left(\frac{1}{3}\right)42$$
$$(1)w = 14$$
$$w = 14$$

*Multiply each side by the multiplicative inverse of the coefficient.*
*1 is the identity element for multiplication.*

*The solution is 14.*

You can use the same steps and a matrix equation to solve a system of equations.

- Write the matrix equation.
- Multiply each side by the inverse of the coefficient matrix.
- The result will be the identity matrix multiplied by the variables.
- Interpret the solution.

Follow these steps in Examples 2 and 3.

**Example 2**

**Use a matrix equation to solve** $\begin{cases} 5x + 3y = -5 \\ 7x + 5y = -11 \end{cases}$.

The matrix equation is $\begin{bmatrix} 5 & 3 \\ 7 & 5 \end{bmatrix} \cdot \begin{bmatrix} x \\ y \end{bmatrix} = \begin{bmatrix} -5 \\ -11 \end{bmatrix}$.

Now find the inverse of the coefficient matrix $\begin{bmatrix} 5 & 3 \\ 7 & 5 \end{bmatrix}$ so you can complete the second step.

The inverse of $\begin{bmatrix} 5 & 3 \\ 7 & 5 \end{bmatrix}$ is $\frac{1}{25 - 21}\begin{bmatrix} 5 & -3 \\ -7 & 5 \end{bmatrix}$ or $\frac{1}{4}\begin{bmatrix} 5 & -3 \\ -7 & 5 \end{bmatrix}$.

Multiply each side of the matrix equation by the inverse matrix. Note the placement of the inverse when multiplying.

$$\frac{1}{4}\begin{bmatrix} 5 & -3 \\ -7 & 5 \end{bmatrix} \cdot \begin{bmatrix} 5 & 3 \\ 7 & 5 \end{bmatrix} \cdot \begin{bmatrix} x \\ y \end{bmatrix} = \frac{1}{4}\begin{bmatrix} 5 & -3 \\ -7 & 5 \end{bmatrix} \cdot \begin{bmatrix} -5 \\ -11 \end{bmatrix}$$

$$\begin{bmatrix} 1 & 0 \\ 0 & 1 \end{bmatrix} \cdot \begin{bmatrix} x \\ y \end{bmatrix} = \frac{1}{4}\begin{bmatrix} 8 \\ -20 \end{bmatrix} \quad \frac{1}{4}\begin{bmatrix} 5 & -3 \\ -7 & 5 \end{bmatrix} \cdot \begin{bmatrix} 5 & 3 \\ 7 & 5 \end{bmatrix} = \begin{bmatrix} 1 & 0 \\ 0 & 1 \end{bmatrix}$$

$$\begin{bmatrix} x \\ y \end{bmatrix} = \begin{bmatrix} 2 \\ -5 \end{bmatrix}$$

The solution is (2, -5).

The graph at the right confirms the solution.

*How else could you check the solution?*

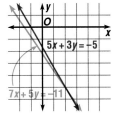

**Example 3** | Use a matrix equation to solve $\begin{cases} 3x - 2y + z = 0 \\ 2x + 3y = 12 \\ y + 4z = -18 \end{cases}$ , if the inverse of the

coefficient matrix is $\dfrac{1}{54}\begin{bmatrix} 12 & 9 & -3 \\ -8 & 12 & 2 \\ 2 & -3 & 13 \end{bmatrix}$.

To help you write the coefficient matrix it may be helpful to write the system so that each equation contains all three variables with the proper coefficients.

$\begin{aligned} 3x - 2y + 1z &= 0 \\ 2x + 3y + 0z &= 12 \\ 0x + 1y + 4z &= -18 \end{aligned}$ ➡ $\begin{bmatrix} 3 & -2 & 1 \\ 2 & 3 & 0 \\ 0 & 1 & 4 \end{bmatrix}$

Write the matrix equation. $\begin{bmatrix} 3 & -2 & 1 \\ 2 & 3 & 0 \\ 0 & 1 & 4 \end{bmatrix} \cdot \begin{bmatrix} x \\ y \\ z \end{bmatrix} = \begin{bmatrix} 0 \\ 12 \\ -18 \end{bmatrix}$

Multiply each side by the inverse and find a solution.

$\dfrac{1}{54}\begin{bmatrix} 12 & 9 & -3 \\ -8 & 12 & 2 \\ 2 & -3 & 13 \end{bmatrix} \cdot \begin{bmatrix} 3 & -2 & 1 \\ 2 & 3 & 0 \\ 0 & 1 & 4 \end{bmatrix} \cdot \begin{bmatrix} x \\ y \\ z \end{bmatrix} = \dfrac{1}{54}\begin{bmatrix} 12 & 9 & -3 \\ -8 & 12 & 2 \\ 2 & -3 & 13 \end{bmatrix} \cdot \begin{bmatrix} 0 \\ 12 \\ -18 \end{bmatrix}$

$\begin{bmatrix} 1 & 0 & 0 \\ 0 & 1 & 0 \\ 0 & 0 & 1 \end{bmatrix} \cdot \begin{bmatrix} x \\ y \\ z \end{bmatrix} = \dfrac{1}{54}\begin{bmatrix} 162 \\ 108 \\ -270 \end{bmatrix}$   *Remember that the product of a matrix and its inverse is the identity matrix.*

$\begin{bmatrix} x \\ y \\ z \end{bmatrix} = \begin{bmatrix} 3 \\ 2 \\ -5 \end{bmatrix}$

**Check:**

| | | |
|---|---|---|
| $3x - 2y + z = 0$ | $2x + 3y = 12$ | $y + 4z = -18$ |
| $3(3) - 2(2) + (-5) \stackrel{?}{=} 0$ | $2(3) + 3(2) \stackrel{?}{=} 12$ | $2 + 4(-5) \stackrel{?}{=} -18$ |
| $0 = 0 \checkmark$ | $12 = 12 \checkmark$ | $-18 = -18 \checkmark$ |

The solution is $(3, 2, -5)$.

You can use matrices to help you solve problems that involve systems of equations. Using matrices often simplifies the process of solving these systems.

## Example 4

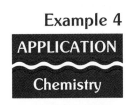

**APPLICATION**

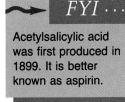

Chemistry

*FYI* $\cdots$

Acetylsalicylic acid was first produced in 1899. It is better known as aspirin.

Giem Nguyen is a chemist who is preparing an acid solution to be used as a cleaner for machine parts. The machine shop needs several batches of 200 mL of solution at a 48% concentration. Giem only has 60% and 40% concentration solutions. The two solutions can be combined to make the 48% solution. How much of each solution should Giem use to make 200 mL of solution?

*EXPLORE*    Let $a$ represent the amount of 60% solution and let $b$ represent the amount of 40% solution.

$a + b = 200$    *The total of the two amounts must be 200 mL.*
Now write an equation that represents the proportions of each solution needed.

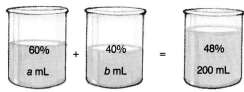

60% $(a)$ + 40% $(b)$ = 48% $(a + b)$    *Each part contributes to the total.*

$0.60a + 0.40b = 0.48a + 0.48b$    *Percents can be written in decimal form.*

$60a + 40b = 48a + 48b$    *Multiply by 100 to remove the decimals.*

$12a - 8b = 0$    *Write the equation in standard form.*

*PLAN*    Write a system of equations. Then write the system as a matrix equation.

$$\begin{cases} a + b = 200 \\ 12a - 8b = 0 \end{cases} \quad \Rightarrow \quad \begin{bmatrix} 1 & 1 \\ 12 & -8 \end{bmatrix} \cdot \begin{bmatrix} a \\ b \end{bmatrix} = \begin{bmatrix} 200 \\ 0 \end{bmatrix}$$

*SOLVE*    To solve the matrix equation, first find the inverse of the coefficient matrix.

$$\frac{1}{ad - bc} \begin{bmatrix} d & -b \\ -c & a \end{bmatrix} \quad \Rightarrow \quad -\frac{1}{20} \begin{bmatrix} -8 & -1 \\ -12 & 1 \end{bmatrix}$$

Now multiply each side of the matrix equation by the inverse and solve.

$$-\frac{1}{20} \begin{bmatrix} -8 & -1 \\ -12 & 1 \end{bmatrix} \cdot \begin{bmatrix} 1 & 1 \\ 12 & -8 \end{bmatrix} \cdot \begin{bmatrix} a \\ b \end{bmatrix} = -\frac{1}{20} \begin{bmatrix} -8 & -1 \\ 12 & 1 \end{bmatrix} \cdot \begin{bmatrix} 200 \\ 0 \end{bmatrix}$$

$$\begin{bmatrix} a \\ b \end{bmatrix} = \begin{bmatrix} 80 \\ 120 \end{bmatrix}$$

This means that 80 mL of the 60% solution is added to 120 mL of the 40% solution to make 200 mL of the 48% solution.

*EXAMINE*    A solution of 48% is closer in acidity to 40% than to 60%. It makes sense that there would be more 40% solution in the mixture than 60% solution. Also, the total amount is 200 mL.

# CHECKING FOR UNDERSTANDING

**Communicating Mathematics**

Read and study the lesson to answer each question.

1. What is the matrix equation for $\begin{cases} 2x - 3y = 8 \\ 7x - 3y = 5 \end{cases}$?

2. What is the product of the coefficient matrix and its inverse?

**Guided Practice**

Write a matrix equation for each system.

3. $\begin{cases} 5a + 2b = -49 \\ 2a + 9b = 5 \end{cases}$

4. $\begin{cases} 2x - 3y + z = 29 \\ x + 4y + 2z = 3 \\ 3x + y - 2z = -3 \end{cases}$

Write each matrix equation as a system of linear equations.

5. $\begin{bmatrix} 5 & 1 \\ 2 & -3 \end{bmatrix} \cdot \begin{bmatrix} x \\ y \end{bmatrix} = \begin{bmatrix} 26 \\ 41 \end{bmatrix}$

6. $\begin{bmatrix} 2 & 1 & -1 \\ 1 & -4 & 3 \\ 6 & -2 & 5 \end{bmatrix} \cdot \begin{bmatrix} m \\ n \\ p \end{bmatrix} = \begin{bmatrix} -7 \\ 5 \\ 9 \end{bmatrix}$

7. Given that the inverse of the coefficient matrix is $\dfrac{1}{6}\begin{bmatrix} 3 & -1 \\ -9 & 5 \end{bmatrix}$, solve the matrix equation $\begin{bmatrix} 5 & 1 \\ 9 & 3 \end{bmatrix} \cdot \begin{bmatrix} x \\ y \end{bmatrix} = \begin{bmatrix} 1 \\ 1 \end{bmatrix}$.

# EXERCISES

**Practice**

Write the system of linear equations represented by each matrix equation.

8. $\begin{bmatrix} 5 & 4 \\ 3 & -5 \end{bmatrix} \begin{bmatrix} x \\ y \end{bmatrix} = \begin{bmatrix} -3 \\ -24 \end{bmatrix}$

9. $\begin{bmatrix} 3 & 1 \\ 4 & -2 \end{bmatrix} \begin{bmatrix} x \\ y \end{bmatrix} = \begin{bmatrix} 13 \\ 24 \end{bmatrix}$

10. $\begin{bmatrix} 2 & 0 & 5 \\ 1 & 8 & 2 \\ 3 & -5 & 7 \end{bmatrix} \begin{bmatrix} x \\ y \\ z \end{bmatrix} = \begin{bmatrix} 1 \\ 2 \\ 3 \end{bmatrix}$

11. $\begin{bmatrix} 1 & -2 & 0 \\ 3 & 1 & 2 \\ 4 & -3 & 3 \end{bmatrix} \begin{bmatrix} x \\ y \\ z \end{bmatrix} = \begin{bmatrix} -8 \\ 9 \\ 1 \end{bmatrix}$

Write a matrix equation for each system of linear equations.

12. $3x - y = 5$
    $2x + 3y = 29$

13. $2x + 5y = 1$
    $3x + 4y = 12$

14. $6a + 9b = 6$
    $4a + 6b = 8$

Use the given inverse matrix $M^{-1}$ to solve each matrix equation.

15. $\begin{bmatrix} 4 & 8 \\ 2 & -3 \end{bmatrix} \cdot \begin{bmatrix} x \\ y \end{bmatrix} = \begin{bmatrix} 7 \\ 0 \end{bmatrix}$ $\qquad M^{-1} = -\dfrac{1}{28}\begin{bmatrix} -3 & -8 \\ -2 & 4 \end{bmatrix}$

**16.** $\begin{bmatrix} 3 & 1 & 1 \\ -6 & 5 & 3 \\ 9 & -2 & -1 \end{bmatrix} \cdot \begin{bmatrix} x \\ y \\ z \end{bmatrix} = \begin{bmatrix} -1 \\ -9 \\ 5 \end{bmatrix}$  $M^{-1} = -\dfrac{1}{9}\begin{bmatrix} 1 & -1 & -2 \\ 21 & -12 & -15 \\ -33 & 15 & 21 \end{bmatrix}$

**17.** $\begin{bmatrix} 1 & 2 & 2 \\ 2 & -1 & 1 \\ 3 & -2 & 3 \end{bmatrix} \cdot \begin{bmatrix} a \\ b \\ c \end{bmatrix} = \begin{bmatrix} 0 \\ -1 \\ -4 \end{bmatrix}$  $M^{-1} = -\dfrac{1}{9}\begin{bmatrix} -1 & -10 & 4 \\ -3 & -3 & 3 \\ -1 & 8 & -5 \end{bmatrix}$

**Solve each system of equations by using a matrix equation.**

**18.** $6a + 2b = 11$
$3a - 8b = 1$

**19.** $4x + 3y = 5$
$8x - 9y = 0$

**Critical Thinking**

**20.** Rosalyn tried to use inverse matrices to solve a system of equations. However she found the determinant of the coefficient matrix to be 0. Can she still use this method? Why or why not? Describe the graph of such a system of equations.

**Applications**

**21. Landscaping** Two trucks have capacities of 10 tons and 12 tons. They made a total of 20 round trips to haul 226 tons of sand to the community park. How many round trips did each truck make?

**22. Metallurgy** To make 20 kg of aluminum alloy with 70% aluminum, a metallurgist wants to use two metals with 55% and 80% aluminum content. How much of each metal should she use?

**Mixed Review**

**23.** Find the inverse of $\begin{bmatrix} 4 & -2 \\ -3 & 6 \end{bmatrix}$. **(Lesson 4-5)**

**24.** Find $M$ if $\begin{bmatrix} 3 & 6 & 1 \\ 2 & -1 & 0 \end{bmatrix} \cdot M = \begin{bmatrix} 3 & 6 & 1 \\ 2 & -1 & 0 \end{bmatrix}$. **(Lesson 4-5)**

**25.** Graph this system of inequalities and name the vertices of the polygon formed. Then find the maximum and minimum values of $f(x, y) = 4x - 3y$ for the region. **(Lesson 3-6)**

$x \leq 5$ $\qquad$ $y \geq -3x$ $\qquad$ $2y \leq x + 7$ $\qquad$ $y \geq x - 4$

**26.** Find the value of $a$ for which the graph of $y = ax - 3$ is perpendicular to the graph of $-6x + 11y = 4$. **(Lesson 2-4)**

**27.** Find the standard form of the equation that passes through $(4, -1)$ and $(-3, 2)$. **(Lesson 2-5)**

**Solve each inequality.** **(Lesson 1-8)**

**28.** $5 < 2x - 9 < 11$

**29.** $|9 - 3t| > 5$

# 4-7 Using Cramer's Rule

**Objective**

After studying this lesson, you should be able to:

- use Cramer's rule to solve a system of linear equations in three variables.

You have learned to solve a system of linear equations in three variables algebraically and by using inverse matrices. In Chapter 3, you learned to solve a system of linear equations in two variables by using Cramer's rule. Now you will learn to use Cramer's rule to solve a system of three equations in three variables.

*Test for Unique Solutions*

The system of equations $\begin{cases} a_1x + b_1y + c_1z = d_1 \\ a_2x + b_2y + c_2z = d_2 \\ a_3x + b_3y + c_3z = d_3 \end{cases}$ has a **unique** solution **if and only if** $\begin{vmatrix} a_1 & b_1 & c_1 \\ a_2 & b_2 & c_2 \\ a_3 & b_3 & c_3 \end{vmatrix} \neq 0.$

To use Cramer's rule on a system of three equations in three variables you follow the same steps as with a system of two equations in two variables. The denominator is the determinant containing the coefficients. The numerators are the same determinant except that the coefficients of the variable for which you are finding a solution are replaced with the constant terms. Study this procedure in the following example.

**Example 1**

Determine whether the system $\begin{cases} a + 2b - c = -7 \\ 2a + 3b + 2c = -3 \\ a - 2b - 2c = 3 \end{cases}$ has a unique solution. If it does, then solve the system using Cramer's rule.

*Each determinant in this example is evaluated by using diagonals.*

$= -6 + 4 + 4 - (-3) - (-4) - (-8)$ or 17

*Since the value of the determinant is not 0, the system has a unique solution.*

$a = \dfrac{\begin{vmatrix} -7 & 2 & -1 \\ -3 & 3 & 2 \\ 3 & -2 & -2 \end{vmatrix}}{17} = \dfrac{42 + 12 + (-6) - (-9) - 28 - 12}{17} = \dfrac{17}{17}$ or 1

$$b = \frac{\begin{vmatrix} 1 & -7 & -1 \\ 2 & -3 & 2 \\ 1 & 3 & -2 \end{vmatrix}}{17}$$

$$c = \frac{\begin{vmatrix} 1 & 2 & -7 \\ 2 & 3 & -3 \\ 1 & -2 & 3 \end{vmatrix}}{17}$$

$$= \frac{6 + (-14) + (-6) - 3 - 6 - 28}{17}$$

$$= \frac{9 + (-6) + 28 - (-21) - 6 - 12}{17}$$

$$= \frac{-51}{17} \text{ or } -3$$

$$= \frac{34}{17} \text{ or } 2$$

The solution is $(1, -3, 2)$.   *Check this solution.*

---

**Example 2**

APPLICATION

Banking

Last year, Kathie Faught invested $48,000, some in stocks, some in bonds, and the remainder in a term account. She earned 4% on the stocks, 7% on the bonds, and 6% on the term account. For the year, she earned a total of $2860. She earned three times as much from the term account as she did from the stocks. How much did she invest in each?

*EXPLORE*   Let $s$ represent the amount in stocks, $b$ represent the amount in bonds, and $t$ represent the amount in the term account.

*PLAN*   Write a system of equations.

$s + b + t = 48000$
$0.04s + 0.07b + 0.06t = 2860$
$0.06t = 3(0.04s)$   or   $0.12s - 0.06t = 0$

*SOLVE*   Solve the system of equations. Use Cramer's rule.
*Each determinant will be evaluated using diagonals.*

$$\begin{vmatrix} 1 & 1 & 1 \\ 0.04 & 0.07 & 0.06 \\ 0.12 & 0 & -0.06 \end{vmatrix} = -0.0042 + 0.0072 - (0.0084 - 0.0024)$$

$$= -0.003$$

*Since the value of the determinant is not zero, the system has a unique solution.*

$$s = \frac{\begin{vmatrix} 48000 & 1 & 1 \\ 2860 & 0.07 & 0.06 \\ 0 & 0 & -0.06 \end{vmatrix}}{-0.003}$$

$$t = \frac{\begin{vmatrix} 1 & 1 & 48000 \\ 0.04 & 0.07 & 2860 \\ 0.12 & 0 & 0 \end{vmatrix}}{-0.003}$$

$$= \frac{-201.6 - (-171.6)}{-0.003} \text{ or } 10,000$$

$$= \frac{343.2 - 403.2}{-0.003} \text{ or } 20,000$$

$$b = \frac{\begin{vmatrix} 1 & 48000 & 1 \\ 0.04 & 2860 & 0.06 \\ 0.12 & 0 & -0.06 \end{vmatrix}}{-0.003} = \frac{174 - 228}{-0.003} \text{ or } 18,000$$

Kathie invested $10,000 in stocks, $18,000 in bonds, and $20,000 in a term account.

EXAMINE  Stocks: 4% of $10,000 = $400   *Three times the amount earned*
         Bonds: 7% of $18,000 = $1260   *in stocks is 3($400) or $1200.*
         Term: 6% of $20,000 = $1200
                              $2860 ✓  The total is correct.

# CHECKING FOR UNDERSTANDING

**Communicating Mathematics**

Read and study the lesson to answer each question.

1. Will Cramer's Rule apply for all systems of linear equations? Explain your answer.

2. If the three linear equations represent three parallel planes, what would be the value of the coefficient determinant?

3. What condition would have to exist for the value of the coefficient determinant to be a fraction?

**Guided Practice**

Write the coefficient determinant for each system of equations.

4. $3a = 5b + 6$
   $a - b = 4$

5. $6t + u = 0$
   $5t - 8u = -19$

6. $3x + 2y = 0$
   $4x - z = 3$
   $3z = -5x$

Name the determinants you would use to solve each system by Cramer's rule. Then solve each system using Cramer's rule.

7. $2x + 4y - z = -6$
   $x - 2y + 3z = 2$
   $x + 2y - 4z = -10$

8. $a - 2b + 3c = -4$
   $2a - b + 4c = -1$
   $2a + 3b + 5c = 1$

# EXERCISES

**Practice**

Determine whether each system of equations has a unique solution.

9. $2a + 4b - c = -6$
   $a - 2b + 3c = 2$
   $a + 2b - 4c = -10$

10. $2x + 2z = 5$
    $3y - 3z = -2$
    $-3x - 2y = 11$

11. $-r + 7s + 2t = 6$
    $2r + s = -11$
    $3r - 6s - 2t = -25$

Solve each system of equations using Cramer's Rule.

12. $4a + b + 3c = 1$
    $2a + c = 3$
    $4a - 6b = 8$

13. $2x - y + 3z = 5$
    $3x + 2y - 5z = 4$
    $x - 4y + 11z = 3$

14. $a + 2b - 3c = -13$
    $2a - b + 3c = 23$
    $3a + b - 3c = -8$

15. $3x - y + 2z = 11$
    $6x - 3y + z = -1$
    $-3x - 2y + 2z = 11$

16. $x + 9y - 2z = 2$
    $-x - 3y + 4z = 1$
    $2x + 3y - 6z = -5$

17. $x + 4y + 3z = 10$
    $2x - 2y + z = 15$
    $x + 2y - 3z = -1$

**Write a system of equations for each problem. Do not solve the problem.**

18. Floyd has 16 coins in pennies, nickels, and dimes. The number of dimes is equal to the sum of the number of pennies and number of nickels. If the total value of the coins is $1.08, how many of each kind does he have?

19. At Burger Heaven, 2 cheeseburgers and 3 orders of fries cost $3.65. A cheeseburger and 2 milkshakes cost $2.47. A cheeseburger, 2 orders of fries, and a milkshake cost $3.01. What is the cost of each item?

20. At Tapeland, T-60 blank tapes cost $3.19, T-90 blank tapes cost $3.89, and T-120 blank tapes cost $4.59. Carl bought 10 blank tapes for $40.30. If he bought twice as many T-90 tapes as T-120 tapes, how many of each tape did he buy?

**Journal**

Write a paragraph to tell what you like and dislike about using Cramer's rule to solve equations.

**Critical Thinking**

21. Study the following problem and determine what effect using Cramer's Rule has when trying to solve it. Does the problem have a unique answer? If so, what is it?

    *Fred Chaves has $1.35 in nickels, dimes, and quarters. If he has the same amount of money in nickels as he has in dimes, how many of each does he have?*

**Applications**

22. **Sports**  At the student-faculty basketball game, Mrs. Winters, Mrs. Gordon, and Mr. Gossell scored a total of 63 points for the faculty team before they collapsed from exhaustion. Mrs. Winters scored twice as many points as Mr. Gossell. Mrs. Gordon scored three points more than the sum of points scored by Mr. Gossell and Mrs. Winters. How many points did each person score?

23. **Retail**  The Yogurt Shoppe sells cones in three sizes: small, 89¢; medium, $1.19; large, $1.39. One day, Kyla Martin sold 52 cones. She sold two more than twice as many mediums as larges. If she sold $58.98 in cones, how many of each size did she sell?

**Mixed Review**

24. Describe $A_{2 \times 4}$.  **(Lesson 4-1)**

25. Graph $y \geq |3x|$.  **(Lesson 2-8)**

26. Solve $\begin{cases} y > x \\ y < x - 3 \end{cases}$ by graphing.  **(Lesson 3-4)**.

27. **Communications**  A certain telephone call to a sports hotline costs $3.38 for the first three minutes and $0.96 for each minute thereafter. What is the cost of a 12-minute phone call?  **(Lesson 1-4)**

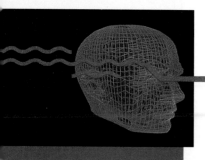

# Technology

## Cramer's Rule

The BASIC program below can be used to solve a system of equations in three variables.

```
5    DIM A[3,4],B[3,4]
10   FOR I=1 TO 3
20   FOR J=1 TO 4
30   READ A[I,J]
40   NEXT J
50   NEXT I
55   DATA 1,2,-1,-7,2,3,2,-3,1,-2,-2,3
60   FOR K=0 TO 3
70   FOR J=1 TO 3
80   FOR I=1 TO 3
90   IF J=K THEN 120
100  LET B[I,J]=A[I,J]
110  GOTO 130
120  LET B[I,J]=A[I,4]
130  NEXT I
140  NEXT J
150  LET M1=B[2,2]*B[3,3]-B[3,2]*B[2,3]
160  LET M2=B[2,1]*B[3,3]-B[3,1]*B[2,3]
170  LET M3=B[2,1]*B[3,2]-B[3,1]*B[2,2]
180  LET D[K]=B[1,1]*M1-B[1,2]*M2+B[1,3]*M3
190  NEXT K
200  IF D[0]=0 THEN 250
210  PRINT "X=";D[1];"/";D[0];"OR";D[1]/D[0]
220  PRINT "Y=";D[2];"/";D[0];"OR";D[2]/D[0]
230  PRINT "Z=";D[3];"/";D[0];"OR";D[3]/D[0]
240  GOTO 260
250  PRINT "NO UNIQUE SOLUTION"
260  END
```

Line 55 in the program shows the coefficients of the system as DATA:

$$x + 2y - z = -7$$
$$2x + 3y + 2z = -3$$
$$x - 2y - 2z = 3$$

In line 10–50, the computer interprets the data as an augmented matrix. In the program, $A(I, J)$ is the element in row $I$ and column $J$ of that matrix.

Lines 60–190 form a loop for evaluating four different $3 \times 3$ determinants. For each determinant, three columns of the augmented matrix are used. The four values are stored as $D(0)$, $D(1)$, $D(2)$, and $D(3)$.

Line 200 checks whether the system has a unique solution by looking at the value of the determinant $D(0)$. If $D(0) = 0$, then the computer will print "NO UNIQUE SOLUTION." Otherwise, a solution will be printed.

The output for this system of equations is:

```
X = 17/17 OR 1
Y = -51/17 OR -3
Z = 34/17 OR 2
```

# EXERCISES

**Change line 55 and run the program for each system of equations.**

1. $2x + 3y + 4z = 4$
   $2x - 8z = -1$
   $4x - 6y + 4z = -1$

2. $5x - y + 2z = 5$
   $2x - 3y + 5z = 1$
   $3x + 2y - 3z = 4$

# 4-8 Using Augmented Matrices

**Objective**

After studying this lesson, you should be able to:

■ solve a system of equations using an augmented matrix.

A system of equations may also be solved using a matrix called an **augmented matrix.** The augmented matrix of a system contains the coefficient matrix with an extra column containing the constant terms. Study how the system below can be written as an augmented matrix.

$$
\begin{aligned}
x - 4y - 2z &= 11 \\
3x + 2y + z &= 5 \\
2x - 4y - 3z &= 19
\end{aligned}
\qquad\Rightarrow\qquad
\begin{bmatrix}
1 & -4 & -2 & 11 \\
3 & 2 & 1 & 5 \\
2 & -4 & -3 & 19
\end{bmatrix}
$$

The system of equations can be solved by manipulating the rows of the matrix rather than the equations themselves. In this way, you perform the same operations that you would in working with the equations, but you do not have to bother writing the variables or worrying about the order in which the terms are written—the organization of the matrix keeps all of this in its proper place.

Suppose you multiplied the first equation by $-2$. The result would be $-2x + 8y + 4z = -22$. The corresponding change in the matrix is that the first row becomes $[-2 \quad 8 \quad 4 \quad -22]$. The result is the same as the result of multiplying the first row by $-2$. This is only one of the row operations you can perform in manipulating an augmented matrix. Here is a summary of the row operations on matrices you can use.

1. **Interchange any two rows.**
2. **Replace any row with a nonzero multiple of that row.**
3. **Replace any row with the sum of that row and a multiple of another row.**

The solution to the system of equations above is $(3, 0.5, -5)$. That is, the three planes meet where $x = 3$, $y = 0.5$, and $z = -5$. Suppose we write these three equations in the form of an augmented matrix.

$$
\begin{aligned}
x &= 3 \\
y &= 0.5 \\
z &= -5
\end{aligned}
\qquad\Rightarrow\qquad
\begin{bmatrix}
1 & 0 & 0 & 3 \\
0 & 1 & 0 & 0.5 \\
0 & 0 & 1 & -5
\end{bmatrix}
$$

Notice that the first three columns form the $3 \times 3$ identity matrix. When doing row operations, your goal should be to find an augmented identity matrix.

Just as there is no one single order of steps to solve a system of equations, there is also no one single group of row operations that arrive at the correct solution. The order in which you solve a system may be different from the way your classmate solves it, but you may both be correct. Study Example 1 to see how the row operations are used.

**Example 1**

Use an augmented matrix to solve $\begin{cases} a + 2b + c = 0 \\ 2a + 5b + 4c = -1. \\ a - b - 9c = -5 \end{cases}$

Write the augmented matrix.

$$\begin{bmatrix} 1 & 2 & 1 & 0 \\ 2 & 5 & 4 & -1 \\ 1 & -1 & -9 & -5 \end{bmatrix}$$

*The first element in row 1 is already 1.*

Multiply row 1 by $-1$ and add to row 3.

$$\begin{bmatrix} 1 & 2 & 1 & 0 \\ 2 & 5 & 4 & -1 \\ 0 & -3 & -10 & -5 \end{bmatrix}$$

*The first element in row 3 is now 0.*

Multiply row 1 by $-2$ and add to row 2.

$$\begin{bmatrix} 1 & 2 & 1 & 0 \\ 0 & 1 & 2 & -1 \\ 0 & -3 & -10 & -5 \end{bmatrix}$$

*The first element in row 2 is now 0, and the second element in row 2 is now 1.*

Multiply row 2 by $-2$ and add to row 1.

$$\begin{bmatrix} 1 & 0 & -3 & 2 \\ 0 & 1 & 2 & -1 \\ 0 & -3 & -10 & -5 \end{bmatrix}$$

*The second element in row 1 is now 0.*

Multiply row 2 by 3 and add to row 3.

$$\begin{bmatrix} 1 & 0 & -3 & 2 \\ 0 & 1 & 2 & -1 \\ 0 & 0 & -4 & -8 \end{bmatrix}$$

*The second element in row 3 is now 0.*

Multiply row 3 by $-\dfrac{1}{4}$.

$$\begin{bmatrix} 1 & 0 & -3 & 2 \\ 0 & 1 & 2 & -1 \\ 0 & 0 & 1 & 2 \end{bmatrix}$$

*The third element in row 3 is now 1.*

Multiply row 3 by 3 and add to row 1.

$$\begin{bmatrix} 1 & 0 & 0 & 8 \\ 0 & 1 & 2 & -1 \\ 0 & 0 & 1 & 2 \end{bmatrix}$$

*The third element in row 1 is now 0.*

Multiply row 3 by $-2$ and add to row 2.

$$\begin{bmatrix} 1 & 0 & 0 & 8 \\ 0 & 1 & 0 & -5 \\ 0 & 0 & 1 & 2 \end{bmatrix}$$

*This matrix contains an augmented identity matrix. Now you can read the solution.*

The last augmented matrix represents $a = 8$, $b = -5$, and $c = 2$. Therefore, the solution is $(8, -5, 2)$.

The process of performing row operations to get the desired matrix is called **reducing a matrix.** The resulting matrix is called a **reduced matrix.**

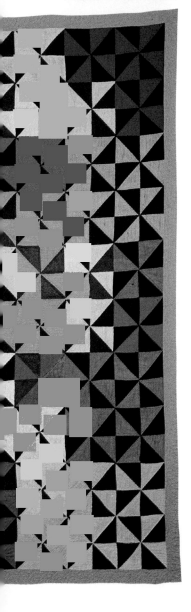

## Example 2

The perimeter of a triangle is 83 inches. The longest side is three times the length of the shortest side and 17 inches more than one-half the sum of the other two sides. Find the length of each side.

*EXPLORE*  Let $a$ = the measure of the longest side of the triangle, $c$ = the measure of the shortest side, and $b$ = the measure of the other side.

*PLAN*  Write equations for all the information you know.

$a + b + c = 83$  *The perimeter is 83 in.*

$a = 3c$  *The longest side is 3 times the length of shortest.*

$a = \dfrac{1}{2}(b + c) + 17$  *The longest side is 17 more than half the sum of the lengths of the other sides.*

*SOLVE*  In order to solve the system, rewrite the equations so that they are all in the form necessary to write an augmented matrix.

$$\begin{array}{rl} a + b + c = 83 \\ a \quad\quad - 3c = 0 \\ 2a - b - c = 34 \end{array} \quad\Rightarrow\quad \begin{bmatrix} 1 & 1 & 1 & 83 \\ 1 & 0 & -3 & 0 \\ 2 & -1 & -1 & 34 \end{bmatrix}$$

After applying row operations on the matrix, we get

$$\begin{bmatrix} 0 & 1 & 0 & 31 \\ 1 & 0 & 0 & 39 \\ 0 & 0 & 1 & 13 \end{bmatrix} \quad\Rightarrow\quad \begin{bmatrix} 1 & 0 & 0 & 39 \\ 0 & 1 & 0 & 31 \\ 0 & 0 & 1 & 13 \end{bmatrix}. \quad \textit{Interchange the first and second rows.}$$

The solution is (39, 31, 13), which means the lengths of the sides of the triangle are 39 inches, 31 inches, and 13 inches.

*EXAMINE*  The sum of the three lengths (39 + 31 + 13) is 83 inches. The longest side, 39, is 3 times 13, the shortest side. Half the sum of 31 and 13 is 22, which is 17 less than 39, the longest side.

As with other methods of solving systems of equations, there is not always a unique solution. However, when using augmented matrices, you can determine what type of solution you have when a solution is not unique. Study the solution of the system shown below.

$$\begin{array}{rl} 2x - y + 4z = 4 \\ x + 2y - 3z = 7 \\ x - 8y + 17z = -13 \end{array} \quad\Rightarrow\quad \begin{bmatrix} 2 & -1 & 4 & 4 \\ 1 & 2 & -3 & 7 \\ 1 & -8 & 17 & -13 \end{bmatrix} \quad\Rightarrow\quad \begin{bmatrix} 1 & 0 & 1 & 3 \\ 0 & 1 & -2 & 2 \\ 0 & 0 & 0 & 0 \end{bmatrix}$$

The row of zeros tells us that the last equation was derived from the first two so that this is a dependent system—there is no unique solution. However, a solution can be written. Let's write the two equations represented in the matrix. Then solve each equation for z.

$$x + z = 3 \qquad\qquad y - 2z = 2$$
$$x = 3 - z \qquad\qquad y = 2 + 2z$$

The solution would be the ordered triple $(3 - z, 2 + 2z, z)$. By choosing values for $z$ you can find points on the line that is the solution to the system.

Let's look at the solution to another system of equations.

$$
\begin{aligned}
x + 5y - 3z &= 4 \\
4x - 3y + 2z &= 4 \\
8x - 6y + 4z &= 14
\end{aligned}
\quad\Rightarrow\quad
\begin{bmatrix}
1 & 5 & -3 & 4 \\
4 & -3 & 2 & 4 \\
8 & -6 & 4 & 14
\end{bmatrix}
\quad\Rightarrow\quad
\begin{bmatrix}
1 & 5 & -3 & 4 \\
0 & -23 & 14 & -12 \\
0 & 0 & 0 & 3
\end{bmatrix}
$$

Notice the last row in the last matrix. This represents the equation $0 = 3$. This cannot be. Therefore the system is inconsistent and no solution exists.

**Example 3**

Describe the solution for the system of equations represented by each reduced augmented matrix.

a. $\begin{bmatrix} 1 & 0 & 7 \\ 0 & 2 & -6 \end{bmatrix}$
b. $\begin{bmatrix} 1 & 0 & 3 & 11 \\ 0 & 2 & 1 & 0 \\ 0 & 0 & 0 & 0 \end{bmatrix}$
c. $\begin{bmatrix} 2 & 0 & 0 & 3 \\ 0 & 0 & 0 & 6 \\ 0 & 5 & 0 & 6 \end{bmatrix}$

a. The solution is $(7, -3)$.

b. The solution is a line containing points $\left(11 - 3z, -\dfrac{1}{2}z, z\right)$.

c. There is no solution.

# CHECKING FOR UNDERSTANDING

**Communicating Mathematics**

Read and study the lesson to answer each question.

1. What is an augmented matrix?

2. What does it mean when an augmented matrix has a row containing all zeros?

3. What does it mean when an augmented matrix has a row that is all zeros except for the last element?

4. What are the row operations that you can use with an augmented matrix?

**5.** Write a system of equations represented by $\begin{bmatrix} 1 & 0 & 3 & -2 \\ 3 & 9 & -2 & -5 \\ -4 & 1 & -7 & 3 \end{bmatrix}$.

**6.** State the row operations you would use to change $\begin{bmatrix} 1 & 3 & 2 \\ 2 & 1 & 7 \end{bmatrix}$ to $\begin{bmatrix} 1 & 0 & \frac{19}{5} \\ 0 & 1 & -\frac{3}{5} \end{bmatrix}$.

**Write an augmented matrix for each system of equations. Then solve each system.**

**7.** $5a - 3b = 7$
$3a + 9b = -3$

**8.** $6x - 7z = 13$
$8y + 2z = 14$
$7x + z = 6$

# EXERCISES

**Practice**

**9.** State the row operations you would use to reduce $\begin{bmatrix} 2 & 4 & 3 \\ -2 & -3 & 1 \end{bmatrix}$.

**Write an augmented matrix for each system of equations. Then solve.**

**10.** $4x + 3y = 10$
$5x - y = 3$

**11.** $7m - 3n = 41$
$2m + 5n = 0$

**12.** $3a - 5b + 2c = 22$
$2a + 3b - c = -9$
$4a + 3b + 3c = 1$

**Describe the solution for the system of equations represented by each reduced augmented matrix.**

**13.** $\begin{bmatrix} 1 & 0 & 3 \\ 0 & 1 & 5 \end{bmatrix}$

**14.** $\begin{bmatrix} 1 & -3 & 5 & -2 \\ 0 & 0 & 0 & 3 \\ 0 & 0 & 1 & 2.6 \end{bmatrix}$

**15.** $\begin{bmatrix} 5 & 0 & 4 & 7 \\ 0 & 1 & -2 & 1 \\ 0 & 0 & 0 & 0 \end{bmatrix}$

**Solve each system of equations using augmented matrices.**

**16.** $6r + s = 9$
$3r + 2s = 0$

**17.** $a + b + c = -2$
$2a - 3b + c = -11$
$-a + 2b - c = 8$

**18.** $2x + y + z = 0$
$3x - 2y - 3z = -21$
$4x + 5y + 3z = -2$

**19.** In triangle $ABC$, the measure of $\angle A$ is twice the measure of $\angle B$. The measure of $\angle C$ exceeds four times the measure of $\angle B$ by 12 degrees. Find the measure of each angle.

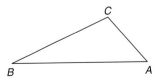

**20.** Solve $\begin{cases} 8m - 3n - 4p = 6 \\ 4m + 9n - 2p = -4 \\ 6m + 12n + 5p = -1 \end{cases}$ using augmented matrices.

**Critical Thinking**

21. You have learned what kind of solution a system has when one row of the reduced matrix contains all zeros. What kind of solution do you think you have when two rows of the three in the augmented matrix contain all zeros? Explain your answer.

**Applications**

22. **Automechanics**  Ann Brauen is inventory manager for a local repair shop. If she orders 6 batteries, 5 cases of spark plugs, and 2 dozen pair of wiper blades, she will pay $830. If she orders 3 batteries, 7 cases of spark plugs, and 4 dozen pair of wiper blades, she will pay $820. If the batteries are $22 less than twice as expensive as a dozen wiper blades, what is the cost of each item on her order?

23. **Restaurant Management**  There are several meals that can be purchased at Frank's Fried Chicken. A salad, 2 rolls, and 2 pieces of chicken costs $3.65. If you order a roll and 3 pieces of chicken, your meal costs $3.20. If a salad is 3 times as expensive as a roll, what is the cost of each item?

**Mixed Review**

24. Write the matrix equation for $\begin{cases} x + 5y + 2z = 10 \\ 3x - 3y + 2z = 2 \\ 2x + 4y - z = -15 \end{cases}$ .  (**Lesson 4-7**)

25. Evaluate $\begin{vmatrix} 4 & 2 \\ 3 & 7 \end{vmatrix}$.  (**Lesson 4-3**)

26. Use Cramer's Rule to solve $\begin{cases} 9a - b = 1 \\ 3a + 2b = 12 \end{cases}$.  (**Lesson 3-3**)

27. Solve $\begin{cases} x + 5y = 14 \\ -2x + 6y = 4 \end{cases}$ by graphing.  (**Lesson 3-1**)

28. What property of real numbers is demonstrated by $x(a + b) = xa + xb$?  (**Lesson 1-2**)

## LANGUAGE CONNECTION

When you need to refer to a specific member of a matrix, a symbol is used to name each element of the matrix. The notation $e_{34}$ denotes the element that is in the third row and fourth column.

What is the first element and last element in a matrix called $B_{m \times n}$?

# Graphing Calculator Exploration: Matrix Row Operations

You can solve a system of linear equations by using the TI-81 calculator and the functions listed on its MATRX menu. Each of the functions is listed below with instructions on the keying procedure after that function has been selected from the MATRX menu. For convenience, suppose your augmented matrix has been entered as matrix [A] .

RowSwap(
**Interchanges two rows.**
- Enter the name of the matrix followed by a comma. *The comma is entered using the* ALPHA *key and the* ⋅ *key.*
- Enter one of the rows you want to interchange followed by a comma.
- Enter the other row you want to interchange followed by ⟮)⟯ .

*To interchange rows 2 and 3 in matrix A:* RowSwap( [A] ⟮,⟯ 2 ⟮,⟯ 3 ⟮)⟯ ENTER

Row+(
**Adds two rows and stores the result in the last row you entered.**
- Enter the name of the matrix followed by a comma.
- Enter the row you want to add, followed by a comma.
- Enter the row you want it added to, followed by ⟮)⟯ .

*To add row 3 to row 1 in matrix A:* Row +( [A] ⟮,⟯ 3 ⟮,⟯ 1 ⟮)⟯ ENTER

*Row(
**Does scalar multiplication on one row.**
- Enter the number you want to multiply by, followed by a comma.
- Enter the name of the matrix followed by a comma.
- Enter the row you want multiplied, followed by ⟮)⟯ .

*To multiply row 3 by –2 in matrix A:* *Row( –2 [A] ⟮,⟯ ⟮,⟯ 3 ⟮)⟯ ENTER

*Row+(
**Multiplies one row and adds the result to another.**
- Enter the number you want to multiply by, followed by a comma.
- Enter the name of the matrix followed by a comma.
- Enter the row you want multiplied, followed by a comma.
- Enter the row you want the result added to, followed by ⟮)⟯ .

*To multiply row 2 by $\frac{1}{2}$ and add it to row 3 in matrix A:*

*Row+( 0.5, ⟮,⟯ [A] ⟮,⟯ 2 ⟮,⟯ 3 ⟮)⟯ ENTER

To perform one operation after another in completely reducing a matrix, use ANS to be your matrix name so the operations will be done on the matrix you just finished.

*To add row 2 of the matrix you just altered to row 1:* Row+( ANS ⟮,⟯ 2 ⟮,⟯ 1 ⟮)⟯ ENTER

Now use your TI-81 to check your answers in Lesson 4-8.

# 4 SUMMARY AND REVIEW

## VOCABULARY

Upon completing this chapter you should be
familiar with the following terms:

| | | | |
|---|---|---|---|
| augmented matrix | **195** | **179** | identity matrix |
| coefficient matrix | **184** | **179** | inverse matrix |
| coordinate matrix | **161** | **156** | matrix |
| Cramer's rule | **190** | **184** | matrix equation |
| determinant | **167** | **156** | matrix logic |
| dilation | **162** | **167** | minor |
| dimension | **161** | **174** | rotation matrix |
| element | **161** | **162** | scalar multiplication |
| expansion by minors | **167** | **163** | translation |

## SKILLS AND CONCEPTS

| OBJECTIVES AND EXAMPLES | REVIEW EXERCISES |
|---|---|

Upon completing this chapter, you should
be able to:

create a matrix and name it using its
dimensions   (**Lesson 4-2**)

$R = [5 \quad -3 \quad 9 \quad 7]$

The matrix above would be named $R_{1 \times 4}$.

Use these exercises to review and prepare
for the chapter test.

**Name each matrix using its dimensions.**

1. $P = \begin{bmatrix} 3 & 2 & 1 \\ -5 & 6 & -3 \end{bmatrix}$   2. $Q = \begin{bmatrix} 6 \\ -2 \\ 1 \end{bmatrix}$

---

perform operations such as addition of
matrices and scalar multiplication of a
matrix   (**Lesson 4-2**)

$2 \begin{bmatrix} 18 & -6 \\ 9 & 21 \end{bmatrix} = \begin{bmatrix} 36 & -12 \\ 18 & 42 \end{bmatrix}$

**Perform the indicated operations.**

3. $3 \begin{bmatrix} 8 & -3 & 2 \\ 4 & 1 & 7 \end{bmatrix}$

4. $2 \begin{bmatrix} 8 & -1 \\ 3 & 4 \end{bmatrix} - 3 \begin{bmatrix} 1 & 6 \\ -2 & -3 \end{bmatrix}$

---

find unknown values in equal matrices
(**Lesson 4-2**)

To find $x$ and $y$ for $\begin{bmatrix} 2x \\ y \end{bmatrix} = \begin{bmatrix} 32 + 6y \\ 7 - x \end{bmatrix}$,
solve the system $2x = 32 + 6y$
and $y = 7 - x$.

5. Find the values of $x$ and $y$ for which the
   equation is true.

   $\begin{bmatrix} 7x \\ x + y \end{bmatrix} = \begin{bmatrix} 5 + 2y \\ 11 \end{bmatrix}$

| OBJECTIVES AND EXAMPLES | REVIEW EXERCISES |
|---|---|

- evaluate the determinant of a 3 × 3 matrix  (**Lesson 4-3**)

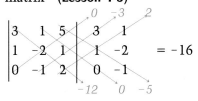

$$\begin{vmatrix} 3 & 1 & 5 \\ 1 & -2 & 1 \\ 0 & -1 & 2 \end{vmatrix} = -16$$

**Evaluate each determinant.**

6. $\begin{vmatrix} 5 & -1 & 2 \\ -6 & -7 & 3 \\ 7 & 0 & 4 \end{vmatrix}$

7. $\begin{vmatrix} 2 & -3 & 1 \\ 0 & 7 & 8 \\ 2 & 1 & 3 \end{vmatrix}$

---

- multiply two matrices and interpret the results  (**Lesson 4-4**)

The product of an $m \times n$ matrix, $A$, and an $n \times r$ matrix, $B$, is the $m \times r$ matrix $AB$. The element in the $i^{\text{th}}$ row and the $j^{\text{th}}$ column of $AB$ is the sum of the products of the corresponding elements in the $i^{\text{th}}$ row of $A$ and the $j^{\text{th}}$ column of $B$.

**Find each product.**

8. $\begin{bmatrix} 2 & -3 \\ 6 & 4 \end{bmatrix} \cdot \begin{bmatrix} 1 & -3 & 4 \\ 3 & 1 & 1 \end{bmatrix}$

9. $[5 \quad -2 \quad 3] \cdot \begin{bmatrix} 2 & 5 \\ -1 & 3 \\ 6 & 4 \end{bmatrix}$

10. $\begin{bmatrix} 4 \\ -1 \\ 3 \end{bmatrix} \cdot \begin{bmatrix} 1 & 0 & 0 \\ 0 & 1 & 0 \\ 0 & 0 & 1 \end{bmatrix}$

---

- write the identity matrix for any given matrix  (**Lesson 4-5**)

The identity matrix is a square matrix which when multiplied by another matrix equals that same matrix.

11. Write the 4 × 4 identity matrix.

12. Write the identity matrix, $I_{6\times6}$.

---

- find the inverse matrix of a 2 × 2 matrix  (**Lesson 4-5**)

Any matrix $M$, $\begin{bmatrix} a & b \\ c & d \end{bmatrix}$, will have an inverse $M^{-1}$ if and only if $\begin{vmatrix} a & b \\ c & d \end{vmatrix} \neq 0$.

Then $M^{-1} = \dfrac{1}{ad - bc}\begin{bmatrix} d & -b \\ -c & a \end{bmatrix}$.

**Find the inverse of each matrix.**

13. $\begin{bmatrix} 8 & 6 \\ 9 & 7 \end{bmatrix}$

14. $\begin{bmatrix} 3 & 2 \\ 4 & -2 \end{bmatrix}$

---

- write a system of linear equations as a matrix and use the inverse to solve the system  (**Lesson 4-6**)

  - Multiply both sides by the inverse of the coefficient matrix.
  - The result will be the identity matrix multiplied by the variables.
  - Interpret the solution.

**Solve each matrix equation.**

15. $\begin{bmatrix} 3 & 2 \\ 1 & -2 \end{bmatrix} \cdot \begin{bmatrix} x \\ y \end{bmatrix} = \begin{bmatrix} 9 \\ 11 \end{bmatrix}$

16. $\begin{bmatrix} 3 & 1 & 1 \\ 2 & 4 & 1 \\ 1 & 3 & 2 \end{bmatrix} \cdot \begin{bmatrix} a \\ b \\ c \end{bmatrix} = \begin{bmatrix} -2 \\ 4 \\ 12 \end{bmatrix}$ if the

inverse is $\dfrac{1}{14}\begin{bmatrix} 5 & 1 & -3 \\ -3 & 5 & -1 \\ 2 & -8 & 10 \end{bmatrix}$.

| OBJECTIVES AND EXAMPLES | REVIEW EXERCISES |
|---|---|

- use Cramer's Rule to solve a system of linear equations in three variables (**Lesson 4-7**)

  - Determine if the system of equations has a unique solution.
  - If the system has a unique solution, then solve using Cramer's Rule.

**Solve each system using Cramer's Rule.**

17. $2a - b - 3c = -20$
    $4a + 2b + c = 6$
    $2a + b - c = -6$
18. $2x - y + 3z = 1$
    $x - y + 4z = 0$
    $3x - 2y + z = -5$

---

- solve a system of equations using an augmented matrix   (**Lesson 4-8**)

  The solution represented by the reduced

  augmented matrix $\begin{bmatrix} 1 & 0 & 0 & 4 \\ 0 & 1 & 0 & 3 \\ 0 & 0 & 1 & \frac{7}{4} \end{bmatrix}$ is

  $\left(4, 3, \frac{7}{4}\right)$.

**Solve using augmented matrices.**

19. $x + 5y = 1$
    $2x - 3y = 15$
20. $a - 2b + c = 6$
    $3a + 2b - c = 0$
    $2a + b - 6c = -2$

## ～～～～ APPLICATIONS AND CONNECTIONS ～～～～

21. Use matrix logic to solve this problem.   (**Lesson 4-1**)
    Alan, Bill, and Cathy each had different lunches. One had soup, one had a sandwich, and one had a salad. Alan did not have a sandwich. Bill did not have soup or a sandwich. What did each person have for lunch?

**Triangle ABC has vertices with coordinates $A(5, -2)$, $B(-3, 4)$, and $C(-2, -3)$. Find the coordinates of the vertices of triangle $A'B'C'$ for each transformation.   (Lesson 4-2)**

22. Triangle $A'B'C'$ has a perimeter three times that of triangle $ABC$.

23. Triangle $A'B'C'$ is translated 4 units right and 1 unit down.

24. Find the area of a triangle with vertices $X(-2, 6)$, $Y(-3, -2)$, and $Z(3, 5)$.
    (**Lesson 4-3**)

25. Triangle $MPQ$ has vertices $M(5, 0)$, $P(-2, 6)$, and $Q(3, -4)$. Find the coordinates of the triangle after it is rotated 90° counterclockwise about the origin.   (**Lesson 4-4**)

26. **Investment Planning**   Maria Hernandez invested $29,000, part in stocks at 4% return, part in bonds at 6% return, and the remainder in a term account at 5.5% return. The amount she invested in stocks was equal to the total amount invested in bonds and term accounts together. If her total annual interest was $1414, how much money did she invest in each item?   (**Lesson 4-7**)

**State the dimensions of each matrix. Then evaluate the determinant of the matrix, if it exists.**

1. $\begin{bmatrix} 2 & -3 & 1 \\ 3 & -1 & 2 \\ 1 & 2 & -3 \end{bmatrix}$

2. $\begin{bmatrix} 7 & -10 & 1 & 4 \\ 6 & 8 & 5 & -1 \end{bmatrix}$

3. $\begin{bmatrix} 1 & 3 & 1 \\ 2 & 1 & -5 \\ 3 & -1 & -4 \end{bmatrix}$

**Perform the indicated operations.**

4. $\begin{bmatrix} 1 & 2 \\ -4 & 3 \\ 5 & 2 \end{bmatrix} \begin{bmatrix} 5 \\ 4 \end{bmatrix}$

5. $\begin{bmatrix} 2 & -4 & 1 \\ 3 & 8 & -2 \end{bmatrix} - 2\begin{bmatrix} 1 & 2 & -4 \\ -2 & 3 & 7 \end{bmatrix}$

6. Solve $\begin{bmatrix} x \\ 2y \end{bmatrix} - 3\begin{bmatrix} y+1 \\ 2x \end{bmatrix} = \begin{bmatrix} 5 \\ -16 \end{bmatrix}$.

7. Find the inverse of $\begin{bmatrix} 5 & -2 \\ 6 & 3 \end{bmatrix}$.

8. Determine whether the system $\begin{cases} r + s + t = 7 \\ 3r - 7s + 2t = 11 \\ -9r + 21s + 3t = -3 \end{cases}$ has a unique solution.

   If it does, find it using Cramer's rule.

9. Solve $\begin{cases} x + 8y = -3 \\ 2x - 6y = -17 \end{cases}$ using a matrix equation.

**Solve each system of equations using augmented matrices.**

10. $6x - y = -15$
    $5x + 2y = -4$

11. $2x - 3y + z = 7$
    $3x - y + 2z = 1$
    $x + 2y - 3z = -14$

**For Exercises 12–14, use △ABC whose vertices have coordinates A(6, 3), B(1, 5), and C(-1, 4).**

12. Translate triangle $ABC$ so that the coordinates of $B'$ are (3, 1). What are the coordinates of $A'$ and $C'$?

13. Find the coordinates of the vertices of a similar triangle whose perimeter is five times that of △ABC.

14. Find the coordinates of the vertices of triangle $ABC$ after it has been rotated counterclockwise 90° about the origin.

**Bonus**
Find the coordinates of a similar triangle whose perimeter is half that of △ABC and has been rotated *clockwise* 90° about the origin.

# College Entrance Exam Preview

The test questions on these pages deal with ratios, proportions, and percents.

**Directions: Choose the one best answer. Write A, B, C, or D.**

1. 40% of 10 inches is how many sixths of 2 feet?

   (A) $\frac{1}{3}$        (B) 1

   (C) 2        (D) 4

2. For nonzero numbers, $a$, $b$, $c$, and $d$, $\frac{a}{b} = \frac{c}{d}$. Which of the following must be true?

   (A) $\frac{a}{b} = \frac{b}{c}$      (B) $\frac{a+b}{b} = \frac{c+b}{d}$

   (C) $\frac{d}{b} = \frac{c}{a}$      (D) $\frac{b}{c+d} = \frac{d}{a+b}$

3. Find the percent of increase if your salary increases from $250 a week to $300.

   (A) $16\frac{2}{3}\%$      (B) 20%

   (C) 22%      (D) 25%

4. If your grade was 90 and is now 75, find the percent of decrease.

   (A) $16\frac{2}{3}\%$      (B) 18%

   (C) 20%      (D) 22%

5. 9 is 6% of what number?

   (A) 100      (B) 120

   (C) 130      (D) 150

6. If $\frac{x}{y} = \frac{5}{6}$, then $18x =$

   (A) $\frac{5y}{3}$    (B) $90y$    (C) $15y$    (D) $\frac{5y}{6}$

7. The price of an item was reduced by 20% then later reduced by 5%. The two reductions were equivalent to the single reduction of

   (A) 15%    (B) 24%    (C) 25%    (D) 75%

8. Ten gallons of gas were added to a tank that had been $\frac{1}{4}$ full. If it is now $\frac{7}{8}$ full, how many gallons does the tank hold?

   (A) 16    (B) 18    (C) 20    (D) 24

9. The ratio of Jean's weight to Jim's weight is 3:4. If Jean gains 30 pounds and Jim does not gain any, the ratio will be 7:8. How much does Jim weigh?

   (A) 60    (B) 170    (C) 180    (D) 240

10. Last year Joe attended one-half the number of sporting events that Jan did. George attended one-third the number that Jan did. If George attended 8 sporting events, how many did Joe attend?

    (A) 1    (B) 4    (C) 8    (D) 12

11. If $\frac{1}{9} = \frac{x}{.45}$, what is the value of $x$?

    (A) 0.05    (B) 0.5    (C) 5    (D) 6

12. In a class of 54 students, 12 are honor students. What part of the class are not honor students?

(A) $\frac{21}{33}$  (B) $\frac{7}{9}$  (C) $\frac{2}{7}$  (D) $\frac{2}{9}$

13. 22% of 440 is 4.4% of

(A) 96.8  (B) 425.92

(C) 220  (D) 2200

14. 75% of $10a$ is $b$. What percent of $2b$ is $a$?

(A) $13\frac{1}{3}$  (B) $6\frac{2}{3}$  (C) $7\frac{1}{2}$  (D) 15

15. John spent $\frac{1}{4}$ of his money on a book and then $\frac{2}{5}$ of the remaining money for lunch. What fractional part of the original amount is left?

(A) $\frac{3}{20}$  (B) $\frac{3}{8}$  (C) $\frac{9}{20}$  (D) $\frac{4}{9}$

16. Write 0.4% as a fraction.

(A) $\frac{1}{250}$  (B) $\frac{2}{125}$

(C) $\frac{1}{25}$  (D) $\frac{2}{5}$

17. $\frac{1}{2}$% of 500 is

(A) 2500  (B) 250

(C) 25  (D) 2.5

18. If $3n = m$, then $\frac{2}{3}m =$

(A) $2n$  (B) $\frac{3}{2}n$  (C) $\frac{2}{3}n$  (D) $\frac{1}{2}n$

19. Ms. Kwan earns a salary of $300 per week plus a 6% commission on her sales. What must her sales be for the week if she earned $345?

(A) $5750  (B) $2070

(C) $750  (D) $207

20. The side of one square is 3 in. and the side of another square is $1\frac{1}{4}$ ft. The ratio of the areas of the two squares is

(A) 1:5  (B) 1:25

(C) 12:5  (D) 144:25

21. What is $t$% of 8?

(A) $\frac{2t}{25}$  (B) $\frac{2}{25t}$  (C) $\frac{25t}{2}$  (D) $\frac{25}{2t}$

# 5 Polynomials

## CHAPTER OBJECTIVES

In this chapter, you will:
- Multiply monomials.
- Represent numbers in scientific notation.
- Factor polynomials.
- Divide polynomials.

Punnett squares can predict the possible types of offspring from two parents with specific traits. If the parents were a pink flower and a red flower, what types of offspring might they produce?

**Punnett Squares**

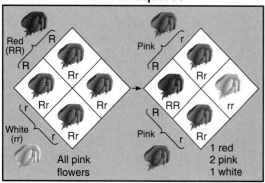

## CAREERS IN GENETIC ENGINEERING

What qualities would you select if you were breeding the perfect melon?
- sweet, flavorful, firm flesh
- early ripening
- easy to see whether ripe or not

How about the perfect tomato?
- full of sun-ripened flavor
- meaty, yet juicy
- easy to grow, harvest, and ship

If you were designing a new strain of cauliflower, how about adding a naturally cheesy flavor? Why not design a broccoli that could be grown in fields irrigated with salt water, or maybe even not irrigated at all. Add resistance to frost and disease too, if you like. Genetic engineers are doing all these things and more right now.

If designer greens don't interest you, how about designer pharmaceuticals? Genetic engineers are now manufacturing insulin and hepatitis-B surface antigen safely and economically, and others are producing new hope for sufferers of AIDS and cystic fibrosis. Still in the planning stages are an artificial pancreas and liver.

A third trend is bioremediation. In this field, genetic engineers design and then produce organisms that eat toxic and hazardous waste.

Genetic engineering, or altering genetic characteristics for a specific purpose, is a young field that grows bigger each year. There may be a career in that field for you.

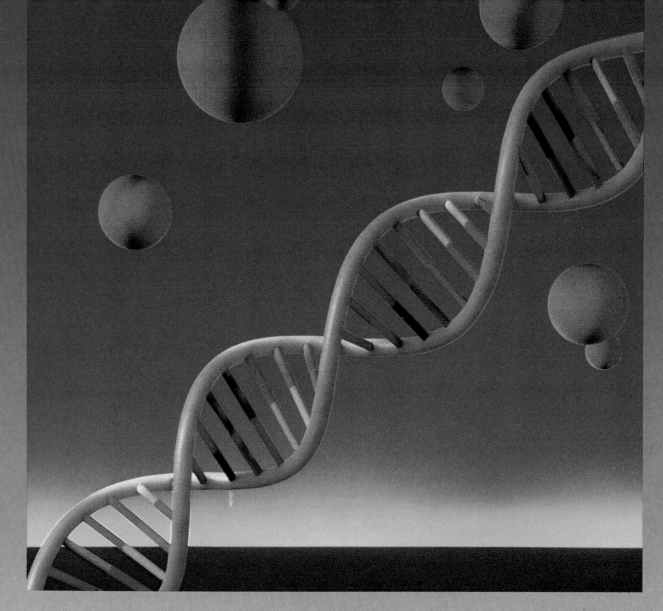

# MORE ABOUT GENETIC ENGINEERING

## Degree Required:

- Bachelor's Degree in Molecular Biology or Chemistry

## Some genetic engineers like:

- working on the cutting edge of technological research and new applications of science
- working in a field that requires creativity
- knowing that their work may benefit others

## Related Math Subjects:

- Advanced Algebra
- Probability/Statistics
- Trigonometry
- Calculus

## Some genetic engineers dislike:

- working long hours indoors
- having to work under pressure
- having to try to keep up with the new developments in the field
- having to meet deadlines

For more information on the various careers in the field of genetic engineering, write to:

Industrial Biotechnology Association
1625 K Street, N.W.
Suite 1100
Washington, D.C. 20006

## 5-1 Monomials

**Objectives**

After studying this lesson, you should be able to:

- multiply monomials and powers of monomials, and
- represent numbers in scientific notation.

**Application**

Stephanie bought some of the supplies for the Wilderness Club's winter weekend getaway. She bought three bags of apples, 4 boxes of granola bars, and 2 packages of recyclable paper plates.

The quantities of items that Stephanie bought can be described using **monomials.** A monomial is an expression that is a number, a variable, or the product of a number and one or more variables. If $a$ is the number of apples in a bag, $3a$ would be a monomial describing the number of apples that Stephanie bought. Some other monomials are $-3$, $z$, $t^4$, and $\frac{2}{3}ab^2$.

Expressions like $\frac{1}{x}$ and $\sqrt{x}$ are not monomials. Monomials cannot contain variables whose exponents can be written as a fraction or as a negative number.

**Constants** are monomials that contain no variables. The numerical factor of a monomial is the **coefficient** of the variable. For example, the coefficient of $g$ in $-4g$ is $-4$. The **degree of a monomial** is the sum of the exponents of its variables. The degree of a nonzero constant is 0. The constant 0 has no degree.

This table summarizes some terms related to monomials.

| Monomial | Coefficient | Variable(s) | Exponent(s) | Degree |
|----------|-------------|-------------|-------------|--------|
| $s$ | 1 | $s$ | 1 | 1 |
| $-7a^2$ | $-7$ | $a$ | 2 | 2 |
| $k^8$ | 1 | $k$ | 8 | 8 |
| $\frac{2}{3}ab^2$ | $\frac{2}{3}$ | $a$ and $b$ | 1, 2 | 3 |

*Remember that $s = 1 \cdot s$.*

If two monomials are the same, or differ only by their numerical coefficients, they are called **like terms.** For example, $5xy^2$ and $12xy^2$ are like terms, but $5x^2y$, $3x^3$, and $12xy^2$ are not.

**Example 1**

**Simplify $4a^3b + 11a^3b - a^3b$.**

$$4a^3b + 11a^3b - a^3b = (4 + 11 - 1)a^3b \qquad \text{\textit{The three terms are like terms.}}$$
$$= 14a^3b$$

**Example 2**

Simplify $(r^3s^2)(r^4s^4)$.

$$(r^3s^2)(r^4s^4) = (r \cdot r \cdot r \cdot s \cdot s)(r \cdot r \cdot r \cdot r \cdot s \cdot s \cdot s \cdot s)$$
$$= r \cdot r \cdot r \cdot r \cdot r \cdot r \cdot r \cdot s \cdot s \cdot s \cdot s \cdot s \cdot s \qquad \text{\textit{7 factors of r and}}$$
$$= r^7 s^6 \qquad\qquad\qquad\qquad\qquad\qquad \text{\textit{6 factors of s}}$$

Example 2 suggests the following property.

| *Multiplying Powers* | **For any real number $a$, and positive integers $m$ and $n$, $a^m \cdot a^n = a^{m+n}$.** |
|---|---|

Let's use this property to find how to raise a power to a power. Try this.

**Example 3**

Simplify $(d^3)^4$.

$$(d^3)^4 = d^3 \cdot d^3 \cdot d^3 \cdot d^3$$
$$= d^{3+3+3+3} \qquad \text{\textit{Multiplying powers}}$$
$$= d^{12}$$

This example suggests the following property.

| *Raising a Power to a Power* | **For any real number $a$, and positive integers $m$ and $n$, $(a^m)^n = a^{mn}$.** |
|---|---|

**Example 4**

Simplify $(xy)^3$.

$$(xy)^3 = (xy)(xy)(xy)$$
$$= x \cdot x \cdot x \cdot y \cdot y \cdot y \qquad \text{\textit{Commutative property}}$$
$$= x^3 y^3$$

Example 4 demonstrates the following property.

| *Finding a Power of a Product* | **For any real numbers $a$, $b$, and positive integer $m$, $(ab)^m = a^m b^m$.** |
|---|---|

We can simplify many kinds of expressions using the properties of exponents and the commutative and associative properties.

**Example 5**

Simplify $(3a^3b)(-5a^2b^2)$.

$$(3a^3b)(-5a^2b^2) = 3 \cdot (-5) \cdot a^3 \cdot a^2 \cdot b \cdot b^2$$
$$= -15 \cdot a^{3+2} \cdot b^{1+2}$$
$$= -15a^5b^3$$

An important use of exponents is **scientific notation.** Very large numbers are often written in scientific notation. Study the following examples.

246,000,000,000

$= 2.46 \times 100,000,000,000$
$= 2.46 \times 10^{11}$   *$10^{11} = 100,000,000,000$*

3,220,000

$= 3.22 \times 1,000,000$
$= 3.22 \times 10^6$   *$10^6 = 1,000,000$*

| *Definition of Scientific Notation* | **A number is in scientific notation when it is in the form $a \times 10^n$, where $1 \le a < 10$ and $n$ is an integer.** |
|---|---|

**Example 6**

Use a calculator to multiply $3.2 \times 10^4$ by $1.9 \times 10^2$. Express the solution in both scientific and decimal notation.
*The exponential shift key may vary depending upon the calculator you are using.*

ENTER:   3.2 [EXP] 4 [×] 1.9 [EXP] 2 [=]   ⎕⎕⎕⎕⎕⎕⎕

*Your calculator may display $3.2 \times 10^4$ as 3.2 + 4.*

The solution is $6.08 \times 10^6$ or 6,080,000.

**Example 7**

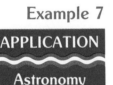
APPLICATION
Astronomy

When a solar flare occurs on the sun, it sends out light waves that travel through space at a speed of $1.08 \times 10^9$ km/h. If a satellite in space detects the flare 2 hours after its occurrence, how far is the satellite from the sun?

$$rt = d \qquad \textit{rate} \times \textit{time} = \textit{distance}$$
$$(1.08 \times 10^9)(2) = d \qquad \textit{Substitute the given values into the formula for distance.}$$
$$2.16 \times 10^9 = d$$

The satellite is $2.16 \times 10^9$ or 2,160,000,000 km from the Sun.

# CHECKING FOR UNDERSTANDING

Communicating
Mathematics

**Read and study the lesson to answer these questions.**

1. Are $4x^2$ and $(4x)^2$ equivalent?

2. Is a negative number raised to the seventh power negative or positive? Explain.

3. In the problem in the introduction, suppose there are $g$ granola bars in a box. Express the number of granola bars that Stephanie bought for the Wilderness Club's getaway as a monomial.

**Guided Practice**

**State whether each expression is a monomial. If it is, name its coefficient and degree.**

4. $3x$

5. $a^2$

6. $4rs + s$

7. $-5ab$

8. $\dfrac{11xy}{7}$

9. $\sqrt{cd}$

10. $5x^3y^2z^4$

11. $0$

12. $\dfrac{3rs}{t}$

**Express each of the following in scientific notation.**

13. 810.4

14. 2100

15. 9,000,000,000

16. 786,500,000

17. 72,100,000

18. 528,000

**Express each of the following in decimal notation.**

19. $4.2 \times 10^4$

20. $2.541 \times 10^2$

21. $5.7 \times 10^1$

22. $4.27 \times 10^1$

23. $3.21 \times 10^6$

24. $7.2 \times 10^4$

# EXERCISES

**Practice**

**Simplify.**

25. $3x + 2x + (-4x)$

26. $4d^3 - d^3 + 2d^3$

27. $4ab^2 - 3ab^2$

28. $3x^2 + 4 - 3x^2$

29. $y^5 \cdot y^7$

30. $b^4 \cdot b^3 \cdot b^2$

31. $8^6 \cdot 8^4 \cdot (8^2)^2$

32. $(y^5)^2$

33. $(3a)^4$

34. $(x^2y^2)^2x^3y^3$

35. $\left(-\dfrac{3}{4}x^2y^3\right)^2\left(\dfrac{8}{9}xy^4\right)$

36. $\left(\dfrac{3}{5}c^2f\right)\left(\dfrac{4}{3}cd\right)^2$

37. $(-4a)(a^2)(-a^3) + 3a^2(a^4)$

38. $2(rk)^2(5rt^2) - k(2rk)(2rt)^2$

39. $(5a)(6a^2b)(3ab^3) + (4a^2)(3b^3)(2a^2b)$

40. $(5mn^2)(m^3n)(-3p^2) + (8np)(3mp)(m^3n^2)$

**Evaluate. Express each answer in both scientific and decimal notation.**

**41.** $(9.5 \times 10^3)^2$

**42.** $(7.2 \times 10^5)(8.1 \times 10^3)$

**43.** $(4.5 \times 10^3)(7.0 \times 10^2)$

**44.** $(2.5 \times 10^2)(1.1 \times 10^2)$

**45.** $(34,000)(0.0056)$

**46.** $(4,300)(0.02)$

**47.** $(3,000)(82,500)$

**48.** $(45,000)(0.0025)$

**49.** $(4.4 \times 10^5) - (3.2 \times 10^5)$

**50.** $1.2 \times 10^3 - 1.2 \times 10^2$

**Critical Thinking**

**51.** The number 64 is a square, a cube, and a sixth power since:
$$64 = 8^2 = 4^3 = 2^6.$$
Find the least integer greater than 1 that is a square, a cube, a sixth, and a ninth power.

**52.** Which is greater, $100^{10}$ or $10^{100}$? Explain your answer.

**Applications**

**53. Astronomy** Moonlight takes 1.25 seconds to reach Earth. If the speed of light is $3.00 \times 10^5$ kilometers per second, how far from Earth is the moon?

**54. Chemistry** The mole is a standard unit of measure in chemistry. One mole of any compound contains $6.02 \times 10^{23}$ molecules. How many molecules are in 19.9 moles of ammonia?

**Mixed Review**

**Solve each system of equations using the augmented matrix method. (Lesson 4-8)**

**55.** $4x - y + z = 6$
$2x + y + 2z = 3$
$3x - 2y + z = 3$

**56.** $x + 3y - 2z = 9$
$-x + 5y + 2z = 31$
$2x - 9z = -32$

**57.** The sum of two numbers is 42. Their difference is 12. What are the two numbers? **(Lesson 3-2)**

**58.** Solve $3x + 4y = 16$ for $y$. **(Lesson 2-2)**

**59.** Solve $|x + 1| \leq 3$. **(Lesson 1-8)**

**60. Sports** It is possible to score 2, 3, 6, or 7 points in a football game. Assuming the game was not a forfeit, are there any total scores less than 50 points that are impossible to make? If so, name them. **(Lesson 1-5)**

**Journal**

Do some research to find some numbers written in scientific notation. Tell what each means and then write the numbers in order from least to greatest.

## 5-2 Dividing Monomials

**Objectives**

After studying this lesson, you should be able to:
- divide monomials, and
- divide expressions written in scientific notation.

**Application**

The spaceprobe *Pioneer 10* was as far from Earth as the planet Pluto in April of 1983. It sent radio signals that traveled at the speed of light back to Earth. If Pluto is $4.58 \times 10^9$ km from Earth and the speed of light is $3.00 \times 10^5$ km per second, how long after *Pioneer 10* sent the signals did the Earth-based tracking stations receive them?

Problems like this one require division of powers. You know that when you multiply powers of the same base you add the exponents. Knowing this, it seems reasonable to expect to subtract exponents when you divide powers. Let's try a problem and see if this is true.

**Example 1**

Simplify $\dfrac{m^8}{m^3}$.

$$\dfrac{m^8}{m^3} = \dfrac{m \cdot m \cdot m \cdot m \cdot m \cdot \overset{1}{\cancel{m}} \cdot \overset{1}{\cancel{m}} \cdot \overset{1}{\cancel{m}}}{\underset{1}{\cancel{m}} \cdot \underset{1}{\cancel{m}} \cdot \underset{1}{\cancel{m}}}$$

*Remember m cannot equal 0.*

$$= m \cdot m \cdot m \cdot m \cdot m$$

*5 or (8 − 3) factors*

$$= m^{8-3} \text{ or } m^5$$

This example suggests that it is true. To divide powers of the same base, you subtract exponents. This is stated more formally below.

**Dividing Powers**

**For any real number $a$, except $a = 0$, and integers $m$ and $n$,**
$$\dfrac{a^m}{a^n} = a^{m-n}.$$

*Why is 0 not an acceptable value for a?*

Let's use this property to do some investigation. Study the two ways of simplifying $\dfrac{n^3}{n^3}$ shown below.

$$\dfrac{n^3}{n^3} = \dfrac{n \cdot n \cdot n}{n \cdot n \cdot n} \qquad \dfrac{n^3}{n^3} = n^{3-3}$$
$$= 1 \qquad\qquad = n^0$$

Since $\dfrac{n^3}{n^3}$ cannot have two values, we can conclude that $n^0 = 1$, where $n$ is not equal to zero. In general, any nonzero number raised to the zero power is equal to 1.

Why is $0^0$ not defined? If we interpret $0^0$ as $0^{m-m}$, then it represents $\dfrac{0^m}{0^m}$. Since $0^m$ is 0, this implies division by zero. Division by zero is not defined, so $0^0$ is also not defined.

**Example 2**

**A chemist has performed an experiment that yields 1.8 × 10²⁴ molecules of ethanol. The mole is the standard unit of measure for the chemical quantity of a substance. There are 6.02 × 10²³ molecules in a mole. How many moles of ethanol did the experiment yield?**

Let $n$ represent the number of moles of ethanol.

number of moles $\times$ number of molecules per mole $=$ number of molecules

$$n(6.02 \times 10^{23}) = 1.8 \times 10^{24}$$

$$n = \frac{1.8 \times 10^{24}}{6.02 \times 10^{23}}$$

$$= \left(\frac{1.8}{6.02}\right) \times \left(\frac{10^{24}}{10^{23}}\right)$$

$$\approx 0.299 \times 10^1 \text{ or } 2.99$$

The experiment yielded about 2.99 moles of ethanol.

Let's do some more investigation with the properties of exponents. Study the two ways of simplifying $\dfrac{t^4}{t^8}$. Assume $t$ is a nonzero real number.

*Method 1*

$$\frac{t^4}{t^8} = \frac{t \cdot t \cdot t \cdot t}{t \cdot t \cdot t \cdot t \cdot t \cdot t \cdot t \cdot t}$$

$$= \frac{1}{t \cdot t \cdot t \cdot t}$$

$$= \frac{1}{t^4}$$

*Method 2*

$$\frac{t^4}{t^8} = t^{4-8}$$

$$= t^{-4}$$

We can conclude that $\dfrac{1}{t^4} = t^{-4}$ since $\dfrac{t^4}{t^8}$ cannot have two values. Find the reciprocal of $\dfrac{t^4}{t^3}$ in two ways. What do you conclude?

*Negative Exponents*

> **For any real number $a$, except $a = 0$, and any integer $n$,**
> $$a^{-n} = \frac{1}{a^n} \text{ and } \frac{1}{a^{-n}} = a^n.$$

You know how to write a very large number in scientific notation. We can also write very small numbers in scientific notation by using negative exponents. Study the following example.

$$0.0064 = 6.4 \times 0.001$$
$$= 6.4 \times \frac{1}{10^3}$$
$$= 6.4 \times 10^{-3}$$

*0.0064 = 6.4 × 0.001*

*0.001 = $\dfrac{1}{1000}$ or $\dfrac{1}{10^3}$*

**Example 3**

Use your calculator to divide $8.4 \times 10^3$ by $2.25 \times 10^{-2}$.

ENTER:    8.4 (EXP) 3 (÷) 2.25 (EXP) 2 (+/-) (=)   ＿ ＿＿＿＿＿＿. ＿＿＿＿

The answer is 373333.3333 or $3.7\overline{3} \times 10^5$.

When you are asked to simplify an expression, write an equivalent form that uses only positive exponents and no parentheses. Each base should appear only once and all fractions should be in simplest form.

**Example 4**

Simplify $\dfrac{(3rs)^2(s^3t)^2}{12s^2t^4}$.

$\dfrac{(3rs)^2(s^3t)^2}{12s^2t^4} = \dfrac{(9r^2s^2)(s^6t^2)}{12s^2t^4}$     *Multiplying powers property*

$= \dfrac{9r^2(s^2 \cdot s^6)t^2}{12s^2t^4}$     *Associative property*

$= \left(\dfrac{9}{12}\right)\left(\dfrac{r^2}{1}\right)\left(\dfrac{s^8}{s^2}\right)\left(\dfrac{t^2}{t^4}\right)$     *Group like terms.*

$= \dfrac{3}{4}r^2s^6t^{-2}$

$= \dfrac{3r^2s^6}{4t^2}$

**Example 5**

Simplify $\dfrac{8^{3n}}{8^{3n-2}}$.

$\dfrac{8^{3n}}{8^{3n-2}} = 8^{3n-(3n-2)}$     *Division of powers*

$= 8^{3n-3n+2}$

$= 8^2$ or $64$

**Example 6**

Simplify $\left(\dfrac{3}{4}\right)^{-3}$.

$\left(\dfrac{3}{4}\right)^{-3} = \left[\left(\dfrac{3}{4}\right)^{-1}\right]^3$     *Use the power property.*

$= \left(\dfrac{4}{3}\right)^3$     $\left(\dfrac{3}{4}\right)^{-1} = \dfrac{1}{\frac{3}{4}} = \dfrac{4}{3}.$

$= \dfrac{4}{3} \cdot \dfrac{4}{3} \cdot \dfrac{4}{3}$

$= \dfrac{4^3}{3^3}$ or $\dfrac{64}{27}$

Example 6 suggests the following properties.

| *Powers of Quotients* | **For any nonzero real numbers $a$ and $b$, and integer $n$,** |
| | $\left(\dfrac{a}{b}\right)^n = \dfrac{a^n}{b^n}$ and $\left(\dfrac{a}{b}\right)^{-n} = \left(\dfrac{b}{a}\right)^n$ or $\dfrac{b^n}{a^n}.$ |

# CHECKING FOR UNDERSTANDING

**Communicating Mathematics**

**Read and study the lesson to answer these questions.**

1. Explain in your own words why $0^0 \neq 1$.
2. Write $7^{-3}$ using a positive exponent.
3. Is $2(x^2y)^3$ in simplest form? Explain your answer.
4. Use the information given in the introductory problem to find the time it took for the Earth-based stations to receive the signals from *Pioneer 10*.

**Guided Practice**

**Simplify. Assume no variable equals 0.**

5. $\dfrac{x^5}{x^3}$

6. $\dfrac{n^4}{n^4}$

7. $\dfrac{r^4}{r}$

8. $\dfrac{t^6}{t^8}$

9. $\dfrac{5y^{10}}{y^{13}}$

10. $\dfrac{1}{m^{-2}}$

11. $\dfrac{3^{-3}}{3^{-2}}$

12. $\left(\dfrac{1}{2}\right)^{-2}$

13. $\left(\dfrac{2}{3}\right)^0$

14. $\left(\dfrac{3}{b}\right)^6$

15. $\left(\dfrac{1}{10}\right)^{-4}$

16. $\left(\dfrac{k}{4}\right)^{-3}$

# EXERCISES

**Practice**

**Simplify. Assume no variable equals 0.**

17. $t^{-2}t^4$

18. $m^{-8}m^3$

19. $\dfrac{12x^8}{4x^3}$

20. $\dfrac{an^6}{n^5}$

21. $\dfrac{-24s^8}{2s^5}$

22. $\dfrac{6mn^2}{3m}$

23. $\dfrac{xy^7}{x^4}$

24. $\dfrac{48a^8}{12a^{11}}$

25. $\dfrac{4z^3}{28z^5}$

26. $\dfrac{-15r^4}{30r^3}$

27. $\dfrac{12b^4}{60b^6}$

28. $\dfrac{2x^{-3}}{6(x^2)^2}$

29. $\dfrac{16b^6c^5}{4b^4c^2}$

30. $\dfrac{8(k^{-2})^2}{4k^{-2}}$

31. $\dfrac{1}{x^0 + y^0}$

32. $\dfrac{-27w^3t^7}{-3w^3t^{12}}$

33. $\dfrac{-15r^5s^2}{5r^5s^{-4}}$

34. $\dfrac{8}{m^0 + n^0}$

35. $\dfrac{-2c^3d^6}{24c^2d^2}$

36. $\dfrac{(3c^2)^2(-d^5)}{-45c^7d^3}$

37. $\dfrac{20a^5b^9}{20ab^7}$

38. $\dfrac{16s^6t^5}{(2s^2t)^2}$

39. $\dfrac{3^{xy+5}}{3^{xy}}$

40. $\dfrac{s^{3x}}{s^{3x-2}}$

41. $\dfrac{5^{2x}}{5^{2x+2}}$

42. $(m^4n^5)^{-2}$

43. $36a^3b^5(12a^2b^2)^{-1}$

44. $3^3x^3y^3(3x)^{-2}$

45. $\left(\dfrac{a}{b^{-1}}\right)^{-1}$

46. $\left(\dfrac{x}{y^{-1}z^2}\right)^{-1}$

47. $\left(\dfrac{1}{5}\right)^{-2} + \left(\dfrac{1}{4}\right)^{-1}$

48. $\dfrac{-15m^5n^8(m^3n^2)}{45m^4n}$

49. $\dfrac{(-2t^3)^2(t^{-2})^{-1}}{(t^2)^{-3}}$

50. $\dfrac{(4x^3y)(4^2x^{-1}y)}{4^3xy^2}$

51. $\left(\dfrac{-3y^4}{2y^2}\right)^{-2}$

52. $\left(\dfrac{1}{2}\right)^{-2} + \left(\dfrac{1}{3}\right)^2$

**Evaluate. Express each answer in both scientific and decimal notation.**

53. $\dfrac{8 \times 10^{-1}}{16 \times 10^{-2}}$

54. $\dfrac{15 \times 10^4}{6 \times 10^{-2}}$

55. $(4.5 \times 10^3)(7.5 \times 10^2)^{-1}$

56. $(6.9 \times 10^3)(1.4 \times 10^3)^{-1}$

57. $\dfrac{0.000000036}{0.00011}$

58. $\dfrac{5,600,000,000}{60,000}$

59. $\dfrac{(84,000,000)(0.00004)}{0.0016}$

60. $\dfrac{(93,000,000)(0.0005)}{0.0015}$

**Critical Thinking**

61. Express the quotient $\dfrac{x + x^2 + x^3 + x^4 + x^5 + x^6 + x^7}{x^{-3} + x^{-4} + x^{-5} + x^{-6} + x^{-7} + x^{-8} + x^{-9}}$ in simplest form. Assume that $x$ is not equal to zero.   *Hint: Simplify the denominator first.*

**Applications**

62. **Astronomy**   Earth has an average distance of $1.496 \times 10^8$ kilometers from the Sun. If light travels $3.00 \times 10^5$ kilometers per second, how long does it take sunlight to reach Earth?

**Journal**

Complete this sentence. "Mathematics is important in science because ___."

63. **Communication**   Television signals travel at the speed of light. If the speed of light is $3.00 \times 10^5$ kilometers per second, how long would it take for signals broadcasting from a television station to reach a house 36 kilometers away?

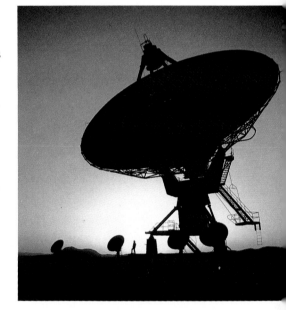

**Mixed Review**

**Evaluate. Express each answer in both scientific and decimal notation. (Lesson 5-1)**

64. $(5 \times 10^6)(9 \times 10^2)$

65. $(1.5 \times 10^3)^4$

66. If $A = \begin{bmatrix} 2 & -3 \\ 1 & 4 \end{bmatrix}$ and $B = \begin{bmatrix} -4 & 0 \\ 2 & 5 \end{bmatrix}$, evaluate $AB$. Write "not defined" if the product does not exist.   (**Lesson 4-4**)

67. In which octant does the point $(5, -1, 9)$ lie?   (**Lesson 3-8**)

68. State the multipliers you would use to eliminate each of the variables from the system $\begin{cases} 3x + 4y = 7 \\ 4x - 3y = 1 \end{cases}$ by addition.   (**Lesson 3-2**)

69. Write the equation $\dfrac{1}{4}x = 2y - 1$ in standard form.   (**Lesson 2-2**)

70. Evaluate $2|-3x| - 9$ if $x = 5$.   (**Lesson 1-6**)

# 5-3 Problem-Solving Strategy: Draw a Diagram

**Objective**

After studying this lesson, you should be able to:

- solve problems by using a diagram.

Many times you write an equation to solve a math problem. Drawing a diagram of the situation described in a problem can be helpful. You can use the diagram to help you write an equation or devise another plan to solve the problem. You will draw diagrams to help you work problems with polynomials in the next lesson.

**Example 1**

**APPLICATION**

**Horticulture**

**The New Bern Garden Club has a square garden in the city park. The garden is divided into five rectangular flower beds that are as long as one side of the square. The perimeter of each rectangle is 60 meters. If the club were to build a fence around the entire garden, how much fencing would they need to buy?**

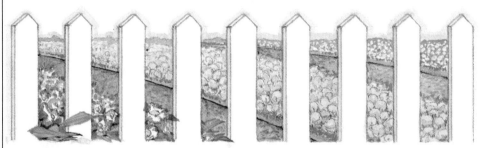

Draw a diagram of the garden to help find a solution.

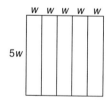

*Let w represent the width of a rectangle. The length of a rectangle is 5w since the figure is a square.*

Now, we can write an equation to represent the perimeter of one rectangle.

$$w + 5w + w + 5w = 60$$
$$12w = 60$$
$$w = 5 \qquad \text{The width of each rectangle is 5 meters.}$$

Each side of the square is $5w$ meters, so the perimeter of the square is $4(5w)$ or 100 meters. The club would need to buy 100 meters of fencing material to enclose the garden.

Example 2

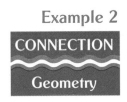

CONNECTION

Geometry

Suppose that you have a square piece of paper on which you draw the largest possible circle. You cut the circle out and discard the leftover scraps of paper. Inside the circle you draw the largest possible square, cut it out, and discard the leftover scraps of paper. How much of the original square do you have left?

Draw a diagram. The shaded region represents the square of paper that is left after all of the scraps have been discarded. If we separate the remaining paper into four congruent triangles as shown by the dotted lines, we can see that one-half of each of the regions remains. So one-half of the original square is left.

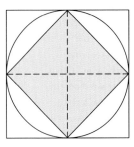

# CHECKING FOR UNDERSTANDING

Communicating Mathematics

Read and study the lesson to answer these questions.

1. How much fencing would the Garden Club have had to buy if the rectangular flower beds had a perimeter of 42 meters?

2. What is the length of a diagonal of the remaining square of paper as described in Example 2?

3. If you were to repeat the process described in Example 2 using the remaining square, how much of the original square would still remain?

Guided Practice

Solve. Draw a diagram.

4. Lee and Susan can wallpaper a 25-square-foot wall in one hour. At that same rate, how long would it take them to wallpaper an area of 5 square feet?

5. Six equilateral triangles are placed together to form a hexagon. The length of a side on one triangle is 2 inches. What is the diameter of the smallest circle that includes all six vertices of the hexagon?

6. Carole was traveling from Asheville to Indianapolis by bus. After half the trip, she fell asleep. When Carole awoke, she had half of the distance she traveled when asleep yet to go. For what fraction of the trip was Carole asleep?

# EXERCISES

**Practice**

### Strategies

Look for a pattern.
Solve a simpler problem.
Act it out.
Guess and check.
Draw a diagram.
Make a chart.
Work backwards.

**Solve. Use any strategy.**

7. How many different acute angles can be traced using the rays in the figure at the right?

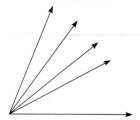

8. You scored 5 points for every correct answer on a math test. For each incorrect answer 2 points were deducted. The test consisted of 15 questions. If you attempted every problem and received a score of 61, how many questions did you answer correctly?

9. Draw an array of dots as shown at the right. Without lifting your pencil, draw four straight lines to pass through all of the dots.

10. Find the digit represented by each different letter in the problem at the right.

$$\begin{array}{r} ABC \\ ABC \\ + ABC \\ \hline BBB \end{array}$$

11. In how many ways can a straight line separate a square into two congruent regions?

12. A parallelogram has consecutive sides with lengths 9 m and 7 m. The measures of the diagonals are integers. How long are the diagonals?

13. Express 96 as the difference of two squares. There are four different possibilities.

14. An isosceles triangle has a base 10 cm long and two sides 13 cm long. A different triangle with two sides 13 cm long has the same area as the first triangle. What is the length of the base of that second triangle?

15. A square barn is 40 feet on each side. The tether for a horse is attached to one corner of the barn as shown at the right. If the rope is 25 feet long, how many square feet of land is the horse able to graze?

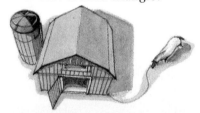

## COOPERATIVE LEARNING ACTIVITY

**Work in groups. Each person in the group must understand the solution and be able to explain it to any person in class.**

Toothpicks are used to lay out a grid like the outlines of the squares on a checkerboard. The grid is *m* toothpicks long and *n* toothpicks wide. How many toothpicks are used?

222 CHAPTER 5 POLYNOMIALS

## 5-4 Polynomials

**Objectives**

After studying this lesson, you should be able to:

- add polynomials,
- subtract polynomials, and
- multiply polynomials.

**Application**

After their project on ecology, the earth science class began a recycling program at Woodfield High School. Tim decided to place a box to collect paper to be recycled next to the library copy machine. The piece of cardboard he chose to make the box was a $y \times y$ inch square. From each corner of the cardboard, he cut a square $x$ inches on a side. Using the diagram, we can see that the area of the remaining cardboard is $y^2 - 4x^2$ square inches.

The expression $y^2 - 4x^2$ is a **polynomial.** A polynomial is a monomial or a sum of monomials. $4 + \dfrac{3}{x}$ is not a polynomial. Why?

*Remember a difference can be written as a sum.*

The monomials that make up a polynomial are called the **terms** of the polynomial. A polynomial with two unlike terms is called a **binomial** and one with three terms is a **trinomial.** The **degree of a polynomial** is the degree of the monomial with greatest degree.

**Example 1**

Find the degree of $8a^3 - 5a^2b^2 - 4ab^2 + b - 1$.

*The terms of polynomials are usually arranged so that the powers of one of the variables are in descending order. In this case, the variable is a.*

$8a^3 - 5a^2b^2 - 4ab^2 + b - 1 = 8a^3 + (-5a^2b^2) + (-4ab^2) + b + (-1)$

$8a^3$ has degree 3.          $-5a^2b^2$ has degree 4.          $-4ab^2$ has degree 3.
$b$ has degree 1.          $-1$ has degree 0.

The highest degree is 4. The degree of the polynomial is 4.

**Example 2**

Simplify $4x^2y - 2xy^3 + y + 6xy^3 - x^2y + 9y$.

$4x^2y - 2xy^3 + y + 6xy^3 - x^2y + 9y$
$= (4x^2y - x^2y) + (-2xy^3 + 6xy^3) + (y + 9y)$
$= (4 - 1)x^2y + (-2 + 6)xy^3 + (1 + 9)y$
$= 3x^2y + 4xy^3 + 10y$

We can simplify a polynomial using the distributive property, as we have just seen, or by adding or subtracting the coefficients of like terms. Study Example 3 to see how to simplify by adding or subtracting coefficients.

**Example 3**

Simplify $(4s^2 + 7st - 2t^2) - (2s^2 + 8st - 5t^2)$.

$$(4s^2 + 7st - 2t^2) - (2s^2 + 8st - 5t^2) = 4s^2 - 2s^2 + 7st - 8st - 2t^2 + 5t^2$$
$$= 2s^2 - st + 3t^2$$

The distributive property is also useful in multiplying monomials.

**Example 4**

Find $5a(5ab^2 + 2a^2b^2 - 9b)$.

$$5a(5ab^2 + 2a^2b^2 - 9b) = 5a \cdot 5ab^2 + 5a \cdot 2a^2b^2 - 5a \cdot 9b$$
$$= 25a^2b^2 + 10a^3b^2 - 45ab$$

**Example 5**

Find $(x + 2)(x + 10)$.

$$\begin{aligned}(x + 2)(x + 10) &= (x + 2)x + (x + 2)10 \quad \textit{Distribute } x + 2.\\ &= (x \cdot x) + (2 \cdot x) + (x \cdot 10) + (2 \cdot 10)\\ &= x^2 + 2x + 10x + 20\\ &= x^2 + 12x + 20\end{aligned}$$

The **FOIL method** is an application of the distributive property that makes multiplying binomials faster. Take another look at Example 5 using the FOIL method.

$$\begin{aligned}(x + 2)(x + 10) &= x \cdot x + x \cdot 10 + 2 \cdot x + 2 \cdot 10\\ &= x^2 \quad + \quad 12x \quad + 20\end{aligned}$$

| *FOIL method of Multiplying Polynomials* | **The product of two binomials is the sum of the products of**<br>**F  the *first* terms,**<br>**O  the *outer* terms,**<br>**I  the *inner* terms, and**<br>**L  the *last* terms.** |
|---|---|

**Example 6**

Use the FOIL method to find $(9a - 3)(a + 4)$.

$$\begin{array}{cccc} & F & O & I & L \end{array}$$
$$\begin{aligned}(9a - 3)(a + 4) &= 9a \cdot a + 9a \cdot 4 + (-3) \cdot a + (-3) \cdot 4\\ &= 9a^2 + 36a - 3a - 12\\ &= 9a^2 + 33a - 12\end{aligned}$$

**Example 7**

Show geometrically that $(a + b)^2 = a^2 + 2ab + b^2$.

Draw a diagram of a square that has sides of length $a + b$.

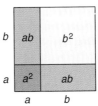

The area of the square can be expressed by $(a + b)^2$ or by $a^2 + 2ab + b^2$.

We can also extend the distributive property to multiply polynomials with more than two terms.

**Example 8**

Find $(2x^2 + 10x - 4)(x - 12)$.

$(2x^2 + 10x - 4)(x - 12)$
$= (2x^2 + 10x - 4)x - (2x^2 + 10x - 4)12$    *Distribute $2x^2 + 10x - 4$.*
$= 2x^2 \cdot x + 10x \cdot x - 4 \cdot x - 2x^2 \cdot 12 - 10x \cdot 12 + 4 \cdot 12$
$= 2x^3 + 10x^2 - 4x - 24x^2 - 120x + 48$
$= 2x^3 - 14x^2 - 124x + 48$

# CHECKING FOR UNDERSTANDING

**Communicating Mathematics**

Read and study the lesson to answer these questions.

1. Explain how the FOIL method and the distributive property are related.

2. Multiply $(2x + 3)$ by $(5x - 8)$ using the FOIL method.

3. Could you modify the FOIL method to find the product in Example 8? If so, how?

**Guided Practice**

Find the degree of each polynomial.

4. $4a^2 + 12ab$

5. $s^2 + 2s + 3$

6. $x^8 + 2x^7 - 3x^4 + 11x^3 + x - 9$

7. $3x^4y^2 - 5x^2y + 3$

8. $3r^5 - 3r^4 - 7r - 5$

9. $m^3 + 2mn^2 + 4n^3$

10. $5xy - 2x^2 - 3y^2$

11. $13xy^7 + 36x^3y^5 - 2x^4y^5 - xy$

Simplify.

12. $(4a + 2) + (2a + 6)$

13. $(5x + 6y) - (3x + 8y)$

14. $(12n^2 - 4n + 8) - (4n^2 - 1)$

15. $3p(p^2 - 2p + 3)$

# EXERCISES

**Practice**   **Find the area of each triangle.**

**16.**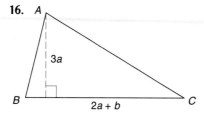

**17.**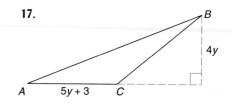

**Simplify.**

**18.** $(9x + 4y) + (7x - 2y)$
**19.** $(-3a + 5b) + (-6a + 8b)$

**20.** $(m^2 + 9m + 3) - (3m^2 + m + 2)$
**21.** $(3a^2 - 5d + 17) - (-a^2 + 5d - 3)$

**22.** $(3y^2 + 5y - 7) + (2y^2 - 7y + 10)$
**23.** $(8r^2 + 5r + 14) - (7r^2 + 6r + 8)$

**24.** $(10n^2 - 3nt + 4t^2) - (3n^2 + 5nt)$
**25.** $(-12y - 6y^2) + (-7y + 6y^2)$

**26.** $(x^3 - 3x^2y + 4xy^2 + y^3) - (7x^3 + x^2y - 9xy^2 + y^3)$

**27.** $4f(gf - bh)$
**28.** $-5mn^2(-3m^2n + 6m^3n - 3m^4n^4)$

**29.** $x^{-4}(x^2 + x - 3)$
**30.** $r^{-3}(r^5 - 2r^3 + r^{-1})$

**31.** $4a^{-1}b^2(a^2b^{-1} + 3a^3b^{-2} + 4^{-2}ab^{-1})$
**32.** $y^2x^{-3}(yx^4 + y^{-1}x^3 + y^{-2}x^2)$

**33.** $(x + 7)(x + 2)$
**34.** $(a + 5)(a - 7)$

**35.** $(s^2 + 5)(s^2 - 4)$
**36.** $(y^2 + y)(y^2 + 5)$

**37.** $(2x + 7)(3x + 5)$
**38.** $(3t - 8)(2t + 7)$

**39.** $(w^2 - 5)(2w^2 + 3)$
**40.** $(2x + 3y)(3x - 5y)$

**41.** $(6p - 5)(7p - 9)$
**42.** $(9y^2 - 1)(y + 2)$

**Find each product.**

**43.** $(a - b)^2$
**44.** $(a - b)(a + b)$

**45.** $(d + 3)^2$
**46.** $(r - 8)^2$

**47.** $(y - 2)^2$
**48.** $(y - 5)(y + 5)$

**49.** $(2p + q^3)^2$
**50.** $(x - 3y)^2$

**51.** $(4m - 3n)^2$
**52.** $(5r - 2)^2$

**53.** $(1 + 4r)^2$
**54.** $(x^3 - y)(x^3 + y)$

**Find the area of each figure.**

**55.**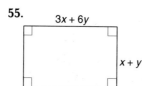

**56.**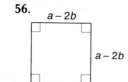

**57.**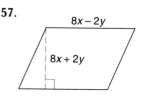

**Find each product.**

**58.** $(a + b)(a^2 - ab + b^2)$

**59.** $(2x - 3)(x^2 - 3x - 8)$

**60.** $(x - y)(x^2 + xy + y^3)$

**61.** $(m - 4)(3m^2 + 5m - 4)$

**62.** $r(r - 2)(r - 3)$

**63.** $(b + 1)(b - 2)(b + 3)$

**64.** $(2x - 3)(x + 1)(3x - 2)$

**65.** $(2a + 1)(a - 2)^2$

**66.** $(a - b)(a^2 + ab + b^2)$

**67.** $(2k + 3)(k^2 - 7k + 21)$

**68.** Write two different polynomials that represent the area of the figure at the right.

**Critical Thinking**

**69.** Recall that the measure of an angle inscribed in a circle is half the measure of its intercepted arc. That is, $m\angle B = \frac{1}{2}m\widehat{ADC}$. Given a circle with inscribed quadrilateral $ABCD$ with the given arc measures, find the ratio of $m\angle A$ to $m\angle B$.

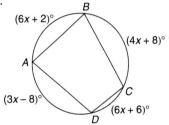

**Applications**

**70. Business** Dawn is writing a computer program to find the salaries of her employees after their annual raise. The percentage of increase is represented by $p$ in the program. Marty's salary is $23,450 now. Write a polynomial to represent Marty's salary in one year and one to represent Marty's salary after three years. Assume that the rate of increase will be the same for the next three years.

**71. Consumer Awareness** The Ready Rentals Car Company rents cars for $19.95 a day with 50 free miles. A charge of $0.25 is assessed for each mile driven over 50 miles. Express the cost of renting a car for one day as a polynomial. Assume that you drive more than 50 miles.

**Mixed Review**

72. **Manufacturing**  The Marysville Metalworks cuts the largest possible circle from a 4-inch square of tin to make the top for a soup can. How much metal do they have for scrap after the circle is cut?  **(Lesson 5-3)**

CONNECTION
Geometry

73. The length of rectangle $ABCD$ is twice its width. The perimeter is 48 meters. If $W$, $X$, $Y$, and $Z$ are the midpoints of the sides of $ABCD$, find the area of parallelogram $WXYZ$.  **(Lesson 5-3)**

74. Find the inverse for the matrix $\begin{bmatrix} 3 & 0 \\ -1 & 5 \end{bmatrix}$.  **(Lesson 4-5)**

75. Use Cramer's Rule to solve the following system of equations.
**(Lesson 3-3)**     $2x - 3y = 0$
                    $6x + 5y = 7$

76. Explain how the graphs of $y = [3x]$ and $y = 3[x]$ differ.  **(Lesson 2-7)**

77. Evaluate $\dfrac{3ab^2 - c^3}{a + c}$ if $a = 3$, $b = 7$, and $c = -2$.  **(Lesson 1-1)**

---

## MID-CHAPTER REVIEW

**Simplify.  (Lesson 5-1)**

1. $5c^2 + 8d^2 - 10c^2$

2. $3d^3 - 4d^3 + d^3$

3. $(4x^2y)^2(3x^2y)$

4. $\left(\dfrac{1}{3}x^2y\right)^3\left(\dfrac{2}{3}xy\right)^2$

5. Evaluate $(32{,}000{,}000)(48{,}000)$. Express the answer in both scientific and decimal notation.  **(Lesson 5-1)**

**Simplify.  (Lesson 5-2)**

6. $\dfrac{18b^4}{6b^2}$

7. $\dfrac{20m^5n^4}{10mn^2}$

8. $\left(\dfrac{1}{4y}\right)^{-1}$

9. $\dfrac{-3w^2t^3}{(4w^2)(6w^3t^4)}$

**Evaluate. Express each answer in both scientific and decimal notation.  (Lesson 5-2)**

10. $\dfrac{42 \times 10^5}{14 \times 10^{-2}}$

11. $\dfrac{682{,}000}{480}$

12. If 2 miles of fence will enclose a square field of 160 acres, how large of a field will 4 miles of fence enclose?  **(Lesson 5-3)**

**Simplify.  (Lesson 5-4)**

13. $(3x - 2y) - (4x + 7y)$

14. $(4d^2 - 2d + 8) - (6d^2 - 7d + 2)$

15. $(3y - 8)(2y + 9)$

16. $(x + 7)(x - 3)$

17. $(m + 2)^2$

18. $(2x - 3)^2$

19. $(x + 2)(x - 3)^2$

20. $(2y - 4)(y^2 + 2y - 7)$

# Factoring

Objective

After studying this lesson, you should be able to:

■ factor polynomials.

You know what it means to express a composite number as the product of its prime factors. We can also express a polynomial as the product of its prime factors. This way of expressing a polynomial is called its factored form. The first step in finding the factored form of a polynomial is to find the greatest common factor, or GCF, of the terms of the polynomial. Let's try factoring $12y^2 + 15y$.

First, write each term as the product of its prime factors.

$12y^2 = 2 \cdot 2 \cdot 3 \cdot y \cdot y$
$15y = 3 \cdot 5 \cdot y$

The greatest common factor of $12y^2$ and $15y$ is $3y$.

The distributive property states that $a(b + c) = ab + ac$. Factoring "undoes" the distribution. That is, $ab + ac = a(b + c)$.

$12y^2 + 15y = (3y \cdot 4y) + (3y \cdot 5)$
$= 3y(4y + 5)$

The factored form of $12y^2 + 15y$ is $3y(4y + 5)$.

A geometric way of interpreting factoring is to think of the area of a rectangle expressed as the product of the length and the width.

$3y$ | Area = $3y(4y + 5)$
or
$12y^2 + 15y$

$4y + 5$

Example 1

CONNECTION

Geometry

**Factor $24a^2b - 18ab^2$. Draw the geometric representation of the expression.**

$24a^2b - 18ab^2 = (2 \cdot 2 \cdot 2 \cdot 3 \cdot a \cdot a \cdot b) - (2 \cdot 3 \cdot 3 \cdot a \cdot b \cdot b)$
$= (6ab \cdot 4a) - (6ab \cdot 3b)$    *6ab is the GCF.*
$= 6ab(4a - 3b)$

$6ab$ | Area = $6ab(4a - 3b)$
or
$24a^2b - 18ab^2$

$4a - 3b$

The rectangle is $6ab$ units wide and $(4a - 3b)$ units long.

Example 2

**Factor $3s^2t + 4st^2 + st^3$.**

$3s^2t + 4st^2 + st^3 = (3 \cdot s \cdot s \cdot t) + (2 \cdot 2 \cdot s \cdot t \cdot t) + (s \cdot t \cdot t \cdot t)$
$= (st \cdot 3s) + (st \cdot 4t) + (st \cdot t^2)$    *st is the GCF.*
$= st(3s + 4t + t^2)$

Sometimes just factoring out the GCF does not result in a completely factored polynomial. In those cases, rearranging and grouping the terms may be helpful.

**Example 3**

Factor $a^2 - 2ab + a - 2b$.

$$
\begin{aligned}
a^2 - 2ab + a - 2b &= (a^2 - 2ab) + (a - 2b) \\
&= a(a - 2b) + 1(a - 2b) \\
&= (a - 2b)(a + 1)
\end{aligned}
$$

*Look at the terms in pairs.*
*Factor out the GCF of each.*
*$(a - 2b)$ is the GCF of the two new terms.*

In the last lesson you learned how to multiply binomials using the FOIL method. Let's see if we can find a way that the process can be reversed to factor a trinomial into its binomial factors.

$$
\begin{aligned}
(x + r)(x + s) &= x \cdot x + x \cdot s + r \cdot x + r \cdot s \\
&= x^2 + (r + s)x + rs
\end{aligned}
$$

Look at the coefficient of the second term, $r + s$. It is the sum of the two constant terms in the original binomials. The constant term in the product is the product of the constants from the original binomials. Now that we know where this coefficient and constant come from, factoring a trinomial like this will be easy.

**Example 4**

Factor $x^2 - 2x - 35$.

We need to find two numbers whose product is $-35$ and whose sum is $-2$. List the pairs of factors of $-35$ and look for the pair that has a sum of $-2$. Make a table to organize our search.

| Factors of $-35$ | Sum of Factors |
|:---:|:---:|
| $-1, 35$ | $34$ |
| $1, -35$ | $-34$ |
| $-5, 7$ | $2$ |
| $5, -7$ | $-2$ |

The two numbers are 5 and $-7$.     *Check this using the FOIL method.*
$x^2 - 2x - 35 = (x + 5)(x - 7)$

**Example 5**

Factor $x^2 - 49$.

We must find two factors of $-49$ that have a sum of 0. The only two that do are 7 and $-7$.

So, $x^2 - 49 = (x + 7)(x - 7)$.

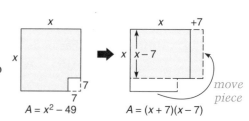

$A = x^2 - 49$     $A = (x + 7)(x - 7)$

Did you notice that the two terms in the original polynomial are perfect squares? The geometric representation is a square with a square cut out. This is a special case called **the difference of two squares.** The difference of two squares is simple to factor.

| Difference of Two Squares | For real numbers $a$ and $b$, $a^2 - b^2 = (a - b)(a + b)$. |
|---|---|

The method we have used on the last two examples works when the coefficient of $x^2$ is 1. Is there an easy way to factor trinomials when the coefficient of $x^2$ is not 1? Try using the FOIL method on two other binomials to see if you can find a rule.

$$(ax + b)(cx + d) = ax \cdot cx + ax \cdot d + b \cdot cx + b \cdot d$$
$$= acx^2 + (ad + bc)x + bd$$

Notice that the product of the *coefficient* of $x^2$ and the *constant* term is *abcd*. The product of the two coefficients of the $x$ term, *bc* and *ad*, is also *abcd*.

**Example 6**

Factor $2x^2 - 11x - 21$.

The product of the coefficient and the constant term is $2 \cdot -21$ or $-42$. So, the two coefficients of $x$ must have a sum of $-11$ and a product of $-42$. They must be 3 and $-14$, since $3 + (-14) = -11$ and $3(-14) = -42$.
*Check to see that this is the only pair that meets the criteria.*

Now, rewrite the expression using $3x$ and $-14x$ in place of $-11x$ and factor by grouping.

$$\begin{aligned} 2x^2 - 11x - 21 &= 2x^2 + (3x - 14x) - 21 & \textit{Substitute } (3x - 14x) \textit{ for } -11x. \\ &= (2x^2 + 3x) + (-14x - 21) & \textit{Group the terms.} \\ &= x(2x + 3) - 7(2x + 3) & \textit{Find the GCF of each group.} \\ &= (2x + 3)(x - 7) & \textit{Distributive property} \end{aligned}$$

**Example 7**

Factor $16r^2 - 24r + 9$.

$$\begin{aligned} 16r^2 - 24r + 9 &= 16r^2 - 12r - 12r + 9 & \textit{-12 and -12 have a sum} \\ &= 4r(4r - 3) - 3(4r - 3) & \textit{of -24 and a product of} \\ &= (4r - 3)(4r - 3) \text{ or } (4r - 3)^2 & \textit{16 } \cdot \textit{ 9 or 144.} \end{aligned}$$

Example 7 shows one of the special cases that occur when factoring. It is a **perfect square trinomial,** which is a square of a binomial. The other special case is **the sum** or **difference of two cubes.** The methods for factoring these special cases are listed on the next page.

| Perfect Square Trinomials | **For any numbers *a* and *b*,** $a^2 + 2ab + b^2 = (a + b)^2$, and $a^2 - 2ab + b^2 = (a - b)^2$. |
|---|---|
| Sum or Difference of Two Cubes | **For any numbers *a* and *b*,** $a^3 + b^3 = (a + b)(a^2 - ab + b^2)$, and $a^3 - b^3 = (a - b)(a^2 + ab + b^2)$. |

**Example 8**

**Factor $x^3 + 64$.**

$$x^3 + 64 = x^3 + (4)^3$$
$$= (x + 4)(x^2 - x \cdot 4 + 4^2)$$
$$= (x + 4)(x^2 - 4x + 16)$$

*Use the pattern*
$a^3 + b^3 = (a + b)(a^2 - ab + b^2)$.

Use the following steps to factor a polynomial.

1. Find the greatest common factor of the terms and factor it out.
2. Check for special products.
   a. If there are *two terms*, look for a difference of two squares, difference of two cubes, or sum of two cubes.
   b. If there are *three terms*, look for a perfect square trinomial.
3. Try other factoring methods.
   a. If there are *three terms*, try the trinomial pattern.
   b. If there are *four or more terms*, try grouping.

**Example 9**

**Factor $3x^3 + 6x^2 - 3x - 6$.**

$$3x^3 + 6x^2 - 3x - 6 = 3(x^3 + 2x^2 - x - 2)$$     *3 is the GCF.*
$$= 3[(x^3 + 2x^2) + (-x - 2)]$$     *Group the terms.*
$$= 3[x^2(x + 2) - 1(x + 2)]$$     *Factor each group.*
$$= 3(x + 2)(x^2 - 1)$$     *Factor out $x + 2$.*
$$= 3(x + 2)(x + 1)(x - 1)$$     *$x^2 - 1$ is the difference of two squares.*

# CHECKING FOR UNDERSTANDING

**Communicating Mathematics**

**Read and study the lesson to answer these questions.**

1. What polynomial is represented by the area of rectangle at the right? What is its factored form?

$2x + 1$

Area = $2x^2 + 7x + 3$    $x + 3$

2. Draw a geometric representation of $(a + 2)^2$.

3. You know how to factor the difference of two squares and the sum or difference of two cubes, but to factor these you must recognize the squares and cubes. Use a calculator to help you make a list of the squares of the integers from 1 to 20 and the cubes of the integers from 1 to 10. Memorize the list so that you will be able to recognize special cases quickly.

4. Group the terms in Example 9 differently and factor. Did you get the same prime factorization?

## Guided Practice

**Factor.**

5. $3s + 3t$  
6. $8a - 2b$  
7. $ab + ac$  
8. $x^2 - x$  
9. $r^2 - 9$  
10. $x^2 - 25$  
11. $100 - m^2$  
12. $y^2 - 81z^2$  
13. $y(3y - 2) + 4k(3y - 2)$  
14. $2x^2 + 6y + 8b$  
15. $9p^2 - 3pq$  
16. $3m(m - 7) + k(m - 7)$

# EXERCISES

## Practice

**Factor.**

17. $-15x^2 - 5x$  
18. $s^2 - 6s + 8$  
19. $2ab(c - d) + 10d(c - d)$  
20. $a^2 + 5a + 6$  
21. $y^2 + 6y + 9$  
22. $a(y - b) - c(y - b)$  
23. $r^2 + 16r + 64$  
24. $x^2 + xy + 3x$  
25. $3a^2 + 6a + 9y$  
26. $a^4 + a^3b + a^2b^2$  
27. $5x^2y - 10xy^2$  
28. $49s^2 - 100$  
29. $x^3 + 8$  
30. $d^3 - 27$  
31. $f^2 - 18f + 81$  
32. $t^2 + 12t + 35$  
33. $s^2 + 12s + 36$  
34. $a^2 + 4ab + 4b^2$  
35. $4x^2 - 9$  
36. $3y^2 + 5y + 2$  
37. $4s^2 - 20s + 21$  
38. $3d^2 - 48$  
39. $x^3 + 2x^2 - 35x$  
40. $4s^2 - 20st + 4t^2$  
41. $p^2 - 4bp + 4b^2$  
42. $6d^2 + 33d - 63$  
43. $f^3 - 1$  
44. $x^4 - 13x^2 + 36$  
45. $(x + y)^2 - \dfrac{1}{4}$  
46. $2r^3 - 16s^3$  
47. $m^2 - k^2 + 6k - 9$  
48. $16y^4 - z^4$  
49. $a + b + 3a^2 - 3b^2$  
50. $a^3b^3 - 27$  
51. $1 - 8m^6$  
52. $4a^2 + 4ab - y^2 + b^2$

## Critical Thinking

53. Factor $a^{2n} - 64$.

54. Factor $x^{3n} - y^{3n}$.

55. Find $\dfrac{3}{a + b}$ if $\left(\dfrac{3}{a + b}\right)^2 - \dfrac{6}{a + b} + 1 = 0$.

## Applications

56. **Manufacturing** A square boat cover is designed as shown at the right. Express the area of the blue region in factored form.

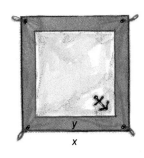

57. **Interior Design** A designer preparing to paint a wall found that its area is $3x^2 - 2x - 5$ square feet. Find the length and width of the rectangular wall. Draw and label a diagram to represent the wall.

**Use the FOIL method to find each product. (Lesson 5-4)**

**58.** $(s + 3)^2$

**59.** $(2x + 4)(7x - 1)$

**60.** State the system of equations represented by the matrix equation
$\begin{bmatrix} 3 & 0 \\ 1 & -2 \end{bmatrix} \begin{bmatrix} x \\ y \end{bmatrix} = \begin{bmatrix} 12 \\ 8 \end{bmatrix}$. **(Lesson 4-7)**

**61.** Find the dimension of matrix $M$ if $A_{3 \times 3} \cdot B_{3 \times 5} = M$. **(Lesson 4-2)**

**62.** Given $f(x, y) = 1.2x - 0.8y$ use a calculator to find the value of $f(0.1, -0.3)$. **(Lesson 3-6)**

**63.** Is $\{(9, 0), (-1, 0), (0, 0), (0, 1), (11, -11)\}$ a function? **(Lesson 2-1)**

**64.** Solve the equation $\frac{3}{4}t + 1 = 10$. **(Lesson 1-3)**

## BIOLOGY CONNECTION

### Genetics and Population Predictions

Reproductive cells contain pairs of chromosomes that contain the genetic code in the form of genes. When the chromosomes split and recombine with other chromosomes, pairs of genes are formed containing codes that may be pure dominant, pure recessive, or hybrid. The Punnett squares shown on page 208 demonstrate these combinations, called genotypes.

Suppose that $p$ represents the ratio of dominant gene $A$ in a population and $q$ represents the ratio of recessive gene $a$ in a population. The next generation is the result of $(p + q)^2$. This results in $p^2$ pure dominant genotypes ($AA$), $q^2$ pure recessive genotypes ($aa$), and $2pq$ hybrid genotypes ($Aa$). Since all members of the population must have at least one recessive gene or dominant gene present in their genotypes, $p + q = 1$.

**Application** In the population of a village, the recessive left-handedness gene ($r$) had a frequency of 1:4, and the dominant right-handedness gene ($R$) had a frequency of 3:4. In the next generation, what would you predict the population genotypes to be?

$p = \frac{3}{4}$, ratio of the dominant gene $\qquad$ $q = \frac{1}{4}$, ratio of the recessive gene

| right-handed (pure) | right-handed (hybrid) | left-handed (pure) |
|---|---|---|
| $p^2 = \left(\frac{3}{4}\right)^2$ or $\frac{9}{16}$ | $2pq = 2\left(\frac{3}{4}\right)\left(\frac{1}{4}\right)$ or $\frac{6}{16}$ | $q^2 = \left(\frac{1}{4}\right)^2$ or $\frac{1}{16}$ |
| 9 out of 16 have $RR$. | 6 out of 16 have $Rr$. | 1 out of 16 have $rr$. |

In the next generation, 15 of 16 would be expected to be right-handed.

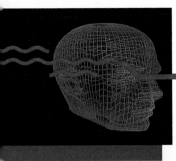

# Technology
## Factoring

The examples below show how to use the *Mathematical Exploration Toolkit (MET)* to factor polynomials. The computer can find common monomial factors, linear factors, and quadratic factors of polynomials in one variable. You will need the CALC commands listed below. Their shortened forms are shown in parentheses. To enter an exponent, you must use the ^ symbol. For example, $x^2$ is entered as $x^2$.

FACTOR (fac)        FSTEPS (fst)        SIMPLIFY (simp)

The FACTOR command will find only the common monomial factor in polynomials with more than one variable. In polynomials of only one variable, it will factor the expression completely. The FSTEPS command displays the steps used to perform the most recent FACTOR command.

**Example 1**   Factor $6x^3y^4 - 9xy^2 + 12xy$.

ENTER:   $6x^3y^4 - 9xy^2 + 12xy$
fac

```
6 x^3 y^4 - 9 x y^2 + 12 x y
3 x y ( 2 x^2 y^3 - 3 y + 4 )
```

**Example 2**   Factor $2p(3q + 6) + 3pq(4p - 2)$.

ENTER:   $2p(3q+6)+3pq(4p-2)$
simp
fac

```
2 p ( 3 q + 6 ) + 3 p q ( 4 p - 2 )
12 p^2 q + 12 p
12 p ( p q + 1 )
```

**Example 3**   Factor $z^7 - 256z^3$.

ENTER:   $z^7 - 256z^3$
fac

```
z^7 - 256 z^3
z^3 ( z - 4 ) ( z + 4 ) ( z^2 + 16 )
```

*Suppose you use* fst *after entering the polynomial.*

```
z^7 - 256 z^3
( z^5 - 16 z^3 ) * ( z^2 + 16 )
( z^4 - 4 z^3 ) * ( ( ( z + 4 ) ( z^2 + 16 ) ) )
( z^3 ) * ( ( z - 4 ) ( ( z + 4 ) ( z^2 + 16 ) )
z^3 ( z - 4 ) ( z + 4 ) ( z^2 + 16 )
```

## EXERCISES

Use FACTOR and FSTEPS to factor each polynomial. Record each step.

1. $96a^3b^3 - 80a^2b + 112ab$

2. $15m(n^2 + 10m) - 20m^2$

3. $9x^4 - 144$

4. $x^9 + 1$

## 5-6 Dividing Polynomials

**Objective**

After studying this lesson, you should be able to:

■ divide polynomials using factoring and long division.

**Application**

Yū Kamin is a genetic engineer working on a vaccine for influenza. The number of people in a small town who catch influenza during an epidemic is estimated to be $n = \dfrac{170t^2}{t^2 + 1}$, where $n$ represents the number of people and $t$ represents the number of weeks from the beginning of the epidemic. After one week of an epidemic, $\dfrac{170(1)^2}{(1)^2 + 1}$ or 85 people would have influenza.

Sometimes it is necessary to divide polynomials by monomials or by other polynomials to solve problems. The properties of exponents that you learned in Lesson 5-2 will be very helpful in this process.

**Example 1**

Divide $25a^4$ by $5a^2$.

$$25a^4 \div 5a^2 = \frac{25a^4}{5a^2} \qquad \text{\textit{A denominator cannot have a value of 0, so } a \neq 0.}$$

$$= \frac{25}{5} \cdot a^{4-2} \qquad \text{\textit{Quotient of powers property}}$$

$$= 5a^2$$

**Example 2**

Simplify $\dfrac{8x^2y^3 - 28x^3y^2}{4xy^2}$.

$$\frac{8x^2y^3 - 28x^3y^2}{4xy^2} = \frac{8x^2y^3}{4xy^2} - \frac{28x^3y^2}{4xy^2} \qquad \text{\textit{x}} \neq 0, \text{\textit{y}} \neq 0$$

$$= \frac{8}{4} \cdot x^{2-1}y^{3-2} - \frac{28}{4} \cdot x^{3-1}y^{2-2}$$

$$= 2xy - 7x^2 \qquad \text{\textit{y}}^{2-2} = \text{\textit{y}}^0 = 1$$

$$= x(2y - 7x)$$

We can use a process similar to long division to divide a polynomial by a polynomial. When doing the division, remember that you must have like terms to add and subtract.

**Example 3**

Divide $9b^2 + 9b - 10$ by $3b - 2$.

$$\begin{array}{r} 3b \phantom{+9b-10} \\ 3b-2 \overline{) 9b^2 + 9b - 10} \\ \underline{9b^2 - 6b} \phantom{-10} \\ 15b - 10 \end{array}$$

➡

$$\begin{array}{r} 3b + 5 \phantom{0} \\ 3b-2 \overline{) 9b^2 + 9b - 10} \\ \underline{9b^2 - 6b} \phantom{-10} \\ 15b - 10 \\ \underline{15b - 10} \\ 0 \end{array}$$

**Example 4**

Simplify $(5m^2 - 34m - 7)(m - 7)^{-1}$. $\quad m \neq 7$

$$(5m^2 - 34m - 7)(m - 7)^{-1} = \frac{5m^2 - 34m - 7}{m - 7}$$

$$= \frac{(5m + 1)(m - 7)}{m - 7} \qquad \textit{Factor the numerator.}$$

$$= 5m + 1 \qquad \textit{Simplify } \frac{m-7}{m-7}.$$

**Example 5**

CONNECTION
Geometry

The area of rectangle $ACED$ is represented by $6x^2 + 38x + 56$. Its width is represented by $2x + 8$. Point $B$ is the midpoint of $AC$. $ABFG$ is a square. Find the length of rectangle $ACED$ and the area of square $ABFG$.

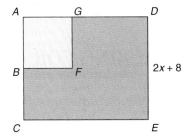

Divide $6x^2 + 38x + 56$ by $2x + 8$ to find the length of the rectangle.

$$\begin{array}{r} 3x + 7 \phantom{0} \\ 2x+8 \overline{) 6x^2 + 38x + 56} \\ \underline{6x^2 + 24x} \phantom{+56} \\ 14x + 56 \\ \underline{14x + 56} \\ 0 \end{array}$$

The length of rectangle $ACED$ is $3x + 7$.

The length of one side of square $ABFG$ is $\dfrac{2x + 8}{2}$ or $x + 4$. The area of the square is $(x + 4)^2$ or $x^2 + 8x + 16$.

**Example 6**

APPLICATION
Health

Simplify the formula for the number of people who catch influenza during an epidemic, $n = \dfrac{170t^2}{t^2 + 1}$, as given in the beginning of the lesson.

$$n = \frac{170t^2}{t^2 + 1}$$

$$= 170t^2 \div (t^2 + 1)$$

➡

$$= 170 - \frac{170}{t^2 + 1}$$

$$\begin{array}{r} 170 \phantom{00000000} \\ t^2+1 \overline{) 170t^2 + 0t + \phantom{0}0} \\ \underline{170t^2 \phantom{+0t} + 170} \\ -170 \end{array}$$

*What happens to the value of n as t becomes very great?*

As you know from long division with real numbers, the divisor is a factor of the dividend if the remainder upon division is zero. The example below uses this fact to prove that a binomial is a factor of the polynomial.

**Example 7**

Show that $4 - n$ is a factor of $n^3 - 6n^2 + 13n - 20$.

$$
\begin{array}{r}
-n^2 + 2n - 5 \\
-n + 4\overline{)\,n^3 - 6n^2 + 13n - 20} \\
\underline{n^3 - 4n^2} \\
-2n^2 + 13n \\
\underline{-2n^2 + 8n} \\
5n - 20 \\
\underline{5n - 20} \\
0
\end{array}
$$

$4 - n = -n + 4$

The remainder is 0, so $4 - n$ is a factor of $n^3 - 6n^2 + 13n - 20$.

# CHECKING FOR UNDERSTANDING

**Communicating Mathematics**

Read and study the lesson to answer these questions.

1. About how many people will have influenza after 6 weeks of an epidemic?

2. Why is $m \neq 7$ in Example 4?

3. What can you conclude about the dividend and divisor if the remainder is zero?

4. When is it easier for you to divide using factoring instead of the long division technique?

**Guided Practice**

Simplify.

5. $\dfrac{5ab^2 - 4ab + 7a^2b}{ab}$

6. $\dfrac{x^3y^2 - x^2y + 2x}{-xy}$

7. $\dfrac{6r^2s^2 + 3rs^2 - 9r^2s}{3rs}$

8. $\dfrac{3(x - 7)^6}{(x - 7)^{10}}$

9. $\dfrac{2(y^2 - 5)^3}{8(y^2 - 5)^5}$

10. $\dfrac{2(t + 3)^4}{10(t + 3)^2}$

11. $\dfrac{c^2 - c - 30}{c - 6}$

12. $\dfrac{m^2 + 8m + 16}{m + 4}$

13. $(w^2 - w^3)(w^2 - 1)^{-1}$

14. $(a^3 - b^3)(a - b)^{-2}$

# EXERCISES

**Simplify.**

**15.** $\dfrac{12pq^3 + 9p^2q^2 - 15p^2q}{3pq}$

**16.** $\dfrac{6m^4n^2 + 4m^2n + 5mn^3}{-mn}$

**17.** $\dfrac{28k^3py - 42kp^2y^2 + 56kp^3y^2}{14kpy}$

**18.** $\dfrac{15r^2s + 23rs^2 + 6s^2}{3rs}$

**19.** $(a^2 - 5a - 84)(a + 7)^{-1}$

**20.** $(2r^2 + 5r - 3) \div (r + 3)$

**21.** $(n^2 - 12n - 45) \div (n + 3)$

**22.** $(6y^2 + 7y - 3)(2y + 3)^{-1}$

**23.** $(2x^2 + x - 16) \div (x - 3)$

**24.** $(8g^2 - 18g - 9)(g - 3)^{-1}$

**25.** $(s^2 + 4s - 16) \div (6 - s)$

**26.** $(8x^2 - 4x + 11)(x + 5)^{-1}$

**27.** $(6y^3 + 13y^2 + y - 2)(3y - 1)^{-1}$

**28.** $(56c^2 - 113c + 59) \div (8c - 7)$

**29.** $(6x^3 + 5x^2 + 9) \div (2x + 3)$

**30.** $(y^3 - 1) \div (y - 1)$

**31.** $(r^3 - 9r^2 + 27r - 28) \div (r - 3)$

**32.** $(m^3 - 7m + 3m^2 - 21)(m^2 - 7)^{-1}$

**33.** $(6x^3 - 5x^2 - 12x - 4) \div (3x + 2)$

**34.** $(x^2 + 4x - 4) \div (x + 2)$

**35.** $(2t^3 - 2t - 3) \div (t - 1)$

**36.** $(v^4 + 4) \div (v^2 - 2v + 2)$

**37.** $(s^3 - 8) \div (s - 2)$

**38.** $(y^4 + 4y^3 + 10y^2 + 12y + 9) \div (y^2 + 2y + 3)$

**39.** $(x^4 - 4x^2 + 12x - 9) \div (x^2 + 2x - 3)$

**40.** Is $3y - 2$ a factor of $6y^3 - y^2 - 5y + 2$?

**41.** One factor of $a^3 - 2a^2 - a + 2$ is $a - 2$. Find the other factors.

**42.** Find the remainder when dividing $x^2 + 3x + 5$ by $x - 2$. If $f(x) = x^2 + 3x + 5$, find $f(2)$. Compare the answers.

**43.** Find the value of $k$ so that the remainder upon dividing $(x^2 + 8x + k)$ by $(x - 4)$ is zero.

**44.** Simplify $\left(\dfrac{y^2 + 2y - 15}{y^2 + 3y - 10}\right)\left(\dfrac{y^2 - 9}{y^2 - 9y + 14}\right)^{-1}$.

**45.** Suppose that the quotient upon dividing one polynomial by another is $3x^2 - x + 32 - \dfrac{121}{x + 4}$. What is the dividend?

**46.** **Entertainment**   A magician asked a member of his audience to choose any number. He said "Multiply your number by 3. Add the sum of your number and 8 to that result. Now divide by the sum of your number and two." The magician announced the final answer without asking the original number. What was the final answer and how did he know what it was?

**47. Manufacturing** A machinist who makes square metal pipes found a formula for the amount of metal she needed to make a pipe. She found that to make a pipe $8x$ inches long she needed $32x^2 + x$ square inches of metal. In figuring the area needed, the machinist allowed some fixed length of metal for overlap of the seam. If the width of the finished pipe will be $x$ inches, how much did the machinist leave for the seam?

Metal needed

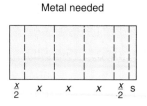

$s$ = width of seam

$\frac{x}{2}$ $\quad x \quad$ $\quad x \quad$ $\quad x \quad$ $\frac{x}{2}$ $\quad s$

Finished Pipe

**Portfolio**

Select some of your work from this chapter that shows how you used a calculator or computer. Place it in your portfolio.

**Mixed Review**

**Factor.** (Lesson 5-5)

**48.** $4a^2 - 16$

**49.** $d^2 - 11d - 26$

**50.** Simplify $\dfrac{1}{2x^0 + y^0}$. (Lesson 5-2)

**51. Manufacturing** Earthly Treasures Inc. makes wood chairs and tables. A chair requires 4 hours of cutting and 4 hours of assembly. A table requires 3 hours of cutting and 2 hours of assembly. Each week there are 40 hours of worker time available in the cutting department and 36 hours of worker time available in the assembly department. If the profits on chairs and tables are $28 and $22 respectively, how many of each item should the company produce for a maximum profit? (Lesson 3-7)

**52.** Determine which numbers in $\{-2, -1, 0, 1, 2\}$ are solutions of $|x| + x = 0$. (Lesson 1-6)

**53.** Name the sets of numbers to which $2.121221222 \ldots$ belongs. (Lesson 1-2)

---

## ~~~~~ HISTORY CONNECTION ~~~~~

What mathematical topics were there before algebra? Historians have relied on cave pictures from ancient civilizations to trace much of the development of our number system. Early man used his fingers to count, but found this inconvenient when numbers exceeded 10. Many caves in North America show piles of stones used as a counting tool. These piles often contain 5 or 10 stones which correspond to the numbers of fingers possible. The Maya Indians of South America had a number system based on 20, which indicates they also used their toes when counting.

# 5-7 Synthetic Division

**Objective**

After studying this lesson, you should be able to:

- divide polynomials using synthetic division.

In Lesson 5-6, you learned to divide a polynomial by another polynomial using long division. A simpler process called **synthetic division** has been devised to divide a polynomial by a binomial. Study the example below to see how the process works.

**Example 1**

**Divide $2x^3 - 7x^2 - 8x + 16$ by $x - 4$.**

Write the terms of the polynomial so that the degrees of the terms are in descending order. Then write just the coefficients as shown at the right.

$$2x^3 - 7x^2 - 8x + 16$$

$$2 \quad -7 \quad -8 \quad 16$$

Write the constant, $r$, of the divisor $x - r$ to the left. In this case, $r = 4$.

$$4 \underline{|\ 2 \quad -7 \quad -8 \quad 16}$$

Now bring the first coefficient down as shown.

$$4 \underline{|\ 2 \quad -7 \quad -8 \quad 16}$$
$$\quad\ \ 2$$

Multiply the first coefficient by $r$ and write the product under the second coefficient.

$$4 \underline{|\ 2 \quad -7 \quad -8 \quad 16}$$
$$\quad\quad\ \ 8$$
$$\quad\ \ 2$$

Add the product and the second coefficient. $-7 + 8 = 1$. Write the sum as shown.

$$4 \underline{|\ 2 \quad -7 \quad -8 \quad 16}$$
$$\quad\quad\ \ 8$$
$$\quad\ \ 2 \quad\ 1$$

Multiply the sum, 1, by $r$. Write the product under the next coefficient.

$$4 \underline{|\ 2 \quad -7 \quad -8 \quad 16}$$
$$\quad\quad\ \ 8 \quad\ 4$$
$$\quad\ \ 2 \quad\ 1$$

Add. $-8 + 4 = -4$. Write the sum as shown.

$$4 \underline{|\ 2 \quad -7 \quad -8 \quad 16}$$
$$\quad\quad\ \ 8 \quad\ 4$$
$$\quad\ \ 2 \quad\ 1 \quad -4$$

Repeat the process. Multiply the sum, $-4$ by $r$ and write the product under the next coefficient.

$$4 \underline{|\ 2 \quad -7 \quad -8 \quad 16}$$
$$\quad\quad\ \ 8 \quad\ 4 \quad -16$$
$$\quad\ \ 2 \quad\ 1 \quad -4$$

Add. $16 + (-16) = 0$. The remainder is 0.

$$4 \underline{|\ 2 \quad -7 \quad -8 \quad 16}$$
$$\quad\quad\ \ 8 \quad\ 4 \quad -16$$
$$\quad\ \ 2 \quad\ 1 \quad -4 \quad | \quad 0$$

Writing the result is easy. The numbers along the bottom row are the coefficients of the powers of $x$ in descending order. Start with the power that is one less than that of the dividend. The result of this division is $2x^2 + x - 4$.

*Check this result. Does $(x - 4)(2x^2 + x - 4) = 2x^3 - 7x^2 - 8x + 16$?*

Let's compare the process of synthetic division to long division. We have used both methods to divide $3x^3 - 8x^2 + 5x - 1$ by $x - 2$. Study the results below.

$$\begin{array}{r|rrrr}
2 & 3 & -8 & 5 & -1 \\
 &  & 6 & -4 & 2 \\
\hline
 & 3 & -2 & 1 & \vert\ 1
\end{array}$$

$$\begin{array}{r}
3x^2 - 2x + 1 \\
x - 2 \overline{)\ 3x^3 - 8x^2 + 5x - 1} \\
\underline{3x^3 - 6x^2\qquad\qquad} \\
-2x^2 + 5x \\
\underline{-2x^2 + 4x} \\
x - 1 \\
\underline{x - 2} \\
1
\end{array}$$

Compare the numbers in the second row of the synthetic division with those that appear in the long division. Why do you think that in synthetic division you add these numbers that you would subtract when using long division?  *Look at the divisors.*

| Example 2 | **Use synthetic division to find $(5s^3 + s^2 - 7) \div (s + 1)$.** |

In synthetic division, every power of the variable must be represented in the dividend. So, $5s^3 + s^2 - 7$ must be written as $5s^3 + s^2 + 0s - 7$.

$$\begin{array}{r|rrrr}
-1 & 5 & 1 & 0 & -7 \\
 &  & -5 & 4 & -4 \\
\hline
 & 5 & -4 & 4 & -11
\end{array}$$   The result is $5s^2 - 4s + 4 - \dfrac{11}{s + 1}$.

To check the result, multiply the divisor, $s + 1$, by the quotient, $5s^2 - 4s + 4$. Then add the remainder, $-11$.

$$\begin{aligned}
(s + 1)(5s^2 - 4s + 4) + (-11) &= s(5s^2 - 4s + 4) + (5s^2 - 4s + 4) + (-11) \\
&= 5s^3 - 4s^2 + 4s + 5s^2 - 4s + 4 - 11 \\
&= 5s^3 + s^2 - 7 \ \checkmark
\end{aligned}$$

It checks because this polynomial is the dividend.

| Example 3 | **Use a calculator to divide $x^3 + 13x^2 - 12x - 8$ by $x + 2$.** |

ENTER: 2 [+/-] [×] 1 [=] [+] 13 [=]   $11$

$$\begin{array}{r|rrr}
-2 & 1 & 13 \\
 &  & -2 \\
\hline
 & 1 & 11 & \vert
\end{array}$$

ENTER: [×] 2 [+/-] [=] [+] 12 [+/-] [=]   $-34$

$$\begin{array}{r|rrr}
-2 & 1 & 13 & 12 \\
 &  & -2 & -22 \\
\hline
 & 1 & 11 & -34\vert
\end{array}$$

ENTER: [×] 2 [+/-] [=] [+] 8 [+/-] [=]   $60$

$$\begin{array}{r|rrrr}
-2 & 1 & 13 & -12 & -8 \\
 &  & -2 & -22 & 68 \\
\hline
 & 1 & 11 & -34 & \vert\ 60
\end{array}$$

The result is $x^2 + 11x - 34 + \dfrac{60}{x + 2}$.

So far we have used synthetic division on problems with divisors with leading coefficients of one. You can also use synthetic division when the divisor has a leading coefficient other than 1. Study the example below.

**Example 4**

*FYI* · · ·

**The Cookie Crate Co. makes cookie tins in several sizes. The largest tin they make is three inches longer than twice its width. It has a volume of $4w^3 + 8w^2 + 3w$ in³, where $w$ represents the width of the tin. If the lid is made separately, how much metal is required to make this cookie tin?**

We need to find the dimensions of the tin to be able to find the surface area. If $w$ represents the width, then the length is $2w + 3$. We can divide the volume by both the width and the length to find the height.

$$\text{height} = \frac{4w^3 + 8w^2 + 3w}{w(2w + 3)} \text{ or } \frac{4w^2 + 8w + 3}{2w + 3}$$

Since synthetic division requires a divisor of the form $x - r$, we must get the divisor in this form before using synthetic division. We can do that by factoring the leading coefficient of the divisor from the divisor and the dividend. Let's rewrite the division and take another look.

$$\text{height} = \frac{2\left(2w^2 + 4w + \frac{3}{2}\right)}{2\left(w + \frac{3}{2}\right)} \text{ or } \frac{2x^2 + 4w + \frac{3}{2}}{w + \frac{3}{2}}$$

*Factor 2 from both the divisor and the dividend. Then simplify the expression.*

Now we can use synthetic division. since the leading coefficient of $w + \frac{3}{2}$ is 1.

The height is $2w + 1$.

$$
\begin{array}{r|rrr}
-\frac{3}{2} & 2 & 4 & \frac{3}{2} \\
 & & -3 & -\frac{3}{2} \\
\hline
 & 2 & 1 & 0
\end{array}
$$

If the lid for the cookie tin is to be made separately, we need to find the sum of the areas of the four sides and the bottom to know how much metal will be needed to make the tin.

| | | | |
|---|---|---|---|
| Area of a long side | $= (2w + 3)(2w + 1)$ | $= 4w^2 + 8w + 3$ |
| Area of a short side | $= w(2w + 1)$ | $= 2w^2 + w$ |
| Area of bottom | $= w(2w + 3)$ | $= 2w^2 + 3w$ |

$$
\begin{aligned}
\text{Total area} &= 2(4w^2 + 8w + 3) + 2(2w^2 + w) + 2w^2 + 3w \\
&= 8w^2 + 16w + 6 + 4w^2 + 2w + 2w^2 + 3w \\
&= 14w^2 + 21w + 6
\end{aligned}
$$

An area of $14w^2 + 21w + 6$ square inches of metal will be needed to make the cookie tin.

**Example 5** | Use synthetic division to find $(2x^3 - x^2 + 5x - 12) \div (2x - 3)$.

$$\frac{2x^3 - x^2 + 5x - 12}{2x - 3} = \frac{2\left(x^3 - \frac{1}{2}x^2 + \frac{5}{2}x - 6\right)}{2\left(x - \frac{3}{2}\right)} \text{ or } \frac{x^3 - \frac{1}{2}x^2 + \frac{5}{2}x - 6}{x - \frac{3}{2}}$$

Now, use synthetic division to divide
$x^3 - \frac{1}{2}x^2 + \frac{5}{2}x - 6$ by $x - \frac{3}{2}$.

$$\frac{3}{2} \begin{array}{|cccc} 1 & -\frac{1}{2} & \frac{5}{2} & -6 \\ & \frac{3}{2} & \frac{3}{2} & 6 \\ \hline 1 & 1 & 4 & 0 \end{array}$$

The result is $x^2 + x + 4$.
*Check the solution.*

# CHECKING FOR UNDERSTANDING

**Communicating Mathematics**

Read and study the lesson to answer these questions.

1. Explain why 2 was factored out of both the divisor and the dividend in Example 5.

2. Why is it necessary to include terms with zero coefficients in the row of numbers for synthetic division?

3. In Example 4, suppose $w = 6.5$. How many square inches of metal are needed to make the box? What is its volume?

**Guided Practice**

Use synthetic division to determine which of the following binomials are factors of $2x^2 - 7x - 4$.

4. $x - 4$       5. $x + 1$       6. $2x - 1$

7. $x - 1$       8. $2x + 1$       9. $2x - 3$

# EXERCISES

**Practice**

Divide using synthetic division.

10. $(3y^3 + 2y^2 - 32y + 2) \div (y - 3)$    11. $(2b^3 + b^2 - 2b + 3) \div (b + 1)$

12. $(2c^3 - 3c^2 + 3c - 4) \div (c - 2)$    13. $(3x^3 - 2x^2 + 2x - 1) \div (x - 1)$

14. $(t^4 - 2t^3 + t^2 - 3t + 2) \div (t - 2)$    15. $(3r^4 - 6r^3 - 2r^2 + r - 6) \div (r + 1)$

16. $(z^4 - 3z^3 - z^2 - 11z - 4) \div (z - 4)$    17. $(2b^3 - 11b^2 + 12b + 9) \div (b - 3)$

18. $(6s^3 - 19s^2 + s + 6) \div (s - 3)$    19. $(x^3 + 2x^2 - 5x - 6) \div (x - 2)$

20. $(x^3 + 3x^2 - 7x + 1) \div (x - 1)$    21. $(n^4 - 8n^3 + 54n + 105) \div (n - 5)$

22. $(2x^4 - 5x^3 + 2x - 3) \div (x - 1)$    23. $(z^5 - 6z^3 + 4z^2 - 3) \div (z - 2)$

24. $(y^4 + 3y^3 + y - 1) \div (y + 3)$    25. $(4s^4 - 5s^2 + 2s + 3) \div (2s - 1)$

26. $(2x^3 - 3x^2 - 8x + 4) \div (2x + 1)$    27. $(4x^4 - 5x^2 - 8x - 10) \div (2x - 3)$

28. $(6j^3 - 28j^2 + 19j + 3) \div (3j - 2)$    29. $(y^5 - 3y^2 - 20) \div (y - 2)$

30. Use synthetic division to find $(3y^3 - 5y - 2) \div (y - 2)$. If $f(y) = 3y^3 - 5y - 2$, find $f(2)$. Compare the remainder from the division to the value of $f(2)$.

31. Use synthetic division to find $(2x^4 - 3x^2 + 1) \div (x + 1)$. If $f(x) = 2x^4 - 3x^2 + 1$, find $f(-1)$. Compare the remainder from the division to the value of $f(-1)$.

**Critical Thinking**

32. Based on your answers to Exercises 30 and 31, how do you think $f(r)$ and the division of $f(r)$ by $x - r$ are related?

**Applications**

33. **Manufacturing**  The volume of a small tin made by The Cookie Crate Co. is $6w^3 - 4w^2 - 16w$ in$^3$, where $w$ represents the width of the tin. If the height of the tin is four inches more than three times the width, what is the length of the tin?

34. **Retail**  The store *Bunches of Boxes and Bags* assembles boxes to package items for mailing. The store manager found that the volume of a box made from a piece of cardboard with a square of length $x$ inches cut from each corner is $4x^3 - 168x^2 + 1728x$ in$^3$. If the piece of cardboard is 48 inches long, how wide is it?

**Computer**

*Press RETURN after each coefficient.*

This BASIC program uses synthetic division to compute the coefficients of the quotient and remainder from a polynomial divided by a linear binomial. Input the degree of the polynomial. Next input the constant $R$ for the divisor $x - R$. Finally, input the coefficients of the polynomial.

```
10   INPUT "DEGREE OF POLYNOMIAL:
     "; N
20   INPUT "CONSTANT R: "; R
30   PRINT "ENTER COEFFICIENTS:"
40   FOR X = 1 TO N + 1
50   INPUT A(X)
60   NEXT X
70   LET B(1) = A(1)
80   PRINT "COEFFICIENTS OF
     QUOTIENT ARE: "
90   PRINT B(1); " ";
100  FOR X = 1 TO N-1
110  B(X + 1) = A(X + 1) + R * B(X)
120  PRINT B(X + 1); " ";
130  NEXT X
140  PRINT
150  PRINT "REMAINDER: "; A(N + 1) +
     R * B(N)
160  END
```

**Enter the program on a computer and use it to perform the following divisions.**

35. $(x^4 + 2x^3 - 7x^2 + 2x - 8) \div (x - 3)$

36. $(x^4 + 8x^3 + 22x^2 + 24x + 9) \div (x + 1)$

37. $(2x^5 + 3x^4 - 6x^3 + 6x^2 - 8x + 3) \div (x + 1)$

38. $(x^5 - 3x^2 - 20) \div (x - 2)$

39. $(x^5 - 15x^3 - 10x^2 + 60x + 72) \div (x + 3)$

**Mixed Review**

**Use long division to simplify each expression.  (Lesson 5-6)**

40. $(x^2 - 7x + 6)(x - 1)^{-1}$

41. $(a^3 + 7a^2 + 10a - 9)(a + 4)^{-1}$

42. Find the dimension of the matrix $M$ if $A_{1 \times 9} \cdot M = B_{1 \times 1}$.  **(Lesson 4-2)**

43. Find the value of the determinant $\begin{vmatrix} 3 & -2 \\ 0 & 1 \end{vmatrix}$.  **(Lesson 4-3)**

44. Write the equation $y = -11x + 2$ in standard form.  **(Lesson 2-2)**

## VOCABULARY

Upon completing this chapter, you should be
familiar with the following terms:

| | | | |
|---|---|---|---|
| binomial | **223** | **210** | monomial |
| coefficient | **210** | **231** | perfect square trinomial |
| constant | **210** | **223** | polynomial |
| degree of monomial | **210** | **212** | scientific notation |
| degree of polynomial | **223** | **231** | sum of cubes |
| difference of cubes | **231** | **241** | synthetic division |
| difference of squares | **231** | **223** | term |
| FOIL | **224** | **223** | trinomial |
| like terms | **210** | | |

## SKILLS AND CONCEPTS

| OBJECTIVES AND EXAMPLES | REVIEW EXERCISES |
|---|---|

Upon completing this chapter, you should
be able to:

Use these exercises to review and prepare
for the chapter test.

■ multiply monomials and powers of
monomials  (**Lesson 5-1**)

$$(x^2y^2)^3(2x)^2 = (x^6y^6)(4x^2)$$
$$= 4x^8y^6$$

**Simplify.**

**1.** $4m + 3m + (-6m)$  **2.** $4d^3 - 7d^3 + 5d^3$

**3.** $y^8(y^5)$  **4.** $x^2(x^4)(x^5)$

**5.** $(3xy^2)^2(4xy)^2$  **6.** $(-3x^2y)^3(2x)^2$

---

■ represent numbers in scientific
notation  (**Lesson 5-1**)

$$(9.1 \times 10^6)(2 \times 10^3)$$
$$= 1.82 \times 10^{10}$$
$$= 18,200,000,000$$

**Evaluate. Express each answer in both
scientific notation and decimal notation.**

**7.** $(3.2 \times 10^5)(4.2 \times 10^9)$

**8.** $(4.7 \times 10^2)(11 \times 10^4)$

---

■ divide polynomials and simplify
expressions containing negative and
zero exponents  (**Lesson 5-2**)

$$(mn^3)^{-3} = \frac{1}{m^3n^9} \qquad \frac{36x^2y}{9xy^2} = \frac{4x}{y}$$

$$a^0 = 1$$

**Simplify.**

**9.** $\dfrac{a^5}{a^3}$  **10.** $\dfrac{5y^{10}}{y^7}$

**11.** $\dfrac{4^{-6}}{4^{-3}}$  **12.** $\left(\dfrac{3}{4}\right)^0$

**13.** $(a^3b^2)^{-2}$  **14.** $\dfrac{25m^2n^3}{5mn}$

| OBJECTIVES AND EXAMPLES | REVIEW EXERCISES |
|---|---|

■ divide expressions written in scientific notation  (**Lesson 5-2**)

$$\frac{24 \times 10^{-2}}{8 \times 10^{3}} = 3 \times 10^{-5}$$
$$= 0.00003$$

**Evaluate. Express each answer in both scientific and decimal notation.**

15. $\dfrac{20 \times 10^5}{10 \times 10^{-2}}$

16. $\dfrac{(34{,}000{,}000)(24{,}000)}{6800}$

---

■ add, subtract, and multiply polynomials  (**Lesson 5-4**)

$(3x + y) + (2x - 4y) = 5x - 3y$

$(x^2 - 4y^2) - (3x^2 - 2y^2)$
$\quad = -2x^2 - 2y^2$

$(3a - b)(a + 2b) = 3a^2 + 6ab - ab - 2b^2$
$\quad\quad\quad\quad\quad = 3a^2 + 5ab - 2b^2$

**Simplify.**

17. $(9x + 2y) - (7x - 3y)$

18. $(4y^2 + 2y - 7) + (6y^2 - 3y + 2)$

19. $(m - 2)(m + 5)$

20. $(2y + 7)(3y - 9)$

21. $(n + 2)(n^2 - 3n + 1)$

22. $(r + 4)(r - 1)^2$

---

■ factor polynomials  (**Lesson 5-5**)

$ab - 2a + b - 2 = a(b - 2) + 1(b - 2)$
$\quad\quad\quad\quad\quad\quad = (a + 1)(b - 2)$

$3a^2 - 48 = 3(a^2 - 16)$
$\quad\quad\quad = 3(a - 4)(a + 4)$

$2x^2 - 5x - 3 = (x - 3)(2x + 1)$

**Factor.**

23. $5a + 5b$     24. $-14x^2 - 7x$

25. $s^2 + 7s + 6$     26. $a(y + 2) - b(y + 2)$

27. $b^3 - 64$     28. $8a^3 + 27$

29. $x^2 + 6x + 9$     30. $4x^2 - 81$

31. $2y^3 - 98y$     32. $y^{2n} - 81$

---

■ divide polynomials using factoring and long division  (**Lesson 5-6**)

$(x^2 + 2) \div (x - 1)$

$$\begin{array}{r} x + 1 \phantom{xxx} \\ x - 1 \overline{)\, x^2 + 0x + 2} \\ \underline{x^2 - x \phantom{xxxx}} \\ x + 2 \\ \underline{x - 1} \\ 3 \end{array}$$

$x + 1 + \dfrac{3}{x - 1}$

**Use division to simplify each expression.**

33. $(6y^2 + 16y + 8) \div (y + 2)$

34. $(x^3 - 2x^2 + 5) \div (x - 3)$

35. $(8s^2 - 2) \div (2s + 1)$

36. $(2d^4 + 2d^3 - 9d^2 - 3d + 9) \div (2d^2 - 3)$

■ divide polynomials using synthetic division   (**Lesson 5-7**)

$(3c^3 + 4c^2 - 2c - 1) \div (c + 2)$

$$
\begin{array}{r|rrrr}
-2 & 3 & 4 & -2 & -1 \\
   &   & -6 & 4 & -4 \\
\hline
   & 3 & -2 & 2 & -5
\end{array}
$$

$3c^2 - 2c + 2 - \dfrac{5}{c + 2}$

**Divide using synthetic division.**

**37.** $(x^3 - 4x^2 + 3x - 7) \div (x - 4)$

**38.** $(n^4 - 16) \div (n + 2)$

**39.** $(6y^3 + 11y^2 + y - 1) \div (2y + 3)$

**40.** $(10a^4 - 11a^3 + a^2 - 3a + 1) \div (2a - 1)$

# ~~~~~ APPLICATIONS AND CONNECTIONS ~~~~~

**41. Astronomy**   A light year is the distance that light travels in one year. If light travels 186,000 miles per second, how many miles are in a light year? Express the answer in both scientific and decimal notation. Assume a 365-day year. (**Lesson 5-1**)

**42. Physics**   Newton's law of gravitation can be used to compute the mass of Earth in grams. His formula applied is:

$$908 = \frac{6.67 \times 10^{-8} \times M}{(6.37 \times 10^8)^2}.$$

If $M$ represents mass, what is the mass of Earth in grams? Express the answer in both scientific and decimal notation. (**Lesson 5-2**)

**43. Travel**   Troy was traveling from Atlanta to Memphis by train. His lunch arrived after one-third of the trip was over. When he finished lunch, he had half the distance traveled before lunch yet to go. Draw a diagram to find the fraction of the trip Troy traveled while he was eating lunch. (**Lesson 5-3**)

**44. Retail**   The Bargain Barn gets a new shipment of merchandise each Monday. When the new shipment comes in, every item in the store is marked down 10%. If the original price of an item is $p$, what will its price be after 3 weeks in the store? (**Lesson 5-4**)

**45. Interior Design**   The amount of carpeting needed for the Leshin's living room is $2x^2 + x - 6$ square yards. They plan to place a new baseboard around the room after the carpeting is installed. The wood for the baseboard is sold according to the length needed. How much wood should they buy for the baseboard?   (**Lesson 5-5**)

**Simplify.**

1. $4y + 7y - 14y$

2. $9a^2 + 4a^2 - 2a^2$

3. $x^4y^3(x^8)$

4. $(4y)^3(2y)^2$

5. $\dfrac{n^4}{n^4}$

6. $\dfrac{y^2(3x)^3}{4x^{-2}}$

7. $\left(\dfrac{4s}{3t}\right)^0$

8. $\dfrac{13^{-6}}{13^{-9}}$

9. $\dfrac{r^2s^{-3}}{s^{-2}t^4}$

**Evaluate. Express each answer in both scientific and decimal notation.**

10. $(7.82 \times 10^3)(934 \times 10^2)$

11. $\dfrac{84{,}000{,}000 \times 0.0013}{0.021}$

12. The Carpet Experts installation team can install carpet in a room that is 10 ft by 10 ft in 1 hour. Draw a diagram and find how long it would take them to do a walk-in closet that is 5 ft by 5 ft.

**Simplify.**

13. $(3x + 2) - (7x - 6)$

14. $(m + 2)(m - 8)$

15. $(4n - 9)(5n + 2)$

16. $(s + 2)(s + 1)^2$

17. $(t - 1)(t^2 - 2t - 6)$

18. $(r - 8)(r + 8)$

**Factor.**

19. $3ab - 9b^2$

20. $y^2 - 7y + 6$

21. $f^2 + 16f + 64$

22. $w^2 - 144$

23. $y^3 - 125$

24. $8x^3 + 1$

25. $x^{2n} - 49$

26. $ay + 2a - by - 2b$

**Use division to simplify each expression.**

27. $\dfrac{5(w + 1)^6}{(w + 1)^3}$

28. $\dfrac{4r^2st + 12rst^2 - 10rt}{2rt}$

29. $(16s^3 - 8s^2 - 40s + 15) \div (2s^2 - 5)$

30. $(x^4 - x^3 + x^2 - 2x - 2) \div (x^2 + 2)$

**Divide using synthetic division.**

31. $(2x^3 + x^2 + 4x - 7) \div (x - 1)$

32. $(m^3 - 2m^2 - 17m + 30) \div (m + 4)$

33. **Manufacturing**  The Owen Container Corporation makes cardboard boxes to sell to manufacturers for transporting products to retail stores. One of their boxes has dimensions $2y$ by $3y$ by $4y$. Write a polynomial to represent the minimum area of cardboard needed to make this closed box if all lengths are in centimeters.

**Bonus**

Simplify $(36x^2 - 84xy + 49y^2 - 16a^2 - 24ab - 9b^2) \div (-18xz + 21yz + 12az + 9bz)$.

# Irrational and Complex Numbers

## CHAPTER OBJECTIVES

In this chapter, you will:

- Simplify radical expressions.
- Simplify expressions with rational exponents.
- Solve equations containing radicals.
- Add, subtract, multiply, and divide complex numbers.

If two stations are 25 miles apart and the range of the radar at the stations is the same, how far must the radar be able to reach?

**Diagram of Instrument Controlled Flight**

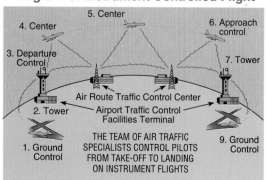

## CAREERS IN AIR TRAFFIC CONTROL

When you were a child, did you ever want to grow up to be an airplane pilot, commanding the wide, blue skies? That's still an attractive career, but you might find it even more appealing to be the person who commands the pilots: the air traffic controller.

Air traffic controllers usually control several planes at once and must be able to make quick decisions about completely different activities. For instance, a small owner-operated plane may be requesting wind and weather information at the same time that a jumbo jet asks for directions on its landing approach. While instructing these pilots, the controller would also be keeping an eye on other planes in his or her assigned airspace, such as those in a holding pattern waiting for permission to land, and making sure they are a safe distance apart.

Can you visualize three-dimensional spaces? Are you articulate enough to give pilots directions quickly and clearly? Do you have a good memory? Are you decisive? Could you pass physical, psychological, and drug-screening tests? Then you just might make a top-notch air traffic controller.

## MORE ABOUT AIR TRAFFIC CONTROL

### Degree Required:

- Bachelor's Degree in General Studies Program, and Intensive Training Program at FAA Academy

### Some air traffic controllers like:

- good opportunities for advancement
- the satisfaction of doing challenging work
- good salaries and benefits

### Related Math Subjects:

- Advanced Algebra
- Trigonometry
- Applied Math

### Some air traffic controllers dislike:

- having to pass an annual physical exam to keep their job
- an irregular work schedule
- working under extremely stressful conditions
- having to keep up with technological advances

For more information about careers in the field of Air Traffic Control, write to:
Air Traffic Control Association
2020 North 14th Street
Arlington, VA 22201

# 6-1 Roots of Real Numbers

**Objectives**

After studying this lesson, you should be able to:
- simplify radicals having various indices, and
- use a calculator to estimate roots of numbers.

**Application**

A carton shaped like a cube has sides 14 inches long. The volume of the carton is $14 \cdot 14 \cdot 14$ or $14^3$ cubic inches ($in^3$). Can you see why we call 14 to the third power 14 *cubed*? Why do you think we call raising a number to the second power *squaring* the number? Why do you suppose that there are not special names for raising a number to any other power?

Raising a number to the *n*th power means using that number as a factor *n* times.

$7^3 = 7 \cdot 7 \cdot 7$ or 343    *7 is used as a factor 3 times, n = 3.*

$2^4 = 2 \cdot 2 \cdot 2 \cdot 2$ or 16    *2 is used as a factor 4 times, n = 4.*

$5^n = \underbrace{5 \cdot 5 \cdot 5 \cdot \ldots \cdot 5}_{n \text{ factors}}$    *5 is used as a factor n times.*

As you know, finding the square root of a number and squaring a number are inverse operations. To find the square root of *n*, you must find a number whose square is *n*. For example, a square root of 25 is 5 since $5^2 = 25$. Since $(-5)^2 = 25$, -5 is also a square root of 25.

| Square Root | **For any real numbers *a* and *b*,** <br> **if $a^2 = b$, then *a* is a square root of *b*.** |
|---|---|

Since finding the square root of a number and squaring a number are inverses, it makes sense that the inverse of raising a number to the *n*th power is finding the **nth root** of the number. For example, to find the cube root of 27, you must find a number whose cube is 27.

$$a^3 = 27 \qquad a \cdot a \cdot a = 27$$

Since $3 \cdot 3 \cdot 3 = 27$, 3 is a cube root of 27.    *Is -3 a cube root of 27? Why or why not?*

| nth Root | **For any real numbers *a* and *b*, and any positive integer *n*,** <br> **if $a^n = b$, then *a* is an nth root of *b*.** |
|---|---|

The symbol $\sqrt[n]{\phantom{x}}$ indicates an $n$th root.

$$\underset{index}{\phantom{x}}\sqrt[n]{512} \leftarrow radicand \qquad \nearrow radical\ sign$$

When no index is given, the radical sign indicates a nonnegative square root.

Some numbers have more than one $n$th root. As we just saw, 25 has two square roots, 5 and $-5$. When there is more than one root, the nonnegative root is called the **principal root.** The symbol $\sqrt[n]{b}$ stands for the principal root. If $n$ is odd and $b$ is negative, there will be no nonnegative root. So, in this case, the principal root is negative.

| | |
|---|---|
| $\sqrt{49} = 7$ | $\sqrt{49}$ indicates the principal square root of 49. |
| $-\sqrt{49} = -7$ | $-\sqrt{49}$ indicates the opposite of the principal square root of 49. |
| $\pm\sqrt{49} = \pm 7$ | $\pm\sqrt{49}$ indicates both square roots of 49. $\pm$ *means positive or negative.* |
| $\sqrt[3]{-64} = -4$ | $\sqrt[3]{-64}$ indicates the principal cube root of $-64$. |
| $-\sqrt[4]{81} = -3$ | $-\sqrt[4]{81}$ indicates the opposite of the principal fourth root of 81. |

The chart below gives a summary of the real $n$th roots of a number $b$.

**The Real $n$th Roots of $b$, $\sqrt[n]{b}$ or $-\sqrt[n]{b}$**

| | $b > 0$ | $b < 0$ | $b = 0$ |
|---|---|---|---|
| $n$ even | one positive root one negative root | no real roots | one real root, 0 |
| $n$ odd | one positive root no negative roots | no positive roots one negative root | |

**Example 1**

Find $\pm\sqrt{81n^2}$.

$$\pm\sqrt{81n^2} = \pm\sqrt{(9n)^2}$$
$$= \pm 9n$$

The square roots of $81n^2$ are $\pm 9n$.

**Example 2**

Find $-\sqrt{(x + 1)^4}$.

$$-\sqrt{(x + 1)^4} = -\sqrt{[(x + 1)^2]^2}$$
$$= -(x + 1)^2$$

The opposite of the principal square root of $(x + 1)^4$ is $-(x + 1)^2$.

**Example 3**

Find $\sqrt[3]{8n^9}$.

$$\sqrt[3]{8n^9} = \sqrt[3]{(2n^3)^3}$$
$$= 2n^3$$

The principal cube root of $8n^9$ is $2n^3$.

**Example 4**

Find $\sqrt[4]{m^4}$.

Since $m^4 = m \cdot m \cdot m \cdot m$, $m$ is a fourth root of $m^4$. Since the index is even, the principal root is nonnegative. However, since $m$ *may* be negative, we must take the absolute value of $m$ to identify the principal root.

$$\sqrt[4]{m^4} = |m|$$

When you find the $n$th root of an even power and an odd power is the result, you must take the absolute value of the result to ensure that the value is nonnegative. If the result is an even power or you find the $n$th root of an odd power, there is no need to take the absolute value. *Why?*

$$\sqrt{(-6)^2} = |-6| \text{ or } 6 \qquad \sqrt{(-4)^6} = |(-4)^3| \text{ or } 64$$

Expressions such as $\sqrt{64}$ and $\sqrt[3]{-\dfrac{1}{8}}$ name rational numbers. There is no need to take the absolute value.

$$\sqrt{64} = 8 \qquad \sqrt[3]{-\frac{1}{8}} = -\frac{1}{2}$$

As you learned in Chapter 1, real numbers that cannot be expressed as terminating or repeating decimals are **irrational numbers.** $\sqrt{2}$ and $\sqrt{3}$ are examples of irrational numbers. Decimal approximations for irrational numbers, such as 3.14 for $\pi$, are often used in applications. You can use a calculator to find decimal approximations. *If you do not have a calculator, refer to the Tables Appendix.*

**Example 5**

Use a calculator to find a decimal approximation for $\sqrt[3]{339}$.

*Method 1* Use the root key. *It may be a second function key.*

ENTER: 339 $\boxed{\sqrt[x]{y}}$ 3 $\boxed{=}$ $6.972682649$

*Method 2* Use the power and reciprocal keys.

ENTER: 339 $\boxed{y^x}$ 3 $\boxed{1/x}$ $\boxed{=}$ $6.972682649$

**Check:**    6.972682649 $\boxed{y^x}$ 3 $\boxed{=}$ $339 \checkmark$

$\sqrt[3]{339}$ is approximately 6.97.

## Example 6

APPLICATION

Manufacturing

Educational Enterprises makes children's toys for use in preschools. One of their most popular items is a set of multicolored blocks designed to teach colors and shapes. The volume of the materials required to make each cubic block is 60 cubic centimeters. What is the surface area that needs to be printed on each block?

Let $s$ represent the length of an edge of a block.

$V = s^3$ — *The volume of a cube is the cube*
$60 = s^3$ — *of the length of an edge.*
$\sqrt[3]{60} = s$ — *Take the cube root of each side.*
$3.91 \approx s$ — *Use a calculator to find an approximation for* $\sqrt[3]{60}$.

The length of an edge of a block is about 3.91 centimeters.

Each face of the block will have an area of $s^2$ square centimeters.

$s^2 = (3.91)^2$ or $15.29$

Since there are six faces, the total surface area to be painted is 6(15.29) or about 91.74 square centimeters.

# CHECKING FOR UNDERSTANDING

**Communicating Mathematics**

Read and study the lesson to answer these questions.

1. Explain why it is not always necessary to take the absolute value of a result to indicate the principal root.

2. Why is $-2n^3$ not a second answer in Example 3?

3. Explain why you can use the inverse key with the power key to find an $n$th root on your calculator.

4. Does $\sqrt[4]{(-x)^4} = x$ no matter what value $x$ represents? Explain.

5. Does $\sqrt[5]{(-x)^5} = x$ no matter what value $x$ represents? Explain.

**Guided Practice**

Simplify.

6. $\sqrt{144}$  7. $-\sqrt{121}$  8. $\sqrt[3]{8}$

9. $\sqrt[3]{y^3}$  10. $\sqrt[4]{16}$  11. $-\sqrt[4]{x^4}$

12. $\sqrt[4]{t^8}$  13. $\sqrt[3]{-125}$  14. $\sqrt[5]{32n^5}$

15. $\sqrt{16a^2b^4}$  16. $\sqrt{(y + 1)^2}$  17. $\sqrt{x^2 + 6x + 9}$

Use a calculator to approximate each value to three decimal places.

18. $\sqrt{3.2}$  19. $\sqrt{55}$  20. $\sqrt[3]{9.8}$

21. $-\sqrt[3]{47}$  22. $-\sqrt[3]{670}$  23. $\sqrt{64}$

# EXERCISES

Use a calculator to find each value to three places. Check your approximation by using the power key.

**24.** $\sqrt{83}$        **25.** $-\sqrt{99}$        **26.** $\sqrt{9.5}$

**27.** $\sqrt[3]{23}$        **28.** $\sqrt[3]{8.1}$        **29.** $-\sqrt[3]{-41}$

Simplify.

**30.** $\pm\sqrt{81}$        **31.** $\sqrt{196}$        **32.** $\sqrt{256}$

**33.** $\sqrt[4]{81}$        **34.** $-\sqrt[3]{27}$        **35.** $\sqrt[3]{-216}$

**36.** $\sqrt[5]{-1}$        **37.** $\sqrt[3]{-1000}$        **38.** $\pm\sqrt{0.49}$

**39.** $\sqrt[3]{0.125}$        **40.** $\sqrt{121n^2}$        **41.** $\sqrt{25y^6}$

**42.** $\sqrt{(3s)^4}$        **43.** $\pm\sqrt{576}$        **44.** $\sqrt{676}$

**45.** $\sqrt{64a^2b^4}$        **46.** $-\sqrt{144b^2c^6}$        **47.** $\sqrt[3]{-8b^3c^3}$

**48.** $\pm\sqrt[3]{27r^3s^3}$        **49.** $\sqrt[3]{64a^6b^3}$        **50.** $\sqrt[4]{625n^8m^4}$

**51.** $\sqrt{(3x+y)^2}$        **52.** $\sqrt{(a+b)^2}$        **53.** $\sqrt[3]{(s+t)^3}$

**54.** $\sqrt[3]{(2x-y)^3}$        **55.** $\sqrt[4]{(r+s)^4}$        **56.** $\sqrt[5]{(2m-3)^5}$

**57.** $\sqrt{x^2+10x+25}$        **58.** $\sqrt{x^2+6x+9}$        **59.** $\sqrt{9a^2+6a+1}$

**60.** $\sqrt{4y^2+12y+9}$        **61.** $\sqrt{s^2-2st+t^2}$        **62.** $\sqrt{4x^2+12xy+9y^2}$

**Critical Thinking**      **63.** Under what conditions is $\sqrt[n]{(-x)^n} = x$?

**Applications**      **64. Physics** The formula for finding the time, $t$, it takes an object dropped from a height of $h$ feet to reach the ground is $t = \sqrt{\dfrac{2h}{g}}$, where $g$ represents the acceleration due to gravity. All objects in free fall near the earth's surface have an acceleration due to gravity of 32 feet per second squared. If a ball is dropped from a window 64 feet high, how long will it take for it to reach the ground?

65. **Aerospace Engineering**   Scientists expect that in future space stations artificial gravity will be created by rotating all or part of the space station. The formula $N = \dfrac{1}{2\pi}\sqrt{\dfrac{a}{r}}$ gives the number of rotations, $N$, required per second to maintain an acceleration of gravity of $a$ meters per second squared on a satellite with a radius of $r$ meters. The acceleration of gravity on Earth is 9.8 m/s². How many rotations per minute will produce an artificial gravity that is equal to half of the gravity on Earth in a space station with a 25 m radius?

**Mixed Review**   **Divide using synthetic division.   (Lesson 5-7)**

66. $(t^3 - 3t + 2) \div (t + 2)$

67. $(5x^3 - 8x^2 + 38x - 18) \div (5x - 3)$

68. Evaluate $(4.5 \times 10^4)(3.33 \times 10^2)$. Express the answer in both scientific and decimal notation.   **(Lesson 5-1)**

69. Solve $\begin{cases} x + y + z = 0 \\ 2x + 4y + z = -1 \\ x - 2y - z = -2 \end{cases}$ using an augmented matrix.   **(Lesson 4-8)**

70. State the dimension of the matrix $\begin{bmatrix} 3 & 2 & 0 \\ -2 & 10 & 6 \end{bmatrix}$. Then find its determinant, if it exists.   **(Lesson 4-1)**

71. In which octant does the point $(3, 9, -2)$ lie?   **(Lesson 3-8)**

72. **Business**   The Burrough's Department Store parking garage charges $1.50 for each hour or fraction of an hour for parking. What type of function does this relationship represent?   **(Lesson 2-7)**

## LANGUAGE CONNECTION

In mathematics, many words have specific definitions. However, when these words are used in everyday language, they frequently have a different meaning. Study each pair of sentences. How do the meanings of the word in boldface differ?

**1a.** Plants receive nourishment and water from their **roots.**
**1b.** The square **roots** of 36 are 6 and -6.

**2a.** This soup tastes **odd.**
**2b.** For any number $a$ and any integer $n$ greater than 1, if $n$ is **odd**, then
   $\sqrt[n]{a^n} = a$.

Write two sentences for each word, one that uses the everyday meaning of the word and one that uses the mathematical meaning.

| | | | |
|---|---|---|---|
| 1. negative | 2. power | 3. rational | 4. coordinate |
| 5. degree | 6. absolute | 7. identity | 8. real |

## 6-2 Products and Quotients of Radicals

**Objectives**

After studying this lesson, you should be able to:
- simplify radical expressions using multiplication and division, and
- rationalize the denominator of a fraction containing a radical expression.

**Application**

A group of graduate students from Purdue University is conducting research on marine life on a small island in the Pacific. An airplane must drop their supplies on the island since there is no clearing large enough for it to land. The airplane is flying at a speed of 484 feet per second at an altitude of 5000 feet. Where should the pilot let the supplies drop so that they land near the camp?

One of the formulas we need to solve this problem is $t = \sqrt{\dfrac{2h}{g}}$, where $t$ represents time, $h$ represents the height of the object when it is dropped, and $g$ represents acceleration due to gravity. Do you think that $t = \dfrac{\sqrt{2} \cdot \sqrt{h}}{\sqrt{g}}$ is an equivalent formula? Let's do some investigation of multiplication and division of radicals to find out.

Try evaluating these roots in two ways. The first way is to find the root and then multiply. The second way is to multiply and then find the root. Then compare the results.

| *Method 1* | *Method 2* |
|---|---|
| $\sqrt{9} \cdot \sqrt{16} = 3 \cdot 4$ or $12$ | $\sqrt{9} \cdot \sqrt{16} = \sqrt{144}$ or $12$ |
| $\sqrt[3]{-8} \cdot \sqrt[3]{27} = -2 \cdot 3$ or $-6$ | $\sqrt[3]{-8} \cdot \sqrt[3]{27} = \sqrt[3]{-216}$ or $-6$ |

The result is the same using either method. These examples demonstrate the following property of radicals.

| | |
|---|---|
| *Product Property of Radicals* | **For any real numbers $a$ and $b$, and any integer $n$, $n > 1$,**<br><br>**1. If $n$ is even, then $\sqrt[n]{ab} = \sqrt[n]{a} \cdot \sqrt[n]{b}$ as long as $a$ and $b$ are both nonnegative, and**<br><br>**2. If $n$ is odd, then $\sqrt[n]{ab} = \sqrt[n]{a} \cdot \sqrt[n]{b}$.** |

When you simplify a square root, first write the prime factorization of the radicand. Then use the product property to isolate the perfect squares. Then simplify each radical.

**Example 1** | Simplify $\sqrt{24a^3b^2}$.

$$\sqrt{24a^3b^2} = \sqrt{2^2 \cdot 2 \cdot 3 \cdot a^2 \cdot a \cdot b^2} \qquad \textit{The prime factorization of 24 is } 2^3 \cdot 3.$$
$$= \sqrt{2^2} \cdot \sqrt{2} \cdot \sqrt{3} \cdot \sqrt{a^2} \cdot \sqrt{a} \cdot \sqrt{b^2} \qquad \textit{Product property of radicals}$$
$$= 2a|b|\sqrt{6a}$$

If $a < 0$, then $\sqrt{a^3}$ has no real roots, so we must assume that $a \geq 0$. Since $a \geq 0$, we don't need to write $\sqrt{a^3} = |a|\sqrt{a}$. However, $b$ could be negative so we must write $\sqrt{b^2}$ as $|b|$.

Simplifying $n$th roots is very similar to simplifying square roots. Find the factors that are $n$th powers and use the product property.

**Example 2** | Simplify $\sqrt[3]{40x^3y^5}$.

$$\sqrt[3]{40x^3y^5} = \sqrt[3]{2^3 \cdot 5 \cdot x^3 \cdot y^3 \cdot y^2} \qquad \textit{Factor into cubes where possible.}$$
$$= \sqrt[3]{2^3} \cdot \sqrt[3]{5} \cdot \sqrt[3]{x^3} \cdot \sqrt[3]{y^3} \cdot \sqrt[3]{y^2} \qquad \textit{Product property of radicals}$$
$$= 2xy\sqrt[3]{5y^2}$$

**Example 3** | Simplify $\sqrt[4]{3n} \cdot \sqrt[4]{5n^7}$.

$$\sqrt[4]{3n} \cdot \sqrt[4]{5n^7} = \sqrt[4]{3n \cdot 5n^7} \qquad \textit{Product property of radicals}$$
$$= \sqrt[4]{15n^8}$$
$$= \sqrt[4]{15} \cdot \sqrt[4]{n^4} \cdot \sqrt[4]{n^4} \qquad \textit{Factor into fourth powers where possible.}$$
$$= n^2\sqrt[4]{15} \qquad \textit{Why aren't absolute values required!}$$

When you multiply rational numbers and radicals, multiply the rationals and the radicals separately and then simplify the product.

**Example 4** | Simplify $3\sqrt{2} \cdot 4\sqrt{10}$.

$$3\sqrt{2} \cdot 4\sqrt{10} = 3 \cdot 4 \cdot \sqrt{2} \cdot \sqrt{10} \qquad \textit{Commutative property of multiplication}$$
$$= 12\sqrt{20}$$
$$= 12 \cdot \sqrt{2^2} \cdot \sqrt{5} \qquad \textit{Product property of radicals}$$
$$= 12 \cdot 2 \cdot \sqrt{5} \text{ or } 24\sqrt{5}$$

Take a look at the division of radicals to see if there is a quotient property similar to the product property of radicals. What do you think we'll discover?

*Method 1*        *Method 2*

$$\frac{\sqrt{81}}{\sqrt{9}} = \frac{9}{3} \text{ or } 3 \qquad\qquad \sqrt{\frac{81}{9}} = \sqrt{9} \text{ or } 3$$

$$\frac{\sqrt[3]{-216}}{\sqrt[3]{27}} = \frac{-6}{3} \text{ or } -2 \qquad\qquad \sqrt[3]{\frac{-216}{27}} = \sqrt[3]{-8} \text{ or } -2$$

Were you right about a quotient property of radicals?

For any real numbers $a$ and $b$, $b \neq 0$, and any integer $n$, $n > 1$,

$$\sqrt[n]{\frac{a}{b}} = \frac{\sqrt[n]{a}}{\sqrt[n]{b}} \text{ if all roots are defined.}$$

**Example 5**

**a. Simplify** $\sqrt[3]{\frac{3}{8}}$.

$$\sqrt[3]{\frac{3}{8}} = \frac{\sqrt[3]{3}}{\sqrt[3]{8}} \qquad \textit{Quotient property of radicals}$$

$$= \frac{\sqrt[3]{3}}{2}$$

**b. Simplify** $\frac{12\sqrt{18}}{4\sqrt{6}}$.

$$\frac{12\sqrt{18}}{4\sqrt{6}} = \frac{12}{4} \cdot \sqrt{\frac{18}{6}}$$

$$= 3\sqrt{3}$$

Fractions are usually written without radicals in the denominator. Radicals are not usually left in fraction form either. The process of eliminating radicals from the denominator or fractions from the radicand is called **rationalizing the denominator.**

To rationalize a denominator, you must multiply the numerator and denominator by a quantity so that the radicand has an exact root. Study the examples below.

If the denominator were $\sqrt{ab^2c^3}$ you would multiply the numerator and the denominator by $\sqrt{ac}$.

*Other multipliers are also possible, but $\sqrt{ac}$ is the simplest multiplier you can use.*

$$\sqrt{ab^2c^3} \cdot \sqrt{ac} = \sqrt{a^2b^2c^4} = a|b|c^2 \qquad \textit{Note that all of the factors in } \sqrt{a^2b^2c^4} \textit{ are perfect squares.}$$

If the denominator were $\sqrt[3]{2mn^2p^4}$, you would multiply the numerator and denominator by $\sqrt[3]{2^2m^2np^2}$.

$$\sqrt[3]{2mn^2p^4} \cdot \sqrt[3]{2^2m^2np^2} = \sqrt[3]{2^3m^3n^3p^6} = 2mnp^2 \qquad \textit{Note that all of the factors in } \sqrt[3]{2^3m^3n^3p^6} \textit{ are perfect cubes.}$$

**Example 6**

**Simplify** $\frac{5}{2\sqrt{3}}$.

$$\frac{5}{2\sqrt{3}} = \frac{5}{2\sqrt{3}} \cdot \frac{\sqrt{3}}{\sqrt{3}} \qquad \textit{Since } \frac{\sqrt{3}}{\sqrt{3}} = 1, \textit{ the value of } \frac{5}{2\sqrt{3}} \textit{ is not changed.}$$

$$= \frac{5\sqrt{3}}{2\sqrt{3} \cdot 3}$$

$$= \frac{5\sqrt{3}}{2\sqrt{3^2}} \text{ or } \frac{5\sqrt{3}}{6}$$

1. The index, $n$, is as small as possible.
2. The radicand contains no factor (other than one) which is the $n$th power of an integer or polynomial.
3. The radicand contains no fractions.
4. No radicals appear in the denominator.

**Example 7**

Simplify $\sqrt[3]{\dfrac{2}{3t}}$.

$$\sqrt[3]{\frac{2}{3t}} = \frac{\sqrt[3]{2}}{\sqrt[3]{3t}} \cdot \frac{\sqrt[3]{3^2t^2}}{\sqrt[3]{3^2t^2}}$$  *Why is $\dfrac{\sqrt[3]{3^2t^2}}{\sqrt[3]{3^2t^2}}$ used to rationalize the denominator?*

$$= \frac{\sqrt[3]{2 \cdot 3^2t^2}}{\sqrt[3]{3^3t^3}} \text{ or } \frac{\sqrt[3]{18t^2}}{3t}$$

**Example 8**

**APPLICATION**

**Physics**

The formula for finding centripetal force, $F_c$, the inward force that must be applied to keep an object moving in a circle, is $F_c = \dfrac{mv^2}{r}$. Let $m$ represent the mass of the object, $v$ represent the velocity, and $r$ represent the radius of the circular path. Solve the formula for the velocity and write the result in simplified form.

$$F_c = \frac{mv^2}{r}$$

$$\frac{F_c r}{m} = v^2$$  *Multiply each side by $\dfrac{r}{m}$.*

$$\sqrt{\frac{F_c r}{m}} = v$$  *Find the square root of each side.*

$$\sqrt{\frac{F_c r}{m}} \cdot \frac{\sqrt{m}}{\sqrt{m}} = v$$  *Rationalize the denominator.*

$$\frac{\sqrt{F_c rm}}{m} = v$$

# CHECKING FOR UNDERSTANDING

**Communicating Mathematics**

Read and study the lesson to answer these questions.

1. What property or properties make the alternative formula that is given in the beginning of the lesson equivalent to the original formula?

2. Why is the product property of radicals for odd indices different than the product property for even indices?

3. Explain why $\dfrac{\sqrt{3x}}{x}$ is in simplified form.

4. Explain how to simplify $\dfrac{4}{\sqrt{2}}$.

## Guided Practice

**Simplify.**

5. $\sqrt{27}$      6. $\sqrt{32}$      7. $\sqrt{98y^4}$

8. $\sqrt{50x^2}$      9. $\sqrt[3]{16}$      10. $\sqrt[4]{48}$

11. $\sqrt{y^3}$      12. $\sqrt{a^5}$      13. $\sqrt[4]{t^5}$

**State the fraction that each radical expression should be multiplied by to rationalize the denominator. Then simplify.**

14. $\dfrac{6}{\sqrt{2}}$      15. $\dfrac{1}{\sqrt{3}}$      16. $\dfrac{1}{\sqrt{x}}$

17. $\dfrac{3}{\sqrt{b}}$      18. $\dfrac{3}{\sqrt[3]{4}}$      19. $\dfrac{7}{\sqrt[3]{9}}$

# EXERCISES

## Practice

**Simplify.**

20. $5\sqrt{50}$      21. $4\sqrt{54}$      22. $\sqrt[3]{32}$

23. $\sqrt[3]{56}$      24. $\sqrt{162}$      25. $\sqrt{675}$

26. $\sqrt{8a^2b^3}$      27. $\sqrt{8x^2y} \cdot \sqrt{2xy}$      28. $\sqrt[4]{81m^4n^5}$

29. $\dfrac{\sqrt{10}}{\sqrt{5}}$      30. $\dfrac{\sqrt{12}}{\sqrt{3}}$      31. $\dfrac{\sqrt{22}}{\sqrt{2}}$

32. $\sqrt[3]{-192}$      33. $6\sqrt{216}$      34. $3\sqrt{242}$

35. $\sqrt[4]{112}$      36. $(-3\sqrt{24})(5\sqrt{20})$      37. $(4\sqrt{18})(2\sqrt{14})$

38. $\sqrt[3]{121} \cdot \sqrt[3]{88}$      39. $\sqrt{3}(\sqrt{6} - 2)$      40. $\sqrt{3x^2z^3} \cdot \sqrt{15x^2z}$

41. $\sqrt[4]{a^5b^3} \cdot \sqrt[4]{81a^3b^2}$      42. $\sqrt{\dfrac{7}{4}}$      43. $\dfrac{\sqrt[3]{81}}{\sqrt[3]{9}}$

44. $\sqrt{\dfrac{1}{3}}$      45. $\sqrt{\dfrac{5}{12a}}$      46. $\sqrt{7}(\sqrt{14} + \sqrt{21})$

47. $\sqrt{a}(\sqrt{b} + \sqrt{ab})$      48. $\sqrt{r}(\sqrt{r} + r\sqrt{s})$      49. $\sqrt[3]{\dfrac{54}{125}}$

50. $\sqrt[4]{\dfrac{5}{16}}$      51. $\sqrt{\dfrac{5}{32x}}$      52. $\sqrt[4]{\dfrac{2}{3}}$

53. Find the radius, $r$, of a sphere whose surface area $S$ is 616 square inches. Use the formula $r = \dfrac{1}{2}\sqrt{\dfrac{S}{\pi}}$

**Critical Thinking**

**54.** Is the statement "All real numbers can be written as a radical." *true* or *false*? Justify your answer.

**Applications**

**55. Manufacturing** Find the length in inches that a pendulum should be for a complete swing to take one second. Use the formula $T = 2\pi\sqrt{\dfrac{L}{384}}$, where $T$ represents time in seconds, and $L$ represents the length of the pendulum.

**56. Physics** Sharon and Anthony dropped a stone from a 150-foot cliff. Find the time, $t$, in seconds that it will take for the stone to reach the ground. Let $t = \dfrac{1}{4}\sqrt{s}$, where $s$ represents the distance in feet the stone will fall.

**Mixed Review**

Simplify. **(Lesson 6-1)**

**57.** $\sqrt{(5b)^4}$

**58.** $-\sqrt{121b^2c^6}$

**59.** Evaluate $(9 \times 10^3)^{-1}(3.5 \times 10^{-2})$. Express the answer in both scientific and decimal notation. **(Lesson 5-2)**

**60.** Can a $3 \times 4$ matrix have an inverse? Explain your answer. **(Lesson 4-5)**

**61.** Find the $x$-, $y$-, and $z$-intercepts for the equation $3x + 6y - 8z = 24$. **(Lesson 3-8)**

**62.** If $h(x) = [3x - 1]$, find $h(-2.1)$. **(Lesson 2-7)**

---

**BIOLOGY CONNECTION**

Nature, seldom content with simple shapes alone, has created all kinds of intricate mathematical designs. The spirals of the shell of a chambered nautilus, for example, is an equiangular spiral. In the cutaway drawing at the right, notice that the outreaching radii form right angles with the chords along the curve and each chord has a value of 1. The Pythagorean Theorem allows us to calculate the measure of the hypotenuse. Notice the pattern of radicals that results with each additional chamber of the nautilus. This pattern was first documented by Jacob Bernoulli (1654–1705). He was so impressed with this spiral that he had it engraved on his tombstone with the inscription *Eadem mutata resurgo* (I shall arise, the same, though changed).

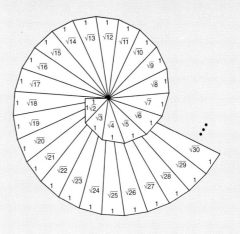

# Computing with Radicals

**Objective**

After studying this lesson, you should be able to:

- add, subtract, multiply, and divide radical expressions.

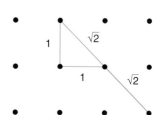

$a^2 + b^2 = c^2$
$1^2 + 1^2 = c^2$
$\pm\sqrt{2} = c$

The geometric dot paper at the right shows the construction of a segment that is $\sqrt{2}$ units long. This can be verified by using the Pythagorean Theorem. The segment in green is also $\sqrt{2}$ units long. Estimate the sum of the two lengths. Based on your estimate, do you think $\sqrt{2} + \sqrt{2} = \sqrt{2 + 2}$ or 2?

As you have just observed, $\sqrt{a} + \sqrt{b} \neq \sqrt{a + b}$ just like $a^2 + b^2 \neq (a + b)^2$. Adding radicals is like adding other monomials. You wouldn't say $x^2 + x^2 = (x + x)^2$. You know that $x^2 + x^2 = 2x^2$ because you add like terms. So you must add like terms with radicals also. Two radical expressions are called **like radical expressions** if both the indices and the radicands are alike. Some examples of like and unlike radical expressions are given below.

$2\sqrt[4]{3}$ and $5\sqrt[4]{3}$ are like expressions.    *Both the indices and radicands are alike.*

$\sqrt[3]{21}$ and $\sqrt[4]{21}$ are not like expressions.    *The indices are not alike.*

$6x\sqrt{5}$ and $6\sqrt{5x}$ are not like expressions.   *The radicands are not alike.*

$\sqrt[4]{7x}$ and $\sqrt[3]{7y}$ are not like expressions.    *Neither the indices nor the radicands are alike.*

To add or subtract radicals, just combine like terms as you do when you add or subtract other monomials.

**Example 1**

**Simplify $4 + 3\sqrt{5} + 7 + 2\sqrt{5}$.**

$$4 + 3\sqrt{5} + 7 + 2\sqrt{5} = (4 + 7) + (3\sqrt{5} + 2\sqrt{5})$$
$$= (4 + 7) + (3 + 2)\sqrt{5} \qquad \textit{$3\sqrt{5}$ and $2\sqrt{5}$ are like expressions.}$$
$$= 11 + 5\sqrt{5}$$

**Example 2**

**Simplify $4\sqrt{27} + 3\sqrt{3} - \sqrt{48}$.**

$$4\sqrt{27} + 3\sqrt{3} - \sqrt{48} = 4\sqrt{3^2 \cdot 3} + 3\sqrt{3} - \sqrt{4^2 \cdot 3} \quad \textit{Simplify each radical.}$$
$$= 4\sqrt{3^2}\sqrt{3} + 3\sqrt{3} - \sqrt{4^2}\sqrt{3}$$
$$= 4 \cdot 3\sqrt{3} + 3\sqrt{3} - 4\sqrt{3}$$
$$= 12\sqrt{3} + 3\sqrt{3} - 4\sqrt{3}$$
$$= 11\sqrt{3} \qquad\qquad \textit{Combine like expressions.}$$

**Example 3**

Simplify $\sqrt[3]{32x} + \sqrt[3]{108x}$.

$$\sqrt[3]{32x} + \sqrt[3]{108x} = \sqrt[3]{2^3 \cdot 4x} + \sqrt[3]{3^3 \cdot 4x} \qquad \textit{Simplify each radical.}$$

$$= \sqrt[3]{2^3} \cdot \sqrt[3]{4x} + \sqrt[3]{3^3} \cdot \sqrt[3]{4x} \qquad \textit{Product of powers property}$$

$$= 2\sqrt[3]{4x} + 3\sqrt[3]{4x}$$

$$= 5\sqrt[3]{4x}$$

**Example 4**

APPLICATION

Sports

*FYI* · · ·

The world record high dive was made by Randal Dickison at Ocean Park, Hong Kong on April 6, 1885. It was 174 feet 8 inches.

Two cliff divers are performing in a diving show. One is diving from a cliff 128 feet high, and the other is diving from a cliff 32 feet high. They start to dive at the same time. Use the formula $t = \sqrt{\dfrac{2s}{g}}$, where $s$ represents the distance in feet, $t$ represents time in seconds, and $g$ represents the acceleration due to gravity, to find how much longer it will take the diver on the higher cliff to enter the water. Assume that the acceleration due to gravity is 32 feet per second squared and that air resistance is not a factor.

Find the time for the diver from a 128-foot cliff to enter the water.

$$t = \sqrt{\frac{2s}{g}}$$

$$= \sqrt{\frac{2(128)}{32}} \qquad \textit{Substitute 128 for s and 32 for g.}$$

$$= \sqrt{8} \text{ or } 2\sqrt{2}$$

The time is $2\sqrt{2}$ seconds.

Now find the difference of the two times.

Find the time for the diver from a 32-foot cliff to enter the water.

$$t = \sqrt{\frac{2s}{g}}$$

$$= \sqrt{\frac{2(32)}{32}} \qquad \textit{Substitute 32 for s and 32 for g.}$$

$$= \sqrt{2}$$

The time is $\sqrt{2}$ seconds.

$$2\sqrt{2} - \sqrt{2} = \sqrt{2} \approx 1.41$$

It will take the diver on the higher cliff about 1.41 seconds longer to enter the water.

We combine radicals like monomials when we add and subtract. We can multiply radicals like binomials using the FOIL method.

**Example 5**

Simplify $(3 + \sqrt{2})(\sqrt{10} + \sqrt{5})$.

$$\begin{array}{cccc} & F & O & I & L \end{array}$$

$$(3 + \sqrt{2})(\sqrt{10} + \sqrt{5}) = 3\sqrt{10} + 3\sqrt{5} + \sqrt{2} \cdot \sqrt{10} + \sqrt{2} \cdot \sqrt{5}$$

$$= 3\sqrt{10} + 3\sqrt{5} + \sqrt{20} \qquad + \sqrt{10}$$

$$= 3\sqrt{10} + 3\sqrt{5} + 2\sqrt{5} \qquad + \sqrt{10}$$

$$= 4\sqrt{10} + 5\sqrt{5}$$

**Example 6**

Simplify $(5 - \sqrt{3})(5 + \sqrt{3})$.

$$(5 - \sqrt{3})(5 + \sqrt{3}) = 25 + 5\sqrt{3} - 5\sqrt{3} - \sqrt{3 \cdot 3}$$
$$= 25 - \sqrt{3^2}$$
$$= 25 - 3$$
$$= 22$$

Binomials, like those in Example 6, of the form $a\sqrt{b} + c\sqrt{d}$ and $a\sqrt{b} - c\sqrt{d}$ where $a$, $b$, $c$, and $d$ are rational numbers called **conjugates** of each other. The product of conjugates is always a rational number. We can use conjugates to rationalize a binomial denominator containing a radical.

**Example 7**

Simplify $\dfrac{2 + \sqrt{5}}{5 - 2\sqrt{5}}$.

$$\frac{2 + \sqrt{5}}{5 - 2\sqrt{5}} = \frac{2 + \sqrt{5}}{5 - 2\sqrt{5}} \cdot \frac{5 + 2\sqrt{5}}{5 + 2\sqrt{5}}$$

*The conjugate of $5 - 2\sqrt{5}$ is $5 + 2\sqrt{5}$.*

$$= \frac{10 + 4\sqrt{5} + 5\sqrt{5} + 2\sqrt{5} \cdot 5}{25 - (2\sqrt{5})^2}$$
$$= \frac{10 + 4\sqrt{5} + 5\sqrt{5} + 10}{25 - 4(5)}$$
$$= \frac{20 + 9\sqrt{5}}{5}$$

# CHECKING FOR UNDERSTANDING

**Communicating Mathematics**

Read and study the lesson to answer these questions.

1. Are $3\sqrt{7}$ and $4\sqrt[3]{7}$ like radical expressions? If not, why not?

2. What is the conjugate of $-4 - 2\sqrt{2}$?

3. Why is the product of conjugates always a rational number?   *Hint: Find the product of the two general conjugates given in the definition of conjugates.*

4. What expression would you multiply $\dfrac{1 - \sqrt{3}}{5 + 2\sqrt{3}}$ by to rationalize the denominator?

**Guided Practice**

Name the conjugate of each expression.

5. $5 - \sqrt{7}$     6. $1 + \sqrt{3}$     7. $\sqrt{3} + \sqrt{10}$

8. $\sqrt{3} + 5$     9. $2 - 2\sqrt{3}$     10. $\sqrt{7} - 3\sqrt{5}$

**Simplify.**

**11.** $5\sqrt{3} - 4\sqrt{3}$

**12.** $7\sqrt{y} - 4\sqrt{y}$

**13.** $8\sqrt[3]{6} + 3\sqrt[3]{6}$

**14.** $\sqrt[5]{3} + 4\sqrt[5]{3}$

**15.** $2\sqrt{2} + \sqrt{8}$

**16.** $\sqrt[3]{40} - 2\sqrt[3]{5}$

# EXERCISES

**Practice**

**Simplify.**

**17.** $-3\sqrt{5} + 5\sqrt{2} + 4\sqrt{20} - 3\sqrt{50}$

**18.** $5\sqrt{2} + 3\sqrt{2} - 8$

**19.** $8\sqrt{3} - 3\sqrt{75}$

**20.** $3\sqrt{7} - 5\sqrt{28}$

**21.** $(3 + \sqrt{5})(4 + \sqrt{5})$

**22.** $(5 + \sqrt{3})(3 - \sqrt{3})$

**23.** $(3x + \sqrt{5y})(3x - \sqrt{5y})$

**24.** $(6 - \sqrt{2})(6 + \sqrt{2})$

**25.** $(4 + \sqrt{3})^2$

**26.** $(a + \sqrt{b})^2$

**27.** $5\sqrt{20} + \sqrt{24} - \sqrt{180} + 7\sqrt{54}$

**28.** $7\sqrt[3]{5t} + 4\sqrt[3]{5t}$

**29.** $\sqrt[3]{54} - \sqrt[3]{128}$

**30.** $8\sqrt[3]{2x} + 3\sqrt[3]{2x} - 8\sqrt[3]{2x}$

**31.** $\sqrt[3]{48} - \sqrt[3]{6}$

**32.** $7\sqrt[3]{2} + 6\sqrt[3]{150}$

**33.** $5\sqrt[3]{135} - 2\sqrt[3]{81}$

**34.** $(5 + \sqrt{2})(3 + \sqrt{2})$

**35.** $(5 + \sqrt{6})(5 - \sqrt{2})$

**36.** $(8 - \sqrt{3})(6 + \sqrt{3})$

**37.** $(7 + \sqrt{11y})(7 - \sqrt{11y})$

**38.** $(\sqrt{3} + \sqrt{5})(\sqrt{12} - \sqrt{5})$

**39.** $\dfrac{1}{3 + \sqrt{5}}$

**40.** $\dfrac{7}{4 - \sqrt{3}}$

**41.** $7\sqrt{24} + \sqrt[3]{24}$

**42.** $\sqrt{98} - \sqrt{72} + \sqrt{32}$

**43.** $7\sqrt[4]{2} + 8\sqrt[4]{2}$

**44.** $\sqrt[4]{x^2} + \sqrt[4]{x^6}$

**45.** $\sqrt[4]{y^4} + \sqrt[3]{y^6} + \sqrt{y^8}$

**46.** $(4\sqrt{5} - 3\sqrt{2})(2\sqrt{5} + 2\sqrt{2})$

**47.** $(m + \sqrt[3]{4})(m^2 - m\sqrt[3]{4} + \sqrt[3]{16})$

**48.** $(2 + \sqrt[3]{s})(4 - 2\sqrt[3]{s} + \sqrt[3]{s^2})$

**49.** $\dfrac{\sqrt{x+1}}{\sqrt{x-1}}$

**50.** $\dfrac{\sqrt{3} + n\sqrt{6}}{4 - \sqrt{n}}$

**51.** $\sqrt{\dfrac{2}{5}} + \sqrt{40} + \sqrt{10}$

**52.** $\sqrt[3]{\dfrac{2}{3}} + \sqrt[3]{144} - \sqrt[3]{243}$

**Critical Thinking**

**53.** As you recall, a set is closed under an operation if the result of performing the operation on any two elements of the set is an element of the set. Is the set of irrational numbers closed under any of the four basic operations?

**Applications**

**54. Automotive Engineering**  An automotive engineer is trying to design a safer car. The maximum force a road can exert on the tires of the car being redesigned is 2000 pounds. What is the maximum velocity in ft/s at which this car can safely round a turn of radius 320 feet? Use the formula $v = \sqrt{\dfrac{F_c r}{100}}$, where $F_c$ is the force the road exerts on the car and $r$ is the radius of the turn.

**55. Sports**  Casey hit a foul-ball straight up over the plate. It reached a height of 112 feet. How long does the catcher have to get ready to catch the ball before it reaches the ground? The formula for the total time is $t = 2\sqrt{\dfrac{2h}{g}}$, where $h$ is the height of the ball and $g$ is the acceleration due to gravity. Assume that the acceleration due to gravity is 32 feet per second squared.

**Mixed Review**

Simplify.  **(Lesson 6-2)**

**56.** $5\sqrt{54}$

**57.** $\sqrt[4]{5m^3n^5} \cdot \sqrt[4]{125m^2n^3}$

**58.** Factor $1 - 8a^3$.  **(Lesson 5-5)**

**59.** Simplify $c^3 \cdot c^2 \cdot c^4$.  **(Lesson 5-1)**

**60.** State the dimension of the matrix $\begin{bmatrix} -1 & 1 & 7 \\ 0 & 4 & 0 \\ 2 & 2 & 3 \end{bmatrix}$. Then evaluate its determinant (if one exists).  **(Lesson 4-1)**

**61.** Find the maximum and minimum values of the function $f(x, y) = x - y$ defined for the polygonal region having vertices with coordinates $(0, 0)$, $(0, 5)$, $(3, 4)$, and $(6, 0)$.  **(Lesson 3-6)**

# Rational Exponents

**Objectives**

After studying this lesson, you should be able to:

■ write expressions with rational exponents in simplest radical form and vice versa, and

■ evaluate expressions in either exponential or radical form.

**Application**

Nina deposited $500 in her account at First Atlanta Bank on March 1. Her account earns 8% interest per year. If she withdraws the money to pay for her college tuition at the beginning of September, how much interest has she earned? The bank uses the formula $A = P(1 + r)^t$ for finding the amount of money ($A$) in a compound interest account at the end of $t$ years if $P$ is the original amount of money deposited and $r$ is the annual interest rate.

Nina had her money in the bank six months. But, $t$ is expressed in years. So substituting $\frac{1}{2}$ for $t$, we must evaluate $A = 500(1 + 0.08)^{\frac{1}{2}}$ to find the amount of interest she earned. But how do we evaluate a fractional exponent? Assume that fractional exponents behave as integral exponents. Then, $5^1 = 5^{\left(\frac{1}{2}\right)\cdot 2}$ or $\left(5^{\frac{1}{2}}\right)^2$.

So, $5^{\frac{1}{2}}$ is a number that when squared equals 5. Since you know that, $(\sqrt{5})^2$ also equals 5, then $5^{\frac{1}{2}} = \sqrt{5}$. This suggests the following definition.

*Definition of $b^{\frac{1}{n}}$*

> **For any real number $b$ and for any integer $n$, $n > 1$,**
> $$b^{\frac{1}{n}} = \sqrt[n]{b}$$
> **except when $b < 0$ and $n$ is even.**

From the definition, we can say that $7^{\frac{1}{4}} = \sqrt[4]{7}$ and $(-8)^{\frac{1}{3}} = \sqrt[3]{-8}$ or $-2$. The expression $(-16)^{\frac{1}{4}}$ is not defined since $-16 < 0$ and 4 is even. Can you tell why we need this restriction?

**Example 1**

**Evaluate $64^{\frac{1}{3}}$.**

Here are two methods of solution.

*Method 1* $\quad 64^{\frac{1}{3}} = \sqrt[3]{64}$ $\qquad$ *Method 2* $\quad 64^{\frac{1}{3}} = (4^3)^{\frac{1}{3}}$

$\qquad\qquad\qquad = \sqrt[3]{4^3}$ $\qquad\qquad\qquad\qquad\qquad = 4^{3\left(\frac{1}{3}\right)}$

$\qquad\qquad\qquad = 4$ $\qquad\qquad\qquad\qquad\qquad\qquad = 4^1$ or 4

**Example 2** | Evaluate $625^{-\frac{1}{4}}$.

*Method 1*

$$625^{-\frac{1}{4}} = \frac{1}{625^{\frac{1}{4}}}$$

$$= \frac{1}{\sqrt[4]{625}} \qquad \text{Definition of } b^{\frac{1}{n}}$$

$$= \frac{1}{\sqrt[4]{5^4}}$$

$$= \frac{1}{5}$$

*Method 2*

$$625^{-\frac{1}{4}} = (5^4)^{-\frac{1}{4}}$$

$$= 5^{4\left(-\frac{1}{4}\right)} \qquad \textit{Raising a power to a power}$$

$$= 5^{-1}$$

$$= \frac{1}{5}$$

How can you evaluate an expression with a fractional exponent in which the numerator is not 1? Study the two methods shown below.

*Method 1*

$$6^{\frac{3}{2}} = \left(6^{\frac{1}{2}}\right)^3 \text{ or } (\sqrt{6})^3$$

*Method 2*

$$6^{\frac{3}{2}} = (6^3)^{\frac{1}{2}} \text{ or } \sqrt{6^3}$$

Since $(\sqrt{6})^3 = \sqrt{6} \cdot \sqrt{6} \cdot \sqrt{6}$ or $\sqrt{216}$ and $\sqrt{6^3} = \sqrt{6 \cdot 6 \cdot 6}$ or $\sqrt{216}$, $(\sqrt{6})^3$ and $\sqrt{6^3}$ have the same value.

| *Definition of Rational Exponents* | **For any nonzero real number $b$, and any integers $m$ and $n$, with $n > 1$,** $$b^{\frac{m}{n}} = \sqrt[n]{b^m} = (\sqrt[n]{b})^m$$ **except when $b < 0$ and $n$ is even.** |
|---|---|

Why does the definition not apply when the index is even and the base is negative? Substitute different values in for $b$ and $n$ to discover the reason.

**Example 3** | Evaluate $8^{\frac{2}{3}}$.

*Method 1*
$$8^{\frac{2}{3}} = (\sqrt[3]{8})^2$$
$$= (2)^2$$
$$= 4$$

*Method 2*
$$8^{\frac{2}{3}} = (2^3)^{\frac{2}{3}}$$
$$= 2^2$$
$$= 4$$

**Example 4** | Evaluate $27^{\frac{1}{3}} \cdot 27^{\frac{4}{3}}$.

*Method 1*
$$27^{\frac{1}{3}} \cdot 27^{\frac{4}{3}} = 27^{\frac{5}{3}}$$
$$= (\sqrt[3]{27})^5$$
$$= (3)^5 \text{ or } 243$$

*Method 2*
$$27^{\frac{1}{3}} \cdot 27^{\frac{4}{3}} = 27^{\frac{5}{3}}$$
$$= (3^3)^{\frac{5}{3}}$$
$$= 3^5 \text{ or } 243$$

As you know, fractions are rational numbers that can also be written as decimal numbers. For example, $\frac{2}{3} = 0.\overline{6}$ and $\frac{3}{4} = 0.75$. We can use a calculator to evaluate expressions with fractional or decimal exponents.

**Example 5**

Use a calculator to evaluate $1331^{\frac{2}{3}}$.

ENTER: 1331 $\boxed{y^x}$ $\boxed{(}$ 2 $\boxed{\div}$ 3 $\boxed{)}$ $\boxed{=}$ $121$

$1331^{\frac{2}{3}} = 121$   *To evaluate an expression with a decimal exponent, enter the decimal after the $\boxed{y^x}$ key.*

When simplifying a radical, find the smallest index possible.

**Example 6**

Express $\sqrt[6]{36}$ in simplest radical form.

In $\sqrt[6]{36}$, the index is 6. Methods 1 and 2 both illustrate using a smaller index.

*Method 1*

$$\sqrt[6]{36} = 36^{\frac{1}{6}}$$
$$= (36^{\frac{1}{2}})^{\frac{1}{3}}$$
$$= (\sqrt{36})^{\frac{1}{3}}$$
$$= 6^{\frac{1}{3}} \text{ or } \sqrt[3]{6}$$

*Method 2*

$$\sqrt[6]{36} = 36^{\frac{1}{6}}$$
$$= (6^2)^{\frac{1}{6}}$$
$$= 6^{\frac{1}{3}} \text{ or } \sqrt[3]{6}$$

**Example 7**

Express $2^{\frac{1}{2}}a^{\frac{2}{3}}b^{\frac{5}{6}}$ in simplest radical form.

$$2^{\frac{1}{2}}a^{\frac{2}{3}}b^{\frac{5}{6}} = 2^{\frac{3}{6}}a^{\frac{4}{6}}b^{\frac{5}{6}}$$   *Rewrite all exponents using the least common denominator, 6.*
$$= (2^3 a^4 b^5)^{\frac{1}{6}}$$
$$= \sqrt[6]{8a^4b^5}$$

**Example 8**

**APPLICATION**

**Music**

The formula for the frequency, $f$, of a note is $f = (2L)^{-1}P^{\frac{1}{2}}m^{-\frac{1}{2}}$, where $L$ is the length of the string, $P$ is the stretching force on the string, and $m$ is the mass of one centimeter of the string. Write the formula in simplest radical form.

$$f = (2L)^{-1}P^{\frac{1}{2}}m^{-\frac{1}{2}}$$
$$= \left(\frac{1}{2L}\right)\sqrt{P}\left(\frac{1}{\sqrt{m}}\right)$$   $(2L)^{-1} = \frac{1}{2L}; \ P^{\frac{1}{2}} = \sqrt{P}; \ m^{-\frac{1}{2}} = \frac{1}{\sqrt{m}}$
$$= \left(\frac{1}{2L}\right)\left(\frac{\sqrt{P}}{\sqrt{m}}\right)$$
$$= \frac{1}{2L}\left(\frac{\sqrt{P}}{\sqrt{m}}\right)\left(\frac{\sqrt{m}}{\sqrt{m}}\right)$$   *Rationalize the denominator.*
$$= \frac{\sqrt{Pm}}{2Lm}$$

# CHECKING FOR UNDERSTANDING

**Communicating Mathematics**

**Read and study the lesson to answer these questions.**

1. Why are fractional exponents not defined when the denominator of the exponent is even and the base is negative?

2. State an expression that is equivalent to $\sqrt[4]{33}$.

3. Is $\sqrt[6]{256}$ in simplest radical form? If not, tell why and write the expression in simplest radical form.

**Guided Practice**

Evaluate.

4. $9^{\frac{3}{2}}$

5. $8^{\frac{2}{3}}$

6. $4^{\frac{3}{2}}$

7. $16^{\frac{3}{4}}$

8. $64^{\frac{5}{6}}$

9. $27^{-\frac{2}{3}}$

10. $\sqrt[3]{8^2}$

11. $343^{\frac{2}{3}}$

12. $\sqrt[4]{81}$

13. $9^{\frac{1}{3}} \cdot 9^{\frac{5}{3}}$

14. $16^{-\frac{3}{2}}$

15. $36^{\frac{3}{4}} \div 36^{\frac{1}{4}}$

Express in simplest radical form.

16. $\sqrt[6]{49}$

17. $\sqrt[4]{36}$

18. $\sqrt[6]{81}$

19. $\sqrt[8]{16}$

20. $\sqrt[4]{25}$

21. $\sqrt[9]{64}$

# EXERCISES

**Practice**

Express using rational exponents.

22. $\sqrt{14}$

23. $\sqrt[3]{17}$

24. $\sqrt[6]{32}$

25. $\sqrt[4]{y}$

26. $\sqrt[3]{m}$

27. $\sqrt{25a^3b^4}$

28. $\sqrt[3]{8x^3y^6}$

29. $\sqrt[4]{27}$

30. $\sqrt[4]{8x^3y^5}$

31. $\sqrt[3]{n^2}$

32. $\sqrt[6]{b^3}$

33. $\sqrt[3]{16a^5b^7}$

Express in simplest radical form.

34. $7^{\frac{1}{2}}$

35. $36^{\frac{1}{4}}$

36. $6^{\frac{1}{3}}$

37. $n^{\frac{3}{4}}$

38. $x^{\frac{3}{2}}y^{\frac{5}{2}}$

39. $2^{\frac{5}{3}}a^{\frac{7}{3}}$

40. $(2m)^{\frac{1}{2}}m^{\frac{1}{2}}$

41. $p^{\frac{5}{2}}q^{\frac{3}{4}}$

42. $4^{\frac{1}{3}}x^{\frac{2}{3}}y^{\frac{4}{3}}$

43. $(3r)^{\frac{2}{5}}s^{\frac{3}{5}}$

44. $x^{\frac{4}{7}}y^{\frac{3}{7}}$

45. $5^{\frac{1}{3}}s^{\frac{2}{3}}t^{\frac{1}{3}}$

Evaluate each expression using a calculator.

**46.** $16^{0.25}$

**47.** $\left(\dfrac{1}{32}\right)^{\frac{1}{5}}$

**48.** $144^{\frac{1}{2}}$

**49.** $\sqrt[4]{256}$

**50.** $25^{2.5}$

**51.** $27^{\frac{4}{3}}$

**52.** $\left(\dfrac{343}{64}\right)^{\frac{1}{3}}$

**53.** $(9^{0.75})^{\frac{2}{3}}$

**54.** $\left(\dfrac{216}{729}\right)^{\frac{2}{3}}$

**55.** $(0.008)^{\frac{1}{3}}$

**56.** $(0.125)^{\frac{2}{3}}$

**57.** $(6^{\frac{2}{3}})^{3}$

Express in simplest radical form.

**58.** $\sqrt[4]{9}$

**59.** $x^{\frac{1}{3}}y^{\frac{1}{2}}$

**60.** $a^{\frac{3}{4}}b^{\frac{1}{3}}c^{\frac{5}{6}}$

**61.** $\sqrt[6]{8}$

**62.** $5a^{\frac{1}{2}}b^{\frac{1}{4}}$

**63.** $x^{\frac{5}{6}}y^{\frac{3}{2}}z^{\frac{7}{3}}$

**64.** $\sqrt[3]{2^5} \cdot \sqrt[4]{2}$

**65.** $\sqrt{3} \cdot \sqrt[4]{3^2}$

**66.** $\sqrt[3]{\sqrt{27}}$

**Critical Thinking**

**67.** Does $\sqrt[n]{\sqrt[m]{b}} = \sqrt[m]{\sqrt[n]{b}}$? Justify your answer.

**Applications**

**68. Music** A technician is tuning a piano. The frequency of the A note above middle C is correctly set at 440 vibrations per second. The

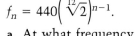

frequency $f_n$ of a note $n$ notes above A should be $f_n = 440\left(\sqrt[12]{2}\right)^{n-1}$.

  **a.** At what frequency should the technician set the A that is one octave, or 12 notes, above the A above middle C?

  **b.** Middle C is nine notes below the A that has frequency 440 vibrations per second. What should the frequency of middle C be?

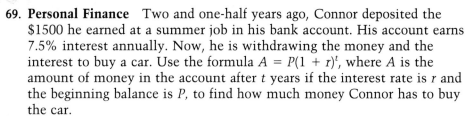

**69. Personal Finance** Two and one-half years ago, Connor deposited the $1500 he earned at a summer job in his bank account. His account earns 7.5% interest annually. Now, he is withdrawing the money and the interest to buy a car. Use the formula $A = P(1 + r)^t$, where $A$ is the amount of money in the account after $t$ years if the interest rate is $r$ and the beginning balance is $P$, to find how much money Connor has to buy the car.

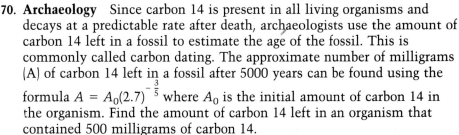

**70. Archaeology** Since carbon 14 is present in all living organisms and decays at a predictable rate after death, archaeologists use the amount of carbon 14 left in a fossil to estimate the age of the fossil. This is commonly called carbon dating. The approximate number of milligrams (A) of carbon 14 left in a fossil after 5000 years can be found using the formula $A = A_0(2.7)^{-\frac{3}{5}}$ where $A_0$ is the initial amount of carbon 14 in the organism. Find the amount of carbon 14 left in an organism that contained 500 milligrams of carbon 14.

**71.** Name the conjugate of the expression $5 + 3\sqrt{3}$. **(Lesson 6-3)**

**72.** Simplify $\sqrt{108} - \sqrt{48} + (\sqrt{3})^3$. **(Lesson 6-3)**

**73.** Use division to simplify $(a^2 - 5ab + 6b^2) \div (a - 3b)$. **(Lesson 5-6)**

**Journal**

Write how you would explain simplifying radical expressions to another student and write the examples you would use.

**74.** Find the inverse of the matrix $\begin{bmatrix} 3 & 1 \\ 2 & -4 \end{bmatrix}$. **(Lesson 4-5)**

**75.** The Worthington Public Library charges a fine of 5¢ per day for overdue books. What will the fine be on a book that is two weeks overdue? **(Lesson 2-2)**

**76.** Simplify the expression $4(5x + 2y) + 9(x - 2y)$. **(Lesson 1-2)**

## MID-CHAPTER REVIEW

**Simplify. (Lesson 6-1)**

**1.** $-\sqrt{81x^2}$

**2.** $\sqrt{a^2 + 14a + 49}$

**3.** $\sqrt[3]{-64x^9}$

**4.** $\sqrt{48m^2n^3}$

**Simplify. (Lesson 6-2)**

**5.** $\sqrt{6}(\sqrt{3} + 5\sqrt{2})$

**6.** $\dfrac{5}{3\sqrt{5}}$

**7.** $\dfrac{12}{\sqrt[3]{4x}}$

**8.** $\sqrt[4]{3b^6r^7} \cdot \sqrt[4]{81b^2r^2}$

**9. Physics** A pebble is dropped from a height of 200 feet. Use the formula $t = \frac{1}{4}\sqrt{s}$, where $t$ is the number of seconds it takes for the pebble to reach the ground and $s$ is the distance in feet it will fall, to find the time it will take for the pebble to reach the ground.

**Simplify. (Lesson 6-3)**

**10.** $2\sqrt{18} + 3\sqrt{8} - 4\sqrt{50}$

**11.** $(5 + \sqrt{3})(7 - 2\sqrt{3})$

**12.** $(11 - \sqrt{7})(11 + \sqrt{7})$

**13.** $\dfrac{1 + \sqrt{3}}{5 - 2\sqrt{3}}$

**14.** Evaluate $8^{\frac{2}{3}} \cdot 9^{\frac{1}{2}}$. **(Lesson 6-4)**

**15.** Express $5^{\frac{2}{3}}x^{\frac{1}{2}}y^{\frac{3}{4}}$ in simplest radical form. **(Lesson 6-4)**

**16.** Express $\sqrt[6]{27a^3b^4c^6}$ using rational exponents. **(Lesson 6-4)**

# 6-5 Problem-Solving Strategy: Identify Subgoals

**Objective**  After studying this lesson, you should be able to:

- solve problems by identifying and achieving subgoals.

Like walking a mile, solving a problem is a series of small steps. If we can identify the steps, or subgoals, that need to be achieved in solving a problem, solving is a simpler process.

**Example**

**Find the sum of the whole numbers from 1 through 200 that are not multiples of 4 or 9. You may use the formula $S = \frac{1}{2}n(n + 1)$ for the sum, $S$, of the integers 1 to $n$.**

The first subgoal we can identify is to find the sum of the integers from 1 to 200. Use the formula provided.

$$S = \frac{1}{2}n(n + 1)$$
$$= \frac{1}{2}(200)(201) \qquad \textit{Replace n with 200.}$$
$$= 20{,}100$$

The second subgoal is to find the sum of the integers from 1 to 200 that are multiples of 4 and the sum of the integers that are multiples of 9.

The sum of the multiples of 4 is given by the following expression.

$$4 + 8 + 12 + \ldots + 200$$
$$= 4(1 + 2 + 3 + \ldots + 50) \qquad \textit{Distributive property}$$

Now use the formula for the sum of the integers 1 to 50.

$$4(1 + 2 + 3 + \ldots + 50)$$
$$= 4\left(\frac{1}{2}(50)(51)\right) \qquad \textit{Replace n with 50.}$$
$$= 5100$$

Use the same technique to find the sum of the multiples of 9.

$$9 + 18 + 27 + \ldots + 198$$
$$= 9(1 + 2 + 3 + \ldots + 22)$$
$$= 9\left(\frac{1}{2}(22)(23)\right) \text{ or } 2277$$

The final subgoal is to find the sum of the numbers between 1 and 100 that are not multiples of 4 or 9.

If we subtract these two sums from the sum of all integers from 1 to 200, the numbers that are multiples of both 4 and 9 will be subtracted twice. So, we must find that sum and add it back so that those numbers are subtracted only once. The multiples of 36 are multiples of both 4 and 9.

Multiples of 36: $36 + 72 + 108 + 144 + 180$ or $540$

Solve the problem: $20{,}100 - 5100 - 2277 + 540 = 13{,}263$

The sum of the whole numbers between 1 and 200 that are not multiples of 4 or 9 is 13,263

## CHECKING FOR UNDERSTANDING

**Communicating Mathematics**

**Read and study the lesson to answer these questions.**

1. Why is it helpful to set subgoals when solving a problem?

2. Why did we need to add the sum of the multiples of 36 in the Example?

3. What is the sum of the integers from 1 through 300 that are not multiples of 4 or 9?

**Guided Practice**

4. Find the value of the expression $\sqrt{\sqrt{\left(\sqrt{\left(\sqrt{\left(\sqrt{2^2}\right)^2}\right)^4}\right)^2}}$. Justify your answer.

## EXERCISES

**Solve. Use any strategy.**

5. How many factors of 2000 are perfect squares?

6. Find the sum $\dfrac{1}{2^1} + \dfrac{1}{2^2} + \dfrac{1}{2^3} + \ldots + \dfrac{1}{2^{10}}$.

7. The proper divisors of a number are the factors of the number that are less than the number. What is the least natural number that has exactly nine proper factors?

8. **Statistics** The average of the ages of the first five presidents at the time of their inauguration is 58 years. If the sum of the ages of the first four presidents at their inauguration is 232, how old was the fifth president when he was inaugurated?

9. **Geometry** If the area of a 14 cm by 14 cm square is increased by 60 cm$^2$, what are the dimensions of the new square?

10. What positive number is equal to its square added to its opposite?

## COOPERATIVE LEARNING ACTIVITY

**Work in groups. Each person in the group must understand the solution and be able to explain it to any person in the class.**

How many zeros appear at the end of the product of the first 100 integers?

# 6-6 Simplifying Expressions with Rational Exponents

**Objective**

After studying this lesson, you should be able to:

■ simplify expressions containing rational exponents.

**Application**

**FYI · · ·**

NASA has launched more than 300 satellites since its start in 1958. Satellites are used for everything from observing weather patterns and providing communication links to monitoring the use of resources.

Aerospace engineers have found that the velocity necessary for a satellite to maintain a circular orbit around Earth is found by the formula

$v = R_e \sqrt{\dfrac{g}{r}}$, where $R_e$ represents the radius of the Earth, $g$ represents the acceleration due to gravity, and $r$ represents the radius of the orbit. Can you rewrite the formula using positive rational exponents? That is, can you simplify the formula?

When you are asked to simplify an expression, you must write the expression with all positive exponents. Furthermore, any exponents in the denominator of a fraction must be positive *integers*.

**Example 1**

*Remember* $\dfrac{1}{5^{\frac{1}{2}}} = \dfrac{1}{\sqrt{5}}$.

Simplify each expression.

**a.** $\dfrac{1}{5^{\frac{1}{2}}}$

$$\dfrac{1}{5^{\frac{1}{2}}} = \dfrac{1}{5^{\frac{1}{2}}} \cdot \dfrac{5^{\frac{1}{2}}}{5^{\frac{1}{2}}} \qquad Why \ \dfrac{5^{\frac{1}{2}}}{5^{\frac{1}{2}}}?$$

$$= \dfrac{5^{\frac{1}{2}}}{5}$$

**b.** $\dfrac{1}{7^{\frac{2}{3}}}$

$$\dfrac{1}{7^{\frac{2}{3}}} = \dfrac{1}{7^{\frac{2}{3}}} \cdot \dfrac{7^{\frac{1}{3}}}{7^{\frac{1}{3}}} \qquad Why \ \dfrac{7^{\frac{1}{3}}}{7^{\frac{1}{3}}}?$$

$$= \dfrac{7^{\frac{1}{3}}}{7}$$

There is more than one way to simplify an expression. Choosing the multiplier carefully may allow you to simplify in fewer steps. Study Example 2 to learn how to choose the correct multiplier.

**Example 2**

Simplify $\dfrac{1}{2^{\frac{3}{2}}}$.

*Method 1*

$$\frac{1}{2^{\frac{3}{2}}} = \frac{1}{2^{\frac{3}{2}}}\left(\frac{2^{\frac{3}{2}}}{2^{\frac{3}{2}}}\right)$$

$$= \frac{2^{\frac{3}{2}}}{2^3} \text{ or } \frac{2^{\frac{1}{2}}}{2^2}$$

*Method 2*

$$\frac{1}{2^{\frac{3}{2}}} = \frac{1}{2^{\frac{3}{2}}}\left(\frac{2^{\frac{1}{2}}}{2^{\frac{1}{2}}}\right)$$

$$= \frac{2^{\frac{1}{2}}}{2^2}$$

*Notice that there are fewer steps in simplifying when the multiplier is in the form $\dfrac{2^{\frac{1}{2}}}{2^{\frac{1}{2}}}$.*

An expression is simplified when it meets all of these conditions:
- it has no negative exponents,
- it has no fractional exponents in the denominator,
- it is not a complex fraction, and
- the index of any remaining radical is the least number possible.

**Example 3**

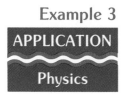

APPLICATION

Physics

A model airplane is fixed on a string so that it flies around in a circle. The designers of the plane would like to find the time it takes for the airplane to make a complete circle. They know that the formula $F_c = m\left(\dfrac{4\pi^2 r}{T^2}\right)$ describes the force required to keep the airplane going in a circle. $m$ represents the mass of the airplane, $r$ represents the radius of the circle, and $T$ represents the time for a revolution. Solve the formula for $T$.

$$F_c = m\left(\frac{4\pi^2 r}{T^2}\right)$$

$$T^2 = m\left(\frac{4\pi^2 r}{F_c}\right) \qquad \textit{Multiply each side by } \frac{T^2}{F_c}.$$

$$T = \sqrt{m\left(\frac{4\pi^2 r}{F_c}\right)} \qquad \textit{Find the square root of each side.}$$

$$= 2\pi\sqrt{\frac{mr}{F_c}} \qquad \textit{Simplify.}$$

$$= \frac{2\pi\sqrt{mrF_c}}{F_c} \qquad \textit{Rationalize the denominator.}$$

**Example 4**

Simplify $\dfrac{x^{\frac{1}{2}} - y^{\frac{1}{2}}}{x^{\frac{1}{2}} + y^{\frac{1}{2}}}$.

$$\frac{x^{\frac{1}{2}} - y^{\frac{1}{2}}}{x^{\frac{1}{2}} + y^{\frac{1}{2}}} = \frac{x^{\frac{1}{2}} - y^{\frac{1}{2}}}{x^{\frac{1}{2}} + y^{\frac{1}{2}}} \cdot \frac{x^{\frac{1}{2}} - y^{\frac{1}{2}}}{x^{\frac{1}{2}} - y^{\frac{1}{2}}} \qquad \textit{The conjugate of } x^{\frac{1}{2}} + y^{\frac{1}{2}} \textit{ is } x^{\frac{1}{2}} - y^{\frac{1}{2}}.$$

$$= \frac{x - 2x^{\frac{1}{2}}y^{\frac{1}{2}} + y}{x - y}$$

# CHECKING FOR UNDERSTANDING

**Communicating Mathematics**

**Read and study the lesson to answer these questions.**

1. Under what conditions does a denominator need to be rationalized?

2. Is the expression $\dfrac{1}{q^{\frac{1}{2}}}$ simplified? If not, why not?

3. Explain your strategy for choosing the multiplier so that rationalizing a denominator will take the fewest steps.

**Guided Practice**

**State a factor that can be used to rationalize the denominator of each expression. Then simplify.**

4. $\dfrac{8}{3^{\frac{1}{2}}}$

5. $\dfrac{16}{4^{\frac{3}{2}}}$

6. $\dfrac{1}{y^{\frac{2}{3}}}$

7. $\dfrac{1}{a^{\frac{1}{3}}}$

8. $x^{-\frac{1}{5}}$

9. $\dfrac{1}{t^{\frac{1}{2}} + 1}$

10. $m^{-\frac{3}{2}}$

11. $\dfrac{q}{q^{\frac{1}{2}} - r^{\frac{1}{2}}}$

12. $\dfrac{a + b}{a^{\frac{1}{2}} + b}$

13. $\dfrac{2}{t^{\frac{3}{2}} + s^{\frac{1}{2}}}$

14. $\dfrac{3}{c^{\frac{3}{2}} + c^{\frac{1}{2}}}$

15. $\dfrac{w + 1}{w - w^{\frac{1}{2}}}$

# EXERCISES

**Practice**

**Simplify.**

16. $x^{-\frac{1}{4}}$

17. $\dfrac{1}{s^{\frac{4}{5}}}$

18. $\dfrac{1}{y^{\frac{2}{5}}}$

19. $t^{-\frac{5}{6}}$

20. $n^{-\frac{3}{2}}$

21. $\dfrac{1}{x^{\frac{1}{2}} + 1}$

22. $\dfrac{p + q}{p^{\frac{1}{2}} + q}$

23. $\dfrac{1}{t^{\frac{3}{2}} + t^{\frac{1}{2}}}$

24. $\dfrac{rt}{r^{\frac{1}{2}} + t^{\frac{1}{2}}}$

25. $\dfrac{24}{6^{\frac{2}{3}}}$

26. $\dfrac{15}{5^{\frac{2}{3}}}$

27. $\dfrac{ab^{\frac{1}{2}}}{c^{\frac{3}{2}}}$

28. $\dfrac{xy}{\sqrt[3]{z}}$

29. $\dfrac{n^{\frac{3}{2}} + 3n^{-\frac{1}{2}}}{n^{\frac{1}{2}}}$

30. $\dfrac{a^{\frac{5}{3}}b + 3a^{-\frac{1}{3}}}{a^{\frac{2}{3}}}$

31. $\dfrac{3x + 4x^2}{x^{-\frac{2}{3}}}$

32. $\left(r^{-\frac{1}{6}}\right)^{-\frac{2}{3}}$

33. $\dfrac{3x}{y^{-\frac{3}{2}} \cdot \sqrt[3]{z}}$

**34.** $\left(y^{\frac{1}{3}}\right)^{-\frac{3}{4}}$

**35.** $\dfrac{r^{\frac{3}{2}}}{r^{\frac{1}{2}}+2}$

**36.** $\dfrac{s^{\frac{1}{2}}+t^{\frac{1}{2}}}{s^{\frac{1}{2}}-t^{\frac{1}{2}}}$

**37.** $\dfrac{b^{\frac{1}{2}}}{b^{\frac{3}{2}}-b^{\frac{1}{2}}}$

**38.** $\dfrac{s^{\frac{1}{2}}+1}{s^{\frac{1}{2}}-1}$

**39.** $\left(\dfrac{x^{-2}y^{-6}}{9}\right)^{-\frac{1}{2}}$

**40.** $\dfrac{8^{\frac{1}{6}}-9^{\frac{1}{4}}}{\sqrt{3}+\sqrt{2}}$

**41.** $\dfrac{a^{-\frac{2}{3}}b^{\frac{1}{2}}}{b^{-\frac{3}{2}}\cdot\sqrt[3]{a}}$

**42.** $\dfrac{x^{\frac{5}{3}}-x^{\frac{1}{3}}y^{\frac{4}{3}}}{x^{\frac{2}{3}}+y^{\frac{2}{3}}}$

**Critical Thinking**

**43.** Evaluate $-\dfrac{4}{9}x^9\left(\dfrac{3}{x^2}-\dfrac{1}{\sqrt[3]{2}}\right)$ when $x=\sqrt[6]{2}$.

**Applications**

**44. Electricity** The formula for finding the total resistance of a parallel circuit with four resistors is

$$R_t = \left(\frac{1}{R_1} + \frac{1}{R_2} + \frac{1}{R_3} + \frac{1}{R_4}\right)^{-1},$$

where $R_t$ is the resistance of resistor $t$. Simplify the formula and find the total resistance for a circuit whose resistors have resistance of 16, 12, 8, and 24 ohms.

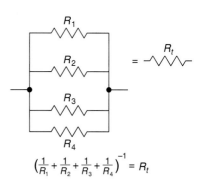

$$\left(\tfrac{1}{R_1}+\tfrac{1}{R_2}+\tfrac{1}{R_3}+\tfrac{1}{R_4}\right)^{-1} = R_t$$

**45. Medicine** A doctor has determined that 20 units of a medication are in a patient's system now. At the end of any hour there is one-third the medication left in his system that was there when the hour began. Assuming that no more injections of medication are made, write an expression for the amount of medication in the patient's system after $t$ hours.

**Mixed Review**

**46.** What fraction of the perfect squares between 0 and 100 are odd? **(Lesson 6-5)**

**47.** How would you write the seventh root of 5 cubed using an exponent? **(Lesson 6-4)**

**48.** Use synthetic division to find $(2x^3 - 2x^2 + 4) \div (x + 1)$. **(Lesson 5-7)**

**49.** Describe the graphs of two linear equations that are dependent. **(Lesson 3-1)**

**50.** Determine the slope of the line passing through the points (9, 0) and (4, -5) **(Lesson 2-4)**

# Solving Equations Containing Radicals

**Objective**

After studying this lesson, you should be able to:

■ solve equations containing radicals.

**Application**

A jeweler is designing a pin in the shape of a parallelogram. The design consists of a square piece of black onyx and a piece of mother-of-pearl on either side. The pieces of mother-of-pearl are isosceles right triangles. A 10-centimeter piece of gold wire is to be placed around the outside of the pin. What are the lengths of the sides of the parallelogram?

Solving some equations involves the use of radical numbers in equations. Let's investigate this type of equation while we find the lengths of the sides of the parallelogram described above.

The drawing at the right models the shape of the pin.
Let $x$ = the length of a side of the square.

Since we know that the triangles are right triangles, we can use the Pythagorean Theorem to find an expression for the length of an end of the parallelogram. *Since the triangle is isosceles, the legs are of equal length.*

$$a^2 + b^2 = c^2$$
$$x^2 + x^2 = c^2$$
$$2x^2 = c^2$$
$$\pm x\sqrt{2} = c$$

Since a measurement is positive, $x\sqrt{2}$ is the length of an end of the parallelogram.

The 10-cm-long gold wire represents the perimeter of the figure. The perimeter is $x + x + x\sqrt{2} + x + x + x\sqrt{2}$ or $4x + 2x\sqrt{2}$. So, $4x + 2x\sqrt{2} = 10$. We must use some of the properties of radical expressions to solve this equation.

$$4x + 2x\sqrt{2} = 10$$
$$x(4 + 2\sqrt{2}) = 10 \qquad \text{\textit{Factor out x.}}$$
$$x = \frac{10}{4 + 2\sqrt{2}} \qquad \text{\textit{Divide each side by } } 4 + 2\sqrt{2}.$$
$$= \left(\frac{10}{4 + 2\sqrt{2}}\right)\left(\frac{4 - 2\sqrt{2}}{4 - 2\sqrt{2}}\right) \qquad \text{\textit{Rationalize the denominator.}}$$
$$= \frac{10 - 5\sqrt{2}}{2}$$

Using a calculator to evaluate this expression, we find that $x$ is about 1.46 cm. So, the sides of the pin are $2x$ or about 2.92 cm and $x\sqrt{2}$ or 2.06 cm long.

**Example 1**

Solve $x + 1 = x\sqrt{2}$.

$$x + 1 = x\sqrt{2}$$
$$1 = x\sqrt{2} - x$$
$$1 = x(\sqrt{2} - 1)$$
$$\frac{1}{\sqrt{2} - 1} = x$$
$$\frac{\sqrt{2} + 1}{\sqrt{2} + 1} \cdot \frac{1}{\sqrt{2} - 1} = x$$
$$\frac{\sqrt{2} + 1}{1} = x$$
$$\sqrt{2} + 1 = x$$

The solution is $\sqrt{2} + 1$.

**Check:**
$$x + 1 = x\sqrt{2}$$
$$(1 + \sqrt{2}) + 1 \stackrel{?}{=} (1 + \sqrt{2})\sqrt{2}$$
$$2 + \sqrt{2} = \sqrt{2} + 2 \checkmark$$

Sometimes variables appear in the radicand. Equations with radicals like this are called **radical equations.** To solve this type of equation, you will need to square (or sometimes cube) each side of the equation to remove the variable from the radical. Solving radical equations in this way sometimes yields **extraneous solutions,** solutions that do not satisfy the original equation. You must check all the possible solutions in the *original* equation and disregard the extraneous solutions.

**Example 2**

Solve $9 + \sqrt{x - 1} = 1$.

We need to remove the radical from this equation. We can achieve this subgoal by isolating the radical on one side of the equation.

$$9 + \sqrt{x - 1} = 1$$
$$\sqrt{x - 1} = -8 \quad \textit{Isolate the radical.}$$
$$x - 1 = 64 \quad \textit{Square each side.}$$
$$x = 65$$

The solution does not check.

**Check:** $9 + \sqrt{x - 1} = 1$
$$9 + \sqrt{65 - 1} \stackrel{?}{=} 1$$
$$9 + \sqrt{64} \stackrel{?}{=} 1$$
$$17 \neq 1$$

The equation has no real solution.

**Example 3**

Solve $5 - \sqrt{b + 2} = 0$.

$$5 - \sqrt{b + 2} = 0$$
$$5 = \sqrt{b + 2}$$
$$25 = b + 2 \quad \textit{Square each side.}$$
$$23 = b$$

The solution is 23.

**Check:** $5 - \sqrt{b + 2} = 0$
$$5 - \sqrt{23 + 2} \stackrel{?}{=} 0$$
$$5 - \sqrt{25} \stackrel{?}{=} 0$$
$$0 = 0 \checkmark$$

**Example 4**

**APPLICATION**

**Fire Fighting**

The Durham City Fire Department is going to buy some new hoses. The hoses they buy must be powerful enough to propel water at least 75 feet into the air. The advertisement for the hose they are considering says that the water flows from the hose at a velocity as high as 72 feet per second. Use the formula $v = \sqrt{2gh}$, where $v$ is the velocity of the water, $g$ is the acceleration due to gravity, and $h$ is the maximum height of the water flow, to determine whether this hose will be suitable. Assume that the acceleration due to gravity is 32 feet per second squared.

$$v = \sqrt{2gh}$$
$$72 = \sqrt{2(32)h} \qquad \textit{Substitute the values into the formula.}$$
$$5184 = 64h \qquad \textit{Square each side.}$$
$$81 = h \qquad \textit{Check this solution.}$$

This hose will propel water up to 81 feet into the air, so it will be suitable.

**Example 5**

Solve $\sqrt[3]{4a - 1} - 3 = 0$.

$$\sqrt[3]{4a - 1} - 3 = 0$$
$$\sqrt[3]{4a - 1} = 3 \qquad \textit{Isolate the radical.}$$
$$4a - 1 = 27 \qquad \textit{Cube each side.}$$
$$4a = 28$$
$$a = 7$$

**Check:**
$$\sqrt[3]{4a - 1} - 3 = 0$$
$$\sqrt[3]{4(7) - 1} - 3 \stackrel{?}{=} 0$$
$$\sqrt[3]{27} - 3 \stackrel{?}{=} 0$$
$$0 = 0 \checkmark$$

The solution is 7.

**Example 6**

Solve $\sqrt{x + 8} - \sqrt{x + 35} = -3$.

$$\sqrt{x + 8} - \sqrt{x + 35} = -3$$
$$\sqrt{x + 8} = \sqrt{x + 35} - 3 \qquad \textit{Isolate one radical.}$$
$$x + 8 = x + 35 - 6\sqrt{x + 35} + 9 \quad \textit{Square each side.}$$
$$-36 = -6\sqrt{x + 35} \qquad \textit{Now isolate the radical again.}$$
$$6 = \sqrt{x + 35} \qquad \textit{Divide each side by } -6.$$
$$36 = x + 35 \qquad \textit{Square each side.}$$
$$x = 1$$

**Check:**
$$\sqrt{x + 8} - \sqrt{x + 35} = -3$$
$$\sqrt{1 + 8} - \sqrt{1 + 35} \stackrel{?}{=} -3$$
$$\sqrt{9} - \sqrt{36} \stackrel{?}{=} -3$$
$$3 - 6 \stackrel{?}{=} -3$$
$$-3 = -3 \checkmark$$

The solution is 1.

**Example 7**

Solve $r = \sqrt[3]{\dfrac{3w}{4\pi d}}$ for $d$.

$$r = \sqrt[3]{\dfrac{3w}{4\pi d}}$$

$$r^3 = \dfrac{3w}{4\pi d} \qquad \textit{Cube each side.}$$

$$r^3 \cdot d = \dfrac{3w}{4\pi d} \cdot d \qquad \textit{Multiply each side by d.}$$

$$\dfrac{r^3 d}{r^3} = \dfrac{3w}{4\pi r^3} \qquad \textit{Divide each side by } r^3.$$

$$d = \dfrac{3w}{4\pi r^3}$$

# CHECKING FOR UNDERSTANDING

**Communicating Mathematics**

**Read and study the lesson to answer these questions.**

1. What is an extraneous solution?

2. Look at Examples 2 and 3. Why do you think Example 2 had an extraneous solution and Example 3 did not?

3. Why do you need to isolate the radical before squaring to remove it?

**Guided Practice**

**Solve each equation. Be sure to check for extraneous solutions.**

4. $\sqrt{x} = 3$

5. $\sqrt{y} = 5$

6. $\sqrt{n} - 8 = 0$

7. $\sqrt{s} - 4 = 0$

8. $\sqrt{2y + 7} = 3$

9. $\sqrt{3m + 7} = 7$

10. $\sqrt[3]{x - 2} = 3$

11. $\sqrt[4]{3w + 7} = 2$

12. $y\sqrt{3} - y = 7$

# EXERCISES

**Practice**

**Solve each equation. Be sure to check for extraneous solutions.**

13. $1 + x\sqrt{2} = 0$

14. $7 + 6n\sqrt{5} = 0$

15. $6 + 2x\sqrt{3} = 0$

16. $2 + 5r\sqrt{10} = 0$

17. $x\sqrt{2} + 3x = 4$

18. $3x + 5 = x\sqrt{3}$

19. $x - x\sqrt{5} = 2$

20. $13 - 3p = p\sqrt{5}$

21. $\sqrt{a - 4} - 3 = 0$

22. $\sqrt{x - 5} - 7 = 0$

Solve each equation. Be sure to check for extraneous solutions.

23. $\sqrt[3]{s + 1} = 2$

24. $\sqrt[3]{n - 1} = 3$

25. $\sqrt[4]{3q} - 2 = 0$

26. $\sqrt[4]{4b} = 3$

27. $\sqrt{2c + 3} - 7 = 0$

28. $\sqrt{3z - 5} - 3 = 1$

29. $\sqrt{4x + 8} + 9 = 11$

30. $\sqrt{1 + 2g} - 6 = -3$

31. $\sqrt{5y + 1} + 6 = 10$

32. $\sqrt[4]{2d + 3} + 5 = 4$

33. $\sqrt{3f + 1} - 2 = 6$

34. $\sqrt[3]{x + 5} + 6 = 4$

35. $\sqrt{y + 5} = \sqrt{2y - 3}$

36. $\sqrt{x - 4} = \sqrt{2x - 3}$

37. $\sqrt{n + 12} - \sqrt{n} = 2$

38. $\sqrt{x + 6} - \sqrt{x} = \sqrt{2}$

39. $\sqrt{y - 5} - \sqrt{y} = 1$

40. $\sqrt{c + 4} = \sqrt{c + 20} - 2$

41. $\sqrt{x - 1} + \sqrt{x + 3} = 5$

42. $\sqrt{4y + 1} - \sqrt{4y - 2} = 3$

43. $\sqrt{y + 1} + \sqrt{y - 3} = 5$

44. $\sqrt{x^2 + 5x} + x + 10 = 0$

45. $\sqrt{x + 12} + 1 = \sqrt{x + 21}$

46. $\sqrt{5y^2 + 7y - 2} - y\sqrt{5} = -4$

Solve for the variable indicated.

47. $y = \sqrt{r^2 + s^2}$ for $r$

48. $t = \sqrt{\dfrac{2s}{g}}$ for $s$

49. $r = \sqrt[3]{\dfrac{2mM}{c}}$ for $c$

50. $m^2 = \sqrt[3]{\dfrac{rp}{g^2}}$ for $p$

**Critical Thinking**

51. For what values of $a$, $b$, and $x$ will the equation $\sqrt{5x - 7} + a = b$ have no real solution?

**Applications**

52. **Air Traffic Control**  The radar used by the air traffic controllers at the Lake County Municipal Airport can receive signals from a circular area of 7854 square miles. Using the formula $A = \pi r^2$, where $A$ represents the area of the circular region and $r$ represents the radius of the region, find the greatest distance from which a signal can be received by the radar.

53. **Energy**  The energy of direct sunlight on a solar cell with an area of one square centimeter is converted into 0.01 watt of electrical energy. Suppose that a square solar cell must deliver 15 watts of energy. What dimensions should the cell have?

54. **Aerospace Engineering** The radius of the orbit of a satellite is found by

$$r = \sqrt[3]{\dfrac{GMt^2}{4\pi^2}},$$ where $t$ represents the time it takes for the satellite to complete one orbit, $G$ represents the constant of universal gravitation, and $M$ is the mass of the central object. Solve the formula for $t$.

**Mixed Review**

Simplify. **(Lesson 6-6)**

55. $\dfrac{rs}{r^{\frac{1}{2}} + r^{\frac{3}{2}}}$

56. $(\sqrt[6]{5}a^{\frac{7}{4}}b^{-\frac{2}{3}})^{12}$

57. Find the degree of the polynomial $a^8 + a^7b + a^6b^2 - a^2b^6 - ab^7 - b^8$. **(Lesson 5-4)**

58. Does the matrix $\begin{bmatrix} 3 & 4 & -2 \\ 0 & 1 & 0 \\ -1 & -1 & 4 \end{bmatrix}$ have an inverse? **(Lesson 4-4)**

59. The Woodward Park High School auditorium seats 150 people. Admission to the spring play is $2.00 for adults and $1.00 for students. The Drama Club has already sold fifty student tickets and the rest are to be sold at the door. How many of each type of ticket should be sold for the Drama Club to earn the maximum amount of money? **(Lesson 3-7)**

60. Is the relation {(0, 0), (1, 0)} a function? Explain your answer. **(Lesson 2-1)**

## HISTORY CONNECTION

**Johann Kepler** (1571–1630) ranks foremost as the mathematician of the sky. He dreamed of a harmony in arithmetic, geometry, and music, and that every planet had its own tune. He described all of this in his 1618 book *Harmonica Mundi* (Harmony of the Worlds). While this may seem comical, he did disclose his third law of planetary motion in this text. This law which states that the square of the time of revolution of each planet is proportional to the cube of its mean distance from the sun, is still used to compare the distances and periods of the planets about the sun. One form of the law can be expressed using rational exponents.

$$\dfrac{T_a}{T_b} = \left(\dfrac{r_a}{r_b}\right)^{\frac{3}{2}},$$ where $T_a$ and $T_b$ are the planets' periods and $r_a$ and $r_b$ are their average distances from the sun.

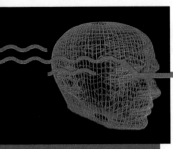

# Technology

## Solving Radical Equations

The *Mathematics Exploration Toolkit (MET)* can be used to solve equations that contain square roots. Some equations can be solved automatically with the SOLVEFOR command. Other equations require you to choose appropriate steps to simplify the equation before the computer can solve it. The computer does not check solutions obtained by squaring both sides of an equation. Therefore, some solutions found by the computer will *not* satisfy the original equation. The CALC commands (and their shortened forms) you can use are listed below.

| | | |
|---|---|---|
| ADD (add) | SUBTRACT (sub) | MULTIPLY (mul) |
| DIVIDE (div) | RAISETO (rai) | SIMPLIFY (simp) |
| STORE (sto) | SUBSTITUTE (subst) | & enters a $\sqrt{\ }$ . |

**Example**    **Solve $\sqrt{x+1} + \sqrt{x-4} = 5$.**

ENTER:  &(x + 3) + &(x − 4) = 5

| ENTER: | |
|---|---|
| &(x + 3) + &(x − 4) = 5 | $\sqrt{x+1}+\sqrt{x-4}=5$ |
| sto r | saves the equation as r |
| rai 2 | $(\sqrt{x+1}+\sqrt{x-4})^2=5^2$ |
| simp, simp | $2\sqrt{x^2-3x-4}+2x-3=25$ |
| sub 2x | $2\sqrt{x^2-3x-4}+2x-3-2x=25-2x$ |
| simp | $2\sqrt{x^2-3x-4}-3=-2x+25$ |
| add 3 | $2\sqrt{x^2-3x-4}-3+3=-2x+25+3$ |
| simp | $2\sqrt{x^2-3x-4}=-2x+28$ |
| rai 2 | $(2\sqrt{x^2-3x-4})^2=(-2x+28)^2$ |
| simp, simp | $4x^2-12x-16=4x^2-112x+784$ |
| sol x | $x=8$ |

Now check the computer's solution.

| ENTER: r | $\sqrt{x+1}+\sqrt{x-4}=5$ |
|---|---|
| subst 8 x | $\sqrt{8+1}+\sqrt{8-4}=5$ |
| simp | $5=5$ |

The solution checks. The solution is $x = 8$.

## EXERCISES

**Use CALC commands to solve each equation. Check each solution.**

**1.** $5x - 6\sqrt{x} + 1 = 0$

**2.** $12 - 3\sqrt{x-5} = 0$

**3.** $\sqrt{4-x} + \sqrt{9-2x} = 0$

**4.** $\sqrt{3x-2} - \sqrt{2x-3} = 1$

## 6-8 Pure Imaginary Numbers

**Objectives**

After studying this lesson, you should be able to:
- simplify radicals containing negative radicands,
- multiply pure imaginary numbers, and
- solve quadratic equations that have pure imaginary solutions.

**FYI ...**

Girolamo Cardano began his career as a doctor and studied, taught, and wrote mathematics as a sideline. He held important positions at The Universities of Pavia and Bologna in Italy, and wrote many works on arithmetic, astronomy, physics, and medicine.

Until the sixteenth century, mathematicians were puzzled by square roots of negative numbers. As you know, some expressions have irrational solutions. For example, the solutions to $x^2 - 5 = 0$ are $\sqrt{5}$ and $-\sqrt{5}$. But the equation $x^2 = -1$ has no solution in the real numbers. This is because the square of a real number is nonnegative. However in 1545, the Italian mathematician Girolamo Cardano published *Ars Magna* in which he began working with what the great mathematician René Descartes later called **imaginary numbers.**

The number $i$ is defined to be a solution to $x^2 = -1$ and is *not* a real number. It is called the **imaginary unit.** Using $i$ as you would any constant, you can define square roots of negative numbers.

Since $i = \sqrt{-1}$, it follows that $i^2 = -1$.

$$(2i)^2 = 2^2 i^2 \text{ or } -4 \implies \sqrt{-4} = \sqrt{4} \cdot \sqrt{-1} \text{ or } 2i$$

$$(i\sqrt{2})^2 = i^2(\sqrt{2})^2 \text{ or } -2 \implies \sqrt{-2} = \sqrt{2} \cdot \sqrt{-1} \text{ or } i\sqrt{2}$$

*To avoid $\sqrt{2}(i)$ being read as $\sqrt{2i}$, write $\sqrt{2}(i)$ as $i\sqrt{2}$.*

| *Definition of Pure Imaginary Numbers* | **For any positive real number $b$,** $$\sqrt{-(b^2)} = \sqrt{b^2} \cdot \sqrt{-1} \text{ or } bi$$ **where $i$ is the imaginary unit, and $bi$ is called a pure imaginary number.** |
|---|---|

*-bi is also a pure imaginary number since -bi = -1(bi).*

**Example 1**

a. Simplify $\sqrt{-25}$.

$$\sqrt{-25} = \sqrt{25} \cdot \sqrt{-1}$$
$$= 5 \cdot \sqrt{-1}$$
$$= 5i$$

b. Simplify $\sqrt{-45}$.

$$\sqrt{-45} = \sqrt{45} \cdot \sqrt{-1}$$
$$= (\sqrt{9} \cdot \sqrt{5})i$$
$$= 3i\sqrt{5}$$

The commutative and associative properties for multiplication hold true for pure imaginary numbers.

**Example 2**

a. Simplify $2i \cdot 7i$.

$$2i \cdot 7i = (2 \cdot 7)(i \cdot i)$$
$$= 14i^2$$
$$= 14(-1) \qquad i^2 = -1$$
$$= -14$$

b. Simplify $\sqrt{-5} \cdot \sqrt{-20}$.

$$\sqrt{-5} \cdot \sqrt{-20} = i\sqrt{5} \cdot i\sqrt{20}$$
$$= i^2\sqrt{100}$$
$$= -1 \cdot 10 \text{ or } -10$$

Simplify the successive powers of $i$. Do you see a pattern?

$i^1 = i$                                   $i^5 = i^4 \cdot i = 1 \cdot i = i$

$i^2 = -1$                                  $i^6 = i^4 \cdot 1^2 = 1 \cdot -1 = -1$

$i^3 = i^2 \cdot i = -1 \cdot i = -i$       $i^7 = 1^1 \cdot i^3 = 1 \cdot (-i) = -i$

$i^4 = i^2 \cdot i^2 = -1 \cdot (-1) = 1$   $i^8 = i^4 \cdot i^4 = 1 \cdot 1 = 1$

**Example 3**

Simplify $i^{13}$.

*Method 1*

$$i^{13} = i^4 \cdot i^4 \cdot i^4 \cdot i^1$$
$$= 1 \cdot 1 \cdot 1 \cdot i$$
$$= i$$

*Method 2*

$$i^{13} = i^{12} \cdot i^1$$
$$= (i^2)^6 \cdot i$$
$$= (-1)^6 \cdot i$$
$$= i$$

**Example 4**

Solve $x^2 + 7 = 0$.

$$x^2 + 7 = 0$$
$$x^2 = -7$$
$$x = \pm\sqrt{-7} \qquad \sqrt{-7} = \sqrt{7} \cdot \sqrt{-1}$$
$$= \pm i\sqrt{7}$$

The solutions, $i\sqrt{7}$ and $-i\sqrt{7}$, are both pure imaginary numbers.

*Check these solutions.*

**Example 5**

**APPLICATION**

**Electronics**

The reactance of an electrical circuit is represented by a pure imaginary number and is found by the formula $X = X_L - X_C$. $X_L$ represents the inductive reactance and $X_C$ represents the capacitive reactance. If the reactance of a circuit is $16i$ ohms and the capacitive reactance is $5i$ ohms, find the inductive reactance.

$$X = X_L - X_C$$
$$16i = X_L - 5i \qquad \textit{Substitute } 16i \textit{ for } X \textit{ and } 5i \textit{ for } X_C.$$
$$21i = X_L$$

The inductive reactance is $21i$ ohms.

# CHECKING FOR UNDERSTANDING

**Communicating Mathematics**

**Read and study the lesson to answer these questions.**

1. What is an imaginary number? Is an imaginary number a real number?
2. Give examples of numbers that are pure imaginary numbers.
3. What is the pattern formed by the powers of $i$?

## Guided Practice

**Simplify.**

**4.** $\sqrt{-49}$

**5.** $\sqrt{-36}$

**6.** $4\sqrt{-3}$

**7.** $\sqrt{-2} \cdot \sqrt{-2}$

**8.** $6\sqrt{-4}$

**9.** $\sqrt{-3} \cdot \sqrt{-3}$

**10.** $\sqrt{-5} \cdot \sqrt{5}$

**11.** $4 \cdot 5i$

**12.** $i^{10}$

# EXERCISES

**Practice**

**Simplify.**

**13.** $\sqrt{-169}$

**14.** $\sqrt{-100}$

**15.** $\sqrt{-50}$

**16.** $\sqrt{-98}$

**17.** $\sqrt{-\dfrac{4}{9}}$

**18.** $\sqrt{-\dfrac{9}{25}}$

**19.** $\sqrt{-\dfrac{1}{5}}$

**20.** $\sqrt{-\dfrac{1}{2}}$

**21.** $i^5$

**22.** $i^{11}$

**23.** $i^{91}$

**24.** $i^{244}$

**25.** $\sqrt{-8} \cdot \sqrt{-2}$

**26.** $\sqrt{-3} \cdot \sqrt{-18}$

**27.** $\sqrt{-14} \cdot \sqrt{-7}$

**28.** $(\sqrt{-5})^2$

**29.** $(\sqrt{-12})^2$

**30.** $(\sqrt{-3})^3$

**31.** $(\sqrt{-4})^3$

**32.** $\sqrt{9} \cdot \sqrt{-9}$

**33.** $\sqrt{3} \cdot \sqrt{-27}$

**34.** $\sqrt{-5} \cdot \sqrt{20}$

**35.** $\sqrt{-8} \cdot \sqrt{6}$

**36.** $(2i)(3i)^2$

**37.** $(-2\sqrt{-8})(3\sqrt{-2})$

**38.** $(4\sqrt{-12})(-2\sqrt{-3})$

**39.** $5i(-2i)^2$

**40.** $(3\sqrt{21})(-2\sqrt{-21})$

**Solve each equation.**

**41.** $a^2 + 16 = 0$

**42.** $x^2 + 64 = 0$

**43.** $m^2 + 121 = 0$

**44.** $n^2 + 169 = 0$

**45.** $3x^2 + 27 = 0$

**46.** $6y^2 = -96$

**47.** $t^2 + 12 = 0$

**48.** $3a^2 + 18 = 0$

**49.** $4x^2 + 5 = 0$

**50.** $5w^2 = -40$

**Critical Thinking**

**51.** Simplify $\sqrt{-81} \cdot \sqrt{-8} + \sqrt[3]{256}$.

**Application**

**52. Electricity** The reactance of an electrical circuit is represented by a pure imaginary number, and is found by the formula $X = X_L - X_C$. $X_L$ represents the inductive reactance and $X_C$ represents the reactance from all capacitors.

**a.** The total reactance for the capacitors in a circuit is $6i$ ohms. If the reactance for the inductions is $14i$ ohms, find the total reactance of the circuit.

**b.** A circuit has a total reactance of $8i$ ohms. If the reactance of the capacitors is $7i$ ohms, find the reactance of the inductions.

**Computer**  53. The BASIC program below simplifies powers of $i$. It uses the pattern of the powers discussed in Exercise 3. Lines 30–60 of the program find the remainder of the power divided by four and then lines 70–130 use that remainder to determine the value of the power.

```
10    PRINT "ENTER THE POWER"
20    INPUT N
30    IF N - 4 * INT(N/4) = THEN 70
40    IF N - 4 * INT(N/4) = 1 THEN 90
50    IF N - 4 * INT(N/4) = 2 THEN 110
60    IF N - 4 * INT(N/4) = 3 THEN 130
70    PRINT "i ^ "; N; " = 1"
80    GOTO 140
90    PRINT "i ^ "; N; " = i"
100   GOTO 140
110   PRINT "i ^ "; N; " = -1"
120   GOTO 140
130   PRINT "i ^ "; N; " = -i"
```

Enter the program and use it to simplify each power of $i$.

a. $i^{17}$  b. $i^{34}$
c. $i^{59}$  d. $i^{92}$
e. $i^{103}$  f. $i^{300}$
g. $i^{997}$  h. $i^{2002}$

**Mixed Review**  Solve each equation.  **(Lesson 6-7)**

54. $2x + 7 = -x\sqrt{2}$

55. $\sqrt{3y^2 + 11y - 5} = y\sqrt{3} + 1$

56. Simplify $\sqrt{(x - 2)^2}$.  **(Lesson 6-1)**

57. Factor $a^3 + 1$.  **(Lesson 5-5)**

58. Naren bought 2 slices of pizza, two cartons of milk, and one chocolate chip cookie for lunch at the Meadow Park High School cafeteria. He spent $2.05 on his lunch. Ted brought a sandwich from home, so he just bought a carton of milk and 3 cookies. His total bill was 75¢. Sarah spent $0.95 on one slice of pizza and a carton of milk. How much each does a slice of pizza, a carton of milk, and a chocolate chip cookie cost? **(Lesson 4-7)**

59. Find the value of the determinant $\begin{vmatrix} 18 & -5 \\ -9 & 11 \end{vmatrix}$.  **(Lesson 3-3)**

60. **Health**  The optimum heart rate is the rate that a person should achieve during exercise for the exercise to be most beneficial. The prediction equation, $r = 0.6(220 - a)$, where $a$ represents age, can be used to find a person's optimum heart rate, $r$. If Peggy is 20 years old, find her optimum heart rate.  **(Lesson 2-2)**

## 6-9 Complex Numbers

**Objective**

After studying this lesson, you should be able to:
- add, subtract, and multiply complex numbers.

*FYI ···*

Research is still being conducted in the field of fractal geometry. So far, fractals have been used to simulate objects in nature, such as clouds and mountains. Their first application to science was to solve the problem of interference on transmission channels like telephone lines.

The picture at the right, called a fractal, was created with the aid of a computer. Many fractal objects are generated using functions that are iterated; that is, repeated over and over. The function is evaluated for some initial value of $x$, then the function is evaluated for the new value. Repeating this process and plotting the points produces interesting and sometimes beautiful pictures. Some of the most exciting fractals are created using this process with a number like $4 + 3i$ as the initial value. Numbers like $4 + 3i$ are called **complex numbers.**

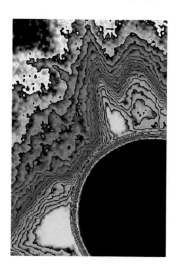

| Definition of a Complex Number | **A complex number is any number that can be written in the form $a + bi$ where $a$ and $b$ are real numbers and $i$ is the imaginary unit. $a$ is called the real part, and $bi$ is called the imaginary part.** |
|---|---|

A real number is also a complex number. For example, $\sqrt{3}$ can be expressed as $\sqrt{3} + 0i$. The imaginary part is 0. A complex number is real only if the imaginary part is zero.

### The Complex Numbers

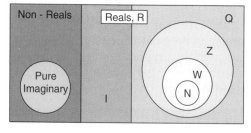

R = reals
I = irrationals
Q = rationals
Z = integers
W = wholes
N = naturals

The diagram at the left shows the relationship among the various sets of numbers that you have studied and the complex numbers.

Two complex numbers are equal if and only if their real parts are equal and their imaginary parts are equal. That is,

$$a + bi = c + di \text{ if and only if } a = c \text{ and } b = d.$$

**Example 1**

Find values for $x$ and $y$ such that $3x + 4yi = 12 + 8i$.

$3x + 4yi = 12 + 8i$

$3x = 12$ and $4y = 8$

$x = 4 \qquad\qquad y = 2$

**Check:** $3x + 4yi = 12 + 8$

$3(4) + 4(2)i \stackrel{?}{=} 12 + 8i$

$12 + 8i = 12 + 8i \checkmark$

To add or subtract complex numbers, we must combine like terms; that is, combine the real parts and combine the imaginary parts.

**Example 2**

> **Simplify $(2 + 5i) + (4 - i)$.**
>
> $(2 + 5i) + (4 - i) = (2 + 4) + (5i - i)$
> $\qquad\qquad\qquad\quad = 6 + 4i$

**Example 3**

> **Simplify $(8 - 2i) - (6 - 4i)$.**
>
> $(8 - 2i) - (6 - 4i) = (8 - 6) + (-2i - (-4i))$
> $\qquad\qquad\qquad\qquad = 2 + 2i$

*The complex plane is also known as the Gaussian plane or an Argand diagram.*

The complex numbers can be graphed on the complex plane, where the horizontal axis represents the real part of the complex number and the vertical axis represents the imaginary part. The complex numbers are represented by the segments whose endpoints are the origin and a point whose coordinates are the real part and the imaginary part of the complex number. Addition of complex numbers can also be represented by graphing. First, graph the two numbers to be added. Then complete the parallelogram that has two sides represented by the segments. The segment from the origin to the fourth vertex of the parallelogram represents the sum of the two original numbers.

**Example 4**

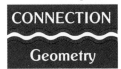

*This way of adding complex numbers assumes the numbers do <u>not</u> lie on the same line.*

> **Graph $-4 + 3i$ and $5 + 2i$ on the complex plane. Find their sum geometrically.**
>
> Graph each complex number using the real part as the $x$-coordinate and the imaginary part as the $y$-coordinate. Connect each point to the origin.
>
> Next, complete the parallelogram. The fourth vertex has coordinates $(1, 5)$ or $1 + 5i$.
>
>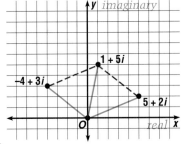
>
> Check by adding algebraically.
> $(-4 + 3i) + (5 + 2i) = (-4 + 5) + (3i + 2i)$
> $\qquad\qquad\qquad\qquad = 1 + (3 + 2)i$
> $\qquad\qquad\qquad\qquad = 1 + 5i \quad$ *It checks.*
>
> *The green segment represents $1 + 5i$.*

You can multiply complex numbers using the FOIL method.

**Example 5**

> **Simplify $(9 - 3i)(2 + 2i)$.**
>
> $\qquad\qquad\qquad\quad F \qquad O \qquad\quad I \qquad\quad L$
> $(9 - 3i)(2 + 2i) = 9 \cdot 2 + 9 \cdot 2i + (-3i) \cdot 2 + (-3i) \cdot 2i$
> $\qquad\qquad\qquad = 18 + 18i - 6i - 6i^2$
> $\qquad\qquad\qquad = (18 + 6) + (18i - 6i) \qquad -6i^2 = 6$
> $\qquad\qquad\qquad = 24 + 12i$

**Example 6** | Simplify $(-2 + 3i)(3 - i)$.

$$
\begin{aligned}
(-2 + 3i)(3 - i) &= (-2) \cdot 3 + (-2) \cdot (-i) + 3i \cdot 3 + 3i \cdot (-i) \\
&= -6 + 2i + 9i - 3i^2 \\
&= (-6 + 3) + (2i + 9i) \quad\quad -3i^2 = -3(-1) = 3 \\
&= -3 + 11i
\end{aligned}
$$

This chart summarizes addition, subtraction, and multiplication of complex numbers.

| For any complex numbers $a + bi$ and $c + di$: |
| :---: |
| $(a + bi) + (c + di) = (a + c) + (b + d)i$ |
| $(a + bi) - (c + di) = (a - c) + (b - d)i$ |
| $(a + bi)(c + di) = (ac - bd) + (ad + bc)i$ |

# CHECKING FOR UNDERSTANDING

**Communicating Mathematics**

**Read and study the lesson to answer these questions.**

1. When are two complex numbers equal?
2. How are the complex and real numbers related?
3. State two phenomena that can be modeled by fractals.
4. What number does the segment from the origin to (4, 2) on the complex plane represent?

**Guided Practice**

**Simplify.**

5. $(4 - i) + (3 + 3i)$
6. $(7 + 2i) + (2 + 8i)$
7. $(5 + 2i) - (2 + 2i)$
8. $(7 - 6i) - (5 - 6i)$
9. $(7 + 3i) + (3 - 3i)$
10. $(2 - 4i) + (2 + 4i)$
11. $4(5 + 3i)$
12. $-6(2 - 3i)$
13. $(2 + 4i)(1 + 3i)$
14. $(1 - 4i)(2 - 3i)$
15. $(4 + i)(4 - i)$
16. $(4 - i)(3 + 2i)$

**Find the values of $x$ and $y$ for which each equation is true.**

17. $x + yi = 2 - 3i$
18. $x - yi = 5 + 6i$
19. $x + 2yi = 3$
20. $x - yi = 4 + 5i$
21. $x - yi = 7 - 2i$
22. $2x + yi = 5i$

# EXERCISES

**Practice**

**Simplify.**

23. $(4 + 2i) + (1 + 3i)$
24. $(2 + 6i) + (4 + 3i)$
25. $(11 + 5i) - (4 + 2i)$
26. $(11 - \sqrt{-3}) - (-4 + \sqrt{-5})$
27. $(8 - 7i) + (-5 - i)$
28. $(5 + \sqrt{-7}) + (-3 + \sqrt{-2})$

**29.** $(-6 - 2i) - (-8 - 3i)$

**30.** $(3 - 11i) - (-5 + 4i)$

**31.** $(1 - 5i\sqrt{3}) + (4 + 2i\sqrt{3})$

**32.** $(8 - 3i\sqrt{5}) + (-3 + 2i\sqrt{5})$

**33.** $3(4 - 5i) - 6(2 - i)$

**34.** $3(-5 - 2i) + 2(-3 + 2i)$

**35.** $(5 + 3i)(6 - i)$

**36.** $(5 + i)(2 - 3i)$

**37.** $(6 - 2i)^2$

**38.** $(2 + i\sqrt{3})^2$

**39.** $(7 - 2i)(4 - 3i)$

**40.** $(7 - i\sqrt{2})(5 + i\sqrt{2})$

**41.** $(3 + 4i)^2$

**42.** $(3 + 2i)^2$

**43.** $(\sqrt{2} + i)(\sqrt{2} - i)$

**44.** $(2 - \sqrt{-3})(2 + \sqrt{-3})$

**45.** $(3 + 2i)(3 - 2i)$

**46.** $(3 + \sqrt{-2})(3 - \sqrt{-2})$

CONNECTION

Geometry

**Graph each addend on the complex plane. Then find their sum geometrically.**

**47.** $(-2 + i) + (4 + 4i)$

**48.** $(9 + 4i) + (3 - 2i)$

**49.** $(-1 - 5i) + (4 + 0i)$

**50.** $(3 + 2i) + (-11 - 5i)$

**Find the values of $x$ and $y$ for which each equation is true.**

**51.** $3x + 2yi = 18 + 7i$

**52.** $3x + 5yi = 6 + 20i$

**53.** $(x - y) + (x + y)i = 2 - 4i$

**54.** $(2x + y) + (x - y)i = 7 - i$

**55.** $(x + 2y) + (2x - y)i = 5 + 5i$

**56.** $(x + 4y) + (2x - 3y)i = 13 + 7i$

**Simplify.**

*Portfolio*

Select an item from this chapter that you feel shows your best work and place it in your portfolio. Explain why you selected it.

**57.** $(1 + 2i)(3 - 4i)(2 + i)$

**58.** $(3 + 3i)(6 - i)(5 + 2i)$

**59.** $(2 - 3i)(7 + 5i)(7 - 5i)$

**60.** $(4 + 3i)(3 + i)(2 - 7i)$

**61.** $(7 - i)(5 + 2i)(4 + 2i)$

**62.** $(5 + i)(9 + 2i)(9 - 2i)$

**63.** Write an expression for the additive inverse of the complex number $a + bi$.

**64.** Show that 0 is the additive identity for the complex numbers.

**65.** Show that 1 is the multiplicative identity for the complex numbers.

**Critical Thinking**

**66.** Under which of the operations, addition, subtraction, or multiplication, are the imaginary numbers closed? Give examples to support your answer.

**Applications**

**67. Fractal Geometry** Suppose the function $f(x) = x^2 - 1$ is to be iterated to produce a fractal. Find the first four points of the iteration if the initial value of $x$ is $(1 + i)$.

**68. Electrical Engineering** The relationship between the flow of electricity, $I$, in a circuit, the resistance to the flow, $Z$, called impedance, and the electromotive force, $E$, called voltage, is given by the formula $E = I \cdot Z$. Electrical engineers use $j$ to represent the imaginary unit. An electrical engineer is designing a circuit that is to have a current of $(6 - j8)$ amps. If impedance of the circuit is $(14 + j8)$ ohms, find the voltage.

**Mixed Review**

69. Simplify $\sqrt{-\dfrac{1}{3}}$. **(Lesson 6-8)**

70. Solve $3a^2 + 24 = 0$. **(Lesson 6-8)**

71. Simplify $\sqrt[4]{5} + 6\sqrt[4]{5} - 2\sqrt[4]{5}$. **(Lesson 6-3)**

72. Divide $(y^4 + 6y^3 - 7y^2 + 7y - 1) \div (y + 3)$ using synthetic division. **(Lesson 5-7)**

73. **Chemistry** The mass of a proton is $1.672 \times 10^{-24}$ grams. If the mass of Earth's moon is $7.35 \times 10^{22}$ kilograms, how many times greater is its mass than that of the proton? **(Lesson 5-2)**

74. Find $\begin{bmatrix} 4 & 5 \\ 2 & -2 \\ 4 & 9 \end{bmatrix} + \begin{bmatrix} -3 & 2 \\ -1 & 4 \\ 4 & 4 \end{bmatrix}$. **(Lesson 4-3)**

75. Solve $\begin{cases} x + y < 8 \\ x + y > 5 \end{cases}$ by graphing. **(Lesson 3-4)**

76. Graph the system $\begin{cases} 3x - y = 4 \\ 9x - 6 = 3y \end{cases}$ and state its solution. Then state whether the system is *consistent and independent, consistent and dependent,* or *inconsistent.* **(Lesson 3-1)**

77. Use the vertical line test to determine if the relation graphed at the right is a function. **(Lesson 2-1)**

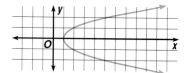

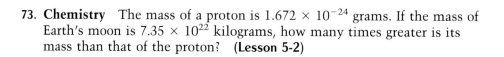

## TECHNOLOGY CONNECTION

### Benoit Mendelbroit

You have probably seen computer-generated images on television or in the movies that looked as if they were real pictures in nature. These images were generated using fractals. Fractal geometry was developed in 1980 by Benoit Mandelbroit, a research mathematician for IBM. Fractal geometry is based on complex numbers and their graphs. With fractal geometry, you can create the "irregular" shapes that appear in nature that you cannot create with traditional geometric shapes such as squares, rectangles, and circles.

# 6-10 Simplifying Expressions Containing Complex Numbers

**Objective**

After studying this lesson, you should be able to:

- simplify rational expressions containing complex numbers in the denominator.

As you know, two radical expressions of the form $a\sqrt{b} + c\sqrt{d}$ and $a\sqrt{b} - c\sqrt{d}$ are conjugates of each other. Since imaginary numbers also involve radicals, numbers of the form $a + bi$ and $a - bi$ are also called conjugates. Recall that the product of two radical conjugates is a rational number. Let's investigate the product of complex conjugates.

**Example 1**

**Find the product of the conjugates $5 + 2i$ and $5 - 2i$.**

$$(5 + 2i)(5 - 2i) = 25 - 10i + 10i - 4i^2 \qquad \textit{Use FOIL.}$$
$$= 25 - 4i^2$$
$$= 25 - (-4) \qquad \textit{4i}^2 = -4$$
$$= 29$$

In this case, the product is rational. Explore the general case.

**Example 2**

**Find $(a + bi)(a - bi)$.**

$$(a + bi)(a - bi) = a^2 - abi + abi - b^2i^2 \qquad \textit{Use FOIL.}$$
$$= a^2 - b^2i^2$$
$$= a^2 - (-b^2) \qquad \textit{b}^2i^2 = -b^2$$
$$= a^2 + b^2$$

Since $a$ and $b$ are real, $a^2 b^2$ will also be real.

To simplify expressions, we must eliminate all radicals in the denominator. Since $i$ is a radical, all imaginary numbers must also be eliminated from denominators.

**Example 3**

**Simplify $\dfrac{9 + 3i}{2i}$.**

$$\frac{9 + 3i}{2i} = \frac{9 + 3i}{2i} \cdot \frac{i}{i} \qquad \textit{Multiply by } \frac{i}{i} \textit{ to remove the i from the denominator.}$$
$$= \frac{9i + 3i^2}{2i^2}$$
$$= \frac{-3 + 9i}{-2} \text{ or } \frac{3 - 9i}{2}$$

We use conjugates of radicals to rationalize the denominators of expressions with radicals in the denominator. We can also use conjugates of complex numbers to rationalize denominators of expressions with complex numbers in the denominator.

**Example 4**

Simplify $\dfrac{5 + i}{1 + 2i}$.

$$\frac{5 + i}{1 + 2i} = \frac{5 + i}{1 + 2i} \cdot \frac{1 - 2i}{1 - 2i} \qquad \textit{1 + 2i and 1 − 2i are conjugates.}$$

$$= \frac{5 - 10i + i - 2i^2}{1 - 4i^2} \qquad \textit{(a + b)(a − b) = a}^2 - b^2$$

$$= \frac{5 - 9i + 2}{1 + 4} \text{ or } \frac{7 - 9i}{5}$$

**Example 5**

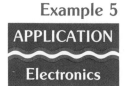
Dr. Sharon Weisman is an electrical engineer designing the electrical circuits for a new office building. There are three basic things to be considered in an electrical circuit: the flow of the electrical current, *I*; the resistance to that flow, *Z*, called impedance; and electromotive force, *E*, called voltage. These quantities are related in the formula $E = I \cdot Z$. The current of the circuit Sharon is designing is to be $(35 - j40)$ amps. Electrical engineers use the letter *j* to represent the imaginary unit. Find the impedance of the circuit if the voltage is to be $(430 - j330)$ volts.

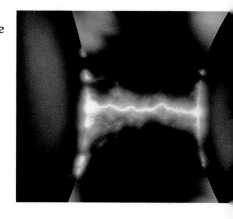

$$E = I \cdot Z$$

$$\frac{E}{I} = Z$$

$$\frac{(430 - j330)}{(35 - j40)} = Z$$

$$\frac{(430 - j330)}{(35 - j40)} \cdot \frac{(35 + j40)}{(35 + j40)} = Z \qquad \textit{Rationalize the denominator.}$$

$$\frac{15{,}050 + j17{,}200 - j11{,}550 - j^2 13{,}200}{1225 - j^2 1600} = Z \qquad \textit{j}^2 = -1$$

$$\frac{28250 + j5650}{2825} = Z$$

$$10 + j2 = Z$$

The impedance will be $(10 + j2)$ ohms.

**Example 6**

Find the multiplicative inverse of $5 - 7i$.

The multiplicative inverse of $5 - 7i$ is $\dfrac{1}{5 - 7i}$.

Now simplify the expression.

$$\dfrac{1}{5 - 7i} = \dfrac{1}{5 - 7i} \cdot \dfrac{5 + 7i}{5 + 7i} \qquad \textit{Multiply by conjugate}$$

$$= \dfrac{5 + 7i}{25 - 49i^2}$$

$$= \dfrac{5 + 7i}{74} \qquad \textit{i}^2 = \textit{-1}$$

**Check:** *The product of a number and its multiplicative inverse is 1.*

$$5 - 7i \cdot \dfrac{5 + 7i}{74} = \dfrac{25 + 49}{74}$$

$$= \dfrac{74}{74} \text{ or } 1 \ \checkmark$$

The inverse of $5 - 7i$ is $\dfrac{5 + 7i}{74}$.

# CHECKING FOR UNDERSTANDING

**Communicating Mathematics**

**Read and study the lesson to answer these questions.**

1. What is the conjugate of the complex number $3 - 2i$?
2. Describe the product of the complex number $a + bi$ and its conjugate.
3. Study Example 6. Describe the multiplicative inverse of a complex number.
4. Does every complex number have a multiplicative inverse?

**Guided Practice**

**Find the conjugate of each complex number.**

5. $4 + i$
6. $1 + 6i$
7. $5 - 4i$
8. $3 - 3i$
9. $5i$
10. $6i$
11. $-10i$
12. $9$
13. $12 - i$

**Show that each pair of numbers are multiplicative inverses of one another.**

14. $2 + 3i;\ \dfrac{2 - 3i}{13}$
15. $5 - 4i;\ \dfrac{5 + 4i}{41}$
16. $6 + 8i;\ \dfrac{3 - 4i}{50}$

# EXERCISES

**Practice** Find the product of each complex number and its conjugate.

**17.** $8 - 2i$

**18.** $3 + 7i$

**19.** $5 - 2i$

**20.** $5$

**21.** $1 + i$

**22.** $12 + 5i$

**23.** $9i$

**24.** $6 + 5i$

**25.** $-10i$

Simplify.

**26.** $\dfrac{4 + 5i}{1 + i}$

**27.** $\dfrac{3 - 2i}{1 - i}$

**28.** $\dfrac{1 + i}{3 + 2i}$

**29.** $\dfrac{11 + i}{2 - i}$

**30.** $\dfrac{3 + 5i}{2i}$

**31.** $\dfrac{1 - i}{4 - 5i}$

**32.** $\dfrac{5 - 6i}{-3i}$

**33.** $\dfrac{2 + i}{5i}$

**34.** $\dfrac{4 - 7i}{-3i}$

**35.** $\dfrac{5}{2 + i}$

**36.** $\dfrac{3}{4 - i}$

**37.** $\dfrac{2}{6 + 5i}$

**38.** $\dfrac{7}{\sqrt{2} - 3i}$

**39.** $\dfrac{4}{\sqrt{3} + 2i}$

**40.** $\dfrac{1 + i\sqrt{2}}{1 - i\sqrt{2}}$

**41.** $\dfrac{\sqrt{3}}{\sqrt{3} - i}$

**42.** $\dfrac{2 + i\sqrt{3}}{2 - i\sqrt{3}}$

**43.** $\dfrac{3 - i\sqrt{5}}{3 + i\sqrt{5}}$

Find the multiplicative inverse of each complex number.

**44.** $2 + i$

**45.** $3 - 4i$

**46.** $7 - 3i$

**47.** $3 + 7i$

**48.** $2 - 5i$

**49.** $10 - 12i$

**50.** $\dfrac{2i}{5 - i}$

**51.** $\dfrac{4i}{3 + i}$

**52.** $\dfrac{-i}{2 - 3i}$

**53.** $\dfrac{i}{8 - 2i}$

**54.** $\dfrac{-3i}{3 + 4i}$

**55.** $a + bi$

Simplify.

**56.** $\dfrac{1 - i\sqrt{3}}{2 - i\sqrt{3}}$

**57.** $\dfrac{2 - i\sqrt{7}}{2 + i\sqrt{7}}$

**58.** $\dfrac{(2 + 3i)^2}{(3 + i)^2}$

**59.** $\dfrac{(3 + 3i)^2}{(1 + i)^2}$

**60.** $\dfrac{(4 + 3i)^2}{(3 - i)^2}$

**61.** $\dfrac{1 - i}{(1 + i)^2}$

**Critical Thinking** **62.** Show that $-\dfrac{1}{2} + \dfrac{1}{2}i\sqrt{3}$ is a cube root of 1. *Hint: Find the cube of the number.*

**63.** Show that $1 + i\sqrt{3}$ is a cube root of $-8$.

**300 CHAPTER 6 IRRATIONAL AND COMPLEX NUMBERS**

**Application**    **64. Electrical Engineering**   In an electrical circuit, the flow of the electrical current, $I$, the impedance, $Z$, and the voltage, $E$, are related by the formula $E = I \cdot Z$.

**Find $I$ given the following values.**

a. $E = (70 + j226)$ volts, $Z = (6 + j8)$ ohms

b. $E = (85 + j110)$ volts, $Z = (3 - j4)$ ohms

c. $E = (60 + j112)$ volts, $Z = (10 - j6)$ ohms

**Find $Z$ given the following values.**

d. $E = (-50 + j100)$ volts, $I = (-6 - j2)$ amps

e. $E = (100 + j10)$ volts, $I = (-8 + j3)$ amps

f. $E = (-70 + j240)$ volts, $I = (-5 + j4)$ amps

**Mixed Review**    **Simplify.   (Lesson 6-9)**

**65.** $(3 + 2i) + (4 + 5i)$                          **66.** $(4 + 3i)(16 - 28i)$

**67.** Express $\sqrt{\sqrt[3]{64n^6}}$ in simplest radical form.   **(Lesson 6-4)**

**68. Chemistry**   Wavelengths of light are measured in Angstroms. An Angstrom is $10^{-8}$ centimeters. The wavelength of cadmium's green line is 5085.8 Angstroms. How many wavelengths of cadmium's green line are there in one meter?   **(Lesson 5-2)**

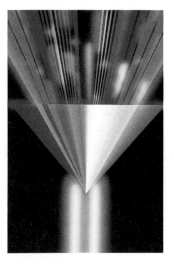

**69.** Determine whether the system
$$\begin{cases} 5a + b + 3c = 3 \\ 5a + b - c = -9 \\ a - b + 5c = 9 \end{cases}$$ has a unique solution. If so, solve the system using
Cramer's Rule.   **(Lesson 4-7)**

**70.** Use the strategy of solving a simpler problem to find the sum of the first 30 positive integers.   **(Lesson 3-5)**

**71.** Draw the graph of $3x + 8y < 11$.   **(Lesson 2-8)**

**72.** Solve the equation $|x + 9| = 22$.   **(Lesson 1-6)**

# SUMMARY AND REVIEW

## VOCABULARY

Upon completing this chapter, you should be
familiar with the following terms:

| | | | |
|---|---|---|---|
| complex number | **292** | **258** | product property of radicals |
| conjugates | **266** | **288** | pure imaginary number |
| extraneous solutions | **282** | **259** | quotient property of radicals |
| imaginary number | **288** | **282** | radical equations |
| imaginary unit | **288** | **253** | radical sign |
| index | **253** | **253** | radicand |
| irrational numbers | **254** | **270** | rational exponents |
| like radical expressions | **264** | **260** | rationalizing the denominator |
| $n$th root | **252** | **252** | square root |
| principal root | **253** | | |

## SKILLS AND CONCEPTS

| OBJECTIVES AND EXAMPLES | REVIEW EXERCISES |
|---|---|

Upon completing this chapter, you should
be able to:

Use these exercises to review and prepare
for the chapter test.

- simplify radicals  (**Lesson 6-1**)

$$\sqrt{36m^4n^6} = \sqrt{(6m^2n^3)^2}$$
$$= 6\,m^2|n^3|$$

**Simplify.**

1. $\sqrt{49x^2}$   2. $\sqrt[3]{-64a^6b^9}$

3. $\sqrt{(3p-5q)^2}$   4. $\sqrt{4n^2+12n+9}$

---

- simplify radical expressions using
  multiplication and division, and
  rationalize the denominator of a fraction
  containing a radical expression
  (**Lesson 6-2**)

$$\sqrt{27a^3} = \sqrt{3^2} \cdot \sqrt{3} \cdot \sqrt{a^2} \cdot \sqrt{a}$$
$$= 3a\sqrt{3a}$$

**Simplify.**

5. $\sqrt{96}$   6. $\sqrt{50x^3y^2}$

7. $\sqrt{6ab} \cdot \sqrt{3a}$   8. $\sqrt[3]{\dfrac{5}{27}}$

9. $\dfrac{15}{2\sqrt{5}}$   10. $\dfrac{4}{\sqrt[4]{2}}$

11. $\sqrt{5}(2\sqrt{10}+3\sqrt{2})$

12. $\sqrt[3]{4}(2\sqrt[3]{4}-5\sqrt[3]{2})$

| OBJECTIVES AND EXAMPLES | REVIEW EXERCISES |
|---|---|

■ add, subtract, multiply, and divide radical expressions   (**Lesson 6-3**)

$$(2 + \sqrt{2})(3 + \sqrt{3})$$
$$= 6 + 2\sqrt{3} + 3\sqrt{2} + \sqrt{6}$$
$$\frac{2 - \sqrt{5}}{1 - 2\sqrt{5}} \cdot \frac{1 + 2\sqrt{5}}{1 + 2\sqrt{5}} = \frac{-8 + 3\sqrt{5}}{1 - 20}$$
$$= \frac{8 - 3\sqrt{5}}{19}$$

**Simplify.**

13. $5 + 2\sqrt{6} - 3\sqrt{6} + 9$

14. $3\sqrt{27} - 5\sqrt{3} + 2\sqrt{48}$

15. $7\sqrt[3]{24x^2} + \sqrt[3]{81x^2}$

16. $(6 + \sqrt{3})(2\sqrt{5} - \sqrt{3})$

17. $(5\sqrt{2} - \sqrt{3})(5\sqrt{2} + \sqrt{3})$

18. $\dfrac{4 - \sqrt{3}}{1 + 2\sqrt{3}}$

---

■ write expressions with rational exponents in simplest radical form and vice versa   (**Lesson 6-4**)

$$\sqrt{36a^2b^3} = 6|a|b^{\frac{3}{2}}$$
$$x^{\frac{4}{5}} = \sqrt[5]{x^4}$$

**Express using rational exponents.**

19. $\sqrt[4]{r^3}$      20. $\sqrt[3]{8m^2n^7}$

**Express in simplest radical form.**

21. $5^{\frac{1}{3}}$      22. $2^{\frac{2}{3}} \cdot x^{\frac{5}{6}} \cdot y^{\frac{1}{2}}$

---

■ evaluate expressions in either exponential or radical form (**Lesson 6-4**)

$$8^{-\frac{1}{3}} = \frac{1}{\sqrt[3]{8}} = \frac{1}{2}$$

**Evaluate.**

23. $125^{\frac{1}{3}}$      24. $4^{-\frac{1}{2}}$

25. $16^{1.25}$      26. $8^{\frac{2}{3}} \cdot 8^{\frac{2}{3}}$

---

■ simplify expressions containing rational exponents   (**Lesson 6-6**)

$$\frac{1}{4^{\frac{3}{2}}} = \frac{1}{4^{\frac{3}{2}}} \cdot \frac{4^{\frac{1}{2}}}{4^{\frac{1}{2}}} = \frac{4^{\frac{1}{2}}}{4^2} = \frac{4^{\frac{1}{2}}}{16}$$

$$\frac{1}{w - y^{\frac{1}{2}}} \cdot \frac{w + y^{\frac{1}{2}}}{w + y^{\frac{1}{2}}} = \frac{w + y^{\frac{1}{2}}}{w^2 - y}$$

**Simplify**

27. $\dfrac{\frac{1}{5^{\frac{1}{3}}}}{}$ i.e. $\dfrac{1}{5^{\frac{1}{3}}}$      28. $\dfrac{2^{\frac{1}{2}}}{2^{\frac{1}{3}}}$

29. $a^{-\frac{3}{4}}$      30. $\dfrac{p^{\frac{1}{2}} - 2q^{\frac{1}{2}}}{p^{\frac{1}{2}} + q^{\frac{1}{2}}}$

---

■ solve equations containing radicals (**Lesson 6-7**)

$$\sqrt{x + 1} - \sqrt{x + 3} = 1$$
$$\sqrt{x + 1} = \sqrt{x + 3} + 1$$
$$x + 1 = x + 3 + 2\sqrt{x + 3} + 1$$
$$\frac{-3}{2} = \sqrt{x + 3}$$
$$\frac{9}{4} = x + 3$$
$$-\frac{3}{4} = x$$

However, this solution does not check so there is no solution.

**Solve each equation.**

31. $a + 3 = a\sqrt{2}$

32. $4x - \sqrt{2} = x\sqrt{3} + 2\sqrt{2}$

33. $5 - \sqrt{3x + 4} = 0$

34. $\sqrt{x + 11} - \sqrt{15 + 2x} = 1$

35. $\sqrt{x - 2} + \sqrt{7x - 6} = 8$

36. $\sqrt[3]{5n + 4} - 4 = 0$

| OBJECTIVES AND EXAMPLES | REVIEW EXERCISES |
|---|---|

■ simplify radicals containing negative radicands and multiply pure imaginary numbers **(Lesson 6-8)**

$$\sqrt{-180} = \sqrt{36} \cdot \sqrt{5} \cdot \sqrt{-1}$$
$$= 6i\sqrt{5}$$

$$2i(i)^2 = 2i(-1) = -2i$$

**Simplify.**

37. $\sqrt{-121}$  38. $\sqrt{-32}$

39. $5i(3i)$  40. $(8i)^3$

---

■ add, subtract, and multiply complex numbers **(Lesson 6-9)**

$$(2 + 3i) + (4 - i) = 6 + 2i$$
$$(13 - 6i) - (10 - 4i) = 3 - 2i$$
$$(1 + 5i)(2 - 3i) = 2 + 7i + 15$$
$$= 17 + 7i$$

**Simplify.**

41. $(8 + 7i) + (13 - 2i)$

42. $(29 - 37i) - (19 + 21i)$

43. $(7 + 6i)(4 - 3i)$

44. $(5 - 7i)^2$

---

■ simplify rational expressions containing complex numbers in the denominator **(Lesson 6-10)**

$$\frac{3 + 4i}{3 - 2i} \cdot \frac{3 + 2i}{3 + 2i} = \frac{1 + 18i}{13}$$

The multiplicative inverse of $2 + 5i$ is
$$\frac{1}{2 + 5i} \text{ or } \frac{2 - 5i}{29}.$$

**Find the conjugate of each complex number.**

45. $8 - 13i$  46. $-3 + 11i$

**Simplify.**

47. $\dfrac{11 + 8i}{2i}$  48. $\dfrac{2 + 3i}{2 - 3i}$

**Find the multiplicative inverse of each complex number.**

49. $8 - i$  50. $12i$

---

# ~~~~ APPLICATIONS AND CONNECTIONS ~~~~

51. **Physics** Find the time, $t$ (in seconds), that it takes for a freely falling object to fall a distance, $s$, of 200 feet. Use the formula $t = \frac{1}{4}\sqrt{s}$. **(Lesson 6-3)**

52. Find the sum of the numbers between 1 and 500 that are not multiples of 2 or 3. You may use the formula $S = \dfrac{n(n + 1)}{2}$ for the sum, $S$, of the integers 1 to $n$. **(Lesson 6-5)**

53. **Energy** A circular solar cell must deliver 18 watts of energy. If each square centimeter of the cell that is in sunlight produces 0.01 watt of energy, how long must the radius of the cell be? **(Lesson 6-7)**

54. **Electronics** The voltage, $E$, (in volts) of an electrical circuit is given by the formula $E = I \cdot Z$, where $I$ represents the current, and $Z$ represents the impedance. Find the voltage of a circuit with current $(20 + j12)$ amps and impedance $(15 + j3)$ ohms. **(Lesson 6-9)**

**Simplify.**

1. $\sqrt{324}$

2. $\sqrt{512}$

3. $\sqrt{169a^3b^2}$

4. $\sqrt[3]{-16y^3}$

5. $\sqrt{9x^2 - 30xy + 25y^2}$

6. $\sqrt{x^2(x-3)^2}$

7. $\sqrt{5a^3} \cdot \sqrt{10ab^3}$

8. $\sqrt{6}(3\sqrt{2} - 2\sqrt{12})$

9. $\sqrt[3]{2}(\sqrt[3]{54} - 3\sqrt[3]{16})$

10. $5\sqrt{8} - 6\sqrt{50} + 4\sqrt{18}$

11. $(5 + \sqrt{3})(7 - 2\sqrt{3})$

12. $(2\sqrt{5} - \sqrt{3})(2\sqrt{5} + \sqrt{3})$

13. $(\sqrt{2} - 3\sqrt{6})(\sqrt{3} + 6)$

14. $\dfrac{6}{\sqrt{3}}$

15. $\dfrac{3}{\sqrt[4]{54}}$

16. $\dfrac{3 - \sqrt{2}}{4 + \sqrt{2}}$

17. What fraction of the odd whole numbers less than 100 are perfect squares?

**Simplify.**

18. $\dfrac{1}{4^{\frac{1}{3}}}$

19. $p^{-\frac{3}{4}}$

20. $\dfrac{3}{a^{-1} + b^{-2}}$

21. $\dfrac{m^{\frac{1}{2}}}{m^{\frac{1}{2}} - n^{\frac{1}{2}}}$

**Solve each equation.**

22. $\sqrt{3t - 2} = 5$

23. $\sqrt{x + 5} + \sqrt{x + 13} = 4$

24. $\sqrt[3]{4n + 9} = 5$

25. $\sqrt{n^2 - 7} = \sqrt{4n + 25}$

26. $x^2 + 100 = 0$

27. $6x^2 + 42 = 0$

**Find the conjugate of each complex number.**

28. $-8 + 5i$

29. $4i$

**Find the multiplicative inverse of each complex number.**

30. $3 - 9i$

31. $11i$

32. Write $5^{\frac{2}{3}}a^{\frac{1}{2}}b^{\frac{3}{8}}$ using a single radical.

33. Write $\sqrt[6]{36x^5y^9z^4}$ in simplified form using rational exponents.

**Bonus**  Find the least positive integer greater than 2 whose cube ends in 8.

# College Entrance Exam Preview

The questions on these pages involve comparing two quantities, one in column A and one in column B. In certain questions, information related to one or both quantities is centered above them. All variables used represent real numbers.

**Directions:**

Write A if the quantity in Column A is greater.

Write B if the quantity in Column B is greater.

Write C if the quantities are equal.

Write D if there is not enough information to determine the relationship.

| Column A | Column B |
|---|---|
| 1. $\dfrac{8a + 12}{2}$ | $4a + 6$ |
| 2. $(-9)^{72}$ | $(-9)^{83}$ |
| 3. $-3 < n < 0$ | |
| $-3(n + n)$ | $(n)(n)(n)$ |
| 4. $n + 11 > 12$ | |
| $3n + 8$ | $16 - 2n$ |
| 5. $a < 0 < b$ | |
| $\dfrac{a}{2}$ | $b^2$ |
| 6. $x^2 > y^2$ | |
| $(y + 1)^2$ | $(x + 1)^2$ |

| Column A | Column B |
|---|---|
| 7. $\frac{1}{2}\%$ of 400 | $0.5 \times 400$ |
| 8. $\dfrac{0.05}{x} = \dfrac{0.2}{0.6}$ | |
| $x$ | $0.2$ |
| 9. $6 > b > -5$ | |
| $\dfrac{b}{5}$ | $\dfrac{5}{b}$ |
| 10. 30% of $r$ is 6 | |
| The percent that 60 is of $r$. | The percent that $r$ is of 6. |
| 11. $0.8 \times 6d$ | $\dfrac{3}{5}$ of $8d$ |
| 12. $23 - 4(2)0$ | $6 + 10(3 - 2)$ |
| 13. $\dfrac{x}{2} = z^2$ | |
| $x$ | $z$ |
| 14. area of a circle with diameter 10 | area of a right triangle with hypotenuse 10 |
| 15. $\dfrac{1}{3} < a < \dfrac{2}{3}$ $\dfrac{2}{3} < b < 1$ | |
| $a + b$ | $a^2 + b^2$ |
| 16. $a$ if $3a < 23$ | $b$ if $23 < 3b$ |

| Column A | Column B |
|---|---|

**17.** The perimeter of the square and rectangle are equal.

| area of the square | area of the rectangle |
|---|---|

## TEST TAKING TIP

If the use of A's and B's in both the column names and answer choices is confusing, change the column names to another pair of letters or numbers, such as $X$ and $Y$ or $I$ or $II$.

Perform any indicated mathematical operations. Change the common information by addition, subtraction, multiplication, or division when it is given as an equation or an inequality.

Remember that the figures are often not drawn to scale and may lead you to make assumptions which are not valid. Try many ways to illustrate the figures given to fit the situations given.

**18.** $\dfrac{x}{y}$ $\qquad$ $\dfrac{y}{x}$

**19.** $a > 0$

$4a - 3a$ $\qquad$ $4 \times 3a$

**20.** the slope of $2x + 3y = 7$ $\qquad$ the slope of $3x - 2y = 7$

**21.** $-x$ $\qquad$ $\dfrac{1}{x}$

**22.** $\begin{bmatrix} 3 & -4 & 5 \\ -2 & 1 & -3 \\ 6 & -5 & -7 \end{bmatrix}$

the signed minor of $e_{23}$ $\qquad$ the signed minor of $e_{32}$

**23.** $5x - 2y = 4$

the $y$-intercept of the line parallel to the given line and containing $(-2, 3)$ $\qquad$ the $y$-intercept of the line perpendicular to given line and containing $(-2, 3)$

**24.** $r < 0$

$\sqrt{r^2}$ $\qquad$ $\sqrt[3]{r^3}$

| Column A | Column B |
|---|---|

**25.** $x + 3y = 5$
$y = 3x + 5$

$x$ $\qquad$ $y$

**26.** $i^x = -i$

$x$ $\qquad$ $6$

# Quadratic Equations

## CHAPTER OBJECTIVES

In this chapter, you will:

- Solve quadratic equations by graphing, factoring, completing the square, and using the quadratic formula.
- Write quadratic equations when two roots of the equation are known.
- Solve nonquadratic equations by methods used to solve quadratic equations.

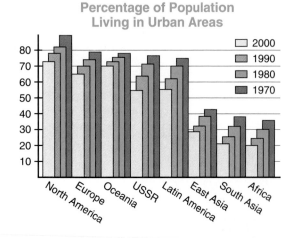

**What trends do you foresee for city planners? Why do you think this trend is occurring?**

Percentage of Population Living in Urban Areas

## CAREERS IN CITY PLANNING

What kinds of cities will we need in the next century? The downtowns of many cities today are degenerating into vacant stores and uninhabited offices and apartments. City planners are replacing these areas with elegant convention centers, glassy hotels and office towers, and festive marketplaces. These features might entice suburban dwellers to leave their small shopping centers and fast food outlets to venture into the city.

What about the next century? Today most of us use our automobile to travel to the market or workplace. The concept of convenience has steered many city planners to expand cities as small, self-sufficient centers. With the increase in the price of oil, this trend will increase. Planners will design areas with apartments, houses, offices, factories, bike trails, lakes, medical facilities, schools, parks, shops and supermarkets, and sports facilities clustered together. In such neighborhoods, people can walk to any local destination and save their cars for long trips.

In designing such environments, city planners will create 3-D mathematical models on their computers. They'll consult with architects, public health officers, economists, and many other specialists. They will become experts in evaluating statistics as they study traffic patterns, weather, and demographics. Will you someday play a part in the design of your own children's hometown?

## MORE ABOUT CITY PLANNING

### Degree Required:

- Bachelor's Degree in City and Community Planning, Environmental Design

### Some city planners like:

- working for the community
- helping to improve the quality of life for citizens
- good salaries and benefits
- variety and challenge in their work

### Related Math Subjects:

- Advanced Algebra
- Trigonometry
- Statistics/Probability
- Economics

### Some city planners dislike:

- keeping up with government rules and regulations and local building codes
- meeting deadlines
- attending meetings after regular work hours and on weekends

For more information on the various careers available in the field of city planning, write to:
American Institute of Certified Planners
1776 Massachusetts Avenue NW
Washington, DC 20036

# Problem-Solving Strategy: Guess-and-Check

### Objective

After studying this lesson, you will be able to:

■ solve problems using the guess-and-check strategy.

### Application

Miss Grusselli owns a ranch in Colorado. She wants to fence off part of the ranch to make a corral for her horses. The fence used for horses is made of sections of wood planking nailed to posts. She has enough lumber to build 16 sections, but she also wants to have the maximum area for her horses. What should the dimensions of the rectangular corral be?

*Let one unit represent each section of wood planking.*

| length (units) | width (units) | perimeter (units) | area (units²) | |
|---|---|---|---|---|
| 7 | 1 | 16 | 7 | |
| 6 | 2 | 16 | 12 | *This is getting closer. Guess again.* |
| 5 | 3 | 16 | 15 | *This is better. Guess again.* |
| 4 | 4 | 16 | 16 | *Better still! Is this it?* |
| 4.5 | 3.5 | 16 | 15.75 | |

The last guess produced an area smaller than the area of the 4 × 4 rectangle. You might try other combinations to convince yourself that the 4 × 4 rectangle will be the figure that has the maximum area of a perimeter of 16 units.

### Example

*You need only divide by primes. What are the first five primes?*

**Find the least prime number greater than 720.**

One way to solve this problem is to check each integer, beginning with 721, for prime divisors until you find a prime number. However, if you use known information, you can save time and effort. Obviously, any even integer greater than 720 is not prime. Since 720 is divisible by 3, every third integer greater than 720 will not prime. Any integer greater than 720 whose last digit is 5 or 0 is not prime. Thus, many possibilities are eliminated.

Try 721.    *Using a calculator is helpful.*
    $721 \div 7 = 103$    *not prime*

Try 727.
| | |
|---|---|
| $727 \div 7 \approx 103.9$ | $727 \div 11 \approx 66.1$ |
| $727 \div 13 \approx 55.9$ | $727 \div 17 \approx 42.8$ |
| $727 \div 19 \approx 38.3$ | $727 \div 23 \approx 31.6$ |
| $727 \div 29 \approx 25.1$ | *Why can you stop after this division?* |

Thus, 727 is the least prime number greater than 720.

# CHECKING FOR UNDERSTANDING

**Communicating Mathematics**

**Read and study the lesson to answer these questions.**

1. In the corral problem, what was the shape of the figure that had the maximum area for a perimeter of 16 units? For any given perimeter, is this type of figure always the rectangle with the maximum area?

2. What is the rule for easily checking if a number is divisible by 3 or by 5?

3. In the example, explain why you can stop dividing after you have tried 29.

**Guided Practice**

4. Rich and Peg raise dogs and puffins. They counted all the heads and got 120. They counted all the feet and got 400. How many dogs and how many puffins do they have?

# EXERCISES

| Strategies |
| --- |
| Look for a pattern. |
| Solve a simpler problem. |
| Act it out. |
| Guess and check. |
| Draw a diagram. |
| Make a chart. |
| Work backwards. |

**Solve. Use any strategy.**

5. Replace each letter with a whole number so that the multiplication problem at the right is correct. Each letter represents a different number.

$$\begin{array}{r} \text{ABCDE} \\ \times 4 \\ \hline \text{EDCBA} \end{array}$$

6. At the conclusion of a committee meeting, a total of 153 handshakes were exchanged. Assuming each person shook hands with everyone else, how many people were at the meeting?

7. With one straight vertical cut and one straight horizontal cut, divide this Greek cross into four pieces that can be assembled in a different way to form a square.

8. What row in the table contains the square of an integer and the cube of a different integer?

| A | 123 | 126 | 164 | 270 | 381 |
| --- | --- | --- | --- | --- | --- |
| B | 52 | 64 | 75 | 81 | 92 |
| C | 120 | 130 | 155 | 166 | 196 |
| D | 320 | 450 | 566 | 678 | 999 |
| E | none of the above | | | | |

---

## ~ COOPERATIVE LEARNING ACTIVITY ~

**Work in groups. Each person in the group must understand the solution and be able to explain it to any person in the class.**

A perfect number is one in which the number equals the sum of its factors, excluding itself. The first perfect number is 6.

The factors of 6 are 1, 2, 3, and 6. Now exclude the 6. The sum of the factors is $1 + 2 + 3$, or 6. Thus, 6 is a perfect number.

Find two other perfect numbers.

# Graphing Calculator Exploration: Quadratic Equations

The graphing calculator is a powerful tool for studying graphs of different equations. In this lesson we will study the graphs of equations of the form $y = ax^2 + bx + c$. These are called **quadratic equations.** The graphs of quadratic equations are called **parabolas.**

**Example 1**

Graph $y = x^2$ using the standard viewing window.

This equation is of the form $y = ax^2 + bx + c$, where $a = 1$, $b = 0$, and $c = 0$.

First, clear the graphics screen. To clear the screen on the Casio fx-7000G, press [SHIFT] [Cls] then [EXE]. To clear the screen on the TI-81, press [Y=] , then use the arrow keys and the [CLEAR] key to select and clear any equations from the $Y=$ list.

Next, make sure that the range parameters are set for the standard viewing window and graph.

*Casio*

ENTER:  [GRAPH] [ALPHA] [X]
        [$x^2$] [EXE]

*TI-81*

ENTER:  [Y=] [X|T] [$x^2$] [GRAPH]

The graph of the equation is shaped like a cup. This is the general shape of a parabola.
*Notice that $a > 0$.*

The standard viewing window allowed us to view a complete graph of $y = x^2$. As you recall from our earlier work on the graphing calculator, a graph is said to be complete if all of the important characteristics of the graph are displayed. In general, the important characteristics of a parabola are the $x$- and $y$-intercepts, a relative maximum or relative minimum point, that is, the top or bottom of the "cup", and the end behavior of the graph. The end behavior refers to the nature of the graph as the $x$ values grow very large or very small.

**Example 2**

**Graph $y = -0.15x^2 + 30$.**

The equation $y = -0.15x^2 + 30$ is in the form $y = ax^2 + bx + c$, where $a = -0.15$, $b = 0$, and $c = 30$.

Let's try graphing in the standard viewing window. Clear the graphics window before graphing.

*Casio*

ENTER: [GRAPH] [(-)] 0.15 [ALPHA] [×] [$x^2$] [+] 30 [EXE]

*TI-81*

ENTER: [Y=] [(-)] 0.15 [X|T] [$x^2$] [+] 30 [GRAPH]

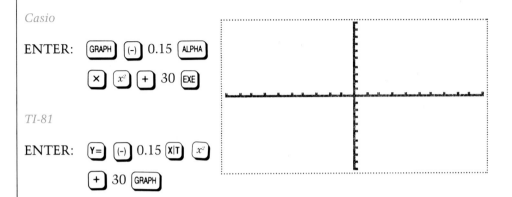

Nothing appears on the graphics screen. We must change the range parameters to be able to view a complete graph. Set the range parameters to [-20, 20] by [-5, 35] with scale factors of 5 for both axes and graph again.

*Recall that [-20, 20] by [-5, 35] with scale factors of 5 for both axes represents Xmin = -20, Xmax = 20, Ymin = -5, and Ymax = 35, with Xscl = 5 and Yscl = 5.*

There is no need to retype the equation after changing the range parameters. Simply press [EXE] on the Casio or [GRAPH] on the TI-81.

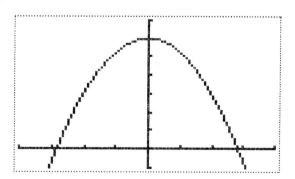

The graph of $y = -0.15x^2 + 30$ is a parabola that opens downward. *Notice that $a < 0$.*

You have solved quadratic equations of the form $0 = ax^2 + bx + c$ by factoring or by using the quadratic formula. You can also solve these equations graphically by using the graphing calculator.

The graph of the quadratic equation $y = ax^2 + bx + c$ represents all of the values of $x$ and $y$ that satisfy the equation. When we solve the equation $0 = ax^2 + bx + c$, we are only interested in the values of $x$ which make the value of the expression $ax^2 + bx + c$ equal to 0. These values are represented by the points at which the graph of the equation crosses the $x$-axis, since the $y$-values of these points are 0. These $x$-intercepts are the **solutions** or **roots** of the quadratic equation.

There are three possible outcomes when solving a quadratic equation. The equation will have either two real solutions, one real solution, or no real solutions. A graph of each of these outcomes is shown below.

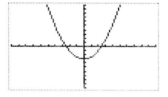    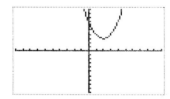

two real solutions          one real solution          no real solutions

**Example 3**

Solve $3x^2 + 6x - 1 = 0$ using the graphing calculator.

Set the range parameters to the standard viewing window and clear the graphics screen before graphing the equation.

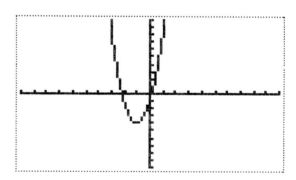

Now use the zoom-in process that we used in the Graphing Calculator Exploration on page 105 to zoom in on each of the $x$-intercepts and determine the solutions. The solutions of this equation are 0.1547 and $-2.1547$, accurate to four decimal places.

It should be noted that solving quadratic equations graphically produces only approximate solutions. While approximations are usually adequate for most applications, if an exact solution is needed the equations should be solved by factoring or using the quadratic formula.

# EXERCISES

Graph $y = x^2 - 6x + 7$ for each set of range parameter values. Then sketch the graph shown on the graphics screen, indicating the scales on each axis.

1. Xmin: -10, Xmax: 10, Xscl: 1, Ymin: -10, Ymax: 10, Yscl: 1

2. Xmin: -1, Xmax: 9, Xscl: 1, Ymin: 0, Ymax: 35, Yscl: 5

3. Xmin: -10, Xmax: 16, Xscl: 2, Ymin: -5, Ymax: 160, Yscl: 10

4. Xmin: -100, Xmax: 100, Xscl: 20, Ymin: -100, Ymax: 2000, Yscl: 400

5. Xmin: -25, Xmax: 25, Xscl: 5, Ymin: -50, Ymax: 1200, Yscl: 200

6. Which of the range parameter values used in Exercises 1 to 5 produced a complete graph of $y = x^2 - 6x + 7$?

7. How many solutions does the equation $0 = x^2 - 6x + 7$ have?

Graph $y = -8x^2 - 1$ for each set of range parameter values. Then sketch the graph shown on the graphics screen, indicating the scales on each axis.

8. Xmin: -10, Xmax: 10, Xscl: 1, Ymin: -10, Ymax: 10, Yscl: 1

9. Xmin: -1, Xmax: 12, Xscl: 1, Ymin: 0, Ymax: 30, Yscl: 5

10. Xmin: -10, Xmax: 20, Xscl: 2, Ymin: -5, Ymax: 100, Yscl: 10

11. Xmin: -20, Xmax: 20, Xscl: 2, Ymin: -100, Ymax: 0, Yscl: 10

12. Xmin: -25, Xmax: 25, Xscl: 5, Ymin: -25, Ymax: 25, Yscl: 5

13. Which of the range parameter values used in Exercises 8 to 12 produced a complete graph of $y = -8x^2 - 1$?

14. How many solutions does the equation $0 = -8x^2 - 1$ have?

Find and state an appropriate viewing window for viewing a completed graph of each function. Then sketch each graph as seen on the graphics screen.

15. $y = 3.2x^2 + 9.2$

16. $y = x^2 - 5x + 6$

17. $y = 6x^2 + 108x + 480$

18. $y = 2x^2 + 9x - 18$

19. $y = 2x^2 - x - 15$

20. $y = 1.3x^2 - 3.8x + 5.1$

Find the solutions of each quadratic equation accurate to four decimal places by using your graphing calculator.

21. $2x^2 - x - 15 = 0$

22. $0.2x^2 - 0.3492x - 0.0738 = 0$

23. $1.2x^2 - 3.6x + 5.8 = 0$

24. $3x^2 + 3.08x - 1.36 = 0$

25. $x^2 + 31.54x + 229.068 = 0$

26. $35x^2 + 66x - 65 = 0$

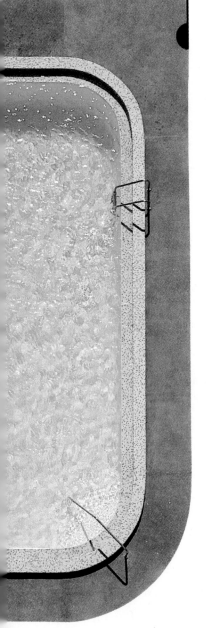

# Solving Quadratic Equations

**Objectives**

After studying this lesson, you should be able to:

- solve quadratic equations by graphing, and
- solve quadratic equations by factoring.

**Application**

Craig Hoffheimer wants to build a swimming pool surrounded by a sidewalk of uniform width. He wants the dimensions of the pool and sidewalk to be 16 meters by 20 meters. The pool has an area of 192 square meters. How wide should the sidewalk be?

*A drawing will help in solving this problem.*

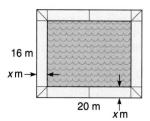

16 m

x m→

20 m

x m

Let $x$ meters be the width of the sidewalk. The length of the pool is $20 - 2x$ meters. The width of the pool is $16 - 2x$ meters.

The area of the pool can be expressed as the product of the length and width.

$$A = \ell w$$
$$= (20 - 2x)(16 - 2x) \qquad \textit{Substitute for } \ell \textit{ and } w.$$
$$= 320 - 72x + 4x^2 \qquad \textit{Use FOIL.}$$
$$= 4x^2 - 72x + 320$$

The area of the pool can be expressed as $4x^2 - 72x + 320$ square meters. Since the area of the pool is 192 square meters, replacing $A$ with 192 results in the equation $192 = 4x^2 - 72x + 320$.

Of course, Craig would probably never use a quadratic equation like $192 = 4x^2 - 72x + 320$ to express the area of a pool. However, in this chapter you will study many other formulas in science and business that involve this type of equation.

A **quadratic equation** is an equation that can be written in the form $ax^2 + bx + c = 0$, where $a \neq 0$. We say that equations like this have a **degree** of 2 since the greatest exponent of the variable is 2. Notice that a quadratic equation only has one variable, and all the exponents are positive.

The values of the variable that satisfy an equation are called **roots** or **solutions** of the equation. There are several methods that you can use to find the roots of a quadratic equation.

One way to determine the roots of a quadratic equation is to graph the related quadratic function. First, the quadratic equation must be written in the general form. The swimming pool equation needs to be rewritten.

$$192 = 4x^2 - 72x + 320$$
$$0 = 4x^2 - 72x + 128 \qquad \textit{Subtract 192 from each side.}$$
$$0 = x^2 - 18x + 32 \qquad \textit{Divide each side by 4.}$$

The related function for this equation is $y = x^2 - 18x + 32$.

To graph this function, first make a table of values that satisfy the function. Then graph these ordered pairs.

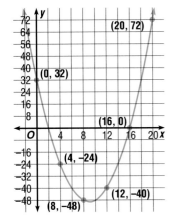

| $x$ | 0 | 4 | 8 | 12 | 16 | 20 |
|---|---|---|---|---|---|---|
| $y$ | 32 | -24 | -48 | -40 | 0 | 72 |

When you graph the ordered pairs, note that the points suggest the shape of a curve called a **parabola**. Notice this parabola crosses the $x$-axis twice and the function has a degree of 2. These intercept points are called the **zeros of a function.** For this function, the two points are $(2, 0)$ and $(16, 0)$. Remember that the quadratic equation is the related quadratic function with $y$ equaling 0. The $x$ values, 2 and 16, are the roots of the equation $0 = x^2 - 18x + 32$.

*You can use your graphing calculator and the TRACE function to find the zeros of the graphed function.*

Now let's solve the swimming pool problem. The solutions to the quadratic equation $0 = x^2 - 18x + 32$ are 2 and 16. The region containing the pool and sidewalk is only 16 meters wide. Therefore, the sidewalk itself cannot be 16 meters wide. The value of $x$ appropriate for this problem is 2. So, the sidewalk should be 2 meters wide on each side of the pool.

Another way of solving a quadratic equation is by **factoring.** The factoring method depends on the **zero product property.**

| *Zero Product Property* | **For any real numbers $a$ and $b$,** **if $ab = 0$, then $a = 0$ or $b = 0$.** |
|---|---|

Let's factor the same equation we solved by graphing.

$$x^2 - 18x + 32 = 0 \qquad \Rightarrow \qquad (x - 16)(x - 2) = 0$$

Now set each factor equal to zero and solve.

$$x - 16 = 0 \qquad x - 2 = 0$$
$$x = 16 \qquad x = 2$$

You see that the same two solutions occur.

When using factoring to solve a real-life problem, you still need to examine each solution carefully to see if it is reasonable for that situation.

**Example 1**

**Solve $x^2 - 2x - 15 = 0$ by graphing. Then verify your solutions by factoring.**

Graph the related quadratic function $y = x^2 - 2x - 15$ by finding and graphing the ordered pairs that satisfy the function. Then graph the parabola suggested by the points.

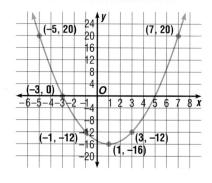

| x | y |
|---|---|
| -7 | 48 |
| -3 | 0 |
| -1 | -12 |
| 1 | -16 |
| 3 | -12 |
| 7 | 20 |

The table tells us one solution of $x^2 - 2x - 15 = 0$ is -3. The graph tells us that the other solution is 5. Let's check our solutions by using factoring.

*To factor, you use the guess-and-check strategy.*

$x^2 - 2x - 15 = 0$
$(x - 5)(x + 3) = 0$      *Factor.*

$x - 5 = 0$   or   $x + 3 = 0$    *Zero product*
     $x = 5$   or      $x = -3$    *property*

The solutions are 5 and -3.

**Check:**   $x^2 - 2x - 15 = 0$
$(5)^2 - 2(5) - 15 \stackrel{?}{=} 0$
$0 = 0$   ✓

$(-3)^2 - 2(-3) - 15 \stackrel{?}{=} 0$
$0 = 0$   ✓

No matter what method you use, you can always check your solutions by substituting the values into the equation and simplifying.

Quadratic equations always have two solutions. Sometimes those two solutions may be the same number.

**Example 2**

**Solve $x^2 + 6x = -9$ by graphing and by factoring.**

First rewrite the equation in quadratic form: $x^2 + 6x + 9 = 0$.

*Notice the graph intersects the x-axis in one point. This means there is only one solution.*

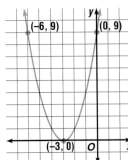

$x^2 + 6x + 9 = 0$
$(x + 3)(x + 3) = 0$

$x + 3 = 0$   or   $x + 3 = 0$
    $x = -3$   or       $x = -3$

Notice that the graph has only one x-intercept, -3. Factoring yields one solution.

The only solution is -3.

## Example 3

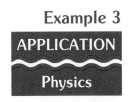

APPLICATION

Physics

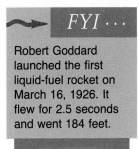

*FYI* · · ·

Robert Goddard launched the first liquid-fuel rocket on March 16, 1926. It flew for 2.5 seconds and went 184 feet.

*The graph shown is the graph of the function that describes the height at any given time. The actual path of the rocket is an entirely different parabola.*

As an object is propelled upwards, gravity pulls it back to Earth. This relationship can be expressed by the formula $s = v_i t - \frac{1}{2}gt^2$, where $s$ is the distance above the starting point, $v_i$ is the initial velocity, $t$ is time elapsed, and $g$ is the acceleration of gravity. Find how long it will take a model rocket propelled into the air at an initial velocity of 80 ft/s to return to ground level, if the acceleration of gravity is 32 ft/s². *Since the rocket is returning to ground level, the distance above ground is 0.*

Substitute the values into the formula and solve the equation. Let's try factoring.

$s = v_i t - \frac{1}{2}gt^2$

$0 = (80)t - \frac{1}{2}(32)t^2$

$0 = 80t - 16t^2$

$0 = 16t(5 - t)$

$16t = 0 \qquad 5 - t = 0$

$t = 0 \qquad 5 = t$

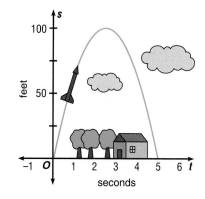

The solution of 0 seconds is the time when the rocket is launched. The solution of 5 seconds is the time elapsed when the rocket returns to the ground. The graph verifies this. So, it will take the rocket 5 seconds to return to the ground.

# CHECKING FOR UNDERSTANDING

**Communicating Mathematics**

Read and study the lesson to answer each question.

1. In the swimming pool problem, what are the dimensions of the pool? Make a scale drawing of the pool and sidewalk.

2. Define each term and explain how they are related.

   **a.** solution      **b.** root      **c.** zero of a function      **d.** $x$-intercept

3. What is the name of the graph of a quadratic function?

4. Can a quadratic equation have more than two solutions? Why or why not?

5. In Example 3, estimate the maximum height of the rocket.

**Guided Practice**

Determine if each equation is a quadratic equation. Write *yes* or *no*.

6. $4x^2 + 7x - 3 = 0$

7. $5y^4 - 7y^2 = 0$

8. $\frac{1}{2}y^2 + \frac{3}{4} = 0$

9. $z^2 + 7z - 3 = z^3$

10. Factor $12m^2 + 25m + 12$.

Determine the solution of each equation from its related graph.

11.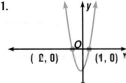
( 2, 0)  (1, 0)

12.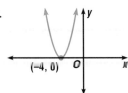
(−4, 0)

13.
(0, 0)  (16, 0)

14. What are the solutions of $(y - 3)(y + 7) = 0$?

# EXERCISES

**Practice**

Solve each equation.

15. $(a + 6)(a + 2) = 0$

16. $z(z - 1)^2 = 0$

17. $(3y + 7)(y + 5) = 0$

18. $(2x + 3)(3x - 1) = 0$

Solve each equation by graphing.

19. $d^2 + 6d + 8 = 0$
20. $z^2 + 4z + 3 = 0$
21. $c^2 + 4c + 4 = 0$
22. $n^2 - 3n = 0$
23. $2w^2 - 3w = 9$
24. $4s^2 - 11s = 3$

Solve each equation by factoring.

25. $y^2 - y = 12$
26. $z^2 - 5z = 0$
27. $p^2 - 12p + 36 = 0$
28. $r^2 + r = 30$
29. $d^2 - 3d = 4$
30. $3c^2 = 5c$
31. $2q^2 + 11q = 21$
32. $18u^2 - 3u = 1$
33. $3t^2 + 4t = 15$
34. $4y^2 = 25$
35. $6r^2 + 7r = 3$
36. $9y^2 + 16 = {}^-24y$

37. **Physics** A tennis ball is shot vertically upward from a launcher with an initial velocity of 64 feet per second. When will the ball return to the ground? How far off the ground will the ball be after 1 second? after 3 seconds?

Solve each equation.

38. $4a^2 - 17a + 4 = 0$

39. $b^2 + 3b = 40$

40. $12m^2 + 25m + 12 = 0$

41. $18n^2 - 3n = 15$

42. $n^3 = 9n$

43. $a^3 = 81a$

44. $35z^3 + 16z^2 = 12z$

45. $18r^3 + 16r = 34r^2$

**CONNECTION**

**Physics**

46. **Physics** Marsha hits a tennis ball upward from the top of a 90-foot-high cliff with an initial velocity of 16 ft/s. Marsha hit the ball when it was 6 feet above the top of the cliff.

   a. Assume that the tennis ball will hit the ground below the cliff. Determine the equation for the height of the ball after $t$ seconds. *Hint: If the initial height is 6 feet, how would you write a distance 90 feet below that height?*

   b. Graph your equation.

   c. Find the number of seconds it will take the ball to hit the ground.

**Critical Thinking**

47. Graph the function $y = x^2 + 2x + 4$. Based on this graph, what are the roots of $0 = x^2 + 2x + 4$? Explain your answer.

**Applications**

48. **City Planning** In the Winton Woods Park, a playground is planned that is 30 m by 20 m. Due to a government grant, the plans are being revised to double the area of the playground by adding strips of the same width to a side and an end of the area to form a rectangle. Find the width of the strips. What are the dimensions of the new playground?

49. **Sports** Plans for the rectangular Pinetown Ice-Skating Rink that is 30 m by 60 m have to be revised because of a budget cut. Strips of equal width will be removed from one end and one side of the plan to create an area of 1000 m². What are the dimensions of the new rink?

**Mixed Review**

50. Find two numbers whose sum is 81 and whose product is 1400. **(Lesson 7-1)**

51. Simplify $(3a)(5a^2b) + (6ab)(10a^2)$. **(Lesson 5-2)**

52. Simplify $\sqrt{3mn^4} \cdot \sqrt{25m^6}$. **(Lesson 6-2)**

53. Solve $\begin{cases} 6x - 2y - 3z = -10 \\ -6x + y + 9z = 3 \\ 8x - 3y = -16 \end{cases}$ . **(Lesson 3-9)**

54. Sam bought 6 deluxe, 5 glazed, and 2 cake doughnuts for $4.00. If he had bought 4 deluxe, 2 glazed, and 7 cake doughnuts, the cost would have been $3.40. A deluxe doughnut costs 5 cents less than twice the cost of a cake doughnut. Find the cost of each type of doughnut. **(Lesson 1-5)**

55. **Electricity** Find the amount of current $I$ (in amperes) produced if the electromotive force $E$ is 1.5 volts, the circuit resistance $R$ is 2.35 ohms, and the resistance $r$ within a battery is 0.15 ohms, using the formula $I = \dfrac{E}{R + r}$. **(Lesson 1-1)**

# 7-3  Completing the Square

**Objective**

After studying this lesson, you should be able to:

■ solve quadratic equations by completing the square.

**Application**

The Finneytown Athletic Association owns a large lot. On a portion of the lot, a square athletic field has been maintained. Each side of the field is 110 feet long. The association would like to equally expand two of the sides of the field for parking spaces so that the entire region has an area of 14,000 ft². What is the width of the parking strips being added?

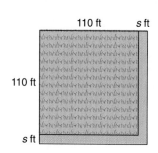

Let $s$ ft represent the width of each parking strip. After expansion, each side of the field would now be $110 + s$ feet long. The equation that expresses the area of the new field, which is still square, is $14,000 = (110 + s)^2$.

Remember that an equation like $100 = y^2$ can be solved by taking the square root of each side. The solutions are $\pm 10$.

The equation $14,000 = (110 + s)^2$ can be solved in this same way.

$$14,000 = (110 + s)^2$$
$$\sqrt{14,000} = \sqrt{(110 + s)^2} \qquad \textit{Take the square root of each side.}$$
$$\pm\sqrt{14,000} = 110 + s$$
$$\pm\sqrt{14,000} - 110 = s$$

Now use your calculator to find approximate values for $s$.

*The $\boxed{\sqrt{x}}$ key may be a second function key.*

ENTER: 14000 $\boxed{\sqrt{x}}$ $\boxed{-}$ 110 $\boxed{=}$ $8.321595662$

ENTER: 14000 $\boxed{\sqrt{x}}$ $\boxed{+/-}$ $\boxed{-}$ 110 $\boxed{=}$ $-228.3215957$

Since our problem asks for a field measurement, the negative solution can be disregarded. The parking strip should be approximately 8.32 feet wide. Check this solution.

Quadratic equations can be solved in the same way as long as one side of the equation contains a perfect square. When the equation does not contain a perfect square, you can use a process called **completing the square** to create a perfect square. Notice the pattern in the following examples.

In a perfect square, there is a relationship between the coefficient of the middle term and the constant term.

$$(x + 7)^2 = x^2 + 14x + 49 \qquad\qquad (x + b)^2 = x^2 + 2bx + b^2$$

$$7 \;=\; \tfrac{1}{2}(14) \to 7^2 = 49 \qquad\qquad b \;=\; \tfrac{1}{2}(2b) \to (b)^2$$

To complete the square in the expression below, you would use the same process. Take half the coefficient of $x$ and square it.

$$x^2 - 8x + \underline{\phantom{~~?~~}}$$

$$\left(-\tfrac{8}{2}\right)^2 \to (-4)^2 \text{ or } 16 \qquad \text{The answer is } x^2 - 8x + 16.$$

*$x^2 - 8x + 16$ is a perfect square trinomial.*

In Example 1, follow through the steps that show how completing the square is used in solving a quadratic equation.

## Example 1

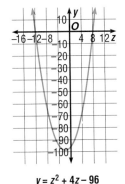

$y = z^2 + 4z - 96$

**Solve $z^2 + 4z = 96$ by completing the square.**

You must find the term that completes the square on the left side. Remember that whatever you add to one side of an equation, you must add to the other side also.

$$z^2 + 4z + ? = 96 + ?$$
$$z^2 + 4z + 4 = 96 + 4 \qquad \left(\tfrac{4}{2}\right)^2 = 4$$
$$(z + 2)^2 = 100 \qquad \text{\textit{Factor the perfect square trinomial.}}$$
$$z + 2 = \pm 10 \qquad \text{\textit{Take the square root of each side.}}$$

$$z + 2 = 10 \quad \text{or} \quad z + 2 = -10$$
$$z = 8 \quad \text{or} \qquad z = -12$$

The solutions are 8 and $-12$.

When the coefficient of the second degree term is not 1, you must first divide the equation by that coefficient before completing the square.

## Example 2

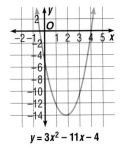

$y = 3x^2 - 11x - 4$

*In Examples 1 and 2, both graphs cross the x-axis twice, so there are two solutions for each equation.*

**Solve $3x^2 - 11x - 4 = 0$ by completing the square.**

$$3x^2 - 11x - 4 = 0$$
$$x^2 - \tfrac{11}{3}x - \tfrac{4}{3} = 0 \qquad \text{\textit{Divide each side by 3.}}$$
$$x^2 - \tfrac{11}{3}x = \tfrac{4}{3} \qquad \text{\textit{Isolate the constant on one side.}}$$
$$x^2 - \tfrac{11}{3}x + \tfrac{121}{36} = \tfrac{4}{3} + \tfrac{121}{36} \qquad \text{\textit{Add} } \left(-\tfrac{11}{3} \div 2\right)^2 \text{\textit{ or }} \tfrac{121}{36} \text{\textit{ to each side.}}$$
$$\left(x - \tfrac{11}{6}\right)^2 = \tfrac{169}{36} \qquad \text{\textit{Factor.}}$$
$$x - \tfrac{11}{6} = \pm\tfrac{13}{6} \qquad \text{\textit{Take the square root of each side.}}$$
$$x = \tfrac{11}{6} \pm \tfrac{13}{6} \qquad \text{\textit{Add} } \tfrac{11}{6} \text{\textit{ to each side.}}$$
$$x = 4 \quad \text{or} \quad x = -\tfrac{1}{3} \qquad \text{The solutions are 4 and } -\tfrac{1}{3}.$$

Not all solutions to quadratic equations are rational numbers. When a solution is irrational, there are two ways you might express the value. The solution written with the radical is the *exact* answer. When you use your calculator to change the expression to an equivalent decimal, the solution is approximate.

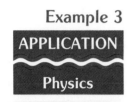

Example 3

APPLICATION

Physics

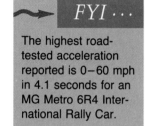

*FYI* · · ·

The highest road-tested acceleration reported is 0–60 mph in 4.1 seconds for an MG Metro 6R4 International Rally Car.

The distance ($s$) an object travels can be computed when the initial speed ($v_i$), time elapsed ($t$), and the rate of constant acceleration ($a$) is known. The formula that relates these factors is $s = v_i t + \frac{1}{2}at^2$. If a car has an initial speed of 20 m/s and a constant acceleration of 2 m/s$^2$, determine the amount of time it takes it to travel 145 m.

$$s = v_i t + \frac{1}{2}at^2$$

$$145 = (20)t + \frac{1}{2}(2)t^2 \qquad \textit{Substitute the known values.}$$

$$145 = 20t + t^2 \qquad \textit{Complete the square by adding } \left[\frac{1}{2}(2b)\right]^2 \textit{ or 100.}$$

$$145 + 100 = t^2 + 20t + 100$$

$$245 = (t + 10)^2 \qquad \textit{Factor.}$$

$$\pm\sqrt{245} = t + 10 \qquad \textit{Take the square root of each side.}$$

$$\pm\sqrt{245} - 10 = t$$

The solutions are $\sqrt{245} - 10$, or about 5.65 seconds and $-\sqrt{245} - 10$, or about $-25.65$ seconds. Since negative time represents time before the car started, the second solution can be disregarded for this problem. The car traveled 145 m in about 5.65 seconds.

# CHECKING FOR UNDERSTANDING

Communicating Mathematics

Read and study the lesson to answer each question.

1. In the athletic field problem, what would be the width of the parking strip if the area of the region were to be 13,000 ft$^2$?

2. Why does a quadratic equation have two solutions but you sometimes may only use one of them to solve a problem?

3. Could you solve the equation in Example 1 by factoring? Explain your answer.

4. Could you solve the equation in Example 1 by graphing? Explain your answer.

Guided Practice

State whether each trinomial is a perfect square. Write *yes* or *no*.

5. $a^2 + 4a + 28$

6. $m^2 - 10m + 25$

7. $a^2 - 3a + \frac{9}{2}$

Find the value of *c* that makes each trinomial a perfect square.

**8.** $x^2 + 2x + c$
**9.** $t^2 + 40t + c$
**10.** $x^2 + 18x + c$
**11.** $r^2 - 9r + c$
**12.** $a^2 - 100a + c$
**13.** $x^2 + 15x + c$

Solve each equation by completing the square.

**14.** $y^2 - 2y = 24$
**15.** $z^2 + 3z = 88$

# EXERCISES

**Practice**

Solve each equation by completing the square.

**16.** $x^2 + 8x - 84 = 0$
**17.** $m^2 + 3m - 180 = 0$
**18.** $n^2 - 8n + 14 = 0$
**19.** $x^2 - 7x + 5 = 0$
**20.** $a^2 - 5a - 10 = 0$
**21.** $t^2 + 3t - 8 = 0$
**22.** $12r^2 - 17r - 5 = 0$
**23.** $b^2 - \frac{3}{4}b + \frac{1}{8} = 0$
**24.** $3t^2 + 4t - 15 = 0$
**25.** $3z^2 - 12z + 4 = 0$
**26.** $6s^2 + 2s + 3 = 0$
**27.** $ax^2 + c = 0$
**28.** $x^2 + bx + c = 0$
**29.** $ax^2 + bx + c = 0$

**Critical Thinking**

**30.** You have learned how to solve quadratic equations using three different methods. How do you know what is the best method to use for any given equation? Is any one method more useful than the other two? Explain your answer.

**Applications**

**31. Hobbies**  Jackie is in charge of building a set for the school play. She wants each rectangular window to have an area of 315 in². She also wants each window to be 6 in. taller than it is wide. What are the dimensions of the window?

**32. Physics**  Michael drives a red sportscar on a race track at an initial velocity of 24 ft/s and begins to accelerate at a constant rate of 8 ft/s².

   **a.** How long will it take him to travel a distance of 100 ft? *Round to the nearest tenth.*

   **b.** How long will it take him to travel a distance of 200 ft? *Round to the nearest tenth.*

   **c.** How long will it take him to travel a distance of 300 ft? *Round to the nearest tenth.*

   **d.** Study your answers in parts a–c. As the distance doubles, does the amount of time double? Explain your answer.

**33. Safety**  Juanita is driving a truck at an initial velocity of 60 ft/s. She sees a stop sign 600 ft ahead of her. If she begins to decelerate at the rate of 3 ft/s², how long will it take her before she stops at the stop sign? *If acceleration is a positive number, what is deceleration?*

**Computer**

**34.** The BASIC program below determines if an equation written in standard form contains a perfect square trinomial. It uses the same process that you use when completing the square to determine if the expression is a perfect square trinomial.

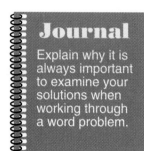
```
10 PRINT "INPUT A, B, and C."     90 IF B^2 = 4*A*C THEN 120
20 INPUT A, B, C                 100 PRINT "NOT A PERFECT
30 PRINT "A", "B", "C",              SQUARE"
   "4*A*C", "B^2"                 110 GOTO 140
40 PRINT A,B,C,4*A*C,B^2          120 PRINT A;"X^2 + ";B;
50 PRINT                              "X + ";C;" = 0"
60 IF A<0 OR C< 0 THEN 100        130 PRINT "(";SQR(A);
70 IF SQR(A) <> INT(SQR(A))            "X + ";SQR(C);")^2 = 0"
   THEN 100                       140 PRINT "X = ";-B/(2*A)
80 IF SQR(C) <> INT(SQR(C))       150 END
   THEN 100
```

**Use the program above to guess-and-check values for $k$ in each expression so that $ax^2 + bx + c$ is a perfect square trinomial.**

**a.** $x^2 - 14x + k$  **b.** $x^2 + kx + 64$  **c.** $4x^2 + kx + 1$

**d.** $kx^2 - 20x + 4$  **e.** $16x^2 + 40x + k$  **f.** $4x^2 - 12x + k$

**Mixed Review**

**35.** Solve $x^2 - 20x = -75$ by factoring.   **(Lesson 7-2)**

**36.** Solve $\sqrt{3w - 2} = 8$.   **(Lesson 6-7)**

**37.** Use long division to show that $a - 5$ is a factor of $a^3 - 5a^2 - 6a + 30$.   **(Lesson 5-6)**

**38.** Solve $\begin{cases} |x + 3| < 4 \\ x + y \geq 2 \end{cases}$ by graphing.   **(Lesson 2-8)**

**39.** Find the next number in the pattern 2, 8, 18, 32, 50, $\underline{\ \ ?\ \ }$.   **(Lesson 2-3)**

---

## ∿∿∿ MID-CHAPTER REVIEW ∿∿∿

**1.** Find the least prime number greater than 840.   **(Lesson 7-1)**

**Solve each equation by graphing.**   **(Lesson 7-2)**

**2.** $x^2 - 2x - 35 = 0$

**3.** $m^2 + 6m = 27$

**Solve each equation by factoring.**   **(Lesson 7-2)**

**4.** $4x^2 - 13x = 12$     **5.** $4t^2 = 25$       **6.** $a^3 = 81a$

**Solve each equation by completing the square.**   **(Lesson 7-3)**

**7.** $4x^2 + 19x - 5 = 0$          **8.** $y^2 + 12y + 4 = 0$

**9. Photography**   The edge of a metallic picture frame forms a rectangle 12 by 16 in. The frame is of uniform width and contains a picture whose area is 164 in². What is the approximate width of the frame?   **(Lesson 7-2)**

# 7-4 The Quadratic Formula and the Discriminant

## Objectives

After studying this lesson, you should be able to:

- solve quadratic equations using the quadratic formula, and
- use the discriminant to determine the nature of the roots of a quadratic equation.

In Lessons 7-2 and 7-3, you learned several ways to solve quadratic equations. Each has its limitations. You might ask, "Isn't there some formula that will work for any quadratic equation?" The answer is "yes" and that formula is called the **quadratic formula.** The formula is derived from solving the general form of a quadratic equation for $x$.

$$ax^2 + bx + c = 0 \quad (a \neq 0)$$ *Start with the general form of a quadratic equation.*

$$x^2 + \frac{b}{a}x + \frac{c}{a} = 0$$ *Divide by a so that the coefficient of $x^2$ is 1.*

$$x^2 + \frac{b}{a}x = -\frac{c}{a}$$ *Subtract $\frac{c}{a}$ from each side.*

*Complete the square by adding*

$$x^2 + \frac{b}{a}x + \left(\frac{b}{2a}\right)^2 = -\frac{c}{a} + \left(\frac{b}{2a}\right)^2$$ $\left(\frac{b}{a} \div 2\right)^2$ *or* $\left(\frac{b}{2a}\right)^2$ *to each side.*

$$\left(x + \frac{b}{2a}\right)^2 = -\frac{c}{a} + \frac{b^2}{4a^2}$$ *Factor the left side.*

$$\left(x + \frac{b}{2a}\right)^2 = \frac{b^2 - 4ac}{4a^2}$$ *Add the fractions on the right side.*

$$\left|x + \frac{b}{2a}\right| = \sqrt{\frac{b^2 - 4ac}{4a^2}}$$ *Take the square root of each side.*

$$x + \frac{b}{2a} = \pm\frac{\sqrt{b^2 - 4ac}}{2a}$$ *Simplify.*

$$x = -\frac{b}{2a} \pm \frac{\sqrt{b^2 - 4ac}}{2a}$$ *Subtract $\frac{b}{2a}$ from each side.*

$$x = \frac{-b \pm \sqrt{b^2 - 4ac}}{2a}$$

This equation is known as the quadratic formula.

| Quadratic Formula | **The solutions of a quadratic equation of the form** $ax^2 + bx + c = 0$ **with** $a \neq 0$ **are given by this formula.** $$x = \frac{-b \pm \sqrt{b^2 - 4ac}}{2a}$$ |
| --- | --- |

A photo editor at a magazine publisher has a 12.5 cm by 8.4 cm photo of a rabbit sitting in the center of a large field of grass. She wishes to get a print that has half the area of the original photo and concentrates on the rabbit. The layout artist needs to be told how much of the grass portion to crop, or cut off. If the same amount is cropped from all edges of the photo, what are the dimensions of the print?

The area of the original photo is (12.5)(8.4) or 105 cm². The problem states that the print has half the area of the original, or $\frac{1}{2}(105)$ cm². The print has a length of 12.5 − 2x cm and a width of 8.4 − 2x cm.

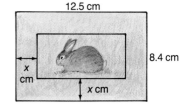

$$A = \ell w$$
$$\frac{1}{2}(105) = (12.5 - 2x)(8.4 - 2x) \qquad \textit{Substitute the known values.}$$
$$52.5 = 105 - 41.8x + 4x^2 \qquad \textit{Multiply.}$$
$$0 = 4x^2 - 41.8x + 52.5 \qquad \textit{Subtract 52.5 from each side.}$$

In the last step, the equation is written in the general form of a quadratic equation. From this form, the values for $a$, $b$, and $c$ can be defined for use in the quadratic formula.

$$x = \frac{-b \pm \sqrt{b^2 - 4ac}}{2a}$$

$$x = \frac{-(-41.8) \pm \sqrt{(-41.8)^2 - 4(4)(52.5)}}{2(4)} \qquad a = 4, b = -41.8, c = 52.5$$

$$x = \frac{41.8 \pm \sqrt{(-41.8)^2 - 4(4)(52.5)}}{8}$$

Now you can use your calculator to help you find approximate solutions. First evaluate the radical part of the formula and store the result in memory.

ENTER: 41.8 [+/-] [x²] [−] 4 [×] 4 [×] 52.5 [=] [√x] [STO]  `30.12042496`

Now find the solutions.

ENTER: [(] 41.8 [+] [RCL] [)] [÷] 8 [=]  `8.99005312`

ENTER: [(] 41.8 [−] [RCL] [)] [÷] 8 [=]  `1.45994688`

The two solutions are approximately 8.99 cm and 1.46 cm. In this problem, 8.99 cm is not a reasonable solution. The new length of the print would be 12.5 − 2(1.46), or 9.58 cm. The new width would be 8.4 − 2(1.46), or 5.48 cm.

**Check:** 9.58(5.48) = 52.4984 $\approx \frac{1}{2}(105)$. ✓

**Example 1**

Solve $t^2 - 3t - 28 = 0$.

This equation is already in general form. So you can use the quadratic formula directly.    $a = 1, b = -3, c = -28$

$$t = \frac{-b \pm \sqrt{b^2 - 4ac}}{2a}$$

$$t = \frac{-(-3) \pm \sqrt{(-3)^2 - 4(1)(-28)}}{2(1)}$$

$$t = \frac{3 \pm \sqrt{121}}{2} \text{ or } \frac{3 \pm 11}{2}$$

$$t = \frac{3 + 11}{2} \text{ or } 7 \quad \text{and} \quad t = \frac{3 - 11}{2} \text{ or } -4.$$

The solutions are 7 and -4.    *Check these solutions.*

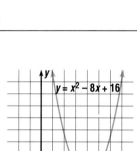

*The related function shows that there are 2 solutions.*

---

**Example 2**

Solve $x^2 - 8x + 16 = 0$.    $a = 1, b = -8, c = 16$

$$x = \frac{-b \pm \sqrt{b^2 - 4ac}}{2a}$$

$$x = \frac{-(-8) \pm \sqrt{(-8)^2 - 4(1)(16)}}{2(1)}$$

$$x = \frac{8 \pm 0}{2}$$

$$x = \frac{8 + 0}{2} \text{ or } 4 \quad \text{and} \quad x = \frac{8 - 0}{2} \text{ or } 4$$

There is one distinct solution, 4.

*The related function shows that there is 1 solution.*

---

**Example 3**

Solve $3p^2 - 5p + 9 = 0$.    $a = 3, b = -5, c = 9$

$$x = \frac{-b \pm \sqrt{b^2 - 4ac}}{2a}$$

$$x = \frac{-(-5) \pm \sqrt{(-5)^2 - 4(3)(9)}}{2(3)}$$

$$x = \frac{5 \pm \sqrt{-83}}{6}$$

Since the radical contains a negative value, the solutions will be imaginary.

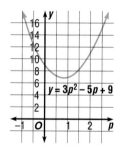

*The related function does not intersect the horizontal axis.*

*The solutions are imaginary.*

The solutions are $\dfrac{5 + i\sqrt{83}}{6}$ and $\dfrac{5 - i\sqrt{83}}{6}$.

*Imaginary solutions appear in conjugate pairs.*

---

These three examples demonstrate a pattern that is useful in determining the nature of the roots of a quadratic equation. In the quadratic formula, the expression under the radical sign, $b^2 - 4ac$, is called the **discriminant.** The discriminant tells the nature of the roots of a quadratic equation.

| Equation | Value of the Discriminant | Roots | Nature of the Roots |
|---|---|---|---|
| Ex. 1 $t^2 - 3t - 28 = 0$ | 121 or $(11)^2$ | 7, -4 | 2 real roots |
| Ex. 2 $x^2 - 8x + 16 = 0$ | 0 | 4 | 1 real root |
| Ex. 3 $3p^2 - 5p + 9 = 0$ | -83 | $\dfrac{5 \pm i\sqrt{83}}{6}$ | 2 imaginary roots |

The chart shows that if the value of the discriminant is a perfect square or 0, the roots are real and rational. Other positive discriminants will yield irrational roots. A negative discriminant means the roots will be imaginary.

**Example 4**

Find the value of the discriminant of each equation and then describe the nature of its roots.

**a.** $2x^2 + x - 3 = 0$

$a = 2, b = 1, c = -3$
$b^2 - 4ac = (1)^2 - 4(2)(-3)$
$\qquad = 1 + 24$
$\qquad = 25$

The value of the discriminant is a positive and a perfect square. So, $2x^2 + x - 3 = 0$ has two real roots and they are rational.

**b.** $x^2 + 8 = 0$

$a = 1, b = 0, c = 8$
$b^2 - 4ac = (0)^2 - 4(1)(8)$
$\qquad = 0 - 32$
$\qquad = -32$

The value of the discriminant is negative. So, $x^2 + 8 = 0$ has two imaginary roots.

# CHECKING FOR UNDERSTANDING

**Communicating Mathematics**

Read and study the lesson to answer each question.

1. In the photo problem, why was 8.99 cm eliminated as a reasonable answer?

2. The two solutions for the photo problem were stated to be *approximately* 8.99 and 1.46. Why was the word approximately used?

3. When the discriminant is a positive number, there will be __?__ __?__ roots.

4. When the discriminant is __?__, there will be one real root.

5. When the discriminant is a negative number, there will be __?__ __?__ roots.

**Guided Practice**

State the values of a, b, and c for each quadratic equation. Then find the value of the discriminant, and solve the equation.

6. $y^2 + 6y + 9 = 0$

7. $a^2 = 16$

8. $5x^2 + 16x + 3 = 0$

9. $6a^2 + 2a + 1 = 0$

10. $-3x^2 + x - 2 = 0$

11. $3a^2 - a + 3 = 0$

# EXERCISES

**Practice**

Find the value of the discriminant for each quadratic equation. Fully describe the nature of its roots. Then solve the equation. Express irrational roots as exact and then use your calculator to give an approximation.

12. $a^2 + 12a + 32 = 0$

13. $y^2 - 4y + 4 = 0$

14. $x^2 - 4x + 1 = 0$

15. $x^2 - 2x - 35 = 0$

16. $x^2 - 10x + 25 = 0$

17. $4x^2 + 8x + 3 = 0$

18. $3x^2 + 11x + 4 = 0$

19. $4y^2 + 16y + 15 = 0$

20. $m^2 - 2m + 5 = 0$

21. $y^2 - 6y + 13 = 0$

22. $c^2 - 12c + 42 = 0$

23. $a^2 = 6a$

24. $3m^2 = 108m$

25. $4x^2 - 8x + 13 = 0$

26. $x^2 - x + 1 = 0$

27. $n^2 + 4n + 29 = 0$

28. $2a^2 - 13a = 7$

29. $a^2 + a - 5 = 0$

30. $11m^2 - 12m = 10$

31. $4a^2 + 3a - 2 = 0$

32. $t^2 - 16t + 4 = 0$

33. $2x^2 + 5x = 9$

**Journal**

Describe which method of solving quadratic equations you prefer to use and why.

**Critical Thinking**

Find three values for $k$ for each equation so that it will have one real root, two real roots, and two imaginary roots.

34. $x^2 + 3x + k = 0$

35. $kx^2 + 3x - 2 = 0$

36. $2x^2 - 5x - k = 0$

**Applications**

37. **Gardening**   The Hillcrest Garden Club wants to double the area of its rectangular rose bed. Strips of the same width will be added to one end and one side to form a rectangle. If the bed is now 17.5 m by 12.2 m, what are the dimensions of the new bed?

38. **Astronomy**   The acceleration due to gravity on the surface of Earth is 9.8 m/s². The acceleration due to gravity on the surface of Mars is 3.7 m/s² and on Venus it is 8.9 m/s². How long will it take a ball thrown upward to return to the ground on each of these planets if the initial velocity is 25 m/s?

**Mixed Review**

39. Solve $x^2 + 14x - 12 = 0$ by completing the square.   (**Lesson 7-3**)

40. Solve $d^2 - 5d - 24 = 0$ by graphing.   (**Lesson 7-2**)

41. Simplify $(4a + 7)(3a - 9)$.   (**Lesson 5-2**)

42. Write $\begin{cases} h + t + u = 10 \\ h - u = 1 \\ h = t + u \end{cases}$   as an augmented matrix.   (**Lesson 4-7**)

43. Given $f(x, y) = 3x + 2y$; find $f(5, -2)$.   (**Lesson 3-6**)

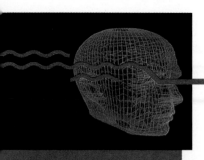

# Technology

## Discriminants and Roots

▶ BASIC
Spreadsheets
Software

```
10 INPUT "ENTER A,B,C OF
      QUADRATIC FORMULA: ";A,B,C
20 LET D = B * B - 4 * A * C
30 IF D < 0 THEN 110
40 IF D > 0 THEN 70
50 PRINT "ONE ROOT: "; -B / (2 * A)
60 GOTO 150
70 LET X = (-B + SQR (D)) / (2 * A)
80 LET Y = (-B - SQR (D)) / (2 * A)
90 PRINT "TWO REAL ROOTS: ";X;", ";Y
100 GOTO 150
110 PRINT "TWO COMPLEX ROOTS: "
120 PRINT -B / (2 * A);"+";SQR
      (-D) / (2 * A);"I"
130 PRINT -B / (2 * A);"-";SQR
      (-D) / (2 * A);"I"
150 PRINT "THE DISCRIMINANT IS ";D
160 END
```

The BASIC program at the left utilizes the discriminant to determine the type of roots an equation has and what those roots are. In order to use the program, you must have the quadratic equation written in standard form.

Example    $3x^2 + 16x = -5$
$3x^2 + 16x + 5 = 0$

$a = 3$, $b = 16$, and $c = 5$

To run the program to solve this equation, type RUN and then enter the values for $a$, $b$, and $c$ when the ? prompt appears on the screen. The screen should print the following information.

```
TWO REAL ROOTS: -.33333333, -5
THE DISCRIMINANT IS 196.
```

# EXERCISES

Use the BASIC program to find the roots of each quadratic equation. Describe the types of roots each equation has. If the roots are real, determine if they are rational or irrational.

**1.** $6n^2 + 8n + 1 = 0$      **2.** $3y^2 + 3y = -2$      **3.** $x^2 = 6x - 13$

Find the value of the discriminant and the roots of each equation. Make a conjecture about the roots and discriminants of each group of equations.

**4.** $m^2 - 4m + 6 = 0$
$m^2 + 4m + 6 = 0$

**5.** $x^2 - 25 = 0$
$2x^2 - 10 = 0$
$16x^2 - 12 = 0$

**6.** $x^2 - 3x - 28 = 0$
$2x^2 - 6x - 56 = 0$
$5x^2 - 15x - 140 = 0$

**7.** $x^2 + 4x + 4 = 0$
$x^2 - 10x + 25 = 0$
$4x^2 + 32x + 64 = 0$

# Sum and Product of Roots

**Objectives**

After studying this lesson, you should be able to:

- find the sum and product of the roots of a quadratic equation,
- find all possible integral roots of a quadratic equation, and
- find a quadratic equation to fit a given condition.

There are times when you may know the roots of a quadratic equation but you don't know the equation itself. For example, suppose the roots of a quadratic equation are $-7$ and $5$, and you want to find the equation.

Remember when you used factoring to solve an equation. You eventually solved two equations set equal to zero to find the two solutions. You can use the process in reverse to find the equation when you know the solutions.

$$
\begin{array}{ll}
x = 5 \qquad\qquad\qquad x = -7 & \textit{Start with the solutions.} \\
\underline{x - 5 = 0 \qquad\qquad\quad x + 7 = 0} & \textit{Find equations set equal to 0.}
\end{array}
$$

$$
\begin{array}{ll}
(x - 5)(x + 7) = 0 & \textit{Use the zero product property.} \\
x^2 + 2x - 35 = 0 & \textit{Find the quadratic equation by} \\
& \textit{multiplying.}
\end{array}
$$

The last step shows the general form of a quadratic equation whose roots are $5$ and $-7$. These roots, their sum, and their product can lead you to the equation in another way. Study the pattern shown below.

$$
\begin{array}{ccccc}
x^2 & + & 2x & + & (-35) = 0 \\
\uparrow & & \uparrow & & \uparrow \\
1 & & \dfrac{-(-7 + 5)}{1} & & \dfrac{-7(5)}{1}
\end{array}
$$

This pattern can be generalized for any quadratic equation by using the roots as defined by the quadratic formula. Let $s_1$ and $s_2$ represent solutions, or roots.

$$
s_1 = \frac{-b + \sqrt{b^2 - 4ac}}{2a} \qquad\qquad s_2 = \frac{-b - \sqrt{b^2 - 4ac}}{2a}
$$

$$
\begin{aligned}
s_1 + s_2 &= \frac{-b + \sqrt{b^2 - 4ac}}{2a} + \frac{-b - \sqrt{b^2 - 4ac}}{2a} \\
&= \frac{-b + (-b) + \sqrt{b^2 - 4ac} - \sqrt{b^2 - 4ac}}{2a} \\
&= \frac{-2b + 0}{2a} \ \text{or} \ -\frac{b}{a}
\end{aligned}
$$

$$
\begin{aligned}
s_1 s_2 &= \frac{-b + \sqrt{b^2 - 4ac}}{2a} \cdot \frac{-b - \sqrt{b^2 - 4ac}}{2a} \\
&= \frac{b^2 - (b^2 - 4ac)}{4a^2} \qquad \textit{Use the distributive property.} \\
&= \frac{b^2 - b^2 + 4ac}{4a^2} \\
&= \frac{4ac}{4a^2} \ \text{or} \ \frac{c}{a}
\end{aligned}
$$

The rule below can help you in checking your solutions to a quadratic equation or in finding the equation when you know the roots.

| Sum and Product of Roots | If the roots of $ax^2 + bx + c = 0$ with $a \neq 0$ are $s_1$ and $s_2$, then $s_1 + s_2 = -\dfrac{b}{a}$ and $s_1 s_2 = \dfrac{c}{a}$. |
|---|---|

**Example 1**

Solve $3x^2 - 16x - 12 = 0$. Then use the sum and product of the roots to check your solution.

$$x = \frac{-b \pm \sqrt{b^2 - 4ac}}{2a}$$

$$= \frac{-(-16) \pm \sqrt{(-16)^2 - 4(3)(-12)}}{2(3)} \qquad \text{In the equation } 3x^2 - 16x - 12 = 0,$$
$$\text{$a = 3$, $b = -16$, and $c = -12$.}$$

$$= \frac{16 \pm \sqrt{400}}{6} \quad \text{or} \quad \frac{16 \pm 20}{6}$$

The roots are $\dfrac{16 + 20}{6}$ and $\dfrac{16 - 20}{6}$ or 6 and $-\dfrac{2}{3}$.

**Check:** *The sum of the roots, $s_1 + s_2$, should be $-\dfrac{b}{a}$ or $\dfrac{16}{3}$.*  *The product of the roots, $s_1 s_2$, should be $\dfrac{c}{a}$ or $-4$.*

$$\left[ 6 + \left( -\frac{2}{3} \right) \right] = \frac{16}{3} \qquad\qquad 6\left( -\frac{2}{3} \right) = -4$$

The solutions are correct.

**Example 2**

Write a quadratic equation that has roots $-\dfrac{5}{4}$ and $\dfrac{16}{5}$.

Find the sum and product of the roots. Express the two fractions with the same denominator.

$$s_1 + s_2 = -\frac{5}{4} + \frac{16}{5} = \frac{39}{20} \quad -\frac{b}{a} \qquad s_1 s_2 = \left( -\frac{5}{4} \right)\left( \frac{16}{5} \right) = -\frac{80}{20} \quad \frac{c}{a}$$

Therefore, $a = 20$, $b = -39$, and $c = -80$.

The equation is $20x^2 - 39x - 80 = 0$.

The method used in Example 2 can also be used with equations whose roots are imaginary.

**Example 3**

Find a quadratic equation that has roots $5 + 2i$ and $5 - 2i$.

$$s_1 + s_2 = (5 + 2i) + (5 - 2i) \qquad\qquad s_1 s_2 = (5 + 2i)(5 - 2i)$$
$$= 10 \qquad -\frac{b}{a} = \frac{10}{1} \qquad\qquad\qquad = 25 + 4 \text{ or } 29 \qquad \frac{c}{a} = \frac{29}{1}$$

Therefore, $a = 1$, $b = -10$, and $c = 29$.
The equation is $x^2 - 10x + 29 = 0$.

---

**Example 4**

APPLICATION

Sports

Claire Homan is practicing her lobs for a tennis tournament. In analyzing her style of play, she must use some physics. Her coach videotapes her practice with a special camera that also records the elapsed time (in seconds) of each frame of film. When reviewing the film, he notes that the ball she lobs is approximately the same distance above the ground after 0.5 second and after 1.5 seconds. At what speed was she hitting the ball? If Claire hits the ball when it is 7 feet in the air, how far above the ground is it at 0.5 second and 1.5 seconds?

You can use the formula you learned in Lesson 7-2 for a free falling object affected by gravity to solve this problem. This formula can be rewritten as a quadratic equation in general form.

$$s = v_i t - \frac{1}{2}gt^2 \quad\Rightarrow\quad 0 = \left(-\frac{1}{2}g\right)t^2 + (v_i)t + (-s)$$
$$\qquad\qquad\qquad\qquad\qquad\quad \uparrow \qquad\quad \uparrow \qquad\quad \uparrow$$
$$\qquad\qquad\qquad\qquad\qquad\quad a \qquad\quad b \qquad\quad c$$

Since the acceleration of gravity is 32 ft/s$^2$, $a = -16$.
The two solutions for $t$ are 0.5 and 1.5.

$$s_1 + s_2 = -\frac{b}{a} \qquad\qquad s_1 \cdot s_2 = \frac{c}{a}$$
$$0.5 + 1.5 = -\frac{b}{-16} \qquad\quad 0.5 \cdot 1.5 = \frac{c}{-16}$$
$$2 = \frac{b}{16} \qquad\qquad\quad 0.75 = \frac{c}{-16}$$
$$32 = b \qquad\qquad\qquad -12 = c$$

The equation is $0 = -16t^2 + 32t - 12$. It can also be written as $12 = 32t - 16t^2$, so that it has the same form as the original equation.

From this form, you can see that $v_i$ is 32. So, she hits the ball at a speed of 32 ft/s. The value of $s$ in this equation is 12, but recall that she hits the ball when it is 7 feet in the air. The actual height of the ball at 0.5 and 1.5 seconds is $12 + 7$ or 19 feet.

*Remember that this graph shows the graph of the formula and* not *the actual path of the ball.*

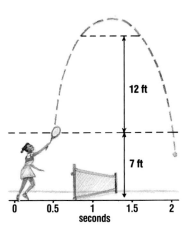

---

You can also use the sum or product to find missing values for $a$, $b$, or $c$ when one of the roots is known.

**Example 5**

Find $k$ such that $-3$ is a root of $x^2 + kx - 24 = 0$.

Let $s_1 = -3$. Solve $s_1 s_2 = \dfrac{c}{a}$ for $s_2$. Then solve for $k$.

$$-3s_2 = \frac{-24}{1} \qquad s_1 + s_2 = -\frac{b}{a}$$
$$s_2 = 8 \qquad\qquad -3 + 8 = -\frac{k}{1}$$
$$\qquad\qquad\qquad -5 = k$$

The value of $k$ is $-5$.  *Check this result.*

# CHECKING FOR UNDERSTANDING

**Communicating Mathematics**

**Read and study the lesson to answer each question.**

1. What property allows you to conclude that if the roots of a quadratic equation are 5 and $-7$, then $(x - 5)(x + 7) = 0$?

2. The product of the roots of a quadratic equation is equal to __?__ expressed in terms of $a$, $b$, and $c$.

3. The sum of the roots of a quadratic equation is equal to __?__ expressed in terms of $a$, $b$, and $c$.

4. The sum of the roots of $5x^2 - 14x - 20 = 0$ would be

    a. $-14$          b. $-\dfrac{14}{5}$          c. $\dfrac{14}{5}$

5. The product of the roots of $5x^2 - 14x - 20 = 0$ would be

    a. $-20$          b. $-4$          c. $4$

**Guided Practice**

**State the sum and the product of the roots of each quadratic equation.**

6. $2x^2 + 8x - 3 = 0$

7. $4x^2 + 3x - 12 = 0$

8. $2x^2 + 7 = 0$

9. $5x^2 = 3$

10. $x^2 + 4x - \dfrac{5}{3} = 0$

11. $3x^2 - \dfrac{x}{5} - \dfrac{4}{5} = 0$

**Solve each equation. Then find the sum and the product of the roots to check your solutions.**

12. $x^2 + x - 6 = 0$

13. $m^2 + 5m + 6 = 0$

14. $s^2 + 5s - 24 = 0$

15. $a^2 - 9a + 20 = 0$

# EXERCISES

**Practice**

Solve each equation. Then find the sum and the product of the roots to check your solutions.

**16.** $x^2 + 6x - 7 = 0$

**17.** $2z^2 - 5z - 3 = 0$

**18.** $y^2 + 5y + 6 = 0$

**19.** $2c^2 - 5c + 1 = 0$

**20.** $6t^2 + 28t - 10 = 0$

**21.** $4a^2 + 21a = 18$

Write a quadratic equation that has the given roots.

**22.** $6, 4$

**23.** $8, -2$

**24.** $6, -6$

**25.** $3, \dfrac{1}{2}$

**26.** $5, \dfrac{2}{3}$

**27.** $-\dfrac{2}{5}, \dfrac{2}{5}$

**28.** $\dfrac{5}{8}, \dfrac{1}{4}$

**29.** $\sqrt{3}, 2\sqrt{3}$

Solve each equation. Then use the sum and the product of the roots to check your solutions.

**30.** $9n^2 - 1 = 0$

**31.** $s^2 - 16 = 0$

**32.** $2x^2 - 7x = 15$

**33.** $15c^2 - 2c - 8 = 0$

**34.** $7s^2 + 5s - 1 = 0$

**35.** $12x^2 + 19x + 4 = 0$

Write a quadratic equation that has the given roots.

**36.** $2 \pm \sqrt{3}$

**37.** $-6i, 6i$

**38.** $\dfrac{5 - 3i}{4}, \dfrac{5 + 3i}{4}$

**39.** $\dfrac{1 + \sqrt{7}}{2}, \dfrac{1 - \sqrt{7}}{2}$

Find $k$ such that the number given is a root of the equation.

**40.** $1; x^2 + kx - 5 = 0$

**41.** $3; x^2 + kx - 21 = 0$

**42.** $3; x^2 + 6x - k = 0$

**43.** $-\dfrac{3}{2}; 2x^2 + kx - 12 = 0$

**Critical Thinking**

**44.** Suppose you were given the equation $x^2 + bx + 12 = 0$ and told that $b$ was an integer. List all the possible integral roots of that equation.

**Applications**

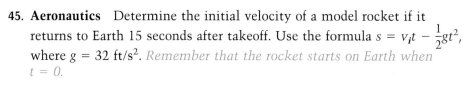

**45. Aeronautics**   Determine the initial velocity of a model rocket if it returns to Earth 15 seconds after takeoff. Use the formula $s = v_i t - \dfrac{1}{2}gt^2$, where $g = 32$ ft/s$^2$. *Remember that the rocket starts on Earth when $t = 0$.*

**46. Horticulture**   Helene Jonson has a rectangular garden 25 ft by 50 ft. She wishes to increase the garden on all sides by an equal amount. The area of the garden will be increased by 400 ft$^2$. By how much will each dimension be increased?

**47. Flight**  Sam shot his model rocket up into the air and counted the seconds it stayed in flight. He noticed that the rocket seemed to be at the same height as the top of a television antenna at 1 and 3 seconds. Find the initial speed of the model rocket and the height of the television antenna.

**Mixed Review**

**48.** Solve $2x^2 - 5x + 4 = 0$ using the quadratic formula.  **(Lesson 7-4)**

**49.** Solve $\sqrt[3]{2x + 1} = 3$.  **(Lesson 6-7)**

**50.** Simplify $\dfrac{(3 + \sqrt{5})}{(1 + \sqrt{2})}$.  **(Lesson 6-2)**

**51.** Factor $ab + 7a + 4b + 28$.  **(Lesson 5-5)**

**52.** Write the identity matrix for a $3 \times 3$ matrix.  **(Lesson 4-4)**

**53. Business**  Derringer Cleaners charges $52 to clean a prom dress. If the equation relating time spent (in hours) to total charge (in dollars) is $C = 12 + 20t$, find the time spent cleaning the dress.  **(Lesson 2-1)**

**54. Geometry**  A piece of wire was cut into two pieces. One was bent into a square and the other into an equilateral triangle. The side of the equilateral triangle has the same whole-number length (in cm) as the side of the square. If the length of the piece of wire is less than 50 cm, find all possible measurements for the sides of the figures.  **(Lesson 1-5)**

---

## HISTORY CONNECTION

Archaeologists in an ancient area of the Middle East have found artifacts that show the Babylonians solved mathematical problems. They completed these calculations in words. Algebraic symbols were not in general usage until the middle of the 17th century A.D. The tablet shown at the right contains one quadratic equation and its solution written sometime between 1900 and 1600 B.C. The tablet explains the solution to a problem about a rectangle whose area is 60 square units.

The solution to the problem, when translated into English, involves a version of a quadratic equation in the form $x^2 + ax = b,\ b > 0$. If $y$ is the length of the rectangle and $x$ is the width, then $y = x + 7$ and $xy = 60$. When solved, we find the rectangle's dimensions are $5 \times 12$ units.

# Quadratic Techniques to Solve Polynomial Equations

**Objectives**

After studying this lesson, you should be able to:

- solve third and fourth degree equations that contain a quadratic factor, and
- solve other nonquadratic equations that can be written in quadratic form.

The problem below demonstrates solving an equation that is not a quadratic equation itself, but contains a factor that is a quadratic. The methods you have learned are used to solve the quadratic part of that equation.

**Connection**

A rectangular prism has a base that is $b$ mm long. The width of the base is 2 mm less than its length. The height is 3 mm greater than the length of the base. The measure of the volume is 6 times the measure of the length of the base. Find the dimensions of the prism.

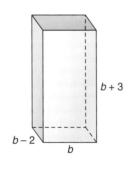

The volume of a prism is the product of the length, width, and height of the prism. That is, $V = \ell wh$.

$\ell = b, w = b - 2, h = b + 3, V = 6b$

$V = \ell wh$

| | |
|---|---|
| $6b = (b)(b - 2)(b + 3)$ | *Substitute values for $\ell$, $w$, $h$.* |
| $6b = b^3 + b^2 - 6b$ | *Multiply.* |
| $0 = b^3 + b^2 - 12b$ | *Subtract $6b$ from each side.* |
| $0 = b(b^2 + b - 12)$ | *Factor out the common factor, $b$.* |
| $0 = b(b - 3)(b + 4)$ | *Factor the quadratic term.* |

Now use the zero product property.

$b = 0$  or  $b + 4 = 0$  or  $b - 3 = 0$
$b = 0$         $b = -4$         $b = 3$

Since this problem deals with measures of a prism, we can eliminate 0 and $^-4$ as reasonable solutions.

$\ell = b = 3$

$w = b - 2 = 3 - 2$ or 1

$h = b + 3 = 3 + 3$ or 6

The dimensions of the prism are 3 mm × 1 mm × 6 mm.

**Check:** The width, $\ell$, is 2 less than 3, the length.
The height, 6, is 3 more than 3, the length.
$V = 3 \cdot 1 \cdot 6$ or 18, which is 6 times the length.

Some equations are not quadratic but can be written in a form that resembles a quadratic equation. For example, the equation $x^4 - 20x^2 + 64 = 0$ can be written as $(x^2)^2 - 20(x^2) + 64 = 0$. Equations that can be written in this way are said to be equations in **quadratic form.**

| | |
|---|---|
| *Definition of Quadratic Form* | **For any numbers *a*, *b*, and *c*, except *a* = 0, an equation that may be written as $a[f(x)]^2 + b[f(x)] + c = 0$, where $f(x)$ is some expression in *x*, is in quadratic form.** |

Once an equation is written in quadratic form, it can be solved by the methods you have aleady learned to use for solving quadratic equations.

**Example 1**

Solve $x^4 - 13x^2 + 36 = 0$.

$$x^4 - 13x^2 + 36 = 0$$
$$(x^2)^2 - 13(x^2) + 36 = 0$$
$$(x^2 - 9)(x^2 - 4) = 0$$
$$(x + 3)(x - 3)(x + 2)(x - 2) = 0$$

$$x + 3 = 0 \qquad x - 3 = 0$$
$$x = -3 \qquad x = 3$$

$$x + 2 = 0 \qquad x - 2 = 0$$
$$x = -2 \qquad x = 2$$

*The graph of $y = x^4 - 13x^2 + 36$ crosses the x-axis 4 times. There will be 4 real solutions.*

The solutions or roots are $-3$, $3$, $-2$, and $2$.

Recall that $(a^m)^n = a^{mn}$ for any positive number *a* and any rational numbers *n* and *m*. This property of exponents that you learned in Chapter 5 is often used when solving equations.

**Example 2**

Solve $x^{\frac{1}{2}} - 6x^{\frac{1}{4}} + 8 = 0$.

$$x^{\frac{1}{2}} - 6x^{\frac{1}{4}} + 8 = 0$$
$$(x^{\frac{1}{4}})^2 - 6(x^{\frac{1}{4}}) + 8 = 0 \quad \textit{f(x) is } x^{\frac{1}{2}}.$$
$$(x^{\frac{1}{4}} - 2)(x^{\frac{1}{4}} - 4) = 0 \quad \textit{Factor to solve for f(x).}$$

$$x^{\frac{1}{4}} - 2 = 0 \quad \text{or} \quad x^{\frac{1}{4}} - 4 = 0$$
$$x^{\frac{1}{4}} = 2 \qquad\qquad x^{\frac{1}{4}} = 4$$
$$(x^{\frac{1}{4}})^4 = (2)^4 \qquad (x^{\frac{1}{4}})^4 = 4^4$$
$$x = 2^4 \text{ or } 16 \qquad x = 4^4 \text{ or } 256$$

**Check:**
$$x^{\frac{1}{2}} - 6x^{\frac{1}{4}} + 8 = 0$$
$$16^{\frac{1}{2}} - 6(16^{\frac{1}{4}}) + 8 \stackrel{?}{=} 0$$
$$4 - 6(2) + 8 \stackrel{?}{=} 0$$
$$0 = 0 \quad \checkmark$$
$$256^{\frac{1}{2}} - 6(256^{\frac{1}{4}}) + 8 \stackrel{?}{=} 0$$
$$16 - 6(4) + 8 \stackrel{?}{=} 0$$
$$0 = 0 \quad \checkmark$$

The solutions are 16 and 256.

**Example 3**

Solve $t^{\frac{2}{3}} = 16$.

$$t^{\frac{2}{3}} = 16$$

$$\left(t^{\frac{2}{3}}\right)^3 = (16)^3 \quad \text{\textit{Cube each side.}}$$

$$t^2 = 16^3 \text{ or } 4^6$$

$$t = \pm(4^6)^{\frac{1}{2}} \quad \text{\textit{Why } \pm?}$$

$$t = \pm 4^3 \text{ or } \pm 64$$

The solutions are 64 and $-64$.

**Check:**

$$t^{\frac{2}{3}} = 16$$
$$64^{\frac{2}{3}} \stackrel{?}{=} 16$$
$$(\sqrt[3]{64})^2 \stackrel{?}{=} 16$$
$$4^2 \stackrel{?}{=} 16$$
$$16 = 16 \quad \checkmark$$

$$t^{\frac{2}{3}} = 16$$
$$(-64)^{\frac{2}{3}} \stackrel{?}{=} 16$$
$$(\sqrt[3]{-64})^2 \stackrel{?}{=} 16$$
$$(-4)^2 \stackrel{?}{=} 16$$
$$16 = 16 \quad \checkmark$$

The quadratic formula can also be used to solve equations that are in quadratic form.

**Example 4**

Solve $x - 7\sqrt{x} - 8 = 0$.

$$x - 7\sqrt{x} - 8 = 0$$

$$(\sqrt{x})^2 - 7(\sqrt{x}) - 8 = 0 \quad \text{\textit{f(x) is } } \sqrt{x}.$$

$$\sqrt{x} = \frac{-b \pm \sqrt{b^2 - 4ac}}{2a}$$

$$= \frac{-(-7) \pm \sqrt{(-7)^2 - 4(1)(-8)}}{2(1)}$$

$$= \frac{7 \pm \sqrt{81}}{2} \quad \text{or} \quad \frac{7 \pm 9}{2}$$

$$\sqrt{x} = 8 \quad \text{or} \quad \sqrt{x} = -1$$

$$x = 64$$

*a = 1,*
*b = -7,*
*c = -8*

**Check:** $\quad x - 7\sqrt{x} - 8 = 0$

$$64 - 7\sqrt{64} - 8 \stackrel{?}{=} 0$$
$$64 - 7 \cdot 8 - 8 \stackrel{?}{=} 0$$
$$0 = 0 \quad \checkmark$$

There is no real number $x$ such that $\sqrt{x} = -1$.
The only solution is 64.

Some cubic equations can be solved using the quadratic formula. First a binomial factor must be found.

**Example 5**

Solve $x^3 - 27 = 0$.

$$x^3 - 27 = 0 \qquad \text{\textit{The left side is the difference of cubes.}}$$

$$(x - 3)(x^2 + 3x + 9) = 0 \qquad \text{\textit{Factor.}}$$

$$x - 3 = 0 \quad \text{or} \quad x^2 + 3x + 9 = 0 \qquad \text{\textit{Zero product property}}$$

$$x = 3 \quad \text{or} \quad x = \frac{-3 \pm \sqrt{(3)^2 - 4(1)(9)}}{2(1)}$$

$$= \frac{-3 \pm \sqrt{-27}}{2}$$

$$= \frac{-3 \pm 3i\sqrt{3}}{2}$$

The solutions are $3$, $\dfrac{-3 + 3i\sqrt{3}}{2}$, and $\dfrac{-3 - 3i\sqrt{3}}{2}$.

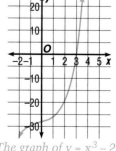

*The graph of $y = x^3 - 27$ crosses the x-axis once. There will be 1 real root.*

# CHECKING FOR UNDERSTANDING

Communicating
Mathematics

**Read and study the lesson to answer each question.**

1. Why were the roots 0 and $-4$ disregarded as possible solutions for measures of a prism?

2. What is the actual volume of the prism in the problem at the beginning of the lesson?

3. In Example 5, three solutions are shown but the graph of the related function crosses the $x$-axis once. Explain this discrepancy.

4. The maximum number of solutions a quadratic equation can have is $\underline{\ ?\ }$.

5. If $pqr = 0$, what property allows us to say that at least one of the three factors is equal to zero?

6. Give two examples of equations that are not quadratic, but can be written in quadratic form.

Guided Practice

**Factor each equation. Is one of the factors a quadratic? Write *yes* or *no*.**

7. $x^4 + 5x^3 + 6x^2 = 0$

8. $x^3 + 10x^2 + 16x = 0$

9. $3m^3 = 2m^2 - 7m$

10. $a^3 = 81a + a^5$

11. $25d^3 + 9d = 30d^2$

12. $16t^4 = 40t^2 - 25t$

**State whether each equation can be written in quadratic form.**

13. $x^4 + 5x^2 + 3 = 0$

14. $6x^4 + 7x = 8$

15. $6n^4 + 8n^2 = 0$

16. $2p + 5\sqrt{p} = 9$

**Solve each equation.**

17. $r^{\frac{1}{3}} = 2$

18. $x^{\frac{1}{4}} = 3$

19. $p^{\frac{3}{2}} - 8 = 0$

20. $k^{\frac{3}{4}} = 27$

# EXERCISES

Practice

**Write each equation so that it contains a quadratic factor or is in quadratic form.**

21. $x - 10x^{\frac{1}{2}} + 25 = 0$

22. $x^{\frac{4}{3}} - 7x^{\frac{2}{3}} + 12 = 0$

23. $z^4 + 6z^3 + 8z^2 = 0$

24. $y^{\frac{1}{2}} - 10y^{\frac{1}{4}} + 16 = 0$

25. $r^{\frac{2}{3}} - 5r^{\frac{1}{3}} + 6 = 0$

26. $s^{\frac{2}{3}} - 9s^{\frac{1}{3}} + 20 = 0$

27. $x^{\frac{1}{2}} + 7x^{\frac{1}{4}} + 12 = 0$

28. $9y^3 + 16y = -24y^2$

**Solve each equation.**

29. $x^4 + 5x^3 + 6x^2 = 0$

30. $x^3 + 10x^2 + 16x = 0$

31. $16x^4 - x^2 = 0$

32. $a^3 = 81a$

**Solve each equation.**

**33.** $s^3 = 8$

**34.** $s^4 = 25$

**35.** $b^4 - 5b^2 + 4 = 0$

**36.** $y^4 - 3y^2 + 2 = 0$

**37.** $a^3 = 125$

**38.** $m - 9\sqrt{m} + 8 = 0$

CONNECTION
Geometry

**39.** A rectangular prism has a width of $w$ inches. The length is 3 inches more than the width. The height is 4 inches more than the width. The measure of the volume of the prism is 32 times the measure of the length. Find the dimensions of the prism and its volume.

**Solve each equation.**

**40.** $r^{\frac{2}{3}} - 12r^{\frac{1}{3}} + 20 = 0$

**41.** $x^{\frac{2}{3}} - 8x^{\frac{1}{3}} + 15 = 0$

**42.** $m - 11m^{\frac{1}{2}} + 30 = 0$

**43.** $y^3 - 16y^{\frac{3}{2}} + 64 = 0$

**44.** $3g^{\frac{2}{3}} - 10g^{\frac{1}{3}} + 8 = 0$

**45.** $3m + m^{\frac{1}{2}} - 2 = 0$

**Critical Thinking**

**46.** How would you solve the equation $|a - 4|^2 - 7|a - 4| = -6$? Write an explanation. Then solve the equation.

**Applications**

**47. Landscaping** A rectangular lawn has dimensions 24 m by 32 m. A walkway of uniform width will be constructed along the inside edges of all four sides of the lawn. The remaining lawn will have an area of 425 m². How wide is the walkway?

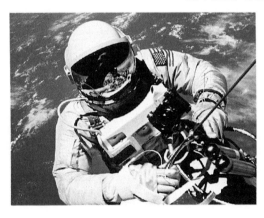

**48. Aeronautics** As an object moves farther from Earth, its weight decreases. The force of gravity decreases with the square of the distance from the center of Earth. The radius $R$ of Earth is about 3960 miles. Let $w_E$ be the weight of a body on Earth, $w_S$ be the weight of a body a certain distance from the center of Earth, and $r$ be the distance of the object above Earth's surface. The formula relating these is $(R + r)^2 = \dfrac{R^2 \cdot w_E}{w_S}$. Determine how far above Earth's surface a 135-pound astronaut is if she weighs 125 pounds in space.

**Mixed Review**

**49.** Solve $x^2 - 3x + 1 = 0$. Then find the sum and the product of the roots to check your solutions. **(Lesson 7-5)**

*Portfolio*

Select an item that shows something new you learned in this chapter and place it in your portfolio.

**50.** Simplify $\sqrt{600}$. **(Lesson 6-1)**

**51.** Use synthetic division to show that $x + 3$ is a factor of $2x^3 + 15x^2 + 22x - 15$. **(Lesson 5-7)**

**52.** Factor $6a^2 + a - 35$. **(Lesson 5-5)**

**53.** Define and give an example of an identity matrix. **(Lesson 4-4)**

**54.** Graph $2x + 3y = 12$. **(Lesson 2-2)**

# SUMMARY AND REVIEW

## VOCABULARY

Upon completing this chapter you should be
familiar with the following terms:

| | | | |
|---|---|---|---|
| completing the square | **322** | **340** | quadratic form |
| degree | **316** | **327** | quadratic formula |
| discriminant | **329** | **316** | roots |
| factoring | **317** | **316** | solutions |
| parabola | **317** | **317** | zero product property |
| quadratic equation | **316** | **317** | zeros of a function |

## SKILLS AND CONCEPTS

| OBJECTIVES AND EXAMPLES | REVIEW EXERCISES |
|---|---|

Upon completing this chapter, you should
be able to:

Use these exercises to review and prepare
for the chapter test.

■ solve problems using the guess-and-
check strategy  **(Lesson 7-1)**

Find two integers whose sum is 8 and
whose product is 16.

Guess: 2, 6     Check: $2(6) = 12$, not 16
Guess: 4, 4     Check: $4(4) = 16$  √

1. Write an 8-digit number using each of
the digits 1, 2, 3, and 4 twice so that the
1s are separated by 1 digit, the 2s are
separated by 2 digits, the 3s are
separated by 3 digits, and the 4s are
separated by 4 digits.

---

■ solve quadratic equations by factoring or
by graphing  **(Lesson 7-2)**

Solve $x^2 + 4x + 3 = 0$.
$(x + 3)(x + 1) = 0$
$x = -3$ and $x = -1$

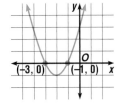

Solve each equation by graphing the related
quadratic function.

2. $g^2 - 4g - 21 = 0$
3. $(x + 7)(2x - 5) = 0$

Solve each equation by factoring.

4. $2x^2 + 5x + 3 = 0$

5. $4p^2 + 9 = 12p$

6. $2x^2 - 8x = 0$

7. $8b^2 + 10b = 3$

■ solve quadratic equations by completing the square   (**Lesson 7-3**)

Solve $x^2 + 6x = 12$.

$x^2 + 6x + \underline{\quad?\quad} = 12 + \underline{\quad?\quad}$
$\phantom{x^2 + }x^2 + 6x + 9 = 12 + 9$
$\phantom{x^2 + 6x + 9}(x + 3)^2 = 21$
$x + 3 = \pm\sqrt{21}$ or $x = -3 \pm\sqrt{21}$

**Solve each equation by completing the square.**

8. $x^2 + 20x + 75 = 0$

9. $x^2 - 5x - 24 = 0$

10. $2t^2 + t - 21 = 0$

11. $r^2 + 4r = 96$

---

■ solve quadratic equations using the quadratic formula   (**Lesson 7-4**)

If $ax^2 + bx + c = 0$, where $a \neq 0$,

$x = \dfrac{-b \pm \sqrt{b^2 - 4ac}}{2a}$.

**Solve each equation by using the quadratic formula.**

12. $3x^2 - 11x + 10 = 0$

13. $2p^2 - 5p + 4 = 0$

14. $4x^2 - 40x + 25 = 0$

15. $y^2 + 3y = 0$

---

■ use the discriminant to determine the nature of the roots of a quadratic equation   (**Lesson 7-4**)

| Discriminant | Description of Roots |
|---|---|
| positive | 2 real roots |
| 0 | 1 real root |
| negative | 2 imaginary roots |

**Find the value of the discriminant of each equation. Describe the roots completely. Then find the roots.**

16. $n^2 = 8n - 16$

17. $7b^2 = 4b$

18. $2y^2 + 6y + 5 = 0$

---

■ find the sum and product of the roots of a quadratic equation   (**Lesson 7-5**)

For $ax^2 + bx + c = 0$ with roots $s_1$ and $s_2$,

$s_1 + s_2 = -\dfrac{b}{a}$   and   $s_1 s_2 = \dfrac{c}{a}$.

**Solve each equation. Find the sum and product of the roots of each quadratic equation.**

19. $x^2 - 12x - 45 = 0$

20. $2m^2 - 10m + 9 = 0$

21. $3s^2 - 11 = 0$

22. $2x^2 = 3 - 3x$

---

■ find the quadratic equation given its roots   (**Lesson 7-5**)

$s_1 = 7 \quad s_2 = 8$
$s_1 + s_2 = \dfrac{15}{1} = -\dfrac{b}{a}$
$s_1 s_2 = \dfrac{56}{1} = \dfrac{c}{a}$

Therefore, $a = 1$, $b = -15$, $c = 56$.
The equation is $x^2 - 15x + 56 = 0$.

**Find a quadratic equation that has the given roots.**

23. $4, -6$

24. $\dfrac{3}{4}, \dfrac{1}{3}$

25. $5 \pm 3i$

26. $2 \pm \sqrt{3}$

■ solve nonquadratic equations using quadratic techniques. (**Lesson 7-6**)

$x^3 - 3x^2 - 54x = 0$
$x(x - 9)(x + 6) = 0$
$x = 0 \quad x = 9 \quad x = -6$

$y - 4\sqrt{y} - 45 = 0$
$(\sqrt{y})^2 - 4(\sqrt{y}) - 45 = 0$
$(\sqrt{y} - 9)(\sqrt{y} + 5) = 0$
$y = 81 \quad \sqrt{y} + 5 = 0$ has no real
$\qquad$ solution.

**Solve each equation.**

27. $3x^3 + 4x^2 - 15x = 0$

28. $m^4 + 3m^3 = 40m^2$

29. $a^3 - 64 = 0$

30. $r + 9\sqrt{r} = -8$

31. $x^4 - 8x^2 + 16 = 0$

32. $x^{\frac{2}{3}} - 9x^{\frac{1}{3}} + 20 = 0$

# APPLICATIONS AND CONNECTIONS

33. **Decorating**  Tonniann and Glenn have a family room with a 9 ft by 12 ft rug. A strip of floor of the same width shows on all sides of the rug. If the area of the room is 270 ft², how wide is the strip?  (**Lesson 7-2**)

34. **Space Exploration**  The Apollo 11 spacecraft propelled the first men to the moon and contained three stages of rockets. The first stage dropped off 2 min 40 s after takeoff and the second stage ignited. The initial velocity of the second stage was 2760 m/s with a constant acceleration of 200 m/s². How long did it take the second stage to travel 7040 m?  (**Lesson 7-3**)

35. **Aeronautics**  A rocket is fired upward from a platform 288 ft above ground level with an initial velocity of 960 ft/s. Find the number of seconds it will take the rocket to reach ground level.  (**Lesson 7-4**)

36. **Geometry**  Sue, Darlene, and Jeanne mow lawns to earn extra money. Jeanne is going to mow one-third of a 100 ft by 120 ft lawn by mowing a strip of uniform width around the outer edge of the lawn. What is the width of the strip?  (**Lesson 7-5**)

**Solve each equation by graphing.**

1. $a^2 + 8a - 33 = 0$

2. $7m^2 - 21m = 0$

3. $5x^2 - 125 = 0$

**Solve each equation by factoring.**

4. $6y^2 - y = 15$

5. $3b^2 + b = 14$

6. $12p^2 - 5p = 3$

**Find the value of the discriminant of each equation. Describe the roots completely. Then solve each equation.**

7. $9a^2 - 30a + 25 = 0$

8. $3x^2 - 4x + 2 = 0$

9. $2x^2 + 3x + 3 = 0$

10. $14r^2 + 35r - 5 = 0$

11. $6t^2 = 2t - 1$

12. $3x^2 - 7x = 20$

13. $n^2 - 4n + 4 = 0$

14. $7k^2 = 4k + 1$

15. $4x^2 = 324$

**Find the sum and the product of the roots for each quadratic equation.**

16. $2z^2 - 3z - 12 = 0$

17. $2x^2 = 3 - 3x$

18. $n + 7 = 4n^2$

**Find a quadratic equation having the given roots.**

19. $0, -3$

20. $\dfrac{4}{3}, \dfrac{2}{3}$

21. $5 + 2i, 5 - 2i$

22. $8$

**Solve each equation.**

23. $p^3 + 8p^2 = 18p$

24. $r^4 - 9r^2 + 18 = 0$

25. $2d + 3\sqrt{d} = 9$

26. $2m^{\frac{2}{3}} - 5m^{\frac{1}{3}} - 12 = 0$

27. $x^{\frac{1}{2}} - 15x^{\frac{1}{4}} + 50 = 0$

28. $x^3 + 27 = 0$

29. The Dolphin Pool Company will build a pool for Sally Wadman having a surface area of 600 ft². Ms. Wadman's pool, along with a deck of uniform width, will have dimensions 30 ft by 40 ft. What will be the width of the deck around the pool?

30. A cannonball is shot upward from the upper deck of a fort with an initial velocity of 192 ft/s. The deck is 32 ft above the ground. How long will it take the cannonball to return to the ground?

**Bonus**

In the figure at the right, the shaded rectangle created by the dotted line is similar to the rectangle with the dimensions 1 by $x$. Write a proportion and solve for $x$ to find the ratio relating the sides of the two rectangles. Round to the nearest thousandth.

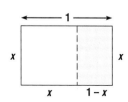

# 8 Quadratic Relations and Functions

## CHAPTER OBJECTIVES

In this chapter, you will:
- Graph quadratic functions.
- Solve problems using quadratic equations.
- Solve quadratic inequalities.

**Which type of crime has escalated in the past twenty years?**

### Crime rate In the United States
Rate per 100,000 population

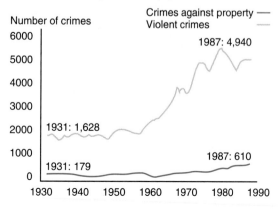

## CAREERS IN FORENSIC SCIENCE

Every 10 seconds, a premise somewhere in the United States is burglarized. Every 22 seconds, a car is stolen. Every 25 minutes, someone is murdered.

If you want to help fight crime, perhaps you'll select the career of criminalist, also known as forensic scientist. A criminalist uses scientific techniques to collect and analyze the physical evidence of crimes and testifies about his or her findings in court.

Criminology is a high-tech field these days. A hundred years ago, a real-life counterpart of the fictional Sherlock Holmes could scrape up a bit of cigarette ash with a trusty penknife and briefly examine it with a hand-held magnifying glass before declaring "Aha! The villain has delivered himself into our hands!" Today, that same bit of ash would have to be analyzed by a spectrophotometer and a gas chromatograph. In the case of blood and other body fluids, the criminalist prepares a DNA profile and does a protein marker analysis to help identify the criminal. Even fingerprints can be lifted by the laser today. It's a demanding job, but one that is filled with mystery, discovery, and satisfaction.

## MORE ABOUT FORENSIC SCIENCE

### Degree Required:

- Bachelor's degree in biochemistry, biophysics, or chemistry, with study in forensic science

### Some criminalists like:

- doing work that is challenging and intellectually stimulating
- the great deal of variety in their work
- good pay

### Related Math Subjects:

- Advanced Algebra
- Trigonometry
- Probability/Statistics
- Calculus

### Some criminalists dislike:

- the stiff competition for jobs
- having to testify in court
- the difficulty to advance beyond a certain level

For more information about the various careers available in the field of Forensic Science, write to:

American Academy of Forensic Sciences
218 East Cache LaPoudre
Colorado Springs, CO 80903

# Graphing Calculator Exploration: Families of Parabolas

In geometry, a collection of related geometric configurations is referred to as a *family*. Likewise, there are families of parabolas. The equations of the parabolas in these families are closely related. In the general form of a quadratic equation, $y = a(x - h)^2 + k$, the parameters are $a$, $h$, and $k$. Changing a parameter in an equation results in a different parabola in the family.

**Example 1**

**Graph the following equations on one set of axes and describe the similarities and differences between the graphs. $y = x^2$, $y = x^2 + 3$, $y = x^2 - 5$.** *The k values for these equations are 0, 3, and -5 respectively.*

Graph each of the equations. Make sure your calculator is set for the standard viewing window. *If you are using a TI-81, make sure all old equations are cleared from the Y = list before graphing.*

*Casio*

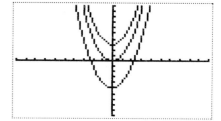

ENTER: GRAPH ALPHA X $x^2$ : GRAPH

ALPHA X $x^2$ + 3 : GRAPH

ALPHA X $x^2$ − 5 EXE

*TI-81*

ENTER: Y= X|T $x^2$ ENTER X|T $x^2$ + 3

ENTER X|T $x^2$ − 5 GRAPH

The graphs are all the same shape, all open upward, and all have their vertex on the y-axis. The graphs have different vertical positions.

Example 1 shows that changing the $k$ value in an equation translates the parabola vertically. If $k > 0$, the parabola is translated $k$ units upward and if $k < 0$, it is translated $k$ units downward. How do you think changing the value of $h$ will change the graphs in a family of parabolas?

**Example 2**

**Graph the following equations on one set of axes and describe the similarities and differences between the graphs. $y = x^2$, $y = (x - 4)^2$, $y = (x + 2)^2$** *The h values for these equations are 0, 4, and -2 respectively.*

Graph the equations. *If you are using a Casio fx-7000G, be sure to clear the graphics screen before graphing.*

*Casio*

ENTER: GRAPH ALPHA X $x^2$ : GRAPH ( ALPHA X − 4 )

$x^2$ : GRAPH ( ALPHA X + 2 ) $x^2$ EXE

*TI-81*

ENTER: [Y=] [X|T] [x²] [ENTER] [(]

[X|T] [−] 4 [)] [x²] [ENTER]

[(] [X|T] [+] 2 [)] [x²]

[GRAPH]

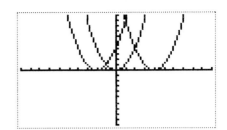

The graphs have the same shape, all open upward, and all have their vertices on the *x*-axis. The graphs have different horizontal positions.

Example 2 demonstrates that changing the *h* value in $y = a(x - h)^2 + k$, translates the graph horizontally. If $h > 0$, the graph translates to the right *h* units and if $h < 0$, the graph translates to the left *h* units.

Changing the value of *a* in the equation $y = a(x - h)^2 + k$ affects the direction of opening and the shape of the graph. If $a > 0$, the graph opens upward and if $a < 0$ the graph opens downward. If $|a| < 1$ the graph is wider than the graph of $y = x^2$ and if $|a| > 1$ the graph is narrower than the graph of $y = x^2$. Graphs of equations with *a* values with the same absolute value, such as $y = 3x^2$ and $y = -3x^2$ have the same shape.

## Example 3

**Graph the following equations on one set of axes and describe the similarities and differences between the graphs $y = x^2$, $y = -x^2$, $y = 0.5x^2$, $y = -2x^2$:**

*Casio*

ENTER: [GRAPH] [ALPHA] [X] [x²] [:]

[GRAPH] [(-)] [ALPHA] [X] [x²]

[:] [GRAPH] 0.5 [ALPHA] [X]

[x²] [:] [GRAPH] [(-)] 2 [ALPHA]

[X] [x²] [EXE]

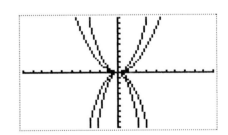

*TI-81*

ENTER: [Y=] [X|T] [x²] [ENTER] [(-)] [X|T]

[x²] [ENTER] 0.5 [X|T] [x²]

[ENTER] [(-)] 2 [X|T] [x²] [GRAPH]

The graphs of $y = x^2$ and $y = 0.5x^2$ open upward, and the graphs of $y = -x^2$ and $y = -2x^2$ open downward. The graphs of $y = x^2$ and $y = -x^2$ have the same shape. The graph of $y = 0.5x^2$ is wider than the graph of $y = x^2$. The graph of $y = -2x^2$ is narrower than the graph of $y = -x^2$, or the graph of $y = x^2$.

# EXERCISES

1. Describe the effect that changing the value of $k$ in an equation of the form $y = a(x - h)^2 + k$ has on the graph of the equation.

2. How does changing the value of $h$ in an equation of the form $y = a(x - h)^2 + k$ affect the graph of the equation?

3. How do the graphs of $y = a(x - h)^2 + k$ and $y = -a(x - h)^2 + k$ compare?

**Use your graphing calculator to graph each pair of equations on the same set of axes. Then compare the graph of the first equation to the graph of the second equation.**

4. $y = (x + 8)^2$, $y = x^2$

5. $y = (x - 5)^2$, $y = x^2$

6. $y = x^2 - 6$, $y = x^2$

7. $y = x^2 + 8$, $y = x^2$

8. $y = (x - 2)^2 + 8$, $y = x^2 + 8$

9. $y = (x - 4)^2 - 7$, $y = (x - 4)^2$

10. $y = (x - 10)^2 + 9$, $y = x^2$

11. $y = (x + 6)^2 - 5$, $y = x^2$

12. $y = (x - 11)^2 - 7$, $y = x^2$

13. $y = (x + 9)^2 + 4$, $y = x^2$

14. $y = 2x^2$, $y = x^2$

15. $y = 0.75x^2$, $y = x^2$

16. $y = 3x^2$, $y = -3x^2$

17. $y = 5x^2 + 1$, $y = x^2 + 1$

18. $y = -\dfrac{5}{3}x^2$, $y = x^2$

19. $y = -0.55x^2 - 10$, $y = x^2 - 10$

20. $y = 0.375(x + 2)^2$, $y = (x + 2)^2$

21. $y = \dfrac{19}{6}(x - 8)^2$, $y = x^2$

22. $y = -7(x - 13)^2$, $y = (x - 13)^2$

23. $y = \dfrac{6}{7}(x + 5)^2$, $y = x^2$

24. $y = 3(x - 9)^2 + 5$, $y = (x - 9)^2 + 5$

25. $y = -0.1(x + 1)^2 + 3$, $y = (x + 1)^2 + 3$

26. $y = \dfrac{5}{2}(x - 3)^2 - 9$, $y = x^2$

27. $y = -\dfrac{3}{4}(x + 12)^2 - 12$, $y = x^2$

**Use the results of Exercises 4–27 to answer each question. Assume that $a$, $h$, and $k$ are real numbers and $a \ne 0$.**

28. How does the graph of $y = (x - h)^2$ compare to the graph of $y = x^2$?

29. How does the graph of $y = x^2 + k$ compare to the graph of $y = x^2$?

30. How does the graph of $y = (x - h)^2 + k$ compare to the graph of $y = (x - h)^2$?

31. How does the graph of $y = (x - h)^2 + k$ compare to the graph of $y = x^2 + k$?

32. How does the graph of $y = (x - h)^2 + k$ compare to the graph of $y = x^2$?

33. How does the graph of $y = ax^2$ compare to the graph of $y = x^2$?

34. How does the graph of $y = ax^2 + k$ compare to the graph of $y = x^2 + k$?

35. How does the graph of $y = a(x - h)^2 + k$ compare to the graph of $y = (x - h)^2 + k$?

36. How does the graph of $y = a(x - h)^2 + k$ compare to the graph of $y = x^2$?

# 8-1 Quadratic Functions

## Objectives

After studying this lesson, you should be able to:

- write functions in quadratic form, and
- identify the quadratic term, the linear term, and the constant term of a quadratic function.

## Application

Mike and Rich are conducting a physics experiment. They will launch a model rocket and use a formula to find the maximum height that the rocket reaches based on the time that it takes for it to land. The rocket they are launching is to have an initial velocity of 160 feet per second. According to the formula, the height of the rocket, $h(t)$, $t$ seconds after it is launched is given by the function $h(t) = 160t - 16t^2$. This function is a quadratic function. You will graph and solve this function in Lesson 8-3.

| Definition of Quadratic Function | A quadratic function is a function described by an equation that can be written in the form $f(x) = ax^2 + bx + c$ where $a \neq 0$. |
|---|---|

In a quadratic function, $ax^2$ is called the **quadratic term**, $bx$ is the **linear term**, and $c$ is the **constant term**.

## Example 1

Write $f(x) = (x + 2)^2 - 6$ in quadratic form. Identify the quadratic term, the linear term, and the constant term.

$$f(x) = (x + 2)^2 - 6 \implies x^2 + 4x + 4 - 6 \quad \text{or} \quad x^2 + 4x - 2$$

The quadratic term is $x^2$, the linear term is $4x$, and the constant term is $-2$.

## Example 2

**APPLICATION**

**Entertainment**

The four theaters in the Studio 25 Cinema have 125 seats each. Based on previous experience, the owner estimates that for each 50¢ increase in the ticket price 20 fewer people will attend a show. Each showing is sold out at the current price of $5.00 a seat. Define a variable and write a quadratic function to describe the Cinema's income after any number of price increases.

*FYI . . .*

The Return of the Jedi holds the record for the highest gross movie ticket sales for one day, $8,440,105.

Let $n$ represent the number of 50¢ price increases. So, $5.00 + 0.50n$ is the ticket price after $n$ increases.
There are $4 \times 125$ or 500 seats in the cinema. $500 - 20n$ is the number of people who will attend a show after $n$ price increases.

*Income = (ticket price)(number of people attending)*

$$
\begin{aligned}
I(n) &= (5.00 + 0.50n)(500 - 20n) \\
&= 2500 - 100n + 250n - 10n^2 \\
&= 2500 + 150n - 10n^2
\end{aligned}
$$

# CHECKING FOR UNDERSTANDING

**Communicating Mathematics**

**Read and study the lesson to answer these questions.**

1. Name the quadratic term, the linear term, and the constant term of the quadratic function found in Example 2.

2. Explain the steps you would use to write $f(x) = 4(x - 1)^2 - 1$ in quadratic form.

3. Is the equation $f(x) = 3(x - 2)^2 + 5$ in quadratic form? If not, write the function in quadratic form.

**Guided Practice**

**State whether each equation describes a quadratic function.**

4. $f(x) = x^2 - 5x + 2$

5. $m(x) = -4x^2 - 6x - 2$

6. $g(x) = 3x + 3$

7. $f(x) = (x - 4)^2$

8. $f(x) = 2(x + 5)^2 + 7$

9. $p(x) = x$

10. $g(x) = -\frac{1}{3}x + \frac{4}{5}$

11. $f(x) = \frac{1}{x^2} + \frac{1}{x} = 1$

**For each function, identify the quadratic term, the linear term, and the constant term.**

12. $f(x) = x^2 + x - 4$

13. $g(x) = 5x^2 - 7x + 2$

14. $g(n) = 3n^2 - 1$

15. $f(n) = \frac{1}{3}n^2 + 4$

16. $n(x) = -4x^2 - 8x - 9$

17. $f(z) = z^2 + 3z$

18. $f(x) = (x + 3)^2$

19. $f(t) = (3t + 1)^2 - 8$

**Journal**

Tell what your favorite sport is and how mathematics is used in it.

# EXERCISES

**Practice**

**Write each function in quadratic form.**

20. $g(x) = (x - 1)^2$

21. $f(x) = (x - 3)^2$

22. $f(m) = (2m - 5)^2$

23. $h(x) = (3x + 2)^2$

24. $f(r) = 3(r - 2)^2$

25. $f(x) = -4(2x - 4)^2$

26. $g(x) = 2(4x + 1)^2$

27. $f(x) = 4(x + 1)^2 + 10$

28. $f(x) = -3(2x + 2)^2 + 6$

29. $g(x) = \frac{1}{6}(6x + 12)^2 + 5$

**Define a variable and write a quadratic function to describe each situation.**

30. the product of two numbers whose sum is 45

**CONNECTION**

**Geometry**

31. the area of a circle in terms of its radius

32. the product of two numbers whose difference is 15

33. the area of an isosceles right triangle in terms of the measure of its legs

34. the area of a rectangle whose perimeter is 30 centimeters

35. the sum of the squares of two numbers whose sum is 10

36. the sum of the squares of two numbers whose difference is 18

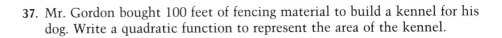

37. Mr. Gordon bought 100 feet of fencing material to build a kennel for his dog. Write a quadratic function to represent the area of the kennel.

CONNECTION
Geometry

38. Cheryl is going to wallpaper her bedroom walls. The perimeter of the room is $4x + 10$ feet. Write a quadratic function to represent the amount of wallpaper in square feet Cheryl will need if the height of the walls is $x$ feet.

**Critical Thinking**

39. Prove that $x^2 - 1$ is always divisible by 8 if $x$ is an odd number.
    *Hint: Substitute $2n + 1$ for $x$, since $2n + 1$ is always odd.*

**Applications**

40. **Business**   Ernie Burton runs a shuttle bus between the airport and the convention center. About 300 passengers ride the shuttle per week. He plans to raise the $8.00 charge. But, based on past experience, he estimates that he will lose 20 passengers per week to the competition for each $1 he increases the fare. Write a quadratic function to describe Mr. Burton's income after he raises the fare.

41. **Entertainment**   The Darien High School winter play brought in $800 last year. This year, the drama teacher intends to raise the $4.00 ticket price. It is estimated that for each 25¢ increase in the price 10 fewer people will come to the play. Write a quadratic function to describe the income after the price increase.

**Mixed Review**

42. Express the equation $x^6 + 3x^3 - 10 = 0$ in quadratic form.   **(Lesson 7-6)**

43. Solve the equation $a - 4a^{\frac{1}{2}} + 3 = 0$.
    **(Lesson 7-6)**

44. Find the product of the complex number $2 + 9i$ and its conjugate.   **(Lesson 6-10)**

45. Evaluate the expression $(0.0016)^{\frac{1}{4}}$.
    **(Lesson 6-4)**

46. Divide $(x^3 - 2.8)$ by $(x + 0.4)$ using synthetic division.   **(Lesson 5-7)**

47. **Physics**   Light from a laser will travel about 300,000 kilometers per second. How many kilometers will it travel in a day?   **(Lesson 5-1)**

48. Use a matrix equation to solve
    $$\begin{cases} 4x + y = -5 \\ x + 4y = 10 \end{cases}.$$   **(Lesson 4-6)**

# Problem-Solving Strategy: Make a Table

**Objective**

After studying this lesson, you should be able to:

■ solve problems by using tables.

Organizing data into a table can be very helpful when solving problems. You can use a table to organize given data and identify missing data, to organize your results, or to look for a pattern.

**Example 1**

**APPLICATION**

**Business**

*FYI* · · ·

The highest bridge is the Royal Gorge over the Arkansas River in Colorado. It is 1053 ft above the water level.

**The toll to cross the bridge over the Rushing River is 50¢. One of the lanes has an automatic gate for exact change only. If the automatic gate will not accept pennies, how many combinations of coins must the gate be programmed to accept?**

The total of the change must always be 50¢. So make a table to show all of the combinations of coins with a total value of 50¢. We'll start with the larger coins and work toward the smaller ones.

| Half-Dollars | Quarters | Dimes | Nickels |
|---|---|---|---|
| 1 | 0 | 0 | 0 |
| 0 | 2 | 0 | 0 |
| 0 | 1 | 2 | 1 |
| 0 | 1 | 1 | 3 |
| 0 | 1 | 0 | 5 |
| 0 | 0 | 5 | 0 |
| 0 | 0 | 4 | 2 |
| 0 | 0 | 3 | 4 |
| 0 | 0 | 2 | 6 |
| 0 | 0 | 1 | 8 |
| 0 | 0 | 0 | 10 |

There are 11 combinations that the gate must be programmed to accept.

**Example 2**

**What is the ones' digit of $7^{100}$?**

Let's make a table and look for a pattern. Use the $y^x$ key on your calculator to find the powers of 7.

| $n$ | $7^n$ | Ones' digit |
|---|---|---|
| 1 | 7 | 7 |
| 2 | 49 | 9 |
| 3 | 343 | 3 |
| 4 | 2401 | 1 |
| 5 | 16,807 | 7 |
| 6 | 117,649 | 9 |
| 7 | 823,543 | 3 |
| 8 | 5,764,801 | 1 |
| 9 | 40,353,607 | 7 |

To find $7^4$,
ENTER: 7 $y^x$ 4 $=$ 2401.

The ones' digits repeat in a cycle of four: 7–9–3–1, 7–9–3–1, . . . . Every fourth power of 7 has a ones' digit of 1, so the ones' digit of $7^{100}$ will be 1.

# CHECKING FOR UNDERSTANDING

## Communicating Mathematics

**Read and study the lesson to answer these questions.**

1. Name some ways you can use a table to help solve a problem.

2. If a driver has 3 dimes, 1 quarter, and 5 pennies, can she use the exact change only toll gate at the Rushing River?

3. What is the ones' digit of $7^{125}$?

4. Can you see a pattern in the tens' digits of the powers of 7?

## Guided Practice

**Make a table to help you answer the following.**

5. If the toll over the Rushing River is raised to 60¢, how many combinations of coins must the gate be programmed to accept?

6. Clay bought a pen at the school book store for 69¢. He gave the clerk $1.00 and received 7 coins in change. What coins might he have received?

7. What is the tens' digit of $11^{10}$?

# EXERCISES

### Strategies

Look for a pattern.
Solve a simpler problem.
Act it out.
Guess and check.
Draw a diagram.
Make a chart.
Work backwards.

**Solve. Use any strategy.**

8. Darren and Emily are playing a game of darts with the target shown at the right. If they each throw three darts in a round, in how many ways could a player score 90 points in one round? Assume that all of the darts hit the target.

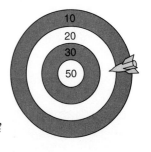

9. What is the remainder of the division $5^{100} \div 7$?

10. A new high school is being built in West Union City. The building will be built around a courtyard with walkways constructed as shown at the right. If the walkways are to be 4 feet wide, how many square feet of sod should the contractor order to cover the grassy area in the courtyard?

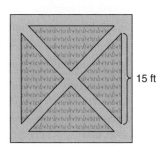

15 ft

11. In how many ways can you write 53 as the sum of two prime numbers?

12. The difference of the squares of two consecutive integers is 1993. What is the sum of the two integers?

13. Find the least positive integer that is divisible by 2, 3, 7, 10, 15, 20, and 21.

14. Three soccer teams, the Hornets, the Warriors, and the Vikings, play each other in the Central State League. The following table shows the results of the games played so far this season. Copy and complete the table and find the score of each game.

| | Games Played | Wins | Losses | Ties | Goals Scored | Goals Against |
|---|---|---|---|---|---|---|
| **Hornets** | 2 | 2 | | | | 1 |
| **Warriors** | 2 | | | 1 | 2 | 4 |
| **Vikings** | 2 | | | | 3 | 7 |

15. Place four coins on your desk with tails up as shown at the right. Choose any three of the coins and turn them over. Continue choosing three coins and turning them over until all of the coins are heads-up. What is the fewest number of such moves that results in all heads-up?

16. Find the digit represented by each different letter in the problem at the right. Each of the digits 1–9 is used exactly once.

$$\begin{array}{r} ABC \\ +\,DEF \\ \hline GHI \end{array}$$

17. The number 25 is a square. When each of its digits is increased by one, the result is also a square, namely 36. There is one four-digit number with the same property. What is it?

18. Use a calculator to find three different ways to express 1000 as the sum of two or more consecutive positive integers.

---

## ～ COOPERATIVE LEARNING ACTIVITY ～

**Work in groups. Each person in the group must understand the solution and be able to explain it to any person in class.**

Arrange nine coins as shown at the right with the center coin tails-up and the rest heads-up. A move consists of turning over the three coins in any row, column, or diagonal. In how many moves can you make all of the coins heads-up?

# Graphing Calculator Exploration:
# Locating the Vertex of a Parabola

The vertex of a parabola is the maximum or minimum point on the parabola. Finding the approximate coordinates of the vertex of a parabola on a graphing calculator is a simple process.

To use the graphing calculator to find the vertex of a parabola, use the tracing and zoom-in features of the calculator, with which you are already familiar. The vertex is located at the maximum point of the parabola if it opens downward and at the minimum point if it opens upward. Once you have graphed the parabola in an appropriate viewing window, use the tracing function to move the cursor close to the vertex. Move the cursor back and forth over the vertex while you watch the $y$-values. Find the point with the highest, or lowest, $y$-value. This is the vertex. Now, use the zoom-in process to determine the coordinates with your desired accuracy.

**Example**

Use the tracing function to determine the coordinates of the vertex of the parabola with equation $y = (x - 2)^2 - 3$.

First graph the parabola.   *Use the standard viewing window.*

*Casio*

ENTER: [GRAPH] [(] [ALPHA] [X] [−] 2
[)] [$x^2$] [−] 3 [EXE]

*TI-81*

ENTER: [Y=] [(] [X|T] [−] 2 [)] [$x^2$]
[−] 3 [GRAPH]

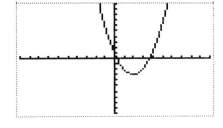

Now use the tracing and zoom-in functions to determine the coordinates of the vertex. The coordinates of the vertex to two decimal places are (2.00, −3.00).   *You may wish to review the Graphing Calculator Exploration on pages 104–107 to refresh your memory on the zoom-in process.*

# EXERCISES

Graph each equation on your graphing calculator. Then use the tracing function to approximate the coordinates of the vertex of the graph to two decimal places.

**1.** $y = 2(x - 10)^2 + 14$

**2.** $y = (x + 8)^2 - 5$

**3.** $y = x^2 - 14x + 70$

**4.** $y = x^2 - 87x - 23$

**5.** $y = 5x^2 - 302x - 321$

**6.** $y = 12x^2 + 18x + 10$

**7.** $y = 0.5x^2 - 37x + 777$

**8.** $y = 10x^2 - 70x + 134$

**9.** $y = \frac{2}{3}x^2 - \frac{3}{4}x + 6$

**10.** $y = 0.5x^2 + 0.29x + 8.9$

# Graphing Quadratic Functions

**Objective**

After studying this lesson, you should be able to:

- graph quadratic equations of the form $y = (x - h)^2 + k$, and identify the vertex and the equation of the axis of symmetry of a parabola.

**Application**

After launching the model rocket described in Lesson 8-1, and measuring the amount of time that the rocket was in flight with a stopwatch at 10 seconds, Mike and Rich graphed the function describing the flight of the rocket. The height of the rocket, $h(t)$, $t$ seconds after being launched is given by the quadratic function $h(t) = 160t - 16t^2$.

First they created a table of values.

| $t$ | $160t - 16t^2$ | $h(t)$ |
|---|---|---|
| 0 | $160(0) - 16(0)^2$ | 0 |
| 1 | $160(1) - 16(1)^2$ | 144 |
| 2 | $160(2) - 16(2)^2$ | 256 |
| 3 | $160(3) - 16(3)^2$ | 336 |
| 4 | $160(4) - 16(4)^2$ | 384 |
| 5 | $160(5) - 16(5)^2$ | 400 |
| 6 | $160(6) - 16(6)^2$ | 384 |
| 7 | $160(7) - 16(7)^2$ | 336 |
| 8 | $160(8) - 16(8)^2$ | 256 |
| 9 | $160(9) - 16(9)^2$ | 144 |
| 10 | $160(10) - 16(10)^2$ | 0 |

Next they plotted the points $(t, h(t))$.

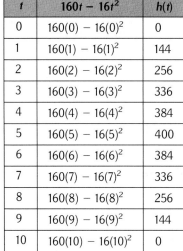

Finally, they connected the points in a smooth curve. The resulting graph is called a **parabola.**

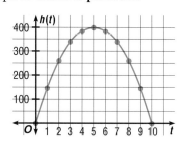

Notice how the height of the rocket is zero again 10 seconds after launching. This means that the rocket has returned to the ground. It appears that the rocket reached its maximum height of 400 feet after 5 seconds. Evaluate $h(t)$ for other values of $t$ between 4 and 6 to check this conclusion.

The graph of any quadratic function is a parabola. All parabolas have an **axis of symmetry.** The axis of symmetry is the line about which the parabola is symmetric. That is, if you could fold the coordinate plane along the axis of symmetry the portions of the parabola on each side of the line would match. The line of symmetry is named by the equation of the line. All parabolas have a **vertex** as well. The vertex is the point of intersection of the parabola and the axis of symmetry.

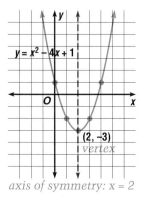

$y = x^2 - 4x + 1$

(2, –3) vertex

axis of symmetry: x = 2

The graphs of $y = x^2$, $y = (x - 2)^2$, $y = x^2 + 3$, and $y = (x - 2)^2 + 3$ are shown below on the same set of axes. Study these graphs.

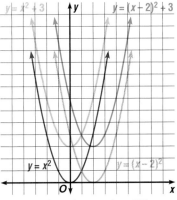

| Equation | Vertex | Axis of Symmetry |
|---|---|---|
| $y = x^2$ | (0, 0) | $x = 0$ |
| $y = (x - 2)^2$ | (2, 0) | $x = 2$ |
| $y = x^2 + 3$ | (0, 3) | $x = 0$ |
| $y = (x - 2)^2 + 3$ | (2, 3) | $x = 2$ |

*Notice that the graphs all have the same shape. The difference is their position.*

We can express the equations for these parabolas in the general form $y = (x - h)^2 + k$. When a quadratic function is written in this form, the values of $h$ and $k$ are easy to determine.

$y = x^2$ ➡ $y = (x - 0)^2 + 0$ ➡ $h = 0, k = 0$

$y = (x - 2)^2$ ➡ $y = (x - 2)^2 + 0$ ➡ $h = 2, k = 0$

$y = x^2 + 3$ ➡ $y = (x - 0)^2 + 3$ ➡ $h = 0, k = 3$

$y = (x - 2)^2 + 3$ ➡ $y = (x - 2)^2 + 3$ ➡ $h = 2, k = 3$

Compare the values for $h$ and $k$ for each equation with the coordinates of the vertex of each graph given above. Also, compare each equation for the axis of symmetry with each value for $h$. What pattern do you notice?

| $y = (x - h)^2 + k$ | |
|---|---|
| Vertex | $(h, k)$ |
| Axis of Symmetry | $x = h$ |

In Chapter 4 you learned that a translation moves a figure on the coordinate plane without changing its size. As the value of $h$ and $k$ change, the graph of $y = (x - h)^2 + k$ is the graph of $y = x^2$ translated $|h|$ units left or right and $|k|$ units up or down. If $h$ is positive, the parabola is translated to the right. If $h$ is negative, it is translated to the left. Likewise, for $k$, the translation is up if $k$ is positive and down if $k$ is negative.

## Example 1

**Name the vertex and the axis of symmetry for the graph of $y = (x + 8)^2 - 1$. How is this graph different from the graph of $y = x^2$?**

Rewrite the equation in the form $y = (x - h)^2 + k$.
$$y = (x - (-8))^2 + (-1)$$
From this equation, $h = -8, k = -1$.

So, the vertex is $(-8, -1)$ and the axis of symmetry is $x = -8$. The shape of the graph is the same as the shape of the graph of $y = x^2$, but it is translated 8 units left and 1 unit down.

**Example 2**

Name the vertex and the axis of symmetry for the graph of $f(x) = (x + 1)^2 + 3$. Then draw the graph.

This function can be rewritten as $f(x) = (x - (-1))^2 + 3$.

So, $h = -1$ and $k = 3$.
The vertex is $(-1, 3)$ and
the axis of symmetry is $x = -1$.

Finding several points on the graph makes graphing easier.

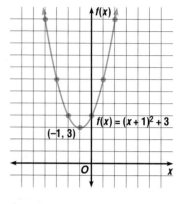

| $x$ | $(x + 1)^2 + 3$ | $f(x)$ |
|-----|-----------------|--------|
| -4  | $(-4 + 1)^2 + 3$ | 12 |
| -3  | $(-3 + 1)^2 + 3$ | 7  |
| -2  | $(-2 + 1)^2 + 3$ | 4  |
| -1  | $(-1 + 1)^2 + 3$ | 3  |
| 0   | $(0 + 1)^2 + 3$  | 4  |
| 1   | $(1 + 1)^2 + 3$  | 7  |
| 2   | $(2 + 1)^2 + 3$  | 12 |

Notice that points with the same $y$-coordinates are the same distance from the axis of symmetry, $x = -1$.

---

**Example 3**

**CONNECTION**

Geometry

Write a function to represent the area of all the rectangles whose length is 2 centimeters more than their width. Graph the function.

Let $w =$ the width of the rectangle and $w + 2 =$ the length.
$$\text{Area} = f(w) = w(w + 2)$$
$$= w^2 + 2w$$

Now rewrite the equation in the form $y = (x - h)^2 + k$. To do this we must complete the square.
$$f(w) = (w^2 + 2w + 1) - 1$$
$$= (w + 1)^2 - 1$$
$$= (w - (-1))^2 + (-1)$$
From this equation, $h = -1$ and $k = -1$.
Therefore, the vertex is $(-1, -1)$ and
the axis of symmetry is $x = -1$.
Find other points on the graph.

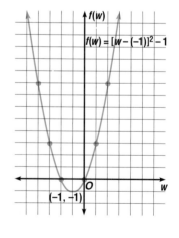

| $w$ | $(w + 1)^2 - 1$ | $f(w)$ |
|-----|-----------------|--------|
| -4  | $(-4 + 1)^2 - 1$ | 8 |
| -3  | $(-3 + 1)^2 - 1$ | 3 |
| -2  | $(-2 + 1)^2 - 1$ | 0 |
| 0   | $(0 + 1)^2 - 1$  | 0 |
| 1   | $(1 + 1)^2 - 1$  | 3 |
| 2   | $(2 + 1)^2 - 1$  | 8 |

Only the points with nonnegative coordinates make sense as measures for the sides of a rectangle.

# CHECKING FOR UNDERSTANDING

**Communicating Mathematics**

Read and study the lesson to answer these questions.

1. Did the model rocket described in the beginning of the lesson reach its maximum height 5 seconds after launch? Justify the answer.

2. The graph of a quadratic function is a ___?___. It is symmetric about a line called the ___?___.

3. Write a quadratic function whose graph has a vertex in the third quadrant.

4. Compare the graphs of the functions $f(x) = (x - 2)^2 + 2$ and $f(x) = (x + 2)^2 + 2$.

5. Write the equation of the function whose graph is shown at the right.

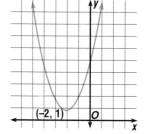

**Guided Practice**

Name the vertex and the axis of symmetry for the graph of each equation.

6. $f(x) = x^2$

7. $g(x) = (x - 4)^2$

8. $f(x) = (x + 6)^2$

9. $h(x) = x^2 + 3$

10. $f(x) = x^2 - 9$

11. $y = (x - 6)^2 + 4$

12. $y = (x + 10)^2 - 7$

13. $y = \left(x - \frac{1}{5}\right)^2 + 1$

14. $f(x) = (x + 1.5)^2 - 3.2$

Write the equation of the quadratic function for each graph.

15.

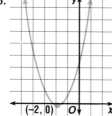

16.

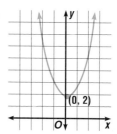

17.

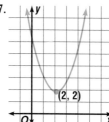

18.

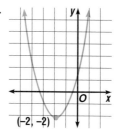

# EXERCISES

**Practice**

Write each equation in the form $f(x) = (x - h)^2 + k$. Then name the vertex and the axis of symmetry for the graph of each function.

19. $f(x) = x^2 - 4x + 4$

20. $f(x) = x^2 + 6x + 9$

21. $f(x) = x^2 + 9$

22. $f(x) = x^2 - 7$

23. $f(x) = x^2 + 10x$

24. $f(x) = x^2 - 12x$

Write each equation in the form $f(x) = (x - h)^2 + k$. Then name the vertex and the axis of symmetry for the graph of each function.

**25.** $f(x) = x^2 + 8x + 20$

**26.** $f(x) = x^2 - 3x + 3$

Graph each equation.

**27.** $y = (x - 5)^2$

**28.** $y = (x + 7)^2$

**29.** $f(x) = x^2 + 6$

**30.** $g(x) = x^2 - 4$

**31.** $y = (x - 3)^2 + 5$

**32.** $f(x) = (x - 8)^2 + 3$

**33.** $g(x) = (x + 2)^2 - 3$

**34.** $y = (x + 4)^2 + 1$

**35.** $f(x) = (x - 1)^2 - 4$

**36.** $y = (x + 11)^2 - 1$

**37.** $y = x^2 + 6x + 2$

**38.** $f(x) = x^2 + 10x + 27$

**39.** $g(x) = x^2 - 2x + 7$

**40.** $y = x^2 + 3x$

**41.** $h(x) = x^2 - 5x$

**42.** $f(x) = x^2 - x - 3$

**Critical Thinking**

Complete the following table for parabolas with equations of the form $y = (x - h)^2 + k$.

| | Axis of Symmetry | Contains the Point | Vertex | Equation of the Parabola |
|---|---|---|---|---|
| **43.** | $x = 0$ | $(2, -1)$ | | |
| **44.** | $x = -2$ | $(-5, 9)$ | | |
| **45.** | $x = -3$ | $(1, 18)$ | | |
| **46.** | $x = 1$ | $(-1, -2)$ | | |

**Applications**

**47. Physics** An arrow is shot upward with an initial velocity of 80 feet per second. The height of the arrow, $h(t)$, in terms of the time since the arrow was released $t$, is $h(t) = 80t - 16t^2$.
a. Draw the graph of the function relating the height of the arrow to the time.
b. How long after the arrow is released does it reach its maximum height? What is that height?

**48. Sports** Shawn hit a foul ball straight up over home plate. The height of the ball over the level of the bat, $h(t)$, is given by the function $h(t) = 48t - 16t^2$, where $t$ is the time in seconds after the ball left the bat.
a. Draw the graph of the function relating the height of the ball over the level of the bat to the time.
b. If the catcher is going to attempt to catch the ball, how long does she have to get ready?
c. Where is the ball when the height found by the function is zero?

**Mixed Review**

**49.** Identify the quadratic term, the linear term, and the constant term of the function $f(x) = 4x^2 - 8x - 2$. **(Lesson 8-1)**

**50.** Find a quadratic equation having 8 and -7 as roots. **(Lesson 7-5)**

**51.** Simplify $\sqrt[3]{2}(3\sqrt[3]{4} + 2\sqrt[3]{32})$. **(Lesson 6-2)**

**52.** Find the degree of the polynomial $16x^4yz + 12x^2y^3z - 24x^3y^2z - 18xy^4z$. **(Lesson 5-4)**

**53.** Determine the slope of the line that passes through the points $(9, 3)$ and $(-3, 9)$. **(Lesson 2-4)**

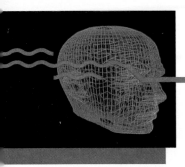

# Technology

## Quadratic Functions

▶ BASIC
Spreadsheets
Software

The BASIC program below can be used to find the vertex, axis of symmetry, and direction of the opening for the graph of any quadratic function. The program also lists six points on the graph, which are three points on either side of the axis of symmetry.

*In order to enter the correct data, the quadratic function must be in the form $f(x) = ax^2 + bx + c$.*

```
10    INPUT "ENTER THE COEFFICIENTS OF THE EQUATION:";
      A,B,C
20    LET H = -B/(2*A)
30    LET K = A*H*H+B*H+C
40    PRINT "THE VERTEX IS (";H;",";K;")."
50    PRINT "THE AXIS OF SYMMETRY IS X = ";H;"."
60    IF A > 0, THEN 80
70    PRINT "THE GRAPH OPENS DOWNWARD.": GOTO 90
80    PRINT "THE GRAPH OPENS UPWARD."
90    PRINT "HERE ARE THE COORDINATES OF SIX ADDITIONAL
      POINTS ON THE GRAPH:"
100   FOR N = H - 3 TO H + 3
110   IF H = INT(H) THEN 130
120   IF N = H - 3 THEN 170
130   LET X = INT(N)
140   IF X = H THEN 170
150   LET Y = A*X*X+B*X+C
160   PRINT "(";X;",";Y;")."
170   NEXT N
180   END
```

Run the program for the function $f(x) = 2x^2 - 5x + 3$.

```
RUN
ENTER THE COEFFICIENTS OF THE EQUATION: 2,-5,3
THE VERTEX IS (1.25,-0.125).
THE AXIS OF SYMMETRY IS X = 1.25.
THE GRAPH OPENS UPWARD.
HERE ARE THE COORDINATES OF SIX ADDITIONAL POINTS ON THE
GRAPH
(-1,10)        (0,3)(1,0)        (2,1)
(3,6)          (4,15)
```

# EXERCISES

Use the program to find the vertex, axis of symmetry, direction of opening, and additional points on the graph of each function. Then graph the function.

1. $f(x) = x^2 - 28x + 186$

2. $f(x) = -x^2 - 16x + 36$

3. $f(x) = -5x^2 + 22x - 3$

4. $f(x) = 10x^2 - 37x - 7$

# Analyzing Graphs of Quadratic Functions

**Objectives**

After studying this lesson, you should be able to:

- graph equations of the form $y = a(x - h)^2 + k$ and identify the vertex, the equation of the axis of symmetry, and the direction of the opening, and
- determine the equation of a parabola from given information about the graph.

**Application**

Athletic Advantage Inc. makes athletic shoes for aerobics and running. Their sales manager has determined that their profit, $P(x)$, on $x$ pairs of shoes can be found by $P(x) = -x^2 + 90x - 500$.

You know that the graph of any quadratic function is a parabola. We've studied graphs of equations of the form $y = (x - h)^2 + k$. Some other quadratic functions can be written in the form $y = a(x - h)^2$. Let's compare the graph of the profit function given above to some other graphs we've already studied.

First, write the equation in the form $y = a(x - h)^2 + k$. We'll do this by completing the square.

$$
\begin{aligned}
P(x) &= -x^2 + 90x - 500 \\
&= -1(x^2 - 90x) - 500 \\
&= -1(x^2 - 90x + 2025) + 2025 - 500 \quad \text{\textit{Add and subtract} } \left(\tfrac{90}{2}\right)^2 \text{\textit{ or 2025 to}} \\
&= -1(x - 45)^2 + 1525 \quad\quad\quad\quad\quad\quad\quad \text{\textit{obtain an equivalent equation.}}
\end{aligned}
$$

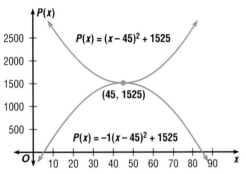

$P(x) = (x - 45)^2 + 1525$

$(45, 1525)$

$P(x) = -1(x - 45)^2 + 1525$

Now graph $P(x) = -1(x - 45)^2 + 1525$ and $P(x) = (x - 45)^2 + 1525$ on the same coordinate plane and compare the graphs.

The graphs have the same vertex and are shaped the same, but the graph of $P(x) = -1(x - 45)^2 + 1525$ opens downward and the graph of $P(x) = (x - 45)^2 + 1525$ opens upward.

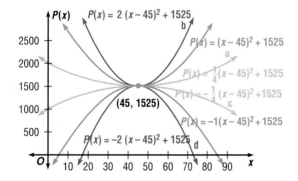

$P(x) = 2(x - 45)^2 + 1525$  b

$P(x) = (x - 45)^2 + 1525$  a

$P(x) = \tfrac{1}{4}(x - 45)^2 + 1525$

$P(x) = -\tfrac{1}{4}(x - 45)^2 + 1525$  c

$(45, 1525)$

$P(x) = -1(x - 45)^2 + 1525$

$P(x) = -2(x - 45)^2 + 1525$  d

Graph the following functions to further investigate the relationships between similar functions. What do you find?

**a.** $P(x) = \dfrac{1}{4}(x - 45)^2 + 1525$ $\quad a = \dfrac{1}{4}$

**b.** $P(x) = 2(x - 45)^2 + 1525$ $\quad a = 2$

**c.** $P(x) = -\dfrac{1}{4}(x - 45)^2 + 1525$ $\quad a = -\dfrac{1}{4}$

**d.** $P(x) = -2(x - 45)^2 + 1525$ $\quad a = -2$

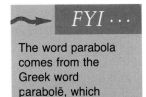
All of the graphs for quadratic functions **a, b, c,** and **d** have the vertex (45, 1525) and the axis of symmetry $x = 45$. When $a$ is negative, the graph opens downward. When $a$ is positive, the graph opens upward. As the value of $|a|$ increases the graph becomes narrower.

The following chart summarizes the characteristics of the graph of $y = a(x - h)^2 + k$.

| $y = a(x - h)^2 + k$ | $a$ is positive | $a$ is negative |
|---|---|---|
| Vertex | $(h, k)$ | $(h, k)$ |
| Axis of Symmetry | $x = h$ | $x = h$ |
| Direction of Opening | upward | downward |
| As the value of $|a|$ increases, the graphs of $y = a(x - h)^2 + k$ narrow. | | |

**Example 1**

**Name the vertex, axis of symmetry, and direction of opening for the graph of $f(x) = -4(x - 9)^2$.**

This function can be written in the form $f(x) = -4(x - 9)^2 + 0$.

From this equation, $a = -4$, $h = 9$, and $k = 0$.

So, the vertex is (9, 0), the axis of symmetry is $x = 9$, and since $a$ is negative, the graph opens downward.

**Example 2**

**Graph $f(x) = -2x^2 - 8x - 1$.**

Write the equation in the form $f(x) = a(x - h)^2 + k$ by completing the square.

$$\begin{aligned} f(x) &= -2x^2 - 8x - 1 \\ &= -2(x^2 + 4x) - 1 \\ &= -2(x^2 + 4x + 4) - 1 - (-2)(4) \\ &= -2(x + 2)^2 + 7 \\ &= -2(x - (-2))^2 + 7 \end{aligned}$$

From the equation, $h = -2$, $k = 7$, and $a = -2$.
So, the vertex is $(-2, 7)$ and
the axis of symmetry is $x = -2$.
Since $a = -2$, the graph opens downward
and is narrower than the graph of
$f(x) = (x + 2)^2 + 7$.

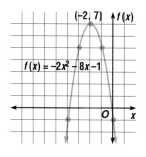

| $x$ | $-2(x + 2)^2 + 7$ | $f(x)$ |
|---|---|---|
| -4 | $-2(-4 + 2)^2 + 7$ | -1 |
| -3 | $-2(-3 + 2)^2 + 7$ | 5 |
| -2 | $-2(-2 + 2)^2 + 7$ | 7 |
| -1 | $-2(-1 + 2)^2 + 7$ | 5 |
| 0 | $-2(0 + 2)^2 + 7$ | -1 |

**Example 3**

**Write the equation of the parabola shown below.**

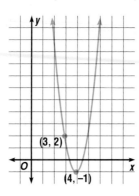

The vertex of the parabola is (4, –1). So $h = 4$, and $k = -1$.

Substitute the values of $h$ and $k$ and the coordinates of one other point on the graph into the general form of the equation and solve for $a$.   *Use (3, 2).*

$$y = a(x - h)^2 + k$$
$$2 = a(3 - 4)^2 + (-1) \quad \textit{Substitute 4 for h, –1 for}$$
$$3 = a(-1)^2 \quad \textit{k, 3 for x, and 2 for y.}$$
$$3 = a$$

The equation of the parabola is $y = 3(x - 4)^2 - 1$ or $y = 3x^2 - 24x + 47$.

---

**Example 4**

APPLICATION

Sports

In the finals of the softball tournament, Jenny hit a homerun to win the game. The ball she hit traveled in a path described by the function $f(x) = -0.004x^2 + x + 4$, where $x$ represents the number of feet the ball has traveled from the plate and $f(x)$ represents the height of the ball in feet. Name the vertex, axis of symmetry, and direction of opening for the graph of this function. Then graph.

First, write the equation in standard form by completing the square.

$$y = -0.004x^2 + x + 4$$
$$= -0.004(x^2 - 250x) + 4$$
$$= -0.004(x^2 - 250x + 15,625) + 4 - (-0.004)(15,625)$$
$$= -0.004(x - 125)^2 + 66.5$$

From this equation, $a = -0.004$, $h = 125$, and $k = 66.5$.
So, the vertex is (125, 66.5), the axis of symmetry is $x = 125$, and since $a$ is negative the graph opens downward.

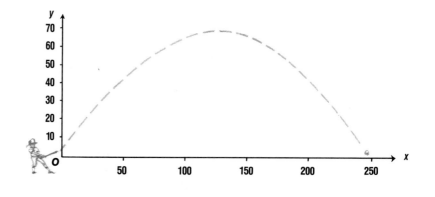

**Example 5**

Graph $f(x) = \frac{1}{3}(x - 1)^2 + 3$.

From this equation, $h = 1$, $k = 3$, and $a = \frac{1}{3}$. Therefore, the vertex is $(1, 3)$ and the axis of symmetry is $x = 1$. Since $a = \frac{1}{3}$, the graph opens upward and is wider than the graph of $f(x) = (x - 1)^2 + 3$.

Find some points on the graph, plot, and connect to form a smooth graph.

| $x$ | $\frac{1}{3}(x - 1)^2 + 3$ | $f(x)$ |
|---|---|---|
| -3 | $\frac{1}{3}(-3 - 1)^2 + 3$ | $8\frac{1}{3}$ |
| -2 | $\frac{1}{3}(-2 - 1)^2 + 3$ | 6 |
| 1 | $\frac{1}{3}(1 - 1)^2 + 3$ | 3 |
| 4 | $\frac{1}{3}(4 - 1)^2 + 3$ | 6 |
| 5 | $\frac{1}{3}(5 - 1)^2 + 3$ | $8\frac{1}{3}$ |

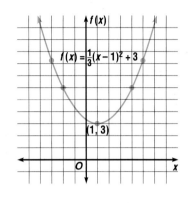

**Example 6**

Write the equation of the parabola that passes through the points $(0, -2)$, $(2, -2)$, and $(3, 4)$.

Each point should satisfy the equation of the parabola. Using the form of the equation $y = ax^2 + bx + c$, find $a$, $b$, and $c$ by substituting the coordinates of the points into the general form of the equation.

$-2 = a(0)^2 + b(0) + c$ ➡ $-2 = c$

$-2 = a(2)^2 + b(2) + c$ ➡ $-2 = 4a + 2b + c$

$4 = a(3)^2 + b(3) + c$ ➡ $4 = 9a + 3b + c$

Now solve the system of equations. From the first equation, $c = -2$, so substitute $-2$ for $c$ in the other two equations.

$-2 = 4a + 2b - 2$ ➡ $0 = 4a + 2b$

$4 = 9a + 3b - 2$ ➡ $6 = 9a + 3b$

Solve using elimination.

$0 = 4a + 2b$   *Multiply by -3.*   $0 = -12a - 6b$

$6 = 9a + 3b$   *Multiply by 2.*   $12 = 18a + 6b$

$$12 = 6a$$
$$2 = a$$

Now substitute 2 for $a$ and solve for $b$.

$0 = 4(2) + 2b$
$-4 = b$

The solution to the system is $(2, -4, -2)$. So the equation of the parabola is $y = 2x^2 - 4x - 2$. *Check this solution by graphing.*

# CHECKING FOR UNDERSTANDING

**Communicating Mathematics**

Read and study the lesson to answer these questions.

1. How can you tell if the graph of a function in the form $y = a(x - h)^2 + k$ will open upward or downward?

2. As $|a|$ decreases what happens to the graph of $y = a(x - h)^2 + k$?

3. Explain the difference between the graph of $y = 2(x - 5)^2 + 3$ and $y = -2(x - 5)^2 + 3$.

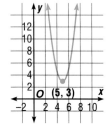

4. The equation $y = 2(x - 5)^2 + 3$ is graphed at the left. Describe the differences between this graph and the graph of $y = \frac{1}{2}(x - 5)^2 + 3$.

**Guided Practice**

Name the vertex, axis of symmetry, and the direction of opening for the graph of each equation.

5. $y = 3(x - 9)^2$

6. $f(x) = -2(x + 3)^2$

7. $y = 5x^2 - 6$

8. $g(x) = -3x^2 + 6$

9. $y = 5(x + 3)^2 - 1$

10. $y = -2(x - 2)^2 - 2$

11. $y = (x + 2)^2 - \frac{4}{3}$

12. $h(x) = 3\left(x - \frac{1}{2}\right)^2 + \frac{1}{4}$

13. Write the equation of a parabola with position 4 units above the parabola with equation $f(x) = -2x^2$.

14. Write the equation of a parabola with position 1 unit to the right and 8 units below the parabola with equation $f(x) = 5x^2$.

# EXERCISES

**Practice**

Write each equation in the form $f(x) = a(x - h)^2 + k$. Then name the vertex, axis of symmetry, and direction of opening for the graph of each equation.

15. $f(x) = x^2 - 4x + 5$

16. $f(x) = -x^2 - 2x + 2$

17. $f(x) = -3x^2 + 12x$

18. $f(x) = 4x^2 + 24x$

19. $f(x) = 3x^2 - 18x + 11$

20. $f(x) = -2x^2 - 20x - 50$

21. $f(x) = -\frac{1}{2}x^2 + 5x - \frac{27}{2}$

22. $f(x) = \frac{1}{3}x^2 - 4x + 15$

**Write the equation of each parabola shown.**

23.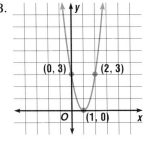

24.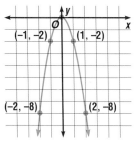

25.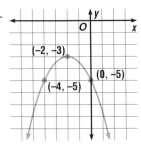

26.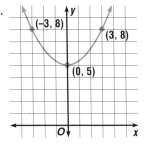

27.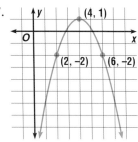

28.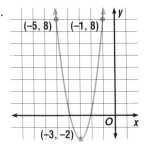

**Write the equation of a parabola that passes through the given points.**

**29.** $(0, 0)$, $(2, 6)$, $(-1, 3)$

**30.** $(1, 6)$, $(-2, 27)$, $(2, 11)$

**31.** $(2, -3)$, $(0, -1)$, $\left(-1, 1\frac{1}{2}\right)$

**32.** $(1, 0)$, $(3, 38)$, $(-2, 48)$

**Graph each equation.**

**33.** $f(x) = 2(x + 3)^2 - 5$

**34.** $f(x) = 3x^2 + 18x + 27$

**35.** $f(x) = \frac{1}{2}(x + 3)^2 - 5$

**36.** $f(x) = \frac{1}{3}(x - 1)^2 + 2$

**37.** $f(x) = x^2 + 6x + 2$

**38.** $f(x) = -2x^2 + 16x - 31$

**39.** $f(x) = -5x^2 - 40x - 80$

**40.** $f(x) = 2x^2 + 8x + 10$

**41.** $f(x) = -9x^2 - 18x - 6$

**42.** $f(x) = -0.25x^2 - 2.5x - 0.25$

**Critical Thinking**

**43.** Given $f(x) = ax^2 + bx + c$ with $a \neq 0$, complete the square and rewrite the equation in the form $f(x) = a(x - h)^2 + k$. State an expression for $h$ and $k$ in terms of $a$, $b$, and $c$.

**Application**

**44. Architecture** The Gateway Arch of the Jefferson National Expansion Memorial in St. Louis is shaped like the parabola whose equation is $f(x) = \frac{1}{315}(-2x^2 + 1260x)$.
   a. Write the equation of the arch in the form $f(x) = a(x - h)^2 + k$.
   b. If the bases of the arch are 630 feet apart, how tall is the arch?

**Mixed Review**    Name the vertex and the axis of symmetry for the graph of each equation.
(Lesson 8-4)

**45.** $f(x) = (x + 2)^2$

**46.** $y = (x - 3)^2 - 11$

**47.** Solve $8s^2 = 200$.   (Lesson 7-2)

**48.** Simplify $(1 - \sqrt{3})^2$.   (Lesson 6-3)

**49.** Simplify $\dfrac{18x^3yz^2 + 27x^2yz + 45x^2y^2z^4}{9xyz}$.   (Lesson 5-6)

**50.** State whether the expression $3xy + x$ is a monomial. If it is, name its coefficient and degree.   (Lesson 5-1)

**51.** Determine whether the system $\begin{cases} 5x + y - z = 0 \\ 2x - y = 7 \\ 3y + 5z = 0 \end{cases}$ has a unique solution.   (Lesson 4-7)

**52.** Solve the system $\begin{cases} x + y > 7 \\ x + y < 10 \end{cases}$ by graphing.   (Lesson 3-4)

---

## MID-CHAPTER REVIEW

**Express each function in quadratic form.**   (Lesson 8-1)

**1.** $g(x) = (x + 2)^2$

**2.** $f(x) = 3(x - 5)^2 + 4$

**Define a variable and write an equation to describe each situation.**   (Lesson 8-1)

**3.** the product of two numbers whose sum is 50

**4.** the area of a rectangle whose perimeter is 42 centimeters

**5.** You want to make a rectangular potholder with the maximum surface on each side. However, you only have 36 inches of piping to put around the edge of the potholder. Make a table to determine the best dimensions for the potholder.   (Lesson 8-2)

**6.** Compare the graphs of $y = (x - 7)^2$ and $y = x^2$.   (Lesson 8-3)

**Name the vertex and axis of symmetry for each equation.**   (Lesson 8-3)

**7.** $f(x) = 3(x - 2)^2 - 4$    **8.** $y = (x + 4)^2$    **9.** $y = x^2 + 4x + 10$    **10.** $g(x) = x^2$

**11.** Write the equation of the parabola that is 2 units to the right of the parabola whose equation is $y = (x + 2)^2 + 3$.   (Lesson 8-4)

**Write each equation in the form $f(x) = a(x - h)^2 + k$. Then name the vertex, axis of symmetry, and direction of opening for the graph of each equation.**   (Lesson 8-4)

**12.** $f(x) = x^2 + 6x + 3$

**13.** $f(x) = -9x^2 - 18x - 10$

# Applications of Quadratic Equations

### Objective

After studying this lesson, you should be able to:

■ solve problems using quadratic equations.

### Application

Ken Virkus has 1200 feet of fencing material to build a pen for his sheep and goats. The pen will be a rectangle with a divider down the middle as shown at the right.

What is the maximum area for the pen? What are the dimensions of the pen with the maximum area?

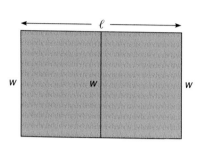

Let $w$ = the width of the pen and $\ell$ represent the length. So, $2\ell + 3w = 1200$. To express the area in one variable, solve for $\ell$ in terms of $w$.

$$2\ell + 3w = 1200$$
$$\ell = \frac{1200 - 3w}{2}$$
$$\ell = 600 - 1.5w$$

The area of the two pens is represented by $A = w(600 - 1.5w)$. Simplify the expression $w(600 - 1.5w)$ and complete the square.

$$
\begin{aligned}
A &= w(600 - 1.5w) \\
&= 600w - 1.5w^2 \\
&= -1.5(w^2 - 400w) \\
&= -1.5(w^2 - 400w + 40{,}000) - (-1.5)(40{,}000) \\
&= -1.5(w - 200)^2 + 60{,}000
\end{aligned}
$$

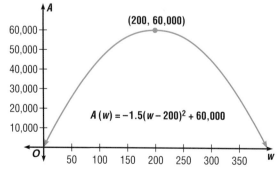

$A(w) = -1.5(w - 200)^2 + 60{,}000$

The graph of this equation is a parabola with vertex (200, 60,000). Since the parabola opens downward, the vertex is a maximum point. The first coordinate of the maximum point represents the width of the rectangular pen. The length is $600 - 1.5(200)$ or 300 feet. The second coordinate of the maximum point represents the maximum area of the pen, 60,000 square feet.

Many problems ask you to find a **maximum** or **minimum value.** If you can write a quadratic function to describe the situation in these problems, you can probably solve it by finding the coordinates of the vertex of the function. The vertex of a parabola that opens downward represents a maximum point of the function. The vertex of a parabola opening upward represents a minimum point of the function.

**Example 1** | **Find two numbers whose difference is 48 and whose product is a minimum.**

*Explore* | Let $n$ represent the lesser number. Then $n + 48$ represents the greater number.

*Plan* | Product $= n(n + 48)$
$= n^2 + 48n$

*Solve* | This is the equation of a quadratic function. Write the equation in the form $y = a(x - h)^2 + k$ to find the vertex. Let $y$ represent the product of the two numbers.

$y = n^2 + 48n$
$= n^2 + 48n + \left(\dfrac{48}{2}\right)^2 - \left(\dfrac{48}{2}\right)^2$   *Complete the square.*
$= (n + 24)^2 - 576$

The vertex is $(-24, -576)$. So, the minimum product, $-576$, occurs when $n$ is $-24$. When $n = -24$, $n + 48 = -24 + 48$ or $24$.

*Examine* | Let's check some other pairs of numbers with a difference of 48 to see that their product is not less than $-576$.

$(-23)(25) = -575 \qquad (-20)(28) = -560 \qquad (-15)(33) = -495$

---

**Example 2**

**APPLICATION**
**Business**

**A ferry transports tourists to the Middle Bass Island on Lake Erie during the summer months. The one-way fare is $6.00 a person and 200 people ride the ferry each day. The owner estimates that for every $0.50 the fare is raised, they will lose 10 customers. What should the fare be for the greatest income for the ferry owner?**

Let $p$ represent the number of $0.50 price increases.
Then $6 + 0.5p$ represents the fare and $200 - 10p$ represents the number of passengers.

    *Income = (number of passengers) × (fare)*
Income $= (200 - 10p) \qquad \times (6 + 0.5p)$
$= 1200 + 100p - 60p - 5p^2$
$= 1200 + 40p - 5p^2$

Complete the square to find the vertex of the parabola.
Income $= 1200 + 40p - 5p^2$
$= -5(p^2 - 8p) + 1200$
$= -5(p^2 - 8p + 16) + 1200 - (-5)(16)$   *Complete the square.*
$= -5(p - 4)^2 + 1280$

The vertex is $(4, 1280)$. So the maximum income, $1280, occurs when the owner raises the fare 4 times. The total increase will be $4(0.50)$ or $2.00. Therefore, a fare of $6.00 + $2.00 or $8.00 will maximize the owner's income.

# CHECKING FOR UNDERSTANDING

**Communicating Mathematics**

**Read and study the lesson to answer these questions.**

1. Describe a situation where someone would want to find the maximum value of a function.

2. Why does the vertex of a parabola represent a maximum value when the parabola opens downward and a minimum value when the parabola opens upward?

3. Does the quadratic function $f(x) = 3(x - 3)^2 + 5$ have a minimum value? If so, where does the minimum value occur?

4. Look back at Example 2. How many price increases should the owner of the ferry make if they would lose 20 people for every $0.50 increase in fare? What would the fare be after that change?

**Guided Practice**

5. Find two numbers whose difference is −20 and whose product is a minimum.

   a. Let $x$ represent the lesser number. Write an expression for the greater number.

   b. Write an expression for the product of the two numbers.

   c. Draw a graph relating the lesser number to the product of the two numbers. Place $x$ on the horizontal axis and the product on the vertical axis.

   d. What are the two numbers?

6. **Business**  The circulation of the Charlotte Arts Council Newsletter is 50,000. Due to increased production costs, the council must increase the current price of 40¢ a copy. According to a recent survey, the circulation of the newsletter will decrease 5000 for each 10¢ increase in price.

   a. Let $p$ represent the number of 10¢ price increases. Write an algebraic expression to describe the decreased circulation after the price increase.

   b. Write an algebraic expression for the increased price per copy.

   c. The Arts Council's income from the newsletter is the product of the number of copies sold and the price per copy. This is a function of the number of price increases. Write an equation to describe the function.

   d. Draw a graph relating the Arts Council's income to the number of price increases. Use the number of price increases for the $x$-axis and the income in dollars for the $y$-axis.

   e. What price per copy will maximize the Arts Council's income from the newsletter?

# EXERCISES

Practice **Solve each problem.**

7. Find two numbers whose sum is 36 and whose product is a maximum.

8. Find two numbers whose difference is 40 and whose product is a minimum.

9. Find two numbers whose sum is 37 and whose product is a maximum.

10. Find two numbers whose difference is 25 and whose product is a minimum.

11. The sum of $x$ and $y$ is $-16$. Find $x$ and $y$ so that $xy$ is a maximum.

12. A rectangle has a perimeter of 40 meters. Find the dimensions of the rectangle with the maximum area.

13. Find the dimensions and maximum area of a rectangle whose perimeter is 24 inches.

Application

14. Melissa Hart plans to put a fence around her rectangular garden. She has 150 feet of fencing material to make the fence. If there is to be a 10 foot opening left for an entrance on one side of the garden, what dimensions should the garden be for maximum area?

15. Marc Jesson has 120 feet of fence to make a rectangular kennel for his dogs. If the house is to be used as one side of the kennel, what would be the length and width for maximum area?

16. A tour bus carries tourists through the historic district of Savannah, Georgia, serving 300 customers a day. The charge is $8.00 per person. The company owner estimates that the company would lose 20 passengers a day for each $1 increase in the fare. What charge would be most profitable for the tour bus company?

17. The Hoosier Commuter Airline transports about 800 passengers a week between Chicago and Fort Wayne. A round-trip ticket is $300. The company executives estimate that for each $5 increase in the ticket price, 10 passengers would be lost to the competition. What ticket price would maximize income for the company?

18. An object is fired upwards from the top of a 200 foot tower at a velocity of 80 feet per second. The height of the object $t$ seconds after firing is given by the formula $h(t) = -16t^2 + 80t + 200$. Find the maximum height reached by the object and the time that that height is reached.

19. The Center Stage Community Theater can seat 500 people. They sold out for every performance last season. They intend to raise the $3.00 admission price for the upcoming season. They estimate that for every $0.20 increase in price, 25 fewer people will attend a performance. What ticket price will maximize the theater's income?

20. It costs the Fresh Air Fan Company $(20x + 1000)$ dollars to produce $x$ ceiling fans. They sell the fans for $(300 - 2x)$ dollars each. How many fans should the company produce each month to maximize their profit?

21. Lian is making a box to collect paper for the school paper drive. She cuts a 5 centimeter square from each corner of a rectangular piece of cardboard and folds the sides up to make the box. If the perimeter of the bottom of the box must be 50 centimeters, what should the length, width, and height of the box be for maximum volume?

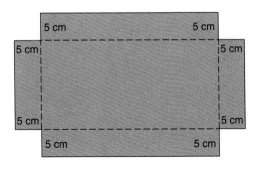

22. A rectangle is inscribed in an isosceles triangle. The base of the triangle is 8 inches and the height is 10 inches. Find the dimensions of the inscribed rectangle with maximum area.

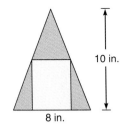

10 in.

8 in.

23. Zach and Amy cut a 32-inch piece of wire into two pieces. Amy bent one piece into a square and Zach bent the other into a rectangle that is 2 inches longer than it is wide. How long should they have cut each piece so that the sum of the areas of the square and the rectangle are minimized?

24. Two hundred people came to last year's winter play at Meadowbrook High School. The ticket price was $2. Since the Drama Club is planning a trip to a Broadway play, they would like to price the tickets so that they earn the maximum amount of money possible. They estimate that for each 25¢ increase in the price, 10 fewer people will come to the play. How much should the tickets be?

25. **Physics**  A ball is thrown straight up with an initial velocity of 64 feet per second. The height of the ball $t$ seconds after it is thrown is given by the formula $h(t) = 64t - 16t^2$. What is the height of the ball after 1.5 seconds? What is its maximum height? After how many seconds will it return to the ground?

26. **Business**  The Images Camera Shop can sell 21 KS-2 Cameras a month at $120 each. The owner estimates that for each $5 decrease in price they could sell three more of these cameras a month. The cameras cost the store $75 each. If the store sells all of the cameras bought each month, what should they charge for a camera to maximize profit? What is the maximum profit?

**Critical Thinking**    27. Find two numbers whose sum is $a$ and whose product is a maximum.

**Computer**    When an object is hurled upward, its height $s$, above the ground is given as a function of time by the formula $s = -16t^2 + v_0t + s_0$. The height is in feet and the time is in seconds. The initial velocity is symbolized by $v_0$ and $s_0$ is the initial height. The BASIC program below finds the height and the time, given the initial velocity and initial height.

```
10   PRINT "ENTER THE INITIAL VELOCITY."
20   INPUT VO
30   PRINT "ENTER THE INITIAL HEIGHT."
40   INPUT SO
50   PRINT
60   PRINT "TIME", "HEIGHT"
70   DEFFNS(T) = -16*T^2 + VO*T + SO
80   LET T = 0
90   PRINT T, FNS(T)
100  LET T = T + 1
110  IF FNS(T) < 0 THEN 130
120  GOTO 90
130  END
```

28. Enter an initial velocity of 256 feet per second and an initial height of 0 feet.

    a. How long will the object be airborne?

    b. What is the greatest height the object will reach?

    c. When will the object be at a height of 768 feet?

29. Suppose a baseball is thrown upward from a height of six feet at an initial velocity of 88 feet per second.

    a. How far off the ground will the ball be after one second?

    b. About how long will the ball be in the air?

    c. At about what time will the ball reach its greatest height?

    d. Approximately what is the greatest height the ball will reach?

**Mixed Review**    30. What is the remainder of the division $8^{100} \div 5$?    (**Lesson 8-2**)

31. Write the equation of a parabola with position 3 units to the right of the parabola with equation $f(x) = x^2$.    (**Lesson 8-4**)

32. Find a value of $c$ that makes $x^2 - 8x + c$ a perfect square. (**Lesson 7-3**)

33. Simplify $(3 + 2i)(4 - i)$.    (**Lesson 6-9**)

34. Factor $9x^2 - 12x + 4$.    (**Lesson 5-6**)

35. State the row operations you would use to obtain a zero in the second column of row one of $\begin{bmatrix} -2 & 1 & 0 \\ 0 & 4 & 12 \end{bmatrix}$.    (**Lesson 4-7**)

# Graphing Quadratic Inequalities

**Objective**

After studying this lesson, you should be able to:

- graph quadratic inequalities.

**Application**

Discovery Audio Inc. makes compact disks. According to their recent sales figures, their profit can be found by the inequality $P(x) \leq -0.003x^2 + 9x - 1500$, where $x$ is the number of CDs sold in a week. The relation is an inequality because it is based on the list price of the CDs, and the company often charges record stores less than list price when they buy in great quantities.

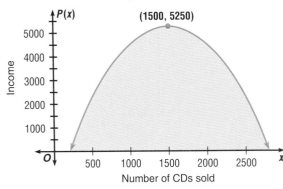

The graph of this inequality is the part of the plane enclosed by the parabola whose equation is $P(x) = -0.003x^2 + 9x - 1500$. The parabola is the **boundary** of each region. Points on the parabola represent the profit if all of the CDs are sold at list price. Points in the interior of the curve represent profits if some of the CDs were sold for a price below list price.

*Notice that the boundary is solid. You can determine whether the boundary is solid or broken the same way that you do with a linear inequality.*

**Example 1**

Graph $y \leq -3x^2 + 12x - 2$.

Complete the square on the right side.

$$y \leq -3x^2 + 12x - 2$$
$$y \leq -3(x^2 - 4x) - 2$$
$$y \leq -3(x^2 - 4x + 4) - 2 - (-3)(4)$$
$$y \leq -3(x - 2)^2 + 10$$

The boundary is a parabola with vertex $(2, 10)$ that opens downward. Test points not on the parabola to see whether points above or below the parabola belong to the graph.

*Region below parabola*

Test $(2, 0)$: $y \leq -3x^2 + 12x - 2$
$$0 \leq -3(2)^2 + 12(2) - 2$$
$$0 \leq -12 + 24 - 2$$
$$0 \leq 10$$

belongs

*Region above parabola*

Test $(5, 0)$: $y \leq -3x^2 + 12x - 2$
$$0 \leq -3(5)^2 + 12(5) - 2$$
$$0 \leq -75 + 60 - 2$$
$$0 \leq -17$$

does not belong

Shade the region below the parabola. Since the inequality contains a "less than or equal" symbol, the boundary is included. So the graph of the parabola itself is solid.

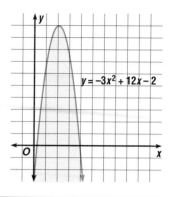

$y = -3x^2 + 12x - 2$

**Example 2**

**A rectangle is 2 inches longer than it is wide. Find the possible dimensions if the area of the rectangle is more than 224 square inches.**

Draw a diagram of the rectangle.

Let $w$ represent the width of the rectangle. The length would be represented by $w + 2$. Thus, the area of the rectangle is $w(w + 2)$.

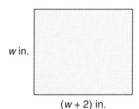

$w$ in.

$(w + 2)$ in.

So, $w(w + 2) > 224$.

$$w(w + 2) > 224$$
$$w^2 + 2w > 224$$
$$w^2 + 2w - 224 > 0$$
$$(w^2 + 2w + 1) - 224 - 1 > 0$$
$$(w + 1)^2 - 225 > 0$$

The boundary of the graph of this relation is a parabola with vertex $(-1, -225)$ that opens upward. Test a point off of the boundary to determine which region should be included in the graph.

Test $(0, 0)$:   $w(w + 2) > 224$
$$0(0 + 2) > 224$$
$$0 > 224$$

Since $(0, 0)$ is above the parabola and it does not satisfy the inequality, the region below the parabola is included in the graph.

The graph includes points that have negative values. Since lengths and areas cannot be negative, these values should be disregarded in determining the solution. So allowing only positive values, the width of the rectangle should be greater than 14 and the length should be greater than 16.

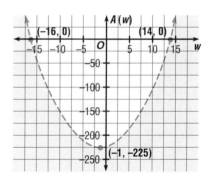

# CHECKING FOR UNDERSTANDING

**Read and study the lesson to answer these questions.**

1. Should the boundary of the graph of $y < 2x^2 + 3x - 5$ be solid or broken? Explain.

2. The equation $y = x^2 - 6x + 5$ is graphed at the right. If you were to graph the inequality $y \geq x^2 - 6x + 5$, would you include the region above or below the parabola? Explain why.

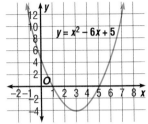

3. Why were the points with negative coordinates disregarded in Example 2?

## Guided Practice

**Determine if the ordered pair is a solution of the given inequality.**

4. $y > x^2$, $(3, 2)$

5. $y > (x - 2)^2$, $(0, 5)$

6. $y \geq x^2 - 16$, $(-3, 0)$

7. $y \leq -3x^2 + 5$, $(2, -7)$

8. $y \leq 2(x + 3)^2$, $(-2, 3)$

9. $y \geq 5x^2 - 6x$, $(1, 1)$

10. $y \geq 4x^2 - 8x - 13$, $(3, -1)$

11. $y < -0.5x^2 + 9x - 2$, $(16, 15)$

# EXERCISES

## Practice

**Draw the graph of each inequality.**

12. $y \leq (x + 3)^2$

13. $y > x^2 + 2x + 1$

14. $y \geq x^2 + 8x + 16$

15. $y > x^2 - 10x + 25$

16. $y \geq x^2 - 49$

17. $y \leq x^2 - 16$

18. $y \geq x^2 + 3x - 18$

19. $y \leq x^2 - 13x + 36$

20. $y > x^2 + x - 30$

21. $y > x^2 - x - 20$

22. $y \geq -x^2 + 6x + 8$

23. $y \leq 2x^2 + x - 3$

24. $y > 4x^2 - 8x + 3$

25. $y \leq -x^2 - 7x + 10$

26. $y < -3x^2 + 5x + 2$

27. $y > -4x^2 - 3x - 6$

**CONNECTION**
**Geometry**

28. A rectangle is 5 centimeters longer than it is wide. Find the possible dimensions if the area of the rectangle is more than 104 square centimeters.

## Critical Thinking

29. Find the intersection of the graphs of $y \geq x^2 - 3$ and $y \leq x^2 + 3$.

## Applications

30. **Forensic Science** A criminalist is investigating a shooting of a police helicopter. A weapon was found with a suspect's fingerprints. The criminalist has deduced that the weapon is capable of firing with an initial velocity of 1100 feet per second, so the height of the bullet $t$ seconds after firing is found by the function $h(t) = -16t^2 + 1100t$. The helicopter was flying at an altitude of 7000 feet at the time it was shot. Tell how you can determine whether it is possible that this gun shot the helicopter.

31. **Sports**  Kristin's homerun hit traveled in a path described by the function $f(x) = -0.003x^2 + x + 4$ where $x$ represents the number of feet the ball has traveled from the plate and $f(x)$ represents the height of the ball. If the scoreboard is 30 feet high and 410 feet from the plate, show that Kristin's homerun ball did not hit the scoreboard.

**Mixed Review**

32. The cruise ship *The Silver Dollar* was rented to take 100 passengers to the Green Mountain Resort. The fare is $5 per person. The owner of the cruise ship has agreed to reduce the fare by 2¢ per passenger for every passenger over 100 in the group. How many passengers will produce a maximum profit for the owner?  **(Lesson 8-5)**

33. Find two numbers whose difference is -40 and whose product is a minimum. **(Lesson 8-5)**

**Journal**
Use graphs of quadratic equations to draw a sketch of a roller coaster ride. Be use to include the equation of each graph.

34. Solve the equation $y^2 - 16y + 10$ using the quadratic formula. **(Lesson 7-4)**

35. Simplify $\sqrt{\dfrac{8}{9}}$.  **(Lesson 6-2)**

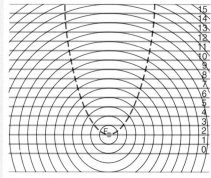

36. **Astronomy**  Venus has an average distance of $1.08 \times 10^8$ kilometers from the Sun. Saturn has an average distance of $1.428 \times 10^9$ kilometers from the Sun. About how much closer to the Sun is Venus? **(Lesson 5-1)**

## HISTORY CONNECTION

**Hypatia** (370–415 A.D.) was the first recorded woman mathematician. Her beauty and talents were legendary at a time when Greek mathematics was neither appreciated nor encouraged. Her lectures at the University of Alexandria were well attended by scholars from Europe, Asia, and Africa. She often spoke about the works of Diophantus and other Greek scholars and wrote a text called *On the Conics of Appollonius*, in which she described graphs such as the parabola.

The diagram at the right illustrates her definition of a parabola as the path of all points the same distance from a fixed point called the focus $F$ and a fixed line. A group of concentric circles is drawn with $F$ as their center. The fixed line is line 0. Lines parallel to line 0 are drawn tangent to each circle. If you connect the points where the circles intersect the parallel lines, the parabola is formed.

# 8-7 Solving Quadratic Inequalities

**Objective**

After studying this lesson, you should be able to:
- solve quadratic inequalities in one variable.

**Application**

The Corporate Commuter Airline, CCA, transports 500 people a week from Atlanta to Washington, D.C. The round trip fare is currently $400. CCA is in a fare war with competing airlines, and company officials estimate that for every $10 decrease in the fare at most 25 passengers will choose them over their competitor. If it costs $144,000 a week to run the airline, how many $10 decreases could the airline make and not lose money?

There are many different changes in the airfare that Corporate Commuter could make that would make the airline remain profitable, since quadratic inequalities have an infinite number of solutions. Let's graph the inequality for the situation described above to find the solution.

First write the inequality. Let $x$ represent the number of $10 price decreases and $y$ represent the profit.

*Profit = (number of passengers)(fare) − cost*     *Since it is "at most 25 passengers," this is an inequality.*

$$y \le (500 + 25x)(400 - 10x) - 144{,}000$$

$$y \le 200{,}000 - 5000x + 10{,}000x - 250x^2 - 144{,}000$$

$$y \le -250x^2 + 5000x + 56{,}000$$

$$y \le -250(x^2 - 20x + 100) + 56{,}000 + 25{,}000 \qquad \textit{Complete the square.}$$

$$y \le -250(x - 10)^2 + 81{,}000$$

Now graph the inequality. It is a parabola with vertex $(10, 81{,}000)$ that opens downward. The axis of symmetry is $x = 10$.

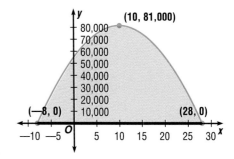

We are looking for all the points where Corporate Commuter is not losing money. That would be where the profit is greater than or equal to 0. Each point on the $x$-axis has a $y$-coordinate of 0. So the points on the $x$-axis that satisfy $y \le -250(x - 10)^2 + 81{,}000$ are solutions to the inequality $0 \le -250(x - 10)^2 + 81{,}000$. Those solutions are $\{x | -8 \le x \le 28\}$.

We could also solve the inequality algebraically.

$0 \le -250x^2 + 5000x + 56{,}000$    *The company is not losing money as long as profit is greater than or equal to 0.*

$0 \le -250(x^2 - 20x - 224)$    *Factor.*

$0 \le -250(x - 28)(x + 8)$

$0 \ge (x - 28)(x + 8)$    *Divide each side by $-250$ and reverse the inequality.*

The product of two numbers is negative if one number is negative and one is positive, so we can write the following.

$(x - 28)$ is negative.              $(x + 8)$ is negative.

$x - 28 \le 0$   and $x + 8 \ge 0$        $x - 28 \ge 0$   and $x + 8 \le 0$

    $x \le 28$ and     $x \ge -8$    or        $x \ge 28$ and     $x \le -8$

       $-8 \le x \le 28$                         never true

*The graphs of $x \ge 28$ and $x \le -8$ never intersect, so $x \ge 28$ and $x \le -8$ can never be true.*

Solving $0 \le -250x^2 + 5000x + 56{,}000$ either graphically or algebraically produces the solution $\{x | -8 \le x \le 28\}$. So, Corporate Commuter would not lose money if they lower their fare between $-8$ and 28 times inclusive.

*If they were to lower the fare $-8$ times, they would actually be increasing the $10 fare 8 times.*

## Example 1

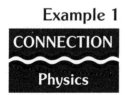

**CONNECTION**

**Physics**

**A ball is thrown upward with an intial velocity of 64 feet per second. The formula $h(t) = 64t - 16t^2$ gives the height of the ball after $t$ seconds. During what time span is the ball in flight?**

The ball is in flight as long as its height is greater than 0. We can find the times that the ball is in flight by solving the inequality $0 < 64t - 16t^2$.

$0 < 64t - 16t^2$

$0 < -16t(t - 4)$

The product of two numbers is positive if both of the numbers are positive or both are negative.

Both factors are positive.      Both factors are negative.

$-16t > 0$ and $t - 4 > 0$     $-16t < 0$ and $t - 4 < 0$

    $t < 0$ and     $t > 4$           $t > 0$ and     $t < 4$

         never true                  $0 < t < 4$

The solution set is $\{t | 0 < t < 4\}$. So the ball is in flight between 0 and 4 seconds after it is thrown.

**Example 2** | **Solve $(x - 7)(x + 2) > 0$.**

Another way that you can solve quadratic inequalities is by using three test points. First solve the equation $(x - 7)(x + 2) = 0$.

$$(x - 7)(x + 2) = 0$$
$$x - 7 = 0 \text{ or } x + 2 = 0$$
$$x = 7 \qquad x = -2 \qquad \textit{These solutions are called \underline{critical points}.}$$

The points 7 and $-2$ separate the $x$-axis into three parts: $x < -2$, $-2 < x < 7$, and $x > 7$.

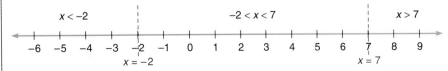

Choose a value from each of the parts and substitute it into $(x - 7)(x + 2) > 0$. Make a table to organize your results.

| Part of $x$-axis | Point, $x$ | $(x - 7)(x + 2)$ | Is $(x - 7)(x + 2) > 0$? |
|---|---|---|---|
| $x < -2$ | $-3$ | $((-3) - 7)((-3) + 2) = 10$ | yes |
| $-2 < x < 7$ | 0 | $(0 - 7)(0 + 2) = -14$ | no |
| $x > 7$ | 8 | $(8 - 7)(8 + 2) = 10$ | yes |

The parts of the $x$-axis whose points are solutions to $(x - 7)(x + 2) > 0$ belong to the solution set. So the solution set is $\{x | x < -2 \text{ or } x > 7\}$.

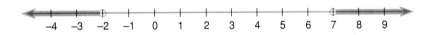

# CHECKING FOR UNDERSTANDING

**Communicating Mathematics**

**Read and study the lesson to answer these questions.**

1. Why might the Corporate Commuter Airline want to charge a fare that will not give maximum profit?

2. What is the solution set for the inequality graphed at the right?

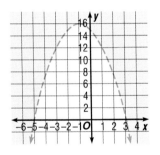

3. What points would you test to find the solution to $(x + 8)(x - 10) > 0$?

**Guided Practice**    State the signs of the factors of each quadratic inequality.

**4.** $(x - 9)(x + 1) < 0$

**5.** $(x + 5)(x - 9) > 0$

**6.** $(x + 11)(x - 1) > 0$

**7.** $(x - 8)(x - 2) \leq 0$

**8.** $(x + 20)(x - 20) < 0$

**9.** $(x + 1)(x - 3) > 0$

**10.** $x^2 + 6x - 27 \leq 0$

**11.** $x^2 + 9x + 20 \leq 0$

**12.** $x^2 - x - 90 > 0$

**13.** $2x^2 + 3x + 1 \leq 0$

State the solution set for each inequality graphed below.

**14.**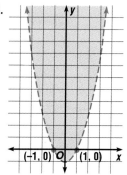

**15.**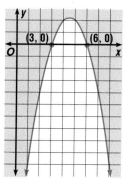

**16.**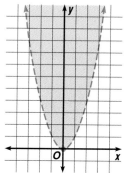

# EXERCISES

**Practice**    Solve each inequality.

**17.** $(x + 2)(x + 9) > 0$

**18.** $(a - 10)(a - 3) \leq 0$

**19.** $(n - 2.5)(n + 3.8) \geq 0$

**20.** $x^2 + 4x - 21 < 0$

**21.** $q^2 + 2q \geq 24$

**22.** $m^2 + m - 6 > 0$

**23.** $2b^2 - b < 6$

**24.** $6s^2 + 5s > 4$

**25.** $p^2 - 4p \leq 5$

**26.** $x^2 - 4x \leq 0$

**27.** $w^2 \geq 2w$

**28.** $c^2 \leq 49$

**29.** $d^2 \geq 3d + 28$

**30.** $b^2 \geq 10b - 25$

**31.** $2x^2 > 25$

**32.** $t^2 + 12t \leq -27$

**33.** $f^2 + 12f + 36 < 0$

**34.** $9v^2 - 6v + 1 \leq 0$

**35.** $8d + d^2 \geq -16$

**36.** $2g^2 - 5g - 3 < 0$

**37.** $-5x - 3x^2 < -2$

**38.** $n^2 \leq 3$

**39.** $4t^2 - 9 < -4t$

**40.** $-2 > -6r - 9r^2$

**Critical Thinking**    Solve each inequality.

**41.** $(x - 1)(x - 3)(x + 4) > 0$

**42.** $(x - 8)(x + 2)(x + 4) \leq 0$

**43.** $(x - 3)(x + 6)(x + 2) < 0$

**44.** $(x + 5)(x + 6)(x + 7) \geq 0$

**45.** $(x + 5)(x + 1)(x - 4)(x - 6) > 0$

**46.** $(x - 2)(x + 2)(x - 1)(x + 3) \geq 0$

47. **Sports** The instant replay facility at the Superdome was moved because it was hit by a high punt kicked by Oakland Raider Ray Guy. The original position of the facility was 90 feet above the playing field. Colin, a high school punter, can kick a football with an initial velocity of 65 feet per second. The height of the football $t$ seconds after he kicks it is found by the function $h(t) = -16t^2 + 65t$. If Colin were to have kicked a football in the Superdome before the instant replay facility was moved, would he have been able to hit it?

48. **Landscaping** Mr. and Mrs. Ortiz are making an English garden in their backyard which they plan to surround with decorative stones. They have enough stones to enclose a rectangular garden with a perimeter of 68 feet. They would like the area of the garden to be at least 240 square feet. What could the width of the garden be?

49. **Business** Judy runs a shuttle bus for the Department of Recreation. She charges $1 per person to ride from the High School to the Community Recreation Center. About 100 students ride the shuttle bus each week. Judy estimates that for every 20¢ she lowers the fare 5 more students would ride each week. If the cost of running the shuttle is $66 a week, how many 20¢ decreases in fare could Judy make without having to subsidize the shuttle bus program?

**Mixed Review**

50. Is $(4, -4)$ a solution to the quadratic inequality $y \le -x^2 + 5x$? **(Lesson 8-6)**

51. Graph $y > x^2 - 7x + 10$. **(Lesson 8-6)**

52. Identify the quadratic term, the linear term, and the constant term of the function $f(x) = x^2 + 3x - 2$. **(Lesson 8-1)**

53. Solve $d^{\frac{1}{3}} = 4$. **(Lesson 7-6)**

54. Simplify $(\sqrt{3a} + \sqrt{2b})(\sqrt{15a} - \sqrt{3b})$. **(Lesson 6-3)**

55. Factor $a^2 - b^2 + 8b - 16$. **(Lesson 5-5)**

56. **Number Theory** The sum of the digits of a three digit number is 13. The tens' digit is 1 less than the ones' digit. The hundreds' digit is 2 less than twice the sum of the ones' and tens' digits. Find the number. **(Lesson 4-4)**

**Portfolio**

Review the items in your portfolio. Make a table of contents of the items, noting why each item was chosen. Replace any items that are no longer appropriate.

## VOCABULARY

Upon completing this chapter, you should be familiar with the following terms:

| | | | |
|---|---|---|---|
| axis of symmetry | **360** | **373** | minimum value |
| boundary | **379** | **360** | parabola |
| constant term | **353** | **353** | quadratic function |
| half plane | **379** | **353** | quadratic term |
| linear term | **353** | **360** | vertex |
| maximum value | **373** | | |

## SKILLS AND CONCEPTS

| OBJECTIVES AND EXAMPLES | REVIEW EXERCISES |
|---|---|

Upon completing this chapter, you should be able to:

Use these exercises to review and prepare for the chapter test.

■ write functions in quadratic form (**Lesson 8-1**)

Express $g(x) = (3x - 1)^2$ in quadratic form.

$$g(x) = 9x^2 - 6x + 1$$

**Express each function in quadratic form.**

1. $f(x) = (x + 2)^2$
2. $h(x) = 2(x - 4)^2 + 6$

■ identify the quadratic term, the linear term, and the constant term of a quadratic function (**Lesson 8-1**)

For $f(x) = 2x^2 + x - 5$, the quadratic term is $2x^2$, the linear term is $x$, and the constant term is $-5$.

**For each function, identify the quadratic term, the linear term, and the constant term.**

3. $f(x) = 3x^2 + 2x - 1$
4. $g(x) = x^2 + 4x - 2$

- graph quadratic equations of the form $y = (x - h)^2 + k$ (**Lesson 8-3**)

Graph $y = (x + 1)^2 - 1$.

The vertex is $(-1, -1)$.
The axis of symmetry is $x = -1$.

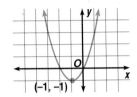

Write the equation of each parabola if its equation has the form $y = (x - h)^2 + k$.

5.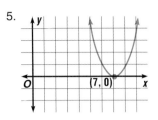

6.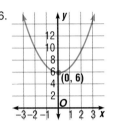

Graph each equation.

7. $y = (x - 3)^2$      8. $f(x) = x^2 + 3$

---

- identify the vertex and the equation of the axis of symmetry of a parabola (**Lesson 8-3**)

The function $f(x) = x^2 - 2x + 3$ written in the form $f(x) = (x - h)^2 + k$ is $f(x) = (x - 1)^2 + 2$. The vertex is $(1, 2)$. The axis of symmetry is $x = 1$.

Write each equation in the form $f(x) = (x - h)^2 + k$. Then name the vertex and the axis of symmetry for the graph of each function.

9. $f(x) = x^2 - 2x + 1$

10. $g(x) = x^2 + 6x + 3$

---

- graph equations of the form $y = a(x - h)^2 + k$ and identify the vertex, the equation of the axis of symmetry, and the direction of the opening (**Lesson 8-4**)

The function $f(x) = 2x^2 - 8x + 9$ written in the form $f(x) = a(x - h)^2 + k$ is $f(x) = 2(x - 2)^2 + 1$. The vertex is $(2, 1)$. The axis of symmetry is $x = 2$. The graph opens upward.

**14.**

Write each equation in the form $f(x) = a(x - h)^2 + k$. Then name the vertex, axis of symmetry, and direction of opening for the graph of each equation.

11. $f(x) = x^2 - 2x + 4$

12. $h(x) = -3x^2 + 18x$

13. $f(x) = -2x^2 - 40x + 10$

14. $g(x) = \frac{1}{3}x^2 + 2x + 7$

---

- solve problems using quadratic equations (**Lesson 8-5**)

If you are asked to find a maximum or minimum value, write a quadratic function to describe the situation by finding the coordinates of the vertex of the functions.

15. Find two numbers whose difference is 5 and whose product is a minimum.

16. **Geometry** Find the dimensions and maximum area of a rectangle whose perimeter is 80 cm.

| OBJECTIVES AND EXAMPLES | REVIEW EXERCISES |
|---|---|

■ graph quadratic inequalities
(Lesson 8-6)

Graph $y \geq (x - 2)^2 + 1$.

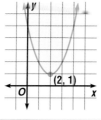

The boundary is a
parabola with vertex
(2, 1) that opens
upward.

**Draw the graph of each inequality.**

17. $y > x^2 + 3x + 4$

18. $y < x^2 + 5x + 6$

19. $y > -2x^2 + 9$

20. $y \geq 3x^2 - 15x + 22$

---

■ solve quadratic inequalities in one
variable. **(Lesson 8-7)**

To solve an inequality, either graph the
inequality or use three test points to
determine the solutions.

**Solve each inequality.**

21. $(x - 10)(x + 1) < 0$

22. $x^2 + 8x - 9 > 0$

23. $6 - 5a - 4a^2 \leq 0$

24. $m^2 - 20m \geq -100$

---

# APPLICATIONS AND CONNECTIONS

25. Make a table to help you find the
remainder of the division $6^{100} \div 5$.
**(Lesson 8-2)**

27. **Business** Quick Cab Co. runs a taxi
cab service between the airport and the
business district. The charge for one-
way service is $11.50. The owners
estimate the 30 passengers will be lost
each day for each $1.50 increase in the
fare. If they currently service 320
passengers per day, what fare should
they charge for maximum
income? **(Lesson 8-5)**

29. **Physics** The Empire State Building is
1250 feet high. If an object is thrown
upward from the top of the building at
an initial velocity of 35 feet per second,
its height $t$ seconds after it is thrown is
given by the function $h(t) = -16t^2 +
35t + 1250$. How long will it be before
the object hits the ground?
**(Lesson 8-5)**

26. **Art** Morgan needs to paint a landscape
for art class. She has 6 feet of framing
material to frame the finished painting.
What should the dimensions of her canvas
be for the painting to be of maximum
area? **(Lesson 8-5)**

28. **Construction** Kipp has 80 feet of
fencing material to make a kennel for
his springer spaniel puppies. The fence
that already runs around his backyard
will be used as one side of the kennel.
What should the dimensions of the
kennel be if Kipp wants to give the
puppies as much area as possible?
What is the area of the completed
kennel? **(Lesson 8-5)**

## Express each function in quadratic form.

**1.** $f(x) = (x + 2)^2 + 4$

**2.** $f(x) = 2(x - 3)^2 + 5$

## Write the equation of each parabola shown.

**3.**

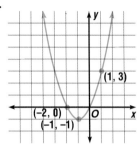

**4.**

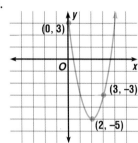

**5.**

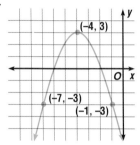

## Graph each equation. Then name the vertex, the axis of symmetry, and the direction of opening for each graph.

**6.** $y = (x + 3)^2$

**7.** $f(x) = -(x - 2)^2$

**8.** $y = (x + 2)^2 + 1$

**9.** $f(x) = -2(x + 4)^2 + 6$

**10.** $g(x) = x^2 + 3x + 6$

**11.** $y = \frac{1}{2}(x + 6)^2 - 3$

**12.** $h(x) = 2x^2 + 8x + 9$

**13.** $y = x^2 - 8x + 9$

**14.** $f(x) = -x^2 - 10x + 10$

**15.** $y = -\frac{2}{3}x^2 + 4x - 3$

**16.** Make a table to help you find the ones digit of $8^{50}$.

**17.** Find two numbers whose difference is $-22$ and whose product is a minimum.

**18. Geometry** A rectangle has a perimeter of 60 cm. Find the dimensions of the rectangle with the maximum area.

**19. Geometry** Corey is building a fence around the vegetable garden with 60 feet of chicken wire. The side of the barn will serve as one side of the fence. Find the dimensions of the garden with the maximum area.

**20. Physics** The initial velocity of a rocket shot straight upward is 40 ft/s. The height of the rocket after $t$ seconds is given by the formula $h(t) = -16t^2 + 40t$. What is the maximum height the rocket reaches? When will it land?

## Draw the graph of each inequality.

**21.** $y < -(x - 2)^2$

**22.** $y \leq x^2 + 6x - 7$

## Solve each inequality.

**23.** $(x + 6)(x - 3) < 0$

**24.** $3x^2 + 2x - 5 \geq 0$

**25.** $a^2 - 8a \geq -16$

**Bonus** Graph all points bounded by the graphs of $y = x^2 - 4$ and $y = -x^2 + 4$.

# College Entrance Exam Preview

The test questions on these pages deal with expression and equations.

**Directions: Choose the one best answer. Write A, B, C, or D.**

1. If $10 - 4y = 18 + 2y$, then what is the value of $3y$?

   (A) $-4$    (B) $-3$    (C) $\dfrac{-4}{3}$    (D) $\dfrac{4}{3}$

2. If it takes 6 hours for 4 people to paint a room, how many hours will it take 5 people, working at the same rate, to paint a room that is the same size?

   (A) $3\dfrac{1}{3}$    (B) $4\dfrac{4}{5}$    (C) $6\dfrac{1}{2}$    (D) $7\dfrac{1}{2}$

3. In a college lecture hall, the number of seats in each row is 25 less than the number of rows. If there are 350 seats in all, find the number of rows.

   (A) 10    (B) 14    (C) 25    (D) 35

4. A grocery store sold 500 candy bars for $255.00. A small candy bar sells for 40¢ and a large bar sells for 65¢. How many of the small bars were sold?

   (A) 55    (B) 77    (C) 220    (D) 280

5. A bottle of type B perfume costs two dollars more than 2 bottles of type C perfume. If the total cost of one bottle of each type is $32, how much more does type B cost then type C?

   (A) 10    (B) 12    (C) 22    (D) 32

6. If $x > 1$, which of the following increases as $x$ increases?

   I. $x - \dfrac{1}{x}$

   II. $\dfrac{1}{x^2 - x}$

   III. $4x^3 - 2x^2$

   (A) I only

   (B) II only

   (C) I and III only

   (D) I, II, and III

7. Evaluate $3x^3 - 2x^2 + x - 1$ if $x = -1$.

   (A) $-7$    (B) $-3$    (C) $-2$    (D) $-1$

8. Eight pencils and five pens cost $5.41, while nine pencils and three pens cost $3.75. What is the cost of a pencil?

   (A) 12¢    (B) 15¢    (C) 38¢    (D) 89¢

9. John sold 3 less than twice the number of pizzas that Sam sold. If Sam sold $x$ pizzas, how many more did John sell?

   (A) $3x - 3$    (B) $2x - 3$

   (C) $x + 3$    (D) $x - 3$

10. What is the value of $(x - y)^3$ if $y = x + 3$?

    (A) $-27$    (B) 0    (C) 6    (D) 9

11. What is the value of $xy$ in the equation $21xy + 77 = 32xy$?

    (A) $\dfrac{-1}{7}$    (B) $\dfrac{1}{7}$    (C) $-7$    (D) 7

12. If $a^3 = 7$, then what is the value of $4a^6$?

    (A) 28    (B) 56    (C) 196    (D) 1372

13. Which is the equation of the line that passes through $(2, -5)$ and is parallel to the graph of $y - 3x = 2$?

    (A) $y = -3x - 11$    (B) $y = 3x - 11$
    (C) $y = 3x - 5$    (D) $y = x - 7$

14. $q * t$ is defined as $q^2 + t^2 + qt$. What is the value of $4 * (-2)$?

    (A) 4    (B) 12    (C) 26    (D) 28

15. What is the value of $c$ if $3x^2 - 12x + c$ is a perfect square?

    (A) 12    (B) 16    (C) 36    (D) 144

16. How many real roots does the equation $2x^2 - 5x - 7 = 0$ have?

    (A) one    (B) two    (C) none
    (D) cannot be determined

17. The number of days in $w$ weeks, $d$ days, and $h$ hours is

    (A) $7w + d + 24h$    (B) $\dfrac{w + d}{7} + 24h$
    (C) $7w + d + \dfrac{h}{24}$    (D) $7w + \dfrac{d - h}{24}$

18. Simplify $\dfrac{\sqrt{108}}{\sqrt{24}}$.

    (A) $\dfrac{3}{\sqrt{2}}$    (B) $\dfrac{6\sqrt{3}}{2\sqrt{6}}$    (C) $\dfrac{\sqrt{3}}{2}$    (D) $\dfrac{3\sqrt{2}}{2}$

19. If $x$, $y$, and $z$ are three odd consecutive integers and $x < y < z$, then $(x - y)(x - z)(y - z) =$

    (A) $-6$    (B) 6    (C) $-16$    (D) 16

20. If $x @ y = x^2 - y^2$ then $3 @ (-2) =$

    (A) 5    (B) 10    (C) 13    (D) 25

21. Jana has scores of 89, 92, 76, and 85 on four tests. What must she score on the next test if she wishes her average to be 86?

    (A) 90    (B) 88    (C) 86    (D) 84

22. A truck and a train leave from the same terminal at 9:35 A.M. If the truck travels at 52 mph and the train travels at 84 mph, how far apart, to the nearest mile, are they at 3:20 P.M.?

    (A) 782    (B) 384    (C) 184    (D) 174

23. The cost of two audio tapes is the same as two-thirds the cost of a compact disk. Mel purchased three audio tapes and two compact disks for $45. What is the cost of two audio tapes?

    (A) $5    (B) $10    (C) $15    (D) $20

24. If $a + b = 7$ and $a^2 - b^2 = -7$, then $b - a =$

    (A) $-1$    (B) 1    (C) 0    (D) 7

# CHAPTER 9

# Conics

## CHAPTER OBJECTIVES

In this chapter, you will:

- Find the distance between two points in a plane.
- Find the midpoint of a segment.
- Identify and graph conic sections from their equations.
- Solve quadratic systems of equations and inequalities.

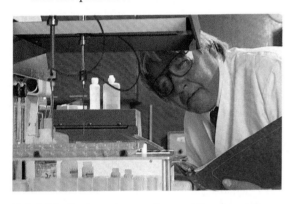

Point A indicates average glucose (blood sugar) level. The red line shows your glucose level after eating an apple. The blue line shows the level after eating a candy bar. Which item will give you more energy over a longer period of time?

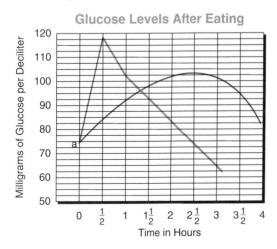

## CAREERS IN MEDICAL TECHNOLOGY

Would you like to be an unseen member of a life-saving team? If you choose this career, you won't be in the spotlight—in fact, the people whose lives you help to save may never see you or know your name. But your special knowledge and careful work will be essential to the doctors and nurses who do see the patients. You'll be a medical technologist.

Medical technologists run many complicated tests. Those who work in large labs may specialize, but if you are a medical technologist in a small lab, you may perform all these functions and more:

- run chemical tests to determine blood sugar levels or cholesterol levels,
- examine tissues to detect the presence of infections or diseases,
- examine blood, tissue, and other body substances with a microscope,
- make cultures of body fluid or tissue samples to determine the presence of viruses, bacteria, fungi, parasites, or other microorganisms, and
- type and cross-match blood samples for transfusions.

No matter what kinds of test you may run, it's always satisfying to know that because of your work the patient's medical picture has become less of a mystery.

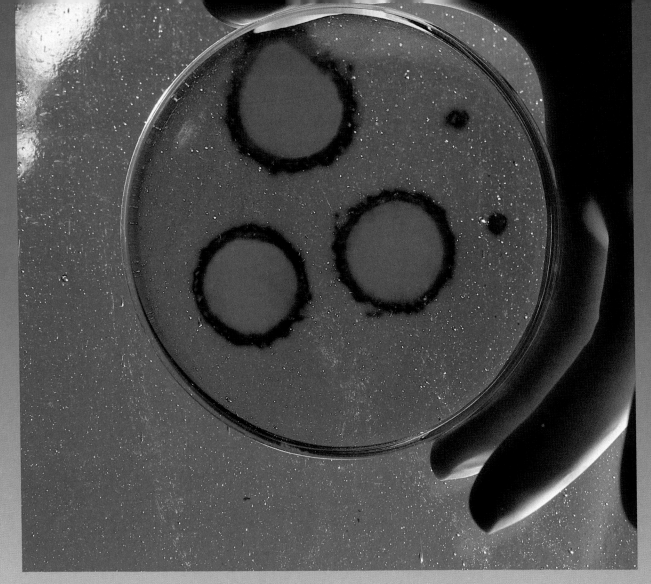

## MORE ABOUT MEDICAL TECHNOLOGY

### Degree Required:

- Bachelor's Degree in one of the Life Sciences

### Some medical technologists like:

- being an important part of the medical field
- the opportunity to work part-time
- working with a minimum of supervision
- being able to specialize in a field of their choice

### Related Math Subjects:

- Advanced Algebra
- Trigonometry
- Statistics
- Probability
- Calculus

### Some medical technologists dislike:

- working in an environment that may have unpleasant odors
- the possibility of infection if cultures are not handled properly
- working nights, weekends, and holidays

For more information on the various careers available in the field of Medical Technology, write to:

American Society for Medical Technology
2021 "L" Street, NW, Suite 400
Washington, D.C. 20036

# The Distance and Midpoint Formulas

**Objectives**

After studying this lesson, you should be able to:

- find the distance between two points in the coordinate plane, and
- find the midpoint of a line segment in the coordinate plane.

**Application**

Mike and Nita are going on a camping vacation to the Great Smoky Mountains National Park. The park appears as B4 on their map. They live in Wilmington, which appears as L1 on the map. If each side of a grid square represents 30 miles, about how far will they have to drive to get to the park?

We can look at the map as a coordinate plane. On a coordinate plane, the location of the park would be represented by the ordered pair (2, 4) instead of B4. The location of Wilmington would be represented by (12, 1) instead of L1. To find the distance between two points on a coordinate plane we can use the Pythagorean Theorem.

First plot the points and draw vertical and horizontal segments from each point to form a right triangle. Then use the Pythagorean Theorem to find the length of the segment between the two points.

*The square of the measure of the hypotenuse of a right triangle equals the sum of the squares of the measures of the other two sides.*

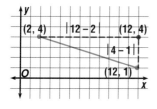

Recall that you can find the distance between two points on a number line using absolute value. That is, the distance between two points whose coordinates are $a$ and $b$ is $|a - b|$ or $|b - a|$. So the length of the horizontal leg is $|12 - 2|$ or 10 units, and the length of the vertical leg is $|4 - 1|$ or 3 units.

$$d^2 = 10^2 + 3^2$$
$$d^2 = 109$$
$$d = \sqrt{109}$$
$$d \approx 10.44 \qquad \textit{Distance is positive.}$$

So Mike and Nita would have to drive about $10.44 \times 30$ or 313.2 miles to get to the park. Of course, the distance we found is in a straight line, and since the roads will not be straight from Wilmington to the park, their actual driving distance will be longer.

Suppose $(x_1, y_1)$ and $(x_2, y_2)$ name two points in the plane. If we form a right triangle by drawing a vertical line through $(x_1, y_1)$ and a horizontal line through $(x_2, y_2)$, the lines will intersect at point $(x_1, y_2)$. *Why?*

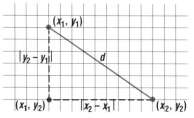

Use the Pythagorean Theorem to find the distance.

$$d^2 = |x_2 - x_1|^2 + |y_2 - y_1|^2$$
$$d^2 = (x_2 - x_1)^2 + (y_2 - y_1)^2 \qquad \text{\textit{Why can } } (x_2 - x_1)^2 \text{ \textit{be substituted for} } |x_2 - x_1|^2?$$
$$d = \sqrt{(x_2 - x_1)^2 + (y_2 - y_1)^2}$$

| *Distance Formula for Two Points in a Plane* | **The distance between two points with coordinates $(x_1, y_1)$ and $(x_2, y_2)$ is given by $d = \sqrt{(x_2 - x_1)^2 + (y_2 - y_1)^2}$.** |
|---|---|

**Example 1**

Use the distance formula to find the distance between (−3, 2) and (9, 4).

$$d = \sqrt{(x_2 - x_1)^2 + (y_2 - y_1)^2}$$
$$= \sqrt{((-3) - 9)^2 + (2 - 4)^2}$$
$$= \sqrt{(-12)^2 + (-2)^2}$$
$$= \sqrt{144 + 4}$$
$$= \sqrt{148} \text{ or } 2\sqrt{37} \qquad \text{The distance is } 2\sqrt{37} \text{ or about } 12.17 \text{ units.}$$

**Example 2**

**CONNECTION**

**Geometry**

**Show that $P(4, 2)$ is the midpoint of the segment from $A(9, 3)$ to $B(−1, 1)$.**

First we must prove that point $P$ is on segment $AB$. It is sufficient to show that segments $AP$ and $AB$ have the same slope. *Why?*

slope of $\overline{AP} = \dfrac{3 - 2}{9 - 4}$       slope of $\overline{AB} = \dfrac{3 - 1}{9 - (-1)}$

$\qquad\qquad = \dfrac{1}{5}$ $\qquad\qquad\qquad\qquad = \dfrac{2}{10}$

$\qquad\qquad\qquad\qquad\qquad\qquad\qquad = \dfrac{1}{5}$

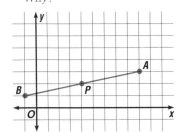

Next we need to show that the distance from $A$ to $P$ equals the distance from $B$ to $P$.

*distance from A to P*        *distance from B to P*

$$d_{\overline{AP}} = \sqrt{(9 - 4)^2 + (3 - 2)^2} \qquad\qquad d_{\overline{BP}} = \sqrt{(-1 - 4)^2 + (1 - 2)^2}$$
$$= \sqrt{5^2 + 1^2} \qquad\qquad\qquad\qquad\qquad = \sqrt{(-5)^2 + (-1)^2}$$
$$= \sqrt{25 + 1} \text{ or } \sqrt{26} \qquad\qquad\qquad = \sqrt{25 + 1} \text{ or } \sqrt{26}$$

Since the distances are equal, $P$ is the midpoint.

| *Midpoint of a Line Segment* | If a line segment has endpoints at $(x_1, y_1)$ and $(x_2, y_2)$ then the midpoint of the line segment has coordinates $\left(\dfrac{x_1 + x_2}{2}, \dfrac{y_1 + y_2}{2}\right)$. |
| --- | --- |

### Example 3

**CONNECTION**

**Geometry**

Find the center of the circle $Q$ with a diameter whose endpoints are $(-3, 10)$ and $(7, -2)$.

The center of the circle will be the midpoint of the diameter.

$$(x, y) = \left(\frac{-3 + 7}{2}, \frac{10 + (-2)}{2}\right)$$

$$= \left(\frac{4}{2}, \frac{8}{2}\right)$$

$$= (2, 4) \qquad \text{The center of circle } Q \text{ is } (2, 4).$$

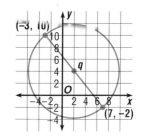

# CHECKING FOR UNDERSTANDING

**Communicating Mathematics**

**Read and study the lesson to answer these questions.**

1. Why were we able to substitute $(x_2 - x_1)^2$ for $|x_2 - x_1|^2$ when we derived the distance formula?

2. Is a point that is equidistant from two points always the midpoint of the segment between them? Explain your answer.

3. Could you find one endpoint of a segment given the other endpoint and the midpoint? If so, demonstrate by finding the endpoint of the segment with one endpoint $(3, 7)$ and midpoint $(0, -2)$.

**Guided Practice**

**The coordinates of two points on the number line are given. Find the distance between each pair of points.**

4. $-8, 9$      5. $-3, 0$     6. $-30, -15$

7. $-12.2, -3.3$     8. $4.4, -8.1$     9. $10\frac{2}{3}, -6\frac{1}{8}$

**Use the distance formula to find the distance between each pair of points.**

10. $(6, 3), (6, -1)$     11. $(9, 5), (4, -7)$     12. $(0, -5), (10, -3)$

# EXERCISES

**Practice**

**Use the distance formula to find the distance between each pair of points.**

13. $(4, 8), (8, 8)$     14. $(0, 5), (-4, -2)$

15. $(9, 0), (6, 7)$     16. $(-4, 9), (1, -3)$

17. $(-4, -10), (-3, -11)$     18. $\left(1, \frac{1}{2}\right), \left(\frac{1}{3}, -2\right)$

19. $(-0.5, 1), (-2.2, -0.3)$     20. $(3, 3), (\sqrt{3}, \sqrt{3})$

21. $(-2\sqrt{7}, 10), (4\sqrt{7}, 8)$     22. $(2\sqrt{3}, 4\sqrt{3}), (2\sqrt{3}, -\sqrt{3})$

**Find the value of *c* such that each pair of points is 5 units apart.**

23. $(7, 2)$, $(3, c)$

24. $(-7, 7)$, $(c, 11)$

25. $(13, 10.1)$, $(9, c)$

26. $(1.2, 5.9)$, $(c, 1.9)$

**Find the midpoint of each line segment whose endpoints are given below.**

27. $(8, 3)$, $(16, 7)$

28. $(5, 9)$, $(12, 18)$

29. $\left(\frac{1}{4}, 3\right)$, $\left(\frac{7}{8}, -\frac{1}{2}\right)$

30. $(-4.3, 2.8)$, $(2.7, 4.9)$

CONNECTION

Geometry

**Solve. Use a drawing.**

31. Find the center of the circle whose diameter has endpoints $(9, 0)$ and $(11, -14)$.

32. Find the perimeter of a quadrilateral with vertices at $(4, 5)$, $(-4, 6)$, $(-5, -8)$, and $(6, 3)$.

33. Triangle *MNO* has vertices $M(3, 5)$, $N(-2, 8)$, and $O(7, -4)$. Find the coordinates of the midpoint of each side.

34. Find the lengths of the diagonals of the parallelogram with vertices at $(-14, 8)$, $(6, 8)$, $(-12, -2)$, and $(8, -2)$.

35. Show that the triangle with vertices $A(-3, 0)$, $B(-1, 4)$ and $C(1, -2)$ is isosceles.

36. Triangle *XYZ* is a right triangle with vertices at $X(0, 1)$, $Y(4, 1)$, and $Z(0, 7)$. Show that the midpoint of the hypotenuse is the same distance from each vertex.

**Journal**

Draw a rectangle on a piece of graph paper. Tell how you could find the point where the diagonals meet by using the formulas in this lesson.

**Critical Thinking**

37. Find the coordinates of a point one-fourth of the distance from $A(11, 2)$ to $B(3, -2)$.

**Applications**

38. **Transportation** A semi-truck traveled 23 miles south on I-71 before turning west on I-70. After traveling 85 miles on I-70, about how far is it from its starting point?

39. **Manufacturing** The hole in a record album needs to be placed in the center of the record. If the record is 12 inches in diameter, and the hole is $\frac{1}{4}$ inch in diameter, how far from the edge should the edge of the hole be placed?

— 12 in. —

**Mixed Review**

**Solve each inequality. (Lesson 8-9)**

40. $(x + 3)(x + 7) > 0$

41. $(a - 1.5)(a + 2.5) \geq 0$

42. Solve the equation $y^4 - y^2 - 30 = 0$. **(Lesson 7-6)**

43. Find values of *x* and *y* for which the sentence $2x + 5yi = 4 + 15i$ is true. **(Lesson 6-9)**

44. Factor $3m^3 + 24p^3$. **(Lesson 5-6)**

## 9-2 Parabolas

**Objectives**

After studying this lesson, you should be able to:

- write equations of parabolas, and
- graph parabolas from given information.

**Application**

*FYI ...*

Drawings showing the reflective properties of a parabola were found among the papers of Leonardo Da Vinci, a gifted scientist and architect as well as a noted sculptor and painter.

The shape of the reflectors in flashlights and searchlights is based on the **parabola.** If you were to cut one of these reflectors down the middle, the cross section would look like the diagram at the right. The light source is placed at a point where the light is reflected in parallel rays. So, the light coming out of the reflector is a straight beam.

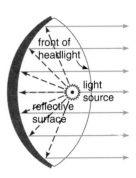

The point where the light source is placed is called the **focus.** A parabola can be defined in terms of the location of its focus and a line called the **directrix.**

*Definition of a Parabola*

> **A parabola is the set of all points in a plane that are the same distance from a given point called the *focus* and a given line called the *directrix*.**

*The directrix is named by the equation of the line.*

The parabola at the right has focus (5, 3) and directrix $y = -1$. We can use the distance formula with the definition of a parabola to find the equation of this parabola. Let $(x, y)$ be a point on the parabola. The distance from this point to the focus must be the same as its distance from the directrix.

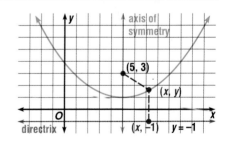

*distance between (x, y) and (5, 3) = distance between (x, y) and (x, −1)*

$$\sqrt{(x - 5)^2 + (y - 3)^2} = \sqrt{(x - x)^2 + (y - (-1))^2}$$
$$(x - 5)^2 + (y - 3)^2 = (x - x)^2 + (y + 1)^2$$
$$(x - 5)^2 + y^2 - 6y + 9 = y^2 + 2y + 1$$
$$(x - 5)^2 + 8 = 8y$$
$$\frac{1}{8}(x - 5)^2 + 1 = y$$

The equation of a parabola with focus (5, 3) and directrix $y = -1$ is $y = \frac{1}{8}(x - 5)^2 + 1$. The axis of symmetry of this parabola is $x = 5$. Notice that the axis of symmetry and the directrix are perpendicular. The point of intersection of the axis of symmetry and the parabola is called the **vertex.** The vertex of this parabola is (5, 1).

The line segment through the focus of a parabola and perpendicular to the axis of symmetry whose endpoints are on the parabola is called the **latus rectum**. In the figure at the right the latus rectum is $\overline{AB}$. The length of the latus rectum of the parabola whose equation is

$y = a(x - h)^2 + k$ is $\left|\dfrac{1}{a}\right|$ units. The endpoints of the latus rectum are $\left|\dfrac{1}{2a}\right|$ units from the focus.

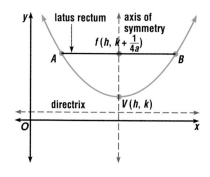

You can write the equation for a parabola in the form $y = a(x - h)^2 + k$ or in the form $x = a(y - k)^2 + h$. Either form gives us valuable information about the graph of the parabola.

| Information about Parabolas | | |
|---|---|---|
| form of equation | $y = a(x - h)^2 + k$ | $x = a(y - k)^2 + h$ |
| axis of symmetry | $x = h$ | $y = k$ |
| vertex | $(h, k)$ | $(h, k)$ |
| focus | $\left(h, k + \dfrac{1}{4a}\right)$ | $\left(h + \dfrac{1}{4a}, k\right)$ |
| directrix | $y = k - \dfrac{1}{4a}$ | $x = h - \dfrac{1}{4a}$ |
| direction of opening | upward if $a > 0$, downward if $a < 0$ | right if $a > 0$, left if $a < 0$ |
| length of latus rectum | $\left|\dfrac{1}{a}\right|$ units | $\left|\dfrac{1}{a}\right|$ units |

**Example 1**

Graph $y = \dfrac{1}{8}(x - 1)^2 + 4$.

Use the chart above to find as much information about the graph as possible. Use the information to graph the parabola.

vertex: $(1, 4)$

axis of symmetry: $x = 1$

focus: $(1, 4 + 2)$ or $(1, 6)$ $\qquad \dfrac{1}{4a} = \dfrac{1}{4\left(\frac{1}{8}\right)}$ or $2$

directrix: $y = 4 - 2$ or $y = 2$

direction of opening: upward since $a > 0$

length of latus rectum: $\left|\dfrac{1}{\frac{1}{8}}\right|$ or $8$ units

**Example 2**

**Graph $6x = y^2 - 6y + 39$.**

First write the equation in the form $x = a(y - k)^2 + h$.

$6x = y^2 - 6y + 39$

$6x = y^2 - 6y + \square \quad \square + 39$  *Complete the square.*

$6x = y^2 - 6y + \left(\dfrac{6}{2}\right)^2 - \left(\dfrac{6}{2}\right)^2 + 39$

$6x = (y - 3)^2 + 30$

$x = \dfrac{1}{6}(y - 3)^2 + 5$

vertex: $(5, 3)$    focus: $\left(6\dfrac{1}{2}, 3\right)$

axis of symmetry: $y = 3$

directrix: $x = 3\dfrac{1}{2}$

direction of opening: right $(a > 0)$

length of

   latus rectum: $\left|\dfrac{1}{\frac{1}{6}}\right|$ or 6 units

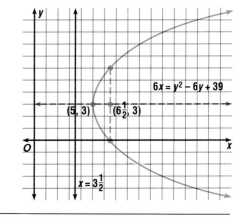

**Example 3**

APPLICATION

Communication

A microphone is placed at the focus of a parabolic reflector to collect sounds for the television broadcast of a football game. The focus of the parabola that is the cross section of the reflector is 5 inches from the vertex. The latus rectum is 20 inches long. Assuming that the focus is at the origin and the parabola opens to the right, write the equation of the cross section.

focus:    $(0, 0)$    *Since the parabola opens to*
vertex:   $(-5, 0)$   *the right, the vertex must*
                      *be to the left of the focus.*

measure of latus rectum $= 20 = \left|\dfrac{1}{a}\right|$, so $a = \dfrac{1}{20}$

Write the equation using the form $x = a(y - k)^2 + h$.

$x = \dfrac{1}{20}y^2 - 5$

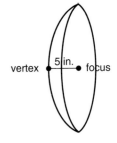

# CHECKING FOR UNDERSTANDING

**Communicating Mathematics**

Read and study the lesson to answer these questions.

1. Name some uses for parabolic reflectors.

2. Describe the relationships between the directrix, the vertex, the axis of symmetry, and the latus rectum.

3. The graph of the parabola whose equation is $x = 2(y + 2)^2 - 1$ is shown at the right. Name the vertex, the axis of symmetry, the focus, and the directrix of the parabola.

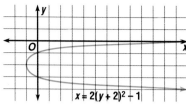

$x = 2(y + 2)^2 - 1$

Find the value of $c$ that makes each trinomial a perfect square.

4. $x^2 - 6x + c$

5. $y^2 + 4y + c$

6. $a^2 - 8a + c$

7. $n^2 - 10n + c$

8. $q^2 + 3q + c$

9. $s^2 - 7s + c$

Express each equation in the form $y = a(x - h)^2 + k$.

10. $12y = x^2$

11. $x^2 = -4y$

12. $y = x^2 + 8x + 20$

13. $y = x^2 + 4x + 1$

Name the vertex, axis of symmetry, focus, directrix, and direction of opening of the parabola whose equation is given. Then find the length of the latus rectum.

14. $y^2 = 6x$

15. $(x + 2)^2 = y - 3$

16. $6x = (y + 2)^2$

17. $x^2 = (y - 1)$

# EXERCISES

Name the vertex, axis of symmetry, focus, directrix, and direction of opening of the parabola whose equation is given. Then find the length of the latus rectum and graph the parabola.

18. $-8y = x^2$

19. $4(y - 2) = (x - 4)^2$

20. $(x + 3)^2 = \frac{1}{4}(y - 2)$

21. $\frac{1}{2}(y + 1) = (x - 8)^2$

22. $4(x - 2) = (y + 3)^2$

23. $(y - 8)^2 = -4(x - 4)$

24. $y = x^2 + 8x + 20$

25. $x = y^2 - 14y + 25$

26. $y = x^2 - 6x + 33$

27. $y = \frac{1}{2}x^2 - 3x + \frac{19}{2}$

28. $y = x^2 + 4x + 1$

29. $x = \frac{1}{4}y^2 - \frac{1}{2}y - 3$

30. $x = 5y^2 - 25y + 60$

31. $y = 3x^2 - 24x + 50$

The focus and directrix of a parabola are given. Write an equation for each parabola. Then draw the graph.

32. $(3, 8)$, $y = 4$

33. $(8, 0)$, $y = 4$

34. $(5, 5)$, $y = -3$

35. $(6, 2)$, $x = 4$

36. $(4, -3)$, $y = 6$

37. $(3, -1)$, $x = -2$

38. $(3, 0)$, $x = -2$

39. $(10, -3)$, $x = 5$

**Write the equation of each parabola described below. Then draw the graph.**

40. vertex (0, 0), focus (0, 3)
41. vertex (5, -1), focus (3, -1)
42. vertex (-7, 4), axis $n = -7$, measure of latus rectum 6, $a < 0$
43. vertex (4, 3), axis of symmetry $y = 3$, measure of latus rectum 4, $a > 0$
44. focus (11, -1), directrix $y = 2$
45. focus (3, 6), directrix $x = 5$
46. focus (7, -7), directrix $x = -2$
47. focus (-1, 2), directrix $y = -1$

**Critical Thinking**

48. There are two parabolas that have vertex (-3, -4) and pass through the point (-1, 4). Find their equations.

**Applications**

49. **Electronics**  The headlights on a car contain parabolic reflectors. A special light bulb with two filaments is used to produce the high and low beams. The filament placed at the focus produces the high beam and low beam is produced by the filament placed off of the focus. If the equation of the parabola that is the cross section of the reflector is $y = \frac{1}{10}x^2$, where should the filament for the high beam be placed?

50. **Engineering**  The shape taken by the supporting cables for a suspension bridge is actually a catenary curve, which very closely approximates a parabola. The cables for a certain bridge are 75 feet above the road at the towers and 10 feet above it at the center of the bridge. Find the equation of the path of the supporting cables if the distance between the towers is 250 feet.

**Mixed Review**

51. Use the distance formula to find the distance between (9, 6) and (8, 0).  **(Lesson 9-1)**

52. Find the midpoint of the segment whose endpoints are (9, 3) and (-6, -8).  **(Lesson 9-1)**

53. **Geometry**  A wire 36 centimeters long is cut into two pieces and each piece is bent into a square. How long should each piece be cut so that the sum of the areas of the two squares is minimal?  **(Lesson 8-5)**

54. Find two consecutive integers whose product is 7832.  **(Lesson 7-2)**

55. Simplify $\dfrac{\frac{3}{4}}{\frac{5}{x}}$.  **(Lesson 6-4)**

56. Divide $(a^4 - 5a^3 - 13a^2 + 53a + 60)$ by $(a + 1)$ using synthetic division.  **(Lesson 5-7)**

57. State whether the expression $-24p^4q$ is a monomial. If it is, name its coefficient and degree.  **(Lesson 5-1)**

## 9-3 Circles

**Objectives**

After studying this lesson, you should be able to:
- write equations of circles, and
- draw a circle having certain properties.

**Application**

Dean and Shina are flying a model airplane on a line. They control the flight of the plane with the 25-foot connecting wire. The plane's path is a circle with radius 25 feet and the controller at the center.

*The measure of a radius is also called a <u>radius</u>.*

A circle is the set of all points in a plane that are equidistant from a given point, called the **center.** Any segment whose endpoints are the center and a point on the circle is a **radius** of the circle.

Let's use the distance formula and the definition of a circle to find the equation of a circle with radius 8 units and center $(4, -5)$. Let $(x, y)$ represent any point on the circle.

*distance between $(x, y)$ and $(4, -5) = 8$*
$$\sqrt{(x - 4)^2 + (y - (-5))^2} = 8 \qquad \textit{Apply the distance formula.}$$
$$(x - 4)^2 + (y - (-5))^2 = 64 \qquad \textit{Square each side.}$$
$$(x - 4)^2 + (y + 5)^2 = 64 \qquad \textit{Simplify.}$$

The equation of the circle with center $(4, -5)$ and radius 8 units is $(x - 4)^2 + (y + 5)^2 = 64$.

| *Equation of a Circle with Center (h, k)* | **The equation of a circle with center $(h, k)$ and radius $r$ units is** $$(x - h)^2 + (y - k)^2 = r^2.$$ |
| --- | --- |

**Example 1**

**APPLICATION**

**Seismology**

An earthquake observation station that is located 30 miles east and 25 miles north of the central station detected that an earthquake occurred in the area. It is estimated that the epicenter was 50 miles away. Write the equation of the set of points that could be the epicenter and graph the solution set.

Assume that the central station is located at the origin. Any point that is 50 miles away from the station could be the epicenter. So the circle whose center is the observation station and whose radius is 50 miles is the solution set. The center of the circle is $(30, 25)$. The equation of the circle is $(x - 30)^2 + (y - 25)^2 = 50^2$.

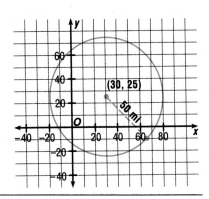

The equation $x^2 + y^2 + 2x - 12y = 35$ also describes a circle. If we were to complete the square for each variable and write the equation in the form $(x - h)^2 + (y - k)^2 = r^2$, the result would be $(x + 1)^2 + (y - 6)^2 = 72$. This circle has center $(-1, 6)$ and radius $\sqrt{72}$ or $6\sqrt{2}$ units.

**Example 2**

Find the center and radius of the circle whose equation is $x^2 + y^2 - 3x + 8y = 20$. Then graph the circle.

$$x^2 + y^2 - 3x + 8y = 20$$
$$x^2 - 3x + \square + y^2 + 8y + \blacksquare = 20 + \square + \blacksquare$$
$$x^2 - 3x + \frac{9}{4} + y^2 + 8y + 16 = 20 + \frac{9}{4} + 16$$
$$\left(x - \frac{3}{2}\right)^2 + (y + 4)^2 = \frac{153}{4}$$

The circle has center $\left(\frac{3}{2}, -4\right)$ and

radius $\sqrt{\frac{153}{4}} = \frac{3\sqrt{17}}{2}$ or about 6.2 units.

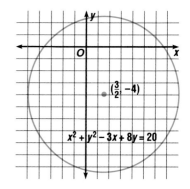

# CHECKING FOR UNDERSTANDING

**Communicating Mathematics**

Read and study the lesson to answer these questions.

1. If the restriction "in a plane" were taken out of the definition of the circle, would the figure change? If so, how?

2. Describe the similarities and differences between the graphs of $(x + 3)^2 + (y - 4)^2 = 16$ and $(x - 3)^2 + (y - 2)^2 = 16$.

3. Concentric circles are circles with the same center, but not necessarily the same radius. Write the equations of two concentric circles. Graph the circles.

4. The circle whose equation is $(x - a)^2 + (y - b)^2 = 25$ has its center in the second quadrant. What do you know about $a$ and $b$?

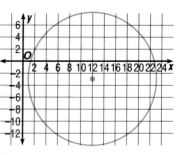

5. Find the equation of the circle graphed at the right. State the center and radius of the circle.

**Guided Practice**

State whether the graph of each equation is a circle or a parabola.

6. $x^2 + y^2 = 36$

7. $x^2 + 4x + 4 = 9y + 27$

8. $y^2 = 6x - 4$

9. $x^2 + y^2 = 7x - 5$

10. $y^2 + y + x^2 = 12 - 3x$

11. $y^2 = 5x$

State the center and radius of each circle whose equation is given.

**12.** $x^2 + y^2 = 49$

**13.** $x^2 + (y - 3)^2 = 36$

**14.** $(x - 4)^2 + (y - 1)^2 = 9$

**15.** $(x + 2)^2 + \left(y - \frac{2}{3}\right)^2 = 8$

**16.** $(x - 4)^2 + y^2 = \dfrac{16}{25}$

**17.** $x^2 + (y + 5)^2 = \dfrac{81}{64}$

Find the center and radius of each circle whose equation is given. Then draw the graph.

**18.** $x^2 + y^2 = 9$

**19.** $(x + 5)^2 + y^2 = 64$

**20.** $(x + 6)^2 + (y - 5)^2 = 16$

**21.** $(x - 3)^2 + (y + 2)^2 = 169$

# EXERCISES

**Practice**

Find the center and radius of each circle whose equation is given. Then draw the graph.

**22.** $x^2 + (y - 4)^2 = 49$

**23.** $x^2 + (y + 2)^2 = 4$

**24.** $(x - 3)^2 + y^2 = 16$

**25.** $x^2 + y^2 = 144$

**26.** $(x - 3)^2 + (y - 1)^2 = 25$

**27.** $(x + 2)^2 + (y - 1)^2 = 81$

**28.** $(x + 3)^2 + (y + 7)^2 = 81$

**29.** $(x - 4)^2 + (y - 9)^2 = 4$

**30.** $x^2 + y^2 + 6x - 2y - 15 = 0$

**31.** $x^2 + y^2 - 18x - 18y + 53 = 0$

**32.** $x^2 + y^2 + 14x + 6y = 23$

**33.** $x^2 + y^2 + 8x - 6y = 0$

**34.** $x^2 - 12x + 84 = -y^2 + 16y$

**35.** $x^2 + y^2 = 4x + 9$

**36.** $x^2 + y^2 - 6y - 16 = 0$

**37.** $3x^2 + 3y^2 + 6y + 9x = 2$

**38.** $x^2 + y^2 + 9x - 8y + 4 = 0$

**39.** $4x^2 + 4y^2 + 36y + 5 = 0$

**40.** $x^2 + y^2 + 4x - 8 = 0$

**41.** $x^2 + y^2 + 2x + 4y - 9 = 0$

**42.** $x^2 + 14x + y^2 + 6y + 50 = 0$

**43.** $x^2 + y^2 + 2x - 10 = 0$

Write an equation for each circle whose center and radius are given.

**44.** (4, 1), 4 units

**45.** (0, 3), 7 inches

**46.** (5, 0), 8 meters

**47.** (-1, -5), 2 cm

**48.** (-8, 7), $\dfrac{1}{2}$ km

**49.** (-3, -9), $\dfrac{5}{6}$ yd

Write the equation of each circle described below.

**50.** The circle has center (4, -2) and passes through (5, 3).

**51.** The circle has center (3, 3) and passes through the origin.

**52.** The endpoints of a diameter of the circle are (2, 5) and (2, -1).

**53.** The endpoints of a diameter of the circle are (-4, 12) and (2, 0).

**54.** The circle has center (-3, 8) and is tangent to the x-axis.

*A circle tangent to a line intersects that line in exactly one point.*

**55.** The center of the circle is (4, -3) and the circle is tangent to the y-axis.

**Critical Thinking**

**56.** Write the equation of a circle that is tangent to both the x- and y-axes.

**57. Air Traffic Control** The control tower for the municipal airport is located at (4, 16) on the county map. The radar used in tracking airplanes can detect a plane up to 13 miles away. Write an equation for the position of the most distant plane that the tower can detect in terms of the county map.

**58. Physics** The model plane described in the beginning of the lesson makes one revolution in 8 seconds. How fast is the model travelling in feet per second? *(Hint: Recall that the formula for the circumference of a circle is $C = 2\pi r$.)*

**59. City Planning** The officials of the city of Westerville are planning the placement of some new fire stations. They want to ensure that all areas of the city are within 5 miles of a fire station. If the city is approximately 10 miles square, will they be able to cover the entire city with two fire stations? *Hint: Draw a diagram.*

**Mixed Review**

**60.** Change the equation $y^2 = 6x$ to the form $x = a(y - k)^2 + h$. **(Lesson 9-2)**

**61.** Write an equation for the parabola with focus (2, 4) and directrix $y = 6$. Then draw the graph. **(Lesson 9-2)**

**62.** Is (3, −5) a solution for the inequality $y \geq x^2 - 16$? **(Lesson 8-7)**

**63.** Express the equation $y^{-6} + 4y^{-3} - 32 = 0$ in quadratic form. **(Lesson 7-6)**

**64.** Solve $3x\sqrt{3} + 2 = 0$. **(Lesson 6-7)**

**65.** Factor $4k^2 + 26k + 30$. **(Lesson 5-5)**

**66.** Find the inverse (if one exists) of the matrix $\begin{bmatrix} 8 & 0 \\ -1 & -4 \end{bmatrix}$. **(Lesson 4-5)**

---

## HISTORY CONNECTION

### The Dogon of Mali

Some 500 to 700 years ago, the astronomer priests of the Dogon of the Republic of Mali in West Africa began accumulating detailed observations and knowledge of the universe that equal modern science. They knew, in detail, of the elliptical paths of the planets, the rings of Saturn, the moons of Jupiter, and the spiral structure of the Milky Way Galaxy. They knew that Earth's moon was barren and described it as dry and dead. Amazingly enough, all these observations were made with the naked eye without the aid of modern tools.

Every 60 years, when the orbits of Jupiter and Saturn converse, the Dogon held a ceremony to the star Sirius. Actually they celebrated Sirius B, a smaller star, that revolves around Sirius A, the brightest star in the sky. They knew that this dwarf star had an elliptical orbit around Sirius A which took 50 years to complete. Modern scientists have not been able to confirm or deny this incredible observation. They know only of Sirius B's existence, which can only be seen with high-powered telescopes.

# Ellipses

**Objectives**

After studying this lesson, you should be able to:

- write equations of ellipses, and
- draw an ellipse having certain properties.

**Application**

*You will read more about this chamber later in this lesson.*

The Statuary Hall in The United States Capitol is shaped like an **ellipse.** In this chamber, a person standing at one of two points called the **foci** of the ellipse can hear someone who is standing on the other **focus point** whispering, even though that point is more than 80 feet away.

| *Definition of Ellipse* | An ellipse is the set of all points in a plane such that the sum of the distances from two given points in the plane, called the *foci*, is constant. |
|---|---|

You can draw an ellipse using two tacks and a piece of string. Push the two tacks into a piece of cardboard. Tie a knot in the piece of string and loop it around the tacks as shown at the right. Place your pencil in the loop and draw around the tacks, making sure to keep the string tight. The foci of your ellipse are the points where the tacks are placed.

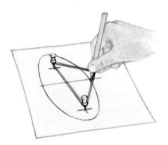

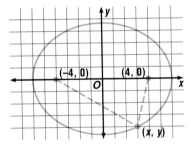

The ellipse at the left has foci $(-4, 0)$ and $(4, 0)$, and the sum of the distances to any point $(x, y)$ on the ellipse to the foci is 12 units. Let's use the distance formula and the definition of an ellipse to find the equation of this ellipse.

Let $(x, y)$ be any point on the ellipse. The sum of the distance between $(x, y)$ and $(-4, 0)$ and the distance between $(x, y)$ and $(4, 0)$ is 12 units.

$$\underset{\substack{\text{distance between} \\ (x, y) \text{ and } (-4, 0)}}{} + \underset{\substack{\text{distance between} \\ (x, y) \text{ and } (4, 0)}}{} = 12$$

$$\sqrt{(x + 4)^2 + y^2} + \sqrt{(x - 4)^2 + y^2} = 12$$

$$\sqrt{(x + 4)^2 + y^2} = 12 - \sqrt{(x - 4)^2 + y^2} \quad \textit{Now square each side.}$$

$$(x + 4)^2 + y^2 = 144 - 24\sqrt{(x - 4)^2 + y^2} + (x - 4)^2 + y^2$$

$$4x - 36 = -6\sqrt{(x - 4)^2 + y^2} \quad \textit{Simplify.}$$

$$16x^2 - 288x + 1296 = 36[(x - 4)^2 + y^2] \quad \textit{Square each side.}$$

$$720 = 20x^2 + 36y^2 \quad \textit{Simplify.}$$

$$1 = \frac{x^2}{36} + \frac{y^2}{20} \quad \textit{Divide by 720.}$$

The equation of the ellipse with foci $(-4, 0)$ and $(4, 0)$ and with 12 units as the sum of the distances from the foci is $\frac{x^2}{36} + \frac{y^2}{20} = 1$.

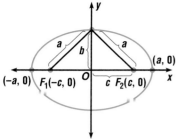

The length of the major axis is 2a. The length of the minor axis is 2b. Notice $a > b$.

There are two axes of symmetry in an ellipse. The points where the ellipse intersects the axes define two line segments whose endpoints lie on the ellipse. The longer segment is called the major axis and the shorter is the minor axis. The foci always lie on the major axis. The intersection of the two axes is the **center** of the ellipse.

Study the ellipse at the left. The sum of the distances from the foci to any point on the ellipse is 2a units. The distance from the center to either focus is $c$ units. By using the Pythagorean Theorem, we can conclude that $b^2 = a^2 - c^2$. If we found the equation of this general ellipse, we would find the standard equation of an ellipse.

**Standard Equation of Ellipse with Center at the Origin**

For ellipses, $a^2 > b^2$

If an ellipse has foci at $(-c, 0)$ and $(c, 0)$ and if the sum of the distances from the foci to any point on the ellipse is 2a units, then the standard equation of the ellipse is $\dfrac{x^2}{a^2} + \dfrac{y^2}{b^2} = 1$, where $b^2 = a^2 - c^2$.

If an ellipse has foci at $(0, -c)$ and $(0, c)$ and if the sum of the distances from the foci to any point on the ellipse is 2a units, then the standard equation of the ellipse is $\dfrac{x^2}{b^2} + \dfrac{y^2}{a^2} = 1$, where $b^2 = a^2 - c^2$.

**Example 1**

**APPLICATION**

**Aeronautics**

A satellite is in an elliptical orbit with the center of Earth at one focus. The major axis of the orbit is 28,900 miles long and the center of the Earth is 8000 miles from the center of the ellipse. Assuming that the center of the ellipse is the origin and the foci lie on the x-axis, write the equation of the path of the satellite.

We know the length of the major axis is 28,900. So we can find $a$.

$$2a = 28,900$$
$$a = 14,450$$

Since Earth is at a focus point and it is 8000 miles from the center, $c = 8000$.

$$b^2 = a^2 - c^2$$
$$b^2 = 14,450^2 - 8000^2$$
$$b^2 = 144,802,500$$

Now we can write the equation.

$$\frac{x^2}{a^2} + \frac{y^2}{b^2} = 1$$
$$\frac{x^2}{(14,450^2)} + \frac{y^2}{144,802,500} = 1$$
$$\frac{x^2}{208,802,500} + \frac{y^2}{144,802,500} = 1$$

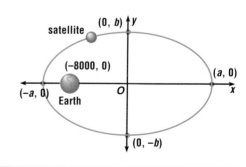

**Example 2** | **Write the equation of the ellipse shown below.**

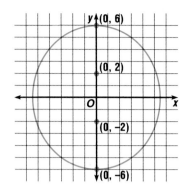

First find the length of the major axis. The distance between $(0, -6)$ and $(0, 6)$ is 12 units.

$$2a = 12$$
$$a = 6$$

Since the foci are at $(0, -2)$ and $(0, 2)$, $c = 2$.

$$b^2 = a^2 - c^2$$
$$b^2 = 6^2 - 2^2$$
$$b^2 = 32$$

Now write the equation.

$$\frac{x^2}{b^2} + \frac{y^2}{a^2} = 1 \qquad \textit{The major axis is vertical.}$$

$$\frac{x^2}{32} + \frac{y^2}{6^2} = 1 \quad \text{or} \quad \frac{x^2}{32} + \frac{y^2}{36} = 1$$

In the equation of an ellipse, $a^2 > b^2$. So it is easy to decide whether the foci are on the $x$-axis or the $y$-axis just by looking at the equation. If $a^2$ is the denominator of the $x^2$ term, the foci are on the $x$-axis. If $a^2$ is the denominator of the $y^2$ term, the foci are on the $y$-axis.

**Example 3** | **Find the foci and the lengths of the major and minor axes of the ellipse whose equation is $4x^2 + y^2 = 16$. Then draw the graph.**

$$4x^2 + y^2 = 16 \qquad \textit{Write the equation in standard form.}$$
$$\frac{x^2}{4} + \frac{y^2}{16} = 1$$

Since $16 > 4$, the foci are on the $y$-axis, $a = 4$, and $b = 2$.

$$b^2 = a^2 - c^2$$
$$4 = 16 - c^2$$
$$c^2 = 12$$
$$c = \sqrt{12} \text{ or } 2\sqrt{3} \qquad 2\sqrt{3} \approx 3.5$$

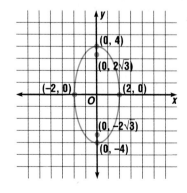

The foci are at $(0, 2\sqrt{3})$ and $(0, -2\sqrt{3})$. The major axis is $2a$ or 8 units long and the minor axis is $2b$ or 4 units long.

An equation in the form $\frac{x^2}{a^2} + \frac{y^2}{b^2} = 1$ or $\frac{x^2}{b^2} + \frac{y^2}{a^2} = 1$ represents an ellipse with center at the origin. If an ellipse has center at a point $(h, k)$, its equation would be similar with $(x - h)$ replacing $x$ and $(y - k)$ replacing $y$.

**Example 4**

Graph $\dfrac{(x + 1)^2}{36} + \dfrac{(y - 2)^2}{9} = 1$.

The graph is the same shape as the graph of $\dfrac{x^2}{36} + \dfrac{y^2}{9} = 1$, but with center translated 1 unit to the left and 2 units up. The center is at $(-1, 2)$ instead of the origin.

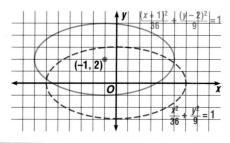

The graph of $2x^2 + y^2 - 4x + 8y - 6 = 0$ is an ellipse. Complete the square for each variable, like you did with equations of circles, to write the equation in standard form.

$$2x^2 + y^2 - 4x + 8y - 6 = 0$$
$$2(x^2 - 2x + \square) + y^2 + 8y + \blacksquare = 6 + 2\square + \blacksquare \quad \textit{Complete the square.}$$
$$2(x^2 - 2x + 1) + y^2 + 8y + 16 = 6 + 2(1) + 16$$
$$2(x - 1)^2 + (y + 4)^2 = 24$$
$$\frac{(x - 1)^2}{12} + \frac{(y + 4)^2}{24} = 1$$

The ellipse has center at $(1, -4)$. Since $24 > 12$, $a^2 = 24$ and $b^2 = 12$ and it has a vertical major axis. So the major axis is $2a$ or $2\sqrt{24}$ (or $4\sqrt{6}$) units long and the minor axis is $2b$ or $2\sqrt{12}$ (or $4\sqrt{3}$) units long.

**Example 5**

Find the center, foci, and lengths of the major and minor axes of the ellipse whose equation is $x^2 + 25y^2 - 8x + 100y + 91 = 0$. Then draw the graph.

$$x^2 + 25y^2 - 8x + 100y + 91 = 0$$
$$x^2 - 8x + \square + 25(y^2 + 4y + \blacksquare) = -91 + \square + 25(\blacksquare) \quad \textit{Complete the square.}$$
$$x^2 - 8x + 16 + 25(y^2 + 4y + 4) = -91 + 16 + 25(4)$$
$$(x - 4)^2 + 25(y + 2)^2 = 25$$
$$\frac{(x - 4)^2}{25} + \frac{(y + 2)^2}{1} = 1$$

The ellipse has center at $(4, -2)$. Since $25 > 1$, $a^2$ must be 25. So $a = 5$, $b = 1$, and the major axis is parallel to the $x$-axis.

$$b^2 = a^2 - c^2$$
$$1 = 25 - c^2$$
$$c^2 = 24$$
$$c = \sqrt{24} \text{ or } 2\sqrt{6} \qquad 2\sqrt{6} \approx 4.9$$

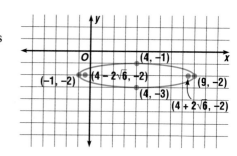

The foci are $2\sqrt{6}$ units to the left and right of the center. The foci are at $(4 + 2\sqrt{6}, -2)$ and $(4 - 2\sqrt{6}, -2)$ or about $(8.9, -2)$ and $(-0.9, -2)$. The length of the major axis is $2a$ or 10 units and the minor axis is $2b$ or 2 units.

# CHECKING FOR UNDERSTANDING

**Communicating Mathematics**

Read and study the lesson to answer these questions.

1. Use the string and tack technique to draw an ellipse with the two foci at one point. What do you observe? What can you conclude?

2. How can you tell which axis is the major axis from the equation of an ellipse?

3. Describe the graph of the equation $\dfrac{x^2}{4} + \dfrac{(y - 2)^2}{1} = 1$.

**Guided Practice**

Name the center of the ellipse with each equation and state whether the major axis is horizontal or vertical.

4. $\dfrac{x^2}{12} + \dfrac{y^2}{8} = 1$

5. $\dfrac{x^2}{5} + \dfrac{y^2}{24} = 1$

6. $\dfrac{x^2}{64} + \dfrac{(y - 3)^2}{36} = 1$

7. $\dfrac{(x - 11)^2}{121} + \dfrac{(y + 8)^2}{144} = 1$

Write the equation of each ellipse in standard form.

8. $8x^2 + 2y^2 = 32$

9. $7x^2 + 3y^2 = 84$

10. $x^2 + 4y^2 + 2x - 24y + 33 = 0$

11. $4x^2 + 9y^2 + 24x - 90y = -225$

12. $6x^2 + 10y^2 + 54x + 20y = 122$

13. $9x^2 + 16y^2 - 54x + 64y + 1 = 0$

Find the center, foci, and lengths of the major and minor axes for each ellipse whose equation is given. Then draw the graph.

14. $\dfrac{x^2}{9} + \dfrac{y^2}{36} = 1$

15. $\dfrac{x^2}{36} + \dfrac{y^2}{16} = 1$

16. $3x^2 + 9y^2 = 27$

17. $16x^2 + 9y^2 = 144$

# EXERCISES

**Practice**

Write the equation of each ellipse.

18.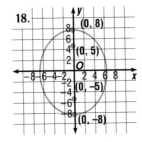

19.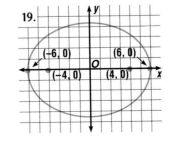

20.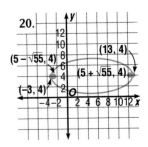

Find the center, foci, and lengths of the major and minor axes for each ellipse whose equation is given. Then draw the graph.

21. $\dfrac{x^2}{25} + \dfrac{y^2}{9} = 1$

22. $\dfrac{x^2}{5} + \dfrac{y^2}{10} = 1$

23. $x^2 + 9y^2 = 9$

24. $36x^2 + 81y^2 = 2916$

25. $27x^2 + 9y^2 = 81$

26. $\dfrac{(x + 9)^2}{36} + \dfrac{(y - 4)^2}{9} = 1$

27. $\dfrac{(x - 8)^2}{4} + \dfrac{(y + 8)^2}{1} = 1$

28. $\dfrac{(x + 2)^2}{20} + \dfrac{(y + 3)^2}{40} = 1$

29. $\dfrac{(x + 8)^2}{121} + \dfrac{(y - 7)^2}{64} = 1$

30. $\dfrac{(x - 4)^2}{16} + \dfrac{(y + 1)^2}{9} = 1$

31. $4x^2 + 9y^2 + 16x - 18y - 11 = 0$

32. $9x^2 + 16y^2 - 18x + 64y = 71$

33. $7x^2 + 3y^2 - 28x - 12y + 19 = 0$

34. $16x^2 + 25y^2 + 32x - 150y = 159$

Write the equation of each ellipse described below.

35. The endpoints of the major axis are at $(0, 10)$ and $(0, -10)$. The foci are at $(0, 8)$ and $(0, -8)$.

36. The foci are at $(12, 0)$ and $(-12, 0)$. The endpoints of the minor axis are at $(0, 5)$ and $(0, -5)$.

37. The major axis is 16 units long and parallel to the $x$-axis. The center is $(5, 4)$ and the minor axis is 9 units long.

38. The endpoints of the major axis are at $(2, 12)$ and $(2, -4)$. The endpoints of the minor axis are at $(4, 4)$ and $(0, 4)$.

39. The major axis is 12 units long and parallel to the $y$-axis. The minor axis is 8 units long and the center is at $(-2, 3)$.

40. The endpoints of the minor axis are at $(-2, 5)$ and $(-2, -1)$. The endpoints of the major axis are at $(-9, 2)$ and $(5, 2)$.

**Critical Thinking**

41. Describe the changes in the shape and size of the ellipse whose equation is $x^2 + 4y^2 = k$ as you increase the value of $k$.

**Applications**

42. **History** The elliptical chamber in the United States Capitol Building is 46 feet wide and 96 feet long.

a. Write an equation to describe the shape of the room. Assume that it is centered at the origin and the major axis is horizontal.

b. John Quincy Adams discovered that he could overhear the conversations being held at the opposing party leader's desk if he stood in a certain spot in the elliptical chamber. Describe the position of the desk and how far away Adams had to stand to overhear.

**Mixed Review**

**43.** Write an equation for the circle whose center is (6, 2) and whose radius is 5 units. **(Lesson 9-3)**

**44.** Write an equation for the circle with center at (1, 5) that passes through the origin. **(Lesson 9-3)**

**45.** Write the equation $f(x) = x^2 + 2x - 2$ in the form $f(x) = a(x - h)^2 + k$. Then name the vertex, axis of symmetry, and direction of opening for the parabola it describes. **(Lesson 8-3)**

**46.** State whether $x^2 + 8x + 64$ is a perfect square. **(Lesson 7-3)**

**47.** Find the product of $7 - 7i$ and its conjugate. **(Lesson 6-10)**

**48.** Rose, Lars, Kris, and Jim were ready to begin playing a game of cards. Lars started shuffling the deck, but dropped the cards. Rose, Kris, and Jim picked up the cards for him. Rose said she had twice as many cards as Jim. If Rose were to give Kris five cards, Kris would have five times as many cards as Jim. Without counting the cards, Lars said he would get a second deck of cards because the deck they were using could not be complete. Find all the possible numbers of cards that might have been in the incomplete deck. **(Lesson 1-5)**

---

## ~~~~~ MID-CHAPTER REVIEW ~~~~~

**Use the distance formula to find the distance between each pair of points.** **(Lesson 9-1)**

**1.** $(-2, 7), (4, -1)$

**2.** $(0, 5), (6, 3)$

**Find the midpoint of the line segment whose endpoints are given below.** **(Lesson 9-1)**

**3.** $(3, -1), (5, -7)$

**4.** $(8, 0), (-5, 12)$

**Name the vertex, axis of symmetry, focus, directrix, and direction of opening of the parabola whose equation is given. Then find the length of the latus rectum and graph.** **(Lesson 9-2)**

**5.** $-12y = x^2$

**6.** $8x = y^2 + 4y + 28$

**The focus and directrix of a parabola are given. Write an equation for each parabola. Then draw the graph.** **(Lesson 9-2)**

**7.** $(3, 5), y = 1$

**8.** $(0, 4), x = 1$

**State whether the graph of each equation is a circle or a parabola.** **(Lesson 9-3)**

**9.** $x^2 + y = 8$

**10.** $x^2 = 16 - y^2$

**State the center and radius of each circle whose equation is given.** **(Lesson 9-3)**

**11.** $x^2 + y^2 = 27$

**12.** $(x + 3)^2 + (y - 1)^2 = 81$

**Write the equation of each ellipse in standard form.** **(Lesson 9-4)**

**13.** $12x^2 + 6y^2 = 168$

**14.** $8x^2 + 4y^2 - 16x - 20y = 7$

**15.** Write the equation of an ellipse whose foci are at (3, 8) and (3, -6). The major axis is 18 units long. **(Lesson 9-4)**

# Hyperbolas

**Objectives**

After studying this lesson, you should be able to:

- write equations of hyperbolas, and
- draw hyperbolas.

**Application**

The orbits of some comets and satellites are shaped like **hyperbolas.** You can see the shape of a hyperbola by placing a table lamp near the wall and observing the shape the light makes. The comets or satellites that have orbits like this follow one branch of the curve and are said to be in an open or escape orbit.

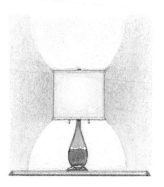

---

*Definition of Hyperbola*

**A hyperbola is the set of all points in a plane such that the absolute value of the difference of the distances from any point on the hyperbola to two given points in the plane, called the *foci*, is constant.**

---

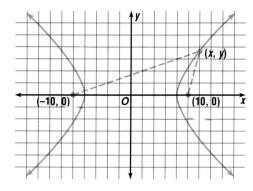

The hyperbola at the left has foci at $(-10, 0)$ and $(10, 0)$. The absolute value of the difference of the distances from the two foci to any point on the hyperbola is 16. Let's use the distance formula and the definition of a hyperbola to find the equation of a hyperbola.

Let $(x, y)$ be any point on the hyperbola. The distance between $(x, y)$ and $(-10, 0)$ *minus* the distance between $(x, y)$ and $(10, 0)$ is $\pm 16$ units.

$$\underset{\substack{\text{distance between}\\(x, y) \text{ and } (-10, 0)}}{} - \underset{\substack{\text{distance between}\\(x, y) \text{ and } (10, 0)}}{} = \pm 16$$

$$\sqrt{(x + 10)^2 + y^2} - \sqrt{(x - 10)^2 + y^2} = \pm 16$$

$$\sqrt{(x + 10)^2 + y^2} = \pm 16 + \sqrt{(x - 10)^2 + y^2} \quad \text{\textit{Add} } \sqrt{(x - 10)^2 + y^2} \text{ \textit{to each side.}}$$

$$(x + 10)^2 + y^2 = 256 \pm 32\sqrt{(x - 10)^2 + y^2} + (x - 10)^2 + y^2 \quad \text{\textit{Square each side.}}$$

$$5x - 32 = \pm 4\sqrt{(x - 10)^2 + y^2} \quad \text{\textit{Simplify.}}$$

$$25x^2 - 320x + 1024 = 16[(x - 10)^2 + y^2] \quad \text{\textit{Square each side.}}$$

$$9x^2 - 16y^2 = 576 \quad \text{\textit{Simplify.}}$$

$$\frac{x^2}{64} - \frac{y^2}{36} = 1 \quad \text{\textit{Divide each side by 576.}}$$

The equation of the hyperbola with foci at $(-10, 0)$ and $(10, 0)$ and with 16 as the absolute value of the difference between the distances from the two foci is $\frac{x^2}{64} - \frac{y^2}{36} = 1$.

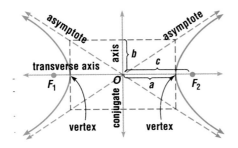

Let's take a close look at the parts of a hyperbola. The midpoint of the segment connecting the foci of a hyperbola is the center of the hyperbola. The point on each branch of the hyperbola that is nearest the center is a **vertex**. As a hyperbola recedes from the center, the branches approach lines called the **asymptotes**.

A hyperbola has many similarities to an ellipse. The distance to the center from a vertex of a hyperbola is $a$ units and the distance from a focus to the center is $c$ units. It has two axes of symmetry. One axis is the **transverse axis**, which is the segment of length $2a$ units whose endpoints are the vertices of the hyperbola. The **conjugate axis** is the segment of length $2b$ units that is perpendicular to the transverse axis at the center. However, the lengths of $a$, $b$, and $c$ are related differently for a hyperbola than for an ellipse. For a hyperbola, $a^2 + b^2 = c^2$.

*Standard Equation of Hyperbola with Center at the Origin*

If a hyperbola has foci at $(-c, 0)$ and $(c, 0)$ and if the absolute value of the difference of the distances from any point on the hyperbola to the two foci is $2a$ units, then the standard equation of the hyperbola is $\dfrac{x^2}{a^2} - \dfrac{y^2}{b^2} = 1$, where $c^2 = a^2 + b^2$.

If a hyperbola has foci at $(0, -c)$ and $(0, c)$ and if the absolute value of the difference of the distances from any point on the hyperbola to the two foci is $2a$ units, then the standard equation of the hyperbola is $\dfrac{y^2}{a^2} - \dfrac{x^2}{b^2} = 1$, where $c^2 = a^2 + b^2$.

**Example 1**

Write the equation of the hyperbola shown below.

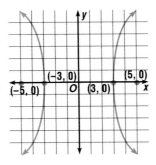

The distance from the center to a vertex is $a$ units and the distance from the center to a focus is $c$ units, so $a = 3$ and $c = 5$.

Use the formula $c^2 = a^2 + b^2$ to find $b^2$.

$$c^2 = a^2 + b^2$$
$$5^2 = 3^2 + b^2$$
$$b^2 = 16$$

The equation is $\dfrac{x^2}{9} - \dfrac{y^2}{16} = 1$.

Sketching the asymptotes of a hyperbola makes graphing it easier. The chart below shows the equations of asymptotes for different hyperbolas.

| Equation of Hyperbola | $\dfrac{x^2}{a^2} - \dfrac{y^2}{b^2} = 1$ | $\dfrac{y^2}{a^2} - \dfrac{x^2}{b^2} = 1$ |
|---|---|---|
| Equations of Asymptotes | $y = \pm\dfrac{b}{a}x$ | $y = \pm\dfrac{a}{b}x$ |
| Transverse Axis | horizontal | vertical |

## Example 2

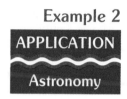

A comet travels along a path that is one branch of the hyperbola whose equation is $\frac{y^2}{225} - \frac{x^2}{400} = 1$. Find the vertices, foci, and the equations of the asymptotes of the hyperbola. Then draw the graph.

The hyperbola has center $(0, 0)$ and a vertical transverse axis. The equations of the asymptotes are $y = \pm\frac{3}{4}x$, since $a = 15$ and $b = 20$. Sketch a $2a$ by $2b$ rectangle and draw the diagonals.

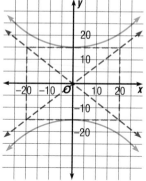

Since $a = 15$, the distance from the center to each vertex is 15 units. The vertices are $(0, 15)$ and $(0, -15)$.

To locate the foci, find the value of $c$.

$$c^2 = a^2 + b^2$$
$$c^2 = 225 + 400$$
$$c^2 = 625$$
$$c = 25$$

So the foci are at $(0, 25)$ and $(0, -25)$.

So far we have discussed equations of hyperbolas centered at the origin. A hyperbola with center $(h, k)$ that is the same shape as one centered at the origin can be described by the standard equation with $(x - h)$ replacing $x$ and $(y - k)$ replacing $y$.

| *Standard Equation of Hyperbola with Center at (h, k)* | The equation of a hyperbola whose center is at $(h, k)$ and with a horizontal transverse axis is $\frac{(x - h)^2}{a^2} - \frac{(y - k)^2}{b^2} = 1$. The equation of a hyperbola whose center is at $(h, k)$ and with a vertical transverse axis is $\frac{(y - k)^2}{a^2} - \frac{(x - h)^2}{b^2} = 1$. |
|---|---|

You can find the slopes of the asymptotes for hyperbolas with centers other than the origin in the same way as you found the slopes for hyperbolas centered at the origin. Remember that the asymptotes always pass through the center of the hyperbola.

## Example 3

Graph $\frac{(x - 1)^2}{9} - \frac{(y + 4)^2}{16} = 1$.

The graph has the same shape as the graph of $\frac{x^2}{9} - \frac{y^2}{16} = 1$, but has center $(1, -4)$.

*The slopes of the asymptotes are $\pm\frac{4}{3}$.*

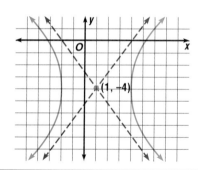

**Example 4**

The graph of $16x^2 - y^2 + 96x + 8y + 112 = 0$ is a hyperbola. Write the equation in standard form and find the vertices, foci, and slopes of the asymptotes of the hyperbola. Then draw the graph.

Complete the square for each variable to write the equation in standard form.

$$16x^2 - y^2 + 96x + 8y + 112 = 0$$
$$16(x^2 + 6x + \square) - (y^2 - 8y + \blacksquare) = -112 + 16\square + (-\blacksquare) \quad \textit{Complete the square.}$$
$$16(x^2 + 6x + 9) - (y^2 - 8y + 16) = -112 + 16(9) - 16$$
$$16(x + 3)^2 - (y - 4)^2 = 16$$
$$\frac{(x + 3)^2}{1} - \frac{(y - 4)^2}{16} = 1$$

The hyperbola has center at $(-3, 4)$ and a horizontal transverse axis. Since $a = 1$ and $b = 4$, the slopes of the asymptotes are 4 and $-4$.

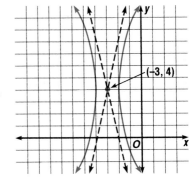

The distance from the center to each vertex is 1 unit, since $a = 1$. So, the vertices are at $(-2, 4)$ and $(-4, 4)$.

To locate the foci, find $c$.

$$c^2 = a^2 + b^2$$
$$c^2 = 1^2 + 4^2$$
$$c = \sqrt{17}$$

The foci are $\sqrt{17}$ units to the left and to the right of the center. So, the foci are at $(-3 + \sqrt{17}, 4)$ and $(-3 - \sqrt{17}, 4)$ or about $(1.1, 4)$ and $(-7.1, 4)$.

# CHECKING FOR UNDERSTANDING

**Communicating Mathematics**

Read and study the lesson to answer these questions.

1. Name some similarities and differences between the hyperbola and the ellipse.

2. The graph of $\dfrac{x^2}{4} + \dfrac{y^2}{1} = 1$ is an ellipse and the graph of $\dfrac{x^2}{4} - \dfrac{y^2}{1} = 1$ is a hyperbola. How can you determine that from their equations?

3. Describe the graph of the equation $\dfrac{(y - 2)^2}{25} - \dfrac{x^2}{4} = 1$.

4. Write the equation of the hyperbola shown at the right. State its vertices, foci, and the slopes of its asymptotes.

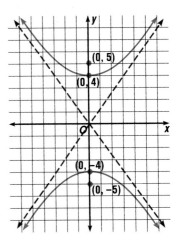

State whether the graph of each function is an ellipse or a hyperbola.

**5.** $\dfrac{x^2}{10} + \dfrac{y^2}{18} = 1$

**6.** $\dfrac{y^2}{1} - \dfrac{x^2}{8} = 1$

**7.** $\dfrac{x^2}{24} - \dfrac{y^2}{36} = 1$

**8.** $\dfrac{y^2}{49} + \dfrac{x^2}{49} = 1$

**9.** $\dfrac{y^2}{20} - \dfrac{x^2}{32} = 1$

**10.** $\dfrac{x^2}{100} + \dfrac{y^2}{25} = 1$

Write the equation of each hyperbola in standard form.

**11.** $2x^2 - y^2 = 10$

**12.** $6y^2 - 34x^2 = 204$

**13.** $(y - 1)^2 - 4(x - 2)^2 = 168$

**14.** $3x^2 - 12y^2 + 45x + 60y = -60$

Find the vertices, foci, and slopes of the asymptotes for each hyperbola whose equation is given. Then draw the graph.

**15.** $\dfrac{x^2}{16} - \dfrac{y^2}{4} = 1$

**16.** $\dfrac{x^2}{6} - \dfrac{y^2}{2} = 1$

**17.** $\dfrac{x^2}{9} - \dfrac{y^2}{4} = 1$

**18.** $x^2 - y^2 = 4$

# EXERCISES

**Practice**   Write the equation of each hyperbola.

**19.**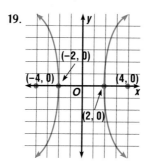

**20.**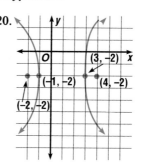

**21.**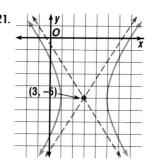

**22.**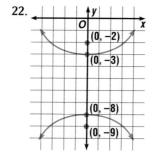

**23.**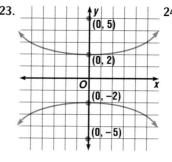

**24.**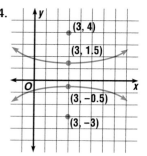

Find the vertices, foci, and slopes of the asymptotes for each hyperbola whose equation is given. Then draw the graph.

**25.** $\dfrac{x^2}{36} - \dfrac{y^2}{1} = 1$

**26.** $\dfrac{y^2}{16} - \dfrac{x^2}{25} = 1$

**27.** $\dfrac{x^2}{9} - \dfrac{y^2}{25} = 1$

**28.** $\dfrac{y^2}{81} - \dfrac{x^2}{25} = 1$

**29.** $\dfrac{x^2}{4} - \dfrac{y^2}{9} = 1$

**30.** $\dfrac{x^2}{81} - \dfrac{y^2}{36} = 1$

**31.** $\dfrac{y^2}{18} - \dfrac{x^2}{20} = 1$  **32.** $\dfrac{x^2}{9} - \dfrac{y^2}{16} = 1$  **33.** $\dfrac{y^2}{100} - \dfrac{x^2}{144} = 1$

**34.** $x^2 - 2y^2 = 2$  **35.** $y^2 = 36 + 4x^2$

**36.** $\dfrac{(x+6)^2}{36} - \dfrac{(y+3)^2}{9} = 1$  **37.** $\dfrac{(y-3)^2}{25} - \dfrac{(x-2)^2}{16} = 1$

**38.** $\dfrac{(y-4)^2}{16} - \dfrac{(x+2)^2}{9} = 1$  **39.** $\dfrac{(x+1)^2}{4} - \dfrac{(y+3)^2}{9} = 1$

**40.** $(x+3)^2 - 4(y-2)^2 = 4$  **41.** $5x^2 - 4y^2 - 40x - 16y = 36$

**42.** $y^2 - 4x^2 - 2y - 16x + 1 = 0$  **43.** $y^2 - 3x^2 + 6y + 6x = 18$

**Write the equation of each hyperbola described below.**

**44.** A hyperbola is centered at the origin with a horizontal transverse axis. The value of $a = 1$ and $b = 4$.

**45.** The center is at $(5, 4)$ and it has a vertical transverse axis. The value of $a = 2$ and $b = 6$.

**46.** The equations of the asymptotes are $3x - 2y = 0$ and $3x + 2y = 0$. It has a horizontal transverse axis and passes through $(2, 0)$.

**An equation of the form $xy = c$ is a hyperbola with the x- and y-axes as asymptotes.**

**47.** Sketch the graph of $xy = 3$.  **48.** Sketch the graph of $xy = -1$.

**Critical Thinking**

**49.** The asymptotes of a hyperbola are given by the equations $5x + 3y = -1$ and $5x - 3y = 11$. Find the equation of the hyperbola if it passes through the point $(4, -2)$.

**Applications**

**50. Chemistry**  Boyle's Law states that if the temperature of a gas is constant, then the pressure exerted by the gas varies inversely as the volume. Symbolically, $PV = k$, where $P$ represents the pressure, $V$ represents the volume, and $k$ is a constant. The constant for a certain gas is 22,000. Graph $PV = k$ for this gas.

**51. Sports**  One lap of the Indianapolis 500 is $2\frac{1}{2}$ miles. Since you know that rate multiplied by time equals distance, you could find the speed at which a car is traveling if you timed one lap. Graph the combinations of times and speeds a car could have for one lap of the Indy 500.

52. Find the center, foci, and lengths of the major and minor axes of the ellipse whose equation is $\frac{x^2}{4} + \frac{y^2}{25} = 1$. Then graph the ellipse. (**Lesson 9-4**)

53. Write the equation of the ellipse whose foci are at (5, 4) and (-3, 4). The major axis is 10 units long. (**Lesson 9-4**)

54. Solve the inequality $x^2 \leq 6$. (**Lesson 8-7**)

55. State the sum and product of the roots of the equation $2x^2 + 4x = 6$. (**Lesson 7-5**)

56. State a factor that can be used to rationalize the denominator of $\dfrac{10}{5^{\frac{2}{3}}}$. (**Lesson 6-6**)

---

## APPLICATION

Most of us are familiar with the mathematical shapes used in architecture. We are used to seeing squares, rectangles, pyramids, and spheres. But some of the more recent architects use designs that are not recognizable. One such example is St. Mary's Cathedral in San Francisco. Its design is called a **hyperbolic paraboloid.** The cathedral was designed by Paul Ryan and John Lee. Pier Nervi of Rome and Pietro Bellaschi of M.I.T. were engineering consultants on the project. When Nervi was asked what Michelangelo would have thought of the cathedral, he replied, "He could not have thought of it. This design comes from geometric theories, not then proven."

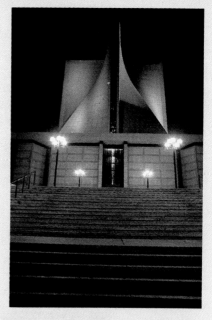

A hyperbolic paraboloid combines a parabola revolved about its axis of symmetry and a three-dimensional hyperbola. The top of St. Mary's is a 2135 cubic foot hyperbolic paraboloid cupola with concrete pylons extending 94 feet into the ground. Each pylon carries a weight of 9 million pounds. The walls are made from 1680 prepounded concrete coffers involving 128 different sizes. The dimensions of the square foundation measure 255 feet by 255 feet. The equation for a hyperbolic paraboloid is $\frac{y^2}{b^2} - \frac{x^2}{a^2} = \frac{z}{c}$, where $a > 0$, $b > 0$, and $c \neq 0$. Since a hyperbolic paraboloid is a three-dimensional figure, $z$ represents values on the third axis.

# 9-6

# Problem-Solving Strategy:
# Use a Model

**Objective**

After studying this lesson, you should be able to:

■ use a simulation to help solve problems.

Sometimes solving a problem directly is impractical. In these cases, it helps to devise a way of modeling, or simulating, a situation to make the problem easier to solve.

**Example**

APPLICATION

Marketing

**HealthWise Cereal has placed contest tickets that are printed with two letters from the name of the cereal in their cereal boxes. To win a $50 prize, you must collect all five different tickets whose letters spell the name of the cereal. The total number of each kind of ticket is the same. How many boxes would you expect to have to buy to win the contest?**

It would be expensive and inconvenient, but you could buy hundreds of boxes of cereal to see how many it took to win. Instead, you could simulate these purchases by assigning each of the tickets a number from 1 to 5. Then roll a die to determine how many boxes you would need to buy. If you roll a 6, you ignore it. You keep rolling until all 5 numbers appear. In the following trials, the first time each number occurs is shown in blue.

|  | Number of boxes to win |
|---|---|
| 5 1 4 3 4 3 5 3 3 1 4 2 | 12 |
| 4 3 2 5 5 4 4 4 3 2 4 4 1 | 13 |
| 1 2 1 5 5 5 4 3 | 8 |
| 3 5 5 4 1 3 1 1 3 1 2 | 11 |
| 4 3 5 1 3 4 3 4 3 4 1 1 3 2 | 14 |

Take an average to find an estimate of the number of boxes you would have to buy.

$$\frac{12 + 13 + 8 + 11 + 14}{5} = 11.6 \approx 12$$

Our estimate is that you would have to buy 12 boxes to win the contest.

*When you simulate a situation in this way, the more trials you conduct, the better your estimation.*

# CHECKING FOR UNDERSTANDING

**Communicating Mathematics**

**Read and study the lesson to answer these questions.**

1. When is it helpful to use simulation to help solve a problem?

2. What is the least number of boxes of HealthWise that you could have to buy to win their contest?

3. Conduct more trials for the Example using your own die. How do your results compare with the results above? with other students' results?

| Strategies |
| --- |
| Look for a pattern. |
| Solve a simpler problem. |
| Act it out. |
| Guess and check. |
| Draw a diagram. |
| Make a chart. |
| Work backwards. |

**4. Statistics**   Miguel had been making only 50% of his free throws all season. During a practice session, Kristin noticed something about Miguel's shooting style and suggested a way he might correct it. Miguel tried the new technique and hit 9 of his next 10 free throws. Did Kristin's suggestion help Miguel's game?

a. With Miguel's old style of shooting, he has a 50% chance of sinking a single shot. Use a coin to simulate 10 free throws. Heads represents a made shot and tails represents a miss. If tossing the coin simulates ten of Miguel's shots, how many would he have made?

b. Use the coin to simulate 10 sets of 10 free throws and average the results. How many does Miguel hit on average?

# EXERCISES

**Solve. Use any strategy.**

5. How many triangles are in the figure at the right?

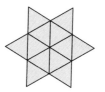

6. The Chimes that ring on The Ohio State University campus ring every fifteen minutes. At one-quarter past the hour, they chime four notes, at half past they chime eight notes, and at one-quarter before the hour they chime twelve notes. On every hour, they chime sixteen notes and the number of the hour. How many notes will be struck by the chimes in one week?

7. What is the 120th odd natural number?

8. Write an expression whose value is 56 using five 4s.

9. Karen was fouled at the final buzzer of the playoff basketball game. She will shoot free throws. She has a success rate of 67% at the free throw line. If her team is behind by one point, do you think she will be able to win the game with these free throws? Do you think she will be able to tie the game?

## COOPERATIVE LEARNING ACTIVITY

**Work in groups. Each person in the group must understand the solution and be able to explain it to any person in class.**

A parking lot attendant keeps the keys for the cars parked in the lot. The spaces in the lot are numbered and the keys are kept in a locked box with numbered pegs to correspond with the spaces. There are twelve cars parked in the lot when the attendant drops the key box and all of the keys are mixed up. If he places the keys on the numbered pegs in random order, how many do you think he will get in the right place?

# Graphing Calculator Exploration: Conic Sections

The parabola, circle, ellipse, and hyperbola are called the **conic sections.** You can use your graphing calculator to graph the conic sections. However, some special care must be taken when graphing the conic sections.

Most of the conic sections are relations, not functions. Since the graphing calculators will only plot functions, we must manipulate the equations before entering them into the calculator for graphing. For example, the equation for a circle, $x^2 + y^2 = 9$, cannot be entered directly, since both types of calculators require that the equation be entered in a "$y =$" format. We must put the equation into this format algebraically before entering it into the calculator.

Graph the circle whose equation is $x^2 + y^2 = 9$.

First put the equation in the "$y =$" format.

$$x^2 + y^2 = 9$$
$$y^2 = 9 - x^2 \qquad \textit{Subtract } x^2 \textit{ from each side.}$$
$$y = \pm\sqrt{9 - x^2} \qquad \textit{Take the square root of each side, including both the positive and negative roots.}$$

Now enter the equations as $y = \sqrt{9 - x^2}$ and $y = -\sqrt{9 - x^2}$ separately, since there is no $\pm$ on the calculator.
*Make sure that the range parameters are set for the standard viewing window.*

*Casio*

ENTER: [GRAPH] [√] [(] 9 [−] [ALPHA]
[X] [$x^2$] [)] [:] [GRAPH] [(-)]
[√] [(] 9 [−] [ALPHA] [X] [$x^2$]
[)] [EXE]

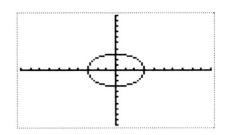

*TI-81*

ENTER: [Y=] [2nd] [√] [(] 9 [−] [X|T]
[$x^2$] [)] [ENTER] [(-)] [2nd] [√]
[(] 9 [−] [X|T] [$x^2$] [)] [GRAPH]

*Sometimes the points near the x-axis or the vertices of a graph will not be graphed by a graphing calculator. These points are included in the graph, but the calculator cannot graph them accurately.*

The graph appears to be an ellipse. The calculator screen must be set so that the units on the $x$- and $y$-axes are equal in length. This adjustment can be made on each of the calculators, so that the graph is not distorted.

*Casio*

ENTER: RANGE SHIFT MCL

RANGE EXE

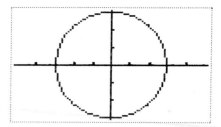

*TI-81*

ENTER: ZOOM 5

The graph now appears as a circle.

Sometimes a complete graph of an equation will not be shown on the square grid provided by the Casio system. In this case, simply reset the range parameters using multiples of the range setting values that the calculator set.

EXAMPLE | **Graph the hyperbola $16x^2 - y^2 + 96x + 8y + 112 = 0$.**

Solve the equation for $y$ by completing the square. The result is

$y = 4 \pm 4 \sqrt{(x + 3)^2 - 1}$.

Now enter the equations to graph the relation.
*Multiply each of the range parameters in the Casio range by 4 to view a complete graph.*

*Casio*

ENTER: GRAPH 4 + 4 √ ( (

ALPHA X + 3 ) $x^2$

− 1 ) : GRAPH 4 − 4

√ ( ( ALPHA X + 3

) $x^2$ − 1 ) EXE

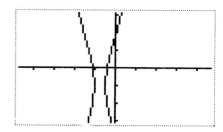

*TI-81*

ENTER: Y= 4 + 4 2nd √ ( ( X|T + 3 ) $x^2$ − 1 ) ENTER

4 − 4 2nd √ ( ( X|T + 3 ) $x^2$ − 1 ) GRAPH

# EXERCISES

Graph each of the following conic sections on your graphing calculator. Sketch the graph that appears on the graphics screen.

1. $x = y^2 + 4y + 28$

2. $(x - 2)^2 + y^2 = 9$

3. $5x^2 + 15y^2 = 225$

4. $9x^2 - 4y^2 - 54x - 40y - 55 = 0$

5. $x^2 + y^2 + 7x - 5 = 0$

6. $y^2 - 20x^2 - 4x - 6y = 36$

# Conic Sections

**Objectives**

After studying this lesson, you should be able to:

- write the equation of a conic section in standard form, and
- identify a conic section from its equation.

**Application**

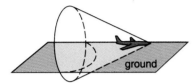
ground

A supersonic jet has a shock wave in the shape of a cone. At the points where the wave hits the ground, a sonic boom is heard. If the jet is traveling parallel to the ground, the sonic boom is heard at points that form one branch of a hyperbola. What shape would the points form if the jet is not traveling parallel to the ground?

By slicing a double cone in different directions, you can form parabolas, circles, ellipses and hyperbolas. For this reason, these curves are called conic sections.

*FYI ...*

Supersonic jets fly faster than the speed of sound. Capt. Charles Yeager became the first person to fly faster than the speed of sound on October 14, 1947.

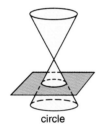
circle

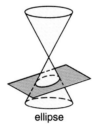
ellipse

parabola

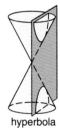
hyperbola

The conic sections can all be described by a quadratic equation.

*Equation of a Conic Section*

**The equation of a conic section can be written in the form $Ax^2 + Bxy + Cy^2 + Dx + Ey + F = 0$ where $A$, $B$, and $C$ are not all zero.**

You can identify the conic section that is represented by a given equation by writing the equation in one of the standard forms you have learned. Study the table below to recall those forms.

| Conic Section | Standard Form of Equation |
|---|---|
| parabola | $y = a(x - h)^2 + k$ <br> or $x = a(y - k)^2 + h$ |
| circle | $(x - h)^2 + (y - k)^2 = r^2$ |
| ellipse | $\dfrac{(x - h)^2}{a^2} + \dfrac{(y - k)^2}{b^2} = 1$ <br> or $\dfrac{(x - h)^2}{b^2} + \dfrac{(y - k)^2}{a^2} = 1$ |

| Conic Section | Standard Form of Equation |
|---|---|
| hyperbola | $\dfrac{(x - h)^2}{a^2} - \dfrac{(y - k)^2}{b^2} = 1$ <br> or $\dfrac{(y - k)^2}{a^2} - \dfrac{(x - h)^2}{b^2} = 1$ <br> or $xy = c$, when $c \neq 0$ |

**Example 1**

Is the graph of $x^2 + y^2 - 8x + 6y + 24 = 0$ a parabola, a circle, an ellipse, or a hyperbola?

Write the equation in standard form.

$$x^2 + y^2 - 8x + 6y + 24 = 0$$
$$x^2 - 8x + \square + y^2 + 6y + \blacksquare = -24 + \square + \blacksquare \quad \textit{Complete the square.}$$
$$x^2 - 8x + 16 + y^2 + 6y + 9 = -24 + 16 + 9$$
$$(x - 4)^2 + (y + 3)^2 = 1 \quad \textit{h is 4, k is -3, and r is 1.}$$

The graph is a circle. It is also an ellipse, with $a$ and $b$ both equal to 1.

**Example 2**

**APPLICATION**

**Aeronautics**

The shock wave generated by a supersonic jet intersects the ground in a curve whose equation is $x^2 + 10x + 5 = 4y^2 + 16y$. What shape is the curve?

$$x^2 + 10x + 5 = 4y^2 + 16y$$
$$x^2 + 10x + \square + 4(y^2 + 4y + \blacksquare) = -5 + \square + (-4)\blacksquare \quad \textit{Complete the square.}$$
$$x^2 + 10x + 25 - 4(y^2 + 4y + 4) = -5 + 25 - 4(4)$$
$$(x + 5)^2 - 4(y + 2)^2 = 4$$
$$\frac{(x + 5)^2}{4} - \frac{(y + 2)^2}{1} = 1 \quad \textit{h is -5, k is -2, a is 2, and b is 1.}$$

The curve is a hyperbola.

You can easily determine the type of conic section represented by an equation of the form $Ax^2 + Bxy + Cy^2 + Dx + Ey + F = 0$ when $B = 0$ by looking at $A$ and $C$.

- If $A = C$, the equation represents a circle.
- If $A$ and $C$ have the same sign and $A \neq C$, the equation represents an ellipse.
- If $A$ and $C$ have opposite signs, the equation represents a hyperbola.
- If $A = 0$ or $C = 0$, but not both, the equation represents a parabola.

# CHECKING FOR UNDERSTANDING

**Communicating Mathematics**

Read and study the lesson to answer these questions.

1. Which curves are conic sections and why are they called conic sections?

2. If a supersonic jet were gaining in altitude, what curve would be formed by the points on the ground where the sonic boom could be heard?

3. Explain why a circle is a special kind of ellipse.

**Guided Practice**

State whether the graph of each equation is a parabola, a circle, an ellipse, or a hyperbola.

**4.** $x^2 + y^2 = 81$

**5.** $y = (x - 3)^2 + 25$

**6.** $x = (y + 4)^2 - 6$

**7.** $\dfrac{x^2}{8} - \dfrac{y^2}{10} = 1$

**8.** $\dfrac{x^2}{6} + \dfrac{y^2}{4} = 1$

**9.** $\dfrac{(x - 4)^2}{9} - \dfrac{(y + 2)^2}{1} = 1$

**10.** $x^2 = 121 - y^2$

**11.** $\dfrac{(y - 7)^2}{3} + \dfrac{(x + 2)^2}{2} = 1$

# EXERCISES

**Practice**

Write the standard form of each equation. State whether the graph of the equation is a parabola, a circle, an ellipse, or a hyperbola. Then graph the equation.

**12.** $4x^2 + 2y^2 = 8$

**13.** $x^2 = 8y$

**14.** $6x^2 + 6y^2 = 162$

**15.** $4y^2 - x^2 + 4 = 0$

**16.** $13x^2 + 13y^2 = 49$

**17.** $y^2 - 2x^2 - 16 = 0$

**18.** $3x^2 + 4y^2 + 8y = 8$

**19.** $y = x^2 + 3x + 1$

**20.** $x + 2 = x^2 + y$

**21.** $\dfrac{(y - 5)^2}{4} - (x + 1)^2 = 4$

**22.** $x^2 - 8y + y^2 + 11 = 0$

**23.** $(y - 4)^2 = 9(x - 4)$

**24.** $3y^2 + 24y - x^2 - 2x + 41 = 0$

**25.** $25y^2 + 9x^2 - 50y - 54x = 119$

**26.** $x^2 + y^2 = x + 2$

**27.** $6x^2 - 24x - 5y^2 - 10y - 11 = 0$

The graph of an equation of the form $Ax^2 + Bxy + Cy^2 + Dx + Ey + F = 0$ is either a conic section or a *degenerated case*. The degenerated cases for the conic sections are stated below. Graph each equation and identify the result.

**28.** $4x^2 - y^2 = 0$

**29.** $4y^2 + 3x^2 + 32y - 6x = -67$

**30.** $x^2 - x = 0$

| Conic | Degenerated Case |
|---|---|
| ellipse or circle | isolated point |
| hyperbola | two intersecting lines |
| parabola | two parallel lines or one line |

**Critical Thinking**

**31.** Graph $\dfrac{x^2}{4} - \dfrac{y^2}{9} = 1$. Then graph $\dfrac{x^2}{4} - \dfrac{y^2}{9} = c$ for several values of $c$ that approach 0. What happens to the graphs as $c$ approaches 0? Describe the graph if $c = 0$.

**Applications**

**32. Astronomy**  In the early 1600s, Johann Kepler studied the orbits of the planets and determined that they are elliptical. It is known now that orbits can take the shape of any of the conic sections. The equations of different orbits are given below. State the shape of each orbit.

a. $x^2 + y^2 = 75{,}000$

b. $y - x^2 = 3x + 5$

c. $x^2 + y^2 - 4x = 9$

d. $x^2 + 5x = y^2 - 6y - 1$

**Computer**

**33.** This BASIC program uses the general form for the equation of a conic section to determine if the graph of the equation is a conic section or a degenerate case.

```
10   PRINT "A*X^2 + C*Y^2 +
     D*X + E*Y + F = 0
20   PRINT "ENTER A, C, D, E, F
25   INPUT A,C,D,E,F
30   IF A <> 0 THEN 60
40   IF D = 0 THEN 100
50   GOTO 120
60   IF C <> 0 THEN 80
70   GOTO 120
80   LET F1 = D^2/(4*A) + E^2/
     (4*C)
90   IF F1 <> F THEN 120
100  PRINT "DEGENERATE
     CASE": GOTO 190
120  IF A = C THEN PRINT
     "CIRCLE": GOTO 190
140  IF A*C > 0 THEN PRINT
     "ELLIPSE": GOTO 190
160  IF A*C < 0 THEN PRINT
     "HYPERBOLA": GOTO 190
180  PRINT "PARABOLA"
190  END
```

**Use the program to determine if each equation is a conic section equation or the degenerate case.**

a. $12x^2 + 36x + 16y^2 + 32y - 5 = 0$

b. $25x^2 - 4y^2 = 100$

c. $x^2 + 12x + y^2 - 8y = -44$

d. $4x^2 + 24x + y^2 - 10y + 45 = 0$

e. $(y + 3)^2 = -12(x - 2)$

f. $16x^2 + 8x + 16y^2 - 32y + 17 = 0$

**Mixed Review**

**34. Probability**  When Kesia and Jeff play darts, Kesia usually wins two out of three games. Jeff read a book on the game and practiced several days before their recent match. Jeff won all three games. Do you think the book and the practice helped Jeff's game?  (**Lesson 9-6**)

**35.** Write the equation of the hyperbola with center $(-2, 2)$ and a vertical transverse axis. $a = 6$ and $c = 10$.  (**Lesson 9-5**)

**36.** Graph the equation $f(x) = 2x^2 + 3$.  (**Lesson 8-3**)

**37.** Solve the equation $6x^2 - 16x - 6 = 0$ using the quadratic formula.  (**Lesson 7-4**)

**38.** Simplify $\sqrt[3]{-27r^3s^3}$.  (**Lesson 6-2**)

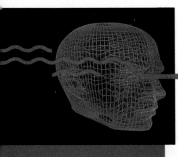

# Technology

## Conic Sections

The *Mathematical Exploration Toolkit (MET)* can be used to graph a system of quadratic equations containing conic sections. From the graphs, you can determine the number of solutions the system has. The CALC commands (and their shortened forms) you can use are listed below.

| | | |
|---|---|---|
| CLEAR f (clr f) | CIRCLE (cir) | COLOR (col) |
| ELLIPSE (ell) | GRAPH (gra) | HYPERBOLA (hyp) |
| SCALE (sca) | | |

The equations of circles, hyperbolas, and ellipses are entered using the values of the graphing parameters given in the general form of each conic section. That is, CIRCLE h k r, ELLIPSE h k a b, and HYPERBOLA h k a b are the commands used, replacing each variable with its numerical value from the general form. A parabola is entered either in the form $y = ax^2 + bx + c$ or $y = (x - h)^2 + k$.

**Example**  Find the number of solutions for the system $\begin{cases} y = (x - 1)^2 \\ (x - 1)^2 + (y - 2)^2 = 4 \end{cases}$.

ENTER:

| | |
|---|---|
| clr f | clears the graphing window |
| sca 10 | sets axes limits at -10 and 10 for x and y |
| y=(x−1)^2 | $y = (x - 1)^2$ |
| col 1 | sets color to cyan (blue) |
| gra | graphs the parabola in blue |
| cir 1 2 2 | $(x - 1)^2 + (y - 2) = 2^2$ |
| col 2 | sets color to magenta (red) |
| gra | graphs the circle in red |

Since the graphs intersect at three points, there are three solutions.

# EXERCISES

**Graph each pair of equations on the same set of axes. Then determine the number of solutions the system has.**

1. $\dfrac{(x - 2)^2}{3^2} + \dfrac{(y + 1)^2}{4^2} = 1$
   $\dfrac{(x + 1)^2}{2^2} - \dfrac{(y - 2)^2}{3^2} = 1$

2. $\dfrac{(x - 2)^2}{9} - \dfrac{(y + 1)^2}{16} = 1$
   $x^2 + (y + 1)^2 = 25$

3. $\dfrac{(x + 1)^2}{4} + \dfrac{(y - 2)^2}{9} = 1$
   $y = -2(2x + 1)^2 + 3$

4. $y = x^2 + 2x - 3$
   $16y^2 - 9x^2 = 144$

**TECHNOLOGY  431**

# Graphing Quadratic Systems

**Objective**

After studying this lesson, you should be able to:

- graph systems of quadratic equations and identify the solution sets.

**Application**

An important concept in chemistry, Boyle's Law, states that if the temperature of a gas is constant, the pressure exerted by the gas varies inversely as the volume. So, $PV = k$, where $P$ represents pressure in kilopascals, $V$ represents volume in cubic decimeters, and $k$ is a constant. A medical technologist provides oxygen for patients with respiratory problems through a tank containing compressed oxygen. The constant for oxygen at 25°C is 504. The volume of the tank is 12 cubic decimeters. What is the pressure of the oxygen in the tank?

We can solve the system of equations described by graphing to find the pressure in the tank. First write the system of equations.

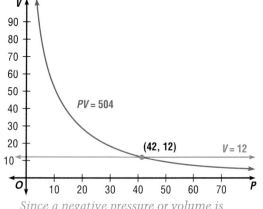

$$PV = k \qquad \textit{Boyle's Law}$$
$$PV = 504 \qquad \textit{The constant for oxygen at 25°C is 504.}$$
$$V = 12 \qquad \textit{The volume of the tank is 12 dm}^3.$$

Now graph.

The point where the line intersects the hyperbola is the solution.

The pressure is 42 kilopascals when the volume is 12 cubic decimeters.

*Since a negative pressure or volume is impossible, we will only consider positive values of P and V.*

If the graphs of a system of equations are a conic section and a straight line, the system will have zero, one, or two solutions.

no solutions

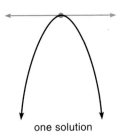

one solution

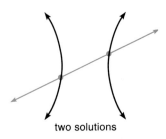

two solutions

**Example 1**

Graph the following system of equations and find its solution.

$$x^2 + y^2 = 9$$
$$x - y = 3$$

The graph of $x^2 + y^2 = 9$ is a circle centered at the origin with a radius of 3 units.

The graph of $x - y = 3$ is a line with slope 1 and $y$-intercept $-3$.

The solutions of the system are $(3, 0)$ and $(0, -3)$. *Check.*

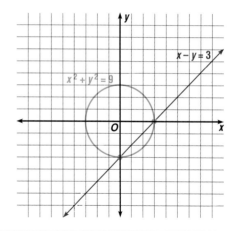

If the graphs of the system of equations are two conic sections, the system will have zero, one, two, three, or four solutions.

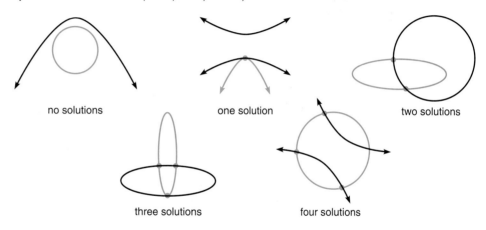

no solutions          one solution          two solutions

three solutions          four solutions

**Example 2**

APPLICATION

Aeronautics

The elliptical orbit of a comet that orbits the Sun is described by the equation $\dfrac{x^2}{9.61} + \dfrac{y^2}{4.84} = 1$. If NASA were to launch a satellite whose orbit would be $\dfrac{x^2}{7.84} + \dfrac{y^2}{5.76} = 1$, would the satellite be in danger of colliding with the comet?

The graph of each equation is an ellipse with center at the origin. The vertices of the comet's orbit are $(0, 2.2)$, $(0, -2.2)$, $(3.1, 0)$, and $(-3.1, 0)$. The vertices of the satellite's orbit are $(0, 2.4)$, $(0, -2.4)$, $(2.8, 0)$, and $(-2.8, 0)$.

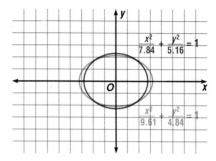

Since the orbits of the comet and the satellite intersect, there is a danger of the two colliding.

**Example 3**

Graph the following system of equations and find its solutions.

$$4y = 25 - x^2$$
$$x^2 - y^2 = 36$$

The graph of $4y = 25 - x^2$ is a parabola with vertex at $(0, 6.25)$. It opens downward and has $x$-intercepts at $(5, 0)$ and $(-5, 0)$.

The graph of $x^2 - y^2 = 36$ is a hyperbola with $x$-intercepts at $(6, 0)$ and $(-6, 0)$. The asymptotes are $y = \pm x$.

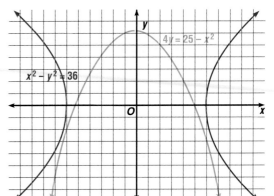

Since the two curves do not intersect, the system has no solutions.

# CHECKING FOR UNDERSTANDING

**Communicating Mathematics**

Read and study the lesson to answer these questions.

1. How many solutions could there be to a system of equations whose graphs are a parabola and a hyperbola? Draw a diagram of each situation.

2. Describe and sketch an example of the graphs of a system of equations with no solutions, if the graph of each equation is a conic section.

3. Could the graphs of two hyperbolas intersect in just one point? If so, draw a sketch of the situation.

4. Use Boyle's Law to find the pressure in a 15 cubic deciliter tank of oxygen. Assume that the temperature is 25°C.

**Guided Practice**

If possible, draw a sketch of the graphs of a system of equations for each situation described below.

5. a hyperbola and a circle that intersect in 2 points

6. two circles that intersect in 3 points

7. two parabolas that intersect in 4 points

8. a hyperbola and an ellipse that do not intersect

Graph each system of equations. Then find the solutions of each system.

9. $x^2 + y^2 = 16$
   $y = 2$

10. $x + y = -7$
    $x^2 + y^2 = 25$

11. $x + y + 1 = 0$
    $(y - 1)^2 = x + 4$

12. $3y = 6 - 5x$
    $\dfrac{x^2}{16} + \dfrac{y^2}{4} = 1$

# EXERCISES

**Practice**

**Graph each system of equations. Then find the solutions of each system.**

**13.** $y = x + 2$
$y = x^2$

**14.** $y = 3x$
$y = x^2 - 4$

**15.** $x^2 + 4y^2 = 20$
$x = y$

**16.** $x^2 - y^2 = 9$
$2y = x - 3$

**17.** $y = 6$
$y^2 = x^2 + 9$

**18.** $y^2 = 100 - x^2$
$x - y = 2$

**19.** $y = 7 - x$
$y^2 + x^2 = 9$

**20.** $y = x - 6$
$\dfrac{x^2}{4} + \dfrac{y^2}{1} = 1$

**21.** $x + 2y = 1$
$x^2 + 4y^2 = 25$

**22.** $x^2 - 4y^2 = 16$
$3x - y = 3$

**23.** $x - y = -2$
$\dfrac{(x-2)^2}{16} + \dfrac{y^2}{16} = 1$

**24.** $y + x^2 = 0$
$x + y = -2$

**25.** $\dfrac{x^2}{36} - \dfrac{y^2}{4} = 1$
$x = y$

**26.** $(x - 1)^2 + 4(y - 1)^2 = 20$
$y = x$

**27.** $5x^2 + y^2 = 30$
$9x^2 - y^2 = -16$

**28.** $3x + 5y = 44$
$\dfrac{(x-3)^2}{25} + \dfrac{(y-4)^2}{9} = 1$

**29.** $x - y = 3$
$(x - 3)^2 + (y + 6)^2 = 36$

**30.** $y = -2x^2$
$x^2 + y^2 = 5$

**31.** $9x^2 + 4y^2 = 36$
$9x^2 - 4y^2 = 36$

**32.** $3x^2 - y^2 = 9$
$x^2 + 2y^2 = 10$

**33.** $y = -x^2 + 3$
$x^2 + 4y^2 = 36$

**34.** $y^2 = x^2 - 25$
$x^2 - y^2 = 7$

**35.** $x^2 + y^2 = 64$
$\dfrac{x^2}{64} + \dfrac{y^2}{1} = 1$

**36.** $y^2 = 16 - x^2$
$x^2 + y^2 = 9$

**37.** $(x - 2)^2 + (y - 2)^2 = 1$
$x^2 = 4 - 4y^2$

**38.** $y^2 = x^2 - 7$
$x^2 + y^2 = 25$

**39.** $y^2 - x^2 = 16$
$x^2 - y^2 = 16$

**40.** $x^2 + 2y^2 = 16$
$y^2 + 2x^2 = 17$

**Critical Thinking**

**41.** Solve the following system of equations by graphing.

$$x - y = -10$$
$$y^2 = 16(x + 10)$$

**Journal**

Use different figures than those used on page 433 to demonstrate the possible number of solutions in a quadratic system. Name the figures you used in each diagram.

**Applications**

**42. Sports** When Greg Norman drives a golf ball, its path is shaped like a parabola. On the 18th fairway, there is a tree that is 60 feet tall.

   **a.** When he drives the ball along the part of the fairway where the tree is, describe three possible heights the ball might reach in terms of the tree.

   **b.** In each situation, how many times during its path will the ball be as high as the tree?

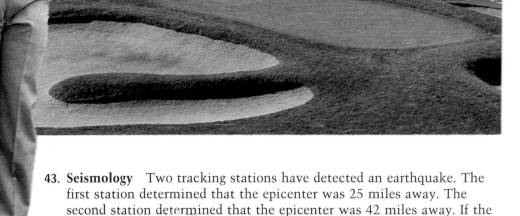

**43. Seismology** Two tracking stations have detected an earthquake. The first station determined that the epicenter was 25 miles away. The second station determined that the epicenter was 42 miles away. If the first station is located at the origin and the second station is 50 miles due east of the first station, where could the epicenter have been?

**Mixed Review**

**44.** State whether the graph of $\dfrac{(x + 3)^2}{1} - \dfrac{(y - 4)^2}{9} = 1$ is a parabola, a circle, an ellipse, or a hyperbola.  **(Lesson 9-7)**

**45.** Write the standard form of the equation $x^2 + 4y^2 = 4$. Graph the equation and state whether the graph is a parabola, a circle, an ellipse, or a hyperbola.  **(Lesson 9-7)**

**46.** State whether the equation $f(x) = -3x^2 - 8x + 7$ describes a quadratic function.  **(Lesson 8-1)**

**47.** Solve the equation $x^2 + 5x - 14 = 0$ by completing the square. **(Lesson 7-3)**

**48.** Simplify $(2p^3 + 7p^2 - 29p + 29) \div (2p - 3)$.  **(Lesson 5-6)**

# Graphing Calculator Exploration: Solving Quadratic Systems

As you know, the graphing calculator is capable of graphing several equations on the screen at one time. You can use this capability with the tracing function to determine the approximate solutions of a system of quadratic equations.

**Example 1**

Graph the following system of equations and find its solutions to two decimal places.

$$4y = 25 - x^2$$
$$x^2 - y^2 = 36$$

First solve each equation for $y$.

$$4y = 25 - x^2 \qquad\qquad x^2 - y^2 = 36$$
$$y = \frac{1}{4}(25 - x^2) \qquad\qquad x^2 - 36 = y^2$$
$$\pm\sqrt{x^2 - 36} = y$$

Now graph the equations. *Make sure your calculator is set for a square viewing window.*

*Casio*

ENTER: [GRAPH] .25 [(] 25 [−] [ALPHA]

[X] [$x^2$] [)] [:] [GRAPH] [√]

[(] [ALPHA] [X] [$x^2$] [−] 36

[)] [:] [GRAPH] [(-)] [√] [(]

[ALPHA] [X] [$x^2$] [−] 36 [)]

[EXE]

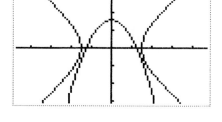

*TI-81*

ENTER: [Y=] .25 [(] 25 [−] [X|T] [$x^2$]

[)] [ENTER] [2nd] [√] [(] [X|T]

[$x^2$] [−] 36 [)] [ENTER] [(-)]

[2nd] [Y-VARS] [▼] [ENTER] [GRAPH]

*The Y-VARS menus allows you to reuse the equations in the Y = list without retyping them.*

There are no intersection points, so there are no real solutions to this system of equations.

**Example 2**

Graph the following system of equations and find its solutions to two decimal places.

$$6x^2 + y^2 = 30$$
$$8x^2 - 2y^2 = 4$$

Solve each equation for $y$.

$$6x^2 + y^2 = 30 \qquad\qquad\qquad 8x^2 - 2y^2 = 4$$
$$y^2 = 30 - 6x^2 \qquad\qquad\qquad 8x^2 - 4 = 2y^2$$
$$y = \pm\sqrt{30 - 6x^2} \qquad\qquad 4x^2 - 2 = y^2$$
$$\qquad\qquad\qquad\qquad\qquad \pm\sqrt{4x^2 - 2} = y$$

Graph the equations.

*Casio*

ENTER:  GRAPH  √  (  30  −  6  ALPHA  X  $x^2$  )  :  GRAPH  (-)  √  (

30  −  6  ALPHA  X  $x^2$  )  :  GRAPH  √  (  4  ALPHA  X  $x^2$

−  2  )  :  GRAPH  (-)  √  (

4  ALPHA  X  $x^2$  −  2  )  EXE

*TI-81*

ENTER:  Y=  2nd  √  (  30  −  6

X|T  $x^2$  )  ENTER  (-)  2nd

Y-VARS  ENTER  ENTER  2nd  √

(  4  X|T  $x^2$  −  2  )

ENTER  (-)  2nd  Y-VARS  ▼  ▼

ENTER  GRAPH

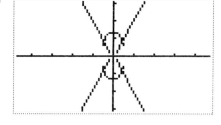

*The points near the axes are part of the graph, but are not drawn by the graphing calculator.*

Now use the zoom-in technique that we used to find the solution of a system of linear equations to find the solutions. The solutions are (1.78, 3.28), (1.78, −3.28), (−1.78, 3.28), and (−1.78, −3.28) to the nearest two decimal places. *Notice the symmetry of these points.*

# EXERCISES

Use your graphing calculator to graph each system of equations and find the solutions to two decimal places.

**1.** $x^2 + y^2 = 25$
$y - x^2 = 1$

**2.** $x^2 + 2y^2 = 10$
$3x^2 - y^2 = 9$

**3.** $x^2 + y^2 = 16$
$x^2 + y^2 = 9$

**4.** $y = x^2 - 6$
$9x^2 - 4y^2 = 36$

**5.** $(x - 1)^2 + y^2 = 9$
$x^2 + 64y^2 = 64$

**6.** $x^2 + 4y^2 = 16$
$(x - 2)^2 + y^2 = 16$

# Solving Quadratic Systems

**Objectives**

After studying this lesson, you should be able to:
- solve systems of equations algebraically, and
- solve systems of inequalities involving quadratics graphically.

You have learned how to solve systems of linear equations algebraically. You can use similar methods to solve systems of quadratic equations.

**Example 1**

Find the solutions of the following system of equations.
$$4x^2 + 4y^2 = 65$$
$$6x - 2y = 5$$

The graphs of the equations are a circle and a line. There are two solutions to the system. *Estimate the solutions from the graph.*

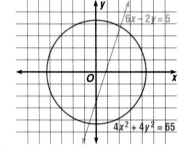

Let's use the substitution method to find the solutions. First, rewrite $6x - 2y = 5$ as $y = 3x - \dfrac{5}{2}$.

$$4x^2 + 4y^2 = 65$$
$$4x^2 + 4\left(3x - \frac{5}{2}\right)^2 = 65 \qquad \textit{Substitute } 3x - \frac{5}{2} \textit{ for y.}$$
$$4x^2 + 4\left(9x^2 - 15x + \frac{25}{4}\right) = 65$$
$$4x^2 + 36x^2 - 60x + 25 = 65$$
$$40x^2 - 60x - 40 = 0$$
$$2x^2 - 3x - 2 = 0$$
$$(2x + 1)(x - 2) = 0$$

$$2x + 1 = 0 \qquad \text{or} \qquad x - 2 = 0 \qquad \textit{Zero product property}$$
$$x = -\frac{1}{2} \qquad\qquad\qquad x = 2$$

Now solve for $y$.

$$y = 3x - \frac{5}{2} \qquad\qquad y = 3x - \frac{5}{2}$$
$$y = 3\left(-\frac{1}{2}\right) - \frac{5}{2} \qquad\qquad y = 3(2) - \frac{5}{2}$$
$$y = -4 \qquad\qquad\qquad y = \frac{7}{2}$$

The solutions are $\left(-\dfrac{1}{2}, -4\right)$ and $\left(2, \dfrac{7}{2}\right)$. *Compare to your estimate. Are these solutions reasonable?*

**Example 2**

Find the solutions to the following system of equations.
$$5x^2 + y^2 = 30$$
$$6x^2 - 2y^2 = 4$$

The graphs of the equations are an ellipse and a hyperbola. The graphs show that there are four solutions to the system.

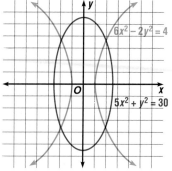

Use the elimination method to solve.

$$\begin{array}{ll} 5x^2 + y^2 = 30 & \textit{Multiply by 2} \\ 6x^2 - 2y^2 = 4 & \end{array} \quad \begin{array}{r} 10x^2 + 2y^2 = 60 \\ 6x^2 - 2y^2 = 4 \\ \hline 16x^2 = 64 \\ x^2 = 4 \\ x = \pm 2 \end{array}$$

Substitute 2 and $-2$ for $x$ to solve for $y$.

$$\begin{array}{ll} 5x^2 + y^2 = 30 & \qquad 5x^2 + y^2 = 30 \\ 5(2)^2 + y^2 = 30 & \qquad 5(-2)^2 + y^2 = 30 \\ y^2 = 10 & \qquad\qquad y^2 = 10 \\ y = \pm\sqrt{10} & \qquad\qquad y = \pm\sqrt{10} \end{array}$$

The solutions are $(2, \sqrt{10})$, $(2, -\sqrt{10})$, $(-2, \sqrt{10})$, and $(-2, -\sqrt{10})$.

**Example 3**

**APPLICATION**

**Aviation**

The radar at the Municipal Airport can detect planes up to 45 miles away. The airport is located at (0, 0) on a map. If a plane is traveling on a straight path given by the equation $2x + y = 12$, show when the plane is within the range of the radar.

The area where the radar can detect the plane can be described by the inequality $x^2 + y^2 \le 45^2$. To find the area where the plane is within the range of the radar, graph both the inequality that describes the range of the radar and the equation that describes the path of the plane and find the intersection of the two graphs.

The graph of $x^2 + y^2 \le 45^2$ is a circle and its interior. This region is shaded on the graph at the right. The graph of $2x + y = 12$ is a line with slope $-2$ and $y$-intercept 12. The intersection of the line and the shaded region is the graph of the area where the plane is in the range of the radar. It is indicated by the thick line segment.

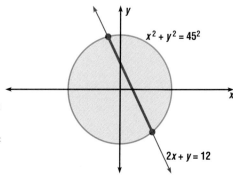

**Example 4**

Solve the following system of inequalities by graphing.
$$x^2 + y^2 < 9$$
$$y < -x^2$$

The graph of $x^2 + y^2 < 9$ is the interior of the circle $x^2 + y^2 = 9$. This region is shaded blue.

The graph of $y < -x^2$ is all of the points within the parabola $y = -x^2$. This region is shaded yellow.

The intersection of these two graphs represents the solutions for the system of inequalities. The points on the curves themselves are not solutions.

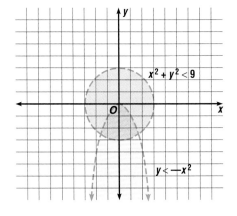

# CHECKING FOR UNDERSTANDING

**Communicating Mathematics**

Read and study the lesson to answer these questions.

1. When would you choose to solve a system of quadratic equations algebraically rather than graphically?

2. If you were to solve a system of quadratic equations and the solution were an imaginary number, what can you conclude about the graphs of the equations of the system?

3. The blue region of the graph at the right represents the solution of $x^2 + y^2 \le 25$ and the yellow region represents the solution of $4y + x^2 \le 25$. What does the green region represent?

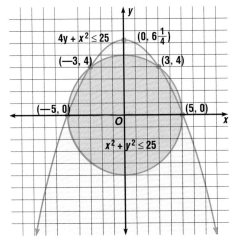

**Guided Practice**

State which method, substitution or elimination, you would use to solve each system of quadratic equations. Then solve each system.

4. $x^2 + y^2 = 25$
   $y - x = 1$

5. $x^2 + 2y^2 = 10$
   $3x^2 - y^2 = 9$

6. $x^2 - y^2 = 25$
   $4y^2 + x^2 = 25$

7. $3x = 4y^2$
   $4y^2 - 2x^2 = 16$

8. $y = 3x^2 + 2$
   $x^2 - 3y^2 = 27$

9. $x^2 + y^2 = 81$
   $x = 2y^2 - 162$

**Write the system of inequalities represented by each graph.**

10.
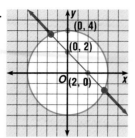
$(0, 4)$
$(0, 2)$
$O$ $(2, 0)$

11.
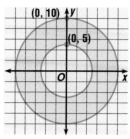
$(0, 10)$
$(0, 5)$
$O$

12.
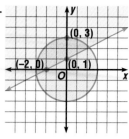
$(0, 3)$
$(-2, 0)$ $(0, 1)$
$O$

# EXERCISES

**Practice**   Solve each system of equations.

**13.** $y = x + 2$
$y = x^2$

**14.** $\dfrac{x^2}{20} + \dfrac{y^2}{5} = 1$
$y = x$

**15.** $y^2 = x^2 - 9$
$2y = x - 3$

**16.** $y = x^2 - 4$
$y = 3x$

**17.** $x + y + 7 = 0$
$x^2 + y^2 = 25$

**18.** $x^2 + 4y^2 = 16$
$5x + 2y = 4$

**19.** $y = 6$
$x^2 - y^2 + 9 = 0$

**20.** $x + 4 = (y - 1)^2$
$x + y + 1 = 0$

**21.** $x + y = 7$
$x^2 + y^2 = 9$

**22.** $x - 2 = y$
$x^2 + y^2 = 100$

**23.** $x^2 + 4y^2 = 4$
$y = x - 6$

**24.** $(x - 2)^2 + y^2 = 16$
$y - x = 2$

**25.** $y = -\dfrac{1}{2}x + \dfrac{1}{2}$
$x^2 + 4y^2 = 25$

**26.** $3x - y = 3$
$x^2 - 4y^2 = 16$

**27.** $x^2 + y = 0$
$x + y = -2$

**28.** $x^2 - 9y^2 = 36$
$x = y$

**29.** $5x^2 + y^2 = 30$
$y^2 - 16 = 9x^2$

**30.** $(x - 3)^2 + (y + 6)^2 = 36$
$x - y = 3$

**31.** $3x + 5y = 44$
$\dfrac{(x - 3)^2}{25} + \dfrac{(y - 4)^2}{9} = 1$

**32.** $x^2 + y^2 = 64$
$x^2 + 64y^2 = 64$

Solve each system of inequalities by graphing.

**33.** $x^2 + y^2 < 25$
$4x^2 - 9y^2 < 36$

**34.** $x^2 + y^2 \geq 49$
$\dfrac{x^2}{16} - \dfrac{y^2}{1} \geq 1$

**35.** $\dfrac{x^2}{25} - \dfrac{y^2}{16} \geq 1$
$y \leq x - 2$

**36.** $x^2 + y^2 \geq 4$
$x^2 + y^2 \leq 36$

**37.** $(y - 3)^2 \geq x + 2$
$x^2 \leq y + 4$

**38.** $9x^2 + y^2 < 81$
$x^2 + y^2 \geq 16$

**39.** $y = x + 3$
$x^2 + y^2 < 25$

**40.** $y^2 < x$
$x^2 - 4y^2 < 16$

**41.** $x^2 + y^2 < 25$
$x + 2y > 1$

**42.** $x + y = 4$
$9x^2 - 4y^2 \geq 36$

**43.** $4x^2 + 9y^2 \geq 36$
$4y^2 + 9x^2 \leq 36$

**44.** $4x^2 + (y - 3)^2 \leq 16$
$x - 2y = -1$

**45.** $(x + 2)^2 + 16(y + 3)^2 \geq 16$
$x + y = 0$

**46.** $x = 2$
$4x^2 + 9y^2 \leq 36$

**47.** $x = 4$
$x^2 + y^2 \geq 16$

**48.** $y = 6$
$y^2 \leq 36 - x^2$

**Critical Thinking**

Solve each system of equations.

**49.** $x + y^2 = 2$
$2y - 2\sqrt{2} = x(\sqrt{2} + 2)$

**50.** $x^2 + y^2 = 1$
$y = 3x + 1$
$x^2 + (y + 1)^2 = 4$

**Applications**

**51. Gardening** Rob and Clarice are going to build a fence around their
vegetable garden. They have 88 feet of fencing
material and plan to have a garden with an
area of 480 square feet. What will the
dimensions of their garden be?

**52. Seismology** Three tracking stations have
detected an earthquake in the area. The first
station is located at the origin on the map.
Each grid on the map represents one square
mile. The second and third tracking stations
are located at (0, 30) and (35, 18) respectively.
The epicenter was 50 miles from the first
station, 40 miles from the second station, and
13 miles from the third station. Where was the
epicenter of the earthquake?

**Mixed Review**

**53.** Graph the following system of equations. Then find the
solutions. **(Lesson 9-8)**
$16x^2 = 4y^2 + 64$
$49x^2 = 4(y^2 - 49)$

**54.** Name the vertex, axis of symmetry and direction of opening of the
parabola whose equation is $f(x) = 4(x - 8)^2$. **(Lesson 9-2)**

**55. Cartography** Edison is located at (9, 3) on the road map. Kettering is
located at (12, 5) on the same map. Each side of a grid on the map
represents 10 miles. Use the distance formula to approximate the
distance between Edison and Kettering. **(Lesson 9-1)**

**56.** Solve the equation $2x^2 + 15x + 7 = 0$. **(Lesson 7-2)**

**57.** Simplify $\dfrac{2 + \sqrt{6}}{2 - \sqrt{6}}$. **(Lesson 6-3)**

**58.** Simplify $(2xy^2)^3 + (2xy^2)^2(6xy^2)$. **(Lesson 5-1)**

# SUMMARY AND REVIEW

## VOCABULARY

Upon completing this chapter, you should be familiar with the following terms:

| | | | |
|---|---|---|---|
| asymptotes | **417** | **400** | focus |
| center | **405** | **416** | hyperbola |
| circle | **405** | **401** | latus rectum |
| conic sections | **407** | **410** | major axis |
| conjugate axis | **417** | **398** | midpoint formula |
| directrix | **400** | **410** | minor axis |
| distance formula | **397** | **400** | parabola |
| ellipse | **409** | **405** | radius |
| foci | **409** | **417** | transverse axis |

## SKILLS AND CONCEPTS

| OBJECTIVES AND EXAMPLES | REVIEW EXERCISES |
|---|---|

Upon completing this chapter, you should be able to:

- use the distance formula to find the distance between a pair of points. **(Lesson 9-1)**

Find the distance between $(3, -2)$ and $(0, 2)$.

$$d = \sqrt{(x_2 - x_1)^2 + (y_2 - y_1)^2}$$
$$= \sqrt{(0 - 3)^2 + [2 - (-2)]^2}$$
$$= \sqrt{25} \text{ or } 5$$

**Find the distance between each pair of points.**

1. $(-8, -7), (-2, -1)$
2. $(3, 6), (7, -8)$
3. $(-2.4, 0.6), (1.7, 0.8)$
4. $(2\sqrt{3}, 4\sqrt{3}), (2\sqrt{3}, -\sqrt{3})$

- use the midpoint formula to find the midpoint of a segment.   **(Lesson 9-1)**

**Find the midpoint of the segment whose endpoints are $(-8, 7)$ and $(2, 15)$.**

$$M = \left(\frac{x_1 + x_2}{2}, \frac{y_1 + y_2}{2}\right)$$
$$= \left(\frac{-8 + 2}{2}, \frac{7 + 15}{2}\right) \text{ or } (-3, 11)$$

**Find the midpoint of each line segment with endpoints having the following coordinates.**

5. $(17, -8), (-13, 1)$
6. $(0.2, 0.6), (0.3, 0.4)$
7. $(5, 2), (-3, 1)$
8. $(2, 2), (\sqrt{2}, \sqrt{2})$

■ graph parabolas.  (**Lesson 9-2**)

Graph $x^2 = 4y$.
Rewrite as $y = \frac{1}{4}x^2$.
vertex: $(0, 0)$
axis of symmetry: $x = 0$
focus: $(0, 1)$
directrix: $y = -1$
direction of opening: up
length of latus rectum: $\left|\dfrac{1}{\frac{1}{4}}\right|$ or 4 units

Name the vertex, axis of symmetry, focus, directrix, and direction of opening of the parabola whose equation is given. Then find the length of the latus rectum and draw the graph.

9. $y^2 = -8x$

10. $(y - 8)^2 = -4(x - 4)$

---

■ graph circles.  (**Lesson 9-3**)

The equation of a circle with radius $r$ and center $(h, k)$ is $(x - h)^2 + (y - k)^2 = r^2$.

Find the center and radius of each circle whose equation is given. Then draw the graph.

11. $x^2 + y^2 = 121$   12. $(x - 3)^2 + (y + 7)^2 = 81$

---

■ graph ellipses.  (**Lesson 9-4**)

An ellipse is the set of all points in a plane such that the sum of the distances from two given points in the plane, called the foci, is constant.

Find the center, foci, and lengths of the major and minor axes for each ellipse whose equation is given. Then draw the graph.

13. $9x^2 + 16y^2 = 144$

14. $\dfrac{(x - 3)^2}{25} + \dfrac{(y + 1)^2}{4} = 1$

---

■ graph hyperbolas. (**Lesson 9-5**)

Graph $\dfrac{x^2}{16} - \dfrac{y^2}{81} = 1$.

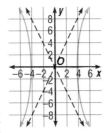

vertices: $(4, 0)$, $(-4, 0)$
foci $(\sqrt{97}, 0)$, $(-\sqrt{97}, 0)$
slopes of asymptotes: $\pm\dfrac{9}{4}$

Find the vertices, foci, and slopes of the asymptotes for each hyperbola whose equation is given. Then draw the graph.

15. $49x^2 - 16y^2 = 784$

16. $25(y + 6)^2 - 20(x - 1)^2 = 500$

---

■ identify a conic section from its equation.  (**Lesson 9-7**)

$Ax^2 + Bxy + Cy^2 + Dx + Ey + F = 0$

| relationship of $A$ and $C$ | graph |
|---|---|
| $A = C$ | circle |
| $A \neq C$, but have same sign | ellipse |
| $A$ and $C$ have opposite signs | hyperbola |
| $A = 0$ or $C = 0$ but not both | parabola |

State whether the graph of each equation is a parabola, a circle, an ellipse, or a hyperbola.

17. $(x - 1)^2 = 4y$

18. $4x^2 + 5y^2 = 20$

19. $3x^2 - 16 = -3y^2$

20. $3y^2 - 7x^2 = 21$

| OBJECTIVES AND EXAMPLES | REVIEW EXERCISES |
|---|---|

■ solve systems of quadratic equations.
(Lesson 9-8 and 9-9)

A system of equations whose graphs are a line and a conic section can have zero, one, or two solutions. A system of equations whose graphs are two conic sections can have zero, one, two, three, or four solutions.

**Graph each system of equations. Then find the solutions of each system.**

**21.** $(x - 2)^2 + y^2 = 16$
$y - x = 2$

**22.** $x^2 - y^2 = 16$
$y^2 - x^2 = 16$

**Solve each system of equations.**

**23.** $x + y = 4$
$y = x^2$

**24.** $x + y = 1$
$x^2 + y^2 = 9$

---

■ solve systems involving quadratic inequalities by graphing. (Lesson 9-9)

Graph the solution set of the following system of inequalities.
$x^2 + y^2 \geq 5$
$2x - 3y = 5$

**Solve each system of inequalities by graphing.**

**25.** $x^2 + y^2 < 25$
$x + y > 5$

**26.** $y \geq x^2 + 4$
$x^2 + y^2 < 49$

---

# APPLICATIONS AND CONNECTIONS

**27. Aviation** An air traffic control tower is located at (12, 25) on a county map. The radar equipment can detect planes up to 48 miles away. Assuming that each side of a grid on the map represents one mile, write an equation for the position of the most distant plane that the tower can detect in terms of the county map. **(Lesson 9-1)**

**29. Aerospace** The path of a comet can be described by the equation $4x^2 - 9y^2 = 36$. Describe this orbit. **(Lesson 9-5)**

**30. Probability** Mr. and Mrs. Porter would like to have two children, a boy and a girl. Use coins to determine how likely it is that they will have a boy and then a girl. **(Lesson 9-6)**

**28. Sports** Elliptipool is a game like pool that is played on the elliptical table. The table has one pocket that is located at one focus of the ellipse. How could a player strike a ball and be guaranteed that the ball will go in the pocket? **(Lesson 9-4)**

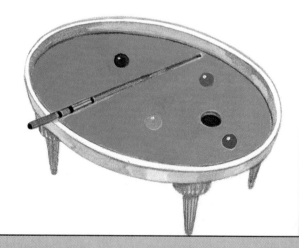

**Use the distance formula to find the distance between each pair of points.**

**1.** $(6, 3), (-6, 0)$

**2.** $(9, -11), (-7, 18)$

**Find the midpoint of each line segment whose endpoints are given below.**

**3.** $(6, 12), (-12, 22)$

**4.** $(-3.2, 2.1), (9.8, -0.6)$

**State whether the graph of each equation is a parabola, a circle, an ellipse, or a hyperbola. Then draw the graph.**

**5.** $y = 3x^2$

**6.** $x^2 + 4x = -(y^2 - 6)$

**7.** $9x^2 + 49y^2 = 441$

**8.** $4x^2 - y^2 = 4$

**9.** $x^2 + 4x + y^2 - 8y = 2$

**10.** $(x + 3)^2 = 8(y + 2)$

**11.** $9x^2 + 9y^2 = 9$

**12.** $y - x^2 = x + 3$

**13.** $2x^2 - 13y^2 + 5 = 0$

**14.** $16(x - 3)^2 + 81(y + 4)^2 = 1296$

**15.** $4x^2 - y^2 = 16$

**16.** $x^2 + 5y^2 = 16$

**17. Statistics** Paul makes two of every three free throws that he attempts. He has been fouled and is to shoot two free throws. Do you think he will make them both?

**Solve each system of equations.**

**18.** $y = -(x + 1)$
$x^2 + y^2 = 25$

**19.** $9x^2 - 16y^2 = 144$
$x^2 + y^2 = 16$

**Solve each system of inequalities by graphing.**

**20.** $x^2 + y < 2$
$x^2 + y^2 < 49$

**21.** $y = 5 - x$
$x^2 + y^2 \geq 49$

**Find the equation for each conic section described below.**

**22.** A parabola has vertex at $(6, -1)$ and focus at $(3, -1)$.

**23. Geometry** A diameter of a circle has endpoints at $(-2, 3)$ and $(4, 5)$.

**24.** An ellipse has center at $(3, 1)$. Its major axis is 12 units long and is parallel to the $y$-axis. Its minor axis is $8\sqrt{2}$ units long.

**25.** The center of a hyperbola is $(2, -4)$. The transverse axis is horizontal and is 6 units long. The conjugate axis is 10 units long.

**Bonus**
**Solve the following system of equations.**
$x^2 + 4y^2 = 4$
$(x - 1)^2 + y^2 = 1$

# Polynomial Functions

## CHAPTER OBJECTIVES

In this chapter, you will:

- Find factors and zeros of polynomials.
- Approximate real zeros of and graph polynomial functions.
- Find the composition of functions.
- Determine the inverse of a function or relation.

The diagram below shows sources of money flowing into a person's life and where it goes. Which of these inflows and outflows do you have? Which ones do you anticipate once you are living on your own?

| INFLOW | OUTFLOW |
|---|---|
| JOBS | HOUSING COSTS |
| GIFTS | FOOD |
| LOANS | CLOTHING & PERSONAL CARE |
| UNEMPLOYMENT COMPENSATION | TRANSPORTATION |
| WELFARE | TAXES |
| INTEREST | INSURANCE PREMIUMS |
| DIVIDENDS | APPLIANCES |
| CAPITAL GAINS | HOUSEHOLD FURNISHINGS |
| SOCIAL SECURITY | MEDICAL COSTS |
| RETIREMENT PENSIONS | CONTRIBUTIONS |
| ANNUITIES | SOCIAL SECURITY PAYMENTS |
| WILLS & TRUSTS | INTEREST PAYMENTS ON DEBTS |
| INSURANCE BENEFITS | OTHER PURCHASES & PAYMENTS |
| ROYALTIES | |
| RENTALS | |

## CAREERS IN FINANCIAL PLANNING

Does money fascinate you? Well, of course. But here's the question that separates future financial planners from everyone else: Would you rather save money and watch it grow than buy things with it? If you can honestly say yes, then financial planning may be for you. A good financial planner sees money as a garden to tend, not a treasure to spend.

Some financial planners work for institutions such as banks, credit unions, savings and loan associations, and finance companies. There they meet with individuals and families. They help their clients analyze cash flow (where it comes from, where it goes), taxes, savings, real estate, retirement plans, planning for college educations, investments, insurance, debts—in short, everything financial. Some people are rather haphazard about these matters. It's the financial planner's job to bring all the loose ends together into one big picture: the client's future financial security.

Other financial planners work for large corporations, setting and implementing financial policy for the firms.

Whatever the level, financial planners watch the money. If you think you could be responsible for the health and growth of major money, then you might want to think seriously about this career.

## MORE ABOUT FINANCIAL PLANNING

### Degree Required:

- Bachelor's Degree in accounting

### Some financial planners like:

- seeing people succeed financially as a result of their advice
- challenge and variety in working with people
- potentially high salary

### Related Math Subjects:

- Algebra
- Advanced Math
- Probability/Statistics

### Some financial planners dislike:

- the competition
- economic uncertainties
- constant changes in federal and state regulations on finance

For more information on the various careers available in the field of Financial Planning, write to:

American Economics Association
1313 21st Avenue, South
Nashville, TN 37212

# Polynomial Functions

**Objectives**

After studying this lesson, you should be able to:
- evaluate polynomial functions, and
- identify general shapes of the graphs of polynomial functions.

**Application**

The number of times that a cricket chirps in an hour is a function of the temperature described by $f(t) = \dfrac{t}{0.2} - 32$, where $f(t)$ is the number of chirps in relation to the temperature $t$ in degrees Celsius. The brightness of the light that fireflies produce is also a function of the temperature. An approximate formula for the intensity of their light, $I(t)$, in lumens is $I(t) = 10 + 0.3t + 0.4t^2 - 0.01t^3$, where $t$ represents the temperature in degrees Celsius. The expression $10 + 0.3t + 0.4t^2 - 0.01t^3$ is a **polynomial in one variable.**

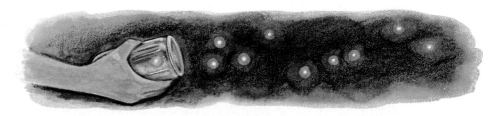

**Definition of Polynomial in One Variable**

A polynomial in one variable, $x$, is an expression of the form $a_0x^n + a_1x^{n-1} + ... + a_{n-2}x^2 + a_{n-1}x + a_n$. The coefficients $a_0$, $a_1$, $a_2$, ..., $a_n$ represent complex numbers (real or imaginary), $a_0$ is not zero, and $n$ represents a nonnegative integer.

**Example 1**

Determine if each expression is a polynomial in one variable.

a. $8x^5 + 6x^4 - 2x^2 + x - 9$

This is a polynomial in one variable, $x$.

b. $8a^2b^4 + 2ab^2 - 1$

This is not a polynomial in one variable. It contains two variables, $a$ and $b$.

c. $x^2 + 2x - \dfrac{1}{x}$

This is not a polynomial in one variable, because the term $\dfrac{1}{x}$ cannot be written in the form $x^n$ with $n$ a nonnegative integer.

d. $10 - 2a + 6a^2$

This is a polynomial in one variable, $a$.

In Chapter 7, we said that the degree of a quadratic equation is 2. The degrees of other polynomials can also be found. The degree of a polynomial in one variable is the greatest exponent of its variable.

9 has degree 0.    *Remember that* $9 = 9x^0$.
$x + 1$ has degree 1.    *Remember* $x = x^1$.
$2x^2 + 4x - 8$ has degree 2.
$7x^6 - 1$ has degree 6.
$a_0x^n + a_1x^{n-1} + ... + a_{n-2}x^2 + a_{n-1}x + a_n$ has degree $n$.

Some of the polynomials in one variable have special names. As you know, a polynomial in one variable of degree 0 is called a **constant.** Polynomials in one variable of degree 1 are **linear expressions,** of degree 2 are **quadratic expressions,** and of degree 3 are **cubic expressions.**

When a polynomial equation is used to represent a function, it is a **polynomial function.** For example, the equation $f(x) = 3x^2 - 8x + 7$ represents a quadratic polynomial function, and the equation $p(x) = 2x^3 + 6x^2 - 5x + 3$ represents a cubic polynomial function. These and other polynomial functions can all be defined by the following general rule.

| *Definition of Polynomial Function* | A polynomial function can be described by an equation of the form $P(x) = a_0x^n + a_1x^{n-1} + ... + a_{n-2}x^2 + a_{n-1}x + a_n$. The coefficients $a_0, a_1, ... , a_{n-1},$ and $a_n$ represent real numbers, $a_0$ is not zero, and $n$ represents a nonnegative integer. |
|---|---|

The graphs of several polynomial functions are shown below. Notice how many times the graph of each function intersects the x-axis. In each case, this is the maximum number of real zeros the function may have.

constant function

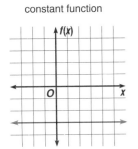

linear function

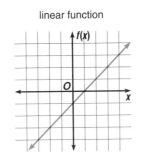

quadratic function

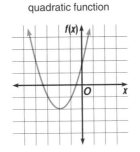

cubic function

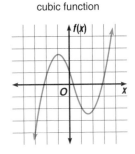

If you know an element in the domain of any polynomial function, you can find the corresponding value in the range. Remember that if $f(x)$ is the function and 4 is an element in the domain, the corresponding element in the range is $f(4)$. Its value is found when the function is evaluated for $x = 4$.

### Example 2

**APPLICATION**

**Biology**

Using the polynomial function for the intensity of the light of a firefly that was given at the beginning of the lesson, find the intensity of the light emitted by a firefly when the temperature is 20°C.

$$I(t) = 10 + 0.3t + 0.4t^2 - 0.01t^3$$
$$I(20) = 10 + 0.3(20) + 0.4(20)^2 - 0.01(20)^3 \quad \text{\textit{Substitute 20 for t.}}$$
$$I(20) = 10 + 6 + 160 - 80 \quad\quad\quad\quad\quad \text{\textit{Evaluate.}}$$
$$I(20) = 96$$

The intensity of the light at 20°C is 96 lumens.

### Example 3

Find $p(a + 1)$ if $p(x) = 4x - x^2 + 2x^3$.

$$p(a + 1) = 4(a + 1) - (a + 1)^2 + 2(a + 1)^3 \quad \text{\textit{Substitute a + 1 for x.}}$$
$$= 4a + 4 - (a^2 + 2a + 1) + 2(a^3 + 3a^2 + 3a + 1)$$
$$= 2a^3 + 5a^2 + 8a + 5$$

### Example 4

Find $4[p(x)] + 3[p(x + 2)]$ if $p(x) = x^3 + 2x^2 - 4$.

$$4[p(x)] + 3[p(x + 2)]$$
$$= 4[x^3 + 2x^2 - 4] + 3[(x + 2)^3 + 2(x + 2)^2 - 4]$$
$$= 4x^3 + 8x^2 - 16 + 3[(x^3 + 6x^2 + 12x + 8) + (2x^2 + 8x + 8) - 4]$$
$$= 4x^3 + 8x^2 - 16 + 3[x^3 + 8x^2 + 20x + 12]$$
$$= 7x^3 + 32x^2 + 60x + 20$$

The coefficient of the term with the highest degree in a polynomial function is called the leading coefficient. The general shapes of the graphs for polynomial functions with positive leading coefficients and degree greater than 0 are shown below.

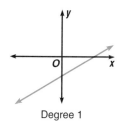

Degree 1

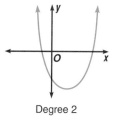

Degree 2

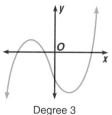

Degree 3

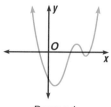

Degree 4

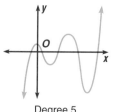

Degree 5

Study the graphs of the odd degree functions. The leftmost points of the graphs of odd degree functions have negative values for $y$. The rightmost points of the graphs of those functions have positive values for $y$.

Now study the graphs of the even degree functions. The leftmost points of the graphs have positive values for $y$ as do the rightmost points of the graphs.

Another pattern can be observed in the graphs of polynomial functions. The simplest polynomial graphs are those with equations in the form $f(x) = x^n$, where $n$ is a positive number. Observe the general shape of even degree polynomial functions and odd degree polynomial functions.

even degree polynomial functions

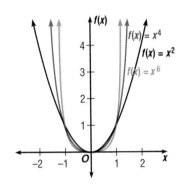

odd degree polynomial functions

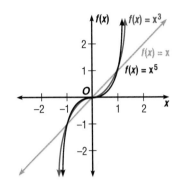

Note that these even degree functions intersect the $x$-axis once. An even degree function may or may not intersect the $x$-axis depending on its location in the coordinate plane. However, an odd degree function will always cross the $x$-axis at least once. Remember that where the graph crosses the $x$-axis is called a zero of the function. On the coordinate plane, these zeros are real numbers.

**Example 5**

Determine if each graph represents an odd degree function or an even degree function. Then state how many real zeros each function has.

a.

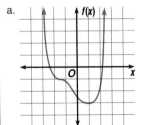

b.

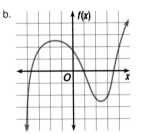

c.

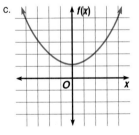

| Graph | leftmost y values | rightmost y values | degree of function | times graph crosses x-axis | number of real zeros |
|---|---|---|---|---|---|
| a. | pos. | pos. | even | 2 | 2 |
| b. | neg. | pos. | odd | 3 | 3 |
| c. | pos. | pos. | even | 0 | 0 |

# CHECKING FOR UNDERSTANDING

**Read and study the lesson to answer these questions.**

1. Write two expressions, one that is a polynomial in one variable and one that is not. Explain why one is a polynomial in one variable and one is not.

2. Is $f(x) = x^3 - 2x + 4$ a polynomial expression or a polynomial function?

3. What would the intensity of the light emitted by a firefly be if the temperature is 15°C?

4. What are some of the characteristics of the graphs of odd degree polynomial functions? How do they differ from the graphs of even degree polynomial functions?

**Guided Practice**

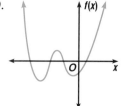

**Determine if each expression is a polynomial in one variable. If it is, state its degree.**

5. $3x^3 - 4x^5 - 6x^2 + 3x$

6. $3a + 5b - 1$

7. $14$

8. $x^2 - 8x + 9$

9. $xy\sqrt{2} + 3$

10. $6n^3 - 4n + 9$

11. $\dfrac{9}{t} + t + 3$

12. $13x^2 - 3x + 14$

13. $x^3 + 5x^2 + x\sqrt{3} + 2$

14. $(8 + 2i)x^2 + (1 - i)x - 8$

**Find $p(2)$ for each function $p(x)$.**

15. $p(x) = 4x + 2$

16. $p(x) = 2x^2 + 6x - 8$

17. $p(x) = 4x^3 - 2x^2 + x - 1$

18. $p(x) = \dfrac{x^2}{4} - 4x + 11$

**Determine whether the degree of the function represented by each graph is even or odd. How many real zeros does each polynomial function have?**

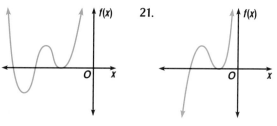

19.     20.     21.

# EXERCISES

**Practice**

**Find $p(2)$ for each function, $p(x)$.**

22. $p(x) = 2x^4 - 3x^3 + 8$

23. $p(x) = -3x^4 + 1$

24. $p(x) = x^5 - x^2$

25. $p(x) = -x^6 + 12$

Find $f(-5)$ for each function $f(x)$.

**26.** $f(x) = 3 - 2x$

**27.** $f(x) = 6x + 9$

**28.** $f(x) = 3x^2$

**29.** $f(x) = x^2 - 2x + 1$

**30.** $f(x) = x^3 + 4x^2 + x + 15$

**31.** $f(x) = x^4 + 10x$

**32.** $f(x) = \dfrac{x^4}{25} - 2$

**33.** $f(x) = 30 - \dfrac{x^3}{6}$

Find $f(x + h)$ for each function $f(x)$.

**34.** $f(x) = x + 1$

**35.** $f(x) = 2x - 3$

**36.** $f(x) = 4x^2$

**37.** $f(x) = x^2 - 2x + 5$

**38.** $f(x) = x^2 - \dfrac{1}{2}x$

**39.** $f(x) = x^3 + 4x$

**40.** $f(x) = \dfrac{4}{3}x^3 - 1$

**41.** $f(x) = 2x^3 - x^2 + 4$

Find $4[p(x)]$ for each function $p(x)$.

**42.** $p(x) = 2x - 4$

**43.** $p(x) = x^2 + 5$

**44.** $p(x) = 6x^3 - 4x^2 + 2$

**45.** $p(x) = \dfrac{x^3}{4} + \dfrac{x^2}{16} - 2$

**Determine whether the degree of the function repeated by each graph is even or odd. How many real zeros does each polynomial function have?**

**46.**

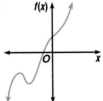

**47.**

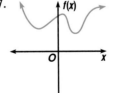

**48.**

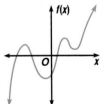

Find $2[f(x + 3)]$ for each function $f(x)$.

**49.** $f(x) = 3x + 8$

**50.** $f(x) = x^2 - 8$

**51.** $f(x) = x^2 + 6x - 18$

**52.** $f(x) = \dfrac{1}{2}x^2 - \dfrac{3}{4}$

Find $2[p(x)] - 3[p(x + 1)]$ for each function $p(x)$.

**53.** $p(x) = 5x - 7$

**54.** $p(x) = x^2 - 7x + 16$

**55.** $p(x) = x^3 + 1$

**56.** $p(x) = (x - 2)^3$

**Critical Thinking**

**57.** A super ball dropped from a height of 36 inches rebounds three-fourths of the distance of the previous bounce. Write an expression to represent the distance traveled by the ball in $n$ bounces. Justify your answer.

**Applications**

**58. Energy** The power generated by a windmill is a function of the speed of the wind. The approximate power is given by the function $P(s) = \dfrac{s^3}{1000}$, where $s$ represents the speed of the wind in kilometers per hour. Find the units of power generated by a windmill when the wind speed is 25 kilometers per hour.

**59. Art** Joyce Cafaro purchases works of art for an art gallery. Two years ago she bought a painting for \$20,000 and last year she bought one for \$35,000. If these paintings appreciate at 14% per year, how much are these two pieces worth now?

**Mixed Review**

**60.** Solve $\begin{cases} x^2 + y^2 = 5 \\ 2x^2 + y = 0 \end{cases}$. **(Lesson 9-9)**

**61.** Solve $\begin{cases} 5x^2 + 5y^2 \geq 25 \\ 2x - 3y = 5 \end{cases}$ by graphing. **(Lesson 9-9)**

**62.** Graph $f(x) = x^2 - 2x + 7$. **(Lesson 8-3)**

**63.** State whether the equation $5x^4 + 6x^3 = 9$ is in quadratic form. **(Lesson 7-6)**

**64.** Find values of $x$ and $y$ for which the equation $x + 2y\boldsymbol{i} = 3$ is true. **(Lesson 6-9)**

**65. Sports** Karl caught three times as many fish as Adam, and ten more fish than Sally. If the total number of fish caught was less than 100, what is the greatest number of fish that Karl could have caught? **(Lesson 2-6)**

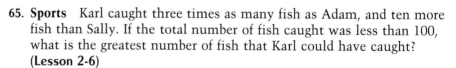

## ~ HISTORY CONNECTION ~

**Karl Friedrich Gauss** (1777–1855) was more than a famous adult mathematician. He displayed his mathematical talents as a child. His father, a gardener and bricklayer, taught Gauss his numbers at an early age. At the age of three, Gauss is said to have found a mathematical error in his father's ledgers of the total of his laborers' salaries. At the age of 10, when he entered school, his teacher asked the students to find the sum of 81297 + 81495 + 81693 + ... + 100899 to overwhelm them with the power of mathematics. Young Gauss stated the correct answer almost at the same time the teacher finished writing the problem on the board. While Gauss was brilliant, he did not like to publicize his thoughts. Many of the mathematical discoveries accredited to Gauss came from his diary, which was found after his death on February 23, 1855.

# Graphing Calculator Exploration: Graphing Polynomial Equations

You can use your graphing calculator to graph polynomial functions and approximate the real zeros of a function. When using your calculator to approximate zeros, it is important to view a complete graph of the function before zooming-in on a certain point. Otherwise, zeros may be overlooked because they were not in the viewing window. Remember that a complete graph of the function shows all the characteristics of the graph, such as all $x$- and $y$-intercepts, relative maximum and minimum points, and the end behavior of the graph.

**Example 1**

Graph the function $f(x) = 3x^3 + 2x^2 - 8x + 7$, so that a complete graph appears in the viewing window. Then approximate each of the real zeros to the nearest hundredth.

Let's try graphing in the standard viewing window.

*Casio*

ENTER: [GRAPH] 3 [ALPHA] [X] [$x^y$] 3

[+] 2 [ALPHA] [X] [$x^2$] [−] 8

[ALPHA] [X] [+] 7 [EXE]

*TI-81*

ENTER: [Y=] 3 [X|T] [^] 3 [+] 2

[X|T] [$x^2$] [−] 8 [X|T] [+]

7 [GRAPH]

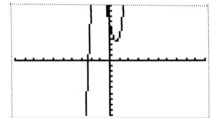

This viewing window will not accommodate a complete graph. Change the viewing window to $[-4, 4]$ by $[-10, 30]$ with a scale factor of 1 for the $x$-axis and 2 for the $y$-axis and graph again by pressing [EXE] on the Casio or [GRAPH] on the TI-81.

This window can accommodate a complete graph.

According to the graph, there is one real zero to this function. There are three zeros for any third degree function, so two of the zeros for this function must be imaginary.

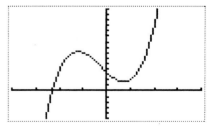

Use the tracing and the zoom-in features of the calculator to approximate the zero. The zero is approximately $-2.28$. *Check this result.*

When using your graphing calculator to approximate real zeros, it is helpful to know that a function with degree $n$ has $n$ zeros. This is a corollary to the Fundamental Theorem of Algebra, which you will study in Lesson 10-4. Since an $n$th degree function has $n$ zeros, a function with degree 5 has 5 zeros. So, if you can see five $x$-intercepts in the viewing window you have all of the zeros in the window. However, these zeros may not all be real. Complex zeros occur in pairs of conjugates, so a fifth degree function may have five, three, or one real zero.

# EXERCISES

**Graph each function so the complete graph is shown. Then sketch the graph of the function and state how many real zeros the function has.**

**1.** $f(x) = x^3$

**2.** $g(x) = 4x^4 + 3$

**3.** $h(x) = 3x^3 + 2x - 1$

**4.** $r(x) = 6x^4 - 3x^2 + 7$

**5.** $m(x) = 7x^5 - 8x^4 + 1$

**6.** $f(x) = x^5 - 9x^4 + 6x^3 + x + 4$

**7.** $g(x) = -4x^4 + 9x + 2$

**8.** $h(x) = 2x^{11} + 6x^5 - 1$

**9.** $f(x) = x^4 + x^3 - 37x^2 - 64x + 84$

**10.** $n(x) = 7x^9 + 4x^7 - 3x^5 + 9x^3 - 1$

**Graph each function so the complete graph is shown. Then approximate each of the real zeros to the nearest hundredth.**

**11.** $f(x) = x^3 - 3$

**12.** $g(x) = x^3 - 5$

**13.** $g(x) = x^3 - 4x + 4$

**14.** $h(x) = -7x^3 - 6x + 1$

**15.** $f(x) = x^3 - 2x^2 + 6$

**16.** $f(x) = x^3 - x^2 + 1$

**17.** $f(x) = x^5 - 6$

**18.** $nx = x^4 + 3x^3 - 4x^2 - 7$

**19.** $m(x) = 3x^4 - x^2 + x - 1$

**20.** $g(x) = x^4 - 4x^3 - 4x^2 + 24x - 6$

**21.** $p(x) = x^4 - x^2 + 6$

**22.** $p(x) = 2x^5 + 3x - 2$

**23.** $r(x) = x^4 - x^2 - 6$

**24.** $f(x) = x^3 + 2x^2 - 3x - 5$

**25.** $g(x) = x^5 - x^3 - x + 1$

**26.** $h(x) = 3x^3 - 16x^2 + 12x + 6$

**27.** $f(x) = x^4 - 4x^2 + 3$

**28.** $c(x) = x^4 - 10x^2 + 21$

**29.** $m(x) = x^5 - 3x^4 + x^3 + 5x^2 - 6x - 1$

**30.** $n(x) = x^4 - 9x^3 + 25x^2 - 24x + 6$

**31.** $q(x) = x^5 + 4x^4 - x^3 - 9x^2 + 3$

**32.** $r(x) = x^5 + 2x^4 - 10x^3 - 20x^2 + 9x + 15$

# 10-2 The Remainder and Factor Theorems

**Objective**

After studying this lesson, you should be able to:

■ find factors of polynomials using the Factor Theorem and synthetic division.

**Application**

Chris Sabo of the Cincinnati Reds hits a high fastball straight up over home plate. The function that describes the height of the ball after $t$ seconds is $h(t) = -16t^2 + 80t + 5$. The graph at the right shows this function. The roots of the function tell at what times the ball is theoretically on the ground. Notice that when $t = 0$, the height of the ball is 5 feet. This is the point at which he hit the ball.

Suppose we find the height of the ball after 4 seconds.

$$h(t) = -16t^2 + 80t + 5$$
$$h(4) = -16(4^2) + 80(4) + 5 \qquad \textit{Replace t with 4.}$$
$$= -256 + 320 + 5$$
$$= 69$$

After 4 seconds, the height of the ball is 69 feet.

Now let's divide the polynomial in the function by $t - 4$, and compare the remainder to $h(4)$.

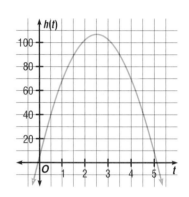

*long division*

$$
\begin{array}{r}
-16t + 16 \\
t - 4 \overline{\smash{\big)}\, -16t^2 + 80t + 5} \\
\underline{-16t^2 + 64t} \\
16t + 5 \\
\underline{16t - 64} \\
69
\end{array}
$$

*synthetic division*

$$
\begin{array}{r|rrr}
4 & -16 & 80 & 5 \\
  &     & -64 & 64 \\
\hline
  & -16 & 16 & 69 \\
\end{array}
$$

Notice that the value of $h(4)$ is the same as the remainder when the polynomial is divided by $t - 4$. This illustrates the **Remainder Theorem.** This theorem states that, for $f(x)$, the value $f(a)$ is equal to the remainder when f(x) is divided by $x - a$.

|  | If a polynomial $f(x)$ is divided by $x - a$, the remainder is the constant $f(a)$, and |
| The Remainder Theorem | $$\text{dividend} = \text{quotient} \cdot \text{divisor} + \text{remainder}$$ $$f(x) = q(x) \cdot (x - a) + f(a)$$ where $q(x)$ is a polynomial with degree one less than the degree of $f(x)$. |

**Example 1**

Let $f(x) = 2x^4 + x^3 - 3x^2 - 5$. Show that $f(2)$ is the remainder when $f(x)$ is divided by $x - 2$.

Use synthetic division to divide by $x - 2$.

```
2 | 2   1   -3    0   -5
  |     4   10   14   28        Long division could also be used.
  ─────────────────────────
    2   5    7   14 | 23
```

The quotient is $2x^3 + 5x^2 + 7x + 14$ with a remainder of 23.

Now find $f(2)$.

$$f(2) = 2(2)^4 + (2)^3 - 3(2)^2 - 5$$
$$= 32 + 8 - 12 - 5 \text{ or } 23$$

Thus, $f(2) = 23$, the same number as the remainder after division by $x - 2$.

As illustrated in Example 1, synthetic division can be used to find the value of a function. Synthetic division, when used to find the value of a function, is often called **synthetic substitution.** This is a very convenient way of finding the value of a function, especially when the degree of the polynomial is greater than two.

**Example 2**

If $f(x) = x^4 - 10x^3 + x^2 - 8x + 1$, find $f(10)$.

When $f(x)$ is divided by $x - 10$, the remainder is $f(10)$.

```
10 | 1  -10    1   -8    1        Check:     Use direct substitution.
   |      10    0   10   20        f(10) = (10)^4 - 10(10)^3 + (10)^2 - 8(10) + 1
   ──────────────────────────
     1    0     1    2 | 21        21 ≟ 10,000 - 10,000 + 100 - 80 + 1
                                   21 = 21  ✓
```

By *synthetic substitution*, $f(10) = 21$.

Consider $f(x) = x^4 + x^3 - 13x^2 - 25x - 12$. If $f(x)$ is divided by $x - 4$, then the remainder is zero. Therefore, 4 is a zero of $f(x)$.

```
4 | 1   1   -13   -25   -12        Check:
  |     4    20    28    12        f(4) = (4)^4 + (4)^3 - 3(4)^2 - 25(4) - 12
  ──────────────────────────
    1   5     7     3 |  0         0 = 256 + 64 - 208 - 100 - 12
                                   0 = 0  ✓
```

From the results of the division and by using the Remainder Theorem, we can make the following statement.

$$\underset{dividend}{x^4 + x^3 - 13x^2 - 25x - 12} = \underset{quotient \cdot divisor}{(x^3 + 5x^2 + 7x + 3)(x - 4)} + \underset{remainder}{0}$$

Since the remainder is zero, $x - 4$ is a factor of $x^4 + x^3 - 13x^2 - 25x - 12$. This illustrates the **Factor Theorem,** which is a special case of the Remainder Theorem.

| *The Factor Theorem* | **The binomial $x - a$ is a factor of the polynomial $f(x)$ if and only if $f(a) = 0$.** |
|---|---|

Suppose you wanted to find the zeros of $f(x) = x^3 + 3x^2 - 6x - 8$. From the graph at the right you find that the graph crosses the x-axis at -4, -1, and 2. These are the zeros of the function. Using these zeros and the zero product property, we can express the polynomial in factored form:

$f(x) = (x + 4)(x + 1)(x - 2)$

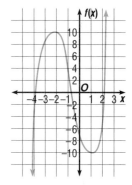

Many polynomial functions are not easily graphed and once graphed, the exact zeros are often difficult to determine. The Factor Theorem can help in finding all the factors of a polynomial. Suppose we wanted to determine if $x + 2$ is a factor of $x^3 - x^2 - 10x - 8$ and, if it is, what the other factors are.

Let $f(x) = x^3 - x^2 - 10x - 8$. If $x + 2$ is a factor of the polynomial, then -2 is a zero. Use the Factor Theorem.

$$\begin{array}{r|rrrr} -2 & 1 & -1 & -10 & -8 \\ & & -2 & 6 & 8 \\ \hline & 1 & -3 & -4 & 0 \end{array}$$

Since the remainder is 0, $x + 2$ is a factor of the polynomial. Further, since $x - 2$ is a factor of the polynomial, it follows that the remainder is 0.

When you divide a polynomial by one of its binomial factors, the quotient is called a **depressed polynomial.** In the polynomial above, $x^3 - x^2 - 10x - 8$ can be factored as $(x + 2)(x^2 - 3x - 4)$. The polynomial $x^2 - 3x - 4$ is the depressed polynomial, which also may be factorable.

$x^2 - 3x - 4 = (x - 4)(x + 1)$

So, $x^3 - x^2 - 10x - 8 = (x + 2)(x - 4)(x + 1)$.

**Example 3**

Show that $x + 2$ is a factor of $x^3 - 2x^2 - 5x + 6$. Then find any remaining factors.

$$
\begin{array}{r|rrrr}
-2 & 1 & -2 & -5 & 6 \\
   &   & -2 & 8 & -6 \\
\hline
   & 1 & -4 & 3 & 0
\end{array}
$$

The remainder is 0, so $x + 2$ is a factor of $x^3 - 2x^2 - 5x + 6$.

So, $x^3 - 2x^2 - 5x + 6 = (x^2 - 4x + 3)(x + 2)$.

Can the depressed polynomial be factored?

$x^2 - 4x + 3 = (x - 3)(x - 1)$

So, $x^3 - 2x^2 - 5x + 6 = (x - 3)(x - 1)(x + 2)$.

The graph of the polynomial function crosses the x-axis at 3, 1, and −2.

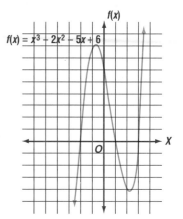

$f(x) = x^3 - 2x^2 - 5x + 6$

# CHECKING FOR UNDERSTANDING

**Communicating Mathematics**

**Read and study the lesson to answer each question.**

1. When the polynomial function $f(x)$ is divided by $x - 4$, the remainder is 12. What is $f(4)$?

2. What is a depressed polynomial?

3. If the divisor is a factor of a polynomial, then what is the remainder after division?

4. In the baseball problem, approximately how long did the catcher have to get under the ball to catch the pop-up for an out?

5. Suppose a depressed polynomial is a quadratic. What methods could you use to determine if the polynomial can be factored?

**Guided Practice**

**State the degree of each polynomial. Then state the degree of the depressed polynomial that would result from dividing the polynomial by one of its binomial factors in the form $x - a$.**

6. $7x^3 - 4x^2 + 3x - 5$

7. $x^5 - 3x^2 + 4$

8. $x^2 - 2x - 9$

9. $x^6 - 3x^5 + 5x^4 + 9$

**Use synthetic substitution to find $g(2)$ for each function $g$.**

10. $g(x) = x^2 - 5$

11. $g(x) = x^3 - 3x^2 + 4x + 8$

12. $g(x) = x^4 - 5x + 2$

13. $g(x) = x^2 - 4x + 4$

# EXERCISES

**Practice**

Divide using synthetic division and write your answer in the form *dividend = quotient · divisor + remainder*. Is the binomial a factor of the polynomial?

**14.** $(x^3 - 4x^2 + 2x - 6) \div (x - 4)$

**15.** $(x^3 - 8x^2 + 2x - 1) \div (x + 1)$

**16.** $(2x^3 + 8x^2 - 3x - 1) \div (x - 2)$

**17.** $(x^4 - 16) \div (x - 2)$

**18.** $(x^3 + 27) \div (x + 3)$

**19.** $(6x^3 + 9x^2 - 6x + 2) \div (x + 2)$

**20.** $(x^3 - 64) \div (x - 4)$

**21.** $(4x^4 - 2x^2 + x + 1) \div (x - 1)$

Use synthetic substitution to find $f(2)$ and $f(-1)$ for each function $f$.

**22.** $f(x) = x^3 - 2x^2 - x + 1$

**23.** $f(x) = x^3 + 2x^2 - 3x + 1$

**24.** $f(x) = 2x^2 - 8x + 6$

**25.** $f(x) = x^3 - 8x^2 - 2x + 5$

**26.** $f(x) = 3x^4 + 8x^2 - 1$

**27.** $f(x) = x^4 + x^3 + x^2 + x + 1$

Given a polynomial and one of its factors, find the remaining factors of the polynomial. Some factors may not be binomials.

**28.** $x^3 + 2x^2 - x - 2;\ x - 1$

**29.** $x^3 - 6x^2 + 11x - 6;\ x - 2$

**30.** $2x^3 + 17x^2 + 23x - 42;\ 2x + 7$

**31.** $x^3 - 3x + 2;\ x - 1$

**32.** $x^3 - x^2 - 5x - 3;\ x + 1$

**33.** $x^4 + 2x^3 - 8x - 16;\ x + 2$

**34.** $8x^4 + 32x^3 + x + 4;\ 2x + 1$

**35.** $16x^5 - 32x^4 - 81x + 162;\ x - 2$

Use the graph of the polynomial function to determine at least one of the binomial factors of the polynomial. Then find all factors of the polynomial.

**36.** $x^5 + x^4 - 3x^3 - 3x^2 - 4x - 4$

**37.** $x^5 + x^4 - x - 1$

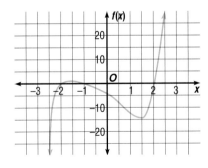
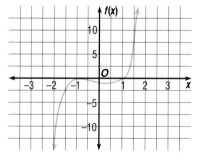

Find values for $k$ so that each remainder is 3.

**38.** $(x^2 + kx - 17) \div (x - 2)$

**39.** $(x^2 - x + k) \div (x - 1)$

**40.** $(x^3 + 4x^2 + x + k) \div (x + 1)$

**41.** $(x^2 + 5x + 7) \div (x + k)$

**42.** Consider the function $f(x) = x^3 + 2x^2 - 5x - 6$.

   **a.** Use synthetic substitution to find the values of $f(-4)$, $f(-2)$, $f(0)$, $f(2)$, and $f(4)$.

   **b.** On a coordinate plane, graph the ordered pairs of the form $(x, f[x])$ you found and connect them to make a smooth curve.

   **c.** How many times does the graph cross the $x$-axis? Does this result agree with what you learned in Lesson 10-1 about the graphs of polynomial functions?

**Applications**

**43. Financial Planning** Jo Phillips, a financial advisor, is helping Ramon's parents develop a plan to save money for his college education. Ramon will start college in six years. According to Ms. Phillips's plan, Ramon's parents will save $1000 each year for the next three years. The fourth and fifth years they will save $1200 each year. The last year before he starts college, they will save $2000.

   **a.** In the formula $A = P(1 + r)^t$, $A =$ the balance, $P =$ the amount invested, $r =$ the interest rate, and $t =$ the number of years the money has been invested. Use this formula to write a polynomial equation to describe the balance of the account when Ramon starts college.

   **b.** Find the balance of the account if the interest rate is 6%.

**44. Physics** A model rocket is shot straight up with an initial velocity of 64 meters per second. The height (in meters) of the rocket after $t$ seconds is given by the function $h(t) = 64t - 4.9t^2$.

   **a.** What do the solutions of the equation represent?

   **b.** How long is the rocket in flight?

**Mixed Review**

**45.** Find $f(2)$ for the function $f(x) = -3x^3 + 2$.　**(Lesson 10-1)**

**46.** Find $3[p(x - 1)]$ for the function $p(x) = x^2 - 4$.　**(Lesson 10-1)**

**47.** Find the center and radius of the circle whose equation is $(x + 4)^2 + y^2 = 49$.　**(Lesson 9-3)**

**48.** Solve the inequality $(y - 12)(y - 5) \le 0$.　**(Lesson 8-7)**

**49.** Solve the equation $169 = 9x^2$.　**(Lesson 7-1)**

**50.** Evaluate $\dfrac{3^0 y + 4y^{-1}}{y^{-\frac{2}{3}}}$ when $y = 8$.　**(Lesson 6-6)**

**51.** Find the product $7 \begin{bmatrix} 4 & 0 \\ 3 & -1 \end{bmatrix}$.　**(Lesson 4-3)**

# 10-3 Problem-Solving Strategy: Combining Strategies

**Objective**

After studying this lesson, you should be able to:

■ solve problems by using more than one strategy.

**Application**

Suppose you wanted to know what is the largest amount of United States coins that you could have and still not be able to make correct change for a $1 bill. There are several ways you could attack this problem.

■ Make a list of all possible change combinations.
■ Make a table to record how many of each type of coin.
■ Work backwards by having change for $1 and then determine what coins you could substitute and/or add but still not have change for $1.
■ Model the problem with real coins, play money, or game pieces.

Whatever method you use, you will discover that two possibilities exist, both of which equal $1.19. They are
1 quarter, 9 dimes, 4 pennies and 3 quarters, 4 dimes, 4 pennies.

Many problems, like the one above, can be solved by using any one of a number of different strategies. Sometimes it takes more than one strategy to solve a problem.

**Example**

> **Mikhail has $3.21 to spend on pens. He knows the price of each pen is greater than 50¢. When he goes to the store, he is able to spend the exact amount of money he has on several pens. How many pens did he purchase? How much did each pen cost if each pen has the same price?**
>
> Let's find what strategies we can use to attack this problem. We could *list all the possibilities* of prices and then find multiples of each price. This strategy would work but it would be very time consuming.
>
> We could *work backwards* from the total $3.21. If he bought 7 pens at 50¢, the total would be $3.50. This exceeds the maximum. So we know the number of pens has to be less than 7. From the problem, we know he bought more than 1 pen. So the answer must be either 2, 3, 4, 5, or 6.
>
> Since we know that Mikhail spent all his money and each pen has the same price, we know that 321 is a multiple of some number. 321 is an odd number so we can *eliminate the possibility* of an even answer. This means the answer could be 3 or 5. 321 is a multiple of 3 and is not a multiple of 5. So 3 is the number of pens Mikhail bought, and by dividing we find that each pen costs $1.07.

# CHECKING FOR UNDERSTANDING

**Communicating Mathematics**

**Read and study the lesson to answer each question.**

1. List all the problem solving strategies you can.

2. Which of these strategies do you think could be combined?

3. Which strategies would you use to solve the U.S. coin problem? Explain why you chose those strategies.

**Guided Practice**

4. Your algebra final exam has 120 questions. Your score is based on 1 point for each correct answer minus one-fourth point for each incorrect answer. Suppose you answered all the questions and got a score of 100 points. How many correct answers did you have?

# EXERCISES

| Strategies |
| --- |
| Look for a pattern. |
| Solve a simpler problem. |
| Act it out. |
| Guess and check. |
| Draw a diagram. |
| Make a chart. |
| Work backwards. |

**Solve. Use any strategy.**

5. There are between 50 and 80 books on the shelf in the computer room. Exactly 20% of the books are on using BASIC and exactly one-seventh of the books are on fractal geometry. If possible, determine how many books there are in all.

6. The square of an integer is a four-digit number whose digits are all even. Find the greatest such number.

7. Darcy is a math major at Indiana University. She is studying for finals with her friends, Janna and Ray at the university library. Janna is an English major and Ray is a sociology major. Ray said, "Isn't it funny that one of us is studying English, one math, and one sociology, but none of us is studying the subject that is our major?" The person studying English said, "So what?" Who is studying each subject?

8. Place addition signs in the left side of the equation below so that a true sentence results.

$$9 \ 8 \ 7 \ 6 \ 5 \ 4 \ 3 \ 2 \ 1 = 99$$

9. Square $ABCD$ has an area of $x$ square units. Each vertex is joined to the midpoint of one opposite side.

   a. What figure is formed in the center of the square?

   b. What is the area of this figure?

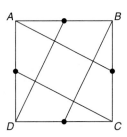

10. After half of the people at a Spanish Club meeting left, one-third of those remaining began to plan the club's Cinco de Mayo (5th of May) celebration. The other 18 people were cleaning up the room. How many people attended that meeting?

11. In the figure at the right, assume all vertical lines are parallel, all angles are right angles, and all horizontal lines are equally distant. What part of the figure is shaded?

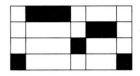

12. You are lost in the wilderness and are trying to get to the nearest town. You know that the people from the nearby village of Nallini are famous for always lying. The people from the village of Yassini always tell the truth. You have come to a fork in the road and there is a person from each of the two villages standing there. You don't know which person is which, so you choose one, point to the right, and ask "If I asked him whether or not this road leads to the nearest town, would he say yes?" "Yes he would" she replied. Should you head right or left at the fork?

13. The desk calendar shown at the right uses four cubes to show the month and the day. The two cubes in the center can be combined to make any date from 01 to 31. The cube on the right has 3 sides printed with the digits 3, 4, and 5. What digits are shown on the other sides and on the other cube?

14. Arrange exactly six straws of the same length to form six congruent triangles.

## COOPERATIVE LEARNING ACTIVITY

**Work in groups. Each person in the group must understand the solution and be able to explain it to any person in class.**

The polynomial equation below has at least one root that is an integer. Factor the polynomial completely and find all the roots of the equation.

$$x^6 + 2x^5 - 26x^4 - 28x^3 + 145x^2 + 26x - 120 = 0$$

# Roots and Zeros

**Objective**

After studying this lesson, you should be able to:

- find the number of positive real zeros, negative real zeros, and complex zeros for a polynomial function.

**Application**

As part of an art class project, Carl is designing a carton for repackaging candy bars that the junior class will sell to raise money for the prom. The volume of the carton must be 120 in³. To hold the correct number of candy bars, the carton must be 3 inches longer than it is wide. The height of the carton should be 2 inches less than the width. Using these restrictions, Carl must determine the dimensions of the carton.

We can find the dimensions of the carton Carl is making by writing a polynomial equation. Then we can use the Factor Theorem and synthetic substitution.

First define each dimension of the carton in terms of the width $w$.

$w$ = width     $w + 3$ = length     $w - 2$ = height

Let $V(w)$ be the function defining the volume.

$$V(w) = (w + 3)(w)(w - 2) \qquad \text{\textit{volume} = lwh}$$
$$120 = w^3 + w^2 - 6w \qquad \textit{Substitute 120 for V(w) and multiply.}$$
$$0 = w^3 + w^2 - 6w - 120 \qquad \textit{Subtract 120 from each side.}$$

We will use a shortened form of synthetic substitution for several values of $w$ to search for the solutions for $0 = w^3 + w^2 - 6w - 120$. The values for $w$ are in the first column of the chart. Beside each value is the last line of the synthetic substitution. Recall that the first three numbers are the coefficients of the depressed polynomial. The last number in each row is the remainder. Study the chart below.

| $w$ | 1 | 1 | $-6$ | $-120$ |
|---|---|---|---|---|
| $-1$ | 1 | 0 | $-6$ | $-114$ |
| 0 | 1 | 1 | $-6$ | $-120$ |
| 1 | 1 | 2 | $-4$ | $-124$ |
| 2 | 1 | 3 | 0 | $-120$ |
| 3 | 1 | 4 | 6 | $-102$ |
| 4 | 1 | 5 | 14 | $-64$ |
| 5 | 1 | 6 | 24 | 0 |
| 6 | 1 | 7 | 36 | 96 |

A remainder of 0 occurs when $r = 5$. This means that $w - 5$ is a factor of the polynomial. The depressed polynomial is $w^2 + 6w + 24$.

The polynomial $w^3 + w^2 - 6w - 120$ can be factored as $(w - 5)(w^2 + 6w + 24)$. The trinomial $w^2 + 6w + 24$ cannot be factored.

$$0 = w^3 + w^2 - 6w - 120$$
$$0 = (w - 5)(w^2 - 6w - 24) \qquad \textit{How do you know } w - 5 \textit{ is a factor?}$$

Use the zero product property.

$w - 5 = 0 \quad$ or $\quad w^2 - 6w + 24 = 0$

$w = 5$

$$w = \frac{-(-6) \pm \sqrt{(-6)^2 - 4(1)(24)}}{2(1)} \qquad \textit{Use the quadratic formula.}$$

$$= \frac{6 \pm \sqrt{-60}}{2}$$

$$= \frac{6 \pm 2i\sqrt{15}}{2} \text{ or } 3 \pm i\sqrt{15}$$

Since imaginary solutions are not applicable to building a carton, use $w = 5$. The dimension of Carl's carton should be 5 inches wide by 5 + 3 or 8 inches long by 5 − 2 or 3 inches high. *Do these dimensions produce the correct volume?*

In Chapter 7, you learned that a **zero** of a function $f(x)$ is any value $a$ such that $f(a) = 0$. This zero is also the **root** or solution of the equation formed when $f(x) = 0$. When the function is graphed, the real zeros of the function will be the $x$-intercepts of the graph. So you see these terms are interrelated.

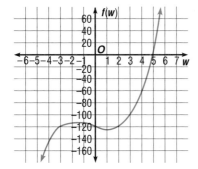

In the equation $w^3 + w^2 - 6w - 120 = 0$, the roots are 5, $3 + i\sqrt{15}$, and $3 - i\sqrt{15}$. One root is real and two are imaginary. The graph of $f(w) = w^3 + w^2 - 6w - 120$, shown at the left, crosses the $x$-axis only once, verifying that there is only one real root.

When you solve a polynomial equation you may have one or more real roots, or no real roots (the roots are imaginary). Remember that real numbers and imaginary numbers belong to the set of complex numbers. So, all polynomial equations have at least one root in the set of complex numbers. This is stated the **Fundamental Theorem of Algebra.**

| | |
|---|---|
| *Fundamental Theorem of Algebra* | **Every polynomial equation with degree greater than zero has at least one root in the set of complex numbers.** |

The following corollary of the Fundamental Theorem of Algebra is an even more powerful tool for problem solving.

| Corollary | **A polynomial equation of the form $P(x) = 0$ of degree $n$ has exactly $n$ roots in the complex numbers.** |
| --- | --- |

**Example 1**

*FYI · · ·*

Karl Friedrich Gauss (1777–1855) is credited with the first proof of the Fundamental Theorem of Algebra.

**Find all roots of $0 = x^3 + 3x^2 - 6x - 8$.**

We can find the roots of the equation by combining some of the strategies listed in Lesson 10-3. Let's list some of the possibilities for roots and then eliminate those that are not. Suppose we begin with integral values from $-5$ to $5$ and use the shortened form of synthetic substitution. Because this polynomial has a degree of 3, the equation has 3 roots. However all of them may not be real.

You can use your calculator to find $f(a)$ quickly. To find $f(-5)$, duplicate the process of synthetic substitution on your calculator.

$-5\underline{\rvert}$  1   3   $-6$   $-8$   is processed by:

ENTER:  5 [+/-] [STO] [×]  1 [+] 3 [=]    *After* [=] *the display is* $-2$.

[×] [RCL] [+]  6 [+/-]  [=]    *After* [=] *the display is* 4.

[×] [RCL] [+]  8 [+/-]  [=] $-28$

The display shown after each [=] gives the second, third, and fourth coefficients of the depressed polynomial. To evaluate $f(x)$ for other values for $x$, simply change the first number entered in the series of keystrokes shown above.

| $r$ | 1 | 3 | $-6$ | $-8$ |
| --- | --- | --- | --- | --- |
| $-5$ | 1 | $-2$ | 4 | $-28$ |
| $-4$ | 1 | $-1$ | $-2$ | 0 |
| $-3$ | 1 | 0 | $-6$ | 10 |
| $-2$ | 1 | 1 | $-8$ | 8 |
| $-1$ | 1 | 2 | $-8$ | 0 |
| 0 | 1 | 3 | $-6$ | $-8$ |
| 1 | 1 | 4 | $-2$ | $-10$ |
| 2 | 1 | 5 | 4 | 0 |

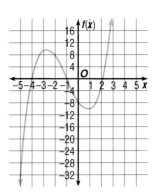

The zeros occur at $x = -4$, $x = -1$, and $x = 2$. The graph of the function verifies that there are three real roots.

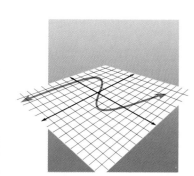

Remember when you solved quadratic equations like $x^2 + 4 = 0$, there were always two imaginary roots. In this case, $2i$ and $-2i$ are the roots. These numbers are conjugate pairs. In any polynomial function, if an imaginary number is a zero of that function, its conjugate is also a zero.

| Complex Conjugates Theorem | Suppose $a$ and $b$ are real numbers with $b \neq 0$. Then, if $a + bi$ is a zero of a polynomial function, then $a - bi$ is also a zero of the function. |
| --- | --- |

**Example 2**

Find all zeros of $f(x) = x^3 - 11x^2 + 40x - 50$ if $3 + i$ is one zero of $f(x)$.

Since $3 + i$ is a zero, $3 - i$ is also a zero. So, both $x - (3 + i)$ and $x - (3 - i)$ are factors of the polynomial $x^3 - 11x^2 + 40x - 50$.

$$f(x) = [x - (3 + i)][x - (3 - i)](\underline{\phantom{?}})$$
$$= (x^2 - 6x + 10)(\underline{\phantom{?}})$$

Since $f(x)$ has a degree of 3, there are three factors. Use division to find the other factor.

$$
\begin{array}{r}
x - 5 \\
x^2 - 6x + 10 \enclose{longdiv}{x^3 - 11x^2 + 40x - 50} \\
\underline{x^3 - 6x^2 + 10x} \\
-5x^2 + 30x - 50 \\
\underline{-5x^2 + 30x - 50} \\
0
\end{array}
$$

Thus, $f(x) = (x^2 - 6x + 10)(x - 5)$.

Since $x - 5$ is a factor, 5 is a zero. The three zeros are $3 + i$, $3 - i$, and 5.

French mathematician René Descartes made more discoveries about the zeros of polynomial functions. His rule of signs is given below.

| Descartes' Rule of Signs | If $P(x)$ is a polynomial whose terms are arranged in descending powers of the variable,<br>■ the number of positive real zeros of $y = P(x)$ is the same as the number of changes in sign of the coefficients of the terms, or is less than this by an even number, and<br>■ the number of negative real zeros of $y = P(x)$ is the same as the number of changes in sign of the coefficients of the terms of $P(-x)$, or is less than this number by an even number. |
| --- | --- |

*When using Descartes' rule of signs, the coefficients are ignored.*

**Example 3**

State the number of positive and negative real zeros for $g(x) = 3x^4 - x^3 + 8x^2 + x - 7$.

Count the number of changes in sign for the coefficients of $g(x)$.

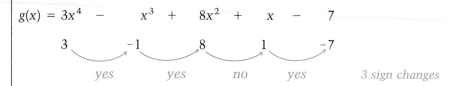

$$g(x) = 3x^4 - x^3 + 8x^2 + x - 7$$

| 3 | −1 | 8 | 1 | −7 |
|---|----|---|---|----|
| yes | yes | no | yes | 3 sign changes |

There are either 3 positive real zeros or 1 (which is 3 − 2) positive real zero.

Now find $g(-x)$ and count the number of changes in signs for its coefficients.

$$g(-x) = 3(-x)^4 - (-x)^3 + 8(-x)^2 + (-x) - 7$$
$$g(-x) = 3x^4 + x^3 + 8x^2 - x - 7$$

| 3 | 1 | 8 | −1 | −7 |
|---|---|---|----|----|
| no | no | yes | no | 1 sign change |

There is exactly 1 negative real zero.

The function $g(x)$ has either 3 or 1 positive real zeros and exactly 1 negative real zero.

---

*Using a graphing calculator or sketching the graph may help in determining the nature of the zeros of a function.*

In Example 3, $g(x)$ has a degree of 4, so it has 4 zeros. Using the information in the example, you can make a chart of the possibilities for these zeros.

| Number of Positive Real Zeros | Number of Negative Real Zeros | Number of Imaginary Zeros | |
|:---:|:---:|:---:|---|
| 3 | 1 | 0 | $3 + 1 + 0 = 4$ |
| 1 | 1 | 2 | $1 + 1 + 2 = 4$ |

**Example 4**

Write the polynomial function of least degree with integral coefficients whose zeros are 6 and $4 - 2i$.

If $4 - 2i$ is a zero, then $4 + 2i$ is also a zero.    *Why?*

Use the zero product property to write a polynomial equation that has these zeros as roots.

$$0 = (x - 6)[x - (4 - 2i)][x - (4 + 2i)]$$    *Why is $x - 6$ a factor?*

Thus, $f(x) = (x - 6)[x - (4 - 2i)][x - (4 + 2i)]$
$$= (x - 6)(x^2 - 8x + 20)$$
$$= x^3 - 14x^2 + 68x - 120$$

# CHECKING FOR UNDERSTANDING

**Communicating Mathematics**

Read and study the lesson to answer these questions.

1. How many zeros does $f(x) = x^4 - 1$ have?

2. If one zero of a polynomial is $2 + i$, name one other zero. What are these zeros called?

3. A polynomial has degree 4 and one positive real zero. Describe the other possible zeros. Justify your answer.

4. Describe how you can use Descartes' Rule of Signs to help you find the zeros of a polynomial function.

**Guided Practice**

Find $f(-x)$ for each function $f(x)$ given.

5. $f(x) = 3x^5 + 7x^2 - 8x + 1$

6. $f(x) = x^4 - 2x^3 + x^2 - 1$

7. $f(x) = 4x^4 - 3x^3 + 2x^2 - x + 1$

8. $f(x) = x^7 - x^3 + 2x - 1$

State the number of positive real zeros, negative real zeros, and imaginary zeros for each function.

9. $f(x) = x^3 + x^2 + x + 1$

10. $f(x) = -x^4 - x^2 - x - 1$

11. $f(x) = x^4 + x^3 - 7x - 1$

12. $f(x) = x^{10} - 1$

# EXERCISES

**Practice**

For each function, state the number of positive real zeros, negative real zeros, and imaginary zeros.

13. $f(x) = 3x^4 + 2x^3 - 3x^2 - 4x + 1$

14. $f(x) = x^4 + x^3 + 2x^2 - 3x - 1$

15. $f(x) = x^3 + 1$

16. $f(x) = x^5 - x^3 - x + 1$

17. $f(x) = x^{10} - x^8 + x^6 - x^4 + x^2 - 1$

18. $f(x) = x^{14} + x^{10} - x^9 + x - 1$

Given a function and one of its zeros, find all of the zeros of the function.

19. $g(x) = x^3 - 6x^2 + 10x - 8;\ 4$

20. $f(x) = x^3 + 2x^2 - 3x + 20;\ -4$

21. $h(x) = 2x^3 - x^2 + 28x + 51;\ -\dfrac{3}{2}$

22. $h(x) = 2x^3 - 17x^2 + 90x - 41;\ \dfrac{1}{2}$

23. $f(x) = 4x^4 + 17x^2 + 4;\ 2i$

24. $f(x) = x^3 + 6x^2 + 21x + 26;\ -2$

25. $f(x) = x^3 - 3x^2 + 9x + 13;\ 2 + 3i$

26. $g(x) = x^3 - 10x^2 + 34x - 40;\ 3 - i$

27. $g(x) = x^4 - 6x^3 + 12x^2 + 6x - 13;\ 3 + 2i$

Write the polynomial function of least degree with integral coefficients that has the given zeros.

28. $2,\ 1 - i$

29. $3,\ 2i$

30. $-1,\ 1,\ 2 - i$

31. $-2 - i,\ 1 + 3i$

32. $4,\ i,\ -1 + i$

33. $3i,\ -2i,\ 1 - i$

**34.** If $f(x) = x^3 + kx^2 - 7x - 15$, find the value of $k$ so that $-2 - i$ is a zero of $f(x)$.

**Applications**

**35. Manufacturing**  The volume of a fudge tin must be 120 cubic centimeters. The tin is 7 centimeters longer than it is wide and six times longer than it is tall. Find the dimensions of the tin.

**36. Manufacturing**  The height of a certain juice can is 4 times the radius of the top of the can. Determine the dimensions of the can if the volume is approximately 17.89 cubic inches.    *Hint: The volume of a right circular cylinder* $= \pi r^2 h.$

**Mixed Review**

**37.** Use synthetic substitution to find $f(3)$ and $f(-2)$ for $f(x) = 2x^2 - 8x + 6.$ **(Lesson 10-2)**

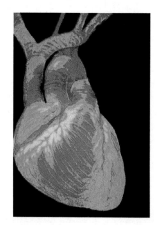

**Journal**

Explain how sketching the graph or using a graphing calculator can help you in determining the number of zeros of a function.

**38.** Write the equation of the parabola with vertex at $(0, 0)$ and focus at $(0, -4)$.   **(Lesson 9-2)**

**39.** Write the function $f(x) = x^2 + 6$ in the form $f(x) = (x - h)^2 + k$. Then name the vertex and axis of symmetry for the graph of the function. **(Lesson 8-2)**

**40. Health**  On the average, an adult heart beats about once every $\frac{5}{6}$ second. If Diane is 18 years old, about how many beats has her heart made? **(Lesson 5-1)**

## ~~~~ MID-CHAPTER REVIEW ~~~~

Find each value if $p(x) = 6x^3 - 3x^2 + 4x - 9$.   **(Lesson 10-1)**

**1.** $p(-3)$  |  **2.** $p(a^2)$  |  **3.** $p(x + 1)$

Given a polynomial and one of its factors, find the remaining factors of the polynomial. **(Lesson 10-2)**

**4.** $x^3 + x^2 - 24x + 36; \; x + 6$  |  **5.** $2x^3 + 13x^2 + x - 70; \; x - 2$

**6.** Four positive two-digit even numbers $xy$ have a fifth power that ends in the digits $xy$. Find them.   **(Lesson 10-3)**

For each function, state the number of positive real zeros, negative real zeros, and imaginary zeros.   **(Lesson 10-4)**

**7.** $f(x) = x^3 - 7x^2 - 6x + 3$  |  **8.** $g(x) = 2x^4 - x^3 - 9x - 12$

Write the polynomial function of least degree with integral coefficients that has the given zeros. **(Lesson 10-4)**

**9.** $1, 1 - i$  |  **10.** $-3, 2, -3 + 2i$

# The Rational Zero Theorem

**Objectives**

After studying this lesson, you should be able to:

- identify all possible rational zeros of a polynomial function using the Rational Zero Theorem, and
- find zeros of polynomial functions.

**Application**

The largest pyramid at Gizeh, Egypt was built by Khufu, a king of the fourth dynasty. The pyramid has a square base and has a volume of 2645 cubic dekameters. The height of the pyramid is approximately 8 dekameters less than the length of a side of the square base. What are the dimensions of the pyramid?    *1 dekameter = 10 meters*

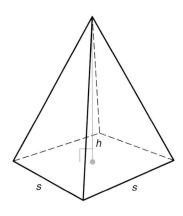

The formula for the volume of a pyramid is $V = \frac{1}{3}Bh$, where $B$ represents the area of the base and $h$ represents the height. Let's set up an equation to find the dimensions of the pyramid.

Let $s$ represent the length of one side of the base of the pyramid. Then the height is $s - 8$.

$$V = \frac{1}{3}Bh$$

$$2645 = \frac{1}{3}(s^2)(s - 8) \qquad V = 2645,\ B = s^2,\ h = s - 8$$

$$7935 = s^3 - 8s^2 \qquad \textit{Multiply each side by 3.}$$

$$0 = s^3 - 8s^2 - 7935$$

We could use synthetic substitution to test possible zeros. But the numbers are so large that we might have to test hundreds of possible zeros before we find one. In situations like this, the **Rational Zero Theorem** can give us some direction in testing possible zeros. This theorem is stated below.

*Rational Zero Theorem*

> Let $f(x) = a_0x^n + a_1x^{n-1} + \ldots + a_{n-1}x + a_n$ represent a polynomial function with integral coefficients. If $\frac{p}{q}$ is a rational number in simplest form and is a zero of $y = f(x)$, then $p$ is a factor of $a_n$ and $q$ is a factor of $a_0$.

A corollary to the Rational Zero Theorem, called the **Integral Zero Theorem,** states that if the coefficients of the polynomial function are integers such that $a_0 = 1$ and $a_n \neq 0$, any rational zeros of the function must be factors of $a_n$.

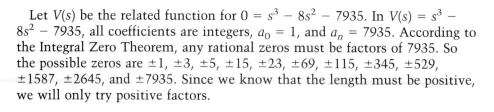

Let $V(s)$ be the related function for $0 = s^3 - 8s^2 - 7935$. In $V(s) = s^3 - 8s^2 - 7935$, all coefficients are integers, $a_0 = 1$, and $a_n = 7935$. According to the Integral Zero Theorem, any rational zeros must be factors of 7935. So the possible zeros are $\pm 1$, $\pm 3$, $\pm 5$, $\pm 15$, $\pm 23$, $\pm 69$, $\pm 115$, $\pm 345$, $\pm 529$, $\pm 1587$, $\pm 2645$, and $\pm 7935$. Since we know that the length must be positive, we will only try positive factors.

According to Descartes' rule of signs, there will be only one positive real root and no negative real roots. We can use synthetic substitution to test for possible zeros and we can stop testing when we find the first zero. Let's make a chart.

Since $s - 8 = h$ and $h$ must be positive, we need only consider values for $s$ that are greater than 8.

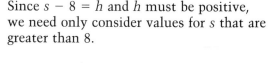

| $s$ | 1 | $-8$ | 0 | $-7935$ |
|---|---|---|---|---|
| 15 | 1 | 7 | 105 | $-6360$ |
| 23 | 1 | 15 | 345 | 0 |

A zero occurs when $s = 23$. Thus, $s - 23$ is a factor of the polynomial and 23 is a root of the equation.

Since Descartes' rule told us there was only one real root, 23 is the only possible measure for the side of the pyramid. The length of one side of the base of the pyramid at Gizeh, Egypt is 23 dekameters, and the height is $23 - 8$ or 15 dekameters.

**Example 1**

**List all of the possible rational zeros of $f(x) = 2x^3 - x^2 - 34x - 56$.**

Since $a_0 \neq 1$, we cannot use the Integral Zero Theorem.

If $\dfrac{p}{q}$ is a rational root, then $p$ is a factor of 56 and $q$ is a factor of 2.

All of the possible values of $p$ are $\pm 1$, $\pm 2$, $\pm 4$, $\pm 7$, $\pm 8$, $\pm 14$, $\pm 28$, and $\pm 56$.

All of the possible values of $q$ are $\pm 1$ and $\pm 2$.

So all of the possible rational zeros are:

$\pm\dfrac{1}{1}$, $\pm\dfrac{1}{2}$, $\pm\dfrac{2}{1}$, $\pm\dfrac{2}{2}$, $\pm\dfrac{4}{1}$, $\pm\dfrac{4}{2}$, $\pm\dfrac{7}{1}$, $\pm\dfrac{7}{2}$, $\pm\dfrac{8}{1}$, $\pm\dfrac{8}{2}$, $\pm\dfrac{14}{1}$, $\pm\dfrac{14}{2}$, $\pm\dfrac{28}{1}$, $\pm\dfrac{28}{2}$, $\pm\dfrac{56}{1}$, and $\pm\dfrac{56}{2}$

or $\pm 1$, $\pm\dfrac{1}{2}$, $\pm 2$, $\pm 4$, $\pm 7$, $\pm\dfrac{7}{2}$, $\pm 8$, $\pm 14$, $\pm 28$, and $\pm 56$.

**Example 2**

The volume of a rectangular solid is 72 cubic units. The width is twice the height and the length is 7 units more than the height. Find the dimensions of the solid.

Let $h$ = the height of the solid.

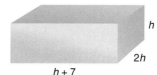

volume = $(2h)(h + 7)(h)$
$$72 = 2h^3 + 14h^2 \qquad \textit{Substitute 72 for volume.}$$
$$0 = 2h^3 + 14h^2 - 72 \qquad \textit{Subtract 72 from each side.}$$
$$0 = h^3 + 7h^2 - 36 \qquad \textit{Divide each side by 2.}$$

All of the possible rational zeros are $\pm 1$, $\pm 2$, $\pm 3$, $\pm 4$, $\pm 6$, $\pm 9$, $\pm 12$, $\pm 18$, and $\pm 36$. Since a measure must be positive and according to Descartes' Rule of Signs there is one positive real zero, we can stop testing possible zeros when we find the first one. Let's make a table and test each possible rational zero.

| $\dfrac{p}{q}$ | 1 | 7 | 0 | $-36$ |
|---|---|---|---|---|
| 1 | 1 | 8 | 8 | $-28$ |
| 2 | 1 | 9 | 18 | 0 |

The zero is 2. So the solid is 4 units by 9 units by 2 units.

---

You have learned many rules to help you determine the number and characteristics of the zeros of a function. Example 3 shows how many of them can be used.

**Example 3**

Find all zeros of $f(x) = 2x^4 - 9x^3 + 2x^2 + 21x - 10$.

- From the corollary to the Fundamental Theorem of Algebra we know there are exactly 4 complex zeros.

- According to Descartes' rule of signs there are either 3 or 1 positive real zeros and exactly 1 negative real zero.

- Now use the Factor Theorem to determine that the possible rational zeros are $\pm\dfrac{1}{2}$, $\pm 1$, $\pm 2$, $\pm\dfrac{5}{2}$, $\pm 5$, and $\pm 10$.

- Use synthetic substitution and a chart to find at least one zero.

| $\dfrac{p}{q}$ | 2 | $-9$ | 2 | 21 | $-10$ | |
|---|---|---|---|---|---|---|
| $\dfrac{1}{2}$ | 2 | $-8$ | $-2$ | 20 | 0 | *One zero is $\dfrac{1}{2}$.* |

The depressed polynomial after division by $x - \frac{1}{2}$ is $2x^3 - 8x^2 - 2x + 20$. Now use a synthetic division chart with this polynomial.

$$
\begin{array}{c|cccc|c}
x & 2 & -8 & -2 & 20 & \\
\hline
1 & 2 & -6 & -8 & & 12 \\
2 & 2 & -4 & -10 & & 0 \quad \text{\textit{Another zero is 2.}}
\end{array}
$$

The depressed polynomial is $2x^2 - 4x - 10$. Use the quadratic formula to find other possible zeros.

$$
\begin{aligned}
x &= \frac{-(-4) \pm \sqrt{(-4)^2 - 4(2)(-10)}}{2(2)} \\
&= \frac{4 \pm \sqrt{96}}{4} \\
&= \frac{4 \pm 4\sqrt{6}}{4} \text{ or } 1 \pm \sqrt{6}
\end{aligned}
$$

The zeros are $\frac{1}{2}$, 2, $1 + \sqrt{6}$, and $1 - \sqrt{6}$. The approximate values of the irrational roots are 3.45 and $-1.45$. So, there are 3 positive zeros and 1 negative zero.

The graph of the function shown at the right crosses the $x$-axis 4 times, confirming that there are 4 real zeros.

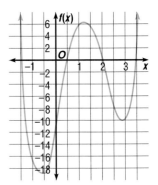

# CHECKING FOR UNDERSTANDING

**Communicating Mathematics**

Read and study the lesson to answer these questions.

1. When can you use the Rational Zero Theorem to determine the possible rational zeros for a polynomial function?

2. Why is it helpful to use the Rational Zero Theorem while finding the zeros of a polynomial function?

3. Describe the possible rational zeros when the leading coefficient, $a_0$, of a polynomial function is one.

4. Write a polynomial function with four possible rational roots.

**Guided Practice**

State all possible rational zeros for each function.

5. $f(x) = x^3 + x + 2$

6. $g(x) = x^3 + 3x^2 - 6x + 5$

7. $f(x) = x^3 + 8x + 6$

8. $h(x) = x^4 - 9x - 3$

9. $p(x) = x^5 - 10$

10. $d(x) = x^3 + 6x^2 - 3x + 1$

11. $g(x) = 3x^3 - 5x^2 - x + 4$

12. $f(x) = 2x^4 + 4x^3 - x^2 - 3$

# EXERCISES

State all possible rational zeros for each function.

**13.** $g(x) = x^3 + 2x^2 - 3x + 5$   **14.** $n(x) = x^3 - 7x + 12$

**15.** $f(x) = x^5 + 6x^3 - 12x + 18$   **16.** $h(x) = 3x^3 - 5x^2 - 11x + 3$

**17.** $p(x) = 6x^3 + 6x^2 - 15x - 2$   **18.** $f(x) = 3x^4 + 15$

**19.** $f(x) = 4x^6 + 6x^4 - 4x^2 - 6$   **20.** $g(x) = 9x^6 - 5x^3 + 27$

Find all the rational zeros for each function.

**21.** $f(x) = x^3 + x^2 - 80x - 300$   **22.** $g(x) = x^3 - x^2 - 34x - 56$

**23.** $g(x) = x^3 - 3x - 2$   **24.** $h(x) = x^4 - 3x^3 + x^2 - 3x$

**25.** $f(x) = x^4 - 3x^3 - 53x^2 - 9x$   **26.** $h(x) = 2x^3 - 11x^2 + 12x + 9$

**27.** $p(x) = 6x^3 + 11x^2 - 3x - 2$   **28.** $g(x) = x^4 + 10x^3 + 33x^2 + 38x + 8$

**29.** $f(x) = x^4 + x^2 - 2$   **30.** $p(x) = x^4 + x^3 - 9x^2 - 17x - 8$

**31.** $g(x) = x^3 + 4x^2 - 3x - 18$   **32.** $f(x) = x^4 - 13x^2 + 36$

**33.** $h(x) = x^3 - x^2 - 40x + 12$   **34.** $g(x) = 48x^4 - 52x^3 + 13x - 3$

**35.** $p(x) = x^5 - 6x^3 + 8x$   **36.** $f(x) = 2x^5 - x^4 - 2x + 1$

**37.** A rectangular solid has a volume of 144 cubic units. The width is twice the height and the length is 2 units more than the width. Find the dimensions of the solid.

Find all zeros of each function.

**38.** $f(x) = 8x^3 - 36x^2 + 22x + 21$

**39.** $g(x) = 6x^3 + 5x^2 - 9x + 2$

**40.** $g(x) = 12x^4 + 4x^3 - 3x^2 - x$

**41.** $p(x) = 6x^4 + 22x^3 + 11x^2 - 38x - 40$

**42.** $h(x) = 5x^4 - 29x^3 + 55x^2 - 28x$

**43.** $g(x) = 9x^5 - 94x^3 + 27x^2 + 40x - 12$

**44.** $p(x) = x^5 - 2x^4 - 12x^3 - 12x^2 - 13x - 10$

**Critical Thinking**

**45.** Suppose $k$ and $2k$ are zeros of $f(x) = x^3 + 4x^2 + 9kx - 90$. Find $k$ and all three zeros of $f(x)$.

**Applications**

**46. Manufacturing**   The volume of a milk carton is 200 cubic inches. The base of the carton is square and the height is 3 inches more than the length of the base. What are the dimensions of the carton?

**47. Food Production**   I.C. Dreams makes ice cream cones shaped like geometric cones. The volume of a cone is about 5.24 cubic inches, and its height is 4 inches more than the radius of the opening of the cone. Determine the dimensions of the cone.   *Hint: The formula for the volume of a cone is $V = \frac{1}{3}\pi r^2 h$.*

48. State the possible number of positive real zeros, negative real zeros, and imaginary zeros of the function $f(x) = 3x^5 - 8x^2 + 1$.   (**Lesson 10-3**)

49. Write the polynomial function of least degree with integral coefficients that has $-2$ and $2 + 3i$ as zeros.   (**Lesson 10-3**)

50. Graph the system $\begin{cases} 4x^2 + 9y^2 = 36 \\ 4x^2 - 9y^2 = 36 \end{cases}$. Then find the solutions.   (**Lesson 9-8**)

51. A possible solution for the system $\begin{cases} 2x - 4y - z = 6 \\ 3x + 4y + 3z = -1 \\ x + y + z = 0 \end{cases}$ is $(1, -1, 0)$. Determine if this ordered triple is a solution.   (**Lesson 3-9**)

---

## LANGUAGE CONNECTION

**Study the mathematical terms used in the statements below.**

$x^3 - 7x + 6$ is a **polynomial expression,** or a **polynomial.** Its factors are $(x - 2)$, $(x - 1)$, and $(x + 3)$.     *When a polynomial is divided by one of its factors, the remainder is zero.*

$x^3 - 7x + 6 = 0$ is a polynomial equation. The **factored form** of the equation is $(x - 2)(x - 1)(x + 3) = 0$. Its **roots,** or **solutions,** are 2, 1, and $-3$.     *When a root of an equation is substituted for the variable, the two sides are equal.*

$f(x) = x^3 - 7x + 6$ is a **polynomial function.** The **zeros** of the function are 2, 1, and $-3$.     *When a zero of a function is substituted for the variable, the result is zero.*

The function $f(x) = x^3 - 7x + 6$ may be expressed as $y = x^3 - 7x + 6$. When a value is substituted for $x$, exactly one value is obtained for $y$. Each ordered pair $(x, y)$ is a **solution** to the equation $y = x^3 - 7x + 6$.     *We do not use the term root for solutions that are ordered pairs.*

**Choose the correct term in parentheses to complete each statement.**

1. $x - 2$ is a *(factor, root)* of the polynomial $x^2 - 4x + 4$.

2. 2 is a *(root, zero)* of the function $f(x) = x^2 - 4x + 4$.

3. $(2, 0)$ is a *(solution, root)* of the equation $y = x^2 - 4x + 4$.

4. 2 is a *(root, zero)* of $x^2 - 4x + 4 = 0$.

5. $y - 6$ is a *(factor, solution)* of $y^2 - 5y - 6$.

6. The zeros of the *(function, expression)* $f(y) = y^2 - 5y - 6$ are 6 and $-1$.

7. The roots of the *(polynomial, equation)* $y^2 - 5y - 6 = 0$ are 6 and $-1$.

8. 7 is a *(factor, zero)* of $f(k) = k - 7$.

9. 7 is a *(solution, factor)* of $k - 7 = 0$.

10. $(3, -4)$ is a *(solution, root)* of $y = k - 7$.

11. $f(k) = k - 7$ is a(n) *(function, expression)*.

12. The factors of the *(equation, expression)* $x^2 - 1$ are $x - 1$ and $x + 1$.

# Graphing Polynomial Functions and Approximating Zeros

**Objectives**

After studying this lesson, you should be able to:

- approximate the real zeros of polynomial functions, and
- graph polynomial functions to find significant points.

**Application**

The space shuttle has an external tank for the fuel that the main engines need for the launch. About eight minutes into the flight, the fuel is gone and the tank is released. This tank is shaped like a capsule, a cylinder with a hemispherical dome at either end. The cylindrical part of the tank has a volume of 1170 cubic meters and a height of 17 meters more than the radius of the tank. What are the dimensions of the tank to the nearest tenth of a meter?

Let's begin searching for the dimensions of the tank by writing a polynomial equation describing the volume of the tank.

Let $r$ = the radius of the tank and $r + 17$ = the height.

*Use the formula for the volume of a cylinder.*

$$V = \pi r^2 h$$
$$V = \pi r^2(r + 17) \qquad \text{\textit{Replace h with r + 17.}}$$
$$1170 = \pi r^3 + 17\pi r^2 \qquad \text{\textit{Replace V with 1170.}}$$
$$0 = \pi r^3 + 17\pi r^2 - 1170 \qquad \text{\textit{Subtract 1170 from each side.}}$$
$$0 = r^3 + 17r^2 - \frac{1170}{\pi} \qquad \text{\textit{Divide each side by }}\pi.$$
$$0 = r^3 + 17r^2 - 372 \qquad \text{\textit{Round to the nearest integer.}}$$

Let $V(r) = r^3 + 17r^2 - 372$. According to Descartes' rule of signs, there is one positive real zero and two or zero negative real zeros. The possible rational zeros are $\pm 1$, $\pm 2$, $\pm 3$, $\pm 4$, $\pm 6$, $\pm 12$, $\pm 31$, $\pm 62$, $\pm 93$, $\pm 124$, $\pm 186$, and $\pm 372$, but upon inspection none of these values is a zero. Let's evaluate some values of $V(r)$. Zero separates the positives from the negatives. If you find a sign change in your list of $V(r)$ values a zero must lie somewhere between those two values.

| $r$ | $V(r)$ | $r$ | $V(r)$ | $r$ | $V(r)$ | $r$ | $V(r)$ |
|---|---|---|---|---|---|---|---|
| -16 | -116 | -10 | 328 | -4 | -164 | 2 | -296 |
| -15 | 78 | -9 | 276 | -3 | -246 | 3 | -192 |
| -14 | 216 | -8 | 204 | -2 | -312 | 4 | -36 |
| -13 | 304 | -7 | 118 | -1 | -356 | 5 | 178 |
| -12 | 348 | -6 | 24 | 0 | -372 | 6 | 456 |
| -11 | 354 | -5 | -72 | 1 | -354 | 7 | 804 |

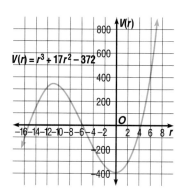

$V(r) = r^3 + 17r^2 - 372$

Use the values from the charts to plot points on a graph. Then sketch the graph of the function $V(r)$ by connecting those points with a smooth curve. The graph will cross the $r$-axis somewhere between the pairs of $r$ values where the corresponding $V(r)$ values change sign. Since $r$-intercepts are zeros of the function, there is a zero between each pair of these $r$ values. This strategy is the **location principle.**

*Note that the function has an odd degree so its leftmost points have negative values for $V(r)$ and the rightmost points have positive values. There are 3 $r$-intercepts, indicating there are 3 real zeros. The zeros are close to $-15$, $-6$, and $4$.*

| *The Location Principle* | **Suppose $y = f(x)$ represents a polynomial function and $a$ and $b$ are two numbers such that $f(a) < 0$ and $f(b) > 0$. Then the function has at least one real zero between $a$ and $b$.** |
|---|---|

Since we would like to approximate the dimensions of the shuttle tank to the nearest tenth of a meter, we will have to repeat the process of evaluating $V(r) = r^3 + 17r^2 - 372$ for successive values of $r$ expressed in tenths. The measure must be positive, so we will evaluate only the $r$ values between 4 and 5. Start with the values closest to 4 since we know that $V(4)$ is closer to 0 than is $V(5)$. Using a calculator will help find these values more easily.

| $r$ | $r^3 + 17r^2 - 372$ | $V(r)$ |
|---|---|---|
| 4 | $4^3 + 17(4)^2 - 372$ | -36 |
| 4.1 | $(4.1)^3 + 17(4.1)^2 - 372$ | -17.309 |
| 4.2 | $(4.2)^3 + 17(4.2)^2 - 372$ | 1.968 |

To evaluate $V(4.1)$:

ENTER: 4.1 $y^x$ 3 $+$ 17

$\times$ 4.1 $x^2$ $-$

372 $=$ $-17.309$

The positive zero of $V(r)$ is between 4.1 and 4.2, but it is closer to 4.2. So, to the nearest tenth, the radius of the cylinder is 4.2 meters. The height of the cylinder is $17 + 4.2$ or 21.2 meters.

**Example 1**

**Approximate the zeros of $f(x) = x^3 + 4x^2 + x - 2$. Then draw the graph.**

According to Descartes' Rule of Signs, there is 1 positive real root and 2 or 0 negative real roots. Let's evaluate several successive values of $x$ to locate the zeros. Then plot the points and connect them to form a smooth graph.

| x | f(x) |
|----|------|
| -4 | -6 |
| -3 | 4 |
| -2 | 4 |
| -1 | 0 |
| 0 | -2 |
| 1 | 4 |

| x | f(x) |
|------|--------|
| -3.5 | 0.625 |
| -3.6 | -0.416 |
| -3.7 | -1.593 |
| 0.5 | -0.375 |
| 0.6 | 0.256 |

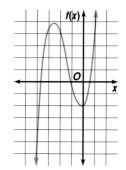

One zero is -1. One other zero lies between -4 and -3 and is approximately -3.6. The other zero lies between 0 and 1 and is approximately 0.6.

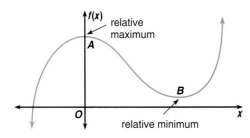

The sideways $s$-shaped graph at the left shows the shape of the graph of a general third degree polynomial function. Point $A$ on the graph is a **relative maximum** of the cubic function since no other nearby points have a greater $y$-coordinate. Likewise, point $B$ is a **relative minimum.** You can use this information to help graph functions that have imaginary zeros.

**Example 2**

**Graph $f(x) = x^3 - 6x - 9$.**

$f(x) = x^3 - 6x - 9$
$f(x) = (x - 3)(x^2 + 3x + 3)$     The function has one real zero, 3.

| x | f(x) |
|------|---------|
| -3 | -18 |
| -2 | -5 |
| -1.5 | -3.375 |
| -1 | -4 |
| 0 | -9 |
| 1 | -14 |
| 1.5 | -14.625 |
| 2 | -13 |
| 3 | 0 |
| 4 | 31 |

*indicates a relative maximum*

*indicates a relative minimum*

← *3 is a zero.*

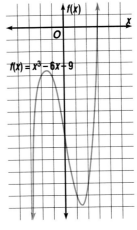

$f(x) = x^3 - 6x - 9$

The graph of this cubic polynomial function has a *sideways* S shape. Thus, it has one relative maximum and one relative minimum. The values of $f(-1.5)$ and $f(1.5)$ were computed to approximate the maximum and minimum more closely.

# CHECKING FOR UNDERSTANDING

**Communicating Mathematics**

**Read and study the lesson to answer these questions.**

1. Why did we not choose one of the negative zeros for the height of the space shuttle tank cylinder?

2. Explain how to use the location principle.

3. Sketch a graph of a third degree polynomial with 1 real zero and two imaginary zeros.

4. State the degree of the polynomial whose graph is shown at the right. Describe the zeros of the function.

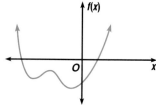

**Guided Practice**

5. Consider the function $f(x) = x^4 - 8x^2 + 10$.

   a. Evaluate $f(x)$ for successive integers between $-4$ and $4$ inclusive.

   b. Between what successive integers do the zeros appear? Approximate those zeros to the nearest tenth.

   c. State the ranges of $x$ values where the values of $f(x)$ are negative and ranges where the values of $f(x)$ are positive.

   d. State the relative maximum(s) and relative minimum(s) if the data you found indicates that there are any.

   e. Graph the function.

# EXERCISES

**Practice**

**Approximate to the nearest tenth the real zeros of each function.**

6. $f(x) = x^3 - 2x^2 + 6$

7. $g(x) = 2x^5 + 3x - 2$

8. $g(x) = x^5 - 6$

9. $h(x) = x^3 + 2x^2 - 3x - 5$

10. $n(x) = x^3 - x^2 + 1$

11. $f(x) = x^3 + 1$

12. $h(x) = 3x^3 - 16x^2 + 12x + 6$

13. $f(x) = x^4 - 4x^2 + 6$

**Graph each function.**

14. $f(x) = x^3$

15. $f(x) = 3x^5$

16. $f(x) = 4x^6$

17. $f(x) = x^3 - x^2 - 4x + 4$

18. $f(x) = x^3 - x$

19. $f(x) = x^3 - x^2 - 8x + 12$

20. $f(x) = x^4 - 81$

21. $f(x) = x^3 + 5$

22. $f(x) = 15x^3 - 16x^2 - x + 2$

23. $f(x) = x^4 - 10x^2 + 9$

24. $f(x) = -x^3 - 13x - 12$

25. $f(x) = -x^3 - 4x^2 - 8x - 8$

**Approximate to the nearest tenth the real zeros of each function. Then use the functional values to graph the function.**

26. $g(x) = x^4 - 9x^3 + 25x^2 - 24x + 6$   27. $h(x) = x^5 + 4x^4 - x^3 - 9x^2 + 3$

28. $f(x) = x^3 - 3x^2 - 2$                   29. $r(x) = x^3 - 3x - 4$

30. $f(x) = x^4 + 7x + 1$                      31. $g(x) = x^5 + x^4 - 2x^3 + 1$

**Critical Thinking**

32. Study the graphs for Exercises 14–31. Make a general statement comparing the graphs of functions of even degree with those of functions of odd degree.

33. The function that represents the volume of a pyramid with a height of the same measure as the side of its square base is $V(s) = \frac{1}{3}s^3$.
   a. Graph the function.
   b. Find the zeros of the function.
   c. Find the maximum and minimum of the function.
   d. Make a conjecture about how all of this data relates.

**Applications**

34. **Aerospace Engineering**   Use the information about the space shuttle fuel tank given in the beginning of the lesson to find the volume of the two hemispheres on the ends of the tank. Then determine the total volume of the tank to the nearest cubic meter.   *Hint: The column of a sphere is given by the formula $V = \frac{4}{3}\pi r^3$.*

35. **Pharmacy**   A syringe is to deliver an injection of 2 cubic centimeters of medication. If the plunger is pulled out two centimeters to have the proper dosage, approximate the radius of the inside of the syringe to the nearest hundredth of a centimeter.   *Hint: The volume of a cylinder is given by the formula $V = \pi r^2 h$.*

**Mixed Review**

36. The Muzik Maze received a shipment of compact discs and cassette tapes in three boxes. The boxes are labeled "compact discs", (CD), "tapes", (T), and "compact discs and tapes" (CDT). None of the boxes are labeled correctly. Steven pulled a compact disc out of the box marked CDT. What is in each box?   **(Lesson 10-6)**

37. Find two numbers whose sum is 60 and whose product is a maximum.   **(Lesson 8-5)**

38. Factor $x^2y^2 - 25x^2$.   **(Lesson 5-5)**

39. State whether $y = \frac{1}{x}$ is a linear function.   **(Lesson 2-2)**

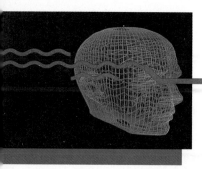

# Technology

## Zeros of Polynomial Functions

Spreadsheet programs can be used to make tables and perform calculations. We can use these capabilities to find approximations for the zeros of polynomial functions.

The spreadsheet program below left is set up to find the values of the function $f(x) = x^3 - 4x + 3$ for values of the domain between $-4$ and $4$. A printout of this program is shown below right. By finding these values, you can use the location principle to determine where zeros occur. Then the spreadsheet can be modified to find function values to get closer approximations of the zeros.

| Polynomial Functions | | |
|---|---|---|
| | **A** | **B** |
| 1 | FUNCTION | F(X) = X³ − 4X + 3 |
| 2 | X | F(X) |
| 3 | -4 | A3^3−4*A3+3 |
| 4 | -3 | A4^3−4*A4+3 |
| 5 | -2 | A5^3−4*A5+3 |
| 6 | -1 | A6^3−4*A6+3 |
| 7 | 0 | A7^3−4*A7+3 |
| 8 | 1 | A8^3−4*A8+3 |
| 9 | 2 | A9^3−4*A9+3 |
| 10 | 3 | A10^3−4*A10+3 |
| 11 | 4 | A11^3−4*A11+3 |

```
= = = = = = = = = = = = = = = = = = = =
        POLYNOMIAL FUNCTIONS
= = = = = = = A = = = = = = B = = = = = =
  1:  FUNCTION  F(X) = X³ − 4X + 3
  2:       X              F(X)
  3:      -4              -45
  4:      -3              -12
  5:      -2               3
  6:      -1               6
  7:       0               3
  8:       1               0
  9:       2               3
 10:       3              18
 11:       4              51
= = = = = = = = = = = = = = = = = = = =
```

According to the location principal, there is a zero between $-3$ and $-2$, 1 is a zero, and there may be a zero between 0 and 1 or between 1 and 2. If you change the values in cells A3 to A11, you can evaluate the function for values in these ranges to determine the zeros more accurately. The zeros of the function are 1 and approximately $-2.303$ and $1.303$.

## EXERCISE

1. Describe how you could modify the spreadsheet program to find values for different functions.

# Composition of Functions

**Objective**

After studying this lesson, you should be able to:

- find the composition of functions.

**Application**

Bill Leshnock is an employee at Pogue's Department Store. As one of his benefits, he receives a 20% employee discount on his purchases at the store. During the Midnight Sale, Bill bought a leather jacket that was on sale for 30% off. The original price of the jacket was $299. How much did Bill pay?

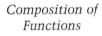

To find how much Bill paid, we will have to first take 30% off of the price of the jacket and then take 20% off of that discounted price. This example illustrates the **composition of functions.**

| *Composition of Functions* | **Suppose $f$ and $g$ are functions such that the range of $g$ is a subset of the domain of $f$. Then the composite function, $f \circ g$, can be described by the equation $[f \circ g](x) = f[g(x)]$.** |
| --- | --- |

*[f ∘ g](x) and f[g(x)] are both read "f of g of x."*

Use the information given about Bill's purchase to find how much he paid for the leather jacket. Let $x$ = the original price.

Let $f(x) = x - 0.2x$     *20% employee discount*

Let $g(x) = x - 0.3x$     *30% off sale price*

$$
\begin{aligned}
[f \circ g](x) &= f[g(x)] \\
&= f(x - 0.3x) && \textit{Substitute } x - 0.3x \textit{ for } g(x). \\
&= f[299 - 0.3(299)] && \textit{Replace } x \textit{ with 299.} \\
&= f(209.30) && \textit{Simplify.} \\
&= 209.30 - 0.2(209.30) && \textit{Evaluate } f \textit{ when } x \textit{ is 239.20.} \\
&= 167.44
\end{aligned}
$$

Bill paid $167.44 for his leather jacket.

**Example 1** | If $f(x) = x^2 + 6$ and $g(x) = 3x - 4$, find $[f \circ g](x)$.

$[f \circ g](x) = f[g(x)]$
$\qquad = f[3x - 4]$     *Substitute 3x − 4 for g(x).*
$\qquad = (3x - 4)^2 + 6$     *Evaluate f when x is 3x − 4.*
$\qquad = 9x^2 - 24x + 22$     *Simplify.*

The graphs of each function are shown at the right.

$f(x)$ is quadratic.
$g(x)$ is linear.
$[f \circ g](x)$ is quadratic.

What type of function would $[g \circ f](x)$ be?

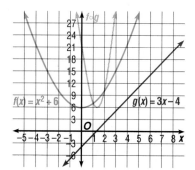

The composition of functions can be shown by mappings. Suppose $f = \{(3, 4), (9, -1), (0, 3)\}$ and $g = \{(4, 9), (3, 3), (-1, 0)\}$. The composition of these functions is shown below.

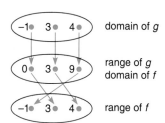

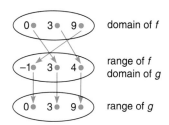

**Example 2** | If $f = \{(2, 5), (8, 4), (-3, 4)\}$ and $g = \{(5, -3), (4, 8)\}$, find $f \circ g$ and $g \circ f$.

$f[g(5)] = f(-3)$       $g[f(2)] = g(5)$       $g[f(-3)] = g(4)$
$\qquad = 4$              $= -3$                $= 8$

$f[g(4)] = f(8)$       $g[f(8)] = g(4)$
$\qquad = 4$              $= 8$

$f \circ g = \{(5, 4), (4, 4)\}$       $g \circ f = \{(2, -3), (8, 8), (-3, 8)\}$

The composition of two functions may not exist. Look back at the definition of the composition of functions. It defines the composition of functions $f$ and $g$, $f \circ g$, when the range of $g$ is a subset of the domain of $f$. If this condition is not met, the composition is not defined.

**Example 3**

If $h = \{(2, 4), (6, 8), (4, 6), (8, 10)\}$ and $k = \{(6, 5), (4, 5), (10, 12), (8, 12)\}$, find $h \circ k$ and $k \circ h$ if they exist.

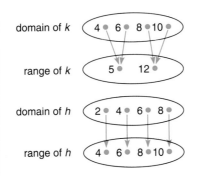

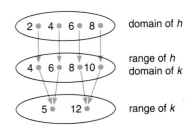

$h \circ k$ does not exist.
The range of $k$ is not a subset of the domain of $h$.

$k \circ h = \{(2, 5), (4, 5), (6, 12), (8, 12)\}$

# CHECKING FOR UNDERSTANDING

**Communicating Mathematics**

**Read and study the lesson to answer these questions.**

1. Find the price Bill Leshnock would pay for a pair of jeans that are on sale 50% off. The regular price of the jeans is $35.99.

2. Would Bill get a greater discount on the jeans if the 20% employee discount was taken before the 50% off sale price?

3. What is meant by the expression $[f \circ g](x)$?

4. Show that if $f(x) = x^2$ and $g(x) = x + 2$, $[f \circ g](x) \neq [g \circ f](x)$.

**Guided Practice**

**For each function, $f$, find $f(3)$, $f(0)$, and $f(-1)$.**

5. $f = \{(3, 9), (4, -1), (0, 6), (-1, 12)\}$

6. $f(x) = x - 5$

7. $f(x) = x^2 + 3x + 2$

8. $f(x) = |x - 3|$

9. $f(x) = x^2 - x^5$

**For each pair of functions find $[f \circ g](2)$ and $[g \circ f](2)$.**

10. $f(x) = x + 4$
    $g(x) = x - 7$

11. $f(x) = x^2 + 1$
    $g(x) = x + 1$

12. $f(x) = x^3 + 4$
    $g(x) = x + 3$

# EXERCISES

**Practice**

For each pair of functions, find $[f \circ g](2)$ and $[g \circ f](2)$.

**13.** $f(x) = x^3$
$g(x) = x^2$

**14.** $f(x) = x$
$g(x) = -x$

**15.** $f = \{(3, 2), (1, -7), (2, 0)\}$
$g = \{(2, 1), (0, 11)\}$

For each pair of functions, find $g[h(x)]$ and $h[g(x)]$.

**16.** $g(x) = x + 4$
$h(x) = x + 9$

**17.** $g(x) = 2x$
$h(x) = 5x$

**18.** $g(x) = x - 1$
$h(x) = x^2$

**19.** $g(x) = -x$
$h(x) = -x$

**20.** $g(x) = x + 1$
$h(x) = x^3$

**21.** $g(x) = |x|$
$h(x) = x - 3$

*Portfolio*

Select some of your work from this chapter that shows how you used a calculator or computer. Place it in your portfolio.

If $f(x) = x^2$, $g(x) = 3x$, and $h(x) = x - 1$, find each value.

**22.** $g[f(1)]$

**23.** $[f \circ h](3)$

**24.** $[h \circ f](3)$

**25.** $[g \circ f](-2)$

**26.** $g[h(-2)]$

**27.** $f[h(-3)]$

**28.** $g[f(x)]$

**29.** $[f \circ g](x)$

**30.** $[f \circ (g \circ h)](x)$

Express $g \circ f$ and $f \circ g$, if they exist, as a set of ordered pairs.

**31.** $f = \{(1, 1), (0, -3)\}$
$g = \{(1, 0), (-3, 1), (2, 1)\}$

**32.** $f = \{(3, 8), (4, 0), (6, 3), (7, -1)\}$
$g = \{(0, 4), (8, 6), (3, 6), (-1, -8)\}$

**Critical Thinking**

**33.** Name two functions $f$ and $g$ such that $f[g(x)] = g[f(x)]$.

**Applications**

**34. Chemistry** While performing an experiment, Joyce needs to record the temperature of a solution at different times. She needs to record the temperature in degrees Kelvin, but only has a thermometer with a Fahrenheit scale. Joyce knows that a Kelvin temperature is 273 degrees greater than a Celsius temperature and that the formula $C = \frac{5}{9}(F - 32)$ converts a Fahrenheit temperature to Celsius. What will she record when the thermometer reads 59°F?

**35. Business** The Clothes Rack adds a 75% markup to the wholesale price of merchandise to determine the selling price. If a jacket that costs $100 wholesale is on sale for 20% off, what would a customer pay for it?

**Mixed Review**

**36.** Graph the function $f(x) = x^6$. **(Lesson 10-6)**

**37.** Approximate to the nearest tenth the real zeros of the function $g(x) = x^4 - 4x^2 + 3$. **(Lesson 10-6)**

**38.** Write an equation for the circle that has a diameter whose endpoints are $(4, -3)$ and $(8, 5)$. **(Lesson 9-3)**

**39.** Evaluate the expression $x(x^2 + 2x + 3)$ if $x = 5$. **(Lesson 1-2)**

# Inverse Functions and Relations

**Objectives**

After studying this lesson, you should be able to:

■ determine the inverse of a function or relation, and
■ graph a function and its inverse.

**Application**

Meredith read about a number game in her Scholastic magazine and decided to try it out on her twin brother Christopher. Christopher was perplexed because his sister could guess what number he was thinking of every time. Suppose Christopher was thinking of the number 47. Here are Meredith's instructions to him.

| Verbal directions | Number | Functional representation |
|---|---|---|
| Think of a number.<br>Double the number.<br>Now add 8. | 47<br>94<br>102 | $f(x) = x$<br>$g(x) = 2[f(x)] = 2x$<br>$h(x) = g(x) + 8 = 2x + 8$ |
| Tell me your final answer and I'll tell you your number. | | |

Not to be outdone by his sister, Christopher tried to figure out this puzzle. In algebra class, the solution dawned on him. Subtraction is the inverse operation of addition and division is the inverse operation of multiplication. She must use these inverses to determine the original number.

The chart below shows the inverse of the puzzle shown above.

| Verbal directions | Number | Functional representation |
|---|---|---|
| Tell me a number.<br>Subtract 8 from the number.<br>Now divide by 2. | 102<br>94<br>47 | $p(x) = x$<br>$r(x) = p(x) - 8 = x - 8$<br>$t(x) = \dfrac{r(x)}{2} = \dfrac{x - 8}{2}$ |

The functions $h(x) = 2x + 8$ and $t(x) = \dfrac{x - 8}{2}$ are **inverse functions.**

| *Definition of Inverse Functions* | **Two functions, $f$ and $g$, are inverse functions if and only if both their compositions are the identity function. That is, $[f \circ g](x) = x$ and $[g \circ f](x) = x$.** |
|---|---|

You can determine if two functions are inverse functions by finding both their compositions. Consider $h(x) = 2x + 8$ and $t(x) = \frac{x - 8}{2}$.

$$[h \circ t](x) = h[t(x)]$$
$$= h\left(\frac{x - 8}{2}\right)$$
$$= 2\left(\frac{x - 8}{2}\right) + 8$$
$$= x$$

$$[t \circ h](x) = t[h(x)]$$
$$= t(2x + 8)$$
$$= \frac{(2x + 8) - 8}{2}$$
$$= x$$

The identity function is $f(x) = x$. Since both $[h \circ t](x)$ and $[t \circ h](x)$ equal $x$, $h(x)$ and $t(x)$ are inverse functions. That is, $h$ is the inverse of $t$ and $t$ is the inverse of $h$. We can write this using this notation.

$$h = t^{-1} \quad \text{and} \quad t = h^{-1}$$

The symbol $h^{-1}$ is read "$h$ inverse" or "the inverse of $h$." By the definition of inverse functions we can write $[h \circ h^{-1}](x) = x$ and $[h^{-1} \circ h](x) = x$. *The $-1$ is <u>not</u> an exponent.*

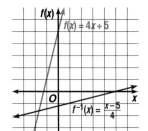

The ordered pairs of inverse functions are related. The functions $f(x) = 4x + 5$ and $f^{-1}(x) = \frac{x - 5}{4}$ are inverse functions.

Evaluate $f(5)$. Then find $f^{-1}[f(5)]$.

$f(5) = 4(5) + 5$ or $25$

The ordered pair $(5, 25)$ belongs to $f$.

$f^{-1}(25) = \frac{(25) - 5}{4}$ or $5$

The ordered pair $(25, 5)$ belongs to $f^{-1}$.

The graphs of the two functions are shown at the left.

So, the inverse of a function can be found by reversing the order of the coordinates of each ordered pair that satisfies the function.

| Property of Inverse Functions | Suppose $f$ and $f^{-1}$ are inverse functions. Then $f(a) = b$ if and only if $f^{-1}(b) = a$. |
| --- | --- |

To find the inverse of a function $f$, you can interchange the variables in the equation, $y = f(x)$. This results in the inverse of the function $f(x)$.

**Example**

Find the inverse of $f(x) = 2x + 1$. Then show that $f$ and its inverse are inverse functions and graph both functions.

Rewrite $f(x)$ as $y = 2x + 1$. Then interchange the variables and solve for $y$.

$x = 2y + 1$    *Interchange variables x and y.*

$x - 1 = 2y$    *Solve for y.*

$\frac{x - 1}{2} = y$    *Notice that the equation defines a function.*

The inverse of $f$ is $f^{-1}(x) = \frac{x - 1}{2}$.

Now show that the compositions of $f$ and $f^{-1}$ are identity functions.

$$[f \circ f^{-1}](x) = f[f^{-1}(x)] \qquad\qquad [f^{-1} \circ f](x) = f^{-1}[f(x)]$$
$$= f\left(\frac{x-1}{2}\right) \qquad\qquad\qquad = f^{-1}(2x+1)$$
$$= 2\left(\frac{x-1}{2}\right) + 1 \qquad\qquad\qquad = \frac{(2x+1)-1}{2}$$
$$= x \qquad\qquad\qquad\qquad\qquad = x$$

The functions are inverses, since both $[f \circ f^{-1}](x)$ and $[f^{-1} \circ f]x$ equal $x$.

Now graph both functions.

Suppose the plane containing the graphs could be folded along the line $g(x) = x$. Then the graphs would coincide.

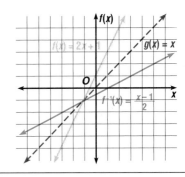

The graphs of a function and its inverse are mirror images, or reflections, of each other with respect to the graph of the identity function. The line whose equation is $g(x) = x$ is the line of symmetry.

In the Example, $f$ and $f^{-1}$ were both functions. However, the inverse of a function is not always a function. For example, the inverse of $f(x) = x^2$ is $f^{-1} = \pm\sqrt{x}$, which is not a function but a relation. You may recall that a relation is a set of ordered pairs. The order of the coordinates in each ordered pair of a relation can be reversed to create a new relation. This new set of ordered pairs and the original relation are **inverse relations.**

| *Definition of Inverse Relations* | **Two relations are inverse relations if and only if whenever one relation contains the element $(a, b)$, the other relation contains the element $(b, a)$.** |
| --- | --- |

# CHECKING FOR UNDERSTANDING

**Communicating Mathematics**

**Read and study the lesson to answer these questions.**

1. How can you find the inverse of a function?

2. Give an example of a function whose inverse is not a function.

3. Are the functions $f(x) = 5x^2 + 6$ and $g(x) = \dfrac{x^2 - 6}{5}$ inverse functions? Explain your answer.

**Guided Practice**

Find the inverse of each relation and determine whether the inverse is a function.

**4.** $\{(3, 2), (4, 2)\}$

**5.** $\{(-1, -2), (-3, -2), (-1, -4), (0, 6)\}$

**6.** $\{(3, 8), (4, -2), (5, -3)\}$

**7.** $\{(2, 4), (-3, 1), (2, 8)\}$

**8.** $\{(1, 3), (1, -1), (1, -3), (1, 1)\}$

**9.** $\{(6, 11), (-2, 7), (0, 3), (-5, 3)\}$

Find the inverse of each function.

**10.** $y = 3x$

**11.** $f(x) = x - 5$

**12.** $y = -3x + 1$

# EXERCISES

**Practice**

Find the inverse of each function.

**13.** $y = 8$

**14.** $g(x) = 4x + 4$

**15.** $f(x) = \frac{1}{2}x + 2$

**16.** $h(x) = x^3$

**17.** $y = x^2 - 9$

**18.** $f(x) = (x - 9)^2$

Graph each function and its inverse.

**19.** $f(x) = x$

**20.** $y = 3x$

**21.** $y = -2x - 1$

**22.** $f(x) = \dfrac{2x - 1}{3}$

**23.** $f(x) = 5x + 3$

**24.** $g(x) = x - 2$

**25.** $g(x) = (x - 4)^2$

**26.** $y = x^2 + 1$

**27.** $y = (x + 2)^2 - 3$

CONNECTION
Geometry

**28.** The vertices of square $ABCD$ form the relation $\{(2, 4), (8, 4), (2, -2), (8, -2)\}$. Find the inverse of this relation and determine if the resulting ordered pairs are also the vertices of a square.

Determine whether each pair of functions are inverse functions.

**29.** $f(x) = x + 6$

$g(x) = x - 6$

**30.** $f(x) = -2x + 3$

$g(x) = 2x - 3$

**31.** $f(x) = 4x - 5$

$g(x) = \dfrac{x + 5}{4}$

**32.** $f(x) = x$

$g(x) = -x$

**33.** $f(x) = \dfrac{x - 2}{3}$

$g(x) = 3x - 2$

**34.** $f(x) = \dfrac{x - 1}{2}$

$g(x) = 2x + 1$

Sketch the graph of the inverse of each relation. Then determine if the inverse is a function.

**35.**

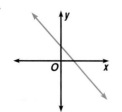

**36.**

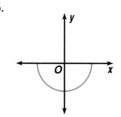

**37.**

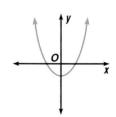

**38.** Find a function that is its own inverse. Can you find more than one?

**39. Business** Sales associates at Egan Electronics earn $7.00 an hour plus a 3% commission on the merchandise they sell. Write a function to describe their income and find how much merchandise they must sell in order to earn $400 in a 40-hour week.

**40. Consumerism** Mr. and Mrs. Pearson bought a stereo on sale from Egan Electronics. They used a $50 gift certificate to help pay for the stereo. The final bill was $453.20 plus tax. If the stereo was on sale 20% off, what is its regular price?

**41.** The DEFFN command in BASIC is used to enter a formula to be evaluated for many numbers. In this program the functions are $f(x) = 2x - 1$ and $g(x) = x^2 + 1$. The computer prints the values of $f(x)$, $g(x)$, $f[g(x)]$, and $g[f(x)]$. You can change lines 10 and 20 to use this program with any pair of functions.

```
10  DEFFNF(X) = 2*X -1
20  DEFFNG(X) = X^2 + 1
30  PRINT "ENTER A NUMBER."
40  INPUT X
50  PRINT "F(X) = ";FNF(X)
60  PRINT "G(X) = ";FNG(X)
70  PRINT "F[G(X)]
        = ";FNF(FNG(X))
80  PRINT "G[F(X)]
        = ";FNG(FNF(X))
90  END
```

Use the BASIC program and a chart to determine if each pair of functions are inverses of each other.

**a.** $f(x) = x + 1$
  $g(x) = x - 1$

**b.** $f(x) = x + 4$
  $g(x) = x - 4$

**c.** $f(x) = \frac{1}{5}(x + 7)$
  $t(x) = {}^-5x + 7$

**42.** If $f(x) = 3x$ and $g(x) = x - 1$, find $f[g(x)]$.  **(Lesson 10-7)**

**43.** If $f = \{(2, 1)(3, 4)(6, -2)\}$ and $g = \{(1, 5), (4, -7), (-2, -3)\}$, express $g \circ f$ and $f \circ g$ as sets of ordered pairs, if they exist.  **(Lesson 10-7)**

**44.** State whether $f(x) = \frac{1}{x^2} + \frac{1}{x} + 1$ describes a quadratic function. Explain your answer.  **(Lesson 8-1)**

**45.** Simplify $\dfrac{x^{\frac{1}{3}}}{x^{\frac{2}{3}} - x^{-\frac{1}{3}}}$.  **(Lesson 6-6)**

**46.** Divide $3a^4 - 2a^3 + 5a^2 - 4a - 2$ by $3a + 1$.  **(Lesson 5-6)**

**Journal**

Make up a number puzzle like the one at the beginning of the lesson. Be sure to try it out on another person to see if it works.

## VOCABULARY

Upon completing this chapter you should be
familiar with the following terms:

| | | | |
|---|---|---|---|
| Complex Conjugates Theorem | **471** | **451** | polynomial function |
| composition of functions | **487** | **450** | polynomial in one variable |
| depressed polynomial | **461** | **475** | Rational Zero Theorem |
| Descartes' rule of signs | **471** | **483** | relative maximum |
| Factor Theorem | **461** | **483** | relative minimum |
| Fundamental Theorem of Algebra | **469** | **459** | Remainder Theorem |
| Integral Zero Theorem | **475** | **469** | root of polynomial equation |
| inverse function | **491** | **460** | synthetic substitution |
| inverse relation | **493** | **469** | zero of polynomial function |
| Location Principle | **482** | | |

## SKILLS AND CONCEPTS

| OBJECTIVES AND EXAMPLES | REVIEW EXERCISES |
|---|---|

Upon completing this chapter, you should
be able to:
- evaluate polynomial functions.
  **(Lesson 10-1)**

Find $p(a + 1)$ if $p(x) = 5x - x^2 + 3x^3$.

$$\begin{aligned}
p(a + 1) &= 5(a + 1) - (a + 1)^2 + \\
&\quad 3(a + 1)^3 \\
&= 5a + 5 - (a^2 + 2a + 1) \\
&\quad + 3(a^3 + 3a^2 + 3a + 1) \\
&= 3a^3 + 8a^2 + 12a + 7
\end{aligned}$$

Use these exercises to review and prepare
for the chapter test.

**Find $p(x - 2)$ for each function $p(x)$.**

**1.** $p(x) = 6x + 3$ \quad **2.** $p(x) = x^2 + 5$

**Find $f(x + h)$ for each function $f(x)$.**

**3.** $f(x) = x^2 - x$ \quad **4.** $f(x) = 2x^3 - 1$

---

- find factors of polynomials using the
  Factor Theorem and synthetic
  division. **(Lesson 10-2)**

The Factor Theorem states that the
binomial $x - a$ is a factor of the
polynomial $f(x)$ if and only if $f(a) = 0$.

**Given a polynomial and one of its factors,
find the remaining factors of the polynomial.
Some factors may not be binomials.**

**5.** $x^3 + 5x^2 + 8x + 4; x + 1$

**6.** $x^4 - 6x^3 + 22x + 15; x + 1$

| OBJECTIVES AND EXAMPLES | REVIEW EXERCISES |
|---|---|

■ find the number of real positive zeros, real negative zeros, and complex zeros for a polynomial function. (**Lesson 10-4**)

State the number of positive real zeros and negative real zeros for $f(x) = 5x^4 + 6x^3 - 8x + 12$.

Since $f(x)$ has 2 sign changes, there are 2 or 0 real positive zeros.

$f(-x) = 5x^4 - 6x^3 + 8x + 12$

Since $f(-x)$ has 2 sign changes, there are 2 or 0 negative real zeros.

**For each function, state the number of positive real zeros, negative real zeros, and complex zeros.**

7. $f(x) = 2x^4 - x^3 + 5x^2 + 3x - 9$

8. $f(x) = 7x^3 + 5x - 1$

9. $f(x) = -4x^4 - x^2 - x - 1$

10. $f(x) = x^4 + x^3 - 7x + 1$

---

■ find zeros of polynomial functions (**Lesson 10-5**)

The Rational Zero Theorem states that if $\frac{p}{q}$ is a rational number in simplest form and a zero of the function $f(x) = a_0x^n + a_1x^{n-1} + ... + a_{n-1}x + a_n$, then $p$ is a factor of $a_n$ and $q$ is a factor of $a_0$.

**Find all the rational zeros for each function.**

11. $f(x) = 2x^3 - 13x^2 + 17x + 12$

12. $g(x) = x^4 + 5x^3 + 15x^2 + 19x + 8$

13. $h(x) = x^3 - 3x^2 - 10x + 24$

14. $f(x) = 2x^3 - 5x^2 - 28x + 15$

---

■ approximate the real zeros of polynomial functions and graph. (**Lesson 10-6**)

The Location Principle states that if $y = f(x)$ represents a polynomial function, and $a$ and $b$ are two numbers such that $f(a) < 0$ and $f(b) > 0$, then the function has at least one zero between $a$ and $b$.

**Approximate to the nearest tenth the real zeros of each function. Then use the functional values to graph the function.**

15. $f(x) = x^3 - x^2 + 1$

16. $g(x) = 4x^3 + x^2 - 11x + 3$

---

■ find the composition of functions. (**Lesson 10-7**)

If $f(x) = x^2 - 2$ and $g(x) = 8x - 1$, find $[f \circ g](x)$ and $[g \circ f](x)$.

$[f \circ g](x) = (8x - 1)^2 - 2$
$= 64x^2 - 16x + 1 - 2$
$= 64x^2 - 16x - 1$
$[g \circ f](x) = 8(x^2 - 2) - 1$
$= 8x^2 - 17$

**For each pair of functions, $f$ and $g$, find $[f \circ g](x)$ and $[g \circ f](x)$.**

17. $f(x) = 2x - 1$     18. $f(x) = x^2 + 2$
    $g(x) = 3x + 4$         $g(x) = x - 3$

19. If $f = \{(2, 1), (-1, 6), (3, 2)\}$ and $g = \{(2, 2), (6, -1), (1, 5)\}$, express $f \circ g$ and $g \circ f$, if they exist, as sets of ordered pairs.

| OBJECTIVES AND EXAMPLES | REVIEW EXERCISES |
|---|---|

■ determine the inverse of a function or relation. **(Lesson 10-8)**

Two functions are inverses if and only if both of the compositions are the identity function. If one function contains the element $(a, b)$, then the other contains $(b, a)$.

**Determine whether each pair of functions are inverse functions.**

**20.** $f(x) = 3x - 4$
$g(x) = \dfrac{x - 4}{3}$

**21.** $f(x) = -2x - 3$
$g(x) = \dfrac{-x - 3}{2}$

# ~~~~~ APPLICATIONS AND CONNECTIONS ~~~~~

**22. Investments**  Four years ago, Mr. and Mrs. Moon bought an apartment building for $200,000. Two years ago, they bought an office building for $500,000. They bought an ocean-front beach house for $100,000 last year.  **(Lesson 10-2)**

  **a.** Assuming that the investments have appreciated at a rate of $x$% per year, write a polynomial function to represent the current worth of their investments.

  **b.** Find the total worth of their investments if they appreciated at a rate of 12% per year.

**24. Manufacturing**  The DrinKone Company makes paper cups that are cone shaped. The volume of a cone is about 7.07 cubic inches. The diameter of the top of the cone is equal to the height of the cone. Determine the dimensions of the cone.  **(Lesson 10-5)**
*Hint: The formula for the volume of a cone is $V = \dfrac{1}{3}\pi r^2 h$*

**25. Geometry**  The volume of a rectangular solid is 2475 cubic units. The length of the box is three units more than twice the width of the box. The height is two units less than the width. Find the dimensions of the box.  **(Lesson 10-4)**

**23. Education**  Craig, Vanessa, Devin, and Anita attend Harding High School. Each will be attending a different school next year. They will attend the DeVry Institute of Technology, Judson College, The Juilliard School, and Case Western Reserve. The student attending DeVry will study electronics and the student attending Juilliard will study the cello. The student attending Judson has not chosen a major and the student attending Case Western Reserve will study chemical engineering. Neither Devin nor Anita can play an instrument. Vanessa will study electronics next year, but Anita has not yet decided what to study. Who will be attending which school?  **(Lesson 10-3)**

**26. Business**  The CD Menagerie adds a 100% markup to the wholesale price of compact discs before placing them in the store for sale. If a CD is on sale for 20% off and the customer pays $12 for it, what is its wholesale price?  **(Lesson 10-7)**

Find $f(3)$ for each function $f$.

**1.** $f(x) = x^3 - 27$

**2.** $f(x) = 2x^4 - 3x^3 + 8$

Find $f(a - 1)$ for each function $f$.

**3.** $f(x) = x^3 - x + 7$

**4.** $f(x) = x^2 - x^3$

Divide using synthetic division and write your answer in the form *dividend* = *quotient* · *divisor* + *remainder*. Is the binomial a factor of the polynomial?

**5.** $(x^2 + 6x - 3) \div (x + 1)$

**6.** $(x^3 + 8x + 1) \div (x + 2)$

**7.** $(x^3 - x^2 - 5x - 3) \div (x + 1)$

**8.** $(x^4 + x^3 + x^2 + x + 1) \div (x + 1)$

For each function, state the number of positive real zeros, negative real zeros, and complex zeros.

**9.** $g(x) = x^3 - x^2 - 14x + 24$

**10.** $f(x) = x^4 + x^3 - 9x^2 - 17x - 8$

Find all the rational zeros for each function.

**11.** $f(x) = x^3 - 3x^2 - 53x - 9$

**12.** $h(x) = 6x^3 + 4x^2 - 14x + 4$

Find all zeros of each function.

**13.** $g(x) = x^3 - 3x - 52$

**14.** $h(x) = 4x^4 + 11x^3 + 10x^2 - 69x - 54$

**15.** Tara, Alan, and Lee each have an after school job. They are a cashier, a pizza delivery person, and a typist. Tara never took typing and cannot drive. Lee works in a busy lawyer's office. Who has what job?

Approximate to the nearest tenth the real zeros of each function. Then use the functional values to graph the function.

**16.** $f(x) = x^3 + 6x^2 + 6x - 4$

**17.** $g(x) = x^3 + 3x^2 - 2x + 1$

If $f(x) = 2x$ and $g(x) = x^2 - 1$, find each value.

**18.** $[f \circ g](4)$

**19.** $[g \circ f](x)$

**20.** Find the inverse of the function $f(x) = x^2 + 2$.

**Bonus**
Suppose $f = \{(0, -3), (2, 5), (-1, 1), (3, 2)\}$, $g = \{(-1, 2), (2, 2), (1, 4), (4, 3), (0, -1)\}$, and $h = \{(4, 2), (1, 0), (-3, 4), (3, -1)\}$. Find $f \circ g \circ h$.

Some test questions on these pages deal with coordinates and geometry. The figures shown may not be drawn to scale.

**Directions: Choose the one best answer. Write A, B, C, or D.**

**1.** What is the length of the diagonal of a square, if the area is $25x^2$?

    (A) $5x$         (B) $5x\sqrt{2}$

    (C) $10x$       (D) $5x^2$

**2.** What percent of rectangle $ACDF$ is shaded?

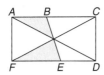

    (A) 25    (B) 30    (C) $33\frac{1}{3}$    (D) 50

**3.** In $1\frac{1}{2}$ hours, the minute hand of a clock rotates through an angle of how many degrees?

    (A) 60°    (B) 180°    (C) 540°    (D) 720°

**4.** The three straight lines intersect at the same point.

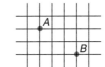

    $a + b =$

    (A) $b + f$       (B) $e + d$

    (C) $f + d$       (D) $c + e$

**5.** Line $p$ is parallel to line $q$.

    $c - a =$

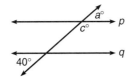

    (A) 0°    (B) 100°    (C) 120°    (D) 150°

**6.**

On line segment $PR$ in the figure above, $PQ = 8$ and $QR = 6$. What is the length of the segment joining the midpoints of segments $PQ$ and $QR$?

    (A) 3    (B) 4    (C) 7    (D) 14

**7.**

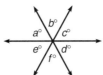

In the triangle above, $4x =$

    (A) 18    (B) 32    (C) 40    (D) 72

**8.**

Above is a section of a graph without the $x$- and $y$-axes. If $A$ has coordinates $(4, 6)$, state the coordinates of $B$.

    (A) $(4, 7)$       (B) $(4, 1)$

    (C) $(9, 2)$       (D) $(7, 4)$

**9.**

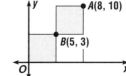

Point $A$ in the figure above has coordinates $(8, 10)$ and point $B$ has coordinates $(5, 3)$. What is the total area of the shaded region?

    (A) 36 square units     (B) 40 square units

    (C) 70 square units     (D) 80 square units

**10.** If $\ell_1 \perp \ell_2$, what is the value of $a + b$?

(A) 45

(B) 90

(C) 110

(D) cannot be determined

**11.**

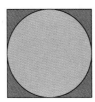

$\overline{QR} \perp \overline{RS}$
$\overline{QR} \perp \overline{PQ}$
$PQ = 10$
$QR = 7$
$RS = 14$

What is the shortest distance from $P$ to $S$?

(A) 24    (B) 25    (C) 28    (D) 31

**12.** What is the area of the triangle with vertices at $(4, -2)$, $(-2, 3)$, and $(4, 3)$?

(A) 60 square units    (B) 30 square units

(C) 15 square units    (D) 7.5 square units

**13.** The area of the circle enclosed in the square is $8\pi$. What is the area of the square?

(A) 64 square units    (B) 32 square units

(C) 16 square units    (D) 8 square units

**14.** The length of a side of a rectangle is $(3x + 4)$ in. The length of the adjacent side is nine less than twice the length of the first side. What is the area of the rectangle in terms of $x$?

(A) $6x^2 - 19x - 36$ in$^2$

(B) $6x^2 + 8x - 12$ in$^2$

(C) $18x^2 + 18x - 8$ in$^2$

(D) $18x^2 + 21x - 4$ in$^2$

---

**15.** The ratio of the circumference of a circle to its radius is

(A) $\pi$    (B) $\frac{\pi}{2}$    (C) $2\pi$    (D) $\frac{2}{\pi}$

# Rational Polynomial Expressions

## CHAPTER OBJECTIVES

In this chapter, you will:

- Simplify rational expressions.
- Solve equations involving rational expressions.
- Graph rational functions.
- Solve problems involving direct, inverse, and joint variation.

Referring to the diagram below, when do you think an outbreak of flu would most handicap the construction team if it results in half of the workers being absent?

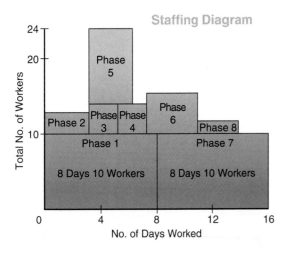

## CAREERS IN CONSTRUCTION MANAGEMENT

The person who manages a construction project has one of the most creative, exciting, and important careers in the world. Before he or she starts, the property is undeveloped (or perhaps holds a structure that will be torn down). When he or she is finished, a new building stands in its place, ready for business.

Construction projects may be small (such as remodeling a home) or huge (such as building an industrial complex or a shopping mall). The construction manager hires and supervises the craftspeople such as electricians, plumbers, clerks, heavy-machinery operators, and engineers. He or she schedules their work and makes sure the needed tools and supplies are on hand for each step. When bad weather or accidents slow down the work, the construction manager reschedules. Obtaining all necessary permits and licenses is another of the manager's responsibilities, too—as well as monitoring compliance with safety codes and other union regulations. The construction manager regularly reviews the blueprints and specifications to be sure the work is right, and at the same time he or she keeps a close eye on the budget to avoid cost overruns. The manager prepares frequent—sometimes daily—progress reports. He or she also reports daily requirements for labor, material, and equipment at the construction site. It's not easy work. But it can be very rewarding.

## MORE ABOUT CONSTRUCTION MANAGEMENT

### Degree Required:

- Bachelor's Degree in construction science

### Some construction managers like:

- seeing the end-product of a project
- solving problems using technology
- high salary and good advancement opportunities

### Related Math Subjects:

- Geometry
- Trigonometry
- Advanced Algebra
- Statistics and Probability

### Some construction managers dislike:

- outdoor conditions they might encounter
- job uncertainty related to economic downturns

For more information on the various careers available in construction management, write to:

Construction Management Association of America
12355 Sunrise Valley Drive
Suite 640
Reston, VA 22091

# Graphing Calculator Exploration: Graphing Rational Functions

A rational function is an equation of the form $f(x) = \frac{p(x)}{q(x)}$, where $p(x)$ and $q(x)$ are polynomial functions and $q(x) \neq 0$. A graphing calculator is a great tool for exploring the graphs of rational functions. These graphs have some features that never appear in the graphs of polynomial functions.

The graphs of rational functions have *breaks in continuity*. This means that, unlike polynomial functions, which can be traced with a pencil never leaving the paper, a rational function is not traceable. Breaks in continuity can appear as vertical asymptotes or as *point discontinuity*. Point discontinuity is like a hole in a graph.

**Example 1**

Graph $y = \frac{x-1}{x-2}$. Use the viewing window [-5, 10] by [-5, 10] with scale factors of 1 for both axes.

*Casio*

ENTER:

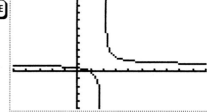

*TI-81*

ENTER: Y= ( X|T − 1 ) ÷ (

X|T − 2 ) GRAPH

There is a break in continuity at $x = 2$. Looking back at the equation, we can see that when $x = 2$, the function is undefined. In this case, the graph has a vertical asymptote with equation $x = 2$. *Sometimes graphing calculators graph this line. This is for clarity, since the line is not part of the graph of the function.*

**Example 2**

Graph $y = \frac{x^2 - 4}{x + 2}$ in the standard viewing window.

*Casio*

ENTER:

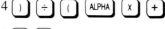

*TI-81*

ENTER: Y= ( X|T $x^2$ − 4 ) ÷ ( X|T + 2 ) GRAPH

The graph looks like a line with a break in continuity at $x = -2$. This occurs because the function $y = \dfrac{x^2 - 4}{x + 2}$ reduces to $y = x - 2$, with the restriction that $x \neq -2$. *You may have to zoom-in to see the break in continuity.*

As you recall, a complete graph of a function includes all of the features of the graph, including the end behavior. The end behavior of some rational functions shows that the graphs approach horizontal asymptotes.

**Example 3**

Graph $y = \dfrac{9x + 1}{3x - 1}$. Use the viewing window $[-5, 5]$ by $[-5, 10]$ with scale factors of 1 on both axes. Then find the equation of the horizontal asymptote.

*Casio*

ENTER: GRAPH ( 9 ALPHA X +

1 ) ÷ ( 3 ALPHA X

– 1 ) EXE

*TI-81*

ENTER: Y= ( 9 X|T + 1 ) ÷

( 3 X|T – 1 ) GRAPH

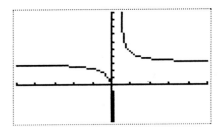

Now trace along the graph and observe the $y$-values as $x$ grows smaller and as $x$ grows larger. The $y$-values approach 3. Thus the equation of the horizontal asymptote is $y = 3$.

# EXERCISES

**Graph each equation on your graphing calculator so the complete graph is shown. Then sketch the graph.**

**1.** $y = \dfrac{1}{x}$

**2.** $y = \dfrac{1}{x^3}$

**3.** $y = \dfrac{-3x}{(x - 1)^2}$

**4.** $y = \dfrac{x^2 - 9}{x + 2}$

**5.** $y = \dfrac{2x}{x^2 + 4}$

**6.** $y = \dfrac{x^2 - 9}{(x + 2)^3}$

**7.** $y = \dfrac{x + 5}{(x^2 - 16)(x + 1)}$

**8.** $y = \dfrac{x - 4}{x^3 + x^2 - 6x}$

**9.** $y = \dfrac{x - 4}{x^4 + 7x^3 - 36x}$

**Use the tracing function on your graphing calculator to determine the equations of the horizontal asymptotes for the graph of each function.**

**10.** $f(x) = \dfrac{22}{x}$

**11.** $f(x) = \dfrac{x - 8}{4x}$

**12.** $f(x) = \dfrac{x}{x^2 - 4}$

**13.** $f(x) = -\dfrac{6}{x^2}$

**14.** $f(x) = -\dfrac{x^2 + 9}{3x^2}$

**15.** $f(x) = \dfrac{5x^2 - 8}{4x^2 + x}$

# Graphing Rational Functions

**Objective**

After studying this lesson you should be able to:

■ locate the asymptotes and sketch the graph of a rational function.

**Application**

Electrical circuits operate according to Ohm's law which states that $\frac{V}{R} = I$.

$V$ = electromotive force (in volts)
$I$ = current (in amperes)
$R$ = resistance of the circuit (in ohms).

The resistance of an electric percolator is 22 ohms. Suppose you wanted to find what current it draws if you know that it has a voltage of 110. One way you could find this is by using the graph that represents Ohm's law if one of the values is known. The graph at the right shows $I = \frac{110}{R}$. The point (22, 5) on the graph shows that when $R = 22$, $I = 5$. The current for the percolator is 5 amperes.

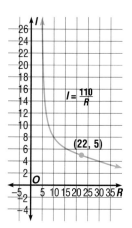

$I = \frac{110}{R}$

(22, 5)

A **rational function** is an equation of the form $f(x) = \frac{p(x)}{q(x)}$, where $p(x)$ and $q(x)$ are polynomial functions and $q(x) \neq 0$. The equation $\frac{V}{R} = I$ is a rational function. Functions such as $f(x) = \frac{x}{x-1}$, $f(x) = \frac{2}{x-2}$, and $f(x) = \frac{4}{(x+2)(x-3)}$ are other examples of rational functions.

The graph of a rational function has two or more branches. The graph above only shows one branch of the function because the negative values would have no real-world meaning in the formula $\frac{V}{R} = I$.

The branches of the graph of a rational function approach lines called **asymptotes.** If the function is not defined when $x = a$, the line with the equation $x = a$ is a vertical asymptote. In the example above, the function would not be defined if $R = 0$. So, the equation of one asymptote is $R = 0$. If the value of the function approaches $b$ as the value of $|x|$ increases, the line with the equation $y = b$ is a horizontal asymptote. In the graph of $I = \frac{110}{R}$, the horizontal asymptote is at $I = 0$, because, as the values of $R$ become greater, the value of $I$ approaches 0.

Before graphing a rational function, it is helpful to graph the asymptotes of the function.

**Example 1**

Graph $f(x) = \dfrac{x}{x-1}$.

If $x = 1$, then $f(x)$ is undefined. Therefore, the equation of a vertical asymptote is $x = 1$. Study the pattern of values of $f(x)$ as the value of $|x|$ increases to determine the equation of the horizontal asymptote.

| x | f(x) |
|------|--------|
| 2 | 2 |
| 5 | 1.25 |
| 10 | 1.111 |
| 100 | 1.010 |
| 1000 | 1.001 |
| 0 | 0 |
| −10 | 0.9090 |
| −100 | 0.9900 |
| −1000 | 0.9990 |

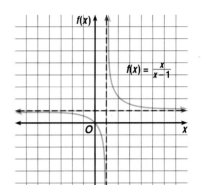

As the value of $|x|$ increases, it appears that the value of the function gets closer and closer to 1. The line with the equation $f(x) = 1$ is a horizontal asymptote of the function.

To graph the function, plot points on either side of each asymptote until the pattern is visible to sketch the graph. *Be sure to test values that are close to the asymptotes.*

Using a calculator can help you find values that pinpoint horizontal asymptotes more easily. The calculator can also help you find values quickly in order to graph ordered pairs of the function.

**Example 2**

Graph $f(x) = \dfrac{6}{(x-3)(x+4)}$.

The vertical asymptotes are the lines whose equations are $x = 3$ and $x = -4$. Use your calculator to find some values of $f(x)$, such as $f(10)$.

ENTER:  6 ÷ ( ( 10 − 3 ) × ( 10 +
4 ) ) = 0.061224489

The same pattern of keystrokes can be used to find other values of $f(x)$ that approach those of the horizontal asymptote.

$f(10) = 0.061224489$
$f(100) = 0.000594766$
$f(1000) = 0.000005994$
$f(-10) = 0.076923076$
$f(-100) = 0.000606796$
$f(-1000) = 0.000006006$

A-11-507b

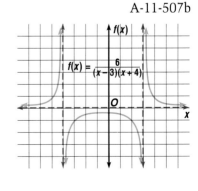

As $|x|$ increases, it appears that the value of the function is approaching 0. The equation of the horizontal asymptote will be $f(x) = 0$. Plot points on either side of each asymptote. Then sketch the graph.

As you saw in Example 2, not all rational functions have graphs that are similar in appearance. Some contain no negative values for $f(x)$.

**Example 3**

Graph $f(x) = \dfrac{2}{(x-2)^2}$

A vertical asymptote is the line whose equation is $x = 2$. Use your calculator to estimate the location of the horizontal asymptote. *Remember any value for x on either side of the asymptote may be used.*

$f(102) = 0.0002$
$f(1002) = 0.000002$
$f(10,002) = 0.00000002$

$f(-98) = 0.0002$
$f(-998) = 0.000002$
$f(-9998) = 0.00000002$

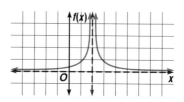

The pattern of values suggests that the equation of the horizontal asymptote will be $f(x) = 0$. Plot points on either side of the vertical asymptote. Then sketch the graph.

# CHECKING FOR UNDERSTANDING

**Communicating Mathematics**

Read and study the lesson to answer each question.

1. What is the equation of the vertical asymptote of the graph of $f(x) = \dfrac{1}{x-3}$?

2. What is the equation of the horizontal asymptote of the graph of $f(x) = \dfrac{1}{x+2}$? Generate a table of values to support your answer.

3. In Example 3, why do you think the values 102, 1002, 10,002, and so on were chosen?

4. In Example 3, why are there no negative values for $f(x)$?

**Guided Practice**

State the equations of the vertical and horizontal asymptotes for each rational function.

5. $f(x) = \dfrac{1}{x-3}$

6. $f(x) = \dfrac{6}{(x-6)^2}$

7. $f(x) = \dfrac{4}{(x-1)(x+5)}$

8. $f(x) = \dfrac{1}{2x}$

# EXERCISES

**Graph each rational function.**

**9.** $f(x) = \dfrac{1}{x}$

**10.** $f(x) = \dfrac{3}{x + 2}$

**11.** $f(x) = \dfrac{x}{x - 2}$

**12.** $f(x) = \dfrac{x - 5}{x + 1}$

**13.** $f(x) = \dfrac{x - 1}{x - 4}$

**14.** $f(x) = \dfrac{-4}{x - 1}$

**15.** $f(x) = \dfrac{x}{x + 1}$

**16.** $f(x) = \dfrac{-2}{(x - 3)^2}$

**17.** $f(x) = \dfrac{4x}{x - 1}$

**18.** $f(x) = \dfrac{-1}{x - 6}$

**19.** $f(x) = \dfrac{1}{(x + 2)^2}$

**20.** $f(x) = \dfrac{8}{(x - 1)(x + 3)}$

**21.** $f(x) = \dfrac{3}{(x - 4)^2}$

**22.** $f(x) = \dfrac{2}{(x - 2)(x + 1)}$

**23.** $f(x) = \dfrac{-5}{(x - 3)(x + 1)}$

**24.** $f(x) = \dfrac{x}{1 - x^2}$

**25.** $f(x) = \dfrac{x}{x^2 - 4}$

**26.** $f(x) = \dfrac{x - 1}{x^2 - 9}$

**Critical Thinking**

**27.** Study your graphs from Exercises 9-26. Make a conjecture on how many asymptotes there will be based on the number of factors in the numerator and denominator of the rational function.

**Applications**

**28. Transportation**   A train travels at one velocity ($V_1$) for a given amount of time ($t_1$) and then another velocity ($V_2$) for a different amount of time ($t_2$). The average velocity is given by the formula $V = \dfrac{V_1 t_1 + V_2 t_2}{t_1 + t_2}$.

   **a.** Draw the graph if $V_1 = 60$ mph, $V_2 = 40$ mph, and $t_2 = 8$ hours.

   **b.** Find $V$ when $t_1 = 9$ hours.

**29. Auto Safety**   If a car hits a tree, the objects in the car (including passengers) keep moving forward until an impact occurs. After impact, objects are repelled. Seatbelts and airbags limit how far you are jolted in this manner. The formula for the velocity you are thrown backward is

$V_f = \dfrac{m_1 - m_2}{m_1 + m_2} \cdot v_i$, where $m_1$ and $m_2$ are the masses of the two objects meeting and $v_i$ is the initial velocity.

   **a.** Graph the function if $m_2 = 7$ kg and $v_i = 5$ m/s.

   **b.** Find the value of $V_f$ when the value of $m_1$ is 5 kg.

**Mixed Review**

**30.** Find the inverse of $f(x) = \dfrac{5x + 2}{2}$.   **(Lesson 10-8)**

**31.** Find the center and radius of a circle whose equation is $x^2 + y^2 = 25$. **(Lesson 9-3)**

**32.** Find $f(-2)$ for $f(x) = 2x^2 + 3x - 5$.   **(Lesson 8-1)**

**33.** Solve $4y^2 - 8y - 7 = 0$ by completing the square. **(Lesson 7-3)**

**34.** Factor $x^2 - x + 3xy - 3y$.   **(Lesson 5-5)**

# Direct, Inverse, and Joint Variation

**Objective**

After studying this lesson, you should be able to:

■ solve problems involving direct, inverse, and joint variation.

**Application**

The LEM (Lunar Exploration Module) used by astronauts to explore the moon's surface during the Apollo space missions weighs about 30,000 pounds on Earth. On the moon, there is less gravity so it weighs less, meaning that less fuel is needed to lift off from the moon's surface. The force of gravity on Earth is about six times as much as that on the moon. The relationship of the two gravities can be expressed in the equation $y = 6x$, where $y$ represents the weight on Earth, and $x$ represents the weight on the moon. The 6 is called the **constant of variation.**

To find out how much the LEM weighs on the moon, you can substitute 30,000 for $y$ and solve the equation.

$$y = 6x$$
$$30,000 = 6x \quad \textit{Substitute 30,000 for y.}$$
$$5000 = x \quad \textit{Divide each side by 6.}$$

The LEM weighs approximately 5000 pounds on the moon.

The relationship described above is an example of a **direct variation.** This means that $y$ is a multiple of $x$. To express a direct variation, we say that $y$ *varies directly as x.*

| *Direct Variation* | **y varies directly as x if there is some constant k such that y = kx. k is called the constant of variation.** |
|---|---|

If you know that $y$ varies directly as $x$ and one set of corresponding values, you can use a proportion to find other unknown values.

$$y_1 = kx_1 \quad \text{and} \quad y_2 = kx_2$$
$$\frac{y_1}{x_1} = k \quad \text{and} \quad \frac{y_2}{x_2} = k$$
$$\text{Therefore, } \frac{y_1}{x_1} = \frac{y_2}{x_2}.$$

Using the properties of equality, you can find many other proportions that relate these same $x$-values and $y$-values.

**Example 1**

If $y$ varies directly as $x$, and $y = 12$ when $x = 15$, find $x$ when $y = 21$.

Use a proportion that relates the values.

$$\frac{y_1}{x_1} = \frac{y_2}{x_2}$$

$$\frac{12}{15} = \frac{21}{x_2} \qquad \text{\textit{Substitute the known values.}}$$

$$12x_2 = (15)(21) \qquad \text{\textit{Cross multiply.}}$$

$$x_2 = 26.25 \qquad \text{\textit{Divide each side by 12.}}$$

The value of $x$ when $y = 21$ is 26.25.

Many quantities are **inversely proportional** or are said to **vary inversely** with each other. For example, if $xy = 12$, as $x$ becomes greater, $y$ becomes less.

| *Inverse Variation* | **$y$ varies inversely as $x$ if there is some constant $k$ such that $xy = k$.** |
|---|---|

Just as with direct variation, a proportion can be used with indirect variation to solve problems where some quantities are known. The following proportion is only one of several that can be formed.

$$x_1y_1 = k \quad \text{and} \quad x_2y_2 = k$$

$$x_1y_1 = x_2y_2 \qquad \text{\textit{Substitution property of equality}}$$

$$\frac{x_1}{y_2} = \frac{x_2}{y_1} \qquad \text{\textit{Divide each side by } } y_1y_2.$$

**Example 2**

If $y$ varies inversely as $x$, and $y = 3$ when $x = 4$, find $y$ when $x = 18$.

$$\frac{x_1}{y_2} = \frac{x_2}{y_1}$$

$$\frac{4}{y_2} = \frac{18}{3} \qquad \text{\textit{Substitute the known values.}}$$

$$18y_2 = 12 \qquad \text{\textit{Cross multiply.}}$$

$$y_2 = \frac{12}{18} \text{ or } \frac{2}{3} \qquad \text{\textit{Divide each side by 18.}}$$

The value of $y$ when $x = 18$ is $\frac{2}{3}$.

Example 3

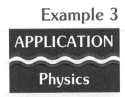

APPLICATION

Physics

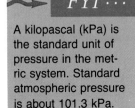

FYI · · ·

A kilopascal (kPa) is the standard unit of pressure in the metric system. Standard atmospheric pressure is about 101.3 kPa.

**The volume of any gas varies inversely with its pressure as long as the temperature remains constant. If a helium-filled balloon has a volume of 3.4 cubic decimeters at a pressure of 120 kilopascals, what is its volume at 101.3 kilopascals?**

Let $V_1$ be the given volume of the helium, 3.4 cubic decimeters
Let $P_1$ be the pressure of the helium at that volume, 120 kilopascals.
Let $P_2$ be the new pressure of the helium, 101.3 kilopascals.

Write a proportion for inversely proportional quantities. Then you can use your calculator to evaluate the proportion.

$$\frac{V_1}{P_2} = \frac{V_2}{P_1} \qquad \textit{Write a proportion.}$$

$$\frac{3.4}{101.3} = \frac{V_2}{120} \qquad \textit{Substitute the known values.}$$

$$\frac{(3.4)(120)}{101.3} = V_2 \qquad \textit{Solve for } V_2.$$

ENTER: ( 3.4 × 120 ) ÷ 101.3 = 4.027640671

The volume of helium at 101.3 kilopascals is about 4.03 cubic decimeters.
*As the pressure decreases, the volume increases.*

Another type of variation is **joint variation.** This type of variation is a direct variation involving three variables as well as a constant.

| *Joint Variation* | **$y$ varies jointly as $x$ and $z$ if there is some number $k$ such that $y = kxz$, where $x \neq 0$ and $z \neq 0$.** |
|---|---|

Example 4

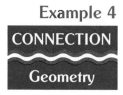

CONNECTION

Geometry

**The area ($A$) of a triangle varies jointly as the base ($b$) and height ($h$). Find the equation of joint variation if $A = 100$, $b = 25$, and $h = 8$.**

Write an equation for joint variation to find the constant of variation.

$A = kbh$     *Let $y = A$, $x = b$, and $z = h$.*

$100 = k(25)(8)$     *Substitute the known values.*

$100 = 200k$     *Multiply.*

$\frac{1}{2} = k$     *Divide each side by 200.*

The equation for the area of a triangle is $A = \frac{1}{2}bh$.

# CHECKING FOR UNDERSTANDING

**Communicating Mathematics**

Read and study the lesson to answer each question.

1. What does $k$ represent in each of the variation equations?

2. In which type of variation would the value of $y$ become less as the value of $x$ becomes greater?

3. Why is joint variation considered to be a direct variation?

**Guided Practice**

State whether each equation represents a direct, inverse, or joint variation. Then name the constant of variation.

4. $xy = -3$

5. $\dfrac{x}{y} = -6$

6. $y = -4x$

7. $\dfrac{x}{2} = y$

8. $y = 3xz$

9. $a = 4b$

10. $y = \dfrac{3}{x}$

11. $a = kcb$

12. $\dfrac{3}{5}a = -\dfrac{5}{4}b$

Write an equation for each statement and then solve the equation.

13. If $y$ varies directly as $x$, and $y = 34$ when $x = 17$, find $y$ when $x = 56$.

14. If $x$ varies inversely as $y$, and $y = 25$ when $x = 5$, find $x$ when $y = 30$.

15. If $y$ varies jointly as $x$ and $z$, and $y = 60$ when $x = 5$ and $z = 4$, find $y$ when $x = 4$ and $z = 10$.

16. If $y$ varies directly as $x$ and inversely as $z$, and $y = 40$ when $x = 20$ and $z = 2$, find $y$ when $x = 60$ and $z = 4$.

# EXERCISES

**Practice**

17. If $y$ varies directly as $x$, and $y = 8$ when $x = 2$, find $y$ when $x = 9$.

18. If $g$ varies directly as $w$, and $g = 10$ when $w = -3$, find $w$ when $g = 4$.

19. If $t$ varies inversely as $r$, and $r = 14$ when $t = -6$, find $r$ when $t = -11$.

20. If $y$ varies inversely as $x$, and $y = \dfrac{1}{5}$ when $x = 9$, find $y$ when $x = -3$.

21. **Cartography**  A map is scaled so that 1 cm represents 15 km. How far apart are two towns if they are 7.9 cm apart on the map?

22. **Manufacturing**  Six feet of steel wire weighs 0.7 kg. How much does 100 ft of the same steel wire weigh?

23. If $y$ varies directly as $x$, and $x = 4$ when $y = 0.5$, find $y$ when $x = 9$.

24. If $y$ varies inversely as $x$, and $y = 1$ when $x = 44$, find $x$ when $y = 40$.

25. Suppose $y$ varies jointly as $x$ and $z$. Find $y$ when $x = 6$ and $z = 8$, if $y = 12$ when $z = 3$ and $x = 4$.

26. If $y$ varies directly as $x$, and $y = -5$ when $x = 0.25$, find $x$ when $y = -7$.

27. If $y$ varies inversely as $x$, and $x = 20$ when $y = 10$, find $x$ when $y = 14$.

CONNECTION

Geometry

28. The area of a parallelogram varies jointly as its base and height. Parallelogram $DUCK$ has a base of 15 meters, a height of 12 meters, and an area of 180 square meters. Find the height of parallelogram $DOVE$ if its area is 1615 square meters and its base is 42.5 meters.

29. The Heavenly Hog has a 10 kg ham that serves 68 people. The Gourmet Diner has a 6 kg ham that serves 44 people. Which eating establishment serves larger portions?

30. If $y$ varies inversely as $x$, and $y = \frac{4}{9}$ when $x = \frac{3}{8}$, find $y$ when $x = \frac{2}{3}$.

31. If $g$ varies directly as $h$, and $g = \frac{3}{4}$ when $h = \frac{2}{5}$, find $g$ when $h = 8$.

32. If $y$ varies jointly as $x$ and $z$, and $y = 25$ when $z = 5$ and $x = 1$, find $y$ when $x = 8$ and $z = 12$.

33. If $y$ varies jointly as $x$ and $z$, and $y = 34$ when $z = 2$ and $x = 17$, find $y$ when $x = 4$ and $z = 8$.

CONNECTION

Geometry

34. The area of a trapezoid varies jointly as the height and the sum of the bases. If the area of a trapezoid is 20 in², its bases are 3 in. and 5 in., and its height is 5 in., find the constant of variation. Then write the general equation for the area of a trapezoid.

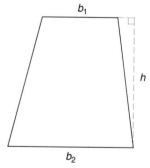

**Critical Thinking**

35. If $y$ varies directly as $x^2$, and $y = 7$ when $x = 9$, find $y$ when $x = 7$.

36. If $y^2$ varies inversely as $x$, and $y = 4$ when $x = 2$, find $y$ when $x = 11$.

**Applications**

37. **Physics** The current ($I$) in an electrical circuit varies inversely with the resistance ($R$) in the circuit.

   a. Use the chart below to write an equation relating the current and resistance.

| $I$ (in amperes) | 0.5 | 1.0 | 1.5 | 2.0 | 2.5 | 3.0 | 4.0 | 5.0 |
|---|---|---|---|---|---|---|---|---|
| $R$ (in ohms) | 12 | 6.0 | 4.0 | 3.0 | 2.4 | 2.0 | 1.5 | 1.2 |

   b. What is the constant of variation?

38. **Auto Mechanics**  When air is pumped into a tire, the pressure required varies inversely as the volume of the air. If the pressure is 30 lb/in$^2$ when the volume is 140 in$^3$, find the pressure when the volume is 100 in$^3$.

39. **Tourism**  In planning a trip you must know that the distance you travel varies jointly as the time and rate of speed. LaDonna Metcalf must travel 396 miles in 8 hours to meet a prospective client. She travels 6 hours at 55 mph. She stopped for a half hour to rest and eat lunch. What is the minimum speed at which she must travel to meet her appointment?

**Mixed Review**

40. Graph $y = \dfrac{-3x}{x-1}$.  **(Lesson 11-1)**

41. Find all rational zeros of $f(a) = a^3 + 2a^2 - 11a - 12$.  **(Lesson 10-4)**

42. Find the solution of $\begin{cases} 3x^2 - y^2 = 26 \\ 2 = y - x \end{cases}$.  **(Lesson 9-9)**

43. Define a variable and write a quadratic function to express the phrase, *the product of two numbers whose difference is 55.*  **(Lesson 7-2)**

44. Simplify $3\sqrt{18} + 8\sqrt{8}$.  **(Lesson 6-3)**

45. Find $3\begin{bmatrix} -4 & 0 & 1 \\ 7 & -2 & 5 \\ 1 & 1 & 4 \end{bmatrix} + \begin{bmatrix} 8 & 0 & 6 \\ -5 & 2 & -1 \\ 4 & -4 & 7 \end{bmatrix}$.  **(Lesson 4-3)**

46. Gladys is three times as old as Maria. In 10 years, Gladys will be twice as old as Maria. What are their ages now?  **(Lesson 3-5)**

## FINE ARTS CONNECTION

Those with a dislike for mathematics rarely see any beauty connected with it. The general population is often unaware of how much mathematics they see and use in everyday life. To help the nonmathematical-minded person relate to some complex mathematical relationships, Helaman R. P. Ferguson at Brigham Young University has found a way to communicate mathematics through sculpture.

The sculpture he calls *Umbilic Torus NC* looks like a twisted bronze ring inscribed with messages from some prehistoric culture. However, if you were to trace your finger along the ring's edge, you would find that your finger is carried around the ring three times before returning to its starting point. The form was created using matrices associated with homogeneous cubic polynomials in three variables.

When asked to explain his art, Ferguson said "Understanding the math would take years of training, but here a person can actually touch a theorem, and get a feel for math."

Umbilic Torus NC

## 11-3 Multiplying and Dividing Rational Expressions

**Objectives**

After studying this lesson, you should be able to:
- simplify rational expressions, and
- simplify complex fractions.

Operations with rational numbers and **rational algebraic expressions** are very similar. A rational number can be expressed as the quotient of two integers. A rational algebraic expression can be expressed as the quotient of two polynomials. In either case, the denominator can never be 0.

$$\frac{2}{3} \quad \frac{415}{100} \quad \frac{-6}{11}$$

*rational numbers*

$$\frac{6}{k} \quad \frac{2x}{x-5} \quad \frac{p^2-25}{p+6}$$

*rational algebraic expressions*

To write a fraction in simplest form, you divide both the numerator and denominator by their greatest common factor (GCF). To simplify a rational algebraic expression, you use similar properties.

**Example 1**

**Simplify** $\dfrac{2x(x-5)}{(x-5)(x^2-1)}$. **Under what conditions is the expression undefined?**

Look for common factors.

$$\frac{2x(x-5)}{(x-5)(x^2-1)} = \frac{2x}{x^2-1} \cdot \frac{x-5}{x-5} \qquad \textit{How is this similar to simplifying } \frac{4}{6}?$$

$$= \frac{2x}{x^2-1} \qquad\qquad \frac{x-5}{x-5} = 1$$

To find when the expression is undefined, completely factor the original denominator.

$$\frac{2x(x-5)}{(x-5)(x^2-1)} = \frac{2x(x-5)}{(x-5)(x-1)(x+1)}$$

The values that would make the denominator equal 0 are 5, 1, or -1. So, the expression is undefined when $x = 5$, $x = 1$, or $x = -1$.

**Example 2**

**Simplify** $\dfrac{z^2w-z^2}{z^3-z^3w}$.

*For what values is the expression undefined?*

First factor the expression to find any common factors.

$$\frac{z^2w-z^2}{z^3-z^3w} = \frac{z^2(w-1)}{z^3(1-w)} \qquad \textit{Factor the GCF of the numerator and then of the denominator.}$$

$$= \frac{z^2(-1)(1-w)}{z^3(1-w)} \qquad w-1 = -1(-w+1) \textit{ or } -1(1-w)$$

$$= -\frac{1}{z} \qquad\qquad \textit{The GCF is } z^2(1-w).$$

Remember that to multiply two fractions you multiply the numerators and then multiply the denominators. To divide two fractions, you multiply by the multiplicative inverse, or reciprocal, of the divisor.

$$\frac{1}{2} \cdot \frac{8}{9} = \frac{1 \cdot \overset{1}{\cancel{2}} \cdot 4}{\underset{1}{\cancel{2}} \cdot 9} \text{ or } \frac{4}{9}$$

$$\frac{1}{3} \div \frac{5}{9} = \frac{1}{3} \cdot \frac{9}{5} = \frac{1 \cdot \overset{1}{\cancel{3}} \cdot 3}{\underset{1}{\cancel{3}} \cdot 5} \text{ or } \frac{3}{5}$$

The same procedures are used for multiplying and dividing rational expressions. These can be generalized by the following rules.

*Multiplying and Dividing Rational Expressions*

> **For all rational expressions $\frac{a}{b}$ and $\frac{c}{d}$,**
>
> $\frac{a}{b} \cdot \frac{c}{d} = \frac{ac}{bd}$, **if $b \neq 0$ and $d \neq 0$, and**
>
> $\frac{a}{b} \div \frac{c}{d} = \frac{a}{b} \cdot \frac{d}{c} = \frac{ad}{bc}$, **if $b \neq 0$, $c \neq 0$, and $d \neq 0$.**

The following examples show how these rules are used with rational expressions.

**Example 3**

Find $\frac{4a}{5b} \cdot \frac{15b}{16a}$. Write the answer in simplest form.

$$\frac{4a}{5b} \cdot \frac{15b}{16a} = \frac{\overset{1}{\cancel{4}} \cdot \overset{1}{\cancel{a}} \cdot \overset{1}{\cancel{5}} \cdot 3 \cdot \overset{1}{\cancel{b}}}{\underset{1}{\cancel{5}} \cdot \underset{1}{\cancel{b}} \cdot \underset{1}{\cancel{4}} \cdot 4 \cdot \underset{1}{\cancel{a}}} \text{ or } \frac{3}{4}$$

*For what values is the expression undefined?*

**Example 4**

Find $\frac{4x^2y}{15a^3b^3} \div \frac{2xy^2}{5ab^3}$. Write the answer in simplest form.

$$\frac{4x^2y}{15a^3b^3} \div \frac{2xy^2}{5ab^3} = \frac{4x^2y}{15a^3b^3} \cdot \frac{5ab^3}{2xy^2}$$ 
*Multiply by the reciprocal of the divisor.*

$$= \frac{\overset{2}{\cancel{4}} \cdot \overset{1}{\cancel{5}} \cdot \overset{x}{\cancel{x^2}} \cdot \overset{1}{\cancel{y}} \cdot \overset{1}{\cancel{a}} \cdot \overset{1}{\cancel{b^3}}}{\underset{3}{\cancel{15}} \cdot \underset{1}{\cancel{2}} \cdot \underset{1}{\cancel{x}} \cdot \underset{y}{\cancel{y^2}} \cdot \underset{a^2}{\cancel{a^3}} \cdot \underset{1}{\cancel{b^3}}}$$ 
*Factor and divide.*

$$= \frac{2x}{3a^2y}$$ 
*Multiply.*

These same steps are followed when the rational expressions contain numerators and denominators that are polynomials.

**Example 5**

Find $\dfrac{x+2}{x+3} \div \dfrac{x^2+x-12}{x^2-9}$. **Write the answer in simplest form.**

$$\dfrac{x+2}{x+3} \div \dfrac{x^2+x-12}{x^2-9} = \dfrac{x+2}{x+3} \cdot \dfrac{x^2-9}{x^2+x-12} \qquad \textit{Multiply by the reciprocal of the divisor.}$$

$$= \dfrac{(x+2)\overset{1}{\cancel{(x+3)}}\overset{1}{\cancel{(x-3)}}}{\underset{1}{\cancel{(x+3)}}(x+4)\underset{1}{\cancel{(x-3)}}} \qquad \textit{Factor and divide.}$$

$$= \dfrac{x+2}{x+4}$$

**Example 6**

The bases of two parallelograms are also the edges of a rectangle C. Parallelogram A has an area of $12x^2 + 2x - 2$ ft$^2$ and height of $2x - 5$ ft. Parallelogram B has an area of $2x^2 - 3x - 5$ ft$^2$ and height of $3x - 1$ ft. Find the area of rectangle C.

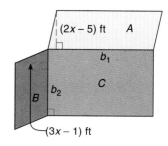

The area of a parallelogram is found by using the formula $A = bh$. Since you know the area and height of each parallelogram, you can find the base measures, $b_1$ and $b_2$, by dividing the area by the height.

$$b_1 = \dfrac{12x^2 + 2x - 2}{2x - 5} \qquad\qquad b_2 = \dfrac{2x^2 - 3x - 5}{3x - 1}$$
$$= \dfrac{(3x-1)(4x+2)}{2x-5} \qquad\qquad = \dfrac{(2x-5)(x+1)}{3x-1}$$

The area of the rectangle is found by $A = bh$. In this case, $b$ and $h$ are the measures of the bases of the parallelograms. The area of rectangle C can be found by multiplying the bases.

$$A = b_1 \cdot b_2$$

$$= \dfrac{\overset{1}{\cancel{(3x-1)}}(4x+2)}{\underset{1}{\cancel{2x-5}}} \cdot \dfrac{\overset{1}{\cancel{(2x-5)}}(x+1)}{\underset{1}{\cancel{3x-1}}}$$

$$= (4x+2)(x+1)$$

$$= 4x^2 + 6x + 2$$

The area of rectangle C is $4x^2 + 6x + 2$ ft$^2$.

A complex fraction is a rational expression whose numerator and/or denominator contains a rational expression. The expressions below are complex fractions.

$$\dfrac{\dfrac{z + 4t}{w}}{6z} \qquad \dfrac{\dfrac{8}{x}}{3 - y} \qquad \dfrac{\dfrac{1}{x} + 3}{\dfrac{2}{x} + 5} \qquad \dfrac{\dfrac{x^2 - 4}{2}}{\dfrac{2 - x}{5}}$$

Remember that a fraction is nothing more than a way to express a division problem. That is, $1 \div 3$ can be expressed as $\dfrac{1}{3}$. So to simplify any complex fraction, rewrite it as a division expression and use the rules for division.

**Example 7**

Simplify $\dfrac{\dfrac{x^2}{x^2 - 25y^2}}{\dfrac{x}{5y - x}}$.

Rewrite the complex fraction as a division expression.

$$\dfrac{\dfrac{x^2}{x^2 - 25y^2}}{\dfrac{x}{5y - x}} = \dfrac{x^2}{x^2 - 25y^2} \div \dfrac{x}{5y - x}$$

$$= \dfrac{x^2}{x^2 - 25y^2} \cdot \dfrac{5y - x}{x} \qquad \textit{Multiply by the reciprocal of the divisor.}$$

$$= \dfrac{\overset{x}{\cancel{x^2}}\,(-1)(\cancel{x - 5y})}{\underset{1}{\cancel{x}}(x + 5y)\underset{1}{\cancel{(x - 5y)}}} \text{ or } \dfrac{-x}{x + 5y}$$

# CHECKING FOR UNDERSTANDING

**Communicating Mathematics**

Read and study the lesson to answer each question.

1. Define greatest common factor.

2. Suppose the numerator of a rational expression is a polynomial and the denominator of a rational expression is a different polynomial. Will factoring the polynomials necessarily provide a way to simplify the expression? Explain your answer.

3. What is the multiplicative inverse of $7a$?

4. What is the multiplicative inverse of $\dfrac{11b}{4c}$?

5. What is the greatest common factor of $(ab - bc)$ and $(3xy + 4tr)$?

**Guided Practice**

Find the GCF of the numerator and denominator for each expression. Then simplify the expression.

6. $\dfrac{42y}{18xy}$

7. $\dfrac{-3x^2y^5}{18x^3y^2}$

8. $\dfrac{42y^2x}{18y^7}$

9. $\dfrac{(-2x^2y)^3}{4x^5y}$

10. $\dfrac{a^3b^2}{(-ab)^3}$

11. $\dfrac{m + 5}{2m + 10}$

12. $\dfrac{(-3t^2u)^3}{(6tu^2)^2}$

13. $\dfrac{4x}{x^2 - x}$

**Write each quotient as a product. Then simplify.**

14. $\dfrac{p^3}{2q} \div \dfrac{-p^2}{4q}$

15. $\dfrac{y^2}{x+2} \div \dfrac{y}{x+2}$

16. $\dfrac{3h}{h+1} \div (h-2)$

# EXERCISES

## Practice

**Simplify each expression.**

17. $\dfrac{3ab}{4ac} \cdot \dfrac{6a^2}{3b^2}$

18. $-\dfrac{3}{5a} \div \left(-\dfrac{9}{15ab}\right)$

19. $\dfrac{3d^3c}{a^4} \div \left(-\dfrac{6dc}{a^5}\right)$

20. $\dfrac{(cd)^3}{a} \cdot \dfrac{ax^2}{xc^2d}$

21. $\left(\dfrac{x^2}{y}\right)^2 \cdot \dfrac{5}{3x}$

22. $\dfrac{5}{m-3} \div \dfrac{10}{m-3}$

23. $\dfrac{(ab)^3}{d^3} \div \dfrac{a^2b^4}{(cd)^4}$

24. $\left(\dfrac{3a^3}{b^2}\right)^3 \cdot \dfrac{4b^2}{3a^7}$

25. $\dfrac{x+y}{a} \div \dfrac{x+y}{a^2}$

26. The area of a triangle can be expressed as $4x^2 - 2x - 6$ square meters. The height of the triangle is $x + 1$ meters. Find the base of the triangle.

**Simplify each expression.**

27. $\dfrac{3x-21}{x^2-49} \div \dfrac{3x}{x^2+7x}$

28. $\dfrac{2x+2}{x^2+5x+6} \div \dfrac{3x+3}{x^2+2x-3}$

29. $-\dfrac{x^2-y^2}{x+y} \cdot \dfrac{1}{x-y}$

30. $\dfrac{a^2+2a-15}{a-3} \div \dfrac{a^2-4}{2}$

31. $\dfrac{y^2-y}{w^2-y^2} \div \dfrac{y^2-2y+1}{1-y}$

32. $\dfrac{a^2-b^2}{2a} \div \dfrac{a-b}{6a}$

33. $\dfrac{(y-2)^2}{(x-4)^2} \cdot \dfrac{x-4}{y-2}$

34. $\dfrac{x^2-y^2}{y^2} \cdot \dfrac{y^3}{y-x}$

35. $\dfrac{x^2+3x-10}{x^2+8x+15} \cdot \dfrac{x^2+5x+6}{x^2+4x+4}$

36. $\dfrac{a^3-b^3}{a+b} \cdot \dfrac{a^2-b^2}{a^2+ab+b^2}$

37. $\dfrac{w^2-11w+24}{w^2-18w+80} \cdot \dfrac{w^2-15w+50}{w^2-9w+20}$

38. $\dfrac{\dfrac{x^2-y^2}{2}}{\dfrac{x-y}{4}}$

39. The lengths of the sides of a right triangle can be expressed as $x + 2$ in., $x + 9$ in., and $x + 10$ in. Find the lengths of the sides.

**Simplify each expression.**

40. $\dfrac{\dfrac{w^2+2w+1}{w+1}}{3}$

41. $\dfrac{\dfrac{5a^2-20}{2a+2}}{\dfrac{10a-20}{4a}}$

42. $\dfrac{\dfrac{2y}{y^2-4}}{\dfrac{3}{y^2-4y+4}}$

43. $\dfrac{\dfrac{p^2+7p}{3p}}{\dfrac{49-p^2}{3p-21}}$

## Critical Thinking

44. Simplify $\dfrac{(a^2-5a+6)^{-1}}{(a-2)^{-2}} \div \dfrac{(a-3)^{-1}}{(a-2)^{-2}}$.

**Applications**

**45. Statistics** After conducting a survey on raising taxes for schools in Worthington, a statistician said the number of women in favor of the tax levy could be expressed by $\dfrac{3 + 10t^2 - 17t}{5t^2 + 4t - 1}$. The number of men in favor of the levy can be expressed by $\dfrac{4t^2 - 9}{3 + 5t + 2t^2}$. Find the ratio of women to men in simplest form. What does this ratio mean? *A ratio is the quotient of two values.*

**46. Demographics** In the United States, the ratio of the number of females to the number of males is almost one. However, in sparsely populated parts of Alaska there are more men than women.

　**a.** If the ratio of men to women in one region of Alaska is expressed as $\dfrac{m^2 + 15m + 54}{m + 6}$ to $\dfrac{m + 9}{3}$, find how many men there are for each woman in that region.

　**b.** Why do you think there are more men than women in these regions?

**Mixed Review**

**47.** If $y$ varies inversely as $x$, and $y = -8$ when $x = 1.5$, find $x$ when $y = -3$. **(Lesson 11-2)**

**48. Real Estate** Budget Realty charges a commission of $4800 on the sale of a $90,000 home. At that rate, how much commission would be charged on the sale of a $219,000 home? **(Lesson 11-2)**

**49.** State whether the graph of $4y^2 - x^2 - 24y + 6x = 11$ is a parabola, a circle, an ellipse, or a hyperbola. **(Lesson 9-7)**

**50.** Find the sum, difference, and product for $6 + i$ and $5 - 2i$. **(Lesson 6-8)**

**51.** Suppose $w$ is a number. Name the most sets of numbers to which it can belong. **(Lesson 6-1)**

---

## ⌇⌇⌇⌇⌇⌇ MID-CHAPTER REVIEW ⌇⌇⌇⌇⌇⌇

**Graph each rational function.** **(Lesson 11-1)**

**1.** $f(x) = \dfrac{2}{x - 5}$　　　　**2.** $f(x) = \dfrac{2}{x^2 - 2x + 1}$　　　　**3.** $f(x) = \dfrac{7}{(x - 2)(x - 3)}$

**4.** If $y$ varies inversely as $x$, and $x = -8$ when $y = -2$, find $x$ when $y = \dfrac{2}{3}$. **(Lesson 11-2)**

**5.** If $m$ varies directly as $n$, and $n = \dfrac{1}{5}$ when $m = 11$, find $m$ when $n = \dfrac{2}{5}$. **(Lesson 11-2)**

**Simplify each expression.** **(Lesson 11-3)**

**6.** $\dfrac{4ab}{2bc} \cdot \dfrac{11a^2b}{5b^2}$　　　　**7.** $\dfrac{7a + 49}{16} \cdot \dfrac{48}{6a + 42}$　　　　**8.** $\dfrac{m^2 + 5m + 4}{6} \div \dfrac{m + 1}{18m + 24}$

# Adding and Subtracting Rational Expressions

**Objectives**

After studying this lesson, you should be able to:
- find the least common denominator of two or more algebraic expressions, and
- add and subtract rational expressions.

**Application**

Ansel Adams (1902–1984) was a famous American photographer known for his style of detailed, focused photos that showed its subjects simply and directly. Most of his photos are in black and white. To take sharp, clear pictures, Adams had to focus the camera precisely. The distance from the object to the lens ($p$) and the distance from the lens to the film ($q$) must be accurately calculated to insure this sharp image. The focal length of the lens is $f$. These measurements are demonstrated in the diagram below.

**Mount Williamson, the Sierra Nevada, from Manzanar, California, 1944.**

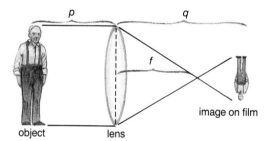

The formula that relates these measures is $\frac{1}{p} + \frac{1}{q} = \frac{1}{f}$.

This formula involves addition of two rational expressions. Remember from arithmetic that to add (or subtract) fractions, they must first be written as equivalent fractions with a common denominator. The least common denominator (LCD) is usually used.

| **Specific case** | | **General case** |
|---|---|---|

$$\frac{2}{3} + \frac{3}{5} = \frac{2 \cdot 5}{3 \cdot 5} + \frac{3 \cdot 3}{5 \cdot 3}$$

*Find equivalent fractions that have a common denominator.*

$$\frac{1}{p} + \frac{1}{q} = \frac{1 \cdot q}{p \cdot q} + \frac{1 \cdot p}{q \cdot p}$$

$$= \frac{10}{15} + \frac{9}{15}$$

$$= \frac{q}{pq} + \frac{p}{pq}$$

*Add the numerators.*

$$= \frac{19}{15}$$

$$= \frac{q + p}{pq}$$

*It is not necessary to rewrite the fraction $\frac{19}{15}$ as the mixed number $1\frac{4}{15}$.*

**Example 1**

Simplify $\dfrac{7x}{13y^2} + \dfrac{4y}{6x^2}$.

$$\dfrac{7x}{13y^2} + \dfrac{4y}{6x^2} = \dfrac{7x}{13y^2} + \dfrac{2y}{3x^2}$$   *Simplify the second term.*

$$= \dfrac{7x(3x^2)}{13y^2(3x^2)} + \dfrac{2y(13y^2)}{3x^2(13y^2)}$$   *The LCD is $39x^2y^2$. Find equivalent fractions that have this denominator.*

$$= \dfrac{21x^3}{39x^2y^2} + \dfrac{26y^3}{39x^2y^2}$$   *Simplify each numerator and denominator.*

$$= \dfrac{21x^3 + 26y^3}{39x^2y^2}$$   *Add the numerators.*

Sometimes the common denominator is not easily recognized, especially when working with algebraic rational expressions. Just as in arithmetic, the LCD must contain all prime factors of each denominator raised to the highest power that occurs in either denominator.

To add $\dfrac{5}{36}$ and $\dfrac{7}{24}$, it is helpful to factor each denominator.

$$36 = 2^2 \cdot 3^2 \qquad 24 = 2^3 \cdot 3$$

The prime factors are 2 and 3. The greatest power of 2 is $2^3$, and the greatest power of 3 is $3^2$. The LCD is $2^3 \cdot 3^2$ or 72.

The LCD for two algebraic rational expressions must also contain each factor of the denominator raised to its highest power.

**Example 2**

Simplify $\dfrac{3y + 1}{2y - 10} + \dfrac{1}{y^2 - 2y - 15}$.

$$\dfrac{3y + 1}{2y - 10} + \dfrac{1}{y^2 - 2y + 15} = \dfrac{3y + 1}{2(y - 5)} + \dfrac{1}{(y - 5)(y + 3)}$$   *The LCD is $2(y - 5)(y + 3)$.*

$$= \dfrac{3y + 1}{2(y - 5)} \cdot \dfrac{y + 3}{y + 3} + \dfrac{1}{(y - 5)(y + 3)} \cdot \dfrac{2}{2}$$

$$= \dfrac{(3y + 1)(y + 3) + 2}{2(y - 5)(y + 3)}$$

$$= \dfrac{3y^2 + 10y + 3 + 2}{2(y - 5)(y + 3)}$$

$$= \dfrac{3y^2 + 10y + 5}{2(y - 5)(y + 3)}$$

In Example 2, you saw that the numerator was expressed as a trinomial, but the denominator was left as a product of factors. When you simplify the numerator, you sometimes discover that the polynomial contains a factor common to the denominator. Thus the rational expression can be further simplified.

**Example 3**

Simplify $\dfrac{w + 12}{4w - 16} - \dfrac{w + 4}{2w - 8}$.

$$\dfrac{w + 12}{4w - 16} - \dfrac{w + 4}{2w - 8} = \dfrac{w + 12}{4(w - 4)} - \dfrac{w + 4}{2(w - 4)}$$   *Factor each expression.*

$$= \dfrac{(w + 12) - (2)(w + 4)}{4(w - 4)}$$   *Since 2 is a factor of 4, it is not necessary to include an extra factor of 2 in the LCD.*

$$= \dfrac{w + 12 - 2w - 8}{4(w - 4)}$$

$$= \dfrac{-w + 4}{4(w - 4)}$$   *Combine like terms in the numerator.*

$$= \dfrac{-1(w - 4)}{4(w - 4)} \text{ or } -\dfrac{1}{4}$$

To simplify a complex fraction that contains sums or differences, you should first express the numerator and denominator as a single rational expression before you divide the two expressions.

**Example 4**

Simplify $\dfrac{\dfrac{1}{x} - \dfrac{1}{y}}{1 + \dfrac{1}{x}}$.

$$\dfrac{\dfrac{1}{x} - \dfrac{1}{y}}{1 + \dfrac{1}{x}} = \dfrac{\dfrac{y}{xy} - \dfrac{x}{xy}}{\dfrac{x}{x} + \dfrac{1}{x}}$$   *In the numerator, the LCD is x y.*
   *In the denominator, the LCD is x.*

$$= \dfrac{\dfrac{y - x}{xy}}{\dfrac{x + 1}{x}}$$   *Simplify the numerator and denominator.*

$$= \dfrac{y - x}{xy} \div \dfrac{x + 1}{x}$$   *Write the complex fraction as a division problem.*

$$= \dfrac{y - x}{xy} \cdot \dfrac{x}{x + 1}$$   *The GCF is x.*

$$= \dfrac{y - x}{y(x + 1)} \text{ or } \dfrac{y - x}{xy + y}$$

# CHECKING FOR UNDERSTANDING

**Communicating Mathematics**

Read and study the lesson to answer each question.

1. What does GCF represent?

2. What does LCD represent?

3. What are equivalent fractions?

4. Two denominators of expressions to be added are $5(a + c)$ and $25a + 25b$. Tell how you would find the LCD. What is it?

**Guided Practice**

Find the LCD for each pair of denominators.

5. 78, 39

6. 12, 27

7. 80, 125

8. $36x^2y$, $20xyz$

9. $x(x - 2)$, $x^2 - 4$

10. $(x + 2)(x + 1)$, $x^2 - 1$

11. $3x + 15$, $x^2 + 2x - 15$

12. $x^2 - 8x$, $y^2 - 8y$

Simplify each expression.

13. $\dfrac{7}{ab} + \dfrac{9}{b}$

14. $\dfrac{11}{10} - \dfrac{7}{2a} - \dfrac{6}{5a}$

15. $3t - 7 + \dfrac{3t + 1}{t - 5}$

# EXERCISES

**Practice**

Simplify each expression.

16. $\dfrac{3a + 2}{a + b} + \dfrac{4}{2a + 2b}$

17. $-\dfrac{18}{9xy} + \dfrac{7}{2x} - \dfrac{2}{3x^2}$

18. $\dfrac{3}{4a} - \dfrac{2}{5a} - \dfrac{1}{2a}$

19. $\dfrac{7}{y - 8} - \dfrac{6}{8 - y}$

20. $\dfrac{x}{x^2 - 9} + \dfrac{1}{2x + 6}$

21. $y - 1 + \dfrac{1}{y - 1}$

22. $3m + 1 - \dfrac{2m}{3m + 1}$

23. $\dfrac{x}{x + 3} - \dfrac{6x}{x^2 - 9}$

24. $\dfrac{3}{a - 2} + \dfrac{2}{a - 3}$

25. $\dfrac{6}{x^2 + 4x + 4} + \dfrac{5}{x + 2}$

26. $\dfrac{8}{2y - 16} - \dfrac{y}{8 - y}$

27. $\dfrac{2a}{3a - 15} + \dfrac{-16a + 20}{3a^2 - 12a - 15}$

28. $\dfrac{5}{x^2 - 3x - 28} + \dfrac{7}{2x - 14}$

29. $\dfrac{x}{x^2 + 2x + 1} - \dfrac{x + 2}{x + 1} - \dfrac{3x}{x + 1}$

30. $\dfrac{m + 3}{m^2 - 6m + 9} - \dfrac{8m - 24}{9 - m^2}$

31. $\dfrac{m^2 + n^2}{m^2 - n^2} + \dfrac{m}{n - m} + \dfrac{n}{m + n}$

32. $\dfrac{x}{x - y} + \dfrac{y}{y^2 - x^2} + \dfrac{2x}{x + y}$

33. $\dfrac{x^2 - 3x + 1}{x^2 - 4} - \dfrac{x^2 + 2x + 4}{2 - x} - \dfrac{x - 4}{x - 2}$

34. $\dfrac{3b - 1}{b^2 - 49} - \dfrac{3b + 2}{14 + 5b - b^2}$

35. $\dfrac{x + 1}{x - 1} + \dfrac{x + 2}{x - 2} + \dfrac{x}{x^2 - 3x + 2}$

36. $\dfrac{(x + y)\left(\dfrac{1}{x} - \dfrac{1}{y}\right)}{(x - y)\left(\dfrac{1}{x} + \dfrac{1}{y}\right)}$

37. $\dfrac{\dfrac{1}{x + 5} + \dfrac{1}{x - 3}}{\dfrac{2x^2 - 3x - 5}{x^2 + 2x - 15}}$

**Simplify each expression.**

**38.** $\dfrac{2x}{x^2 + 7x + 10} + \dfrac{x - 1}{x^2 - 25}$   **39.** $\dfrac{3x}{4x^2 - 1} + \dfrac{5}{2x^2 - x} + \dfrac{2x + 1}{2x^2 + 5x + 2}$

**Applications**

**40. Photography**   Refer to the lens formula at the beginning of the lesson. Usha Smith has a camera with a focal length of 10 cm. When the lens is 12 cm from the film, the camera is focused to take a picture of her dog. How far from the lens is the dog?

**41. Sports**   Ricki can throw a football at a speed of $\dfrac{4}{x + 2}$ mph. Juan can throw a football at a speed of $\dfrac{7}{x - 3}$ mph. How much greater is Ricki's speed than Juan's?

**42. Auto Racing**   A Ferrari has a top speed of 189 mph. A Lamborghini can travel at a speed of $\dfrac{x}{x - 3}$ mph. If the Lamborghini is faster than the Ferrari, express the difference as a simplified rational expression.

**43. Landscaping**   A mad mathematics professor called the hardware store to find out how much fence he would need to outline his pentagon-shaped garden. He gave them the measures of each side as $\dfrac{1}{x}, \dfrac{2}{x - 2}, \dfrac{3}{x}, \dfrac{4}{x - 2}$, and $\dfrac{x}{x - 2}$. The manager of the store panicked until he found an algebra student to help him out. What was the total of the five sides of the garden?

**Mixed Review**

**44.** Simplify $\dfrac{\dfrac{3x + 5}{3x + 1} - 2}{3 + \dfrac{3x}{1 - 2x}}$.   **(Lesson 11-3)**

**45.** If $f(x) = 3x^2$ and $g(x) = x^2 - 1$, find $[f \circ g](2)$.   **(Lesson 10-7)**

**46.** Solve $x^2 - 8x + 7 \geq 0$.   **(Lesson 8-7)**

**47.** Solve $\sqrt{z + 12} - \sqrt{z} = 2$.   **(Lesson 6-7)**

**48.** Graph $r(x) = |-4x|$.   **(Lesson 2-7)**

**49. Travel**   The distance a car moves along the street in one revolution of the tires is directly proportional to the diameter of the tire. Mrs. Witmer's car travels about 88 inches in one revolution of the tires. Find the approximate diameter of the tire.   **(Lesson 11-2)**

## 11-5 Solving Rational Equations

**Objective**

After studying this lesson, you should be able to:
- solve rational equations.

An equation that contains one or more rational expressions is called a **rational equation.** It is easiest to solve a rational equation if the fractions are eliminated. This can be done by multiplying each side of the equation by the least common denominator (LCD). Remember that when you multiply each side by the LCD, each term on each side must be multiplied by the LCD.

**Application**

Remember that the formula that relates the focal length $f$ of a lens, the distance $p$ from the lens to the object, and the distance $q$ from the lens to the image on the film is $\frac{1}{p} + \frac{1}{q} = \frac{1}{f}$. Find $q$ if $p = 45$ cm and $f = 5$ cm.

$$\frac{1}{p} + \frac{1}{q} = \frac{1}{f}$$

$$\frac{1}{45} + \frac{1}{q} = \frac{1}{5}$$

$$45q\left(\frac{1}{45} + \frac{1}{q}\right) = 45q\left(\frac{1}{5}\right)$$

$$45q\left(\frac{1}{45}\right) + 45q\left(\frac{1}{q}\right) = \frac{45q}{5}$$

$$q + 45 = 9q$$

$$45 = 8q$$

$$\frac{45}{8} = q$$

The value of $q$ is $\frac{45}{8}$ or 5.625 cm.

**Check:**

$$\frac{1}{p} + \frac{1}{q} = \frac{1}{f}$$

$$\frac{1}{45} + \frac{1}{\frac{45}{8}} \stackrel{?}{=} \frac{1}{5}$$

$$\frac{1}{45} + \left(1 \cdot \frac{8}{45}\right) \stackrel{?}{=} \frac{1}{5}$$

$$\frac{1}{45} + \frac{8}{45} \stackrel{?}{=} \frac{1}{5}$$

$$\frac{1}{5} = \frac{1}{5} \checkmark$$

In an equation, the LCD must be a common denominator for *all* denominators in the equation. This may involve finding the LCD for three or more denominators.

**Example 1**

Solve $\frac{9}{28} + \frac{3}{z + 2} = \frac{3}{4}$.

The LCD for the three denominators is $28(z + 2)$.

$$\frac{9}{28} + \frac{3}{z + 2} = \frac{3}{4}$$

$$28(z + 2)\left(\frac{9}{28} + \frac{3}{z + 2}\right) = 28(z + 2)\left(\frac{3}{4}\right)$$

$$\frac{9}{28}(28)(z + 2) + \left(\frac{3}{z + 2}\right)(28)(z + 2) = \left(\frac{3}{4}\right)(28)(z + 2)$$

$$(9z + 18) + (84) = 21z + 42$$

$$60 = 12z$$

$$5 = z$$

**Check:**

$$\frac{9}{28} + \frac{3}{z + 2} = \frac{3}{4}$$

$$\frac{9}{28} + \frac{3}{5 + 2} \stackrel{?}{=} \frac{3}{4}$$

$$\frac{9}{28} + \frac{3}{7} \stackrel{?}{=} \frac{3}{4}$$

$$\frac{9}{28} + \frac{12}{28} \stackrel{?}{=} \frac{3}{4}$$

$$\frac{3}{4} = \frac{3}{4} \checkmark$$

Remember that a rational expression is undefined when the value for a variable results in a denominator of zero. When solving rational equations, you should watch for solutions that would produce a denominator of zero. These values must be excluded from your list of solutions. This is one reason that checking your solutions in the *original* equation is so important.

**Example 2**

Solve $\dfrac{7}{m-3} = \dfrac{m+4}{m-3}$.

$$\dfrac{7}{m-3} = \dfrac{m+4}{m-3}$$

$$(m-3)\left(\dfrac{7}{m-3}\right) = (m-3)\left(\dfrac{m+4}{m-3}\right) \qquad \textit{The LCD is } m-3.$$

$$7 = m + 4$$

$$3 = m$$

When you check your solution, you find that 3 produces a zero in the denominator. This value is not a solution. Since there are no other solutions to choose from, this equation has no solution.

**Example 3**

Solve $r + \dfrac{r^2-5}{r^2-1} = \dfrac{r^2+r+2}{r+1}$.

$$r + \dfrac{r^2-5}{r^2-1} = \dfrac{r^2+r+2}{r+1}$$

$$(r^2-1)\left(r + \dfrac{r^2-5}{r^2-1}\right) = (r^2-1)\left(\dfrac{r^2+r+2}{r+1}\right) \qquad \textit{The LCD is } (r^2-1).$$

$$(r^2-1)r + (r^2-1)\left(\dfrac{r^2-5}{r^2-1}\right) = (r-1)(r+1)\left(\dfrac{r^2+r+2}{r+1}\right)$$

$$r^3 - r + r^2 - 5 = (r-1)(r^2+r+2)$$

$$r^3 + r^2 - r - 5 = r^3 + r - 2$$

$$r^2 - 2r - 3 = 0$$

$$(r-3)(r+1) = 0$$

$$r - 3 = 0 \quad \text{or} \quad r + 1 = 0$$

$$r = 3 \qquad\qquad r = -1$$

**Check:** $\quad r + \dfrac{r^2-5}{r^2-1} = \dfrac{r^2+r+2}{r+1}$

When you check the value of $-1$, you get a zero for the denominator. So, $-1$ must be eliminated from your list of solutions.

$$3 + \dfrac{3^2-5}{3^2-1} \stackrel{?}{=} \dfrac{3^2+3+2}{3+1}$$

$$3 + \dfrac{4}{8} \stackrel{?}{=} \dfrac{14}{4}$$

$$\dfrac{7}{2} = \dfrac{7}{2} \checkmark$$

The solution is 3.

# CHECKING FOR UNDERSTANDING

**Communicating Mathematics**

Read and study the lesson to answer each question.

1. Explain why you multiply each side of an equation by the least common denominator as the first step in solving a rational equation.

2. Explain why the equation $x + \dfrac{1}{x - 1} = 1 + \dfrac{1}{x - 1}$ has no solution.

3. Explain how to find the LCD of a rational equation.

**Guided Practice**

Find the LCD for each equation. State what values should be excluded as solutions.

4. $\dfrac{1}{5} = \dfrac{2}{10y}$

5. $\dfrac{1}{x} + \dfrac{1}{2} = \dfrac{2}{x}$

6. $\dfrac{9}{x + 5} = \dfrac{6}{x - 3}$

7. $\dfrac{3m}{2 + m} - \dfrac{5}{7} = 4$

8. $\dfrac{1 - b}{1 + b} = \dfrac{2b}{2b + 3}$

9. $\dfrac{6}{x} = \dfrac{9}{x^2}$

Solve each equation.

10. $x + 3 = \dfrac{4}{x}$

11. $r^2 + \dfrac{17r}{6} = \dfrac{1}{2}$

12. $\dfrac{2y}{3} - \dfrac{y + 3}{6} = 2$

# EXERCISES

**Practice**

Solve each equation. Check your solutions.

13. $\dfrac{y + 1}{3} + \dfrac{y - 1}{3} = \dfrac{4}{3}$

14. $\dfrac{2y - 5}{6} - \dfrac{y - 5}{4} = \dfrac{3}{4}$

15. $\dfrac{5 + 7p}{8} - \dfrac{3(5 + p)}{10} = 2$

16. $8 - \dfrac{2 - 5x}{4} = \dfrac{4x + 9}{3}$

17. $x + 5 = \dfrac{6}{x}$

18. $\dfrac{1}{y^2 - 1} = \dfrac{2}{y^2 + y - 2}$

19. $x + \dfrac{12}{x} - 8 = 0$

20. $\dfrac{5}{6} - \dfrac{2m}{2m + 3} = \dfrac{19}{6}$

21. $\dfrac{1}{9} + \dfrac{1}{2a} = \dfrac{1}{a^2}$

22. $\dfrac{1}{1 - x} = 1 - \dfrac{x}{x - 1}$

23. $\dfrac{2p}{2p + 3} - \dfrac{2p}{2p - 3} = 1$

24. $\dfrac{4}{x - 2} - \dfrac{x + 6}{x + 1} = 1$

25. $\dfrac{x - 4}{x - 2} = \dfrac{x - 2}{x + 2} + \dfrac{1}{x - 2}$

26. $\dfrac{x - 3}{2x} = \dfrac{x - 2}{2x + 1} - \dfrac{1}{2}$

27. $\dfrac{12}{x^2 - 16} - \dfrac{24}{x - 4} = 3$

28. $\dfrac{6}{a - 7} = \dfrac{a - 49}{a^2 - 7a} + \dfrac{1}{a}$

**Solve each equation. Check your solutions.**

**29.** $\dfrac{2}{y+2} - \dfrac{y}{2-y} = \dfrac{y^2+4}{y^2-4}$

**30.** $\dfrac{t+4}{t} + \dfrac{3}{t-4} = \dfrac{-16}{t^2-4t}$

**31.** $\dfrac{x+3}{x+2} = 2 - \dfrac{3}{x^2+5x+6}$

**32.** $\dfrac{x}{x^2-1} + \dfrac{2}{x+1} = \dfrac{1}{2x-2}$

**Critical Thinking**

**33.** Find the values of $A$ and $B$, if $\dfrac{A}{z+2} + \dfrac{B}{2z-3} = \dfrac{5z-11}{2z^2+z-6}$.

**Applications**

**34. Travel**  During one-fourth of the time it took him to travel from Denver to Cheyenne, Alvin drove in a snowstorm at a speed of 40 mph. He drove the rest of the time at a speed of 65 mph. What was his average speed for the entire trip?

**35. Statistics**  A number $x$ is said to be the harmonic mean of $y$ and $z$ if $\dfrac{1}{x}$ is the average of $\dfrac{1}{y}$ and $\dfrac{1}{z}$.

  **a.** Find $y$ if $x = 8$ and $z = 20$.

  **b.** Find $x$ if $y = 5$ and $z = 8$.

**Journal**

Tell why it is necessary to check your solutions when solving a rational equation. Give an example in your explanation.

**Mixed Review**

**Simplify each expression.**  **(Lesson 11-4)**

**36.** $\dfrac{7x}{13y^2} + \dfrac{4x}{13y^2}$

**37.** $\dfrac{3}{x} + 4$

**38.** $\dfrac{9}{4a} + \dfrac{-7}{5b}$

**39.** Find the equations of the asymptotes of the graph of $f(x) = \dfrac{-2x}{(x-1)(x-4)}$. **(Lesson 11-1)**

**Find the solutions of each system of equations.**  **(Lesson 9-9)**

**40.** $x^2 + y^2 = 25$
$\qquad x = 2$

**41.** $x - 2 = -2y$
$\qquad y - 1 = 2x + x^2$

**42. Geometry**  Find the coordinates of the midpoint of $\overline{AB}$ with $A(5, 5)$ and $B(\sqrt{5}, \sqrt{5})$. **(Lesson 9-1)**

**43.** Graph $y < x^2 + x - 1$. **(Lesson 8-6)**

**44.** Simplify $(6 + \sqrt{3})(7 - \sqrt{2})$. **(Lesson 6-3)**

**45. Statistics**  The table below shows the number of miles driven per week by 10 people and the amount of fuel each used for that week. **(Lesson 2-6)**

| Miles Driven | 120 | 322 | 250 | 300 | 350 | 135 | 50 | 150 | 180 | 70 |
|---|---|---|---|---|---|---|---|---|---|---|
| Fuel Used (gallons) | 7.5 | 14 | 11 | 10 | 10 | 3.5 | 2.3 | 5 | 6.2 | 3.8 |

  **a.** Draw a scatter plot and find a prediction equation.

  **b.** Estimate how much fuel a person who drives 280 miles a week would use.

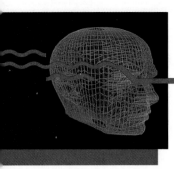

# Technology

## Rational Expressions

The *Mathematical Exploration Toolkit (MET)* can be used to simplify rational expressions. The CALC commands (and their shortened forms) you can use are listed below.

FACTOR (fac)           REDUCE (red)
SIMPLIFY (simp)      $\wedge$ *is used to enter an exponent.*

The SIMPLIFY command combines rational expressions. Once combined, use the FACTOR command to factor the numerator and denominator. In this form, the REDUCE command can be used to divide out any factors common to the numerator and denominator.

**Example 1**   **Simplify $\dfrac{x^2-1}{x-1}$.**

ENTER:   $(x\wedge 2 - 1)/(x - 1)$

          fac

          red

$$\dfrac{x^2-1}{x-1}$$
$$\dfrac{(x-1)(x+1)}{x-1}$$
$$x+1$$

**Example 2**   **Simplify $\dfrac{y^2-4}{y^2} \div \dfrac{y+2}{y}$.**

ENTER:   $((y\wedge 2 - 4)/y\wedge 2)/((y + 2)/y)$

          simp

          fac, fac

          red

$$\dfrac{y^2-4}{y^2} \cdot \dfrac{y}{y+2}$$
$$\dfrac{y^3-4y}{y^3\,2y^2}$$
$$\dfrac{y(y-2)(y+2)}{yy(y+2)}$$
$$\dfrac{y-2}{y}$$

# EXERCISES

**Use CALC to simplify each rational expression.**

1. $\dfrac{a^3-8}{a^2-4}$

2. $\dfrac{x^2-9}{x^2-6x+9}$

3. $\dfrac{x^3-2x^2}{x^4-x^2}$

4. $\dfrac{2y^2-y-1}{y^2-1} \cdot \dfrac{y+1}{2y^2+y}$

5. $\dfrac{x}{x-1} - \dfrac{x-1}{x} - \dfrac{1}{x^2-x}$

## 11-6 Problem-Solving Strategy: Organizing Data

**Objective**

After studying this lesson, you should be able to:

■ solve problems by organizing data.

**Application**

Two machines at the Vernon Wilson Chocolate Company can produce the same number of pounds of chocolate in one hour. Machine A produces 1.5-ounce pieces of candy, and machine B produces pieces twice that weight. If machine A produces 128 pieces of candy in one hour, how many pieces does machine B produce?

Before trying to write an equation to solve this problem, first draw an illustration of the situation and then organize the data you have in the problem by labeling it on the illustration.

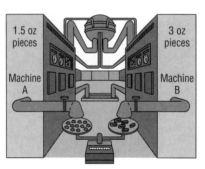

Equal weight in one hour

Machine A → 128 pieces weighing 1.5 ounces each or 192 ounces

Machine B → ___?___ pieces weighing 2(1.5) ounces each or 3(?) ounces

We know the total output of machine B by weight is the same as machine A. So if we let $x$ equal the number of pieces from machine B, we can write an equation.

*output of machine A = output of machine B*
$$192 = 3x$$
$$64 = x$$

Machine B produces 64 pieces in an hour.

There are many ways to organize the data in a given problem. You might use tables, different types of graphs, or diagrams. Your choice of display depends on the type of problem and your approach to solving it.

APPLICATION

Commerce

FYI · · ·

OPEC stands for
Organization of
Petroleum Exporting
Countries. It includes
Nigeria, Saudi Arabia,
Venezuela, Ecuador,
Gabon, Iran, Iraq,
Kuwait, Libya,
Qatar, United Arab
Emirates, and
Indonesia.

**The sources of oil (in thousands of 42-gallon barrels) imported daily by the United States in 1989 are listed below. What part of the total amount of oil is imported from each source?**

| *OPEC countries* | | *Non-OPEC countries* | |
|---|---|---|---|
| Nigeria | 809 | Canada | 910 |
| Saudi Arabia | 1224 | Mexico | 763 |
| Venezuela | 867 | United Kingdom | 217 |
| Other OPEC | 772 | Other | 569 |

A circle graph is an excellent way to represent information that needs to be analyzed as parts of the whole. To show each source, you must find the total amount of oil imported and then calculate the percent each group represents.

The total is 6131 thousand barrels daily. Use your calculator to find the percent by dividing the number for each source by 6131.

For example, to find Nigeria's percent,

ENTER: 809 ÷ 6131 = $0.131952373$

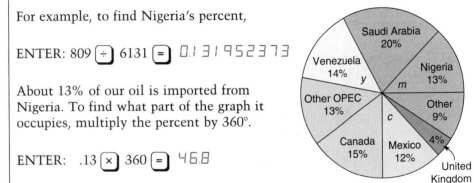

About 13% of our oil is imported from Nigeria. To find what part of the graph it occupies, multiply the percent by 360°.

ENTER: .13 × 360 = $46.8$

The central angle for Nigeria should be about 47°.

When completed, the graph should resemble the one above.

# CHECKING FOR UNDERSTANDING

**Communicating Mathematics**

Read and study the lesson to answer each question.

1. Name some different ways to display data.

2. How does drawing a sketch of the situation help you to organize your data?

3. Name as many different types of graphs as you can that might be used to display data.

**Guided Practice**

4. You are given the temperatures recorded every 10 minutes for a given day. What type of graph could you use to show this information?

5. Dr. Xenon must make a 30% saline solution. He has one bottle of 60% saline solution and a bottle of 20% solution. He needs 280 mL of the 30% solution. Make a sketch to demonstrate the data in this problem.

# EXERCISES

**Solve. Use any strategy.**

6. Write the digits 1 through 8 in the squares so that no two consecutive numbers are next to each other either vertically, horizontally, or diagonally.

7. **Probability**  The sum of the last four digits of Dan's telephone number is 6, and none of the digits is 0. If the probability of guessing Dan's phone number is $\frac{1}{n}$, where $n$ is the number of possible numbers, what is the probability of guessing Dan's number on the first try?

8. Fifteen balls are numbered 1 through 15 and arranged in a triangular format so that the three balls in the middle have a sum of 39 and the five balls on each side of the triangle have a sum of 39. Copy the sketch at the right and number each ball.

9. In the ancient land of Radico, a decree went out that each farmer should buy 100 head of livestock using exactly 100 radicands. In those days, cows cost 10 radicands each, pigs cost 3 radicands each, and geese cost 0.5 radicands each. Every farmer had to buy at least one of each type of livestock. What combination met with the decree?

10. **Geometry**  The view of a rectangular box at the right shows 3 faces of the box. The areas of those faces are 24 in$^2$, 32 in$^2$, and 48 in$^2$. Find the volume of the box.

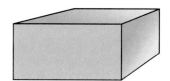

## COOPERATIVE LEARNING ACTIVITY

**Work in groups. Each person in the group must understand the solution and be able to explain it to any person in the class.**

Refer to Exercise 5. Share your sketches of the problem and then find a solution to the problem using any method you wish.

# 11-7 Applications of Rational Equations

**Objective**

After studying this lesson, you should be able to:
- use rational expressions to solve problems.

**Application**

The Delaware Demolition Company wants to build a brick wall to hide the area where they store wrecked cars from public view. One bricklayer can build this wall in 5 days. Another bricklayer can do the job in 4 days. The Delaware Demolition Company decides to hire them both to work together. How long will it take the two bricklayers to finish this wall?

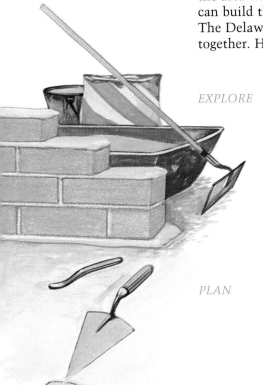

*EXPLORE*  In one day, bricklayer A can complete $\frac{1}{5}$ of the wall. In 2 days, he can complete $\frac{1}{5} \cdot 2$ or $\frac{2}{5}$ of the wall. In $t$ days, he can complete $\frac{1}{5} \cdot t$ or $\frac{t}{5}$ of the wall.

In one day, bricklayer B can complete $\frac{1}{4}$ of the wall. Using the same pattern as above, in $t$ days she can complete $\frac{1}{4} \cdot t$ or $\frac{t}{4}$ of the wall.

*PLAN*  In $t$ days, bricklayer A can complete $\frac{t}{5}$ of the wall and, in that same time, bricklayer B can complete $\frac{t}{4}$ of the wall. Together, they can complete the entire wall.

$$bricklayer\ A\ +\ bricklayer\ B\ =\ entire\ wall$$
$$\frac{t}{5}\ \ +\ \ \frac{t}{4}\ \ =\ \ 1$$

*SOLVE*

$$\frac{t}{5} + \frac{t}{4} = 1$$

$$20\left(\frac{t}{5} + \frac{t}{4}\right) = 20(1) \qquad \textit{Multiply by the LCD.}$$

$$4t + 5t = 20 \qquad \textit{Distributive property}$$

$$9t = 20 \qquad \textit{Combine like terms.}$$

$$t = \frac{20}{9} \quad \text{or} \quad 2\frac{2}{9} \qquad \textit{Divide each side by 9.}$$

The wall can be built in $2\frac{2}{9}$ days.

*EXAMINE*  Bricklayer A builds $\frac{1}{5}t$ or $\frac{1}{5}\left(\frac{20}{9}\right)$ of the wall. $\qquad \frac{1}{5}\left(\frac{20}{9}\right) = \frac{4}{9}$

Bricklayer B builds $\frac{1}{4}t$ or $\frac{1}{4}\left(\frac{20}{9}\right)$ of the wall. $\qquad \frac{1}{4}\left(\frac{20}{9}\right) = \frac{5}{9}$

Since $\frac{4}{9} + \frac{5}{9} = \frac{9}{9}$ or 1, the solution checks.

Sometimes it is helpful to make a drawing when organizing the data given in the problem. In this way you can evaluate how to write an equation that helps you solve the problem.

**Example**

A car travels 300 km in the same time that a freight train travels 200 km. The speed of the car is 20 km/h more than the speed of the train. Find the speed of the car and the speed of the train.

*EXPLORE*  You know the distances both vehicles traveled. You also know that both vehicles traveled the same amount of time, and that the speed of the car is 20 km/h faster than that of the train.

time = t

20 km/h faster

300 km

time = t

200 km

*PLAN*  The formula that relates distance, time, and rate is $d = rt$. Since both vehicles travel the same time, you want to rewrite the formula in terms of $t$. That is, $\frac{d}{r} = t$. Then you can equate the formulas for the two vehicles in terms of $t$.

Let $r$ be the speed of the train. The car goes 20 km/h faster so the speed of the car would be $r + 20$.

$$\underbrace{car's\ time}_{} \quad = \quad \underbrace{train's\ time}_{}$$

$$\frac{distance}{rate} = \frac{300}{r + 20} \qquad \frac{distance}{rate} = \frac{200}{r}$$

*SOLVE*

$$\frac{300}{r + 20} = \frac{200}{r}$$

$$\overset{1}{r(r + 20)}\left(\frac{300}{r + 20}\right) = \overset{1}{r(r + 20)}\left(\frac{200}{r}\right) \qquad \textit{Multiply each side by the LCD.}$$

$$300r = 200r + 4000$$

$$100r = 4000 \qquad \textit{Subtract 200r from each side.}$$

$$r = 40 \qquad \textit{Divide by 100.}$$

The speed of the freight train is 40 km/h, and the speed of the car is 40 + 20 or 60 km/h.

*EXAMINE*    At 40 km/h, it takes the train $\frac{200}{40}$ or 5 hours to travel 200 km.

At 60 km/h, it takes the car $\frac{300}{60}$ or 5 hours to travel 300 km. Both vehicles travel their given distances in the same amount of time.

# CHECKING FOR UNDERSTANDING

**Communicating Mathematics**

**Read and study the lesson to answer each question.**

1. What is the purpose of each step of the 4-step problem-solving plan?

2. If a girl can mow a lawn in 3 hours, what part of the lawn can she mow in one hour?

3. In the bricklayer problem, why was the sum of the rational expression equal to 1?

**Guided Practice**

**Write an expression or equation for each problem. Then solve.**

4. A pipe will fill a swimming pool in 12 hours. What part of the pool will be filled after 5 hours?

5. A tank can be filled in 6 hours and drained in 18 hours. If the drain is left open, how many hours will it take to fill an empty tank?

6. If Joe can travel at a speed of 12 mph on his bicycle, how far can he ride in 5 hours?

7. Salina drove 524 miles in 9 hours. What was her average speed per hour?

# EXERCISES

**Applications**

**Solve each problem.**

8. **Zoology**   A panda can eat the leaves from a certain bamboo stem in 14 minutes. Together, two pandas could eat the leaves from that same stem in 9 minutes. How long would it have taken the second panda to eat the leaves from the stem by itself?

9. **Oil Refining**   An empty oil tank can be filled by pipeline in 10 hours. The tank can be emptied in 20 hours by opening a valve. If the valve is opened while the pipeline is filling the tank, how long will it take for the tank to fill?

10. **Construction**  A painter works on a job for 10 days and is then joined by an associate. Together they finish the job in 6 more days. The associate could have done the job alone in 30 days. How long would it have taken the painter to do the job alone?

11. **Number Theory**  The ratio of 4 less than a number to 26 more than that number is 1 to 3. What is the number?

12. **Number Theory**  Five times the multiplicative inverse of a number is added to the number and the result is 10.5 What is the number?

13. **Aviation**  A plane flies 2000 miles from Chicago to Los Angeles with a 50 mph tail wind in $3\frac{1}{3}$ hours. Returning against the same wind, it takes 4 hours. What is the speed of the plane if there is no wind?

14. **Navigation**  The speed of the current in the Puget Sound is 5 mph. A barge travels with the current 26 miles and returns in $10\frac{2}{3}$ hours. What is its speed in still water?

15. **Trucking**  Two trucks can carry loads of coal in a ratio of 5 to 2. The smaller truck has a capacity 3 tons less than that of the larger truck. What is the capacity of the larger truck?

16. **Banking**  The simple interest for one year on a sum of money is $108. Suppose the interest rate is increased by 2%. Then $450 less than the original sum could be invested and yield the same annual interest.

   a. How much is the original sum invested?

   b. At what rate was the money invested?

17. **Number Theory**  The denominator of a fraction is 1 less than twice the numerator. If 7 is added to both the numerator and denominator, the resulting fraction has a value of $\frac{7}{10}$. Find the original fraction.

18. **Chemistry**  A chemist needs to make 1000 mL of a 30% alcohol solution by mixing 25% and 55% solutions. How much of each should she use?

19. **Banking**  A sum of money is invested for one year at 8% simple interest rate. If the sum of money is increased by $200 and the interest rate is lowered to 6%, then the annual interest is increased by $1.

   a. How much is the original sum of money invested?

   b. What was the original amount of interest?

**20.** Pipe A can fill a tank in 4 hours and pipe B can fill the tank in 3 hours. With the tank empty, pipe A is turned on, and one hour later, pipe B is turned on. How long will pipe B run before the tank is full?

**21.** Pipe C can fill an empty tank in 6 hours and pipe D can empty the tank in 2 hours. With the tank empty, pipe C is turned on, and pipe D is left open. How long will it take to fill the tank? What is wrong with this problem?

**22.** The Greek mathematician Euclid, who lived about 300 B.C., is credited with developing a method for finding the greatest common factor of two integers. This method is called the Euclidean Algorithm. For example, to find the GCF of 232 and 136, you use a process involving division by the remainder of a previous division. Study the following calculations.

$$dividend = quotient \cdot divisor + remainder$$

$$232 = 1 \cdot 136 + 96$$
$$136 = 1 \cdot 96 + 40$$
$$96 = 2 \cdot 40 + 16$$
$$40 = 2 \cdot 16 + 8$$
$$16 = 2 \cdot 8 + 0$$

8 is the GCF of 232 and 136.

*Portfolio*

Place your favorite word problem from this chapter in your portfolio with a note explaining why it is your favorite.

The BASIC program at the right performs the Euclidean Algorithm. Use the program to find the GCF for each pair of integers.

a. 187, 221

b. 182, 1690

c. 4807, 5083

d. 1078, 1547

e. 41, 3

f. 199, 24

g. 766, 424

h. 197, 37

```
10   PRINT "ENTER THE GREATER NUMBER"
20   INPUT X
30   PRINT "ENTER THE LESSER NUMBER"
40   INPUT Y
50   IF INT (X / Y) = X / Y THEN 150
60   PRINT "DIVIDEND = QUOTIENT *
     DIVISOR + REMAINDER"
70   LET Q = INT (X / Y)
80   LET R = X - (Q * Y)
90   PRINT X;"=";Q;"*";Y;"+";R
100  IF R = 0 THEN 150
110  LET X = Y
120  LET Y = R
140  GOTO 70
150  PRINT Y;"IS THE GREATEST
     COMMON FACTOR."
180  END
```

**Simplify each expression. (Lessons 11-3, 11-4)**

**23.** $\dfrac{x^2 - y^2}{x + y} \cdot \dfrac{1}{x - y}$

**24.** $\dfrac{4}{3a} - \dfrac{7}{5a} + \dfrac{1}{2a}$

**25.** $\dfrac{13x^2}{40y} \div \dfrac{26x^2}{70y^3}$

**26.** Write the simplest polynomial function with integral coefficients if two of its zeros are $-2$ and $4 + 3i$. **(Lesson 10-3)**

**27.** Find the distance between points $X(3, 6)$ and $Y(7, -8)$. **(Lesson 9-1)**

**28.** Use long division to find $(10s^3 - 3s^2 - 31s - 6) \div (2s + 3)$. **(Lesson 5-6)**

# SUMMARY AND REVIEW

## VOCABULARY

Upon completing this chapter you should be
familiar with the following terms:

| | | | |
|---|---|---|---|
| asymptote | **506** | **512** | joint variation |
| constant of variation | **510** | **516** | rational algebraic expression |
| direct variation | **510** | **527** | rational equation |
| inverse variation | **511** | **506** | rational function |

## SKILLS AND CONCEPTS

### OBJECTIVES AND EXAMPLES

Upon completing this chapter, you should
be able to:

- sketch the graph of a rational function
  (**Lesson 11-1**)

  Find the values for which the function is
  undefined to locate the vertical
  asymptotes. Then plot points on either
  side of the asymptotes.
  For $\dfrac{5}{(x-3)(x)} = y$, the vertical asymptotes
  have equations $x = 3$ and $x = 0$.

- solve problems involving direct, inverse,
  and joint variation   (**Lesson 11-2**)

  direct: $y = kx$
  inverse: $xy = k$
  joint: $y = kxz$

### REVIEW EXERCISES

Use these exercises to review and prepare
for the chapter test.

**Graph each rational function.**

1. $f(x) = \dfrac{4}{x-2}$   2. $f(x) = \dfrac{x}{x+3}$

3. $f(x) = \dfrac{2}{x}$   4. $f(x) = \dfrac{1}{2-x^2}$

5. $f(x) = \dfrac{5}{(x+1)(x-3)}$

6. If $y$ varies directly as $x$, and $y = 21$
   when $x = 7$, find $x$ when $y = -5$.

7. If $y$ varies inversely as $x$, and $y = 9$
   when $x = 2.5$, find $y$ when $x = -0.6$.

8. If $y$ varies directly as $x$, and $x = 28$
   when $y = 18$, find $x$ when $y = 63$.

9. If $y$ varies inversely as $x$, and $x = 28$
   when $y = 18$, find $x$ when $y = 63$.

10. If $y$ varies jointly as $x$ and $z$, and $x = 2$
    and $z = 4$ when $y = 16$, find $y$ when
    $x = 5$ and $z = 8$.

11. If $y$ varies jointly as $x$ and $z$, and $x = 4$
    and $z = 2$ when $y = 25$, find $x$ when
    $y = 12$ and $z = 20$.

■ multiply and divide rational expressions
(**Lesson 11-3**)

**Simplify each expression.**

$$\frac{3x}{2y} \cdot \frac{8y^3}{6x^2} = \frac{\overset{1}{\cancel{3}} \cdot x \cdot \overset{4}{\cancel{8}} \cdot \overset{y^2}{\cancel{y^3}}}{\underset{1}{\cancel{2}} \cdot y \cdot \underset{2}{\cancel{6}} \cdot \underset{x}{\cancel{x^2}}} = \frac{2y^2}{x}$$

**12.** $\dfrac{-4ab}{21c} \cdot \dfrac{14c^2}{22a^2}$

**13.** $\dfrac{y-2}{a-x} \cdot (a-3)$

$$\frac{x^2-4}{x^2-9} \div \frac{x+2}{x-3} = \frac{x^2-4}{x^2-9} \cdot \frac{x-3}{x+2}$$

**14.** $\dfrac{x+y}{a} \div \dfrac{x+y}{a^2}$

**15.** $\dfrac{a^2-b^2}{6b} \div \dfrac{a+b}{36b^2}$

$$= \frac{(x+2)(x-2)(x-3)}{(x+3)(x-3)(x+2)}$$

$$= \frac{x-2}{x+3}$$

**16.** $\dfrac{y^2-y-12}{y+2} \div \dfrac{y-4}{y^2-4y-12}$

■ simplify complex fractions
(**Lesson 11-3**)

**Simplify each expression.**

Rewrite as a division expression.

**17.** $\dfrac{\dfrac{1}{n^2-6n+9}}{\dfrac{n+3}{2n^2-18}}$

**18.** $\dfrac{\dfrac{x^2+7x+10}{x+2}}{\dfrac{x^2+2x-15}{x+2}}$

$$\frac{\dfrac{1}{x}}{\dfrac{2x}{17}} = \frac{1}{x} \div \frac{2x}{17}$$

$$= \frac{1}{x} \cdot \frac{17}{2x} = \frac{17}{2x^2}$$

■ add and subtract rational expression
(**Lesson 11-4**)

**Simplify each expression.**

Rewrite all rational expressions as
equivalent fractions with a common
denominator.

**19.** $\dfrac{-9}{4a} + \dfrac{7}{3b}$

**20.** $\dfrac{x+2}{x-5} + 6$

**21.** $\dfrac{x-1}{x^2-1} + \dfrac{2}{5x+5}$

**22.** $\dfrac{7}{y} - \dfrac{2}{3y}$

$$\frac{3x}{x-y} + \frac{4x}{y-x} = \frac{3x}{x-y} + \frac{-4x}{x-y} = \frac{-x}{x-y}$$

**23.** $\dfrac{7}{y-2} - \dfrac{11}{2-y}$

**24.** $\dfrac{14}{x+y} - \dfrac{9}{y^2-x^2}$

**25.** $\dfrac{\dfrac{5x}{4}}{\dfrac{6x}{5}} + \dfrac{\dfrac{2x}{ab}}{\dfrac{3x}{a}}$

**26.** $\dfrac{\dfrac{2a+4}{a}}{6+\dfrac{2}{a^2}}$

■ solve rational equations. (**Lesson 11-5**)

$$\frac{1}{x-1} + \frac{2}{x} = 0$$

Multiply each side of the equation by the common denominator of the equation.

$$x(x-1)\left[\frac{1}{x-1} + \frac{2}{x}\right] = x(x-1)(0)$$

Simplify the products.

$$\frac{x(x-1)}{x-1} + \frac{2x(x-1)}{x} = 0$$

Solve the equation.

$$x + 2(x-1) = 0$$
$$x + 2x - 2 = 0$$
$$3x = 2$$
$$x = \frac{2}{3}$$

**Solve each equation.**

27. $\dfrac{3}{y} + \dfrac{7}{y} = 9$

28. $1 + \dfrac{5}{y-1} = \dfrac{7}{6}$

29. $\dfrac{3x+2}{4} = \dfrac{9}{4} - \dfrac{3-2x}{6}$

30. $\dfrac{1}{r^2-1} = \dfrac{2}{r^2+r-2}$

31. $\dfrac{x}{x^2-1} + \dfrac{2}{x+1} = 1 + \dfrac{1}{2x-2}$

# APPLICATIONS AND CONNECTIONS

32. **Meteorology** The temperature at noon on Monday was 54°F. On Tuesday, it was 33°F. The temperatures at noon on Wednesday through Sunday were 60°, 44°, 32°, 54°, and 78°, respectively. Make a drawing to organize this data and tell where the greatest change from day to day occurred. (**Lesson 11-6**)

34. **Construction** Mike Welch can paint his house in 15 hours. His friend Joe can paint the house in 20 hours. If they work together, how long will it take them to paint the house? (**Lesson 11-7**)

35. **Number Theory** One integer is 2 less than another integer. Three times the reciprocal of the lesser integer plus five times the reciprocal of the greater integer is $\dfrac{7}{8}$. What are the two integers? (**Lesson 11-7**)

33. **Physics** The intensity of illumination on a surface varies inversely as the square of the distance from the light source. A surface is 12 meters from a light source. How far must the surface be from the source to receive twice as much illumination? (**Lesson 11-7**)

**Simplify each expression.**

1. $\dfrac{7ab}{9c} \cdot \dfrac{81c^2}{91a^2b}$

2. $\dfrac{a^2 - ab}{3a} \div \dfrac{a - b}{15b^2}$

3. $\dfrac{7}{5a} - \dfrac{10}{3ab}$

4. $\dfrac{6}{x - 5} + 7a$

5. $\dfrac{x^2 - y^2}{a^2 - b^2} \cdot \dfrac{a + b}{x - y}$

6. $\dfrac{x - y}{a - b} - \dfrac{x + y}{a + b}$

7. $\dfrac{x^2 - 2x + 1}{y - 5} \div \dfrac{x - 1}{y^2 - 25}$

8. $\dfrac{x + 2}{x - 1} + \dfrac{6}{7x - 7}$

9. $\dfrac{y}{y - 9} - \dfrac{-9}{9 - y}$

10. $\dfrac{\dfrac{x^2 - 1}{x^2 - 3x - 10}}{\dfrac{x^2 - 12x + 35}{x^2 + 3x + 2}}$

11. $\dfrac{\dfrac{1}{x} - \dfrac{1}{2x}}{\dfrac{2}{x} + \dfrac{4}{3x}}$

12. $\dfrac{\dfrac{2}{x - 4} + \dfrac{5}{x + 1}}{\dfrac{3x}{x^2 - 3x - 4}}$

**Solve each equation.**

13. $a - \dfrac{5}{a} = 4$

14. $\dfrac{3}{x} + \dfrac{x}{x + 2} = \dfrac{-2}{x + 2}$

15. $\dfrac{y}{y - 3} + \dfrac{6}{y + 3} = 1$

**Graph each rational function.**

16. $y = \dfrac{3x}{x + 2}$

17. $y = \dfrac{2}{x - 1}$

18. $y = \dfrac{3}{x} - \dfrac{7}{x}$

19. Jesse Cruz can wordprocess 75 pages of manuscript in 8 hours. Tedra Szatro can produce the same number of pages in 13 hours. If they work together, how long will it take to process 75 pages?

20. Two barrels have capacities in the ratio of 7 to 4. The larger barrel holds 12 gallons less than two smaller barrels. How much will each barrel hold?

21. Suppose $y$ varies directly as $x$. If $y = 10$, then $x = -3$. Find $y$ when $x = 20$.

22. Suppose $y$ varies inversely as $x$. If $y = 9$, then $x = -\dfrac{2}{3}$. Find $x$ when $y = -7$.

23. Suppose $y$ varies jointly as $x$ and $z$. If $y = 45$ when $x = 3$ and $z = 5$, find $y$ when $x = 2$ and $z = 4$.

24. Suppose $y$ varies jointly as $x$ and $z$. If $x = 10$ when $y = 250$ and $z = 5$, find $x$ when $y = 2.5$ and $z = 4.5$.

25. The city swimming pool can be filled from two sources, a well or city water. The pipe for the city water fills the pool in 6 hours. The pipe from the well fills the pool in 10 hours. How long will it take the pool to fill if both sources are piped in at the same time? Make an illustration of the problem. Then solve.

**Bonus**

Simplify. $\dfrac{\dfrac{9x^2 - 12x + 4}{6x^2 - 13x + 6}}{\dfrac{6x^2 + 13x + 6}{4x^2 - 9}} \div \dfrac{\dfrac{9x^2 - 4}{6x^2 - 5x - 6}}{\dfrac{6x^2 + 5x - 6}{4x^2 - 12x + 9}}$

# Exponential and Logarithmic Functions

## CHAPTER OBJECTIVES

In this chapter, you will:

- Solve equations involving logarithmic and exponential functions.
- Find common and natural logarithms of numbers.
- Solve problems using estimation.
- Use logarithms to solve problems.

If the ant is $\frac{3}{8}$ inch long, estimate the size of the microchip.

## CAREERS IN SYSTEMS ANALYSIS

The world depends on computers today, and it depends on them far more than anyone could have believed possible even thirty years ago. Research, business, government, education, art, sports, entertainment—almost every company in every field now has a computer system to help it function and grow. Thus it follows that the career of designing computer systems is a powerful one.

A systems analyst must understand the technical language of computer programming well enough to explain to the programmers what is needed. But he or she must also understand the workings of business. The systems analyst analyzes an organization's needs and then designs and creates a system of workers, information, materials, and computer technology that will meet those needs. Ideally, the machines and the people using them should work together like an artist's hand and mind. It costs a great deal to reorganize a company to the newest technology. If the system has not been designed correctly, there is little the computer programmer or the computer operator can do to make it meet the company's needs.

Some of today's systems analysts have an even bigger opportunity before them: the design and creation of robots. Robots are computers that can operate machines without direction from human operators. It's a challenging kind of system to analyze.

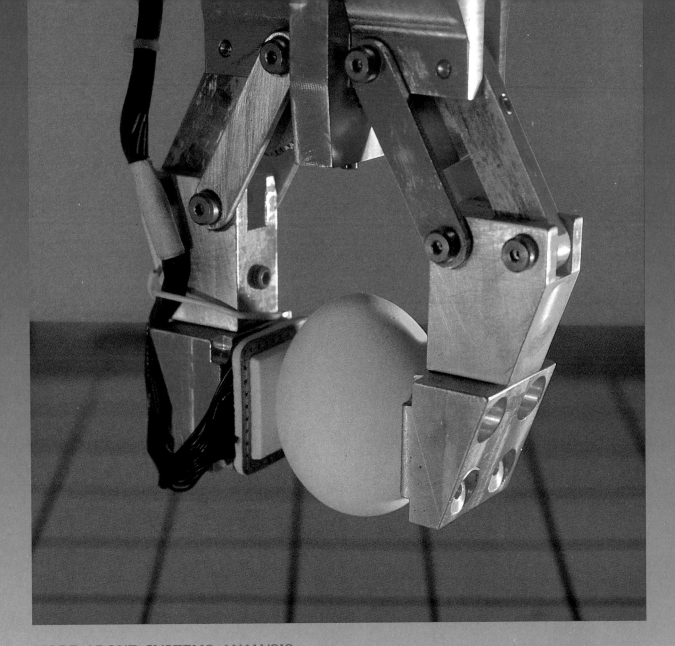

## MORE ABOUT SYSTEMS ANALYSIS

### Degree Required

- Bachelor's degree in Computer Science

### Some systems analysts like:

- challenge and variety in their work
- good employment opportunities
- good salaries
- interaction with people in solving problems

### Related Math Subjects:

- Advanced Algebra
- Geometry
- Statistics/Probability

### Some systems analysts dislike:

- the need to update their knowledge with advances in technology
- working overtime to meet deadlines
- frustrating problems in computer programming

For more information on the various careers available in the field of Systems Analysis, write to:

Association of the Institute for Certification of Computer Professionals
2200 East Devon Avenue
Suite 268
Des Plaines, IL  60018

# 12-1 Real Exponents and Exponential Functions

**Objective**

After studying this lesson, you should be able to:

■ simplify expressions and solve equations involving real exponents.

**Application**

*This bacteria is often referred to as e. coli.*

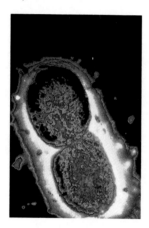

Mitosis is a process of cell duplication in which one cell divides into two. The *escherichia coli* is one of the fastest growing bacteria. It can reproduce itself in 15 minutes. If you begin with one escherichia coli cell, how many cells will there be in one hour?

Of course, in one hour there are four 15-minute intervals. After the first 15-minute interval, the cell will divide and form 2 cells. After the second 15-minute interval, each of the two cells will divide to form a total of 4 cells. After the third and fourth 15-minute intervals, 8 and 16 cells, respectively, will result.

This pattern can be summarized in the table at the right.

The total number of cells, *y*, can be expressed as a function of time where *x* is the number of 15-minute intervals. This function, $y = 2^x$, is an **exponential function.**

| 15-minute Intervals | Total Number of Cells | Pattern |
|---|---|---|
| 0 | 1 | $2^0$ |
| 1 | 2 | $2^1$ |
| 2 | 4 | $2^2$ |
| 3 | 8 | $2^3$ |
| 4 | 16 | $2^4$ |
| . | . | . |
| . | . | . |
| . | . | . |
| $x$ | $y$ | $2^x$ |

Let's take a close look at the graph of $y = 2^x$, where $x$ is a rational number. Make a table of values to help draw the curve. Of course, negative values for $x$ have no meaning in mitosis. But, they must be included to show rational values for $x$.

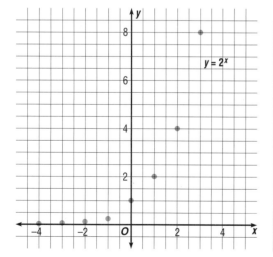

| $x$ | $2^x$ or $y$ | $y$ (approx.) |
|---|---|---|
| $-4$ | $2^{-4} = \frac{1}{16}$ | 0.06 |
| $-3$ | $2^{-3} = \frac{1}{8}$ | 0.12 |
| $-2$ | $2^{-2} = \frac{1}{4}$ | 0.25 |
| $-1$ | $2^{-1} = \frac{1}{2}$ | 0.5 |
| $-\frac{1}{2}$ | $2^{-\frac{1}{2}} = \frac{1}{2}\sqrt{2}$ | 0.7 |
| 0 | $2^0 = 1$ | 1 |

| $x$ | $2^x$ or $y$ | $y$ (approx.) |
|---|---|---|
| $\frac{1}{2}$ | $2^{\frac{1}{2}} = \sqrt{2}$ | 1.4 |
| 1 | $2^1 = 2$ | 2 |
| $\frac{3}{2}$ | $2^{\frac{3}{2}} = 2\sqrt{2}$ | 2.8 |
| 2 | $2^2 = 4$ | 4 |
| $\frac{5}{2}$ | $2^{\frac{5}{2}} = 4\sqrt{2}$ | 5.7 |
| 3 | $2^3 = 8$ | 8 |

Since $2^x$ has not been defined when $x$ is irrational, there are "holes" in the graph of $y = 2^x$. We can expand the domain of $y = 2^x$ to include the rational numbers and the irrational numbers. This will include all real numbers.

Consider the expression $2^{\sqrt{2}}$. Since $1.4 < \sqrt{2} < 1.5$, it makes sense that $2^{1.4} < 2^{\sqrt{2}} < 2^{1.5}$. By selecting closer approximations for $\sqrt{2}$, closer approximations for $2^{\sqrt{2}}$ will result.

$$2^{1.4} < 2^{\sqrt{2}} < 2^{1.5}$$
$$2^{1.41} < 2^{\sqrt{2}} < 2^{1.42}$$
$$2^{1.414} < 2^{\sqrt{2}} < 2^{1.415}$$
$$2^{1.4142} < 2^{\sqrt{2}} < 2^{1.4143}$$
$$2^{1.41421} < 2^{\sqrt{2}} < 2^{1.41422}$$

So, approximate values for $2^x$ when $x$ is irrational can be found by using rational approximations for $x$.

| | |
|---|---|
| *Definition of Irrational Exponents* | **If $x$ is an irrational number and $a > 0$, then $a^x$ is the real number between $a^{x_1}$ and $a^{x_2}$, for all possible choices of rational numbers $x_1$ and $x_2$ such that $x_1 < x < x_2$.** |

Now since $2^x$ is defined when $x$ is an irrational number, the domain of $y = 2^x$ is the set of all real numbers. There are no "holes" in the graph now. The graph is a smooth curve. You could use an accurate graph of $y = 2^x$ to estimate the value of $2^x$ when $x$ is any real number.

**Example 1**

APPLICATION

Biology

Use the graph of $y = 2^x$ shown below to find the number of cells present in a sample after $\sqrt{3}$ fifteen-minute intervals. Then check your estimation using a calculator.

The value of $x$ is $\sqrt{3}$ and $1.7 < \sqrt{3} < 1.8$.

From the graph, the value of $y$ is approximately 3.3.

To check:

ENTER: 2 $\boxed{y^x}$ 3 $\boxed{\sqrt{x}}$ $\boxed{=}$

3.321 99609

The calculator verifies our estimate from the graph. After $\sqrt{3}$ fifteen-minute intervals, there would be about 3 cells.

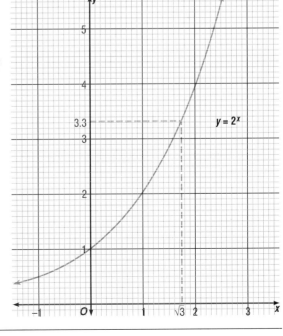

All of the properties of rational exponents apply to real exponents as well.

**Example 2**

Simplify $5^{\sqrt{2}} \cdot 5^{\sqrt{3}}$.

$5^{\sqrt{2}} \cdot 5^{\sqrt{3}} = 5^{\sqrt{2} + \sqrt{3}}$    *Recall the product of powers property, $a^m \cdot a^n = a^{m+n}$.*
*Use your calculator to check this result.*

**Example 3**

Simplify $(6^{\sqrt{5}})^{\sqrt{2}}$.

$(6^{\sqrt{5}})^{\sqrt{2}} = 6^{\sqrt{5} \cdot \sqrt{2}}$
$\qquad\quad = 6^{\sqrt{10}}$    *Recall the power of a power property, $(a^m)^n = a^{mn}$.*

---

*Definition of*
*Exponential Function*

An equation of the form $y = a^x$, where $a > 0$ and $a \neq 1$, is called an exponential function.

---

Several exponential functions have been graphed at the right. Compare the graphs of functions where $a > 1$ to those where $a < 1$. Notice that when $a > 1$ the value of $y$ increases as the value of $x$ increases. When $a < 1$, the value of $y$ decreases as the value of $x$ increases.

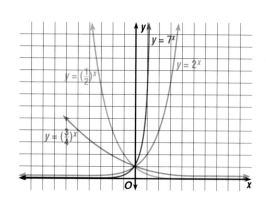

The following property is very useful when solving equations involving exponential functions.

---

*Property of Equality*
*for Exponential*
*Functions*

Suppose $a$ is a positive number other than 1. Then $a^{x_1} = a^{x_2}$ if and only if $x_1 = x_2$.

---

Notice that $a$ cannot equal 1. Since $1^x = 1$ for any real number, $1^{x_1} = 1^{x_2}$ for any choice of $x_1$ and $x_2$.

**Example 4**

Solve $9^3 = 3^{x^2}$ for x.

$$9^3 = 3^{x^2}$$

$$(3^2)^3 = 3^{x^2}$$     *Write each term with the same base; in this case, 3.*

$$3^6 = 3^{x^2}$$     $(a^b)^c = a^{bc}$

$$6 = x^2$$     *Property of equality for exponential functions*

$$\pm\sqrt{6} = x$$

**Check:** $9^3 = 3^{x^2}$

$$9^3 \stackrel{?}{=} 3^{(\pm\sqrt{6})^2}$$

$$9^3 \stackrel{?}{=} 3^6$$

$$729 = 729 \checkmark$$

# CHECKING FOR UNDERSTANDING

**Communicating Mathematics**

Read and study the lesson to answer these questions.

1. Name three ways that you can find an approximate value for $2^x$ when $x$ is an irrational number.

2. Use each of the three methods that you named to approximate $2^{\sqrt{5}}$. Which do you think is the most accurate? Why?

3. Why is 1 excluded as a value for $a$ in the property of equality for exponential functions?

**Guided Practice**

Use the graph of $y = 2^x$ on page 547 or a calculator to approximate each expression to the nearest tenth.

4. $2^{\sqrt{5}}$        5. $2^{0.7}$        6. $2^{-0.3}$

7. $2^{1.1}$        8. $2^{-1}$        9. $8^{\sqrt{2}}$

Use the rule of exponents to simplify each expression.

10. $7^{\sqrt{3}} \cdot 7^{2\sqrt{3}}$        11. $(2^{\sqrt{3}})^{\sqrt{3}}$        12. $3(2^{\sqrt{2}})(2^{-\sqrt{2}})$

Solve each equation.

13. $5^x = 5^{-3}$        14. $6^x = 216$        15. $7^y = \dfrac{1}{49}$

16. $10^x = 0.001$        17. $2^{2x} = \dfrac{1}{8}$        18. $\left(\dfrac{1}{5}\right)^{b-3} = 125$

# EXERCISES

**Practice**

Use the graph of $y = 2^x$ on page 547 or a calculator to evaluate each expression to the nearest tenth.

19. $2^{-1.1}$        20. $4^{-0.3}$        21. $16^{0.4}$

22. $8^{-0.3}$        23. $2^{\sqrt[3]{2}}$        24. $2^{\sqrt[3]{7}}$

Simplify each expression.

25. $(3^{\sqrt{8}})^{\sqrt{2}}$        26. $5^{\sqrt{3}} \cdot 5^{\sqrt{27}}$        27. $64^{\sqrt{7}} \div 2^{\sqrt{7}}$

28. $(y^{\sqrt{3}})^{\sqrt{27}}$        29. $(x^{\sqrt{2}})^{\sqrt{8}}$        30. $(m^{\sqrt{2}} + n^{\sqrt{2}})^2$

**Solve each equation.**

**31.** $3^y = 3^{3y+1}$

**32.** $5^{3y+4} = 5^y$

**33.** $3^x = 9^{x+1}$

**34.** $2^5 = 2^{2x-1}$

**35.** $8^{r-1} = 16^{3r}$

**36.** $2^{x+3} = \dfrac{1}{16}$

**37.** $9^{3y} = 27^{y+2}$

**38.** $\dfrac{1}{27} = 3^{x-5}$

**39.** $\left(\dfrac{1}{3}\right)^q = 3^{q-6}$

**40.** $25^{2m} = 125^{m-3}$

**41.** $2^{2n-1} = 8^{n+7}$

**42.** $2^{x^2+1} = 32$

**43.** $4^{x-1} = 8^x$

**44.** $36^x = 6^{x^2-3}$

**45.** $9^{x^2-2x} = 27^{x^2+1}$

**Graph each equation.**

**46.** $y = 4^x$

**47.** $y = 3^x$

**48.** $y = \left(\dfrac{1}{4}\right)^x$

**Critical Thinking**

**49.** Compare the graphs of $y = 4^x$ and $y = \left(\dfrac{1}{4}\right)^x$. What do you notice?

**Applications**

**50. Communication**  Sally's office has a system to let people know when the department will have a meeting. Sally calls three people. Then those three people each call three other people, and so on until the whole department is notified. If it takes ten minutes for a person to call three people and all calls are completed within 30 minutes, how many people will be notified in the last round?

**51. Finance**  Graham's grandparents started a savings account for him when he was born. They invested $100 in an account that pays 8% interest compounded annually.
  **a.** Write an exponential equation to express the amount of money in the account on Graham's xth birthday.
  **b.** How much is in the account on his 16th birthday?

**Mixed Review**

**52.** How much of a 60% saline solution must be added to a 35% saline solution to get 600 milliliters of a 50% solution?  **(Lesson 11-7)**

**53.** Use synthetic substitution to find $f(3)$ and $f(-2)$ for the function $f(x) = x^3 + 8x + 1$.  **(Lesson 10-2)**

**54.** Is $(4, 0)$ a solution for the inequality $y \le x^2 - 7$? Explain your answer.  **(Lesson 8-6)**

**55. Geometry**  The length of a rectangle is 2 units more than its width. The area of the rectangle is 1763 square units. What are the dimensions of the rectangle?  **(Lesson 7-4)**

**56.** Simplify $\dfrac{9x^{-\frac{4}{3}} - 4y^{-2}}{3x^{-\frac{2}{3}} + 2y^{-1}}$.  **(Lesson 6-6)**

**57.** Factor $b(3b - 2y) - (3b - 2y)$.  **(Lesson 5-5)**

**Journal**

Describe how the graph of an exponential function differs from the other curved graphs you have studied.

# Graphing Calculator Exploration:
## Graphing Exponential and Logarithmic Functions

You can draw graphs of exponential functions and logarithmic functions on your graphing calculator. As you know, an exponential function is a function of the form $y = a^x$, where $a > 0$ and $a \neq 1$. A logarithmic function is a function of the form $x = a^y$, where $a > 0$ and $a \neq 1$. It is denoted $y = \log_a x$, and is the inverse function of an exponential function.

**Example 1**

Graph the set of exponential equations, $y = 2^x$ and $y = \left(\dfrac{1}{2}\right)^x$, on the same set of axes. Use the viewing window [−5, 5] by [0, 10] with scale factors of 1 on both axes.

Graph the equations.  *Be sure to set your range parameters before graphing.*

*Casio*

ENTER: [GRAPH] 2 [$x^y$] [ALPHA] [X] [:] [GRAPH]

.5 [$x^y$] [ALPHA] [X] [EXE]

*TI-81*

ENTER: [Y=] 2 [^] [X|T] [ENTER] .5 [^] [X|T] [GRAPH]

The set of functions has an axis of symmetry, the y-axis. The graphs have the point (0, 1) in common, and both have the x-axis as a horizontal asymptote.  *Do you think that these characteristics are common to all exponential functions? Why?*

The Casio graphing calculator has the graph of the logarithmic function $y = \log_{10} x$ built in. Try pressing [GRAPH] [LOG] [EXE] to view the graph.

**Example 2**

Graph the set of equations, $y = 2^x$ and $y = \log_2 x$, on the same set of axes. Use the viewing window [−2, 10] by [−2, 10] with scale factors of 1 on both axes.

To graph logarithmic functions with bases other than 10, you must use the change of base formula, $\log_a x = \dfrac{\log_{10} x}{\log_{10} a}$. You will use this formula more extensively in Lesson 12-7.

*Casio*

ENTER:

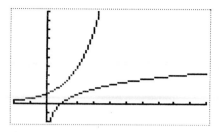

*TI-81*

ENTER:

Notice that the graphs are reflections of each other over the line $y = x$.

---

**Example 3**

**Solve $4^{2x-5} = 3^x$.**

There are two methods that you could use with your graphing calculator to find the solution to this equation. The first method is to graph the equations, $y = 4^{2x-5}$ and $y = 3^x$, and find the $x$-coordinate of the point of intersection. The second method is to graph the equation $y = 4^{2x-5} - 3^x$, and find the $x$-intercept. Let's try the second method. Use the viewing window $[-1, 5]$ by $[-35, 10]$.

*Casio*

ENTER:

*TI-81*

ENTER:

Now use the zoom-in process to estimate the solution. The solution is approximately 4.14  *Check this solution using the first method.*

# EXERCISES

Graph each equation on your graphing calculator so the complete graph is shown. Then sketch the graph.

1. $y = 10^x$       2. $y = 12.5^x$      3. $y = 0.1^x$

4. $y = \log_{20} x$      5. $y = \log_5 x$      6. $y = \log_{0.2} x$

Solve each equation using your graphing calculator.

7. $0.5^x = 75$      8. $25^{2x-1} = 512$      9. $10^{2x} = 17^{1-x}$

10. $0.3^{5x+1} = 0.7^{x-1}$      11. $5^x = 3^{x+2}$      12. $3.32^{3x-2} = 0.5$

13. $6^{x-1} = 10^x$      14. $2^x = x^2$      15. $3^x = x^3$

16. $0.35^{6-x} = 28$      17. $0.75^{x+3} = 4^{2x-1}$      18. $3^{4x-7} = 4^{2x+3}$

## 12-2 Logarithms and Logarithmic Functions

**Objectives**

After studying this lesson, you should be able to:
- write exponential equations in logarithmic form and vice versa,
- evaluate logarithmic expressions, and
- solve equations involving logarithmic functions.

**Application**

The strength of an earthquake is measured using the Richter scale, which is based on powers of 10. The Richter scale is a **logarithmic scale.** This means that a 3 on the scale is 10 times stronger than a 2 and 100 times stronger than a 1. An earthquake that measured 6.9 on the Richter scale hit the San Francisco Bay area on October 17, 1989. This earthquake was $10^{6.9}$ or 7,943,280 times stronger than the weakest earthquake that a seismograph can detect.

*FYI* · · ·

The 1989 earthquake started a wave of long-distance phone calls to the area. About 140 million calls were made the next day, the most ever for a single day.

The two tables below show two related exponential equations. You will recognize the equation in the table on the left as an exponential function.

Given the exponent, $x$, compute the power of 2, $y$.

| $x$ | $2^x = y$ | $y$ |
|---|---|---|
| $-1$ | $2^{-1} = y$ | ? |
| 2 | $2^2 = y$ | ? |
| 3 | $2^3 = y$ | ? |
| 6 | $2^6 = y$ | ? |

Given the power of 2, $x$, compute the exponent, $y$.

| $y$ | $2^y = x$ | $x$ |
|---|---|---|
| ? | $2^y = \dfrac{1}{2}$ | $\dfrac{1}{2}$ |
| ? | $2^y = 4$ | 4 |
| ? | $2^y = 8$ | 8 |
| ? | $2^y = 64$ | 64 |

In the relation shown in the table at the right, $2^y = x$, the exponent $y$ is called the **logarithm**, base 2, of $x$. This relation is written as $\log_2 x = y$. This equation is read "the log base 2 of $x$ is equal to $y$." The logarithm corresponds to the exponent. Study the diagram below.

**Exponential Equation**          **Logarithmic Equation**

$$n = b^p \qquad\qquad p = \log_b n$$

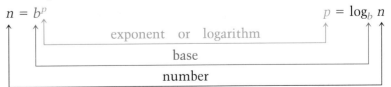

exponent   or   logarithm

base

number

| *Definition of Logarithm* | Suppose $b > 0$ and $b \neq 1$. For $n > 0$, there is a number $p$ such that $\log_b n = p$ if and only if $b^p = n$. |
|---|---|

| Exponential Equation | Logarithmic Equation |
|---|---|
| **a.** $5^2 = 25$ | $\log_5 25 = 2$ |
| **b.** $10^5 = 100{,}000$ | $\log_{10} 100000 = 5$ |
| **c.** $8^0 = 1$ | $\log_8 1 = 0$ |
| **d.** $2^{-4} = \dfrac{1}{16}$ | $\log_2 \dfrac{1}{16} = -4$ |
| **e.** $9^{\frac{1}{2}} = 3$ | $\log_9 3 = \dfrac{1}{2}$ |

The chart at the left shows some equivalent exponential and logarithmic equations.

Examples 1 and 2 show how to find the value of a variable in a logarithmic equation $\log_b x = y$ when values for two of the variables are known.

**Example 1**

**a. Solve $\log_3 243 = y$ for $y$.**

$\log_3 243 = y$

$3^y = 243$    *Definition of logarithm*

$3^y = 3^5$

$y = 5$    *Property of equality for exponential functions*

**b. Solve $\log_9 x = -3$ for $x$.**

$\log_9 x = -3$

$9^{-3} = x$    *Definition of logarithm*

$x = \dfrac{1}{729}$

**Example 2**

APPLICATION

Communication

The superintendent of Fairborn School District has set up a telephone network to notify each faculty or staff member in the event of a school cancellation. She has estimated that each set of calls can be made in ten minutes and that the whole faculty and staff can be contacted in forty minutes. If 256 calls are made in the last round, how many people will each person call?

Since $40 \div 10 = 4$, there are 4 rounds of calls.

$\log_x 256 = 4$    *The network contacts 256 people in the last round.*

$x^4 = 256$    *Definition of logarithm*

$x = \sqrt[4]{256}$

$x = 4$    *Since $b > 0$, $-4$ is not a solution.*

*How many staff members does the superintendent have?*

Each person contacts 4 people.

Let's look at the graphs of an exponential function and its corresponding logarithmic relation. The domain of the relation $y = \log_2 x$ is the set of all positive reals. The range is the set of all real numbers.

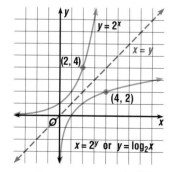

*For every point $(a, b)$ on $y = 2^x$, there is a point on $y = \log_2 x$ with coordinates $(b, a)$.*

$y = 2^x$

| $x$ | $-4$ | $-3$ | $-2$ | $-1$ | $0$ | $1$ | $2$ | $3$ |
|---|---|---|---|---|---|---|---|---|
| $y$ | $\dfrac{1}{16}$ | $\dfrac{1}{8}$ | $\dfrac{1}{4}$ | $\dfrac{1}{2}$ | $1$ | $2$ | $4$ | $8$ |

$2^y = x$   or   $y = \log_2 x$

| $x$ | $\dfrac{1}{16}$ | $\dfrac{1}{8}$ | $\dfrac{1}{4}$ | $\dfrac{1}{2}$ | $1$ | $2$ | $4$ | $8$ |
|---|---|---|---|---|---|---|---|---|
| $y$ | $-4$ | $-3$ | $-2$ | $-1$ | $0$ | $1$ | $2$ | $3$ |

*The $x$ and $y$ values are reversed.*

Notice that the graphs are reflections of each other along the line $y = x$. The relations are inverses of each other. Using the vertical line test, you can see that no vertical line can intersect the graph of $y = \log_2 x$ in more than one place, so $y = \log_2 x$ is a **logarithmic function.**

| *Definition of Logarithmic Function* | **An equation of the form $y = \log_b x$, where $b > 0$ and $b \neq 1$, is called a logarithmic function.** |
|---|---|

Since the exponential function and the logaritnmic function are inverses of each other, their composites are the identity function. Let $f(x) = \log_b x$ and $g(x) = b^x$. For $f(x)$ and $g(x)$ to be inverses, it must be true that $f(g(x)) = x$ and $g(f(x)) = x$.

$$f(g(x)) = x \qquad\qquad g(f(x)) = x$$
$$f(b^x) = x \qquad\qquad g(\log_b x) = x$$
$$\log_b b^x = x \qquad\qquad b^{\log_b x} = x$$

**Example 3**

Evaluate each expression.

a. $\log_8 8^4$
   $\log_8 8^4 = 4 \qquad \log_b b^x = x$

b. $6^{(\log_6 (3x-1))}$
   $6^{(\log_6 (3x-1))} = 3x - 1 \qquad b^{\log_b x} = x$

A property similar to the property of equality for exponential functions applies to the logarithmic functions.

| *Property of Equality for Logarithmic Functions* | **Suppose $b > 0$ and $b \neq 1$. Then $\log_b x_1 = \log_b x_2$ *if and only if* $x_1 = x_2$.** |
|---|---|

**Example 4**

Solve each of the following.

a. $\log_3 (3x - 6) = \log_3 (2x + 1)$

$$\log_3 (3x - 6) = \log_3 (2x + 1)$$
$$3x - 6 = 2x + 1 \quad \textit{Property of equality for logarithmic functions}$$
$$x = 7$$

b. $\log_6 (3x - 1) = \log_6 (2x + 4)$

$$\log_6 (3x - 1) = \log_6 (2x + 4)$$
$$3x - 1 = 2x + 4 \quad \textit{Property of equality for logarithmic functions}$$
$$x = 5$$

# CHECKING FOR UNDERSTANDING

**Communicating Mathematics**

Read and study the lesson to answer these questions.

1. In the equation $2^y = x$, what is the exponent $y$ called?

2. Give an example of a logarithmic function.

3. The principal of Fairborn High School needed to contact each member of the Booster Association. She estimated that by using the network method to have each person call 4 people in ten minutes, each member could be contacted in one-half hour. How many members will be contacted in the last round of calls?

4. In your own words, write a definition of a logarithm.

5. How much stronger is an earthquake with a Richter scale rating of 7 than an aftershock with a rating of 4?

**Guided Practice**

Write each equation in logarithmic form.

6. $2^3 = 8$

7. $10^3 = 1000$

8. $6^{-3} = \dfrac{1}{216}$

Write each equation in exponential form.

9. $\log_2 64 = 6$

10. $\log_{10} 0.01 = -2$

11. $\log_9 27 = \dfrac{3}{2}$

Solve each equation.

12. $\log_8 y = -2$

13. $\log_b 81 = 4$

14. $\log_{12} \dfrac{1}{12} = x$

15. $\log_b 64 = 6$

16. $\log_{10} \dfrac{1}{1000} = x$

17. $\log_4 y = 4$

# EXERCISES

**Practice**

Evaluate each expression.

18. $\log_{10} 10000$

19. $\log_4 16$

20. $\log_{13} 169$

21. $\log_8 \dfrac{1}{64}$

22. $\log_3 \dfrac{1}{243}$

23. $\log_{25} 5$

Solve each equation.

24. $\log_{\frac{1}{2}} 8 = x$

25. $\log_3 729 = x$

26. $\log_b 81 = 2$

27. $\log_3 y = 2$

28. $\log_5 x = -1$

29. $\log_b 18 = 1$

30. $\log_b 81 = 4$

31. $\log_5 x = -2$

32. $\log_3 (4 + y) = \log_3 (2y)$

33. $\log_5 (2x - 3) = \log_5 (x + 2)$
34. $\log_8 (3y - 1) = \log_8 (y + 4)$
35. $\log_9 (5x - 1) = \log_9 (3x + 7)$
36. $\log_{10} (x - 1)^2 = \log_{10} 0.01$
37. $\log_2 (x^2 + 6x) = \log_2 (x - 4)$
38. $\log_{12} (7x - 3) = \log_{12} (5 - x^2)$
39. $\log_{10} (x^2 + 36) = \log_{10} 100$
40. $\log_9 (x^2 + 9x) = \log_9 10$

**Graph each pair of equations on the same set of axes.**

41. $y = \log_3 x$ and $y = 3^x$
42. $y = \log_{\frac{1}{2}} x$ and $y = \left(\frac{1}{2}\right)^x$
43. $y = 10^x$ and $y = \log_{10} x$
44. $y = 4^x$ and $y = \log_4 x$

**Show that each statement is true.**

45. $\log_4 4 + \log_4 16 = \log_4 64$
46. $\log_4 16 = 2 \log_4 4$
47. $\log_2 8 \cdot \log_8 2 = 1$
48. $\log_{10} [\log_3 (\log_4 64)] = 0$

**Critical Thinking**

49. Show that $f(x) = \log_b x$ and $g(x) = b^x$ are inverse functions.

**Applications**

50. **Chemistry** The pH of a solution is a measure of its acidity. A low pH indicates an acidic solution, and a high pH indicates a basic solution. Neutral water has a pH of 7. The pH of a solution is related to the concentration of hydrogen ions by the formula $\text{pH} = \log_{10} \frac{1}{H^+}$, where $H^+$ represents the hydrogen atoms in gram atoms per liter. If the pH of a solution is 4, what is the concentration of hydrogen ions?

51. **Biology** An amoeba divides into two amoebas once every hour. How long would it take for a single amoeba to become a colony of 4096 amoebas?

**Mixed Review**

52. Simplify the expression $11^{\sqrt{5}} \cdot 11^{\sqrt{45}}$. **(Lesson 12-1)**

53. Find the greatest common factor of the numerator and the denominator of $\frac{3x + 15}{9x + 45}$. Then simplify the expression. **(Lesson 11-1)**

54. State the number of positive real zeros, negative real zeros, and complex zeros for the function $f(x) = -x^3 + x^2 - x + 1$. **(Lesson 10-4)**

55. Find the vertices, foci, and slopes of the asymptotes for the hyperbola whose equation is $36y^2 - 81x^2 = 2916$. **(Lesson 9-5)**

56. State the sum and product of the roots of the quadratic equation $x^2 - 18x - 120 = 0$. **(Lesson 7-5)**

57. Find the product of $3 - 7i$ and its conjugate. **(Lesson 6-10)**

58. The sum of the digits of a three-digit number is 22. The tens digit exceeds the hundreds digit by 1. When the digits are reversed, the new number is 297 greater than the original number. Find the number. **(Lesson 3-9)**

# 12-3 Properties of Logarithms

**Objective**

After studying this lesson, you should be able to:

- solve equations or simplify and evaluate expressions using properties of logarithms.

**Application**

*FYI* · · ·

A large portion of the acid rain is a result of the burning of fossil fuels. Finding other sources of energy will help solve the problem.

Acid rain is a serious problem in many parts of the world. It raises the pH level of lakes and streams, killing fish and other waterlife. The pH scale measures acidity and is a logarithmic scale. Lower pH numbers represent acids, and higher pH numbers represent bases. Acid rain has a pH of about 4.2, while neutral water has a pH of 7. So, acid rain is about 630 times more acidic than neutral water, since $10^{7-4.2} \approx 630$. The pH scale is an example of a logarithmic scale.

Since logarithms are exponents, the properties of logarithms can be derived from the properties of exponents that you already know.

The product of powers is found by adding exponents. So it makes sense that the logarithm of a product is found by adding the logarithms.

**a.** $\log_3 (9 \cdot 27)$
$= \log_3 (3^2 \cdot 3^3)$
$= \log_3 (3^{2+3})$
$= 2 + 3$

**b.** $\log_3 9 + \log_3 27$
$= \log_3 3^2 + \log_3 3^3$
$= 2 + 3$

So, $\log_3 (9 \cdot 27) = \log_3 9 + \log_3 27$. This illustrates the following property.

| *Product Property of Logarithms* | **For all positive numbers *m*, *n* and *b*, where *b* ≠ 1,** $\log_b mn = \log_b m + \log_b n.$ |
| --- | --- |

To prove this property, let $b^x = m$ and $b^y = n$.
Then $\log_b m = x$ and $\log_b n = y$.

$$b^x b^y = mn$$
$$b^{x+y} = mn \qquad \textit{Property of exponents}$$
$$\log_b b^{x+y} = \log_b mn \qquad \textit{Property of equality for logarithmic functions}$$
$$x + y = \log_b mn \qquad \textit{Definition of inverse functions}$$
$$\log_b m + \log_b n = \log_b mn \qquad \textit{Substitution property}$$

**Example 1**

Given $\log_2 5 = 2.322$, find each logarithm.

**a.** $\log_2 20 = \log_2 (2^2 \cdot 5)$
$= \log_2 2^2 + \log_2 5$
$= 2 + 2.322 \text{ or } 4.322$

**b.** $\log_2 25 = \log_2 (5 \cdot 5)$
$= \log_2 5 + \log_2 5$
$= 2.322 + 2.322 \text{ or } 4.644$

To find the quotient of powers, you subtract the exponents. So, to find the logarithm of a quotient, you subtract logarithms.

**a.** $\log_3 \left(\dfrac{81}{27}\right) = \log_3 \left(\dfrac{3^4}{3^3}\right)$
$= \log_3 (3^{4-3})$
$= 4 - 3$

**b.** $\log_3 81 - \log_3 27 = \log_3 3^4 - \log_3 3^3$
$= 4 - 3$

So, $\log_3 \left(\dfrac{81}{27}\right) = \log_3 81 - \log_3 27$. This illustrates the following property.

| *Quotient Property of Logarithms* | **For all positive numbers $m$, $n$, and $b$, where $b \neq 1$,** $$\log_b \frac{m}{n} = \log_b m - \log_b n.$$ |
|---|---|

To prove this property, let $b^x = m$ and $b^y = n$.
Then $\log_b m = x$ and $\log_b n = y$.

$$\frac{b^x}{b^y} = \frac{m}{n}$$

$$b^{x-y} = \frac{m}{n} \qquad \textit{Property of exponents}$$

$$\log_b b^{x-y} = \log_b \frac{m}{n} \qquad \textit{Property of equality for logarithmic functions}$$

$$x - y = \log_b \frac{m}{n} \qquad \textit{Definition of inverse functions}$$

$$\log_b m - \log_b n = \log_b \frac{m}{n} \qquad \textit{Substitution property}$$

The following example illustrates the quotient property of equality for logarithms.

**Example 2**

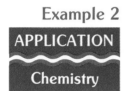

**The pH of a solution is related to the number of gram atoms of hydrogen ions, $H^+$, by the formula pH $= \log_{10} \dfrac{1}{H^+}$. If the pH level of a lake is 5, how much more acidic is it than neutral water, that has a pH of 7?**

The pH of neutral water is 7. Since $\log_{10} \left(\dfrac{1}{10^{-7}}\right) = 7$, there are $10^{-7}$ gram atoms of hydrogen ions in a liter of neutral water. Similarly, since $\log_{10} \left(\dfrac{1}{10^{-5}}\right) = 5$, there $10^{-5}$ gram atoms of hydrogen ions in a liter of the lake water. If we divide these two numbers, we will find how much stronger the lake water is than neutral water.

$$x = \frac{10^{-5}}{10^{-7}}$$

$$\log_{10} x = \log_{10} \frac{10^{-5}}{10^{-7}}$$

$$\log_{10} x = \log_{10} 10^{-5} - \log_{10} 10^{-7}$$

$$\log_{10} x = -5 - (-7)$$

$$\log_{10} x = 2$$

$$x = 100$$

**Check:** $x = \dfrac{10^{-5}}{10^{-7}}$
$= 10^{-5-(-7)}$
$= 10^2$
$= 100 \checkmark$

So the lake water is 100 times more acidic than the neutral water.

**Example 3** | Given $\log_{12} 9 = 0.884$ and $\log_{12} 18 = 1.163$, find each logarithm.

**a.** $\log_{12}\left(\dfrac{3}{4}\right)$                 **b.** $\log_{12} 2$

$$\log_{12}\left(\frac{3}{4}\right) = \log_{12}\left(\frac{9}{12}\right)$$
$$= \log_{12} 9 - \log_{12} 12$$
$$= 0.884 - 1$$
$$= -0.116$$

$$\log_{12} 2 = \log_{12}\left(\frac{18}{9}\right)$$
$$= \log_{12} 18 - \log_{12} 9$$
$$= 1.163 - 0.884$$
$$= 0.279$$

The quotient property can be used to solve equations involving logarithms.

**Example 4** | Solve $\log_5 4 + \log_5 x = \log_5 36$.

$$\log_5 4 + \log_5 x = \log_5 36$$
$$\log_5 x = \log_5 36 - \log_5 4$$
$$\log_5 x = \log_5\left(\frac{36}{4}\right) \qquad \textit{Quotient property of logarithms}$$
$$\log_5 x = \log_5 9$$
$$x = 9 \qquad \textit{Property of equality for logarithmic functions}$$

The power of a power is found by multiplying the two exponents. This suggests that the logarithm of a power is found by multiplying the logarithm and the exponent.

**a.** $\log_3 9^4$
$$= \log_3 (3^2)^4$$
$$= \log_3 3^{2 \cdot 4}$$
$$= 2 \cdot 4$$

**b.** $4 \log_3 9$
$$= (\log_3 9) \cdot 4$$
$$= (\log_3 3^2) \cdot 4$$
$$= 2 \cdot 4$$

So, $\log_3 9^4 = 4 \log_3 9$. This suggests the following property.

| *Power Property of Logarithms* | **For any real number $p$ and positive numbers $m$ and $b$ where $b \neq 1$, $\log_b m^p = p \cdot \log_b m$.** |
| --- | --- |

In exercise 3, you will prove this property. Examples 5 and 6 illustrate how to use this property to solve equations involving logarithms.

**Example 5** | Solve $2 \log_6 4 - \dfrac{1}{3} \log_6 8 = \log_6 x$.

$$2 \log_6 4 - \frac{1}{3} \log_6 8 = \log_6 x$$
$$\log_6 4^2 - \log_6 8^{\frac{1}{3}} = \log_6 x \qquad \textit{Power property of logarithms}$$
$$\log_6 16 - \log_6 2 = \log_6 x$$
$$\log_6\left(\frac{16}{2}\right) = \log_6 x \qquad \textit{Quotient property of logarithms}$$
$$\log_6 8 = \log_6 x$$
$$8 = x \qquad \textit{Property of equality for logarithmic functions}$$

**Example 6**

Solve $\log_4 (x + 2) + \log_4 (x - 4) = 2$.

$$\log_4 (x + 2) + \log_4 (x - 4) = 2$$

$$\log_4 (x + 2)(x - 4) = 2 \qquad \textit{Product property of logarithms}$$

$$(x + 2)(x - 4) = 4^2 \qquad \textit{Definition of logarithm}$$

$$x^2 - 2x - 8 = 16$$

$$x^2 - 2x - 24 = 0$$

$$(x - 6)(x + 4) = 0$$

$$x - 6 = 0 \quad \text{or} \quad x + 4 = 0$$

$$x = 6 \quad \text{or} \quad x = -4$$

**Check:** $\log_4 (x + 2) + \log_4 (x - 4) = 2 \qquad\qquad \log_4 (x + 2) + \log_4 (x - 4) = 2$

$\log_4 (6 + 2) + \log_4 (6 - 4) = 2 \qquad\qquad \log_4 (-4 + 2) + \log_4 (-4 - 4) = 2$

$\log_4 8 + \log_4 2 = 2 \qquad\qquad\qquad\qquad \log_4 (-2) + \log_4 (-8) = 2$

$\log_4 16 = 2 \qquad\qquad\qquad\qquad$ *Both $\log_4 (-2)$ and $\log_4 (-8)$ are*

$2 = 2 \checkmark \qquad\qquad\qquad\qquad$ *undefined, so $-4$ is not a solution.*

The only solution is 6.

# CHECKING FOR UNDERSTANDING

**Communicating Mathematics**

**Read and study the lesson to answer these questions.**

1. Does the more acidic of two solutions have a higher or lower pH?

2. Show that the acid rain described at the beginning of the lesson is about 630 times as acidic as neutral water.

3. Prove the power property of logarithms.

**Guided Practice**

**Express each logarithm as the sum or difference of simpler logarithmic expressions.**

4. $\log_8 x^3 y$
5. $\log_5 (xy)^2$
6. $\log_4 \dfrac{ab}{c}$
7. $\log_2 rt^{\frac{1}{2}}$

**Evaluate each expression.**

8. $8^{\log_8 3 \,+\, \log_8 2}$
9. $10^{4 \log_{10} 2}$
10. $9^{\log_9 12 \,-\, \log_9 4}$

**Solve each equation.**

11. $\log_3 7 + \log_3 x = \log_3 14$
12. $\log_2 10 - \log_2 t = \log_2 2$

# EXERCISES

**Practice**

**Use $\log_2 3 = 1.585$ and $\log_2 7 = 2.807$ to evaluate each expression.**

13. $\log_2 49$
14. $\log_2 27$
15. $\log_2 \dfrac{7}{3}$

16. $\log_2 36$
17. $\log_2 0.75$
18. $\log_2 48$

19. $\log_2 108$
20. $\log_2 \dfrac{36}{49}$
21. $\log_2 \dfrac{7}{16}$

**Solve each equation.**

**22.** $\log_2 3 + \log_2 7 = \log_2 x$

**23.** $\log_3 56 - \log_3 8 = \log_3 x$

**24.** $\log_3 14 + \log_3 y = \log_3 42$

**25.** $\log_9 x = \frac{1}{2} \log_9 144 - \frac{1}{3} \log_9 8$

**26.** $\log_{10} m = \frac{1}{2} \log_{10} 81$

**27.** $\log_7 m = \frac{1}{3} \log_7 64 + \frac{1}{2} \log_7 121$

**28.** $\log_{10} 7 + \log_{10} (n - 2) = \log_{10} 6n$

**29.** $3 \log_{10} x = \log_{10} 27$

**30.** $\log_{10} (m + 3) - \log_{10} m = \log_{10} 4$

**31.** $2 \log_3 x + \log_3 0.1 = \log_3 5 + \log_3 2$

**32.** $3 \log_5 x - \log_5 4 = \log_5 16$

**33.** $\log_2 15 + \log_2 14 - \log_2 105 = \log_2 x$

**34.** $\log_{10} y + \log_{10} (y + 21) = 2$

**35.** $\log_4 (x + 3) + \log_4 (x - 3) = 2$

**36.** $\log_2 (y + 2) - 1 = \log_2 (y - 2)$

**37.** $\log_8 (n + 1) - \log_8 n = \log_8 4$

**Solve for a.**

**38.** $\log_n a = \log_n (y + 3) - \log_n 3$

**39.** $\log_b 2a - \log_b x^3 = \log_b x$

**40.** $\log_x a^2 + 5 \log_x y = \log_x a$

**41.** $\log_b 4 + 2 \log_b a = 2 \log_b (n + 1)$

**Critical Thinking**

**42.** Explain why 1 is excluded from the possible values for $b$ in each of the three properties you learned in this lesson.

**Applications**

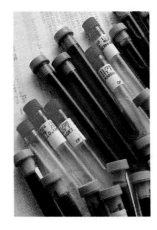

**43. Medicine** The pH of a person's blood can be found by using the Henderson-Hasselbach formula. The formula is pH $= 6.1 + \log_{10} \left( \frac{B}{C} \right)$, where $B$ represents the concentration of bicarbonate, which is a base, and $C$ represents the concentration of carbonic acid, which is acidic. Most people have a blood pH of about 7.4.

a. Rewrite the formula to eliminate the logarithm of a quotient.

b. As you know, a pH of 7 is neutral and lower pH numbers represent acidic solutions. The pH levels higher than 7 represent basic solutions. Is blood normally an acid, a base, or neutral?

c. If you have a scientific calculator, it will find logarithms base 10. Use your calculator to find the pH of a persons blood if the concentration of bicarbonate is 25 and the concentration of carbonic acid is 2.

**Mixed Review**

**44.** Solve the equation $\log_{10} (3n) = \log_{10} (n + 2)$. **(Lesson 12-2)**

**45.** If $f = \{(0, 2), (1, 1), (3, 3), (4, -1)\}$ and $g = \{(0, 0), (1, 1), (2, 4), (3, 9)\}$, find $f \circ g$ and $g \circ f$ if they exist. **(Lesson 10-7)**

**46.** Use the quadratic formula to solve $x^2 - 8x + 22 = 0$. **(Lesson 7-4)**

**47.** Factor $3pq + 3ps + q^2 - s^2$. **(Lesson 5-5)**

**48.** Solve the inequality $|x + 3| + |x - 3| > 8$. **(Lesson 1-8)**

# 12-4 Common Logarithms

**Objectives**

After studying this lesson, you should be able to:

- identify the characteristic and the mantissa of a logarithm, and
- find common logarithms and antilogarithms.

Logarithms can make certain computations easier. When logarithms are used, multiplication changes to addition and division changes to subtraction. For this reason, logarithms were used quite extensively for calculations before calculators became readily available. The most useful logarithms are base 10, since our number system is base 10. Base 10 logarithms are called **common logarithms.** These are usually written without the subscript 10, so $\log_{10} x$ is written as $\log x$.

Let's find the common logarithm of a number and explore common logarithms. Expressing a number in scientific notation is helpful when working with common logarithms.

**Example 1**

**Given that log 8.1 = 0.9085, find the value of log 81,000.**

$$81,000 = 8.1 \times 10^4 \quad \textit{Express 81,000 using scientific notation.}$$

$$\begin{aligned}
\log 81,000 &= \log (8.1 \times 10^4) \\
&= \log 8.1 + \log 10^4 \\
&= 0.9085 + 4 \\
&= 4.9085
\end{aligned}$$

The log of 81,000 is 4.9085. *Verify this with your calculator.*
*Is $10^{4.9085} = 81,000$?*

Now look closely at the result of Example 1, 4.9085. It, like all common logarithms, is made up of two parts, the **characteristic** and the **mantissa.** The mantissa is the logarithm of a number between 1 and 10, in the case of Example 1 that number is 0.9085. The characteristic is the exponent of ten that is used when the number is expressed in scientific notation. In Example 1, the characteristic is 4.

$$\log 81,000 = 4.\underbrace{9085}$$

*characteristic    mantissa*

*The mantissa is the decimal post of a logarithm.*

The mantissa is usually expressed as a positive number. This makes using a table of logarithms much easier. So to avoid negative mantissas, we rewrite the negative mantissa as the difference of a positive number and an integer, usually 10.

## Example 2

**Use your calculator to find log 0.0027. Write the result with a positive mantissa.** *If you do not have a scientific calculator, see the Appendix.*

The [LOG] key is used to find common logarithms.

ENTER: 0.0027 [LOG] $-2.5686362$

The value of log 0.0027 is approximately $-2.5686$.

To write the logarithm with a positive mantissa, add and subtract 10.

$(-2.5686 + 10) - 10 = 7.4314 - 10$   *10 − 10 = 0 and a + 0 = a.*

The characteristic of the logarithm is $7 - 10$ or $-3$. Thus, the mantissa is 0.4314.

---

Sometimes an application of logarithms requires that you use the inverse of logarithms, exponentiation. When you are given a logarithm and asked to find the number, you are finding the **antilogarithm.** That is, if log $x = a$, then $x = $ antilog $a$.

## Example 3

**Find the antilogarithm of 2.579 using a calculator.**

To find an antilogarithm on your calculator, use the [$10^x$] key. *If your calculator does not have a [$10^x$] key, use the [INV] key and then the [LOG] key.*

ENTER: 2.579 [$10^x$] [=] $379.31498$

The antilogarithm of 2.579 is approximately 379.

---

## Example 4

APPLICATION

Physics

**The intensity of sound in decibels, $I_d$, is related to the intensity of sound in watts per square meter, $I$, by the formula $I_d = 10 \log \dfrac{I}{10^{-12}}$. Sound waves whose intensities reach more than 1 W/m² can cause damage to a person's hearing. If the sound at a recent concert reached 110 decibels, were the listeners in danger of causing damage to their hearing?**

Use antilogarithms to find the value of $I$ in the equation $I_d = 10 \log \dfrac{I}{10^{-12}}$.

$$I_d = 10 \log \frac{I}{10^{-12}}$$

$$110 = 10 \log \frac{I}{10^{-12}} \qquad \text{\textit{Substitute 110 for } } I_d.$$

$$11 = \log \frac{I}{10^{-12}} \qquad \text{\textit{Divide each side by 10.}}$$

$$\text{antilog } 11 = \text{antilog}\left(\log \frac{I}{10^{-12}}\right)$$ *Take the antilogarithm of each side.*

$$10^{11} = \frac{I}{10^{-12}}$$ *Definition of antilogarithm*

$$10^{11} \cdot 10^{-12} = I$$ *Multiply each side by $10^{-12}$.*

$$0.1 = I$$

The intensity of the music was 0.1 $W/m^2$, which is less than 1, so there was no danger of causing damage to their hearing.

# CHECKING FOR UNDERSTANDING

**Communicating Mathematics**

**Read and study the lesson to answer these questions.**

1. What function are you performing when you take the antilogarithm of a value, such as $x$?

2. What base does the calculator ⌊LOG⌋ key use? What are these logarithms called?

3. For what values do you think the logarithms are programmed into calculators so that the value of any function using logarithms can be found?

**Guided Practice**

**If log 573 = 2.7582, find each number.**

4. characteristic of log 573

5. log 5.73

**If log 0.023 = −1.6383, find each number.**

6. mantissa of log 0.023

7. antilog 0.3617

# EXERCISES

**Practice**

**Use a calculator to find the logarithm of each number, rounded to four decimal places. Then state the characteristic.**

8. 53.7

9. 800.2

10. 2.25

11. 0.057

12. 4.322

13. 0.295

**Use a calculator to find the antilogarithm of each logarithm, rounded to four decimal places.**

14. 0.2586

15. 2.2249

16. 1.0024

17. −0.2586

18. −2.0112

19. −1.9725

20. 8.1342 − 10

21. 2.2675 − 3

22. 4.9243

Use a calculator to find the logarithm of each number, rounded to four decimal places. Then state the characteristic and the mantissa.

**23.** 98.29        **24.** 13.54        **25.** 827.1

**26.** $10^{-7}$        **27.** 5.4265        **28.** 0.265

**29.** 0.005        **30.** 6678        **31.** 1000

**Critical Thinking**

**32.** Try to find $(-3)^3$ on your calculator. Some calculators will say "ERROR", even though $(-3)^3 = -27$. Can you think of a reason why they might do this?

**Applications**

**33. Seismology** As you know, the Richter scale is a logarithmic scale. An earthquake that measures 6 on the Richter scale is $10^6$ times as intense as the weakest earthquake perceptible by a seismograph.
   **a.** How much more intense is an earthquake that measures 5.3 on the Richter scale than the weakest perceptible earthquake?
   **b.** How much more intense was the San Francisco earthquake of 1989, a 6.9 on the Richter scale, than its strongest after shock, a 4.3 on the Richter scale?

**Mixed Review**

**34.** Solve the equation $2 \log_6 3 + 3 \log_6 2 = \log_6 x$. **(Lesson 12-3)**

**35.** If $y$ varies inversely as $x$ and when $y = 5$, $x = -2$, then find $x$ when $y = 15$. **(Lesson 11-2)**

**36.** Find all zeros for the function $f(x) = 2x^3 + 9x^2 - 20x - 75$. **(Lesson 10-3)**

**37.** State the dimension and evaluate the determinant (if one exists) for the matrix $\begin{bmatrix} 5 & 12 \\ 0 & -4 \end{bmatrix}$. **(Lesson 4-1)**

## ～～～ MID-CHAPTER REVIEW ～～～

Solve each equation. **(Lesson 12-1)**

**1.** $12^5 = 12^{2x+1}$      **2.** $3^{x+3} = \dfrac{1}{81}$      **3.** $5^{x^2-3} = 25^x$

Evaluate each expression. **(Lesson 12-2)**

**4.** $\log_{10} 100$      **5.** $\log_{36} 6$      **6.** $\log_5 \dfrac{1}{125}$

Solve each equation. **(Lesson 12-3)**

**7.** $\log_6 8 + \log_6 3 = \log_6 x$    **8.** $3 \log_2 x = \log_2 512$    **9.** $\log_2 (9x + 5) - \log_2 (x^2 - 1) = 2$

Use a calculator to find each value, round to four decimal places. **(Lesson 12-4)**

**10.** log 229        **11.** antilog −0.0951        **12.** antilog 0.7935 − 3

**13. Seismology** The earthquake that occurred in San Francisco in 1989 registered a 6.9 on the Richter scale. In 1906, an earthquake occurred in San Francisco that is estimated to have registered an 8.3 on the Richter scale. How much more intense was the 1906 earthquake? **(Lesson 12-4)**

## 12-5 Natural Logarithms

**Objective**

After studying this lesson, you should be able to:
- find natural logarithms of numbers.

**Application**

*FYI* · · ·

7% interest compounded continuously yields an interest rate of about $7\frac{1}{4}$% per year.

Sean's grandparents opened a savings account for him when he was born. They placed $1000 in an account that paid 7% interest compounded continuously. The amount of money in the account, $A$, after $t$ years is found by using the formula $A = Pe^{rt}$, where $P$ represents the amount of principal, $r$ represents the annual interest rate, and $e$ is a special irrational number.

Sean is now 16 years old and would like to buy a used car that costs $2500. Does he have enough money in his savings account to buy it? *This problem will be solved in Example 3.*

The number $e$, used in the interest formula, is used extensively in science and mathematics. It is an irrational number whose value is approximately 2.718. $e$ is the base for the **natural logarithms,** which is abbreviated **ln.** So, the natural logarithm of $e$ is 1. All of the properties of logarithms that you have learned apply to the natural logarithms as well. The key marked ⓛⓝ on your calculator is the natural logarithm key.

**Example 1**

**Use your calculator to find ln 3.965.**

ENTER: 3.965 ⓛⓝ  $1.3775059$

The natural logarithm of 3.695 is approximately 1.3775.

You can take antilogarithms of natural logarithms as well. The symbol for the antilogarithm of $x$ is antiln $x$.

**Example 2**

**Find each value.**

**a.** Find $x$ if ln $x$ = 3.9824.

$\ln x = 3.9824$
$x = \text{antiln } 3.9824$

ENTER: 3.9824 ⓔˣ  $53.645629$

*If your calculator has no* ⓔˣ *key, use the* ⓘⓝⓥ *key and then the* ⓛⓝ *key.*

So $x$ is approximately 53.6456.

**b.** Find $e$ if ln $e$ = 1.

$\ln e = 1$
$e = \text{antiln } 1$

ENTER: 1 ⓔˣ  $2.7182818$

So $e$ is approximately 2.7183.

The exponential equations in the following applications contain the number $e$. Equations involving $e$ are easier to solve using natural logarithms, than using common logarithms since $\ln e = 1$.

**Example 3**

APPLICATION

Finance

**Use the formula $A = Pe^{rt}$, to determine whether Sean can buy a used car costing \$2500 with the \$1000 investment his grandparents made for him 16 years ago.**

Use the formula $A = Pe^{rt}$.

$A = Pe^{rt}$

$A = (1000)e^{(0.07)(16)}$     *principal = \$1000, rate = 7%, time = 16 years*

$A = 1000e^{1.12}$

$\ln A = \ln (1000e^{1.12})$     *Take the natural logarithm of each side.*

$\ln A = \ln 1000 + (1.12) \ln e$     *Power and product properties of logarithms*

$\ln A = \ln 1000 + 1.12$     *Since e is the base for natural logarithms, $\ln e = 1$.*

$\ln A = 8.0277553$

$A = 3064.85$     *Take the antiln of each side.*

Sean has \$3064.85 in his account. This is more than enough to buy the car.

Natural logarithms can also be used to determine the amount you need to invest now in order to have a certain amount later.

**Example 4**

APPLICATION

Finance

**Danica is saving money to go on a trip to Europe after her college graduation. She will finish college six years from now. If the six-year certificate of deposit that she buys now pays 8% interest compounded continuously, how much should she invest now in order to have \$3000 for the trip?**

Again, use the formula $A = Pe^{rt}$.

$A = Pe^{rt}$

$3000 = Pe^{(0.08)(6)}$

$3000 = Pe^{0.48}$

$\ln 3000 = \ln (Pe^{0.48})$

$\ln 3000 = \ln P + (0.48) \ln e$

$\ln 3000 = \ln P + 0.48$

$7.5263676 \approx \ln P$

$1856.35 \approx P$

Danica should invest \$1856.35 to earn enough for the trip.

# CHECKING FOR UNDERSTANDING

**Communicating Mathematics**

Read and study the lesson to answer these questions.

1. What is the base of the natural logarithms?

2. How much would Sean have in his savings account if the interest rate had been 7.25%?

3. When should you choose to use natural logarithms instead of common logarithms to solve a problem?

**Guided Practice**

Use your calculator to find each value, rounded to four decimal places.

4. ln 2.58

5. ln 4.28

6. ln 48.987

7. antiln 0.4253

8. antiln 1.7015

9. antiln $-0.7876$

# EXERCISES

**Practice**

Use your calculator to find each value, rounded to four decimal places.

10. ln 9.45

11. ln 7.21

12. ln 1.42

13. antiln 0.469

14. antiln 0

15. antiln 2.2289

16. ln 56.9

17. ln 0.543

18. ln 65

19. ln $e$

20. antiln 3.56

21. antiln 0.52

22. antiln 0.288

23. antiln $-1.679$

24. ln 1000

Solve each equation.

25. $2000 = 5e^{0.045x}$

26. $2 = e^{5k}$

27. $\ln 3.6 = \ln (e^{0.031t})$

28. $65 = e^{6n}$

29. $25 = e^{0.075y}$

30. $\ln 40.5 = \ln (e^{0.21t})$

**Critical Thinking**

31. The great Swiss mathematician Leonhard Euler, for whom the number $e$ is named, defined $e$ as the sum of the series $1 + \dfrac{1}{1} + \dfrac{1}{1 \cdot 2} + \dfrac{1}{1 \cdot 2 \cdot 3} + \dfrac{1}{1 \cdot 2 \cdot 3 \cdot 4} + \ldots$ . Calculate $e$ to the nearest ten-thousandth using this series.

**Applications**

32. **Finance** Mr. and Mrs. Grauser invested $500 at 6.5% compounded continuously.
   a. Find the value of the investment after 7 years.
   b. When will the Grausers' investment be tripled?

33. **Chemistry** Radium 226 decomposes radioactively. The amount of a radioactive substance present after $t$ years is found by the formula $y = ne^{kt}$, where $n$ is the initial amount of the substance and $k$ is a constant. It takes 1800 years for half of a sample of Radium 226 to decompose.
   a. Use 100 grams as the original amount to find the constant $k$ for this substance.
   b. How much of a 1-gram sample of Radium 226 will remain after 10,000 years?

**Mixed Review**
34. Use a calculator to find the common logarithm of 349.948, rounded to four decimal places. Then state the characteristic and the mantissa. (**Lesson 12-4**)

35. Add $\dfrac{3}{x-2} + \dfrac{2}{x-3}$. (**Lesson 11-4**)

36. Find the solutions of the system $\begin{aligned} 2y^2 &= 10 - x^2 \\ 3x^2 - 9 &= y^2. \end{aligned}$ (**Lesson 9-9**)

37. Solve the inequality $x^2 \le 36$. (**Lesson 8-7**)

38. Carol and Frank want to buy some new living room furniture. A sofa, love seat, and coffee table cost $1230. The sofa costs twice as much as the love seat. The sofa and the coffee table cost $880. What are the prices of each piece of furniture? (**Lesson 3-9**)

## ~~~~~ HISTORY CONNECTION ~~~~~

Logarithms were first described by **John Napier** (1550–1617 A.D.), a Scottish laird. This Baron of Murchiston managed his large estate, while pursuing writing and mathematics as a hobby. He first used the words "artifical number" to describe his system of exponential values. Later, when he published his findings in *Descripto* (1619), he used the word logarithm which he formed from the Greek words *logos* (meaning ratio) and *arithmos* (meaning number).

A Salvilian professor of geometry at Oxford University, **Henry Briggs** (1561–1639), read Napier's work and was enthusiastic in studying this further. It was Briggs who first used the terms *mantissa* and *characteristic*. The early logarithm tables printed both the mantissa and characteristic. It was not until the mid-18th century that tables containing only the mantissas were widely accepted.

John Napier

# 12-6 Problem-Solving Strategy: Using Estimation

**Objective**

After studying this lesson, you should be able to:

■ solve problems using estimation.

Sometimes solving a problem is a long process involving many steps. Occasionally we make a miscalculation, take a wrong turn, or even mis-key on a calculator and get an incorrect answer. This is why it is very helpful to estimate the solution to a problem before performing the calculations. Then if our solution is not close to the estimate, we know that we have made a mistake.

**Example 1**

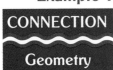

CONNECTION

Geometry

**Find the approximate area of a regular hexagon whose sides are 29 cm long.**

A regular hexagon can be inscribed in a circle. The figure shows the hexagon inscribed in a circle.

The diagonals of the hexagon form equilateral triangles, so the radius of the circle is also 29 cm.

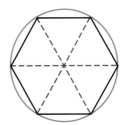

You can estimate that the size of the hexagon is a little less than the area of the circle, which is found using the formula $A = \pi r^2$. Use your calculator to find the area of the circle.

ENTER: $\boxed{\pi}$ $\boxed{\times}$ 29 $\boxed{x^y}$ $\boxed{=}$ $\mathtt{2642.079422}$

The area of the circle is about 2642 cm². The area of the hexagon would be close to but less than 2642 cm². The actual area of the hexagon is $\frac{2523\sqrt{3}}{2}$ cm², which is about 2185 cm².

**Example 2**

APPLICATION

Finance

**Ted invested \$9300 in a two year certificate of deposit (CD) that has an annual yield of 9.57%. How much interest did he earn from the CD?**

Ted will earn 9.57% each year. Since 10% of \$9300 is \$930, an estimate of Ted's earnings is $2 \times 930$ or \$1860. *Will the actual value be greater or less than \$1860?*

$I = Prt$         *Use the equation for simple interest.*

$I = (9300)(0.0957)(2)$

$I = 1780.02$

Ted will earn \$1780.02 in interest. This is a reasonable answer according to our estimate.

## Example 3

**APPLICATION**

Aerospace

A satellite requires 10 watts of power to operate its equipment for one year. A power supply aboard the satellite generates power according to the equation $P = 50e^{-\frac{t}{250}}$, where $P$ represents the power in watts and $t$ represents the time in days. How much power will the power supply generate in one year?

Since the satellite requires 10 watts of power in a year, this is probably a good estimate of the power to be generated in one year.

$$P = 50e^{-\frac{t}{250}}$$

$$P = 50e^{-\frac{365}{250}} \qquad \textit{There are 365 days in a year.}$$

$$\ln P = \ln 50 + \left(-\frac{365}{250}\right) \ln e$$

$$\ln P = 3.9120 - \frac{365}{250} \qquad \textit{In e = 1}$$

$$\ln P = 2.4520230$$

$$P = 11.611814$$

The power supply generates about 11.6 watts of energy in a year. This is a reasonable answer according to our estimate.

# CHECKING FOR UNDERSTANDING

**Communicating Mathematics**

Read and study the lesson to answer these questions.

1. When might you use estimation in every day life?

2. When might you use estimation in solving mathematics problems?

3. Where would you look for a mistake if your estimate is very different than your solution?

**Guided Practice**

Solve each problem using estimation.

4. Give an estimate of the area of a square with a side 9.39 inches long. Then find the actual area and compare your solution to your estimate.

5. A roll of wallpaper will cover 75 square feet. The room Mary wishes to wallpaper is 9 feet wide and 12 feet long with a standard 8 foot ceiling.
   a. Estimate how many rolls of wallpaper she will need for the job.
   b. Find the actual number of rolls Mary will need.
   c. If she had bought the wallpaper according to your estimate, would she have bought the right amount?

# EXERCISES

## Strategies

Look for a pattern.
Solve a simpler problem.
Act it out.
Guess and check.
Draw a diagram.
Make a chart.
Work backwards.

**Solve. Use any strategy.**

6. The sign in front of an Alaskan bank displays both Fahrenheit and Celsius temperatures. At what temperature are the readings the same?

7. How can two fathers and two sons divide twenty-one $1 bills evenly among them? Each must receive an equal number of bills.

8. A beautiful 9 × 12 meter rug was damaged by the moving company. After the damaged part was cut out, a 1 × 8 meter rectangular hole resulted in the very center of the rug. Using straight cuts, cut the remaining rug into two parts that, when sewn together, will form a square.

9. The following pattern contains all the digits from 0 through 9. Discover the pattern and complete the sequence.
8, 5, 4, 9, 1, $\underline{\ ?\ }$, $\underline{\ ?\ }$, $\underline{\ ?\ }$, $\underline{\ ?\ }$, $\underline{\ ?\ }$

10. Keshia bought an eighteen month certificate of deposit with an annual yield of 8.79%. If the original investment was $1800, approximately how much interest did she earn?

11. Trace the square at the right. Then show at least five ways to divide the square into four congruent parts.

12. Use the following clues to find what year the first Super Bowl was played.
   (1) No digit is an 8.
   (2) The hundreds digit is 3 more than the tens digit.
   (3) The sum of the digits is 23.

## COOPERATIVE LEARNING ACTIVITY

**Work in groups. Each person in the group must understand the solution and be able to explain it to any person in class.**

Leon is helping his younger sister with her homework. She is learning to simplify fractions and has simplified the fraction $\frac{154}{253}$ to $\frac{14}{23}$ by canceling out the two middle digits. Her procedure was wrong but the answer is correct. After Leon showed his sister how to simplify fractions correctly, he tried to find other fractions that would simplify correctly by illegally canceling. How many can you find? Can you generalize the form of these fractions?

## 12-7 Exponential Equations

**Objective**

After studying this lesson, you should be able to:

- solve equations with variable exponents using logarithms.

**Application**

Jess Burgess is a systems analyst for an investment corporation. He is designing a computer program that will help financial planners analyze various investments and, as a result, choose wise investments for their clients. One of the things the computer program must be able to do is to find what the value of an investment that earns a given interest rate will be at some time in the future. The formula for finding that value is

$A = P\left(1 + \dfrac{r}{n}\right)^{nt}$, where $A$

represents the value of the investment in the future, $P$ is the original investment, $r$ is the annual interest rate, $t$ is the number of years of the investment, and $n$ is the number of times the interest is compounded each year.

The value of an investment at some future date is one example of an **exponential equation.** Exponential equations are equations in which the variables appear as exponents. These equations can be solved using the property of equality for logarithmic functions.

**Example 1**

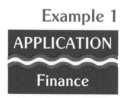

**APPLICATION**

**Finance**

Use the compound interest formula, $A = P\left(1 + \dfrac{r}{n}\right)^{nt}$, to find how long it would take for an investment of \$2500 to triple if it is invested in an account that earns 6% interest compounded quarterly.

$$A = P\left(1 + \frac{r}{n}\right)^{nt} \qquad \textit{Compound interest formula}$$

$$7500 = 2500\left(1 + \frac{0.06}{4}\right)^{4t} \qquad \textit{P is \$2500, r is 0.06, n is 4,}$$
$$\textit{and A is 3P or \$7500.}$$

$$3 = (1.015)^{4t} \qquad \textit{Divide each side by 2500.}$$

$$\log 3 = \log (1.015)^{4t} \qquad \textit{Property of equality for logarithmic functions}$$

$$\log 3 = 4t \log (1.015) \qquad \textit{Power property of logarithms}$$

$$\frac{\log 3}{4 \log (1.015)} = t$$

$$t \approx 18.4472$$

The investment of \$2500 would triple in a little less than $18\frac{1}{2}$ years.

**Example 2**

Solve $4^x = 24$.

$$4^x = 24$$

$$\log 4^x = \log 24 \qquad \textit{Property of equality for logarithmic functions}$$

$$x \log 4 = \log 24 \qquad \textit{Power property of logarithms}$$

$$x = \frac{\log 24}{\log 4}$$

$$x = \frac{1.3802}{0.6021}$$

$$x \approx 2.2923$$

The solution is approximately 2.2923.   *Check this result.*

**Example 3**

Solve $7^{x-2} = 5^{3-x}$.

$$7^{x-2} = 5^{3-x}$$

$$\log 7^{x-2} = \log 5^{3-x} \qquad \textit{Property of equality for logarithmic functions}$$

$$(x - 2) \log 7 = (3 - x) \log 5 \qquad \textit{Power property of logarithms}$$

$$x \log 7 - 2 \log 7 = 3 \log 5 - x \log 5 \qquad \textit{Distributive property}$$

$$x \log 7 + x \log 5 = 3 \log 5 + 2 \log 7$$

$$x(\log 7 + \log 5) = 3 \log 5 + 2 \log 7 \qquad \textit{Distributive property}$$

$$x = \frac{3 \log 5 + 2 \log 7}{\log 7 + \log 5} \qquad \textit{Use a calculator to perform calculations.}$$

$$x = \frac{3(0.6990) + 2(0.8451)}{0.8451 + 0.6990}$$

$$x \approx 2.4527$$

The solution is approximately 2.4527.   *Check this result.*

It is possible to evaluate expressions involving logarithms with different bases. Since your calculator isn't programmed with all of the possible bases for logarithms, the **change of base formula** is very helpful.

| | |
|---|---|
| *Change of Base Formula* | **For all positive numbers $a$, $b$ and $n$, where $a \neq 1$ and $b \neq 1$,** $$\log_a n = \frac{\log_b n}{\log_b a}.$$ |

**Example 4**

Express each logarithm in terms of common logarithms. Then find its value.

**a.** $\log_4 22$                                **b.** $\log_{12} 95$

$$\log_a n = \frac{\log_b n}{\log_b a} \quad \textit{Change of base formula}$$

$$\log_4 22 = \frac{\log 22}{\log 4} \quad \begin{array}{l} a = 4, n = 22, \\ b = 10 \end{array}$$

$$= \frac{1.3424}{0.6021}$$

$$\approx 2.2295$$

The value of $\log_4 22$ is approximately 2.2295.

$$\log_a n = \frac{\log_b n}{\log_b a} \quad \textit{Change of base formula}$$

$$\log_{12} 95 = \frac{\log 95}{\log 12} \quad \begin{array}{l} a = 12, n = 95, \\ b = 10 \end{array}$$

$$= \frac{1.9777}{1.0792}$$

$$\approx 1.8326$$

The value of $\log_{12} 95$ is approximately 1.8326.

# CHECKING FOR UNDERSTANDING

**Communicating Mathematics**

**Read and study the lesson to answer these questions.**

1. Is $x^4 = 256$ an exponential equation? If not, why not?

2. When might the change of base formula come in handy?

3. Could you use the change of base formula to express a logarithm in terms of natural logarithms? If not, why not?

**Guided Practice**

**State $x$ in terms of common logarithms. Then find the value of $x$.**

4. $6^x = 72$         5. $8^x = 100$         6. $9^{2x} = 144$

7. $x = \log_4 169$       8. $x = \log_6 90$       9. $3^x = \sqrt{34}$

10. $2^x = 5\sqrt{2}$        11. $3^{-x} = 22$        12. $2^{-x} = \sqrt{7}$

# EXERCISES

**Practice**

**Approximate each logarithm to three decimal places.**

13. $\log_5 15$            14. $\log_8 72$            15. $\log_{12} 169$

16. $\log_4 100$          17. $\log_{12} 15$          18. $\log_2 36$

19. $\log_{15} 5$            20. $\log_9 108$          21. $\log_{11} 104$

**Use logarithms to solve each equation.**

22. $8^x = 45$             23. $2^x = 27$           24. $2.1^{x-5} = 9.32$

25. $7.6^{a-2} = 41.7$      26. $5^{x+2} = 15.3$      27. $x = \log_4 51.6$

28. $9^{x-4} = 6.28$        29. $x = \log_{20} 1000$    30. $25^{x^2} = 50$

31. $6^{x^2-2} = 48$         32. $4.3^{3x+1} = 78.5$     33. $2.7^{x^2-1} = 52.3$

**34.** $5^{x-1} = 3^x$

**35.** $12^{x-4} = 4^{2-x}$

**36.** $7^{x-2} = 5^x$

**37.** $5^{2x} = 9^{x-1}$

**38.** $2^{2x+3} = 3^{3x}$

**39.** $2^{3y} = 3^{y+1}$

**40.** $2^{5x-1} = 3^{2x+1}$

**41.** $4^{5y-6} = 3^{2y+5}$

**42.** $5^{4y+1} = 32^{2y}$

**43.** $24^{3x} = 6^{2x+1}$

**44.** $2^n = \sqrt{3^{n-2}}$

**45.** $\sqrt[3]{4^{x-1}} = 6^{x-2}$

**Critical Thinking**

**46.** Let $x$ be any real number and $a$, $b$, and $n$ be positive real numbers where $a \neq 1$ and $b \neq 1$. Show that if $x = \log_a n$, then $x = \dfrac{\log_b n}{\log_b a}$.

**Applications**

**47. Finance**  Nalani saved $500 of the money she earned working at the Dairy Dream last summer. She deposited the money in a certificate of deposit that earns 8.75% interest compounded monthly. If she rolls over the CD at the same rate each year, when will Nalani's CD have a balance of $800?

**48. Education**  Dwain withdrew all of the $2500 in his savings account to pay the tuition for his first semester at college. The account had earned 12% interest compounded monthly, and no withdrawals or additional deposits were made.
  **a.** If Dwain's original deposit was $1250, how long ago did he open the account?
  **b.** If Dwain's original deposit was $1500, how long ago did he open the account?

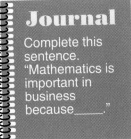
**49. Business**  The T.C. Company has a savings plan for their employees. The employee makes an initial contribution of $1000 and the company pays 8% interest compounded quarterly.
  **a.** If an employee participating in the plan withdraws the balance of the account after five years, how much will the company have paid into the account?
  **b.** If an employee participating in the plan withdraws the balance of the account after thirty-five years, how much will the company have paid into the account?

**Mixed Review**

**50.** Solve the equation $\ln 9.5 = \ln (e^{0.2x})$. **(Lesson 12-5)**

**51.** Write the equation $f(x) = x^2 + 16x + 67$ in the form $f(x) = (x - h)^2 + k$. Then name the vertex and axis of symmetry for the graph of the function. **(Lesson 8-2)**

**52.** State the sum and product of the roots of the equation $-5x^2 + x + 3 = 0$. **(Lesson 7-5)**

**53. Physics** The formula for the time, $T$, in seconds that it takes for a pendulum to make a complete swing (back and forth) is $T = 2\pi\sqrt{\dfrac{L}{32}}$, where $L$ represents the length of the pendulum. Find the time for a complete swing of a pendulum whose length is 6 feet. *If your calculator does not have a $\pi$ key, use 3.14 to approximate $\pi$.* **(Lesson 6-2)**

**54.** Simplify $\dfrac{-3w^6t^7}{(-27w^3t^2)(wt)^2}$. **(Lesson 5-2)**

**55.** State the domain and range of the relation $\{(9, 3), (-8, 3), (1, -3), (0, 0)\}$. Then state if the relation is a function. **(Lesson 2-1)**

**56. Sports** The Wildcats play 84 games this season. It is now midseason and they have won 30 games. To win at least 60% of *all* of their games, how many of the remaining games must they win? **(Lesson 1-7)**

## ∼∼∼ BIOLOGY CONNECTION ∼∼∼

Many problems involving estimation can be solved using **interpolation** and **extrapolation.** Interpolation is used to estimate a value between two known values. Extrapolation is used to estimate a value that is either greater than or less than two known values. Both use proportions to find the missing value. Environmentalists often use these methods to estimate wildlife populations.

The deer population in Deercreek Park is known for 1987 and 1992. The deer population in 1991 can be estimated using interpolation while the deer population for 1994 can be estimated using extrapolation.

| Year | Population | Year | Population |
|---|---|---|---|
| +3 ⌈ 1987 ⌉ +5 | +x ⌈ 4320 ⌉ | +5 ⌈ 1987 ⌉ | 4320 ⌉ +2460 |
| → 1991 | → unknown +2460 | → 1992 +11 | 6780← +y |
| 1992 ← | 6780 ← | 1994 ← | unknown← |

$$\frac{3}{5} = \frac{x}{2460}$$
$$x = 1476$$

$$\frac{11}{5} = \frac{y}{2460}$$
$$y = 5412$$

The deer population in 1991 is about 1476 more than 4320, or 5796 deer. The deer population for 1994 would be about 5412 more than 4320, or 9732 deer. *You could also round the numbers to have another estimate.*

**578 CHAPTER 12 EXPONENTIAL AND LOGARITHMIC FUNCTIONS**

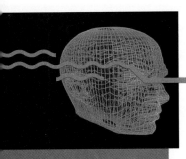

# Technology

## Compound Interest

You can use a spreadsheet to calculate how your funds are increasing due to interest that compounds. Use the formula $A = P\left(1 + \dfrac{r}{n}\right)^{nt}$, where $A$ is the amount in the account after $t$ years, with an initial ivestment of $P$ dollars at an interest rate $r$, if the interest is compounded $n$ times per year.

The spreadsheet below left is set up to find the value of an investment after 1 to 20 years. The principal is entered in cell B1, the number of times the interest is compounded per year is entered in cell B2, and the rate is entered in B3. Cells B5 to B24 contian a variation of the compound interest formula that will compute the balance in the account after successive years.

Below right is a partial printout from the spreadsheet on compound interest. It shows the amount accumulated in an account that bears 5.5% interest compounded monthly if the principal is $9500.

| COMPOUND INTEREST | | |
|---|---|---|
| | **A** | **B** |
| 1 | PRINCIPAL | |
| 2 | TIMES COMPOUNDED | |
| 3 | RATE | |
| 4 | YEAR | BALANCE |
| 5 | 1 | B1*(1 + B3/B2)† (B3*A5) |
| 6 | 2 | B1*(1 + B3/B2)† (B3*A6) |
| 24 | 20 | B1*(1 + B3/B2)† (B3*A24) |

```
= = = = = = = = = = = = = = = = = = =
      COMPOUND INTEREST

= = = = = = A = = = = = = B = = =

  1:     PRINCIPAL      9500

  2:       TIMES         12
        COMPOUNDED

  3:       RATE         0.055

  4:       YEAR        BALANCE

  5:        1         10035.87

  6:        2         10601.98

 24:       20         28467.94
```

# EXERCISES

1. Describe how you could modify the spreadsheet program to print 50 years of account balances.

2. Describe how you could modify the spreadsheet program to print comparisons of the amount accumulated in accounts of two different interest rates.

# Applications of Logarithms

**Objective**

After studying this lesson, you should be able to:

- use logarithms to solve problems.

**Application**

Bacteria reproduce by mitosis, a process by which a cell divides to produce two identical cells. Their growth can be described by the **general formula for growth and decay.** This is the exponential function $y = ne^{kt}$, where $y$ is the final amount, $n$ is the initial amount, $k$ is a constant and $t$ is the time. The *escherichia coli*, or *e. coli*, can reproduce in 15 minutes. If you began with just one bacterium, how many would there be in four hours?

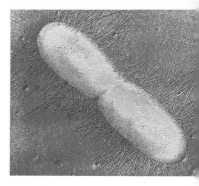

Use the general formula for growth and decay to find the number of bacteria present after four hours. First we must find the constant $k$ when $t$ is given in minutes. The constant will be positive since the number of bacteria is increasing. If the number were decreasing, as with radioactive decay, the value of $k$ would be negative.

$$y = ne^{kt}$$
$$2 = 1e^{k(15)} \quad \text{\textit{One bacterium can produce two in 15 minutes.}}$$
$$2 = e^{15k}$$
$$\ln 2 = \ln e^{15k} \quad \text{\textit{Take the natural logarithm of each side.}}$$
$$\ln 2 = 15k \ln e \quad \text{\textit{Power property of logarithms}}$$
$$\ln 2 = 15k \quad \text{\textit{ln e = 1}}$$
$$\frac{\ln 2}{15} = k$$
$$\frac{0.6931}{15} = k$$
$$0.0462 \approx k \quad \text{\textit{The constant is approximately 0.0462.}}$$

Now apply the exponential growth and decay formula to find the number of bacteria present after four hours.

$$y = ne^{kt}$$
$$y = 1e^{(0.0462)(240)} \quad \text{\textit{There are 240 minutes in four hours.}}$$
$$y = e^{11.088} \quad \text{\textit{Simplify.}}$$
$$\ln y = \ln e^{11.088} \quad \text{\textit{Take the natural logarithm of each side.}}$$
$$\ln y = 11.088 \ln e \quad \text{\textit{Power property of logarithms}}$$
$$\ln y = 11.088 \quad \text{\textit{Take the antiln of each side.}}$$
$$y \approx 65382$$

There will be approximately 65,382 bacteria in four hours.

The general formula for growth and decay also describes the amount of decaying materials left after time. Study the example below.

**Example 1**

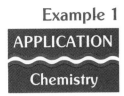

APPLICATION

Chemistry

**Radioactive isotopes decay with time. In 9 years, just half of the mass of a 20-gram sample of an isotope remains. This period of time is called the half-life of the isotope. Find the constant $k$ for this isotope when $t$ is given in years.**

$$y = ne^{kt} \qquad \textit{Since this problem involves decay, k will be negative.}$$
$$10 = 20e^{k(9)} \qquad \textit{Substitute 10 for y, 20 for n, and 9 for t.}$$
$$0.5 = e^{9k}$$
$$\ln 0.5 = \ln e^{9k}$$
$$\ln 0.5 = 9k \ln e$$
$$\ln 0.5 = 9k$$
$$\frac{\ln 0.5}{9} = k$$
$$k \approx -0.0770$$

The value of the constant for this isotope is approximately $-0.0770$.

Certain assets, such as cars, houses, and business equipment, depreciate or appreciate with time. The formula $V_n = P(1 + r)^n$, where $V_n$ is the new value, $P$ is the initial value, $r$ is the fixed rate of appreciation or depreciation, and $n$ is the number of years, can be used to compute the value of an asset. The value of $r$ for a depreciating asset will be negative, and the value of $r$ for an appreciating asset will be positive.

**Example 2**

APPLICATION

Business

**Zeller Industries bought a piece of weaving equipment for $50,000. It is expected to depreciate at a steady rate of 10% each year. When will the value have depreciated to $25,000?**

$$V_n = P(1 + r)^n$$
$$25000 = 50000(1 - 0.10)^n \qquad \textit{Substitute 25,000 for } V_n.$$
$$0.5 = 0.9^n \qquad \textit{50,000 for P,}$$
$$\log 0.5 = \log 0.9^n \qquad \textit{and } -.10 \textit{ for r.}$$
$$\log 0.5 = n \log 0.9$$
$$\frac{\log 0.5}{\log 0.9} = n$$
$$n \approx 6.58$$

The value of the equipment will be $25,000 in about $6\frac{1}{2}$ years.

You have studied compound interest where the interest is compounded at various intervals. When interest is compounded continuously, the formula for finding the amount in the account, $A$, after $t$ years is $A = Pe^{rt}$, where $P$ is the initial investment, and $r$ is the annual interest rate.

### Example 3

**APPLICATION**

**Finance**

*FYI···*

You can estimate how long it will take for an investment to double by dividing the interest rate into 72. For example, an investment that pays 8% will double in about 9 years. This is known as the "Rule of 72."

The Saver's Club at Citizen's Fidelity Bank promises to double your money in $8\frac{1}{2}$ years. Assuming that the interest is compounded continuously, what is the interest rate?

$$A = Pe^{rt}$$

$$2 = 1e^{r(8.5)} \qquad \text{\textit{Substitute 2 for A, 1 for P since the amount is doubled.}}$$

$$2 = e^{8.5r}$$

$$\ln 2 = \ln e^{8.5r} \qquad \text{\textit{Take the natural logarithm of each side.}}$$

$$\ln 2 = 8.5r \ln e \qquad \text{\textit{ln } a^b = b \text{ ln } a}$$

$$\ln 2 = 8.5r$$

$$\frac{\ln 2}{8.5} = r$$

$$r \approx 0.0815$$

The interest rate is 8.15%.

# CHECKING FOR UNDERSTANDING

**Communicating Mathematics**

**Read and study the lesson to answer these questions.**

1. Describe a situation where the constant $k$ in the formula for growth and decay is positive and one where $k$ is negative. Describe the situation if the value of $k$ is zero.

2. The value of houses usually increase with time. Which formula would you use to find the value of a house that has appreciated 15% each year?

3. You are considering opening a savings account. Lake County Savings offers an account earning 8% interest compounded continuously and Beneficial Bank offers an account earning 8% compounded daily.

   a. Which account should you choose?

   b. Are there any factors other than interest that you should consider when you choose between the two banks?

Solve for each variable.

**4.** $40 = 200e^{7k}$

**5.** $50000 = 25000(1 + 0.1)^n$

**6.** $200 = 100e^{.06t}$

**7.** $2500 = 4e^{.58t}$

# EXERCISES

**Practice**

**Solve each problem.**

**8.** For a certain strain of bacteria, $k$ is 0.775 when $t$ is measured in hours. How long will it take 2 bacteria to increase to 1000 bacteria?

**9.** Jackson deposited $100 in a savings account that pays 6% interest compounded continuously. He just withdrew the entire balance of $200. How long ago did he open the account?

**10.** The constant $k$ for a radioactive substance is $-0.08042$ when $t$ is measured in years. In how many years will a 250-gram sample reduce to 50 grams?

**11.** A culture of a certain bacteria will grow from 500 to 4000 bacteria in 90 minutes. Find the constant $k$ for this bacteria if $k$ is in hours.

**12.** Charlotte invested $1000 in a certificate of deposit three years ago. The CD is now worth $1276. Assuming that the interest was compounded continuously, what was the interest rate?

**13.** A piece of office equipment valued at $25,000 depreciates at a steady rate of 10% annually. In how many years will it be worth $5000?

**14.** Keith has saved $2000 to buy a synthesizer that will cost about $2500. If he has the money in an account paying 7.25% compounded continuously, when will Keith be able to buy the synthesizer?

**15.** Suppose you deposited $10 in a savings account that pays 8% interest compounded continuously.
   **a.** In how many years will the account have a balance of $100?
   **b.** In how many years will the balance be $1000?

**16.** The Holub's bought a condominium for $63,000. Assuming that its value will appreciate 8% a year, how much will the condo be worth in five years when the Holub's are ready to move?

**17.** Ten years ago, Cathy's mother bought a new car for $6000. Cathy is now going to buy the car for $600. Assuming a steady rate of depreciation what was the annual rate of depreciation?

18. A radioactive substance decays according to the equation $A = A_0 \times 10^{-0.024t}$, where $t$ is in hours. Find the half-life of the substance, that is when $A = 0.5A_0$.

19. Colin has saved $500 of the money he earned working at Carousel Music Store. If he spends 10% of the money each week, after how many weeks will he have less than $1? *(Hint: Use $V_n = P(1 + r)^n$, where n is the number of weeks.)*

20. A piece of machinery valued at $2,500 depreciates at a steady rate of 10% yearly. The owner of the business will replace the equipment when its value has depreciated to $500. In how many years will the equipment be replaced?

## Critical Thinking

21. Compare the formulas for exponential growth, $y = ne^{kt}$, and for continuously compounded interest, $A = Pe^{rt}$. Explain how the formulas are related.

## Applications

22. **Electronics**  The output in watts of a power supply is given by $w = 50e^{-0.004t}$, where $t$ is the time in days. In how many days will the power output be reduced to 20 watts?

23. **Chemistry**  Radium-226 decomposes radioactively. Its half-life, that is the time that it takes for half of the sample to decompose, is 1800 years. Find the constant $k$ for this compound. Use 100 grams as the original amount.

24. **Real Estate**  Ten years ago, Mr. and Mrs. Oon bought a house for $49,000. Their home is now worth $120,000. Assuming a steady rate of growth, what was the annual rate of appreciation?

## Computer

This BASIC program determines the number of payments needed to accumulate an amount of money, given the amount of each payment, the annual interest rate, and the number of payments per year. The formula

$$S = R\left[\frac{(1 + I)^N - 1}{I}\right] \text{ finds}$$

the amount accumulated when payments are made at regular intervals and interest is compounded at the end of each payment period.

```
10  INPUT "ENTER THE AMOUNT TO BE
    ACCUMULATED: $"; S
20  INPUT "ENTER THE AMOUNT OF
    EACH PAYMENT: $"; R
30  INPUT "ENTER THE ANNUAL
    INTEREST RATE AS A PERCENT: ";
    IR
40  INPUT "ENTER THE NUMBER OF
    PAYMENTS PER YEAR: "; PAY
50  LET I = IR/(PAY*100):PRINT
60  LET N = LOG(S*I/R+1)/LOG(1+I)
70  IF N = INT(N+1) THEN 90
80  LET N = INT(N+1)
90  PRINT N; "PAYMENTS ARE NEEDED
    TO ACCUMULATE $"; S
100 END
```

In the formula, $R$ represents the amount of each payment, $I$ represents the annual interest rate, $N$ represents the number of payments, and $S$ represents the amount of money accumulated immediately after the final payment. In the program, the formula is solved for $N$.

**Use the program on the previous page to determine the number of payments needed to accumulate the indicated amount of money, given the amount of each payment, the annual interest rate, and the number of payments each year.**

25. $4000; $300; 9%; 2

26. $7500; $400; 8.5%; 4

27. $8995; $156; 9.25%; 12

28. $14,600; $195; 8.75%; 12

29. $96,000; $850; 9.65%; 12

30. $90,000; $425; 9.65%; 26

31. Each month, Michael deposits $100 into an account that earns 8.75% compounded monthly. How many months will it take him to accumulate $3000 in his account?

32. Every three months, Guillermo invests $300 in an account that pays 8.3% interest compounded quarterly. Joyce invests $98 each month in her account, which pays 9.3% interest compounded monthly. Who will accumulate $5000 more quickly?

**Mixed Review**

33. Approximate $\log_7 12$ to three decimal places. **(Lesson 12-7)**

34. State all possible rational zeros for the function $f(x) = 3x^4 - 5x^2 + 4$. **(Lesson 10-5)**

35. For the function $f(x) = (x + 9)^2$, identify the quadratic term, the linear term, and the constant term. **(Lesson 8-1)**

36. **Photography** Shina Murakami is a professional photographer. She has a photograph that is 4 inches wide and 6 inches long. She wishes to make a print of the photograph for a competition. The area of the new print is to be five times the area of the original. If Ms. Murakami is going to add the same amount to the length and the width of the photograph, what will the dimensions of the new print be? **(Lesson 7-4)**

37. **Physics** Metal expands and contracts with changes in temperature. The change in the length of steel per degree Celsius is given by the constant $11 \times 10^{-6}$. Over a period of time, the temperature of a steel bridge 200 meters long varies by 70° Celsius. What is the change in the length of the bridge in centimeters? **(Lesson 5-1)**

38. Solve the inequality $3x + 1 < x + 5$. **(Lesson 1-7)**

## VOCABULARY

Upon completing this chapter, you should be
familiar with the following terms:

| | | | |
|---|---|---|---|
| antilogarithm | **564** | **546** | exponential function |
| characteristic | **563** | **553** | logariths |
| common logarithm | **563** | **555** | logarithmic function |
| $e$ | **567** | **563** | mantissa |
| exponential equation | **574** | **567** | natural logarithm |

# SKILLS AND CONCEPTS

| OBJECTIVES AND EXAMPLES | REVIEW EXERCISES |
|---|---|

Upon completing this chapter, you should
be able to:

Use these exercises to review and prepare
for the chapter test.

■ simplify expressions and solve equations
involving real exponents  **(Lesson 12-1)**

**Simplify each expression.**

Simplify $(9^{\sqrt{5}})^{\sqrt{5}}$.
$(9^{\sqrt{5}})^{\sqrt{5}} = 9^{\sqrt{5} \cdot \sqrt{5}}$
$\qquad = 9^5$

Solve $9^{3r} = 27^{r-2}$.

$$9^{3r} = 27^{r-2}$$
$$(3^2)^{3r} = (3^3)^{r-2}$$
$$3^{6r} = 3^{3r-6}$$
$$6r = 3r - 6$$
$$r = -2$$

**1.** $3^{\sqrt{2}} \, 3^{\sqrt{2}}$

**2.** $(9^{\sqrt{2}})^{\sqrt{2}}$

**3.** $\dfrac{49^{\sqrt{2}}}{7^{\sqrt{12}}}$

**4.** $(x^{\sqrt{5}})^{\sqrt{20}}$

**Solve each equation.**

**5.** $2^{6x} = 4^{5x+2}$

**6.** $(\sqrt{3})^{n+1} = 9^{n-1}$

**7.** $49^{3p+1} = 7^{2p-5}$

**8.** $9^{x^2} = 27^{x^2-2}$

■ write exponential equations in
logarithmic form and vice versa.
**(Lesson 12-2)**

**Write each equation in logarithmic form.**

Write $3^3 = 27$ in logarithmic form.
$$3^3 = 27$$
$$3 = \log_3 27$$

Write $\log_4 64 = 3$ in exponential form.
$$\log_4 64 = 3$$
$$64 = 4^3$$

**9.** $7^3 = 343$

**10.** $5^{-2} = \dfrac{1}{25}$

**11.** $4^0 = 1$

**12.** $4^{\frac{3}{2}} = 8$

**Write each equation in exponential form.**

**13.** $\log_4 64 = 3$

**14.** $\log_8 2 = \dfrac{1}{3}$

**15.** $\log_6 \dfrac{1}{36} = -2$

**16.** $\log_6 1 = 0$

| OBJECTIVES AND EXAMPLES | REVIEW EXERCISES |
|---|---|

■ evaluate logarithmic expressions and solve logarithmic equations. (**Lesson 12-2**)

Evaluate $\log_3 3^5$.
$$\log_3 3^5 = x$$
$$3^x = 3^5$$
$$x = 5$$

Solve $\log_b 16 = 4$.
$$b^4 = 16$$
$$b = \sqrt[4]{16}$$
$$b = 2$$

**Evaluate each expression.**

**17.** $\log_5 5^7$   **18.** $6^{\log_6 7}$

**19.** $\log_n n^3$   **20.** $n^{\log_n 3}$

**Solve each equation.**

**21.** $\log_b 9 = 2$   **22.** $\log_b 9 = \dfrac{1}{2}$

**23.** $\log_{16} 2 = x$   **24.** $\log_4 x = -\dfrac{1}{2}$

---

■ solve equations involving logarithmic functions.   (**Lesson 12-2**)

For $b > 0$ and $b \neq 1$, $\log_b x_1 = \log_b x_2$ if and only if $x_1 = x_2$.

Solve $\log_3 10 = \log_3 (2x)$.
$$10 = 2x$$
$$5 = x$$

**Solve each equation.**

**25.** $\log_6 12 = \log_6 (5x - 3)$

**26.** $\log_4 (1 - 2x) = \log_4 (x + 10)$

**27.** $\log_7 (x^2 + x) = \log_7 12$

**28.** $\log_2 (x - 1)^2 = \log_2 7$

---

■ solve equations or simplify and evaluate expressions using the product, quotient, and power properties.   (**Lesson 12-3**)

For all positive numbers $m$, $n$ and $b$, where $b \neq 1$.

$$\log_b mn = \log_b m + \log_b n$$

$$\log_b \frac{m}{n} = \log_b m - \log_b n$$

$$\log_b m^n = n \log_b m$$

**Use $\log_9 7 = 0.8856$ and $\log_9 4 = 0.6309$ to evaluate each expression.**

**29.** $\log_9 28$   **30.** $\log_9 49$

**31.** $\log_9 144$   **32.** $\log_9 15.75$

**Solve each equation.**

**33.** $\log_3 x - \log_3 4 = \log_3 12$

**34.** $\log_2 y = \dfrac{1}{3} \log_2 27$

**35.** $\log_5 7 + \dfrac{1}{2} \log_5 4 = \log_5 x$

**36.** $2 \log_2 x - \log_2 (x + 3) = 2$

---

■ identify the characteristic and mantissa of a logarithm.   (**Lesson 12-4**)

$\log 4,300 \approx 3.6336$

3 is the characteristic.
0.6335 is the mantissa.

**If log 36.2 = 1.5587, find each number.**

**37.** the characteristic of $\log 36.2$

**38.** the mantissa of $\log 36.2$

**If log 0.00927 = −2.0329, find each number.**

**39.** the characteristic of $\log 0.00927$

**40.** the mantissa of $\log 0.00927$

| OBJECTIVES AND EXAMPLES | REVIEW EXERCISES |
|---|---|

■ find common logarithms and antilogarithms. (**Lesson 12-4**)

Use the $\boxed{\text{LOG}}$ key of your calculator to find common logarithms, the $\boxed{10^x}$ key to find the antilogarithm.

**Find the logarithm of each number**

**41.** 0.003141      **42.** 50,030

**Find the antilogarithm of each logarithm.**

**43.** 0.5574 − 4      **44.** 2.9876

---

■ find natural logarithms of numbers. (**Lesson 12-5**)

Use the $\boxed{\text{LN}}$ key of your calculator to find common logarithms, the $\boxed{e^x}$ key to find the antilogarithm.

**Find each value.**

**45.** ln 2.3      **46.** ln 9.25

**47.** antiln 1.9755      **48.** antiln 2.246

---

■ solve equations with variable exponents using logarithms. (**Lesson 12-7**)

$$3^{x-4} = 5^{x-1}$$
$$\log 3^{x-4} = \log 5^{x-1}$$
$$(x - 4)\log 3 = (x - 1)(\log 5)$$
$$x \log 3 - 4 \log 3 = x \log 5 - \log 5$$
$$x = \frac{4 \log 3 - \log 5}{\log 3 - \log 5}$$
$$x \approx -5.4502$$

**Use logarithms to solve each equation.**

**49.** $2^x = 53$      **50.** $\log_4 11.2 = x$

**51.** $2.3^{x^2} = 66.6$      **52.** $3^{4x-7} = 4^{2x+3}$

**53.** $\sqrt{3^b} = 2^{b+1}$      **54.** $6^{3y} = 8^{y-3}$

**55.** $300 = 20e^{5t}$

**56.** $500 = P(1 + 0.2)^5$

# APPLICATIONS AND CONNECTIONS

**57. Business** A car valued at $14,000 depreciates 18% a year. After how many years will the value have depreciated to $1000? Estimate the answer and then check your estimate against the answer you obtain using $V = P(1 + r)^n$. (**Lesson 12-6**)

**58. Science** For a certain strain of bacteria, $k$ is 0.872 when $t$ is measured in days. How long will it take 9 bacteria to increase to 738 bacteria? (**Lesson 12-8**)

**59. Finance** If $200 is invested at 6% annual interest compounded continuously when will the investment be worth $300? Use $A = Pe^{rt}$. (**Lesson 12-8**)

**60. Science** A bacteria culture grows from 400 to 5000 bacteria in 2 hours. Find the constant $k$ for the growth formula $y = ne^{kt}$ where $t$ is in hours. (**Lesson 12-8**)

# CHAPTER 12 TEST

Write each equation in logarithmic form.

**1.** $6^4 = 1296$

**2.** $3^7 = 2187$

Write each equation in exponential form.

**3.** $\log_5 625 = 4$

**4.** $\log_8 16 = \dfrac{4}{3}$

Evaluate each expression.

**5.** $\log_{12} 12^2$

**6.** $4^{\log_4 3}$

**7.** $\log_b b^{1.6}$

Solve each equation.

**8.** $9^x = 3^{3x-2}$

**9.** $27^{2p+1} = 3^{4p-1}$

**10.** $\log_m 144 = -2$

**11.** $\log_2 128 = y$

**12.** $\log_3 x - 2\log_3 2 = 3\log_3 3$

**13.** $\log_{\sqrt{7}} x = 4$

**14.** $\log_5(8r - 7) = \log_5(r^2 + 5)$

**15.** $\log_9(x + 4) + \log_9(x - 4) = 1$

Use $\log_4 7 = 1.4037$ and $\log_4 3 = 0.7925$ to evaluate each expression.

**16.** $\log_4 21$

**17.** $\log_4 9$

**18.** $\log_4 36$

**19.** $\log_4 \dfrac{7}{12}$

Find each value, rounded to four decimal places.

**20.** $\log 769{,}000$

**21.** $\log 0.00535$

**22.** $\ln 9.6$

**23.** antilog $-3.1649$

**24.** antilog $6.3337$

**25.** antiln $0.4055$

Use logarithms to solve each equation.

**26.** $7.6^{x-1} = 431$

**27.** $\log_4 37 = x$

**28.** $3^x = 5^{x-1}$

**29.** $\sqrt{2^{b-4}} = 6^b$

**30.** $4^{2x-3} = 9^{x+2}$

**31.** $45.9 = e^{0.75t}$

**32. Finance** Suppose that a pilgrim ancestor of Jenny Chambers deposited $10 in a savings account at Provident Savings Bank. The annual interest rate was 4% compounded continuously. The account is now worth $75,000. How long ago was the account started? (Use $A = Pe^{rt}$.)

**33. Biology** A certain culture of bacteria will grow from 500 to 4000 bacteria in 1.5 hours. Find the constant $k$ for the growth formula. (Use $y = ne^{kt}$.)

**Bonus** Solve $\log x^2 = (\log x)^2$.

The questions on these pages deal with rational expressions and radicals.

**Directions: Choose the one best answer. Write A, B, C, or D.**

1. If $\frac{x}{y} = z$ and $y = z$, find $y$ in terms of $x$.

   (A) $y$

   (B) $\pm\sqrt{y}$

   (C) $\pm\sqrt{x}$

   (D) $\pm\sqrt{xz}$

2. If the average of $x$ and $y$ equals the average of $x$, $y$, and $z$, then express $z$ in terms of $x$ and $y$.

   (A) $x + y$

   (B) $2(x + y)$

   (C) $\frac{x + y}{2}$

   (D) $\frac{x + y}{3}$

3. If the product of a number and $2b$ is increased by $y$, the result is $p$. Find the number in terms of $b$, $y$, and $p$.

   (A) $2by - p$

   (B) $\frac{2b}{y - p}$

   (C) $\frac{p - y}{2b}$

   (D) $\frac{y - b}{2b}$

4. If $\frac{1}{p} = \sqrt{0.25}$, then $p^2 =$

   (A) $0.25$

   (B) $4$

   (C) $25$

   (D) $400$

5. Of the following numbers, which is the greatest?

   (A) $\frac{1}{3\sqrt{3}}$

   (B) $\frac{1}{3}$

   (C) $\frac{\sqrt{3}}{3}$

   (D) $\sqrt{3}$

6. If $\frac{a + b}{a} = \frac{5}{4}$, then $\frac{b}{a} =$

   (A) $\frac{1}{4}$

   (B) $\frac{5}{4}$

   (C) $\frac{7}{4}$

   (D) $\frac{9}{4}$

7. Of the following, which is the closest to the value of $\frac{65.9 \times 0.49}{3.3}$?

   (A) $10$

   (B) $80$

   (C) $100$

   (D) $450$

8. The reciprocal of $\frac{5}{b - 1} + \frac{3}{b}$ is

   (A) $\frac{b^2 - b}{15}$

   (B) $\frac{b - 1}{2}$

   (C) $\frac{b^2 - b}{8b - 3}$

   (D) $\frac{2b - 1}{8}$

9. Simplify $\dfrac{1 \div \frac{1}{b}}{\frac{1}{b}}$.

   (A) $1$

   (B) $\frac{1}{b}$

   (C) $b$

   (D) $b^2$

10. If $4b - 3a = 0$, then what is the value of $\frac{16b^2}{a^2}$?

    (A) $\frac{1}{9}$

    (B) $9$

    (C) $16$

    (D) $\frac{256}{9}$

11. If $xyz = 8$ and $y = z$, then $x =$

    (A) $y^2$

    (B) $\frac{8}{y^2}$

    (C) $8y^2$

    (D) $\frac{1}{y^2}$

12. If $\frac{2b}{5a} = 12$, then $\frac{2b - 10a}{5a} =$

    (A) $5$

    (B) $10$

    (C) $14$

    (D) $24$

**13.** If $\frac{x}{6} + 4 = 1$, the value of $\frac{x}{3}$ is

(A) $-36$    (B) $-18$

(C) $-6$    (D) $6$

**14.** If $3 + \frac{d}{4} = \frac{81}{2}$, then $d =$

(A) $10$    (B) $80$

(C) $100$    (D) $150$

**15.** Betty can mow the lawn in $x$ hours. Ted can mow the same lawn in $y$ hours. If they work together, how long will it take them to mow the lawn?

(A) $\frac{x + y}{2}$    (B) $\frac{1}{x + y}$

(C) $\frac{xy}{x + y}$    (D) $\frac{x + y}{xy}$

**16.** Mike saves $c$ cents per week. In $n$ weeks how much money, in terms of dollars, will he have saved?

(A) $nc$    (B) $100nc$

(C) $\frac{nc}{100}$    (D) $\frac{n + c}{100}$

**17.** If $\frac{a}{b} = c$, then $\log c =$

(A) $\frac{\log a}{\log b}$

(B) $\log a - \log b$

(C) $\log (a - b)$

(D) $\frac{\log a}{b}$

**18.** Which of the following is not a real number?

(A) $\sqrt{(-3)^2}$    (B) $\sqrt[3]{(-3)^3}$

(C) $\sqrt{-(3)^2}$    (D) $\sqrt[3]{-(3)^3}$

**19.** Find the value of $(5\sqrt{3})^2$.

(A) $75$    (B) $225$    (C) $10\sqrt{6}$    (D) $150$

**20.** A man owns $\frac{1}{4}$ of a business. He sells half of his share for $12,000. What is the total value of the business?

(A) $1500    (B) $48,000

(C) $96,000    (D) $108,000

**21.** If a runner can cover 2 miles in 25 minutes, what is the rate in miles per hour?

(A) $5$    (B) $4.8$    (C) $3$    (D) $0.8$

**22.** If $x^2 + y^2 = 15$ and $(x + y)^2 = 35$, then $xy =$

(A) $5$    (B) $10$    (C) $20$    (D) $40$

**23.** If $5y + 1$ is an odd integer, what is the next consecutive odd integer?

(A) $3y + 1$    (B) $5y + 3$

(C) $7y + 1$    (D) $7y + 3$

# 13

# Sequences and Series

## CHAPTER OBJECTIVES

In this chapter you will:

- Find terms in arithmetic and geometric sequences.
- Find sums of arithmetic and geometric series.
- Use the Binomial Theorem to find terms of a binomial expansion.

**Survey at least 20 people about how they treat their illnesses. How do your results compare with those in the chart?**

### How Americans Handle Everyday Health Problems

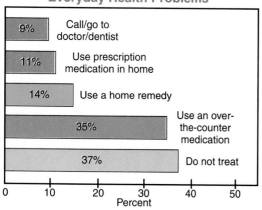

| | |
|---|---|
| 9% | Call/go to doctor/dentist |
| 11% | Use prescription medication in home |
| 14% | Use a home remedy |
| 35% | Use an over-the-counter medication |
| 37% | Do not treat |

0   10   20   30   40   50
Percent

## CAREERS IN PHARMACOLOGY

Today's pharmacists don't do as much actual grinding, measuring, and mixing of ingredients to form powders, tablets, capsules, ointments, and solutions as they did in the early 1900's. Most medicines are produced by pharmaceutical companies.

Today's pharmacists now advise customers and physicians about the proper selection and use of medicines. There are so many pharmaceutical drugs on the market that most physicians are familiar with only a few hundred of them. Pharmacists must know all of them. This means they must be aware of all possible drug interactions and side effects. Computerized records help pharmacists keep track of their customers so that they can advise them more efficiently.

The job opportunities for a pharmacist may be as varied as the cities in which they work. A small town may only have one drugstore in which the pharmacist is the owner. This pharmacist may have to be on call to fill prescriptions in case of an emergency. Larger pharmacies may employ several pharmacists. Their duties may include supervision, inventory of counter merchandise, and even public health testing.

## MORE ABOUT PHARMACOLOGY

### Degree Required:

- Bachelor's Degree in Pharmacy, and a license from the state board

### Some pharmacists like:

- pleasant working conditions
- working with people
- variety of specialties available to them

### Related Math Subjects:

- Advanced Algebra
- Trigonometry
- Statistics/Probability

### Some pharmacists dislike:

- having to work nights, weekends, and holidays
- the responsibility of being 100% accurate and being knowledgeable of all current developments

For more information on the various careers available in the field of Pharmacology, write to:

American Pharmaceutical Association and Academy of Pharmacy Practice and Management
2215 Constitution Avenue NW
Washington, DC 20037

# 13-1 Problem-Solving Strategy: Look for a Pattern

**Objective**  After studying this lesson, you should be able to:
- find the next number in a sequence.

**Application**

The top view of a pine cone shows an example of a pattern often found in nature. The number of clockwise spirals is 13, and the number of counterclockwise spirals is 8. These are two numbers in a famous pattern of numbers called the **Fibonacci sequence,** named after its discoverer, Leonardo Fibonacci, who presented it in 1201. A sequence is an ordered set of numbers related mathematically. The first numbers of the Fibonacci sequence are shown below. The symbol ... means that the sequence continues indefinitely.

1, 1, 2, 3, 5, 8, 13, 21, 34, 55, 89, 144, ...

After the two first numbers, each number in the sequence is the sum of the two numbers that precede it. That is, 3 = 2 + 1, 34 = 21 + 13, and so on.

**Example 1**  **Find the next three terms in the sequence 6, 10, 15, 21, 28, ...**

One way to find a pattern in a sequence is to find the difference of consecutive terms.

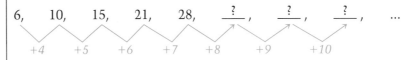

6,　　10,　　15,　　21,　　28,　　$\underline{?}$,　　$\underline{?}$,　　$\underline{?}$,　　...

　　+4　　+5　　+6　　+7　　+8　　+9　　+10

The difference increases by 1 for each term. The next three terms of this sequence are 36, 45, and 55.

**Example 2**  **Find the next three terms in the sequence 1, 2, 6, 24, 120 ...**

The pattern in this sequence is formed by multiplication. Each successive multiplier increases by 1.

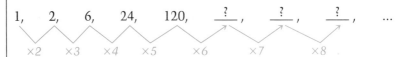

1,　　2,　　6,　　24,　　120,　　$\underline{?}$,　　$\underline{?}$,　　$\underline{?}$,　　...

　　×2　　×3　　×4　　×5　　×6　　×7　　×8

The next three terms of this sequence are 720, 5040, and 40,320.

# CHECKING FOR UNDERSTANDING

**Communicating Mathematics**

**Read and study the lesson to answer each question.**

1. In your own words, write a definition of *sequence*.
2. What operations can be used to form the pattern in a sequence?
3. For whom is the Fibonacci sequence named?
4. Find other examples (besides the spirals of a pine cone) of the Fibonacci sequence in nature.

**Guided Practice**

5. Find the first 20 terms of the Fibonacci sequence.
6. Observe the pattern in $4 \times 6 = 24$, $14 \times 16 = 224$, $24 \times 26 = 624$, and $34 \times 36 = 1224$. Without using a calculator or multiplication, write the product of $124 \times 126$.

# EXERCISES

| Strategies |
| --- |
| Look for a pattern. |
| Solve a simpler problem. |
| Act it out. |
| Guess and check. |
| Draw a diagram. |
| Make a chart. |
| Work backwards. |

**Solve. Use any strategy.**

7. Find $n$ if $n(n - 1)(n - 2)(n - 3)(n - 4) = 95,040$.
8. The first fifty natural numbers are multiplied together. How many zeros appear as the last digits of the product?
9. A bicycle is on sale at a 25% discount. It is reduced another 10% because of a coupon and another 3% for paying cash. These discounts are taken one after the other. If the final price of the bicycle is $195.89, what was the original price?
10. Trace the figure at the right without tracing over any segments you have already drawn. *Tracings which intersect at a point are allowed.*

11. All the sides of a rectangle *ABCD* have measures that are whole numbers. Find the greatest area possible for the rectangle if the perimeter is 42 cm.

## ~ COOPERATIVE LEARNING ACTIVITY ~

**Work in groups. Each person in the group must understand the solution and be able to explain it to any person in class.**

The middle sapphire on a string of 33 sapphires is the largest and most expensive. Starting from one end, each sapphire is worth $100 more than the one before it up to and including the middle sapphire. From the other end, each sapphire is worth $150 more than the one before it, up to and including the middle sapphire. The string of sapphires is worth $65,000. What is the value of the middle sapphire?

# Arithmetic Sequences

**Objectives**

After studying this lesson, you should be able to:

- find the *n*th term of an arithmetic sequence,
- find the position of a given term in an arithmetic sequence, and
- find arithmetic means.

**Application**

Emerson Fittipaldi is a professional race car driver from Brazil. As he drives out of a curve, he enters the straightaway at 119.9 mph. While on the straightaway, he steadily increases his speed by 78.3 mph. After 9 seconds, his speed is 198.2 mph. The chart below shows how his speed increased each second after entering the straightaway.

| Number of seconds | 0 | 1 | 2 | 3 | 4 | 5 | 6 | 7 | 8 | 9 |
|---|---|---|---|---|---|---|---|---|---|---|
| Speed in mph | 119.9 | 128.6 | 137.3 | 146.0 | 154.7 | 163.4 | 172.1 | 180.8 | 189.5 | 198.2 |

The graph shows the information from the table. If the points on the graph were connected, what kind of figure would you have? What type of equation does this represent?

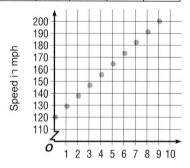

In the previous lesson, you learned to look for patterns in a list of numbers. The points in the graph represent the speeds at successive seconds. This set of numbers is another example of a *sequence*. Each number in a sequence is called a **term.** The first term is symbolized by $a_1$, the second term by $a_2$, and so on to $a_n$, the *n*th term.

The sequence shown in the chart above contains ten terms. Therefore, $a_1 = 119.9$ and $a_{10} = 198.2$. Each term of the sequence can be found by adding 8.7 to the previous term. A sequence of this type is called an **arithmetic sequence.** The number added to find the next term of an arithmetic sequence is called the **common difference** and is symbolized by the variable *d*.

| *Definition of Arithmetic Sequence* | **An arithmetic sequence is a sequence in which each term, after the first, is found by adding a constant, called the common difference, to the previous term.** |
|---|---|

To find the next terms in an arithmetic sequence, first find the common difference $d$ by subtracting any term from its succeeding term. Then add the common difference to the last term to find successive terms.

**Example 1**

**Find the next four terms of the arithmetic sequence 33, 39, 45, ....**

Find the common difference $d$ by subtracting two consecutive terms.

$39 - 33 = 6$ and $45 - 39 = 6$     So, $d = 6$.

Now add 6 to the last term of the sequence, and then continue adding until the next four terms are found.

$45 + 6 = 51$
$51 + 6 = 57$
$57 + 6 = 63$
$63 + 6 = 69$

The next four terms of the sequence are 51, 57, 63, and 69.

There is a pattern in the way terms of an arithmetic sequence are formed. A formula to find any term of an arithmetic sequence can be found if you know the first term and the common difference. This type of formula is known as a **recursive formula.** Recursive means that each succeeding term is formulated from one or more previous terms. Let's use the terms of the sequence in Example 1.

| $a_1$ | $a_2$ | $a_3$ | $a_4$ | $a_5$ | ... | $a_n$ |
|-------|-------|-------|-------|-------|-----|-------|
| 33 | 39 | 45 | 51 | 57 | ... | $a_n$ |
| $33 + 0(6)$ | $33 + 1(6)$ | $33 + 2(6)$ | $33 + 3(6)$ | $33 + 4(6)$ | ... | $33 + (n-1)(6)$ |
| $a_1 + 0 \cdot d$ | $a_1 + 1 \cdot d$ | $a_1 + 2 \cdot d$ | $a_1 + 3 \cdot d$ | $a_1 + 4 \cdot d$ | ... | $a_1 + (n-1)d$ |

*Why is the expression for $a_n$ equal to $33 + (n-1)6$?*

The following formula generalizes this pattern for any sequence.

| *Formula for the nth Term of an Arithmetic Sequence* | **The $n$th term, $a_n$, of an arithmetic sequence with first term $a_1$ and common difference $d$ is given by** $$a_n = a_1 + (n-1)d.$$ |
|---|---|

**Example 2**

Suppose a race car driver increases her speed at a constant rate. What will her speed be after 15 seconds, if her initial speed is 85 mph and her rate of acceleration is 4.5 mph per second?

$a_1 = 85$ and $d = 4.5$       *Why does $a_1 = 85$ and $d = 4.5$?*

Find $a_{16}$ using $a_n = a_1 + (n - 1)d$.     *Why do you need to find $a_{16}$?*

$a_n = a_1 + (n - 1)d$
$a_{16} = 85 + (16 - 1)(4.5)$
$a_{16} = 67.5$ or $152.5$       Her speed will be $152.5$ mph.

---

**Example 3**

**APPLICATION**

**Flight**

On a lengthy hot-air balloon flight, Hugo Furst floated into a warm air mass and began losing altitude. Because the air was too warm, he could not use his propane burner effectively to give him enough lift to clear a power line tower in front of him. So he threw an empty propane tank overboard to lighten the load. The distance the tank falls can be found using $d = 16t^2$, where $d$ is the distance in feet and $t$ is time elapsed in seconds. What would be the distance the tank falls *during* the 5th second?

| Time elapsed (seconds) | Total distance fallen (ft) | Distance fallen since the last second (ft) | Difference from second to second |
|:---:|:---:|:---:|:---:|
| 1 | $16(1)^2 = 16$ | $16 - 0 = 16$ | |
| 2 | $16(2)^2 = 64$ | $64 - 16 = 48$ | $48 - 16 = 32$ |
| 3 | $16(3)^2 = 144$ | $144 - 64 = 80$ | $80 - 48 = 32$ |
| 4 | $16(4)^2 = 256$ | $256 - 144 = 112$ | $112 - 80 = 32$ |

The distances fallen in each second form the arithmetic sequence 16, 48, 80, .... The last column of the table tells that the common difference is 32. The first term is 16. Now use the formula for the $n$th term to find the distance fallen during the 5th second.

$a_n = a_1 + (n - 1)d$
$a_{10} = 16 + (5 - 1)(32)$

Now use your calculator to help you find this distance.

ENTER:   16 $\boxed{+}$ $\boxed{(}$ 5 $\boxed{-}$ 1 $\boxed{)}$ $\boxed{\times}$ 32 $\boxed{=}$

The propane tank fell 144 feet during the 5th second.

---

Sometimes you may know two terms of a sequence but they are not consecutive terms of that sequence. The terms between any two nonconsecutive terms of an arithmetic sequence are called **arithmetic means.** In the sequence below, 30, 39, and 48 are the three arithmetic means between 21 and 57.

12, 21, 30, 39, 48, 57, 66, 75, ...

**Example 4**

> **Find the four arithmetic means between 19 and 54.**
>
> You can use the $n$th term formula to find the common difference. In the sequence 19, ____, ____, ____, ____, 54, 19 is $a_1$ and 54 is $a_6$.
>
> $a_n = a_1 + (n - 1)d$
> $54 = 19 + (6 - 1)d$    *Substitute the known values.*
> $54 = 19 + 5d$
> $35 = 5d$
> $7 = d$
>
> Now use the value of $d$ to find the four arithmetic means.
>
> $19 + 7 = 26$    $26 + 7 = 33$    $33 + 7 = 40$    $40 + 7 = 47$
>
> The arithmetic means are 26, 33, 40, and 47.

# CHECKING FOR UNDERSTANDING

**Communicating Mathematics**

Read and study the lesson to answer each question.

1. How would you determine if a list of numbers is an arithmetic sequence?

2. What is a common difference?

3. What is the formula for finding the $n$th term of an arithmetic sequence and what does each part of the formula represent?

4. What do you call the terms between any two nonconsecutive terms of an arithmetic sequence?

**Guided Practice**

Name the first five terms of each arithmetic sequence described.

5. $a_1 = 4, d = 3$       6. $a_1 = 7, d = 5$       7. $a_1 = 16, d = -2$

8. $a_1 = 38, d = -4$      9. $a_1 = \frac{3}{4}, d = -\frac{1}{4}$      10. $a_1 = \frac{3}{8}, d = \frac{5}{8}$

Name the next four terms of each arithmetic sequence.

11. 5, 9, 13, ...        12. 11, 14, 17, ...       13. 2, -3, -8, ...

14. 21, 15, 9, ...       15. $\frac{1}{2}, \frac{3}{2}, \frac{5}{2}, $ ...       16. -5.4, -1.4, 2.6, ...

# EXERCISES

**Find the *n*th term of each arithmetic sequence.**

**17.** $a_1 = -1$, $d = -10$, $n = 25$      **18.** $a_1 = -3$, $d = -9$, $n = 11$

**19.** $a_1 = 7$, $d = 3$, $n = 14$      **20.** $a_1 = -7$, $d = 3$, $n = 17$

**21.** $a_1 = 2$, $d = \dfrac{1}{2}$, $n = 8$      **22.** $a_1 = \dfrac{3}{4}$, $d = -\dfrac{5}{4}$, $n = 13$

**23.** $a_1 = 20$, $d = 4$, $n = 100$      **24.** $a_1 = 13$, $d = 3$, $n = 101$

**Complete each statement.**

**25.** 124 is the $\underline{\ ?\ }$ th term of $-2$, 5, 12, ....

**26.** 142 is the $\underline{\ ?\ }$ th term of $-3$, 2, 7, ....

**27.** $-28$ is the $\underline{\ ?\ }$ th term of 7, 2, $-3$, ....

**28.** $-\dfrac{17}{4}$ is the $\underline{\ ?\ }$ th term of $2\dfrac{1}{4}$, 2, $1\dfrac{3}{4}$, ....

**29.** Let $A = 24$, $B = 36$, and $C =$ the arithmetic mean of $A$ and $B$. Graph $A$, $B$, and $C$ on a number line. What is the relationship of $C$ to segment $AB$?

**Find the indicated term in each arithmetic sequence.**

**30.** $a_{12}$ for $-17$, $-13$, $-9$, ...      **31.** $a_{21}$ for 10, 7, 4, ...

**32.** $a_{32}$ for 4, 7, 10, 13, ...      **33.** $a_{10}$ for 8, 3, $-2$, ...

**Find the missing terms in each arithmetic sequence. Then graph each sequence using the *x*-axis for the number of the term and the *y*-axis for the term itself.**

**34.** 55, $\underline{\ ?\ }$, $\underline{\ ?\ }$, $\underline{\ ?\ }$, 115      **35.** $-8$, $\underline{\ ?\ }$, $\underline{\ ?\ }$, 3

**36.** $-10$, $\underline{\ ?\ }$, $\underline{\ ?\ }$, $\underline{\ ?\ }$, $\underline{\ ?\ }$, 2      **37.** 2, $\underline{\ ?\ }$, $\underline{\ ?\ }$, $\underline{\ ?\ }$, $\underline{\ ?\ }$, $\underline{\ ?\ }$, 20

**38.** $\underline{\ ?\ }$, $-6$, $\underline{\ ?\ }$, $\underline{\ ?\ }$, 15, $\underline{\ ?\ }$      **39.** $\underline{\ ?\ }$, 49, $\underline{\ ?\ }$, $\underline{\ ?\ }$, 28

**40.** The last term of an arithmetic sequence is 207, the common difference is 3, and the number of terms is 14. What is the first term of the sequence?

**41.** The third term of an arithmetic sequence is 14 and the ninth term is $-1$. Find the first four terms of the sequence.

**42.** The fifth term of an arithmetic sequence is 19 and the 11th term is 43. Find the first term and the 87th term.

**43.** Find three numbers that have a sum of 27, a product of 288, and form an arithmetic sequence.

**44.** Use an arithmetic sequence to find how many multiples of seven are between 11 and 391. *Hint: What is the least multiple in that range? What is the greatest multiple?*

**45.** In the Fibonacci sequence from Lesson 13-1, you learned that each term is found by adding a number to the previous numbers.

    **a.** Is the Fibonacci sequence an arithmetic sequence? Explain your answer.

    **b.** Let $F_1 = 1$, $F_2 = 1$, $F_3 = 2$, $F_4 = 3$, and so on to identify each term of the Fibonacci sequence. Write an expression for $F_n$ in terms of the preceding two terms.

**Applications**

**46. Skydiving** During a free fall, a skydiver falls 16 feet in the first second, 48 feet in the second second, and 80 feet in the third second. If she continues to fall at this rate, how many feet will she fall during the 8th second?

**47. Aeronautics** A rocket rises 20 feet in the first second, 60 feet in the second second, and 100 feet in the third second. If it continues at this rate, how many feet will it rise in the 20th second?

**Mixed Review**

**48.** Find the next three terms of the following pattern $a$, $ac$, $ace$, $\underline{\ ?\ }$, $\underline{\ ?\ }$, $\underline{\ ?\ }$. **(Lesson 13-1)**

**49.** Evaluate $\log_8 8^5$. **(Lesson 12-2)**

**50.** Simplify $\dfrac{3x^2 - 5x + 2}{2x^2 - 5x - 3} + \dfrac{x^2 + x - 2}{2x^2 - x - 3}$. **(Lesson 11-3)**

**51.** Find $g[f(-2)]$ if $f(x) = 3x^2 + 12x - 5$ and $g(x) = x^3 + 5x^2 - 4x + 12$. **(Lesson 10-7)**

**52.** Graph $y \le -x^2 + 8x - 11$. **(Lesson 8-6)**

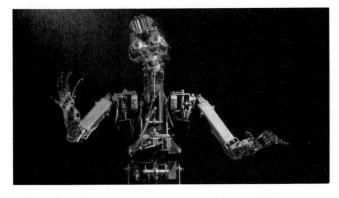

**Journal**

Describe the most challenging thing you have learned in this course thus far. Tell why you think it is most challenging.

**53. Manufacturing** The Grasco Company makes widgets and gadgets. At least 500 widgets and 700 gadgets are needed to meet minimum, daily demands. The machinery can produce no more than 1200 widgets and 1400 gadgets per day. The combined number of widgets and gadgets that the packaging department can handle is 2300 per day. **(Lesson 2-6)**

    **a.** If the company sells widgets for 40¢ each and gadgets for 50¢ each, how many of each type should be produced for maximum daily income?

    **b.** What is the maximum daily income?

# 13-3 Arithmetic Series

**Objectives**

After studying this lesson, you should be able to:

- find sums of arithmetic series and find specific terms in the series, and
- use sigma notation to express the sum.

**Application**

Skydivers fall greater distances each second they free fall. In Lesson 13-2, you learned that these free-fall distances form an arithmetic sequence.

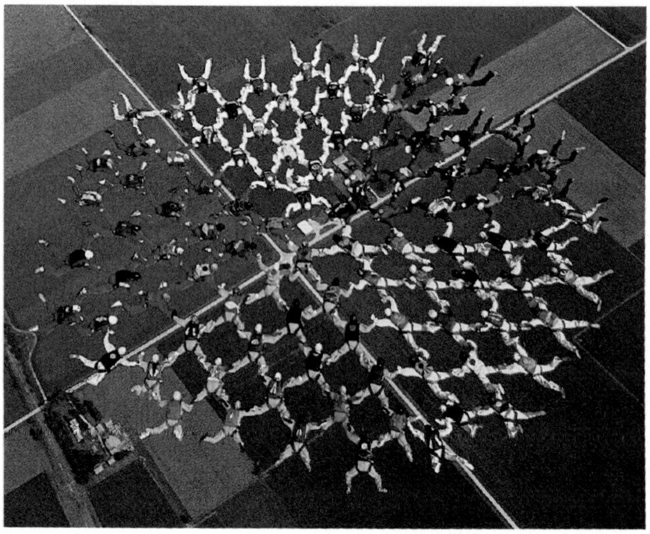

16, 48, 80, 112, 144, 176, 208, ...

To find out what the total distance fallen by the skydiver is, you would add the terms in the sequence.

16 + 48 + 80 + 112 + 144 + 176 + 208 + ...

The indicated sum of the terms of a sequence is called a **series.** The series shown above is an **arithmetic series.**

The list below shows examples of arithmetic sequences and their corresponding arithmetic series.

| Arithmetic Sequence | Arithmetic Series |
| --- | --- |
| 3, 6, 9, 12, 15 | 3 + 6 + 9 + 12 + 15 |
| −8, −2, 4 | −8 + (−2) + 4 |
| $\frac{4}{5}, \frac{8}{5}, \frac{12}{5}, \frac{16}{5}$ | $\frac{4}{5} + \frac{8}{5} + \frac{12}{5} + \frac{16}{5}$ |
| $a_1, a_2, a_3, a_4, \ldots, a_n$ | $a_1 + a_2 + a_3 + a_4 + \cdots + a_n$ |

**FYI ...**

During a jump, skydivers usually create a controlled free-fall by spreading their arms and legs. Even in a controlled free-fall, they may reach a speed of 146 ft/s or about 98 mph.

The symbol $S_n$ is used to represent the sum of the first $n$ terms of a series. For example, $S_3$ means the sum of the first three terms of a series. In the series 3 + 6 + 9 + 12 + 15, $S_3$ would be 3 + 6 + 9 or 18.

If a series has a large number of terms, it is not convenient to list all the terms and then find their sum. To develop a general formula for the sum of any arithmetic series, let's consider the series of skydiving distances.

16 + 48 + 80 + 112 + 144 + 176 + 208

Suppose we write $S_7$ in two different orders and find the sum.

$$
\begin{array}{rccccccccccccc}
S_7 = & 16 + & 48 + & 80 + & 112 + & 144 + & 176 + & 208 \\
+\ S_7 = & 208 + & 176 + & 144 + & 112 + & 80 + & 48 + & 16 \\
\hline
2 \cdot S_7 = & 224 + & 224 + & 224 + & 224 + & 224 + & 224 + & 224
\end{array}
$$

*7 sums of 224*

$$2 \cdot S_7 = 7(224)$$
$$S_7 = \frac{7}{2}(224) \quad \textit{Divide each side by 2.}$$

Now let's analyze what these numbers represent in terms of $S_n$. In the equation $S_7 = \frac{7}{2}(224)$, 7 represents $n$, and 224 represents the sum of the first and last terms, $a_1 + a_n$. Thus, we can replace the equation with the formula $S_n = \frac{n}{2}(a_1 + a_n)$. This formula can be used to find the sum of any arithmetic series.

---

*Sum of an Arithmetic Series*

**The sum $S_n$ of the first $n$ terms of an arithmetic series is given by $S_n = \frac{n}{2}(a_1 + a_n)$.**

---

**Example 1**

**Find the sum of the first 100 positive integers.**

In this series, $a_1 = 1$ and $a_n = a_{100} = 100$.

$$S_n = \frac{n}{2}(a_1 + a_n)$$

$$S_{100} = \frac{100}{2}(1 + 100) \quad \textit{Substitute the known values.}$$

$$= 50(101) \quad \text{or} \quad 5050$$

The sum of the first 100 positive integers is 5050.

---

In Lesson 13-2, you learned that in an arithmetic sequence, $a_n = a_1 + (n - 1)d$. Using this formula and substitution gives us another version of the formula for the sum of an arithmetic sequence.

$$S_n \div \frac{n}{2}(a_1 + a_n)$$

$$= \frac{n}{2}\{a_1 + [a_1 + (n - 1)d]\} \quad \textit{Substitute } a_1 + (n - 1)d \textit{ for } a_n.$$

$$= \frac{n}{2}[2a_1 + (n - 1)d]$$

You can use this formula when you don't know the value of the last term.

**Example 2**

Find the sum of the first 50 terms of an arithmetic series where $a_1 = 5$ and $d = 25$.

The series is $5 + 30 + 55 + 80 + \ldots$    *Use the formula for $S_n$.*

$S_n = \frac{n}{2}[2a_1 + (n - 1)d]$

$\quad = \frac{50}{2}[2(5) + (50 - 1)25]$    *Substitute the known values.*

$\quad = 30{,}875$

**Example 3**

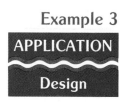

A 24-hour Redimarket has its evening employees do most of the restocking of shelves and arranging of displays. At Thanksgiving the store creates large displays of those items most frequently bought at that time. Su Makita has the job of arranging the canned yams. He designs a display with 15 rows with each row having one less can than the row below it. If the bottom row has 27 cans, how many cans are in the display?

$S_n = \frac{n}{2}[2a_1 + (n - 1)d]$    $a_1 = 27$, $d = -1$, and $n = 15$

$\quad = \frac{15}{2}[2(27) + (15 - 1)(-1)]$

$\quad = 300$

There are 300 cans of yams in the display.

It is sometimes necessary to use both sum formulas to solve a problem. You must analyze the information you are given and then decide which formula to use first.

**Example 4**

Find the first three terms of an arithmetic series where $a_1 = 17$, $a_n = 101$, and $S_n = 472$.

First, find $n$.                 Next, find $d$.

$S_n = \frac{n}{2}(a_1 + a_n)$          $a_n = a_1 + (n - 1)d$

$\phantom{S_n}$                 $101 = 17 + (8 - 1)d$

$472 = \frac{n}{2}(17 + 101)$          $84 = 7d$

$472 = 59n$                 $12 = d$

$\quad n = 8$

Now determine $a_2$ and $a_3$.

$a_2 = 17 + 12$ or $29$     $a_3 = 29 + 12$ or $41$

The first three terms are 17, 29, and 41.

Writing out a series is often time-consuming and lengthy. To simplify this, mathematicians use a more concise notation called **sigma** or **summation notation.**

$2 + 4 + 6 + 8 + \cdots + 20$ can be expressed as $\sum\limits_{n=1}^{10} 2n$.

$\sum\limits_{n=1}^{10} 2n$ *is read the summation from 1 to 10 of 2n.*

When using sigma notation, the variable defined below the $\Sigma$ (sigma) is called the **index of summation.** The upper number is the upper limit of the index. To generate the terms of the series, successively replace the index of summation with each value of $n$, in this series, 1, 2, 3, and so on through 10.

**Example 5**

Write the terms of $\sum\limits_{k=2}^{6} (3k + 2)$ and find the sum.

In this series, replace $k$ with 2 and then 3, 4, 5, and 6 to find the terms.

$$\sum\limits_{k=2}^{6} (3k + 2) = \overset{k=2}{3(2) + 2} + \overset{k=3}{3(3) + 2} + \overset{k=4}{3(4) + 2} + \overset{k=5}{3(5) + 2} + \overset{k=6}{3(6) + 2}$$
$$= 8 + 11 + 14 + 17 + 20$$
$$= 70$$

The sum is 70.

Just as a polynomial can be expressed in more than one form, the summation of a series can be expressed in different ways. The summation of the series in Example 5 is expressed as $\sum\limits_{k=2}^{6} (3k + 2)$. It can also be expressed as $\sum\limits_{k=2}^{6} [5 + 3(k - 1)]$. *Verify that these two expressions are equivalent.*

# CHECKING FOR UNDERSTANDING

**Communicating Mathematics**

Read and study the lesson to answer each question.

1. The indicated sum of terms of a sequence is called a __?__ .
2. The sum of a series can be written in __?__ or __?__ notation.
3. What is the purpose of the index of summation?

**Guided Practice**

Find the sum of each series.

4. $4 + 7 + 10 + 13 + 16 + 19 + 22 + 25$
5. $1 + 5 + 9 + 13 + 17 + 21 + 25 + 29$

Find $S_n$ for each arithmetic series described.

6. $a_1 = 2$, $a_n = 200$, $n = 100$

7. $a_1 = 5$, $a_n = 100$, $n = 200$

8. $a_1 = 4$, $n = 15$, $d = 3$

9. $a_1 = 50$, $n = 20$, $d = -4$

10. $-3 + (-7) + (-11) + \ldots + a_{10}$

11. $9 + 11 + 13 + 15 + \ldots$ for $n = 12$

12. the sum of the greatest 100 negative integers

13. the sum of the first 100 positive even integers

# EXERCISES

**Practice**     Find $S_n$ for each arithmetic series described.

14. $a_1 = 3$, $a_n = -38$, $n = 8$

15. $a_1 = 85$, $n = 21$, $a_n = 25$

16. $a_1 = 34$, $n = 9$, $a_n = 2$

17. $a_1 = 76$, $n = 16$, $a_n = 31$

18. $a_1 = 4$, $d = -1$, $n = 7$

19. $a_1 = 5$, $d = \dfrac{1}{2}$, $n = 13$

Find the sum of each arithmetic series.

20. $6 + 12 + 18 + \ldots + 96$

21. $34 + 30 + 26 + \ldots + 2$

22. $7 + 14 + 21 + 28 + \ldots + 98$

23. $10 + 4 + (-2) + \ldots + (-50)$

24. $\displaystyle\sum_{n=1}^{25} 2n$

25. $\displaystyle\sum_{r=3}^{6} (r + 2)$

26. $\displaystyle\sum_{n=1}^{30} (2n - 1)$

27. $\displaystyle\sum_{n=21}^{75} (2n + 5)$

28. $\displaystyle\sum_{n=10}^{50} (3n - 1)$

29. $\displaystyle\sum_{j=1}^{6} (24 - 9j)$

Find $S_n$ for each arithmetic series described.

30. $a_1 = 91$, $d = -4$, $a_n = 15$

31. $d = 5$, $n = 16$, $a_n = 72$

32. $d = -4$, $n = 9$, $a_n = 27$

33. $a_1 = -2$, $d = \dfrac{1}{2}$, $a_n = 5$

Find the first three terms of each arithmetic series.

34. $a_1 = 7$, $a_n = 139$, $S_n = 876$

35. $n = 14$, $a_n = 53$, $S_n = 378$

36. $n = 21$, $a_n = 78$, $S_n = 1008$

37. $a_1 = 6$, $a_n = 306$, $S_n = 1716$

**Critical Thinking**     38. Evaluate $\displaystyle\sum_{a=3}^{6} (a - 2)^2$ and $\displaystyle\sum_{a=1}^{4} a^2$. What do you notice?

39. Write an argument to prove or disprove the statement
$$2 \sum_{k=3}^{7} k^2 = \sum_{k=3}^{7} 2k^2.$$

**Applications**

**40. Consumerism** A pile of fireplace logs at the Garden Shop has 10 logs on the top layer, 11 logs in the next layer, and so on. The pile contains 13 layers. Write an expression that represents how many logs are in the pile, and then evaluate that expression.

**41. Education** Milford High School is planning a pep rally for their basketball team that is headed for the state tournament. The school wants to invite parents as well as students. Milford Auditorium has 21 seats in the first row. Each of the other rows has one more seat than the one in front of it and there are 30 rows of seats. If they anticipate 1200 people will come to the pep rally, will there be a seat for everyone? Justify your answer.

**Mixed Review**

**42.** Find the 57th term of the arithmetic sequence 6, 15, 24, 33, .... **(Lesson 13-2)**

**43.** Solve $\log_5 (3x + 7) = \log_5 (x^2 - 4x - 1)$. **(Lesson 12-7)**

**44.** Find $p(-4)$ if $p(x) = 3x^2 - 7x + 1$. **(Lesson 8-1)**

**45.** Solve $\begin{cases} x + 2y - 3z = -13 \\ 2x - y + 3z = 23 \\ 3x + y - 3z = -8 \end{cases}$ using augmented matrices. **(Lesson 4-8)**

**46. Geometry** In isosceles triangle $ABC$, the length of each leg is 5 cm less than three times the length of its base. Find its dimensions if the perimeter of the triangle is 88 cm. **(Lesson 1-6)**

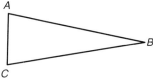

---

## MID-CHAPTER REVIEW

**1.** If the first 27 terms in the sequence 3, 33, 333, 3333, 33,333, ... are added together, what is the digit in the hundreds place of their sum? **(Lesson 13-1)**

**2.** The Lucas sequence was developed from the Fibonacci sequence. In the Lucas sequence, $L_1 = F_1$ and $L_n = F_{n+1} + F_{n-1}$ for $n \geq 2$. Find the first 8 terms of the Lucas sequence.

**3.** Find the 12th term in the arithmetic sequence $\frac{3}{4}, \frac{3}{2}, \frac{9}{4}, \ldots$. **(Lesson 13-2)**

**4.** Find the 100th term in an arithmetic sequence where $a_1 = 20$ and $d = 4$. **(Lesson 13-2)**

**5.** Find $S_n$ for $d = 5$, $n = 16$, and $a_n = 72$. **(Lesson 13-3)**

**6.** Find $S_n$ for $a_1 = 9$, $d = -6$, and $n = 14$. **(Lesson 13-3)**

**7. Personal Finance** Louise McCall retired as a secretary at the lumberyard after 30 years of employment. If her salary was $4500 the first year, and she received an $820 raise at the end of each year of service, what was her total salary for those 30 years? **(Lesson 13-3)**

## 13-4 Geometric Sequences

**Objectives**

After studying this lesson, you should be able to:
- find the *n*th term of a geometric sequence,
- find the position of a given term in a geometric sequence, and
- find geometric means.

**Application**

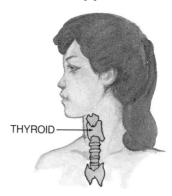

THYROID

Iodine-131 is used medically as a tracer isotope in monitoring the activity of the thyroid gland. A patient is given a compound containing the radioactive iodine. A physician uses a Geiger-Mueller counter to monitor the activity of the iodine in the thyroid. The amount of iodine retained by this gland is a measure of its ability to function.

Iodine-131 has a half-life of about 8 days. That means that approximately every 8 days, half the mass of iodine decays into another element. Then in the next 8 days, half of the remaining iodine decays, and so on.

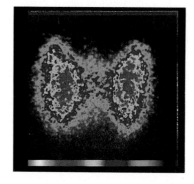

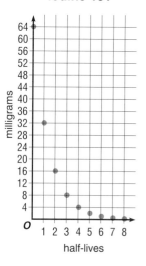

**Iodine-131**

milligrams vs. half-lives

Suppose a container held a mass of 64 milligrams of iodine-131. The graph shows the remaining mass of iodine after each half-life. What type of figure do these points suggest?

The pattern of masses forms a sequence of numbers known as a **geometric sequence.** The terms in this example are 64, 32, 16, 8, 4, 2, 1, 0.5, ...   *What would be the next number in the sequence? How did you find it?*

| *Definition of Geometric Sequence* | **A geometric sequence is one in which each term after the first is found by multiplying the previous term by a constant called the common ratio.** |
|---|---|

In any geometric sequence, the **common ratio *r*** is found by dividing any term by the previous term.

**Example 1**

Find the next two terms of the geometric sequence 4, 12, 36, ....

To find the common ratio, find the quotient of any two consecutive terms.

$\dfrac{12}{4} = 3 \qquad \dfrac{36}{12} = 3$   The common ratio is 3.

The fourth term is $36(3)$ or $108$.

The fifth term is $108(3)$ or $324$.

The next two terms of the sequence are 108 and 324.

---

Successive terms of a geometric sequence are usually expressed as the product of $r$ and the previous term. Thus, a geometric sequence is also a recursive sequence. Since each succeeding term contains a factor of $r$, each term can be expressed as a product of $a_1$ and a power of $r$. Notice each pattern of terms in the table below.

| the sequence | $a_1$ | $a_2$ | $a_3$ | $a_4$ | ... | $a_n$ |
|---|---|---|---|---|---|---|
| terms expressed using the previous term and the common ratio | $a_1$ | $a_1 r$ | $a_2 r$ | $a_3 r$ | ... | $a_{n-1} r$ |
| terms expressed using the first term and the common ratio | $a_1$ | $a_1 r$ | $a_1 r^2$ | $a_1 r^3$ | ... | $a_1 r^{n-1}$ |

The $n$th term, $a_n$, in the last column can be expressed as either $a_{n-1} r$ or $a_1 r^{n-1}$.

---

*Formula for the nth term of a Geometric Sequence*

**The $n$th term, $a_n$, of a geometric sequence with first term $a_1$ and common ratio $r$ is given by either formula.**
$$a_n = a_{n-1} r \qquad \text{or} \qquad a_n = a_1 r^{n-1}$$

---

**Example 2**

Write the first five terms of a geometric sequence in which $a_1 = 5$ and $r = 2$.

Since the first term is given, write each term using the formula $a_n = a_1 r^{n-1}$.

| $a_1$ | $a_2$ | $a_3$ | $a_4$ | $a_5$ |
|---|---|---|---|---|
| 5 | $5(2^{2-1})$ | $5(2^{3-1})$ | $5(2^{4-1})$ | $5(2^{5-1})$ |
| 5 | $5(2^1)$ | $5(2^2)$ | $5(2^3)$ | $5(2^4)$ |
| 5 | 10 | 20 | 40 | 80 |

The first five terms of the sequence are 5, 10, 20, 40, and 80.

**Example 3**

Find the seventh term, $a_7$, of a geometric sequence in which $a_3 = 96$ and $r = 4$.

*Method 1*

The general form of the third term of a sequence is $a_1r^2$.

Find $a_1$.          Then find $a_7$.

$$a_3 = a_1r^2 \qquad a_n = a_1r^{n-1}$$
$$96 = a_1(4^2) \qquad a_7 = a_1r^{7-1}$$
$$\frac{96}{16} = a_1 \qquad\quad = 6(4^6)$$
$$6 = a_1 \qquad\qquad = 24{,}576$$

*Method 2*

Begin with the third term and use $r$ to find each successive term.

$$a_4 = 96 \cdot 4 = 384 \quad {\scriptstyle a_4 = a_3 \cdot r}$$
$$a_5 = 384 \cdot 4 = 1536$$
$$a_6 = 1536 \cdot 4 = 6144$$
$$a_7 = 24{,}576$$

The seventh term is 24,576.

**Example 4**

APPLICATION

Finance

Fred read about a "foolproof" way to become a millionaire. You save 1¢ on the first day. Then each day thereafter, save double the amount you saved the day before. Find the amount he should save on the 20th day of his plan.

In this sequence, $a_1 = 1$. Since the amount is twice that of the day before, $r = 2$.

$$a_n = a_1r^{n-1}$$
$$a_{20} = 1 \cdot 2^{20-1}$$
$$a_{20} = 524{,}288$$

*Is this foolproof way to become a millionaire a practical one?*

On the 20th day, Fred should save 524,288¢ or $5242.88.

You learned that the missing terms between two nonconsecutive terms in an arithmetic sequence were called arithmetic means. Likewise, the missing term or terms between two nonconsecutive terms in a geometric sequence are called **geometric means.** In the sequence, 3, 12, 48, 192, 768, ..., the three geometric means between 3 and 768 are 12, 48, and 192. You can use the common ratio to find missing geometric means in a given sequence.

**Example 5**

**Find the two geometric means between 81 and 3. The sequence is 81, ____ , ____ , 3, ....**

Use the general formula for the $n$th term to find the value of $r$. Since $a_1 = 81$, $a_4 = 3$, and $n = 4$, $a_n = a_1 r^{n-1}$ becomes $a_4 = a_1 \cdot r^3$ or $3 = 81r^3$.

$3 = 81 \cdot r^3$

$\dfrac{1}{27} = r^3$    *Divide each side by 81.*

$\sqrt[3]{\dfrac{1}{27}} = \sqrt[3]{r^3}$    *Take the cube root of each side.*

$\dfrac{1}{3} = r$         $\dfrac{1}{27} = \dfrac{1^3}{3^3}$

$a_2 = 81\left(\dfrac{1}{3}\right)$ or $27$         $a_3 = 81\left(\dfrac{1}{3}\right)^2$ or $9$

The missing geometric means are 27 and 9.

Sometimes there is more than one way to find the missing geometric means for a sequence. If only two terms of the sequence are given, you may find that these terms actually are parts of two different sequences.

**Example 6**

**Find the three geometric means between 6 and 96. Then graph the sequence using the $x$-axis for the number of the term and the $y$-axis for the term itself.**

Approach the problem in the same way as you did in Example 5.

$a_5 = a_1 r^{5-1}$   $a_n = a_1 r^{n-1}$

$96 = 6 \cdot r^4$   *Substitute the known values.*

$16 = r^4$      *Divide each side by 6.*

$\pm 2 = r$      $2^4 = 16$ and $(-2)^4 = 16$

Since there are two values for $r$, two sequences exist.

| $r$ | $a_1$ | $a_2$ | $a_3$ | $a_4$ | $a_5$ |
|---|---|---|---|---|---|
| $r = 2$ | 6 | 6(2) or 12 | $6(2^2)$ or 24 | $6(2^3)$ or 48 | 96 |
| $r = -2$ | 6 | 6(-2) or -12 | $6(-2)^2$ or 24 | $6(-2)^3$ or -48 | 96 |

The geometric means could be either 12, 24, and 48 or $-12$, 24, and $-48$.

Each sequence forms a different graph. Each graph represents a different polynomial which reflects the values of $a_1$ and $r$.

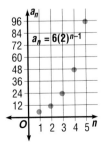

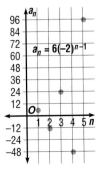

## Example 7

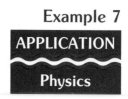

A vacuum pump removes $\frac{1}{5}$ of the air from a sealed container on each stroke of its piston. What percent of the air remains after five strokes of the piston?

Let 1 represent the original amount of air. After the first stroke, $1 - \frac{1}{5}$ or $\frac{4}{5}$ of the air remains. The second strike removes $\frac{1}{5}$ of the remaining air. Thus the amount that remains after two strokes is $\frac{4}{5}\left(1 - \frac{1}{5}\right) = \frac{4}{5} \cdot \frac{4}{5}$ or $\frac{16}{25}$. This pattern can be expressed as a geometric sequence.

| number of strokes | 0 | 1 | 2 | 3 | 4 | 5 |
|---|---|---|---|---|---|---|
| sequence | 1, | $\frac{4}{5}$, | $\frac{16}{25}$, | ? ___, | ? ___, | ? ___ |
| term | $a_1$ | $a_2$ | $a_3$ | $a_4$ | $a_5$ | $a_6$ |

Now use the formula $a_n = a_1 r^{n-1}$ to find $a_6$, the amount of air left after five strokes.

$a_n = a_1 \cdot r^{n-1}$

$a_6 = 1 \cdot \left(\frac{4}{5}\right)^5$ or $\frac{4^5}{5^5}$   *Substitute 1 for $a_1$, 6 for n, and $\frac{4}{5}$ for r.*

$= \frac{1024}{3125}$ or about 0.328

About 32.8% of the air remains after 5 strokes.

# CHECKING FOR UNDERSTANDING

**Communicating Mathematics**

Read and study the lesson to answer each question.

1. What do you call the constant that, when multiplied by the previous term, yields the next term of the sequence?

2. Why is a geometric sequence recursive?

3. What are the terms between any two nonconsecutive terms of a geometric sequence called?

4. What is a half-life?

**Guided Practice**

Determine whether each sequence is geometric. If so, find the common ratio.

5. 4, 20, 100, 500

6. 2, 4, 6, 8

7. $\frac{3}{2}, \frac{9}{4}, \frac{27}{8}, \frac{81}{16}$

8. 7, 14, 21, 28

9. 1, 4, 9, 16, 25

10. 9, 6, 4, $\frac{8}{3}$

11. Find the first four terms of the geometric sequence in which $a_1 = 3$ and $r = -2$.

# EXERCISES

**Practice**

Find the next two terms of each geometric sequence.

**12.** 90, 30, 10, ...

**13.** 2, 6, 18, ...

**14.** 20, 30, 45, ...

**15.** 729, 243, 81, ...

**16.** $\dfrac{1}{27}, \dfrac{1}{9}, \dfrac{1}{3}, \ldots$

**17.** $-\dfrac{1}{4}, \dfrac{1}{2}, -1, \ldots$

Find the first four terms of each geometric sequence described.

**18.** $a_1 = 3, r = -2$

**19.** $a_1 = 27, r = -\dfrac{1}{3}$

**20.** $a_1 = 12, r = \dfrac{1}{2}$

**21.** Graph the first eight terms of the sequence, $-1, 2, -4, 8, \ldots$ Use the $x$-axis to represent $n$ and the $y$-axis to represent $a_n$.

Find the $n$th term of each geometric sequence described.

**22.** $a_1 = 4, n = 3, r = 5$

**23.** $a_1 = 2, n = 5, r = 2$

**24.** $a_1 = 7, n = 4, r = 2$

**25.** $a_1 = 243, n = 5, r = -\dfrac{1}{3}$

**26.** $a_3 = 32, n = 6, r = -\dfrac{1}{2}$

**27.** $a_4 = 16, n = 8, r = \dfrac{1}{2}$

Find the missing geometric means. Then graph each sequence, using the $x$-axis for the number of the term and the $y$-axis for the term itself.

**28.** 3, ——, ——, ——, 48

**29.** 1, ——, ——, 8

**30.** 8, ——, ——, ——, ——, $\dfrac{1}{4}$

**31.** 3, ——, 75

**32.** 5, ——, ——, ——, 80

**33.** 7, ——, ——, ——, 112

**34.** ——, ——, $-12$, ——, ——, 96

**35.** ——, ——, ——, 24, ——, ——, ——, 384

Write the formula for the $n$th term for each graphed sequence.

**36.**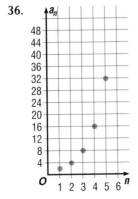

**37.**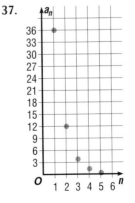

**38.**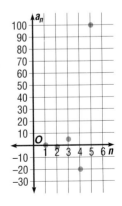

**Critical Thinking**

**39.** Write an argument to show that $a_n = a_{n-1}r$ and $a_n = a_1 r^{n-1}$ are equivalent.

**Applications**

**40. Physics**   A vacuum pump removes $\frac{1}{20}$ of the air in a sealed jar on each stroke of its pistons. How many strokes of the piston are required to remove 99% of the air from the jar?

  **a.** What methods could you use to solve the problem?

  **b.** Find the solution.

**41. Aeronautics**   A vacuum pump removes $\frac{1}{10}$ of the air from a space capsule on each stroke of its piston. What percent of the air remains after 10 strokes of the piston?

**42. Demographics**   The population of Sunville increases by 10% each year. It currently has a population of 20,000.

  **a.** Predict its population to the nearest 100 people 5 years from now.

  **b.** At this rate, in how many years will the population be at least 100,000?

**43. Cinematography**   To produce a vanishing effect on film, an animated character's image is filmed at 50%. Then each reduced image is filmed at 50% again. If this process of filming at 50% is repeated 8 more times, what is the area of the resulting image in comparison to the original image?

**Mixed Review**

**44.** Find $S_n$ for an arithmetic series in which $a_1 = 11$, $a_n = 44$, and $n = 23$.   **(Lesson 13-3)**

**45.** Find all rational zeros of $f(x) = x^3 - 4x^2 - 11x + 30$.   **(Lesson 10-3)**

**46.** Find $f(m + 2)$ if $f(x) = 2x^3 - 5x^2$.   **(Lesson 8-1)**

**47.** Factor $a^4 - 16b^8$.   **(Lesson 5-5)**

**48.** Graph $3x - 4y = 12$.   **(Lesson 2-4)**

**49. Geometry**   The formula for the area of a regular hexagon is $A = \frac{1}{2}ap$, where $a$ is the measure of the apothem and $p$ is the perimeter of the hexagon. Find the area of a hexagon whose side is 6 inches long and whose apothem is 1.73 times the length of the side.   **(Lesson 1-1)**

---

## ~~~~ FINE ARTS CONNECTION ~~~~

Music and mathematics are closely related. Have you ever wondered why a harp or pipe organ has a particular shape? Stringed instruments, like pianos and harps, and those formed from columns of air, like pipe organs, reflect the shape of an exponential curve. The length of each string or pipe is a term in an exponential sequence. The pitch of a string is determined by its thickness and tautness. The pitch of a pipe is determined by its thickness and diameter.

# Geometric Series

After studying this lesson, you should be able to:

- find sums of geometric series, and find specific terms in the series, and
- use sigma notation to express the sum.

**Application**

Fred decided to try the foolproof way to become a millionaire by saving 1¢ on the first day and doubling the amount saved each day. On the first day, he placed a penny in a large glass bottle. The next day he placed 2 cents in the bottle, and so on. After a week, he wondered how much he had saved. How could he compute his savings without keeping extensive records?

The amounts saved in the first week were 1¢, 2¢, 4¢, 8¢, 16¢, 32¢, and 64¢. To find the total of this geometric sequence, you add these numbers. The sum of the terms of a geometric sequence is called a **geometric series.**

You can develop a formula for finding the sum of a geometric series. Let $S_7$ represent the sum of the first seven terms of the geometric series. In the series above, $a_1 = 1$ and $r = 2$.

$$S_7 = 1 + 2 + 4 + 8 + 16 + 32 + 64 \qquad \textit{Now multiply each side by 2.}$$
$$-\quad 2S_7 = \qquad 2 + 4 + 8 + 16 + 32 + 64 + 128 \qquad \textit{Align the terms.}$$
$$\overline{S_7 - 2S_7 = 1 + 0 + 0 + 0 + \quad 0 + \quad 0 + \quad 0 + (-128)} \qquad \textit{Subtract the two equations}$$

$$(1 - 2)S_7 = 1 - 128 \qquad \textit{Factor out } S_7.$$
$$S_7 = \frac{1 - 128}{1 - 2} \text{ or } 127 \qquad \textit{Divide each side by } 1 - 2.$$

Fred saved $1.27.

In Fred's problem, you can relate the equation $S_7 = \dfrac{1 - 128}{1 - 2}$ to terms of the series. Note that 1 is $a_1$, 128 is $a_7$, and 2 is $r$. If we substitute these values, the solution resembles a general formula.

$$S_7 = \frac{a_1 - a_8}{1 - r} \quad \text{or} \quad S_7 = \frac{a_1 - a_1 r^7}{1 - r} \qquad \textit{Remember, } a_8 = a_1 r^{8-1}.$$

Notice that we rewrote $a_8$ in terms of $a_1$ and $r$ so that the formula contained as few different variables as possible. This can be generalized for the $n$th term.

---

**Sum of a Geometric Series**

The sum, $S_n$, of the first $n$ terms of a geometric series is given by the following formula.

$$S_n = \frac{a_1 - a_1 r^n}{1 - r} \quad \text{or } S_n = \frac{a_1(1 - r^n)}{1 - r} \text{ where } r \neq 1$$

---

The general formula for the sum of a geometric series limits the value of $r$. When you have a geometric series in which $r = 1$, the denominator would be zero, so $S_n$ is found by the formula $S_n = n \cdot a_1$.

### Example 1

Find the sum of the first five terms of a geometric series for which $a_1 = 4$ and $r = -3$.

Use the formula for the sum of a geometric series.

$$S_n = \frac{a_1 - a_1 r^n}{1 - r} \quad \blacktriangleright \quad S_5 = \frac{4 - 4(-3)^5}{1 - (-3)}$$

$$= \frac{976}{4} \text{ or } 244$$

The sum of the first five terms is 244.

### Example 2

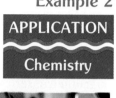

**APPLICATION**

**Chemistry**

The half-life of plutonium (Pu-236) is 2.85 years. Use a calculator to find how much of a 100 gram sample of Pu-236 is left after 10 half-lives.

To find the amount left, use the sum of a series to determine the amount that decays. Then subtract that amount from the initial amount, 100 grams. After the first half-life, 0.5(100) or 50 grams has decayed. So, $a_1 = 50$ and $r = 0.5$ in the series describing the total amount decayed.

$$S_n = \frac{a_1 - a_1 r^n}{1 - r} = \frac{50 - 50(0.5)^{10}}{1 - 0.5} \qquad a_1 = 50,\ r = 0.5,\ n = 10$$

ENTER: $\boxed{(}\ 50\ \boxed{-}\ 50\ \boxed{\times}\ 0.5\ \boxed{y^x}\ 10\ \boxed{)}\ \boxed{\div}$

$\boxed{(}\ 1\ \boxed{-}\ 0.5\ \boxed{)}\ \boxed{=}\quad 99.90234375$

Approximately $100 - 99.9$ or 0.1 gram of Pu-236 is left after 10 half-lives.

Another form of the formula for $S_n$ can be developed and used when you don't know how many terms are in the series.

$$a_n = a_1 r^{n-1}$$
$$a_n \cdot r = a_1 r^{n-1} \cdot r \qquad \textit{Multiply each side by r.}$$
$$a_n r = a_1 r^n \qquad\qquad \textit{a}_1 r^{n-1} \cdot r = a_1 \cdot r^{n-1+1}$$

Thus, we can substitute $a_n r$ into any formula containing $a_1 r^n$.

$$S_n = \frac{a_1 - a_1 r^n}{1 - r} \qquad \textit{Sum of a geometric series}$$

$$S_n = \frac{a_1 - a_n r}{1 - r} \qquad \textit{Substitute } a_n r \textit{ for } a_1 r^n. \qquad \textit{Remember, } r \neq 1.$$

### Example 3

Find the sum of a geometric series for which $a_1 = 48$, $a_n = 3$, and $r = -\frac{1}{2}$.

Since we do not know the value of $n$, use $S_n = \frac{a_1 - a_n r}{1 - r}$.

$$S_n = \frac{a_1 - a_n r}{1 - r} = \frac{48 - 3\left(-\frac{1}{2}\right)}{1 - \left(-\frac{1}{2}\right)} \text{ or } 33$$

The sum is 33.

**Example 4**

**Find $a_1$ in a geometric series where $S_7 = 3279$ and $r = 3$.**

Use the sum formula and substitute in all values you know. Since we have $S_7$, we know $n = 7$.

$$S_7 = \frac{a_1 - a_1 r^7}{1 - r} \qquad S_n = \frac{a_1 - a_1 r^n}{1 - r}$$

$$3279 = \frac{a_1 - a_1 \cdot 3^7}{1 - 3} \qquad r = 3,\ S_7 = 3279$$

$$3279 = \frac{a_1(1 - 2187)}{-2} \qquad a_1 \text{ is a common factor.}$$

$$\frac{3279(-2)}{-2186} = a_1 \qquad \text{Solve for } a_1.$$

$$a_1 = 3$$

In Lesson 13-3, you learned that sigma notation can be used to express an arithmetic series. The same is true for a geometric series.

**Example 5**

**Write the terms of $\displaystyle\sum_{j=1}^{5} 2(4)^{j-1}$ and find the sum.**

$$\sum_{j=1}^{5} 2(4)^{j-1} = (2 \cdot 4^{1-1}) + (2 \cdot 4^{2-1}) + (2 \cdot 4^{3-1}) + (2 \cdot 4^{4-1}) + (2 \cdot 4^{5-1})$$
$$= 2(1) + 2(4) + 2(16) + 2(64) + 2(256)$$
$$= 2 + 8 + 32 + 128 + 512 \text{ or } 682$$

The sum is 682.

In Example 5, $a_1 = 2$, $n = 5$, and $r = 4$. The sum could also have been found by using the sum formula. So, $S_5 = \dfrac{2(1 - 4^5)}{1 - 4}$ or 682.

**Example 6**

**Use sigma notation to express $1 - 3 + 9 - 27 + 81 - 243$.**

This is the sum of the first six terms of a geometric series, so $n = 6$. The value of $r$ is $-3$ and $a_1 = 1$. Use the formula for the $n$th term of a sequence to find a general form of each term.

$$a_n = a_1 r^{n-1}$$
$$a_n = 1(-3)^{n-1} \qquad a_1 = 1,\ r = -3$$

The sum of the terms from $a_1$ to $a_6$ is expressed in sigma notation as $\displaystyle\sum_{n=1}^{6} (-3)^{n-1}$.

# CHECKING FOR UNDERSTANDING

**Communicating Mathematics**

Read and study the lesson to answer each question.

1. A geometric __?__ is the sum of the terms of a geometric sequence.

2. The sum of a geometric series can be found by two formulas, __?__ and __?__ .

3. The formula for the sum of a geometric series states that $r \neq 1$. Why is this stipulation included?

4. Suppose the first and fifth terms of a geometric series are known and you are asked to find the three geometric means. How is it possible to get two correct sets of answers?

**Guided Practice**

For each geometric series, state the first term, the common ratio, the last term, and the number of terms.

5. $9 - 18 + 36 - 72$

6. $3 + 1.5 + 0.75 + 0.375$

7. $a_1 + 8 + 32 + 128 + a_5$

8. $a_1 + 6 - 3 + \dfrac{3}{2} + a_5$

9. In a certain geometric series, $a_1 = 20$, $a_4 = -\dfrac{5}{16}$, and the number of terms is 5. Find the common ratio and the last term.

10. In a certain geometric series, $a_2 = 6$ and $a_4 = 24$. Find the first term, the common ratio, and the sixth term.

Find the sum of each geometric series.

11. $7 + 7 + 7 + \ldots$ to 9 terms

12. $2 + (-6) + 18 + \ldots$ to 6 terms

13. $3 + 6 + 12 + \ldots$ to 6 terms

14. $8 + 4 + 2 + \ldots$ to 6 terms

15. $\dfrac{1}{9} - \dfrac{1}{3} + 1 - \ldots$ to 5 terms

16. $1296 - 216 + 36 - \ldots$ to 5 terms

# EXERCISES

**Practice**

Find the sum of each geometric series described.

17. $16 + 16 + 16 + \ldots$ to 11 terms.

18. $75 + 15 + 3 + \ldots$ to 10 terms.

19. $a_1 = 5$, $r = 3$, $n = 12$

20. $a_1 = 256$, $r = 0.75$, $n = 9$

21. $a_1 = 7$, $r = 2$, $n = 14$

22. $a_1 = 12$, $a_5 = 972$, $r = -3$

23. $a_1 = 16$, $r = -\dfrac{1}{2}$, $n = 10$

24. $a_1 = 625$, $r = \dfrac{2}{5}$, $n = 8$

25. $a_1 = 1$, $a_5 = \dfrac{1}{16}$, $r = -\dfrac{1}{2}$

26. $a_1 = 243$, $r = -\dfrac{2}{3}$, $n = 5$

27. $a_1 = 343$, $a_4 = -1$, $r = -\dfrac{1}{7}$

28. $a_1 = 625$, $a_5 = 81$, $r = \dfrac{3}{5}$

29. $a_1 = 125$, $a_5 = \dfrac{1}{5}$, $r = \dfrac{1}{5}$

30. $a_1 = 4$, $a_6 = \dfrac{1}{8}$, $r = \dfrac{1}{2}$

**31.** $a_2 = 1.5$, $a_5 = 0.1875$, $n = 9$      **32.** $a_3 = \dfrac{3}{4}$, $a_6 = \dfrac{3}{32}$, $n = 6$

**33.** $a_3 = \dfrac{5}{4}$, $a_4 = -\dfrac{5}{16}$, $n = 6$      **34.** $a_2 = -12$, $a_5 = -324$, $n = 10$

**Use sigma notation to express each series.**

**35.** $2 - 6 + 18 - 54 + 162 - 486$

**36.** $243 - 162 + 108 - 72 + 48 - 32$

**Find $a_1$ for each geometric series described.**

**37.** $S_n = 244$, $r = -3$, $n = 5$      **38.** $S_n = 32$, $r = 2$, $n = 6$

**39.** $a_n = 324$, $r = 3$, $S_n = 484$      **40.** $S_n = 635$, $a_n = 320$, $r = 2$

**41.** $S_n = 15.75$, $r = 0.5$, $a_n = 0.25$      **42.** $S_n = 1022$, $r = 2$, $n = 9$

**Critical Thinking**

**43.** Use the formula for the $n$th term of a geometric sequence to show that, when $r = 1$, the sum of a geometric series is $S_n = n \cdot a_1$.

**Applications**

**44. Recreation**   One minute after it is released, a hot air balloon rose 80 feet. In each succeeding minute the balloon rose only 60% as far as it rose in the previous minute.

   **a.** Write a geometric sequence that describes the rise of the hot air balloon during the first 5 minutes.

   **b.** How far will the balloon rise in 6 minutes?

**45. Communications**   The teaching staff of Fairmeadow High School informs its members of school cancellation by telephone. The principal calls 2 teachers, each of whom in turn calls 2 other teachers, and so on. In order to inform the entire staff, 6 rounds of calls are made. Counting the principal, find how many people are on staff at Fairmeadow High.

**46. Meteorology**   A 5-day rain caused the Olentangy River to rise. After the first day, the river rose one inch. Each day the rise in the river tripled. How much had the river risen after 5 days?

**Mixed Review**

**47.** Find the missing terms of the geometric sequence $\underline{\ ?\ }$, $\underline{\ ?\ }$, 3, 9, 27. **(Lesson 13-4)**

**48. Biology**   The number of a certain type of bacteria can increase from 80 to 164 in 3 hours. Find the approximate value of $k$ in the growth formula. **(Lesson 12-9)**

**49.** Danny and Sonia run a lawn service during the summer. Sonia can cut the lawns of all customers in 15 hours. Danny can do the same in 12 hours. Sonia works alone for 8 hours, after which Danny joins her. How long will it take both of them to finish the job? **(Lesson 8-5)**

**50.** What are the values of $n$ and $m$ if $A_{n \times m}$ has an inverse? **(Lesson 4-4)**

**51.** Solve $\begin{cases} 3x - 7y = -1 \\ 3x + 7y = 13 \end{cases}$ using Cramer's rule. **(Lesson 3-3)**

# 13-6 Infinite Geometric Series

**Objective**

After studying this lesson, you should be able to:

- find the sum of an infinite geometric series.

**Application**

The first swing of a pendulum measures 25 cm. The lengths of the successive swings of the pendulum form the geometric sequence 25, 20, 16, 12.8, ....

Suppose this pendulum continues to swing back and forth indefinitely. Then the sequence shown above becomes an infinite geometric sequence.

The total distance the pendulum travels can be expressed as the **infinite geometric series**

$$25 + 20 + 16 + 12.8 + ....$$

In this series, $a_1 = 25$ and $r = \dfrac{20}{25}$ or 0.8. So the series can be expressed as

$$25 + 25(0.8)^1 + 25(0.8)^2 + 25(0.8)^3 + 25(0.8)^4 + 25(0.8)^5 + 25(0.8)^6 + ....$$

Use your calculator to *look for a pattern* in the values of $0.8^n$ as $n$ increases.

$$0.8^1 = 0.8 \qquad 0.8^{10} \approx 0.1073742 \qquad 0.8^{50} \approx 0.0000143 \qquad 0.8^{70} \approx 0.0000002$$

Use your calculator to *look for a pattern* as to how this affects $a_n$.

$$a_1 = 25 \qquad a_{10} \approx 3.3554432 \qquad a_{50} \approx 0.000446 \qquad a_{70} \approx 0.00000514$$

Use your calculator to *look for a pattern* in the values of $S_n$ as $n$ increases.

$$S_1 = 25 \qquad S_{10} = \frac{25 - 25(0.8^{10})}{0.2} \qquad S_{50} = \frac{25 - 25(0.8^{50})}{0.2} \qquad S_{70} = \frac{25 - 25(0.8^{70})}{0.2}$$
$$\approx 111.57823 \qquad\qquad \approx 124.99822 \qquad\qquad \approx 124.99998$$

*Remember that $|r| < 1$ means $-1 < r < 1$.*

The sum $S_n$ appears to approach 125 as $n$ becomes infinitely great. In an infinite geometric series where $|r| < 1$, as the value of $n$ increases, the value of each term approaches zero. This implies that the sum converges to some number. This can be analyzed by observing how this affects the sum formula for a geometric series.

$$S_n = \frac{a_1 - a_1 r^n}{1 - r}$$ as $n$ approaches infinity and $r^n$ approaches 0, $S_n = \dfrac{a_1}{1 - r}$.

This is known as the sum of an infinite geometric series.

---

*Sum of an Infinite Geometric Series*

**The sum, $S$, of an infinite geometric series where $-1 < r < 1$ is given by the following formula.**

$$S = \frac{a_1}{1 - r}$$

An infinite geometric series in which $|r| > 1$ does not have a sum. For example, consider the series $1 + 2 + 4 + 8 + 16 + ...$, where $a_1 = 1$ and $r = 2$. The terms of this series keep increasing. So the sum becomes greater with each additional term and never approaches any particular number.

**Example 1**

Find the total distance traveled by the pendulum as it comes to rest.

$S = 25 + 20 + 16 + 12.8 + ...$, with $r = 0.8$ and $a_1 = 25$

$$S = \frac{a_1}{1 - r} = \frac{25}{1 - 0.8} \text{ or } 125 \qquad a_1 = 25, r = 0.8$$

The pendulum travels 125 cm.

*How does this compare with the pattern on page 620?*

**Example 2**

Find the sum of the infinite geometric series $\dfrac{4}{3} - \dfrac{2}{3} + \dfrac{1}{3} - \dfrac{1}{6} + ....$

To find the value of $r$, divide any term by its preceding term.

$$r = -\frac{2}{3} \div \frac{4}{3}$$

$$= -\frac{2}{3} \cdot \frac{3}{4} \text{ or } -\frac{1}{2}$$

Since $|r| < 1$, you can use the formula $S = \dfrac{a_1}{1 - r}$.

$$S = \frac{\frac{4}{3}}{1 - \left(-\frac{1}{2}\right)} = \frac{\frac{4}{3}}{\frac{3}{2}} \text{ or } \frac{8}{9} \qquad \text{The sum is } \frac{8}{9}.$$

The sum of an infinite geometric series can be used to express a repeating decimal as a rational number in the form of $\dfrac{a}{b}$. Remember that repeating decimals such as $0.\overline{1}$ and $0.0\overline{43}$ represent 0.11111... and 0.0434343..., respectively.

**Example 3**

Express $0.\overline{1}$ as a rational number in the form $\dfrac{a}{b}$.

First express $0.\overline{1}$ as an infinite geometric series.

$0.\overline{1} = 0.1 + 0.01 + 0.001 + 0.0001 + ...$

Each term is $\dfrac{1}{10}$ of the preceding term, so $r = 0.1$. Also $a_1 = 0.1$.

$$S = \frac{a_1}{1 - r}$$

$$= \frac{0.1}{1 - 0.1} \qquad a_1 = 0.1, r = 0.1$$

$$= \frac{0.1}{0.9} \text{ or } \frac{1}{9} \qquad\qquad \text{The decimal } 0.\overline{1} \text{ can be expressed as } \frac{1}{9}.$$

The sum of an infinite geometric series can also be written using sigma notation. Since it is impossible to count the number of terms in an infinite series, the upper limit of the index is written as ∞, which is the symbol for **infinity**.

**Example 4**

Evaluate $\displaystyle\sum_{n=1}^{\infty} 36\left(-\frac{1}{3}\right)^{n-1}$.

$$\sum_{n=1}^{\infty} 36\left(-\frac{1}{3}\right)^{n-1} = 36\left(-\frac{1}{3}\right)^{1-1} + 36\left(-\frac{1}{3}\right)^{2-1} + 36\left(-\frac{1}{3}\right)^{3-1} + \ldots$$

$$= 36 - 12 + 4 - \frac{4}{3} + \frac{4}{9} + \ldots$$

Therefore, $a_1 = 36$ and $r = -\frac{1}{3}$. Now use the formula $s = \frac{a_1}{1-r}$.

$$S = \frac{36}{1-\left(-\frac{1}{3}\right)} = \frac{36}{\frac{4}{3}} \text{ or } 27$$

Therefore, $\displaystyle\sum_{n=1}^{\infty} 36\left(-\frac{1}{3}\right)^{n-1} = 27$.

**Example 5**

**APPLICATION**

**Physics**

To test its elasticity, a rubber ball is dropped into a 30-foot hollow tube that is calibrated so that the scientist can measure the height of each subsequent bounce. The scientist found that on each bounce, the ball rises to a height $\frac{2}{5}$ the height of the previous bounce. How far will the ball travel before it stops bouncing?

A diagram of the situation is helpful in exploring this problem.

This situation involves the sum of two infinite geometric series, one that contains the measures of the upward bounces and one that contains the measures of the downward bounces.

| Down | Up | Down | Up | Down |
|------|----|------|----|------|
| 30 | 12 | 12 | 4.8 | 4.8 |

Let $S_D$ be the downward sum or $30 + 12 + 4.8 + 1.92 + \ldots$.
Let $S_U$ be the upward sum or $12 + 4.8 + 1.92 + \ldots$.

$$S_D = \frac{30}{1 - 0.4} \qquad\qquad S_U = \frac{12}{1 - 0.4}$$

$$= \frac{30}{0.6} \text{ or } 50 \qquad\qquad = \frac{12}{0.6} \text{ or } 20$$

The ball will travel $50 + 20$ or 70 feet before it stops bouncing.

# CHECKING FOR UNDERSTANDING

**Communicating Mathematics**

Read and study the lesson to answer each question.

1. Why are the series in this lesson called infinite geometric series?

2. Refer to the patterns of numbers shown on page 620.
   a. Describe the value of $(0.8)^n$ as $n$ increases.
   b. Describe the value $a_n$ approaches as $n$ increases.
   c. Describe the value $S_n$ approaches as $n$ increases.

3. What does the symbol $\infty$ represent?

4. Under what condition can you find the sum of an infinite geometric series?

**Guided Practice**

Find $a_1$ and $r$ for each series. Then find the sum, if it exists.

5. $12 + 3 + \dfrac{3}{4} + \dfrac{3}{16} + \ldots$

6. $1 - 3 + 9 - 27 + \ldots$

7. $1 - \dfrac{1}{3} + \dfrac{1}{9} - \dfrac{1}{27} + \ldots$

8. $\dfrac{1}{2} + \dfrac{1}{3} + \dfrac{2}{9} + \dfrac{4}{27} + \ldots$

9. $48 + 16 + \dfrac{16}{3} + \dfrac{16}{9} + \ldots$

10. $1 + \dfrac{3}{2} + \dfrac{9}{4} + \dfrac{27}{8} + \ldots$

Use the sum of an infinite geometric series to express each decimal as a rational number in the form $\dfrac{a}{b}$.

11. $0.\overline{7}$

12. $0.\overline{73}$

13. $0.1\overline{52}$

14. $0.9\overline{3}$

# EXERCISES

**Practice**

Find the sum of each infinite geometric series, if it exists.

15. $a_1 = 7, r = -\dfrac{3}{4}$

16. $a_1 = 6, r = \dfrac{11}{12}$

17. $a_1 = 18, r = -\dfrac{2}{7}$

18. $\dfrac{1}{3} + \dfrac{1}{9} + \dfrac{1}{27} + \ldots$

19. $9 + 6 + 4 + \ldots$

20. $2 + 6 + 18 + \ldots$

21. $\dfrac{3}{4} + \dfrac{1}{2} + \dfrac{1}{3} + \ldots$

22. $1 - \dfrac{1}{4} + \dfrac{1}{16} - \ldots$

23. $12 - 4 + \dfrac{4}{3} - \ldots$

24. $a_1 = 27, r = -\dfrac{4}{5}$

25. $10 - \dfrac{5}{2} + \dfrac{5}{8} - \ldots$

26. $12 + 6 + 3 + \ldots$

27. $3 - 9 + 27 - \ldots$

28. $3 - 2 + \dfrac{4}{3} - \ldots$

29. $10 - 1 + 0.1 - \ldots$

Express each decimal as rational number in the form $\dfrac{a}{b}$.

30. $0.\overline{9}$

31. $0.3\overline{1}$

32. $0.\overline{410}$

33. $0.4\overline{5}$

**34. Physics** The end of a swinging pendulum 90 cm long travels 50 cm on its first swing. Each succeeding swing is 0.9 as long as the preceding one.

   **a.** Express the distance as an infinite geometric series.

   **b.** How far will the pendulum travel before coming to rest?

**Find the first three terms of each infinite geometric series.**

**35.** $S = 16$, $r = \dfrac{3}{4}$        **36.** $S = 28$, $r = -\dfrac{2}{7}$        **37.** $S = \dfrac{27}{4}$, $r = -\dfrac{1}{3}$

**Critical Thinking**

**38.** Explain how $S_n = \dfrac{a_1 - a_1 r^n}{1 - r}$ becomes $S = \dfrac{a_1}{1 - r}$ when $-1 < r < 1$ and $n$ approaches infinity. Give several real number examples to demonstrate your explanation.

**Applications**

**39. Aviation** A hot-air balloon rises 80 feet in the first minute of flight. If in each succeeding minute the balloon rises only 90% as far as in the previous minute, what will be its maximum altitude if it is allowed to rise without limit?

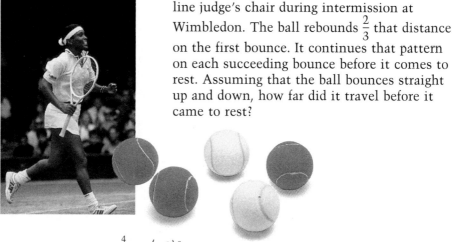

**40. Sports** A tennis ball fell 60 inches from the line judge's chair during intermission at Wimbledon. The ball rebounds $\dfrac{2}{3}$ that distance on the first bounce. It continues that pattern on each succeeding bounce before it comes to rest. Assuming that the ball bounces straight up and down, how far did it travel before it came to rest?

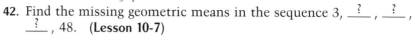

**Mixed Review**

**41.** Find the sum of $\displaystyle\sum_{s=1}^{4} 24\left(-\dfrac{1}{2}\right)^{s}$. **(Lesson 13-5)**

**42.** Find the missing geometric means in the sequence 3, __?__ , __?__ , __?__ , 48. **(Lesson 10-7)**

**43.** If $f(x) = 2x + 3$, $g(x) = x - 1$, and $h(x) = x^2 + 4$, find $[g \circ h \circ f](2)$. **(Lesson 10-7)**

**44. Retailing** A salesman receives $25 for every vacuum cleaner he sells. If he sells more than 10 vacuum cleaners, he will receive an additional $1.75 for each successive sale until he is paid a maximum of $46 per vacuum cleaner. How many must he sell to reach this maximum? **(Lesson 3-6)**

**45.** Solve $2x - 4(x + 2) = -2x - 8$. **(Lesson 1-3)**

**Journal**

Look up the word series in a dictionary. Describe the different definitions you find.

## 13-7 The Binomial Theorem

**Objectives**

After studying this lesson, you should be able to:

- expand powers of binomials using Pascal's triangle and the Binomial Theorem, and
- find specific terms of the binomial expansion.

You have observed patterns in both geometric and arithmetic sequences. Another pattern can be observed when binomials, such as $(a + b)$, are raised to successive powers. The patterns of the coefficients and the exponents of the terms can be described in the **Binomial Theorem.** But first, let us observe some of the patterns that appear when $(a + b)^n$ is expanded for $n = 0, 1, 2, 3,$ and 4.

$$(a + b)^0 = 1a^0b^0$$
$$(a + b)^1 = 1a^1b^0 + 1a^0b^1$$
$$(a + b)^2 = 1a^2b^0 + 2a^1b^1 + 1a^0b^2$$
$$(a + b)^3 = 1a^3b^0 + 3a^2b^1 + 3a^1b^2 + 1a^0b^3$$
$$(a + b)^4 = 1a^4b^0 + 4a^3b^1 + 6a^2b^2 + 4a^1b^3 + 1a^0b^4$$

*$a + b \neq 0$*

*Note that the coefficients are one. What is the pattern of the exponents of a and b in each line?*

**FYI** ...

Blaise Pascal (1623–1662) is best known for his accomplishments in science. Pascal's law is the principle used to operate hydraulic pumps, air compressors, and vacuum pumps.

Look at the exponents in each term of the expansion of $(a + b)^4$. Notice that the sum of the exponents is 4, the exponent of the binomial. Look at the number of terms in each expression. Notice the number of terms is one more than the exponent of the binomial. Here is a list of patterns seen in the expansion of $(a + b)^n$.

1. The exponent of $(a + b)^n$ is the exponent of $a$ in the first term and the exponent of $b$ in the last term.

2. In successive terms, the exponent of $a$ decreases by one. It is $n$ in the first term and zero in the last term.

3. In successive terms, the exponent of $b$ increases by one. It is zero in the first term and $n$ in the last term.

4. The sum of the exponents of each term is $n$.

5. The coefficients are symmetric. They increase at the beginning and decrease at the end of the expansion.

The coefficients form a pattern that is often displayed in a triangular formation. This is known as **Pascal's triangle.** Notice that each new row is formed by starting and ending with 1. Then each coefficient is the sum of the pair of coefficients above it in the previous row.

| | | | | | | | | | | | | |
|---|---|---|---|---|---|---|---|---|---|---|---|---|
| $(a + b)^0$ | | | | | | 1 | | | | | | |
| $(a + b)^1$ | | | | | 1 | | 1 | | | | | |
| $(a + b)^2$ | | | | 1 | | 2 | | 1 | | | | |
| $(a + b)^3$ | | | 1 | | 3 | | 3 | | 1 | | | |
| $(a + b)^4$ | | 1 | | 4 | | 6 | | 4 | | 1 | | |
| $(a + b)^5$ | 1 | | 5 | | 10 | | 10 | | 5 | | 1 | |
| $(a + b)^6$ | 1 | 6 | | 15 | | 20 | | 15 | | 6 | | 1 |

**Example 1**

Use the pattern in Pascal's triangle to write $(a + b)^7$ in expanded form.

The next line of Pascal's triangle is

> 1  7  21  35  35  21  7  1.    *($(a + b)^7$ has 8 terms.*

$(a + b)^7$
$$= 1a^7b^0 + 7a^6b^1 + 21a^5b^2 + 35a^4b^3 + 35a^3b^4 + 21a^2b^5 + 7a^1b^6 + 1a^0b^7$$
$$= a^7 + 7a^6b + 21a^5b^2 + 35a^4b^3 + 35a^3b^4 + 21a^2b^5 + 7ab^6 + b^7$$

Another way to show the coefficients is by writing them in terms of the previous term, as you would in a sequence.

| | | | | | |
|---|---|---|---|---|---|
| $(a + b)^0$ | | | 1 | | |
| $(a + b)^1$ | | 1 | $\frac{1}{1}$ | | |
| $(a + b)^2$ | 1 | $\frac{2}{1}$ | $\frac{2 \cdot 1}{1 \cdot 2}$ | | |
| $(a + b)^3$ | 1 | $\frac{3}{1}$ | $\frac{3 \cdot 2}{1 \cdot 2}$ | $\frac{3 \cdot 2 \cdot 1}{1 \cdot 2 \cdot 3}$ | |
| $(a + b)^4$  1 | $\frac{4}{1}$ | $\frac{4 \cdot 3}{1 \cdot 2}$ | $\frac{4 \cdot 3 \cdot 2}{1 \cdot 2 \cdot 3}$ | $\frac{4 \cdot 3 \cdot 2 \cdot 1}{1 \cdot 2 \cdot 3 \cdot 4}$ | |

*Eliminate common factors that are shown in color. The coefficients are symmetrical.*

This sequence pattern provides the coefficients of a binomial expansion without writing the previous rows of coefficients. This pattern is summarized in the **Binomial Theorem.**

*The Binomial Theorem*

**If $n$ is a positive integer, then**
$$(a + b)^n = 1a^nb^0 + \frac{n}{1}a^{n-1}b^1 + \frac{n(n-1)}{1 \cdot 2}a^{n-2}b^2 + \ldots + 1a^0b^n.$$

**Example 2**

Use the Binomial Theorem to express $(m + n)^8$ in expanded form.

The expansion will have nine terms. Find the first five terms, using the Binomial Theorem and then use symmetry to find the remaining terms.

$(m + n)^8$
$$= m^8n^0 + \frac{8}{1}m^7n^1 + \frac{8 \cdot 7}{1 \cdot 2}m^6n^2 + \frac{8 \cdot 7 \cdot 6}{1 \cdot 2 \cdot 3}m^5n^3 + \frac{8 \cdot 7 \cdot 6 \cdot 5}{1 \cdot 2 \cdot 3 \cdot 4}m^4n^4 + \ldots$$
$$= m^8 + 8m^7n + 28m^6n^2 + 56m^5n^3 + 70m^4n^4 + \ldots$$
$$= m^8 + 8m^7n + 28m^6n^2 + 56m^5n^3 + 70m^4n^4 + 56m^3n^5 +$$
$$28m^2n^6 + 8mn^7 + n^8$$

*Note that in terms having the same coefficients the exponents are reversed, as in $28m^6n^2$ and $28m^2n^6$.*

The patterns in the products of the coefficients in Example 2 are parts of special products called factorials. The product $4 \cdot 3 \cdot 2 \cdot 1$ can be expressed as $4!$ and is read *4 factorial*. By definition, $0! = 1$. For all positive values of $n$, $n! = n(n - 1)(n - 2)(n - 3) \ldots (1)$.

**Example 3**

Evaluate $\dfrac{10!}{4!6!}$.

$\dfrac{10!}{4!6!} = \dfrac{10 \cdot 9 \cdot 8 \cdot 7 \cdot 6 \cdot 5 \cdot 4 \cdot 3 \cdot 2 \cdot 1}{4 \cdot 3 \cdot 2 \cdot 1 \cdot 6 \cdot 5 \cdot 4 \cdot 3 \cdot 2 \cdot 1}$

*Note that $10! = 10 \cdot 9 \cdot 8 \cdot 7 \cdot 6!$.*

*So $\dfrac{10!}{4!6!} = \dfrac{10 \cdot 9 \cdot 8 \cdot 7 \cdot 6!}{4!6!}$ or $\dfrac{10 \cdot 9 \cdot 8 \cdot 7}{4 \cdot 3 \cdot 2 \cdot 1}$*

$= \dfrac{10 \cdot 9 \cdot 8 \cdot 7}{4 \cdot 3 \cdot 2 \cdot 1}$ or $210$

You can also evaluate this expression by using the $\boxed{x!}$ key on your calculator.

*The $\boxed{x!}$ key may be the second function of another calculator key.*

ENTER:  10 $\boxed{x!}$ $\boxed{\div}$ $\boxed{(}$ 4 $\boxed{x!}$ $\boxed{x}$ 6 $\boxed{x!}$ $\boxed{)}$ $\boxed{=}$  $\mathit{210}$

In the Example 2, notice that products, such as $\dfrac{8 \cdot 7 \cdot 6}{1 \cdot 2 \cdot 3}$, can be written as a quotient of factorials. In this case, $\dfrac{8 \cdot 7 \cdot 6}{1 \cdot 2 \cdot 3} = \dfrac{8!}{3!5!}$. Using this same method, we can rewrite any binomial expansion using factorials.

$$(m + n)^8 = \dfrac{8!}{0!8!}m^8 + \dfrac{8!}{1!7!}m^7n + \dfrac{8!}{2!6!}m^6n^2 + \dfrac{8!}{3!5!}m^5n^3 + \dfrac{8!}{4!4!}m^4n^4 +$$
$$\dfrac{8!}{5!3!}m^3n^5 + \dfrac{8!}{6!2!}m^2n^6 + \dfrac{8!}{7!1!}m^1n^7 + \dfrac{8!}{8!0!}n^8$$

This same pattern can be used to write the expansion using sigma notation.

$$(m + n)^8 = \sum_{k=0}^{8} \dfrac{8!}{k!(8 - k)!}m^{8-k}n^k$$

The Binomial Theorem can also be written both in factorial notation and in sigma notation.

$$(a + b)^n = \dfrac{n!}{0!(n - 0)!}a^n + \dfrac{n!}{1!(n - 1)!}a^{n-1}b^1 + \dfrac{n!}{2!(n - 2)!}a^{n-2}b^2 + \ldots$$
$$= \sum_{k=0}^{n} \dfrac{n!}{k!(n - k)!}a^{n-k}b^k$$

## Example 4

**Express $(2s + t)^4$ using sigma notation. Then write the expansion.**

$$(2s + t)^4 = \sum_{k=0}^{4} \frac{4!}{k!(4-k)!} (2s)^{4-k}t^k \qquad \textit{Now construct each term.}$$

$$= \frac{4!}{0!(4-0)!}(2s)^{4-0}t^0 + \frac{4!}{1!(4-1)!}(2s)^{4-1}t^1 + \frac{4!}{2!(4-2)!}(2s)^{4-2}t^2 +$$

$$\frac{4!}{3!(4-3)!}(2s)^{4-3}t^3 + \frac{4!}{4!(4-4)!}(2s)^{4-4}t^4$$

$$= \frac{4\cdot 3\cdot 2\cdot 1}{1\cdot 4\cdot 3\cdot 2\cdot 1}(2s)^4 + \frac{4\cdot 3\cdot 2\cdot 1}{1\cdot 3\cdot 2\cdot 1}(2s)^3 t + \frac{4\cdot 3\cdot 2\cdot 1}{2\cdot 1\cdot 2\cdot 1}(2s)^2 t^2 +$$

$$\frac{4\cdot 3\cdot 2\cdot 1}{3\cdot 2\cdot 1\cdot 1}(2s)t^3 + \frac{4\cdot 3\cdot 2\cdot 1}{4\cdot 3\cdot 2\cdot 1\cdot 1}t^4$$

$$= 16s^4 + 32s^3 t + 24s^2 t^2 + 8st^3 + t^4$$

Sometimes a particular term in the expansion of a binomial is needed. Notice that in the sigma notation form of the Binomial Theorem, $k$ is 0 for the first term, 1 for the second term, and so on. In general, the value of $k$ is always one less than the number of the term you are seeking.

## Example 5

**Find the sixth term of $(p + q)^{12}$.**

First use the Binomial Theorem to write the general form of the expansion.

$$(p + q)^{12} = \sum_{k=0}^{12} \frac{12!}{k!(12-k)!} p^{12-k}q^k$$

*In the sixth term, $k = 5$ since $k$ starts at 0.*

The sixth term, $\dfrac{12!}{5!(12-5)!} p^{12-5}q^5$, is $\dfrac{12\cdot 11\cdot 10\cdot 9\cdot 8}{5\cdot 4\cdot 3\cdot 2\cdot 1} p^7q^5$ or $792p^7q^5$.

The Binomial Theorem can also be used to compute probability.

## Example 6

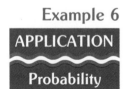

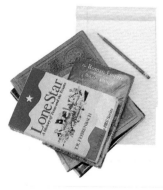

**Sharon didn't study for her U.S. history quiz so she had to guess at all 10 true/false questions. What is the probability that exactly half of her answers are correct?**

Let $p$ be the probability that true is the correct answer and $q$ be the probability that false is the correct answer. Since there are 10 questions, we can use the Binomial Theorem to find any term in the expansion of $(p + q)^{10}$.

$$(p + q)^{10} = \sum_{k=0}^{10} \frac{10!}{k!(10-k)!} p^{10-k}q^k$$

Getting half the answers correct means that Sharon would get exactly 5 correct and 5 incorrect. So, the probability can be computed using the term where $k = 5$, the sixth term.

$$\frac{10!}{5!(10-5)!} p^5q^5 \text{ or } 252p^5q^5$$

Each question has two possible answers. Only one is correct. So the probability that *true* is the correct answer is 1 out of 2 or $\frac{1}{2}$. Likewise, the probability that *false* is the correct answer is also $\frac{1}{2}$. Evaluate the expression for the sixth term if $p = \frac{1}{2}$ and $q = \frac{1}{2}$.

$$252p^5q^5 \quad \blacktriangleright \quad 252\left(\frac{1}{2}\right)^5\left(\frac{1}{2}\right)^5 = 252\left(\frac{1}{32}\right)\left(\frac{1}{32}\right) \text{ or } \frac{63}{256}$$

The probability that Sharon guesses exactly 5 answers correctly is $\frac{63}{256}$ or about 25%.

# CHECKING FOR UNDERSTANDING

**Communicating Mathematics**

Read and study the lesson to answer each question.

1. Explain how to form additional rows of Pascal's triangle.
2. Without writing the expansion, tell how many terms are in the expansion of $(w + z)^{12}$.
3. In your own words, explain how the Binomial Theorem works.
4. If you guess at the answer to a true/false question, what is the probability you will guess correctly?

**Guided Practice**

Use your calculator to evaluate each expression.

5. $9!$
6. $12!$
7. $\dfrac{10!}{8!}$
8. $\dfrac{31!}{28!}$
9. $\dfrac{6!}{3!}$
10. $\dfrac{10!}{4!6!}$

State the number of terms in the expansion of each expression. Then find the fourth term of that expansion.

11. $(a + 3)^4$
12. $(k + m)^7$
13. $(b - z)^5$

# EXERCISES

**Practice**

Expand each binomial.

14. $(r + s)^6$
15. $(y + p)^7$
16. $(x - y)^3$
17. $(r - m)^6$
18. $(2m + y)^5$
19. $(3r + y)^4$
20. $(2b + x)^6$
21. $(2x + 3y)^4$
22. $(3x - 2y)^5$
23. $(2m - 3)^6$
24. $\left(2 + \dfrac{x}{2}\right)^6$
25. $\left(\dfrac{y}{3} + 3\right)^6$

Find the indicated term of each expansion.

26. seventh term of $(x - y)^{15}$
27. fifth term of $(x + y)^7$
28. fourth term of $(2x + 3y)^9$
29. eighth term of $(3a - 5b)^{11}$

**Simplify.**

**30.** $\dfrac{k!}{(k-1)!}$      **31.** $\dfrac{(k+3)!}{(k+2)!}$      **32.** $\dfrac{3!4(k-3)!}{(k-2)!}$      **33.** $(k+1)!(k+2)$

**Critical Thinking**

**34.** Ball bearings fall down a chute toward a tray. As they fall, they branch out. At each branch, there is an equal chance to go either way.

    **a.** If 16 ball bearings go through the chute, how many will be in the tray for each branch?

    **b.** If 64 ball bearings go through the chute, how many will be in the tray for each branch?

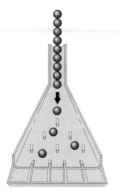

**Applications**

**35. Sports** In the high school football playoffs, Kyle had a success rate of 2 out of 3 passes completed in the first quarter. What is the likelihood of exactly 5 completions in six attempts?

**36. Cooking** In cooking class, Sonja had a success rate of 4 out of 5 for her souffles. One out of 5 collapse. For a party for her parents, she is preparing 6 souffles. What is the likelihood of exactly 4 of them being successes?

**Computer**

**37.** The BASIC program at the right generates the line of coefficients in Pascal's triangle for $(a+b)^n$. You must input the value of $n$ when running the program.

**Use the program to express each binomial in expanded form.**

    **a.** $(a+b)^4$      **b.** $(a+b)^{12}$

    **c.** $(x-y)^6$      **d.** $(x-y)^{10}$

```
  1   PRINT "ENTER THE VALUE
        OF N"
  5   INPUT Y
 10   FOR N = 0 TO Y
 20   FOR R = 0 TO N
 30   LET C = 1
 40   IF N < N - R + 1 THEN 80
 50   FOR X = N TO N - R + 1
        STEP - 1
 60   LET C = C*X/(N - X + 1)
 70   NEXT X
 80   PRINT C;" ";
 90   NEXT R
100   PRINT
110   NEXT N
120   END
```

**Mixed Review**

**38.** Find the sum of the infinite series $\dfrac{2}{3} + \dfrac{1}{3} + \dfrac{1}{6} + \ldots$. **(Lesson 13-6)**

**39. Finance** Suppose a pilgrim ancestor of Kevin White left $150 in a savings account and interest was compounded continuously at 4%. If the account is now worth $24,000,000, how long ago was the account started? Use $A = Pe^{rt}$. **(Lesson 12-8)**

**40.** Simplify $\dfrac{1}{2x-3} - \dfrac{1}{3x-2}$. **(Lesson 11-4)**

**41.** For $f(x) = 7x^5 + 4x^4 - 3x^3 - 2x^2 + 7x + 1$, state the number of positive real zeros, negative real zeros, and imaginary zeros. **(Lesson 10-3)**

*Portfolio*

Select an item from your work in this chapter that shows your creativity and place it in your portfolio.

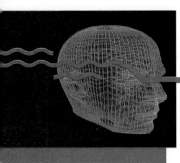

# Technology

## Amortization

**BASIC**
▶ **Spreadsheets**
**Software**

Mortgages for homes and other consumer loans are repaid in a series of equal payments made over a period of time. This is called **amortization.** The money from each of the payments is divided into an interest payment and a principal payment. The interest due for the period since the last payment is paid first, and then the balance of the payment goes toward reducing the principal. The spreadsheet program below constructs an amortization schedule for a two-year loan.

| AMORTIZATION SCHEDULE | | | |
|---|---|---|---|
| | A | B | C | D |
| 1 | | LOAN AMOUNT | PAYMENT | INTEREST RATE |
| 2 | | | | |
| 3 | PAYMENT | INTEREST PAID | PRINCIPAL PAID | BALANCE DUE |
| 4 | 0 | 0 | 0 | B2 |
| 5 | 1 | D2/12*D4 | C2-B5 | D4-C5 |
| 6 | 2 | D2/12*D5 | C2-B6 | D5-C6 |
| 28 | 24 | D2/12*D27 | D27 | $0 |

Sondra is taking out a two-year loan to buy a used car. The loan amount is $2500, the monthly payment is $115.94, and the interest rate is 10.5%. A portion of the amortization schedule for her loan is shown below.

```
= = = = = = = = = = = = = = = = = = = = = = = = = = = = = = = = = = = = = = = =
                  AMORTIZATION SCHEDULE
= = = = A = = = = = = B = = = = = = = = = C = = = = = = = = = D = = = =
 1              LOAN AMOUNT       PAYMENT        INTEREST RATE
 2                2500            115.94              .105
 3    PAYMENT   INTEREST PAID   PRINCIPAL PAID     BALANCE DUE
 4       0            0               0              2500
 5       1          21.88           94.06          2405.94
 6       2          21.05           94.89          2311.05
28      24           1.01          114.92             $0
```

# EXERCISES

1. Explain why the amount of interest paid per months grows smaller.

2. Explain how you could modify the spreadsheet program to make an amortization table for a 30-year mortgage.

# SUMMARY AND REVIEW

## VOCABULARY

Upon completing this chapter you should be
familiar with the following terms:

| | | | |
|---|---|---|---|
| arithmetic means | **599** | **615** | geometric series |
| arithmetic sequence | **596** | **605** | index of summation |
| arithmetic series | **602** | **620** | infinite geometric series |
| Binomial Theorem | **625** | **622** | infinity ($\infty$) |
| common difference | **596** | **597** | $n$th term |
| common ratio | **608** | **625** | Pascal's triangle |
| Fibonacci sequence | **594** | **597** | recursive formula |
| geometric means | **610** | **605** | sigma (or summation) notation ($\Sigma$) |
| geometric sequence | **608** | **596** | term |

## SKILLS AND CONCEPTS

### OBJECTIVES AND EXAMPLES

Upon completing this chapter, you should
be able to:

- find the $n$th term of an arithmetic
  sequence.   (**Lesson 13-2**)

  The $n$th term of an arithmetic sequence
  with first term $a_1$ and common difference
  $d$ is given by $a_n = a_1 + (n - 1)d$.

- find the position of a term in an
  arithmetic sequence.   (**Lesson 13-2**)

  $-3$ is what term of 7, 5, 3 ...?
  $-3 = 7 + (n - 1)(-2)$
  $6 = n$      $-3$ is the 6th term.

- find the arithmetic means of an
  arithmetic sequence.   (**Lesson 13-2**)

  Find the two arithmetic means between
  4 and 25.
  $25 = 4 + (4 - 1)d$      So, $7 = d$.
  $4 + 7 = 11$      $11 + 7 = 18$
  The arithmetic means are 11 and 18.

### REVIEW EXERCISES

Use these exercises to review and prepare
for the chapter test.

**Find the indicated term in each arithmetic
sequence.**

1. $a_5$ for 6, 14, 22, ...   2. $a_{22}$ for $-5$, 2, 9, ...
3. $a_1 = 5$, $d = -2$,   4. $a_1 = -2$,
   $n = 9$              $d = -3$, $n = 15$

**Complete each statement.**

5. 72 is the __?__ th term of $-5$, 2, 9, ...
6. $-37$ is the __?__ th term of 1, $-1$, $-3$, $-5$, ...
7. 49 is the __?__ th term of 4, 9, 14, ...

**Find the arithmetic means in each sequence.**

8. 12, ___ , ___ , 4

9. $-7$, ___ , ___ , ___ , 9

10. ___ , 6, ___ , ___ , $-3$, ___

■ find the sums of arithmetic series and find specific terms.   (**Lesson 13-3**)

The sum $S_n$ of the first $n$ terms of an arithmetic series is given by

$S_n = \frac{n}{2}(a_1 + a_n)$.

**Find $S_n$ for each arithmetic series.**

11. $a_1 = 12$, $a_n = 117$, $n = 36$

12. $4 + 10 + 16 + \ldots + 106$

13. Evaluate $\sum_{n=2}^{29} (3n + 1)$.

14. Find the first three terms of an arithmetic series if $a_1 = 3$, $a_n = 24$, and $S_n = 108$.

---

■ find the $n$th term of a geometric sequence and any missing geometric means.   (**Lesson 13-4**)

The $n$th term $a_n$ of a geometric sequence with first term $a_1$ and common ratio $r$ is given by $a_n = a_{n-1}r$ or $a_n = a_1 r^{n-1}$.

**Find the indicated term in each geometric sequence.**

15. $a_6$ for $\frac{2}{3}, \frac{4}{3}, \frac{8}{3}, \ldots$

16. $a_5$ if $a_1 = 7$ and $r = 3$

17. the next two terms of $4, -12, 36, \ldots$

18. Find the geometric means of $7.5$, ___, ___, ___, $120$.

---

■ find sums of geometric series. (**Lesson 13-5**)

The sum, $S_n$, of the first $n$ terms of a geometric series is given by $S_n = \frac{a_1 - a_1 r^n}{1 - r}$ or $S_n = \frac{a_1(1 - r^n)}{1 - r}$; $r \neq 1$.

**Find the sum of each geometric series described.**

19. $a_1 = 12$, $r = 3$, $n = 5$

20. $a_1 = 625$, $a_n = 16$, $r = \frac{2}{5}$

21. $a_1 = 4$, $r = -\frac{1}{2}$, $n = 6$

---

■ find the first term of a described geometric series.   (**Lesson 13-5**)

Find $a_1$ if $S_5 = 2.75$ and $r = -2$.

$\frac{11}{4} = \frac{a_1[1 - (-2)^5]}{1 - (-2)}$

$\frac{11}{4} = \frac{a_1(33)}{3}$

$\frac{1}{4} = a_1$

**For the geometric series, find $a_1$.**

22. $S_n = 1031$, $r = \frac{2}{5}$, $n = 5$

23. $S_n = 30$, $n = 4$, $r = -2$

24. $S_n = -61$, $n = 5$, $r = -1$

---

■ find the sum of an infinite geometric series.   (**Lesson 13-6**)

The sum, $S$, of an infinite geometric series where $-1 < r < 1$ is given by

$S = \frac{a_1}{1 - r}$.

**Find the sum of each infinite geometric series, if it exists.**

25. $a_1 = -2$ and $r = -\frac{5}{8}$

26. $\frac{1}{8} - \frac{3}{16} + \frac{9}{32} - \frac{27}{64} + \ldots$

27. Evaluate $\sum_{n=1}^{\infty} \frac{1}{2}\left(\frac{1}{3}\right)^{n-1}$.

28. Express $0.\overline{09}$ as a rational number in the form $\frac{a}{b}$.

■ expand powers of binomials using Pascal's triangle and the Binomial Theorem. **(Lesson 13-7)**

If $n$ is a positive integer, then

$$(a + b)^n = \sum_{k=0}^{n} \frac{n!}{k!(n-k)!} a^{n-k}b^k$$

**Expand each binomial.**

**29.** $(a + b)^3$      **30.** $(x - 2)^4$

**31.** $(3r + s)^5$

■ find specific terms of the binomial expansion. **(Lesson 13-7)**

The general form of the expansion of $(c + d)^{11}$ is

$$\sum_{k=0}^{11} \frac{11!}{k!(11-k)!} a^{11-k}b^k$$

**Find the indicated term.**

**32.** fourth term of $(x + 2y)^6$

**33.** second term of $(4x - 5)^{10}$

# APPLICATIONS AND CONNECTIONS

**34. Number Theory** The Leibniz series is $\frac{\pi}{4} = 1 + \left(-\frac{1}{3}\right) + \frac{1}{5} + \left(-\frac{1}{7}\right) + \frac{1}{9} + \dots + \frac{(-1)^{n-1}}{2n-1} + \dots$ . Use the first 10 terms of the series and your calculator to find a decimal approximation of $\pi$. **(Lesson 13-1)**

**35. Physics** A rocket rises 40 feet in the first second, 60 feet in the second second, and 80 feet in the third second. If it continues to rise at this rate, how many feet will it rise in the 10th second? **(Lesson 13-2)**

**36. Recreation** One minute after it is released, a gas-filled balloon rises 100 feet. In each succeeding minute the balloon rises only 50% as far as it rose in the previous minute. How far will the balloon rise in 5 minutes? **(Lesson 13-3)**

**37. Business** Mr. Olsen invested in computer equipment worth $900,000. The equipment depreciates at the rate of 25% per year of the previous year's value. What will be the value of his equipment at the end of four years? **(Lesson 13-5)**

**38. Fine Arts** A layered sculpture is arranged so that there are 5 diamonds on the top design layer, 7 diamonds on the second layer, 9 diamonds on the third layer, and so on. How many diamonds are on the twentieth layer? **(Lesson 13-5)**

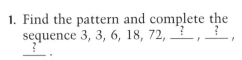
1. Find the pattern and complete the sequence 3, 3, 6, 18, 72, ___?___ , ___?___ , ___?___ .

2. How many integers between 26 and 415 are multiples of 9?

3. Find the next four terms of the arithmetic sequence 42, 37, 32, ....

4. Find the next two terms of the geometric sequence $\frac{1}{81}, \frac{1}{27}, \frac{1}{9}, \ldots$

5. Find the 27th term of an arithmetic sequence if $a_1 = 2$ and $d = 6$.

6. Find the sixth term of a geometric sequence if $a_1 = 5$ and $r = -2$.

7. Find the sum of the arithmetic series where $a_1 = 7$, $n = 31$, and $a_n = 127$.

8. Find the sum of the geometric series where $a_1 = 125$, $r = \frac{2}{5}$, and $n = 4$.

9. Find the three arithmetic means between $-4$ and 16.

10. Find the two geometric means between 7 and 189.

**Find each sum.**

11. $\displaystyle\sum_{k=3}^{15} (14 - 2k)$

12. $\displaystyle\sum_{n=1}^{\infty} \frac{1}{3}(-2)^{n-1}$

13. Find the sum of the series $91 + 85 + 79 + \ldots + (-29)$.

14. Find the sum of the geometric series $12 - 6 + 3 - \frac{3}{2} + \ldots$

15. Express $0.3\overline{2}$ as a rational number in the form $\frac{a}{b}$.

16. Expand $(2s - 3t)^5$.

17. Find the third term of $(x + y)^8$

18. **Nature**   The counterclockwise and clockwise spirals of a sunflower can be described by the sixth and seventh terms of the Fibonacci sequence. What are these two terms?

19. **Physics**   A vacuum pump removes $\frac{1}{7}$ of the air from a jar on each stroke of its piston. What percent of the air remains after four strokes of the piston?

20. **Design**   A landscaper is designing a wall of white brick and red brick. The pattern starts with 20 red bricks on the bottom row. Each row above it contains 3 fewer red bricks than the preceding row. If the top row contains no red bricks, how many rows are there and how many red bricks were used?

**Bonus**
A side of an equilateral triangle is 20 inches long. The midpoints of its sides are joined to form a smaller equilateral triangle. If this process is continued infinitely, find the sum of the perimeters of the triangles.

# 14

# Statistics

## CHAPTER OBJECTIVES

In this chapter you will:

- make bar graphs, line graphs, circle graphs, line plots, stem-and-leaf plots, and box-and-whisker plots.
- find the median, mean, mode, range, quartiles, interquartile range, standard deviation, and outliers of sets of data.
- solve problems involving normally distributed data.

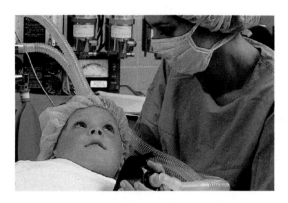

Anesthetics remain in the body long after a patient is awake. What concentration of this anesthetic should be kept in the patient's bloodstream during surgery?

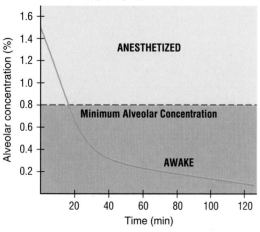

## CAREERS IN ANESTHESIOLOGY

What kind of doctor spends the most time in the operating room? It isn't surgeons; it isn't even obstetricians. It's anesthesiologists. The doctor who gives the patient anesthesia ("relief from pain") really runs the operating room.

An anesthesiologist must be both a doctor with a broad knowledge of surgery and a clinical pharmacologist. More than any other doctor, the anesthesiologist must be a master of *pharmacodynamics*, the science dealing with reactions between drugs and living systems. For surgery to proceed safely, the patient must be in no pain or shock with all systems operating normally. However, a surgical operation is a violent physical invasion. The body could be excused for reacting as though attacked. The job of the anesthesiologist is to prevent such a reaction.

In most surgeries, he or she renders the patient unconscious as well as anesthetized. That means the anesthesiologist is then responsible for the patient's breathing, heartbeat, and all other life functions. He or she must thoroughly know the patient's medical history, the operation and its risks. Calculating and giving correct dosages of anesthetics is key to surgical success. In fact, it is the anesthesiologist who is legally responsible for the anesthetics given. That underlies the life-or-death consequences of mathematical accuracy.

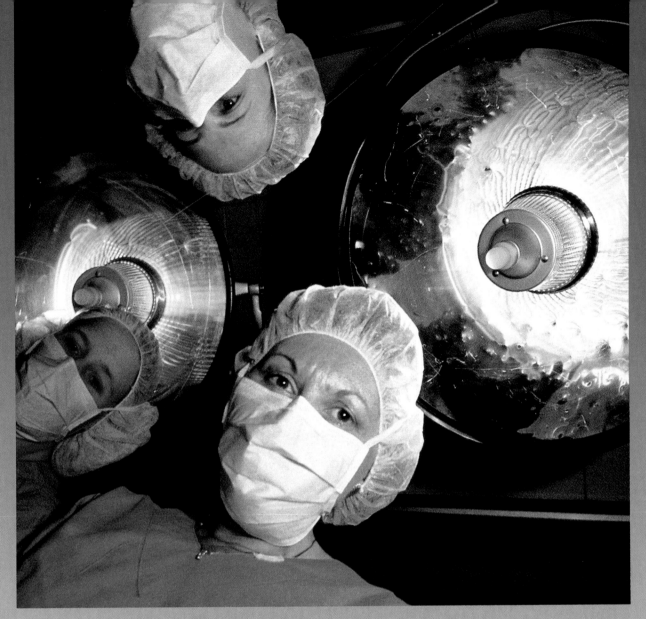

# MORE ABOUT ANESTHESIOLOGY

## Degree Required:
- Medical Doctor plus five years of medical residency

## Some anesthesiologists like:
- good working conditions
- the high salaries and prestige
- the satisfaction they derive from helping others

## Related Math Subjects:
- Advanced Algebra
- Trigonometry
- Geometry
- Statistics/Probability

## Some anesthesiologists dislike:
- being responsible for the life of a patient
- the long preparation to enter the field
- having to be available at all times

For more information on the various careers available in the field of anesthesiology, write to:

American Society of Anesthesiologists
515 Busse Highway
Park Ridge, IL 60068

# 14-1 Problem-Solving Strategy: Make a Graph

**Objective**

After studying this lesson, you should be able to:

■ solve problems by organizing data and making graphs.

**Application**

Data is easier to read and interpret when it is organized. One way to organize data is by using tables. The following table shows the changes in population in some cities from the 1980 census to the 1990 census.

| City | 1990 population | Change from 1980 |
|------|-----------------|------------------|
| Boise, ID | 123,059 | +20.4% |
| Charleston, WV | 56,012 | −12.4% |
| Detroit, MI | 970,156 | −19.4% |
| Kansas City, MO | 427,799 | −4.5% |
| Louisville, KY | 265,660 | −11.1% |
| Newark, NJ | 260,097 | −27.3% |
| Oklahoma City, OK | 441,154 | +9.2% |
| Portland, ME | 64,084 | +4.1% |
| Sioux Falls, SD | 100,281 | +23.3% |
| Wilmington, DE | 70,278 | +0.1% |

You can use the information in the table to quickly answer questions like:

■ Which city had the greatest percent change in population?   *Newark, NJ*
■ Which city had very little change in population?   *Wilmington, DE*
■ Which city has the greatest population?   *Detroit, MI*

The census, which provided the information for this table, is a survey that polls all residents of the United States. Some surveys only poll a **sample** of a population. For a survey to correctly represent the larger group, a representative sample that correctly represents the group should be taken randomly.

The information gathered from a survey can be organized in a table or in different types of graphs. Study the graphs in the following examples.

Bar graphs are useful for showing how quantities compare. Circle graphs are best for showing how parts are related to the whole.

## Example 1

APPLICATION

Sports

**Use the data in the table to draw a bar graph and a circle graph.**

**Injuries to students 5–24 years old in organized sports programs (1988–1989)**

| Sport | Football | Basketball | Soccer | Baseball | Gymnastics |
|---|---|---|---|---|---|
| Number of Injuries | 83,378 | 75,565 | 22,415 | 19,106 | 11,904 |

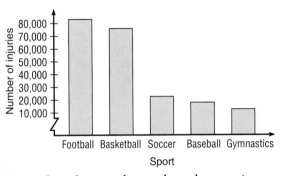

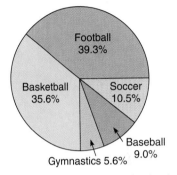

To make a bar graph, graph each quantity on the horizontal axis with the vertical axis showing the range.

To make a circle graph, find the percentage of the whole that each quantity represents. Then find the number of degrees each central angle should have.

Line graphs are usually used for showing changes over time. They make it easy to identify trends.

## Example 2

APPLICATION

Education

**Use the data in the table to draw a line graph.**

**High School Graduates (in thousands)**

| Year | 1960 | 1965 | 1970 | 1975 | 1980 | 1985 | 1990 |
|---|---|---|---|---|---|---|---|
| Number | 1864 | 2665 | 2896 | 3140 | 3058 | 2683 | 2793 |

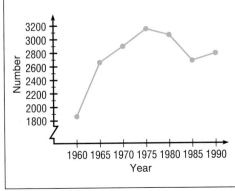

To make a line graph, plot each ordered pair on the graph. Then connect successive points with lines.

# CHECKING FOR UNDERSTANDING

**Communicating Mathematics**

Read and study the lesson to answer these questions.

1. Describe a situation where you would use each type of graph—bar, circle, and line.

2. The circle graph at the right represents the cost of a $4 paperback book broken down by type. How much of the $4 goes to the bookstore?

3. Would a daytime telephone poll be the best method for taking a survey on career choices? Why or why not?

4. Look through a newspaper or magazine to find different types of graphs. What type of graphs are used most often? Do you think the type of graph they chose is the best one for the information?

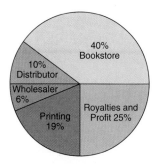

**Guided Practice**

5. Use the data in the table to answer each of the following. The figures are given in millions of people.

| Employment Status | 1965 | 1975 | 1985 |
|---|---|---|---|
| Employed in nonagricultural industries | 66.7 | 82.4 | 104.0 |
| Employed in agriculture | 4.4 | 3.4 | 3.2 |
| Unemployed | 3.4 | 7.9 | 8.3 |
| Total civilian labor force | 74.5 | 93.7 | 115.5 |

a. How many people were employed in agriculture in 1975?

b. What percentage of the civilian labor force was employed in nonagricultural industries in 1985?

c. Make a circle graph to show the breakdown of the civilian labor force in 1965.

# EXERCISES

Solve. Use any strategy.

6. Jack and Kara are running laps around the Bradley Community Park track in opposite directions. Jack makes one lap every 55 seconds. He meets Kara every 30 seconds. How fast can Kara run one lap?

7. Simplify $\dfrac{3^{x+3} - 3(3^x)}{3(3^{x+2})}$.

**Strategies**

Look for a pattern.
Solve a simpler problem.
Act it out.
Guess and check.
Draw a diagram.
Make a chart.
Work backwards.

8. The Jackson Carton Co. makes cardboard cartons. Their medium size carton is 20% longer and 30% wider than their smallest carton. The cartons are the same height. How much greater is the volume of the medium size carton?

9. Find the real roots of $|x - 2|^2 - 2|x - 2| = 8$.

10. The average monthly automobile production in thousands in the United States for 1982 to 1988 is shown in the table below. Make a bar graph and a line graph of the data.

| Year | 1984 | 1985 | 1986 | 1987 | 1988 | 1989 | 1990 |
|------|------|------|------|------|------|------|------|
| Production | 647.8 | 682.1 | 652.4 | 591.6 | 592.6 | 568.6 | 506.4 |

11. How many times do the hands of a clock cross in one day?

12. The numbers of presidents of different political affiliations are listed in the table below. Make a circle graph of this data.

| Political Party | Democratic | Democratic-Republican | Federalist | Republican | Union | Whig |
|-----------------|------------|-----------------------|------------|------------|-------|------|
| Number of Presidents | 14 | 4 | 2 | 17 | 1 | 4 |

13. A palindrome is a word or number that is the same when read forwards and backwards. Some examples are TOOT, RADAR, 232, and 11899811. List all the two-, three-, and four-digit palindromes that are prime numbers.

14. How many divisions are *necessary* to determine whether or not a number is prime?

## COOPERATIVE LEARNING ACTIVITY

**Work in groups. Each person in the group must understand the solution and be able to explain it to any person in class.**

Start at the 0 at the top left-hand corner and move one square to a 1, then move two squares to a 2, then three squares to a 3, and so on. Make a path to the 8 in the bottom right-hand corner without revisiting any square. You may move only horizontally or vertically; diagonal moves are not allowed.

| 0 | 1 | 3 | 2 | 5 | 3 | 5 | 6 |
|---|---|---|---|---|---|---|---|
| 1 | 3 | 2 | 4 | 5 | 4 | 6 | 7 |
| 2 | 3 | 4 | 3 | 4 | 6 | 7 | 5 |
| 3 | 4 | 5 | 3 | 7 | 7 | 6 | 7 |
| 4 | 5 | 6 | 5 | 4 | 5 | 7 | 6 |
| 7 | 6 | 5 | 4 | 5 | 7 | 5 | 7 |
| 6 | 5 | 4 | 5 | 6 | 4 | 5 | 6 |
| 7 | 6 | 6 | 7 | 4 | 6 | 7 | ⑧ |

## 14-2 Line Plots and Stem-and-Leaf Plots

**Objectives**

After studying this lesson, you should be able to:

- represent data using line plots and stem-and-leaf plots, and
- read and interpret data from line plots and stem-and-leaf plots.

Numerical data is often organized and displayed using either **line plots** or **stem-and-leaf plots.** In a line plot, data is recorded and displayed using a number line.

You can make a line plot of the enrollments of thirty small American colleges and universities as follows.

| School | Enrollment (in hundreds) | School | Enrollment (in hundreds) |
|---|---|---|---|
| Amherst College | 16 | Northwestern University | 74 |
| Berea College | 15 | University of Notre Dame | 75 |
| Brown University | 58 | Oberlin College | 29 |
| Carnegie-Mellon University | 43 | Princeton University | 46 |
| Case Western Reserve University | 30 | Purdue University | 29 |
| University of Chicago | 33 | Rice University | 26 |
| Columbia University | 29 | University of Rochester | 48 |
| Dartmouth College | 37 | Smith College | 27 |
| Duke University | 59 | Southern Methodist University | 57 |
| Emory University | 52 | Stanford University | 65 |
| Georgetown University | 59 | Vanderbilt University | 52 |
| Grinnell College | 13 | Wake Forest University | 35 |
| Harvard University | 66 | Wellesley College | 22 |
| Johns Hopkins University | 27 | Williams College | 20 |
| Massachusetts Institute of Technology | 42 | Yale University | 52 |

The data range from 13 to 75. The number line must be drawn long enough to contain all of these values. An "x" is used to indicate the enrollment of a school.

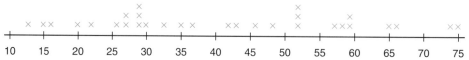

When making a stem-and-leaf plot, each piece of data is separated into two parts that are used to form the stem and leaf for the piece of data. The data is usually organized into two columns. The column on the left side is the stem, which usually consists of the digits in the greatest common place value used in all the data. For example, if the greatest common place value is hundreds, then the stem of 580 is 5 and the stem of 1293 is 12. The column on the right contains the leaves, which are 80 for 580 and 93 for 1293.

## Example 1

APPLICATION

Education

**Make a stem-and-leaf plot of the enrollments of the colleges and universities listed on the previous page.**

The data ranges from 13 to 75 and the greatest common place value of all the data is tens. So, the stems are the numbers from 1 to 7. To plot the number 16, use 1 as the stem and 6 as the leaf. We can organize the data in a stem-and-leaf plot in two different ways. The plot on the left shows the leaves in order of the given data and the plot on the right shows the leaves in numerical order.

| Stem | Leaf |
|------|------|
| 1 | 6 5 3 |
| 2 | 9 7 9 9 6 7 2 0 |
| 3 | 0 3 7 5 |
| 4 | 3 2 6 8 |
| 5 | 8 9 2 9 7 2 2 |
| 6 | 6 5    *6 | 5 represents 65 hundred students.* |
| 7 | 4 5 |

| Stem | Leaf |
|------|------|
| 1 | 3 5 6 |
| 2 | 0 2 6 7 7 9 9 9 |
| 3 | 0 3 5 7 |
| 4 | 2 3 6 8 |
| 5 | 2 2 2 7 8 9 9 |
| 6 | 5 6 |
| 7 | 4 5 |

Sometimes it is a good idea to break a stem into two parts so that the data is organized more conveniently.

## Example 2

APPLICATION

Consumerism

**Maria was browsing through the Shoe Shoppe catalog. The catalog contains 29 pairs of shoes that can be ordered through the mail. The prices are $53, $42, $49, $38, $39, $48, $37, $48, $37, $39, $58, $59, $32, $50, $59, $37, $36, $30, $40, $33, $30, $45, $40, $30, $35, $48, $37, $48, and $50. Make a stem-and-leaf plot of the shoe prices.**

Since the data ranges from 30 to 59, the stems can be 3, 4, and 5. If we make the plot this way however, the stems 3 and 4 will have a great number of leaves. So, in order to make the stem-and-leaf plot easier to interpret, we will break each stem into two parts to represent the data.

Let 3 | represent prices from $30 to $34, and
● | represent prices from $35 to $39.

| Stem | Leaf |
|------|------|
| 3 | 0 0 0 2 3 |
| ● | 5 6 7 7 7 7 8 9 9 |
| 4 | 0 0 2 |
| ● | 5 8 8 8 8 9 |
| 5 | 0 0 3 |
| ● | 8 9 9    *5 | 0 represents $50.* |

*The leaves have been ordered from least to greatest.*

Sometimes data values have more than two digits. In these cases, it is sometimes necessary to round or truncate each piece of data before it can be plotted in a stem-and-leaf plot. This way each leaf will have only one digit. For example, if the greatest common place value is thousands, then the leaf for 1573 is 573. If 1573 is rounded to 1600, then the stem is 1 and the leaf is 6. If 1573 is truncated to 1500, then the stem is 1 and the leaf is 5.

Example 3

**APPLICATION**

**Sports**

The winners of the men's springboard diving competition at the Olympic Games are listed below. Make a stem-and-leaf plot of the rounded and of the truncated values of the points they received to win the competition.

| Year | Winner | Points |
|------|--------|--------|
| 1932 | Michael Galitzen | 161.38 |
| 1936 | Richard Degener | 163.57 |
| 1948 | Bruce Harlan | 163.64 |
| 1952 | David Browning | 205.59 |
| 1956 | Robert Clotworthy | 159.56 |
| 1960 | Gary Tobian | 170.00 |
| 1964 | Ken Sitzberger | 159.90 |

| Year | Winner | Points |
|------|--------|--------|
| 1968 | Bernard Wrightson | 170.15 |
| 1972 | Vladimir Vasin | 594.09 |
| 1976 | Phil Boggs | 619.05 |
| 1980 | Alexandr Portnov | 905.02 |
| 1984 | Greg Louganis | 754.41 |
| 1988 | Greg Louganis | 730.80 |
| 1992 | Mark Lenzi | 676.53 |

*Round to the nearest 10. So, using rounded data, 7 | 3 represents 725 to 734 points.*

*Truncate after the ten. So, using truncated data, 7 | 3 represents 730 to 739 points.*

| Stem | Leaf |
|------|------|
| 1 | 6 6 6 6 6 7 7 |
| 2 | 1 |
| 3 | |
| 4 | |
| 5 | 9 |
| 6 | 2 8 |
| 7 | 3 5 |
| 8 | |
| 9 | 1 |

| Stem | Leaf |
|------|------|
| 1 | 5 5 6 6 6 7 7 |
| 2 | 0 |
| 3 | |
| 4 | |
| 5 | 9 |
| 6 | 1 7 |
| 7 | 3 5 |
| 8 | |
| 9 | 0 |

Example 4

**APPLICATION**

**Meteorology**

The average monthly temperatures in selected cities for January and July are given in the table below. Make a back-to-back stem-and-leaf plot of the rounded temperatures of the cities.

| City | January Temp. | July Temp. |
|------|--------------|-----------|
| Baton Rouge, LA | 50.8 | 82.1 |
| Caribou, ME | 10.7 | 65.1 |
| Charlotte, NC | 40.5 | 78.5 |
| Chicago, IL | 21.4 | 73.0 |
| Dallas, TX | 44.0 | 86.3 |
| Denver, CO | 29.5 | 73.4 |

| City | January Temp. | July Temp. |
|------|--------------|-----------|
| Indianapolis, IN | 26.0 | 75.1 |
| Jacksonville, FL | 53.2 | 81.3 |
| Juneau, AK | 21.8 | 55.7 |
| Roswell, NM | 41.4 | 77.7 |
| San Diego, CA | 56.8 | 70.3 |
| Tulsa, OK | 35.2 | 83.2 |

| January Temperatures | Stem | July Temperatures |
|---------------------|------|-------------------|
| 1 | 1 | |
| 1 2 6 | 2 | |
| 0 5 | 3 | |
| 1 1 4 | 4 | |
| 1 3 7 | 5 | 6 |
| | 6 | 5 |
| | 7 | 0 3 3 5 8 9 |
| | 8 | 1 2 3 6 |

*6 | 2 | represents 26°F.*

*| 8 | 1 represents 81°F.*

# CHECKING FOR UNDERSTANDING

**Communicating Mathematics**

**Read and study the lesson to answer these questions.**

1. Describe a situation in which a line plot is an appropriate way to display data.

2. Compare the two stem-and-leaf plots made in Example 3. How different are the two? Is one more accurate than the other? Explain your answer.

3. Name some collections of data for which a back-to-back stem-and-leaf plot is an appropriate way to display the data.

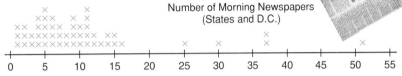

**Guided Practice**

4. Use the line plot below to answer each question.

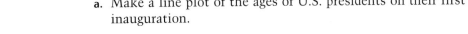

Number of Morning Newspapers
(States and D.C.)

a. What is the greatest number of newspapers in one state?

b. What is the least number of newspapers in one state?

c. How many states have 10 to 20 newspapers?

d. How many newspapers do most states have?

5. Use the stem-and-leaf plot at the right to answer each question.

a. What was the highest winning score?

b. What was the lowest winning score?

c. In 1986, Jane Geddes won the U.S. Women's Open Golf Championship with a score of 287. How many winners won with scores lower than Ms. Geddes'?

d. What scores did more of the winners earn than any other?

**Winning Scores of U.S. Women's Open 1970–1992**

| Stem | Leaf |
|------|------|
| 27 | |
| ● | 7 8 9 |
| 28 | 0 0 0 3 3 4 4 |
| ● | 5 7 7 8 9 |
| 29 | 0 0 0 2 2 |
| ● | 5 5 9 |

*29 | 0 represents a score of 290.*

# EXERCISES

**Practice**

6. Each number below represents the age of a U.S. president on his first inauguration.

| 57 | 61 | 57 | 57 | 58 | 57 | 61 | 54 | 68 | 51 | 49 |
|----|----|----|----|----|----|----|----|----|----|----|
| 50 | 48 | 65 | 52 | 56 | 46 | 54 | 49 | 50 | 47 | 55 |
| 54 | 42 | 51 | 56 | 55 | 51 | 54 | 51 | 60 | 62 | 43 |
| 56 | 61 | 52 | 69 | 64 | 64 | 55 | 55 | | | |

a. Make a line plot of the ages of U.S. presidents on their first inauguration.

**b.** What was the age of the oldest president on his first inauguration?

**c.** What was the age of the youngest president on his first inauguration?

**d.** What is the difference in the ages of the youngest and oldest presidents?

**e.** How many ages are given?

**f.** Which age(s) occur most frequently?

**g.** How many presidents were in their 60s when they were first inaugurated?

**7.** The following table gives scoring information for the top 15 scoring players in the National Basketball Association for 1991–92.

| Player, Team | Games | Points | Average |
|---|---|---|---|
| Michael Jordan, Chicago | 80 | 2404 | 30.1 |
| Karl Malone, Utah | 81 | 2272 | 28.0 |
| Chris Mullin, Golden State | 81 | 2074 | 25.6 |
| Clyde Drexler, Portland | 76 | 1903 | 25.0 |
| Patrick Ewing, New York | 82 | 1970 | 24.0 |
| Tim Hardaway, Golden State | 81 | 1893 | 23.4 |
| David Robinson, San Antonio | 68 | 1578 | 23.2 |
| Charles Barkley, Philadelphia | 75 | 1730 | 23.1 |
| Mitch Richmond, Sacramento | 80 | 1803 | 22.5 |
| Glen Rice, Miami | 79 | 1765 | 22.3 |
| Ricky Pierce, Seattle | 78 | 1690 | 21.7 |
| Hakeem Olajuwon, Houston | 70 | 1510 | 21.6 |
| Brad Dougherty, Cleveland | 73 | 1566 | 21.5 |
| Scottie Pippen, Chicago | 82 | 1720 | 21.0 |
| Reggie Lewis, Boston | 82 | 1703 | 20.8 |

**a.** Make a stem-and-leaf plot of the number of games played by each player.

**b.** Who played the most games?

**c.** Who played the least games?

**d.** How many more games did the player with the most games play than the player with the least games?

**e.** How many games did most of the players play?

**f.** How many players played more than 80 games?

**g.** Make a stem-and-leaf plot of the number of points scored by each player. Round scores to the nearest ten.

**h.** Who scored the most points?

**i.** Who scored the fewest points?

**j.** How many more points did the highest scoring player make than the lowest scoring player?

**k.** How many players scored less than 2000 points?

**l.** Make a stem-and-leaf plot of the average number of points scored per game. Round the number of points to the nearest whole number.

m. Who had the highest per game average?

n. Who had the lowest per game average?

o. What is the difference between the averages of the players with the highest and lowest average?

p. What is the most frequently occurring average?

q. How many players averaged more than 25 points per game?

8. The prize money earned in 1992 by the top five women and men tennis players is listed in the table below.

| Player | Prize Money |
| --- | --- |
| Monica Seles | $1,732,352 |
| Aranxa Sanchez Vicario | 1,093,155 |
| Steffi Graf | 1,068,026 |
| Gabriela Sabatini | 824,065 |
| Natalia Zvereva | 519,144 |

| Player | Prize Money |
| --- | --- |
| Jim Courier | $1,550,045 |
| Stefan Edberg | 1,367,029 |
| Pete Sampras | 1,131,372 |
| Andre Agassi | 1,000,484 |
| Petr Korda | 793,823 |

a. Make a back-to-back stem-and-leaf plot of the prize money for women and men.

b. What information about the relationship between the prize money available to women and men does it present?

**Critical Thinking**

9. Record the heights, to the nearest inch, of the students in your class. Make a stem-and-leaf plot and a line plot of the information. Also make a back-to-back stem-and-leaf plot comparing the heights of male and female students. What observations can you make?

**Applications**

10. **Sports** Use the information about tennis players' earnings given above. Make a line plot. Who earned the most money?

11. **History** Use the information about the ages of the presidents given in Exercise 6 to make a stem-and-leaf plot. In what age range were most of the presidents when they were inaugurated?

**Mixed Review**

12. Use the information about tennis players' earnings given above to make a circle graph showing how much of the prize money was won by each player. **(Lesson 14-1)**

**Journal**

Describe the patterns you might see when using a line plot or stem-and-leaf plot. Include some graphs in your description.

13. Name the first five terms of the arithmetic sequence described by $a_1 = 10$, $d = 2.2$ **(Lesson 13-1)**

14. Evaluate the expression $9^{\log_9 2}$. **(Lesson 12-2)**

15. Use synthetic substitution to find $f(3)$ and $f(-2)$ if $f(x) = x^4 + x^3 + x^2 + x + 1$. **(Lesson 10-2)**

16. Express $f(x) = 5(3x - 2)^2 + 4$ in quadratic form. **(Lesson 8-1)**

17. The sum of Kari's age and her mother's age is 52. Kari's mother is 20 years older than Kari. How old is each? **(Lesson 3-2)**

# Central Tendency: Median, Mode, and Mean

**Objectives**

After studying this lesson, you should be able to:

- find the median, mode, and mean of sets of data, and
- use the median, mode, and mean to interpret data.

**Application**

The table below shows the gross incomes for 1992 of some of the most successful entertainers or entertainment groups. What income is the most representative of the gross incomes listed?

| Entertainer | Gross Income (in millions) | Entertainer | Gross Income (in millions) |
|---|---|---|---|
| Kevin Costner | $21 | Eddie Murphy | $24 |
| Robin Williams | 20 | Jack Nicholson | 14 |
| Bill Cosby | 42 | Arnold Schwarzenegger | 28 |
| Michael Douglas | 14 | U2 | 27 |
| Madonna | 24 | Charles Schultz | 24 |
| Julio Iglesias | 25 | David Copperfield | 20 |
| Mel Gibson | 22 | Arsenio Hall | 12 |
| Guns N' Roses | 26 | Steven Spielberg | 30 |
| Michael Jackson | 26 | Tom Cruise | 18 |
| Steven King | 15 | Sylvester Stallone | 16 |
| Prince | 35 | Hammer | 16 |
| Garth Brooks | 24 | Oprah Winfrey | 46 |

The average income is a value that is most representative of the entire set of incomes. The most commonly used averages are the **median, mode,** and **mean.** In this application, the most representative value is the value in the middle of the group.

---

*Definition of Median, Mode, and Mean*

> The median of a set of data is the middle value. If there are two middle values, it is the value halfway between.
> The mode of a set of data is the most frequent value. Some sets of data have multiple modes and others have no mode.
> The mean of a set of data is the sum of all the values divided by the number of values.

---

Let's investigate the information on the entertainers' incomes.

**median:** To find the median, begin by arranging the data values in order from least to greatest.

*The range of the data is 46 − 12, or 34.*

12  14  14  15  16  16  18  20  20  21  22  24

24  24  24  25  26  26  27  28  30  35  42  46

Since there is an even number of values, we must find the number halfway between the two middle values. The middle values are 24 and 24. So, the median is $\frac{24 + 24}{2}$ or $24 million.

**mode:** To find the mode of the entertainers' incomes, look for the most frequently occurring value. Since 24 occurs 4 times, the mode is 24 million dollars.

*Some calculators have a $\boxed{\bar{x}}$ key that calculates the mean.*

**mean:** The first step in finding the mean is to add all of the values. Then divide the sum by the number of values, which is 24.

$$\frac{\text{sum of values}}{\text{number of values}} = \frac{569}{24} \text{ or approximately 23.7 million dollars.}$$

As you can see, the median, mode, and mean are not always the same number. In this case, the median and the mode are the same and are greater than the mean. This means that a few people have incomes that are less than the rest. The median is the most representative average of the incomes.

Extreme values are those data values that are vastly different from the central group of data values. Every data value affects the value of the mean, so when extreme values are included in a set of data, the mean may become less representative of the set. However, the values of the median and the mode are not affected by extreme values in the set.

## Example 1

APPLICATION

Business

The prices of some new cars are listed below. Determine the mean of the values in the left column and of the values in the right column. Determine the extent to which each mean is representative of the data.

| Subaru Justy | $7576 | Toyota Tercel | $8245 |
|---|---|---|---|
| Hyundai Excel | 7417 | Ford Escort | 8617 |
| Pontiac Sunbird | 9970 | Chevrolet Cavalier | 8882 |
| Nissan Sentra | 9458 | Geo Storm | 8677 |

$$\text{mean} = \frac{7576 + 7417 + 9970 + 9458}{4}$$
$$= \frac{34,421}{4} \text{ or } \$8605.25$$

$$\text{mean} = \frac{8245 + 8617 + 8882 + 8677}{4}$$
$$= \frac{34,421}{4} \text{ or } \$8605.25$$

This mean is not close to any one of the four data values. In this case, it is not a representative value.

There are no extreme values in this set. In this case, the mean is representative of the data.

In some sets of data any average, the mean, median, or mode, is representative of the data.

**Example 2**

The Parents' Association of the Plain City School District is going to give a pizza party for the graduating senior class. They purchased 6 large cheese pizzas for $6.95 each, 15 large pepperoni pizzas for $8.25 each, 10 large sausage and pepper pizzas for $9.50 each, and 14 large vegetable pizzas for $11.95 each. Find the median, mode, and mean of the cost of the pizzas for the party.

| Arrange the prices in order from least to greatest. | $ 6.95 | 6 pizzas |
|---|---|---|
| | 8.25 | 15 pizzas |
| | 9.50 | 10 pizzas |
| | 11.95 | 14 pizzas |

Since this set of data has 45 values, there is one middle value. The 23rd value is the middle, so the median is $9.50.

Most of the pizzas the Parents' Association bought were $8.25, so this is the mode of the data.

Use your calculator to find the mean.

ENTER: ( 6 × 6.95 + 15 × 8.25 + 10 × 9.50 + 14 × 11.95 ) ÷ 45 = $9.505555556$

The mean is about $9.51.

In some applications, there is no mode and it may be impossible to compute the mean.

**Example 3**

There are seven problem-solving teams in Mr. George's Algebra 2 class. They were given a problem to solve, and their times for solving the problem are listed below. What is the average amount of time it takes a problem-solving team in Mr. George's class to solve a problem?

| Students | Time |
|---|---|
| Marsha and Mike | 12 |
| Monica and Ethan | 14 |
| Becky and Kristin | 15 |
| Matt and Estevan | 16 |
| Byron and Kelly | 18 |
| Sam and Emily | 19 |
| Eric and Lydia | never finished |

Since there are seven teams, the median is the time for the team with the fourth fastest time. That is Matt and Estevan. The median time is 16 minutes.

None of the times occur more than once, so there is no mode.

Because there is no time recorded for Eric and Lydia, we cannot determine the mean of the times.

Since no mode or mean can be determined, the median, 16 minutes, is the average time for the teams to solve a problem.

# CHECKING FOR UNDERSTANDING

**Communicating Mathematics**

**Read and study the lesson to answer these questions.**

1. Of the entertainers listed in the table at the beginning of the lesson, who had the greatest gross income in 1992? Who had the lowest?

2. What is the difference between the greatest income and the least income of the entertainers?

3. In your own words, tell the difference between the median, mode, and the mean.

4. Describe some situations in which the median or the mode would be more representative of the data than the mean.

**Guided Practice**

**Find the median, mode, and mean for each set of data.**

5. {0, 2, 4, 4, 5}
6. {2, 4, 6, 8, 10}
7. {9, 9, 9, 9, 9, 9}
8. {4, 1, 2, 1, 1}
9. {2, 56, 8, 43, 44}
10. {239, 299, 318, 399, 399}

# EXERCISES

**Practice**

**Find the median, mode, and mean for each set of data.**

11. {4.8, 5.7, 2.1, 2.1, 4.8, 2.1}
12. {216, 399, 219, 179, 180, 399}
13. {11, 10, 13, 12, 12, 13, 15}
14. {80, 50, 65, 55, 70, 65, 75, 50}
15. {100, 45, 105, 98, 97, 101}
16. {2.0, 2.2, 2.1, 2.2, 2.4, 2.2, 2.3}

17. A die was tossed 25 times with the results shown below. Find the median, mode, and mean for the tosses.

| 5 | 6 | 1 | 3 | 5 | 6 | 1 | 6 | 6 | 6 | 3 | 4 | 5 |
| 6 | 1 | 2 | 4 | 2 | 1 | 4 | 2 | 5 | 4 | 4 | 1 | |

18. The height in feet of 20 of the highest mountains in the world are given below. Find the median, mode, and mean for the heights.

| 26,504 | 26,041 | 26,400 | 26,750 | 26,810 |
| 29,108 | 26,470 | 26,360 | 26,090 | 26,000 |
| 29,064 | 26,291 | 25,910 | 25,895 | 28,208 |
| 25,925 | 27,890 | 27,790 | 26,760 | 26,660 |

19. Find the median, mode, and mean of the hourly wages of 200 employees. One hundred earn $5.00 per hour, ten earn $6.25 per hour, ten earn $7.75 per hour, twenty earn $4.50 per hour, and sixty earn $5.90 per hour.

20. The stem-and-leaf plot at the right shows the points scored by the winning teams in the first 25 Super Bowls. Find the median, mode, and mean for the scores.

| Stem | Leaf |
|------|------|
| 1 | 4 6 6 6 |
| 2 | 0 0 1 3 4 4 6 7 7 7 |
| 3 | 1 2 3 5 5 8 8 9 |
| 4 | 2 6 |
| 5 | 5 |

*1 | 4 represents 14 points.*

**21.** The union and the company executives are currently negotiating a raise in salaries for all of the MicroTech employees. Three of the employees have salaries of $300,000 each. However, a majority of the employees have salaries of about $30,000 per year.

   **a.** You are a vice-president and would like to show that the current salaries are reasonable. Would you quote the median, mode, or mean as the "average" salary to justify your claim? Why?

   **b.** You are the union representative for your department and maintain that a pay raise is in order. Which of the median, mode, or mean would you quote to justify your claim? Why?

**Critical Thinking**

**22.** Write the formula for the mean of a set of data using the summation symbol.   *Hint: Let $x_i$ represent the ith data value.*

**Applications**

**23. Health**   When physicians say that the typical adult female requires 44 grams of protein per day to maintain good health, do you think they are using the median, the mode, or the mean? Why?

**24. Business**   The back-to-back stem-and-leaf plot below shows the median weekly incomes of male and female workers in various occupations. Find the median, mode, and mean of the male workers' incomes and of the female workers' incomes. Compare the results.

| Males | Stem | Females |
|------:|:----:|:--------|
|       |  2   | 0 1 4 4 |
| 5 6   |  ●   | 6 7 9   |
| 0     |  3   | 1       |
| 7     |  ●   | 5 6     |
| 0 2 2 3 | 4  | 1 4     |
| 5 8   |  ●   | 9       |
| 1     |  5   |         |
|       |  ●   |         |
|       |  6   |         |
| 6 7   |  ●   |         |

|2| 0 *represents $200 per week.*

**Mixed Review**

**25. Cartography**   A map is scaled so that 1 inch represents 25 miles. How far apart are two towns if they are 8.2 inches apart on the map? **(Lesson 11-2)**

**26.** Write the equation of the parabola whose focus is (0, 3) and whose directrix is $y = -1$.   **(Lesson 9-2)**

**27.** Find the product $6\begin{bmatrix} 5 & 6 \\ 0 & -2 \end{bmatrix}$.   **(Lesson 4-3)**

**28.** Solve $\left| x - \dfrac{7}{3} \right| = 6$.   **(Lesson 1-6)**

# Variation: Range, Interquartile Range, and Outliers

**Objectives**

After studying this lesson, you should be able to:

- find the range and interquartile range for a set of data, and
- determine if any values in a set of data are outliers.

**Application**

If the record low and high temperatures in each of the fifty states were all the same, there would be no point in studying about how the record temperatures vary. However, values in a set of data usually vary. Study the table of record high and low temperatures below. The variation within a set of data is called **dispersion**.

| State | Low | High | State | Low | High | State | Low | High |
|---|---|---|---|---|---|---|---|---|
| AL | −27 | 112 | LA | −16 | 114 | OH | −39 | 113 |
| AK | −80 | 100 | ME | −48 | 105 | OK | −27 | 120 |
| AZ | −40 | 127 | MD | −40 | 109 | OR | −54 | 119 |
| AR | −29 | 120 | MA | −34 | 107 | PA | −42 | 111 |
| CA | −45 | 134 | MI | −51 | 112 | RI | −23 | 104 |
| CO | −60 | 118 | MN | −59 | 114 | SC | −20 | 111 |
| CT | −32 | 105 | MS | −19 | 115 | SD | −58 | 120 |
| DE | −17 | 110 | MO | −40 | 118 | TN | −32 | 113 |
| FL | −2 | 109 | MT | −70 | 117 | TX | −23 | 120 |
| GA | −17 | 113 | NE | −47 | 118 | UT | −50 | 116 |
| HI | 14 | 100 | NV | −50 | 122 | VT | −50 | 105 |
| ID | −60 | 118 | NH | −46 | 106 | VA | −29 | 110 |
| IL | −35 | 117 | NJ | −34 | 110 | WA | −48 | 118 |
| IN | −35 | 116 | NM | −50 | 116 | WV | −37 | 112 |
| IA | −47 | 118 | NY | −52 | 108 | WI | −54 | 114 |
| KS | −40 | 121 | NC | −29 | 109 | WY | −63 | 114 |
| KY | −34 | 114 | ND | −60 | 121 | | | |

There are several ways to measure variation. The simplest is called the **range**.

| *Definition of Range* | **The range of a set of data is the difference between the greatest and least values in the set.** |
|---|---|

The greatest record high temperature is 134° in California, and the least record high temperature is 100° in Alaska. So the range of the record high temperatures is $134 - 100$ or 34°.

Because the range is the difference between the greatest and least values in a set of data, it is affected by extreme values. In these cases it is not a good measure of variation.

Another commonly used measure of variation is called the **interquartile range.** The interquartile range is the difference between the upper and lower **quartiles.** Quartiles are the values in a set of data that separate the data into four equal parts. The median is one of the quartiles. It separates the data into two equal parts. The remaining two quartiles are the medians of two parts into which the median separates a set of data. The quartile that is less than the median is called the **lower quartile** (LQ) and the quartile that is greater than the median is called the **upper quartile** (UQ).

### Example 1

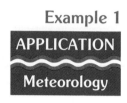

APPLICATION

Meteorology

**Find the interquartile range of the record low temperatures of the United States.**

First, place the temperatures in ascending order.

−80 −70 −63 −60 −60 −60 −59 −58 −54 −54 −52 −51 −50 −50 −50 −50
−48 −48 −47 −47 −46 −45 −42 −40 −40 −40 −40 −39 −37 −35 −35 −34
−34 −34 −32 −32 −29 −29 −29 −27 −27 −23 −23 −20 −19 −17 −17 −16
−2   14

The median is the value halfway between the 25th value, −40, and the 26th value, which is also −40. So the median is −40°.

The median separates the set of data into two sets that each contain 25 values. The median itself is not in either of these sets. The 13th value in each of these sets will be the quartile. The 13th value of the set that is below the median is −50, and the 13th value of the set that is above the median is −29°.

The lower quartile is  50 and the upper quartile is −29. So the interquartile range is −29 − (−50) or 21°.

Notice that the interquartile range in Example 1 is very small as compared to the range of the entire set of data. Since the interquartile range represents the range in which 50% of the data fall, this means that the data are clustered closely around the median of the set. Since most of the data are close to the median, values like −80° and 14° seem rather extreme. Very extreme values are called **outliers.**

*Definition of Outlier*

> **An outlier is any value in a set of data that is at least 1.5 interquartile ranges beyond the upper or lower quartile.**

### Example 2

APPLICATION

Meteorology

**Find any outliers in the record low temperatures.**

The interquartile range of the record low temperatures is 21°. Find the values that are 1.5 interquartile ranges beyond the upper and lower quartiles. The upper quartile is −29 and the lower quartile is −50.

$$-29 + 1.5(21) = 2.5° \qquad\qquad -50 - 1.5(21) = -81.5°$$

Any record low temperature that is above 2.5° or below −81.5° is an outlier. So, the only outlier is 14°, the record low for Hawaii.

**Example 3**

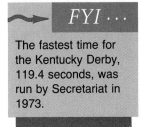

*FYI · · ·*

The fastest time for the Kentucky Derby, 119.4 seconds, was run by Secretariat in 1973.

The stem-and-leaf plot below represents the winning times at the Kentucky Derby from 1965 to 1990. Find the interquartile range of the times and determine if there are any outliers.

There are 26 leaves in this plot. The median is between the 13th and 14th times. Thus, the lower quartile is the 7th time, 121.8, and the upper quartile is the 21st time, 122.4.

| Stem | Leaf |
|------|------|
| 119 | 4 |
| 120 | 2 6 |
| 121 | 2 2 6 8 8 9 |
| 122 | 0 0 0 0 0 2 2 2 4 4 4 4 4 8 |
| 123 | 2 4 4 |
| 124 | 0 |

*120 | 6 represents a time of 120.6.*

The interquartile range is 122.4 − 121.8 or 0.6 seconds.

$$122.4 + 1.5(0.6) = 123.3 \qquad 121.8 - 1.5(0.6) = 120.9$$

Any time above 123.3 or below 120.9 is an outlier. So, 123.4, 123.4, 124.0, 120.2, 120.6, and 119.4 are outliers.

# CHECKING FOR UNDERSTANDING

**Communicating Mathematics**

Read and study the lesson to answer these questions.

1. What is the range of the record low temperatures for the United States?

2. Which has the greater variation, the record high temperatures or the record low temperatures?

3. In your own words, describe an outlier.

4. Give an example of a set of data that is greatly varied and a set that is not greatly varied.

**Guided Practice**

Find the range for each set of data.

5. {250, 275, 325, 300, 200, 225, 175}

6. {48, 36, 40, 37, 29, 45, 38, 51, 47, 38}

7. {15, 13, 19, 7, 82, 8, 3, 22, 19, 31, 12, 9}

8.
| Stem | Leaf |
|------|------|
| 4 | 1 3 9 |
| 5 | 2 3 6 9 |
| 6 | 4 4 5 |
| 7 | 2 4 7 |

*4 | 1 represents 41.*

9.
| Stem | Leaf |
|------|------|
| 3 | 0 0 1 2 4 |
| ● | 5 6 6 6 8 9 |
| 4 | 1 1 3 4 4 |
| ● | 5 5 6 |

*3 | 4 represents 34.*

10–14. Find the quartiles and the interquartile range for each set of data in Exercises 5–9.

# EXERCISES

**Practice**  **Find the range, quartiles, and interquartile range for each set of data.**

**15.** {4, 1, 3, 7, 7, 5, 4, 1, 8, 20, 2, 11, 7, 7, 1}

**16.** {51, 57, 49, 47, 23, 82, 49, 47, 54, 58}

**17.** {1055, 1075, 1095, 1125, 1005, 975, 1125, 1100, 1145, 1025, 1075}

**18.** {71, 81, 65, 95, 85, 59, 88, 66, 53, 75, 96, 57, 63, 76, 64, 82, 98, 65}

**19.**
| Stem | Leaf |
|---|---|
| 0 | 1 1 8 8 9 |
| 1 | 4 5 5 7 7 7 9 |
| 2 | 1 4 4 5 8 8 8 8 9 |
| 3 | 0 1 3 6 6 7 8 |
| 4 | 4 5 6 9 |

2 | 4 represents 240.

**20.**
| Stem | Leaf |
|---|---|
| 4 | 0 |
| 5 | 7 9 |
| 6 | 0 4 5 6 6 7 8 8 |
| 7 | 0 1 1 2 2 3 |
| 8 | 0 5 9 |

5 | 9 represents 5.9.

**21–26.** Find any outliers in each set of data in Exercises 15–20.

**27.** The prices of several zoom lenses available for use with a single-lens reflex camera are listed below.

$396  $290  $350  $298  $239  $200  $150  $235  $265  $175
$230  $150  $140  $275  $500  $275  $180  $350  $130  $180

Find the range, quartiles, and interquartile range of the prices. Then identify any outliers.

**Critical Thinking**  **28.** Give an example of a set of data with a small interquartile range and a large range. Does the set have any outliers?

**Applications**  **29. Business**  The sales of the 15 largest American businesses are given below. Find the range, quartiles, and interquartile range for the sales figures. Then determine if there are any outliers.

| Company | Sales (in billions) | Company | Sales (in billions) |
|---|---|---|---|
| Amoco | 21 | IBM | 60 |
| Chevron | 25 | Mobil | 48 |
| Chrysler | 35 | Occidental Petroleum | 19 |
| Du Pont | 33 | Philip Morris | 26 |
| Exxon | 80 | Proctor and Gamble | 19 |
| Ford Motor | 92 | Shell Oil | 21 |
| General Electric | 49 | Texaco | 34 |
| General Motors | 121 | | |

**30. Consumerism**  The prices for several bike helmets are given below. Find the range, quartiles, and interquartile range for the prices. Then determine if there are any outliers.

$51  $40  $58  $60  $30  $45  $66
$40  $87  $65  $41  $60  $40  $35
$47  $49  $54  $50  $52  $47  $39

**Mixed Review**

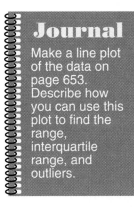

**Journal**

Make a line plot of the data on page 653. Describe how you can use this plot to find the range, interquartile range, and outliers.

31. Find the median, mode, and mean of the prices of the bike helmets given in Exercise 30. **(Lesson 14-3)**

32. State the first term, the common ratio, the last term, and the number of terms for the geometric series $2 + 6 + 18 + 54 + 162$. **(Lesson 13-5)**

33. Find the inverse of the function $f(x) = 3x - 4$. **(Lesson 10-8)**

34. Solve $x\sqrt{5} + x = 3$. **(Lesson 6-7)**

35. Simplify $3a^3b^2(-2ab^2 + 4a^2b - 7a)$. **(Lesson 5-4)**

36. **Business** The Friendly Fix-It Company charges $35 for any in-home repair. In addition, the technician charges $10 an hour after the first half-hour. How much will an in-home repair of $t$ hours cost? **(Lesson 2-5)**

# MID-CHAPTER REVIEW

**Use the data in the table below for Exercises 1–7.**

### Funding for Public Education 1980 to 1987 (in thousands)

| School Year | Federal | State | Local | Total |
|---|---|---|---|---|
| 1980–81 | $ 9,768,262 | $50,182,659 | $45,998,166 | $105,949,087 |
| 1981–82 | 8,186,466 | 52,436,435 | 49,568,356 | 110,191,257 |
| 1982–83 | 8,339,990 | 56,282,157 | 52,875,354 | 117,497,502 |
| 1983–84 | 8,576,547 | 60,232,981 | 57,245,892 | 126,055,419 |
| 1984–85 | 9,105,569 | 67,168,684 | 61,020,425 | 137,294,678 |
| 1985–86 | 9,956,009 | 73,673,174 | 65,375,698 | 149,004,882 |
| 1986–87 | 10,145,899 | 79,022,572 | 69,659,003 | 158,827,473 |

1. Make a line graph of the total funding for public education from 1980 to 1987. **(Lesson 14-1)**

2. Make a bar graph to show the state funding of public education from 1980 to 1987. **(Lesson 14-1)**

3. Make a circle graph to show the sources of funding for public education for the 1985–86 school year. **(Lesson 14-1)**

4. Make a back-to-back stem-and-leaf plot of the rounded and truncated values of local funding for public education from 1980 to 1987. **(Lesson 14-2)**

5. Make a line plot of the values of state funding for public education from 1980 to 1987. **(Lesson 14-3)**

6. Find the median, mode, and mean values of the federal funding for public education from 1980 to 1987. **(Lesson 14-3)**

7. Find the range, quartiles, and interquartile range of the total funding for public education from 1980 to 1987. Then determine if there are any outliers. **(Lesson 14-4)**

# Box-and-Whisker Plots

**Objective**

After studying this lesson, you should be able to:

- represent data using box-and-whisker plots.

Numerical data can be represented using a **box-and-whisker plot.** In a box-and-whisker plot, the quartiles and the extreme values of a set of data are displayed using a number line. The nutritional information for some sandwiches sold by leading fast food restaurants are listed below.

| Sandwich | Calories per serving | Calories per Ounce | Sandwich | Calories per serving | Calories per Ounce |
|----------|---------------------|-------------------|----------|---------------------|-------------------|
| Arby's Ham and Cheese | 380 | 69.1 | Hardee's Big Cheese | 495 | 82.5 |
| Arby's Roast Beef | 350 | 70 | Hardee's Big Deluxe | 675 | 75.8 |
| Arby's Super Roast Beef | 620 | 63.6 | McDonald's Quarter Pounder | 418 | 73.3 |
| Burger King Hamburger | 310 | 75.8 | McDonald's Quarter Pounder with Cheese | 518 | 76.7 |
| Burger King Cheeseburger | 360 | 78.4 | McDonald's Big Mac | 541 | 83.3 |
| Burger King Whopper | 760 | 72.8 | Roy Roger's Hamburger | 425 | 113.3 |
| Dairy Queen's Hamburger | 360 | 69.2 | Roy Roger's Cheeseburger | 475 | 126.7 |
| Dairy Queen's Cheeseburger | 410 | 71.9 | Wendy's Hamburger | 472 | 67.4 |
| Friendly's Big Beef Hamburger | 420 | 57.7 | Wendy's Cheeseburger | 577 | 68.7 |
| Friendly's Big Beef Cheeseburger | 480 | 61.5 | Wendy's Double Hamburger | 797 | 67 |

To make a box-and-whisker plot of the calories, first arrange the values in ascending order and find the quartiles.

310  350  360  360  380  410  418  420  425  472
475  480  495  518  541  577  620  675  760  797

The median is halfway between the 10th and 11th values.

$$\frac{472 + 475}{2} = 473.5$$

The lower quartile is between the 5th value, 380, and the 6th value, 410. The upper quartile is between the 15th value, 541, and the 16th value, 577.

lower quartile: $\frac{380 + 410}{2} = 395$          upper quartile: $\frac{541 + 577}{2} = 559$

The extreme values are the least value, 310, and the greatest value, 797.

*Abbreviations are often used for statistical terms:*
*LQ: lower quartile*
*UQ: upper quartile*
*LV: least value*
*GV: greatest value*

To make the box-and-whisker plot, draw a number line and plot the quartiles, the median, and the extreme values.

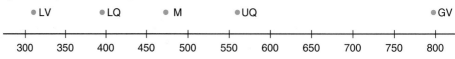

Draw a box to designate the interquartile range and mark the median by drawing a segment containing its point in the box. Draw segments (whiskers) connecting the lower quartile to the least value and the upper quartile to the greatest value.

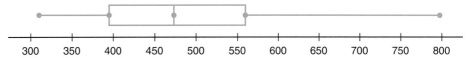

Box-and-whisker plots can be drawn horizontally, as above, or vertically. The plot of the calories of the sandwiches has been drawn vertically at the right. Notice that the box contains 50% of the data and each whisker contains 25% of the data.

The interquartile range for the calories in the sandwiches was 559 − 395 or 164 calories. Since each calorie count was above 395 − 1.5(164) or 149 calories and below 559 + 1.5(164) or 805 calories, there were no outliers. However, outliers can be displayed on a box-and-whisker plot. Each outlier is represented by a point only, and the whisker is extended only to the last value of the data that is not an outlier.

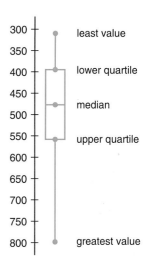

### Example 1

**APPLICATION**

**Sports**

*FYI* ...

The largest of the major league stadiums is Cleveland Stadium, home of the Cleveland Indians. Fenway Park, home of the Boston Red Sox, is the smallest major league stadium.

**The seating capacities of the major league baseball stadiums are listed below. Find any outliers in the data. Then draw a box-and-whisker plot.**

| 34,182 | 39,600 | 40,625 | 43,508 | 44,087 | 45,000 | 49,219 | 52,003 |
| 52,392 | 52,416 | 53,000 | 53,192 | 54,017 | 54,224 | 55,300 | 55,833 |
| 56,000 | 57,545 | 58,000 | 58,150 | 58,433 | 58,727 | 59,149 | 64,538 |
| 64,593 | 74,483 |

There are 26 values. The median is halfway between the 13th and 14th values. So the median is $\frac{54017 + 54224}{2}$ or 54,120.5. The lower quartile is the 7th value, 49,219, and the upper quartile is the 20th value, 58,150.

The interquartile range is 58,150 − 49,219 or 8931.
$$49,219 − 1.5(8931) = 35,822.5$$
$$58,150 + 1.5(8931) = 71,546.5$$
The values 34,182 and 74,483 are outliers.

Draw a number line and plot the quartiles and outliers. Also plot 39,600 and 64,593, since these are the last data values that are not outliers. Extend the whiskers to these points and leave the outliers as single points.

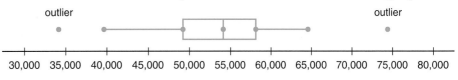

# CHECKING FOR UNDERSTANDING

**Communicating Mathematics**

**Read and study the lesson to answer these questions.**

1. Which type of sandwich listed in the table on page 658 had the highest number of calories per ounce?

2. The box in a box-and-whisker plot is separated by a line indicating the median. What portion of the data is represented by each part of the box?

3. What can you tell about a set of data from a box-and-whisker plot?

4. If the box of a box plot was very long, what could you tell about the set of data?

**Guided Practice**

5. Use box-and-whisker plot I to answer each question.

   I.

   a. What is the range of the data?

   b. What is the median of the data?

   c. What percent of the data is greater than 28?

   d. Between what two values of the data is the middle 50% of the data?

6. Use box-and-whisker plot II to answer each question.

   II.

   a. What percent of the data is less than 76?

   b. What percent of the data is less than 92?

   c. What percent of the data is greater than 64 and less than 92?

   d. Under what conditions would a set of data have this type of box-and-whisker plot?

7. Use box-and-whisker plot III to answer each question.

   III.

   a. What is the range of the data?

   b. What values of the data are outliers?

   c. What percent of the data is greater than 560?

   d. What percent of the data is less than 580?

# EXERCISES

**Practice**

8. The number of calories in a regular serving of french fries at different restaurants are listed below. Make a box-and-whisker plot of the data.

| Restaurant | Calories | Restaurant | Calories |
|------------|----------|------------|----------|
| Burger Chef | 250 | Hardee's | 239 |
| Burger King | 240 | McDonald's | 211 |
| Carl's Jr. | 220 | Roy Rogers | 240 |
| Dairy Queen | 200 | Wendy's | 327 |
| Friendly's | 125 | | |

9. Make a box-and-whisker plot of the calories per ounce of the sandwiches listed on page 658.

10. The number of medals won by the top 16 countries at the 1992 Summer Olympics are given below. Make a box-and-whisker plot of the data.

| Country | Number of Medals | Country | Number of Medals |
|---|---|---|---|
| Unified Team | 112 | Australia | 27 |
| United States | 108 | Spain | 22 |
| Germany | 82 | Japan | 22 |
| China | 54 | Britain | 20 |
| Cuba | 31 | Italy | 19 |
| Hungary | 30 | Poland | 19 |
| South Korea | 29 | Canada | 18 |
| France | 29 | Romania | 18 |

11. The amount of sodium, in milligrams, per serving of certain brands of peanut butter is given below. Make a box-and-whisker plot of the data.

195 210 180 225 255 225 195 225 203 225 195 195 188 191
210 233 225 248 225 210 240 180 225 240 180 225 240 240
195 189 178 255 225 225 225 194 210 225 195 188 205

12. Juanita's bowling scores for each week of her summer bowling league are given below.

153 167 154 172 167 166 201 158 166 134 163 167 188
187 144 154 176 170 129 139 190 221 165 160 171 170
149 168 197 161 166 165 169 200 204 151 178 161 162

a. Make a box-and-whisker plot of the data.

b. Make a stem-and-leaf plot of the same data.

c. Find the mode and median using the stem-and-leaf plot. How do these averages compare with the information from your box-and-whisker plot?

13. The table below shows the median ages of men and women at the time of their first marriage for the decades of 1890 through 1990.

| Year | Men | Women | Year | Men | Women |
|---|---|---|---|---|---|
| 1890 | 26.1 | 22.0 | 1950 | 22.8 | 20.3 |
| 1900 | 25.9 | 21.9 | 1960 | 22.8 | 20.3 |
| 1910 | 25.1 | 21.6 | 1970 | 23.2 | 20.8 |
| 1920 | 24.6 | 21.2 | 1980 | 24.7 | 22.0 |
| 1930 | 24.3 | 21.3 | 1990 | 26.2 | 25.1 |
| 1940 | 24.3 | 21.5 | | | |

a. Make a box-and-whisker plot for the men's and women's ages.

b. Compare the two plots.

c. Write a paragraph analyzing the trend suggested by the data. Include any reasons you might think exist for this trend.

**14.** Describe a set of data for which the box-and-whisker plot has no whiskers.

**15.** **Consumerism** The prices of 14 video cameras are listed below. Make a box-and-whisker plot of the prices.
$877  $819  $1100  $1450  $812  $973  $1399
$890  $1409  $949  $900  $775  $1299  $1399

**16.** **Sports** The number of touchdown passes made by the top 20 quarterbacks in the National Football League through the 1991 season are shown in the table below. Make a box-and-whisker plot of the data.

| Player | Passes | Player | Passes |
|---|---|---|---|
| Fran Tarkenton | 342 | John Brodie | 214 |
| Johnny Unitas | 290 | Terry Bradshaw | 212 |
| Dan Marino | 266 | Jim Hart | 209 |
| Sonny Jergensen | 255 | Roman Gabriel | 201 |
| Dan Fouts | 254 | Ken Anderson | 197 |
| John Hadl | 244 | Norm Snead | 196 |
| Joe Montana | 242 | Bobby Layne | 196 |
| Y. A. Tittle | 242 | Joe Ferguson | 196 |
| Len Dawson | 239 | Dave Krieg | 195 |
| George Blanda | 236 | Ken Stabler | 194 |

**17.** Use the information in the table for Exercise 16 to find the range, quartiles, and interquartile range of the number of passes for touchdowns made by the top 20 quarterbacks. Identify any outliers. **(Lesson 14-4)**

**18.** Write the expression $\sum_{n=0}^{5} 4^{n-2}$ in expanded form and find the sum.

**(Lesson 13-5)**

**19.** Solve $\log_{10}(y-1) + \log_{10}(y+2) = \log_7 7$. **(Lesson 12-3)**

**20.** Simplify $\dfrac{x^2 - y^2}{x} \div \dfrac{y-x}{x^2}$. **(Lesson 11-3)**

**21.** Find the center and radius of the circle whose equation is $(x+8)^2 + (y-3)^2 = 25$. Then draw the graph. **(Lesson 9-3)**

## CHALLENGE

Sequence A and Sequence B are both arithmetic sequences that have infinitely many terms in common. Find three of these common terms.

A = 2, 14, 26, . . .          B = 1, 8, 15, . . .

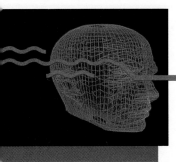

# Technology

## Statistical Graphs

The table below shows the 1990 populations (in millions) of the ten largest cities in the world and the predicted populations for the year 2000.

| City | Tokyo | Mexico City | Sao Paulo | Seoul | New York | Osaka | Bombay | Calcutta | Buenos Aires | Rio de Janeiro |
|------|-------|-------------|-----------|-------|----------|-------|--------|----------|--------------|----------------|
| 1990 | 27.0 | 20.2 | 18.1 | 16.3 | 14.6 | 13.8 | 11.8 | 11.7 | 11.5 | 11.4 |
| 2000 | 30.0 | 27.9 | 25.4 | 22.0 | 14.6 | 14.3 | 15.4 | 14.1 | 12.9 | 14.2 |

There are many ways this data could be displayed. *Data Insights* software provides a quick way to display this data using your computer. You must first enter each group of data in a column shown on the screen. Then return to the main menu. The PLOT option allows you to select what type of display you wish to use. It also allows you to customize your graph with headings and your choice of numerical increments for the axes.

Four types of displays available are line plot, histogram, stem-and-leaf plot, and box plot (box-and-whisker plot). All of these, except the line plot, allow you to graph more than one set of data. When graphing box plots, you may also choose vertical or horizontal displays.

The graphs below are a histogram and a vertical box plot of the data in the table. *Data #1* are the 1990 figures, and *Data#2* are the 2000 figures.

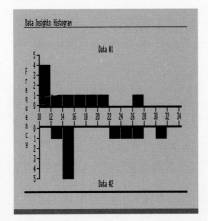
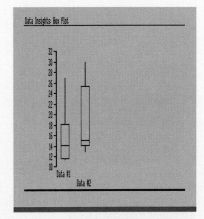

# EXERCISES

**Use the Data Insights software and the data from the table above to create each of the following graphs.**

1. line plot of Data #1.

2. line plot of Data #2

3. horizontal box plot of both sets of data

4. stem-and-leaf plot of both sets of data

# Variation: Standard Deviation

**Objective**

After studying this lesson, you should be able to:
- find the standard deviation for a set of data.

**Standard deviation** is the most commonly used measure of variation. It is used more often than the other measures of variation we discussed in Lesson 14-4, since it is closely tied to the normal distribution, which you will study in Lesson 14-7. Standard deviation is the average measure of how much each value in a set of data differs from the mean. Standard deviation is often symbolized by SD or the lower case Greek sigma, $\sigma$.

*Definition of Standard Deviation*

> From a set of data with $n$ values, if $x_i$ represents a value such that $i$ and $n$ are positive integers and $1 \le i \le n$, and $\bar{x}$ represents the mean, then the standard deviation can be found as follows.
>
> $$SD = \sqrt{\frac{\sum_{i=1}^{n}(x_i - \bar{x})^2}{n}}$$

To find the standard deviation of a set of data, follow these steps.
1. Find the mean.
2. Find the difference between each value in the set of data and the mean.
3. Square each difference.
4. Find the mean of the squares.
5. Take the principal square root of this mean.

**Example 1**

**Ramon is studying the effects of different fertilizers on tree growth. The trees he planted last year are now 49 cm, 54 cm, 61 cm, 49 cm, 54 cm, 51 cm, 56 cm, and 58 cm tall. Find the standard deviation of the heights.**

First find the mean height of the trees.

$$\text{mean height} = \frac{49 + 54 + 61 + 49 + 54 + 51 + 56 + 58}{8} \text{ or } 54 \text{ cm} \quad \text{Thus, } \bar{x} = 54.$$

Then find the standard deviation.

$$SD = \sqrt{\frac{\sum_{i=1}^{n}(x_i - \bar{x})^2}{n}} = \sqrt{\frac{(-5)^2 + (0)^2 + (7)^2 + (-5)^2 + (0)^2 + (-3)^2 + (2)^2 + (4)^2}{8}}$$

$$= \sqrt{\frac{25 + 0 + 49 + 25 + 0 + 9 + 4 + 16}{8}}$$

$$= \sqrt{\frac{128}{8}} = \sqrt{16} \text{ or } 4 \quad \text{The standard deviation is 4 cm.}$$

Example 2

**APPLICATION**

**Entertainment**

**The rental revenues for the top 16 videos of 1991 are given below. Use your calculator to find the standard deviation of the revenues.**

| Title | Revenue (in millions) | Title | Revenue (in millions) |
|---|---|---|---|
| The Addams Family | $55 | Hot Shots! | $33 |
| Backdraft | 40 | The Naked Gun 2½ | 44 |
| Beauty and the Beast | 39 | Robin Hood: Prince of Thieves | 86 |
| Cape Fear | 32 | The Silence of the Lambs | 60 |
| City Slickers | 61 | Sleeping with the Enemy | 46 |
| Dances with Wolves | 53 | Star Trek VI: The Undiscovered Country | 32 |
| Home Alone | 60 | Teenage Mutant Ninja Turtles II | 42 |
| Hook | 40 | Terminator 2 | 112 |

You can use a calculator to perform the arithmetic needed to find the standard deviation. However, your calculator may have a statistics mode that simplifies this calculation. Press $\boxed{\text{MODE}}$ and $\boxed{\text{STAT}}$ to put your calculator in statistics mode. *The statistics functions are often second key functions.*

Enter the data by entering each number and then pressing the $\boxed{\Sigma+}$ key.

ENTER: 55 $\boxed{\Sigma+}$ 40 $\boxed{\Sigma+}$ 39 $\boxed{\Sigma+}$ . . . 112 $\boxed{\Sigma+}$   *Wrong entries can be deleted by using $\boxed{\Sigma-}$ .*

Pressing $\boxed{n}$ tells you how many entries you have.

To find the standard deviation, press $\boxed{\sigma n}$ .   $20.60406134$

The standard deviation is about $20 million dollars.

It is very important to keep the mean of a set in mind when studying the standard deviation of a set of data. For example, suppose that a company that sells video equipment found that the standard deviation of monthly prices for their equipment sold in the last two years was $50. If the mean of the prices over the last two years was $200, then the standard deviation indicates a great deal of variation. If the mean of the prices was $800, then the standard deviation indicates very little variation.

# CHECKING FOR UNDERSTANDING

**Communicating Mathematics**

**Read and study the lesson to answer these questions.**

1. In your own words, explain standard deviation.

2. A recent survey shows that the mean number of full-time employees in law enforcement in large cities is 2881 with a standard deviation of 515. What does the standard deviation say about the variation within the number of employees in the cities?

3. If the mean number of full-time employees in law enforcement in large cities had been 2881 with a standard deviation of 50, what would you say about the variation of the number of employees in the cities?

4. Use the information in Example 2 to find the standard deviation of the rental revenues of the top 10 videos for 1991. Do these revenues vary more or less than the revenues of the top 16?

**Guided Practice**    **Find the mean and the standard deviation for each set of data.**

5. {11, 7, 2, 4, 1}

6. {4, 6, 14, 8, 2, 2}

7. {300, 200, 225, 175, 325, 275, 250}

8. {29, 45, 38, 51, 47, 39, 37, 40, 36, 48}

9. {81, 95, 79, 85, 82, 90, 63, 84, 84, 85, 80, 72, 82, 85, 86}

# EXERCISES

**Practice**    **Find the standard deviation for each set of data.**

10. {5, 7, 3, 4, 2, 4, 4, 4, 4, 5, 4, 3, 4, 3, 4} **1.1**

11. {1145, 1100, 1125, 1050, 1175, 835, 1075, 1095}

12.
| Stem | Leaf |
|------|------|
| 3 | 0 0 1 2 4 |
| ● | 5 6 6 6 8 9 |
| 4 | 1 1 3 4 4 |
| ● | 5 5 6 7 |

*3 | 4 represents 3.4.*

13.
| Stem | Leaf |
|------|------|
| 4 | 1 3 9 |
| 5 | 2 3 6 9 |
| 6 | 4 4 5 7 8 |
| 7 | 2 4 7 |

*5 | 2 represents 52.*

14. The weights in pounds of the starting players for three area high schools' football teams are given below.

West High:    160, 180, 190, 200, 210, 170, 250, 220, 180, 200, 240
Ridgemont:   160, 190, 210, 230, 240, 220, 150, 190, 210, 160, 240
Grandview:   250, 170, 205, 220, 185, 215, 205, 210, 205, 185, 170

a. Find the standard deviation for the weights of the players on the West High team.

b. Find the standard deviation for the weights of the players on the Ridgemont team.

c. Find the standard deviation for the weights of the players on the Grandview team.

d. Which of the teams has the most variation in weights?

**15.** Corporate Car Leasing leases cars to companies for use by employees. The mileages (miles per gallon) for the cars leased to three of their customers are listed below.

Parcels To Go:  22, 14, 33, 11, 25, 11, 22, 14, 36, 35, 28, 20, 36, 15, 21, 12, 22, 10

Selby Sales:  32, 16, 22, 24, 23, 13, 23, 31, 15, 21, 24, 27, 30, 21, 12, 24

T.C. Industries:  23, 28, 16, 30, 12, 22, 11, 33, 25, 28, 21, 25, 16, 30, 12, 29, 18, 24, 13, 25

a. Find the standard deviation for the mileages for Parcels To Go.

b. Find the standard deviation for the mileages for Selby Sales.

c. Find the standard deviation for the mileages for T.C. Industries.

d. Which of the companies had the least variation in mileages?

**16.** The average points scored by the leading scorer in the NBA from 1960 to 1989 are given below. Make a stem-and-leaf plot of the data. Then find the mode, mean, quartiles, interquartile range, standard deviation, and any outliers in the data.

| 37.9 | 38.4 | 50.4 | 44.8 | 36.5 | 34.7 | 33.5 | 35.6 | 27.1 | 28.4 |
| 31.2 | 31.7 | 34.8 | 34.0 | 30.6 | 34.5 | 31.1 | 31.1 | 27.2 | 29.6 |
| 33.1 | 30.7 | 32.3 | 28.4 | 30.6 | 32.9 | 30.3 | 37.1 | 35.0 | 32.5 |

**Critical Thinking**

**17.** Under what circumstances would the standard deviation of a set of data equal zero?

**Applications**

**18. Business**  The hourly wages of eight employees of the Sequoia Insurance Company are $4.45, $5.50, $5.50, $6.30, $7.80, $11.00, $12.20, and $17.20. Find the mean and the standard deviation of the wages of these employees.

**19. Astronomy**  An astronomer made ten measurements of the angular distance between two stars. The measurements were 11.20°, 11.17°, 10.92°, 11.06°, 11.19°, 10.97°, 11.09°, 11.05°, 11.22°, and 11.03°. Find the mean and the standard deviation of the measurements of the distance between the two stars.

**Mixed Review**

**20.** Solve $2.7^x = 52.3$ using logarithms.  (**Lesson 12-7**)

**21.** Find all rational zeros of the function $f(x) = x^4 - 6x^3 - 3x^2 - 24x - 28$. (**Lesson 10-5**)

**22.** Solve $x^2 - 7x - 8 \le 0$.  (**Lesson 8-7**)

**23. Geometry**  Find the length and width of a rectangle if its perimeter is 44 units and its area is 117 square units.  (**Lesson 7-4**)

# The Normal Distribution

**Objective**

After studying this lesson, you should be able to:

- solve problems involving normally distributed data.

**Application**

*FYI …*

The gross national product of a nation is the total value of the goods and services produced in the nation during a specified period of time.

One way of analyzing data is to consider how frequently each value occurs. The table below shows the frequencies of the percentages of their gross national product that different industrialized countries spend on education in a year. The bar graph on the right shows the frequencies of the percentages in the table.

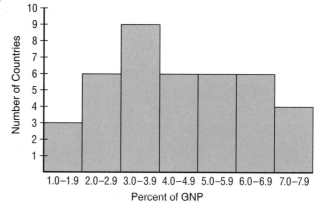

| Percent of GNP | Number of Countries |
|----------------|---------------------|
| 1.0–1.9 | 3 |
| 2.0–2.9 | 6 |
| 3.0–3.9 | 9 |
| 4.0–4.9 | 6 |
| 5.0–5.9 | 6 |
| 6.0–6.9 | 6 |
| 7.0–7.9 | 4 |

The bar graph shows a **frequency distribution** of the scores. That is, it shows how they are spread out over the range of 1.0% to 7.9%. A bar graph, like this one, that shows a frequency distribution is called a **histogram.**

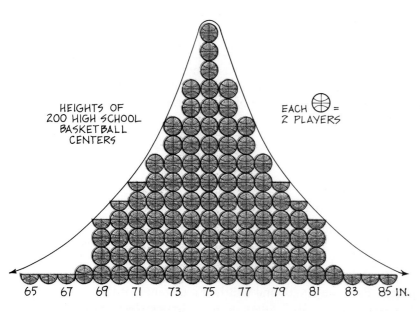

HEIGHTS OF 200 HIGH SCHOOL BASKETBALL CENTERS

EACH ⊕ = 2 PLAYERS

Curves are often used to show frequency distributions, especially when the distribution contains a large number of values. While the curves may be of any shape, many distributions have graphs shaped like the one at the left. A distribution with this type of graph is called a **normal distribution.**

The curve of the graph of a normal distribution is symmetric and is often called a **bell curve.** The shape of the curve indicates that the frequencies in a normal distribution are concentrated around the center portion of the distribution. What does this tell you about the mean?

Normal distributions have these properties:

1. The graph is maximized at the mean.

2. About 68% of the values are within one standard deviation from the mean. *Of the 68%, 34% are greater than the mean and 34% are less.*

3. About 95% of the values are within two standard deviations from the mean. *Of the 95%, 47.5% are greater than the mean and 47.5% are less.*

4. About 99% of the values are within three standard deviations from the mean. *Of the 99%, 49.5% are greater than the mean and 49.5% are less.*

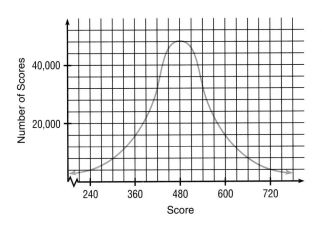

Suppose the scores on the mathematics component of the Scholastic Aptitude Test (SAT) of 1,000,000 students are recorded and the frequency of those scores is normally distributed. If the mean score is 480 and the standard deviation is 90, then the graph at the left approximates the curve for the frequency distribution of the scores.

As shown by the graph, the mean is the most frequent score. Of the 1,000,000 students, about 680,000 scored between 390 and 570 points. About 950,000 students scored between 300 and 660 points. About 990,000 students scored between 210 and 750 points.

Normal distributions occur quite frequently. In addition to test scores, the number of errors made by typists and the lengths of newborn babies can be represented by normal distributions. In all of these cases, the number of data must be sufficiently large for the distribution to be normal.

## Example 1

APPLICATION
Medicine

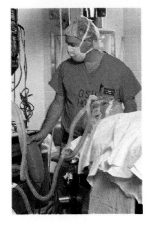

The correct number of milligrams of anesthetic an anesthesiologist must administer to a patient is normally distributed. The mean is 100 milligrams and the standard deviation is 20 milligrams. Of a sample of 200 patients, about how many people require less than 80 milligrams of anesthetic for a response?

This frequency distribution is shown by the following curve. The percentages represent the percentages of patients requiring the dosage within the given interval.

The percentage of people requiring less than 80 milligrams of anesthetic is 0.5% + 2% + 13.5% or 16%.

$200 \times 16\% = 32$

So about 32 of the 200 patients require less than 80 milligrams of anesthetic.

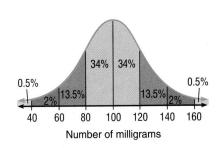

Example 2

**APPLICATION**

**Health**

The lengths of babies born at the hospital in Neenah, Wisconsin in the last year were normally distributed. The mean length was 20.4 inches, and the standard deviation was 0.8 inches. What percentage of babies born in the hospital were born between 18.8 and 21.2 inches long?

Of the babies born at the hospital last year, 68% had lengths between 19.6 and 21.2 inches. Another 13.5% had lengths between 18.8 and 19.6 inches. Therefore, 68% + 13.5% or 81.5% of the babies were between 18.8 and 21.2 inches long.

# CHECKING FOR UNDERSTANDING

**Communicating Mathematics**

Read and study the lesson to answer these questions.

1. In your own words, define histogram.

2. On the graph of the frequencies of percentages of gross national product spent on education, the percentages are graphed on the __?__ axis and the frequencies are graphed on the __?__ axis.

3. Where is the concentration of values in a normal distribution?

4. Recently, Mrs. Sung gave a test in her trigonometry class. The scores were normally distributed with a mean of 85% and a standard deviation of 3%. What percentage of her class would you expect to have scored between 79% and 91% on the test?

5. A frequency distribution like the one graphed at the right is said to be skewed with a long tail to the right. Describe some situations when you would expect data to be distributed in this way.

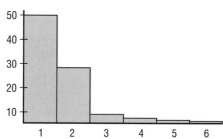

**Guided Practice**

6. Suppose 500 items are normally distributed.

   a. How many items are within one standard deviation from the mean?

   b. How many items are within two standard deviations from the mean?

   c. How many items are within three standard deviations from the mean?

   d. How many items are within one standard deviation less than the mean?

   e. How many items are within two standard deviations greater than the mean?

# EXERCISES

**Practice**

7. The diameters of metal fittings made by a machine are normally distributed. The mean diameter is 7.5 cm, and the standard deviation is 0.5 cm.

   a. What percentage of the fittings have diameters between 7.0 cm and 8.0 cm?

   b. What percentage of the fittings have diameters between 7.5 cm and 8.0 cm?

   c. What percentage of the fittings have diameters greater than 6.5 cm?

   d. Of 100 fittings, how many will have a diameter between 6.0 cm and 8.5 cm?

8. The lifetimes of 10,000 light bulbs are normally distributed. The mean lifetime is 300 days, and the standard deviation is 40 days.

   a. How many light bulbs will last between 260 and 340 days?

   b. How many light bulbs will last between 220 and 380 days?

   c. How many light bulbs will last less than 300 days?

   d. How many light bulbs will last more than 300 days?

   e. How many light bulbs will last more than 380 days?

   f. How many light bulbs will last less than 180 days?

9. Jalisa and Wes were conducting an experiment in statistics. They asked each student in their mathematics classes to toss a fair coin 100 times and record the number of heads they obtained. They found that the number of heads was almost distributed normally. The mean number of heads was 50 and the standard deviation was 5.

   a. What percentage of the students who tossed the coins obtained less than 50 heads?

   b. What percentage of the students who tossed the coins obtained between 35 and 45 heads?

   c. What percentage of the students who tossed the coins obtained between 40 and 60 heads?

10. The weights of boxes of cereal filled by a machine are normally distributed. The mean weight is 510 grams with a standard deviation of 4 grams.

   a. Of 1000 boxes, how many weigh at least 510 grams?

   b. Of 1000 boxes, how many weigh between 502 and 514 grams?

   c. A machine at the end of the production line checks the weight of the boxes before they are shipped to the stores. If a box weighs more than 522 grams, it is sent back to the beginning of the line to be emptied and reused. Of 1000 boxes, how many will be sent back for this reason?

   d. The machine that checks the weight of the boxes sends back boxes that weigh less than 502 grams. Of 1000 boxes, how many will be sent back for this reason?

**Critical Thinking**

11. The scores for all American students taking the Scholastic Aptitude Test are normally distributed. Would you expect the scores of the students in your Algebra 2 class to be normally distributed? Explain your answer.

**Applications**

12. **Education** The table below shows the number of male and female students who scored within a particular range on the verbal component of the Scholastic Aptitude Test.

| Score | Number of Males | Number of Females |
|---|---|---|
| 200–249 | 20,993 | 26,576 |
| 250–299 | 35,900 | 45,677 |
| 300–349 | 61,662 | 76,402 |
| 350–399 | 83,187 | 97,328 |
| 400–449 | 97,269 | 108,582 |
| 450–499 | 85,904 | 89,576 |
| 500–549 | 68,284 | 67,544 |
| 550–599 | 44,932 | 41,513 |
| 600–649 | 25,139 | 21,404 |
| 650–699 | 14,974 | 11,364 |
| 700–749 | 5,500 | 3,986 |
| 750–800 | 621 | 365 |

a. Construct a histogram for the scores of male students.

b. Construct a histogram for the scores of female students.

c. Which of the two sets of data appears to be closer to a normal distribution?

d. Do you think that the scores for all students would be closer to a normal distribution than either of the separate sets? Explain your answer.

**Computer**

13. A claim appears in a consumers magazine that 30% of American households have at least one personal computer. Laura Chenault believes that this percentage is incorrect in her neighborhood. To prove her point, she conducted a survey in her neighborhood. She went to 20 households and asked if there was a personal computer at home. Eight of them answered *YES* and the rest answered *NO*.

The BASIC program at the top of the next page can be used to analyze the data to see if it provides enough evidence to accept or reject the claim that 30% of American households have at least one personal computer. This claim is what statisticians call a **null hypothesis.** The program results are reliable if the hypothetical percentages of *YES* and *NO* answers are greater than 5.

$N$ = the number of people surveyed.     $C$ = the number of YES answers.
$P$ = the percentage of American households with a personal computer.
$Q$ = the percentage of American households without a personal computer.
$Z$ = number of standard deviations from the mean to a YES answer.

```
10   READ P,Q,Z                          120  IF ABS(C-X1)>Z*SD THEN 150
20   DATA 0,3,0,7,1,96                   130  PRINT "ACCEPT THE NULL
30   INPUT "WHAT IS THE NUMBER                HYPOTHESIS SINCE";C2; "<= "
     OF PEOPLE SURVEYED?";N                   ;C; "< = ";C1;","
40   INPUT "WHAT IS THE NUMBER           140  GOTO 160
     OF YES ANSWERS?";C                  150  PRINT "REJECT THE NULL
50   LET X1 = N*P                             HYPOTHESIS SINCE ";C;" > ";
60   LET X2 = N*Q                             C1;" OR ";C;" > ";C2","
70   IF X1 < = 5 THEN 155                151  GOTO 160
80   IF X2 < = 5 THEN 155                155  PRINT "RESULTS NOT RELIABLE."
90   LET SD = SQR(X1*Q)                  160  END
100  LET C1 = X1 + Z*SD
110  LET C2 = X1 - Z*SD
```

**Use the program to answer each question.**

a. If $N = 30$ and $C = 17$, is the null hypothesis rejected or accepted?

b. What would happen if you ran the program for $N = 10$?

c. A claim appears in a national newspaper that 45% of American homes have at least one VCR. To verify these statistics, a local paper interviewed 55 households and found only 30 of them had a VCR. Is the national claim correct for this small town? *Change line 20 to DATA 0.45, 0.55, 1.96.*

**Mixed Review**

14. Find the eighth, ninth, and tenth terms of the sequence for which $a_n = 2n + 1$. **(Lesson 13-2)**

15. Solve $\log_5 (4x) = \log_5 (x^2 - 5)$. **(Lesson 12-3)**

16. Simplify $\dfrac{a + 1}{a^2 - 1} + \dfrac{2}{4a - 4}$. **(Lesson 11-4)**

17. Write the standard form of the equation $8x^2 - 16x + 4y^2 - 20y = 7$. Graph the equation and state whether the graph of the equation is a parabola, a circle, an ellipse, or a hyperbola. **(Lesson 9-7)**

## APPLICATION

You should always carefully examine statistical data represented in magazines, advertising brochures, and newspapers. Data can be misrepresented in order to lead you to a desired conclusion. For example, a group wants to sponsor a rock concert in a public park. They conduct a survey and find that 8 out of 10 people surveyed approve of the concert. The city council approves the concert based on these results. However, upon scrutiny, you may find that the sample was composed of teenagers and those who do not live in the immediate area of the park. Thus the survey was not representative of everyone's views.

Also be aware of misleading graphs. A change in the scale of one of the axes can make a line graph look more dramatic. The choice of graph may make accurate interpretations more difficult. Look at the graph at the right. Is it a reliable graph? Explain your answer.

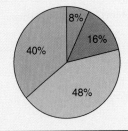

## VOCABULARY

Upon completing this chapter you should be familiar with the following terms:

| | | | |
|---|---|---|---|
| bell curve | **668** | **648** | mode |
| box-and-whisker plot | **658** | **668** | normal distribution |
| frequency distribution | **668** | **654** | outlier |
| histogram | **668** | **654** | quartile |
| interquartile range | **654** | **653** | range |
| line plot | **642** | **638** | sample |
| mean | **648** | **664** | standard deviation |
| median | **648** | **642** | stem-and-leaf plot |

## SKILLS AND CONCEPTS

| OBJECTIVES AND EXAMPLES | REVIEW EXERCISES |
|---|---|

Upon completing this chapter, you should be able to:

■ represent data using line plots and stem-and-leaf plots.  (**Lesson 14-2**)

**The prices of twelve toasters are listed below. Make a line plot and a stem-and-leaf plot of the prices.**

$40   $45   $38   $40   $47   $39
$18   $18   $14   $25   $27   $25

```
              ×          ×
        ×  ×    ×  ×           ×
+++++++++++++++++++++++++++×××++++×+×+++++
10        20        30        40        50
Stem | Leaf       Prices of toasters
```

| Stem | Leaf |
|---|---|
| 1 | 4 8 8 |
| 2 | 5 5 7 |
| 3 | 8 9 |
| 4 | 0 0 5 7 |

*2 | 5 represents $25.*

**Use these exercises to review and prepare for the chapter test.**

The table shows the average number of goals allowed by the top goaltenders in professional hockey.

| Player | Avg | Player | Avg |
|---|---|---|---|
| Roy | 2.53 | Vernon | 3.13 |
| Luit | 2.53 | Essensa | 3.15 |
| Lemelin | 2.81 | Ranford | 3.19 |
| Puppa | 2.89 | Casey | 3.22 |
| Moog | 2.89 | Beaupre | 3.22 |
| Richter | 3.00 | Wamsley | 3.26 |
| Cloutier | 3.09 | Malarchuk | 3.35 |

1. Make a line plot for the goals allowed per game averages for the goaltenders.

2. Make a stem-and-leaf plot of the goals allowed per game averages for the goaltenders.

## OBJECTIVES AND EXAMPLES

■ find the median, mode, and mean of a set of data. **(Lesson 14-3)**

**Find the median, mode, and mean of the set {66, 67, 68, 69, 70, 73, 74, 76, 78, 78, 84}.**

median: 6th value = 73    mode = 78

mean = $\dfrac{66 + 67 + \ldots + 78 + 84}{11}$ or 73

---

■ find the range, interquartile range, and outliers for a set of data. **(Lesson 14-4)**

**Find the range, interquartile range, and any outliers for the scores in the stem-and-leaf plot below.**

| Stem | Leaf |
|------|------|
| 6 | 0 1 2 |
| 7 | 2 2 8 |
| 8 | 0 2 3 |
| 9 | 9 |

greatest value = 99
least value = 60
range = 39

median = 6th score, 78
lower quartile = 3rd score, 62
upper quartile = 9th score, 83
interquartile range = 83 − 62 or 21
There are no outliers.

---

■ represent data using a box-and-whisker plot. **(Lesson 14-5)**

**Use the information in the stem-and-leaf plot above to make a box-and-whisker plot.**

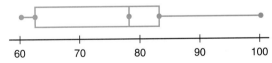

---

■ find the standard deviation of a set of data. **(Lesson 14-6)**

To find the standard deviation of a set of data:
1. Find the mean.
2. Find the difference between each value in the set of data and the mean.
3. Square each difference.
4. Find the mean of the squares.
5. Take the principal square root of this mean.

## REVIEW EXERCISES

**Find the median, mode, and mean for each set of data.**

3. {7.1, 5.0, 2.7, 9.1, 8.1, 6.3, 8.5}

4. {2.1, 4.8, 2.1, 5.7, 2.1, 4.8, 2.1}

5. {20,270, 29,002, 14,255, 18,700, 28,146, 22,835, 21,201, 15,781, 25,263, 19,344}

6. {3, 6, 7, 7, 7, 7, 9, 10, 10, 10, 10, 10, 13, 14, 14, 16, 17, 17, 19, 20, 21, 31}

---

**Find the range, quartiles, interquartile range, and any outliers for each set of data.**

7. {90, 92, 78, 93, 79, 85, 89, 88, 84, 86}

8. {0.4, 0.2, 0.5, 0.9, 0.3, 0.4, 0.5, 1.9, 0.5, 0.7, 0.8, 0.6, 0.2, 0.1, 0.4}

9. {100, 99, 93, 94, 96, 94, 95, 101, 109, 108, 104, 106, 125, 100, 104, 98, 19}

10. {10, 50, 90, 40, 60, 40, 50, 90, 0}

11.

| Stem | Leaf |
|------|------|
| 1 | 1 1 2 2 2 |
| 2 | 2 5 7 7 |
| 3 | 0 3 5 6 8 |
| 4 | 0 0 1 1 2 5 |
| 5 | 0 2 6 8 |

*4 | 0 represents 40.*

---

12. The value in millions of U.S. dollars of the exports of 12 industrialized countries for 1989 are listed below. Make a box-and-whisker plot of the data, labeling any outliers.

| | | |
|---|---|---|
| 101,261 | 116,013 | 178,846 | 343,195 |
| 138,503 | 275,173 | 109,212 | 152,447 |
| 364,080 | 107,877 | 44,424 | 32,206 |

---

**Find the standard deviation for each set of data.**

13. {5.2, 5.7, 6.0, 5.6, 2.4}

14. {13.7, 15.0, 13.7, 16.9, 13.6, 14.3, 14.8, 14.8, 15.1, 15.4, 14.9}

15. {2490, 3700, 4300, 2370, 4730, 3105, 4056, 2905, 2130, 2930, 2770, 3320, 3384, 4750, 3500, 4078}

16. {797, 6481, 117, 3750, 500, 3842, 1748, 3359, 1375, 3662, 1876, 801, 200}

- solve problems involving normally distributed data.

**A Normal Distribution**

34% | 34%

0.5%          0.5%

13.5%     13.5%

2%     2%

34  36  38  40  42  44  46

Number of milligrams

The number of hours of TV watched weekly by 3000 families is normally distributed. The mean is 22 hours and the standard deviation is 7.5 hours.

17. How many families watch at least 22 hours of TV per week?

18. How many families watch TV between 7 and 29.5 hours per week?

19. What percentage of the 3000 families watch more than 37 hours of TV a week?

# ~~~~~ APPLICATIONS AND CONNECTIONS ~~~~~

20. **Ecology** The weights in tons of different types of waste found in a landfill are given in the table below.

| Type | Weight | Type | Weight |
| --- | --- | --- | --- |
| plastic | 2,000 | metal | 2,125 |
| yard debris | 4,400 | glass | 1,750 |
| food waste | 1,850 | other | 2,875 |
| paper | 10,000 | | |

a. Draw a circle graph to show the composition of the landfill. (**Lesson 14-1**)

b. Draw a bar graph to show how the weights of the types of waste compare. (**Lesson 14-1**)

22. **Safety** The numbers of job-related injuries at a construction site for each month of 1990 are listed below.

10  13  15  39  21  24
19  16  39  17  23  25

a. Make a stem-and-leaf plot of the numbers of injuries. (**Lesson 14-2**)

b. Make a line plot of the numbers of injuries. (**Lesson 14-2**)

c. Find the median, mode, and mean of the numbers of injuries. (**Lesson 14-3**)

21. **Education** The numbers of books in the libraries of the top 12 college or university libraries in the United States are listed below.

11,781,270   8,718,619   7,561,615
7,366,672    6,237,521   6,066,136
5,976,588    5,894,135   5,753,147
5,144,830    5,063,051   4,908,985

a. Find the median, mode, and mean of the numbers of books. (**Lesson 14-3**)

b. Find the range, quartiles, interquartile range, and any outliers. (**Lesson 14-4**)

23. **Demographics** The monthly incomes of 10,000 workers in Gahanna are distributed normally. Suppose the mean monthly income is $1250 and the standard deviation is $250.

a. How many workers earn more than $1500 per month? (**Lesson 14-7**)

b. How many workers earn less than $750 per month? (**Lesson 14-7**)

c. What percentage of the 10,000 workers earn between $500 and $1750 a month? (**Lesson 14-7**)

d. What percentage of the workers earn less than $1750 a month? (**Lesson 14-7**)

The following high temperatures were recorded during a cold spell in Cleveland lasting thirty-eight days. Use this data for Exercises 1–9.

29° 26° 17° 12° 5° 4° 25° 17° 23° 18° 13° 6° 25° 20° 27° 22° 26° 30° 31°
2° 12° 27° 16° 27° 16° 30° 6° 16° 5° 0° 5° 29° 18° 16° 22° 29° 8° 23°

1. Draw a circle graph to show how many days the temperature was 0° to 10°, 11° to 20°, 21° to 30°, and above 30°.

2. Draw a line plot of the temperatures.

3. What is the range of the temperatures?

4. Find the median, mode, and mean of the temperatures.

5. Make a stem-and-leaf plot of the temperatures.

6. Find the quartiles and the interquartile range of the temperatures.

7. Find any outliers in the temperatures.

8. Make a box-and-whisker plot of the temperatures.

9. Find the standard deviation of the temperatures.

10. Find the median, mode, and mean of the hourly wages of 200 workers. One hundred earn $5.00 an hour, ten earn $5.75 an hour, ten earn $6.75 an hour, twenty earn $4.50 an hour, and sixty earn $5.25 an hour.

Connie is buying a pair of headphones to use with her stereo receiver. The prices of the 21 different types of stereo headphones sold at the Stereo Studio are $100, $150, $75, $79, $149, $120, $80, $70, $400, $190, $50, $80, $148, $40, $85, $60, $160, $90, $90, $125, and $120.

11. Make a stem-and-leaf plot of the prices of the headphones.

12. Find the median, mode, and mean of the prices of the headphones.

13. Find the range, quartiles, interquartile range, and any outliers of the prices.

14. Make a box-and-whisker plot of the prices.

15. Find the standard deviation of the prices.

The scores on a college entrance exam are normally distributed. The mean score is 510, and the standard deviation is 80.

16. Of the 50,000 people who took the exam, how many scored above 510?

17. Of the 50,000 people who took the exam, how many scored between 430 and 590?

18. What percentage of the people who took the exam scored below 670?

19. A student must score above 750 on the college entrance exam to qualify for a full scholarship at Carleton University. What percentage of the people who took the exam will qualify?

20. **Real Estate** Ten homes were sold in Oak Hills in January. Of these, one sold for $225,000, three sold for $100,000, and six sold for $85,000. Find the median, mode, and mean of the prices of these homes. Which "average" would you quote if you were an area home owner?

**Bonus** The square of the standard deviation is called the variance. Find the variance of the prices of the headphones, given in Exercises 11–15.

The questions on these pages involve comparing two quantities, one in Column A and one in Column B. In certain questions, information related to one or both quantities is centered above them. All variables used stand for real numbers.

**Directions:**

Write **A** if the quantity in Column A is greater.

Write **B** if the quantity in Column B is greater.

Write **C** if the quantities are equal.

Write **D** if there is not enough information to determine the relationship.

| Column A | Column B |
|---|---|

**1.** $\quad 0.4 \qquad\qquad\qquad \sqrt{0.4}$

**2.**
$$0 < x < 7$$
$$0 < y < 9$$
$\qquad y \qquad\qquad\qquad\qquad x$

**3.**
$$n < 0$$
$$b < 0$$
$\qquad n + b \qquad\qquad\qquad n - b$

**4.** one tenth of the product of the first positive ten integers $\qquad$ the product of the first positive nine integers

**5.** $\quad 12\%$ of 1600 $\qquad\qquad 16.5\%$ of 1200

**6.** $\qquad\qquad y \neq 0$
$\qquad \dfrac{y}{3} \qquad\qquad\qquad\qquad \dfrac{3}{y}$

| Column A | Column B |
|---|---|

**7.** $\quad \begin{array}{c} \frac{2}{3} \\ \frac{5}{7} \end{array} \qquad\qquad\qquad \dfrac{6}{17}$

**8.** $\qquad$ Let $^*y$ denote the least integer equal to or greater than $y$.
$\qquad ^*0.4 \qquad\qquad\qquad\qquad ^*1.0$

**9.** $\qquad\qquad b = d + 1$
$\qquad$ the average of $\qquad\qquad$ the average of
$\qquad a$, $b$, and $c \qquad\qquad\qquad a$, $c$, and $d$

**10.** $\qquad\qquad c$ and $d$ are integers greater than 1.
$\qquad [1 + (-c)]^d \qquad\qquad\qquad (-c)^d$

**11.** $\qquad\qquad b < 0$
$\qquad \dfrac{b^9}{b^4} \qquad\qquad\qquad\qquad \dfrac{b^{10}}{b^5}$

**12.** $\qquad\qquad \dfrac{5}{b} = 2;\ 5 = \dfrac{2}{c}$
$\qquad c + \dfrac{1}{3} \qquad\qquad\qquad b + \dfrac{11}{6}$

**13.** $\qquad\qquad \dfrac{4}{k} < 0$
$\qquad \dfrac{-1}{k} \qquad\qquad\qquad\qquad 4k$

**14.** $\qquad\qquad \dfrac{1}{k} < 0$
$\qquad \dfrac{-1}{k} \qquad\qquad\qquad\qquad k$

| Column A | Column B |
|---|---|

**15.** $\dfrac{3^4 + 3^5}{3^4}$ $\qquad$ $\dfrac{3^2 + 3^3}{3^2}$

---

**16.** $n! = n(n - 1)(n - 2) \ldots (2)1$

$\dfrac{8!}{2!6!}$ $\qquad\qquad$ $4!$

---

**17.** $2 + (-4) + 8 + \ldots$

sum of the $\qquad$ sum of the
first 6 terms $\qquad$ first 7 terms

---

**18.** $y$ if $y = 3^x$ $\qquad$ $y$ if $y = 3^{-x}$

---

**19.**

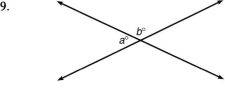

$a$ $\qquad\qquad\qquad$ $180 - b$

---

**20.** $\sqrt{0.2}$ $\qquad\qquad$ $(0.2)^2$

---

**21.** $a > b > c > d > 0$
$a - d$ $\qquad\qquad$ $b - c$

---

**22.**

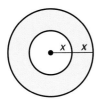

the area of the $\qquad$ the area of the
the inner circle $\qquad$ shaded region

---

**23.** the number of $\qquad$ 5 points
points needed
to divide a
segment into
five parts

| Column A | Column B |
|---|---|

**24.**

$A$ is the midpoint of $\overline{RB}$.
$B$ is the midpoint of $\overline{AS}$.

$RA$ $\qquad\qquad\qquad$ $SB$

# 15 Probability

## CHAPTER OBJECTIVES

In this chapter you will:

- Find ways of grouping objects by combinations and permutations.
- Find the probability of an event.
- Predict events based on probability.

Explain why you think the world's population has grown more since 1990 than it did in the previous century.

### Timing of Each Additional Billion of World Population

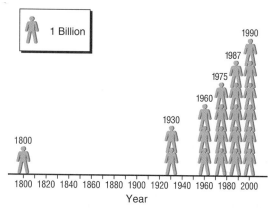

## CAREERS IN ENVIRONMENTALISM

It's obvious: If we don't clean up after ourselves, our grandchildren will all live in one big toxic landfill. There won't be many plants or animals there, either (except for gulls and rats).

Everyone pretty much agrees about that. But it's another thing altogether to accept the responsibility to clean up a river, replant a forest, or detoxify a swamp. Each person thinks, *"It's too late to save this area. What can one person do? I'm only a student—the adults ought to do it. That's what they pay taxes for."*

No one can tell another person how much time and energy to give to the environmental movement. But if you want to make your living helping Earth recover from years of abuse, consider becoming an environmentalist.

Do you wonder about your drinking water, your air, your garden soil? Do you worry about oil spills and hazardous wastes? Does your street need more shade trees? Is your city's air full of smog? You may ask, *"What can I do about it?"*

Here are only some of the specializations available: agricultural engineer, biologist, communicator, consultant, ecologist, educator, environmental health scientist, environmental engineer, fisheries conservationist, forester, health physicist, industrial hygienist, landuse planner, public health doctor, range manager, recreationist, social scientist, soil conservationist, wildlife conservationist. Take your pick.

## MORE ABOUT ENVIRONMENTALISM

**Degree Required:**

- Bachelor's Degree in Biology, Chemistry, or Environmental Science

**Some environmentalists like:**

- traveling around the world
- satisfaction that their work is helping to improve the quality of life and preserve the planet
- good salaries

**Related Math Subjects:**

- Advanced Algebra
- Trigonometry
- Probability/Statistics

**Some environmentalists dislike:**

- working with dangerous organisms and toxic substances
- working irregular hours
- pressure from political groups and industry

For more information on the various careers available in the field of Environmentalism, write to:

Institute of Environmental Sciences
940 East Northwest Highway
Mount Prospect, IL 60056

## 15-1 Problem-Solving Strategy: Using Models

**Objective**

After studying this lesson, you should be able to:

■ solve problems by using models.

Sometimes a problem is easier to solve if you act out the situation instead of trying to figure out which mathematical equation or formula to use. When you act things out, you can use models to represent different aspects of the problem you are solving.

**Example**

Suppose Grisky's Ice Cream Shoppe has 4 flavors of ice cream, 3 sauces, and 3 additional toppings. You are assigned the task of making every kind of sundae possible. Each sundae must have 2 scoops of 1 flavor of ice cream, 1 sauce, and 1 additional topping. How many different sundaes would you make?

One way to model this problem would be to use cutouts of circles to represent ice cream, squares to represent sauces, and triangles to represent the toppings. Use the initials of the items to label each figure.

Now use the items to physically build each sundae. Have your partner record each different type you create. When all possibilities are exhausted, count your results. You should have 36 different sundaes.

## CHECKING FOR UNDERSTANDING

**Communicating Mathematics**

**Read and study the lesson to answer each question.**

1. Describe another way besides using models to determine the number of different sundaes that could be made.

2. Discuss disadvantages of using models to solve problems.

3. Find the most likely sum you will roll each time two dice are thrown.

    **a.** Let *x* be the number on the first die and *y* be the number on the second die. What are all the possibilities for (*x, y*)?

    **b.** What is the smallest sum possible on each roll?

    **c.** What is the greatest sum possible on each roll?

    **d.** What is the most likely sum rolled? Why?

# EXERCISES

### Strategies

Look for a pattern.
Solve a simpler problem.
Act it out.
Guess and check.
Draw a diagram.
Make a chart.
Work backwards.

**Solve. Use any strategy.**

4. Three cruise ships leave New York for France on the same day. The round trip takes the first ship 12 days, the second ship 16 days, and the third ship 20 days. How many days will elapse before all three ships will leave New York again on the same day if none of the ships sits idle?

5. In the land of Id, there are 3 coins. Each has a different value (17¢, 36¢, or 55¢), a different color (black, silver, or gold), and a different size (small, medium, or large). A randomly selected coin is either a black coin, small, or worth 17¢. Another coin is either gold, large, or worth 36¢. The large coin is worth more than the silver coin. What color, size, and value are each of the coins?

6. The box at the right is made of 64 cubes. Two striped paths are painted on it to look like a ribbon. How many small cubes have no paint on them?

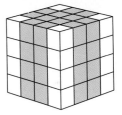

7. Determine the maximum and minimum perimeter of a figure made up of 12 square tiles.

8. George can walk around a circular nature trail in 40 minutes. Amanda walks on the same trail starting at the same time but goes in the opposite direction. She meets George every 15 minutes. How long does it take Amanda to walk around the trail?

### Journal

Describe other topics in this book where making a model helps to solve the problem.

## COOPERATIVE LEARNING ACTIVITY

**Work in groups. Each person in the group must understand the solution and be able to explain it to any person in class.**

There are 3 different geometry books and 3 different algebra books to be arranged in a row on a shelf. How many different ways can the six books be arranged if no two algebra books can be side by side?

## 15-2 The Counting Principle

**Objective**

After studying this lesson, you should be able to:
- solve problems using the Basic Counting Principle.

**Application**

Jolie Martis is going to buy a new automobile. She has already chosen the make and model of the car but she still has three more decisions to make.

1. Does she want standard or automatic transmission?
2. Does she want to have a cassette player or does she want a compact disc player?
3. Does she want a silver, red, or white exterior?

These three decisions are called **independent events** since one decision does not affect the others. The **tree diagram** shown below illustrates all the different choices Jolie has in making her final three decisions.

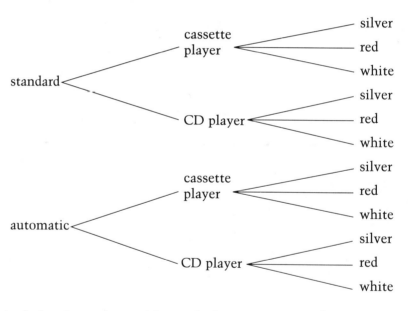

One of Jolie's choices is a red car with standard transmission and cassette player. There are 11 other choices, making a total of 12.

*Remember that each decision is an independent event.*

You can find the total number of choices that Jolie has without drawing a diagram.

| Choices: | standard/automatic | cassette/CD player | silver/white/red |
|---|---|---|---|
| Number of choices: | 2 | 2 | 3 |

The total number of choices can be found by multiplying the number of choices for each decision. Thus, the total number of choices is $2 \cdot 2 \cdot 3$ or 12. This application is an example of the **Basic Counting Principle.**

| | |
|---|---|
| *Basic Counting Principle* | **Suppose an event can occur in $p$ different ways. Another event can occur in $q$ different ways. There are $p \cdot q$ ways both events can occur.** |

This principle can be extended to any number of events.

**Example 1**

**How many different 3-letter patterns can be formed using the letters $x$, $y$, and $z$, if a letter can be used more than once?**

Since each choice of letter is not affected by the previous choice, these are *independent events*.

| Letter: | 1st | 2nd | 3rd |
|---|---|---|---|
| Number of choices: | 3 | 3 | 3 |

There are $3 \cdot 3 \cdot 3$ or 27 possible patterns.

Some applications involve **dependent events.** That is, the number of choices for one event *does affect* other events.

**Example 2**

**How many different 3-letter patterns can be formed using the letters $x$, $y$, and $z$, if each letter is used exactly once?**

After the first letter is chosen, it cannot be chosen again. So there are only two choices for the second letter. Likewise, after the second choice is made, there is only one choice for the third letter. These are *dependent events*.

| Letter: | 1st | 2nd | 3rd |
|---|---|---|---|
| Number of choices: | 3 | 2 | 1 |

There are $3 \cdot 2 \cdot 1$ or 6 patterns. Note that $3 \cdot 2 \cdot 1 = 3!$.

**Example 3**

**How many 7-digit phone numbers can begin with the prefix 890?**

Since each digit can be used any number of times, there are 10 choices for each of the last four digits of the phone number. These are independent events.

| Digit in phone number: | 4th | 5th | 6th | 7th |
|---|---|---|---|---|
| Number of choices: | 10 | 10 | 10 | 10 |

There are $10 \cdot 10 \cdot 10 \cdot 10 = 10^4$ or 10,000 phone numbers.

| Example 4 | Suppose five points in a plane represent towns that are connected by roads. Starting at any one town, how many different routes are there so that you visit each town exactly once? |
|---|---|

Let's name the points A, B, C, D, and E. Since each town you visit limits the number of towns left to visit, these are dependent events.

$5 \cdot 4 \cdot 3 \cdot 2 \cdot 1 = 5!$

| Points: | 1st | 2nd | 3rd | 4th | 5th |
|---|---|---|---|---|---|
| Number of choices: | 5 | 4 | 3 | 2 | 1 |

There are $5 \cdot 4 \cdot 3 \cdot 2 \cdot 1$ or 120 routes you could take.

# CHECKING FOR UNDERSTANDING

**Communicating Mathematics**

Read and study the lesson to answer each question.

1. Describe the difference between independent and dependent events.
2. Give two examples of independent events.
3. Give two examples of dependent events.
4. Are the results of tossing a coin several times independent events or dependent events?

**Guided Practice**

5. Tell whether the events are independent or dependent.
   a. selecting a mystery book and a history book at the library
   b. drawing cards from a deck to form a 5-card hand
   c. selecting the color and model of a new automobile

6. How many different batting orders does a baseball team of nine players have if the pitcher bats last?

7. At Columbus High School, Darrin is taking six different classes. Assuming that each of these classes is offered each period, how many different schedules might he have?

# EXERCISES

**Practice**

Tell whether the events are independent or dependent.

8. choosing the color and size of a pair of pants
9. choosing a president, secretary, and treasurer for the Pep Club
10. choosing five numbers in a bingo game
11. choosing the winner and loser of a chess game
12. Each of five people guess the total number of runs in a baseball game. They write down the guess without telling what it is.
13. The numerals 0 through 9 are written on pieces of paper and placed in a jar. Three of them are selected one after the other without replacing any of the pieces of paper.

**Solve each problem.**

14. The letters g, h, j, k, and l are to be used to form 5-letter passwords for an office security system. How many passwords can be formed if the letters can be used more than once in any password?

15. A store has 15 sofas, 12 lamps, and 10 tables at half price. How many different combinations of a sofa, a lamp, and a table can be sold at half price?

**Draw a tree diagram to illustrate all the possibilities.**

16. the possibilities for boys and girls in a family with two children

17. the possibilities for boys and girls in a family with three children

**Solve each problem.**

18. A license plate must have two letters (not I or O) followed by three digits. The last digit cannot be zero. How many possible plates are there?

19. There are five roads from Albany to Briscoe, six from Briscoe to Chadwick, and three from Chadwick to Dover. How many different routes are there from Albany to Dover via Briscoe and Chadwick?

20. For a particular model of car, a car dealer offers 6 versions of that model, 18 body colors, and 7 upholstery colors. How many different possibilities are available for that model?

21. How many ways can six different books be arranged on a shelf?

22. Three different colored dice are tossed. How many distinct outcomes can occur?

23. How many ways can six books be arranged on a shelf if one of the books is a dictionary and it must be on an end?

24. Using the letters from the word *equation,* how many 5-letter patterns can be formed in which q is followed immediately by u?

25. Consider the letters a, e, i, o, r, s, and t.

   a. How many different 4-letter patterns can be formed from these letters if no letter occurs more than once?

   b. How many of these patterns begin with a vowel and end with a consonant?

26. How many 5-digit numbers exist between 65,000 and 69,999 if no digit is to be repeated in each number?

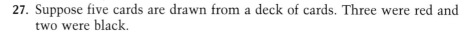

**Critical Thinking**

*In this text, when referring to a deck of cards, we mean a standard deck of 52 cards.*

27. Suppose five cards are drawn from a deck of cards. Three were red and two were black.

   a. How many possibilities are there for this hand?

   b. Suppose exactly one of the black cards is a face card. Now how many possibilities are there?   *A face card is a jack, queen, or king.*

**Applications**

28. **Dining**   A Chinese restaurant offers a special price for customers who dine before 6:30 P.M. This offer includes an appetizer, a soup, and an entree all for $6.95. There are 4 choices of appetizers, 3 soups, and 5 entrees. How many different meals are available under this offer?

29. **Transportation**   Four ferry boats run round trips between Harrod and Lafayette.

   a. How many different ways can a traveler make a round trip?

   b. How many different ways can a traveler make a round trip, by riding a different ferry on the return trip?

**Mixed Review**

30. Suppose you roll 2 dice six times. What is the most probable total of all six rolls? Why?   **(Lesson 15-1)**

31. Find the 27th term of $3 + 6 + 12 + 24 \ldots$ . Express using exponents. **(Lesson 13-5)**

32. A piece of farm equipment valued at $60,000 depreciates 10% per year by the fixed rate method. After how many years will the value have depreciated to $45,000?   **(Lesson 11-7)**

33. Graph $y = \dfrac{3x}{x + 2}$.   **(Lesson 9-5)**

34. Find $[-6.2] + [4.3] + [-2.87] + [0.5]$.   **(Lesson 2-7)**

**Journal**

Suppose you forgot how the counting principle works. How could you figure out a problem without it?

## HISTORY CONNECTION

   Your skill at winning many kinds of games often relies on your ability to count the possible outcomes given certain choices. This skill of gaming can be traced to the early history of mankind.

   An early form of our six-faced die has been found in Assyrian and Sumerian archaeological sites. It is made from the astragalus, which is the bone above the heel bone in sheep and deer. Babylonian and early Egyptian sites (about 3600 B.C.) have contained polished astragali and other colored markers used in game playing. However, these astragali are not uniform in shape and weight. So, a likelihood of a given roll was unpredictable. Perhaps this is one reason that there was not development of probability theory (unlike other fields of mathematics) in the early civilizations.

# Linear Permutations

**Objective**

After studying this lesson, you should be able to:
- solve problems involving permutations.

**Application**

Before the beginning of the game, the girls basketball team waits for the introduction of the starters. The five starters sit in a row of 12 chairs while their teammates stand around them. How many different ways, or arrangements, are there for the five starters to sit in the 12 chairs?

After the first starting player chooses one of the 12 chairs, the second starter only has 11 choices left, and so on, until the fifth starter is seated. The number of choices for each player is affected by the choice of the previous player, so these are dependent events.

| Player: | 1st | 2nd | 3rd | 4th | 5th |
|---|---|---|---|---|---|
| Number of choices: | 12 | 11 | 10 | 9 | 8 |

There are $12 \cdot 11 \cdot 10 \cdot 9 \cdot 8$ or 95,040 seating arrangements possible.

The five starters were seated in a certain order. When a group of people or objects are arranged in a certain order, the arrangement is called a **permutation.** In a permutation, the *order* of the objects is very important. The arrangement of objects in a line is called a **linear permutation.**

Notice that $12 \cdot 11 \cdot 10 \cdot 9 \cdot 8$ is part of 12!. We can write an equivalent expression in terms of 12!.

$$12 \cdot 11 \cdot 10 \cdot 9 \cdot 8 = 12 \cdot 11 \cdot 10 \cdot 9 \cdot 8 \cdot \frac{7 \cdot 6 \cdot 5 \cdot 4 \cdot 3 \cdot 2 \cdot 1}{7 \cdot 6 \cdot 5 \cdot 4 \cdot 3 \cdot 2 \cdot 1}$$

$$= \frac{12 \cdot 11 \cdot 10 \cdot 9 \cdot 8 \cdot 7 \cdot 6 \cdot 5 \cdot 4 \cdot 3 \cdot 2 \cdot 1}{7 \cdot 6 \cdot 5 \cdot 4 \cdot 3 \cdot 2 \cdot 1} \text{ or } \frac{12!}{7!}$$

*In $P(12, 5)$, $n = 12$ and $r = 5$. Also, $P(12, 5) = \frac{12!}{(12 - 5)!}$.*

Note that the denominator of $\frac{12!}{7!}$ is the same as $(12 - 5)!$.

*$P(n, r)$ can also be written $_nP_r$.*

The number of ways to arrange 12 things taken 5 at a time is written as $P(12, 5)$. Thus, $P(n, r)$ is read "$n$ objects taken $r$ at a time" and is defined in the following manner.

**Definition of
$P(n, r)$**

**The number of permutations of *n* objects taken *r* at a time is defined as follows.**

$$P(n, r) = \frac{n!}{(n - r)!}$$

## Example 1

**APPLICATION**

**Photography**

**The eight high school cheerleaders are to have their pictures taken for the Eagle, their yearbook. How many different ways can three cheerleaders be chosen and lined up for each action shot?**

Find the number of permutations of 8 people, taken 3 at a time.

$$P(n, r) = \frac{n!}{(n - r)!}$$

$$P(8, 3) = \frac{8!}{(8 - 3)!} \qquad n = 8, r = 3$$

$$= \frac{8 \cdot 7 \cdot 6 \cdot \cancel{5} \cdot \cancel{4} \cdot \cancel{3} \cdot \cancel{2} \cdot \cancel{1}}{\cancel{5} \cdot \cancel{4} \cdot \cancel{3} \cdot \cancel{2} \cdot \cancel{1}} \text{ or } 336$$

There are 336 possible ways three cheerleaders can be lined up for an action shot.

In Example 1, you may have observed that the factors of $(n - r)!$ are contained in $n!$. Instead of writing all the factors of each term, you could also have evaluated the expression in the following way.

$$\frac{8!}{(8 - 3)!} = \frac{8 \cdot 7 \cdot 6 \cdot \cancel{5!}}{\cancel{5!}} \qquad \frac{5!}{5!} = 1 \qquad \textit{Remember, by}$$
$$= 8 \cdot 7 \cdot 6 \text{ or } 336 \qquad\qquad\qquad \textit{definition, 0! = 1.}$$

## Example 2

**How many ways can 4 algebra books, 3 chemistry books, and 5 history books be arranged on a shelf if the books are ordered according to subject?**

First consider how many different ways the books in each subject can be arranged.

The algebra books can be arranged in $P(4, 4)$ or 4! different ways.
The chemistry books can be arranged in $P(3, 3)$ or 3! different ways.
The history books can be arranged in $P(5, 5)$ or 5! different ways.

Now consider how many ways the 3 subjects can be arranged. There are 3 subjects, so $P(3, 3) = 3!$ ways.

The total number of ways the books can be arranged is the product of these four permutations.

$4! \cdot 3! \cdot 5! \cdot 3!$ or 103,680 ways

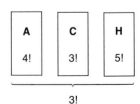

How many different arrangements can be made from the letters of the word *free*? The four letters can be arranged in $P(4, 4)$ or $4!$ ways. However, some of these 24 arrangements look the same. If we labeled the *e*'s as $e_1$ and $e_2$, then $e_1fre_2$ is different from $e_2fre_1$. However, if you drop the subscripts, the two arrangements are indistinguishable. That is, they look the same. Whenever you have items that cannot be distinguished from one another, you must account for this in your final count of possible permutations.

The two *e*'s can be arranged in $P(2, 2)$ or $2!$ ways. To find the number of different arrangements divide $P(4, 4)$ by $2!$.

$$\frac{P(4, 4)}{P(2, 2)} = \frac{4!}{2!}$$

$$= \frac{4 \cdot 3 \cdot 2!}{2!} \text{ or } 12 \qquad \text{There are 12 ways to arrange the letters.}$$

When some objects are alike, use the following rule to find the number of permutations of those objects.

| Permutations with Repetitions | **The number of permutations of $n$ objects of which $p$ are alike and $q$ are alike is** $$\frac{n!}{p!q!}$$ |
| --- | --- |

This rule can be extended for any number of objects that are repeated.

**Example 3**

How many 7-letter patterns can be formed from the letters of *benzene?*

Find the number of permutations of 7 letters of which 3 are *e*'s and 2 are *n*'s. You must divide $7!$ by both $3!$ and $2!$.

$$\frac{7!}{3!2!} = \frac{7 \cdot 6 \cdot 5 \cdot 4 \cdot 3!}{3! \cdot 2 \cdot 1} \text{ or } 420$$

There are 420 7-letter patterns.

# CHECKING FOR UNDERSTANDING

**Communicating Mathematics**

Read and study the lesson to answer each question.

1. What is a permutation?

2. What is the value of $P(5, 5)$?

3. Write an expression for the number of ways 7 books can be arranged in a row taken 3 books at a time.

**Determine whether each statement is *true* or *false*.**

**4.** $5! - 3! = 2!$

**5.** $6 \cdot 5! = 6!$

**6.** $\dfrac{6!}{3!} = 2!$

**7.** $(6 - 3)! = 6! - 3!$

**How many different ways can the letters of each word be arranged?**

**8.** FLOWER

**9.** STUDY

**10.** POP

# EXERCISES

**Practice** **Determine whether each statement is *true* or *false*.**

**11.** $\dfrac{6!}{8!} \cdot \dfrac{8!}{6!} = 1$

**12.** $3! + 4! = 5 \cdot 3!$

**13.** $\dfrac{6!}{30} = 4!$

**14.** $\dfrac{P(9, 9)}{9!} = 1$

**15.** $\dfrac{3!}{3} = \dfrac{2!}{2}$

**16.** $1!2!3!2! = 4!$

**How many different ways can the letters of each word be arranged?**

**17.** SEE

**18.** PEGGY

**19.** LEVEL

**20.** MISSISSIPPI

**21.** ALASKA

**22.** ALGEBRA

**23.** PARALLEL

**24.** ESSENTIAL

**25.** PERPENDICULAR

**Evaluate each expression.**

**26.** $\dfrac{P(10, 3)}{P(5, 3)}$

**27.** $\dfrac{P(6, 4)}{P(5, 3)}$

**28.** $\dfrac{P(5, 3)}{P(8, 5)P(5, 5)}$

**29.** $\dfrac{P(6, 3)P(4, 2)}{P(5, 2)}$

**Solve each problem.**

**30.** Don has 5 pennies, 3 nickels, and 4 dimes. The coins of each denomination are indistinguishable. How many ways can he arrange the coins in a row?

**31.** How many 6-digit numbers can be made using the digits from 833,284?

**32.** Five algebra and four geometry books are to be arranged on a shelf. How many ways can they be arranged if all the algebra books must be together?

**33.** How many ways can 4 nickels and 5 dimes be distributed among 9 children if each is to receive one coin?

**34.** There are 4 green boxes, 1 red box, and 1 blue box to be stacked one on top of the other. How many ways can they be stacked if the red box and the blue box are to be separated?

**Find $n$ in each equation.**

**35.** $P(n, 4) = 3[P(n, 3)]$

**36.** $n[P(5, 3)] = P(7, 5)$

**37.** $7[P(n, 5)] = P(n, 3) \cdot P(9, 3)$

**38.** $P(n, 4) = 40[P(n - 1, 2)]$

**39.** $208P(n, 2) = P(16, 4)$

**40.** $9P(n, 5) = P(n, 3) \cdot P(9, 3)$

**Critical Thinking**

**41.** In the junior class at Grenada Hills High School, the following statistics are true.

115 students take algebra 2 and 99 students take chemistry.
49 students take Spanish.
15 students take algebra 2 and chemistry.
20 students take algebra 2 and Spanish.
22 students take chemistry and Spanish.
12 students take algebra 2, chemistry, and Spanish.
31 students are not taking any of these three courses.

How many students are in the junior class?

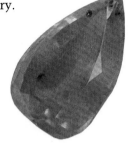

**Applications**

**42. Gemology** Madame Estelle designs jewelry. She is designing a bracelet that will contain a gem in each link of the bracelet. She has 8 emeralds, 5 rubies, and 3 diamonds. The gems in each type of stone are indistinguishable from one another. How many different bracelet designs are possible?

**43. Statistics** Nine scores received on a test were 82, 91, 75, 83, 64, 83, 77, 91, and 75.

**a.** In how many different orders might the scores be recorded?

**b.** What is the average of these test scores?

**44. Communication** There are 3 identical red flags and 5 identical white flags that are used to send signals. All 8 flags are arranged in a row. How many signals can be given?

**Mixed Review**

**45.** How many ways can you have 50¢ using at least 1 quarter? **(Lesson 15-2)**

**46.** State the stems that would be used to make a stem-and-leaf plot for {63, 57, 49, 52, 64, 31, 27, 82, 47, 61}. **(Lesson 14-2)**

**47.** Solve $\log_3(x + 6) = 2 \log_3 x$. **(Lesson 12-3)**

**48. Travel** A plane flies from Baltimore to Omaha with a tailwind of 120 km/h in 2.5 hours. On the return trip flying into the same wind takes 3.75 hours. In still air, the plane would have the same speed both ways. Find the distance from Baltimore to Omaha. **(Lesson 11-7)**

**49.** Solve the system $\begin{cases} x + 2y = 11 \\ x - 4y = 2 \end{cases}$. **(Lesson 3-2)**

**50.** If $a = 2$, $b = -6$, and $c = 3$, find the value of $d$ if $a^3b^2 + 4ac + 2d \geq 6c^2 - 4ab$. **(Lesson 1-7)**

# Circular Permutations

**Objective**

After studying this lesson, you should be able to:

■ solve problems involving circular permutations.

**Application**

Lee's Catering offers party trays that are made up of six items. Mrs. Sitkowski orders a tray with cheese, pickles, ham, salami, turkey, and roast beef. How many different ways can these items be arranged on the tray if each item must fit in its own section?

Let these letters represent the items.

| | |
|---|---|
| c = cheese | p = pickles |
| h = ham | s = salami |
| t = turkey | r = roast beef |

Think of each tray as a circle. Three possible arrangements of the various items are shown below.

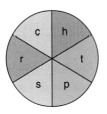 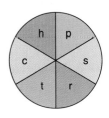 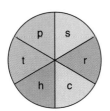

How does the arrangement change as the first tray is turned? Which arrangement is *really* different from the other two?

When 6 different objects are placed in a line, there are 6! or 720 arrangements of the 6 objects taken 6 at a time. However, when the 6 objects are arranged in a circle, some of the arrangements are alike. These arrangements fall into groups of 6. Once an arrangement is determined, the other five members of the group are formed by rotating the circle. Then you can rearrange the items and another group of 6 is formed. Thus, the total number of *really* different arrangements of 6 objects around a circle is $\frac{1}{6}$ of the total number of arrangements in a line.

$$\frac{1}{6} \cdot 6! = \frac{6 \cdot 5 \cdot 4 \cdot 3 \cdot 2 \cdot 1}{6}$$

$$= 5 \cdot 4 \cdot 3 \cdot 2 \cdot 1$$

$$= 5! \text{ or } 120 \qquad \text{\textit{Note that } 5! = (6 - 1)!.}$$

There are $(6 - 1)!$ arrangements of 6 objects in a circle. This is known as a **circular permutation**.

<table>
<tr><td>*Circular Permutations*</td><td>If *n* distinct objects are arranged in a circle, then there are $\frac{n!}{n}$ or $(n - 1)!$ permutations of the objects around the circle.</td></tr>
</table>

**Example 1**

**Five students are assigned to each project group. One group goes to the library to begin their research. If they sit at a round table, how many different seating arrangements are possible?**

Since there are 5 people, there will be $(5 - 1)!$ or $4!$ arrangements.

$(5 - 1)! = 4!$

$\qquad = 4 \cdot 3 \cdot 2 \cdot 1$ or 24

There are 24 seating arrangements possible for the group.

Suppose *n* objects are in a circular arrangement, but the position of the objects is related to a fixed point. Rotating the circle will relate a different object to the fixed point and will make a new arrangement of the objects. Because of this fixed point, the permutations are now considered linear. The number of permutations for a circular permutation with a fixed point is *n*!. Consider the following contrasting situations.

I. Let each circle represent a table and the labeled points be the people at that table. Let the arrow represent the seat next to the door. How many arrangements are possible?

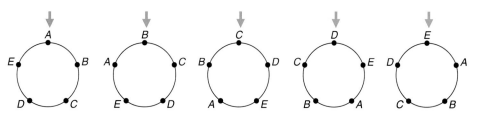

These arrangements can be considered different because in each one, a different person sits next to the door. Thus, there are $P(5, 5)$ or $5!$ arrangements relative to a fixed point.

$5! = 5 \cdot 4 \cdot 3 \cdot 2 \cdot 1$ or 120

II. Suppose three keys are placed on a key ring. How many different arrangements are possible?

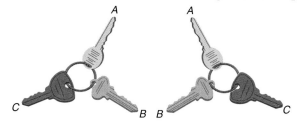

It appears that there are at most $(3 - 1)!$ or 2 different arrangements of keys on the ring.

But what happens if the first key ring arrangement is turned over?

When the key ring is turned over, the first arrangement becomes the second arrangement. Then there is really only one arrangement of the three keys. These two arrangements are **reflections** of each other. As a result, there are only half as many arrangements when reflections are possible.

$$\frac{(3-1)!}{2} = \frac{2 \cdot 1}{2} \text{ or } 1$$

Reflections also occur in linear arrangements. Suppose Lyle holds a multicolored flag up for Rosa to see. From his perspective the flag is red, white, blue, and yellow, in order from left to right. However, from Rosa's view, it is yellow, blue, white, and red, from left to right.

If you were asked how many possible arrangements of these colors there were for the flag, your answer would be half of 4! or 12 arrangements.

**Example 2**

Six charms are to be placed on a bracelet.
**a. How many different ways can they be placed if the bracelet has no clasp?**

a. Since there is no fixed starting point for the arrangements, this is a circular permutation. However, since the bracelet can be turned over, it is also a reflection. So, the number of arrangements will be half that of a circular permutation.

$$\begin{aligned} \frac{(6-1)!}{2} &= \frac{5!}{2} \\ &= \frac{5 \cdot 4 \cdot 3 \cdot 2 \cdot 1}{2} \text{ or } 60 \end{aligned}$$

There are 60 different ways to arrange the charms.

**b. How many different ways can they be placed if the bracelet has a clasp?**

b. Since there is a clasp, this is treated as a linear permutation. However, it is still reflective.

$$\frac{6!}{2} = \frac{6 \cdot 5 \cdot 4 \cdot 3 \cdot 2 \cdot 1}{2} \text{ or } 360$$

There are 360 different ways to arrange the charms.

# CHECKING FOR UNDERSTANDING

**Communicating Mathematics**

Read and study the lesson to answer each question.

1. How does a circular permutation differ from a linear permutation?
2. Explain how a reflection affects arrangements.
3. Under what condition is a circular arrangement treated as a linear permutation?

**Guided Practice**

Determine whether each arrangement is linear or circular. Then determine if it is also a reflection. Then find the number of arrangements.

4. 8 charms on a bracelet that has no clasp
5. a football huddle of 11 players
6. placing 6 coins in a circle on a table
7. 10 beads on a necklace with a clasp

# EXERCISES

**Practice**

Determine whether each arrangement is linear or circular. Then determine if it is also a reflection.

8. a pearl necklace that is open
9. four people seated around a square table relative to each other
10. people seated around a square table relative to one chair
11. a baseball team's batting order

Evaluate each expression.

12. $\dfrac{P(6, 4) \cdot P(5, 2)}{5!}$

13. $\dfrac{P(8, 3) \cdot P(5, 4)}{P(6, 6)}$

14. $\dfrac{P(12, 6)}{P(12, 3) \cdot P(8, 2)}$

Solve each problem.

15. How many ways can 6 keys be arranged on a key ring?
16. How many ways can 6 campers be arranged around a campfire?
17. How many ways can 8 charms be arranged on a bracelet that has no clasp?
18. How many ways can 4 councilmen and 4 councilwomen be seated alternately at a round table?
19. How many ways can 5 hors d'oeuvres be arranged around a plate?
20. How many ways can 6 points be labeled $A$ through $F$ on the circle at the right, relative to its $x$-intercept?

CONNECTION

Geometry

21. How many ways can 5 dinner guests be seated around a circular table if the only two married guests are seated next to each other?

22. A necklace has 20 links and no clasp. How many ways can 6 different colored beads be added to the necklace if each bead requires one link?

**Critical Thinking**

23. **Construction** On a square city block, there are 8 lots on each side of the street between cross-streets. If a builder is to build any one of five basic designed homes on each lot on the block, how many different arrangements could exist? Write your answer in exponential form.

**Applications**

24. **Design** Forty pearls are strung in a circle. Twenty-eight are white and 12 are black. How many ways can the pearls be strung if a clasp is used?

25. **Electricity** Eight switches are connected on a circuit so that if any one or more of the switches are closed, the light will go on. How many combinations exist of open and closed switches that will permit the light to go on?

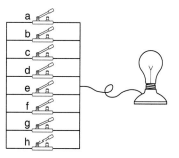

26. **Environmentalism** Sam Marshall is a world environmentalist who keeps video tapes of his research in different countries. On one circular carousel he has six tapes of South America, four tapes of Africa, and two tapes of Australia. How many different ways can he arrange those tapes on the carousel?

**Mixed Review**

27. How many ways can 5 books be placed on a shelf? **(Lesson 15-3)**

28. Write $\frac{2}{3} + \frac{3}{5} + \frac{4}{7} + \frac{5}{9} + \cdots + \frac{21}{41}$ using sigma notation. **(Lesson 13-3)**

29. Tell the nature of the roots of $0 = x^3 - x^2 + 2x - 3$. **(Lesson 10-4)**

30. Solve $35 = 3b^2 - 8b$. Then use the sum and product of the roots to check your solution. **(Lesson 7-5)**

31. Use synthetic division to find $\dfrac{x^3 - 4x^2 - 7x + 10}{x - 1}$. **(Lesson 5-7)**

32. Discuss the relationships of the slopes of parallel lines compared with the slopes of perpendicular lines. **(Lesson 2-4)**

## 15-5 Combinations

**Objective**

After studying this lesson, you should be able to:
- solve problems involving combinations.

**Application**

Suppose six girls from a group of nine are chosen to start the championship volleyball game. In this case, the order in which the girls are chosen is not important. Such a selection is called a **combination.**

The combination of nine objects or people taken six at a time is written as $C(9, 6)$. You know that six objects can be arranged in 6! ways. These arrangements are eliminated with finding the number of combinations because order does not matter.

$$C(9, 6) = \frac{P(9, 6)}{6!}$$

$$= \frac{9!}{(9 - 6)! \cdot 6!} \qquad P(9, 6) = \frac{9!}{(9 - 6)!}$$

$$= \frac{9!}{3! \cdot 6!} \text{ or } 84$$

The six starters can be chosen from the nine team members in 84 ways.

In the solution above, notice that 3! and $(9 - 6)!$ are equivalent. This suggests the following definition.

| *Definition of $C(n, r)$* | **The number of combinations of $n$ distinct objects taken $r$ at a time is defined as follows.** $$C(n, r) = \frac{n!}{(n - r)!r!}$$ |
| --- | --- |

*$C(n, r)$ can also be written $_nC_r$.*

The basic difference between a permutation and a combination is whether order is considered (as in permutation) or not (as in a combination).

**Example 1**

The U.S. history class is preparing for its final exam. From a class of 18 males and 12 females, how many study groups of 2 males and 3 females can be formed?

Order is not considered. There are two questions to be considered.
- How many ways can 2 males be chosen from 18?
- How many ways can 3 females be chosen from 12?

The answer is the product of these two combinations, $C(18, 2)$ and $C(12, 3)$.

You can use your calculator to evaluate this expression.

$$C(18, 2) \cdot C(12, 3) = \frac{18!}{(18 - 2)!2!} \cdot \frac{12!}{(12 - 3)!3!} \text{ or } \frac{18! \, 12!}{16! \, 2! \, 9! \, 3!}$$

ENTER: 18 $\boxed{x!}$ $\boxed{\times}$ 12 $\boxed{x!}$ $\boxed{\div}$ $\boxed{(}$ 16 $\boxed{x!}$ $\boxed{\times}$ 2 $\boxed{x!}$ $\boxed{\times}$

9 $\boxed{x!}$ $\boxed{\times}$ 3 $\boxed{x!}$ $\boxed{)}$ $\boxed{=}$ `33660`

There are 33,660 possible combinations of 2 males and 3 females in a group.

## Example 2

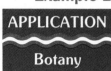

APPLICATION

Botany

**A bucket at Fireside Florists contains 8 red carnations, 5 white daisies, and 4 blue carnations. How many bouquets can be chosen so that each bouquet has 2 red carnations, 1 white daisy, and 2 blue carnations?**

This involves the product of three combinations—one for each type of flower.

$C(8, 2)$     Select 2 of the 8 red carnations.
$C(5, 1)$     Select 1 of the 5 white daisies.
$C(4, 2)$     Select 2 of the 4 blue carnations.

$$C(8, 2) \cdot C(5, 1) \cdot C(4, 2) = \frac{8!}{6!2!} \cdot \frac{5!}{4!1!} \cdot \frac{4!}{2!2!}$$
$$= \frac{8 \cdot 7 \cdot 6!}{6! \cdot 2} \cdot \frac{5 \cdot 4!}{4!} \cdot \frac{4 \cdot 3 \cdot 2!}{2 \cdot 2!}$$
$$= 28 \cdot 5 \cdot 6 \text{ or } 840$$

There are 840 bouquets that could be made.

An important skill in problem solving is being able to distinguish between permutations and combinations.

## Example 3

CONNECTION

Geometry

*A decagon is a polygon with 10 vertices and 10 sides. A diagonal is a segment that connects any two nonconsecutive vertices of a polygon.*

**Find the total number of diagonals that can be drawn in a decagon.**

Each diagonal has two endpoints. Suppose one has endpoints $A$ and $E$. Then segments $AE$ and $EA$ are the same. Thus, order is not important. The combination of 10 points taken two at a time gives the total number of segments connecting those 10 points.

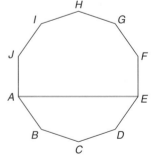

$$C(10, 2) = \frac{10!}{8!2!} \text{ or } 45$$

However, 45 is not our answer. Since 10 of the segments connecting the points are sides of the decagon, you must subtract 10 from 45, the number of combinations.

The total number of diagonals in a decagon is $45 - 10$ or 35.

Some applications can involve *both* permutations and combinations.

**Example 4**

> **From a deck of 52 cards, how many ways can 5 cards be drawn so that 3 are of one suit and 2 are of another?**
>
> First consider how many ways 2 suits can be chosen from 4. Since a different number of cards is being selected from each suit, order is important. Then consider the combinations possible with each suit.
>
> $P(4, 2)$      Select 2 suits from 4 suits.
> $C(13, 3)$    Select 3 cards from 13 cards in one suit.
> $C(13, 2)$    Select 2 cards from 13 cards in the other suit.
>
> $$P(4, 2) \cdot C(13, 3) \cdot C(13, 2) = \frac{4!}{2!} \cdot \frac{13!}{10!3!} \cdot \frac{13!}{11!2!}$$
> $$= 12 \cdot 286 \cdot 78 \text{ or } 267{,}696$$
>
> There are 267,696 ways to draw the cards.

# CHECKING FOR UNDERSTANDING

**Communicating Mathematics**

**Read and study the lesson to answer each question.**

1. Describe the difference between a permutation and a combination.

2. A starting baseball team is formed from a group of 12 players. Explain why this might be considered a permutation.

3. A student approaches you and asks you for the permutation to your permutation lock. You look at them with confusion and then say, "You mean what's the combination to my combination lock?" The student insists upon using the word permutation instead of combination. Who is correct mathematically?

**Guided Practice**

**Determine whether each situation involves a permutation or a combination.**

4. 5 books on a shelf

5. a rummy hand of 7 cards

6. a seating chart in a classroom

7. finding the diagonals of a polygon

**Evaluate each expression.**

8. $C(5, 3)$

9. $C(12, 5)$

10. $C(3, 2) \cdot C(8, 3)$

11. $\dfrac{C(10, 3)}{C(5, 2)}$

# EXERCISES

**Practice**  **Determine whether each situation involves a permutation or a combination.**

12. 3-letter patterns, chosen from the letters in the word *algebra*

13. a hand of 5 cards from a deck of cards

14. a team of 5 people chosen from a group of 12 people

**Determine whether each situation involves a permutation or a combination.**

15. the answers on a true/false test

16. the batting order of the Los Angeles Dodgers

17. putting students in assigned seats

18. 8 guests seated around a table for dinner

19. a 2-man, 2-woman subcommittee of the Campaign Funds committee that has 8 men and 7 women

**Evaluate each expression.**

20. $C(7, 2)$      21. $C(8, 3)$      22. $C(8, 5) \cdot C(7, 3)$   23. $C(24, 21)$

**On a circle are nine points randomly placed. In how many different ways can you form each polygon?**

24. triangle      25. quadrilateral      26. pentagon

27. hexagon      28. octagon      29. decagon

**Solve each problem.**

30. You are required to read 5 books from a list of 12 great American novels. How many different groups can be selected?

31. There are 85 telephones in the editorial department of 'TEEN magazine. How many 2-way connections can be made among the office phones?

32. How many starting baseball teams of 9 members can be formed from a bench of 14 multitalented players?

33. There are 27 people in an algebra class, but only 25 computers in the computer lab, so each student must take turns going to the lab. How many different groups of 25 can the teacher send to the lab?

34. Suppose there are 8 points in a plane such that no three points are collinear. How many distinct triangles could be formed with 3 of these points as vertices?

35. Consider a deck of cards.

   a. How many different 5-card hands can have 5 cards of the same suit?

   b. How many different 4-card hands can have each card from a different suit?

**Find the value of *n*.**

36. $C(n, 8) = C(n, 3)$         37. $C(n, 12) = C(30, 18)$

38. $C(n, 5) = C(n, 7)$         39. $C(14, 3) = C(n, 11)$

**A bag contains 9 blue, 4 red, and 6 white chips. How many ways can 5 chips be selected to meet each condition?**

40. all white                     41. all blue

42. 2 red, 2 white, 1 blue      43. all red

44. 2 are blue             45. 2 one color, 3 another color

**From a group of 8 juniors and 10 seniors, a committee of 5 is to be formed to discuss plans for the spring dance. How many committees can be formed given each condition?**

46. all juniors            47. 3 juniors, 2 seniors

48. 1 junior, 4 seniors     49. all seniors

**Critical Thinking**

50. Prove $P(n, r) = r! \cdot C(n, r)$.

**Applications**

51. **Sports**  How many different baseball teams of nine players can Gardena High School put on the field if they have four players that can only pitch and the remaining twelve players can play any of the other eight positions?

52. **Statistics**  The California state lottery is called "6/53," meaning a winner must select six different correct numbers from the numbers 1 through 53.

    a. How many different ways can six numbers be selected?

    b. Suppose you had to select the numbers in a specific order. What would be the number of possibilities?

    c. Refer to your answers in parts **a** and **b**. Describe each situation as a permutation or a combination. Explain.

53. **Science Fiction**  The planets Laffaglonia and Madaglonia have been at war for centuries. At the Laffaglonia spy-training school, 10 agents are being trained for a secret mission. What the Laffaglonians don't know is that 2 of the 10 are actually Madaglonian counterspies. If 5 of the agents are chosen at random for a secret mission, how many groups will have at least one Madaglonian counterspy?

54. How many ways can five people be seated around a circular table relative to each other? **(Lesson 15-4)**

55. **Business** A piece of machinery valued at $75,000 depreciates at a steady rate of 8% yearly. When will the value be $15,000? Use $V_n = P(1 + r)^n$. **(Lesson 13-7)**

56. **Finance** If $200 is invested at 8% interest compounded continuously, when will the investment double? Use $A = Pe^{rt}$. **(Lesson 12-7)**

57. Graph $4x^2 + 25y^2 = 100$ and $4x^2 - 25y^2 = 100$. What kind of graph does each equation have? **(Lesson 9-7)**

58. Solve $\begin{cases} 2x - y - 5z = 3 \\ x + 4y - 2z = 3 \\ 5x + 3y + 2z = 1 \end{cases}$ using matrices. **(Lesson 4-7)**

59. Find the next three terms in this pattern.
    $a, a^2, a^2b, a^2b^2, a^2b^2c, \underline{\ ?\ }, \underline{\ ?\ }, \underline{\ ?\ }$ **(Lesson 2-3)**

60. Find all integral ordered pairs $(x, y)$ such that $xy = 64$. **(Lesson 1-5)**

## MID-CHAPTER REVIEW

1. Six books are to be arranged on a shelf in your classroom. **(Lesson 15-1)**
   a. Name some models you could use to represent these books and how to arrange them.
   b. How many ways can the books be arranged if the first and last must always be the same two books?

2. At the Burger Bungalow, you can order your hamburger with or without cheese, with or without onions or pickles, and either rare, medium, or well-done. **(Lesson 15-2)**
   a. What type of pictorial representation could you use to calculate the number of choices?
   b. How many different hamburgers are possible?

3. **Government** How many ways can the 100 United States senators seat themselves in a 100-seat auditorium if there are no restrictions? *Write your answer in factorial form.* **(Lesson 15-3)**

4. In Kentucky, license plate numbers are composed of three letters followed by a dash and three numbers. The letters I and O are never used. How many license plates are possible? **(Lesson 15-3)**

5. Find the number of ways Sam, Renee, Julie, Henri, and Denzel can sit around a table if Denzel and Julie insist upon sitting together. **(Lesson 15-4)**

6. In bridge, all 52 cards in a deck are dealt among four players. How many different hands are possible? *Write your answer in factorial form.* **(Lesson 15-5)**

## 15-6 Probability

**Objective**

After studying this lesson, you should be able to:

■ find the probability of an event and determine the odds of success or failure.

When a coin is tossed, only two outcomes are possible—*heads* or *tails*. The desired outcome is called a **success.** Any other outcome is referred to as a **failure.** The likelihood of a success or of a failure is called the **probability** of the event.

| Probability of Success and of Failure | **If an event can succeed in $s$ ways and fail in $f$ ways, then the probabilities of success, $P(s)$, and of failure, $P(f)$, are as follows.** $$P(s) = \frac{s}{s+f} \qquad P(f) = \frac{f}{s+f}$$ |
|---|---|

*What does the sum of s and f represent?*

If the event cannot succeed, $P(s) = 0$ and $P(f) = 1$. If the event cannot fail, $P(s) = 1$ and $P(f) = 0$. Thus, $P(s)$ and $P(f)$ are always between 0 and 1, inclusive. In fact, the sum of $P(s)$ and $P(f)$ is always 1. Thus, they are called *complements*. So, if $P(s) = \frac{1}{3}$, then $P(f) = 1 - \frac{1}{3}$ or $\frac{2}{3}$.

**Example 1**

**A box contains 5 blue pencils and 4 white pencils. If one pencil is chosen at random, what is the probability that it is blue?**

*The term at random means that an outcome is chosen without any preference.*

The probability of selecting a blue pencil is written $P(blue\ pencil)$.

There are 5 ways to select a blue pencil from the box and 4 ways not to select a blue pencil from the box. The total number of possible selections is 5 + 4, or 9.

$$P(blue\ pencil) = \frac{s}{s+f} \qquad \text{\textit{Replace s with 5 and f with 4.}}$$

$$= \frac{5}{5+4} \text{ or } \frac{5}{9}$$

The probability of selecting a blue pencil is $\frac{5}{9}$ or about 0.556.

The counting methods you have studied for permutations and combinations are often used in determining probability.

**Example 2**

> There are 6 men and 3 women on city council. A committee of 2 is to be selected at random to study the city's plans for park expansion. What is the probability that the 2 selected are women?
>
> Since a committee is being formed, order is not important. So find the probability using combinations.
>
> There are $C(3, 2)$ ways to choose 2 women from 3 women. So, $C(3, 2) = s$. There are $C(9, 2)$ ways to form a committee of 2 from a group of 9 people. So, $s + f = C(9, 2)$.
>
> $$P(\text{two women}) = \frac{C(3, 2)}{C(9, 2)} \qquad P(s) = \frac{s}{s + f}$$
>
> $$= \frac{\dfrac{3!}{1!2!}}{\dfrac{9!}{7!2!}} \qquad C(n, r) = \frac{n!}{(n - r)!r!}$$
>
> $$= \frac{3}{36} \text{ or } \frac{1}{12}$$
>
> The probability of selecting 2 women for the committee is $\frac{1}{12}$ or about 0.083.   *0.083 = 8.3%*

You may have heard someone say, "The odds of that happening are 50/50." **Odds** is the term for the ratio of successes to failures.

**Definition of Odds**

> The odds of the successful outcome of an event is expressed as the ratio of the number of ways it can succeed to the number of ways it can fail.
>
> **Odds = number of successes:number of failures or $\dfrac{s}{f}$**

**Example 3**

> What are the odds of rolling a die and getting a 3?
>
> Remember that when rolling a die there are 6 possible outcomes. The number 3 is only on one face of the die. Each of the other five faces have a number other than 3.
>
> Odds of rolling a 3 $= \dfrac{1}{5}$    *1 outcome (3) is a success.*
> *5 outcomes (1, 2, 4, 5, 6) are failures.*
>
> The odds of rolling a 3 are 1 to 5.

Sometimes when figuring the odds, you must first find out the total number of possible outcomes. This can involve finding permutations and combinations.

**Example 4**

Suppose Amar draws 5 cards from a deck of 52 cards. What are the odds that the first 4 cards drawn will be of one suit and the 5th card will be of another suit?

First, find how many 5-card hands meet these conditions.

$P(4, 2)$  Select 2 suits among 4. Order is important since different numbers of cards are to come from each suit.

$C(13, 4)$  Select 4 cards from a suit containing 13 cards.

$C(13, 1)$  Select 1 card from the other suit.

Now use the counting principle.

$$P(4, 2) \cdot C(13, 4) \cdot C(13, 1) = \frac{4!}{2!} \cdot \frac{13!}{9!4!} \cdot \frac{13!}{12!1!} \text{ or } 111{,}540$$

So, the number of successes is 111,540.

Now find the total number of possible 5-card hands.

$$C(52, 5) = \frac{52!}{47!5!} \text{ or } 2{,}598{,}960$$

The number of 5-card hands that do not meet the conditions is 2,598,960 − 111,540 or 2,487,420. So, the number of failures is 2,487,420.

The odds of a 5-card hand with the first 4 cards of one suit and the 5th card of another suit are $\frac{111{,}540}{2{,}487{,}420}$ or $\frac{143}{3189}$. This is about $\frac{1}{22}$ or 0.045.

# CHECKING FOR UNDERSTANDING

**Communicating Mathematics**

Read and study the lesson to answer each question.

1. What is meant by $P(s)$?

2. What is the value of $P(s) + P(f)$? Justify your result.

3. If $P(f) = \frac{4}{5}$, what is $P(s)$?

4. Can $P(s) = \frac{3}{2}$? Explain your answer.

5. Can the odds of an event occurring be $\frac{3}{2}$? Explain your answer.

**Guided Practice**

State the odds of an event occurring given the probability of the event.

6. $\frac{2}{3}$

7. $\frac{5}{9}$

8. $\frac{1}{2}$

9. $\frac{3}{7}$

State the probability of an event occurring given the odds of the event.

10. $\frac{3}{5}$        11. $\frac{4}{3}$        12. $\frac{8}{1}$        13. $\frac{1}{5}$

# EXERCISES

**Practice**    State the odds of an event occurring given the probability of the event.

14. $\frac{3}{4}$        15. $\frac{1}{7}$        16. $\frac{5}{8}$        17. $\frac{7}{15}$

State the probability of an event occurring given the odds of the event.

18. $\frac{3}{7}$        19. $\frac{5}{1}$        20. $\frac{6}{5}$        21. $\frac{1}{1}$

Solve each problem.

22. The odds are 6-to-1 that the crosstown rivals will win in the championship football game on Friday night. What is the probability that they will win?

23. The probability of Kellyn getting an A on her final exam is $\frac{3}{4}$. What are the odds that she will *not* get an A?

A canister contains 20 pieces of candy: 5 strawberry flavored, 9 watermelon flavored, and 6 mint flavored. Two are selected at random. Find each probability.

24. P(*2 strawberry*)             25. P(*2 watermelon*)

26. P(*2 mint*)             27. P(*1 strawberry and 1 mint*)

There are 5 frozen juice bars and 8 frozen yogurt bars in the freezer. Dana reaches in the freezer and grabs 2 without looking. Find the probability of each selection. Then find the odds of that selection.

28. P(*2 juice bars*)    29. P(*2 yogurt bars*)    30. P(*1 of each kind of bar*)

Tommie's bank contains 7 pennies, 4 nickels, and 5 dimes. His parents tell him that he can spend the first three coins that he can shake out of the bank. Find each probability.

31. P(*all pennies*)             32. P(*all nickels*)

33. P(*1 dime, 2 nickels*)         34. P(*2 pennies, 1 dime*)

35. P(*1 dime, 1 nickel, 1 penny*)      36. P(*2 dimes, 1 quarter*)

**Suppose you select 2 letters at random from the word** *algebra*. **Find each probability.**

**37.** $P$(*selecting 2 consonants*)          **38.** $P$(*selecting 2 vowels*)

**39.** $P$(*selecting 1 vowel and 1 consonant*)

**From a deck of 52 playing cards, 5 cards are dealt.**

**40.** What are the odds of each event occurring?

    **a.** all aces          **b.** all face cards          **c.** all from one suit

**41.** What are the odds against each event occurring?

    **a.** all hearts          **b.** all face cards          **c.** all from one suit

    **d.** the first 3 from one suit and the last 2 from another

**Critical Thinking**

**42.** A red die and a green die are tossed. What is the probability that the number showing on the green die is greater than the number showing on the red die?

**Applications**

**43. Finance**   The state of Ohio has a Super Lotto drawing twice a week in which 6 numbers out of 46 are drawn at random. The proceeds from the lottery help to finance education in the state. What is the probability of winning the Super Lotto?

**44. Sports**   A dart board is designed as shown at the right. A dart is thrown and hits the board. Find each probability.   *Hint: What is the area of each ring?*

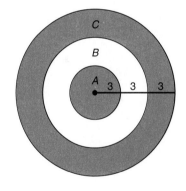

    **a.** $P$(*landing in the A ring*)

    **b.** $P$(*landing in the B ring*)

    **c.** $P$(*landing in the C ring*)

**Mixed Review**

**45.** The Groveport Community Children's Chorus now has six altos and eight sopranos. For the songs they are performing at the spring concert, they need two alto soloists and two soprano soloists. How many ways can these four soloists be selected at random?   **(Lesson 15-5)**

**46.** Expand $(4r + s)^5$.   **(Lesson 13-7)**

**47.** Solve $\log_7 y = 4$.   **(Lesson 12-3)**

**48.** If $f(x) = 2x + 3$ and $g(x) = x^2 + 1$, find $f[g(x)]$.   **(Lesson 10-7)**

**49.** Graph $2x^2 - x - 6 < y$.   **(Lesson 8-6)**

**50.** Solve $6x^2 + 7x - 3 = 0$ by completing the square.   **(Lesson 7-3)**

**51.** Simplify $\left(-\dfrac{1}{2} + \dfrac{i\sqrt{3}}{2}\right)^3$.   **(Lesson 6-10)**

# Multiplying Probabilities

**Objective**

After studying this lesson, you should be able to:

- find the probability of two or more independent or dependent events.

Suppose you toss a red die and a blue die. The probability that the red die shows a 2 is $\frac{1}{6}$. The probability that the blue die shows a 2 is $\frac{1}{6}$. By using the Counting Principle, the probability that both dice show 2s is $\frac{1}{6} \cdot \frac{1}{6}$ or $\frac{1}{36}$.

Since the outcome of tossing the red die does not affect the outcome of tossing the blue die, the events are independent.

| *Probability of Two Independent Events* | **If two events, _A_ and _B_, are independent, then the probability of both events occurring is found as follows.**<br>$$P(A \text{ and } B) = P(A) \cdot P(B)$$ |
|---|---|

**Example 1**

Suppose you spin the spinner three times. Find the probability of each outcome.

**a.** $A$, then $A$, then $A$     **b.** $A$, then $B$, then $C$     **c.** $B$, then $C$, then $C$

Each of these outcomes is independent since what you spin on one turn does not affect the results of other turns.

**a.** $P(A,A,A) = P(A) \cdot P(A) \cdot P(A)$

$$= \frac{3}{6} \cdot \frac{3}{6} \cdot \frac{3}{6}$$

$$= \frac{1}{2} \cdot \frac{1}{2} \cdot \frac{1}{2} \text{ or } \frac{1}{8}$$

**b.** $P(A,B,C) = P(A) \cdot P(B) \cdot P(C)$

$$= \frac{3}{6} \cdot \frac{2}{6} \cdot \frac{1}{6}$$

$$= \frac{1}{2} \cdot \frac{1}{3} \cdot \frac{1}{6} \text{ or } \frac{1}{36}$$

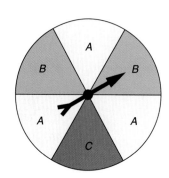

**c.** $P(B,C,C) = P(B) \cdot P(C) \cdot P(C)$

$$= \frac{2}{6} \cdot \frac{1}{6} \cdot \frac{1}{6}$$

$$= \frac{1}{3} \cdot \frac{1}{6} \cdot \frac{1}{6} \text{ or } \frac{1}{108}$$

Example 2

**APPLICATION**

**Entertainment**

The high school chorus often sings selections from a songbook called *Great American Tunes*. Even though they contain the same songs, 5 of the songbooks have red covers and 4 have blue. One student selects a songbook at random from the shelf, looks up a composer, and places the book back on the shelf. A second student does the same thing. What is the probability that both students looked at songbooks with red covers?

The events are independent since the first book is placed back on the shelf. The outcome of the second selection is not affected by the results of the first selection.

$P(both\ red) = P(red) \cdot P(red)$

$$= \frac{5}{9} \cdot \frac{5}{9}\ or\ \frac{25}{81} \qquad \frac{25}{81} \approx 0.309,\ which\ is\ a\ little\ less\ than\ \frac{1}{3}.$$

In the example above, what is the probability of selecting 2 red songbooks if the first selection is *not* put back on the shelf? These events are *dependent* because the outcome of the first selection does affect the outcome of the second selection. Suppose the first selection is red.

| first selection | second selection | |
|---|---|---|
| $P(red) = \dfrac{5}{9}$ | $P(red) = \dfrac{4}{8}$ | *Notice that when the red book is removed, there is not only one less red book but one less book on the shelf.* |

$P(both\ red) = P(red) \cdot P(red\ following\ red)$

$$= \frac{5}{9} \cdot \frac{4}{8}\ or\ \frac{5}{18}$$

The probability of selecting 2 red songbooks is $\frac{5}{18}$ or about 0.278. This is an application of the probability of two dependent events.

| *Probability of Two Dependent Events* | **If two events, A and B, are dependent, then the probability of both events occurring is found as follows.** $$P(A\ and\ B) = P(A) \cdot P(B\ following\ A)$$ |
|---|---|

**Example 3**

There are 9 pennies, 7 dimes, and 5 nickels in an antique coin collection. Suppose two coins are to be selected at random from the collection without replacing the first one. Find the probability of each event.

Because the coins are not replaced, these events are dependent. Thus, $P(A\ and\ B) = P(A) \cdot P(B\ following\ A)$. Let $p$ = penny, $d$ = dime, and $n$ = nickel.

**a.** a penny, then a dime

$P(p,\ then\ d) = P(p) \cdot P(d\ following\ p)$

$$P(p,\ then\ d) = \frac{9}{21} \cdot \frac{7}{20}\ or\ \frac{3}{20}$$

The probability is $\frac{3}{20}$ or 0.15.

**b.** two nickels

$P(n,\ then\ n) = P(n) \cdot P(n\ following\ n)$

$$P(n,\ then\ n) = \frac{5}{21} \cdot \frac{4}{20}\ or\ \frac{1}{21}$$

The probability is $\frac{1}{21}$ or about 0.048.

| Example 4 | From a deck of 52 cards, an 8, a 9, and then another 8 are selected in that order. |
|---|---|

**a.** First find the probability of this event occurring if the cards are replaced after each selection.

When the cards are replaced, the events are independent.

$$P(8, 9, 8) = P(8) \cdot P(9) \cdot P(8)$$

$$P(8, 9, 8) = \frac{4}{52} \cdot \frac{4}{52} \cdot \frac{4}{52} \text{ or } \frac{1}{2197}$$

The probability is $\frac{1}{2197}$ or about 0.0005.

**b.** Find the probability of the event occurring if the cards are not replaced.

When the cards are not replaced, the events are dependent.

$$P(8, 9, 8) = P(8) \cdot P(9 \text{ following } 8) \cdot P(8 \text{ following } 9 \text{ following } 8)$$

$$P(8, 9, 8) = \frac{4}{52} \cdot \frac{4}{51} \cdot \frac{3}{50} \text{ or } \frac{2}{5525}$$

The probability is $\frac{2}{5525}$ or about 0.0004.

# CHECKING FOR UNDERSTANDING

**Communicating Mathematics**

**Read and study the lesson to answer each question.**

1. What is the difference between independent events and dependent events?

2. What effect does replacing the item before the second choice have on finding the probability of the event?

3. In a given situation of selecting items, is the probability greater or less if replacement does not occur?

**Guided Practice**

**Determine if each event is independent or dependent. Then find the probability.**

4. A bag contains 5 red, 3 green, and 8 blue marbles. Three are selected in sequence without replacement. What is the probability of selecting a red, a green, and a blue in that order?

5. There are 4 glasses of iced tea and 3 glasses of lemonade on the counter. Bill drinks two of them at random. What is the probability that he drank 2 glasses of iced tea?

# EXECISES

**Practice**

**Determine if each event is independent or dependent. Then find the probability.**

6. Monique came home from school to find a bowl of 5 apricots and 4 plums on the table. She decides to have a snack. First she selects one and then puts it back. She then selects another. What is the probability both selections were apricots?

7. When Josh plays Sven on his video game, the odds are 3 to 2 that he will win. What is the probability that he will win the next four games?

**The Scrabble® tiles A, B, E, I, J, K, and M are placed face down in the lid of the game and mixed up. Two tiles are chosen at random. Find each probability.**

8. *P(selecting 2 vowels)*, if no replacement occurs

9. *P(selecting 2 vowels)*, if replacement occurs

10. *P(selecting the same letter twice)*, if no replacement occurs

**Christine helps her dad do the dishes. There are 5 bowls, 5 glasses, and 6 plates sitting ready to be washed. She accidentally knocks two items off the counter and breaks them. Find each probability.**

11. *P(breaking 2 plates)*      12. *P(breaking 2 bowls)*

13. *P(breaking a bowl, then a glass)*      14. *P(breaking a bowl and a glass)*

**Two dice are tossed. Find each probability.**

15. *P(two 3s)*      16. *P(no 3s)*

17. *P(3 and 4)*      18. *P(3 and any other number)*

19. *P(2 numbers alike)*      20. *P(2 different numbers)*

21. A jar contains 5 peanut butter cookies, 3 caramel delights, and 7 lemon cookies. If 3 cookies are selected in succession, find the probability of selecting one of each if:

     a. no cookies are replaced.      b. each cookie is replaced.

22. A box contains 8 blue markers, 7 black markers, and 3 red markers. Three are chosen one after the other. Find the probability that each is a different color if:

     a. no replacement occurs.      b. replacement occurs each time.

**Ping-Pong™ balls numbered consecutively 1 to 100 are placed in a large hopper. Five balls are drawn at random. Find each probability.**

23. *P(selecting all odd numbers)*, if replacement occurs

24. *P(selecting all odd numbers)*, if no replacement occurs

25. *P(selecting 5 consecutive numbers)*, if no replacement occurs

**Find each probability if 13 cards are drawn from a deck of 52 playing cards, and no replacement occurs.**

26. *P(all diamonds)*  27. *P(all one suit)*

28. *P(all red cards)*  29. *P(all face cards)*

**Critical Thinking**

30. A bent coin has $P(heads) = \frac{1}{3}$ and $P(tails) = \frac{2}{3}$. What type of game rules could you devise so that the probability of success and the probability of failure are equal?

**Applications**

31. **Sports**  Three darts are thrown at the dart board shown at the right.

   a. Find the probability of all three darts landing in the center ring.

   b. Find the probability that each dart lands in a different ring.

   c. Find the probability that one dart is in the second ring and two are in the outer ring.

32. **School**  Students in geometry class are practicing constructions. The classroom tool box contains 20 compasses. 12 of them have red pencils, 5 have blue pencils, and 3 have yellow pencils. Find the probability of picking two compasses, one with a yellow pencil and one with a red pencil. None of the compasses are put back in the box.

**Mixed Review**

33. One die is blue and the other is white. If the dice are rolled, find the probability that the blue die shows an odd number and the white die shows an even number.  **(Lesson 15-6)**

34. Simplify $\dfrac{2z^2}{z^2 - 2z - 35} - \dfrac{z + 7}{z + 5}$.  **(Lesson 11-4)**

35. **Geometry**  Find the midpoints of the sides of a triangle with vertices $R(0, 4)$, $S(2, 6)$, and $T(4, -2)$.  **(Lesson 9-1)**

36. Write the matrix $I_{2 \times 2}$.  **(Lesson 4-4)**

## 15-8 Adding Probabilities

**Objective**

After studying this lesson, you should be able to:
- find the probability of mutually exclusive events or inclusive events.

Suppose a card is drawn from a standard deck of 52 cards. What is the probability of drawing an ace or a king? Since no card is both an ace and a king, the events are said to be **mutually exclusive.** That is, the two events cannot occur at the same time. The probability of two mutually exclusive events can be found by adding their individual probabilities.

$$P(\text{ace or king}) = P(\text{ace}) + P(\text{king})$$
$$= \frac{4}{52} + \frac{4}{52}$$
$$= \frac{8}{52} \text{ or } \frac{2}{13}$$

The probability of drawing an ace or a king is $\frac{2}{13}$ or about 0.154.

| *Probability of Mutually Exclusive Events* | **The probability of one or the other of two mutually exclusive events, $A$ and $B$, occurring is the sum of their probabilities.** $P(A \text{ or } B) = P(A) + P(B)$ |
|---|---|

Remember some of the special cases you had to account for when dealing with permutations. When some of the permutations were the same as a previous permutation, you had to adjust the formula to account for that duplication.

When two events are **inclusive events,** they can occur at the same time. The formula for the probability of inclusive events is an adjustment to the formula for mutually exclusive events. What is the probability of drawing an ace or a red card from a deck of cards?

| $P(\text{ace})$ | $P(\text{red card})$ | $P(\text{red ace})$ |
|---|---|---|
| $\frac{4}{52}$ | $\frac{26}{52}$ | $\frac{2}{52}$ |
| *1 ace in each suit* | *hearts and diamonds* | *ace of hearts and ace of diamonds* |

The probability of drawing a red ace is counted twice, one for an ace and once for a red card. To find the correct probability you must subtract $P(\text{red ace})$ from the sum of $P(\text{aces})$ and $P(\text{red cards})$.

$$P(\text{ace or red card}) = \frac{4}{52} + \frac{26}{52} - \frac{2}{52} \text{ or } \frac{7}{13}$$

The probability of drawing an ace or a red card is $\frac{7}{13}$, or about 0.538.

| *Probability of Inclusive Events* | **The probability of one or the other of two inclusive events, *A* and *B*, occurring is the sum of the individual probabilities decreased by the probability of both occurring.**<br>$$P(A \text{ or } B) = P(A) + P(B) - P(A \text{ and } B)$$ |
|---|---|

**Example 1**

Leroy has 6 nickels, 4 pennies, and 3 dimes in his pocket. He takes one coin from his pocket at random. What is the probability it is a penny or a nickel?

These are mutually exclusive events since a coin cannot be a penny *and* a nickel. Find the sum of the individual probabilities.

$P(penny \text{ or } nickel) = P(penny) + P(nickel)$

$$P(penny \text{ or } nickel) = \frac{4}{13} + \frac{6}{13} \text{ or } \frac{10}{13}$$

The probability of selecting a penny or a nickel is $\frac{10}{13}$ or about 0.769.

**Example 2**

A card is selected from a deck of 52 cards. What is the probability it is a red card or a face card?

Since some red cards are also face cards, these events are inclusive.

$P(red \text{ or } face \text{ } card) = P(red) + P(face \text{ } card) - P(red \text{ } face \text{ } cards)$

$$P(red \text{ or } face \text{ } card) = \frac{26}{52} + \frac{12}{52} - \frac{6}{52} \text{ or } \frac{8}{13}$$

The probability of selecting a red card or a face card is $\frac{8}{13}$ or about 0.615.

**Example 3**

There are 6 women and 7 men on the committee for city park enhancement. A subcommittee of 5 members is being selected at random to study the feasibility of redoing the landscaping in one of the parks. What is the probability that the committee will have at least 3 women?

*At least 3 women* means that the committee may have 3, 4, or 5 women. It is not possible to select a group of 3, a group of 4, and a group of 5 all in the same 5-member committee. The events are mutually exclusive.

$P(at \text{ } least \text{ } 3 \text{ } women) = P(3 \text{ } women) + P(4 \text{ } women) + P(5 \text{ } women)$

$\qquad\qquad$ *3 women, 2 men* $\quad$ *4 women, 1 man* $\quad$ *P(5 women, 0 men)*

$$= \frac{C(6, 3) \cdot C(7, 2)}{C(13, 5)} + \frac{C(6, 4) \cdot C(7, 1)}{C(13, 5)} + \frac{C(6, 5) \cdot C(7, 0)}{C(13, 5)}$$

$$= \frac{140}{429} + \frac{35}{429} + \frac{2}{429} \text{ or } \frac{59}{143}$$

The probability of at least 3 women on the committee is $\frac{59}{143}$ or about 0.413.

# CHECKING FOR UNDERSTANDING

**Communicating Mathematics**

**Read and study the lesson to answer each question.**

1. Give two examples of mutually exclusive events.

2. Give two examples of inclusive events.

3. Suppose you wanted to select a number at random from the numbers 1 to 100. What events must you consider in calculating the probability that the number is even or a multiple of 7?

4. Draw a Venn diagram to illustrate the events in Example 2.

**Guided Practice**

**Determine if each event is inclusive or mutually exclusive. Then find the probability.**

5. A hopper contains balls numbered consecutively from 1 to 10. A ball is chosen at random and a die is rolled. What is the probability of getting a 2 on only one of them?

6. Two cards are drawn from a standard deck of cards. What is the probability that the two cards are both kings or are both queens?

# EXERCISES

**Practice**

**Determine if each event is inclusive or mutually exclusive. Then find the probability.**

7. Mrs. Martell has 15 photos of relatives in her wallet. Five are pictures of her children, 3 are pictures of her sisters, and 7 are pictures of her grandchildren. She selects three photos at random. What is the probability that she has selected 3 photos of her children or 3 photos of her grandchildren?

8. Five coins are dropped onto the floor. What is the probability that at least three of them land heads-up?

9. Two cards are drawn from a deck of cards. What is the probability of having drawn a black card or an ace?

10. In homeroom, 3 of the 16 girls have red hair and 2 of the 15 boys have red hair. What is the probability of selecting a boy or a red-haired person as homeroom representative to student council?

Ken has 11 coasters in a kitchen drawer. Six are cork and 5 are plastic. He selects three at random to use in the family room. Find each probability.

**11.** *P(all 3 cork or all 3 plastic)*
**12.** *P(exactly 2 plastic)*
**13.** *P(at least 2 plastic)*
**14.** *P(at least 2 cork)*

Two cards are drawn from a deck of cards. Find each probability.

**15.** *P(both black or both face cards)*
**16.** *P(both aces or both face cards)*
**17.** *P(both aces or both red)*
**18.** *P(both either red or an ace)*

The lunchroom workers are counting the money in the cash registers after lunch on Tuesday. Seven coins fall from the counter to the floor. Find each probability.

**19.** *P(landing 3 heads or 2 tails)*
**20.** *P(landing at least 5 heads)*
**21.** *P(landing 3 heads or 3 tails)*
**22.** *P(landing all heads or all tails)*

Six men and eight women arrive at the same time at a no-appointment-needed hair salon. There are six stylists waiting to serve them. Find the probability of each group of six being served first.

**23.** *P(all men or all women)*

**24.** *P(5 men or 5 women)*

**25.** *P(4 men or 4 women)*

**26.** *P(at least 3 men)*

The numbers 1 through 25 are written on Ping-Pong™ balls and placed in one hopper. The numbers 20 through 40 are also written on Ping-Pong™ balls and placed in a different hopper. One ball is chosen at random from each spinning hopper. Find each probability.

**27.** *P(each is a 20)*
**28.** *P(neither is a 20)*
**29.** *P(at least one is a 22)*
**30.** *P(each is greater than 10)*

**Critical Thinking**

**31.** Three urns sit on a table. In the gold urn are 2 red, 3 white, and 1 blue marble. In the silver urn are 5 red, 2 white, and 7 blue marbles. In the ceramic urn are 6 red and 4 white marbles. After an urn is chosen at random, one marble is selected. Suppose it is blue.

a. What is the probability that it came from the ceramic urn?

b. What is the probability that it came from the gold urn? *Hint: Think of the probability of blue from the gold urn in relation to the probability of blue from gold, or silver, or ceramic.*

**Applications**

**32. Sports** The Baltimore Oriole pitching staff has 4 left-handers and 7 right-handers. If 2 are selected at random to warm up, what is the probability that at least one of them is a left-hander?

33. **Business** At an income-tax information center, there are 8 phone operators on duty during the first shift from 7:00 A.M.–3:00 P.M., 12 operators on duty during the second shift from 3:00 P.M.–11:00 P.M., and 3 operators on duty during the third shift from 11:00 P.M.–7:00 A.M. Three operators are selected at random to be on a committee to revise procedures. What is the probability that all work the first shift or all work the third shift?

**Computer**

34. The BASIC program below can be used to find probabilities of choosing a sample from each of two populations without replacement. Such probabilities are also called *hypergeometric*.

**Portfolio**

Place your favorite word problem from this chapter in your portfolio with a note explaining why it is your favorite.

```
10 INPUT "ENTER THE NUMBER      120 PRINT "THE PROBABILITY
   IN GROUP A. ";A                OF ";X; "FROM GROUP A
20 INPUT "ENTER THE NUMBER        AND ";Y; "FROM GROUP B
   IN GROUP B. ";B                 IS ";P;"."
30 INPUT "ENTER THE NUMBER      130 END
   CHOSEN FROM GROUP A. ";X     200 LET F1=1: F2=1: F3=1
40 INPUT "ENTER THE NUMBER      210 FOR I=1 TO C1
   CHOSEN FROM GROUP B. ";Y     220 LET F1=F1*I
50 LET C1=A: C2=X: C3=A-X       230 NEXT I
60 GOSUB 200: LET Y1=C          240 FOR I=1 TO C2
70 LET C1=B: C2=Y: C3=B-Y       250 LET F2=F2*I
80 GOSUB 200: LET Y2=C          260 NEXT I
90 LET C1=A+B: C2=X+Y: C3=A     270 FOR I=1 TO C3
   +B-C2                        280 LET F3=F3*I
100 GOSUB 200: LET Y3=C         290 NEXT I
110 LET P = Y1*Y2/Y3            300 LET C=F1/(F2*F3)
                                310 RETURN
```

**Use this program to find each probability.**

a. A bag contains 4 blue and 5 red marbles. Five marbles are drawn at random. Find *P(3 blue, 2 red)*, *P(4 blue, 1 red)*, and *P(all red)*.

b. The math club has 6 boys and 12 girls. Three were selected at random to enter a contest. Find *P(2 boys, 1 girl)* and *P(1 boy, 2 girls)*.

c. For every 11 computer disks, one is defective. If 10 disks are chosen at random, find *P(1 defective disk)* and *P(no defective disks)*.

**Mixed Review**

35. Mr. Sunami has a corporate board meeting to attend. He has 4 suits and 5 shirts. He selects one suit and one shirt at random to wear to the meeting. What is the probability of each random outfit? **(Lesson 15-7)**

36. From a group of 8 men and 7 women, how many different committees of 3 men and 4 women can be formed? **(Lesson 15-1)**

37. Find the missing geometric means in the sequence 6, $\underline{\ ?\ }$, $\underline{\ ?\ }$, 93.75. **(Lesson 13-4)**

38. Solve $\log_6 (x + 5) + \log_6 (x - 4) = 2$. **(Lesson 12-3)**

39. If $p = 0.5ans$, find $p$ if $a = 10.4$, $n = 6$, and $s = 12$. **(Lesson 1-1)**

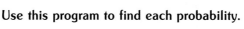

# 15-9 Simulation and Binomial Experiments

**Objectives**

After studying this lesson, you should be able to:
- use simulation to solve various probability problems, and
- use binomial experiments to find probabilities.

**Application**

Brett normally makes 2 out of every 3 free throws she attempts in a basketball game. In other words, the probability that Brett will make a free throw is $\frac{2}{3}$. Suppose Brett attempts 4 free throws during Friday night's game. What is the probability that she will make 3 free throws and miss only one?

Let $S$ stand for scoring when she attempts a free throw. Let $M$ stand for missing when she attempts a free throw.

The possible ways of scoring on 3 free throws and missing 1 free throw are shown at the right. This shows the combination of 4 things (free throws) taken three at a time (scores), or $C(4, 3)$.

| M | S | S | S |
|---|---|---|---|
| S | M | S | S |
| S | S | M | S |
| S | S | S | M |

The terms of the binomial expansion of $(S + M)^4$ can be used to find the probabilities of each combination of scores and misses.
$$(S + M)^4 = S^4 + 4S^3M + 6S^2M^2 + 4SM^3 + M^4$$

| term | meaning | coefficient of term |
|------|---------|---------------------|
| $S^4$ | 1 way to score all 4 times | $C(4, 4) = 1$ |
| $4S^3M$ | 4 ways to score 3 times and miss 1 time | $C(4, 3) = 4$ |
| $6S^2M^2$ | 6 ways to score 2 times and miss 1 time | $C(4, 2) = 6$ |
| $4SM^3$ | 4 ways to score 1 time and miss 3 times | $C(4, 1) = 4$ |
| $M^4$ | 1 way to miss all 4 times | $C(4, 0) = 1$ |

The probability that Brett scores on a free throw is $\frac{2}{3}$. So, the probability that she misses is $\frac{1}{3}$. To find the probability of scoring 3 out of 4 free throws, substitute $\frac{2}{3}$ for $S$ and $\frac{1}{3}$ for $M$ in the term $4S^3M$.

$$4S^3M = 4\left(\frac{2}{3}\right)^3\left(\frac{1}{3}\right) \text{ or } \frac{32}{81}, \text{ which is about } 0.395.$$

Problems that can be solved using binomial expansion are called **binomial experiments**.

| Conditions of a Binomial Experiment | **A binomial experiment exists *if and only if* these conditions occur.** <br> ▪ **There are exactly two possible outcomes for any trial.** <br> ▪ **There is a fixed number of trials.** <br> ▪ **The trials are independent.** <br> ▪ **The probability of each trial is the same.** |
| --- | --- |

**Example 1**

When Marty came home from school, he emptied his pockets onto his desk. There were 5 coins in his pocket. What is the probability that 3 coins landed heads and 2 coins landed tails?

There are only two possible outcomes, heads ($H$) or tails ($T$). The tossing of 5 coins are independent events. For each toss of a coin the probability is the same, and 5 coins are involved. This is a binomial experiment.

When $(H + T)^5$ is expanded, the term $H^3T^2$ represents 3 heads and 2 tails. The coefficient of $H^3T^2$ is $C(5, 3)$.

$P(3 \text{ heads, 2 tails}) = C(5, 3)H^3T^2$    *Replace H with P(H) and T with P(T).*

$$= \frac{5 \cdot 4}{2 \cdot 1}\left(\frac{1}{2}\right)^3\left(\frac{1}{2}\right)^2 \text{ or } \frac{5}{16} \quad \textit{Both P(H) and P(T) equal } \frac{1}{2}.$$

The probability of 3 heads and 2 tails is $\frac{5}{16}$ or about 0.313.

**Example 2**

To practice for a jigsaw puzzle competition, Chad and Rashad put together 7 jigsaw puzzles. The probability that Chad puts in the last piece of a puzzle is $\frac{1}{5}$. The probability that Rashad puts in the last piece is $\frac{4}{5}$. What is the probability that Chad will put in the last piece of at least 3 puzzles?

There are only two possible outcomes for whoever puts in the last piece: Chad ($C$) or Rashad ($R$). Look at the binomial expansion of $(C + R)^7$.

$$(C + R)^7 = C^7 + 7C^6R + 21C^5R^2 + 35C^4R^3 + 35C^3R^4 + 21C^2R^5 + 7CR^6 + R^7$$

The probability of Chad putting in the last piece of at least 3 puzzles equals the sum of the probabilities of putting in the last piece of 3, 4, 5, 6, or all 7 puzzles.

$P(\text{Chad putting in the last piece of at least 3 puzzles})$
$$= C^7 + 7C^6R + 21C^5R^2 + 35C^4R^3 + 35C^3R^4$$

$$= \left(\frac{1}{5}\right)^7 + 7\left(\frac{1}{5}\right)^6\left(\frac{4}{5}\right) + 21\left(\frac{1}{5}\right)^5\left(\frac{4}{5}\right)^2 + 35\left(\frac{1}{5}\right)^4\left(\frac{4}{5}\right)^3 + 35\left(\frac{1}{5}\right)^3\left(\frac{4}{5}\right)^4$$

$$= \frac{1}{78,125} + \frac{28}{78,125} + \frac{336}{78,125} + \frac{2240}{78,125} + \frac{8960}{78,125} \text{ or } \frac{2313}{15,625}$$

The probability of Chad putting in the last piece of at least 3 puzzles is $\frac{2313}{15,625}$, or about 0.148.

*You first learned about simulation in Lesson 9-6.*

Another type of experimental method for finding probability is **simulation.** In simulation, a device is used to model the event, and you observe how the model responds to the conditions listed in a given problem. This process saves long and difficult explorations or samplings.

**Example 3**

A new family is moving in next door. We know that there are three children in the family, but have not seen any of them. What is the probability that there is at least one girl in the family?

$P(boy) = \frac{1}{2}$ and $P(girl) = \frac{1}{2}$. When tossing a coin, $P(H) = \frac{1}{2}$ and $P(T) = \frac{1}{2}$, so we could use coins to simulate the three children in the family.

Let $H$ be boys and $T$ be girls. Now toss 3 coins and record your results. Twenty tosses of the coins reveal these results.

| | | | | | |
|---|---|---|---|---|---|
| TTH | **HHH** | HTT | TTT | THT | *The trials in blue are the ones that represent at least one girl.* |
| TTH | HHT | HTH | THH | HTT | |
| THH | TTH | **HHH** | **HHH** | HTT | |
| **HHH** | THH | HHT | HTT | TTT | |

In our simulation, 16 of 20 trials yielded at least one T, or at least one girl. Therefore, based on this simulation, the probability of at least one girl in the family is $\frac{16}{20}$ or about 0.8.

*The theoretical probability is 0.875.*

In Example 3, other devices could be used to create the simulation. For example, you could roll three dice and let odd numbers be one gender and even numbers be the other. Another way to use the dice is to let the numbers 1, 2, and 3 represent boys and 4, 5, and 6 represent girls.

# CHECKING FOR UNDERSTANDING

**Communicating Mathematics**

Read and study the lesson to answer each question.

1. Can a binomial experiment represent dependent events?

2. Name the conditions that must be satisfied for a problem to be classified as a binomial experiment.

3. In a binomial experiment, if $P(s) = n$, what is $P(f)$?

4. Name some objects that could be used to simulate a given situation.

**Guided Practice**

**Determine if each situation represents a binomial experiment or not. Solve those that represent a binomial experiment.**

5. What is the probability of 2 heads and 1 tail if Angie tosses a coin 3 times?

6. What is the probability of Sergio drawing 4 aces from a deck of cards for each condition?

   **a.** He replaces the card each time.    **b.** He does not replace the card.

7. Eight red sour balls, 4 green sour balls, and 6 purple sour balls are placed in a paper bag. Two are selected with replacement after the first selection. Find each probability.

   **a.** both red          **b.** both green          **c.** both purple
   **d.** 1 red, 1 green    **e.** 1 red, 1 purple     **f.** 1 green, 1 purple

# EXERCISES

**Practice**

**Find each probability if a coin is tossed four times.**

**8.** *P(no heads)*          **9.** *P(2 heads, 2 tails)*          **10.** *P(3 or more heads)*

**Find each probability if a die is tossed five times.**

**11.** *P(only one 4)*          **12.** *P(at least three 4s)*          **13.** *P(no more than two 4s)*

Sandra Wilder carries tubes of lipstick in a bag in her purse. The probability of pulling out the color she wants is $\frac{1}{3}$. If she uses her lipstick 4 times in a day, find each probability.

**14.** *P(never the correct lipstick)*
**15.** *P(at least 3 times correct)*
**16.** *P(no more than 3 times correct)*

Mark Clingan guesses at all 10 true/false questions on his sociology test. Find each probability.

**17.** *P(7 correct)*          **18.** *P(at least 6 correct)*          **19.** *P(all wrong)*

Jojo MacMahon plays for the Pickerington Badgers baseball team. He is now batting 0.200 (meaning 200 hits in 1000 times at bat). Find each probability for the next 5 times at bat.

**20.** *P(exactly 3 hits)*          **21.** *P(at least 4 hits)*          **22.** *P(at least 2 hits)*

**Find each probability if three coins are tossed.**

**23.** *P(3 tails)*                    **24.** *P(3 heads)*
**25.** *P(at least 2 tails)*          **26.** *P(exactly 2 heads)*

If a thumbtack is dropped, the probability of its landing point up is 0.4. Mrs. Wilson drops 10 tacks while putting up the lunch menus for the next week on the bulletin board. Find each probability.

**27.** *P(all point up)*     **28.** *P(exactly 3 point up)*   **29.** *P(at least 6 point up)*

**Critical Thinking**

**30.** Mikel Fatur is a quarterback on the junior varsity squad at Monroe High School. In his freshman season, he has completed $\frac{2}{3}$ of his passes. Assume he will do the same in his sophomore year. Use simulation to find the probability of completing at least 6 of 10 passes for the entire game if he has already completed 4 of 5 passes in the first half.

**Applications**

**31. Traffic Control** The probability that a signal at Darby Avenue is green is $\frac{3}{5}$. What is the probability that exactly 3 of the next seven cars will have to stop?

**32. Baseball** Four of every 7 pitches thrown by Elias Ramos are strikes. What is the probability that 4 of the next 5 pitches will be strikes? Write your answer as a percent.

**33. Skeet Shooting** Skeet shooting, also called trapshooting, involves a person shooting at clay discs, called clay pigeons, propelled into the air by a machine. Harold usually hits 9 out of 10 clay pigeons. If he shoots 12 times, find each probability.

    **a.** *P(all misses)*

    **b.** *P(exactly 7 hits)*

    **c.** *P(all hits)*

    **d.** *P(at least 10 hits)*

**Mixed Review**

**34.** There are 8 girls and 8 boys on the faculty advisory committee. Three are juniors. Find the probability of selecting a boy or a girl from the committee who is not a junior. **(Lesson 15-8)**

**35. Physics** A ball dropped 120 feet bounces $\frac{2}{3}$ of the height from which it fell on each bounce. How far will it travel before coming to rest? **(Lesson 13-6)**

**36.** Solve $9^{3y} = 27^{y-1}$. **(Lesson 12-7)**

**37. Number Theory** A fraction has a value of $\frac{6}{7}$. If 1 is added to its numerator, its value is $\frac{7}{8}$. Find the original fraction. **(Lesson 11-7)**

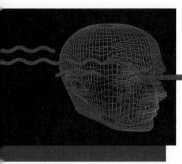

# Technology

## Coin Toss Simulation

▶ BASIC
Spreadsheets
Software

```
 10 LET T2 = 0: LET T3 = 0
 20 LET H2 = 0: LET H3 = 0
 30 FOR I = 1 TO 100
 40 LET T = 0: LET H = 0
 50 FOR N = 1 TO 3
 60 LET R = RND (1)
 70 IF R < 0.5 THEN 110
 80 PRINT "TAIL";
 90 LET T = T + 1
100 GOTO 130
110 PRINT "HEAD";
120 LET H = H + 1
130 NEXT N
134 PRINT
140 IF H = 3 THEN 180
150 IF T = 3 THEN 200
160 IF H = 2 AND T = 1 THEN 220
170 IF H = 1 AND T = 2 THEN 240
180 LET H3 = H3 + 1
190 GOTO 250
200 LET T3 = T3 + 1
210 GOTO 250
220 LET H2 = H2 + 1
230 GOTO 250
240 LET T2 = T2 + 1
250 NEXT I
254 PRINT
260 PRINT "HHH: ";H3
270 PRINT "TTT: ";T3
280 PRINT "HHT: ";H2
290 PRINT "HTT: ";T2
300 END
```

Probability is used to predict the outcome in games of chance involving tossing coins, rolling dice, selecting cards, or winning sweepstakes. A computer program can be used to do the actual counting in simulated situations where large samples are needed.

The BASIC program at the left simulates tossing three coins, prints the outcomes, and keeps totals of the possible outcomes of the 100 samples. Each time the program is run, the totals may be different due to the random selection of values in line 60.

The random numbers selected by the computer lie between 0 and 1. The program defines heads as numbers less than 0.5 and tails as numbers greater than or equal to 0.5.

The output shows each toss of the coin and lists the totals for the four possible combinations: 3 heads; 3 tails; 2 heads and 1 tail; or 1 head and 2 tails.

**Example:**

| | |
|---|---|
| HHH: | 11 |
| TTT: | 19 |
| HHT: | 38 |
| HTT: | 32 |

Enter the program and run it several times. Notice the change in the totals. When you use the program for the exercises below, it may take the computer several minutes to complete the large samples.

## EXERCISES

1. Run the program once. Write the actual percentage for each combination.
2. Calculate the expected probability for each combination.
3. Change the program to increase the sample size to 1000. Delete lines 80 and 110. Then run the program. Find the probability for each combination. Compare these to your results in Exercise 2.
4. Make a conjecture about the reliability of your probabilities in relation to the size of the sample.

# 15 SUMMARY AND REVIEW

## VOCABULARY

Upon completing this chapter you should be
familiar with the following terms:

| | | | |
|---|---|---|---|
| Basic Counting Principle | 684 | 689 | linear permutation |
| binomial experiment | 720 | 715 | mutually exclusive events |
| circular permutation | 694 | 706 | odds |
| combination | 699 | 689 | permutation |
| dependent events | 685 | 705 | probability |
| failure | 705 | 722 | simulation |
| inclusive events | 715 | 705 | success |
| independent events | 684 | 684 | tree diagram |

## SKILLS AND CONCEPTS

| OBJECTIVES AND EXAMPLES | REVIEW EXERCISES |
|---|---|

Upon completing this chapter, you should
be able to:
- solve problems using the Basic Counting
  Principle  **(Lesson 15-2)**

  How many 3-letter patterns are there if

  **a.** repetition is allowed?

  $26 \cdot 26 \cdot 26 = 17{,}576$ patterns

  **b.** repetition is not allowed?

  $26 \cdot 25 \cdot 24 = 15{,}600$ patterns

**Use these exercises to review and prepare
for the chapter test.**

1. Using the digits 0, 1, 2, 3, and 4, how
   many 3-digit patterns can be formed if
   the numbers can be used more than
   once?

2. Using the digits 5, 6, 7, 8, and 9, how
   many 3-digit patterns can be formed if
   each number can only be used once?

---

- solve problems involving
  permutations  **(Lesson 15-3)**

  Find the number of permutations of

  **a.** 9 things taken 3 at a time.

  $P(9, 3) = \dfrac{9!}{(9 - 3)!} = \dfrac{9!}{6!}$ or 504

  **b.** the letters in MISSOURI.
  There are 2 Is and 2 Ss in 8 letters.

  $\dfrac{n!}{p!q!} = \dfrac{8!}{2!2!} = 10{,}080$

**Evaluate each expression.**

3. $\dfrac{P(7, 3)}{P(5, 2)}$

4. $\dfrac{P(8, 5)}{P(5, 3)}$

**On a shelf are 8 mystery and 7 romance
novels. How many ways can they be arranged
for each situation?**

5. all mysteries together.

6. all mysteries together, all romances
   together.

| OBJECTIVES AND EXAMPLES | REVIEW EXERCISES |
|---|---|

■ solve problems involving circular permutations **(Lesson 15-4)**

Find the number of ways to arrange 5 fruits around a plate.
$(n - 1)! = (5 - 1)! = 4!$ or 24 ways

Find the number of ways to arrange 6 large wooden beads on an elastic cord tied into a circle. (a reflection)

$\frac{(n - 1)!}{2} = \frac{(6 - 1)!}{2}$ or 60 ways

7. How many ways can 8 people be seated at a round table?

8. How many ways can 10 charms be placed on a bracelet that has a clasp?

9. Art is to make a flag with 5 differently colored rectangles arranged side by side. How many different ways could these rectangles be arranged?

■ solve problems involving combinations **(Lesson 15-5)**

Seven cards are drawn from a deck of 52 cards. In how many ways can 4 of one suit and 3 of another suit be drawn?

$P(4, 2) \cdot C(13, 4) \cdot C(13, 3)$

$= \frac{4!}{2!} \cdot \frac{13!}{4!9!} \cdot \frac{13!}{2!10!}$ or 2,453,880

There are 2,453,880 such hands.

10. How many baseball teams can be formed from 15 players if only 3 pitch while the others play the remaining 8 positions?

11. From a deck of 52 cards, how many different 4-card hands exist?

■ find the probability and odds of an event **(Lesson 15-6)**

Find the probability of selecting a green marble if a bag holds 14 marbles, 6 of which are green.

$P(green) = \frac{6}{14} = \frac{3}{7} \approx 0.429$

$odds = \frac{6}{8}$ or $\frac{3}{4}$

12. A card is selected from a deck of 52 cards. What is the probability that it is a queen? What are the odds?

13. A bag contains 6 red and 2 white marbles. If two marbles are selected, what is the probability that one is red and the other is white?

■ find the probability of independent or dependent events **(Lesson 15-7)**

Four green marbles and six red marbles are in a bag. Find the probability of drawing a green marble and then a red marble, if the marbles are not replaced.

$P(1g, 1r) = P(g) \cdot P(r\ following\ g)$

$= \frac{4}{10} \cdot \frac{6}{9}$ or $\frac{4}{15}$

14. In his pocket, Jose has 5 dimes, 7 nickels, and 4 pennies. He selects 4 coins. What is the probability that he has 2 dimes and 2 pennies?

15. Ben has 6 blue socks and 4 black socks in a drawer. One dark morning he pulls out 2 socks. What is the probability that he has 2 black socks?

- find the probability of mutually exclusive events   (**Lesson 15-8**)

What is the probability of drawing a heart or a club from a deck of cards?

$P(heart \text{ or } club) = P(heart) + P(club)$

$$= \frac{13}{52} + \frac{13}{52} \text{ or } \frac{1}{2}$$

16. From a deck of 52 cards, one card is selected. What is the probability that it is an ace or a face card?

17. In the numbers 1 through 20, what is the probability of selecting a number at random that is a multiple of 5 or a multiple of 7?

---

- find the probability of inclusive events (**Lesson 15-8**)

What is the probability of drawing a red card or a queen from a deck of cards?

$P(R \text{ or } Q) = P(R) + P(Q) - P(RQ)$

$$= \frac{26}{52} + \frac{4}{52} - \frac{2}{52} \text{ or } \frac{7}{13}$$

18. If a letter is selected at random from the alphabet, what is the probability that it is a letter from the words CAT or SKATE?

19. If a card is selected from a deck of cards, find the probability that it is not red or not a face card.

---

- solve a problem by simulation or binomial experiments   (**Lesson 15-1, Lesson 15-9**)

Six coins are tossed. Find $P(4H, 2T)$.

$(H + T)^6 = H^6 + 6H^5T + 15H^4T^2 +$
$\quad 20H^3T^3 + 15H^2T^4 + 6HT^5 + T^6$

$15H^4T^2 = 15\left(\frac{1}{2}\right)^4\left(\frac{1}{2}\right)^2 = \frac{15}{64}$

20. Four coins are tossed. What is the probability that they show 3 heads and 1 tail?

21. A die is tossed 5 times. What is the probability of at least two 3s?

---

# APPLICATIONS AND CONCEPTS

22. **Geometry**   Find the number of diagonals in a polygon that has 20 sides.   (**Lesson 15-1**)

23. **Tourism**   A taxi in The Netherlands can hold 4 passengers safely. A party of 6 people wish to travel by taxi to the Hague. How many different groups can occupy the first taxi if it will be full? (**Lesson 15-5**)

**Evaluate each expression.**

1. $P(6, 4)$
2. $P(8, 3)$
3. $C(8, 3)$
4. $C(6, 4)$

**Solve each problem.**

5. In a row are 8 chairs. How many ways can 5 people be seated?

6. From 8 shirts, 6 pairs of slacks, and 4 jackets, how many different outfits can be made?

7. How many ways can the letters from the word *television* be arranged?

8. How many different basketball teams could be formed from a group of 12 girls?

9. How many ways can 11 books be arranged on a shelf?

10. How many ways can 6 keys be placed on a key ring?

11. Nine points are placed on a circle. How many triangles can be formed using these points, three at a time, as vertices?

12. From a group of 4 men and 5 women, a committee of 3 is to be formed. What is the probability that it will have 2 men and 1 woman?

13. A red die and a green die are tossed. What is the probability that the red will show even and the green will show a number greater than four?

14. From a deck of cards, what is the probability of selecting a 4 followed by a 7 if no replacement occurs?

15. A letter is drawn at random from the letters *A, B, C, D, E,* and *F.* What is the probability the letter is a vowel?

16. A state is chosen at random from the United States. What is the probability that the state is one of the five smallest states in population?

17. **Geometry** There are 9 points in a plane such that no three of the points lie on the same line. How many different segments connect the points?

18. **Transportation** A fleet of limousines is composed of 6 white limousines and 4 black limousines. A company wishes to rent 5 of them. How many ways can 2 white and 3 black limousines be selected?

19. Five bent coins are tossed. The probability of heads is $\frac{2}{3}$ for each of them. What is the probability that no more than 2 will show heads?

20. While shooting arrows, William Tell can hit an apple 9 out of 10 times. What is the probability that he will hit it exactly 4 out of the next 7 times?

**Bonus** Determine all real values of $w$ for which the statement
$C(w + 1, 2) = 9 \cdot C(w, 1)$ is true.

# 16 Trigonometric Functions

## CHAPTER OBJECTIVES

In this chapter, you will:
- Find the values of trigonometric functions.
- Solve problems involving right triangles using right triangle trigonometry.
- Solve triangles using the law of sines and the law of cosines.

The blueprint of a unit in a multi-family complex is shown below with its scale. Find the approximate size of each room.

## CAREERS IN ARCHITECTURE

I f you could design your idea of the perfect house, what would it look like? How about designing the perfect office building, apartment building, school, hospital, factory, or airport terminal?

Think about the task of drawing up the plans for any of those buildings. You'd be relating its complex functions to space, color, texture, warmth, and light. It's not a career for everyone. Architecture requires a mind where art and science can meet to create a constructed human environment.

Of course, an architect usually isn't working for himself or herself. There is a client who must be satisfied and whose budget must be met. And don't forget to follow local building codes, zoning laws, fire regulations, handicapped-access ordinances, and all other applicable laws. They mostly deal with the functional aspects of the building: that includes heating, air-conditioning, ventilation, electricity, plumbing, and security. But there's more—Will there be a centralized intercom system? How about centralized vacuum cleaning? And what about outdoors—Will the architect also create the landscaping plan? How about outdoor lights? Will there be fences or walls? Is there ample parking?

Architecture is a rewarding career. It is art that can be lived in and used every day. And some of it will last for centuries to come.

## MORE ABOUT ARCHITECTURE

### Degree Required:

- Bachelor's Degree in architecture plus three years experience or a masters degree in architecture plus two years experience

### Some architects like:

- working with people
- solving problems
- the opportunity to be creative
- creating something that will last

### Related Math Subjects:

- Advanced Algebra
- Geometry
- Trigonometry
- Calculus
- Statistics/Probability

### Some architects dislike:

- working long hours
- having to redo plans to please a client
- working with difficult or unpleasant people
- meeting deadlines

For more information on the various careers available in the field of Architecture, write to:

American Institute of Architects
1735 New York Avenue
Washington, D.C. 20006

# Angles and the Unit Circle

**Objectives**

After studying this lesson, you should be able to:
- change radian measure to degree measure and vice versa, and
- identify coterminal angles.

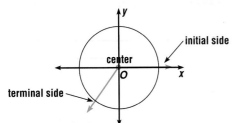

In the coordinate plane at the left, a circle is centered at the origin and two rays extend from the center. These rays form an angle. One ray, called the **initial side** of the angle, is fixed along the positive x-axis. The other ray, called the **terminal side** of the angle, can rotate about the center. An angle positioned like this, with its initial side along the x-axis and its vertex at the origin, is said to be in **standard position.**

The measures of angles in standard position whose terminal sides rotate counterclockwise are positive.

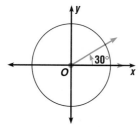 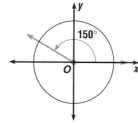 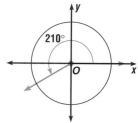

The measures of angles in standard position whose terminal sides rotate clockwise are negative.

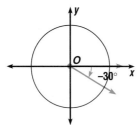 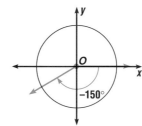 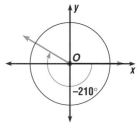

**FYI · · ·**

The word *trigonometry* comes from the Greek *trigōnon-metria* meaning *angle measure*. The Greeks developed trigonometric ratios in an attempt to track the planets and stars.

When terminal sides rotate, they may sometimes make one or more complete revolutions about the center of the circle. An angle whose terminal side has made exactly one revolution about the center has a measure of 360°.

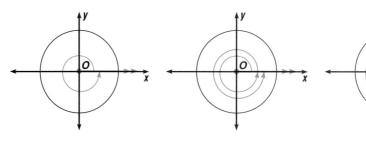

one revolution
360°

two revolutions
360° • 2 or 720°

three revolutions
360° • 3 or 1,080°

Angles in standard position that have the same terminal side are called **coterminal angles.** For instance, 30°, 390°, and 750° are coterminal angles. The measures of coterminal angles always differ by an integral multiple of 360°.  *Why?*

A **unit circle** is a circle centered at the origin whose radius is 1 unit long. Form an angle in standard position so that the rays of the angle intercept an arc 1 unit long. The measure of this angle is defined to be 1 **radian.** While degrees are the most common unit of angle measure, radians can also be used to describe the measure of an angle.

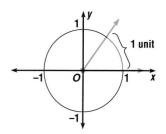

The circumference of any circle is $2\pi r$ where $r$ is the radius measure. So the circumference of a unit circle is $2\pi(1)$ or $2\pi$ units. An angle representing one complete revolution of the circle measures $2\pi$ radians. Since we know that one complete revolution is 360°, the following proportion illustrates the relationship between radians and degrees.

$$\frac{\text{degree measure}}{360} = \frac{\text{radian measure}}{2\pi}$$

If you know the degree measure of an angle and you need to find the radian measure, multiply the number of degrees by $\frac{\pi}{180}$.     $\frac{2\pi}{360} = \frac{\pi}{180}$

**Example 1**

Change each degree measure to radian measure.

**a.** 60°

$$60 \cdot \frac{\pi}{180} = \frac{60\pi}{180} \text{ or } \frac{\pi}{3}$$

**b.** 270°

$$270 \cdot \frac{\pi}{180} = \frac{270\pi}{180} \text{ or } \frac{3\pi}{2}$$

**c.** −45°

$$-45 \cdot \frac{\pi}{180} = -\frac{45\pi}{180} \text{ or } -\frac{\pi}{4}$$

If you know the radian measure of an angle and you need to find the degree measure, multiply the number of radians by $\frac{180}{\pi}$.

**Example 2**

Change each radian measure to degree measure.

**a.** $\frac{\pi}{12}$

$$\frac{\pi}{12} \cdot \frac{180}{\pi} = \left(\frac{180\pi}{12\pi}\right)^{\circ}$$
$$= 15°$$

**b.** $-\pi$

$$-\pi \cdot \frac{180}{\pi} = \left(-\frac{180\pi}{\pi}\right)^{\circ}$$
$$= -180°$$

**c.** 1

$$1 \cdot \frac{180}{\pi} = \left(\frac{180}{\pi}\right)^{\circ}$$
$$\approx 57.30°$$

## Example 3

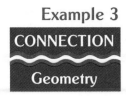

**CONNECTION**

**Geometry**

The numbers on the face of a clock divide the circular face into twelve congruent sectors. Find the degree measure and the radian measure of the angle between any two consecutive numbers on a clock.

Since the face is divided into twelve congruent pieces, each angle must measure $\frac{1}{12}$ of the total measure of the circle. Since a clock face represents one complete revolution, the total measure is 360° or $2\pi$ radians.

$$
\begin{array}{cc}
\textit{degrees} & \textit{radians} \\
\dfrac{\text{total measure}}{12} = \dfrac{360°}{12} & \dfrac{\text{total measure}}{12} = \dfrac{2\pi}{12} \\
= 30° & = \dfrac{\pi}{6} \text{ radians}
\end{array}
$$

The measure of the angle between any two consecutive numbers on a clock face is 30° or $\frac{\pi}{6}$ radians.

*Convert 30° to radians to check that these measures are equivalent.*

# CHECKING FOR UNDERSTANDING

**Communicating Mathematics**

**Read and study the lesson to answer these questions.**

1. An angle with its vertex at the origin and its initial side along the positive $x$-axis is said to be in __?__.

2. Describe coterminal angles and give an example of two angles that are coterminal.

3. Define unit circle.

4. Find the degree measure and the radian measure of the angle formed by the hands of a clock when it is exactly 5 o'clock.

**Guided Practice**

**Suppose angles with the following measures are in standard position. For each angle, name the quadrant that contains the terminal side.**

5. $\dfrac{4\pi}{7}$       6. $\dfrac{5\pi}{3}$       7. $-\dfrac{12\pi}{5}$

8. 97°       9. $-330°$       10. $-115°$

**Determine whether each pair of angles is coterminal.**

11. $\dfrac{\pi}{2}, \dfrac{3\pi}{2}$       12. $0, 8\pi$       13. $\dfrac{2\pi}{5}, \dfrac{4\pi}{5}$

14. $75°, 435°$       15. $20°, 2000°$       16. $-24°, 384°$

# EXERCISES

**Practice**

**Determine whether each pair of angles is coterminal.**

17. $\dfrac{11\pi}{3}, \dfrac{16\pi}{3}$       18. $\dfrac{\pi}{8}, \dfrac{33\pi}{8}$       19. $-\dfrac{5\pi}{9}, \dfrac{5\pi}{9}$

20. $440°, 80°$       21. $-11°, -371°$       22. $-39°, 681°$

**Change each degree measure to radian measure.**

**23.** $180°$          **24.** $-90°$          **25.** $45°$

**26.** $540°$          **27.** $-225°$          **28.** $315°$

**29.** $135°$          **30.** $-210°$          **31.** $-120°$

**Change each radian measure to degree measure.**

**32.** $\dfrac{2\pi}{3}$          **33.** $\pi$          **34.** $\dfrac{5\pi}{6}$

**35.** $-\dfrac{7\pi}{4}$          **36.** $-\dfrac{8\pi}{3}$          **37.** $\dfrac{9\pi}{4}$

**38.** $5$          **39.** $-2\dfrac{1}{3}$          **40.** $-6\dfrac{1}{2}$

**Critical Thinking**

**41.** Explain why the measure of coterminal angles differ by a multiple of $360°$. How would the radian measure of coterminal angles differ?

**Application**

**42. Sports**   A scuba diver uses a compass to navigate a straight line. The compass is aligned to the diver and a point toward which the diver wishes to travel. The ring on the compass, called a bezel, is turned so that the index needle is aligned over the compass needle. Now as long as the diver swims so that the compass needle aligns with the index needle, she will be swimming in a straight line to the desired location. If she wants to return to her starting point, she simply swims in the direction that is $180°$ from her original direction. This direction is called the reciprocal heading.

**a.** Carita has been swimming so that her compass points toward $18°$. Where should the needle point when she swims back to the dock where she started?

**b.** On a certain dive, Carita found a formation that she wished to show her friend Dan on their next dive. When she saw the formation, she was swimming back to the dock so that her compass pointed to $234°$. When she and Dan return to the formation, what should the heading on the compass be?

**Computer** Coterminal angles are those angles whose measures differ by a multiple of 360°. If the degree measure of an angle is entered into this program, the coterminal angle between 0° and 360° will be printed. The BASIC program below uses loops to add or subtract 360° until the angle is between 0° and 360°.

```
10   PRINT "ENTER A DEGREE MEASURE."
20   INPUT A
30   IF A < 360 THEN 60
40   LET A = A - 360
50   GOTO 30
60   IF A > = 0 THEN 90
70   LET A = A + 360
80   GOTO 60
90   PRINT "ITS COTERMINAL ANGLE
     MEASURES ";A;" DEGREES."
100  END
```

**Find the coterminal angle between 0° and 360° for each of the following.**

**43.** $720°$        **44.** $1373.56°$        **45.** $-981°$

**State whether each pair of angles are coterminal.**

**46.** $-59°, 661°$        **47.** $29°, 9361°$        **48.** $49°, 1129°$

**49.** How should lines 30, 40, and 70 be modified if the angle measure is given in radians?

**Mixed Review**

**50. Probability** Chris guessed on all 10 questions on a true-false test. What is the probability that all of the guesses were correct? **(Lesson 15-9)**

**51.** Find the sum of the geometric series for which $a_1 = 5$, $r = 4$, and $n = 8$. **(Lesson 13-5)**

**52.** Show that $\log_3 27 + \log_3 3 = \log_3 81$. **(Lesson 12-2)**

**53.** Graph the equation $f(x) = -x^2 - 4x - 10$. **(Lesson 8-3)**

**54.** Simplify $5^{-3}b^3x^4y^{-1}$. **(Lesson 5-2)**

---

## CHALLENGE

**If $(a, b)$ is on the unit circle, prove that each of the following points is also on the unit circle.**

**a.** $(a, -b)$        **b.** $(b, a)$        **c.** $(b, -a)$

# Sine and Cosine Functions

**Objectives**

After studying this lesson, you should be able to:
- find the least possible angle that is coterminal to a given angle, and
- find the values of expressions involving sine and cosine.

**Application**

The waves shown at the right represent the fluctuations in energy produced by musical tones. An instrument called an oscilloscope converted the sound waves to electrical impulses and displayed the curves on the screen. A pure musical tone produces a curve that is the graph of the **sine function.** The sine function is one of the functions that can be defined in terms of the unit circle.

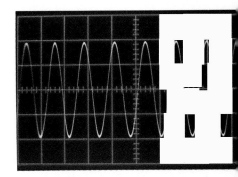

Consider an angle in standard position. Let the Greek letter θ (theta) stand for the measurement of the angle. The terminal side of this angle intersects the unit circle at a particular point. The x-coordinate of this point is called **cosine θ.** The y-coordinate of this point is called **sine θ.** Sine is abbreviated sin and cosine is abbreviated cos.

*Definition of Sine and Cosine*

**Let θ stand for the measurement of an angle in standard position. Let (x, y) represent the coordinates of the point where the terminal side intersects the unit circle. Then**

$$\cos \theta = x \qquad \text{and} \qquad \sin \theta = y.$$

**Example 1**

**Find sin 45°.**

Consider the right triangle formed by two sides and a diagonal of a square. One side of this triangle is part of the initial side of a 45° angle, and the hypotenuse is along the terminal side.

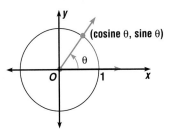

*The triangle formed is an isosceles triangle.*

$$\sqrt{s^2 + s^2} = 1 \qquad \begin{array}{l} \textit{Pythagorean Theorem} \\ c = 1,\ a = s,\ b = s \end{array}$$

$$\sqrt{2s^2} = 1$$

$$s\sqrt{2} = 1$$

$$s = \frac{1}{\sqrt{2}} \text{ or } \frac{\sqrt{2}}{2} \qquad \textit{The length of each side is } \frac{\sqrt{2}}{2} \textit{ or about 0.707 units.}$$

The coordinates of the point labeled (x, y) are $\left(\frac{\sqrt{2}}{2}, \frac{\sqrt{2}}{2}\right)$. So, $\sin 45° = \frac{\sqrt{2}}{2}$.

## Example 2

**Find sin 210°.**

Study the graph at the right. The dashed line cuts the *x*-axis and the terminal side of the angle to form a 30°–60° right triangle. In a 30°–60°-right triangle, the side opposite the 30° angle is half the length of the hypotenuse. The length of the radius of the circle is 1 unit. Thus, *s* = 1 unit and $\frac{s}{2} = \frac{1}{2}$ unit.

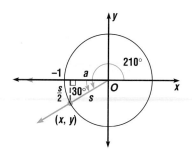

The *y*-coordinate of the point (*x*, *y*) is $-\frac{1}{2}$. So, sin 210° = $-\frac{1}{2}$. *Use the Pythagorean Theorem to check this result.*

The table at the right lists the signs of the sine and cosine functions for angles in standard position with terminal sides in each of the four quadrants. Recall that the coordinates of the point where the terminal side intersects the unit circle are (cos θ, sin θ).

| Quadrant II | Quadrant I |
|---|---|
| cos θ  − | cos θ  + |
| sin θ  + | sin θ  + |

| Quadrant III | Quadrant IV |
|---|---|
| cos θ  − | cos θ  + |
| sin θ  − | sin θ  − |

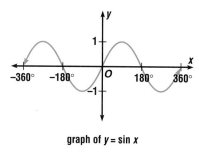

graph of *y* = sin *x*

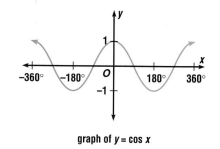

graph of *y* = cos *x*

You could use the technique used in the examples to complete the following tables. *You should memorize the frequently-used values listed in these tables.*

| degrees | 0 | 30 | 45 | 60 | 90 | 120 | 135 | 150 | 180 | 210 | 225 | 240 | 270 | 300 | 315 | 330 | 360 |
|---|---|---|---|---|---|---|---|---|---|---|---|---|---|---|---|---|---|
| radians | 0 | $\frac{\pi}{6}$ | $\frac{\pi}{4}$ | $\frac{\pi}{3}$ | $\frac{\pi}{2}$ | $\frac{2\pi}{3}$ | $\frac{3\pi}{4}$ | $\frac{5\pi}{6}$ | $\pi$ | $\frac{7\pi}{6}$ | $\frac{5\pi}{4}$ | $\frac{4\pi}{3}$ | $\frac{3\pi}{2}$ | $\frac{5\pi}{3}$ | $\frac{7\pi}{4}$ | $\frac{11\pi}{6}$ | $2\pi$ |
| sin θ | 0 | $\frac{1}{2}$ | $\frac{\sqrt{2}}{2}$ | $\frac{\sqrt{3}}{2}$ | 1 | $\frac{\sqrt{3}}{2}$ | $\frac{\sqrt{2}}{2}$ | $\frac{1}{2}$ | 0 | $-\frac{1}{2}$ | $-\frac{\sqrt{2}}{2}$ | $-\frac{\sqrt{3}}{2}$ | $-1$ | $-\frac{\sqrt{3}}{2}$ | $-\frac{\sqrt{2}}{2}$ | $-\frac{1}{2}$ | 0 |
| cos θ | 1 | $\frac{\sqrt{3}}{2}$ | $\frac{\sqrt{2}}{2}$ | $\frac{1}{2}$ | 0 | $-\frac{1}{2}$ | $-\frac{\sqrt{2}}{2}$ | $-\frac{\sqrt{3}}{2}$ | $-1$ | $-\frac{\sqrt{3}}{2}$ | $-\frac{\sqrt{2}}{2}$ | $-\frac{1}{2}$ | 0 | $\frac{1}{2}$ | $\frac{\sqrt{2}}{2}$ | $\frac{\sqrt{3}}{2}$ | 1 |

The next chart contains the same information for angles from 360° to 720°. Compare the information in the two charts. As you can see, the values of sin θ and cos θ are the same for two angles that are coterminal.

| degrees | 360 | 390 | 405 | 420 | 450 | 480 | 495 | 510 | 540 | 570 | 585 | 600 | 630 | 660 | 675 | 690 | 720 |
|---|---|---|---|---|---|---|---|---|---|---|---|---|---|---|---|---|---|
| radians | $2\pi$ | $\dfrac{13\pi}{6}$ | $\dfrac{9\pi}{4}$ | $\dfrac{7\pi}{3}$ | $\dfrac{5\pi}{2}$ | $\dfrac{8\pi}{3}$ | $\dfrac{11\pi}{4}$ | $\dfrac{17\pi}{6}$ | $3\pi$ | $\dfrac{19\pi}{6}$ | $\dfrac{13\pi}{4}$ | $\dfrac{10\pi}{3}$ | $\dfrac{7\pi}{2}$ | $\dfrac{11\pi}{3}$ | $\dfrac{15\pi}{4}$ | $\dfrac{23\pi}{6}$ | $4\pi$ |
| $\sin\theta$ | 0 | $\dfrac{1}{2}$ | $\dfrac{\sqrt{2}}{2}$ | $\dfrac{\sqrt{3}}{2}$ | 1 | $\dfrac{\sqrt{3}}{2}$ | $\dfrac{\sqrt{2}}{2}$ | $\dfrac{1}{2}$ | 0 | $-\dfrac{1}{2}$ | $-\dfrac{\sqrt{2}}{2}$ | $-\dfrac{\sqrt{3}}{2}$ | $-1$ | $-\dfrac{\sqrt{3}}{2}$ | $-\dfrac{\sqrt{2}}{2}$ | $-\dfrac{1}{2}$ | 0 |
| $\cos\theta$ | 1 | $\dfrac{\sqrt{3}}{2}$ | $\dfrac{\sqrt{2}}{2}$ | $\dfrac{1}{2}$ | 0 | $-\dfrac{1}{2}$ | $-\dfrac{\sqrt{2}}{2}$ | $-\dfrac{\sqrt{3}}{2}$ | $-1$ | $-\dfrac{\sqrt{3}}{2}$ | $-\dfrac{\sqrt{2}}{2}$ | $-\dfrac{1}{2}$ | 0 | $\dfrac{1}{2}$ | $\dfrac{\sqrt{2}}{2}$ | $\dfrac{\sqrt{3}}{2}$ | 1 |

Every 360°, or $2\pi$ radians, represents one complete revolution of the terminal side. As you can see by comparing the table above with the one on page 738, for every 360° or $2\pi$ radians, the sine and cosine functions repeat their values. So, we can say that the sine and cosine functions are **periodic.** Each has a **period** of 360° or $2\pi$ radians.

| *Definition of* *Periodic Function* | **A function is called periodic if there is a number $a$ such that $f(x) = f(x + a)$ for all $x$ in the domain of the function. The least positive value of $a$ for which $f(x) = f(x + a)$ is called the period of the function.** |
|---|---|

For the sine and cosine functions, $\cos(x + 360°) = \cos x$ and $\sin(x + 2\pi) = \sin x$. What are the values of $\cos(x + 2\pi)$ and $\sin(x + 360°)$? How can a negative angle be written as a positive angle?

**Example 3**

Find the value of each function.

a. $\sin 420°$

$\sin 420° = \sin(60 + 360)°$

$\qquad\ \ = \sin 60°$

$\qquad\ \ = \dfrac{\sqrt{3}}{2}$

b. $\cos\left(-\dfrac{3\pi}{4}\right)$

$\cos\left(-\dfrac{3\pi}{4}\right) = \cos\left(-\dfrac{3\pi}{4} + 2\pi\right)$

$\qquad\qquad\ \ = \cos\left(\dfrac{5\pi}{4}\right)$ or $\cos 225°$

$\qquad\qquad\ \ = -\dfrac{\sqrt{2}}{2}$

# CHECKING FOR UNDERSTANDING

**Communicating Mathematics**

**Read and study the lesson to answer these questions.**

1. The terminal side of an angle with measure $\theta$ intersects the unit circle at a particular point.

   a. What is the $x$-coordinate of the point called?

   b. What is the $y$-coordinate called?

2. What are the signs of the sine and cosine values for an angle whose terminal side lies in the third quadrant?

3. How are angles with measures of $-45°$ and $315°$ related?

State whether the value of each function is positive or negative.

**4.** $\sin 200°$      **5.** $\cos 320°$      **6.** $\sin(-135°)$

**7.** $\cos 405°$      **8.** $\cos(-225°)$      **9.** $\sin 445°$

Find each value. Use the tables on pages 738 and 739 or the properties of unit circles, and 30°–60° right triangles and 45°–45° right triangles.

**10.** $\sin \dfrac{4\pi}{3}$      **11.** $\cos\left(-\dfrac{7\pi}{3}\right)$      **12.** $\sin\left(-\dfrac{\pi}{6}\right)$

**13.** $\cos\left(-\dfrac{3}{4}\pi\right)$      **14.** $\sin(-240°)$      **15.** $\sin 660°$

# EXERCISES

For each of the following, find the least positive angle measurement that is coterminal.

**16.** $680°$      **17.** $-70°$      **18.** $1020°$

**19.** $-450°$      **20.** $3\pi$      **21.** $\dfrac{9\pi}{2}$

**22.** $-\dfrac{\pi}{4}$      **23.** $\dfrac{27\pi}{4}$      **24.** $-760°$

Find each value.

**25.** $\sin 45°$      **26.** $\cos 150°$      **27.** $\cos \dfrac{11}{3}\pi$

**28.** $\sin\left(-\dfrac{5}{3}\pi\right)$      **29.** $\sin \dfrac{3\pi}{2}$      **30.** $\cos \dfrac{7}{4}\pi$

**31.** $\sin(-180°)$      **32.** $\cos\left(-\dfrac{7}{4}\pi\right)$      **33.** $\cos(-60°)$

**34.** $\sin 300°$      **35.** $\sin\left(-\dfrac{\pi}{6}\right)$      **36.** $\sin \dfrac{4}{3}\pi$

**37.** $4(\sin 30°)(\cos 60°)$      **38.** $\dfrac{\sin 30° + \cos 60°}{2}$

**39.** $\dfrac{4 \sin 300° + 2 \cos 30°}{3}$      **40.** $\sin 30° + \sin 60°$

**41.** $(\sin 60°)^2 + (\cos 60°)^2$      **42.** $8(\sin 120°)(\cos 120°)$

**43.** Use the method shown in Example 2 to prove that $\sin 60° = \dfrac{\sqrt{3}}{2}$.

**44. Sports**   Reba and Kyle are flying a kite. The kite string forms a 45° angle with the ground and the kite is 50 meters above the ground.
   **a.** Draw a diagram of the triangle formed by the kite string and the ground.
   **b.** Find the approximate length of the kite string.

**45. Construction** A support for the roof of a house is shown at the right. The center piece is 3 feet long and forms 60° angles with each of the braces. If each of the braces are 6 feet long, find the cosine of 60°.

6 ft  60° 60°  6 ft

3 ft

**Mixed Review**

**46. Statistics** During a cold spell lasting 43 days, the following high temperatures (in degrees Fahrenheit) were recorded in Chicago.

```
26  17  12   5   4  25  17  23  13   6  25  19  27  22  26
20  31  24  12  27  16  27  16  30   7  31  16   5  29  18
16  22  29   8  31  13  24   5  -7  20  29  18  12
```

Find the median, mode, and mean of the temperatures.   **(Lesson 14-3)**

**47.** Jane and Pedro are stacking cans of mixed fruit in a display at the end of the grocery aisle. The top layer of the stack of cans has 1 can, the second layer has 4 cans, the third layer has 9 cans, and so on. How many cans are in the stack if it has 12 layers?   **(Lesson 13-1)**

**48.** Simplify $\dfrac{\dfrac{3x}{4x-1}}{1+\dfrac{3x}{x-1}}$.   **(Lesson 11-4)**

**49.** Find all zeros for the function $f(x) = x^3 - 2x^2 - 13x - 10$. **(Lesson 10-4)**

**50.** Simplify $(9 + 6i) - (3 + 2i)$.   **(Lesson 6-9)**

**51.** Graph $4y - x \le 6$.   **(Lesson 2-8)**

## APPLICATIONS IN MUSIC

As you know, periodic functions are those functions whose values repeat after a certain period. The graphs of these functions are repetitive. Music is also periodic in nature. As you saw on the oscillogram on page 737, the graph produced by a musical tone closely resembles a sine curve. Hence, the sine curve is called pure or simple by musicians.

Graphs of musical tones have the same general shape, but different **frequencies.** The frequency of a curve is the number of cycles, or repetitions, in a given interval. The musical notes with higher frequencies are the higher pitched notes.

The vibrations of the string at the right show how musical notes are produced. The lowest frequency is called a fundamental and its multiples are called harmonics. Doubling the frequency raises the pitch one octave. So, the second harmonic is one octave above the fundamental. The fourth harmonic is one octave above the second harmonic and two octaves above the fundamental.

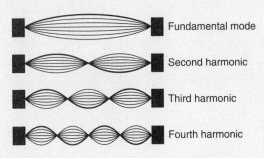

Fundamental mode

Second harmonic

Third harmonic

Fourth harmonic

# Other Trigonometric Functions

**Objective**

After studying this lesson, you should be able to:

■ find the values of other trigonometric functions.

Other trigonometric functions are defined using sine and cosine.

*Definition of Tangent (tan), Cotangent (cot), Secant (sec), and Cosecant (csc)*

**Let θ stand for the measurement of an angle in standard position on the unit circle. Then the following equations hold true.**

**If cos θ ≠ 0, tan θ = $\dfrac{\sin θ}{\cos θ}$ and sec θ = $\dfrac{1}{\cos θ}$.**

**If sin θ ≠ 0, cot θ = $\dfrac{\cos θ}{\sin θ}$ and csc θ = $\dfrac{1}{\sin θ}$.**

**Example 1**

*The tables on pages 738 and 739 give values for sin θ and cos θ.*

Find each of the following.

**a. tan 135°**

$$\tan 135° = \frac{\sin 135°}{\cos 135°} \quad \begin{array}{l}\textit{Definition}\\ \textit{of tangent}\end{array}$$

$$= \frac{\dfrac{\sqrt{2}}{2}}{-\dfrac{\sqrt{2}}{2}}$$

$$= -1$$

**b. sec 3.5π**

$$\sec 3.5π = \frac{1}{\cos 3.5π} \quad \begin{array}{l}\textit{Definition}\\ \textit{of secant}\end{array}$$

$$= \frac{1}{0}$$

sec 3.5π is undefined.

*Division by zero is undefined.*

**c. cot $\dfrac{5π}{2}$**

$$\cot \frac{5π}{2} = \frac{\cos \dfrac{5π}{2}}{\sin \dfrac{5π}{2}} \quad \begin{array}{l}\textit{Definition}\\ \textit{of cotangent}\end{array}$$

$$= \frac{0}{1}$$

$$= 0$$

**d. csc (−150°)**

$$\csc (-150°) = \frac{1}{\sin (-150°)} \quad \begin{array}{l}\textit{Definition}\\ \textit{of cosecant}\end{array}$$

$$= \frac{1}{\sin (210°)}$$

$$= \frac{1}{-\dfrac{1}{2}}$$

$$= -\frac{2}{1} \text{ or } -2$$

**Example 2**

**APPLICATION**

**Surveying**

A surveyor finds that the angle between two merging roads is 120°. Find the value of each of the trigonometric functions for this angle.

$$\sin 120° = \frac{\sqrt{3}}{2}$$

$$\csc 120° = \frac{1}{\sin 120°}$$
$$= \frac{1}{\frac{\sqrt{3}}{2}}$$
$$= \frac{2}{\sqrt{3}} \text{ or } \frac{2\sqrt{3}}{3}$$

$$\cos 120° = -\frac{1}{2}$$

$$\sec 120° = \frac{1}{\cos 120°}$$
$$= \frac{1}{-\frac{1}{2}}$$
$$= -\frac{2}{1} \text{ or } -2$$

$$\tan 120° = \frac{\sin 120°}{\cos 120°}$$
$$= \frac{\frac{\sqrt{3}}{2}}{-\frac{1}{2}}$$
$$= -\sqrt{3}$$

$$\cot 120° = \frac{\cos 120°}{\sin 120°}$$
$$= \frac{-\frac{1}{2}}{\frac{\sqrt{3}}{2}}$$
$$= -\frac{1}{\sqrt{3}} \text{ or } -\frac{\sqrt{3}}{3}$$

# CHECKING FOR UNDERSTANDING

**Communicating Mathematics**

Read and study the lesson to answer these questions.

1. If $\sin \theta = \frac{3}{5}$ and $\cos \theta = \frac{4}{5}$, find the other trigonometric functions for $\theta$.

2. For any angle $\theta$, $-1 \leq \sin \theta \leq 1$. What can you say about the range for $\csc \theta$?

3. Two functions are called reciprocal functions if their product is 1. For example, $y = \sin x$ and $y = \csc x$ are reciprocal functions since $\csc x = \frac{1}{\sin x}$ and $(\sin x)\left(\frac{1}{\sin x}\right) = 1$. What are the reciprocal functions for $y = \cos x$ and for $y = \tan x$?

**Guided Practice**

State the values of $0° \leq \theta \leq 360°$ for which each expression is not defined.

4. $\sin \theta$
5. $\tan \theta$
6. $\cos \theta$
7. $\cot \theta$
8. $\csc \theta$
9. $\sec \theta$

Find the value of each of the trigonometric functions for each angle measure.

10. $60°$
11. $-225°$
12. $\frac{3\pi}{2}$

# EXERCISES

**Practice**

Find each value.

**13.** cot 135°

**14.** csc 45°

**15.** sec 300°

**16.** $\tan\left(-\dfrac{\pi}{3}\right)$

**17.** csc (−210°)

**18.** $\tan\dfrac{7\pi}{6}$

**19.** cot (−60°)

**20.** sec 240°

**21.** $\cot\left(-\dfrac{\pi}{6}\right)$

**22.** $\csc\left(-\dfrac{\pi}{6}\right)$

**23.** sec (−120°)

**24.** cot 210°

**25.** $\csc\dfrac{\pi}{2}$

**26.** $\tan\dfrac{9}{4}\pi$

**27.** $\tan\left(-\dfrac{5}{6}\pi\right)$

**28.** $\cot\dfrac{7\pi}{4}$

**29.** tan (−300°)

**30.** cot 540°

**31.** sec (−30°)

**32.** tan 405°

**33.** csc 180°

**34.** sec 390°

**35.** cot (−600°)

**36.** $\csc\left(-\dfrac{7}{6}\pi\right)$

**37.** $\dfrac{\tan 300°}{\csc 540°}$

**38.** $\tan\dfrac{13\pi}{4} - \sec\pi$

**39.** $\csc\left(-\dfrac{5}{2}\pi\right)\sec\left(-\dfrac{11}{6}\pi\right)$

**Critical Thinking**

**40.** Angles with measures whose sum is 90° are called complementary angles. Explain how the values of the trigonometric functions for complementary angles are related.    *Hint: Find the values of the trigonometric functions for angles of 30° and 60° and compare.*

**Applications**

**41. Navigation**   A ship's instruments indicate that it is headed on a course that forms an angle of 12.7° with the equator. After the ship has traveled 205 miles, it will still be just 45 miles from the equator. Assuming that sin 12.7° = $\dfrac{9}{41}$ and cos 12.7° = $\dfrac{40}{41}$, find the values of the other trigonometric functions of an angle with degree measure 12.7°.

**42. Sports**   The sail at the right is in the shape of a right triangle. The bottom edge of the sail, called the foot, is 60 inches long. The edge of the sail that attaches to the mast, called the luff, is 144 inches long. The angle at the bottom of the sail measures approximately 67.4°. Find the value of each of the trigonometric functions for this angle.

67.4°

**Mixed Review**

43. Shelby has 7 pennies, 4 dimes, 2 quarters, and 1 nickel in her wallet. She chooses one at random. What is the probability that it is a penny or a nickel? **(Lesson 15-8)**

44. State the first term, the common ratio, the last term, and the number of terms for the geometric series described by $a_1 = 12$, $a_4 = \frac{3}{2}$, $n = 6$. **(Lesson 13-5)**

45. Find the solution of the following system of equations. $4x^2 + 9y^2 = 36$
   **(Lesson 9-9)** $\qquad\qquad\qquad\qquad\qquad\qquad\qquad 4x^2 - 9y^2 = 36$

46. Solve $\sqrt[3]{3y - 1} - 2 = 0$. **(Lesson 6-7)**

47. **Geometry** The perimeter of a triangle is 56 feet. The length of the longest side is three times the difference between the lengths of the other two sides. The longest side is also twice as long as the shortest side. Find the lengths of the three sides. **(Lesson 3-9)**

48. Determine the slope of a line that passes through the points (6, 8) and (10, 22). **(Lesson 2-4)**

---

## APPLICATIONS IN OPTICS

Scientists use refractometers to measure the change in direction, or bending, of light as it moves from one medium to another. The measurement is called the *index of refraction*. Refractometers are used in gemology to help identify different stones. Chemists use refractometers to determine the composition of unknown solutions or substances.

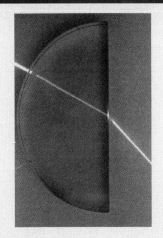

A refractometer uses the basic principles of **Snell's Law.** If $n$ represents the index of refraction, $I$ represents the angle of incidence, and $r$ represents the angle of refraction, then the equation below states their relationship as light passes from a vacuum into another medium.

$n = \dfrac{\sin I}{\sin r}$ $\qquad 0° < I \le 90°$, $0° < r \le 90°$

**EXAMPLE** A beam of light moves from a vacuum to glass. If the angle of refraction is 30° and the index of refraction is $\sqrt{2}$, what is the angle of incidence?

$$n = \frac{\sin I}{\sin r}$$

$$\sqrt{2} = \frac{\sin I}{\sin 30°} \qquad \textit{Substitute } \sqrt{2} \textit{ for n and 30° for r.}$$

$$\sqrt{2} = \frac{\sin I}{\frac{1}{2}} \qquad \textit{sin } 30° = \frac{1}{2}$$

$$\frac{\sqrt{2}}{2} = \sin I$$

$$45° = I \qquad \text{The angle of incidence is 45°.}$$

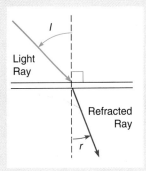

# Inverse Trigonometric Functions

**Objective**

After studying this lesson, you should be able to:

■ find the values of expressions involving trigonometric functions.

**Application**

Phyllis Washington is an architect working on the design for a new home. She knows that the support for the roof will be shaped like two right triangles. Each right triangle will have one leg 8 feet long and a hypotenuse 16 feet long.

The sine of angle θ will be $\frac{1}{2}$. What will the measure of angle θ be? *You will solve this problem in Example 1.*

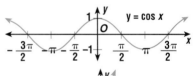

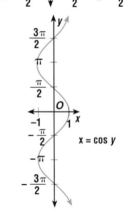

Sometimes you know the value of some trigonometric function for an angle and need to find the measure of the angle. The concept of inverse functions can be applied to find the inverse of trigonometric functions.

In Lesson 10-8, you learned that the inverse of a function is the relation in which all of the values of *x* and *y* are reversed. The graphs of the *y* = cos *x* function and its inverse, *x* = cos *y*, are shown at the left. Notice that the inverse is not a function, since it fails the vertical line test. None of the inverses of the trigonometric functions are functions.

We must restrict the domain of the trigonometric functions so that the inverses are functions. The values in these restricted domains are called **principal values.** Capital letters are used to distinguish trigonometric functions with restricted domains from the usual trigonometric functions.

*Definitions of the Cosine, Sine, and Tangent Functions*

> $y = \text{Cos } x$ *if and only if* $y = \cos x$ *and* $0 \le x \le \pi.$
>
> $y = \text{Sin } x$ *if and only if* $y = \sin x$ *and* $-\dfrac{\pi}{2} \le x \le \dfrac{\pi}{2}.$
>
> $y = \text{Tan } x$ *if and only if* $y = \tan x$ *and* $-\dfrac{\pi}{2} \le x \le \dfrac{\pi}{2}.$

The inverse of the Cosine function is called the **Arccosine function.** It is symbolized by **Cos$^{-1}$** or **Arccos.**

| | |
|---|---|
| *Definition of*<br>*Inverse Cosine* | **Given $y = \text{Cos } x$, the inverse Cosine function is defined by the following equation.**<br>$y = \text{Cos}^{-1} x$ **or** $y = \text{Arccos } x$ |

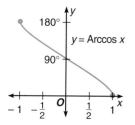

The Arccosine function has the following characteristics.
- Its domain is the set of real numbers from –1 to 1.
- Its range is the set of angle measurements from 0 to $\pi$.
- $\text{Cos } x = y$ if and only if $\text{Cos}^{-1} y = x$.
- $(\text{Cos}^{-1} \circ \text{Cos})(x) = (\text{Cos} \circ \text{Cos}^{-1})(x) = x$     *Recall function composition from Lesson 10-7.*

The definitions of the Arcsine and Arctangent functions are similar to the definition of the Arccosine function.

| | |
|---|---|
| *Definition of*<br>*Inverse Sine*<br>*and Inverse Tangent* | **Given $y = \text{Sin } x$, the inverse Sine function is defined by the following equation.**<br>$$y = \text{Sin}^{-1} x \text{ or } y = \text{Arcsin } x$$<br>**Given $y = \text{Tan } x$, the inverse Tangent function is defined by the following equation.**<br>$$y = \text{Tan}^{-1} x \text{ or } y = \text{Arctan } x$$ |

The expressions in each row are equivalent.

| | |
|---|---|
| $y = \text{Sin } x$ | $x = \text{Sin}^{-1} y$ or $x = \text{Arcsin } y$ |
| $y = \text{Cos } x$ | $x = \text{Cos}^{-1} y$ or $x = \text{Arccos } y$ |
| $y = \text{Tan } x$ | $x = \text{Tan}^{-1} y$ or $x = \text{Arctan } y$ |

Now, use the definitions of the inverse functions to solve the application presented on page 746.

**Example 1**

**The support for a roof is shaped like two right triangles each with one leg 8 feet long and the hypotenuse 16 feet long. The sine of angle $\theta$ will be $\frac{1}{2}$. What will $\theta$ be?**

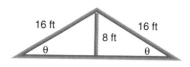

We know the sine of the angle and need to find the measurement of the angle. Use the Arcsine function to find the angle whose sine is $\frac{1}{2}$.

$$\theta = \text{Arcsin } \frac{1}{2}$$
$$\text{Sin } \theta = \frac{1}{2}$$
$$\theta = \frac{\pi}{6} \qquad \theta \text{ is } \frac{\pi}{6} \text{ since it must be that } -\frac{\pi}{2} \leq \theta \leq \frac{\pi}{2}.$$

So the angle at the edge of the roof is $\frac{\pi}{6}$ radians or 30°.

**Example 2** | Find $\cos\left(\text{Arcsin } \dfrac{\sqrt{2}}{2}\right)$.

Arcsin $\dfrac{\sqrt{2}}{2}$ is the angle whose sine is $\dfrac{\sqrt{2}}{2}$, so let $\theta = $ Arcsin $\dfrac{\sqrt{2}}{2}$. Then Sin $\theta$ $= \dfrac{\sqrt{2}}{2}$ and $\theta = \dfrac{\pi}{4}$. Therefore, Arcsin $\dfrac{\sqrt{2}}{2} = \dfrac{\pi}{4}$. *Why is $\theta$ not $\dfrac{3\pi}{4}$?*

$$\cos\left(\text{Arcsin } \dfrac{2\sqrt{2}}{2}\right) = \cos\dfrac{\pi}{4} \quad \text{\textit{Substitute} } \dfrac{\pi}{4} \text{ \textit{for Arcsin} } \dfrac{\sqrt{2}}{2}.$$
$$= \dfrac{\sqrt{2}}{2}$$

In the examples so far, the principal value of each inverse trigonometric function was known to be the value of a trigonometric function for some angle. Sometimes it is *not* known to which angle the principal value corresponds. In such a case, a diagram is almost a necessity.

**Example 3** | Find $\tan\left(\text{Sin}^{-1}\left(\dfrac{5}{13}\right)\right)$.

Let $\theta = \text{Sin}^{-1}\left(\dfrac{5}{13}\right)$.

Draw a diagram using the unit circle with angle $\theta$ so that its sine is $\dfrac{5}{13}$.

Substitute $\theta$ for $\text{Sin}^{-1}\left(\dfrac{5}{13}\right)$ in the original expression to obtain $\tan \theta$. So, now we need to find $\tan \theta$. To do this we need to find $\cos \theta$.

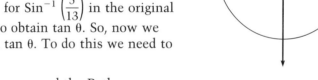

From the diagram and the Pythagorean Theorem, we see that $(\cos \theta)^2 + \left(\dfrac{5}{13}\right)^2 = 1$.

$$(\cos \theta)^2 + \left(\dfrac{5}{13}\right)^2 = 1 \qquad \text{\textit{Solve for} } \cos \theta.$$
$$\cos^2 \theta = 1 - \dfrac{25}{169}$$
$$\cos^2 \theta = \dfrac{144}{169}$$
$$\cos \theta = \dfrac{12}{13} \qquad \text{\textit{Take the square root of each side.}}$$

Now find $\tan \theta$.

$$\tan \theta = \dfrac{\sin \theta}{\cos \theta} \qquad \text{\textit{Definition of tangent.}}$$
$$= \dfrac{\dfrac{5}{13}}{\dfrac{12}{13}} \text{ or } \dfrac{5}{12} \qquad \text{Therefore, } \tan\left(\text{Sin}^{-1}\left(\dfrac{5}{13}\right)\right) = \dfrac{5}{12}.$$

# CHECKING FOR UNDERSTANDING

**Communicating Mathematics**

Read and study the lesson to answer these questions.

1. How are $y = \text{Sin } x$ and $y = \text{Arcsin } x$ related?

2. Why must we restrict the domains of the trigonometric functions before finding the inverse functions?

3. How do you know when the domain of a trigonometric function is restricted?

**Guided Practice**

Write each of the following in the form of an inverse function.

4. $a = \cos b$

5. $\sin y = x$

6. $\tan \alpha = \beta$

7. $\sin 30° = \dfrac{1}{2}$

8. $2 \cos 45° = y$

9. $-\dfrac{4}{3} = \tan x$

Find each value.

10. $\sin \left( \text{Cos}^{-1} \left( \dfrac{2}{3} \right) \right)$

11. $\cos \left( \text{Cos}^{-1} \dfrac{4}{5} \right)$

12. $\cos \left( \text{Cos}^{-1} \dfrac{1}{2} \right)$

# EXERCISES

**Practice**

Find each value.

13. $\text{Cos}^{-1} \left( -\dfrac{1}{2} \right)$

14. $\sin \left( \text{Sin}^{-1} \dfrac{1}{2} \right)$

15. $\text{Sin}^{-1} \left( \cos \dfrac{\pi}{2} \right)$

16. $\text{Tan}^{-1} (-1)$

17. $\text{Sin} \dfrac{\pi}{6}$

18. $\text{Sin}^{-1} 1$

19. $\tan \left( \text{Cos}^{-1} \dfrac{6}{7} \right)$

20. $\cot \left( \text{Sin}^{-1} \dfrac{5}{6} \right)$

21. $\cot \left( \text{Sin}^{-1} \dfrac{7}{9} \right)$

22. $\sin \left( \text{Arctan} \dfrac{\sqrt{3}}{3} \right)$

23. $\cos \left( \text{Arcsin} \dfrac{3}{5} \right)$

24. $\tan (\text{Arctan } 3)$

25. $\text{Sin}^{-1} \left( \tan \dfrac{\pi}{4} \right)$

26. $\text{Arctan } \sqrt{3}$

27. $\text{Arccos} \dfrac{\sqrt{3}}{2}$

28. $\cos (\text{Tan}^{-1} \sqrt{3})$

29. $\cos \left[ \text{Arcsin} \left( -\dfrac{1}{2} \right) \right]$

30. $\cos (\text{Tan}^{-1} 1)$

31. $\cos \left[ \text{Cos}^{-1} \left( -\dfrac{\sqrt{2}}{2} \right) - \dfrac{\pi}{2} \right]$

32. $\sin \left( 2 \text{ Sin}^{-1} \dfrac{1}{2} \right)$

33. $\sin \left( 2 \text{ Cos}^{-1} \dfrac{3}{5} \right)$

34. $\tan \left[ \text{Cos}^{-1} \left( -\dfrac{3}{5} \right) \right]$

35. $\sin \left( \text{Sin}^{-1} \dfrac{\sqrt{3}}{2} \right)$

36. $\sin [\text{Arctan} (-\sqrt{3})]$

37. $\text{Cos}^{-1} \left( \sin \dfrac{3\pi}{2} \right)$

38. $\text{Tan}^{-1} \left( \cos \dfrac{3\pi}{2} \right)$

39. $\text{Sin}^{-1} \left( \cos \dfrac{\pi}{6} \right)$

**Critical Thinking**

40. Prove that $(\text{Cos}^{-1} \circ \text{Cos})(x) = (\text{Cos} \circ \text{Cos}^{-1})(x) = x$.

**Applications**

**41. Navigation** The Western Princess sailed due east 24 miles before turning north. When the Princess became disabled and radioed for help, the rescue boat needed to know the fastest route to her. The navigator of the Princess found that the fastest route for the rescue boat would be 48 miles. The cosine of the angle at which the rescue boat should sail is $\frac{1}{2}$.

a. Draw a diagram that represents the situation.

b. Find the angle at which the rescue boat should travel to aid the Western Princess.

**42. Travel** Michelle and her family took a vacation to Oregon. As Michelle observed Mount Hood from a distance, the sine of the angle at which she looked up was $\frac{40}{41}$. Find the tangent of this angle.

**Mixed Review**

**43. Statistics** The stem-and-leaf plot at the right represents the number of passengers served by certain airports in 1988. Find the range, quartiles, and interquartile range for the data. Then find any outliers. **(Lesson 14-4)**

| Stem | Leaf |
|------|------|
| 1 | 6 6 7 7 8 8 8 9 |
| 2 | 0 0 2 2 2 7 7 9 9 |
| 3 | 1 5 |
| 4 | 0 1 5 |
| 5 | 3 |

$3 \mid 1 = 31,000,000$

**44.** Write the expression $\sum\limits_{k=1}^{10} (2 + k)$ in expanded form and find the sum. **(Lesson 13-3)**

**45. Biology** Bacteria of a certain strain can grow from 80 to 164 bacteria in 3 hours. Find $k$ for the growth formula for this strain. **(Lesson 12-8)**
*Hint: The exponential growth formula is $y = ne^{kt}$.*

**46. Travel** The time it takes to drive a certain distance varies inversely as to the rate of speed. Golda and John drove at 55 miles per hour for 4 hours to visit some friends in another city. Because of bad weather, they can only drive 45 miles per hour on the way home. How long will it take them to get home? **(Lesson 11-2)**

**47. Health** Ty's heart rate is usually 120 beats per minute when he runs. If he runs for 2 hours every day, about how many beats will his heart make during two weeks worth of exercise sessions? Express the answer in both decimal and scientific notation. **(Lesson 5-1)**

**48.** Find $AA$ if $A = \begin{bmatrix} 2 & 7 \\ 0 & -1 \end{bmatrix}$. Write "not defined" if the product does not exist. **(Lesson 4-4)**

# 16-5 Finding Values for Trigonometric Functions

**Objective**

After studying this lesson, you should be able to:

■ use a calculator to find values of trigonometric functions.

**Application**

Ships and airplanes measure distance in nautical miles. A nautical mile is equal to one **minute,** or $\frac{1}{60}$ of a degree of arc length on a meridian. The actual length varies slightly since Earth is not a perfect sphere. The formula 1 nautical mile equals $(6077 - 31 \cos 2\theta)$ feet, where $\theta$ is the latitude in degrees will give the approximate length of a nautical mile in a certain latitude. You will find the length of a nautical mile along the 30th parallel in Example 4.

**Example 1**

**Find sin 67° using your calculator.**    *Set your calculator in degree mode.*

ENTER:  67 (SIN) $0.9205048535$

The sine of 67° is approximately 0.9205.

**Example 2**

**Find x in each of the following using your calculator.**

**a. If sin x = 0.7590, find x.**

ENTER:  0.7590 (SIN⁻¹) $49.37611923$

*The* (SIN⁻¹) *key may be a second function key on your calculator.*

Therefore, x is approximatley 49.3761°.    *Round to four decimal places.*

**b. Suppose csc x = 4.2024. Find the value of x to the nearest minute.**

First use the second function and sin keys to find the decimal value of x.

$\csc x = \frac{1}{\sin x}$, $\sin x = \frac{1}{4.2024}$

ENTER:  4.2024 (¹/ₓ) (SIN⁻¹) $13.76612556$

Now subtract the whole number part and multiply the decimal portion by 60 to find the minutes.
ENTER:  (−) 13 (=) (×) 60 (=) $45.96753346$

Therefore, x is approximately 13°46′.

**Example 3**

**Find each of the following.**

**a.** $\cot\left(\dfrac{\pi}{12}\right)$ — *Set your calculator in radian mode.*

$\cot x = \dfrac{1}{\tan x}$

ENTER: [π] [÷] 12 [=] [TAN] [¹/ₓ] 3.732050808

$\cot\left(\dfrac{\pi}{12}\right)$ is approximately 3.7321. — *Round to four decimal places.*

**b.** $\tan 67°38'$ — *Make sure your calculator is in degree mode.*

Minutes must be changed to decimal form before you can use the calculator to find the trigonometric value. This is done by dividing the minutes by 60.

ENTER: 67 [+] 38 [÷] 60 [=] [TAN] 2.430193842

So, $\tan 67°38'$ is approximately 2.4302.

Now solve the application presented on page 751.

**Example 4**

APPLICATION

Navigation

*FYI* · · ·

The 30th parallel runs through New Orleans; Cairo, Egypt; and Chongqing, China.

**Use the formula 1 nautical mile = (6077 − 31 cos 2θ), where θ is the latitude in degrees to find the length of a nautical mile on the 30th parallel.**

1 nautical mile = 6077 − 31 cos 2θ — *Formula for the length of a nautical mile.*
　　　　　　　 = 6077 − 31 cos 2(30) — *Substitute 30 for θ.*
　　　　　　　 = 6077 − 31 cos 60

Use your calculator to find the value of the expression.

ENTER: 6077 [−] 60 [COS] [×] 31 [=] 6061.5

Therefore, one nautical mile is about 6061.5 feet on the 30th parallel. This is $\dfrac{6061.5}{5280}$ or about 1.15 miles.

Your calculator can perform multiple functions with trigonometric values.

**Example 5**

**Find** $\tan\left(\text{Cos}^{-1}\left(\dfrac{\sqrt{2}}{2}\right)\right)$.

ENTER: 2 [√] [÷] 2 [=] [COS⁻¹] [TAN] 1

Therefore, $\tan\left(\text{Cos}^{-1}\left(\dfrac{\sqrt{2}}{2}\right)\right)$ is 1. — *Check this result by drawing a diagram.*

# CHECKING FOR UNDERSTANDING

**Communicating Mathematics**

Read and study the lesson to answer these questions.

1. How can you use a calculator to find the trigonometric value of a measurement that is given in degrees and minutes?

2. Write the key sequence you would use to find the value of cot 48° on your calculator. Find the value of cot 48°.

3. How can you find the minutes of a degree measurement found on a calculator?

**Guided Practice**

Use a calculator to find each value. Round your answers to four decimal places.

4. $\sin 42°$
5. $\cos 81°$
6. $\tan 5°$
7. $\cos 42°20'$
8. $\sin 3°10'$
9. $\tan 89°50'$

Use a calculator to find the value of $x$, in degrees, for each trigonometric function. Round your answers to four decimal places.

10. $\sin x = 0.7364$
11. $\cos x = 0.9912$
12. $\tan x = 0.5923$
13. $\sec x = 2.7504$
14. $\csc x = 1.9735$
15. $\cot x = 3.1376$

# EXERCISES

**Practice**

Use a calculator to find each value. Round your answers to four decimal places.

16. $\csc 75°$
17. $\sin 730°$
18. $\tan -90°$

19. $\cos 5\pi$
20. $\cot \dfrac{-3\pi}{5}$
21. $\sin \dfrac{7\pi}{3}$

22. $\tan \dfrac{-11\pi}{8}$
23. $\sec \dfrac{-8\pi}{5}$
24. $\cot 250°$

25. $\sec 600°$
26. $\sec 540°$
27. $\csc 7\pi$

28. $\csc (-890°)$
29. $(\sin 95°)(\tan 37°)$

30. $\left(\sin \dfrac{\pi}{3}\right)\left(\cos \dfrac{\pi}{8}\right)$
31. $\dfrac{3(\sin 50°) + 9(\cos 10°)}{\tan 290°}$

32. $\text{Sin}^{-1} \dfrac{15}{16}$
33. $\text{Cos}^{-1} 0.89$

34. $\text{Tan}^{-1} \dfrac{3}{2}$
35. $\dfrac{3 \tan \dfrac{\pi}{6}}{2 \sin \dfrac{\pi}{3} - \sec 3\pi}$

36. $\tan \left(\text{Cos}^{-1} \dfrac{\sqrt{3}}{3}\right)$
37. $\sin [\text{Tan}^{-1} (-\sqrt{6})]$

**Critical Thinking**

**38.** Some calculators allow you to work in a unit of angle measure called **gradians** or grads. There are 400 gradians in a circle.

   **a.** Find the trigonometric values for an angle that measures 100 grads.

   **b.** What is the measure of this angle in degrees and in radians?

**Applications**

**39. Physics**   The maximum height that a projected object reaches is given by the formula $H = \dfrac{V_o{}^2 \sin^2 \theta}{2g}$, where $V_o$ represents the initial velocity, $\theta$ represents the degree measure of the angle which the path of the object makes with the ground, and $g$ represents the acceleration due to gravity. An arrow is released at an angle of 65° to the ground and with an initial velocity of 100 feet per second. If the acceleration due to gravity is 32 feet per second, what is the maximum height reached by the arrow?

**40. Electronics**   The power $P$ in watts absorbed by an AC circuit is given by the formula $P = IV \cos \theta$, where $I$ is the current in amps, $V$ is the voltage, and $\theta$ is the measure of the phase angle. Find the power absorbed by a circuit if its current is 2 amps, its voltage is 120 volts, and its phase angle is 70°.

**41. Optics**   Rainbows that result from light passing through a prism are caused by the refraction of light. When the light passes from one medium, the air, into the other, glass, the light ray is bent and the rainbow results. According to Snell's Law, the angle at which the light ray approaches the prism, called the angle of incidence, and the angle at which the angle is bent, called the angle of refraction, are related by the formula $2 \sin I = 3 \sin r$. $I$ is the measure of the angle of incidence and $r$ is the measure of the angle of refraction. Find the angle of refraction for this prism in degrees and minutes if the angle of incidence is 60°.

**Mixed Review**

**42. Sports**   Suppose Lynn and Maria swim 8 one-lap races. The probability that Lynn wins a race is $\dfrac{2}{3}$ and that Maria wins is $\dfrac{1}{3}$. What is the probability that Maria will win at least 3 of the races?   **(Lesson 15-9)**

**43. Statistics**   The points scored by the winning team in each National Football League game over a three-day period were 17, 24, 23, 30, 21, 30, 24, 20, 34, 24, 23, 30, 40, and 31. Find the median, mode, mean, and standard deviation of the scores.   **(Lesson 14-6)**

**Journal**

Write two different sequences of keystrokes you could use to find cot 80° on your calculator. Are there others?

44. Express $\log_3 35$ in terms of common logarithms. Then use a calculator to find its value.   (**Lesson 12-7**)

45. Simplify the expression $\dfrac{\dfrac{x}{x^2 - 16}}{\dfrac{4x - 4}{x^2 - 2x - 8}}$.   (**Lesson 11-3**)

46. Solve $\dfrac{1}{3}|6x + 5| = 7$.   (**Lesson 1-6**)

---

## ~~~ MID-CHAPTER REVIEW ~~~

**Change each degree measure to radian measure.   (Lesson 16-1)**

1. $90°$

2. $150°$

3. $-135°$

**Change each radian measure to degree measure.   (Lesson 16-1)**

4. $\dfrac{3\pi}{2}$

5. $-\dfrac{7\pi}{4}$

6. $2$

**For each of the following, find the least positive angle measurement that is coterminal. (Lesson 16-2)**

7. $420°$

8. $1400°$

9. $-\dfrac{8\pi}{9}$

**Find each value.   (Lesson 16-3)**

10. $\cos 150°$

11. $\tan 900°$

12. $\csc \dfrac{4}{3}\pi$

13. $\sec \dfrac{9}{4}\pi$

14. $\dfrac{\tan(-135°)}{\sec 270°}$

15. $\left(\cot\left(-\dfrac{3\pi}{4}\right)\right)\left(\sec\left(-\dfrac{\pi}{4}\right)\right)$

**Find each of the following.   (Lesson 16-4)**

16. $\text{Arctan } 1$

17. $\text{Sin}^{-1}(-1)$

18. $\text{Cos } 45°$

19. $\text{Sin}^{-1}\left(-\dfrac{1}{2}\right)$

20. $\text{Tan}^{-1}\dfrac{\sqrt{3}}{3}$

21. $\text{Tan}\left(-\dfrac{\pi}{4}\right)$

**Use a calculator to find each value. Round your answers to four decimal places.   (Lesson 16-5)**

22. $\csc 85°$

23. $\sin -530°$

24. $\dfrac{2(\sin 75°) + 6(\tan 12°)}{\cos(-25°)}$

25. **Physics**   The formula $R = \dfrac{V_o{}^2 \sin 2\theta}{g}$ gives the range of an object shot at an initial velocity of $V_o$ feet per second at an angle of $\theta°$ to the ground. $g$ represents the acceleration due to gravity, which is 32 feet per second. Find the angle at which a projectile was shot if its initial velocity was 128 feet per second and its range was 512 feet.   (**Lesson 16-5**)

# Solving Right Triangles

**Objectives**

After studying this lesson, you should be able to:
- use right triangles to find trigonometric values, and
- solve problems involving right triangles using right triangle trigonometry.

**Application**

A car is traveling 500 meters along a raised exit ramp, which makes a 3° angle with the ground. How far did the car rise on the 500-meter ramp?

Trigonometry can be used to solve problems like this where you need to find a missing measure of a triangle. *You will solve this problem in Example 3.*

Consider the right triangle below.

The **hypotenuse** of the triangle is side $\overline{AB}$. Its length is $c$ units.

The side **opposite** angle $A$ is side $\overline{BC}$. Its length is $a$ units.

The side **adjacent** to angle $A$ is side $\overline{AC}$. Its length is $b$ units.

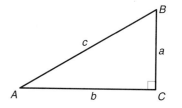

For convenience of notation, we refer to the angle with vertex at $A$ as angle $A$ and use $A$ to stand for its measurement. Similarly, we refer to angle $B$ and its measurement as $B$, and angle $C$ and its measurement as $C$.

Suppose a unit circle is drawn with its center at vertex $A$ as shown below.

The triangle in the interior of the unit circle is similar to triangle $ABC$. So, the measures of the corresponding sides of the triangle are proportional. That is,
$$\frac{1}{c} = \frac{x}{b} = \frac{y}{a}.$$

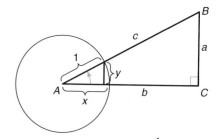

*SOH-CAH-TOA is a helpful mnemonic device for remembering the first three equations.*

$sin = \dfrac{opposite}{hypotenuse}$

$cos = \dfrac{adjacent}{hypotenuse}$

$tan = \dfrac{opposite}{adjacent}$

In this figure, $\sin A = y$. Since the two triangles are similar, $\dfrac{y}{a} = \dfrac{1}{c}$. So, $y = \dfrac{a}{c}$ and, by substitution, $\sin A = \dfrac{a}{c}$. Using this figure, we can define the trigonometric values in the following way.

$$\sin A = \frac{a}{c} \qquad \cos A = \frac{b}{c} \qquad \tan A = \frac{a}{b}$$

$$\csc A = \frac{c}{a} \qquad \sec A = \frac{c}{b} \qquad \cot A = \frac{b}{a}$$

**Example 1**

Find the sine, cosine, tangent, cosecant, secant, and cotangent of angle A. Round each value to four decimal places.

$\sin A = \dfrac{9}{15}$ or 0.6000     $\csc A = \dfrac{15}{9}$ or 1.6667

$\cos A = \dfrac{12}{15}$ or 0.8000     $\sec A = \dfrac{15}{12}$ or 1.2500

$\tan A = \dfrac{9}{12}$ or 0.7500     $\cot A = \dfrac{12}{9}$ or 1.3333

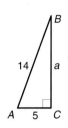

**Example 2**

Find sin A. Round your answer to four decimal places.

$a^2 + 5^2 = 14^2$     *Use the Pythagorean Theorem to find the value of a.*

$a^2 = 171$

$a = \sqrt{171}$

$\sin A = \dfrac{opposite}{hypotenuse}$

$= \dfrac{\sqrt{171}}{14}$

Now use a calculator to find the decimal value.

ENTER: 171 $\boxed{\sqrt{}}$ $\boxed{\div}$ 14 $\boxed{=}$  *0.9340497736*

sin A is approximately 0.9340.

Consider two special right triangles. Triangle *ABC* is an isosceles right triangle. Assume that the congruent sides are each 1 unit long. This is a 45°-45°-right triangle. Find the length of the hypotenuse.

$1^2 + 1^2 = x^2$     *Use the Pythagorean Theorem.*

$2 = x^2$

$\sqrt{2} = x$     The hypotenuse is $\sqrt{2}$ units long.

Now, write the trigonometric values.

$\sin 45° = \dfrac{1}{\sqrt{2}}$ or $\dfrac{\sqrt{2}}{2}$     $\cos 45° = \dfrac{1}{\sqrt{2}}$ or $\dfrac{\sqrt{2}}{2}$     $\tan 45° = \dfrac{1}{1}$ or 1

$\csc 45° = \dfrac{\sqrt{2}}{1}$ or $\sqrt{2}$     $\sec 45° = \dfrac{\sqrt{2}}{1}$ or $\sqrt{2}$     $\cot 45° = \dfrac{1}{1}$ or 1

Triangle *DEG* is an equilateral triangle. Assume that each side is 2 units long. The altitude $\overline{EF}$ forms a 30°-60°-right triangle. Since altitude $\overline{EF}$ is the perpendicular bisector of side $\overline{DG}$, the length of side $\overline{DF}$ is 1. Find the length of altitude $\overline{EF}$.

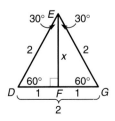

$x^2 + 1^2 = 2^2$     *Pythagorean Theorem*

$x^2 = 3$     *Subtract 1 from each side.*

$x = \sqrt{3}$     Altitude $\overline{EF}$ is $\sqrt{3}$ units long.

Now write the trigonometric values.

$$\sin 30° = \frac{1}{2} \qquad \cos 30° = \frac{\sqrt{3}}{2} \qquad \tan 30° = \frac{1}{\sqrt{3}} \text{ or } \frac{\sqrt{3}}{3}$$

$$\csc 30° = \frac{2}{1} \text{ or } 2 \qquad \sec 30° = \frac{2}{\sqrt{3}} \text{ or } \frac{2\sqrt{3}}{3} \qquad \cot 30° = \frac{\sqrt{3}}{1} \text{ or } \sqrt{3}$$

$$\sin 60° = \frac{\sqrt{3}}{2} \qquad \cos 60° = \frac{1}{2} \qquad \tan 60° = \frac{\sqrt{3}}{1} \text{ or } \sqrt{3}$$

$$\csc 60° = \frac{2}{\sqrt{3}} \text{ or } \frac{2\sqrt{3}}{3} \qquad \sec 60° = \frac{2}{1} \text{ or } 2 \qquad \cot 60° = \frac{1}{\sqrt{3}} \text{ or } \frac{\sqrt{3}}{3}$$

Trigonometric functions can be used to solve problems involving right triangles. Example 3 illustrates the solution to the application presented on page 756.

## Example 3

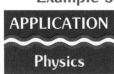

**Physics**

A car is traveling 500 meters along a raised exit ramp, whose angle to the ground has a measurement of 3°. How far did it rise during this distance? Round the answer to the nearest tenth of a meter.

Draw a triangle to represent the situation.

$$\sin 3° = \frac{x}{500} \qquad sin = \frac{opposite}{hypotenuse}$$

$$0.0523 \approx \frac{x}{500}$$

$$26.2 \approx x$$

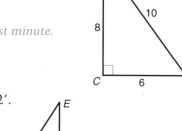

The car will rise about 26.2 meters as it travels 500 meters on the exit ramp.

Finding all of the measures of the sides and the angles in a right triangle is called *solving the triangle.*

## Example 4

Solve each right triangle. Round measures of sides to the nearest tenth and measures of angles to the nearest minute.

**a.** $\triangle ABC$

$$\cos A = \frac{6}{10} \qquad cos = \frac{adjacent}{hypotenuse}$$

$$\cos A = 0.6000$$

$$A = 53°08' \qquad \textit{Round to the nearest minute.}$$

$$53°08' + B \approx 90°$$

$$B \approx 36°52'$$

Therefore, $A \approx 53°08'$ and $B \approx 36°52'$.

**b.** $\triangle DEF$

$$\frac{d}{18} = \sin 57° \qquad \frac{e}{18} = \cos 57°$$

$$\frac{d}{18} \approx 0.8387 \qquad \frac{e}{18} \approx 0.5446$$

$$d \approx 15.1 \qquad e \approx 9.8$$

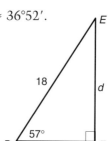

$57° + E = 90°$    *Angles D and E are complementary.*
$E = 41°$
Therefore, $d \approx 15.1$, $e \approx 9.8$, and $E = 41°$.

**c.** $\triangle ABC$

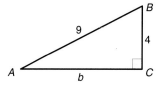

$4^2 + b^2 = 9^2$    *Since $\triangle ABC$ is a right triangle,*
$16 + b^2 = 81$    *use the Pythagorean Theorem.*
$b^2 = 65$
$b \approx 8.1$    *Round to the nearest tenth.*

$\sin A = \dfrac{4}{9} \approx 0.4444$    $\cos B = \dfrac{4}{9} \approx 0.4444$    Therefore, $b \approx 8.1$,
$A \approx 26°23'$    $B \approx 63°37'$    $A \approx 26°23'$, and
$B \approx 63°37'$.

# CHECKING FOR UNDERSTANDING

**Communicating Mathematics**

Read and study the lesson to answer these questions.

1. What does it mean to solve a triangle?

2. If $C$ is the right angle in right triangle $ABC$, what can we say about the measures of angles $A$ and $B$?

3. If angles $A$ and $B$ are complementary angles, what can we say about $\sin A$ and $\cos B$?

**Guided Practice**

Use the triangle below to state equations that would enable you to solve each problem. Then solve. Round measures of sides to the nearest tenth and measures of angles to the nearest minute.

4. If $A = 15°$ and $c = 37$, find $a$.

5. If $A = 76°$ and $a = 13$, find $b$.

6. If $A = 49°13'$ and $a = 10$, find $c$.

7. If $a = 21.2$ and $A = 71°13'$, find $b$.

8. If $a = 13$ and $B = 16°$, find $c$.

9. If $A = 19°07'$ and $b = 11$, find $c$.

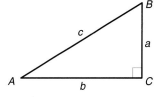

# EXERCISES

**Practice**

Solve each triangle described. Round measures of sides to the nearest hundredth and measures of angles to the nearest minute.

10. $c = 10$, $a = 8$
11. $c = 13$, $a = 12$
12. $a = 2$, $b = 7$
13. $c = 21$, $b = 18$
14. $a = 11$, $b = 21$
15. $b = 6$, $c = 13$
16. $A = 63°$, $a = 9.7$
17. $A = 16°$, $c = 14$
18. $A = 37°15'$, $b = 11$
19. $B = 42°10'$, $a = 9$
20. $B = 64°$, $c = 19.2$
21. $B = 83°$, $b = \sqrt{31}$
22. $a = 33$, $B = 33°$
23. $c = 6$, $B = 13°$
24. $b = 42$, $A = 77°$
25. $a = 9$, $B = 49°$
26. $b = 22$, $A = 22°22'$
27. $a = 44$, $B = 44°44'$

**28.** $B = 18°$, $a = \sqrt{15}$     **29.** $A = 55°55'$, $c = 16$     **30.** $A = 45°$, $c = 7\sqrt{2}$

**31.** $c = 25$, $A = 15°$     **32.** $B = 30°$, $b = 11$     **33.** $\tan A = \dfrac{7}{8}$, $a = 7$

**34.** $a = 7$, $A = 27°$     **35.** $\tan B = \dfrac{8}{6}$, $b = 8$     **36.** $\sin A = \dfrac{1}{3}$, $a = 5$

**37.** Solve the isosceles right triangle shown below. $CA = CB$.

**38.** Solve the equilateral triangle shown below.

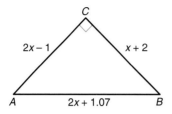
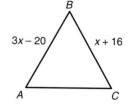

**Critical Thinking**

**39.** Can you solve a right triangle if you are given only the length of one side and the measure of one of the acute angles? Justify your answer.

**Applications**

**40.** **Civil Engineering** The bearing of a road between two points, $A$ and $B$, is the measure of the positive angle with vertex at point $A$ that is measured clockwise from point $B$. Find the bearing of a road that runs directly from $A$ to $B$ if $B$ is 3 miles north and 1.7 miles east of point $A$.

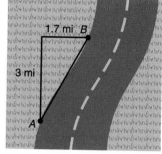

**41.** **Travel** Tom visited the Washington Monument on his summer vacation. He stood 1200 feet away from the base of the monument to take a picture and had to look up at an angle of $24°49'$ to see the top. Approximately how tall is the Washington Monument?

**Mixed Review**

**42.** How many different outfits can be made from 7 pairs of pants, 3 shirts, and 4 pairs of shoes? **(Lesson 15-1)**

**43.** Write $\displaystyle\sum_{n=1}^{5} 3^{n-1}$ in expanded form and find the sum. **(Lesson 13-5)**

**44.** **Finance** Luis' grandfather invested $150 at 5% interest compounded quarterly. When the account was recently given to Luis, it contained $6230. How long ago did Luis' grandfather invest the $150? **(Lesson 12-7)**

**45.** **Chemistry** How much of a 35% salt solution must be added to a 10% salt solution to get 250 milliliters of a 20% solution? **(Lesson 11-7)**

**46.** Find the value of the discriminant for the quadratic equation $y^2 - 8y + 12 = 0$. Describe the nature of the roots. If the roots are real, tell whether they are rational or irrational. Then solve the equation. **(Lesson 7-4)**

**47.** Margaret mixed $r$ cups of cashews with $s$ cups of peanuts. This gave her 12 cups of nuts with one part cashews to three parts peanuts. Find $r$ and $s$. **(Lesson 3-2)**

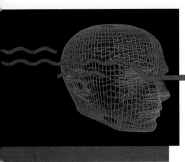

# Technology

## Solving Triangles

The BASIC program below determines the number of solutions of a triangle given the lengths of two sides and the degree measure of the angle opposite one of them.

```
10    INPUT A,A1,B1
20    LET B = B1*SIN (A*3.1415927 / 180)
30    PRINT "B SIN A= ";B
40    IF A > = 90 THEN 110
50    IF A1 < B THEN 120
60    IF A1 = B THEN 130
70    IF A1 < B1 THEN 100
80    IF A1 > B1 THEN 130
100   PRINT "TWO SOLUTIONS EXIST." : GOTO 140
110   IF A1 > B1 THEN 130
120   PRINT "NO SOLUTION EXISTS.": GOTO 140
130   PRINT "ONE SOLUTION EXISTS."
140   END
```

In the program, $A$ represents the degree measure of the angle, $A1$ represents the measure of one side of the triangle, and $B1$ represents the measure of the other side.

It is important that you enter the data correctly when prompted.

- First enter the degree measure of the angle.
- Next enter the measure of the side opposite that angle.
- Then enter the measure of the other side.

The data must be entered in this order to obtain a correct output.

For a triangle in which $A = 26°$, $a = 3$, and $b = 7$, there is no solution, as shown by the output at the right.

```
] RUN
?26,3,7
B SIN A=3.06859807
NO SOLUTION EXISTS.
```

# EXERCISES

**Use the program to determine the number of solutions. Also state the value of** $b \sin A$ **rounded to four decimal places.**

**1.** $A = 103°$, $a = 5$, $b = 2$

**2.** $A = 175°$, $a = 19$, $b = 2$

**3.** $A = 77°$, $a = 1$, $b = 5$

**4.** $A = 130°$, $a = 8$, $b = 10$

**5.** $A = 38.62°$, $a = 15$, $b = 16$

**6.** $A = 10.524°$, $a = 1$, $b = 2$

**7.** $A = 53°20'$, $a = 9$, $b = 6$

**8.** $A = 9°55'$, $a = 3$, $b = 6$

## 16-7 Applications of Right Triangles

**Objective**

After studying this lesson, you should be able to:

■ solve word problems using right triangle trigonometry.

**Application**

Terry Jacobs is standing on the deck of an aircraft carrier observing the approach of an airplane. The angle formed by her line of sight to the airplane and a horizontal is called the **angle of elevation.** The angle formed by the line of sight from the airplane pilot to Ms. Jacobs and a horizontal is called the **angle of depression.** The angles of elevation and depression are alternate interior angles, and since the lines of sight are parallel, they have equal measures.

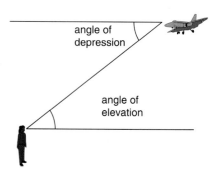

There are many applications of angles of depression and angles of elevation. The following examples illustrate some of these applications.

**Example 1**

**APPLICATION**

**Forestry**

Tami is standing 30 feet from the base of a tree. She measured the angle to the top of the tree with an astrolabe, which is a sighting device used to measure angles. As Tami looks to the top of the tree, the measurement of the angle of elevation is 65°. Find the height of the tree to the nearest tenth of a foot.

Let $h$ represent the height of the tree.

$$\frac{h}{30} = \tan 65°$$

$$\frac{h}{30} \approx 2.1445$$

$$h \approx 64.335$$

The tree is about 64.3 feet tall.

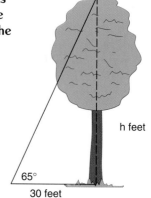

## Example 2

**APPLICATION**

Navigation

*FYI · · ·*

The seventh wonder of the world was the Pharos (lighthouse) of Alexandria. It was built by Sostratus of Chidus during the third century B.C. on the island of Pharos off the coast of Egypt.

**The top of a lighthouse is 120 meters above sea level. From the top of the lighthouse, the measurement of the angle of depression to a boat on the ocean is 43°. How far is the boat from the foot of the lighthouse?**

Let $x$ represent the distance from the lighthouse.

$$\tan 43° = \frac{120}{x} \qquad tan = \frac{opposite}{adjacent}$$

$$x = \frac{120}{\tan 43°}$$

$$x \approx \frac{120}{0.9325}$$

$$x \approx 128.69$$

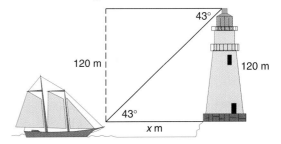

The boat is about 129 meters from the lighthouse.

## Example 3

**APPLICATION**

Astronomy

**Eratosthenes, an astronomer who lived in Greece in the third century B.C., is credited with providing the first accurate measure of Earth's circumference. He found that at noon on the day of the summer solstice, the sun was directly over the city of Syene. At the same time, in Alexandria, which is north of Syene, the sun was 7°12′ south of being directly overhead. If the distance between the two cities was 5000 stadia, find Eratosthenes' measure of the circumference of Earth.**

Draw a diagram.

Since there are 360° in a full rotation around Earth, the following proportion can be written.

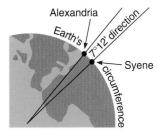

$$\frac{360°}{7°12'} = \frac{c}{5000}$$
$$(7°12')c = (360°)5000$$
$$c = \frac{(360°)5000}{7°12'}$$
$$c = \frac{(360°)5000}{7.2°} \qquad 7°12' = 7\frac{12°}{60} \text{ or } 7.2°$$
$$c = 250,000$$

Eratosthenes' measure of Earth's circumference was 250,000 stadia or 24,661 miles. This is only 158 miles less than the currently accepted value. *The stadium (singular form of stadia) is an ancient unit of measurement equal to about 0.098 mile.*

Example 4

APPLICATION

Broadcasting

Vonda and Bill are standing 100 feet apart and in a straight line with the WWV television tower. The angle of elevation from Bill to the tower is 30° and the angle of elevation from Vonda is 20°. Find the height of the television tower to the nearest foot.

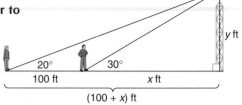

$$\tan 30° = \frac{y}{x}$$

$x \tan 30° = y$   *Solve for y.*

$x(0.5774) = y$   *tan 30° = 0.5774*

$$\tan 20° = \frac{y}{x + 100}$$   *Solve for y.*

$(\tan 20°)(x + 100) = y$

$0.3640(x + 100) = y$   *tan 20° = 0.3640*

$0.3640x + 36.40 = y$

$0.5774x = 0.3640x + 36.40$

$0.2134x = 36.40$

$x \approx 170.6$   *Round to the nearest tenth.*

$$\frac{y}{170.6} \approx \tan 30°$$   *tan 30° = $\frac{y}{x}$*

$y \approx 98.5$   *Round to the nearest tenth.*

The antenna is about 99 feet tall.

# CHECKING FOR UNDERSTANDING

**Communicating Mathematics**

**Read and study the lesson to answer these questions.**

1. What is the angle of elevation?

2. How is the angle of depression related to the angle of elevation?

**Guided Practice**

**Use the triangle to state the equation that would enable you to solve each problem. Then solve.**

3. If $c = 16$ and $a = 7$, find $b$.

4. If $b = 10$ and $c = 20$, find $a$.

5. If $a = 7$ and $b = 12$, find $A$.

6. If $a = b$ and $c = 12$, find $B$.

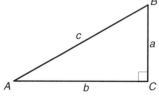

# EXERCISES

For each triangle, find sin A, cos A, and tan A. Round your answers to four decimal places.

**7.**

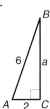

**8.**

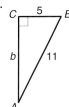

**9.**

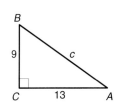

Solve each problem. Round measures of lengths to the nearest hundredth and measures of angles to the nearest minute.

10. The flagpole in front of Stevenson High School casts a shadow 40 feet long when the measurement of the angle of elevation to the sun is 31°20′. How tall is the flagpole?

11. Aaron is standing 300 meters from the base of a radio tower. According to his astrolabe, the measurement of the angle of elevation to the top of the tower is 40°. How high is the tower?

12. According to the pilot's instruments, the measurement of the angle of depression of an aircraft carrier from a plane 1000 feet above the water is 63°18′. How far is the plane from the carrier?

13. Dana is flying a kite to which the angle of elevation is 70°. The string on the kite is 65 meters long. How far is the kite above the ground?

14. A tree was broken in a recent storm. The top of the tree touches the ground 13 meters from the base. The top of the tree makes an angle of 29° with the ground. How tall was the tree before it was broken?

15. Naren Thomas is an architect designing a new parking garage for the city. The floors of the garage are to be 20 feet apart. The exit ramps between each pair of floors are to be 120 feet long. What is the measurement of the angle of elevation of each ramp?

16. A railroad track rises 10 feet for every 400 feet along the track. What is the measurement of the angle the track forms with the horizontal?

17. Curtis and Cindy are observing the Washington Monument from $\frac{1}{4}$ mile away. The monument is 555 feet tall. What is the angle of elevation from Curtis and Cindy to the top of the monument? (1 mile = 5280 feet)

18. The diagram at the right shows square *ABCD*. The midpoint of side $\overline{AD}$ is *E*. Find the values of *x*, *y*, and *z* to the nearest minute.

19. Tabina and Russ are standing in a straight line with the base of a building. The measurement of the angle of elevation to the top of the building from the point where Tabina is standing is 38°20′. From the point where Russ is standing, 50 feet closer to the building, the measurement of the angle of elevation is 45°. How tall is the building?

20. Two apartment buildings are separated by an alley. As Beth looks out of her window 60 feet above the ground, she uses her astrolabe to find that the measurement of the angle of depression of the second building is 50°. She also finds that the angle of elevation to the top of the second building is 40°. About how tall is the second building?

21. A pendulum 50 centimeters long is moved 40° from the vertical. How far did the tip of the pendulum rise?

22. Olivia and Jason are standing 200 feet apart and in line with the base of a flagpole. The measurement of the angle of elevation of the top of the pole from Olivia is 30° and from Jason is 60°. How far is the flagpole from each of them?

CONNECTION

Geometry

23. The isosceles triangle $RST$ at the right has base $\overline{TS}$ measuring 10 centimeters and base angles each measuring 39°. Find the length of the altitude $\overline{QR}$.

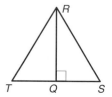

**Critical Thinking**

24. Prove that the angle of elevation is congruent to its corresponding angle of depression.

**Applications**

25. **Transportation** A train travels 5000 meters along a track whose angle of elevation has a measurement of 2°. How much did it rise during this distance?

26. **Broadcasting** A television antenna sits atop a building. From a point 200 feet from the base of the building, the measurement of the angle of elevation of the top of the antenna is 80°. The measurement of the angle of elevation of the bottom of the antenna from the same point is 75°. How tall is the antenna?

27. **Travel** A ship sails due north from its home port for 90 kilometers. It then turns east and sails for 40 kilometers before turning north again to sail for 70 kilometers. How far is the ship from its port?

**Journal**

Make up a problem that uses right triangles and solve it.

**Mixed Review**

28. Find the value of $n$ if $C(n, 5) = C(n, 15)$. (**Lesson 15-4**)

29. Solve $3^5 = 3^{2n-1}$ for $n$. (**Lesson 12-1**)

30. Find all zeros of the function $g(x) = 24x^4 - 94x^3 + 61x^2 + 21x - 18$. (**Lesson 10-4**)

## 16-8 Law of Sines

**Objective**

After studying this lesson, you should be able to:

■ solve triangles and problems using the Law of Sines.

**Application**

A ship is sighted at sea from two observation points on the coastline which are 30 miles apart. The angle between the coastline and the line between the ship and the first observation point measures 34°. The angle between the coastline and the line between the ship and the second observation point measures 45°34′. How far is the ship from the second observation point?

You can use trigonometric functions to solve problems like this one which involve triangles that are *not* right triangles.
*You will solve this problem in Example 3.*

Consider $\triangle ABC$ with height $h$ units and sides with lengths $a$ units, $b$ units, and $c$ units. The area of this triangle is given by the equation area $= \frac{1}{2}bh$. Also, $\sin A = \frac{h}{c}$ or $h = c \sin A$. By combining these equations, we can find a new formula for the area of the triangle.

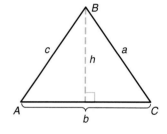

area $= \frac{1}{2}bh = \frac{1}{2}b(c \sin A)$     *h = c sin A*

You can find two other formulas for the area of the triangle in a similar way.

$$\text{area} = \frac{1}{2}ac \sin B \qquad \text{area} = \frac{1}{2}ab \sin C$$

All of these formulas represent the area of the same triangle. So, the following must be true.

$$\frac{1}{2}bc \sin A = \frac{1}{2}ac \sin B = \frac{1}{2}ab \sin C$$

The **Law of Sines** is obtained by dividing each of expression above by $\frac{1}{2}abc$.

$$\frac{\sin A}{a} = \frac{\sin B}{b} = \frac{\sin C}{c}$$

*Law of Sines*

Let $\triangle ABC$ be any triangle with $a$, $b$, and $c$ representing the measures of sides opposite angles with measurements $A$, $B$, and $C$ respectively. Then,
$$\frac{\sin A}{a} = \frac{\sin B}{b} = \frac{\sin C}{c}.$$

**Example 1**

Find the area of $\triangle ABC$ if $a = 11$, $b = 13$, and $C = 31°10'$.

$\text{area} = \frac{1}{2}ab \sin C$

$\approx \frac{1}{2}(11)(13) \sin 31°10'$     *sin 31°10' ≈ 0.5175*

$\approx 37.001$

To the nearest whole unit, the area is 37 square units.

**Example 2**

Use the Law of Sines to solve the triangle. Round when needed.

$\dfrac{\sin B}{b} = \dfrac{\sin A}{a}$     *Law of Sines*

$\dfrac{\sin B}{79} = \dfrac{\sin 45°}{83}$

$\sin B = \dfrac{79 \sin 45°}{83} \approx 0.6730$

$B \approx 42°18'$     *Round to the nearest minute.*

$45° + 42°18' + C \approx 180°$     *The sum of the angle*
$\qquad\qquad\quad C \approx 92°42'$     *measures in a triangle is 180°.*

$\dfrac{\sin 92°42'}{c} = \dfrac{\sin 45°}{83}$

$c = \dfrac{83 \sin 92°42'}{\sin 45°}$

$c \approx 117.3$     *Round to the nearest tenth.*

*Example 3 is the solution to the application on page 767.*

Therefore, $B \approx 42°18'$, $C \approx 92°42'$, and $c \approx 117.25$.

**Example 3**

APPLICATION

Navigation

A ship is sighted at sea from two observation points on the coastline which are 30 miles apart. The angle between the coastline and the line between the ship and the first observation point measures 34°. The angle between the coastline and the line between the ship and the second observation point measures 45°34'. How far is the ship from the second observation point?

Draw a diagram.

First, find the measure of the third angle.

$34° + 45°34' + C = 180°$
$\qquad\qquad\qquad C = 100°26'$

Now, use the Law of Sines to find $x$.

$\dfrac{\sin 100°26'}{30} = \dfrac{\sin 34°}{x}$

$x = \dfrac{30 \sin 34°}{\sin 100°26'}$

$x = 17.06$

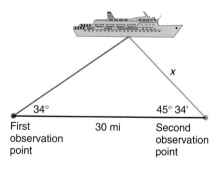

The ship is approximately 17 miles from the second observation point.

# CHECKING FOR UNDERSTANDING

**Communicating Mathematics**

Read and study the lesson to answer these questions.

1. One formula for the area of a triangle is $A = \frac{1}{2}bh$. State another formula for the area of a triangle.

2. For what kind of triangles does the Law of Sines hold true?

3. How far is the ship described in Example 3 from the first observation point?

**Guided Practice**

State an equation that would enable you to find the area of each triangle. Then find the area to the nearest tenth.

4. $a = 15$, $b = 20$, $C = 63°$

5. $b = 10$, $c = 17$, $A = 46°$

6. $a = 6$, $c = 4$, $B = 52°$

7. $a = 15$, $b = 30$, $C = 90°$

For each triangle ABC described, state an equation that would enable you to find each value. Then find the value.

8. If $a = 20$, $A = 40°$, and $B = 60°$, find $b$.

9. If $b = 10$, $a = 14$, and $A = 50°$, find $B$.

10. If $b = 2.8$, $A = 53°$, and $B = 61°$, find $a$.

11. If $c = 12$, $b = 16$, and $B = 42°$, find $C$.

# EXERCISES

**Practice**

Find the area of each triangle described below.

12. $a = 15$, $b = 22$, $C = 90°$

13. $a = 12$, $b = 12$, $C = 50°$

14. $a = 11.5$, $c = 14$, $B = 20°$

15. $a = 11$, $c = 5$, $B = 50°6'$

16. $b = 4$, $c = 19$, $A = 73°24'$

17. $a = 9.4$, $c = 13.5$, $B = 95°$

Solve each triangle described below.

18. $A = 40°$, $B = 60°$, $c = 20$

19. $a = 8$, $A = 49°$, $B = 57°$

20. $a = 80$, $b = 70$, $A = 83°10'$

21. $B = 70°$, $C = 58°$, $a = 84$

22. $A = 30°$, $C = 70°$, $c = 8$

23. $c = 17$, $b = 15$, $C = 64°40'$

24. $a = 14$, $b = 7.5$, $A = 103°$

25. $a = 23$, $A = 73°25'$, $C = 24°30'$

26. $b = 8$, $B = 36°36'$, $C = 119°$

27. $A = 105°$, $a = 18$, $b = 14$

28. An isosceles triangle has a base of 22 centimeters and exactly one angle measuring 36°. Find its perimeter. *Hint: The two angles opposite the congruent sides of an isosceles triangle are congruent.*

29. The longest side of a triangle is 34 yards. Two angles of the triangle are 40° and 65°. Find the length of the other two sides.

30. A triangular lot faces two streets that meet at an angle measuring 85°. The sides of the lot facing the streets are each 160 feet in length. Find the perimeter of the lot.

31. Points $X$ and $Y$ are on opposite sides of a valley. Point $C$ is 60 kilometers from point $X$. Angle $YXC$ is 108° and angle $YCX$ is 35°. Find the width of the valley.

32. A flower bed is in the shape of an obtuse triangle. One angle is 45° and the opposite side is 28 feet long. The longest side is 36 feet long. Find the measures of the remaining angles and side.

**Critical Thinking**

33. Prove that the Law of Sines holds true for right triangles.

**Applications**

34. **Surveying** A building 60 feet tall is on the top of a hill. A surveyor stands at a point on the hill and finds that the angle of elevation to the top of the building has measurement 42° and to the bottom of the building has measurement 18°.

   a. Draw a diagram to represent this situation.

   b. How far is the surveyor from the base of the building?

35. **Aviation** Two planes left John F. Kennedy Airport at the same time. Each plane is flying at a speed of 110 miles per hour. One plane flew in the direction 60° east of north and the other flew in the direction of 40° east of south.

   a. Draw a diagram to represent this situation.

   b. How far apart are the planes after three hours?

**Mixed Review**

36. **Statistics** The number of years of life expected at birth for women in certain countries is given below. Make a box-and-whisker plot of the data, labeling any outliers. **(Lesson 14-5)**
   77.2   76.8   76.0   74.3   77.5   78.8   78.4   75.4   77.5
   76.0   73.7   75.6   77.2   79.5   79.5   75.0   72.9   76.2
   79.9   79.6   74.0   77.6   75.6   75.9   73.2

37. Expand the binomial $(x + 4)^6$. **(Lesson 13-7)**

38. **Biology** For a certain strain of bacteria, $k$ in the exponential growth formula is 0.782 when $t$ is measured in hours. How long will it take 10 bacteria to increase to 500 bacteria? *Hint: The exponential growth formula is $y = ne^{kt}$.* **(Lesson 12-8)**

39. If $f(x) = x + 1$ and $g(x) = x^2$, find $f[g(x)]$ and $g[f(x)]$. **(Lesson 10-7)**

40. **Geometry** The formula for the area of a trapezoid is $A = \frac{h}{2}(b_1 + b_2)$, where $A$ represents the measure of the area, $h$ represents the measure of the altitude, and $b_1$ and $b_2$ represent the measures of the bases. Find the measure of the area of the trapezoid whose height is 8 cm and whose bases measure 12 cm and 20 cm. **(Lesson 1-1)**

# 16-9 Problem-Solving Strategy: Examine the Solution

**Objective**

After studying this lesson, you should be able to:

■ determine, from a given set of information, the number of possible solutions and solve the triangle if solutions do exist.

When solving a triangle, you must analyze the data you are given in order to determine whether there is a solution or not.

When the lengths of two sides of a triangle and the measurement of the angle opposite one of them are given, a single solution does not always exist. In such a case, one of the following will be true.

1. No triangle exists.
2. Exactly one triangle exists.
3. Two triangles exist.

In other words, there may be no solutions, one solution, or two solutions.

Suppose you are given *a*, *b*, and *A*. First consider the case where $A < 90°$.

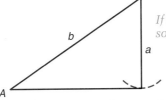

*If a = b sin A, the solution is a right triangle.*

*If a = b sin A, one solution exists.*

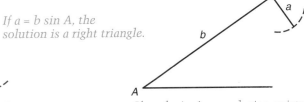

*If a < b sin A, no solution exists.*

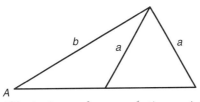

*If b sin A < a < b, two solutions exist.*

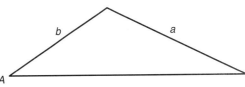

*If a > b, one solution exists.*

Consider the case where $A ≥ 90°$.

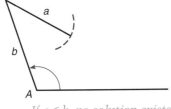

*If a ≤ b, no solution exists.*

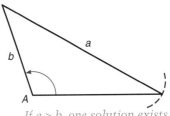

*If a > b, one solution exists.*

**Example 1**

Solve the triangle where $A = 40°$, $b = 12$, and $a = 5$.

$$b \sin A = 12 \sin 40°$$
$$\approx 12(0.6428)$$
$$\approx 7.7136$$

Since $40° < 90°$ and $5 < 7.7136$, no solution exists.
*Check this solution by trying to draw this triangle.*

**Example 2**

Solve the triangle where $A = 58°$, $a = 26$, and $b = 29$.

$$b \sin A = 29 \sin 58°$$
$$\approx 29(0.8480)$$
$$\approx 24.5920$$

Since $58° < 90°$ and $24.5920 < 26 < 29$, there are two solutions.

$$\frac{\sin 58°}{26} = \frac{\sin B}{29} \qquad \text{Use the Law of Sines.}$$

$$\sin B = \frac{29 \sin 58°}{26} \qquad \text{sin } 58° \approx 0.8480$$

$$\approx 0.9458$$

$B \approx 71°3'$ or $108°57'$    *Round to the nearest minute.*

Solution I: $m\angle B \approx 71°3'$     Solution II: $m\angle B \approx 108°57'$

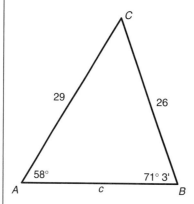   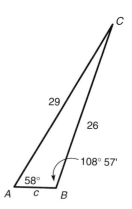

$58° + 71°3' + C \approx 180°$        $58° + 108°57' + C \approx 180°$
$\qquad\qquad C \approx 50°57'$           $\qquad\qquad\qquad C \approx 13°3'$

$c \approx \dfrac{26 \sin 50°57'}{\sin 58°}$  *sin 50°57' ≈ 0.7875*    $c \approx \dfrac{26 \sin 13°3'}{\sin 58°}$  *sin 13°3' ≈ 0.2258*

$\approx 24.1$  *Round to the nearest tenth.*    $\approx 7.0$  *Round to the nearest tenth.*

One solution is $B \approx 71°3'$,          Another solution is $B \approx 108°57'$,
$C \approx 50°57'$, and $c \approx 24.1$.         $C \approx 13°3'$, and $c \approx 7.0$.

**Example 3**

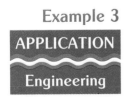

APPLICATION

Engineering

Kira was given an assignment to draw and then construct a triangular model of three steel girders for her engineering class. Two of the girders measured 7 cm and 6 cm, and the angle opposite the 7 cm girder had to be 30°. Could she construct the triangle? If so, how long did the third girder have to be?

$a = 7$, $b = 6$, and $A = 30°$.

Since $30° < 90°$ and $7 > 6$, one solution exists. There is one way that Kira can construct the triangle.

$\dfrac{\sin 30°}{7} = \dfrac{\sin B}{6}$   *Use the Law of Sines.*

$\sin B = \dfrac{6 \sin 30°}{7}$   *sin 30° = 0.5000.*

$\sin B \approx 0.4286$

$B \approx 25°23'$   *Round to the nearest minute.*

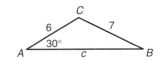

$30° + 25°23' + C \approx 180°$
$C \approx 124°37'$

$\dfrac{\sin 30°}{7} = \dfrac{\sin 124°37'}{c}$   *sin 124°37' ≈ 0.8230*

$c = \dfrac{7 \sin 124°37'}{\sin 30°} \approx 11.5$   *Round to the nearest tenth.*

Kira should make a triangle with $B \approx 25°23'$, $C \approx 124°37'$, and $c \approx 11.5$.

# CHECKING FOR UNDERSTANDING

**Communicating Mathematics**

**Read and study the lesson to answer these questions.**

1. When you are given the lengths of two sides of a triangle and the measurement of the angle opposite one of them, how many possible solutions to the triangle are there?

2. If you are given $a$, $b$, and $A$ for a triangle and $A > 90°$, how many solutions are possible for this triangle?

3. You are given $a$, $b$, and $A$ for a triangle. $a \leq b$ and $A < 90°$. What can you say about this triangle?

**Guided Practice**

**Determine the number of possible solutions for each triangle $ABC$ described. If a solution exists, solve the triangle.**

4. $a = 6$, $b = 8$, $A = 150°$

5. $a = 6$, $b = 8$, $A = 36°52'$

6. $a = 64$, $c = 90$, $C = 98°$

7. $a = 12$, $b = 19$, $A = 57°$

8. $a = 9$, $b = 20$, $A = 31°$

9. $a = 12$, $b = 14$, $A = 90°$

10. $a = 125$, $b = 150$, $A = 25°$

11. $A = 40°$, $b = 16$, $a = 10$

12. $a = 18$, $b = 20$, $A = 120°$

13. $A = 40°$, $b = 10$, $a = 8$

# EXERCISES

Solve. Use any method.

14. Kirby needed to draw a triangle for his geometry class. He made one side 40 mm long. Another side was 32 mm long with the angle opposite measuring 48°19′. What was the length of the third side?

15. Simplify $\left[ \left(\frac{1}{2}\right)^{-1} + \left(\frac{1}{3}\right)^{-1} + \left(\frac{1}{4}\right)^{-1} + \left(\frac{1}{5}\right)^{-1} \right]^{-1}$.

16. Donna changed the three on a die to a five. On another die, she changed the one to a three. What is the probability of rolling a sum of 6 with these two dice?

17. The Cox family is making a triangular deck in their backyard. Two of the pieces of lumber they have to make the deck are 20 feet and 15 feet. If the angle opposite the 15-foot piece is to be 61°, can they construct the deck? If so, how long must the third piece of lumber be?

18. Paul, Eric, and Garnet are playing a card game. They have a rule that when a player loses a hand, he must subtract enough points from his score to double each of the other player's scores. First Paul loses a hand, then Eric, and then Garnet. Each player now has 8 points. Who lost the most points?

19. Sarah rolls two dice and finds the product of the numbers on the faces. What is the most likely product?

20. The shapes below can be folded to make dice. Show how to fill in the missing numbers so that the numbers on opposite faces have a sum of 7.

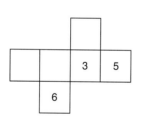 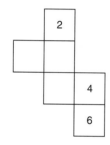 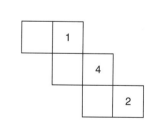

21. Insert operational symbols, +, −, ×, or ÷, and parentheses if necessary to make a true equation from 6 3 6 2 4 = 5.

## COOPERATIVE LEARNING ACTIVITY

**Work in groups. Each person in the group must understand the solution and be able to explain it to any person in class.**

Construct a triangle with three altitudes of length 3 cm, 5 cm, and 6 cm. Can you construct a triangle whose altitudes are 2 cm, 6 cm, and 9 cm? Explain your answer.

# 16-10 Law of Cosines

**Objective**

After studying this lesson, you should be able to:

- solve triangles and problems using the Law of Cosines.

**Application**

A pilot is flying from Chicago, Illinois to Columbus, Ohio, a distance of 300 miles. She starts her flight 15° off course and flies on this course for 75 miles. How far is she from Columbus?

Problems like this, where two sides of a triangle and an included angle are given, cannot be solved using the Law of Sines. You will use another formula, the **Law of Cosines** to solve this problem in Example 2.

Consider $\triangle ABC$ with height $h$ units and sides with lengths $a$ units, $b$ units, and $c$ units. Suppose segment $AD$ is $x$ units long. Then segment $DC$ is $(b - x)$ units long.

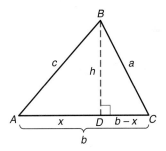

How are $A$, $a$, $b$, and $c$ related?

$$a^2 = (b - x)^2 + h^2 \qquad \textit{Use the Pythagorean Theorem for } \triangle BDC.$$
$$= b^2 - 2bx + x^2 + h^2 \qquad \textit{Expand } (b - x)^2.$$
$$= b^2 - 2bx + c^2 \qquad \textit{In } \triangle ADB, c^2 = x^2 + h^2.$$
$$= b^2 - 2b(c \cos A) + c^2 \qquad \cos A = \frac{x}{c}, \textit{ so } x = c \cos A.$$
$$= b^2 + c^2 - 2bc \cos A$$

You can find two other formulas relating the lengths of sides to the cosine of $B$ and $C$ in a similar way. All three formulas, can be summarized as follows.

**Law of Cosines**

Let $\triangle ABC$ be any triangle with $a$, $b$, and $c$ representing the measures of sides opposite angles with measurement $A$, $B$, and $C$ respectively. Then, the following equations hold true.
$$a^2 = b^2 + c^2 - 2bc \cos A$$
$$b^2 = a^2 + c^2 - 2ac \cos B$$
$$c^2 = a^2 + b^2 - 2ab \cos C$$

You can use the Law of Cosines to solve a triangle in the following cases.

1. To find the length of the third side of any triangle if the lengths of two sides and the measurement of the included angle are given.

2. To find the measurement of an angle of a triangle if the lengths of the three sides are given.

**Example 1**

Solve each triangle. Round lengths to the nearest hundredth and angle measures to the nearest minute.

**a.** $A = 51°$, $b = 40$, $c = 45$

First, determine $a$ using the Law of Cosines.

$a^2 = b^2 + c^2 - 2bc \cos A$  *Use the Law of Cosines.*
$a^2 = 40^2 + 45^2 - 2(40)(45) \cos 51°$  *$A = 51°, b = 40, c = 45$*
$a^2 = 1359.45$
$\phantom{a^2}a \approx 36.87$

Next, use the Law of Sines to determine the measure of a second angle.

$$\frac{\sin A}{a} = \frac{\sin B}{b} \quad \textit{Use the Law of Sines.}$$

$$\frac{\sin 51°}{36.87} \approx \frac{\sin B}{40} \quad \textit{a} \approx 36.87, A = 51°, b = 40$$

$$\sin B \approx \frac{40 \sin 51°}{36.87} \quad \textit{sin } 51° \approx 0.7771$$

$$\sin B \approx 0.8431$$

$$B \approx 57°28' \quad \textit{Round to the nearest minute.}$$

Finally, determine the measure of the third angle.

$51° + 57°28' + C \approx 180°$
$\phantom{51° + 57°28' +} C \approx 71°32'$

Therefore, $a \approx 36.87$, $B \approx 57°28'$, and $C \approx 71°32'$.

**b.** $a = 5$, $b = 6$, and $c = 7$

First, use the Law of Cosines to find the measure of an angle.

$$a^2 = b^2 + c^2 - 2bc \cos A \quad \textit{Use the Law of Cosines.}$$
$$5^2 = 6^2 + 7^2 - 2(6)(7) \cos A \quad \textit{a = 5, b = 6, c = 7}$$
$$2(6)(7) \cos A = 6^2 + 7^2 - 5^2$$
$$\cos A = \frac{6^2 + 7^2 - 5^2}{2(6)(7)}$$
$$\cos A \approx 0.7143$$
$$A \approx 44°25' \quad \textit{Round to the nearest minute.}$$

Next, use the Law of Sines to determine the measure of another angle.

$$\frac{\sin A}{a} = \frac{\sin B}{b}$$  *Use the Law of Sines.*

$$\frac{\sin 44°25'}{5} \approx \frac{\sin B}{6}$$  *a = 5, b = 6, A ≈ 44°25'*

$$\sin B \approx \frac{6 \sin 44°25'}{5}$$

$$\sin B \approx 0.8398$$

$$B \approx 57°7'$$  *Round to the nearest minute.*

Finally, determine the measure of the third angle.

$$44°25' + 57°7' + C \approx 180°$$
$$C \approx 78°28'$$

Therefore, $A \approx 44°25'$, $B \approx 57°7'$, and $C \approx 78°28'$.

**Example 2**

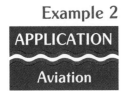

**A pilot is flying from Chicago to Columbus, a distance of 300 miles. She starts her flight 15° off course and flies on this course for 75 miles. How far is she from Columbus?**  *This is the application presented on page 775.*

First, draw a diagram that represents the situation.

Since you know the measure of two sides and the included angle, use the Law of Cosines.

$A = 15°$, $b = 300$ miles, and $c = 75$ miles.

$$a^2 = b^2 + c^2 - 2bc \cos A$$
$$a^2 = 300^2 + 75^2 - 2(300)(75) \cos 15°$$  *Use the Law of Cosines.*

$$a^2 \approx 52{,}158.34$$
$$a \approx 228.38$$  The pilot is about 228 miles from Columbus.

# CHECKING FOR UNDERSTANDING

**Communicating Mathematics**

**Read and study the lesson to answer these questions.**

1. What measures can you find using the Law of Cosines in a triangle if the lengths of two sides and the measurement of the included angle are given?

2. What measures can you find using the Law of Cosines in a triangle if the lengths of all three sides are given?

3. How do you know when to use the Law of Cosines and when to use the Law of Sines to solve a triangle?

4. By what angle will the pilot in Example 2 have to correct her path to reach Columbus?

Determine whether the Law of Sines or the Law of Cosines should be used first to solve each triangle described below. Then solve each triangle.

**5.** $a = 10$, $A = 40°$, $c = 8$

**6.** $A = 40°$, $b = 6$, $c = 7$

**7.** $a = 14$, $c = 21$, $B = 60°$

**8.** $a = 14$, $b = 15$, $c = 16$

**9.** $A = 40°$, $C = 70°$, $c = 14$

**10.** $a = 11$, $b = 10.5$, $C = 35°$

**11.** $a = 11$, $b = 17$, $B = 42°58'$

**12.** $A = 56°$, $C = 22°34'$, $c = 12.2$

# EXERCISES

**Practice** Solve each triangle described below.

**13.** $A = 35°$, $b = 16$, $c = 19$

**14.** $a = 140$, $b = 185$, $c = 166$

**15.** $a = 5$, $b = 12$, $c = 13$

**16.** $a = 20$, $c = 24$, $B = 47°$

**17.** $a = 21.5$, $b = 13$, $C = 38°20'$

**18.** $A = 40°$, $B = 59°$, $c = 14$

**19.** $a = 51$, $c = 61$, $B = 19°$

**20.** $a = 13.7$, $A = 25°26'$, $B = 78°$

**21.** $a = 11$, $b = 13$, $c = 15$

**22.** $a = 345$, $b = 648$, $c = 442$

**23.** $c = 10.3$, $a = 21.5$, $b = 16.71$

**24.** $A = 28°50'$, $b = 5$, $c = 4.9$

**25.** $A = 29°$, $b = 7.6$, $c = 14.1$

**26.** $a = 8$, $b = 24$, $c = 18$

**27.** Two sides of a triangular plot of land have lengths of 400 feet and 600 feet. The measurement of the angle between those sides is 46°20'. Find the perimeter and the area of the plot.

**28.** The sides of a triangle are 6.8 cm, 8.4 cm, and 4.9 cm. Find the measure of the smallest angle.

**29.** The sides of a parallelogram are 55 cm and 71 cm. Find the length of each diagonal if the larger angle measures 106°.

**30.** Circle $Q$ has a radius of 15 cm. Two radii, $\overline{QA}$ and $\overline{QB}$, form an angle of 123°. Find the length of chord $\overline{AB}$.

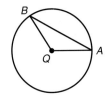

**31.** The sides of a triangular lot are 50 meters, 70 meters, and 85 meters. Find the measure of the angle opposite the shortest side.

**Critical Thinking** **32.** A triangle has angles that measure 48°, 79°, and 53°. Can you find the measures of the sides? If not, why not?

**33. Navigation**  Two ships, the *Indiana* and the *Hoggett Bay,* left San Francisco at noon. The *Indiana* traveled 40° north of west at a speed of 20 knots. The *Hoggett Bay* traveled 10° west of south at a speed of 15 knots. How far apart will they be at 11:00 P.M.? (1 knot = 1 nautical mile per hour)

**34. Broadcasting**  A 40-foot television antenna stands on top of a building. From a point on the ground, the angles of elevation to the top and the bottom of the antenna measure 56° and 42° respectively. How tall is the building?

**35. Aviation**  Ed Gallagher flew his plane 1200 kilometers north before turning 15° clockwise. He flew 850 kilometers in that direction and then landed. How far is Mr. Gallagher from his starting point?

**36. Navigation**  A ship at sea is 70 miles from one radio transmitter and 130 miles from another. The measurement of the angle between the signals is 130°. How far apart are the transmitters?

**Mixed Review**

**37. Statistics**  The estimated populations for 1990 of selected metropolitan areas in the U.S. are listed in the table below. Make a back-to-back stem-and-leaf plot of the rounded and truncated values of the data. **(Lesson 14-2)**

| Capital City | Pop. (thousands) | Capital City | Pop. (thousands) |
|---|---|---|---|
| Albany, NY | 874 | Honolulu, HI | 836 |
| Austin, TX | 782 | Jackson, MS | 395 |
| Baton Rouge, LA | 528 | Little Rock, AR | 513 |
| Boulder, CO | 225 | Madison, WI | 367 |
| Charleston, WV | 250 | Montgomery, AL | 293 |
| Des Moines, IA | 393 | Oklahoma City, OK | 959 |
| Harrisburg, PA | 588 | Salem, OR | 278 |
| Hartford, CT | 768 | Sarasota, FL | 278 |

*Portfolio*

Select some of your work from this chapter that shows how you used a calculator or computer. Place it in your portfolio.

**38.** Solve $\log_x \sqrt{5} = \frac{1}{4}$.  **(Lesson 12-2)**

**39.** Simplify $\dfrac{x^2 + 49}{x^2 - 49} + \dfrac{x}{7 - x} + \dfrac{7}{x + 7}$.  **(Lesson 11-4)**

**40.** Find the inverse of the function $g(x) = (x - 4)^2$.  **(Lesson 10-8)**

# SUMMARY AND REVIEW

## VOCABULARY

Upon completing this chapter, you should be
familiar with the following terms:

| | | | |
|---|---|---|---|
| adjacent side | **756** | **767** | Law of Sines |
| angle of depression | **762** | **751** | minutes |
| angle of elevation | **762** | **756** | opposite side |
| Arccosine | **747** | **739** | period |
| Arcsine | **747** | **739** | periodic function |
| Arctangent | **747** | **746** | principal value |
| cosecant | **742** | **733** | radian |
| cosine | **736** | **742** | secant |
| cotangent | **742** | **736** | sine |
| coterminal | **733** | **732** | standard position |
| gradian | **754** | **742** | tangent |
| hypotenuse | **756** | **732** | terminal side |
| initial side | **732** | **733** | unit circle |
| Law of Cosines | **775** | | |

## SKILLS AND CONCEPTS

| OBJECTIVES AND EXAMPLES | REVIEW EXERCISES |
|---|---|

Upon completing this chapter, you should
be able to:

Use these exercises to review and prepare
for the chapter test.

■ change radian measure to degree measure
and vice versa.  (**Lesson 16-1**)

**Change each degree measure to radians.**

Change the degree measure 240° to
radians.

**1.** 255°  **2.** −315°

$$240 \cdot \frac{\pi}{180} = \frac{240\pi}{180} \text{ or } \frac{4\pi}{3}$$

**3.** 270°  **4.** 120°

Change the radian measure $-\frac{4\pi}{3}$ to
degrees.

**Change each radian measure to degrees.**

$$-\frac{4\pi}{3} \cdot \frac{180}{\pi} = \left(-\frac{720\pi}{3\pi}\right)^\circ \text{ or } -240°$$

**5.** $\frac{\pi}{3}$  **6.** $\frac{7\pi}{4}$

**7.** $-\frac{5\pi}{12}$  **8.** $\frac{4}{3}$

■ evaluate expressions involving sine and cosine. **(Lesson 16-2)**

**Find sin 570°.**

$$\sin 570° = \sin (210 + 360)°$$
$$= \sin 210°$$
$$= -\frac{1}{2}$$

**Find each value.**

9. $\cos 210°$

10. $\cos 3\pi$

11. $\sin (-150°)$

12. $\sin \frac{5}{4}\pi$

13. $\cos (-135°)$

14. $\cos 300°$

15. $(\sin 30°)^2 + (\cos 30°)^2$

16. $(\sin 45°)(\sin 225°)$

---

■ find the values of other trigonometric functions. **(Lesson 16-3)**

**Find tan 150°.**

$$\tan 150° = \frac{\sin 150°}{\cos 150°}$$
$$= \frac{\frac{1}{2}}{-\frac{\sqrt{3}}{2}}$$
$$= -\frac{1}{\sqrt{3}} \text{ or } -\frac{\sqrt{3}}{3}$$

**Find each value.**

17. $\csc 135°$

18. $\csc \pi$

19. $\sec (-30°)$

20. $\cot \frac{7}{6}\pi$

21. $\tan 120°$

22. $\sec (-60°)$

---

■ find the value of expressions involving trigonometric functions. **(Lesson 16-4)**

**Find $Cos^{-1} \left(-\frac{\sqrt{3}}{2}\right)$.**

$$\theta = Cos^{-1} \left(-\frac{\sqrt{3}}{2}\right)$$
$$Cos\ \theta = -\frac{\sqrt{3}}{2}$$
$$\theta = \frac{5\pi}{6}$$

**Find each value.**

23. $Sin^{-1} (-1)$

24. $Cos^{-1} \left(\frac{\sqrt{3}}{2}\right)$

25. $Tan^{-1} \sqrt{3}$

26. $Sin^{-1} \left(\tan \frac{\pi}{4}\right)$

27. $\cos (Sin^{-1} 1)$

28. $\sin \left(2\ Sin^{-1} \frac{1}{2}\right)$

---

■ use a calculator to find values of trigonometric functions. **(Lesson 16-5)**

**If sin x = 0.5346, find x.**

ENTER:

0.5346 $\boxed{\text{SIN}^{-1}}$ ヨ∂.ヨ।Ь ∂ВВЧ∂

Therefore, $x$ is approximately 32.3167°.

**Use a calculator to find each value. Round your answers to four decimal places.**

29. $\cos x = 0.7924$

30. $\csc x = 1.3729$

31. $\left(\cos \frac{2\pi}{3}\right)\left(\sin \frac{\pi}{4}\right)$

32. $\tan [Cos^{-1} 0.8 + Sin^{-1} (-0.4)]$

■ solve problems involving right triangles using right triangle trigonometry. (**Lesson 16-6**)

Solve the right triangle.

$$\frac{a}{14} = \sin 42° \qquad \frac{b}{14} = \cos 42°$$

$$a \approx 9.4 \qquad b \approx 10.4$$

$$B = 90° - 42° \text{ or } 48°$$

**Solve △ABC (with right angle C). Round lengths to the nearest tenth and angle measures to the nearest minute.**

33. If $c = 16$ and $a = 7$, find $b$.

34. If $a = 7$ and $b = 12$, find $A$.

35. If $b = 10$ and $c = 20$, find $a$.

36. If $a = b$ and $c = 12$, find $B$.

37. If $A = 25°$ and $c = 6$, find $b$.

38. If $a = 1$ and $b = 3$, find $A$.

■ solve triangles using the Law of Sines. (**Lesson 16-8**)

According to the Law of Sines,

$$\frac{\sin A}{a} = \frac{\sin B}{b} = \frac{\sin C}{c}.$$

**Use the Law of Sines to solve each triangle described below.**

39. $A = 83°10'$, $a = 80$, $b = 10$

40. $A = 50°$, $b = 12$, $a = 10$

41. $B = 46°$, $C = 83°$, $b = 65$

42. $A = 45°$, $B = 30°$, $b = 20$

■ solve triangles using the Law of Cosines. (**Lesson 16-10**)

According to the Law of Cosines,
$$a^2 = b^2 + c^2 - 2bc \cos A$$
$$b^2 = a^2 + c^2 - 2ac \cos B$$
$$c^2 = a^2 + b^2 - 2ab \cos C$$

**Use the Law of Cosines to solve each triangle described below.**

43. $C = 65°$, $a = 4$, $b = 7$

44. $b = 2$, $c = 5$, $A = 60°$

45. $a = 6$, $b = 7$, $C = 40°$

46. $B = 24°$, $a = 42$, $c = 6.5$

# ～～～～ APPLICATIONS AND CONNECTIONS ～～～～

47. **Aviation**  A pilot 3000 feet above the ocean notes the measurement of the angle of depression to a ship is 42°. How far is the plane from the ship? (**Lesson 16-7**)

48. **Geometry**  When Tracy stands 50 meters from the school's flagpole, the angle of elevation to the top measures 48°. How tall is the pole? (**Lesson 16-7**)

49. **Firefighting**  A firefighter needs to use a 14-meter ladder to enter a window which is 13.5 meters above the ground. How far from the base of the building should the ladder be placed? (**Lesson 16-7**)

50. **Engineering**  An engineering student is assigned to construct a triangular model of three steel girders. Two of the girders measured 20 cm and 15 cm, and the angle opposite the 15 cm girder had to be 61°. Could the model be made? If so, how long is the third girder? (**Lesson 16-9**)

**Change each degree measure to radians.**

1. $-45°$
2. $275°$
3. $330°$
4. $-600°$

**Change each radian measure to degrees.**

5. $-\dfrac{\pi}{6}$
6. $\dfrac{11}{2}\pi$
7. $-\dfrac{7\pi}{4}$
8. $-2\dfrac{1}{3}$

**Find each value.**

9. $\cos(-120°)$
10. $\sin\dfrac{7}{4}\pi$
11. $\cos\dfrac{3\pi}{4}$
12. $\sin 390°$

13. $\tan 135°$
14. $\cot 300°$
15. $\sec\left(-\dfrac{7}{6}\pi\right)$
16. $\csc\dfrac{5\pi}{6}$

17. $\mathrm{Sin}^{-1}\left(-\dfrac{\sqrt{3}}{2}\right)$
18. $\mathrm{Arctan}\ 1$
19. $\mathrm{Cos}^{-1}(\sin-60°)$
20. $\sin 2\left(\mathrm{Arccos}\ \dfrac{1}{2}\right)$

**Find the value of x to the nearest minute.**

21. $\sin x = 0.6712$
22. $\cos x = 0.1389$
23. $\csc x = 1.2955$
24. $\tan x = 0.8999$

**Solve each right triangle. Round lengths to the nearest tenth and angle measures to the nearest minute.**

25. $a = 7, A = 49°$
26. $a = 7, c = 16$
27. $B = 75°, b = 6$
28. $A = 22°, c = 8$

29. Determine the number of possible solutions for a triangle where $A = 40°$, $b = 10$ and $a = 14$. If a solution exists, solve the triangle.

30. **Firefighting** A firefighter leaned a 32-foot ladder against a building. The top of the ladder touches the building 26 feet above the ground. What is the measurement of the angle formed between the ladder and the ground?

31. **Aviation** A plane flew 1000 kilometers north. Then it changed direction by turning 20° clockwise and flew for another 700 kilometers. How far was the plane from its starting point?

32. **Geology** From the top of a cliff, a geologist spots a dried river bed. The measurement of the angle of depression to the river bed is 70°. The cliff is 50 meters high. How far is the river bed from the base of the cliff?

33. **Geometry** The longest side of a triangle is 23 yards. Two of the angles are 23° and 49°. Find the measures of the other two sides and the remaining angle.

**Bonus** Suppose $\sin x = 0.5660$ and the terminal side of angle $x$ is in the second quadrant. Find the value of $x$ to the nearest minute.

# College Entrance Exam Preview

The test questions on these pages deal with a variety of concepts from arithmetic, algebra, and geometry.

**Directions: Choose the one best answer. Write A, B, C, or D.**

1. The length of a rectangle is 1 unit and the width is $w$. If the width is increased by 3 units, by how many units will the perimeter be increased?

   (A) 4          (B) 6

   (C) $2w + 3$      (D) $2w + 6$

2. $O$ is the center of the circle below. $\overline{MO}$ is perpendicular to $\overline{NO}$ and the area of triangle $MON$ is 40. What is the area of circle $O$?

   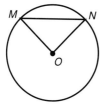

   (A) $80\pi$   (B) $40\pi$   (C) $20\pi$   (D) $160\pi$

3. The distance between point $X(4, 0)$ and $Y$ is 8. The coordinates of point $Y$ could be any of the following except

   (A) $(-4, 0)$          (B) $(0, 4\sqrt{3})$

   (C) $(4, 8)$           (D) $(8, 0)$

4. $x^2 + y^2 = 16$
   $xy = 8$
   $(x + y)^2 = ?$

   (A) 32   (B) 22   (C) 16   (D) 24

5. If $x = -6$ and $\dfrac{1}{2y} = -12$, what is the value of $y$ in terms of $x$?

   (A) $x - 6$          (B) $2x - 6$

   (C) $\dfrac{1}{2x}$           (D) $\dfrac{1}{4x}$

6. 5% of what number is 81?

   (A) 405            (B) 1620

   (C) 16,200          (D) 0.405

7. The area of triangle $ABC$ is 16 square units. Angle $C$ is 45°. What is $AC$?

   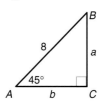

   (A) $16\sqrt{2}$          (B) $4\sqrt{2}$

   (C) $8\sqrt{2}$          (D) $24\sqrt{2}$

8. A person travels five miles north, fifteen miles west, and fifteen miles north. How far (to the nearest mile) is he from the starting point?

   (A) 35   (B) 30   (C) 30   (D) 25

9. If $n^2 - 4 = -3n$, what is the value of $\left(n + \dfrac{3}{2}\right)^2$?

   (A) 5   (B) $6\dfrac{1}{4}$   (C) 11

   (D) cannot be determined

10. If $x^2 = 25$, then $2^{x-1}$ could equal

    (A) 2   (B) 4   (C) 8   (D) 16

11. If $\frac{x}{6} > x$, which could be a value for $x$?

(A) $-6$ (B) 0
(C) 5 (D) 6

12. If $0 < n < 1$, which of the following increases as $n$ increases?

I. $1 - n^2$
II. $n - 1$
III. $\frac{1}{n^2}$

(A) II only
(B) III only
(C) I and III only
(D) II and III only

13. If $-a < 0 < -b$, which of the following is true?

(A) $0 < b < 0$
(B) $a < 0 < b$
(C) $b < 0 < a$
(D) $0 < b < a$

14. If $b$ is an odd integer greater than one, which of the following must be an odd integer?

(A) $b^3 - 1$
(B) $b^3 - b^2$
(C) $1 + b^3$
(D) $\frac{3b - 3}{b - 1}$

15. If $k$ is any odd integer and $x = 6k$, then $\frac{x}{2}$ will always be

(A) odd (B) even
(C) positive (D) negative

16. If $a + b = 8$ and $3b - 4 = -13a$, then what is the value of $a$?

(A) $\frac{-13}{3}$ (B) $-3$
(C) $-2$ (D) 10

**TEST TAKING TIP**

You can prepare for taking a college entrance exam by working through practice tests such as this. The more you work with the material in a format similar to the actual test, the better you become in the art of test taking. Do not wait until the night before taking a college entrance exam to review. Allow yourself plenty of time in the weeks before taking a test to review the basic skills and formulas which are tested. Find your weaknesses so that you can seek help.

When taking a college entrance exam work quickly through the problems, skipping those problems which take more than the average amount of time allotted for doing a problem. Mark these in the test booklet and come back to them after you have completed the problems which you could easily do. Take time out after several problems to freshen your mind.

17. If $x > y$ and $z < 0$, which of the following are true?

I. $xz < yz$
II. $x + z > y + z$
III. $x - z < y - z$

(A) I only
(B) II only
(C) I and II only
(D) I, II, and III

# 17 Trigonometric Graphs, Identities, and Equations

## CHAPTER OBJECTIVES

In this chapter, you will:

- Draw graphs of trigonometric functions.
- Use trigonometric identities.
- Solve trigonometric equations.
- Work with complex numbers in polar form.

Water flows down into deep water tables like (a). But when the water table is high as in (b) the water flows toward bodies of water. What might affect the level of the water table?

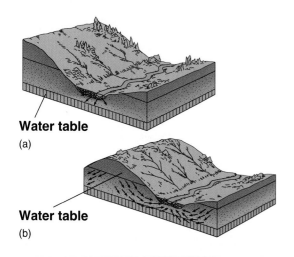

**Water table**

(a)

**Water table**

(b)

## CAREERS IN GEOLOGY

Would you like a career that required you to travel to far-away sites by helicopter or jeep? Would you enjoy covering large areas on foot with a backpack full of rock samples? Or would you like to work overseas or even at sea?

The career: geologist. The goal: Earth. Geologists study it at close range by studying bits of it—rocks —in their labs. They study Earth at long range with satellites, seismographs, sonar, and other instruments.

A geologist might work in the petroleum industry exploring for oil and natural gas. He or she might advise engineers who construct dams, tunnels, and highways. Another field is environmental protection. Geologists can monitor groundwater quality, manage and clean up toxic wastes. Or they can research how human activities affect Earth's atmosphere, oceans, rivers, and land.

A geologist can spend a lifetime studying time. How old is the moon? When did certain creatures live and die to become the fossils we see today? When did certain mountain ranges rise and fall, drying up some seas and creating others? One of the latest techniques for dating objects uses the isotope chlorine-36. Another uses Earth's axis of rotation, which has changed its orientation several times since Earth was formed.

## MORE ABOUT GEOLOGY

### Degree Required:

- Bachelor's Degree in geology or geophysics

### Related Math Subjects:

- Advanced Algebra
- Geometry
- Trigonometry
- Calculus
- Statistics/Probability

For more information on the various careers available in Geology, write to:

American Institute of Professional Geologists
7828 Vance Drive
Arvada, CO 80003

### Some geologists like:

- solving problems
- the variety of their work
- traveling
- working with others
- good salaries

### Some geologists dislike:

- the possibility of injury while on the job
- being far away from home and family for long periods of time
- the primitive living conditions in the field
- the physically demanding nature of the work while at a site under investigation

# Graphing Calculator Exploration: Graphing Trigonometric Functions

Graphing calculators are capable of graphing trigonometric functions in both degrees and radians. In this lesson, we will use degrees.

**Example 1**

Graph the function $y = \sin x$.

The Casio fx-7000G has the sine function as a built-in function. It will automatically set the viewing window for viewing a complete graph of the function. *Make sure your calculator is in degree mode.*

*Casio*

ENTER: [GRAPH] [SIN] [EXE]

For the TI-81, the viewing window must be set. Use the window [-360, 360] by [-1.6, 1.6] with a scale factor of 180 for the x-axis and 0.5 for the y-axis.

*TI-81*

ENTER: [Y=] [SIN] [X|T] [GRAPH]

Notice that the y-values of the graph are between -1 and 1. The **amplitude** of the graph of a periodic function is half the absolute value of the difference between its maximum and minimum values. So the amplitude of this graph is 1. The period of this function is 360°. *How could you graph this function in radians?*

**Example 2**

Graph $y = 2 \cos 3x$. Use the viewing window [-360, 360] by [-2.6, 2.6] with a scale factor of 180 for the x-axis and 1 for the y-axis. State the amplitude and period of the graph.

*Casio*

ENTER: [GRAPH] 2 [COS] [(] 3 [ALPHA] [X]
[)] [EXE]

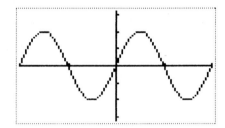

*TI-81*

ENTER: [Y=] 2 [COS] [(] 3 [X|T] [)] [GRAPH]

The amplitude of this graph is 2. The period is 120°. *Zoom-in to check this result.*

As Example 2 demonstrates, the amplitude of the function $y = a \sin bx$ or $y = a \cos bx$ is $|a|$ and the period is $\dfrac{360°}{|b|}$.

**Example 3**

Graph $y = \sin(x + 90°)$. Use the viewing window $[-360, 360]$ by $[-1.6, 1.6]$ with a scale factor of 180 for the x-axis and 0.5 for the y-axis. How does this graph compare to the graph of $y = \sin x$?

*Casio*

ENTER:

90 ⃞ ⃞EXE⃞

*TI-81*

ENTER: ⃞Y=⃞ ⃞SIN⃞ ⃞(⃞ ⃞X|T⃞ ⃞+⃞ 90 ⃞)⃞

⃞GRAPH⃞

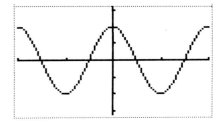

This graph is shaped the same as the graph of $y = \sin x$, except it is shifted to the left by 90°. This is called a 90° **phase shift.**

You can also graph the other trigonometric functions on the graphing calculator.

**Example 4**

Graph $y = \csc x$. Use the viewing window $[-360, 360]$ by $[-5, 5]$ with a scale factor of 180 on the x-axis and 1 on the y-axis. State the period of the graph.

$$\csc x = \frac{1}{\sin x}$$

*Casio*

ENTER:

⃞EXE⃞

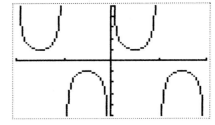

*The vertical asymptotes may appear on the viewing screen.*

*TI-81*

ENTER: ⃞Y=⃞ 1 ⃞÷⃞ ⃞SIN⃞ ⃞X|T⃞ ⃞GRAPH⃞

The period of this graph is 360°. This graph has no amplitude, since there are no maximum or minimum values for the function.

# EXERCISES

Graph each equation on your graphing calculator so the complete graph is shown. Then sketch each graph.

**1.** $y = \cos x$        **2.** $y = \tan(-x)$        **3.** $y = -2 \sin 3x$

**4.** $y = \csc x$        **5.** $y = \cot x$        **6.** $y = \tan(x - 180°)$

**7.** $y = 2 \cos \frac{1}{2}x$        **8.** $y = 3 \sin \frac{2}{3}x$        **9.** $y = 5 \csc(-2x)$

**10.** $y = 3 \tan(-30x)$        **11.** $y = -9 \cot 3.5x$        **12.** $y = 15 \cos 15x$

**13.** $y = 12 \cos(x + 45°)$        **14.** $y = 4 \tan(90° - x)$

**15.** $y = 0.5 \csc(2x - 90°)$        **16.** $y = 0.2 \sec(360° - 3x)$

# Graphs of Trigonometric Functions

**Objectives**

After studying this lesson, you should be able to:

- graph trigonometric functions, and
- find the amplitude and period for variations of the sine and cosine functions.

**Application**

*FYI* ···

The force that a spring exerts increases linearly with the amount it is stretched or compressed. This relationship, which shows that such a spring oscillates in harmonic motion, is called Hooke's Law.

If a weight is attached to a spring and the weight is pushed up, or pulled down, and released, it tends to rise and fall alternately. The weight is said to be oscillating in harmonic motion. If the position of the weight, $y$, is graphed over time, $t$, the result is the graph of a sine or cosine curve.

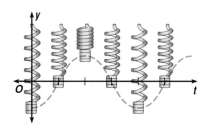

To graph the sine or cosine function, use the horizontal axis for the values of $\theta$ expressed in either degrees or radians. Use the vertical axis for values of $\sin \theta$ or $\cos \theta$. Ordered pairs for these points are of the form $(\theta, \sin \theta)$ and $(\theta, \cos \theta)$.

| $\theta$ | 0 | 30 | 45 | 60 | 90 | 120 | 135 | 150 | 180 | 210 | 225 | 240 | 270 | 300 | 315 | 330 | 360 |
|---|---|---|---|---|---|---|---|---|---|---|---|---|---|---|---|---|---|
| $\sin \theta$ | 0 | $\frac{1}{2}$ | $\frac{\sqrt{2}}{2}$ | $\frac{\sqrt{3}}{2}$ | 1 | $\frac{\sqrt{3}}{2}$ | $\frac{\sqrt{2}}{2}$ | $\frac{1}{2}$ | 0 | $-\frac{1}{2}$ | $-\frac{\sqrt{2}}{2}$ | $-\frac{\sqrt{3}}{2}$ | $-1$ | $-\frac{\sqrt{3}}{2}$ | $-\frac{\sqrt{2}}{2}$ | $-\frac{1}{2}$ | 0 |
| nearest tenth | 0 | 0.5 | 0.7 | 0.9 | 1 | 0.9 | 0.7 | 0.5 | 0 | $-0.5$ | $-0.7$ | $-0.9$ | $-1$ | $-0.9$ | $-0.7$ | $-0.5$ | 0 |
| $\cos \theta$ | 1 | $\frac{\sqrt{3}}{2}$ | $\frac{\sqrt{2}}{2}$ | $\frac{1}{2}$ | 0 | $-\frac{1}{2}$ | $-\frac{\sqrt{2}}{2}$ | $-\frac{\sqrt{3}}{2}$ | $-1$ | $-\frac{\sqrt{3}}{2}$ | $-\frac{\sqrt{2}}{2}$ | $-\frac{1}{2}$ | 0 | $\frac{1}{2}$ | $\frac{\sqrt{2}}{2}$ | $\frac{\sqrt{3}}{2}$ | 1 |
| nearest tenth | 1 | 0.9 | 0.7 | 0.5 | 0 | $-0.5$ | $-0.7$ | $-0.9$ | $-1$ | $-0.9$ | $-0.7$ | $-0.5$ | 0 | 0.5 | 0.7 | 0.9 | 1 |

After plotting several points, complete the graphs of $y = \sin \theta$ and $y = \cos \theta$ by connecting the points with a smooth, continuous curve.

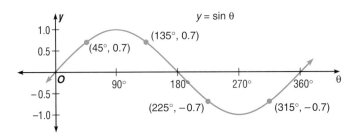

*Negative values of $\theta$ would be represented to the left of zero.*

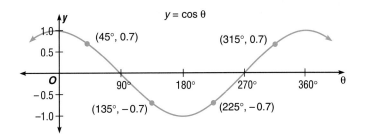

As you recall from your studies of the sine and cosine functions in Chapter 16, both of these functions have a **period** of 360° or $2\pi$ radians. That is, the graph of both functions repeat themselves every 360° or $2\pi$ radians. The following example is a variation of the sine function that has a period less than 360°.

**Example 1**

**Graph $y = \sin 2\theta$. State the period.**

First, complete a table of values.

| $\theta$ | 0° | 15° | 30° | 45° | 60° | 75° | 90° | 105° | 120° | 135° | 150° | 165° | 180° |
|---|---|---|---|---|---|---|---|---|---|---|---|---|---|
| $2\theta$ | 0° | 30° | 60° | 90° | 120° | 150° | 180° | 210° | 240° | 270° | 300° | 330° | 360° |
| $\sin 2\theta$ | 0 | $\frac{1}{2}$ | $\frac{\sqrt{3}}{2}$ | 1 | $\frac{\sqrt{3}}{2}$ | $\frac{1}{2}$ | 0 | $-\frac{1}{2}$ | $-\frac{\sqrt{3}}{2}$ | $-1$ | $-\frac{\sqrt{3}}{2}$ | $-\frac{1}{2}$ | 0 |

Then plot the points given by the ordered pairs of the form $(\theta, y)$, and connect to form a curve.

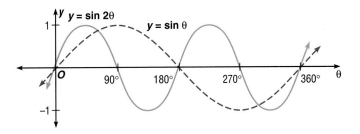

The graph of $y = \sin 2\theta$ repeats every 180° or $\pi$ radians. Therefore, the period of $y = \sin 2\theta$ is 180° or $\pi$ radians.  *Notice that the period is $\frac{360°}{2}$ or $\frac{2\pi}{2}$.*

The trigonometric function you graphed in Example 1 has a maximum value of 1 and a minimum value of -1. The **amplitude** of the graph of a periodic function is the absolute value of half the difference between its maximum value and its minimum value. So in Example 1, the amplitude of the graph is $\left| \dfrac{1 - (-1)}{2} \right|$ or 1. The graphs in the following examples have amplitudes other than 1.

Example 2

Graph $y = \frac{1}{2} \cos \theta$. State the amplitude.

| $\theta$ | 0° | 30° | 60° | 90° | 120° | 150° | 180° | 210° | 240° | 270° | 300° | 330° | 360° |
|---|---|---|---|---|---|---|---|---|---|---|---|---|---|
| $\cos \theta$ | 1 | $\frac{\sqrt{3}}{2}$ | $\frac{1}{2}$ | 0 | $-\frac{1}{2}$ | $-\frac{\sqrt{3}}{2}$ | $-1$ | $-\frac{\sqrt{3}}{2}$ | $-\frac{1}{2}$ | 0 | $\frac{1}{2}$ | $\frac{\sqrt{3}}{2}$ | 1 |
| $\frac{1}{2} \cos \theta$ | $\frac{1}{2}$ | $\frac{\sqrt{3}}{4}$ | $\frac{1}{4}$ | 0 | $-\frac{1}{4}$ | $-\frac{\sqrt{3}}{4}$ | $-\frac{1}{2}$ | $-\frac{\sqrt{3}}{4}$ | $-\frac{1}{4}$ | 0 | $\frac{1}{4}$ | $\frac{\sqrt{3}}{4}$ | $\frac{1}{2}$ |

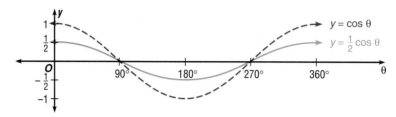

The amplitude of $y = \frac{1}{2} \cos \theta$ is $\frac{1}{2}$.  *Notice that the amplitude is $\left|\frac{1}{2}\right|$ or $\frac{1}{2}$.*

There are many applications of trigonometry in the real world. The motion of a weight on a spring is one such application.

Example 3

The motion of a weight on a certain kind of spring can be described by the equation $y = 3 \cos 2\pi t$, where $y$ is the number of inches the weight is from its equilibrium position, when $y = 0$, after $t$ seconds.

**a.** Graph the function and state the amplitude and the period.

| $t$ seconds | 0 | $\frac{1}{12}$ | $\frac{1}{6}$ | $\frac{1}{4}$ | $\frac{1}{3}$ | $\frac{5}{12}$ | $\frac{1}{2}$ | $\frac{7}{12}$ | $\frac{2}{3}$ | $\frac{3}{4}$ | $\frac{5}{6}$ | $\frac{11}{12}$ | 1 |
|---|---|---|---|---|---|---|---|---|---|---|---|---|---|
| $2\pi t$ radians | 0 | $\frac{\pi}{6}$ | $\frac{\pi}{3}$ | $\frac{\pi}{2}$ | $\frac{2\pi}{3}$ | $\frac{5\pi}{6}$ | $\pi$ | $\frac{7\pi}{6}$ | $\frac{4\pi}{3}$ | $\frac{3\pi}{2}$ | $\frac{5\pi}{3}$ | $\frac{11\pi}{6}$ | $2\pi$ |
| $\cos 2\pi t$ | 1 | $\frac{\sqrt{3}}{2}$ | $\frac{1}{2}$ | 0 | $-\frac{1}{2}$ | $-\frac{\sqrt{3}}{2}$ | $-1$ | $-\frac{\sqrt{3}}{2}$ | $-\frac{1}{2}$ | 0 | $\frac{1}{2}$ | $\frac{\sqrt{3}}{2}$ | 1 |
| $3 \cos 2\pi t$ | 3 | $\frac{3\sqrt{3}}{2}$ | $\frac{3}{2}$ | 0 | $-\frac{3}{2}$ | $-\frac{3\sqrt{3}}{2}$ | $-3$ | $-\frac{3\sqrt{3}}{2}$ | $-\frac{3}{2}$ | 0 | $\frac{3}{2}$ | $\frac{3\sqrt{3}}{2}$ | 3 |

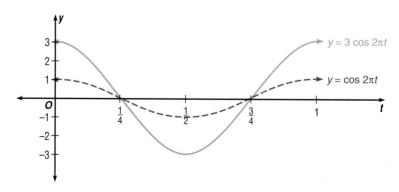

*Notice that the amplitude is $|3|$ or 3 and the period is $\frac{2\pi}{|2\pi|}$ or 1.*

The amplitude of the graph of $y = 3 \cos 2\pi t$ is 3 and the period is 1.

**b. Find the position of the weight after 6 seconds.**

Find $y$ when $t = 6$.   $y = 3 \cos 2\pi t$

$\qquad\qquad\qquad\qquad = 3 \cos 2\pi(6)$   *Substitute 6 for t.*

$\qquad\qquad\qquad\qquad = 3 \cos 12\pi = 3(1) \text{ or } 3$   *$\cos 12\pi = 1$*

After 6 seconds, the weight will be 3 in. above the equilibrium position.

From these examples, we can make the following generalizations.

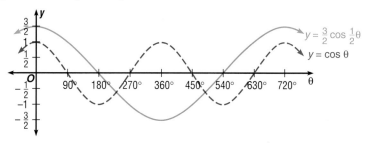

| *Amplitudes and Periods* | **For functions of the form $y = a \sin b\theta$ and $y = a \cos b\theta$ the amplitude is $\lvert a \rvert$ and the period is $\dfrac{360°}{\lvert b \rvert}$ or $\dfrac{2\pi}{\lvert b \rvert}$.** |
| --- | --- |

**Example 4**

State the amplitude and period of $y = \dfrac{3}{2} \cos \dfrac{1}{2}\theta$. Then draw the graph.

The amplitude is $\left\lvert \dfrac{3}{2} \right\rvert$ or $\dfrac{3}{2}$. The period is $\dfrac{360°}{\left\lvert \dfrac{1}{2} \right\rvert}$ or $720°$.   *$720° = 4\pi$ radians*

The graph has a shape like $y = \cos \theta$.

The other trigonometric functions can be graphed as well. Make a table of values and graph $y = \tan \theta$.   *nd = not defined*

| $\theta$ | 0° | 30° | 45° | 60° | 90° | 120° | 135° | 150° | 180° | 210° | 225° | 240° | 270° | 300° | 315° | 330° | 360° |
| --- | --- | --- | --- | --- | --- | --- | --- | --- | --- | --- | --- | --- | --- | --- | --- | --- | --- |
| $\tan \theta$ | 0 | $\dfrac{\sqrt{3}}{3}$ | 1 | $\sqrt{3}$ | nd | $-\sqrt{3}$ | $-1$ | $-\dfrac{\sqrt{3}}{3}$ | 0 | $\dfrac{\sqrt{3}}{3}$ | 1 | $\sqrt{3}$ | nd | $-\sqrt{3}$ | $-1$ | $-\dfrac{\sqrt{3}}{3}$ | 0 |

The tangent function, $y = \tan \theta$, is not defined for 90°, 270°, and so on. We say that it is not defined for $90° + k \cdot 180°$, where $k$ is an integer. The graph is separated by vertical asymptotes, indicated by dashed lines.

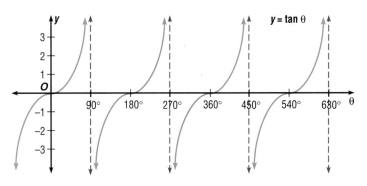

The period of the tangent function is 180° or $\pi$ radians. Since the tangent function is not defined for $90° + k \cdot 180°$, it has no amplitude.

The graphs of the secant, cosecant, and cotangent functions are shown below. Compare them to the graphs of the cosine, sine, and tangent functions, which are shown in red.

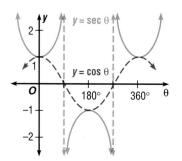
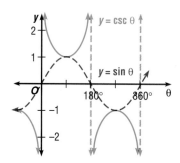
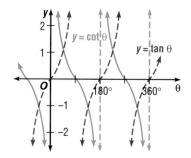

Notice that the periods of the secant and cosecant functions are 360° or $2\pi$ radians. The period of the cotangent function is 180° or $\pi$ radians. What are the amplitudes of the secant, cosecant, and cotangent functions?

**Example 5**

Graph $y = -\dfrac{1}{2} \csc 2\theta$.

*nd = not defined*

| $\theta$ | 0° | 15° | 30° | 45° | 60° | 75° | 90° | 105° | 120° | 135° | 150° | 165° | 180° |
|---|---|---|---|---|---|---|---|---|---|---|---|---|---|
| $2\theta$ | 0° | 30° | 60° | 90° | 120° | 150° | 180° | 210° | 240° | 270° | 300° | 330° | 360° |
| $-\dfrac{1}{2}\csc 2\theta$ | nd | $-1$ | $-\dfrac{\sqrt{3}}{3}$ | $-\dfrac{1}{2}$ | $\dfrac{\sqrt{3}}{3}$ | $-1$ | nd | 1 | $\dfrac{\sqrt{3}}{3}$ | $\dfrac{1}{2}$ | $\dfrac{\sqrt{3}}{3}$ | 1 | nd |

The period is $\dfrac{360°}{|2|}$ or 180°. This function is not defined for 0°, 90°, 180°, and so on. Thus, it has no amplitude.

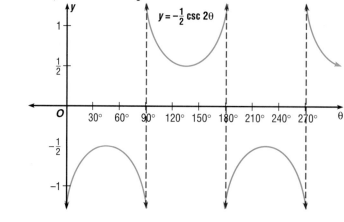

# CHECKING FOR UNDERSTANDING

**Communicating Mathematics**

Read and study the lesson to answer these questions.

1. What does it mean to say that the period of a function is 180°?

2. In your own words, explain what the amplitude of a function is.

3. Judging from the results of Example 5, how can you find the amplitude and the period of $y = a \csc b\theta$?

**Guided Practice**

State the amplitude and period of each function.

4. $y = \sin \theta$

5. $y = 3 \sin \theta$

6. $y = \frac{1}{2} \cos \theta$

7. $y = 3 \cos \frac{1}{2}\theta$

8. $y = \frac{2}{3} \cos \theta$

9. $y = 6 \sin \frac{2}{3}\theta$

10. $y = -2 \sin \theta$

11. $y = -3 \sin \frac{2}{3}\theta$

12. $y = 4 \sin \frac{1}{2}\theta$

13. $y = -\frac{1}{2} \cos \frac{3}{4}\theta$

14. $3y = 2 \sin \frac{1}{2}\theta$

15. $y = -6 \sin 2\theta$

State the period of each function.

16. $y = \sec 3\theta$

17. $y = 4 \sec \theta$

18. $y = 3 \tan \theta$

19. $y = \cot 5\theta$

20. $y = \cot \frac{1}{3}\theta$

21. $y = \tan 3\theta$

22. $y = \tan \frac{1}{2}\theta$

23. $y = \csc 2\theta$

24. $y = \csc \frac{3}{4}\theta$

25. $y = \frac{1}{2} \sec \frac{1}{2}\theta$

26. $y = 6 \cot 2\theta$

27. $y = \frac{3}{4} \csc \frac{2}{3}\theta$

# EXERCISES

**Practice**

Graph each function.

28. $y = \sin \theta$

29. $y = 3 \sin \theta$

30. $y = \frac{1}{2} \cos \theta$

31. $y = \frac{2}{3} \cos \theta$

32. $y = \cos 3\theta$

33. $y = \sin 4\theta$

34. $y = 5 \sin \theta$

35. $y = \cos 2\theta$

36. $y = -2 \sin \theta$

37. $y = \cot \theta$

38. $y = 3 \sec \theta$

39. $y = \sec 3\theta$

40. $y = \csc \frac{1}{3}\theta$

41. $y = 2 \sec \theta$

42. $y = \frac{1}{3} \sec \theta$

43. $y = 2 \tan \theta$

44. $y = \csc 2\theta$

45. $y = 4 \sin \frac{1}{2}\theta$

46. $y = 6 \sin \frac{2}{3}\theta$

47. $y = 3 \cos \frac{1}{2}\theta$

48. $y = 4 \cos \frac{3}{4}\theta$

49. $y = -3 \sin \frac{2}{3}\theta$

50. $y = -6 \sin 2\theta$

51. $y = 2 \sin \frac{1}{5}\theta$

52. $y = -\cot \theta$

53. $y = 3 \csc \frac{1}{2}\theta$

54. $y = -\frac{1}{2} \cot 2\theta$

55. $y = \frac{1}{2} \tan \theta$

56. $3y = 2 \sin \frac{1}{2}\theta$

57. $\frac{3}{4}y = \frac{2}{3} \sin \frac{3}{5}\theta$

58. $\frac{1}{2}y = 3 \sin 2\theta$

59. $y = -\frac{1}{2} \cos \frac{3}{4}\theta$

60. $\frac{1}{2}y = 5 \csc 3\theta$

**Critical Thinking**

**61.** Graph and state the equation involving sine that has an amplitude of 3 and a period of 540°.

**Application**

**62. Zoology**  In predator-prey systems, the number of predators and the number of prey tend to vary periodically. In a certain region with coyotes as predators and rabbits as prey, the rabbit population $R$ varied according to the equation $R = 1000 + 250 \sin \frac{\pi t}{2}$, with the time, $t$, in years since January 1, 1960.

    **a.** Graph the function for the number of rabbits.

    **b.** What was the population of rabbits on January 1, 1960?

    **c.** What is the maximum rabbit population? On what date was the maximum population of rabbits first reached?

    **d.** What is the minimum rabbit population? On what date was the minimum population of rabbits first reached?

**Mixed Review**

**63.** Determine whether the pair of angles whose measurements are 434° and 794° are coterminal.  **(Lesson 16-1)**

**64. Statistics**  Find the median, mode, and mean of {90, 89, 83, 95, 99, 100, 82, 81, 87, 94, 97, 98}.  **(Lesson 14-3)**

**65.** Use sigma notation to express $6 + 9 + 12 + 15 + 18 + 21$.
**(Lesson 13-2)**

**66.** Simplify $(4^{\sqrt{3}})^{\sqrt{2}}$.  **(Lesson 12-1)**

**67.** Find all zeros of the function $f(x) = 4x^4 - 35x^3 + 78x^2 + 28x - 165$.
**(Lesson 10-4)**

**68.** Simplify $(6 + i)(6 - i)$.  **(Lesson 6-9)**

**69. Business**  The Bright Idea Company makes floor lamps and table lamps, each of which must be assembled and packed. The time an employee takes to assemble a floor lamp is 18 minutes and to assemble a table lamp is 12 minutes. It takes 1 minute to package a floor lamp and 2 minutes to package a table lamp. Each employee can spend 240 hours assembling lamps and 20 hours packing lamps.

    **a.** If a floor lamp sells for $85 and a table lamp sells for $75, how many of each should the company make to maximize their revenue?

    **b.** What is the maximum weekly revenue?  **(Lesson 3-7)**

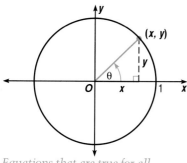

## 17-2 Trigonometric Identities

**Objective**

After studying this lesson, you should be able to:

- use trigonometric identities to simplify and/or evaluate expressions.

Let $\theta$ be the measurement of an angle in standard position. Let $(x, y)$ be the coordinates of the point of intersection of the terminal side of the angle and the unit circle. Then, as you know, $\cos \theta = x$ and $\sin \theta = y$.

The equation for the unit circle is $x^2 + y^2 = 1$. By substituting $\cos \theta$ for $x$ and $\sin \theta$ for $y$, we have the following equation.

$$(\cos \theta)^2 + (\sin \theta)^2 = 1$$

This equation is usually written as follows.

$$\cos^2 \theta + \sin^2 \theta = 1$$

*Equations that are true for all values of the variables for which they are defined are called identities.*

An equation like $\cos^2 \theta + \sin^2 \theta = 1$ is called an **identity** because it is true for all values of $\theta$. Some other trigonometric identities are given below.

---

*Basic Trigonometric Identities*

**The following trigonometric identities hold true for all values of $\theta$ except those for which either side of the equation is undefined.**
$$\cos^2 \theta + \sin^2 \theta = 1$$
**The following two identities result when each side of the equation $\cos^2 \theta + \sin^2 \theta = 1$ is divided by $\sin^2 \theta$ or by $\cos^2 \theta$.**
$$\tan^2 \theta + 1 = \sec^2 \theta$$
$$\cot^2 \theta + 1 = \csc^2 \theta$$

---

The steps in the following example show the development of a basic trigonometric identity. The expression on one side of the equation is transformed into the exact form of the expression on the other side.

**Example 1**

Show that $1 + \tan^2 \theta = \sec^2 \theta$.

$$
\begin{aligned}
1 + \tan^2 \theta &= 1 + \left(\frac{\sin \theta}{\cos \theta}\right)^2 \quad \text{\textit{Definition of} } \tan \theta \\
&= 1 + \frac{\sin^2 \theta}{\cos^2 \theta} \\
&= \frac{\cos^2 \theta}{\cos^2 \theta} + \frac{\sin^2 \theta}{\cos^2 \theta} \quad \text{\textit{1}} = \frac{\cos^2 \theta}{\cos^2 \theta} \\
&= \frac{\cos^2 \theta + \sin^2 \theta}{\cos^2 \theta} \\
&= \frac{1}{\cos^2 \theta} \text{ or } \sec^2 \theta \quad \text{Therefore, } 1 + \tan^2 \theta = \sec^2 \theta.
\end{aligned}
$$

You can use trigonometric identities to find values of trigonometric functions.

**Example 2**

Find $\csc \beta$ if $\cot \beta = \dfrac{3}{5}$ and $180° \le \beta \le 270°$.

$$\csc^2 \beta = 1 + \cot^2 \beta \qquad \textit{Trigonometric identity: } 1 + \cot^2 \theta = \csc^2 \theta$$

$$= 1 + \left(\dfrac{3}{5}\right)^2 \qquad \textit{Substitute } \dfrac{3}{5} \textit{ for } \cot \beta.$$

$$= \dfrac{34}{25} \qquad \textit{Take the square root of each side.}$$

$$\csc \beta = -\sqrt{\dfrac{34}{25}} \qquad \textit{csc } \beta \textit{ is negative for values of } \beta \textit{ between } 180° \textit{ and } 270°.$$

$$= -\dfrac{\sqrt{34}}{5}$$

**Example 3**

If $\tan A = 4$, find $\cos A$. Assume that angle $A$ is acute.

$$\sec^2 A = 1 + \tan^2 A \qquad \textit{Trigonometric identity}$$

$$= 1 + (4)^2$$

$$= 17$$

$$\left(\dfrac{1}{\cos^2 A}\right) = 17 \qquad \sec A = \dfrac{1}{\cos A}$$

$$\cos^2 A = \dfrac{1}{17}$$

$$\cos A = \sqrt{\dfrac{1}{17}} \qquad \textit{cos } A \textit{ is positive for acute angles.}$$

$$= \dfrac{\sqrt{17}}{17}$$

Trigonometric identities can also be used to simplify expressions containing trigonometric functions. Simplifying an expression that contains trigonometric functions means that the expression is written as a numerical value or in terms of a single trigonometric function, if possible.

**Example 4**

Simplify $\dfrac{1}{1 + \sin x} + \dfrac{1}{1 - \sin x}$.

$$\dfrac{1}{1 + \sin x} + \dfrac{1}{1 - \sin x} = \dfrac{(1 - \sin x) + (1 + \sin x)}{(1 + \sin x)(1 - \sin x)}$$

$$= \dfrac{2}{1 - \sin^2 x}$$

$$= \dfrac{2}{\cos^2 x} \qquad \sin^2 x + \cos^2 x = 1$$

$$= 2 \sec^2 x \qquad \sec x = \dfrac{1}{\cos x}$$

# CHECKING FOR UNDERSTANDING

**Communicating Mathematics**

**Read and study the lesson to answer these questions.**

1. What do the coordinates $(x, y)$ represent on the unit circle?

2. What does it mean to simplify an expression containing trigonometric functions?

3. When $\sin \theta < 0$ and $\cos \theta < 0$, in which quadrant does the terminal side of $\theta$ lie?

4. In which quadrants might the angle whose measurement is $\beta$ terminate if $\tan \beta = -\dfrac{4}{3}$?

**Guided Practice**

**Simplify each expression.**

5. $\tan \theta \cos^2 \theta$

6. $\dfrac{\sin^2 \theta + \cos^2 \theta}{\sin^2 \theta}$

7. $\csc^2 \theta - \cot^2 \theta$

8. $\csc^2 \gamma - \cot^2 \gamma$

9. $\cos \alpha \csc \alpha$

10. $\tan x \csc x$

11. $\dfrac{\cos x \csc x}{\tan x}$

12. $\dfrac{\tan x}{\sin x}$

13. $\sin \theta \cot \theta$

14. $\dfrac{1 - \sin^2 \alpha}{\sin^2 \alpha}$

15. $\dfrac{1 + \tan^2 x}{1 + \cot^2 x}$

16. $\dfrac{\tan \beta}{\cot \beta}$

**Solve for values of $\theta$ between 0° and 90°.**

17. If $\cot \theta = 2$, find $\tan \theta$.

18. If $\sin \theta = \dfrac{4}{5}$, find $\cos \theta$.

19. If $\cos \theta = \dfrac{2}{3}$, find $\sin \theta$.

20. If $\cos \theta = \dfrac{2}{3}$, find $\csc \theta$.

# EXERCISES

**Practice**

**Solve for values of $\theta$ between 0° and 90°.**

21. If $\cos \theta = \dfrac{4}{5}$, find $\tan \theta$.

22. If $\sin \theta = \dfrac{1}{2}$, find $\cos \theta$.

23. If $\sin \theta = \dfrac{3}{4}$, find $\sec \theta$.

24. If $\tan \theta = 4$, find $\sin \theta$.

**Solve for values of $\theta$ between 90° and 180°.**

25. If $\sin \theta = \dfrac{1}{2}$, find $\tan \theta$.

26. If $\cos \theta = -\dfrac{3}{5}$, find $\csc \theta$.

27. If $\tan \theta = -2$, find $\sec \theta$.

28. If $\sin \theta = \dfrac{3}{5}$, find $\cos \theta$.

**Solve for values of $\theta$ between 180° and 270°.**

29. If $\sec \theta = -3$, find $\tan \theta$.

30. If $\cos \theta = -\dfrac{3}{5}$, find $\csc \theta$.

31. If $\sin \theta = -\dfrac{1}{2}$, find $\cos \theta$.

32. If $\cot \theta = \dfrac{1}{4}$, find $\csc \theta$.

**Solve for values of θ between 270° and 360°.**

33. If $\tan \theta = -1$, find $\sec \theta$.

34. If $\cos \theta = \dfrac{5}{13}$, find $\sin \theta$.

35. If $\csc \theta = -\dfrac{5}{3}$, find $\cos \theta$.

36. If $\sec \theta = \dfrac{5}{3}$, find $\cos \theta$.

**Simplify each expression.**

37. $\sec^2 \theta - 1$

38. $\csc \alpha \cos \alpha \tan \alpha$

39. $\tan \beta \cot \beta$

40. $\sin x + \cos x \tan x$

41. $\dfrac{1}{\sin^2 \theta} - \dfrac{\cos^2 \theta}{\sin^2 \theta}$

42. $\sin \beta (1 + \cot^2 \beta)$

43. $2(\csc^2 \theta - \cot^2 \theta)$

44. $\dfrac{\tan^2 \theta - \sin^2 \theta}{\tan^2 \theta \sin^2 \theta}$

**Show that each equation is an identity.**

45. $1 + \cot^2 \theta = \csc^2 \theta$

46. $\sin x \sec x = \tan x$

47. $\dfrac{\sec \theta}{\csc \theta} = \tan \theta$

48. $\sec \alpha - \cos \alpha = \sin \alpha \tan \alpha$

**Critical Thinking**

49. If $\tan \beta = \dfrac{3}{4}$, find $\dfrac{\sin \beta \sec \beta}{\cot \beta}$.

**Applications**

50. **Surveying**  A surveyor found that the angle between the line from his position to a building and the line from his position to a road had a sine value of $\dfrac{5}{9}$.

   a. Find the value of the cosine of this angle. Assume that the angle is acute.

   b. Find the value of the cotangent of this angle. Assume that the angle is obtuse.

   c. Find the value of the secant of this angle. Assume that the angle is obtuse.

**Mixed Review**

51. **Entertainment**  Jeanine is writing a word puzzle for the school newspaper. She is scrambling words for the readers to unscramble to find the answer to the riddle. How many different ways could Jeanine scramble the word HOMECOMING?  **(Lesson 15-2)**

52. State the excluded values of $x$ if $\dfrac{2x + 1}{3x} - \dfrac{x - 1}{x} = \dfrac{1}{2x + 3}$. Then solve and check.  **(Lesson 11-5)**

53. Write an equation for the circle whose center is $(6, 0)$ and whose radius is 6 inches.  **(Lesson 9-3)**

54. **Horticulture**  Bill Taylor's garden has a fence along one side. He wishes to fence in the other three sides with 200 feet of fencing. Write a quadratic function to describe the area of the garden.  **(Lesson 8-1)**

55. Simplify $(8b^2 - 4b + 1) \div (2b - 1)$.  **(Lesson 5-7)**

56. State the domain and range of the relation $\{(0, 4), (8, -8), (8, 3), (0, -12), (0, 0)\}$. Then state if the relation is a function.  **(Lesson 2-1)**

# Graphing Calculator Exploration:
# Verifying Trigonometric Identities

You can use your graphing calculator to determine whether an equation is a trigonometric identity. Recall that equations that are *true* for all values of the variable for which they are defined are called *identities*.

The expressions on each side of the equals sign can be graphed as two different functions to verify identities. For example, if you were trying to verify that $\sec^2 x + \csc^2 x = \sec^2 x \csc^2 x$ is an identity, you would graph $y = \sec^2 x + \csc^2 x$ and $y = \sec^2 x \csc^2 x$. If the graphs of the two functions don't match, then the equation is not an identity. If the graphs do coincide, then the equation *may* be an identity. The equation must be verified algebraically to be sure that it is an identity.

**Example 1**

Use your graphing calculator to determine whether $\sec^2 x + \csc^2 x = \sec^2 x \csc^2 x$ may be an identity.

Graph the equations $y = \sec^2 x + \csc^2 x$ and $y = \sec^2 x \csc^2 x$. Use the viewing window $[-360, 360]$ by $[-10, 10]$ with a scale factor of 180 for the $x$-axis and 1 for the $y$-axis.

Recall that $\sec x = \dfrac{1}{\cos x}$ and $\csc x = \dfrac{1}{\sin x}$.

*Casio*

ENTER:
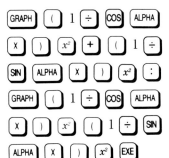

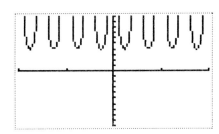

*TI-81*

ENTER:

Since the graphs of the two functions coincide, the equation *may* be an identity. *Graph the functions again in the opposite order to be sure that the graphs match.*

**Example 2**

Use a graphing calculator to determine whether $\tan \frac{x}{2} = \frac{2 \sin x}{1 + 2 \cos x}$ may be an identity.

Graph the equations $y = \tan \frac{x}{2}$ and $y = \frac{2 \sin x}{1 + 2 \cos x}$. Use the viewing window $[-180, 540]$ by $[-6, 6]$ with a scale factor of 180 for the $x$-axis and 2 for the $y$-axis.

*Casio*

ENTER: [GRAPH] [TAN] [(] [ALPHA] [X] [÷]

2 [)] [:] [GRAPH] [(] 2 [SIN]

[ALPHA] [X] [)] [÷] [(] 1 [+]

2 [COS] [ALPHA] [X] [)] [EXE]

*TI-81*

ENTER: [Y=] [TAN] [(] [X|T] [÷] 2 [)]

[ENTER] [(] 2 [SIN] [X|T] [)] [÷]

[(] 1 [+] 2 [COS] [X|T] [)]

[GRAPH]

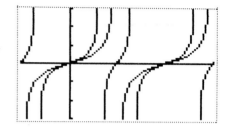

*Vertical asymptotes may appear on the viewing screen.*

The graphs of the functions do not match, so the equation is not an identity.

# EXERCISES

Use your graphing calculator to determine whether each equation *may* be an identity. Write *yes* or *no*.

**1.** $\dfrac{1 + \sin x}{\cos x} = \dfrac{\cos x}{1 - \sin x}$

**2.** $\cot x + \tan x = \csc x \cot x$

**3.** $\tan x \, (\cot x + \tan x) = \sec^2 x$

**4.** $\cos^2 x + \tan^2 x \cos^2 x = 1$

**5.** $\sin (90° - x) = \cos x$

**6.** $\cot^2 x \sec^2 x = 1 + \cot^2 x$

**7.** $\sin x + \cos x \tan x = 2 \sin x$

**8.** $\cos x \sin x \tan x = 1$

**9.** $\csc x - \cos x = \sin x \tan x$

**10.** $\dfrac{1}{\sec x} + \dfrac{1}{\csc x} = 1$

**11.** $\sin (x - 90°) = \cos x$

**12.** $\dfrac{1}{\sin^2 x} + \dfrac{1}{\cos^2 x} = 1$

**13.** $\dfrac{1 + \tan x}{1 + \cot x} = \dfrac{\sin x}{\cos x}$

**14.** $\dfrac{\sec x}{\sin x} - \dfrac{\sin x}{\cos x} = \cot x$

**15.** $\dfrac{\cos x}{\sec x - 1} + \dfrac{\cos x}{\sec x + 1} = 2 \cot^2 x$

**16.** $\dfrac{\sin x}{1 + \cos x} = \dfrac{1 - \cos x}{\sin x}$

**17.** $\dfrac{\csc^2 x}{\csc x - 1} = \dfrac{1 + \sin x}{\sin x}$

**18.** $\dfrac{\tan x}{1 + \tan x} = \dfrac{\sin x}{\sin x + \cos x}$

**19.** $\cos 3x + 1 = 2 \cos^2 x$

**20.** $\cos 2x + 2 \sin^2 x = 1$

# Verifying Trigonometric Identities

**Objective**

After studying this lesson, you should be able to:

- verify trigonometric identities using various methods.

**Application**

The formula for the height of a projected object is $h = \dfrac{v_0^2 \sin^2 \theta}{2g}$, where $\theta$ is the measure of the angle of the elevation of the path of the object, $v_0$ is the initial velocity of the

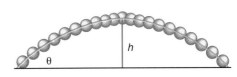

object, and $g$ is the acceleration due to gravity. Is $h = \dfrac{v_0^2 \tan^2 \theta}{2g \sec^2 \theta}$ an equivalent form for the height of the object? *You will solve this problem in Example 2.*

You can use the basic trigonometric identities and the definitions of the trigonometric functions to verify other identities. For example, suppose you wish to know if $\tan \theta(\cot \theta + \tan \theta) = \sec^2 \theta$ is an identity. It is not sufficient to try some value of $\theta$ and conclude that the statement is true for all values of $\theta$ if it is true for that one. To verify that an equation is an identity, the general case must be considered.

Verifying an identity is like checking the solution to an equation. You do not know if the expressions on each side of the equation are equal. That is what you are trying to verify. So, you must simplify one or both sides of the sentence *separately* until they are the same. Often, it is easier to work with only one side of the sentence. You may choose either side.

$\tan \theta(\cot \theta + \tan \theta) \overset{?}{=} \sec^2 \theta$    *Simplify the left side only.*

$\tan \theta \left( \dfrac{1}{\tan \theta} + \tan \theta \right) \overset{?}{=} \sec^2 \theta$    *$\cot \theta = \dfrac{1}{\tan \theta}$*

$1 + \tan^2 \theta \overset{?}{=} \sec^2 \theta$    *Distributive property*

$\sec^2 \theta = \sec^2 \theta$    *$1 + \tan^2 \theta = \sec^2 \theta$*

Thus, $\tan \theta(\cot \theta + \tan \theta) = \sec^2 \theta$ is an identity.

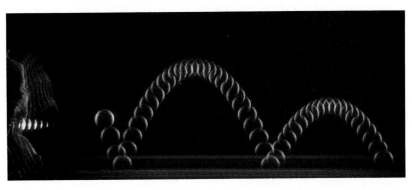

**Example 1**

Verify $\dfrac{\cos^2 x}{1 - \sin x} = 1 + \sin x$.

Notice that if the denominator, $1 - \sin x$, is multiplied by $1 + \sin x$, the result is $1 - \sin^2 x$, which equals $\cos^2 x$.

$$\dfrac{\cos^2 x}{1 - \sin x} \overset{?}{=} 1 + \sin x$$

$$\dfrac{\cos^2 x}{1 - \sin x} \cdot \dfrac{1 + \sin x}{1 + \sin x} \overset{?}{=} 1 + \sin x \qquad \textit{Multiply the numerator and denominator by 1 + sin x.}$$

$$\dfrac{\cos^2 x(1 + \sin x)}{1 - \sin^2 x} \overset{?}{=} 1 + \sin x \qquad \textit{Simplify.}$$

$$\dfrac{\cos^2 x(1 + \sin x)}{\cos^2 x} \overset{?}{=} 1 + \sin x \qquad \textit{cos}^2\textit{ x = 1 − sin}^2\textit{ x}$$

$$1 + \sin x = 1 + \sin x$$

The identity could also be verified by working from the right side. The first step would be to multiply $1 + \sin x$ by $\dfrac{1 - \sin x}{1 - \sin x}$.

---

**Example 2**

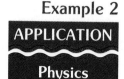

**APPLICATION**

**Physics**

The formula for the height of a projected object is $h = \dfrac{v_o^2 \sin^2 \theta}{2g}$.

Is $h = \dfrac{v_o^2 \tan^2 \theta}{2g \sec^2 \theta}$ an equivalent formula for the height of the object?

$$\dfrac{v_o^2 \sin^2 \theta}{2g} \overset{?}{=} \dfrac{v_o^2 \tan^2 \theta}{2g \sec^2 \theta} \qquad \textit{Simplify the right side.}$$

$$\dfrac{v_o^2 \sin^2 \theta}{2g} \overset{?}{=} \dfrac{v_o^2 \left(\dfrac{\sin^2 \theta}{\cos^2 \theta}\right)}{2g\left(\dfrac{1}{\cos^2 \theta}\right)} \qquad \textit{tan}^2\textit{ } \theta = \dfrac{\textit{sin}^2\textit{ } \theta}{\textit{cos}^2\textit{ } \theta}, \textit{ sec}^2\textit{ } \theta = \dfrac{1}{\textit{cos}^2\textit{ } \theta}$$

$$\dfrac{v_o^2 \sin^2 \theta}{2g} \overset{?}{=} \dfrac{\dfrac{v_o^2 \sin^2 \theta}{\cos^2 \theta}}{\dfrac{2g}{\cos^2 \theta}} \qquad \textit{Simplify.}$$

$$\dfrac{v_o^2 \sin^2 \theta}{2g} = \dfrac{v_o^2 \sin^2 \theta}{2g} \qquad \textit{Multiply the numerator and denominator of the right side by cos}^2\textit{ } \theta.$$

So, the formulas are equivalent.

---

The following suggestions may be helpful as you verify trigonometric identities. Study the examples to see how these suggestions can be used to verify an identity.

- Start with the more complicated side of the equation. Transform the expression into the form of the simpler side.

    or

    Work with each side of the equation at the same time. Transform each expression separately into the same form.

- Substitute one or more basic trigonometric identities to simplify the expression.
- Try factoring or multiplying to simplify the expression.
- Multiply both the numerator and the denominator by the same trigonometric expression.

**Example 3**

Verify that $1 - \cot^4 \beta = 2 \csc^2 \beta - \csc^4 \beta$.

$$1 - \cot^4 \beta \stackrel{?}{=} 2 \csc^2 \beta - \csc^4 \beta$$
$$(1 - \cot^2 \beta)(1 + \cot^2 \beta) \stackrel{?}{=} \csc^2 \beta \, (2 - \csc^2 \beta) \qquad \textit{Factor each side.}$$
$$[1 - (\csc^2 \beta - 1)][\csc^2 \beta] \stackrel{?}{=} (2 - \csc^2 \beta)(\csc^2 \beta) \qquad \textit{1 + cot}^2 \, \beta = \csc^2 \beta$$
$$(2 - \csc^2 \beta)(\csc^2 \beta) = (2 - \csc^2 \beta)(\csc^2 \beta) \qquad \textit{Simplify.}$$

Thus, the identity is verified.

# CHECKING FOR UNDERSTANDING

**Communicating Mathematics**

Read and study the lesson to answer these questions.

1. Describe the different methods you can use to verify trigonometric identities.

2. Verify the identity in Example 2 by simplifying each side separately into the same form.

**Guided Practice**

Verify that each of the following is an identity.

3. $\sin \theta \sec \theta \cot \theta = 1$

4. $\tan^2 x \cos^2 x = 1 - \cos^2 x$

5. $\csc y \sec y = \cot y + \tan y$

6. $\tan \alpha \sin \alpha \cos \alpha \csc^2 \alpha = 1$

# EXERCISES

**Practice**

Verify that each of the following is an identity.

7. $\sec^2 x - \tan^2 x = \tan x \cot x$

8. $\dfrac{1}{\sec^2 \theta} + \dfrac{1}{\csc^2 \theta} = 1$

9. $\dfrac{\sec \alpha}{\sin \alpha} - \dfrac{\sin \alpha}{\cos \alpha} = \cot \alpha$

10. $\tan^2 \theta - \sin^2 \theta = \tan^2 \theta \sin^2 \theta$

11. $\dfrac{\sin \alpha}{1 - \cos \alpha} + \dfrac{1 - \cos \alpha}{\sin \alpha} = 2 \csc \alpha$

12. $\dfrac{\sec \beta + \csc \beta}{1 + \tan \beta} = \csc \beta$

13. $\dfrac{1 - \cos x}{\sin x} = \dfrac{\sin x}{1 + \cos x}$

14. $\dfrac{\sin \theta}{\sec \theta} = \dfrac{1}{\tan \theta + \cot \theta}$

15. $\dfrac{1 - \cos x}{1 + \cos x} = (\csc x - \cot x)^2$

16. $\dfrac{\sec \theta + 1}{\tan \theta} = \dfrac{\tan \theta}{\sec \theta - 1}$

17. $\dfrac{\cot \theta + \csc \theta}{\sin \theta + \tan \theta} = \cot \theta \csc \theta$

18. $\cos^2 x + \tan^2 x \cos^2 x = 1$

19. $\dfrac{1 - 2 \cos^2 \beta}{\sin \beta \cos \beta} = \tan \beta - \cot \beta$

20. $\dfrac{1 + \tan^2 \theta}{\csc^2 \theta} = \tan^2 \theta$

**Verify that each of the following is an identity.**

21. $\dfrac{\cos y}{1 + \sin y} + \dfrac{\cos y}{1 - \sin y} = 2 \sec y$

22. $\dfrac{1 + \sin x}{\sin x} = \dfrac{\cot^2 x}{\csc x - 1}$

23. $\cot x(\cot x + \tan x) = \csc^2 x$

24. $\cos^4 \theta - \sin^4 \theta = \cos^2 \theta - \sin^2 \theta$

25. $\dfrac{1 + \tan \gamma}{1 + \cot \gamma} = \dfrac{\sin \gamma}{\cos \gamma}$

26. $\dfrac{\tan^2 x}{\sec x - 1} = 1 + \dfrac{1}{\cos x}$

27. $1 + \sec^2 x \sin^2 x = \sec^2 x$

28. $\sin \theta + \cos \theta = \dfrac{1 + \tan \theta}{\sec \theta}$

**Critical Thinking**

29. Create a trigonometric identity. Explain the method you used to do this. Then trade with another student and verify each other's identities.

**Applications**

30. **Optics**  The illumination, $E$, in footcandles on a surface that is $R$ feet from a source of light with intensity $I$ candelas is $E = \dfrac{I \cos \theta}{R^2}$, where $\theta$ is the measure of the angle between the direction of the light and a line perpendicular to the surface being illuminated.

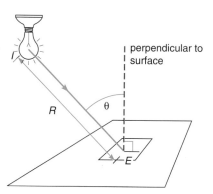

a. Show that the formula $E = \dfrac{I \cot \theta}{R^2 \csc \theta}$ is equivalent to the one given above.

b. Find the illumination on a surface that is 10 feet from a source of light that emits 12 candelas. Angle $\theta$ measures $0°$.

c. An illumination of about 20 footcandles is recommended for reading. How far away should a 75-watt light of intensity 90 candelas be placed if angle $\theta$ measures $30°$?

**Mixed Review**

31. **Probability**  Leshia wins 3 out of every 4 games of chess that she plays. If she played 7 games with Edward, what is the probability that Leshia won all of the games?  **(Lesson 15-9)**

32. **Statistics**  The cost per cup of several different types of juices are given below. Make a stem-and-leaf plot of the costs.  **(Lesson 14-2)**

11¢   9¢   6¢   9¢   12¢   4¢   4¢   8¢   6¢   4¢   16¢
12¢   11¢   19¢   7¢   19¢   7¢   19¢   4¢   6¢   5¢   6¢

33. Write an expression for $\log_5 82$ in terms of common logarithms. Then approximate the logarithm to three decimal places.  **(Lesson 12-7)**

34. Find the sum and the product of the roots of the quadratic equation $2x^2 + 10x + 8 = 0$. Then solve the equation. **(Lesson 7-5)**

35. Find the transpose and the inverse of $\begin{bmatrix} -10 & 3 \\ 0 & -2 \end{bmatrix}$. **(Lesson 4-4)**

# Problem-Solving Strategy: Working Backwards

**Objective**

After studying this lesson, you should be able to:

- solve problems using the strategy of working backwards.

Most problems are given to you with a set of conditions. Then you must find a solution. However, in some cases, like verifying a trigonometric identity, it is faster to determine how the problem ends and then work backwards rather than start from the beginning to find the solution.

**Example 1**

**Find the sum of the reciprocals of two numbers whose sum is 2 and whose product is 3.**

Let $x$ and $y$ be the numbers.
We could set up these equations.

$$x + y = 2$$
$$xy = 3$$

But, solving the system of equations is complicated. Rather than using this approach, work backwards. The desired outcome is $\frac{1}{x} + \frac{1}{y}$.

$$\frac{1}{x} + \frac{1}{y} = \frac{y}{xy} + \frac{x}{xy} \qquad \textit{The LCD is xy.}$$
$$= \frac{x + y}{xy}$$

Looking back to our two original equations, we can see that $x + y = 2$ and $xy = 3$. So, $\frac{x + y}{xy} = \frac{2}{3}$.

The sum of the reciprocals is $\frac{2}{3}$.

**Example 2**

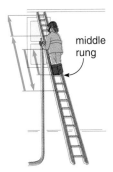

**A fire fighter spraying water on a fire stood on the middle rung of a ladder. The smoke lessened, so she moved up 3 rungs. It got too hot, so she backed down 5 rungs. Later, she went up 7 rungs and stayed until the fire was out. Then, she climbed the remaining 4 rungs and went into the building. How many rungs does the ladder have?**

We know that the fire fighter was on the middle rung in the beginning.

She climbed up 3, backed up 5, went up 7, then up 4 more. So she went up $3 + (-5) + 7 + 4$ or 9 rungs.

middle rung

Since she had to climb 9 rungs to enter the building and she was on the middle rung, there are 19 rungs on the ladder.

# CHECKING FOR UNDERSTANDING

**Communicating Mathematics**

Read and study the lesson to answer these questions.

1. When can you use the strategy of working backwards?

2. How many rungs would the ladder used by the fire fighter in Example 2 have had if she had been standing three-fourths of the way up the ladder?

3. Explain how the strategy of working backwards can be used when you are verifying a trigonometric identity.

**Guided Practice**

Solve each problem by working backwards.

4. If the sum of two numbers is 4 and the product of the numbers is 7, find the sum of the squares of the reciprocals of these numbers.

5. A pirate found a treasure chest containing silver coins. He buried half of them and gave half of the remaining coins to his mother. If he was left with 4550 coins, how many were in the chest when he found it?

6. After cashing her paycheck, Caroline paid her father back the $15 she had borrowed. She then spent half of the remaining money on clothes, and then spent half of what remained on a concert ticket. She bought a cassette tape for $7.45 and had $10.25 left. What was the amount of Caroline's paycheck?

**Exercises**

Use any strategy.

**Strategies**

Look for a pattern.
Solve a simpler problem.
Act it out.
Guess and check.
Draw a diagram.
Make a chart.
Work backwards.

7. Juana collects postcards from the places that she has visited. If she counts the postcards by twos, threes, fours, fives, or sixes, there is always one left over. If she counts the postcards by sevens, none remain. What is the fewest number of postcards Juana could have?

8. Find the value of $\sqrt{11 + \sqrt{72}} + \sqrt{11 - \sqrt{72}}$.

9. Kevin and Madeline are starting to play a game. They each roll one die to determine who will go first. What is the most likely product of the numbers on the dice? Justify your answer.

10. How many positive integers less than 500 have an odd number of positive integral factors?

11. If the sum of two numbers is 2 and the product of the same two numbers is 3, find the sum of the squares of the reciprocals of these numbers.

**12.** A customer bought a magazine from a bookstore for $3.50 and paid the cashier with a $10 bill. Since the cash register did not contain enough small bills to give the customer change, the bookstore manager went next door to the record store to get change. After the customer left, the record store manager came in to say that the $10 bill is counterfeit. The bookstore manager apologized and exchanged the counterfeit bill for a good one. How much money did the bookstore lose?

**13.** Mathematically combine six 5s so that 200 results.

**14.** Solve for $x$ if $(\log x)^2 = \log (x^2)$.

**15.** The fifth power of a two-digit number, $x$, ends in $x$. Find $x$.

**16.** What fraction of the area of rectangle $WXYZ$ is shaded?

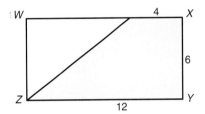

**17.** Tim used to collect model cars. He decides to give them away. First, he gave half of them plus half a car to his younger brother, Brad. Then he gives half of what is left plus half a car more to his friend Tammy. Tim has one car left, which he gives to Aaron. How many cars did Tim start with? (Assume that no car is cut in half.)

## COOPERATIVE LEARNING ACTIVITY

**Work in groups. Each person in the group must understand the solution and be able to explain it to any person in class.**

Arrange the numbers 1, 2, 2, 3, 4, 6, 6, and 12 in place of the letters $a$, $b$, $c$, $d$, $e$, $f$, $g$, and $h$ so that $abc = cde = efg = gha$.

| a | b | c |
|---|---|---|
| h |   | d |
| g | f | e |

# Sum and Difference of Angles Formulas

**Objective**

After studying this lesson, you should be able to:

- find values of sine and cosine involving sum and difference formulas, and
- verify identities using the sum and difference formulas.

**Application**

A geologist surveys a rectangular piece of land to determine whether it is suitable for development. He stands at one corner of the piece of property and measures the angle between one side of the lot and the line from his position to the opposite corner of the lot as 30°. He then measures the angle between that line and the line to the point in the property where a river crosses the property line as 45°. If the geologist stands 100 yards from the opposite corner of the property, how far is he from the point where the river crosses the property line? *You will solve this problem in Example 5.*

It is often helpful to use formulas for the trigonometric values of the difference or sum of two angles. For example, you could find sin 15° by evaluating sin (45 − 30)°.

The figure at the right shows two angles, α and β, in standard position on the unit circle.

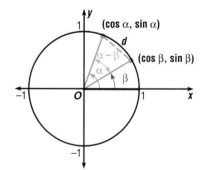

Use the distance formula to find $d$, where $(x_1, y_1) = (\cos \alpha, \sin \alpha)$ and $(x_2, y_2) = (\cos \beta, \sin \beta)$.

$$d = \sqrt{(\cos \alpha - \cos \beta)^2 + (\sin \alpha - \sin \beta)^2}$$

$$d^2 = (\cos \alpha - \cos \beta)^2 + (\sin \alpha - \sin \beta)^2$$

$$d^2 = (\cos^2 \alpha - 2 \cos \alpha \cos \beta + \cos^2 \beta) + (\sin^2 \alpha - 2 \sin \alpha \sin \beta + \sin^2 \beta)$$

$$d^2 = \cos^2 \alpha + \sin^2 \alpha + \cos^2 \beta + \sin^2 \beta - 2 \cos \alpha \cos \beta - 2 \sin \alpha \sin \beta$$

$$d^2 = 1 + 1 - 2 \cos \alpha \cos \beta - 2 \sin \alpha \sin \beta \qquad \text{\textit{sin}}^2 \alpha + \cos^2 \alpha = 1 \text{ and}$$
$$\qquad\qquad\qquad\qquad\qquad\qquad\qquad\qquad\qquad\quad sin^2 \beta + cos^2 \beta = 1$$

$$d^2 = 2 - 2 \cos \alpha \cos \beta - 2 \sin \alpha \sin \beta$$

Now find the value of $d^2$ when the angle having measure $\alpha - \beta$ is in standard position on the unit circle, as shown in the figure below.

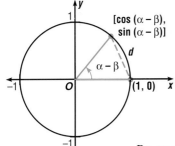

$$d = \sqrt{[\cos (\alpha - \beta) - 1]^2 + [\sin (\alpha - \beta) - 0]^2}$$
$$\begin{aligned} d^2 &= [\cos (\alpha - \beta) - 1]^2 + [\sin (\alpha - \beta) - 0]^2 \\ &= [\cos^2 (\alpha - \beta) - 2 \cos (\alpha - \beta) + 1] + \sin^2 (\alpha - \beta) \\ &= \cos^2 (\alpha - \beta) + \sin^2 (\alpha - \beta) - 2 \cos (\alpha - \beta) + 1 \\ &= \qquad\qquad 1 \qquad\qquad - 2 \cos (\alpha - \beta) + 1 \\ &= 2 - 2 \cos (\alpha - \beta) \end{aligned}$$

By equating the two expressions for $d^2$, it is possible to find a formula for $\cos (\alpha - \beta)$.

$$\begin{aligned} d^2 &= d^2 \\ 2 - 2 \cos (\alpha - \beta) &= 2 - 2 \cos \alpha \cos \beta - 2 \sin \alpha \sin \beta \\ -1 + \cos (\alpha - \beta) &= -1 + \cos \alpha \cos \beta + \sin \alpha \sin \beta \qquad \textit{Divide each side by } -2. \\ \cos (\alpha - \beta) &= \cos \alpha \cos \beta + \sin \alpha \sin \beta \qquad \textit{Add 1 to each side.} \end{aligned}$$

Use the formula for $\cos (\alpha - \beta)$ to find a formula for $\cos (\alpha + \beta)$.

$$\begin{aligned} \cos (\alpha + \beta) &= \cos [\alpha - (-\beta)] \\ &= \cos \alpha \cos (-\beta) + \sin \alpha \sin (-\beta) \\ &= \cos \alpha \cos \beta - \sin \alpha \sin \beta \qquad \textit{cos } (-\beta) = \cos \beta; \sin (-\beta) = -\sin \beta \end{aligned}$$

**Example 1**

Use the formula for $\cos (\alpha - \beta)$ to find $\cos (90° - \theta)$.

$$\begin{aligned} \cos (90° - \theta) &= \cos 90° \cos \theta + \sin 90° \sin \theta \\ &= 0 \cdot \cos \theta + 1 \cdot \sin \theta \\ &= \sin \theta \end{aligned}$$

**Example 2**

Use the formula for $\cos (90° - \theta)$ from Example 1 to find $\sin (90° - \gamma)$.

$$\begin{aligned} \sin (90° - \gamma) &= \cos [90° - (90° - \gamma)] \qquad \textit{Substitute } (90° - \gamma) \textit{ for } \theta \\ &= \cos (90° - 90° + \gamma) \qquad\quad \textit{in } \sin \theta = \cos (90° - \theta). \\ &= \cos \gamma \end{aligned}$$

**Example 3**

Use the formulas for $\cos (90° - \theta)$ and $\sin (90° - \theta)$ from Examples 1 and 2 and the formula for $\cos (\alpha - \beta)$ to find $\sin (\alpha - \beta)$.

$$\begin{aligned} \sin (\alpha - \beta) &= \cos [90° - (\alpha - \beta)] \quad \textit{Substitute } (\alpha - \beta) \textit{ for } \theta \textit{ in } \sin \theta = \cos (90° - \theta). \\ &= \cos [(90° - \alpha) + \beta)] \qquad \textit{Distributive and associative properties} \\ &= \cos (90° - \alpha) \cos \beta - \sin (90° - \alpha) \sin \beta \\ &= \sin \alpha \cos \beta - \cos \alpha \sin \beta \end{aligned}$$

We can state the sum and difference identities as follows.

| *Sum and Difference of Angles Formulas* | **The following identities hold true for all values of $\alpha$ and $\beta$.**<br>$\cos (\alpha \pm \beta) = \cos \alpha \cos \beta \mp \sin \alpha \sin \beta$<br>$\sin (\alpha \pm \beta) = \sin \alpha \cos \beta \pm \cos \alpha \sin \beta$ |
| --- | --- |

The following examples show how to evaluate expressions using the sum and difference formulas.

### Example 4

Evaluate $\cos 105°$.

$\cos 105° = \cos (60° + 45°)$
$= \cos 60° \cos 45° - \sin 60° \sin 45°$
$= \dfrac{1}{2} \cdot \dfrac{\sqrt{2}}{2} - \dfrac{\sqrt{3}}{2} \cdot \dfrac{\sqrt{2}}{2}$ or $\dfrac{\sqrt{2} - \sqrt{6}}{4}$

$\cos (\alpha + \beta) =$
$\cos \alpha \cos \beta - \sin \alpha \sin \beta$
$\dfrac{\sqrt{2} - \sqrt{6}}{4} \approx -0.2588$

### Example 5

**APPLICATION**

**Geology**

As shown in the diagram below, a geologist measures the angle between one side of a rectangular lot and the line from his position to the opposite corner of the lot as 30°. He then measures the angle between that line and the line to the point in the property where a river crosses the property line as 45°. If the geologist stands 100 yards from the opposite corner of the property, how far is he from the point where the river crosses the property line?

First find $x$.

$\sin 30° = \dfrac{x}{100}$

$100\left(\dfrac{1}{2}\right) = x$

$50 = x$

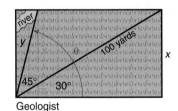

Geologist

Now, find $\sin \theta$ expressed as $\sin (30° + 45°)$.

$\sin (30° + 45°) = \sin 30° \cos 45° + \cos 30° \sin 45°$
$= \dfrac{1}{2} \cdot \dfrac{\sqrt{2}}{2} + \dfrac{\sqrt{3}}{2} \cdot \dfrac{\sqrt{2}}{2}$ or $\dfrac{\sqrt{2} + \sqrt{6}}{4}$

$\sin \theta = \dfrac{50}{y}$

$y = \dfrac{50}{\dfrac{\sqrt{2} + \sqrt{6}}{4}}$ *Substitute* $\dfrac{\sqrt{2} + \sqrt{6}}{4}$ *for* $\sin \theta$.

$y = \dfrac{50}{\dfrac{\sqrt{2} + \sqrt{6}}{4}} \cdot \dfrac{\sqrt{2} - \sqrt{6}}{\sqrt{2} - \sqrt{6}}$

$y = \dfrac{50 (\sqrt{2} - \sqrt{6})}{\dfrac{2 - 6}{4}} = 50\sqrt{6} - 50\sqrt{2}$ or about 51.76

The geologist stands approximately 51.76 yards from the point where the river crosses the property line.

# CHECKING FOR UNDERSTANDING

**Communicating Mathematics**

Read and study the lesson to answer these questions.

1. Is it true that $\sin (x + y) = \sin x + \sin y$? Justify your answer.

2. Describe a method to find the exact value for $\cos 15°$. Then find the value.

**Guided Practice**

Express each angle measurement in terms of sums or differences of 30°, 45°, 60°, and 90° or their multiples.

3. $-15°$      4. $165°$      5. $75°$

6. $285°$      7. $255°$      8. $345°$

# EXERCISES

**Practice**

Evaluate each expression.

9. $\sin 75°$      10. $\sin 285°$      11. $\cos 75°$

12. $\sin 165°$      13. $\cos 195°$      14. $\sin 105°$

15. $\cos 255°$      16. $\cos 345°$      17. $\cos 165°$

18. $\sin 40° \cos 20° + \cos 40° \sin 20°$      19. $\sin 65° \cos 35° - \cos 65° \sin 35°$

20. $\cos 25° \cos 5° - \sin 25° \sin 5°$      21. $\cos 80° \cos 20° + \sin 80° \sin 20°$

Verify that each of the following is an identity.

22. $\cos (270° - \theta) = -\sin \theta$      23. $\sin (270° - \theta) = -\cos \theta$

24. $\sin (180° + \theta) = -\sin \theta$      25. $\sin (90° + \theta) = \cos \theta$

26. $\cos (180° + \theta) = -\cos \theta$      27. $\cos (90° + \theta) = -\sin \theta$

28. $\sin \left(\theta + \dfrac{\pi}{3}\right) - \cos \left(\theta + \dfrac{\pi}{6}\right) = \sin \theta$

29. $\sin (x + y) \sin (x - y) = \sin^2 x - \sin^2 y$

30. $\sin (60° + \theta) + \sin (60° - \theta) = \sqrt{3} \cos \theta$

31. $\sin (x + 30°) + \cos (x + 60°) = \cos x$

Use the identity $\tan (\alpha - \beta) = \dfrac{\tan \alpha - \tan \beta}{1 + \tan \alpha \tan \beta}$ to evaluate each expression.

32. $\tan (315° - 120°)$      33. $\tan (225° - 120°)$

34. $\tan (30° + 30°)$      35. $\tan 195°$

**Critical Thinking**

36. Use the formulas for $\sin (\alpha + \beta)$ and $\cos (\alpha + \beta)$ to derive the formula for $\tan (\alpha + \beta)$. *Hint: Divide all terms of the expression by $\cos \alpha \cos \beta$.*

**37. Geology**  A geologist stands on a ledge and finds that the angle of depression to a river's surface is 12°. The angle of depression to the river bed below the surface is 13°. The geologist is 1500 feet from the river's surface.

    **a.** Write an expression for the sine of the angle between the line from the geologist to the river's surface and the line from the geologist to the river bed.

    **b.** How far above the river bed is the surface of the water?

**Mixed Review**

**38.** Find the least positive angle measurement that is coterminal to $-120°$. **(Lesson 16-1)**

**39.** Find the sum of the geometric series described by $a_1 = 2$, $a_9 = 512$, $n = 10$. **(Lesson 13-5)**

**40.** Find $\dfrac{2x^6y^3}{a^3b^7} \div \dfrac{4x^2y}{ab^3}$. **(Lesson 11-1)**

**41.** If $f(x) = x^2 + 8$ and $g(x) = x - 3$, find $[f \circ g](3)$ and $[g \circ f](3)$. **(Lesson 10-7)**

**42.** Write the equation of a parabola with position 3 units to the right of the parabola with equation $f(x) = x^2$. **(Lesson 8-3)**

**43.** State the property shown by the statement "If $7 = n$, then $n = 7$." **(Lesson 1-3)**

## MID-CHAPTER REVIEW

**State the amplitude and period of each function.  (Lesson 17-1)**

**1.** $y = \sin 4\theta$        **2.** $y = 5 \cos \theta$        **3.** $y = 4 \cos \dfrac{3}{4}\theta$

**State the period of each function.  (Lesson 17-1)**

**4.** $y = \csc 4\theta$        **5.** $y = 4 \tan \theta$        **6.** $y = \cot \dfrac{1}{5}\theta$

**Solve for values between 90° and 180°.  (Lesson 17-2)**

**7.** If $\cot \theta = -\dfrac{3}{5}$, find $\csc \theta$.        **8.** If $\tan \alpha = -\dfrac{3}{5}$, find $\cos \alpha$.

**Verify that each of the following is an identity.  (Lesson 17-3)**

**9.** $\cot \beta = \cos \beta \csc \beta$        **10.** $1 = \dfrac{1}{\sec^2 \alpha - \sec^2 \alpha \sin^2 \alpha}$

**11. Finance**  Martin is selling tickets to a fundraiser. He sold 4 to family members, then gave half of what he had left to his sister for her to sell. He sold twelve more to coworkers and had 15 tickets left to sell. How many tickets did Martin have in all?  **(Lesson 17-4)**

**Evaluate each expression.  (Lesson 17-5)**

**12.** $\sin 195°$        **13.** $\cos 285°$        **14.** $\sin -75°$

# Double-Angle and Half-Angle Formulas

**Objectives**

After studying this lesson, you should be able to:

- find values of sine and cosine involving half and double angles, and
- verify identities using half and double angle formulas.

You can use the formula for $\sin(\alpha + \beta)$ to find the sine of twice an angle $\theta$ and the formula for $\cos(\alpha + \beta)$ to find the cosine of twice an angle $\theta$. Let $\theta$ represent the measure of the angle.

$$\sin 2\theta = \sin(\theta + \theta) \qquad\qquad \cos 2\theta = \cos(\theta + \theta)$$
$$= \sin\theta\cos\theta + \cos\theta\sin\theta \qquad = \cos\theta\cos\theta - \sin\theta\sin\theta$$
$$= 2\sin\theta\cos\theta \qquad\qquad\qquad = \cos^2\theta - \sin^2\theta$$

You can find alternate forms for $\cos 2\theta$ by making substitutions into the expression $\cos^2\theta - \sin^2\theta$.

$$\cos^2\theta - \sin^2\theta = (1 - \sin^2\theta) - \sin^2\theta$$
$$= 1 - 2\sin^2\theta \qquad \textit{Substitute } 1 - \sin^2\theta \textit{ for } \cos^2\theta.$$
$$\cos^2\theta - \sin^2\theta = \cos^2\theta - (1 - \cos^2\theta)$$
$$= 2\cos^2\theta - 1 \qquad \textit{Substitute } 1 - \cos^2\theta \textit{ for } \sin^2\theta.$$

These formulas are called the **double-angle formulas.**

*Double-Angle Formulas*

| The following identities hold true for all values of $\theta$. |
| --- |
| $\sin 2\theta = 2\sin\theta\cos\theta$ $\qquad$ $\cos 2\theta = \cos^2\theta - \sin^2\theta$ |
| $\cos 2\theta = 1 - 2\sin^2\theta$ |
| $\cos 2\theta = 2\cos^2\theta - 1$ |

**Example 1**

Suppose $x$ is between $90°$ and $180°$ and $\sin x = \frac{1}{2}$. Find $\cos 2x$.

Use the identity $\cos 2x = 1 - 2\sin^2 x$.

$$\cos 2x = 1 - 2\sin^2 x$$
$$= 1 - 2\left(\frac{1}{2}\right)^2 \qquad \textit{Substitute } \frac{1}{2} \textit{ for } \sin x.$$
$$= \frac{1}{2}$$

The value of $\cos 2x$ is $\frac{1}{2}$.

**Example 2**

The range of a projected object is the distance that it travels from the point where it is released. In the absence of air resistance, a projectile released at an angle of elevation, θ, with an initial velocity of $v_o$ has a range of

$R = \dfrac{v_o^2}{g} \sin 2\theta$, where $g$ is the acceleration due to gravity. Find the range of a projectile with an initial velocity of 88 feet per second if $\sin \theta = \dfrac{3}{5}$ and $\cos \theta = \dfrac{4}{5}$. The acceleration due to gravity is 32 feet per second squared.

Use the double-angle formula to find sin 2θ.

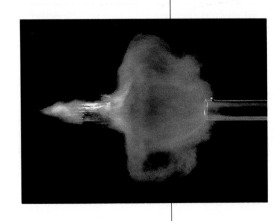

$$\sin 2\theta = 2 \sin \theta \cos \theta \qquad \textit{Double-angle formula}$$

$$= 2\left(\frac{3}{5}\right)\left(\frac{4}{5}\right) \qquad \textit{Substitute } \tfrac{3}{5} \textit{ for sin θ and } \tfrac{4}{5} \textit{ for cos θ.}$$

$$= \frac{24}{25}$$

Now find the range.

$$R = \frac{v_o^2}{g} \sin 2\theta \qquad \textit{Use the formula for range.}$$

$$= \frac{(88)^2}{32}\left(\frac{24}{25}\right) \qquad \textit{Substitute 88 for } v_o, \textit{ 32 for g, and } \tfrac{24}{25} \textit{ for sin 2θ.}$$

$$= 232.32$$

The range of this projectile is 232.32 feet.

---

You can also derive formulas to find the cosine and sine of half a given angle. Let α represent the measure of this angle.

Find $\cos \dfrac{\alpha}{2}$.

$$2 \cos^2 \theta - 1 = \cos 2\theta$$

$$2 \cos^2 \frac{\alpha}{2} - 1 = \cos \alpha$$

$$\cos^2 \frac{\alpha}{2} = \frac{1 + \cos \alpha}{2}$$

$$\cos \frac{\alpha}{2} = \pm\sqrt{\frac{1 + \cos \alpha}{2}}$$

Find $\sin \dfrac{\alpha}{2}$.

$$1 - 2 \sin^2 \theta = \cos 2\theta$$

$$1 - 2 \sin^2 \frac{\alpha}{2} = \cos \alpha$$

$$\sin^2 \frac{\alpha}{2} = \frac{1 - \cos \alpha}{2}$$

$$\sin \frac{\alpha}{2} = \pm\sqrt{\frac{1 - \cos \alpha}{2}}$$

*Use double-angle formulas.*

*Substitute α for 2θ and $\frac{\alpha}{2}$ for θ.*

*Solve for the squared term.*

*Take the square root of each side.*

These are called the **half-angle** formulas.

| *Half-Angle Formulas* | **The following identities hold true for all values of $\alpha$.** |
| --- | --- |
| | $$\cos \frac{\alpha}{2} = \pm\sqrt{\frac{1 + \cos \alpha}{2}} \qquad \sin \frac{\alpha}{2} = \pm\sqrt{\frac{1 - \cos \alpha}{2}}$$ |

**Example 3**

Suppose $\sin x = -\dfrac{9}{41}$ and $x$ is in the fourth quadrant. Find $\cos \dfrac{x}{2}$.

Since $\cos \dfrac{x}{2} = \pm\sqrt{\dfrac{1 + \cos x}{2}}$, we must find $\cos x$ first.

Use $\cos^2 x + \sin^2 x = 1$.

$$\cos^2 x + \sin^2 x = 1$$

$$\cos^2 x + \left(-\frac{9}{41}\right)^2 = 1 \qquad \textit{Substitute } -\frac{9}{41} \textit{ for sin x.}$$

$$\cos^2 x = 1 - \frac{81}{1681}$$

$$\cos^2 x = \frac{1600}{1681}$$

$$\cos x = \pm\frac{40}{41}$$

Since $x$ is in the fourth quadrant, $\cos x = \dfrac{40}{41}$.

$$\cos \frac{x}{2} = \pm\sqrt{\frac{1 + \cos x}{2}} \qquad \textit{Half-angle formula}$$

$$= \pm\sqrt{\frac{1 + \dfrac{40}{41}}{2}} \qquad \textit{Substitute } \frac{40}{41} \textit{ for cos x.}$$

$$= \pm\sqrt{\frac{81}{82}} \text{ or } \pm\frac{9\sqrt{82}}{82}$$

Since $x$ is in the fourth quadrant, $x$ is between $270°$ and $360°$. Thus, $\dfrac{x}{2}$ is between $135°$ and $180°$, and $\cos \dfrac{x}{2}$ is negative. The solution is $-\sqrt{\dfrac{81}{82}}$ or $-\dfrac{9\sqrt{82}}{82}$ or about $-0.9939$.

# CHECKING FOR UNDERSTANDING

**Communicating Mathematics**

**Read and study the lesson to answer these questions.**

1. Use the information from Example 1 to find the value of $\sin 2x$ if $x$ is between $90°$ and $180°$ and $\sin x = \dfrac{1}{2}$.

2. Explain how to find $\sin x$ if $2x$ is in the third quadrant.

3. Use the information from Example 3 to find $\sin \dfrac{x}{2}$ if $\sin x = -\dfrac{9}{41}$ and $x$ is in the fourth quadrant.

4. If $x$ is in the first quadrant, in which quadrant does the terminal side for $2x$ lie?

5. $x$ is a second quadrant angle. In which quadrant does the terminal side for $2x$ lie?

6. If $x$ is a fourth quadrant angle, in which quadrant does the terminal side for $2x$ lie?

7. $x$ is a third quadrant angle. In which quadrant does the terminal side for $2x$ lie?

8. $x$ is a first quadrant angle. In which quadrant does the terminal side for $\frac{x}{2}$ lie?

9. If $x$ is an angle whose terminal side lies in the second quadrant, in which quadrant does the terminal side for $\frac{x}{2}$ lie?

Find $\sin 2x$, $\cos 2x$, $\sin \frac{x}{2}$, and $\cos \frac{x}{2}$ for each of the following.

10. $\sin x = \frac{5}{13}$, $x$ is in the second quadrant

11. $\cos x = \frac{1}{5}$, $x$ is in the fourth quadrant

# EXERCISES

Find $\sin 2x$, $\cos 2x$, $\sin \frac{x}{2}$, and $\cos \frac{x}{2}$ for each of the following, given the quadrant in which the terminal side of $x$ lies.

12. $\cos x = \frac{3}{5}$, first quadrant

13. $\sin x = \frac{4}{5}$, second quadrant

14. $\cos x = -\frac{2}{3}$, third quadrant

15. $\cos x = -\frac{1}{3}$, third quadrant

16. $\sin x = -\frac{3}{4}$, fourth quadrant

17. $\sin x = -\frac{3}{5}$, third quadrant

18. $\sin x = -\frac{1}{4}$, third quadrant

19. $\cos x = -\frac{1}{3}$, second quadrant

Evaluate each expression using the half-angle formulas.

20. $\sin 105°$

21. $\cos \frac{\pi}{8}$

22. $\sin 22\frac{1}{2}°$

23. $\sin 195°$

24. $\cos \frac{19\pi}{12}$

25. $\sin \frac{7\pi}{8}$

Verify that each of the following is an identity.

26. $(\sin x + \cos x)^2 = 1 + \sin 2x$

27. $\cos^2 2x + 4 \sin^2 x \cos^2 x = 1$

28. $\sin 2x = 2 \cot x \sin^2 x$

29. $\sin^2 \theta = \frac{1}{2}(1 - \cos 2\theta)$

The formula for the product of complex numbers can be used to find the square of a complex number.

$$[r(\cos \theta + i \sin \theta)]^2 = [r(\cos \theta + i \sin \theta)] \cdot [r(\cos \theta + i \sin \theta)]$$
$$= r^2[\cos (\theta + \theta) + i \sin (\theta + \theta)]$$
$$= r^2(\cos 2\theta + i \sin 2\theta)$$

The formula for the product of complex numbers can be used to find any power of complex numbers. This is known as DeMoivre's Theorem.

*DeMoivre is pronounced D'Mwov.*

| DeMoivre's Theorem | $[r(\cos \theta + i \sin \theta)]^n = r^n(\cos n\theta + i \sin n\theta)$ |
|---|---|

**Example 4**

Find $(2\sqrt{3} + 2i)^4$ using polar coordinates.

First write $2\sqrt{3} + 2i$ in polar form. Let $x = 2\sqrt{3}$ and $y = 2$.

$$r = \sqrt{(2\sqrt{3})^2 + (2)^2} \qquad \theta = \text{Arctan} \frac{2}{2\sqrt{3}} \quad \textit{Notice } x > 0.$$

$$= \sqrt{16} \text{ or } 4 \qquad\qquad = \text{Arctan} \frac{\sqrt{3}}{3}$$

$$= \frac{\pi}{6}$$

Since $r = 4$ and $\theta = \frac{\pi}{6}$, $2\sqrt{3} + 2i = 4\left(\cos \frac{\pi}{6} + i \sin \frac{\pi}{6}\right)$.

Next, use DeMoivre's Theorem to find the fourth power of the complex number in polar form.

$$(2\sqrt{3} + 2i)^4 = \left[ 4\left(\cos \frac{\pi}{6} + i \sin \frac{\pi}{6}\right) \right]^4$$

$$= (4)^4\left(\cos \frac{4\pi}{6} + i \sin \frac{4\pi}{6}\right)$$

$$= 256\left(\cos \frac{2\pi}{3} + i \sin \frac{2\pi}{3}\right)$$

Finally, write the result in rectangular form.

$$256\left(\cos \frac{2\pi}{3} + i \sin \frac{2\pi}{3}\right) = 256\left(-\frac{1}{2} + i\frac{\sqrt{3}}{2}\right)$$

$$= -128 + 128i\sqrt{3}$$

So, $(2\sqrt{3} + 2i)^4 = -128 + 128i\sqrt{3}$.

# CHECKING FOR UNDERSTANDING

**Communicating Mathematics**

Read and study the lesson to answer these questions.

1. Describe how you can convert $a + bi$ to polar form.

2. Write $1 - i$ in polar form.

3. Find the product of $6\left(\cos \frac{\pi}{6} + i \sin \frac{\pi}{6}\right)$ and $2\left(\cos \frac{2\pi}{3} + i \sin \frac{2\pi}{3}\right)$.

**Guided Practice**

Write each number in polar form.

4. $1 - i$

5. $7i$

6. $-2 + 2i$

Write each number in rectangular form.

7. $2(\cos 0 + i \sin 0)$

8. $3(\cos \pi + i \sin \pi)$

9. $\cos \frac{\pi}{2} + i \sin \frac{\pi}{2}$

State the polar coordinates of each point.

10.

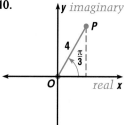

11.

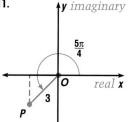

12.

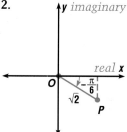

# EXERCISES

**Practice**

Write each number in polar form.

13. $1 + i$

14. $-3 - 3i$

15. $3i$

16. $-5 - i$

17. $-2 + 5i$

18. $2\sqrt{3} - 3i$

Write each number in rectangular form.

19. $\sqrt{2}\left(\cos \frac{5\pi}{4} + i \sin \frac{5\pi}{4}\right)$

20. $6\left(\cos \frac{3\pi}{2} + i \sin \frac{3\pi}{2}\right)$

21. $12\left(\cos \frac{5\pi}{3} + i \sin \frac{5\pi}{3}\right)$

22. $2(\cos 3 + i \sin 3)$

Find each product. Then express the result in rectangular form.

23. $8\left(\cos \frac{3\pi}{4} + i \sin \frac{3\pi}{4}\right) \cdot 2\left(\cos \frac{5\pi}{4} + i \sin \frac{5\pi}{4}\right)$

24. $3\left(\cos \frac{7\pi}{6} + i \sin \frac{7\pi}{6}\right) \cdot 6\left(\cos \frac{\pi}{6} + i \sin \frac{\pi}{6}\right)$

25. $5\left(\cos \frac{3}{4}\pi + i \sin \frac{3}{4}\pi\right) \cdot 2\left(\cos \frac{2}{3}\pi + i \sin \frac{2}{3}\pi\right)$

26. $\frac{1}{3}\left(\cos \frac{7\pi}{8} + i \sin \frac{7\pi}{8}\right) \cdot 3\sqrt{3}\left(\cos \left(-\frac{\pi}{4}\right) + i \sin \left(-\frac{\pi}{4}\right)\right)$

27. $(1 + i)(-1 - i)$

28. $(\sqrt{3} + i)(-2 + 2i)$

29. $(2 - 2i)(1 - i)$

30. $(8 - 2i)(3 + 5i)$

**Find each power. Express the result in rectangular form.**

**31.** $[3(\cos \pi + i \sin \pi)]^3$

**32.** $\left[2\left(\cos \dfrac{\pi}{2} + i \sin \dfrac{\pi}{2}\right)\right]^5$

**33.** $\left[2\left(\cos \dfrac{\pi}{4} + i \sin \dfrac{\pi}{4}\right)\right]^5$

**34.** $\left(\cos \dfrac{7\pi}{6} + i \sin \dfrac{7\pi}{6}\right)^3$

**35.** $(-3 + 3i)^3$

**36.** $(3 + 4i)^4$

**Critical Thinking**

**37.** Use DeMoivre's Theorem to find the square root of *i*.

**Applications**

**38. Biology** The spiral of a chambered nautilus is a logarithmic spiral. You can graph the shape of a nautilus on a polar coordinate system. For integral values of *n*, plot the points $(1 + 0.1n, 10n°)$ on polar graph paper. Connect the points with a smooth curve.

**39. Entertainment** The groove in a record is a spiral like the chambered nautilus. Look at the graph you drew for Exercise 38. The loops of the spiral become shorter as they approach the center. Does the needle on the record player travel along the groove faster or slower as it approaches the center of the record? *Hint: The turntable makes the same number of revolutions per minute every minute that the record plays.*

**Mixed Review**

**40. Sports** Patty is practicing her free throws. She knows that the rim of the basket is 10 feet above the floor. From the spot on the floor where she is standing, the angle of elevation to the rim is 33°33′. Find the distance from Patty's feet to the rim. **(Lesson 16-6)**

**41. Probability** What is the probability that you can toss a fair coin five times and get five heads? **(Lesson 15-9)**

**42.** A local business would like to find a word to represent the last four digits in their telephone number. Each of the digits in their number has three corresponding letters on the telephone dial. How many letter combinations can they make? **(Lesson 15-1)**

**43. Statistics** Two dice were thrown 18 times with the following sums. Find the median, mode, and mean of the sums. **(Lesson 14-3)**
8  11  10  8  8  7  10  3  5  9  10  8  2  9  5  2  3  7

**44. Demographics** The population of Grove City increases by approximately 3% each year. If its population is now 15,000, approximately what will its population be in ten years? **(Lesson 13-5)**

**45.** Write an expression for $\log_{12} 50$ in terms of common logarithms. Then find the value of the expression. **(Lesson 12-4)**

**46.** Find the vertices and foci of the hyperbola whose equation is $25x^2 - 4y^2 = 100$. **(Lesson 9-5)**

# SUMMARY AND REVIEW

## VOCABULARY

Upon completing this chapter you should be familiar with the following terms:

| | | | |
|---|---|---|---|
| amplitude | **791** | **829** | polar form |
| DeMoivre's Theorem | **831** | **829** | pole |
| identity | **797** | **829** | rectangular form |
| period | **791** | **822** | trigonometric equations |
| polar coordinates | **829** | **829** | trigonometric form |

## SKILLS AND CONCEPTS

| OBJECTIVES AND EXAMPLES | REVIEW EXERCISES |
|---|---|

Upon completing this chapter, you should be able to:

■ find the amplitude and period for variations of the sine and cosine functions, and graph them.   **(Lesson 17-1)**

State the amplitude and the period of the function $y = 2 \cos \theta$. Then graph.

For $y = a \sin b\theta$ or $y = a \cos b\theta$, amplitude $= |a| = |2|$ or 2, and

period $= \dfrac{360°}{|b|} = \dfrac{360°}{|1|}$ or 360°.

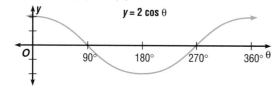

$y = 2 \cos \theta$

Use these exercises to prepare for the chapter test.

**State the period and amplitude for each function. Then graph the function.**

**1.** $y = \sin x$

**2.** $y = -\dfrac{1}{2} \cos \theta$

**3.** $y = 4 \sin 2\theta$

**4.** $y = \sin \dfrac{1}{2}\theta$

**State the period of each function. Then graph.**

**5.** $y = 5 \sec x$

**6.** $y = \tan 4\theta$

**7.** $y = 2 \cot 6\theta$

**8.** $y = \dfrac{1}{2} \csc \dfrac{2}{3}\theta$

■ use trigonometric identities to simplify and/or evaluate expressions.
**(Lesson 17-2)**

Basic Trigonometric Identities

$\sin^2 \theta + \cos^2 \theta = 1$
$1 + \tan^2 \theta = \sec^2 \theta$
$1 + \cot^2 \theta = \csc^2 \theta$

**Solve for values of θ between 90° and 180°.**

**9.** If $\sin \theta = \dfrac{1}{2}$, find $\cos \theta$.

**10.** If $\cot \theta = -\dfrac{1}{4}$, find $\csc \theta$.

**Solve for values of θ between 270° and 360°.**

**11.** If $\csc \theta = -\dfrac{5}{3}$, find $\cot \theta$.

**12.** If $\sin \theta = -\dfrac{1}{2}$, find $\sec \theta$.

- verify trigonometric identities.
  (Lesson 17-3)

  Verify $\tan x + \cot x = \sec x \csc x$.

  $$\tan x + \cot x \overset{?}{=} \sec x \csc x$$

  $$\frac{\sin x}{\cos x} + \frac{\cos x}{\sin x} \overset{?}{=} \sec x \csc x$$

  $$\frac{\sin^2 x + \cos^2 x}{\cos x \sin x} \overset{?}{=} \sec x \csc x$$

  $$\frac{1}{\cos x \sin x} \overset{?}{=} \sec x \csc x$$

  $$\sec x \csc x = \sec x \csc x$$

**Verify that each of the following is an identity.**

13. $\sin^4 x - \cos^4 x = \sin^2 x - \cos^2 x$

14. $\dfrac{\sin \theta}{\tan \theta} + \dfrac{\cos \theta}{\cot \theta} = \cos \theta + \sin \theta$

15. $\dfrac{\sin \theta}{1 - \cos \theta} = \csc \theta + \cot \theta$

16. $\cot^2 \theta \sec^2 \theta = 1 + \cot^2 \theta$

17. $\sec x(\sec x - \cos x) = \tan^2 x$

18. $\dfrac{\cos x}{\csc x} - \dfrac{\sin x}{\cos x} = -\sin^2 x \tan x$

19. $\dfrac{\csc \theta + 1}{\cot \theta} = \dfrac{\cot \theta}{\csc \theta - 1}$

---

- evaluate expressions and verify trigonometric identities using the sum and difference formulas.   (Lesson 17-5)

  Sum and Difference Formulas

  $\cos(\alpha \pm \beta) = \cos \alpha \cos \beta \mp \sin \alpha \sin \beta$
  $\sin(\alpha \pm \beta) = \sin \alpha \cos \beta \pm \cos \alpha \sin \beta$

**Evaluate each expression.**

20. $\sin 105°$

21. $\cos 15°$

22. $\cos 285°$

23. $\sin 195°$

**Verify that each of the following is an identity.**

24. $\cos(90° - \theta) = \sin \theta$

25. $\cos(x + y) + \cos(x - y) = 2 \cos x \cos y$

26. $\cos(60° + \theta) + \cos(60° - \theta) = \cos \theta$

---

- evaluate expressions and verify trigonometric identities using the half- and double-angle formulas.
  (Lesson 17-6)

  Double-Angle Formulas

  $\sin 2\theta = 2 \sin \theta \cos \theta$
  $\cos 2\theta = \cos^2 \theta - \sin^2 \theta$
  $\cos 2\theta = 1 - 2 \sin^2 \theta$
  $\cos 2\theta = 2 \cos^2 \theta - 1$

  Half-Angle Formulas

  $$\cos \frac{\alpha}{2} = \pm \sqrt{\frac{1 + \cos \alpha}{2}}$$

  $$\sin \frac{\alpha}{2} = \pm \sqrt{\frac{1 - \cos \alpha}{2}}$$

27. If $\sin x = -\dfrac{3}{5}$ and $x$ is in the third quadrant, find $\sin 2x$.

28. If $\sin x = \dfrac{1}{4}$ and $x$ is in the first quadrant, find $\cos 2x$.

29. If $\cos x = \dfrac{2}{5}$ and $\cos 2x = -\dfrac{17}{25}$, find $\sin x$.

30. If $\cos x = \dfrac{1}{6}$ and $x$ is in the first quadrant, find $\sin \dfrac{x}{2}$.

31. Verify $(\sin x - \cos x)^2 = 1 - \sin 2x$.

■ solve trigonometric equations.
(Lesson 17-7)

Solve $\sin 2\theta + \sin \theta = 0$ if $0° \le \theta < 360°$.

$$\sin 2\theta + \sin \theta = 0$$
$$2 \sin \theta \cos \theta + \sin \theta = 0$$
$$\sin \theta (2 \cos \theta + 1) = 0$$

$\sin \theta = 0$    or    $2 \cos \theta + 1 = 0$

$\theta = 0°$ or $180°$      $\cos \theta = -\dfrac{1}{2}$

$\theta = 120°$ or $240°$

**Find all solutions if $0° \le \theta < 360°$.**

32. $2 \cos^2 x + \sin^2 x = 2 \cos x$

33. $\cos x = 1 - \sin x$

34. $2 \sin 2x = 1$

**Solve each equation for all values of $\theta$ if $\theta$ is measured in radians.**

35. $6 \sin^2 \theta - 5 \sin \theta - 4 = 0$

36. $\cos 2\theta \sin \theta = 1$

37. $\sin \dfrac{\theta}{2} + \cos \dfrac{\theta}{2} = \sqrt{2}$

■ work with complex numbers in
rectangular and polar form.
(Lesson 17-8)

Find $(2 + 2i)^4$ using polar coordinates.

$r = \sqrt{(2)^2 + (2)^2}$     $\theta = \text{Arctan } \dfrac{2}{2}$

$= 2\sqrt{2}$         $= \dfrac{\pi}{4}$

$(2 + 2i)^4 = \left[ 2\sqrt{2}\left(\cos \dfrac{\pi}{4} + i \sin \dfrac{\pi}{4}\right)\right]^4$

$= (2\sqrt{2})^4\left(\cos \dfrac{4\pi}{4} + i \sin \dfrac{4\pi}{4}\right)$

$= 64(\cos \pi + i \sin \pi)$

$= 64(-1 + 0i)$

$= -64$

**Express each complex number in polar form.**

38. $-6i$               39. $-2 + 2i\sqrt{3}$

**Express each complex number in rectangular form.**

40. $4\left(\cos \dfrac{5\pi}{6} + i \sin \dfrac{5\pi}{6}\right)$

41. $8\left(\cos \dfrac{7\pi}{4} + i \sin \dfrac{7\pi}{4}\right)$

**Find each product. Express the result in rectangular form.**

42. $2\left(\cos \dfrac{\pi}{3} + i \sin \dfrac{\pi}{3}\right) \cdot 4\left(\cos \dfrac{\pi}{3} + i \sin \dfrac{\pi}{3}\right)$

43. $(2 + 2i)^8$        44. $(-2 - 2i\sqrt{3})^3$

# APPLICATIONS AND CONNECTIONS

45. **Surveying** A surveyor found that the angle between the line from her position to a building site and the line from her position to a road has a sine value of $\dfrac{4}{5}$. Find the cosine and the cotangent of this angle. **(Lesson 17-2)**

46. **Finance** Tim received his paycheck Tuesday morning. He put half of the money in his savings account. He then gave $10 to a local charity. He spent one-fourth of his remaining pay on pizza with friends and $12 was left. How much was Tim's paycheck? **(Lesson 17-4)**

# 17 TEST

State the amplitude and period of each function. Then graph.

**1.** $y = 2 \sin 2x$

**2.** $y = 2 \cos \frac{1}{5}\theta$

**3.** $y = \frac{3}{4} \cos \frac{2}{3}x$

State the period of each function. Then graph.

**4.** $y = \csc 6\theta$

**5.** $y = 3 \tan \frac{1}{3}\theta$

**6.** $y = \frac{4}{3} \cot \frac{1}{2}\theta$

Solve each of the following for values of $\theta$ between 180° and 270°.

**7.** If $\sin \theta = -\frac{1}{2}$, find $\tan \theta$.

**8.** If $\cot \theta = \frac{3}{4}$, find $\sec \theta$.

Verify that each of the following is an identity.

**9.** $\dfrac{\cos x}{1 - \sin^2 x} = \sec x$

**10.** $\dfrac{\sec x}{\sin x} - \dfrac{\sin x}{\cos x} = \cot x$

**11.** $\dfrac{1 + \tan^2 \theta}{\cos^2 \theta} = \sec^4 \theta$

**12. Entertainment**   On a game show, all the contestants start with the same number of points and they are awarded points for questions answered correctly and lose points for questions answered incorrectly. Kim answered four 20-point questions correctly, then two 50-point questions incorrectly. In her final round question, Kim doubled her score and won the game with 460 points. How many points does each player have at the start of the game?

Evaluate each expression.

**13.** $\cos 165°$

**14.** $\sin 255°$

**15.** $2 \sin 75° \cos 75°$

**16.** If $x$ is in the first quadrant and $\cos x = \frac{3}{4}$, find $\sin \frac{x}{2}$.

**17.** If $\cos 2x = \frac{7}{9}$ and $\sin x = \frac{1}{3}$, find $\cos x$.

Find all solutions if $0° \le x < 360°$.

**18.** $\sec x = 1 + \tan x$

**19.** $\cos 2x + \sin x = 1$

**20.** $2 \sin x \cos x - \sin x = 0$

Solve each equation for all values of $\theta$ if $\theta$ is measured in radians.

**21.** $\sin \frac{\theta}{2} + \cos \theta = 1$

**22.** $3 \tan^2 \theta - \sqrt{3} \tan \theta = 0$

**23.** Express $-4 + 4i$ in polar form.

**24.** Express $2\left(\cos \frac{\pi}{3} + i \sin \frac{\pi}{3}\right)$ in rectangular form.

**25.** Find the product $4\left(\cos \frac{3\pi}{2} + i \sin \frac{3\pi}{2}\right) \cdot 3\left(\cos \frac{\pi}{4} + i \sin \frac{\pi}{4}\right)$. Express the answer in rectangular form.

**Bonus**  Find $(1 - i)^8$ by DeMoivre's Theorem.

# APPENDIX: USING TABLES

Tables of common logarithms and values of trigonometric functions are provided for use in case a scientific calculator is not available. This guide will show you how to use these tables to find common logarithms, antilogarithms, and values of trigonometric functions and inverse trigonometric functions.

## How To Use Logarithmic Tables

You can use the logarithmic tables to find the common logarithm of any number. The values in the table have been rounded to the nearest ten-thousandth.

To find the logarithm of a number greater than or equal to 1 but less than 10, read across the row that contains the first two digits of the number and down the column that contains the third digit. For example, if you need to find the logarithm of 1.23 read across the row labeled 12 and down the column labeled 3. The common logarithm of 1.23 is 0.0899.

**Common Logarithms of Numbers**

| $n$ | 0 | 1 | 2 | 3 | 4 |
|-----|------|------|------|------|------|
| 10 | 0000 | 0043 | 0086 | 0128 | 0170 |
| 11 | 0414 | 0453 | 0492 | 0531 | 0569 |
| 12 | 0792 | 0828 | 0864 | 0899 | 0934 |

$$\log 1.23 = 0.0899$$

To find the logarithm of a number that is not between 1 and 10, use the table to find the mantissa and use scientific notation to supply the characteristic.

**Example 1**

**Find log 395.**

First, write the number in scientific notation.

$$395 = 3.95 \times 10^2$$

The characteristic is the power of 10. In this case, the characteristic is 2. Now use the table to find the logarithm of the 3.95, this is the mantissa.

$$\log 3.95 = 0.5966 \qquad \textit{The mantissa is 0.5966.}$$

The logarithm of 395 is 2.5966.

You can also use the table of logarithms to find antilogarithms of numbers by simply reversing the process for finding a logarithm.

**Example 2**

> **Find antilog 3.5821.**
>
> Begin by separating the logarithm into the characteristic and the mantissa.
>
> $$\text{characteristic} = 3 \qquad \text{mantissa} = 0.5821$$
>
> Then use the table to find the number for which 0.5821 is the logarithm. It is located in the row labeled 38 and the column labeled 2.
>
>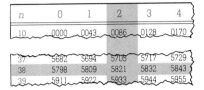
>
> $$\begin{aligned} \text{antilog } 3.5821 &= (\text{antilog } 0.5821) \times 10^3 \\ &= 38.2 \times 10^3 \\ &= 3820 \end{aligned}$$
>
> *log 3.82 = 0.5821*
>
> The antilog of 3.5821 is 3820.

The table of logarithms includes mantissas of numbers with three significant digits. Sometimes you may need to find logarithms more of numbers with more digits. You will need to use **interpolation** to use the table to find these logarithms.

**Example 3**

> **Approximate the value of log 1.327.**
>
> The logarithm of 1.327 is between the logarithms of 1.32 and 1.33. Find these values in the table and write a proportion to find log 1.327.
>
>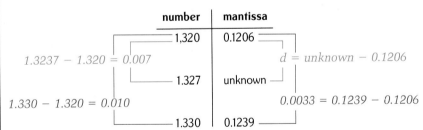
>
> $$\frac{0.007}{0.010} = \frac{d}{0.0033}$$
>
> $$0.00231 = d$$
> $$0.0023 = d \qquad \textit{Round to the nearest ten-thousandth, like the values in the table.}$$
>
> Since the values in the table are increasing, add $d$ to the mantissa of 1.320.
>
> $$\begin{aligned} \log 1.327 &= \log 1.320 + d \\ &= 0.1206 + 0.0023 \\ &= 0.1229 \qquad \text{The logarithm of 1.327 is approximately 0.1229.} \end{aligned}$$

You can use the method of interpolation to find an antilogarithm that cannot be obtained directly from the table as well.

# How To Use Trigonometric Tables

You can use the trigonometric tables to find decimal approximations for the values of trigonometric functions. Values are listed for angle measurements between 0° and 90° in intervals of ten minutes.

Angle measurements from 0°00' to 45°00' are listed on the left-hand side of the tables. When finding a value for an angle in this range, use the column titles at the top of the table. Angle measurements from 45°00' to 90°00' are listed on the right-hand side of the tables. When finding a value for an angle in this range, use the column titles at the bottom on the table.

**Example 1**

Find cos 27°10'.

Read across the row labeled 27°10' and down the column labeled Cos. *Remember to use the column titles at the top of the table for an angle between 0° and 45°.*

**Values of Trigonometric Functions**

| Angle | Sin | Cos | Tan | Cot | Sec | Csc | |
|-------|------|------|------|------|------|------|-----|
| 27°00' | 0.4540 | 0.8910 | 0.5095 | 1.963 | 1.122 | 2.203 | 63°00' |
| 10' | 0.4566 | 0.8897 | 0.5132 | 1.949 | 1.124 | 2.190 | 50' |
| 20' | 0.4592 | 0.8884 | 0.5169 | 1.935 | 1.126 | 2.178 | 40' |
| 30' | 0.4617 | 0.8870 | 0.5206 | 1.921 | 1.127 | 2.166 | 30' |
| 40' | 0.4643 | 0.8857 | 0.5243 | 1.907 | 1.129 | 2.154 | 20' |
| 50' | 0.4669 | 0.8843 | 0.5280 | 1.894 | 1.131 | 2.142 | 10' |
| 28°00' | 0.4695 | 0.8829 | | | | | |

The value of cos 27°10' is 0.8897

**Example 2**

Find tan 54°30'.

Read across the row labeled 54°30' and up the column labeled Tan. *Remember to use the column titles at the bottom of the table for an angle between 45° and 90°.*

| | | | | | | 1.736 | 50' |
|------|--------|--------|--------|--------|--------|--------|--------|
| 20' | 0.5783 | 0.8158 | 0.7089 | 1.411 | 1.226 | 1.729 | 40' |
| 30' | 0.5807 | 0.8141 | 0.7133 | 1.402 | 1.228 | 1.722 | 30' |
| 40' | 0.5831 | 0.8124 | 0.7177 | 1.393 | 1.231 | 1.715 | 20' |
| 50' | 0.5854 | 0.8107 | 0.7221 | 1.385 | 1.233 | 1.708 | 10' |
| 36°00' | 0.5878 | 0.8090 | 0.7265 | 1.376 | 1.236 | 1.701 | 54°00' |
| | Cos | Sin | Cot | Tan | Csc | Sec | Angle |

The value of tan 54°30' is 1.402.

You can use the sum and difference formulas and the half- and double-angle formulas for sine and cosine to find the trigonometric values of angles whose measurements are not listed in the table. For example, you can find sin 153° by using the difference formula for and sin (180° − 27°).

You can also use interpolation to find approximations of trigonometric values for angle measurements that are not listed in the tables.

# COMMON LOGARITHMS OF NUMBERS

| n | 0 | 1 | 2 | 3 | 4 | 5 | 6 | 7 | 8 | 9 |
|---|---|---|---|---|---|---|---|---|---|---|
| 10 | 0000 | 0043 | 0086 | 0128 | 0170 | 0212 | 0253 | 0294 | 0334 | 0374 |
| 11 | 0414 | 0453 | 0492 | 0531 | 0569 | 0607 | 0645 | 0682 | 0719 | 0755 |
| 12 | 0792 | 0828 | 0864 | 0899 | 0934 | 0969 | 1004 | 1038 | 1072 | 1106 |
| 13 | 1139 | 1173 | 1206 | 1239 | 1271 | 1303 | 1335 | 1367 | 1399 | 1430 |
| 14 | 1461 | 1492 | 1523 | 1553 | 1584 | 1614 | 1644 | 1673 | 1703 | 1732 |
| 15 | 1761 | 1790 | 1818 | 1847 | 1875 | 1903 | 1931 | 1959 | 1987 | 2014 |
| 16 | 2041 | 2068 | 2095 | 2122 | 2148 | 2175 | 2201 | 2227 | 2253 | 2279 |
| 17 | 2304 | 2330 | 2355 | 2380 | 2405 | 2430 | 2455 | 2480 | 2504 | 2529 |
| 18 | 2553 | 2577 | 2601 | 2625 | 2648 | 2672 | 2695 | 2718 | 2742 | 2765 |
| 19 | 2788 | 2810 | 2833 | 2856 | 2878 | 2900 | 2923 | 2945 | 2967 | 2989 |
| 20 | 3010 | 3032 | 3054 | 3075 | 3096 | 3118 | 3139 | 3160 | 3181 | 3201 |
| 21 | 3222 | 3243 | 3263 | 3284 | 3304 | 3324 | 3345 | 3365 | 3385 | 3404 |
| 22 | 3424 | 3444 | 3464 | 3483 | 3502 | 3522 | 3541 | 3560 | 3579 | 3598 |
| 23 | 3617 | 3636 | 3655 | 3674 | 3692 | 3711 | 3729 | 3747 | 3766 | 3784 |
| 24 | 3802 | 3820 | 3838 | 3856 | 3874 | 3892 | 3909 | 3927 | 3945 | 3962 |
| 25 | 3979 | 3997 | 4014 | 4031 | 4048 | 4065 | 4082 | 4099 | 4116 | 4133 |
| 26 | 4150 | 4166 | 4183 | 4200 | 4216 | 4232 | 4249 | 4265 | 4281 | 4298 |
| 27 | 4314 | 4330 | 4346 | 4362 | 4378 | 4393 | 4409 | 4425 | 4440 | 4456 |
| 28 | 4472 | 4487 | 4502 | 4518 | 4533 | 4548 | 4564 | 4579 | 4594 | 4609 |
| 29 | 4624 | 4639 | 4654 | 4669 | 4683 | 4698 | 4713 | 4728 | 4742 | 4757 |
| 30 | 4771 | 4786 | 4800 | 4814 | 4829 | 4843 | 4857 | 4871 | 4886 | 4900 |
| 31 | 4914 | 4928 | 4942 | 4955 | 4969 | 4983 | 4997 | 5011 | 5024 | 5038 |
| 32 | 5051 | 5065 | 5079 | 5092 | 5105 | 5119 | 5132 | 5145 | 5159 | 5172 |
| 33 | 5185 | 5198 | 5211 | 5224 | 5237 | 5250 | 5263 | 5276 | 5289 | 5302 |
| 34 | 5315 | 5328 | 5340 | 5353 | 5366 | 5378 | 5391 | 5403 | 5416 | 5428 |
| 35 | 5441 | 5453 | 5465 | 5478 | 5490 | 5502 | 5514 | 5527 | 5539 | 5551 |
| 36 | 5563 | 5575 | 5587 | 5599 | 5611 | 5623 | 5635 | 5647 | 5658 | 5670 |
| 37 | 5682 | 5694 | 5705 | 5717 | 5729 | 5740 | 5752 | 5763 | 5775 | 5786 |
| 38 | 5798 | 5809 | 5821 | 5832 | 5843 | 5855 | 5866 | 5877 | 5888 | 5899 |
| 39 | 5911 | 5922 | 5933 | 5944 | 5955 | 5966 | 5977 | 5988 | 5999 | 6010 |
| 40 | 6021 | 6031 | 6042 | 6053 | 6064 | 6075 | 6085 | 6096 | 6107 | 6117 |
| 41 | 6128 | 6138 | 6149 | 6160 | 6170 | 6180 | 6191 | 6201 | 6212 | 6222 |
| 42 | 6232 | 6243 | 6253 | 6263 | 6274 | 6284 | 6294 | 6304 | 6314 | 6325 |
| 43 | 6335 | 6345 | 6355 | 6365 | 6375 | 6385 | 6395 | 6405 | 6415 | 6425 |
| 44 | 6435 | 6444 | 6454 | 6464 | 6474 | 6484 | 6493 | 6503 | 6513 | 6522 |
| 45 | 6532 | 6542 | 6551 | 6561 | 6571 | 6580 | 6590 | 6599 | 6609 | 6618 |
| 46 | 6628 | 6637 | 6646 | 6656 | 6665 | 6675 | 6684 | 6693 | 6702 | 6712 |
| 47 | 6721 | 6730 | 6739 | 6749 | 6758 | 6767 | 6776 | 6785 | 6794 | 6803 |
| 48 | 6812 | 6821 | 6830 | 6839 | 6848 | 6857 | 6866 | 6875 | 6884 | 6893 |
| 49 | 6902 | 6911 | 6920 | 6928 | 6937 | 6946 | 6955 | 6964 | 6972 | 6981 |
| 50 | 6990 | 6998 | 7007 | 7016 | 7024 | 7033 | 7042 | 7050 | 7059 | 7067 |
| 51 | 7076 | 7084 | 7093 | 7101 | 7110 | 7118 | 7126 | 7135 | 7143 | 7152 |
| 52 | 7160 | 7168 | 7177 | 7185 | 7193 | 7202 | 7210 | 7218 | 7226 | 7235 |
| 53 | 7243 | 7251 | 7259 | 7267 | 7275 | 7284 | 7292 | 7300 | 7308 | 7316 |
| 54 | 7324 | 7332 | 7340 | 7348 | 7356 | 7364 | 7372 | 7380 | 7388 | 7396 |

*The values given are mantissas correct to four decimal places. For example, log 5.42 = 0.7340*

# COMMON LOGARITHMS OF NUMBERS

| n | 0 | 1 | 2 | 3 | 4 | 5 | 6 | 7 | 8 | 9 |
|---|---|---|---|---|---|---|---|---|---|---|
| 55 | 7404 | 7412 | 7419 | 7427 | 7435 | 7443 | 7451 | 7459 | 7466 | 7474 |
| 56 | 7482 | 7490 | 7497 | 7505 | 7513 | 7520 | 7528 | 7536 | 7543 | 7551 |
| 57 | 7559 | 7566 | 7574 | 7582 | 7589 | 7597 | 7604 | 7612 | 7619 | 7627 |
| 58 | 7634 | 7642 | 7649 | 7657 | 7664 | 7672 | 7679 | 7686 | 7694 | 7701 |
| 59 | 7709 | 7716 | 7723 | 7731 | 7738 | 7745 | 7752 | 7760 | 7767 | 7774 |
| 60 | 7782 | 7789 | 7796 | 7803 | 7810 | 7818 | 7825 | 7832 | 7839 | 7846 |
| 61 | 7853 | 7860 | 7868 | 7875 | 7882 | 7889 | 7896 | 7903 | 7910 | 7917 |
| 62 | 7924 | 7931 | 7938 | 7945 | 7952 | 7959 | 7966 | 7973 | 7980 | 7987 |
| 63 | 7993 | 8000 | 8007 | 8014 | 8021 | 8028 | 8035 | 8041 | 8048 | 8055 |
| 64 | 8062 | 8069 | 8075 | 8082 | 8089 | 8096 | 8102 | 8109 | 8116 | 8122 |
| 65 | 8129 | 8136 | 8142 | 8149 | 8156 | 8162 | 8169 | 8176 | 8182 | 8189 |
| 66 | 8195 | 8202 | 8209 | 8215 | 8222 | 8228 | 8235 | 8241 | 8248 | 8254 |
| 67 | 8261 | 8267 | 8274 | 8280 | 8287 | 8293 | 8299 | 8306 | 8312 | 8319 |
| 68 | 8325 | 8331 | 8338 | 8344 | 8351 | 8357 | 8363 | 8370 | 8376 | 8382 |
| 69 | 8388 | 8395 | 8401 | 8407 | 8414 | 8420 | 8426 | 8432 | 8439 | 8445 |
| 70 | 8451 | 8457 | 8463 | 8470 | 8476 | 8482 | 8488 | 8494 | 8500 | 8506 |
| 71 | 8513 | 8519 | 8525 | 8531 | 8537 | 8543 | 8549 | 8555 | 8561 | 8567 |
| 72 | 8573 | 8579 | 8585 | 8591 | 8597 | 8603 | 8609 | 8615 | 8621 | 8627 |
| 73 | 8633 | 8639 | 8645 | 8651 | 8657 | 8663 | 8669 | 8675 | 8681 | 8686 |
| 74 | 8692 | 8698 | 8704 | 8710 | 8716 | 8722 | 8727 | 8733 | 8739 | 8745 |
| 75 | 8751 | 8756 | 8762 | 8768 | 8774 | 8779 | 8785 | 8791 | 8797 | 8802 |
| 76 | 8808 | 8814 | 8820 | 8825 | 8831 | 8837 | 8842 | 8848 | 8854 | 8859 |
| 77 | 8865 | 8871 | 8876 | 8882 | 8887 | 8893 | 8899 | 8904 | 8910 | 8915 |
| 78 | 8921 | 8927 | 8932 | 8938 | 8943 | 8949 | 8954 | 8960 | 8965 | 8971 |
| 79 | 8976 | 8982 | 8987 | 8993 | 8998 | 9004 | 9009 | 9015 | 9020 | 9025 |
| 80 | 9031 | 9036 | 9042 | 9047 | 9053 | 9058 | 9063 | 9069 | 9074 | 9079 |
| 81 | 9085 | 9090 | 9096 | 9101 | 9106 | 9112 | 9117 | 9122 | 9128 | 9133 |
| 82 | 9138 | 9143 | 9149 | 9154 | 9159 | 9165 | 9170 | 9175 | 9180 | 9186 |
| 83 | 9191 | 9196 | 9201 | 9206 | 9212 | 9217 | 9222 | 9227 | 9232 | 9238 |
| 84 | 9243 | 9248 | 9253 | 9258 | 9263 | 9269 | 9274 | 9279 | 9284 | 9289 |
| 85 | 9294 | 9299 | 9304 | 9309 | 9315 | 9320 | 9325 | 9330 | 9335 | 9340 |
| 86 | 9345 | 9350 | 9355 | 9360 | 9365 | 9370 | 9375 | 9380 | 9385 | 9390 |
| 87 | 9395 | 9400 | 9405 | 9410 | 9415 | 9420 | 9425 | 9430 | 9435 | 9440 |
| 88 | 9445 | 9450 | 9455 | 9460 | 9465 | 9469 | 9474 | 9479 | 9484 | 9489 |
| 89 | 9494 | 9499 | 9504 | 9509 | 9513 | 9518 | 9523 | 9528 | 9533 | 9538 |
| 90 | 9542 | 9547 | 9552 | 9557 | 9562 | 9566 | 9571 | 9576 | 9581 | 9586 |
| 91 | 9590 | 9595 | 9600 | 9605 | 9609 | 9614 | 9619 | 9624 | 9628 | 9633 |
| 92 | 9638 | 9643 | 9647 | 9652 | 9657 | 9661 | 9666 | 9671 | 9675 | 9680 |
| 93 | 9685 | 9689 | 9694 | 9699 | 9703 | 9708 | 9713 | 9717 | 9722 | 9727 |
| 94 | 9731 | 9736 | 9741 | 9745 | 9750 | 9754 | 9759 | 9763 | 9768 | 9773 |
| 95 | 9777 | 9782 | 9786 | 9791 | 9795 | 9800 | 9805 | 9809 | 9814 | 9818 |
| 96 | 9823 | 9827 | 9832 | 9836 | 9841 | 9845 | 9850 | 9854 | 9859 | 9863 |
| 97 | 9868 | 9872 | 9877 | 9881 | 9886 | 9890 | 9894 | 9899 | 9903 | 9908 |
| 98 | 9912 | 9917 | 9921 | 9926 | 9930 | 9934 | 9939 | 9943 | 9948 | 9952 |
| 99 | 9956 | 9961 | 9965 | 9969 | 9974 | 9978 | 9983 | 9987 | 9991 | 9996 |

# VALUES OF TRIGONOMETRIC FUNCTIONS

| Angle | Sin | Cos | Tan | Cot | Sec | Csc | |
|---|---|---|---|---|---|---|---|
| 0°00′ | 0.0000 | 1.0000 | 0.0000 | — | 1.000 | — | 90°00′ |
| 10′ | 0.0029 | 1.0000 | 0.0029 | 343.8 | 1.000 | 343.8 | 50′ |
| 20′ | 0.0058 | 1.0000 | 0.0058 | 171.9 | 1.000 | 171.9 | 40′ |
| 30′ | 0.0087 | 1.0000 | 0.0087 | 114.6 | 1.000 | 114.6 | 30′ |
| 40′ | 0.0116 | 0.9999 | 0.0116 | 85.94 | 1.000 | 85.95 | 20′ |
| 50′ | 0.0145 | 0.9999 | 0.0145 | 68.75 | 1.000 | 68.76 | 10′ |
| 1°00′ | 0.0175 | 0.9998 | 0.0175 | 57.29 | 1.000 | 57.30 | 89°00′ |
| 10′ | 0.0204 | 0.9998 | 0.0204 | 49.10 | 1.000 | 49.11 | 50′ |
| 20′ | 0.0233 | 0.9997 | 0.0233 | 42.96 | 1.000 | 42.98 | 40′ |
| 30′ | 0.0262 | 0.9997 | 0.0262 | 38.19 | 1.000 | 38.20 | 30′ |
| 40′ | 0.0291 | 0.9996 | 0.0291 | 34.37 | 1.000 | 34.38 | 20′ |
| 50′ | 0.0320 | 0.9995 | 0.0320 | 31.24 | 1.001 | 31.26 | 10′ |
| 2°00′ | 0.0349 | 0.9994 | 0.0349 | 28.64 | 1.001 | 28.65 | 88°00′ |
| 10′ | 0.0378 | 0.9993 | 0.0378 | 26.43 | 1.001 | 26.45 | 50′ |
| 20′ | 0.0407 | 0.9992 | 0.0407 | 24.54 | 1.001 | 24.56 | 40′ |
| 30′ | 0.0436 | 0.9990 | 0.0437 | 22.90 | 1.001 | 22.93 | 30′ |
| 40′ | 0.0465 | 0.9989 | 0.0466 | 21.47 | 1.001 | 21.49 | 20′ |
| 50′ | 0.0494 | 0.9988 | 0.0495 | 20.21 | 1.001 | 20.23 | 10′ |
| 3°00′ | 0.0523 | 0.9986 | 0.0524 | 19.08 | 1.001 | 19.11 | 87°00′ |
| 10′ | 0.0552 | 0.9985 | 0.0553 | 18.07 | 1.002 | 18.10 | 50′ |
| 20′ | 0.0581 | 0.9983 | 0.0582 | 17.17 | 1.002 | 17.20 | 40′ |
| 30′ | 0.0610 | 0.9981 | 0.0612 | 16.35 | 1.002 | 16.38 | 30′ |
| 40′ | 0.0640 | 0.9980 | 0.0641 | 15.60 | 1.002 | 15.64 | 20′ |
| 50′ | 0.0669 | 0.9978 | 0.0670 | 14.92 | 1.002 | 14.96 | 10′ |
| 4°00′ | 0.0698 | 0.9976 | 0.0699 | 14.30 | 1.002 | 14.34 | 86°00′ |
| 10′ | 0.0727 | 0.9974 | 0.0729 | 13.73 | 1.003 | 13.76 | 50′ |
| 20′ | 0.0756 | 0.9971 | 0.0758 | 13.20 | 1.003 | 13.23 | 40′ |
| 30′ | 0.0785 | 0.9969 | 0.0787 | 12.71 | 1.003 | 12.75 | 30′ |
| 40′ | 0.0814 | 0.9967 | 0.0816 | 12.25 | 1.003 | 12.29 | 20′ |
| 50′ | 0.0843 | 0.9964 | 0.0846 | 11.83 | 1.004 | 11.87 | 10′ |
| 5°00′ | 0.0872 | 0.9962 | 0.0875 | 11.43 | 1.004 | 11.47 | 85°00′ |
| 10′ | 0.0901 | 0.9959 | 0.0904 | 11.06 | 1.004 | 11.10 | 50′ |
| 20′ | 0.0929 | 0.9957 | 0.0934 | 10.71 | 1.004 | 10.76 | 40′ |
| 30′ | 0.0958 | 0.9954 | 0.0963 | 10.39 | 1.005 | 10.43 | 30′ |
| 40′ | 0.0987 | 0.9951 | 0.0992 | 10.08 | 1.005 | 10.13 | 20′ |
| 50′ | 0.1016 | 0.9948 | 0.1022 | 9.788 | 1.005 | 9.839 | 10′ |
| 6°00′ | 0.1045 | 0.9945 | 0.1051 | 9.514 | 1.006 | 9.567 | 84°00′ |
| 10′ | 0.1074 | 0.9942 | 0.1080 | 9.255 | 1.006 | 9.309 | 50′ |
| 20′ | 0.1103 | 0.9939 | 0.1110 | 9.010 | 1.006 | 9.065 | 40′ |
| 30′ | 0.1132 | 0.9936 | 0.1139 | 8.777 | 1.006 | 8.834 | 30′ |
| 40′ | 0.1161 | 0.9932 | 0.1169 | 8.556 | 1.007 | 8.614 | 20′ |
| 50′ | 0.1190 | 0.9929 | 0.1198 | 8.345 | 1.007 | 8.405 | 10′ |
| 7°00′ | 0.1219 | 0.9925 | 0.1228 | 8.144 | 1.008 | 8.206 | 83°00′ |
| 10′ | 0.1248 | 0.9922 | 0.1257 | 7.953 | 1.008 | 8.016 | 50′ |
| 20′ | 0.1276 | 0.9918 | 0.1287 | 7.770 | 1.008 | 7.834 | 40′ |
| 30′ | 0.1305 | 0.9914 | 0.1317 | 7.596 | 1.009 | 7.661 | 30′ |
| 40′ | 0.1334 | 0.9911 | 0.1346 | 7.429 | 1.009 | 7.496 | 20′ |
| 50′ | 0.1363 | 0.9907 | 0.1376 | 7.269 | 1.009 | 7.337 | 10′ |
| 8°00′ | 0.1392 | 0.9903 | 0.1405 | 7.115 | 1.010 | 7.185 | 82°00′ |
| 10′ | 0.1421 | 0.9899 | 0.1435 | 6.968 | 1.010 | 7.040 | 50′ |
| 20′ | 0.1449 | 0.9894 | 0.1465 | 6.827 | 1.011 | 6.900 | 40′ |
| 30′ | 0.1478 | 0.9890 | 0.1495 | 6.691 | 1.011 | 6.765 | 30′ |
| 40′ | 0.1507 | 0.9886 | 0.1524 | 6.561 | 1.012 | 6.636 | 20′ |
| 50′ | 0.1536 | 0.9881 | 0.1554 | 6.435 | 1.012 | 6.512 | 10′ |
| 9°00′ | 0.1564 | 0.9877 | 0.1584 | 6.314 | 1.012 | 6.392 | 81°00′ |
| | Cos | Sin | Cot | Tan | Csc | Sec | Angle |

*For the values of cos, sin, tan, and so on for angles greater than 45°, use the angle measures listed on the right and the functions on the bottom. For example, cos 81° = 0.1564.*

# VALUES OF TRIGONOMETRIC FUNCTIONS

| Angle | Sin | Cos | Tan | Cot | Sec | Csc | |
|---|---|---|---|---|---|---|---|
| 9°00′ | 0.1564 | 0.9877 | 0.1584 | 6.314 | 1.012 | 6.392 | 81°00′ |
| 10′ | 0.1593 | 0.9872 | 0.1614 | 6.197 | 1.013 | 6.277 | 50′ |
| 20′ | 0.1622 | 0.9868 | 0.1644 | 6.084 | 1.013 | 6.166 | 40′ |
| 30′ | 0.1650 | 0.9863 | 0.1673 | 5.976 | 1.014 | 6.059 | 30′ |
| 40′ | 0.1679 | 0.9858 | 0.1703 | 5.871 | 1.014 | 5.955 | 20′ |
| 50′ | 0.1708 | 0.9853 | 0.1733 | 5.769 | 1.015 | 5.855 | 10′ |
| 10°00′ | 0.1736 | 0.9848 | 0.1763 | 5.671 | 1.015 | 5.759 | 80°00′ |
| 10′ | 0.1765 | 0.9843 | 0.1793 | 5.576 | 1.016 | 5.665 | 50′ |
| 20′ | 0.1794 | 0.9838 | 0.1823 | 5.485 | 1.016 | 5.575 | 40′ |
| 30′ | 0.1822 | 0.9833 | 0.1853 | 5.396 | 1.017 | 5.487 | 30′ |
| 40′ | 0.1851 | 0.9827 | 0.1883 | 5.309 | 1.018 | 5.403 | 20′ |
| 50′ | 0.1880 | 0.9822 | 0.1914 | 5.226 | 1.018 | 5.320 | 10′ |
| 11°00′ | 0.1908 | 0.9816 | 0.1944 | 5.145 | 1.019 | 5.241 | 79°00′ |
| 10′ | 0.1937 | 0.9811 | 0.1974 | 5.066 | 1.019 | 5.164 | 50′ |
| 20′ | 0.1965 | 0.9805 | 0.2004 | 4.989 | 1.020 | 5.089 | 40′ |
| 30′ | 0.1994 | 0.9799 | 0.2035 | 4.915 | 1.020 | 5.016 | 30′ |
| 40′ | 0.2022 | 0.9793 | 0.2065 | 4.843 | 1.021 | 4.945 | 20′ |
| 50′ | 0.2051 | 0.9787 | 0.2095 | 4.773 | 1.022 | 4.876 | 10′ |
| 12°00′ | 0.2079 | 0.9781 | 0.2126 | 4.705 | 1.022 | 4.810 | 78°00′ |
| 10′ | 0.2108 | 0.9775 | 0.2156 | 4.638 | 1.023 | 4.745 | 50′ |
| 20′ | 0.2136 | 0.9769 | 0.2186 | 4.574 | 1.024 | 4.682 | 40′ |
| 30′ | 0.2164 | 0.9763 | 0.2217 | 4.511 | 1.024 | 4.620 | 30′ |
| 40′ | 0.2193 | 0.9757 | 0.2247 | 4.449 | 1.025 | 4.560 | 20′ |
| 50′ | 0.2221 | 0.9750 | 0.2278 | 4.390 | 1.026 | 4.502 | 10′ |
| 13°00′ | 0.2250 | 0.9744 | 0.2309 | 4.331 | 1.026 | 4.445 | 77°00′ |
| 10′ | 0.2278 | 0.9737 | 0.2339 | 4.275 | 1.027 | 4.390 | 50′ |
| 20′ | 0.2306 | 0.9730 | 0.2370 | 4.219 | 1.028 | 4.336 | 40′ |
| 30′ | 0.2334 | 0.9724 | 0.2401 | 4.165 | 1.028 | 4.284 | 30′ |
| 40′ | 0.2363 | 0.9717 | 0.2432 | 4.113 | 1.029 | 4.232 | 20′ |
| 50′ | 0.2391 | 0.9710 | 0.2462 | 4.061 | 1.030 | 4.182 | 10′ |
| 14°00′ | 0.2419 | 0.9703 | 0.2493 | 4.011 | 1.031 | 4.134 | 76°00′ |
| 10′ | 0.2447 | 0.9696 | 0.2524 | 3.962 | 1.031 | 4.086 | 50′ |
| 20′ | 0.2476 | 0.9689 | 0.2555 | 3.914 | 1.032 | 4.039 | 40′ |
| 30′ | 0.2504 | 0.9681 | 0.2586 | 3.867 | 1.033 | 3.994 | 30′ |
| 40′ | 0.2532 | 0.9674 | 0.2617 | 3.821 | 1.034 | 3.950 | 20′ |
| 50′ | 0.2560 | 0.9667 | 0.2648 | 3.776 | 1.034 | 3.906 | 10′ |
| 15°00′ | 0.2588 | 0.9659 | 0.2679 | 3.732 | 1.035 | 3.864 | 75°00′ |
| 10′ | 0.2616 | 0.9652 | 0.2711 | 3.689 | 1.036 | 3.822 | 50′ |
| 20′ | 0.2644 | 0.9644 | 0.2742 | 3.647 | 1.037 | 3.782 | 40′ |
| 30′ | 0.2672 | 0.9636 | 0.2773 | 3.606 | 1.038 | 3.742 | 30′ |
| 40′ | 0.2700 | 0.9628 | 0.2805 | 3.566 | 1.039 | 3.703 | 20′ |
| 50′ | 0.2728 | 0.9621 | 0.2836 | 3.526 | 1.039 | 3.665 | 10′ |
| 16°00′ | 0.2756 | 0.9613 | 0.2867 | 3.487 | 1.040 | 3.628 | 74°00′ |
| 10′ | 0.2784 | 0.9605 | 0.2899 | 3.450 | 1.041 | 3.592 | 50′ |
| 20′ | 0.2812 | 0.9596 | 0.2931 | 3.412 | 1.042 | 3.556 | 40′ |
| 30′ | 0.2840 | 0.9588 | 0.2962 | 3.376 | 1.043 | 3.521 | 30′ |
| 40′ | 0.2868 | 0.9580 | 0.2994 | 3.340 | 1.044 | 3.487 | 20′ |
| 50′ | 0.2896 | 0.9572 | 0.3026 | 3.305 | 1.045 | 3.453 | 10′ |
| 17°00′ | 0.2924 | 0.9563 | 0.3057 | 3.271 | 1.046 | 3.420 | 73°00′ |
| 10′ | 0.2952 | 0.9555 | 0.3089 | 3.237 | 1.047 | 3.388 | 50′ |
| 20′ | 0.2979 | 0.9546 | 0.3121 | 3.204 | 1.048 | 3.356 | 40′ |
| 30′ | 0.3007 | 0.9537 | 0.3153 | 3.172 | 1.049 | 3.326 | 30′ |
| 40′ | 0.3035 | 0.9528 | 0.3185 | 3.140 | 1.049 | 3.295 | 20′ |
| 50′ | 0.3062 | 0.9520 | 0.3217 | 3.108 | 1.050 | 3.265 | 10′ |
| 18°00′ | 0.3090 | 0.9511 | 0.3249 | 3.078 | 1.051 | 3.236 | 72°00′ |
| | Cos | Sin | Cot | Tan | Csc | Sec | Angle |

# VALUES OF TRIGONOMETRIC FUNCTIONS

| Angle | Sin | Cos | Tan | Cot | Sec | Csc | |
|---|---|---|---|---|---|---|---|
| 18°00′ | 0.3090 | 0.9511 | 0.3249 | 3.078 | 1.051 | 3.236 | 72°00′ |
| 10′ | 0.3118 | 0.9502 | 0.3281 | 3.047 | 1.052 | 3.207 | 50′ |
| 20′ | 0.3145 | 0.9492 | 0.3314 | 3.018 | 1.053 | 3.179 | 40′ |
| 30′ | 0.3173 | 0.9483 | 0.3346 | 2.989 | 1.054 | 3.152 | 30′ |
| 40′ | 0.3201 | 0.9474 | 0.3378 | 2.960 | 1.056 | 3.124 | 20′ |
| 50′ | 0.3228 | 0.9465 | 0.3411 | 2.932 | 1.057 | 3.098 | 10′ |
| 19°00′ | 0.3256 | 0.9455 | 0.3443 | 2.904 | 1.058 | 3.072 | 71°00′ |
| 10′ | 0.3283 | 0.9446 | 0.3476 | 2.877 | 1.059 | 3.046 | 50′ |
| 20′ | 0.3311 | 0.9436 | 0.3508 | 2.850 | 1.060 | 3.021 | 40′ |
| 30′ | 0.3338 | 0.9426 | 0.3541 | 2.824 | 1.061 | 2.996 | 30′ |
| 40′ | 0.3365 | 0.9417 | 0.3574 | 2.798 | 1.062 | 2.971 | 20′ |
| 50′ | 0.3393 | 0.9407 | 0.3607 | 2.773 | 1.063 | 2.947 | 10′ |
| 20°00′ | 0.3420 | 0.9397 | 0.3640 | 2.747 | 1.064 | 2.924 | 70°00′ |
| 10′ | 0.3448 | 0.9387 | 0.3673 | 2.723 | 1.065 | 2.901 | 50′ |
| 20′ | 0.3475 | 0.9377 | 0.3706 | 2.699 | 1.066 | 2.878 | 40′ |
| 30′ | 0.3502 | 0.9367 | 0.3739 | 2.675 | 1.068 | 2.855 | 30′ |
| 40′ | 0.3529 | 0.9356 | 0.3772 | 2.651 | 1.069 | 2.833 | 20′ |
| 50′ | 0.3557 | 0.9346 | 0.3805 | 2.628 | 1.070 | 2.812 | 10′ |
| 21°00′ | 0.3584 | 0.9336 | 0.3839 | 2.605 | 1.071 | 2.790 | 69°00′ |
| 10′ | 0.3611 | 0.9325 | 0.3872 | 2.583 | 1.072 | 2.769 | 50′ |
| 20′ | 0.3638 | 0.9315 | 0.3906 | 2.560 | 1.074 | 2.749 | 40′ |
| 30′ | 0.3665 | 0.9304 | 0.3939 | 2.539 | 1.075 | 2.729 | 30′ |
| 40′ | 0.3692 | 0.9293 | 0.3973 | 2.517 | 1.076 | 2.709 | 20′ |
| 50′ | 0.3719 | 0.9283 | 0.4006 | 2.496 | 1.077 | 2.689 | 10′ |
| 22°00′ | 0.3746 | 0.9272 | 0.4040 | 2.475 | 1.079 | 2.669 | 68°00′ |
| 10′ | 0.3773 | 0.9261 | 0.4074 | 2.455 | 1.080 | 2.650 | 50′ |
| 20′ | 0.3800 | 0.9250 | 0.4108 | 2.434 | 1.081 | 2.632 | 40′ |
| 30′ | 0.3827 | 0.9239 | 0.4142 | 2.414 | 1.082 | 2.613 | 30′ |
| 40′ | 0.3854 | 0.9228 | 0.4176 | 2.394 | 1.084 | 2.595 | 20′ |
| 50′ | 0.3881 | 0.9216 | 0.4210 | 2.375 | 1.085 | 2.577 | 10′ |
| 23°00′ | 0.3907 | 0.9205 | 0.4245 | 2.356 | 1.086 | 2.559 | 67°00′ |
| 10′ | 0.3934 | 0.9194 | 0.4279 | 2.337 | 1.088 | 2.542 | 50′ |
| 20′ | 0.3961 | 0.9182 | 0.4314 | 2.318 | 1.089 | 2.525 | 40′ |
| 30′ | 0.3987 | 0.9171 | 0.4348 | 2.300 | 1.090 | 2.508 | 30′ |
| 40′ | 0.4014 | 0.9159 | 0.4383 | 2.282 | 1.092 | 2.491 | 20′ |
| 50′ | 0.4041 | 0.9147 | 0.4417 | 2.264 | 1.093 | 2.475 | 10′ |
| 24°00′ | 0.4067 | 0.9135 | 0.4452 | 2.246 | 1.095 | 2.459 | 66°00′ |
| 10′ | 0.4094 | 0.9124 | 0.4487 | 2.229 | 1.096 | 2.443 | 50′ |
| 20′ | 0.4120 | 0.9112 | 0.4522 | 2.211 | 1.097 | 2.427 | 40′ |
| 30′ | 0.4147 | 0.9100 | 0.4557 | 2.194 | 1.099 | 2.411 | 30′ |
| 40′ | 0.4173 | 0.9088 | 0.4592 | 2.177 | 1.100 | 2.396 | 20′ |
| 50′ | 0.4200 | 0.9075 | 0.4628 | 2.161 | 1.102 | 2.381 | 10′ |
| 25°00′ | 0.4226 | 0.9063 | 0.4663 | 2.145 | 1.103 | 2.366 | 65°00′ |
| 10′ | 0.4253 | 0.9051 | 0.4699 | 2.128 | 1.105 | 2.352 | 50′ |
| 20′ | 0.4279 | 0.9038 | 0.4734 | 2.112 | 1.106 | 2.337 | 40′ |
| 30′ | 0.4305 | 0.9026 | 0.4770 | 2.097 | 1.108 | 2.323 | 30′ |
| 40′ | 0.4331 | 0.9013 | 0.4806 | 2.081 | 1.109 | 2.309 | 20′ |
| 50′ | 0.4358 | 0.9001 | 0.4841 | 2.066 | 1.111 | 2.295 | 10′ |
| 26°00′ | 0.4384 | 0.8988 | 0.4877 | 2.050 | 1.113 | 2.281 | 64°00′ |
| 10′ | 0.4410 | 0.8975 | 0.4913 | 2.035 | 1.114 | 2.268 | 50′ |
| 20′ | 0.4436 | 0.8962 | 0.4950 | 2.020 | 1.116 | 2.254 | 40′ |
| 30′ | 0.4462 | 0.8949 | 0.4986 | 2.006 | 1.117 | 2.241 | 30′ |
| 40′ | 0.4488 | 0.8936 | 0.5022 | 1.991 | 1.119 | 2.228 | 20′ |
| 50′ | 0.4514 | 0.8923 | 0.5059 | 1.977 | 1.121 | 2.215 | 10′ |
| 27°00′ | 0.4540 | 0.8910 | 0.5095 | 1.963 | 1.122 | 2.203 | 63°00′ |
| | Cos | Sin | Cot | Tan | Csc | Sec | Angle |

# VALUES OF TRIGONOMETRIC FUNCTIONS

| Angle | Sin | Cos | Tan | Cot | Sec | Csc | |
|---|---|---|---|---|---|---|---|
| 27°00′ | 0.4540 | 0.8910 | 0.5095 | 1.963 | 1.122 | 2.203 | 63°00′ |
| 10′ | 0.4566 | 0.8897 | 0.5132 | 1.949 | 1.124 | 2.190 | 50′ |
| 20′ | 0.4592 | 0.8884 | 0.5169 | 1.935 | 1.126 | 2.178 | 40′ |
| 30′ | 0.4617 | 0.8870 | 0.5206 | 1.921 | 1.127 | 2.166 | 30′ |
| 40′ | 0.4643 | 0.8857 | 0.5243 | 1.907 | 1.129 | 2.154 | 20′ |
| 50′ | 0.4669 | 0.8843 | 0.5280 | 1.894 | 1.131 | 2.142 | 10′ |
| 28°00′ | 0.4695 | 0.8829 | 0.5317 | 1.881 | 1.133 | 2.130 | 62°00′ |
| 10′ | 0.4720 | 0.8816 | 0.5354 | 1.868 | 1.134 | 2.118 | 50′ |
| 20′ | 0.4746 | 0.8802 | 0.5392 | 1.855 | 1.136 | 2.107 | 40′ |
| 30′ | 0.4772 | 0.8788 | 0.5430 | 1.842 | 1.138 | 2.096 | 30′ |
| 40′ | 0.4797 | 0.8774 | 0.5467 | 1.829 | 1.140 | 2.085 | 20′ |
| 50′ | 0.4823 | 0.8760 | 0.5505 | 1.816 | 1.142 | 2.074 | 10′ |
| 29°00′ | 0.4848 | 0.8746 | 0.5543 | 1.804 | 1.143 | 2.063 | 61°00′ |
| 10′ | 0.4874 | 0.8732 | 0.5581 | 1.792 | 1.145 | 2.052 | 50′ |
| 20′ | 0.4899 | 0.8718 | 0.5619 | 1.780 | 1.147 | 2.041 | 40′ |
| 30′ | 0.4924 | 0.8704 | 0.5658 | 1.767 | 1.149 | 2.031 | 30′ |
| 40′ | 0.4950 | 0.8689 | 0.5696 | 1.756 | 1.151 | 2.020 | 20′ |
| 50′ | 0.4975 | 0.8675 | 0.5735 | 1.744 | 1.153 | 2.010 | 10′ |
| 30°00′ | 0.5000 | 0.8660 | 0.5774 | 1.732 | 1.155 | 2.000 | 60°00′ |
| 10′ | 0.5025 | 0.8646 | 0.5812 | 1.720 | 1.157 | 1.990 | 50′ |
| 20′ | 0.5050 | 0.8631 | 0.5851 | 1.709 | 1.159 | 1.980 | 40′ |
| 30′ | 0.5075 | 0.8616 | 0.5890 | 1.698 | 1.161 | 1.970 | 30′ |
| 40′ | 0.5100 | 0.8601 | 0.5930 | 1.686 | 1.163 | 1.961 | 20′ |
| 50′ | 0.5125 | 0.8587 | 0.5969 | 1.675 | 1.165 | 1.951 | 10′ |
| 31°00′ | 0.5150 | 0.8572 | 0.6009 | 1.664 | 1.167 | 1.942 | 59°00′ |
| 10′ | 0.5175 | 0.8557 | 0.6048 | 1.653 | 1.169 | 1.932 | 50′ |
| 20′ | 0.5200 | 0.8542 | 0.6088 | 1.643 | 1.171 | 1.923 | 40′ |
| 30′ | 0.5225 | 0.8526 | 0.6128 | 1.632 | 1.173 | 1.914 | 30′ |
| 40′ | 0.5250 | 0.8511 | 0.6168 | 1.621 | 1.175 | 1.905 | 20′ |
| 50′ | 0.5275 | 0.8496 | 0.6208 | 1.611 | 1.177 | 1.896 | 10′ |
| 32°00′ | 0.5299 | 0.8480 | 0.6249 | 1.600 | 1.179 | 1.887 | 58°00′ |
| 10′ | 0.5324 | 0.8465 | 0.6289 | 1.590 | 1.181 | 1.878 | 50′ |
| 20′ | 0.5348 | 0.8450 | 0.6330 | 1.580 | 1.184 | 1.870 | 40′ |
| 30′ | 0.5373 | 0.8434 | 0.6371 | 1.570 | 1.186 | 1.861 | 30′ |
| 40′ | 0.5398 | 0.8418 | 0.6412 | 1.560 | 1.188 | 1.853 | 20′ |
| 50′ | 0.5422 | 0.8403 | 0.6453 | 1.550 | 1.190 | 1.844 | 10′ |
| 33°00′ | 0.5446 | 0.8387 | 0.6494 | 1.540 | 1.192 | 1.836 | 57°00′ |
| 10′ | 0.5471 | 0.8371 | 0.6536 | 1.530 | 1.195 | 1.828 | 50′ |
| 20′ | 0.5495 | 0.8355 | 0.6577 | 1.520 | 1.197 | 1.820 | 40′ |
| 30′ | 0.5519 | 0.8339 | 0.6619 | 1.511 | 1.199 | 1.812 | 30′ |
| 40′ | 0.5544 | 0.8323 | 0.6661 | 1.501 | 1.202 | 1.804 | 20′ |
| 50′ | 0.5568 | 0.8307 | 0.6703 | 1.492 | 1.204 | 1.796 | 10′ |
| 34°00′ | 0.5592 | 0.8290 | 0.6745 | 1.483 | 1.206 | 1.788 | 56°00′ |
| 10′ | 0.5616 | 0.8274 | 0.6787 | 1.473 | 1.209 | 1.781 | 50′ |
| 20′ | 0.5640 | 0.8258 | 0.6830 | 1.464 | 1.211 | 1.773 | 40′ |
| 30′ | 0.5664 | 0.8241 | 0.6873 | 1.455 | 1.213 | 1.766 | 30′ |
| 40′ | 0.5688 | 0.8225 | 0.6916 | 1.446 | 1.216 | 1.758 | 20′ |
| 50′ | 0.5712 | 0.8208 | 0.6959 | 1.437 | 1.218 | 1.751 | 10′ |
| 35°00′ | 0.5736 | 0.8192 | 0.7002 | 1.428 | 1.221 | 1.743 | 55°00′ |
| 10′ | 0.5760 | 0.8175 | 0.7046 | 1.419 | 1.223 | 1.736 | 50′ |
| 20′ | 0.5783 | 0.8158 | 0.7089 | 1.411 | 1.226 | 1.729 | 40′ |
| 30′ | 0.5807 | 0.8141 | 0.7133 | 1.402 | 1.228 | 1.722 | 30′ |
| 40′ | 0.5831 | 0.8124 | 0.7177 | 1.393 | 1.231 | 1.715 | 20′ |
| 50′ | 0.5854 | 0.8107 | 0.7221 | 1.385 | 1.233 | 1.708 | 10′ |
| 36°00′ | 0.5878 | 0.8090 | 0.7265 | 1.376 | 1.236 | 1.701 | 54°00′ |
| | Cos | Sin | Cot | Tan | Csc | Sec | Angle |

# VALUES OF TRIGONOMETRIC FUNCTIONS

| Angle | Sin | Cos | Tan | Cot | Sec | Csc | |
|---|---|---|---|---|---|---|---|
| 36°00′ | 0.5878 | 0.8090 | 0.7265 | 1.376 | 1.236 | 1.701 | 54°00′ |
| 10′ | 0.5901 | 0.8073 | 0.7310 | 1.368 | 1.239 | 1.695 | 50′ |
| 20′ | 0.5925 | 0.8056 | 0.7355 | 1.360 | 1.241 | 1.688 | 40′ |
| 30′ | 0.5948 | 0.8039 | 0.7400 | 1.351 | 1.244 | 1.681 | 30′ |
| 40′ | 0.5972 | 0.8021 | 0.7445 | 1.343 | 1.247 | 1.675 | 20′ |
| 50′ | 0.5995 | 0.8004 | 0.7490 | 1.335 | 1.249 | 1.668 | 10′ |
| 37°00′ | 0.6018 | 0.7986 | 0.7536 | 1.327 | 1.252 | 1.662 | 53°00′ |
| 10′ | 0.6041 | 0.7969 | 0.7581 | 1.319 | 1.255 | 1.655 | 50′ |
| 20′ | 0.6065 | 0.7951 | 0.7627 | 1.311 | 1.258 | 1.649 | 40′ |
| 30′ | 0.6088 | 0.7934 | 0.7673 | 1.303 | 1.260 | 1.643 | 30′ |
| 40′ | 0.6111 | 0.7916 | 0.7720 | 1.295 | 1.263 | 1.636 | 20′ |
| 50′ | 0.6134 | 0.7898 | 0.7766 | 1.288 | 1.266 | 1.630 | 10′ |
| 38°00′ | 0.6157 | 0.7880 | 0.7813 | 1.280 | 1.269 | 1.624 | 52°00′ |
| 10′ | 0.6180 | 0.7862 | 0.7860 | 1.272 | 1.272 | 1.618 | 50′ |
| 20′ | 0.6202 | 0.7844 | 0.7907 | 1.265 | 1.275 | 1.612 | 40′ |
| 30′ | 0.6225 | 0.7826 | 0.7954 | 1.257 | 1.278 | 1.606 | 30′ |
| 40′ | 0.6248 | 0.7808 | 0.8002 | 1.250 | 1.281 | 1.601 | 20′ |
| 50′ | 0.6271 | 0.7790 | 0.8050 | 1.242 | 1.284 | 1.595 | 10′ |
| 39°00′ | 0.6293 | 0.7771 | 0.8098 | 1.235 | 1.287 | 1.589 | 51°00′ |
| 10′ | 0.6316 | 0.7753 | 0.8146 | 1.228 | 1.290 | 1.583 | 50′ |
| 20′ | 0.6338 | 0.7735 | 0.8195 | 1.220 | 1.293 | 1.578 | 40′ |
| 30′ | 0.6361 | 0.7716 | 0.8243 | 1.213 | 1.296 | 1.572 | 30′ |
| 40′ | 0.6383 | 0.7698 | 0.8292 | 1.206 | 1.299 | 1.567 | 20′ |
| 50′ | 0.6406 | 0.7679 | 0.8342 | 1.199 | 1.302 | 1.561 | 10′ |
| 40°00′ | 0.6428 | 0.7660 | 0.8391 | 1.192 | 1.305 | 1.556 | 50°00′ |
| 10′ | 0.6450 | 0.7642 | 0.8441 | 1.185 | 1.309 | 1.550 | 50′ |
| 20′ | 0.6472 | 0.7623 | 0.8491 | 1.178 | 1.312 | 1.545 | 40′ |
| 30′ | 0.6494 | 0.7604 | 0.8541 | 1.171 | 1.315 | 1.540 | 30′ |
| 40′ | 0.6517 | 0.7585 | 0.8591 | 1.164 | 1.318 | 1.535 | 20′ |
| 50′ | 0.6539 | 0.7566 | 0.8642 | 1.157 | 1.322 | 1.529 | 10′ |
| 41°00′ | 0.6561 | 0.7547 | 0.8693 | 1.150 | 1.325 | 1.524 | 49°00′ |
| 10′ | 0.6583 | 0.7528 | 0.8744 | 1.144 | 1.328 | 1.519 | 50′ |
| 20′ | 0.6604 | 0.7509 | 0.8796 | 1.137 | 1.332 | 1.514 | 40′ |
| 30′ | 0.6626 | 0.7490 | 0.8847 | 1.130 | 1.335 | 1.509 | 30′ |
| 40′ | 0.6648 | 0.7470 | 0.8899 | 1.124 | 1.339 | 1.504 | 20′ |
| 50′ | 0.6670 | 0.7451 | 0.8952 | 1.117 | 1.342 | 1.499 | 10′ |
| 42°00′ | 0.6691 | 0.7431 | 0.9004 | 1.111 | 1.346 | 1.494 | 48°00′ |
| 10′ | 0.6713 | 0.7412 | 0.9057 | 1.104 | 1.349 | 1.490 | 50′ |
| 20′ | 0.6734 | 0.7392 | 0.9110 | 1.098 | 1.353 | 1.485 | 40′ |
| 30′ | 0.6756 | 0.7373 | 0.9163 | 1.091 | 1.356 | 1.480 | 30′ |
| 40′ | 0.6777 | 0.7353 | 0.9217 | 1.085 | 1.360 | 1.476 | 20′ |
| 50′ | 0.6799 | 0.7333 | 0.9271 | 1.079 | 1.364 | 1.471 | 10′ |
| 43°00′ | 0.6820 | 0.7314 | 0.9325 | 1.072 | 1.367 | 1.466 | 47°00′ |
| 10′ | 0.6841 | 0.7294 | 0.9380 | 1.066 | 1.371 | 1.462 | 50′ |
| 20′ | 0.6862 | 0.7274 | 0.9435 | 1.060 | 1.375 | 1.457 | 40′ |
| 30′ | 0.6884 | 0.7254 | 0.9490 | 1.054 | 1.379 | 1.453 | 30′ |
| 40′ | 0.6905 | 0.7234 | 0.9545 | 1.048 | 1.382 | 1.448 | 20′ |
| 50′ | 0.6926 | 0.7214 | 0.9601 | 1.042 | 1.386 | 1.444 | 10′ |
| 44°00′ | 0.6947 | 0.7193 | 0.9657 | 1.036 | 1.390 | 1.440 | 46°00′ |
| 10′ | 0.6967 | 0.7173 | 0.9713 | 1.030 | 1.394 | 1.435 | 50′ |
| 20′ | 0.6988 | 0.7153 | 0.9770 | 1.024 | 1.398 | 1.431 | 40′ |
| 30′ | 0.7009 | 0.7133 | 0.9827 | 1.018 | 1.402 | 1.427 | 30′ |
| 40′ | 0.7030 | 0.7112 | 0.9884 | 1.012 | 1.406 | 1.423 | 20′ |
| 50′ | 0.7050 | 0.7092 | 0.9942 | 1.006 | 1.410 | 1.418 | 10′ |
| 45°00′ | 0.7071 | 0.7071 | 1.000 | 1.000 | 1.414 | 1.414 | 45°00′ |
| | Cos | Sin | Cot | Tan | Csc | Sec | Angle |

# GLOSSARY

**absolute value**  The absolute value of a number is the number of units that it is from zero on the number line.  (31)

**additive identity**  Zero is the additive identity. The sum of any number and zero is identical to the original number.  (14)

**additive inverse**  If the sum of two numbers is zero, they are called additive inverses or opposites of each other.  (14)

**amplitude**  For functions of the form $y = a \sin b\theta$ and $y = a \cos b\theta$, the amplitude is $|a|$.  (793)

**angle of depression**  An angle of depression is the angle formed by a horizontal line and the line of sight to an object at a lower level.  (762)

**angle of elevation**  The angle of elevation is the angle formed by a horizontal line, and the line of sight to an object at a higher level.  (762)

**antilogarithm**  If $\log x = a$, then $x = $ antilog $a$.  (564)

**Arccosine**  Given $y = \text{Cos } x$, the inverse cosine function is defined by $y = \text{Cos}^{-1} x$ or $y = \text{Arccos } x$.  (746)

**Arcsine**  Given $y = \text{Sin } x$, the inverse sine function is defined by $y = \text{Sin}^{-1} x$ or $y = \text{Arc sin } x$.  (747)

**Arctangent**  Given $y = \text{Tan } x$, the inverse tangent function is defined by $y = \text{Tan}^{-1} x$ or $y = \text{Arctan } x$.  (747)

**arithmetic means**  The terms between any two nonconsecutive terms of an arithmetic sequence are called arithmetic means.  (598)

**arithmetic sequence**  An arithmetic sequence is a sequence in which the difference between any two consecutive terms is the same.  (596)

**arithmetic series**  The indicated sum of the terms of an arithmetic sequence is called an arithmetic series.  (602)

**associativity**  The way you group, or associate, three or more numbers does not change their sum or their product.  (14)

**asymptote**  Asymptotes are lines that a curve approaches.  (417)

**augmented matrix**  An augmented matrix is a matrix representation of a system of equations. Each row of the matrix corresponds to an equation in the system. Each column corresponds to the coefficients of a given variable or the constant term.  (195)

**axis of symmetry**  An axis of symmetry is the line about which a figure is symmetric.  (360)

**bar graph**  A bar graph shows how specific quantities compare to one another.  (639)

**binomial**  A polynomial with two unlike terms is a binomial.  (223)

**boundary**  A boundary is a line or curve that separates a graph into two parts.  (379)

**box and whisker plot**  In a box and whisker plot, the quartiles and extreme values of a set of data are displayed using a number line.  (658)

**characteristic**  The characteristic is the power of 10 by which that number is multiplied when the number is expressed in scientific notation.  (563)

**circle**  A circle is a set of points in a plane each of which is the same distance from a given point. The given distance is the radius of the circle and the given point is the center of the circle.  (405)

**circle graph**  A circle graph shows how parts are related to the whole.  (639)

**circular permutations**  If $n$ objects are arranged in a circle, then there are $\dfrac{n!}{n}$ or $(n - 1)!$ permutations of the $n$ objects around the circle.  (695)

**coefficient** The numerical factor of a monomial is the coefficient. (210)

**coefficient matrix** A coefficient matrix is a matrix representation of the coefficients of the variables in a system of equations. Each row of the matrix corresponds to an equation in the system. Each column corresponds to the coefficients of a given variable. (183)

**combination** The number of combinations of $n$ objects, taken $r$ at a time, is defined as
$$C(n,r)=\frac{n!}{(n-r)!r!}\ (699)$$

**common difference** The common difference of an arithmetic sequence is the constant that is the difference between successive terms. (596)

**common logarithms** Common logarithms are logarithms to base 10. (563)

**common ratio** The common ratio of a geometric sequence is the constant that is the ratio of successive terms. (608)

**commutativity** The order in which two numbers are added or multiplied does not change their sum or product. That is, for all numbers $a$ and $b$, $a + b = b + a$ and $a \cdot b = b \cdot a$. (14)

**complex fraction** A complex rational expression, also called a complex fraction, is an expression whose numerator or denominator, or both, contain rational expressions. (519)

**complex number** A complex number is any number that can be written in the form $a + bi$ where $a$ and $b$ are real numbers and $i$ is the imaginary unit. $a$ is the real part and $bi$ is the imaginary part. (292)

**composition of functions** Given functions $f$ and $g$, the composite function $f \circ g$ can be described by $[f \circ g](x) = f[g(x)]$. (487)

**conic section** A conic section is a curve formed by slicing a hollow double cone with a plane. The equation of a conic section can be written in the form $Ax^2 + Bxy + Cy^2 + Dx + Ey + F = 0$ where $A$, $B$, and $C$ are not all zero. (427)

**conjugate axis** The conjugate axis of a hyperbola is the segment perpendicular to the transverse axis at its center. (417)

**conjugates 1.** Binomials of the form $a + b\sqrt{c}$ and $a - b\sqrt{c}$ are conjugates of each other. (266) **2.** Complex conjugates arecomplex numbers of the form $a + bi$ and $a - bi$. (297)

**consistent and dependent system** A system of equations where the graphs of the equations are the same line is called a consistent and dependent system. There is an infinite number of solutions to this system of equations. (109)

**consistent and independent system** A system of equations that has one ordered pair as its solution is a consistent and independent system. (109)

**constant** A monomial that contains no variable is a constant. (210)

**constant function** A constant function is a function of the form $f(x) = b$ where the slope is zero. (62)

**constant of variation** The constant $k$ in either of the equations $y = kx$ or $y = \dfrac{k}{x}$ is called the constant of variation. (510)

**coordinate matrix** A matrix containing coordinates of a geometric figure is called a coordinate matrix. (161)

**coordinate plane** The plane determined by the perpendicular axes is called the coordinate plane. (52)

**coordinates** Each point in the coordinate plane corresponds to an ordered pair of numbers called its coordinates. (52)

**cosecant** Let $\theta$ stand for the measurement of an angle in standard position on the unit circle. Then the following equation holds true: $\csc \theta = \dfrac{1}{\sin \theta}$ (742)

**cosine** Let $\theta$ stand for the measurement of an angle in standard position on the unit circle. Let $(x, y)$ represent the point where the terminal side intersects the unit circle. Then the following equation holds: $\cos \theta = x$ (737)

**cotangent** Let $\theta$ stand for the measurement of an angle in standard position on the unit circle. Then the following equation holds true: $\cot \theta = \dfrac{\cos \theta}{\sin \theta}$ (742)

**coterminal angles** Angles in standard position that have the same terminal side are called coterminal angles. (733)

**cubic expression** A cubic expression is a polynomial in one variable of degree 3. (451)

**degree of monomial** The degree of a monomial is the sum of the exponents of its variables. (210)

**degree of polynomial** The degree of a polynomial is the degree of the monomial of greatest degree. (223)

**dependent events** Two events are dependent when the outcome of the first event affects the outcome of the second event. (685)

**depressed polynomial** A polynomial whose degree is less than the original polynomial is called a depressed polynomial. It is the result of factoring out a factor of the original polynomial. (461)

**determinant** A determinant is a square array of numbers having a numerical value. (118)

**dilation** A dilation is a transformation in which size is altered. (162)

**dimension** In a matrix consisting of $n$ rows and $m$ columns, the matrix is said to have dimension $n \times m$ (read "$n$ by $m$"). (161)

**directrix** See parabola (400)

**direct variation** A direct variation is a linear function described by $y = mx$ or $f(x) = mx$ where $m \neq 0$. (87)

**discriminant** In the quadratic formula, the expression under the radical sign, $b^2 - 4ac$, is called the discriminant. (329)

**distributive property** For all numbers $a$, $b$, and $c$,
$a(b + c) = ab + ac$ and
$(b + c)a = ba + ca$. (15)

**domain** The domain is the set of all first coordinates of the ordered pairs of a relation (53)

**element** An element of a matrix is any value in the array of values. (161)

**ellipse** An ellipse is the set of all points in a plane such that the sum of the distances from two given points in the plane, called the foci, is constant. (409)

**equation** A statement of equality between two mathematical expressions is called an equation. (18)

**expansion by minors** Expansion by minors is a method that can be used to find the value of any third or higher order determinant. (167)

**exponential equation** An equation in which the variables appear as exponents is called an exponential equation. (575)

**exponential function** An equation in the form $y = a^x$, where $a > 0$ and $a \neq 1$, is called an exponential function. (548)

**factorial** If $n$ is a positive integer, the expression $n!$ ($n$ factorial) is defined as $n! = n(n - 1)(n - 2) \cdots (1)$ (627)

**Fibonacci sequence** A Fibonacci sequence is a special sequence often found in nature. It is named after its discoverer, Leonardo Fibonacci. (594)

**focus (foci)** See parabola (400), ellipse (409), or hyperbola. (416)

**frequency distribution** A frequency distribution shows how data are spread out. (668)

**function** A function is a relation in which each element of the domain is paired with exactly one element of the range. (53)

**geometric means** The terms between any two nonconsecutive terms of a geometric sequence are called geometric means. (610)

**geometric sequence** A geometric sequence is a sequence in which each term after the first is the product of the preceding term and the common ratio. (608)

**geometric series** The indicated sum of the terms of a geometric sequence is called a geometric series. (615)

**greatest integer** The greatest integer of $x$ is written $[x]$ and means the greatest integer *not* greater than $x$. (88)

**half-plane**   A half-plane is the region of a graph on one side of a boundary.   (379)

**histogram**   A histogram is a bar graph that shows a frequency distribution.   (668)

**hyperbola**   A hyperbola is the set of all points in a plane such that the absolute value of the difference of the distances from any point on the hyperbola to two given points in the plane, called the foci, is constant.   (416)

**hypotenuse**   The hypotenuse is the side opposite the right angle in a right triangle.   (756)

**identity**   An identity is an equation that is true for all values of the variable for which both sides of the equation are defined.   (797)

**identity function**   An identity function is a linear function described by $y = x$ or $f(x) = x$.   (87)

**identity matrix for multiplication**   The identity matrix, $I$, for multiplication is a square matrix with a 1 for every element of the principal diagonal and a 0 in all other positions.   (179)

**imaginary number**   An imaginary number is a complex number of the form $a + bi$, where $b \neq 0$. The imaginary unit $i$ is defined by $i^2 = -1$.   (288)

**inclusive events**   Two events are inclusive if the outcomes of the events may be the same.   (715)

**inconsistent system**   An inconsistent system is a system of equations where the graph of the equations is parallel lines. There is no solution to this system of equations.   (109)

**independent events**   Two events are independent if the outcome of one event does not affect the outcome of the other event.   (684)

**index of summation**   An index of summation is a variable used with the summation symbol ($\Sigma$).   (605)

**infinite geometric series**   An infinite geometric series is the indicated sum of the terms of an infinite geometric sequence.   (620)

**integers (Z)**   The set of numbers {. . ., $-3$, $-2$, $-1$, $0$, $1$, $2$, $3$, . . .}.   (13)

**interquartile range**   The interquartile range of a set of data is the difference between the upper and lower quartiles of the set.   (654)

**inverse functions**   Two polynomial functions $f$ and $g$ are inverse functions if and only if both their compositions are the identity function.   (491)

**inverse matrix**   For a matrix $A$, with a nonzero determinant, the inverse matrix, $A^{-1}$, of $A$ is that matrix with the property $$A \cdot A^{-1} = A^{-1} \cdot A = I \quad (179)$$

**inverse relations**   Two relations are inverse relations if and only if whenever one relation contains the element $(a, b)$, the other relation contains the element $(b, a)$.   (493)

**inverse variation**   A rational equation in two variables of the form $xy = k$, where $k$ is a constant, is called an inverse variation. The constant $k$ is called the constant of variation, and $y$ is said to vary inversely as $x$.   (511)

**irrational numbers (I)**   Irrational numbers are real numbers that cannot be written as terminating or repeating decimals.   (13)

**joint variation**   If $y = kxy$ where $k$ is a constant, $x \neq 0$, and $y \neq 0$, $y$ is said to vary jointly as $x$ and $z$. (512)

**latus rectum**   A latus rectum is the line segment through the focus of a parabola perpendicular to its axis of symmetry with endpoints on the parabola.   (401)

**like terms**   Two monomials that are the same or differ only by their coefficients are called like terms.   (210)

**linear equation**  A linear equation is an equation whose graph is a straight line. (60)

**linear expression**  A linear expression is a polynomial in one variable of degree 1. (451)

**linear function**  A linear function can be defined by $f(x) = mx + b$ where $m$ and $b$ are real numbers. Any function whose ordered pairs satisfy a linear equation in two variables is a linear function.  (62)

**linear permutation**  The arrangement of $n$ objects in a certain linear order is called a linear permutation. The number of linear permutations of $n$ objects, taken $r$ at a time, is defined as $P(n, \ r) = \dfrac{n!}{(n-r)!}$. (689)

**linear programming**  Linear programming is a method for finding the maximum or the minimum value of a function in two variables subject to given constraints on the variables.  (130)

**line graph**  A line graph shows trends or changes.  (639)

**line plot**  In a line plot, data are recorded and displayed using a number line  (642)

**logarithm**  Suppose $b > 0$ and $b \neq 1$. Then for $n > 0$, there is a number $p$ such that $\log_b n = p$ if and only if $b^p = n$.  (553)

**logarithmic function**  An equation of the form $y = \log_b x$ where $b > 0$ and $b \neq 1$ is called a logarithmic function.  (555)

**lower quartile**  The quartile that is less than the median is called the lower quartile.  (654)

**major axis**  The major axis is the longer of the two line segments that form the axes of symmetry for an ellipse.  (410)

**mantissa**  The mantissa is the logarithm of a number between 1 and 10.  (563)

**mapping**  A mapping illustrates how each element in the domain of a relation is paired with an element in the range. (53)

**matrix**  A matrix is a rectangular arrangement of terms in rows and columns enclosed in brackets or large parentheses. (161)

**matrix equation**  The matrix equation form of a system of equations is an equation of the form $AX = C$. $A$ is the coefficient matrix for the system. $X$ is the column matrix consisting of the variables of the system. $C$ is the column matrix consisting of the constant terms of the system. (184)

**mean**  The mean of a set of data is the sum of all the values divided by the number of values.  (648)

**median**  The median of a set of data is the middle value. If there are two middle values, it is the value halfway between. (648)

**minor**  A minor is the determinant formed when the row and column containing the element are deleted.  (167)

**minor axes**  The minor axis is the shorter of the two line segments that form the axes of symmetry for an ellipse.  (410)

**minute**  A minute is a unit of angle measure that is $\dfrac{1}{60}$ of a degree.  (751)

**mode**  The mode of a set of data is the most frequent value.  (648)

**monomial**  A monomial is an expression that is a number, a variable, or the product of a number and one or more variables.  (210)

**multiplicative identity**  One is the multiplicative identity. The product of any number and one is identical to the original number.  (14)

**multiplicative inverses**  If the product of two numbers is one, they are called multiplicative inverses or reciprocals of each other.  (14)

**mutually exclusive events**  Two events are mutually exclusive if their outcomes can never be the same.  (715)

**natural logarithms**  Natural logarithms are logarithms to the base e.  (568)

**natural numbers (N)**  The set of numbers {1, 2, 3, 4 . . . .}  (13)

**normal distribution** Normal distributions have bell-shaped, symmetric graphs. About 68% of the items are within one standard deviation from the mean. About 95% of the items are within two standard deviations from the mean. About 99% of the items are within three standard deviations from the mean. (668)

**nth root** For any numbers $a$ and $b$, and any positive integer $n$, if $a^n = b$, then $a$ is an $n$th root of $b$. (252)

**nth term** The $n$th term of an arithmetic sequence with the first term $a_1$ and common difference $d$ is given by the following equation.
$$a_n = a_1 + (n - 1)d \quad (597)$$
The $n$th term of a geometric sequence with first term $a$, and common ratio $r$ is given by the following equation.
$$a_n = a_1 r^{n-1} \quad (608)$$

**octant** Three mutually perpendicular planes separate space into eight regions, each called an octant. (139)

**odds** The odds of the successful outcome of an event is expressed as the ratio of the number of ways it can succeed to the number of ways it can fail.

$$\text{Odds} = \text{the ratio of } s \text{ to } f \text{ or } \frac{s}{f} \quad (706)$$

**open sentence** Sentences with variables to be replaced are called open sentences. (18)

**ordered pair** Points in a plane can be located by using ordered pairs of real numbers. The ordered pair are the coordinates of the point. (52)

**ordered triple** **1.** The solution to an equation in three variables is called an ordered triple. (139) **2.** Each point in space corresponds to three numbers called an ordered triple. (139)

**origin** The origin is the point on the coordinate plane whose coordinates are (0, 0). (52)

**outlier** An outlier is any value in a set of data that is at least 1.5 interquartile ranges beyond the upper or lower quartile. (654)

**parabola** **1.** The general shape of the graph of a quadratic function is called a parabola. (317) **2.** A parabola is the set of all points that are the same distance from a given point and a given line. The point is called the focus. The line is called the directrix. (400)

**parallel lines** In a plane, lines with the same slope are called parallel lines. Also, vertical lines are parallel. (68)

**Pascal's triangle** Pascal's triangle is the pyramid formation of the coefficients of binomial expansions. (625)

**period** For a function $f$, the least positive value of $a$ for which $f(x) = f(x + a)$ is the period of the function. (739)

**periodic function** A function $f$ is called periodic if there is a number $a$ such that $f(x) = f(x + a)$. (739)

**permutation** A permutation is the arrangement of things in a certain order. (689)

**perpendicular lines** Two nonvertical lines are perpendicular if and only if the product of their slopes is $-1$. Any vertical line is perpendicular to any horizontal line. (69)

**polar coordinates** The polar coordinates of a point $P$ are written in the form $(r, \theta)$, where $r$ is the distance from the pole to point $P$ and $\theta$ is the measure of an angle which has the polar axis as its initial side and $\overrightarrow{OP}$ as its terminal side. (829)

**polar form** The polar form of the complex number $x + yi$ is $r(\cos \theta + i \sin \theta)$. (829)

**polynomial** A polynomial is a monomial or a sum of monomials. (223)

**polynomial function** A polynomial equation in the form $p(x) = a_n x^n + a_{n-1} x^{n-1} + \cdots + a_1 x + a_0$ is a polynomial function. The coefficients $a_0, a_1, a_2, \ldots, a_{n-1}, a_n$ are real numbers, $a_0$ is not zero, and $n$ is a nonnegative integer. (451)

**polynomial in one variable** A polynomial in one variable, $x$, is an expression of the form $a_0 x^n + a_1 x^{n-1} + \cdots + a_{n-2} x^2 + a_{n-1} x + a_n$. The coefficients $a_0, a_1, a_2$ are real numbers, $a_0$ is not zero, and $n$ represents a nonnegative integer. (450)

**prediction equation** The equation of the line suggested by the dots on a scatter plot is a prediction equation. (80)

**principal root** The principal root is the nonnegative root. If there is no nonnegative root, the principal root is negative. (253)

**principal values** The values in the domain of the functions like Cosine, Sine, and Tangent are the principal values. (746)

**probability** If an event can succeed in $s$ ways and fail in $f$ ways, then the probabilities of success $P(s)$ and of failure $P(f)$ are

$$P(s) = \frac{s}{s+f} \text{ and } P(f) = \frac{f}{s+f} \quad (705)$$

**pure imaginary number** For any positive real number, $b$, $\sqrt{-(b^2)} = \sqrt{b^2}\sqrt{-1}$ or $b\textbf{\textit{i}}$ where $\textbf{\textit{i}}$ is a number whose square is $-1$. $b\textbf{\textit{i}}$ is called a pure imaginary number. (288)

**quadrants** Two perpendicular number lines separate the plane into four parts called quadrants. (52)

**quadratic equation** Any equation that can be written in the form $ax^2 + bx + c = 0$, where $a$, $b$, and $c$ are real numbers and $a \neq 0$, is a quadratic equation. (316)

**quadratic expression** A quadratic expression is a polynomial in one variable of degree 2. (451)

**quadratic form** For any numbers $a$, $b$, and $c$, except $a = 0$, an equation that may be written as $a[f(x)]^2 + b[f(x)] + c = 0$, where $f(x)$ is some expression in $x$, is in quadratic form. (340)

**quadratic formula** The solutions of a quadratic equation of the form $ax^2 + bx + c = 0$ with $a \neq 0$ are given by the quadratic formula.

$$x = \frac{-b \pm \sqrt{b^2 - 4ac}}{2a} \quad (327)$$

**quadratic function** A quadratic function is a function described by an equation of the form $f(x) = ax^2 + bx + c$ where $a \neq 0$. (353)

**quartiles** In a set of data, quartiles are the values that separate the data into four equal parts. (654)

**radian** A radian is an angle that intercepts an arc whose length is 1 unit. (733)

**radical equations** Radical equations are equations containing variables in the radicands. (282)

**radius** The radius of a circle is the distance from the center to any point on the circle. (405)

**range** **1.** The range is a set of all second coordinates of the ordered pairs of a relation. (53) **2.** The range of a set of data is the difference between the greatest and least values in the set. (653)

**rational equation** A rational equation is an equation that contains one or more rational expressions. (527)

**rational exponents** For any nonzero number $b$, and any integers $m$ and $n$, $n > 1$

$$b^{\frac{m}{n}} = \sqrt[n]{b^m} = (\sqrt[n]{b})^m$$

except when $\sqrt[n]{b}$ does not represent a real number. (270)

**rational function** A rational function is an equation of the form $f(x) = \frac{p(x)}{q(x)}$ where $p(x)$ and $q(x)$ are polynomial functions and $q(x) \neq 0$. (506)

**rational numbers** (**Q**) Rational numbers are numbers that can be expressed in the form $\frac{m}{n}$ where $m$ and $n$ are integers and $n$ is not zero. (13)

**real numbers** (**R**) Irrational numbers together with rational numbers form the set of real numbers. (13)

**reciprocal** The reciprocal of a number is its multiplicative inverse. (14)

**rectangular form** The rectangular form of a complex number is $a + b\textbf{\textit{i}}$. (829)

**relation** A relation is a set of ordered pairs. (53)

**root** A root is a solution of an equation. (316)

**rotation**  A rotation is a transformation in which the plane containing the image is rotated around a fixed point.  (174)

**scalar multiplication**  In scalar multiplication of a matrix, each element of the matrix is multiplied by a constant to form a new matrix.  (162)

**scatter plot**  A scatter plot shows visually the nature of a relationship, both shape and closeness.  (80)

**scientific notation**  A number is expressed in scientific notation when it is in the form $a \times 10^n$ where $1 \le a < 10$ and $n$ is an integer.  (212)

**secant**  Let $\theta$ stand for the measurement of an angle in standard position on the unit circle. Then the following equation holds true: $\sec \theta = \dfrac{1}{\cos \theta}$  (742)

**sequence**  A sequence is a set of numbers in a specific order.  (596)

**series**  The indicated sum of the terms of a sequence is a series.  (602)

**sigma notation**  The $\Sigma$ symbol is used to indicate a sum of a series.  (605)

**sine**  Let $\theta$ stand for the measurement of an angle in standard position on the unit circle. Let $(x, y)$ represent the point where the terminal side intersects the unit circle. Then the following equation holds. $\sin \theta = y$  (737)

**slope**  The slope of a line described by $f(x) = mx + b$ is $m$. Slope is also given by $m = \dfrac{y_2 - y_1}{x_2 - x_1}$  (66)

**solution set**  A solution set is the set of all replacements for variables that make an open sentence true.  (32)

**square matrix**  A matrix that has the same number of rows as columns is called a square matrix.  (161)

**square root**  For any numbers $a$ and $b$, if $a^2 = b$, then $a$ is a square root of $b$. (252)

**standard deviation**  From a set of data with $n$ values, if $x_i$ represents a value such that $1 \le i \le n$, and $\overline{x}$ represents the mean, then the standard deviation is $\sqrt{\dfrac{\sum\limits_{i=1}^{n} (x_i - \overline{x})^2}{n}}$.  (664)

**standard form**  The standard form of a linear equation is $Ax + By = C$ where $A$, $B$, and $C$ are real numbers and $A$ and $B$ are not both zero.  (60)

**standard position**  An angle with its vertex at the origin and its initial side along the positive $x$-axis is said to be in standard position.  (732)

**stem and leaf plot**  In a stem and leaf plot, each piece of data is separated into two numbers that are used to form the stem and leaf. The data are organized into two columns. The column on the left is the stem and the column on the right is the leaf.  (642)

**sum of a geometric series**  The sum, $S_n$, of the first $n$ terms of a geometric series is given by the following formula.
$$S_n = \frac{a_1 - a_1 r^n}{1 - r} \text{ or } S_n = \frac{a_1(1 - r^n)}{1 - r} \text{ where } r \ne 1 \quad (615)$$

**sum of an arithmetic series**  The sum, $S_n$, of the first $n$ terms of an arithmetic series is given by the following formula.
$$S_n = \frac{n}{2}(a_1 + a_n) \text{ or}$$
$$S_n = \frac{n}{2}[2a, + (n - 1)d] \quad (603)$$

**sum of an infinite geometric series**  The sum of an infinite geometric series is given by the formula
$$S = \frac{a_1}{1 - r} \text{ where } -1 < r < 1. \quad (620)$$

**synthetic division**  A shortcut method used to divide polynomials by binomials is called synthetic division.  (241)

**synthetic substitution**  Synthetic substitution is the process of using the Remainder Theorem and synthetic division to find the value of a function.  (460)

**system of equations**  A set of equations with the same variables is a system of equations.  (108)

**system of inequalities**  A set of inequalities with the same variables is a system of inequalities.  (122)

**tangent**   Let $\theta$ stand for the measurement of an angle in standard position on the unit circle. Then the following equation holds true: $\tan \theta = \dfrac{\sin \theta}{\cos \theta}$   (742)

**term**   **1.** Each monomial in a polynomial is called a term.   (223) **2.** A term is each number in a sequence.   (596)

**trace**   A trace is the line formed by the intersection of a plane with one of the three coordinate planes.   (141)

**translation**   A translation is a transformation in which a figure is moved from one location to another without changing its orientation, size, or shape.   (163)

**transverse axis**   The line segment of a hyperbola of length $2a$ that has its endpoints at the vertices is called the transverse axis.   (417)

**tree diagram**   A tree diagram is a diagram used to show the total number of possible outcomes.   (684)

**trichotomy property**   For any two numbers $a$ and $b$, one of the following statements is true.
$$a < b, \; a = b, \; a > b \quad (36)$$

**trigonometric equation**   A trigonometric equation is an equation involving one or more trigonometric functions which is true for some, but not all, values of the variable.   (822)

**trigonometric form**   The trigonometric form of the complex number $x + yi$ is $r(\cos\theta + i\sin\theta)$.   (829)

**trinomial**   A polynomial with three unlike terms is called a trinomial.   (223)

**unit circle**   A unit circle has a radius of 1 unit.   (733)

**upper quartile**   The quartile that is greater than the median is called the upper quartile.   (654)

**variable**   A variable is a symbol that represents an unknown quantity.   (18)

**vertex of hyperbola**   The point on each branch of the hyperbola nearest the center is called a vertex of the hyperbola.   (417)

**vertex of parabola**   The point of intersection of the parabola and its axis of symmetry is called the vertex of the parabola.   (360)

**vertical line test**   If any vertical line drawn on the graph of a relation passes through no more than one point of that graph, then the relation is a function.   (54)

**whole numbers (W)**   The set of numbers {0, 1, 2, 3, . . .}.   (13)

**x-intercept**   An $x$-intercept is the value of $x$ when the value of the function or $y$ is zero.   (69)

**y-intercept**   A $y$-intercept is the value of a function when $x$ is 0.   (69)

**Z**

**zero exponent**   For any number $a$, except $a = 0$, $a^0 = 1$.   (215)

**zero of function**   For any polynomial function $f(x)$, if $f(a) = 0$, then $a$ is a zero of the function.   (317)

**zero product property**   For any real numbers $a$ and $b$, if $ab = 0$ then $a = 0$ or $b = 0$.   (317)

# SELECTED ANSWERS

## CHAPTER 1 EQUATIONS AND INEQUALITIES

### Pages 10–12 Lesson 1-1

**5.** $-3$ **7.** 54 **9.** 2 **11.** $-8$ **13.** 37
**15.** $A = (y + 5)(y - 5)$ cm$^2$ **17.** 21 **19.** 41 **21.** 4
**23.** 10 **25.** 11 **27.** 26 **29.** 0.09 **31.** $-63$
**33.** 272.16 **35.** 0 **37.** 0.75 **39.** \$737
**41.** \$39,548.57 **43.** 93.6 **45.** $53\frac{1}{4}$ **49.** 754 gallons
**51. a.** 24.85 ft$^2$ **b.** 2614.68 ft$^2$

### Pages 16–17 Lesson 1-2

**7.** Q, R **9.** I, R **11.** N, W, Z, Q, R **13.** Q, R
**15.** Z, Q, R **17.** N, W, Z, Q, R **19.** commutative $\times$
**21.** commutative $+$ **23.** 1; N, W, Z, Q, R **25.** $-9$;
Z, Q, R **27.** $-24$; Z, Q, R **29.** $\frac{3}{2}$, 1.5; Q, R
**31.** true **33.** false; $\frac{3}{2}$, 1.5 **35.** true **37.** true
**39.** additive inverse **41.** commutative $+$
**43.** multiplicative inverse **45.** additive identity
**47.** $10a + 2b$ **49.** $12 + 20a$ **51.** $\frac{2}{3}a + \frac{5}{2}b$
**53.** $4.4m - 2.9n$ **57a.** no **57b.** yes **57c.** The
difference of two such numbers is always divisible by
9. **58.** 0 **59.** $-2$ **60.** 94 **61.** \$17,400

### Page 21 Lesson 1-3

**5.** Reflexive property of equality **7.** Addition
property of equality **9.** Multiplication property of
equality **11.** 3 **13.** $\frac{5}{4}$ **15.** $-4$ **17.** $-\frac{21}{8}$, $-2.625$
**19.** 2 **21.** $\frac{7}{4}$, 1.75 **23.** $-\frac{2}{3}$, $-0.667$ **25.** $-6$
**27.** all reals **29.** 1 **33.** $t = \dfrac{A - p}{pr}$ **35.** $\sqrt{3}$; I, R
**36.** commutative $+$ **37.** 44.4

### Pages 26–28 Lesson 1-4

**5.** $5x - 4$ **7.** $3 - 2n$ **9.** $6x^2$ **11.** $4(8 + x)$
**13.** $8 + 4n$ **15.** \$7 **17.** 118 **19.** 18 years old
**21.** Mrs. Gampp, 43; Brittany, 19 **23.** \$428 **25.** 4
adults, 20 students **27.** 3 hours **31.** add 6 inches to
the length and 3 inches to the width **32.** Reflexive
property of equality **33.** Substitution property
**34.** $4x + 8$ **35.** $-3.5$

### Page 28 Mid-Chapter Review

**1.** 39 **2.** 83 **3.** $-5$ **4.** $-79$ **5.** \$97.71 **6.** I, R
**7.** Q, R **8.** N, W, Z, Q, R **9.** distributive
**10.** associative **11.** 25.75 **12.** no solution
**13.** $3n + 9$ **14.** $(n + 2)^2$ **15.** 168 **16.** 13

### Page 30 Lesson 1-5

**5.** WWWWLL, WWWLLW, WLWLWW, WWWLWL,
WWLWLW, LWWLWW, WWLWWL, WLWWLW,
WLLWWW, WLWWWL, LWWWLW, LWLWWW,
LWWWWL, WWLLWW, LLWWWW **7.** Assume
that your monthly salary is \$1000 to start with and
the first cut or raise will occur in January. Your final
salary will be \$990 a month with either option.
However, if you take the cut first, your salary for
January will be \$900. If you take the raise first, you
salary for January will be \$1100. So, if you take the
cut first, your total income for the year will be \$900
$+ 11(\$990)$ or \$11,790. If you take the raise first, your
total income for the year will be \$1100 $+ 11(\$990)$ or
\$11,990. Taking the raise first gives you more total
income for the year. **9.** 36 students

### Pages 34–35 Lesson 1-6

**5.** 16 **7.** $-16$ **9.** 5 **11.** 2, $-2$ **13.** $-2$, $-1$, 0
**15.** no solution **17.** 13, $-25$ **19.** 15, $-7$
**21.** 6, $-18$ **23.** 5, $-13$ **25.** 10.5, $-19.5$ **27.** $-\frac{7}{2}$
**29.** 5, $-\frac{19}{3}$ **31.** 3.5, 10.5 **33.** 0, $-\frac{10}{3}$ **35.** 1
**37.** 1, $-9$ **39.** no solution **43.** 1.05 and 0.95 pounds
**45.** 139, 193, 319, 391, 913, 931 **46.** E, A, AH; E,
AH, A; A, E, AH; A, AH, E; AH, E, A; AH, A, E
**47.** $x + x^2$ **48.** 7; N, W, Z, Q, R

### Pages 39–40 Lesson 1-7

**5.**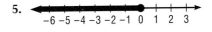

**7.** 

**9.** 

**11.** $\{x|x \geq 5\}$ **13.** $\{s|s < 6\}$ **15.** $\{t|t \leq -8\}$
**17.** $\{y|y > 17.6\}$ **19.** $\{x|x < 0\}$ **21.** $\{z|z \leq 5\}$
**23.** $\{x|x > 2.25\}$ **25.** $\{x|x \leq 3.5\}$ **27.** $\{m|m \geq 1\}$
**29.** $\{y|y < -1\}$ **31.** $\{x|x \geq 423\}$ **33.** $\{x|x \leq 1.\overline{6}\}$
**35.** $\{x|x \leq -1.425\}$ **37.** no solution **39.** \$60.37
**41.** 20 games **43.** 81 **45.** \$75,000; \$56,250
**47.** $9.999 \leq d \leq 10.001$ **49.** 11 **50.** no solution
**51.** $-\frac{15}{2}$ **52.** $2x + 17y$ **53.** $\frac{3}{2}$

**5.** $|x| < 7$     **7.** $|x| > 11$

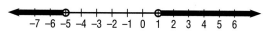

**9.** $|x| < 6$   **11.** $|x| \le 9$   **13.** $|x| \le 6$

**15.** $\{x | x < -5 \text{ or } x > 1\}$

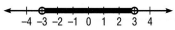

**19.** $\{x | -3 < x < 3\}$

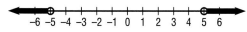

**23.** $\{x | x < -5 \text{ or } x > 5\}$

**27.** $\varnothing$

**31.** $\{x | -30 < x < 60\}$

**35.** $\{x | -1 < x < 4\}$

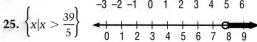

**39.** $\{x | -2 \le x \le 7\}$

**43.** $\{x | x \ge 0\}$

**49a.** maximum = 7.322 cm; minimum = 7.318 cm
**49b.** Tolerance = 0.03 cm   **50.** My best friend is taller than I am; My best friend and I are the same height; My best friend is shorter than I am.
**51.** 21.25   **52.** 3264   **53.** passenger train, 89 mph; express train, 99 mph   **54.** additive identity   **55.** 8%

---

**1.** 92   **3.** $\frac{13}{12}$   **5.** 6,400 feet   **7.** Q, R   **9.** I, R
**11.** additive identity   **13.** $2a - 4b$   **15.** $-\frac{33}{13}$   **17.** 8
**19.** length 21 ft, width 11 ft   **21.** no solution

**23.** $\{x | x < 6\}$

**25.** $\left\{ x \left| x > \frac{39}{5} \right. \right\}$

**27.** $\{x | x < 0 \text{ or } x \ge 1\}$

**29.** all reals

**31.** 64   **33.** from 5.25 to 6 gallons

---

# CHAPTER 2   LINEAR RELATIONS AND FUNCTIONS

**7.** quadrant III   **9.** no quadrant, $y$-axis   **11.** relation
**13.** D = {4, 8, -10}, R = {3, 4, 8, -2}; no

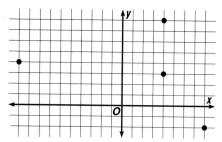

**17.** 4   **21.** $\frac{7}{10}$   **23.** -1   **25.** undefined   **27.** not a function
**29.** not a function   **31.** function   **33.** 2423
**35.** 41.3776   **37.** (-3, -3)   **39.** -12   **41.** $\frac{a^2 + 3a - 10}{a + 2}$
**43.** no

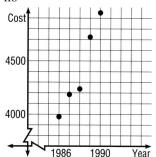

**45.** $\{x | x < -3 \text{ or } x > 9\}$   **46.** $\{y | -8 < y < 6\}$   **47.** 11 posters   **48.** N, W, Z, Q, R   **49.** -84.96   **50.** -9800

---

**5.** no; it contains an exponent greater than 1.   **7.** yes
**9.** yes   **11.** $3x - y = 2$   **13.** $x = 10$
**15.** $3x - y = -12$   **17.** $y = 4.5 - 2x$

**19.**

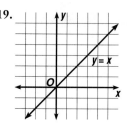

**21.** $m = \frac{n + 3}{-5}$   **23.** $r = \frac{144 - 9t}{8}$

**25.**

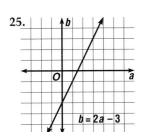

$b = 2a - 3$

**29.**

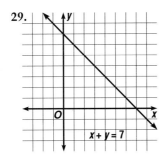

$x + y = 7$

**33.**

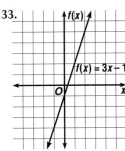

$f(x) = 3x - 1$

**39.** \$102   **41.** It is not a function because one of the *x*-values is paired with two different *y*-values.

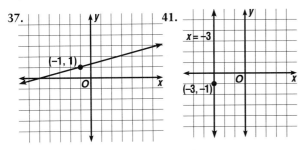

**42.** 9   **43.** 24 ways   **44.** 10.5 inches   **45.** $p + 5$

**Page 65   Lesson 2-3**

**5.** 36   **7.** 30   **9.** \$686

**Pages 70–73   Lesson 2-4**

**9.** 2, −2, 1   **11.** 2, none, 0   **13.** 2, 3, $-\dfrac{2}{3}$   **15.** $-\dfrac{5}{2}$
**17.** $\dfrac{3}{5}$   **19.** 8   **21.** falls   **23.** rises   **25.** rises
**27.** *y*: 9, *x*: $-\dfrac{3}{2}$   **29.** *y*: −2, *x*: none   **31.** *y*: −6, *x*: $\dfrac{6}{5}$
**33.** *y*: −2, *x*: 2   **35.** *y*: 5, *x*: 3

**37.**

**41.**

$x = -3$

(−1, 1)

(−3, −1)

**47.** 65 mph   **48.** 50   **49.** 5   **50.** $\dfrac{1}{2}$   **51.** distributive property

**Page 72   Mid-Chapter Review**

**1a.** D = {0, 5, 10, 15, 20, 25, 30, 35, 40},
R = {30, 27, 16, 9, 4, 1, −2, −4, −5}

**1b.** yes

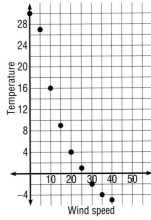

**2.** 19   **3.** $y = \dfrac{7 - 2x}{4}$   **4.** $s = 21 - \dfrac{21}{8}t$
**5.** $c = \dfrac{3}{4}a + \dfrac{1}{2}b - 6$   **6.** no

**7.**

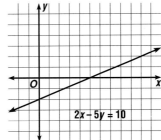

$2x - 5y = 10$

**8.** 135

**9.** $m = \dfrac{2}{3}$, *x*: 3, *y*: −2

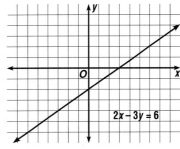

$2x - 3y = 6$

**10.** $m = -\dfrac{2}{3}$, *x*: 4.5, *y*: 3

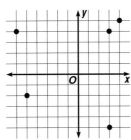

$2x + 3y = 9$

**Pages 76–79 Lesson 2-5**

**7.** $3x + y = -4$ **9.** $y = 2.5x$ **11.** $y = 0$
**13.** $m = -\dfrac{3}{4}$, $y$: $-3$ **15.** $m = -0.2$, $y$: $-6$
**17.** $m = -\dfrac{3}{5}$, $y$: 6 **19.** $2x + 3y = 12$ **21.** $y = 3x - 6$
**23.** no slope-intercept form **25.** $y = -\dfrac{4}{5}x - \dfrac{7}{5}$,
$4x + 5y = -7$ **27.** $y = -\dfrac{5}{2}x + 16$, $5x + 2y = 32$
**29.** $y = \dfrac{3}{2}x$, $3x - 2y = 0$ **31.** $y = \dfrac{3}{4}x - \dfrac{1}{4}$,
$3x - 4y = 1$ **33.** $y = -1$, $y = -1$ **35.** $y = \dfrac{1}{3}x + \dfrac{2}{3}$
**37.** $y = -\dfrac{1}{15}x - \dfrac{23}{5}$, $x + 15y = -69$ **39.** 3 **41.** $\dfrac{5}{7}$
**43.** 15,963.5 lb/in² **45.** $d = 6000 - 75x$
**47a.** $m = -2$, $y$: 2; $y = -2x + 2$ **47b.** $m = \dfrac{4}{3}$, $y$: $5\dfrac{2}{3}$,
$y = \dfrac{4}{3}x + 5\dfrac{2}{3}$ **47c.** $m = \dfrac{4}{3}$, $y$: $-\dfrac{2}{3}$; $y = \dfrac{4}{3}x - \dfrac{2}{3}$
**47d.** $m = \dfrac{1}{2}$, $y$: $-3\dfrac{1}{2}$, $y = \dfrac{1}{2}x - 3\dfrac{1}{2}$ **48.** $\dfrac{2}{3}$ **49.** $x$: 4,
$y$: $-6$ **50.** 0 **51.** $\{x|x > -7\}$ **52.** 24 ways

**Pages 83–84 Lesson 2-6**

**5.** 3.0 cm **7a.** 400 **7b.** 2,000,000 **7c.** $S = 400A +$ 2,000,000 **7d.** $6,000,000 **7e.** $35,000

**9a.**

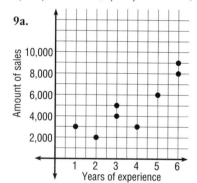

**9b.** $y = 1333x$
**9c.** $10,666
**9d.** $0
**9e.** about 5.5 years

**13a.** $y = -50x + 226.5$ **13b.** 31,500 bushels
**13c.** $4.02

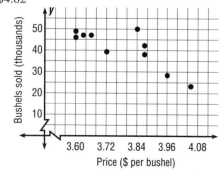

**15.** $y + x = 6$ **16.** $-2$

**17.**

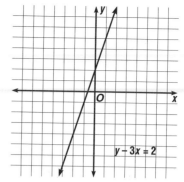

**18.** 152 **19.** $-9x + 14y$

**Pages 89–90 Lesson 2-7**

**7.** G **9.** A **11.** D **13.** identity **15.** greatest integer function **17.** $-5$ **19.** 2

**21.**

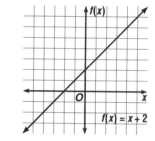

**25.**

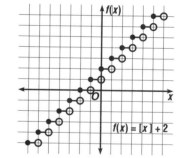

**29.** same graph translated 1 unit down

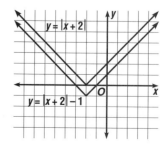

**33.** same shape graph reflected over x-axis

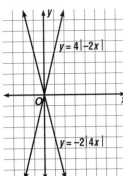

**35.** If $a > 0$, the two graphs are the same. If $a < 0$, the graphs are reflected images over the x-axis.

**11.**

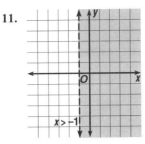

$x > -1$

**37.**

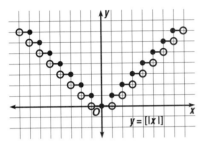

$y = [|x|]$

**15.**

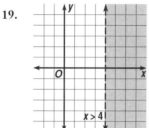

$y > 5x - 3$

**39.**

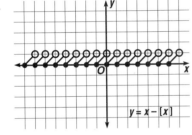

$y = x - [x]$

**19.**

$x > 4$

**43.** step function

**44a.** $1200y = -x + 40.3$

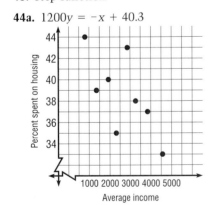

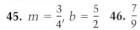

**44b.** about 37.8%   **44c.** about $1725

**45.** $m = \dfrac{3}{4}$, $b = \dfrac{5}{2}$   **46.** $\dfrac{7}{9}$

**Pages 94–95   Lesson 2-8**

**7.** $(0, 0), (2, -3)$   **9.** $(-1, 2)$

**23.**

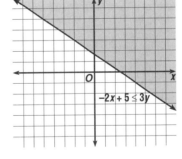

$-2x + 5 \le 3y$

**27.**

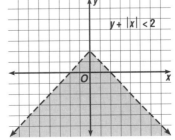

$y + |x| < 2$

**31.**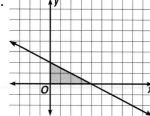

**35.**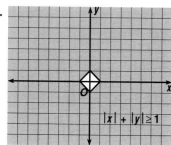

**40.**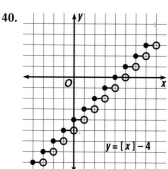

**41.** $a^2 - 6a + 14$

**42.** all reals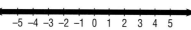

**43.** 1000

**7.** yes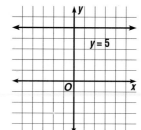

**9.** yes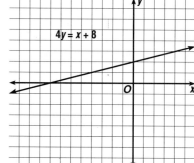

**11.** $-\dfrac{3}{8}$          **15.** parallel

**13.** perpendicular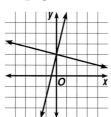

**17.** $y = -3x - 1$; $3x + y = -1$  **19.** $y = \dfrac{5}{2}x + 5$; $5x - 2y = -10$  **21.** $y = \dfrac{2}{3}x - \dfrac{1}{3}$; $2x - 3y = 1$

**23.** 70.3 kg

## Pages 96–98   Chapter 2 Summary and Review

**1.** D = {−3, 3, 4, 6}, R = {−6, −3, −2, 2}; yes          **3.** D = {1, 2, 3, −3.5}, R = {−4.5, 4, 4.5}; yes

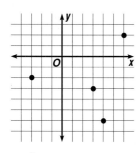   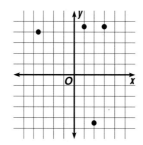

**5.** $12a^2 - 4a - 1$

**25.** 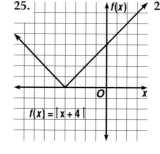   **29.**

**33.** $c = 10 + 0.15x$; yes   **35.** $s = 5.5h - 10$, \$166

# CHAPTER 3 SYSTEMS OF EQUATIONS AND INEQUALITIES

**Pages 110–111   Lesson 3-1**

**5.** consistent and independent, $\left(\frac{11}{7}, \frac{1}{7}\right)$

**7.** no solutions: inconsistent

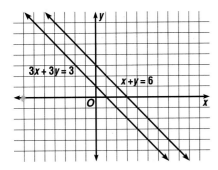

**11.** (2, −6); consistent and independent

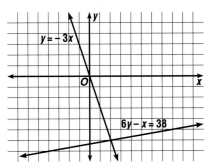

**15.** no solutions; inconsistent

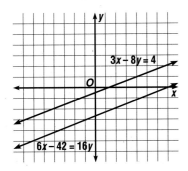

**19.** (5, 3); consistent and independent

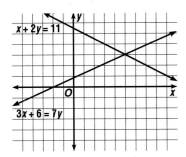

**23.** no solutions; inconsistent

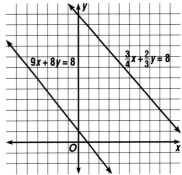

**25.** $a = 3$, $b = 8$   **27.** $a = -4$, $b = 5$   **29.** 2 of the $30 drums and 5 of the $20 drums   **31.** (0, 0), (−1, −3)

**32.**

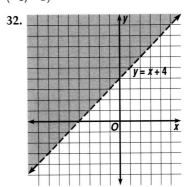

**33.** −1   **34.** D = {9, 2, 1}, R = {3, −7, 1}; yes
**35.** $\{x|x > 12\}$

**Pages 114–116   Lesson 3-2**

**5.** $x$: first by 3 and second by −2; $y$: first by 4 and second by 3; (2, 1)   **7.** $x$: first by −2 and second by 3; $y$: first by −5 and second by 4; (−2, 3)   **9.** $x$: first by −3; $y$: first by 7 and second by 8; (4, 1)   **11.** $\left(\frac{4}{3}, \frac{2}{3}\right)$
**13.** (3, −5)   **15.** (4, −1)   **17.** (−9, −7)   **19.** (−5, 6)
**21.** $\left(\frac{1}{4}, \frac{2}{3}\right)$   **23.** (5.25, 0.75)   **25.** (3, 3)   **27.** $\left(-1, \frac{2}{3}\right)$
**29.** $\left(\frac{21}{4}, -\frac{27}{4}\right)$   **31.** $\left(\frac{10}{3}, \frac{8}{3}\right)$   **33.** (10, 25)
**35.** (−5, −2), (4, 4), (−2, −8), (1, 10)   **37.** 15 inches by 28 inches   **39a.** 11 years ago; 1 and 5   **39b.** 2 and 20; 7 years ago   **39c.** 28 and 56; 11 years from now
**40.** (−2, −4)   **41.** (0, 2)   **42.** $y = -2x + 7$   **43.** 3

**Pages 120–121   Lesson 3-3**

**5.** 0   **7.** 1   **9.** 20   **11.** −13

**13.** $x = \dfrac{\begin{vmatrix} 3 & -1 \\ 5 & 2 \end{vmatrix}}{\begin{vmatrix} 4 & -1 \\ 3 & 2 \end{vmatrix}}$, $y = \dfrac{\begin{vmatrix} 4 & 3 \\ 3 & 5 \end{vmatrix}}{\begin{vmatrix} 4 & -1 \\ 3 & 2 \end{vmatrix}}$; (1, 1)

**15.** $x = \dfrac{\begin{vmatrix} 8 & -2 \\ 21 & -5 \end{vmatrix}}{\begin{vmatrix} 1 & -2 \\ 3 & -5 \end{vmatrix}}$, $y = \dfrac{\begin{vmatrix} 1 & 8 \\ 3 & 21 \end{vmatrix}}{\begin{vmatrix} 1 & -2 \\ 3 & -5 \end{vmatrix}}$; $(2, -3)$

**17.** $s = \dfrac{\begin{vmatrix} 6 & 1 \\ 2 & -1 \end{vmatrix}}{\begin{vmatrix} 1 & 1 \\ 1 & -1 \end{vmatrix}}$, $t = \dfrac{\begin{vmatrix} 1 & 6 \\ 1 & 2 \end{vmatrix}}{\begin{vmatrix} 1 & 1 \\ 1 & -1 \end{vmatrix}}$; $(4, 2)$

**19.** $-18$  **21.** $17$  **23.** $-31.38$  **25.** $(5, 1)$  **27.** $(3, -1)$
**29.** $\left(2, \dfrac{13}{8}\right)$  **31.** $(-3, 1)$  **33.** $\left(\dfrac{2}{3}, 0\right)$  **35.** $(0.75, 0.5)$

**37.** $(6, 9)$  **39.** $\left(\dfrac{9}{64}, \dfrac{49}{16}\right)$  **41.** half inch, 4¢; quarter
inch, 3¢  **43.** $(-1, 2)$  **44.** $(-4, -4)$  **45.** $-4$
**46.** $D = \{-11, 0, 1, 3, 9, 12\}$, $R = \{-6, -4, -3, 0, 1, 7, 8\}$;
no  **47.** $17, -1$

## Pages 124–125   Lesson 3-4

**5.** no  **7.** no  **9.** $(3, 1), (-3, -1), (2, 1), (1, 2), (-1, -2)$
**11.** $(3, 1), (2, 1), (1, 2)$  **13.** $(2, 1), (1, 2), (-1, -2)$

**15.**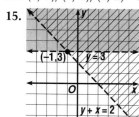

**19.**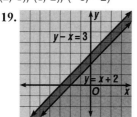

**23.**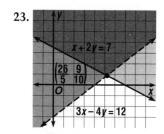

**27.**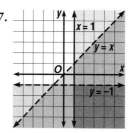

**31.**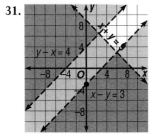

**35.** $y < 0$ or $x < 0$  **36.** $(7, -2)$  **37.** $(4.27, -5.11)$
**38.** $83$  **39.** $x + 3y = -54$  **40.** $p \le 400$

## Page 125   Mid-Chapter Review

**1.** $(2, 1)$  **2.** $(1, 3)$  **3.** $\left(\dfrac{10}{3}, \dfrac{1}{6}\right)$  **4.** $(3, 3)$  **5.** $\left(\dfrac{13}{5}, -\dfrac{22}{5}\right)$

**6.** $34$  **7.** $(3, 1)$  **8.** $\left(-\dfrac{1}{7}, -\dfrac{17}{7}\right)$

**9.**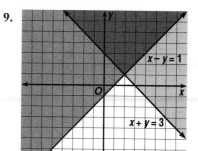

**10.**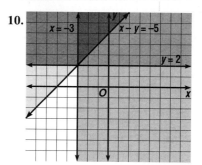

## Pages 127–128   Lesson 3-5

**5.** 34th  **7.** 1027 pages  **9.**
**11.** 0  **13.** no; 3

## Pages 132–133   Lesson 3-6

**5.** $14$  **7.** $-4$  **9.** max: $f(0, 8) = 24$; min: $f(0, 0) = 0$
**11.** max: $f(4, 0) = 2$; min: $f(0, 8) = -12$  **13.** $18$
**15.** $17$  **17.** $1$
**19.** vertices: $(1, 2), (1, 4), (5, 8), (5, 2)$;
max: $f(5, 2) = 11$; min: $(1, 4) = -5$

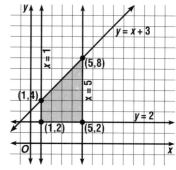

**23.** vertices: $(0, 1), (1, 3), (6, 3), (10, 1)$;
max: $f(1, 3) = 8$; min: $f(10, 1), = -7$

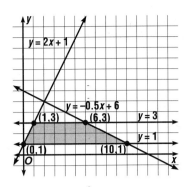

**27.** vertices: $(-4.08, -2)$, $(-1.58, 1)$, $(-0.4, 1)$, $(-0.4, -2)$; max: $f(-0.4, 1) = 0.4$; min: $f(-4.08, -2) = -20.32$ **29.** vertices : $(1, 0)$, $(0, 1)$, $(0, 5)$, $(3, 0)$, $(3, 5)$; max: $f(3, 5) = 56$; min: $f(1, 0) = 12$ **31.** 100 acres of corn, 150 acres of soybeans

**33.**

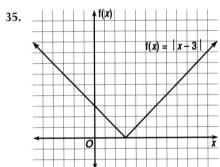

**35.**

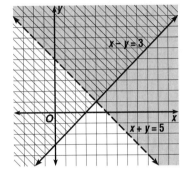

**36.** $(8, 0)$ and $(0, -2)$ **37.** 90

**Pages 136–138 Lesson 3-7**

**5.** $0 \le a \le 9$; $b \ge 18$

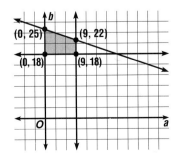

**9.** $P(0, 18) = 360$, $P(0, 25) = 500$, $P(9, 22) = 530$, $P(9, 18) = 450$ **11.** 530 pounds **13.** 30 footballs, 0 basketballs **15a.** $a \ge 0$, $b \ge 0$, $4a + b \le 32$, $a + 6b \le 54$ **15b.** 14 gallons $(6A, 8B)$ **17a.** 30 jean jackets, 10 leather jackets **17b.** 10 jean jackets, 20 leather jackets

**20.** vertices: $(0, 3)$, $(0, 6)$, $(2, 5)$, $(1, 3)$; max: $f(1, 3) = -3$; min: $f(0, 6) = -12$

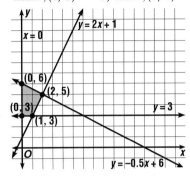

**21.** vertices: $(0, 60)$, $(0, 70)$, $(10, 70)$, $(50, 30)$, $(50, 10)$; max: $f(50, 30) = 290$; min: $f(0, 60) = 180$

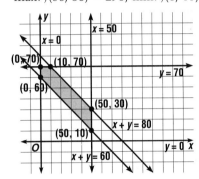

**22.** no **23.** no solution **24.** Reflexive property of equality

**Pages 142–143 Lesson 3-8**

**5.** 4 **7.** 2 **9.** $x$: 1, $y$: $-5$, $z$: 2 **11.** $x$: 10, $y$: 6, $z$: 15 **13.** 4 **15.** 7 **17.** 3

**19.** $x$: $\frac{5}{2}$, $y$: $-10$, $z$: 5; **23.** $x$: $-\frac{12}{5}$, $y$: $\frac{3}{2}$, $z$: none

$4x - y = 10$,     $5x = -12$,
$2x + z = 5$,     $2y = 3$,
$-y + 2z = 10$    $5x - 8y = -12$

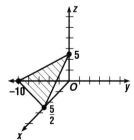

**25.** $12x + 2y - 3z = 6$  **27.** $2x - 8y - z = 2$
**29.** $5x - 2z = -10$  **31a.** 1200 cubic inches
**31b.** 680 square inches  **31c.** It is eight times greater.
**33a.** $c \geq 0,\ r \geq 0,\ c + r \leq 275,\ \dfrac{c}{9} + \dfrac{c}{12} \leq 25$
**33b.** 225 acres of cotton and 0 acres of corn, for a
profit of \$5625  **34.** $(0, 0), (1, -2),$ and $(-3, 1)$
**35.** $\dfrac{32}{7}$  **36.** all reals

**27.** $x$: 4, $y$: 4, $z$: $-2$;
$x + y = 4$;
$x - 2z = 4$;
$y - 2z = 4$
**29.** $(-4, 2, 1)$  **31.** 99
**33.** 24 multiple-choice
and 6 essay

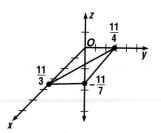

---

### Pages 147–149  Lesson 3-9

**5.** yes  **7.** no  **9.** yes  **11.** $(-9, 2, 4)$  **13.** $(0, 1, 2)$
**15.** $(5, 3, 7)$  **17.** $(-8, 13, -5)$  **19.** $(0, 2, -1)$
**21.** $(-2, 1, 3)$  **23.** $(4, -8, 3)$  **25.** $\left(\dfrac{2}{3}, 1, \dfrac{3}{2}\right)$
**27.** $\left(\dfrac{1}{2}, \dfrac{1}{3}, \dfrac{1}{4}\right)$  **29.** 2, 1, 3  **33.** pizza–\$1.05, salad–
\$1.75, soda–\$0.50  **35.** $x$: $-8$, $y$: 4, $z$: $-6$  **36.** $x$: 4,
$y$: 6, $z$: $-12$  **37.** $(1, -2)$  **38.** $c = 25(t - 0.5) + 35$
**39.** $y = -\dfrac{3}{2}x - \dfrac{1}{2}$

### Pages 150–152  Chapter 3 Summary and Review

**1.** infinite; consistent
and dependent

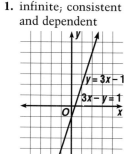

**5.** $(6.25, 1.75)$
**7.** $\left(\dfrac{14}{5}, \dfrac{4}{5}\right)$
**9.** $-2$
**11.** $(7, 2)$

**13.**

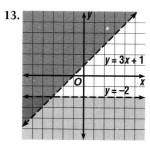

**17.** vertices: $(0, 0), (0, 3),$
$(2, 0), \left(\dfrac{3}{2}, \dfrac{3}{2}\right)$;
max: $f(0, 3) = 12$;
min: $f(0, 0) = 0$

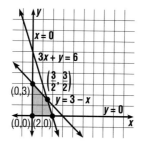

**19.** 2  **21.** none
**23.** 1  **25.** 3

---

## CHAPTER 4  MATRICES

### Pages 157–158  Lesson 4-1

**3.** Carol is a teacher and does sculpture. Dena is a
lawyer and fixes cars. Rae is a doctor and likes
gardening.  **5.** 16: A, B, C, D, E, H, I, K, M, O, T, U,
V, W, X, Y  **7a.** yes  **7b.** yes  **7c.** no  **7d.** yes
**9.** rabbit, duck

### Pages 164–166  Lesson 4-2

**7.** $V_{1 \times 3}$; $[14\ \ -10\ \ -52]$  **9.** $x = 2\dfrac{1}{2}$, $y = 1$, $z = 3$
**11.** $A_{3 \times 2} = \begin{bmatrix} 8 & -12 \\ 16 & 4 \\ 0 & 12 \end{bmatrix}$  **13.** $C_{1 \times 3} = [10\ \ -4\ \ 5]$
**15.** $x = 1$, $y = 2$  **17.** $x = 5$, $y = 3$, $z = 2$  **19.** $x = 2$,
$y = -5$  **21.** $\begin{bmatrix} 7 & 0 & 23 \\ 10 & 15 & -15 \end{bmatrix}$  **23.** $\begin{bmatrix} 28 \\ -16 \\ 3 \end{bmatrix}$
**25a.** $\begin{bmatrix} 4 & 4 & 4 & 4 \\ -2 & -2 & -2 & -2 \end{bmatrix}$  **25b.** $B'(10, -1), T'(1, -7),$
$U'(7, 3)$  **27.** $x = 2$, $y = 7$, $z = -2$  **29.** $x = 3$, $y = 5$,
$m = 10$, $r = 2$  **31.** $M(0, 0)$; $N(-3, 10)$; $P(-8, 6)$;
$Q(-10, 4)$  **33a.** $M = \begin{bmatrix} 120 & 97 & 64 & 75 \\ 80 & 59 & 36 & 60 \\ 72 & 84 & 29 & 48 \end{bmatrix}$,
$T = \begin{bmatrix} 112 & 87 & 56 & 74 \\ 84 & 65 & 39 & 70 \\ 88 & 98 & 43 & 60 \end{bmatrix}$, $S = \begin{bmatrix} 232 & 184 & 120 & 149 \\ 164 & 124 & 75 & 130 \\ 160 & 182 & 72 & 108 \end{bmatrix}$
**33b.** 106.25 pounds  **34.** $2x - 7y = -16$  **35.** $1\dfrac{2}{9}$
**36.** $\left\{a | a \geq \dfrac{4}{3}\right\}$  **37.** $-139$

### Pages 170–172  Lesson 4-3

**5.** 31  **7.** no determinant  **9.** $x = \pm 7$
**11.** $A = \dfrac{1}{2}\begin{vmatrix} 4 & -5 & 1 \\ 3 & 8 & 1 \\ -2 & 3 & 1 \end{vmatrix}$  **13.** 87  **15.** no determinant
**17.** $x = -\dfrac{14}{3}$ or 2  **19.** 24  **21.** $-60$  **23.** $x = 4, -1$
**25.** $x = \dfrac{5}{3}, -4$  **27.** If the area $= 0$, the points are
collinear.  **31.** 5  **32.** $\begin{bmatrix} -2 & 9 & 22 \\ 20 & 12 & -1 \end{bmatrix}$

**33.** $\begin{bmatrix} -28 & 20 & -44 \\ 8 & -16 & 36 \end{bmatrix}$   **34.**

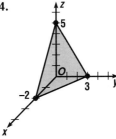

**35.** 15, 9   **36.** $5x - 6y = 30$

## Pages 175–178   Lesson 4-4

**7.** $4 \times 2$   **9.** not defined   **11.** $\begin{bmatrix} 5 & -2 \\ 1 & 2 \end{bmatrix}$   **13.** $1 \times 2$

**15.** not defined   **17.** $\begin{bmatrix} 15 & -8 & -10 \\ -7 & 23 & 16 \end{bmatrix}$   **19.** $[-1, 33]$

**21.** $[-18]$   **23.** not possible to evaluate
**25.** $D'(1, -1), E'(1, -4), F'(3, -4), G'(3, -1)$   **27.** not
defined   **29.** $\begin{bmatrix} -39 & 9 \\ 5 & 16 \end{bmatrix}$   **31.** $(-5, 3), (7, 2), (4, -1)$
**33.** $w = 1, y = 0, x = 0, z = 1$; the same matrix you
began with
**35.** nickels  $243.20     **37.** $-68$   **38.** $(-6, -1)$
     dimes    $199.80
     quarters  $521.95
            $964.95

**39.** 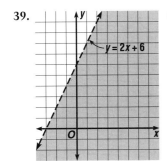   **40.**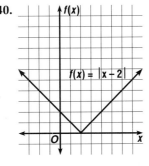

**41.** $164, $123

## Pages 182–183   Lesson 4-5

**5.** yes   **7.** no, det = 0   **9.** $-\dfrac{1}{3}\begin{bmatrix} 1 & -2 \\ -2 & 1 \end{bmatrix}$

**11.** $\dfrac{1}{32}\begin{bmatrix} 1 & 5 \\ -6 & 2 \end{bmatrix}$   **13.** $\dfrac{1}{7}\begin{bmatrix} 1 & -1 \\ 4 & 3 \end{bmatrix}$   **15.** no inverse
exists   **17.** true   **19.** false
**23.** $\begin{bmatrix} -5 & -2 & 15 & -15 \\ 21 & -6 & -63 & 27 \\ -11 & 18 & 33 & 23 \end{bmatrix}$

**24.**

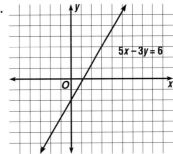

**25.** $\dfrac{9}{2}, \dfrac{4}{9}, -2$   **26.** 18 arrangements

## Page 183   Mid-Chapter Review

**1.** Sue/Lou, Liz/Matt, Jane/Bob   **2.** $M'(-1, 8),$
$A'(-2, -3), T'(-7, 2), H'(-11, 8)$   **3.** $\begin{bmatrix} 4 & -0.5 \\ 0 & -1.5 \end{bmatrix}$

**4.** $\begin{bmatrix} 4 & 1 \\ 0 & -8 \\ -2 & -16 \end{bmatrix}$   **5.** $-119$   **6.** 30 square units

**7.** $\begin{bmatrix} -15 & 6 \\ 19 & 43 \\ -6 & -24 \end{bmatrix}$   **8.** cannot be evaluated

**9.** $-\dfrac{1}{17}\begin{bmatrix} 1 & -5 \\ -3 & -2 \end{bmatrix}$

## Pages 188–189   Lesson 4-6

**3.** $\begin{bmatrix} 5 & 2 \\ 2 & 9 \end{bmatrix} \cdot \begin{bmatrix} a \\ b \end{bmatrix} = \begin{bmatrix} -49 \\ 5 \end{bmatrix}$   **5.** $5x + y = 26$
                                   $2x - 3y = 41$
**7.** $\left(\dfrac{1}{3}, -\dfrac{2}{3}\right)$   **9.** $3x + y = 13$   **11.**     $x - 2y = -8$
                       $4x - 2y = 24$      $3x + y + 2z = 9$
                                       $4x - 3y + 3z = 1$
**13.** $\begin{bmatrix} 2 & 5 \\ 3 & 4 \end{bmatrix} \cdot \begin{bmatrix} x \\ y \end{bmatrix} = \begin{bmatrix} 1 \\ 12 \end{bmatrix}$   **15.** $\left(\dfrac{3}{4}, \dfrac{1}{2}\right)$
**17.** $\left(\dfrac{2}{3}, 1, -\dfrac{4}{3}\right)$   **19.** $\left(\dfrac{3}{4}, \dfrac{2}{3}\right)$   **21.** 7 trips for the 10-ton
truck and 13 for the 12-ton truck
**23.** $\dfrac{1}{18}\begin{bmatrix} 6 & 2 \\ 3 & 4 \end{bmatrix}$   **24.** $\begin{bmatrix} 1 & 0 & 0 \\ 0 & 1 & 0 \\ 0 & 0 & 1 \end{bmatrix}$   **25.** $(1, -3), (-1, 3),$
$(5, 6), (5, 1)$; $17; -13$   **26.** $-\dfrac{11}{6}$   **27.** $3x + 7y = 5$
**28.** $\{x \mid 7 < x < 10\}$   **29.** $\left\{t \mid t < \dfrac{4}{3} \text{ or } t > \dfrac{14}{3}\right\}$

## Pages 192–193   Lesson 4-7

**5.** $\begin{vmatrix} 6 & 1 \\ 5 & -8 \end{vmatrix}$

**7.** $x = \dfrac{\begin{vmatrix} -6 & 4 & -1 \\ 2 & -2 & 3 \\ -10 & 2 & -4 \end{vmatrix}}{\begin{vmatrix} 2 & 4 & -1 \\ 1 & -2 & 3 \\ 1 & 2 & -4 \end{vmatrix}}, \quad y = \dfrac{\begin{vmatrix} 2 & -6 & -1 \\ 1 & 2 & 3 \\ 1 & -10 & -4 \end{vmatrix}}{\begin{vmatrix} 2 & 4 & -1 \\ 1 & -2 & 3 \\ 1 & 2 & -4 \end{vmatrix}},$

$z = \dfrac{\begin{vmatrix} 2 & 4 & -6 \\ 1 & -2 & 2 \\ 1 & 2 & -10 \end{vmatrix}}{\begin{vmatrix} 2 & 4 & -1 \\ 1 & -2 & 3 \\ 1 & 2 & -4 \end{vmatrix}}; \left(-3, \dfrac{1}{2}, 2\right)$  **9.** yes  **11.** no

**13.** no unique solution  **15.** $\left(-\dfrac{1}{3}, 2, 7\right)$  **17.** $\left(6, -\dfrac{1}{2}, 2\right)$

**19.**  $\quad 2c + 3f = 3.65$
$\quad\quad c + 2m = 2.47$
$\quad c + 2f + m = 3.01$

**23.** 17 small, 24 medium, and 11 large  **24.** a matrix with 2 rows and 4 columns

**25.** 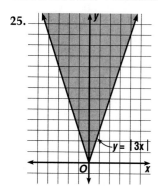  **26.**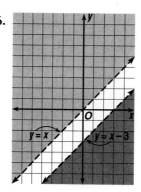

**27.** $12.02

**Pages 199–200   Lesson 4-8**

**5.**  $\quad x + 3z = -2$
$\quad 3x + 9y - 2z = -5$
$\quad -4x + y - 7z = 3$

**7.** $\begin{bmatrix} 5 & -3 & 7 \\ 3 & 9 & 3 \end{bmatrix}; \left(1, -\dfrac{2}{3}\right)$

**11.** $\begin{bmatrix} 7 & -3 & 41 \\ 2 & 5 & 0 \end{bmatrix}; (5, -2)$  **13.** unique  **15.** an equation in two variables  **17.** $(-1, 2, -3)$  **19.** 48°, 24°, 108°  **23.** chicken, $0.95 per piece; salad, $1.05; roll, $0.35  **24.** $\begin{bmatrix} 1 & 5 & 2 \\ 3 & -3 & 2 \\ 2 & 4 & -1 \end{bmatrix} \cdot \begin{bmatrix} x \\ y \\ z \end{bmatrix} = \begin{bmatrix} 10 \\ 2 \\ -15 \end{bmatrix}$

**25.** 34  **26.** $\left(\dfrac{2}{3}, 5\right)$  **27.** $(4, 2)$  **28.** Distributive property

**Pages 202–204   Chapter 4 Summary and Review**

**1.** $P_{2 \times 3}$  **3.** $\begin{bmatrix} 24 & -9 & 6 \\ 12 & 3 & 21 \end{bmatrix}$  **5.** $(3, 8)$  **7.** $-36$

**9.** $[30 \quad 31]$  **11.** $\begin{bmatrix} 1 & 0 & 0 & 0 \\ 0 & 1 & 0 & 0 \\ 0 & 0 & 1 & 0 \\ 0 & 0 & 0 & 1 \end{bmatrix}$  **13.** $\dfrac{1}{2}\begin{bmatrix} 7 & -6 \\ -9 & 8 \end{bmatrix}$

**15.** $(5, -3)$  **17.** $\left(-\dfrac{1}{2}, 1, 6\right)$  **19.** $(6, -1)$  **21.** Alan: soup, Bill: salad, Cathy: sandwich  **23.** $A'(9, -3)$, $B'(1, 3)$, $C'(2, -4)$  **25.** $M'(0, 5)$, $P'(-6, -2)$, $Q'(4, 3)$

## CHAPTER 5   POLYNOMIALS

**Pages 213–214   Lesson 5-1**

**5.** yes, 1, 2  **7.** yes, $-5$, 2  **9.** no  **11.** yes, 0, none  **13.** $8.104 \times 10^2$  **15.** $9 \times 10^9$  **17.** $7.21 \times 10^7$  **19.** 42,000  **21.** 57  **23.** 3,210,000  **25.** $x$  **27.** $ab^2$  **29.** $y^{12}$  **31.** $8^{14}$  **33.** $81a^4$  **35.** $\dfrac{1}{2}x^5y^{10}$  **37.** $7a^6$  **39.** $114a^4b^4$  **41.** $9.025 \times 10^7$; 90,250,000  **43.** $3.15 \times 10^6$; 3,150,000  **45.** $1.904 \times 10^2$; 190.4  **47.** $2.475 \times 10^8$; 247,500,000  **49.** $1.2 \times 10^5$; 120,000  **53.** $3.75 \times 10^5$ kilometers  **55.** $(2, 1, -1)$  **56.** $(2, 5, 4)$  **57.** 27, 15  **58.** $y = 4 - \dfrac{3}{4}x$  **59.** $\{x|4 \le x \le 2\}$  **60.** yes; 1

**Pages 218–219   Lesson 5-2**

**5.** $x^2$  **7.** $r^3$  **9.** $\dfrac{5}{y^3}$  **11.** $\dfrac{1}{3}$  **13.** 1  **15.** 10,000  **17.** $t^2$  **19.** $3x^5$  **21.** $-12s^3$  **23.** $\dfrac{y^7}{x^3}$  **25.** $\dfrac{1}{7z^2}$  **27.** $\dfrac{1}{5b^2}$  **29.** $4b^2c^3$  **31.** $\dfrac{1}{2}$  **33.** $-3s^6$  **35.** $-\dfrac{cd^4}{12}$  **37.** $a^4b^2$  **39.** 243  **41.** $\dfrac{1}{25}$  **43.** $3ab^3$  **45.** $\dfrac{1}{ab}$  **47.** 29  **49.** $4t^{14}$  **51.** $\dfrac{4}{9y^4}$  **53.** $5 \times 10^0$; 5  **55.** $6 \times 10^0$; 6  **57.** $3.\overline{27} \times 10^{-4}$; $0.0003\overline{27}$  **59.** $2.1 \times 10^6$; 2,100,000  **63.** $1.2 \times 10^{-4}$ or 0.00012 seconds  **64.** $4.5 \times 10^9$; 4,500,000,000  **65.** $5.0625 \times 10^{12}$; 5,062,500,000,000  **66.** $\begin{bmatrix} -14 & -15 \\ 4 & 20 \end{bmatrix}$  **67.** 2  **68.** one answer is $x$: first by 4, second by $-3$; $y$: first by 3, second by 4.  **69.** $x - 8y = -4$  **70.** 21

**Pages 221–222   Lesson 5-3**

**5.** 4 inches  **7.** 10  **9.**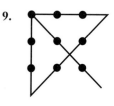
**11.** Any line in the plane of the square that passes through the point of intersection of the diagonals of the square separates it into two congruent parts, so there are infinitely many ways.  **13.** $10^2 - 2^2$, $11^2 - 5^2$, $14^2 - 10^2$, $25^2 - 23^2$  **15.** $\approx 1473$ square feet

**5.** 2   **7.** 6   **9.** 3   **11.** 9   **13.** $2x - 2y$
**15.** $3p^3 - 6p^2 + 9p$   **17.** $10y^2 + 6y$   **19.** $-9a + 13b$

**21.** $4a^2 - 10d + 20$   **23.** $r^2 - r + 6$   **25.** $-19y$
**27.** $4gf^2 - 4fbh$   **29.** $\dfrac{1}{x^2} + \dfrac{1}{x^3} - \dfrac{3}{x^4}$   **31.** $12a^2 +$

$4ab + \dfrac{b}{4}$   **33.** $x^2 + 9x + 14$   **35.** $s^4 + s^2 - 20$
**37.** $6x^2 + 31x + 35$   **39.** $2w^4 - 7w^2 - 15$
**41.** $42p^2 - 89p + 45$   **43.** $a^2 - 2ab + b^2$   **45.** $d^2 +$
$6d + 9$   **47.** $y^2 - 4y + 4$   **49.** $4p^2 + 4pq^3 + q^6$
**51.** $16m^2 - 24mn + 9n^2$   **53.** $1 + 8r + 16r^2$
**55.** $3x^2 + 9xy + 6y^2$   **57.** $64x^2 - 4y^2$   **59.** $2x^3 -$
$9x^2 - 7x + 24$   **61.** $3m^3 - 7m^2 - 24m + 16$
**63.** $b^3 + 2b^2 - 5b - 6$   **65.** $2a^3 - 7a^2 + 4a + 4$
**67.** $2k^3 - 11k^2 + 21k + 63$   **71.** $19.95 +$
$0.25\,(m - 50)$   **72.** $16 - 4\pi$ or about 3.4 square

inches   **73.** 64 square meters   **74.** $\begin{bmatrix} \frac{1}{3} & 0 \\ \frac{1}{15} & \frac{1}{5} \end{bmatrix}$

**75.** $\left(\dfrac{3}{4}, \dfrac{1}{2}\right)$   **76.** The graph of $y = 3[x]$ jumps by threes
at intervals of one unit. The graph of $y = [3x]$ jumps
by ones at intervals of $\dfrac{1}{3}$ unit.   **77.** 449

### Page 228   Mid-Chapter Review

**1.** $8d^2 - 5c^2$   **2.** 0   **3.** $48x^6y^3$   **4.** $\dfrac{4}{243}x^8y^5$
**5.** $1.536 \times 10^{12}$; 1,536,000,000,000   **6.** $3b^2$
**7.** $2m^4n^2$   **8.** $4y$   **9.** $-\dfrac{1}{8w^3t}$   **10.** $3 \times 10^7$; 30,000,000
**11.** $1.4208\overline{3} \times 10^3$, $1420.8\overline{3}$   **12.** 640 acres
**13.** $-x - 9y$   **14.** $-2d^2 + 5d + 6$
**15.** $6y^2 + 11y - 72$   **16.** $x^2 + 4x - 21$
**17.** $m^2 + 4m + 4$   **18.** $4x^2 - 12x + 9$
**19.** $x^3 - 4x^2 - 3x + 18$   **20.** $2y^3 - 22y + 28$

### Pages 233–234   Lesson 5-5

**5.** $3(s + t)$   **7.** $a(b + c)$   **9.** $(r + 3)(r - 3)$
**11.** $(10 + m)(10 - m)$   **13.** $(3y - 2)(y + 4k)$
**15.** $3p(3p - q)$   **17.** $-5x(3x + 1)$
**19.** $2(c - d)(ab + 5d)$   **21.** $(y + 3)^2$   **23.** $(r + 8)^2$
**25.** $3(a^2 + 2a + 3y)$   **27.** $5xy(x - 2y)$
**29.** $(x + 2)(x^2 - 2x + 4)$   **31.** $(f - 9)^2$   **33.** $(s + 6)^2$
**35.** $(2x - 3)(2x + 3)$   **37.** $(2s - 7)(2s - 3)$
**39.** $x(x + 7)(x - 5)$   **41.** $(p - 2b)^2$
**43.** $(f - 1)(f^2 + f + 1)$   **45.** $\left(x + y - \dfrac{1}{2}\right)\left(x + y + \dfrac{1}{2}\right)$
**47.** $(m - k + 3)(m + k - 3)$
**49.** $(a + b)(1 + 3a - 3b)$
**51.** $(1 - 2m^2)(1 + 2m^2 + 4m^4)$
**57.** $(3x - 5)(x + 1)$;

$x + 1$ □

$3x - 5$

**58.** $s^2 + 6s + 9$   **59.** $14x^2 + 26x - 4$   **60.** $3x = 12$;
$x - 2y = 8$   **61.** $M_{3\times5}$   **62.** $f(0.1, -0.3) = 0.36$
**63.** no   **64.** $t = 12$

### Pages 238–240   Lesson 5-6

**5.** $5b - 4 + 7a$   **7.** $2rs + s - 3r$   **9.** $\dfrac{1}{4(y^2 - 5)^2}$
**11.** $c + 5$   **13.** $-\dfrac{w^2}{w + 1}$   **15.** $4q^2 + 3pq - 5p$
**17.** $2k^2 - 3py + 4p^2y$   **19.** $a - 12$   **21.** $n - 15$
**23.** $2x + 7 + \dfrac{5}{x - 3}$   **25.** $-s - 10 + \dfrac{44}{6 - s}$
**27.** $2y^2 + 5y + 2$   **29.** $3x^2 - 2x + 3$
**31.** $r^2 - 6r + 9 - \dfrac{1}{r - 3}$   **33.** $2x^2 - 3x - 2$
**35.** $2t^2 + 2t - \dfrac{3}{t - 1}$   **37.** $s^2 + 2s + 4$
**39.** $x^2 - 2x + 3$   **41.** $a + 1, a - 1$
**43.** $-48$   **47.** $\dfrac{1}{8}$ inch   **48.** $4(a - 2)(a + 2)$
**49.** $(d - 13)(d + 2)$   **50.** $\dfrac{1}{3}$   **51.** 7 chairs, 4 tables
**52.** $-2, -1, 0$   **53.** I, R

### Pages 244–245   Lesson 5-7

**5.** no   **7.** no   **9.** no   **11.** $2b^2 - b - 1 + \dfrac{4}{b + 1}$
**13.** $3x^2 + x + 3 + \dfrac{2}{x - 1}$   **15.** $3r^3 - 9r^2 + 7r - 6$   **17.**
$2b^2 - 5b - 3$   **19.** $x^2 + 4x + 3$
**21.** $n^3 - 3n^2 - 15n - 21$
**23.** $z^4 + 2z^3 - 2z^2 - \dfrac{3}{z - 2}$
**25.** $2s^3 + s^2 - 2s + \dfrac{3}{2s - 1}$
**27.** $2x^3 + 3x^2 + 2x - 1 - \dfrac{13}{2x - 3}$
**29.** $y^4 + 2y^3 + 4y^2 + 5y + 10$   **31.** Both are 0
**33.** $2w - 4$ inches   **35.** $x^3 + 5x^2 + 8x + 26 + \dfrac{70}{x - 3}$
**37.** $2x^4 + x^3 - 7x^2 + 13x - 21 + \dfrac{24}{x + 1}$
**39.** $x^4 - 3x^3 - 6x^2 + 8x + 36 - \dfrac{36}{x + 3}$   **40.** $x - 6$
**41.** $a^2 + 3a - 2 - \dfrac{1}{a + 4}$   **42.** $9 \times 1$   **43.** 3
**44.** $11x + y = 2$

### Pages 246–247   Chapter 5 Summary and Review

**1.** $m$   **3.** $y^{13}$   **5.** $144x^4y^6$   **7.** $1.344 \times 10^{15}$;
1,344,000,000,000,000   **9.** $a^2$   **11.** $\dfrac{1}{64}$   **13.** $\dfrac{1}{a^6b^4}$
**15.** $2 \times 10^7$; 20,000,000   **17.** $2x + 5y$
**19.** $m^2 + 3m - 10$   **21.** $n^3 - n^2 - 5n + 2$
**23.** $5(a + b)$   **25.** $(s + 1)(s + 6)$   **27.** $(b - 4)$
$(b^2 + 4b + 16)$   **29.** $(x + 3)^2$   **31.** $2y(y - 7)(y + 7)$
**33.** $6y + 4$   **35.** $4s - 2$   **37.** $x^2 + 3 + \dfrac{5}{x - 4}$

**39.** $3y^2 + y - 1 + \dfrac{2}{2y + 3}$  **41.** $5.865696 \times 10^{12}$ or 5,865,696,000,000 miles  **43.** one-half  **45.** $(0.9)^3 p$ or $0.729\,p$

## CHAPTER 6  IRRATIONAL AND COMPLEX NUMBERS

### Pages 255–257   Lesson 6-1
**7.** $-11$  **9.** $y$  **11.** $-|x|$  **13.** $-5$  **15.** $4|a|b^2$
**17.** $|x + 3|$  **19.** $7.416$  **21.** $-3.609$  **23.** $8.000$
**25.** $-9.950$  **27.** $2.844$  **29.** $3.448$  **31.** $14$  **33.** $3$
**35.** $-6$  **37.** $-10$  **39.** $0.5$  **41.** $5|y^3|$  **43.** $\pm 24$
**45.** $8|a|b^2$  **47.** $-2bc$  **49.** $4a^2b$  **51.** $|3x + y|$
**53.** $s + t$  **55.** $|r + s|$  **57.** $|x + 5|$  **59.** $|3a + 1|$
**61.** $|s - t|$  **65.** about 4.2 rotations per minute
**66.** $t^2 - 2t + 1$  **67.** $x^2 - x + 7 + \dfrac{3}{5x - 3}$
**68.** $1.4985 \times 10^7$; 14,985,000  **69.** $(-1, 0, 1)$
**70.** $2 \times 3$, no determinant  **71.** $4$  **72.** a step function

### Pages 262–263   Lesson 6-2
**5.** $3\sqrt{3}$  **7.** $7y^2\sqrt{2}$  **9.** $2\sqrt[3]{2}$  **11.** $y\sqrt{y}$  **13.** $t\sqrt[4]{t}$
**15.** $\dfrac{\sqrt{3}}{\sqrt{3}}; \dfrac{\sqrt{3}}{3}$  **17.** $\dfrac{\sqrt{b}}{\sqrt{b}}; \dfrac{3\sqrt{b}}{b}$  **19.** $\dfrac{\sqrt[3]{3}}{\sqrt[3]{3}}; \dfrac{7\sqrt[3]{3}}{3}$
**21.** $12\sqrt{6}$  **23.** $2\sqrt[3]{7}$  **25.** $15\sqrt{3}$  **27.** $4xy\sqrt{x}$
**29.** $\sqrt{2}$  **31.** $\sqrt[4]{11}$  **33.** $36\sqrt{6}$  **35.** $2\sqrt[4]{7}$  **37.** $48\sqrt{7}$
**39.** $3\sqrt{2} - 2\sqrt{3}$  **41.** $3a^2b\sqrt[6]{b}$  **43.** $\sqrt[3]{9}$  **45.** $\dfrac{\sqrt{15a}}{6a}$
**47.** $\sqrt{ab} + a\sqrt{b}$  **49.** $\dfrac{3\sqrt[3]{2}}{5}$  **51.** $\dfrac{\sqrt{10x}}{8x}$  **53.** 7 inches
**55.** 9.73 inches  **57.** $25b^2$  **58.** $-11|bc^3|$
**59.** $3.\overline{8} \times 10^{-6}$; $0.0000038\overline{8}$  **60.** No, only square matrices have inverses.  **61.** $x = 8, y = 4, z = -3$
**62.** $-8$

### Pages 266–268   Lesson 6-3
**5.** $5 + \sqrt{7}$  **7.** $\sqrt{3} - \sqrt{10}$  **9.** $2 + 2\sqrt{3}$  **11.** $\sqrt{3}$
**13.** $11\sqrt[3]{6}$  **15.** $4\sqrt{2}$  **17.** $5\sqrt{5} - 10\sqrt{2}$  **19.** $-7\sqrt{3}$
**21.** $17 + 7\sqrt{5}$  **23.** $9x^2 - 5y$  **25.** $19 + 8\sqrt{3}$
**27.** $4\sqrt{5} + 23\sqrt{6}$  **29.** $-\sqrt[3]{2}$  **31.** $\sqrt[3]{6}$
**33.** $15\sqrt[3]{5} - 6\sqrt[3]{3}$  **35.** $25 - 5\sqrt{2} + 5\sqrt{6} - 2\sqrt{3}$
**37.** $49 - 11y$  **39.** $\dfrac{3 - \sqrt{5}}{4}$  **41.** $14\sqrt{6} + 2\sqrt[3]{3}$
**43.** $15\sqrt[4]{2}$  **45.** $|y| + y^2 + y^4$  **47.** $m^3 + 4$
**49.** $\dfrac{\sqrt{x^2 - 1}}{x - 1}$  **51.** $\dfrac{16\sqrt{10}}{5}$  **55.** $2\sqrt{7}$ or 5.3 seconds
**56.** $15\sqrt{6}$  **57.** $5mn^2\sqrt[4]{m}$
**58.** $(1 - 2a)(1 + 2a + 4a^2)$  **59.** $c^9$  **60.** $3 \times 3$; $-68$
**61.** maximum: 6 at $(6, 0)$; minimum: $-5$ at $(0, 5)$

### Pages 272–274   Lesson 6-4
**5.** $4$  **7.** $8$  **9.** $\dfrac{1}{9}$  **11.** $49$  **13.** $81$  **15.** $6$  **17.** $\sqrt{6}$
**19.** $\sqrt{2}$  **21.** $\sqrt[3]{4}$  **23.** $17^{\frac{1}{3}}$  **25.** $y^{\frac{1}{4}}$  **27.** $5a^2 b^{\frac{3}{2}}$
**29.** $27^{\frac{1}{4}}$  **31.** $n^{\frac{2}{3}}$  **33.** $16^{\frac{1}{3}} a^{\frac{5}{3}} b^{\frac{7}{3}}$  **35.** $\sqrt{6}$  **37.** $\sqrt[4]{n^3}$
**39.** $2a^2\sqrt[3]{4a}$  **41.** $p^2\sqrt[4]{p^2 q^3}$  **43.** $\sqrt[5]{9r^2 s^3}$
**45.** $\sqrt[3]{5s^2 t}$  **47.** $\dfrac{1}{2}$  **49.** $4$  **51.** $81$  **53.** $3$  **55.** $0.2$
**57.** $36$  **59.** $\sqrt[6]{x^2 y^3}$  **61.** $\sqrt{2}$  **63.** $yz^2\sqrt[6]{x^5 y^3 z^2}$
**65.** $3\sqrt[6]{3}$  **69.** $\$1797.27$  **71.** $5 - 3\sqrt{3}$  **72.** $5\sqrt{3}$
**73.** $a - 2b$  **74.** $\dfrac{1}{14}\begin{bmatrix} 4 & 1 \\ 2 & -3 \end{bmatrix}$  **75.** 70¢
**76.** $29x - 10y$

### Page 274   Mid-Chapter Review
**1.** $-9|x|$  **2.** $|a + 7|$  **3.** $-4x^3$  **4.** $4|m|n\sqrt{3n}$
**5.** $3\sqrt{2} + 10\sqrt{3}$  **6.** $\dfrac{\sqrt{5}}{3}$  **7.** $\dfrac{6\sqrt[3]{2x^2}}{x}$  **8.** $3b^2 r^2 \sqrt[4]{3r}$
**9.** 3.53 seconds  **10.** $-8\sqrt{2}$  **11.** $29 - 3\sqrt{3}$  **12.** $114$
**13.** $\dfrac{11 + 7\sqrt{3}}{13}$  **14.** $12$  **15.** $\sqrt[12]{5^8 x^6 y^9}$  **16.** $3^{\frac{1}{2}} a^{\frac{2}{3}} b^{\frac{1}{3}}|c|$

### Page 276   Lesson 6-5
**5.** $6$  **7.** $48$  **9.** 16 cm by 16 cm

### Pages 279–280   Lesson 6-6
**5.** $\dfrac{4^{\frac{1}{2}}}{4^{\frac{1}{2}}}$, 2  **7.** $\dfrac{a^{\frac{2}{3}}}{a^{\frac{2}{3}}}$, $\dfrac{a^{\frac{2}{3}}}{a}$  **9.** $\dfrac{t^{\frac{1}{2}} - 1}{t^{\frac{1}{2}} - 1}$, $\dfrac{t^{\frac{1}{2}} - 1}{t - 1}$  **11.** $\dfrac{q^{\frac{1}{2}} + r^{\frac{1}{2}}}{q^{\frac{1}{2}} + r^{\frac{1}{2}}}$,
$\dfrac{q^{\frac{3}{2}} + qr^{\frac{1}{2}}}{q - r}$  **13.** $\dfrac{t^{\frac{3}{2}} - s^{\frac{1}{2}}}{t^{\frac{3}{2}} - s^{\frac{1}{2}}}$, $\dfrac{2t^{\frac{3}{2}} - 2s^{\frac{1}{2}}}{t^3 - s}$  **15.** $\dfrac{w + w^{\frac{1}{2}}}{w + w^{\frac{1}{2}}}$,
$\dfrac{w^2 + w^{\frac{3}{2}} + w + w^{\frac{1}{2}}}{w^2 - w}$  **17.** $\dfrac{s^{\frac{5}{5}}}{s}$  **19.** $\dfrac{t^{\frac{1}{6}}}{t}$  **21.** $\dfrac{x^{\frac{1}{2}} - 1}{x - 1}$
**23.** $\dfrac{t^{\frac{1}{2}}}{t(t + 1)}$  **25.** $4 \cdot 6^{\frac{1}{3}}$  **27.** $\dfrac{ab^{\frac{1}{2}} c^{\frac{1}{2}}}{c^2}$  **29.** $\dfrac{n^2 + 3}{n}$
**31.** $3x^{\frac{5}{3}} + 4x^{\frac{8}{3}}$  **33.** $\dfrac{3xy^{\frac{3}{2}} z^{\frac{2}{3}}}{z}$  **35.** $\dfrac{r^2 - 2r^{\frac{3}{2}}}{r - 4}$  **37.** $\dfrac{1}{b^{\frac{3}{3}} - 1}$
**39.** $3xy^3$  **41.** $\dfrac{b^2}{a}$  **45.** $\dfrac{20}{3^t}$ units  **46.** $\dfrac{1}{2}$  **47.** $5^{\frac{3}{7}}$
**48.** $2x^2 - 4x + 4$  **49.** They are the same line.
**50.** $1$

### Pages 284–286   Lesson 6-7
**5.** $25$  **7.** $16$  **9.** $14$  **11.** $3$  **13.** $-\dfrac{\sqrt{2}}{2}$  **15.** $-\sqrt{3}$
**17.** $\dfrac{12 - 4\sqrt{2}}{7}$  **19.** $\dfrac{1 + \sqrt{5}}{-2}$  **21.** $13$  **23.** $7$  **25.** $5\dfrac{1}{3}$

27. 23  29. −1  31. 3  33. 21  35. 8  37. 4
39. no solution  41. 5.41  43. 7.41  45. 4
47. $r = \pm\sqrt{y^2 - s^2}$  49. $c = \dfrac{2mM}{r^3}$  53. $10\sqrt{15}$ or
about 38.73 centimeters on each side
55. $\dfrac{r^{\frac{1}{2}}s}{1+r}$  56. $\dfrac{25a^{21}}{b^8}$  57. 8  58. yes  59. 100 adults
and 50 students  60. Yes, each member of the domain is paired with exactly one member of the range.

**Pages 290–291  Lesson 6-8**
5. $6i$  7. $-2$  9. $-3$  11. $20i$  13. $13i$  15. $5i\sqrt{2}$
17. $\dfrac{2}{3}i$  19. $\dfrac{i\sqrt{5}}{5}$  21. $i$  23. $-i$  25. $-4$  27. $-7\sqrt{2}$
29. $-12$  31. $-8i$  33. $9i$  35. $4i\sqrt{3}$  37. 24
39. $-20i$  41. $\pm4i$  43. $\pm11i$  45. $\pm3i$  47. $\pm2i\sqrt{3}$
49. $\pm i\dfrac{\sqrt{5}}{2}$  53a. $i$  53b. $-1$  53c. $-i$  53d. 1
53e. $-i$  53f. 1  53g. $i$  53h. $-1$  54. $\dfrac{7\sqrt{2} - 14}{2}$
55. $\dfrac{6(11 + 2\sqrt{3})}{109}$  56. $|x - 2|$  57. $(a + 1)(a^2 - a + 1)$
58. 65¢, 30¢, 15¢  59. 153  60. 120

**Pages 294–296  Lesson 6-9**
5. $7 + 2i$  7. 3  9. 10  11. $20 + 12i$  13. $-10 + 10i$
15. 17  17. $x = 2, y = -3$  19. $x = 3, y = 0$
21. $x = 7, y = 2$  23. $5 + 5i$  25. $7 + 3i$  27. $3 - 8i$
29. $2 + i$  31. $5 - 3i\sqrt{3}$  33. $-9i$  35. $33 + 13i$
37. $32 - 24i$  39. $22 - 29i$  41. $-7 + 24i$  43. 3
45. 13

47. $2 + 5i$   49. $3 - 5i$

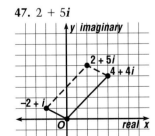
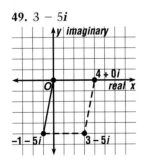

51. $x = 6, y = \dfrac{7}{2}$  53. $x = -1, y = -3$  55. $x = 3,$
$y = 1$  57. $20 + 15i$  59. $148 - 222i$
61. $130 + 110i$  63. $-a - bi$  65. $(a + bi) \cdot$
$1 = (a \cdot 1) + (bi \cdot 1) = a + bi$  67. $(1 + i, -1 + 2i),$
$(-1 + 2i, -4 - 4i), (-4 - 4i, -1 + 32i),$
$(-1, + 32i, -1024 - 64i)$  69. $\dfrac{i\sqrt{3}}{3}$  70. $\pm2i\sqrt{2}$

71. $5\sqrt[4]{5}$  72. $y^3 + 3y^2 - 16y + 55 - \dfrac{166}{y + 3}$
73. approximately $4.4 \times 10^{46}$ times  74. $\begin{bmatrix} 1 & 7 \\ 1 & 2 \\ 8 & 13 \end{bmatrix}$
75. 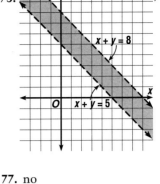  76. no solution, inconsistent

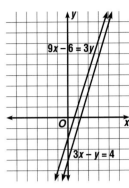

77. no

**Pages 299-301  Lesson 6-10**
5. $4 - i$  7. $5 + 4i$  9. $-5i$  11. $10i$  13. $12 + i$
15. $5 - 4i \cdot \dfrac{5 + 4i}{41} = \dfrac{25 - 16i^2}{41} = \dfrac{25 + 16}{41} = 1$
17. 68  19. 29  21. 2  23. 81  25. 100  27. $\dfrac{5 + i}{2}$
29. $\dfrac{21 + 13i}{5}$  31. $\dfrac{9 + i}{41}$  33. $\dfrac{1 - 2i}{5}$  35. $2 - i$
37. $\dfrac{12 - 10i}{61}$  39. $\dfrac{4\sqrt{3} - 8i}{7}$  41. $\dfrac{3 + i\sqrt{3}}{4}$  43. $\dfrac{2 - 3i\sqrt{5}}{7}$
45. $\dfrac{3 + 4i}{25}$  47. $\dfrac{3 - 7i}{58}$  49. $\dfrac{5 + 6i}{122}$  51. $\dfrac{1 - 3i}{4}$
53. $-2 - 8i$  55. $\dfrac{a - bi}{a^2 + b^2}$  57. $\dfrac{-3 - 4i\sqrt{7}}{11}$  59. 9
61. $\dfrac{-1 - i}{2}$  65. $7 + 7i$  66. $148 - 64i$  67. 2n
68. $1.9663 \times 10^6$  69. yes, $(-2, 4, 3)$  70. 465

71. 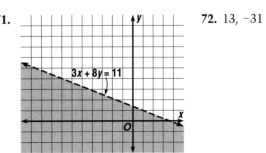  72. 13, −31

3x + 8y = 11

**Pages 302-304  Chapter 6 Summary and Review**
1. $7|x|$  3. $|3p - 5q|$  5. $4\sqrt{6}$  7. $3|a|\sqrt{2b}$  9. $\dfrac{3\sqrt{5}}{2}$
11. $10\sqrt{2} + 3\sqrt{10}$  13. $14 - \sqrt{6}$  15. $17\sqrt[3]{3x^2}$

**17.** 47  **19.** $r^{\frac{3}{4}}$  **21.** $\sqrt[3]{5}$  **23.** 5  **25.** 32  **27.** $\dfrac{5^{\frac{2}{3}}}{5}$

**29.** $\dfrac{a^{\frac{1}{4}}}{a}$  **31.** $3+3\sqrt{2}$  **33.** 7  **35.** 6  **37.** $11i$

**39.** $-15$  **41.** $21+5i$  **43.** $46+3i$  **45.** $8+13i$

**47.** $\dfrac{8-11i}{2}$  **49.** $\dfrac{8+i}{65}$  **51.** $\dfrac{5\sqrt{2}}{2}$ or about 3.54 seconds

**53.** about 23.94 cm

## CHAPTER 7   QUADRATIC EQUATIONS

### Page 311   Lesson 7-1
**5.** $A = 2, B = 1, C = 9, D = 7, E = 8$

**7.**

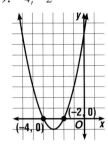

### Pages 319–321   Lesson 7-2
**7.** no  **9.** no  **11.** $-2, 1$  **13.** $0, 16$  **15.** $-6, -2$
**17.** $-\dfrac{7}{3}, -5$

**19.** $-4, -2$  **23.** $-1.5, 3$

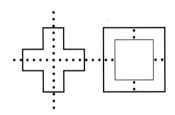

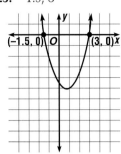

**25.** $4, -3$  **27.** 6  **29.** $4, -1$  **31.** $\dfrac{3}{2}, -7$  **33.** $\dfrac{5}{3}, -3$

**35.** $\dfrac{1}{3}, -\dfrac{3}{2}$  **37.** 4 seconds, 48 ft, 48 ft  **39.** $5, -8$

**41.** $-\dfrac{5}{6}, 1$  **43.** $0, 9, -9$  **45.** $0, 1, \dfrac{8}{9}$

**49.** 20 m × 50 m  **50.** 56, 25  **51.** $75a^3b$

**52.** $5m^3n^2\sqrt{3m}$  **53.** $\left(\dfrac{1}{4}, 6, -\dfrac{1}{6}\right)$  **54.** deluxe, 35¢; glazed, 30¢; cake, 20¢  **55.** 0.6 amperes

### Pages 324–326   Lesson 7-3
**5.** no  **7.** no  **9.** 400  **11.** $\dfrac{81}{4}$  **13.** $\dfrac{225}{4}$  **15.** $-11, 8$
**17.** $-15, 12$  **19.** $\dfrac{7\pm\sqrt{29}}{2}$  **21.** $\dfrac{-3\pm\sqrt{41}}{2}$  **23.** $\dfrac{1}{2}, \dfrac{1}{4}$
**25.** $2\pm\dfrac{2\sqrt{6}}{3}$  **27.** $\pm\dfrac{i\sqrt{ac}}{a}$  **29.** $\dfrac{-b\pm\sqrt{b^2-4ac}}{2a}$
**31.** $21\times 15$ in.  **33.** 20 seconds  **35.** $15, 5$  **36.** 22
**37.**

$$
\begin{array}{r}
a^2 - 6 \\
a - 5 \overline{)\,a^3 - 5a^2 - 6a + 30\,} \\
\underline{a^3 - 5a^2\phantom{aaaaaaaaaaa}} \\
-6a + 30 \\
\underline{-6a + 30} \\
0
\end{array}
$$

**38.**

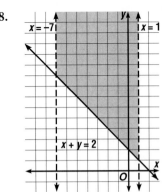

**39.** 72

### Page 326   Mid-Chapter Review
**1.** 853  **2.** $7, -5$  **3.** $-9, 3$  **4.** $-\dfrac{3}{4}, 4$  **5.** $\pm\dfrac{5}{2}$

**6.** $0, \pm9$  **7.** $\dfrac{1}{4}, -5$  **8.** $-6\pm4\sqrt{2}$  **9.** 0.52 in.

### Pages 330–331   Lesson 7-4
**7.** $a = 1, b = 0, c = -16$; 64; $\pm 4$  **9.** $a = 6, b = 2, c = 1$; $-20$; $\dfrac{-1\pm i\sqrt{5}}{6}$  **11.** $a = 3, b = -1, c = 3$; $-35$; $\dfrac{1\pm i\sqrt{35}}{6}$  **13.** 0; 1 R, Q; 2  **15.** 144; 2 R, Q; 7, $-5$

**17.** 16; 2 R, Q; $-\dfrac{3}{2}, -\dfrac{1}{2}$  **19.** 16; 2 R, Q; $-\dfrac{3}{2}, -\dfrac{5}{2}$
**21.** $-16$; 2 Im; $3\pm 2i$  **23.** 36; 2 R, Q; 0, 6
**25.** $-144$; 2 Im; $\dfrac{2\pm 3i}{2}$  **27.** $-100$; 2 Im $-2\pm 5i$
**29.** 21; 2 R, I; $\dfrac{-1\pm\sqrt{21}}{2}$; 1.79, $-2.79$  **31.** 41; 2 R, I; $\dfrac{-3\pm\sqrt{41}}{8}$; 0.43, $-1.18$  **33.** 97; 2 R, I; $\dfrac{-5\pm\sqrt{97}}{4}$; 1.21, $-3.71$  **37.** 18.2 m × 23.5 m  **39.** $-7\pm\sqrt{61}$

**40.** 8, −3

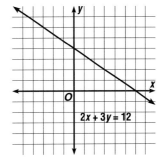

**41.** $12a^2 - 15a - 63$

**42.** $\begin{bmatrix} 1 & 1 & 1 & 10 \\ 1 & 0 & -1 & 1 \\ 1 & -1 & -1 & 0 \end{bmatrix}$

**43.** 11

**54.**

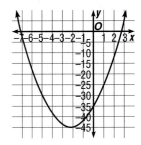

$2x + 3y = 12$

---

### Pages 336–338   Lesson 7-5

**7.** $-\dfrac{3}{4}$, −3   **9.** 0, $-\dfrac{3}{5}$   **11.** $\dfrac{1}{15}$, $-\dfrac{4}{15}$   **13.** −3, −2

**15.** 5, 4   **17.** 3, $-\dfrac{1}{2}$   **19.** $\dfrac{5 \pm \sqrt{17}}{4}$   **21.** $\dfrac{3}{4}$, −6

**23.** $x^2 - 6x - 16 = 0$   **25.** $2x^2 - 7x + 3 = 0$

**27.** $25x^2 - 4 = 0$   **29.** $x^2 - (3\sqrt{3})x + 6 = 0$   **31.** ±4   **33.** $\dfrac{4}{5}$, $-\dfrac{2}{3}$   **35.** $-\dfrac{1}{4}$, $-\dfrac{4}{3}$   **37.** $x^2 + 36 = 0$

**39.** $2x^2 - 2x - 3 = 0$   **41.** 4   **43.** −5   **45.** 240 ft/s

**47.** 64 ft/s, 48 ft   **48.** $\dfrac{5 \pm i\sqrt{7}}{4}$   **49.** 13

**50.** $-3 + 3\sqrt{2} - \sqrt{5} + \sqrt{10}$   **51.** $(b + 7)(a + 4)$

**52.** $\begin{bmatrix} 1 & 0 & 0 \\ 0 & 1 & 0 \\ 0 & 0 & 1 \end{bmatrix}$   **53.** 2 hours   **54.** 1 cm, 2 cm, 3 cm, 4 cm, 5 cm, 6 cm, 7 cm

---

### Pages 342–343   Lesson 7-6

**7.** yes   **9.** yes   **11.** yes   **13.** yes   **15.** yes   **17.** 8

**19.** 4   **21.** $(x^{\frac{1}{2}})^2 - 10(x^{\frac{1}{2}}) + 25 = 0$

**23.** $z^2 (z^2 + 6z + 8) = 0$   **25.** $(r^{\frac{1}{3}})^2 - 5(r^{\frac{1}{3}}) + 6 = 0$

**27.** $(x^{\frac{1}{4}})^2 + 7(x^{\frac{1}{4}}) + 12 = 0$   **29.** 0, −3, −2   **31.** 0, $\pm\dfrac{1}{4}$

**33.** $-1 \pm i\sqrt{3}$, 2   **35.** ±2, ±1   **37.** 5, $\dfrac{-5 \pm 5i\sqrt{3}}{2}$

**39.** $w = 4$ in., $l = 7$ in., $h = 8$ in., $V = 224$ in.$^3$

**41.** 27, 125   **43.** 4   **45.** $\dfrac{4}{9}$   **47.** 3.5 m

**49.** $\dfrac{3 \pm \sqrt{5}}{2}$   **50.** $10\sqrt{6}$   **51.** $\begin{array}{r|rrrr} -3 & 2 & 15 & 22 & -15 \\ & & -6 & -27 & 15 \\ \hline & 2 & 9 & -5 & 0 \end{array}$

**52.** $(3a - 7)(2a + 5)$   **53.** $A$ matrix $I$ such that $A \cdot I = A$. $I_{3 \times 3} = \begin{bmatrix} 1 & 0 & 0 \\ 0 & 1 & 0 \\ 0 & 0 & 1 \end{bmatrix}$

---

### Pages 344–346   Chapter 7 Summary and Review

**1.** 41312432   **3.** $-7, \dfrac{5}{2}$

**5.** $\dfrac{3}{2}$   **7.** $\dfrac{1}{4}$, $-\dfrac{3}{2}$   **9.** −3, 8   **11.** −12, 8   **13.** $\dfrac{5 \pm i\sqrt{7}}{4}$

**15.** −3, 0   **17.** 16, 2 real roots; rational; $\dfrac{4}{7}$, 0

**19.** −3, 15; 12, −45   **21.** $\dfrac{\pm\sqrt{33}}{3}$; 0, $-\dfrac{11}{3}$

**23.** $x^2 + 2x - 24 = 0$   **25.** $x^2 - 10x + 34 = 0$

**27.** $\dfrac{5}{3}$, −3, 0   **29.** 4, $-2 \pm 2i\sqrt{3}$   **31.** 2, −2

**33.** 3 feet   **35.** 60.3 seconds

---

## CHAPTER 8   QUADRATIC RELATIONS AND FUNCTIONS

### Pages 354–355   Lesson 8-1

**5.** yes   **7.** yes   **9.** no   **11.** no   **13.** $5x^2$; $-7x$; 2

**15.** $\dfrac{1}{3}n^2$; 0, 4   **17.** $z^2$; $3z$; 0   **19.** $9t^2$; $6t$; −7

**21.** $f(x) = x^2 - 6x + 9$   **23.** $h(x) = 9x^2 + 12x + 4$

**25.** $f(x) = -16x^2 + 64x - 64$   **27.** $f(x) = 4x^2 + 8x + 14$   **29.** $g(x) = 6x^2 + 24x + 29$   **31.** $r =$ the radius; area $= \pi r^2$   **33.** $s =$ the measure of the length of a leg; area $= \dfrac{1}{2}s^2$   **35.** $x =$ one of the numbers; sum $= 2x^2 - 20x + 100$   **37.** $w =$ the width of the kennel; area $= 50w - w^2$

**41.** $I(p) = 800 + 10p - 2.5p^2$   **42.** $1(x^3)^2 + 3(x^3) - 10 = 0$   **43.** 1 or 9   **44.** 85   **45.** 0.2   **46.** $x^2 - 0.4x + 0.16 - \dfrac{2.864}{x + 0.4}$   **47.** $2.592 \times 10^{10}$ km/day

**48.** $(-2, 3)$

**Pages 357–358  Lesson 8-2**

**5.** 15  **7.** 0  **9.** Remainders of successive powers of 5 divided by 7 repeat in a cycle of six: 5, 4, 6, 2, 3, 1, . . . So $5^{100} \div 7$ will have a remainder of 2.  **11.** 0
**13.** 420  **15.** 4  **17.** 2025

**Pages 363–364  Lesson 8-3**

**7.** $(4, 0)$, $x = 4$  **9.** $(0, 3)$, $x = 0$  **11.** $(6, 4)$, $x = 6$
**13.** $\left(\frac{1}{5}, 1\right)$, $x = \frac{1}{5}$  **15.** $y = [x - (-2)]^2 + 0$
**17.** $y = (x - 2)^2 + 2$  **19.** $f(x) = (x - 2)^2$, $(2, 0)$, $x = 2$  **21.** $f(x) = (x - 0)^2 + 9$, $(0, 9)$, $x = 0$
**23.** $f(x) = [x - (-5)]^2 - 25$, $(-5, -25)$, $x = -5$
**25.** $f(x) = [x - (-4)]^2 + 4$, $(-4, 4)$, $x = -4$

**27.** ID: AW-SA-363a
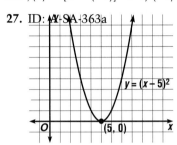

**31.**
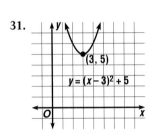

**35.**
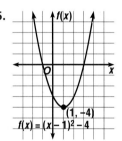

**39.**
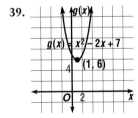

**43.** $(0, -5)$; $y = x^2 - 5$  **45.** $(-3, 2)$; $y = (x + 3)^2 + 2$

**47a.**
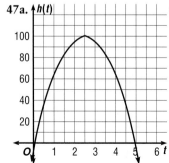

**47b.** 2.5 seconds; 100 feet

**49.** $4x^2$; $-8x$; $-2$  **50.** $x^2 - x - 56 = 0$  **51.** 14
**52.** 6  **53.** $-\frac{1}{2}$

**Pages 370–372  Lesson 8-4**

**5.** $(9, 0)$, $x = 9$, up  **7.** $(0, -6)$, $x = 0$, up
**9.** $(-3, -1)$, $x = -3$, up  **11.** $\left(-2, -\frac{4}{3}\right)$, $x = -2$, up
**13.** $f(x) = -2x^2 + 4$  **15.** $f(x) = (x - 2)^2 + 1$; $(2, 1)$; $x = 2$; up  **17.** $f(x) = -3(x - 2)^2 + 12$; $(2, 12)$; $x = 2$; down  **19.** $f(x) = 3(x - 3)^2 - 16$; $(3, -16)$; $x = 3$; up
**21.** $f(x) = -\frac{1}{2}(x - 5)^2 - 1$; $(5, -1)$; $x = 5$; down
**23.** $y = 3(x - 1)^2$  **25.** $y = -\frac{1}{2}(x + 2)^2 - 3$
**27.** $y = -\frac{3}{4}(x - 4)^2 + 1$  **29.** $y = 2x^2 - x$
**31.** $y = \frac{1}{2}x^2 - 2x - 1$  **33.**
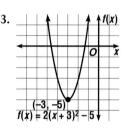

**37.**
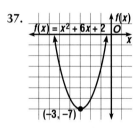

**41.**
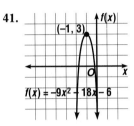

**45.** $(-2, 0)$, $x = -2$  **46.** $(3, -11)$, $x = 3$  **47.** 5, $-5$
**48.** $4 - 2\sqrt{3}$  **49.** $2x^2z + 3x + 5xyz$  **50.** no
**51.** yes  **52.**
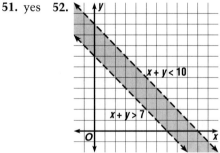

**Page 372  Mid-Chapter Review**

**1.** $g(x) = x^2 + 4x + 4$  **2.** $f(x) = 3x^2 - 30x + 79$
**3.** Let $n$ = one of the numbers; product = $50n - n^2$
**4.** Let $l$ = the length; area = $21l - l^2$  **5.** 9 in. × 9 in.  **6.** They have the same shape, but the graph of $y = (x - 7)^2$ is shifted 7 units to the right.
**7.** $(2, -4)$, $x = 2$  **8.** $(-4, 0)$, $x = -4$  **9.** $(-2, 6)$, $x = -2$  **10.** $(0, 0)$, $x = 0$  **11.** $y = x^2 + 3$
**12.** $f(x) = (x + 3)^2 - 6$, $(-3, -6)$, $x = -3$, up
**13.** $f(x) = -9(x + 1)^2 - 1$, $(-1, -1)$, $x = -1$, down

**Pages 375–378   Lesson 8-5**

**5a.** $x + 20$   **5b.** $x(x + 20)$ or $x^2 + 20x$

**5c.**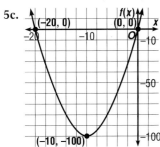

**5d.** $-10$ and $10$   **7.** $18, 18$   **9.** $\dfrac{37}{2}, \dfrac{37}{2}$   **11.** $-8, -8$
**13.** 6 in. by 6 in.; 36 sq. in.   **15.** $60 \text{ ft} \times 30 \text{ ft}$
**17.** \$350   **19.** \$3.50   **21.** 12.5 cm by 12.5 cm by
5 cm   **23.** 16 inches   **25.** 60 ft; 64 ft; 4 seconds
**29a.** 78 feet   **29b.** between 5 and 6 seconds
**29c.** about 3 seconds   **29d.** 126 feet   **30.** 1
**31.** $f(x) = (x - 3)^2$   **32.** 16   **33.** $14 + 5i$
**34.** $(3x - 2)^2$

**Pages 381–382   Lesson 8-6**

**5.** yes   **7.** yes   **9.** yes   **11.** no

**13.** 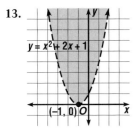   **17.**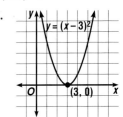

**21.** 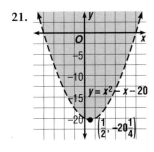   **25.**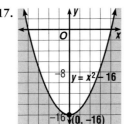

**31.** Substitute values for $x$ and $f(x)$. $30 \neq -90.3$. The
ball hits the ground about 337 ft from the plate.
**32.** 175   **33.** $-20, 20$   **34.** $8 \pm 3\sqrt{6}$   **35.** $\dfrac{2\sqrt{2}}{3}$
**36.** $1.32 \times 10^9$ kilometers

**Pages 386–387   Lesson 8-7**

**5.** both $> 0$ or both $< 0$   **7.** one $\geq 0$ and one $\leq 0$
**9.** both $> 0$ or both $< 0$   **11.** one $\geq 0$ and one $\leq 0$
**13.** one $\geq 0$ and one $\leq 0$   **15.** $\{x \mid x \leq 3 \text{ or } x \geq 6\}$

**17.** $\{x \mid x > -2 \text{ or } x < -9\}$   **19.** $\{n \mid n \geq 2.5 \text{ or } n \leq -3.8\}$
**21.** $\{q \mid q \geq 4 \text{ or } q \leq -6\}$   **23.** $\left\{b \mid -\dfrac{3}{2} < b < 2\right\}$
**25.** $\{p \mid -1 \leq p \leq 5\}$   **27.** $\{w \mid w \leq 0 \text{ or } w \geq 2\}$
**29.** $\{d \mid d \leq -4 \text{ or } d \geq 7\}$
**31.** $\left\{x \mid x > \dfrac{5\sqrt{2}}{2} \text{ or } x < \dfrac{-5\sqrt{2}}{2}\right\}$   **33.** $\varnothing$   **35.** {all reals}
**37.** $\left\{x \mid x > \dfrac{1}{3} \text{ or } x < -2\right\}$
**39.** $\left\{t \mid \dfrac{-1 - \sqrt{10}}{2} < t < \dfrac{-1 + \sqrt{10}}{2}\right\}$   **47.** no
**49.** Let $n = $ number of decreases; $-17 \leq n \leq 2$

**50.** yes   **51.**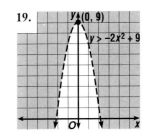

**52.** $x^2$; $3x$; $-2$   **53.** 64   **54.** $3a\sqrt{5} - 3\sqrt{ab} + \sqrt{30ab} - b\sqrt{6}$   **55.** $(a - b + 4)(a + b - 4)$
**56.** 823

**Pages 388–390   Chapter 8 Summary and Review**
**1.** $f(x) = x^2 + 4x + 4$   **3.** $3x^2$; $2x$; $-1$

**5.** $y = (x - 7)^2$   **7.**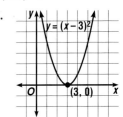

**9.** $f(x) = (x - 1)^2$, $(1, 0)$, $x = 1$
**11.** $f(x) = (x - 1)^2 + 3$, $(1, 3)$, $x = 1$, up
**13.** $f(x) = -2(x + 10)^2 + 210$, $(-10, 210)$, $x = -10$,
down   **15.** $\dfrac{5}{2}, -\dfrac{5}{2}$

**17.** 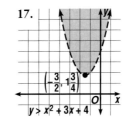   **19.**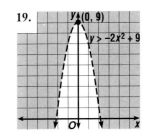

**21.** $\{x|-1 < x < 10\}$  **23.** $\left\{a\middle|a \geq \dfrac{3}{4} \text{ or } a \leq -2\right\}$  **25.** 1
**27.** \$13.75  **29.** 10 seconds

## CHAPTER 9  THE DISTANCE AND MIDPOINT FORMULAS

### Pages 398–399  Lesson 9-1
**5.** 3 units  **7.** 8.9 units  **9.** $16\dfrac{19}{24}$ units  **11.** 13 units
**13.** 4 units  **15.** $\sqrt{58}$ units  **17.** $\sqrt{2}$ units
**19.** $\sqrt{4.58}$ units  **21.** 16 units  **23.** 5 or $-1$  **25.** 7.1
or 13.1  **27.** (12, 5)  **29.** $\left(\dfrac{9}{16}, \dfrac{5}{4}\right)$  **31.** (10, $-7$)

**33.** midpoint of $\overline{MN} = \left(\dfrac{1}{2}, \dfrac{13}{2}\right)$

midpoint of $\overline{MO} = \left(5, \dfrac{1}{2}\right)$

midpoint of $\overline{NO} = \left(\dfrac{5}{2}, 2\right)$

**35.** $AB = \sqrt{(-3 - (-1))^2 + (0 - 4)^2}$
$= \sqrt{(-2)^2 + (-4)^2}$
$= \sqrt{4 + 16}$
$= \sqrt{20}$
$AC = \sqrt{(-3 - 1)^2 + (0 - (-2))^2}$
$= \sqrt{(-4)^2 + 2^2}$
$= \sqrt{16 + 4}$
$= \sqrt{20}$

**39.** $5\dfrac{7}{8}$ inches  **40.** $\{x|x > -3 \text{ or } x < -7\}$
**41.** $\{a|a \geq 1.5 \text{ or } a \leq -2.5\}$  **42.** $\pm i\sqrt{5}, \pm\sqrt{6}$
**43.** $x = 2, y = 3$  **44.** $3(m + 2p)(m^2 - 2mp + 4p^2)$
**45.** $4x - y = 11; 5y = 25$

### Pages 403–404  Lesson 9-2
**5.** 4  **7.** 25  **9.** $\dfrac{49}{4}$  **11.** $y = -\dfrac{1}{4}x^2$
**13.** $y = (x + 2)^2 - 3$
**15.** $(-2, 3); x = -2; \left(-2, 3\dfrac{1}{4}\right); y = 2\dfrac{3}{4};$ up; 1

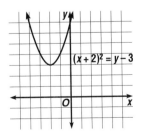

**19.** $(4, 2); x = 4; (4, 3); y = 1;$ up; 4

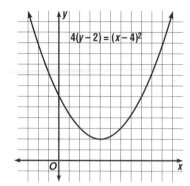

**23.** $(4, 8); y = 8; (3, 8);$
$x = 5,$ left; 4

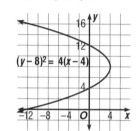

**27.** $(3, 5); x = 3; \left(3, 5\dfrac{1}{2}\right);$
$y = 4\dfrac{1}{2};$ up; 2

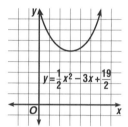

**31.** $(4, 2); x = 4; \left(4, 2\dfrac{1}{12}\right);$
$y = 1\dfrac{11}{12};$ up; $\dfrac{1}{3}$

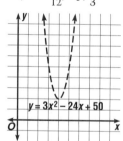

**35.** $x = \dfrac{1}{4}(y - 2)^2 + 5$

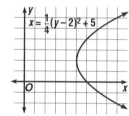

**39.** $x = \dfrac{1}{10}(y + 3)^2 + 7\dfrac{1}{2}$

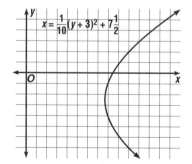

**43.** $x = \frac{1}{4}(y - 3)^2 + 4$    **47.** $y = \frac{1}{6}(x + 1)^2 + \frac{1}{2}$

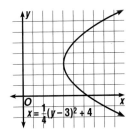

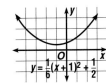

**49.** $\left(0, 2\frac{1}{2}\right)$   **51.** $\sqrt{37}$ units   **52.** $\left(\frac{3}{2}, -\frac{5}{2}\right)$   **53.** 18 cm

**54.** 88 and 89 or $-88$ and $-89$   **55.** $\frac{3x^{\frac{1}{5}}}{x}$

**56.** $a^3 - 6a^2 - 7a + 60$   **57.** yes, $-24$, 5

## Pages 406–408   Lesson 9-3

**7.** parabola   **9.** circle   **11.** parabola   **13.** (0, 3), 6

**15.** $\left(-2, \frac{2}{3}\right)$, $2\sqrt{2}$   **17.** $(0, -5)$, $\frac{9}{8}$

**19.** $(-5, 0)$, 8

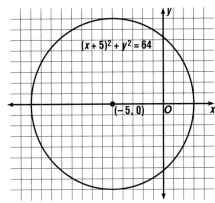

**23.** $(0, -2)$, 2     **27.** $(-2, 1)$, 9

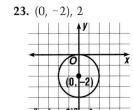

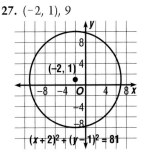

**31.** $(9, 9)$, $\sqrt{109}$

**35.** $(2, 0)$, $\sqrt{13}$

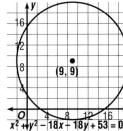

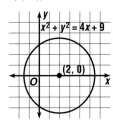

**39.** $\left(0, -\frac{9}{2}\right)$, $\sqrt{19}$     **43.** $(-1, 0)$, $\sqrt{11}$

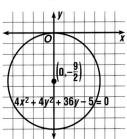

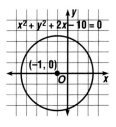

**45.** $x^2 + (y - 3)^2 = 49$   **47.** $(x + 1)^2 + (y + 5)^2 = 4$

**49.** $(x + 3)^2 + (y + 9)^2 = \frac{25}{36}$   **51.** $(x - 3)^2 +$ $(y - 3)^2 = 18$   **53.** $(x + 1)^2 + (y - 6)^2 = 45$

**55.** $(x - 4)^2 + (y + 3)^2 = 16$   **57.** $(x - 4)^2 +$ $(y - 16)^2 = 169$   **59.** no   **60.** $x = \frac{1}{6}y^2$

**61.** $y = -\frac{1}{4}(x - 2)^2 + 5$       **62.** yes

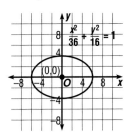

**63.** $1(y^{-3})^2 + 4(y^{-3}) - 32 = 0$   **64.** $\frac{-2\sqrt{3}}{9}$

**65.** $2(2k + 3)(k + 5)$   **66.** $\begin{bmatrix} \frac{1}{8} & 0 \\ -\frac{1}{32} & -\frac{1}{4} \end{bmatrix}$

## Pages 413–415   Lesson 9-4

**5.** $(0, 0)$, V   **7.** $(11, -8)$, V   **9.** $\frac{x^2}{12} + \frac{y^2}{28} = 1$

**11.** $\frac{(x + 3)^2}{9} + \frac{(y - 5)^2}{4} = 1$   **13.** $\frac{(x - 3)^2}{16} + \frac{(y + 2)^2}{9} = 1$

**15.** $(0, 0)$; $(\pm 2\sqrt{5}, 0)$;    **17.** $(0, 0)$; $(0, \pm\sqrt{7})$; 12, 8           8, 6

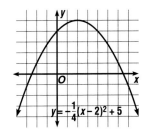

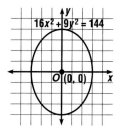

**19.** $\frac{x^2}{36} + \frac{y^2}{20} = 1$

**21.** $(0, 0)$; $(\pm 4, 0)$; 10, 6     **25.** $(0, 0)$; $(0, \pm\sqrt{6})$; 6, $2\sqrt{3}$

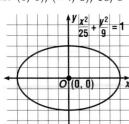

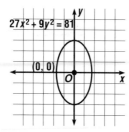

**29.** $(-8, 7)$; $(-8 \pm\sqrt{57}, 7)$; 22, 16

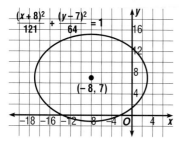

**33.** $(2, 2)$; $(2, 4)$, $(2, 0)$; $2\sqrt{7}$, $2\sqrt{3}$

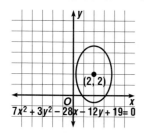

**35.** $\dfrac{x^2}{36} + \dfrac{y^2}{100} = 1$    **37.** $\dfrac{(x-5)^2}{64} + \dfrac{4(y-4)^2}{81} = 1$

**39.** $\dfrac{(x+2)^2}{16} + \dfrac{(y-3)^2}{36} = 1$    **43.** $(x-6)^2 +$
$(y-2)^2 = 25$   **44.** $(x-1)^2 + (x-5)^2 = 26$
**45.** $f(x) = (x+1)^2 - 3$; $(-1, -3)$; $x = -1$; up   **46.** no
**47.** 98    **48.** 19, 27, 35, 43, 51

**Page 415   Mid-Chapter Review**
**1.** 10 units   **2.** $2\sqrt{10}$ units   **3.** $(4, -4)$   **4.** $\left(\dfrac{3}{2}, 6\right)$
**5.** $(0, 0)$; $x = 0$; $(0, -3)$; $y = 3$; down; 12   **6.** $(3, -2)$;
$y = -2$; $(5, -2)$; $x = 1$; right; 8
**7.** $y = \dfrac{1}{8}(x-3)^2 + 3$      **8.** $x = -\dfrac{1}{2}(y-4)^2 + \dfrac{1}{2}$

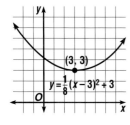

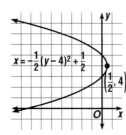

**9.** parabola   **10.** circle   **11.** $(0, 0)$, $3\sqrt{3}$ units
**12.** $(-3, 1)$, 9 units   **13.** $\dfrac{x^2}{14} + \dfrac{y^2}{28} = 1$
**14.** $\dfrac{(x-1)^2}{5} + \dfrac{(y-2.5)^2}{10} = 1$   **15.** $\dfrac{(x-3)^2}{32} + \dfrac{(y-1)^2}{81} = 1$

**Pages 420–422   Lesson 9-5**
**5.** ellipse   **7.** hyperbola   **9.** hyperbola
**11.** $\dfrac{x^2}{5} - \dfrac{y^2}{10} = 1$   **13.** $\dfrac{(y-1)^2}{168} - \dfrac{(x-2)^2}{42} = 1$
**15.** $(\pm 4, 0)$; $(\pm 2\sqrt{5}, 0)$;   **17.** $(\pm 3, 0)$; $(\pm\sqrt{13}, 0)$;
$\pm\dfrac{1}{2}$            $\pm\dfrac{2}{3}$

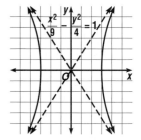

**19.** $\dfrac{x^2}{4} - \dfrac{y^2}{12} = 1$   **21.** $\dfrac{(x-3)^2}{4} - \dfrac{(y+5)^2}{9} = 1$
**23.** $\dfrac{y^2}{4} - \dfrac{x^2}{21} = 1$
**25.** $(\pm 6, 0)$; $(\pm\sqrt{37}, 0)$;   **29.** $(\pm 2, 0)$; $(\pm\sqrt{13}, 0)$;
$\pm\dfrac{1}{6}$            $\pm\dfrac{3}{2}$

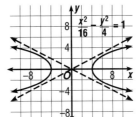

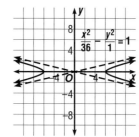
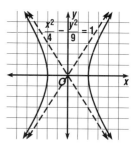

**33.** $(0, \pm 10)$; $(0, \pm 2\sqrt{61})$;   **37.** $(2, -2)$, $(2, 8)$;
$\pm\dfrac{5}{6}$            $(2, 3 \pm\sqrt{41})$; $\pm\dfrac{5}{4}$

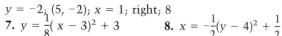
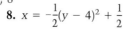

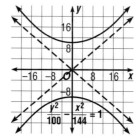
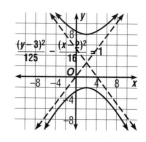

**41.** $(4 \pm 2\sqrt{5}, -2)$; $(4 \pm 3\sqrt{5}, -2)$; $\pm\dfrac{\sqrt{5}}{2}$

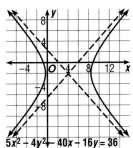

$5x^2 - 4y^2 - 40x - 16y = 36$

**45.** $\dfrac{(y-4)^2}{4} - \dfrac{(x-5)^2}{36} = 1$

**47.**

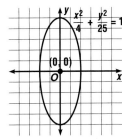

$xy = 3$

**51.**

$rt = 2.5$

**52.** $(0, 0)$; $(0, \pm\sqrt{21})$; 10 units, 4 units

$\dfrac{x^2}{4} + \dfrac{y^2}{25} = 1$

$(0, 0)$

**53.** $\dfrac{(x-1)^2}{25} + \dfrac{(y-4)^2}{9} = 1$    **54.** $\{x \mid -\sqrt{6} \le x \le \sqrt{6}\}$

**55.** $-2, -3$    **56.** $\dfrac{5^{\frac{1}{3}}}{5^{\frac{1}{3}}}$

**Page 424   Lesson 9-6**

**5.** 20   **7.** 239

**Pages 429–431   Lesson 9-7**

**5.** parabola   **7.** hyperbola   **9.** hyperbola   **11.** ellipse

**13.** $y = \dfrac{1}{8}x^2$; parabola

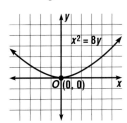

$x^2 = 8y$

$O(0, 0)$

**17.** $\dfrac{y^2}{16} - \dfrac{x^2}{8} = 1$; hyperbola

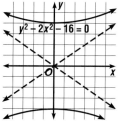

$y^2 - 2x^2 - 16 = 0$

**21.** $\dfrac{(y-5)^2}{16} - \dfrac{(x+1)^2}{4} = 1$; hyperbola

**25.** $\dfrac{(x-3)^2}{25} + \dfrac{(y-1)^2}{9} = 1$; ellipse

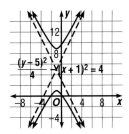

$\dfrac{(y-5)^2}{4} - (x+1)^2 = 4$

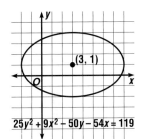

$(3, 1)$

$25y^2 + 9x^2 - 50y - 54x = 119$

**29.** isolated point

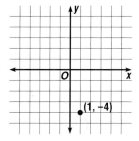

$(1, -4)$

**33a.** ellipse   **33b.** hyperbola   **33c.** circle
**33d.** ellipse   **33e.** parabola   **33f.** degenerated case
**35.** $\dfrac{(y-2)^2}{36} - \dfrac{(x+2)^2}{64} = 1$

**36.**

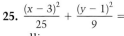

$(0, 3)$
$f(x) = 2x^2 + 3$

**37.** $3, -\dfrac{1}{3}$   **38.** $-3rs$

**Pages 434–436  Lesson 9-8**

**5.**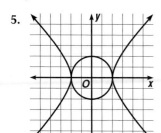

**9.** $(\pm 2\sqrt{3}, 2)$

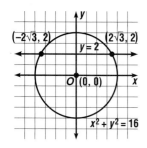

**13.** $(-1, 1), (2, 4)$

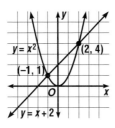

**17.** $(\pm 5.2, 6)$

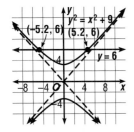

**21.** $(4, -1.5), (-3, 2)$

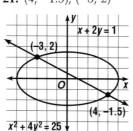

**25.** no solution

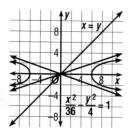

**29.** $(3, 0), (-3, -6)$

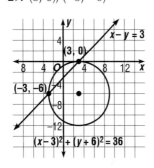

**33.** $(0.3)$ $(\pm 2.4, -2.8)$

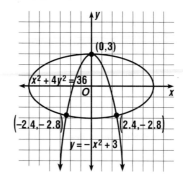

**37.** no solution

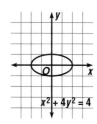

**43.** approximately $(13.6, 21)$ or $(13.6, -21)$: that is 13.6 miles east and 21 miles north or south

**44.** hyperbola

**45.** $\dfrac{x^2}{4} + \dfrac{y^2}{1} = 1$; ellipse

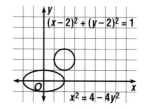

**46.** yes  **47.** $-7, 2$  **48.** $p^2 + 5p - 7 + \dfrac{8}{2p - 3}$

**Pages 441–443  Lesson 9-9**

**5.** elimination; $(2, \pm\sqrt{3}), (-2, \pm\sqrt{3})$

**7.** substitution; no solutions  **9.** substitution; $(0, 9)$, $(0, -9), \left(-\dfrac{1}{2}, \pm\dfrac{\sqrt{323}}{2}\right)$  **11.** $x^2 + y^2 \geq 25$, $x^2 + y^2 \leq 100$  **13.** $(2, 4), (-1, 1)$  **15.** $(3, 0)$, $(-5, -4)$  **17.** $(-4, -3), (-3, -4)$  **19.** $(\pm 3\sqrt{3}, 6)$

**21.** no solutions  **23.** no solutions  **25.** $\left(4, -\dfrac{3}{2}\right)$, $(-3, 2)$  **27.** $(-1, -1), (2, -4)$  **29.** $(1, \pm 5), (-1, \pm 5)$

**31.** $(3, 7), (8, 4)$

**33.**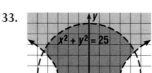

**37.**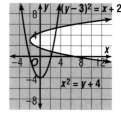

**41.**

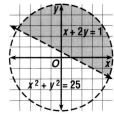

**45.**

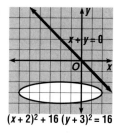

$(x + 2)^2 + 16 (y + 3)^2 = 16$

**51.** 20 ft by 24 ft  **53.** no solution  **54.** $(8, 0)$; $x = 8$; up  **55.** 36 miles  **56.** $-7, -\dfrac{1}{2}$
**57.** $-5 - 2\sqrt{6}$  **58.** $32x^3 y^6$

## Pages 444–446  Chapter 9 Summary and Review

**1.** $6\sqrt{2}$  **3.** $\sqrt{16.85}$  **5.** $\left(2, -\dfrac{7}{2}\right)$  **7.** $\left(1, \dfrac{3}{2}\right)$
**9.** $(0, 0)$; $y = 0$; $(-2, 0)$;  **13.** $(0, 0)$; $(\pm\sqrt{7}, 0)$; 8, 6 $x = 2$; left; 8

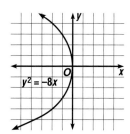
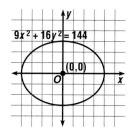

**17.** parabola  **19.** circle

**21.** $(-2, 0), (2, 4)$  **23.** $(1.6, 2.4),$ $(-2.6, 6.6)$

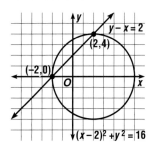

**25.**

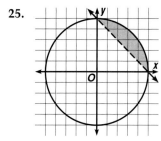

**27.** $(x - 12)^2 + (y - 25)^2 = 2304$  **29.** a hyperbola with foci $(\pm\sqrt{13}, 0)$, vertices $(\pm 3, 0)$, and asymptotes $y = \pm\dfrac{3}{2}$

# CHAPTER 10  POLYNOMIAL FUNCTIONS

## Pages 454–456  Lesson 10-1

**5.** yes, 5  **7.** yes, 0  **9.** no  **11.** no  **13.** yes, 3
**15.** 10  **17.** 25  **19.** even, 4  **21.** odd, 2  **23.** $-47$
**25.** $-52$  **27.** $-21$  **29.** 36  **31.** 575  **33.** $\dfrac{305}{6}$
**35.** $2x + 2h - 3$  **37.** $x^2 + 2xh + h^2 - 2x - 2h + 5$
**39.** $x^3 + 3hx^2 + 3h^2x + h^3 + 4x + 4h$
**41.** $2x^3 + 6hx^2 + 6h^2x + 2h^3 - x^2 - 2hx - h^2 + 4$
**43.** $4x^2 + 20$  **45.** $x^3 + \dfrac{x^2}{4} - 8$  **47.** even, 0
**49.** $6x + 34$  **51.** $2x^2 + 24x + 18$  **53.** $-5x - 8$
**55.** $-x^3 - 9x^2 - 9x - 4$  **59.** \$65,892
**60.** $(1, -2), (-1, -2)$
**61.**

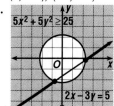

**62.**

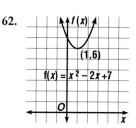

**63.** no  **64.** $x = 3, y = 0$; $x = 0, y = -\dfrac{3}{2}i$  **65.** 45 fish

## Pages 462–464  Lesson 10-2

**7.** 5, 4  **9.** 6, 5  **11.** 12  **13.** 0
**15.** $(x^3 - 8x^2 + 2x - 1) =$ $(x^2 - 9x + 11)(x + 1) - 12$; no
**17.** $(x^4 - 16) = (x^3 + 2x^2 + 4x + 8)(x - 2) + 0$; yes
**19.** $(6x^3 + 9x^2 - 6x + 2) =$ $(6x^2 - 3x)(x + 2) + 2$; no
**21.** $(4x^4 - 2x^2 + x + 1) =$ $(4x^3 + 4x^2 + 2x + 3)(x - 1) + 4$; no  **23.** 11, 5
**25.** $-23, -2$  **27.** 31, 1  **29.** $x - 3, x - 1$  **31.** $x - 1,$ $x + 2$  **33.** $x - 2, x^2 + 2x + 4$  **35.** $2x - 3, 2x + 3,$ $4x^2 + 9$  **37.** $(x + 1)^2(x - 1)(x^2 + 1)$  **39.** 3  **41.** 1, 4
**43a.** $A = 1000(1 + r)^6 + 1000(1 + r)^5 + 1000(1 + r)^4 + 1200(1 + r)^3 + 1200(1 + r)^2 + 2000(1 + r)$
**43b.** \$8916.76  **45.** $-22$  **46.** $3x^2 - 6x - 9$
**47.** $(-4, 0)$; 7  **48.** $\{y | 5 \le y \le 12\}$
**49.** $\dfrac{13}{3}, -\dfrac{13}{3}$  **50.** 34  **51.** $\begin{bmatrix} 28 & 0 \\ 21 & -7 \end{bmatrix}$

## Pages 466–467  Lesson 10-3

**5.** 70  **7.** Darcy - English, Janna - sociology, Ray - Math  **9a.** a square  **9b.** $\dfrac{x}{5}$ square units
**11.** one-fourth  **13.** cube 1: 0, 1, 2, 6, 7, 8; cube 2: 1, 2, 0, 3, 4, 5. Nine can be formed by an upside-down 6.

## Pages 473–474  Lesson 10-4

**5.** $f(-x) = -3x^5 + 7x^2 + 8x + 1$
**7.** $f(-x) = 4x^4 + 3x^3 + 2x^2 + x + 1$  **9.** 0; 3 or 1; 0 or 2  **11.** 1; 3 or 1; 0 or 2  **13.** 2 or 0; 2 or 0; 0, 2, or 4  **15.** 0; 1; 2  **17.** 5, 3, or 1; 5, 3, or 1; 0, 2, 4, 6,

**19.** $4, 1 + i, 1 - i$  **21.** $-\dfrac{3}{2}, 1 + 4i, 1 - 4i$

**23.** $2i, -2i, \dfrac{i}{2}, -\dfrac{i}{2}$  **25.** $2 + 3i, 2 - 3i, -1$  **27.** $3 - 2i$, $3 + 2i, -1, 1$  **29.** $f(x) = x^3 - 3x^2 + 4x - 12$
**31.** $f(x) = x^4 + 2x^3 + 7x^2 + 30x + 50$  **33.** $f(x) = x^6 - 2x^5 + 15x^4 - 26x^3 + 62x^2 - 72x + 72$
**35.** 5 cm by 12 cm by 2 cm  **37.** $0, 30$  **38.** $y = -\dfrac{x^2}{16}$
**39.** $f(x) = x^2 + 6$; $(0, 6), x = 0$  **40.** $681,177,600$ or $6.811776 \times 10^8$

### Pages 474  Mid-Chapter Review

**1.** $-210$  **2.** $6a^6 - 3a^4 + 4a^2 - 9$  **3.** $6x^3 + 15x^2 + 16x - 2$  **4.** $x - 3, x - 2$  **5.** $2x + 7, x + 5$  **6.** $24$, $32, 68, 76$  **7.** 2 or 0; 1; 0 or 2  **8.** 1; 1; 2  **9.** $f(x) = x^3 - 3x^2 + 4x - 2$  **10.** $f(x) = x^4 + 7x^3 + 13x^2 - 23x - 78$

### Pages 478–480  Lesson 10-5

**5.** $\pm 1, \pm 2$  **7.** $\pm 1, \pm 2, \pm 3, \pm 6$  **9.** $\pm 1, \pm 2, \pm 5, \pm 10$
**11.** $\pm 1, \pm 2, \pm 4, \pm\dfrac{1}{3}, \pm\dfrac{2}{3}, \pm\dfrac{4}{3}$  **13.** $\pm 1, \pm 5$  **15.** $\pm 1,$

$\pm 2, \pm 3, \pm 6, \pm 9, \pm 18$  **17.** $\pm 1, \pm 2, \pm\dfrac{1}{2}, \pm\dfrac{1}{3}, \pm\dfrac{1}{6}, \pm\dfrac{2}{3}$
**19.** $\pm 1, \pm\dfrac{1}{2}, \pm\dfrac{1}{4}, \pm 2, \pm 3, \pm\dfrac{3}{2}, \pm\dfrac{3}{4}, \pm 6$  **21.** $-5, -6, 10$
**23.** $-1, -1, 2$  **25.** $0, 9$  **27.** $\dfrac{1}{2}, -\dfrac{1}{3}, -2$  **29.** $1, -1$
**31.** $2, -3, -3$  **33.** $-6$  **35.** $0, 2, -2$  **37.** 6 units by 8
units by 3 units  **39.** $\dfrac{2}{3}, \dfrac{-3 \pm \sqrt{17}}{4}$  **41.** $-2, \dfrac{4}{3}, \dfrac{-3 \pm i}{2}$
**43.** $3, \dfrac{2}{3}, -\dfrac{2}{3}, \dfrac{-3 \pm \sqrt{13}}{2}$
**47.** radius $= 1$ in. and height $= 5$ in.  **48.** 2 or 0; 1; 2
or 4  **49.** $f(x) = x^3 - 2x^2 + 5x + 26$
**50.** $(\pm 3, 0)$  **51.** yes

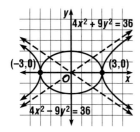

### Pages 484–485  Lesson 10-6

**5a.** $f(-4) = 138, f(-3) = 19, f(-2) = -6, f(-1) = 3$, $f(0) = 10, f(1) = 3, f(2) = -6, f(3) = 19, f(4) = 138$
**5b.** $-3$ and $-2$, $-2$ and $-1$, 1 and 2, 2 and 3; $-2.6$, $-1.2, 1.2, 2.6$  **5c.** negative between $-2.5$ and $-1.2$
and between 1.2 and 2.5; positive between $-1.2$ and
1.2, less than $-2.5$ and greater than 2.5
**5d.** minimums at 2 and $-2$, maximum at 0

**5e.**

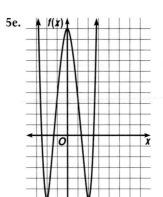

**7.** 0.6  **9.** 1.6, $-1.3$, $-2.4$  **11.** $-1$  **13.** no real zeros

**15.**

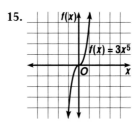

**19.**

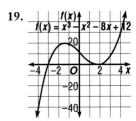

**23.**

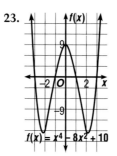

**27.** $-3.6, -1.6, -0.7, 0.6, 1.3$

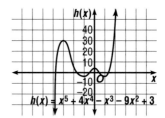

**31.** $-2.0$

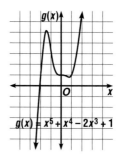

**33b.** 0 **33c.** There are none. **33d.** The ends of an even degree graph both point up or down and the ends of an odd degree function point in opposite directions. **35.** 0.56 centimeters **36.** CDT - compact disks; CD - tapes; T - compact disks and tapes **37.** 30 and 30 **38.** $x^2(y + 5)(y - 5)$ **39.** no

**Pages 489–490 Lesson 10-7**

**5.** 9, 6, 12 **7.** 20, 2, 0 **9.** $-234, 0, 2$ **11.** 10, 6 **13.** 64, 64 **15.** $-7, 11$ **17.** $10x, 10x$ **19.** $x, x$ **21.** $|x - 3|, |x| - 3$ **23.** 4 **25.** 12 **27.** 16 **29.** $9x^2$ **31.** $g \circ f = \{(1, 0), (0, 1)\}$ **35.** \$140

**36.**

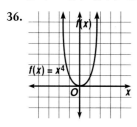

**37.** 1, $-1$, 1.7, $-1.7$ **38.** $(x - 6)^2 + (y - 1)^2 = 20$ **39.** 190

**Pages 494–495 Lesson 10-8**

**5.** $\{(-2, -1), (-2, -3), (-4, -1), (6, 0)\}$, no **7.** $\{(4, 2), (1, -3), (8, 2)\}$, yes **9.** $\{(11, 6), (7, -2), (3, 0), (3, -5)\}$, no **11.** $f^{-1}(x) = x + 5$ **13.** $x = 8$ **15.** $f^{-1}(x) = 2x - 4$ **17.** $y = \pm\sqrt{x + 9}$

**19.**

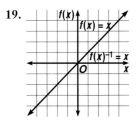

**23.**

**27.**

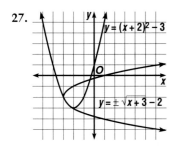

**29.** yes **31.** yes **33.** no **35.** yes **37.** no **39.** $I(m) = 280 + 0.03m$; \$4000 **41a.** yes **41b.** yes **41c.** no **42.** $3x - 3$ **43.** $g \circ f = \{(2, 5), (3, -7), (6, -3)\}$; $f \circ g$ does not exist **44.** no, because of the variable in the denominator

**45.** $\dfrac{x^{\frac{2}{3}}}{x - 1}$ **46.** $a^3 - a^2 + 2a - 2$

**Pages 496–498 Chapter 10 Summary and Review**

**1.** $6x - 9$ **3.** $x^2 + 2xh + h^2 - x - h$ **5.** $x + 2$, $x + 2$ **7.** 3 or 1; 1; 2 or 0 **9.** 0; 0 or 2; 4 or 2 **11.** $-\dfrac{1}{2}$, 3, 4 **13.** $-3, 2, 4$

**15.** 0.8 See graph at right.

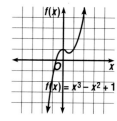

**17.** $[f \circ g](x) = 6x + 7$, $[g \circ f](x) = 6x + 1$ **19.** $f \circ g$ does not exist, $g \circ f = \{(2, 5), (-1, -1), (3, 2)\}$ **21.** yes **23.** Craig: Julliard; Vanessa: DeVry; Devin: Case Western Reserve; Anita: Judson **25.** 9 by 11 by 25 units

## CHAPTER 11 RATIONAL POLYNOMIAL EXPRESSIONS

**Pages 508–509 Lesson 11-1**

**5.** $x = 3, y = 0$ **7.** $x = 1, x = -5, y = 0$

**9.**

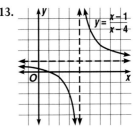

**13.**

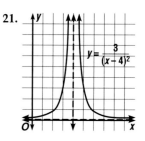

**17.**

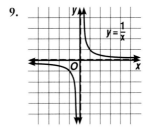

**21.**

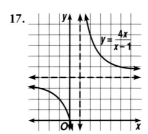

**25.**

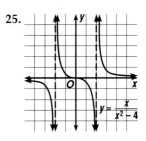

**29a.**

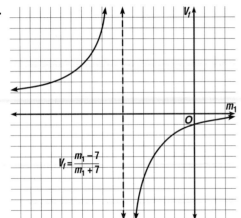

**29b.** $V_f \approx -0.83$ m/s  **30.** $f^{-1}(x) = \dfrac{2x - 2}{5}$  **31.** $(0, 0)$, 5

**32.** $-3$  **33.** $1 \pm \dfrac{\sqrt{11}}{2}$  **34.** $(x + 3y)(x - 1)$

**Pages 513–515  Lesson 11-2**

**5.** direct, $-\dfrac{1}{6}$  **7.** direct, $\dfrac{1}{2}$  **9.** direct, 4  **11.** joint, $k$

**13.** $y = 112$  **15.** $y = 120$  **17.** $y = 36$  **19.** $r = \dfrac{84}{11}$

**21.** 118.5 km  **23.** 1.125  **25.** $y = 48$  **27.** $x = \dfrac{100}{7}$

**29.** Heavenly Hog  **31.** $g = 15$  **33.** $y = 32$

**37a.** $I = \dfrac{6}{R}$  **37b.** 6  **39.** 44 mph

**40.**

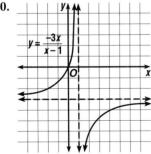

**41.** $-1, 3, -4$  **42.** $(-3, -1)$, $(5, 7)$

**43.** $f(x) = x(x - 55)$  **44.** $25\sqrt{2}$  **45.** $\begin{bmatrix} -4 & 0 & 9 \\ 16 & -4 & 14 \\ 7 & -1 & 19 \end{bmatrix}$

**46.** Gladys 30, Maria 10

**Pages 519–521  Lesson 11-3**

**7.** $3x^2y^2$, $-\dfrac{y^3}{6x}$  **9.** $4x^5y$, $-2xy^2$  **11.** $(m + 5)$, $\dfrac{1}{2}$

**13.** $x$, $\dfrac{4}{x - 1}$  **15.** $y$  **17.** $\dfrac{3a^2}{2bc}$  **19.** $-\dfrac{ad^2}{2}$  **21.** $\dfrac{5x^3}{3y^2}$

**23.** $\dfrac{ac^4d}{b}$  **25.** $a$  **27.** 1  **29.** $-1$  **31.** $-\dfrac{y}{w^2 - y^2}$

**33.** $\dfrac{y - 2}{x - 4}$  **35.** $\dfrac{x - 2}{x + 2}$  **37.** $\dfrac{w - 3}{w - 4}$  **39.** 5, 12, 13 inches

**41.** $\dfrac{a(a + 2)}{a + 1}$  **43.** $-1$  **45.** 1; An equal number of men and women favor the tax.  **47.** $x = 4$  **48.** \$11,680
**49.** hyperbola  **50.** $11 - i$, $1 + 3i$, $32 - 7i$
**51.** Complex, Real, Rational, Integer, Whole, Natural

**Page 521  Mid-Chapter Review**

**1.**

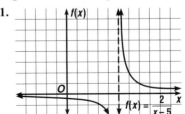

**3.**

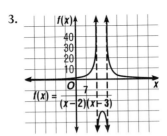

**4.** $x = 24$  **5.** $m = 22$  **6.** $\dfrac{22a^3}{5bc}$  **7.** $\dfrac{7}{2}$  **8.** $(m + 4)(3m + 4)$

**Pages 525–526  Lesson 11-4**

**5.** 78  **7.** 2000  **9.** $x(x - 2)(x + 2)$

**11.** $3(x + 5)(x - 3)$  **13.** $\dfrac{7 + 9a}{ab}$  **15.** $\dfrac{3t^2 - 19t + 36}{t - 5}$

**17.** $\dfrac{-12x + 21xy - 4y}{6x^2y}$  **19.** $\dfrac{13}{y - 8}$  **21.** $\dfrac{y^2 - 2y + 2}{y - 1}$

**23.** $\dfrac{x(x - 9)}{(x + 3)(x - 3)}$  **25.** $\dfrac{5x + 16}{(x + 2)^2}$  **27.** $\dfrac{2(a - 2)}{3(a + 1)}$

**29.** $\dfrac{-4x^2 - 5x - 2}{(x + 1)^2}$  **31.** 0  **33.** $\dfrac{x^3 + 4x^2 + 7x + 17}{x^2 - 4}$

**35.** $\dfrac{2x^2 + x - 4}{(x - 1)(x - 2)}$  **37.** $\dfrac{2}{2x - 5}$

**41.** $\dfrac{-3x - 26}{(x + 2)(x - 3)}$ mph

**43.** $\dfrac{x^2 + 10x - 8}{x(x - 2)}$  **44.** $\dfrac{1 - 2x}{3x + 1}$

**45.** 27

**46.** $\{x | x \le 1 \text{ or } x \ge 7\}$  **47.** 4

**48.** See graph at right.

**49.** about 28 in.

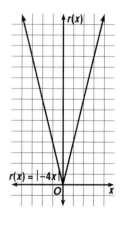

**Pages 529–530   Lesson 11-5**

**5.** $2x, 0$   **7.** $7(2 + m), -2$   **9.** $x^2, 0$   **11.** $-3, \frac{1}{6}$   **13.** 2

**15.** 5   **17.** $-6, 1$   **19.** 2, 6   **21.** $-6, \frac{3}{2}$   **23.** $\frac{-3 + 3\sqrt{2}}{2}$
**25.** 14   **27.** $-6, -2$   **29.** $\varnothing$   **31.** $0, -4$   **35a.** 5
**35b.** 6.15   **36.** $\frac{11x}{13y^2}$   **37.** $\frac{3 + 4x}{x}$   **38.** $\frac{45b - 28a}{20ab}$
**39.** $x = 1, x = 4, y = 0$   **40.** $(2, \pm\sqrt{21})$
**41.** $(0, 1), \left(-\frac{5}{2}, \frac{9}{4}\right)$   **42.** $\left(\frac{5 + \sqrt{5}}{2}, \frac{5 + \sqrt{5}}{2}\right)$

**43.**

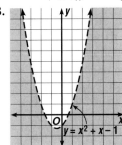

**44.** $42 - 6\sqrt{2} + 7\sqrt{3} - \sqrt{6}$
**45a.** Sample: $d = 25f$   **45b.** about 11 gallons

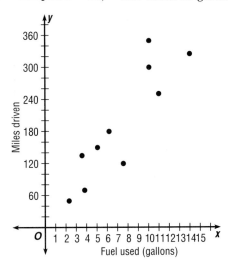

**Page 534   Lesson 11-6**

**5.**

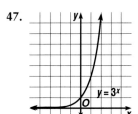

**7.** $\frac{1}{10}$   **9.** 5 cows, 1 pig, 94 geese

**Pages 537–539   Lesson 11-7**

**5.** $\frac{x}{6} - \frac{x}{18} = 1$   **7.** $r = \frac{524}{9}$   **9.** 20 h   **11.** 19
**13.** 550 mph   **15.** 5 tons   **17.** $\frac{7}{13}$   **19a.** \$550

**19b.** \$44   **23.** 1   **24.** $\frac{13}{30a}$   **25.** $\frac{7y^2}{8}$   **26.** $f(x) = x^3 -$
$6x^2 + 9x + 50$   **27.** $2\sqrt{53}$   **28.** $5s^2 - 9s - 2$

**Pages 540–542   Chapter 11 Summary and Review**

**1.**

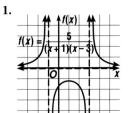

**5.**

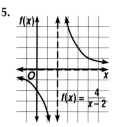

**7.** $y = -37.5$   **9.** $x = 8$   **11.** $x = 0.192$
**13.** $\frac{ay - 2a - 3y + 6}{a - x}$   **15.** $6b(a - b)$   **17.** $\frac{2}{n - 3}$
**19.** $\frac{28a - 27b}{12ab}$   **21.** $\frac{7}{5(x + 1)}$   **23.** $\frac{18}{y - 2}$   **25.** $\frac{25b + 16}{24b}$
**27.** $\frac{10}{9}$   **29.** 3   **31.** $\frac{3}{2}$   **33.** $6\sqrt{2}$ meters   **35.** 8, 10

## CHAPTER 12   EXPONENTIAL AND LOGARITHMIC FUNCTIONS

**Pages 549–550   Lesson 12-1**

**5.** 1.6   **7.** 2.1   **9.** 18.9   **11.** $2^3$   **13.** $-3$   **15.** $-2$   **17.**
$-\frac{3}{2}$   **19.** 0.5   **21.** 3.0   **23.** 2.4   **25.** 81   **27.** $2^{5\sqrt{7}}$

**29.** $x^4$   **31.** $-\frac{1}{2}$   **33.** $-2$   **35.** $-\frac{1}{3}$   **37.** 2   **39.** 3
**41.** $-22$   **43.** $-2$   **45.** $-1, -3$

**47.**

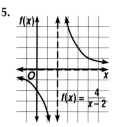

**51a.** $y = 100 \times (1.08)^x$   **51b.** \$342.59   **52.** 360 mL
**53.** 52, $-23$   **54.** yes; $0 \le 4^2 - 7$   **55.** 41 by 43 units
**56.** $\frac{3x^{\frac{1}{3}}y - 2x}{xy}$   **57.** $(b - 1)(3b - 2y)$

**Pages 556–557   Lesson 12-2**
**7.** $\log_{10} 1000 = 3$   **9.** $2^6 = 64$   **11.** $9^{\frac{3}{2}} = 27$   **13.** 3
**15.** 2   **17.** 256   **19.** 2   **21.** $-2$   **23.** $\frac{1}{2}$   **25.** 6   **27.** 9

**29.** 18   **31.** $\frac{1}{25}$   **33.** 5   **35.** 4   **37.** no solution
**39.** $\pm 8$

**41.**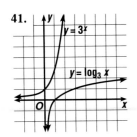

**43.**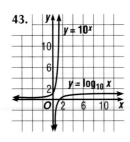

**45.** $\log_4 4 + \log_4 16 \overset{?}{=} \log_4 64$ **47.** $\log_2 8 \cdot \log_8 2 \overset{?}{=} 1$
   $1 + 2 = 3$    $3 \cdot \dfrac{1}{3} = 1$

**51.** 12 hours   **52.** $11^{4\sqrt{5}}$   **53.** $x + 5; \dfrac{1}{3}$   **54.** 3 or 1; 0;

0 or 2   **55.** $(0, \pm 9); (0, \pm 3\sqrt{13}; \pm\dfrac{3}{2}$   **56.** 18, -120

**57.** 58   **58.** 679

## Pages 561–562   Lesson 12-3

**5.** $2\log_5 x + 2\log_5 y$   **7.** $\log_2 r + \dfrac{1}{2}\log_2 t$   **9.** 16
**11.** 2   **13.** 5.614   **15.** 1.222   **17.** $-0.415$   **19.** 6.755
**21.** $-1.193$   **23.** 7   **25.** 6   **27.** 44   **29.** 3   **31.** 10
**33.** 2   **35.** 5   **37.** $\dfrac{1}{3}$   **39.** $\dfrac{x^4}{2}$   **41.** $\dfrac{1}{2}[n + 1]$
**43a.** $\text{pH} = 6.1 + \log_{10} B - \log_{10} C$   **43b.** a very
weak base   **43c.** 7.197   **44.** 1   **45.** neither exist
**46.** $4 \pm i\sqrt{6}$   **47.** $(q + s)(3p + q - s)$
**48.** $\{x|x > 4 \text{ or } x < -4\}$

## Pages 565–566   Lesson 12-4

**5.** 0.7582   **7.** 2.3   **9.** 2.9032; 2   **11.** $-1.2441; -2$
**13.** $-0.5302; -1$   **15.** 167.8   **17.** 0.5513   **19.** 0.0107
**21.** 0.1851   **23.** 1.9925, 1, 0.9925   **25.** 2.9156, 2,
0.9156   **27.** 0.7345, 0, 0.7345   **29.** $-2.3010, -3,$
0.6990   **31.** 3, 3, 0   **33a.** $10^{5.3}$ or about 199,526
times more intense   **33b.** $10^{6.9-4.3} = 10^{2.6}$ or about
398 times stronger   **34.** 72   **35.** $-\dfrac{2}{3}$   **36.** $3, -5, -\dfrac{5}{2}$
**37.** $2 \times 2; -20$

## Page 566   Mid-Chapter Review

**1.** 2   **2.** $-7$   **3.** $3, -1$   **4.** 2   **5.** $\dfrac{1}{2}$   **6.** $-3$   **7.** 24
**8.** 8   **9.** 3   **10.** 2.3598   **11.** 0.8033   **12.** 0.0062
**13.** $10^{8.3-6.9} = 10^{1.4}$ or about 25 times more intense

## Pages 569–570   Lesson 12-5

**5.** 1.454   **7.** 1.530   **9.** 0.4549   **11.** 1.976   **13.** 1.598
**15.** 9.290   **17.** $-0.6106$   **19.** 1   **21.** 1.682
**23.** 0.1866   **25.** 133.1   **27.** 41.32   **29.** 42.92
**33a.** $-0.000385$   **33b.** approximately 0.0213 grams
**34.** 2.5440; 2; 0.5440   **35.** $\dfrac{5x - 13}{(x - 2)(x - 3)}$
**36.** $(2, \pm\sqrt{3}), (-2, \pm\sqrt{3})$   **37.** $\{x|-6 \le x \le 6\}$
**38.** sofa–$700, love seat–$350, coffee table–$180

## Pages 572–573   Lesson 12-6

**5a.** 5 rolls   **5b.** 4.48 rolls   **5c.** Yes, since she can't
buy 4.48 rolls, she must buy 5.   **7.** grandfather,
father, and son each receive $7   **9.** 7, 6, 3, 2, 0
(alphabetical order)
**11.** sample answers are:

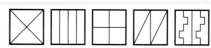

## Pages 576–578   Lesson 12-7

**5.** $\dfrac{\log 100}{\log 8}$; 2.2146   **7.** $\dfrac{\log 169}{\log 4}$; 3.7004

**9.** $\dfrac{\log 34}{2 \log 3}$; 1.6049   **11.** $\dfrac{-\log 22}{\log 3}$; $-2.8136$   **13.** 1.683
**15.** 2.064   **17.** 1.090   **19.** 0.594   **21.** 1.937
**23.** 4.7549   **25.** 3.8394   **27.** 2.8446   **29.** 2.3059
**31.** $\pm 2.0397$   **33.** $\pm 2.2325$   **35.** 3.2838   **37.** $-2.1507$
**39.** 1.1201   **41.** 2.9172   **43.** 0.3011   **45.** 2.3475
**47.** 5 years, 5 months   **49a.** $485.95
**49b.** $14,996.47   **50.** 11.26   **51.** $f(x) = (x + 8)^2 + 3$;
$(-8, 3); x = -8$   **52.** $\dfrac{1}{5}, -\dfrac{3}{5}$   **53.** 2.72 seconds

**54.** $\dfrac{wt^3}{9}$   **55.** domain: $\{-8, 0, 1, 9\}$; range: $\{-3, 0, 3\}$;
yes   **56.** 21 games

## Pages 583–585   Lesson 12-8

**5.** 7.27   **7.** 11.1   **9.** about 11 years, 6 months
**11.** 1.3863   **13.** 15.28 years   **15a.** 28.78 years
**15b.** 57.56 years   **17.** 20.56%   **19.** 59 weeks
**23.** $-0.000385$   **25.** 11   **27.** 48   **29.** 81   **31.** 28
**33.** 1.277   **34.** $\pm 1, \pm 2, \pm 4, \pm\dfrac{1}{3}, \pm\dfrac{2}{3}, \pm\dfrac{4}{3}$   **35.** $x^2$,
$18x, 81$   **36.** 10 in. by 12 in.   **37.** 15.4 cm
**38.** $\{x|x < 2\}$

## Pages 586–588   Chapter 12 Summary and Review

**1.** $3^{2\sqrt{2}}$   **3.** $7^{2\sqrt{2}-2\sqrt{3}}$   **5.** $-1$   **7.** $-\dfrac{7}{4}$   **9.** $\log_7 343 = 3$

**11.** $\log_4 1 = 0$   **13.** $4^3 = 64$   **15.** $6^{-2} = \dfrac{1}{36}$   **17.** 7

**19.** 3   **21.** 3   **23.** $\dfrac{1}{4}$   **25.** 3   **27.** $-4, 3$   **29.** 1.5165
**31.** 2.2618   **33.** 48   **35.** 14   **37.** 1   **39.** $-3$
**41.** $-2.5029$   **43.** 0.0003609   **45.** 0.8329   **47.** 7.21
**49.** 5.7286   **51.** $\pm 2.2452$   **53.** $-4.8188$   **55.** 0.5416
**57.** 13.30 years   **59.** 6.76 years

## CHAPTER 13   SEQUENCES AND SERIES

## Page 595   Lesson 13-1

**5.** 1, 1, 2, 3, 5, 8, 13, 21, 34, 55, 89, 144, 233, 377,
610, 987, 1597, 2584, 4181, 6765   **7.** 12   **9.** $299.18
**11.** 110 cm$^2$

**Pages 599–601  Lesson 13-2**

**5.** 4, 7, 10, 13, 16  **7.** 16, 14, 12, 10, 8  **9.** $\frac{3}{4}, \frac{1}{2}, \frac{1}{4}$,

0, $-\frac{1}{4}$  **11.** 17, 21, 25, 29  **13.** $-13, -18, -23, -28$

**15.** $\frac{7}{2}, \frac{9}{2}, \frac{11}{2}, \frac{13}{2}$  **17.** $-241$  **19.** 46  **21.** $\frac{11}{2}$  **23.** 416
**25.** 19  **27.** 8  **29.** 30, the midpoint  **31.** $-50$
**33.** $-37$
**35.** $-\frac{13}{3}, -\frac{2}{3}$

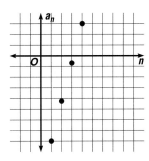

**39.** 56, 42, 35

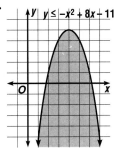

**41.** 19, 16.5, 14, 11.5  **43.** 2, 9, 16  **47.** 780 ft
**48.** aceg, acegi, acegik  **49.** 5
**50.** $\frac{(3x-2)(x+1)(2x-3)}{(x+2)(x-3)(2x+1)}$  **51.** $-3388$

**52.**

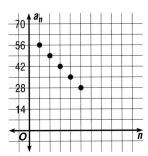

**53a.** 900 W, 1400 G  **53b.** $1060

**Pages 605–607  Lesson 13-3**

**5.** 120  **7.** 10,500  **9.** 240  **11.** 240  **13.** 10,100
**15.** 1155  **17.** 856  **19.** 104  **21.** 162  **23.** $-220$
**25.** 26  **27.** 5555  **29.** $-45$  **31.** 552  **33.** $\frac{45}{2}$
**35.** 1, 5, 9  **37.** 6, 36, 66  **41.** No, because the
auditorium only has 1065 seats.  **42.** 510
**43.** $x = -1, 8$  **44.** 77  **45.** (3, 1, 6)  **46.** 37, 37, 14

**1.** 3  **2.** 1, 3, 4, 7, 11, 18, 29, 47  **3.** 9  **4.** 416
**5.** 552  **6.** $-420$  **7.** $491,700

**Pages 612–614  Lesson 13-4**

**5.** yes, 5  **7.** yes, $\frac{3}{2}$  **9.** no  **11.** 3, $-6$, 12, $-24$
**13.** 54, 162  **15.** 27, 9  **17.** 2, $-4$  **19.** 27, $-9$, 3, $-1$

**21.**

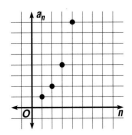

**23.** 32  **25.** 3  **27.** 1

**29.** 2, 4

**33.** 14, 28, 56 or $-14$, 28, $-56$

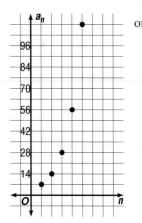 or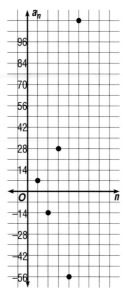

**37.** $a_n = 36\left(\dfrac{1}{3}\right)^{n-1}$  **41.** 34.9%  **43.** $\dfrac{1}{1024}$ or approximately 0.098% of the original  **44.** 632.5
**45.** $-3$, 2, 5  **46.** $2m^3 + 7m^2 + 4m - 4$
**47.** $(a^2 + 4b^4)(a - 2b^2)(a + 2b^2)$

**48.**

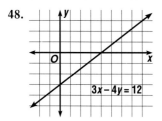

**49.** 186.84 square inches

**Pages 618–619  Lesson 13-5**
**5.** 9; $-2$; $-72$; 4  **7.** 2; 4; 512; 5  **9.** $-\dfrac{1}{4}$, $\dfrac{5}{64}$  **11.** 63

**13.** 189  **15.** $\dfrac{61}{9}$  **17.** 176  **19.** 1,328,600

**21.** 114,681  **23.** 10.66  **25.** $\dfrac{11}{16}$  **27.** 300  **29.** $\dfrac{781}{5}$

**31.** 5.99  **33.** $\dfrac{4095}{256}$  **35.** $\displaystyle\sum_{n=1}^{6} 2(-3)^{n-1}$  **37.** 4  **39.** 4

**41.** 8  **45.** 127  **47.** $\dfrac{1}{3}$, 1  **48.** 0.2392  **49.** 3 hours 7 minutes  **50.** $n = m$  **51.** (2, 1)

**Pages 623–624  Lesson 13-6**
**5.** 12, $\dfrac{1}{4}$, 16  **7.** 1, $-\dfrac{1}{3}$, $\dfrac{3}{4}$  **9.** 48, $\dfrac{1}{3}$, 72  **11.** $\dfrac{7}{9}$  **13.** $\dfrac{152}{999}$

**15.** 4  **17.** 14  **19.** 27  **21.** $\dfrac{9}{4}$  **23.** 9  **25.** 8

**27.** does not exist  **29.** $\dfrac{100}{11}$  **31.** $\dfrac{31}{99}$  **33.** $\dfrac{41}{90}$

**35.** $4 + 3 + \dfrac{9}{4}$  **37.** $9 - 3 + 1$  **39.** 800 feet
**41.** $-7.5$  **42.** $\pm 6$, 12, $\pm 24$  **43.** 52  **44.** 22  **45.** all reals

**Pages 629–630  Lesson 13-7**
**5.** 362, 880  **7.** 90  **9.** 120  **11.** 5, $108a$  **13.** 6, $-10b^2z^3$  **15.** $y^7 + 7y^6p + 21y^5p^2 + 35y^4p^3 + 35y^3p^4 + 21y^2p^5 + 7yp^6 + p^7$  **17.** $r^6 - 6r^5m + 15r^4m^2 - 20r^3m^3 + 15r^2m^4 - 6rm^5 + m^6$
**19.** $81r^4 + 108r^3y + 54r^2y^2 + 12ry^3 + y^4$
**21.** $16x^4 + 96x^3y + 216x^2y^2 + 216xy^3 + 81y^4$
**23.** $64m^6 - 576m^5 + 2160m^4 - 4320m^3 + 4860m^2 - 2916m + 729$  **25.** $\dfrac{1}{729}y^6 + \dfrac{2}{27}y^5 + \dfrac{5}{3}y^4 + 20y^3 + 135y^2 + 486y + 729$  **27.** $35x^3y^4$
**29.** $-2{,}088{,}281{,}250a^4b^7$  **31.** $k + 3$  **33.** $(k + 2)!$
**35.** $26\dfrac{1}{3}$%  **37a.** 1, 4, 6, 4, 1  **37b.** 1, 12, 66, 220, 495, 792, 924, 792, 495, 220, 66, 12, 1  **37c.** 1, $-6$, 15, $-20$, 15, $-6$, 1  **37d.** 1, $-10$, 45, $-120$, 210, $-252$, 210, $-120$, 45, $-10$, 1  **38.** $\dfrac{4}{3}$  **39.** about 300 years
**40.** $\dfrac{x + 1}{(2x - 3)(3x - 2)}$  **41.** 2 or 0 positive real; 3 or 1 negative real; 0, 2, or 4 imaginary  **42.** 24 hours

**Pages 632–634  Chapter 13 Summary and Review**
**1.** 38  **3.** $-11$  **5.** 12  **7.** 10  **9.** $-3$, 1, 5  **11.** 2322
**13.** 1330  **15.** $\dfrac{64}{3}$  **17.** $-108$, 324  **19.** 1452  **21.** $\dfrac{21}{8}$

**23.** $-6$  **25.** $-\dfrac{16}{13}$  **27.** $\dfrac{3}{4}$  **29.** $a^3 + 3a^2b + 3ab^2 + b^3$
**31.** $243r^5 + 405r^4s + 270r^3s^2 + 90r^2s^3 + 15rs^4 + s^5$
**33.** $-13{,}107{,}200x^9$  **35.** 220 feet  **37.** \$284,766

## CHAPTER 14  STATISTICS

**Pages 640–641  Lesson 14-1**
**5a.** 3,400,000  **5b.** 90%

**5c.**

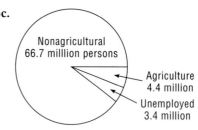

Civilian Labor Force, 1965

**7.** $\dfrac{8}{9}$  **9.** 6, $-2$  **11.** 24 times  **13.** 11, 101, 131, 151, 181, 191, 313, 353, 373, 383, 727, 757, 787, 797, 919, and 929

**Pages 645–647  Lesson 14-2**
**5a.** 299  **5b.** 277  **5c.** 11  **5d.** 280, 290

**7a.**

| Stem | Leaf |
|------|------|
| 6 | 8 |
| 7 | 0  3 |
| ● | 5  6  8  9 |
| 8 | 0  0  1  1  1  2  2  2 |
| ● | 7 \| 3 represents 73 games. |

**7b.** Ewing, Pippen, and Lewis
**7c.** Robinson **7d.** 14 **7e.** 81 or 82 **7f.** 6

**9g.**

| Stem | Leaf |
|------|------|
| 16 | 1 |
| 17 | 3  6 |
| 18 | 2  3 |
| 19 | |
| 20 | 3  4  9 |
| 21 | 0  2  8  8 |
| 22 | 5 |
| 23 | 3 |
| 24 | 16 \| 1 represents 1610 points. |
| 25 | |
| 26 | 3 |

**9h.** Jordan **9i.** Tripucka **9j.** 1027 **9k.** 5
**9l.**

| Stem | Leaf |
|------|------|
| 2 | 3  3  3  3  3 |
| ● | 5  6  6  6  7  7  7  8  9 |
| 3 | 3 |

2 \| 3 represents 23 points.
**9m.** Jordan **9n.** Johnson and McHale **9o.** 10
**9p.** 22.5 and 26.5 **9q.** 9

**11.**

| Stem | Leaf |
|------|------|
| 4 | 2  3 |
| ● | 6  6  7  8  9  9 |
| 5 | 0  0  1  1  1  1  2  2  4  4  4  4 |
| ● | 5  5  5  5  6  6  6  7  7  7  7  8 |
| 6 | 0  1  1  1  2  4  4 |
| ● | 5  8  9    4 \| 2 represents 42 years old. |

Most of the presidents were 50–59.

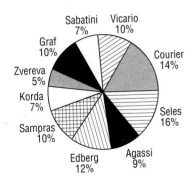

**13.** 10, 12.2, 14.4, 16.6, 18.8 **14.** 2 **15.** 121, 11
**16.** $f(x) = 45x^2 - 60x + 24$ **17.** 36 and 16

**Pages 651–652   Lesson 14-3**
**5.** 4; 4; 3 **7.** 9; 9; 9 **9.** 43; no mode; 30.6
**11.** 3.45; 2.1; 3.6 **13.** 12; 12 and 13; about 12.3

**15.** 99; no mode; 91 **17.** 4; 6; about 3.7 **19.** $5.00;
$5.00; $5.42 **21a.** mean; It is higher. **21b.** mode; It
is lower and is what most employees make. It reflects
the most representative worker. **25.** 205 miles

**26.** $y = \dfrac{1}{8}x^2 + 1$ **27.** $\begin{bmatrix} 30 & 36 \\ 0 & -12 \end{bmatrix}$ **28.** $-\dfrac{11}{3}, \dfrac{25}{3}$

**Pages 655–657   Lesson 14-4**
**5.** 150 **7.** 79 **9.** 16 **11.** 37, 39, 47; 10 **13.** 50.5,
59, 68.5; 18 **15.** 19; 2, 5, 7; 5 **17.** 170; 1025, 1075,
1125; 100 **19.** 480; 160, 265, 345; 185 **21.** 20
**23.** none **25.** none **27.** $370; $177.50, $237, $294;
$116.50; $500 **29.** 102; 21, 34, 60; 39; 121 **31.** 49;
40; 50.3 **32.** 2; 3; 162; 5 **33.** $f^{-1}(x) = \dfrac{x+4}{3}$

**34.** $\dfrac{3\sqrt{5}-3}{4}$ **35.** $-6a^4b^4 + 12a^5b^3 - 21a^4b^2$

**36.** $c = 10\left(t - \dfrac{1}{2}\right) + 35$

**Page 657   Mid-Chapter Review**

**1.**
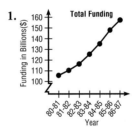

**3.** Sources of Funding for 1985-1986
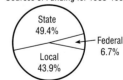

**4.**

| rounded | stem | truncated |
|---------|------|-----------|
| 6 | 4 | 5  9 |
| 7  3  0 | 5 | 2  7 |
| 5  1 | 6 | 1  5  9 |
| 0 | 7 | |

Using rounded data,          Using truncated data,
1\|6\| represents            \|6\|1 represents
$60,500,000,000 to           $61,000,000,000 to
$61,499,999,000.             $61,999,999,000.

**5.**

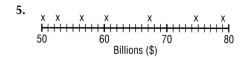

**6.** $9,105,569,000; no mode; $9,154,106,000
**7.** $52,878,386,000; $110,191,257,000,
$126,055,419,000, $149,004,882,000; $38,813,625,000;
none

**21.** $(-8, 3)$; 5

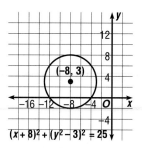

$(x + 8)^2 + (y - 3)^2 = 25$

**Pages 660–662   Lesson 14-5**

**5a.** 20   **5b.** 26   **5c.** 25%   **5d.** 21 and 28   **7a.** 210
**7b.** 670, 700   **7c.** 50%   **7d.** 75%

**9.**

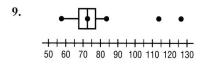

LV = 57.7, LQ = 63.05, M = 72.35,
UQ = 77.55, GV = 83.3,
outliers: 113.3, 126.7

**11.**

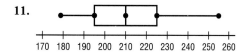

LV = 178, LQ = 195, M = 210, UQ = 225
GV = 83.3, outliers: none

**13a.**

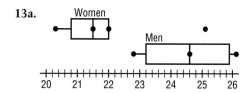

Men:   LV = 22.8        Women:   LV = 20.3
           LQ = 23.2                     LQ = 20.8
           M = 24.6                       M = 21.5
           UQ = 25.9                     UQ = 22.0
           GV = 26.2                     GV = 22.0
           outliers: none               outliers: 25.1

**13b.** Men marry at a later age than women.

**15.**

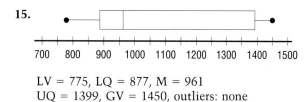

LV = 775, LQ = 877, M = 961
UQ = 1399, GV = 1450, outliers: none

**17.** 148; 196.5, 225, 249; 52.5; 342   **18.** $\frac{1}{16} +$
$\frac{1}{4} + 1 + 4 + 16 + 64; 85\frac{5}{16}$   **19.** 3   **20.** $-x^2 - xy$

**Pages 666–667   Lesson 14-6**

**5.** 5; 3.6   **7.** 250; 50   **9.** 82.2; 7.1   **11.** 97.9
**13.** 10.6   **15a.** 8.8   **15b.** 5.9   **15c.** 6.7
**15d.** Selby Sales   **19.** 11.09°; 0.097°   **20.** 3.9839
**21.** 7, $-1$   **22.** $\{x|-1 \le x \le 8\}$   **23.** 9 by 13 units

**Pages 670–673   Lesson 14-7**

**7a.** 68%   **7b.** 34%   **7c.** 97.5%   **7d.** 97   **9a.** 50%
**9b.** 15.5%   **9c.** 95%   **13a.** rejected
**13b.** X1 = 10 * 0.3 = 3
             X2 = 10 * 0.7 = 7
             Since X1 is less than 5, the procedure is stopped.
**13c.** yes   **14.** 17, 19, 21   **15.** 5   **16.** $\frac{3}{2(a - 1)}$

**17.** $\frac{(x + 1)^2}{5} + \frac{(y - 2.5)^2}{10} = 1$; ellipse

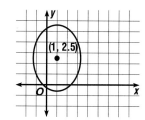

**Pages 674–676   Chapter 14 Summary and Review**

**1.**

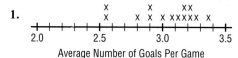

Average Number of Goals Per Game

**3.** 7.1; none; about 6.7   **5.** 20,735.5; none; 21,479.7
**7.** 15; 84, 87, 90; 6; no outliers   **9.** 106; 94.5, 100,
105; 10.5; 19, 125   **11.** 47; 23.5, 35.5, 41.5; 18; no
outliers   **13.** 1.3   **15.** 784.6   **17.** 1500   **19.** 2.5%
**21a.** 6,021,362; no mode; 6,706,047.4
**21b.** 6,872,285; 5,448,988.5, 6,021,362, 7,464,143.5;
2,015,155; 11,781,270   **23a.** 1600   **23b.** 250
**23c.** 97%   **23d.** 97.5%

# CHAPTER 15 PROBABILITY

**Page 683   Lesson 15-1**

**3a.** (1, 1), (1, 2), (1, 3), (1, 4), (1, 5), (1, 6), (2, 1), (2, 2), (2, 3), (2, 4), (2, 5), (2, 6), (3, 1), (3, 2), (3, 3), (3, 4), (3, 5), (3, 6), (4, 1), (4, 2), (4, 3), (4, 4), (4, 5), (4, 6), (5, 1), (5, 2), (5, 3), (5, 4), (5, 5), (5, 6), (6, 1), (6, 2), (6, 3), (6, 4), (6, 5), (6, 6)   **3b.** 2   **3c.** 12
**3d.** 7, because more combinations for 7 exist than for any other number between 2 and 12.   **5.** 17¢ gold medium, 36¢ silver small, 55¢ black large
**7.** minimum: 14 units, maximum: 26 units

**Pages 686–688   Lesson 15-2**

**5a.** independent   **5b.** dependent   **5c.** dependent or independent, based upon the availability of each color for a particular model   **7.** 6! or 720 schedules
**9.** dependent, if a person can only hold one office
**11.** dependent   **13.** dependent   **15.** 1800 combinations

**17.**

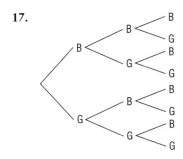

**19.** 90 routes   **21.** 720 ways   **23.** 240 ways
**25a.** 840 patterns   **25b.** 240 patterns
**27a.** 10,140,000   **27b.** 1,872,000   **29a.** 16 ways
**29b.** 12 ways   **30.** $6 \times 7$ or 42, since 7 is the most probable total each time the dice are rolled
**31.** $3(2)^{26}$   **32.** 2.73 years

**33.**                                  **34.** $-6$

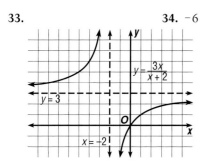

**Pages 692–693   Lesson 15-3**

**5.** true   **7.** false   **9.** 120   **11.** true   **13.** true
**15.** false   **17.** 3   **19.** 30   **21.** 120   **23.** 3360
**25.** 778,377,600   **27.** 6   **29.** 72   **31.** 180   **33.** 126
**35.** 6   **37.** 12   **39.** 15   **43a.** 45,360

**43b.** about 80.1   **45.** 13 ways   **46.** 2, 3, 4, 5, 6, 7, 8
**47.** 3   **48.** 1800 km   **49.** $\left(8, \dfrac{3}{2}\right)$   **50.** $d \geq -105$

**Pages 697–698   Lesson 15-4**

**5.** circular, not reflection, 3,628,800   **7.** linear, reflection, 1,814,400   **9.** circular, not reflection
**11.** linear, not reflection   **13.** 56   **15.** 60   **17.** 2520
**19.** 24   **21.** 12   **23.** $5^{28}$   **25.** 255 ways   **27.** 120

**28.** $\displaystyle\sum_{n=1}^{20} \dfrac{n+1}{2n+1}$   **29.** 3 positive real roots, or 1

positive real root and 2 imaginary roots   **30.** $5, -\dfrac{7}{3}$

**31.** $x^2 - 3x - 10$   **32.** Parallel lines have equal slopes, whereas the slopes of perpendicular lines are negative reciprocals of each other.

**Pages 701–704   Lesson 15-5**

**5.** combination   **7.** combination   **9.** 792   **11.** 12
**13.** combination   **15.** permutation   **17.** permutation
**19.** combination   **21.** 56   **23.** 2024   **25.** 126
**27.** 84   **29.** none   **31.** 3570   **33.** 351   **35a.** 5148
**35b.** 28,561   **37.** 30   **39.** 14   **41.** 126   **43.** 0
**45.** 2808   **47.** 2520   **49.** 252   **51.** 1980 teams
**53.** 196   **54.** 24 ways   **55.** 19.3 years   **56.** about 8.7 years

**57.** ellipse, hyperbola

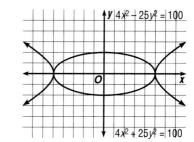

**58.** $\left(\dfrac{1}{5}, \dfrac{2}{5}, -\dfrac{3}{5}\right)$   **59.** $a^2b^2c^2$, $a^2b^2c^2d$, $a^2b^2c^2d^2$
**60.** (1, 64), (2, 32), (4, 16), (8, 8), (16, 4), (32, 2), (64, 1), (−1, −64), (−2, −32), (−4, −16), (−8, −8), (−16, −4), (−32, −2), (−64, −1)

**Page 704   Mid-Chapter Review**

**1a.** squares, pencils, buttons—arranged in a straight line   **1b.** 24   **2a.** tree diagram   **2b.** 24   **3.** 100!
**4.** 13,824,000   **5.** 3!2! or 12 ways   **6.** $\dfrac{52!}{13!39!}$

**Pages 707–709   Lesson 15-6**

**7.** $\dfrac{5}{4}$   **9.** $\dfrac{3}{4}$   **11.** $\dfrac{4}{7}$   **13.** $\dfrac{1}{6}$   **15.** $\dfrac{1}{6}$   **17.** $\dfrac{7}{8}$   **19.** $\dfrac{5}{6}$   **21.** $\dfrac{1}{2}$

**23.** $\dfrac{1}{3}$   **25.** $\dfrac{18}{95} \approx 0.189$   **27.** $\dfrac{3}{19} \approx 0.158$

**29.** $\dfrac{14}{39} \approx 0.359, \dfrac{14}{25}$   **31.** $\dfrac{1}{16} \approx 0.063$   **33.** $\dfrac{3}{56} \approx 0.054$

**35.** $\frac{1}{4} = 0.25$  **37.** $\frac{2}{7} \approx 0.286$  **39.** $\frac{4}{7} \approx 0.571$
**41a.** 66,607:33  **41b.** 108,257:33  **41c.** 16,627:33
**43.** $\frac{1}{9,366,819}$  **45.** 420  **46.** $1024r^5 + 1280r^4s +$
$640r^3s^2 + 160r^2s^3 + 20rs^4 + s^5$  **47.** 2401
**48.** $2x^2 + 5$

**49.** 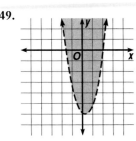  **50.** $-\frac{3}{2}, \frac{1}{3}$  **51.** 1

## Pages 712–714  Lesson 15-7
**5.** dependent, $\frac{2}{7}$  **7.** independent, $\frac{81}{625}$  **9.** $\frac{9}{49} \approx 0.184$

**11.** $\frac{1}{8} = 0.125$  **13.** $\frac{5}{48} \approx 0.104$  **15.** $\frac{1}{36} \approx 0.028$

**17.** $\frac{1}{36} \approx 0.028$  **19.** $\frac{1}{6} \approx 0.167$  **21a.** $\frac{1}{26} \approx 0.038$

**21b.** $\frac{7}{225} \approx 0.031$  **23.** $\frac{1}{32} \approx 0.031$  **25.** $\frac{1}{94,109,400} \approx$

$1.06 \times 10^{-8}$  **27.** $\frac{4}{635,013,559,600} \approx 6.3 \times 10^{-12}$

**29.** 0  **31a.** $\frac{1}{729}$  **31b.** $\frac{5}{243}$  **31c.** $\frac{25}{243}$  **33.** $\frac{1}{4}$

**34.** $\frac{z^2 + 49}{(z + 5)(z - 7)}$  **35.** (1, 5), (2, 1), (3, 2)

**36.** $\begin{bmatrix} 1 & 0 \\ 0 & 1 \end{bmatrix}$

## Pages 717–719  Lesson 15-8
**5.** inclusive, $\frac{7}{30} \approx 0.233$  **7.** exclusive, $\frac{9}{91} \approx 0.099$

**9.** inclusive, $\frac{175}{221} \approx 0.79$  **11.** $\frac{2}{11} \approx 0.181$

**13.** $\frac{14}{33} \approx 0.424$  **15.** $\frac{188}{663} \approx 0.284$  **17.** $\frac{55}{221} \approx 0.249$

**19.** $\frac{7}{16} \approx 0.438$  **21.** $\frac{35}{64} \approx 0.547$  **23.** $\frac{29}{3003} \approx 0.010$

**25.** $\frac{70}{143} \approx 0.490$  **27.** $\frac{1}{525} \approx 0.002$  **29.** $\frac{46}{525} \approx 0.088$

**33.** $\frac{57}{1771} \approx 0.032$  **35.** $\frac{1}{20}$  **36.** 1960  **37.** 15, 37.5
**38.** 7  **39.** 374.4

## Pages 723–724  Lesson 15-9
**5.** binomial, $\frac{3}{8}$  **7a.** binomial, $\frac{16}{81}$  **7b.** binomial, $\frac{4}{81}$

**7c.** binomial, $\frac{1}{9}$  **7d.** not binomial  **7e.** not binomial

**7f.** not binomial  **9.** $\frac{3}{8} = 0.375$  **11.** $\frac{3125}{7776} \approx 0.402$

**13.** $\frac{625}{648} \approx 0.965$  **15.** $\frac{1}{9} \approx 0.111$  **17.** $\frac{15}{128} \approx 0.117$

**19.** $\frac{1}{1024} \approx 0.001$  **21.** $\frac{21}{3125} \approx 0.007$  **23.** $\frac{1}{8} = 0.125$

**25.** $\frac{1}{2} = 0.5$  **27.** 0.0001049  **29.** 0.1662386

**31.** $\frac{4536}{15,625} \approx 0.290$  **33a.** $\left(\frac{1}{10}\right)^{12} = 1.0 \times 10^{-12}$

**33b.** 0.0037881  **33c.** $\left(\frac{9}{10}\right)^{12} \approx 0.2824295$

**33d.** 0.8891300  **34.** $\frac{13}{16}$  **35.** 600 ft  **36.** $-1$  **37.** $\frac{48}{56}$

## Pages 726–728  Chapter 15 Summary and Review
**1.** 125 patterns  **3.** 10.5  **5.** 1,625,702,400  **7.** 5040
**9.** 60 ways  **11.** 270,725  **13.** $\frac{3}{7} \approx 0.429$

**15.** $\frac{2}{15} \approx 0.133$  **17.** $\frac{3}{10} = 0.3$  **19.** $\frac{23}{26} \approx 0.885$

**21.** $\frac{763}{3888} \approx 0.196$  **23.** 15 ways

## CHAPTER 16  TRIGONOMETRIC FUNCTIONS

### Pages 734–736  Lesson 16-1
**5.** II  **7.** IV  **9.** I  **11.** no  **13.** no  **15.** no  **17.** no
**19.** no  **21.** yes  **23.** $\pi$  **25.** $\frac{\pi}{4}$  **27.** $-\frac{5\pi}{4}$  **29.** $\frac{3\pi}{4}$

**31.** $-\frac{2\pi}{3}$  **33.** 180°  **35.** −315°  **37.** 405°

**39.** $\frac{-420°}{\pi} \approx -133.69°$  **43.** 0°  **45.** 99°  **47.** no
**49.** 30 IF A < 2 * 3.1415927 THEN 70; 40 LET A = A − 2 * 3.1415927; 70 LET A = A + 2 × 3.1415927
**50.** $\frac{1}{1024} \approx 0.001$  **51.** 109,225
**52.** $\log_3 27 + \log_3 3 \stackrel{?}{=} \log_3 81$
  $3 \quad + \quad 1 \quad = 4 \checkmark$

**53.** 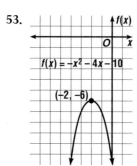  **54.** $\frac{b^3x^4}{5^3y}$

$f(x) = -x^2 - 4x - 10$
(−2, −6)

**Pages 740–741  Lesson 16-2**

**5.** +  **7.** +  **9.** +  **11.** $\frac{1}{2}$  **13.** $-\frac{\sqrt{2}}{2}$  **15.** $-\frac{\sqrt{3}}{2}$

**17.** 290°  **19.** 270°  **21.** $\frac{\pi}{2}$  **23.** $\frac{3\pi}{4}$  **25.** $\frac{\sqrt{2}}{2}$  **27.** $\frac{1}{2}$

**29.** $-1$  **31.** 0  **33.** $\frac{1}{2}$  **35.** $-\frac{1}{2}$  **37.** 1  **39.** $-\frac{\sqrt{3}}{3}$

**41.** 1  **45.** $\frac{1}{2}$  **46.** 19; 16; 18.5  **47.** 650 cans

**48.** $\frac{3x(x-1)}{(4x-1)^2}$  **49.** $-1, -2, 5$  **50.** $6 + 4i$

**51.**

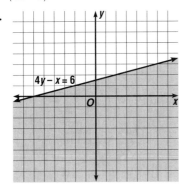

4y − x = 6

**Pages 743–745  Lesson 16-3**

**5.** 90°, 270°  **7.** 0°, 180°, 360°  **9.** 90°, 270°  **11.** sin $-225° = \frac{\sqrt{2}}{2}$; cos $-225° = -\frac{\sqrt{2}}{2}$; tan $-225° = -1$; sec $-225° = -\sqrt{2}$; cot $-225° = -1$; csc $-225° = \sqrt{2}$

**13.** $-1$  **15.** 2  **17.** 2  **19.** $-\frac{\sqrt{3}}{3}$  **21.** $-\sqrt{3}$  **23.** $-2$

**25.** 1  **27.** $\frac{\sqrt{3}}{3}$  **29.** $\sqrt{3}$  **31.** $\frac{2\sqrt{3}}{3}$  **33.** undefined

**35.** $-\frac{\sqrt{3}}{3}$  **37.** undefined  **39.** $-\frac{2\sqrt{3}}{3}$  **41.** tan 12.7° = $\frac{9}{40}$; sec 12.7° = $\frac{41}{40}$; cot 12.7° = $\frac{40}{9}$; csc 12.7° = $\frac{41}{9}$

**43.** $\frac{4}{7}$  **44.** 12; $\frac{1}{2}$, $\frac{3}{8}$; 6  **45.** $(\pm 3, 0)$  **46.** 3  **47.** 24 ft, 20 ft, 12 ft  **48.** $\frac{7}{2}$

**Pages 749–750  Lesson 16-4**

**5.** $y = \text{Sin}^{-1} x$  **7.** $30° = \text{Sin}^{-1} \frac{1}{2}$  **9.** $x = \text{Tan}^{-1}\left(-\frac{4}{3}\right)$

**11.** $\frac{4}{5}$  **13.** 120°  **15.** 0°  **17.** $\frac{1}{2}$  **19.** $\frac{\sqrt{13}}{6}$  **21.** $\frac{4\sqrt{2}}{7}$

**23.** $\frac{4}{5}$  **25.** $\frac{\pi}{2}$ or 90°  **27.** 30°  **29.** $\frac{\sqrt{3}}{2}$  **31.** $\frac{\sqrt{2}}{2}$

**33.** $\frac{24}{25}$  **35.** $\frac{\sqrt{3}}{2}$  **37.** $\pi$ or 180°  **39.** $\frac{\pi}{3}$ or 60°

**41a.**

48 miles  cos $A = \frac{1}{2}$  24 miles  A

**41b.** 60° north of east  **43.** 37,000,000; 18,000,000, 22,000,000, 31,000,000, 13,000,000; 53,000,000
**44.** $3 + 4 + 5 + 6 + 7 + 8 + 9 + 10 + 11 + 12$; 75
**45.** 0.2392  **46.** $4\frac{8}{9}$ hours  **47.** 201,600; $2.016 \times 10^5$

**48.** $\begin{bmatrix} 4 & 7 \\ 0 & 1 \end{bmatrix}$

**Pages 753–755  Lesson 16-5**

**5.** 0.1564  **7.** 07392  **9.** 343.8  **11.** 7.6067
**13.** 68.6796  **15.** 17.6779  **17.** 0.1736  **19.** $-1$
**21.** 0.8660  **23.** 3.2361  **25.** $-2$  **27.** undefined
**29.** 0.7507  **31.** $-4.0624$  **33.** 27.1268°  **35.** 0.6340
**37.** $-0.9258$  **39.** 128.3 feet  **41.** 35°16'  **42.** $\frac{3489}{6561}$ or $\frac{1163}{2187}$; approximately 0.532  **43.** 24; 24 and 30; 26.5; approximately 6 points  **44.** $\log_3 35 = \frac{\log 35}{\log 3}$; approximately 3.2362  **45.** $\frac{x(x+2)}{4(x-1)(x+4)}$  **46.** $-\frac{13}{3}, \frac{8}{3}$

**Page 755  Mid-Chapter Review**

**1.** $\frac{\pi}{2}$  **2.** $\frac{5\pi}{6}$  **3.** $-\frac{3\pi}{4}$  **4.** 270°  **5.** $-315°$

**6.** $\frac{360°}{\pi} \approx 114.59°$  **7.** 60°  **8.** 320°  **9.** $\frac{10\pi}{9}$  **10.** $-\frac{\sqrt{3}}{2}$

**11.** 0  **12.** $-\frac{2\sqrt{3}}{3}$  **13.** $\sqrt{2}$  **14.** undefined  **15.** $\sqrt{2}$

**16.** 45°  **17.** $-90°$  **18.** $\frac{\sqrt{2}}{2}$  **19.** $-30°$  **20.** 30°
**21.** $-1$  **22.** 1.0038  **23.** $-0.1736$  **24.** 3.5387
**25.** 45°

**Pages 759–760  Lesson 16-6**

**5.** tan 76° = $\frac{13}{b}$; 3.2  **7.** tan 71°13' = $\frac{21.2}{b}$; 7.2

**9.** cos 19°07' = $\frac{11}{c}$; 11.6  **11.** $b = 5$, $A = 67°23'$, $B = 22°37'$  **13.** $a = 10.82$, $A = 31°$, $B = 59°$
**15.** $a = 11.53$, $A = 62°31'$, $B = 27°29'$  **17.** $a = 3.86$, $b = 13.46$, $B = 74°$  **19.** $A = 47°50'$, $c = 12.14$, $b = 8.15$  **21.** $A = 7°$, $a = 0.68$, $c = 5.61$
**23.** $A = 77°$, $a = 5.85$, $b = 1.35$  **25.** $A = 41°$, $b = 10.35$, $c = 13.72$  **27.** $A = 45°16'$, $c = 61.94$, $b = 43.59$  **29.** $B = 34°5'$, $a = 13.25$, $b = 8.97$
**31.** $B = 75°$, $a = 6.47$, $b = 24.15$  **33.** $A = 41°11'$, $B = 48°49'$, $b = 8$, $c = 10.63$  **35.** $B = 53°8'$, $A = 36°52'$, $a = 6$, $c = 10$  **37.** $x = 3$, $a = b = 5$, $c = 7.07$, $A = B = 45°$  **41.** 555 feet  **42.** 84

**43.** $1 + 3 + 9 + 27 + 81$; 121   **44.** 75 years ago
**45.** 100 ml   **46.** 16; 2 real-rational; 6, 2   **47.** $r = 3$,
$s = 9$

## Pages 764–766   Lesson 16-7

**3.** $7^2 + b^2 = 16^2$; 14.4   **5.** $\tan A = \dfrac{7}{12}$; 30°15′
**7.** $\sin A = 0.9428$, $\cos A = 0.3333$, $\tan A = 2.8284$
**9.** $\sin A = 0.5692$, $\cos A = 0.8222$, $\tan A = 0.6923$
**11.** 251.73 m   **13.** 61.08 m   **15.** 9°36′   **17.** 22°48′
**19.** 188.89 ft   **21.** 11.7 m   **23.** 4.05 cm
**25.** 174.50 m   **27.** 164.92 km   **28.** 20   **29.** 3   **30.** 3,
$\dfrac{2}{3}$, $\dfrac{3}{4}$, $-\dfrac{1}{2}$

## Pages 769–770   Lesson 16-8

**5.** area $= \dfrac{1}{2}(10)(17) \sin 46°$; 61.1   **7.** area $= \dfrac{1}{2}(15)(30)$

$\sin 90°$; 225   **9.** $\dfrac{\sin 50°}{14} = \dfrac{\sin B}{10}$; 33°10′

**11.** $\dfrac{\sin 42°}{16} = \dfrac{\sin C}{12}$; 30°7′   **13.** 55.155   **15.** 21.097
**17.** 63.2086   **19.** $C = 74°$, $b = 8.89$, $c = 10.19$
**21.** $A = 52°$, $b = 100.17$, $c = 90.40$   **23.** $B = 52°53'$,
$A = 62°27'$, $a = 16.68$   **25.** $B = 82°5'$, $b = 23.77$,
$c = 9.95$   **27.** $B = 48°42'$, $C = 26°18'$, $c = 8.3$
**29.** 31.9 yd and 22.6 yd   **31.** 57.2 km

**35a.**   **35b.** 424.24 miles

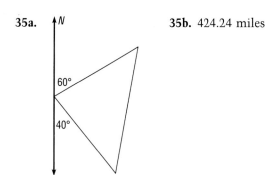

**36.**

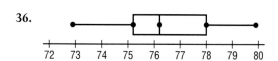

median: 76.2          outliers: none
lower quartile: 75.2   greatest value: 79.9
upper quartile: 78     least value: 72.9
interquartile range: 2.8

**37.** $x^6 + 24x^5 + 240x^4 + 1280x^3 + 3840x^2 +$
$6144x + 4096$   **38.** 5.003 hours   **39.** $x^2 + 1$; $x^2 + 2x$
$+ 1$   **40.** 128 square cm

## Pages 773–774   Lesson 16-9

**5.** 2; $B = 53°7'$, $C = 90°1'$, $c = 10$; $B = 126°53'$,
$C = 16°15'$, $c = 2.8$   **7.** None   **9.** None   **11.** None
**13.** 2; $B = 53°28'$, $C = 86°32'$, $c = 12.4$; $B = 126°32'$,
$C = 13°28'$, $c = 2.9$   **15.** $\dfrac{1}{14}$   **17.** no solution   **19.** 6 or 12
**21.** A sample answer is $6 + 3 - 6 - 2 + 4 = 5$.

## Pages 778–779   Lesson 16-10

**5.** law of sines; $b = 14.7$, $B = 109°3'$, $C = 30°57'$
**7.** law of cosines; $b = 18.5$, $A = 40°57'$, $C = 79°3'$
**9.** law of sines; $a = 9.6$, $b = 14$, $B = 70°$   **11.** law of
sines; $c = 23.3$, $A = 26°10'$, $C = 110°52'$   **13.** $a = 10.91$,
$B = 57°16'$, $C = 87°44'$   **15.** $A = 22°37'$, $B = 67°23'$,
$C = 90°$   **17.** Two solutions: $C = 13.88$, $A = 73°54'$,
$B = 67°46'$; $C = 13.88$, $A = 106°6'$, $B = 35°31'$
**19.** $b = 20.95$, $A = 52°25'$, $C = 108°35'$
**21.** $A = 45°34'$, $B = 57°34'$, $C = 76°52'$
**23.** $A = 102°55'$, $B = 49°15'$, $C = 27°50'$
**25.** $a = 8.31$, $B = 26°19'$, $C = 124°41'$   **27.** 1434.26 ft;
86,804.28 ft$^2$   **29.** 76.90 cm, 101.09 cm   **31.** 36°1′
**33.** 334.6 nautical miles   **35.** 2033 km

**37.**

| | Rounded | | | | Stem | Truncated | | | |
|---|---|---|---|---|---|---|---|---|---|
| 9 | 8 | 8 | 5 | 3 | 2 | 2 | 5 | 7 7 9 |
| | | | 9 | 7 | 3 | 6 | 9 | 9 |
| | | | | 0 | 4 | | | |
| | | 9 | 3 | 1 | 5 | 1 | 2 | 8 |
| | | | | | 6 | | | |
| | | | 8 | 7 | 7 | 6 | 8 | |
| | | | 7 | 4 | 8 | 3 | 7 | |
| | | | | 6 | 9 | 5 | | |

Using rounded data, 7 |3| represents 365,000 to
374,000 people.

Using truncated data, |3| 6 represents 360,000 to
369,000 people.

**38.** 25   **39.** 0   **40.** $y = \pm\sqrt{x} + 4$

## Pages 780–782   Chapter 16 Summary and Review

**1.** $\dfrac{17\pi}{12}$   **3.** $\dfrac{3\pi}{2}$   **5.** 60°   **7.** $-75°$   **9.** $-\dfrac{\sqrt{3}}{2}$   **11.** $-\dfrac{1}{2}$
**13.** $-\dfrac{\sqrt{2}}{2}$   **15.** 1   **17.** $\sqrt{2}$   **19.** $\dfrac{2\sqrt{3}}{3}$   **21.** $-\sqrt{3}$
**23.** $-90°$   **25.** 60°   **27.** 0   **29.** 37.5896°   **31.** $-0.3536$
**33.** 14.4   **35.** 17.3   **37.** 5.4   **39.** $B = 7°8'$, $C = 89°42'$,
$c = 80.6$   **41.** $A = 51°$, $a = 70.22$, $c = 89.69$
**43.** $c = 6.43$, $A = 34°19'$, $B = 80°41'$   **45.** $c = 4.54$,
$A = 58°9'$, $B = 81°51'$   **47.** 4483.48 ft   **49.** 3.7 m

# CHAPTER 17 TRIGONOMETRIC GRAPHS, IDENTITIES, AND EQUATIONS

**Pages 795–796   Lesson 17-1**

**5.** 3, $2\pi$   **7.** 3, $4\pi$   **9.** 6, $3\pi$   **11.** 3, $3\pi$   **13.** $\dfrac{1}{2}$, $\dfrac{8\pi}{3}$

**15.** 6, $\pi$   **17.** $2\pi$   **19.** $\dfrac{\pi}{5}$   **21.** $\dfrac{\pi}{3}$   **23.** $\pi$   **25.** $4\pi$
**27.** $3\pi$

**29.** 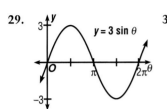   **33.**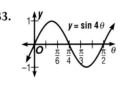

**37.**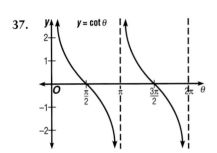

**41.**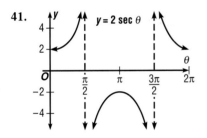

**45.**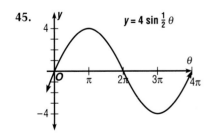

**49.**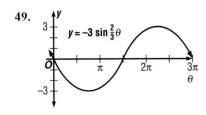

**53.**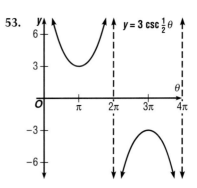

**57.**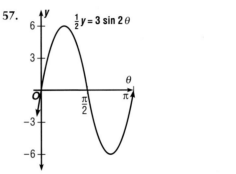

**61a.**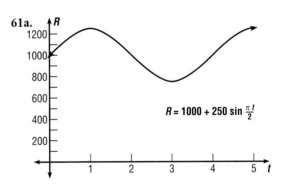

**63.** yes   **64.** 92; no mode; 91.25

**65.** A sample answer is $\displaystyle\sum_{n=1}^{6} 3 + 3n$.   **66.** $4^{\sqrt{6}}$

**67.** 3, $-\dfrac{5}{4}$, $\dfrac{7 \pm \sqrt{5}}{2}$   **68.** 37   **69a.** 600 floor lamps, 300 table lamps   **69b.** \$73,500

**Pages 799–800   Lesson 17-2**

**5.** $\sin\theta\cos\theta$   **7.** 1   **9.** $\cot\alpha$   **11.** $\cot^2 x$   **13.** $\cos\theta$
**15.** $\tan^2 x$   **17.** $\dfrac{1}{2}$   **19.** $\dfrac{\sqrt{5}}{3}$   **21.** $\dfrac{3}{4}$   **23.** $\dfrac{4\sqrt{7}}{7}$

**25.** $-\dfrac{\sqrt{3}}{3}$   **27.** $-\sqrt{5}$   **29.** $2\sqrt{2}$   **31.** $-\dfrac{\sqrt{3}}{2}$   **33.** $\sqrt{2}$

**35.** $\dfrac{4}{5}$   **37.** $\tan^2\theta$   **39.** 1   **41.** 1   **43.** 2

**45.** $1 + \cot^2 \theta \overset{?}{=} \csc^2 \theta$

$$\dfrac{\sin^2 \theta + \cos^2 \theta}{\sin^2 \theta} \overset{?}{=} \csc^2 \theta$$

$$\dfrac{1}{\sin^2 \theta} \overset{?}{=} \csc^2 \theta$$

$$\csc^2 \theta = \csc^2 \theta$$

**47.** $\dfrac{\sec \theta}{\csc \theta} \overset{?}{=} \tan \theta$

$$\dfrac{\dfrac{1}{\cos \theta}}{\dfrac{1}{\sin \theta}} \overset{?}{=} \tan \theta$$

$$\dfrac{1}{\cos \theta} \cdot \dfrac{\sin \theta}{1} \overset{?}{=} \tan \theta$$

$$\dfrac{\sin \theta}{\cos \theta} \overset{?}{=} \tan \theta$$

$$\tan \theta = \tan \theta$$

**51.** 907,200  **52.** $0, -\dfrac{3}{2}; -2, 3$  **53.** $(x - 6)^2 +$
$y^2 = 36$  **54.** $200x - 2x^2$  **55.** $4b + \dfrac{1}{2b - 1}$
**56.** domain: $\{0, 8\}$; range: $\{-12, -8, 0, 3, 4\}$; no

**Pages 805–806  Lesson 17-3**

**3.** $\sin \theta \sec \theta \cot \theta \overset{?}{=} 1$

$$\sin \theta \cdot \dfrac{1}{\cos \theta} \cdot \dfrac{1}{\tan \theta} \overset{?}{=} 1$$

$$\dfrac{\sin \theta}{\cos \theta} \cdot \dfrac{1}{\tan \theta} \overset{?}{=} 1$$

$$\tan \theta \cdot \dfrac{1}{\tan \theta} \overset{?}{=} 1$$

$$1 = 1$$

**5.** $\csc y \sec y \overset{?}{=} \cot y + \tan y$

$$\dfrac{1}{\sin y} \cdot \dfrac{1}{\cos y} \overset{?}{=} \dfrac{\cos y}{\sin y} + \dfrac{\sin y}{\cos y}$$

$$\dfrac{1}{\sin y} \cdot \dfrac{1}{\cos y} \overset{?}{=} \dfrac{\cos^2 y + \sin^2 y}{\sin y \cos y}$$

$$\dfrac{1}{\sin y \cos y} = \dfrac{1}{\sin y \cos y}$$

**7.** $\sec^2 x - \tan^2 x \overset{?}{=} \tan x \cot x$

$$1 \overset{?}{=} \tan x \cdot \dfrac{1}{\tan x}$$

$$1 \overset{?}{=} \dfrac{\tan x}{\tan x}$$

$$1 = 1$$

**9.** $\dfrac{\sec \alpha}{\sin \alpha} - \dfrac{\sin \alpha}{\cos \alpha} \overset{?}{=} \cot \alpha$

$$\dfrac{1}{\cos \alpha \sin \alpha} - \dfrac{\sin \alpha}{\cos \alpha} \overset{?}{=} \cot \alpha$$

$$\dfrac{1 - \sin^2 \alpha}{\cos \alpha \sin \alpha} \overset{?}{=} \cot \alpha$$

$$\dfrac{\cos^2 \alpha}{\cos \alpha \sin \alpha} \overset{?}{=} \cot \alpha$$

$$\dfrac{\cos \alpha}{\sin \alpha} \overset{?}{=} \cot \alpha$$

$$\cot \alpha = \cot \alpha$$

**11.** $\dfrac{\sin \alpha}{1 - \cos \alpha} + \dfrac{1 - \cos \alpha}{\sin \alpha} \overset{?}{=} 2 \csc \alpha$

$$\dfrac{\sin^2 \alpha + (1 - \cos \alpha)^2}{(1 - \cos \alpha)(\sin \alpha)} \overset{?}{=} \dfrac{2}{\sin \alpha}$$

$$\dfrac{\sin^2 \alpha + 1 - 2 \cos \alpha + \cos^2 \alpha}{(1 - \cos \alpha) \sin \alpha} \overset{?}{=} \dfrac{2}{\sin \alpha}$$

$$\dfrac{2 - 2 \cos \alpha}{(1 - \cos \alpha) \sin \alpha} \overset{?}{=} \dfrac{2}{\sin \alpha}$$

$$\dfrac{2(1 - \cos \alpha)}{(1 - \cos \alpha) \sin \alpha} \overset{?}{=} \dfrac{2}{\sin \alpha}$$

$$\dfrac{2}{\sin \alpha} = \dfrac{2}{\sin \alpha}$$

**13.** $\dfrac{1 - \cos x}{\sin x} \overset{?}{=} \dfrac{\sin x}{1 + \cos x}$

$$\dfrac{1 - \cos x}{\sin x} \overset{?}{=} \dfrac{\sin x(1 - \cos x)}{(1 + \cos x)(1 - \cos x)}$$

$$\dfrac{1 - \cos x}{\sin x} \overset{?}{=} \dfrac{\sin x(1 - \cos x)}{1 - \cos^2 x}$$

$$\dfrac{1 - \cos x}{\sin x} \overset{?}{=} \dfrac{\sin x(1 - \cos x)}{\sin^2 x}$$

$$\dfrac{1 - \cos x}{\sin x} = \dfrac{1 - \cos x}{\sin x}$$

**15.** $\dfrac{1 - \cos x}{1 + \cos x} \overset{?}{=} (\csc x - \cot x)^2$

$$\dfrac{1 - \cos x}{1 + \cos x} \overset{?}{=} \left(\dfrac{1}{\sin x} - \dfrac{\cos x}{\sin x}\right)^2$$

$$\dfrac{1 - \cos x}{1 + \cos x} \overset{?}{=} \left(\dfrac{1 - \cos x}{\sin x}\right)^2$$

$$\dfrac{1 - \cos x}{1 + \cos x} \overset{?}{=} \dfrac{(1 - \cos x)^2}{\sin^2 x}$$

$$\dfrac{1 - \cos x}{1 + \cos x} \overset{?}{=} \dfrac{(1 - \cos x)^2}{1 - \cos^2 x}$$

$$\dfrac{1 - \cos x}{1 + \cos x} \overset{?}{=} \dfrac{(1 - \cos x)(1 - \cos x)}{(1 - \cos x)(1 + \cos x)}$$

$$\dfrac{1 - \cos x}{1 + \cos x} = \dfrac{1 - \cos x}{1 + \cos x}$$

**17.** $\dfrac{\cot \theta + \csc \theta}{\sin \theta + \tan \theta} \overset{?}{=} \cot \theta \csc \theta$

$$\dfrac{\dfrac{\cos \theta}{\sin \theta} + \dfrac{1}{\sin \theta}}{\sin \theta + \dfrac{\sin \theta}{\cos \theta}} \overset{?}{=} \dfrac{\cos \theta}{\sin \theta} \cdot \dfrac{1}{\sin \theta}$$

$$\dfrac{\dfrac{\cos \theta + 1}{\sin \theta}}{\dfrac{\sin \theta \cos \theta + \sin \theta}{\cos \theta}} \overset{?}{=} \dfrac{\cos \theta}{\sin \theta} \cdot \dfrac{1}{\sin \theta}$$

$$\dfrac{\dfrac{\cos \theta + 1}{\sin \theta}}{\dfrac{\sin \theta(\cos \theta + 1)}{\cos \theta}} \overset{?}{=} \dfrac{\cos \theta}{\sin \theta} \cdot \dfrac{1}{\sin \theta}$$

$$\dfrac{\cos \theta + 1}{\sin \theta} \cdot \dfrac{\cos \theta}{\sin \theta(\cos \theta + 1)} \overset{?}{=} \dfrac{\cos \theta}{\sin^2 \theta}$$

$$\dfrac{\cos \theta}{\sin^2 \theta} = \dfrac{\cos \theta}{\sin^2 \theta}$$

**19.** $\dfrac{1-2\cos^2\beta}{\sin\beta\cos\beta} \overset{?}{=} \tan\beta - \cot\beta$

$\dfrac{1-2\cos^2\beta}{\sin\beta\cos\beta} \overset{?}{=} \dfrac{\sin\beta}{\cos\beta} - \dfrac{\cos\beta}{\sin\beta}$

$\dfrac{1-2\cos^2\beta}{\sin\beta\cos\beta} \overset{?}{=} \dfrac{\sin^2\beta - \cos^2\beta}{\cos\beta\sin\beta}$

$\dfrac{1-2\cos^2\beta}{\sin\beta\cos\beta} \overset{?}{=} \dfrac{1 - \cos^2\beta - \cos^2\beta}{\cos\beta\sin\beta}$

$\dfrac{1-2\cos^2\beta}{\sin\beta\cos\beta} = \dfrac{1-2\cos^2\beta}{\sin\beta\cos\beta}$

**21.**

$\dfrac{\cos y}{1+\sin y} + \dfrac{\cos y}{1-\sin y} \overset{?}{=} 2\sec y$

$\dfrac{\cos y(1-\sin y) + \cos y(1+\sin y)}{(1+\sin y)(1-\sin y)} \overset{?}{=} \dfrac{2}{\cos y}$

$\dfrac{\cos y - \cos y \sin y + \cos y + \cos y \sin y}{1 - \sin^2 y} \overset{?}{=} \dfrac{2}{\cos y}$

$\dfrac{2\cos y}{\cos^2 y} \overset{?}{=} \dfrac{2}{\cos y}$

$\dfrac{2}{\cos y} = \dfrac{2}{\cos y}$

**23.** $\cot x(\cot x + \tan x) \overset{?}{=} \csc^2 x$

$\cot^2 x + \tan x \cot x \overset{?}{=} \csc^2 x$

$\cot^2 x + \tan x \cdot \dfrac{1}{\tan x} \overset{?}{=} \csc^2 x$

$\cot^2 x + 1 \overset{?}{=} \csc^2 x$

$\csc^2 x = \csc^2 x$

**25.**

$\dfrac{1+\tan\gamma}{1+\cot\gamma} \overset{?}{=} \dfrac{\sin\gamma}{\cos\gamma}$

$\dfrac{1+\dfrac{\sin\gamma}{\cos\gamma}}{1+\dfrac{\cos\gamma}{\sin\gamma}} \overset{?}{=} \dfrac{\sin\gamma}{\cos\gamma}$

$\dfrac{\dfrac{\cos\gamma + \sin\gamma}{\cos\gamma}}{\dfrac{\sin\gamma + \cos\gamma}{\sin\gamma}} \overset{?}{=} \dfrac{\sin\gamma}{\cos\gamma}$

$\dfrac{\cos\gamma + \sin\gamma}{\cos\gamma} \cdot \dfrac{\sin\gamma}{\cos\gamma + \sin\gamma} \overset{?}{=} \dfrac{\sin\gamma}{\cos\gamma}$

$\dfrac{\sin\gamma}{\cos\gamma} = \dfrac{\sin\gamma}{\cos\gamma}$

**27.** $1 + \sec^2 x \sin^2 x \overset{?}{=} \sec^2 x$

$1 + \dfrac{1}{\cos^2 x} \cdot \dfrac{\sin^2 x}{1} \overset{?}{=} \sec^2 x$

$1 + \dfrac{\sin^2 x}{\cos^2 x} \overset{?}{=} \sec^2 x$

$1 + \tan^2 x \overset{?}{=} \sec^2 x$

$\sec^2 x = \sec^2 x$

**31.** $\dfrac{2187}{16,384}$ or about $0.133$

**32.** 

| Stem | Leaf |
|------|------|
| 0 | 4 4 4 4 |
| ● | 5 6 6 6 6 7 7 8 9 9 |
| 1 | 1 1 2 2 |
| ● | 6 9 9 9 |

$1 \mid 1$ represents 11¢.

**33.** $\dfrac{\log 82}{\log 5}$; 2.738  **34.** $-5, 4; -1, -4$

**35.** $\begin{bmatrix} -10 & 0 \\ 3 & -2 \end{bmatrix}$; $\begin{bmatrix} -\dfrac{1}{10} & -\dfrac{3}{20} \\ 0 & -\dfrac{1}{2} \end{bmatrix}$

### Pages 808–809  Lesson 17-4

**5.** 18,200  **7.** 301  **9.** 6 or 12  **11.** $-\dfrac{2}{9}$  **13.** A sample answer is $5[5(5+5)-(5+5)]$.  **15.** 24, 25, 32, 43, 49, 51, 57, 68, 75, 76, 93, or 99  **17.** 7

### Pages 813–814  Lesson 17-5

**3.** $30° - 45°$  **5.** $30° + 45°$  **7.** $225° + 30°$

**9.** $\dfrac{\sqrt{6}+\sqrt{2}}{4}$  **11.** $\dfrac{\sqrt{6}-\sqrt{2}}{4}$  **13.** $\dfrac{-\sqrt{6}-\sqrt{2}}{4}$

**15.** $\dfrac{\sqrt{2}-\sqrt{6}}{4}$  **17.** $\dfrac{-\sqrt{6}-\sqrt{2}}{4}$  **19.** $\dfrac{1}{2}$  **21.** $\dfrac{1}{2}$

**23.**

$\sin(270° - \theta) \overset{?}{=} -\cos\theta$

$\sin 270° \cos\theta - \cos 270° \sin\theta \overset{?}{=} -\cos\theta$

$-1 \cdot \cos\theta - 0 \cdot \sin\theta \overset{?}{=} -\cos\theta$

$-\cos\theta = -\cos\theta$

**25.**

$\sin(90° + \theta) \overset{?}{=} \cos\theta$

$\sin 90° \cos\theta + \cos 90° \sin\theta \overset{?}{=} \cos\theta$

$1 \cdot \cos\theta + 0 \cdot \sin\theta \overset{?}{=} \cos\theta$

$\cos\theta = \cos\theta$

**27.**

$\cos(90° + \theta) \overset{?}{=} -\sin\theta$

$\cos 90° \cos\theta - \sin 90° \sin\theta \overset{?}{=} -\sin\theta$

$0 \cdot \cos\theta - 1 \cdot \sin\theta \overset{?}{=} -\sin\theta$

$-\sin\theta = -\sin\theta$

**33.** $-2 - \sqrt{3}$  **35.** $2 - \sqrt{3}$  **37a.** $\sin 1° = \sin 13° \cos 12° - \cos 13° \sin 12°$  **37b.** 27.5 feet

**38.** 240°  **39.** 2046  **40.** $\dfrac{x^4 y^2}{2a^2 b^4}$  **41.** 8, 14

**42.** $f(x) = (x-3)^2$  **43.** symmetric property of equality

### Page 814  Mid-Chapter Review

**1.** $1, \dfrac{\pi}{2}$ or 90°  **2.** $5, 2\pi$ or 360°  **3.** $4, \dfrac{8\pi}{3}$ or 480°

**4.** $\dfrac{\pi}{2}$ or 90°  **5.** $\pi$ or 180°  **6.** $5\pi$ or 900°  **7.** $\dfrac{\sqrt{34}}{5}$

**8.** $-\dfrac{5\sqrt{34}}{34}$

**9.** $\cot\beta \overset{?}{=} \cos\beta \csc\beta$

$\cot\beta \overset{?}{=} \cos\beta \left(\dfrac{1}{\sin\beta}\right)$

$\cot\beta \overset{?}{=} \dfrac{\cos\beta}{\sin\beta}$

$\cot\beta = \cot\beta$

**10.** $1 \overset{?}{=} \dfrac{1}{\sec^2\alpha - \sec^2\alpha \sin^2\alpha}$    $1 \overset{?}{=} \dfrac{1}{1}$

$1 \overset{?}{=} \dfrac{1}{\sec^2\alpha(1-\sin^2\alpha)}$    $1 = 1$

$1 \overset{?}{=} \dfrac{1}{\dfrac{1}{\cos^2\alpha}(\cos^2\alpha)}$

11. 58  12. $\frac{\sqrt{2}-\sqrt{6}}{4}$  13. $\frac{\sqrt{6}-\sqrt{2}}{4}$  14. $\frac{-\sqrt{6}-\sqrt{2}}{4}$

**Pages 818–819  Lesson 17-6**

5. III or IV  7. I or II  9. I  11. $-\frac{4\sqrt{6}}{25}, -\frac{23}{25}, \frac{\sqrt{10}}{5},$

$-\frac{\sqrt{15}}{5}$  13. $-\frac{24}{25}, -\frac{7}{25}, \frac{2\sqrt{5}}{5}, \frac{\sqrt{5}}{5}$  15. $\frac{4\sqrt{2}}{9}, -\frac{7}{9}, \frac{\sqrt{6}}{3},$

$-\frac{\sqrt{3}}{3}$  17. $\frac{24}{25}, \frac{7}{25}, \frac{3\sqrt{10}}{10}, \frac{\sqrt{10}}{10}$  19. $-\frac{4\sqrt{2}}{9}, -\frac{7}{9}, \frac{\sqrt{6}}{3}, \frac{\sqrt{3}}{3}$

21. $\frac{\sqrt{2+\sqrt{2}}}{2}$  23. $-\frac{\sqrt{2-\sqrt{3}}}{2}$  25. $\frac{\sqrt{2-\sqrt{2}}}{2}$

27. $\cos^2 2x + 4\sin^2 x\cos^2 x \stackrel{?}{=} 1$
$\cos^2 2x + \sin^2 2x \stackrel{?}{=} 1$
$1 = 1$

29. $\sin^2\theta \stackrel{?}{=} \frac{1}{2}(1-\cos 2\theta)$

$\sin^2\theta \stackrel{?}{=} \frac{1}{2}[1-(1-2\sin^2\theta)]$

$\sin^2\theta \stackrel{?}{=} \frac{1}{2}(2\sin^2\theta)$

$\sin^2\theta = \sin^2\theta$

31. $\sin^4 x - \cos^4 x \stackrel{?}{=} 2\sin^2 x - 1$
$(\sin^2 x - \cos^2 x)(\sin^2 x + \cos^2 x) \stackrel{?}{=} 2\sin^2 x - 1$
$(\sin^2 x - \cos^2 x)\cdot 1 \stackrel{?}{=} 2\sin^2 x - 1$
$[\sin^2 x - (1-\sin^2 x)] \stackrel{?}{=} 2\sin^2 x - 1$
$\sin^2 x - 1 + \sin^2 x \stackrel{?}{=} 2\sin^2 x - 1$
$2\sin^2 x - 1 = 2\sin^2 x - 1$

33. $2\cos^2\frac{x}{2} \stackrel{?}{=} 1 + \cos x$  35a. 270.6 feet

$2\left(\pm\sqrt{\frac{1+\cos x}{2}}\right)^2 \stackrel{?}{=} 1 + \cos x$

$2\left(\frac{1+\cos x}{2}\right) \stackrel{?}{=} 1 + \cos x$

$1 + \cos x = 1 + \cos x$

35b. 270.6 feet  35c. 45°; The greatest value of the sine function is 1 and sin 2θ = 1 when θ = 45°.
36. 90°
37.

| Stem | Leaf | | | | | | |
|---|---|---|---|---|---|---|---|
| 4 | | | | | | | |
| ● | 6 | 6 | | | | | |
| 5 | 2 | | | | | | |
| ● | 5 | 5 | 5 | 8 | 8 | 9 | 9 |
| 6 | 0 | 0 | 0 | 2 | | | |
| ● | 5 | 5 | 9 | 9 | | | |
| 7 | 0 | 0 | | | | | |
| ● | 6 | 8 | | | | | |

6 | 0 represents $60.

38. 10, 14, 18, 22  39. 5  40. inverse; 10

**Pages 825–827  Lesson 17-7**

5. 2; 150°, 210°  7. 2; 0°, 180°  9. 2; about 108°26′, about 288°26′  11. 0  13. 0  15. 8; 0°, 45°, 90°, 135°, 180°, 225°, 270°, 315°  17. 60°, 120°, 240°, 300°
19. 90°, 270°  21. 30°, 150°  23. $\frac{7\pi}{6}, \frac{11\pi}{6}$  25. $\frac{4\pi}{3}, \frac{5\pi}{3}$

27. $\frac{\pi}{2}, \frac{7\pi}{6}, \frac{11\pi}{6}$  29. 45° + n · 180°  31. 0° + n · 120°
33. 90° + n · 360°, 180° + n · 360°  35. 90° + n · 180°, 120° + n · 360°, 240° + n · 360°  37. 0° + n · 180°, 90° + n · 360°  39. $\frac{7\pi}{6} + 2n\pi, \frac{11\pi}{6} + 2n\pi$

41. $\frac{\pi}{6} + n\pi, \frac{5\pi}{6} + n\pi$  43. $\frac{2\pi}{3} + 2n\pi, \frac{4\pi}{3} + 2n\pi$

45. $\frac{\pi}{3} + 2n\pi, \frac{5\pi}{3} + 2n\pi$  47. $0 + n\pi, \frac{\pi}{6} + 2n\pi,$

$\frac{5\pi}{6} + 2n\pi$  49a. −2 feet from the equilibrium point
49b. 0.5 + n seconds, where n is any integer  49c. 2 seconds  51a. 60°  51b. 90°, 270°  51c. no solution
51d. −360°, −225°, −180°, −45°, 0°  51e. 0°, 30°, 150°, 180°, 360°, 390°, 510°, 540°, 720°  52. $\frac{1}{2}$

53. $\frac{25}{144} \approx 0.174$  54. 6.5; 0; 7.4  55. 950
56. $\log_3 27 + \log_3 3 \stackrel{?}{=} \log_3 81$
    $3 + 1 = 4$  57. x = 0

**Pages 832–833  Lesson 17-8**

5. $7\left(\cos\frac{\pi}{2} + i\sin\frac{\pi}{2}\right)$  7. 2  9. i

11. $3\left(\cos\frac{5\pi}{4} + i\sin\frac{5\pi}{4}\right)$  13. $\sqrt{2}\left(\cos\frac{\pi}{4} + i\sin\frac{\pi}{4}\right)$

15. $3\left(\cos\frac{\pi}{2} + i\sin\frac{\pi}{2}\right)$

17. $\sqrt{29}(\cos 1.95 + i\sin 1.95)$  19. $-1 - i$
21. $6 - 6i\sqrt{3}$  23. $16(\cos 2\pi + i\sin 2\pi); 16$
25. $10\left(\cos\frac{17\pi}{12} + i\sin\frac{17\pi}{12}\right); -2.59 - 9.66i$  27. $-2i$
29. $-4i$  31. $-27$  33. $-16\sqrt{2} - 16i\sqrt{2}$
35. $54 + 54i$  39. The needle travels slower as it approaches the center of the record.  40. 18.09 feet or 18 feet 1 inch  41. $\frac{1}{32}$ or 0.03125  42. 81  43. 8;

8; 6.94  44. 20,159 people  45. $\frac{\log 50}{\log 12}$; 1.5743

46. $(\pm 2, 0); (\pm\sqrt{29}, 0)$

**Pages 834–836  Chapter 17 Summary and Review**

1. 1, 2π

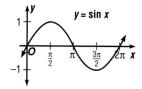

**5.** $2\pi$

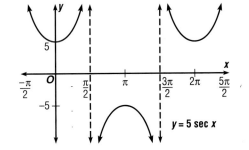

$y = 5 \sec x$

**9.** $-\dfrac{\sqrt{3}}{2}$  **11.** $-\dfrac{4}{3}$

**13.**
$$\sin^4 x - \cos^4 x \overset{?}{=} \sin^2 x - \cos^2 x$$
$$(\sin^2 x - \cos^2 x)(\sin^2 x + \cos^2 x) \overset{?}{=} \sin^2 x - \cos^2 x$$
$$(\sin^2 x - \cos^2 x) \cdot 1 \overset{?}{=} \sin^2 x - \cos^2 x$$
$$\sin^2 x - \cos^2 x = \sin^2 x - \cos^2 x$$

**15.** $\dfrac{\sin\theta}{1 - \cos\theta} \overset{?}{=} \csc\theta + \cot\theta$

$$\dfrac{\sin\theta}{1 - \cos\theta} \overset{?}{=} \dfrac{1}{\sin\theta} + \dfrac{\cos\theta}{\sin\theta}$$

$$\dfrac{\sin\theta}{1 - \cos\theta} \overset{?}{=} \dfrac{1 + \cos\theta}{\sin\theta}$$

$$\dfrac{\sin\theta}{1 - \cos\theta} \overset{?}{=} \dfrac{(1 + \cos\theta)(1 - \cos\theta)}{\sin\theta(1 - \cos\theta)}$$

$$\dfrac{\sin\theta}{1 - \cos\theta} \overset{?}{=} \dfrac{1 - \cos^2\theta}{\sin\theta(1 - \cos\theta)}$$

$$\dfrac{\sin\theta}{1 - \cos\theta} \overset{?}{=} \dfrac{\sin^2\theta}{\sin\theta(1 - \cos\theta)}$$

$$\dfrac{\sin\theta}{1 - \cos\theta} = \dfrac{\sin\theta}{1 - \cos\theta}$$

**17.**
$$\sec x(\sec x - \cos x) \overset{?}{=} \tan^2 x$$
$$\sec^2 x - 1 \overset{?}{=} \tan^2 x$$
$$\tan^2 x = \tan^2 x$$

**19.**
$$\dfrac{\csc\theta + 1}{\cot\theta} \overset{?}{=} \dfrac{\cot\theta}{\csc\theta - 1}$$
$$\dfrac{(\csc\theta + 1)(\csc\theta - 1)}{\cot\theta(\csc\theta - 1)} \overset{?}{=} \dfrac{\cot\theta}{\csc\theta - 1}$$
$$\dfrac{\csc^2\theta - 1}{\cot\theta(\csc\theta - 1)} \overset{?}{=} \dfrac{\cot\theta}{\csc\theta - 1}$$
$$\dfrac{\cot^2\theta}{\cot\theta(\csc\theta - 1)} \overset{?}{=} \dfrac{\cot\theta}{\csc\theta - 1}$$
$$\dfrac{\cot\theta}{\csc\theta - 1} = \dfrac{\cot\theta}{\csc\theta - 1}$$

**21.** $\dfrac{\sqrt{6} + \sqrt{2}}{4}$  **23.** $\dfrac{\sqrt{2} - \sqrt{6}}{4}$  **25.** $2\cos x \cos y$

**27.** $\dfrac{24}{25}$  **29.** $\dfrac{\sqrt{21}}{5}$

**31.**
$$(\sin x - \cos x)^2 \overset{?}{=} 1 - \sin 2x$$
$$\sin^2 x - 2\sin x \cos x + \cos^2 x \overset{?}{=} 1 - \sin 2x$$
$$1 - 2\sin x \cos x \overset{?}{=} 1 - \sin 2x$$
$$1 - \sin 2x = 1 - \sin 2x$$

**33.** $0°, 90°$  **35.** $\dfrac{7\pi}{6} + 2n\pi, \dfrac{11\pi}{6} + 2n\pi$  **37.** $\dfrac{\pi}{2} + 2n\pi$

**39.** $4\left(\cos\dfrac{2\pi}{3} + i\sin\dfrac{2\pi}{3}\right)$  **41.** $4\sqrt{2} - 4i\sqrt{2}$

**43.** $4096$

# INDEX

simplifying, 15, 17, 47, 516-520, 531
verbal, 24
Extraneous solutions, 282, 284-285
Extrapolation, 578

Factorials, 627-630, 634
Factors
  greatest common, 229-232, 516, 519-520
  of polynomials, 229-233, 235, 238, 247, 317-320, 323-324, 339, 344, 461-463, 468-469, 471-472, 480
  prime, 523
  quadratic, 339, 342, 346
Factor Theorem, 461, 478, 496
Failures, 705-707
Families of equations, 68
Feasible regions, 129-130
Ferguson, Helaman R. P., 515
Fibonacci, Leonardo, 594
Fibonacci sequence, 594-595
Foci
  of ellipses, 409-415, 445
  of hyperbolas, 416-421, 445
  of parabolas, 400-404, 445
FOIL method, 224-225, 265, 293, 297, 316
Fractals, 292, 295-296
Fractions
  adding, 522
  complex, 519
  decimals to, 621, 623
  dividing, 517
  multiplying, 517
  reciprocals, 517-519
  in simplest form, 516
Franklin, Benjamin, 18
Frequencies, 741
Frequency distributions, 668-673
  normal, 664, 668-672, 676
  skewed, 670
Functions, 53-63, 66-79, 96-98
  absolute value, 87-90
  Arccosine, 746-747, 749, 751-753, 781
  Arcsine, 747-749, 751, 781
  Arctangent, 747, 749, 753
  composition of, 487-498
  constant, 62, 87, 89, 451
  cosecant, 742-744, 751, 753, 756-758, 794-795, 797-800, 805-806, 834-835

cosine, 737-743, 746-751, 753-759, 775-779, 781-782, 790-795, 797-800, 804-806, 810-836
cotangent, 742-744, 752-753, 756-758, 794-795, 797-800, 803, 805-806, 834-835
cubic, 451-452, 484
direct variation, 87, 89
even degree, 452-455
exponential, 546-552
greatest integer, 88-90
identity, 87, 492-493, 498, 555
inverse, 491-495, 498, 551, 555, 557, 746-753
linear, 62-63, 66-87, 91-94, 96-98, 108-121, 451-452
logarithmic, 551-552, 554-557, 581-584, 587
mappings, 488-489
odd degrees, 452-455, 483
periodic, 739, 741
polynomial, 451-464, 468-499, 506
quadratic, 317-319, 353-355, 360-378, 388-389, 451-452
rational, 504-509, 540
secant, 742-744, 753, 756-758, 794-795, 797-800, 803-805, 834-835
sine, 737-743, 746-751, 753-759, 767-770, 772-773, 776-778, 781-782, 790-791, 793-800, 803-805, 810-836
step, 88-90
tangent, 742-744, 746-749, 752-758, 762-764, 781, 793-795, 797-800, 803-805, 813-814, 834-835
vertical line test for, 54, 56
zeros of, 317, 453-455, 461, 469-486, 497
Fundamental Theorem of Algebra, 458, 469-470, 477

General formula for growth and decay, 581
Geometric means, 610-611, 614
Geometric sequences, 608-614, 633
  common ratios, 608-614, 633
  geometric means, 610-611, 614
  recursive formulas for, 609, 611, 633
Geometric series, 615-624, 633
  infinite, 620-624, 633
Geometry
  angles, 49, 199, 227, 732-740, 756

area, 12, 169, 171-172, 223, 225-226, 228-229, 237, 285, 310, 316, 354-355, 362, 376, 380-381, 512, 514, 518, 572-573, 767-769
circles, 285, 382, 398-399, 405-407, 733-734
cones, 20-21
convex polygons, 64, 128
cubes, 252, 255
cylinders, 482
decagons, 700
diagonals, 64-65, 700
hyperbolic paraboloids, 422
midpoints, 397-399, 444
parallel lines, 68, 75-77, 97, 382
parallelograms, 12, 78, 281, 514, 518
perimeters, 24, 26, 63, 112, 148, 220, 281, 310, 376
perpendicular lines, 68-70, 76-77, 97
prisms, 343, 468, 478
pyramids, 476
Pythagorean Theorem, 263-264, 281, 396-397, 410, 737-738, 748, 757, 775, 829
rectangles, 57, 69, 71, 112, 229, 237, 310, 316, 354-355, 362, 376, 380-381
spheres, 9
squares, 223, 225, 572-573
trapezoids, 12, 514
triangles, 12, 26, 148, 169, 171-172, 199, 512, 737-738, 756, 756-779, 782
volume, 9, 20-21, 252, 255, 343, 468, 476, 478, 482, 484, 486
Goddard, Robert, 319
Gradians, 754
Graphing Calculator Explorations
  conic sections, 425-426
  evaluating expressions, 22-23
  families of parabolas, 350-352
  graphing exponential and logarithmic functions, 551-552
  graphing linear equations, 58-59
  graphing polynomial equations and approximating real zeros, 457-458
  graphing rational functions, 504-505
  graphing systems of equations, 104-107
  lines of regression, 85-86
  locating the vertex of a parabola, 359
  matrices, 159-160
  matrix row operations, 201
  quadratic equations, 312-315
  solving quadratic systems, 437-438

## V

## W

## X

## Y

## Z

# PHOTO CREDITS

**Cover,** G. R. Beechler

**v(t)**, Michael A. Kellar/The Stock Market, **(b)**, C. Garoutte/Tom Stack & Assoc; **vi**, Aaron Haupt; **vii(t)**, Randy Duchaine/The Stock Market, **(b)**, Erik Simonsen/The Image Bank; **viii(t)**, Michel Tcherevkoff/The Image Bank, **(b)**, Hartman-Dewitt/Comstock; **ix(t)**, Rob Atkins/The Image Bank, **(b)**, Larry Hamill; **x**, Aaron Haupt; **xi(t)**, Will McIntyre/Photo Researchers, **(b)**, Kunio Owaki/The Stock Market; **xii(t)**, Thomas Kitchin/Tom Stack & Assoc, **(b)**, Hank Morgan/Rainbow; **xiii(t)**, Charles West/The Stock Market, **(b)**, Robert J. Herke/The Image Bank; **xiv**, Aaron Haupt; **xv(t)**, Galen Rowell/Mountain Light, **(b)**, Grant V. Faint/The Image Bank; **xvi**, Galen Rowell/Mountain Light; **6**, Doug Martin; **7**, Michael A. Kellar/The Stock Market; **8**, Skip Comer; **10**, Ted Rice; **12**, Ralph Cowan/Tony Stone Images; **14**, Doug Martin; **17**, Skip Comer; **18**, Elaine Shay; **21**, Monica V. Brown; **22**, Elaine Shay; **24**, Skip Comer; **27**, Elaine Shay; **28**, Skip Comer; **29**, Tony Stone Images; **30**, Skip Comer; **34**, Doug Martin; **35**, Bettmann Archive; **36**, C. J. Zimmerman/FPG; **38**, Skip Comer; **40**, John Biever/Sports Illustrated © TIME, Inc; **45**, Gary Guisinger/Photo Researchers; **50**, Marty Snyderman; **51**, C. Garoutte/Tom Stack & Assoc; **52**, H. P. Merton/The Stock Market; **53**, Skip Comer; **57**, Mike Surowiak/Tony Stone Images; **61**, Four By Five/Superstock; **63**, Larry Hamill; **65**, Elaine Shay; **75**, Stephen Frink/The Stock Market; **78(t)**, Skip Comer, **(b)**, David Dennis; **79**, The Bettmann Archive; **81,82**, Skip Comer; **84**, Andy Sacks/Tony Stone Images; **87**, The Telegraph Color Library/FPG; **90**, Ronald C. Modra/Sports Illustrated; **92**, David Sailors/The Stock Market; **94**, Skip Comer; **95**, Andrew Sacks/Tony Stone Images; **102**, Doug Martin; **103**, R. Duchain/The Stock Market; **109,112**, Skip Comer; **116**, Doug Martin; **122**, SuperStock; **125**, Skip Comer; **126**, Doug Martin; **129**, The Telegraph Colour Library/FPG; **131,133**, Skip Comer; **137(t)**, Lewis Portnoy/The Stock Market, **(b)**, Skip Comer; **139**, Seth Goltzer/The Stock Market; **143**, Frank Cezus/Tony Stone Images; **149**, Skip Comer; **154**, John Madera/The Stock Market; **155**, Erik Simonsen/The Image Bank; **156**, Chris Jones/The Stock Market; **157**, Elaine Shay; **166**, Skip Comer; **167**, Jeff Zaruba/The Stock Market; **172**, Connie Geocaris/Tony Stone Images; **173**, Skip Comer; **174**, David L. Perry; **187,191**, Skip Comer; **193**, George Anderson; **200**, Skip Comer; **204**, Elaine Shay; **208**, Matt Meadows/Peter Arnold, Inc; **209**, Michel Tcherevkoff/Image Bank; **210**, D. Stoecklein/The Stock Market; **212**, California Institute of Technology; **214**, Daniele Pellegrini/Photo Researchers; **215**, NASA; **216**, Paul Silverman/Fundamental Photographs; **219**, Bob Thomason/Tony Stone Images; **221**, Elaine Shay; **223,236**, Doug Martin; **240,243**, Skip Comer; **248**, Lee Foster/FPG; **250**, Doug Martin; **251**, Hartman-Dewitt/Comstock; **255(l)**, Doug Martin, **(r)**, Ken Frick; **256**, Allen B. Smith/Tom Stack & Assoc; **258**, George H. Matchneer; **261**, Stock Concepts; **263**, file photo; **265**, Roy Morsch/The Stock Market; **268**, Hank Morgan/Science Source/Photo Researchers; **276**, file photo; **277**, NASA; **280**, Elaine Shay; **281,275**, Doug Martin; **286**, NASA; **291**, Al Satterwhite/The Image Bank; **292**, Digital Art/Westlight; **295**, Michael Melford/Image Bank; **296(t)**, Hansen Planetarium, **(b)**, Courtesy IBM; **298**, Michael Melford/Image Bank; **301**, Dale O'Dell/Comstock; **308**, Doug Martin; **309**, Rob Atkins/The Image Bank; **310**, Denver Bryan/Comstock; **311**, Johnny Johnson; **316**, Hank Morgan/Rainbow; **320**, Tim Courlas; **321**, Henri Georgi/Comstock; **323**, Skip Comer; **325**, Steven E. Sutton/DUOMO; **328**, Skip Comer; **331**, NASA; **337**, Doug Martin; **338**, Yale Babylonian Collection; **343,346**, NASA; **348**, Doug Martin; **349**, Larry Hamill; **353**, Doug Martin; **355**, Hank Morgan/Science Scource/Photo Researchers; **358**, Elaine Shay; **364**, Skip Comer; **366**, Tim Courlas; **371**, Hartman-Dewitt/Comstock; **373**, A. Schmidecker/FPG; **376**, Matt Bradley/Tom Stack & Assoc; **379(t)**, Skip Comer, **(b)**, Gary Gladstone/The Image Bank; **381**, Doug Martin; **382**, Harvey Lloyd/Stock Market; **383**, Doug Martin; **387**, Gary Bumgarner/Tony Stone Images; **390**, Alan Carey; **394**, Doug Martin; **395**, Will & Deni McIntyre/Photo Researchers; **400**, Nadia MacKenzie/Tony Stone Images; **402**, David M. Dennis; **405**, John Turner/FPG; **408**, Russ Kinne/Comstock; **410**, Don Carrol/The Image Bank; **414**, Jack Zehrt/FPG; **418**, K. Iwasaki/The Stock Market; **421**, Richard Dole/DUOMO; **422**, Gerald L. French/The Photo File; **424**, Doug Martin; **428**, Charles Palek/Tom Stack & Assoc; **432**, Doug Martin; **434**, Brian Parker/Tom Stack & Assoc; **436(l)**, Paul J. Sutton/DUOMO, **(r)**, David Madison/DUOMO; **440**, Jim Brown/The Stock Market; **443**, Ted Mahieu/The Stock Market; **448**, Doug Martin; **449**, Kunio Owaki/The Stock Market; **452**, E. R. Degginger; **454(t)**, Robert & Linda Mitchell, **(b)**, Robert & Linda Mitchell; **456**, Elaine Shay; **459(t)**, Bryan Yablonski/DUOMO, **(b)**, file photo; **462**, Al Tielemans/DUOMO; **464**, Doug Martin; **465**, Elaine Shay; **467(t)**, Richard Steedman/The Stock Market, **(b)**, Luis Villota/The Stock Market; **468**, Doug Martin; **469**, Elaine Shay; **474**, Howard Sochurek/The Stock Market; **476(t)**, Smithsonian Institution, **(b)**, Paul Steel/The Stock Market; **479**, S. Barry O'Rourke/The Stock Market; **481**, Andy Canfield/The Image Bank; **485(t)**, Hartman-Dewitt/Comstock, **(b)**, Tim Courlas; **487,489**, Tim Courlas; **490**, Doug Martin; **491,495**, Tim Courlas; **498**, Skip Comer; **502**, Larry Hamill; **503**, Thomas Kitchin/Tom Stack & Assoc; **506**, Skip Comer; **509**, National Highway Traffic Safety Administration; **510**, NASA; **512**, Skip Comer; **514**, Elaine Shay; **515(t)**, Ray Mathis/The Stock Market, **(b)**, UMBILLIC TORUS NC by Sculptor/Mathematician Helaman Ferguson, photo credit: Terry Clough; **518**, Ceoffrey Gove/The Image Bank; **521**, John Kelly/The Image Bank; **522**, Ansel Adams/Ansel Adams Publication Rights Trust/Mount Williamson/Sierra Nevada; **526**, Franco Fontana/The Image Bank; **530**, Joanna McCarthy/The Image Bank; **532,534**, Skip Comer; **536**, Don Landwehrle/The Image Bank; **537**, Jany Sauvanet/Photo Researchers; **538**, Walter Bibikow/The Image Bank; **542**, Tim Courlas; **544(t)**, Doug Martin, **(b)**, Antonio Luiz Hamdan/The Image Bank; **545**, Hank

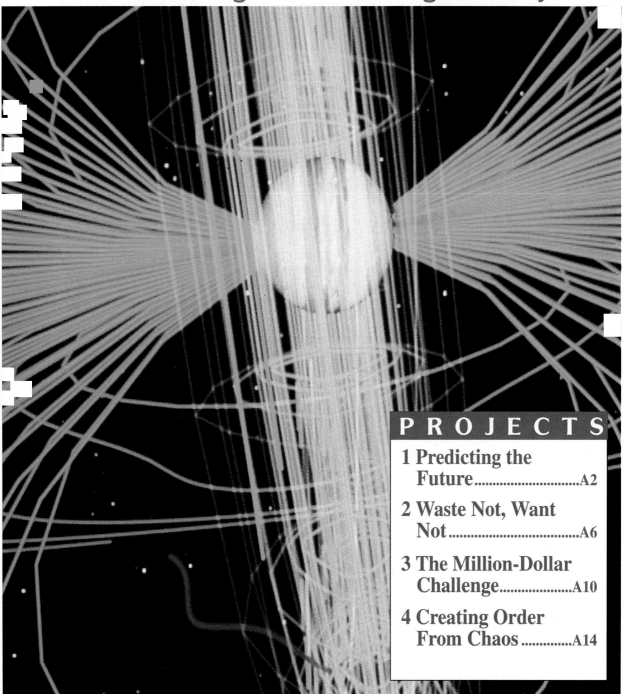

# GLENCOE

# EXTENDED PROJECTS

## for Merrill Algebra 2 with Trigonometry

# 1 Predicting the Future

The better we understand nature, the more accurately we are able to predict its behavior. Ancient peoples had no understanding of the movements of the sun and the moon, so predicting such events with accuracy was impossible. As a result, they believed that eclipses occurred when their gods were angry. When astronomers discovered the causes of eclipses and calculated the orbits of the sun and the moon, they were able to predict the timing of eclipses with pinpoint accuracy.

Human nature, too, is better understood today than ever before. Corporations routinely take the findings of sociologists and psychologists and use them in advertising campaigns to persuade people to purchase their products.

Despite great advances in science, however, much of nature remains a mystery. Making accurate predictions about these areas is as chancy as predicting eclipses was for ancient peoples. Will there be a hurricane? What parts of the world will be affected? How will people vote in the presidential election? Will the stock market go up or down?

In an effort to make predictions about natural occurrences that are not fully understood, some people search for *predictors*, events or facts which, for no obvious reason, seem to correlate with the occurrences. For example, geologist Ruth Simon has discovered that cockroaches become more active during the hours leading up to a major earthquake. Political analysts have noted that the candidate with the longer name has won nearly 75% of the presidential elections in American history.

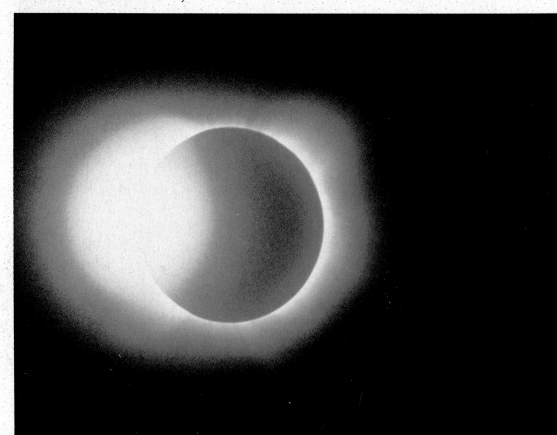

# Find a Predictor

In this project, you will attempt to find a predictor that can be used to predict the behavior of some natural phenomenon. You will analyze several possible predictors to find the one with the best promise. Then you will use the one you chose to make a prediction about the future.

# Getting Started

Follow these steps to carry out your project.

- Choose a phenomenon that interests you whose behavior cannot be predicted with total accuracy. It could be something that occurs in the natural world (for example, weather patterns, changes in sizes of wildlife populations, or appearances of comets). Or it could involve human nature (for example, voting patterns, success of relief pitchers in baseball, or consumer preferences in fashions, films, or music). Be sure that there is plenty of data available on the phenomenon you choose and that the phenomenon is not completely arbitrary in its behavior.

- Collect data on the phenomenon. Study it, searching for patterns, and discuss your findings with your group.

- Brainstorm with your group to identify at least four possible predictors of the phenomenon. Spend a good deal of time searching for promising predictors. Those you settle on may have a seemingly close connection with the phenomenon or, like the "last name" predictor for presidents, may seem unrelated to it.

- Collect data on your predictors.

- Analyze your predictors. Your analysis should include some of the methods discussed in Chapters 1-4. You may wish to draw scatter plots and find prediction equations. You can use a graphing calculator to draw lines of regression or a computer to draw median-fit lines. After you complete your analysis, choose the predictor you judge to have the most promise.

- Use your predictor to predict the future behavior of the phenomenon you have studied.

- Write a report summarizing the work of your group. Explain how you chose your predictors and how you analyzed them to choose the most promising one. Describe the results of your prediction. Give reasons for any discrepancies between predicted and actual results.

# Extensions

1. Find and express in algebraic language a function relating your predictor and the phenomenon you studied in this project.

2. Describe possible connections between your predictor and the phenomenon that may explain the success of your predictor.

3. Find examples in history where deepening understanding of nature allowed people to predict the behavior of phenomenon previously believed to be arbitrary and unpredictable.

# Culminating Activities

Show what you have learned in this project by completing one of the following activities.

1. Make an oral presentation to the class describing your results in this project.

2. Conduct a classroom debate on this topic: "Mathematics Can Be Used to Illuminate All of the Phenomena of Nature."

3. Outline a method for finding predictors of little understood phenomena.

# Waste Not, Want Not

A manufacturer uses rectangular pieces of sheet metal with squares removed from the corners as patterns for open-topped wastebaskets.

The sides are folded up to produce the rectangular-prism-shaped wastebaskets.

The volume and surface area of a wastebasket depend on the size of the corners that are removed. Suppose a design engineer begins with a 12-by-20 inch rectangle. Removal of 3-inch squares results in a 14-by-6-by-3 inch wastebasket. This wastebasket would have a volume of 252 cubic inches and a surface area of 288 square inches.

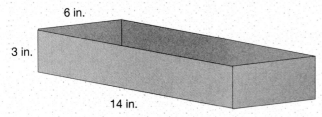

Removal of 4-inch squares results in a 12-by-4-by-4 inch wastebasket with a volume of 192 cubic inches and a surface area of 224 square inches.

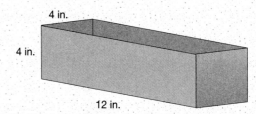

Two important constraints have been placed on the design engineer: wastebaskets should have a large volume to attract customers, but as small a surface area as possible, to cut down on the cost of material.

# Design a Wastebasket

A design engineer has been asked to answer the following questions. Given a rectangular piece of sheet metal measuring $m$ inches by $n$ inches:

1. How can it be cut to obtain a wastebasket with the maximum volume? What is the surface area of the wastebasket?

2. How can it be cut to obtain a particular desired volume? If there are several ways to cut it, which one uses the least material?

3. What are the relationships between the volume and surface area of a wastebasket as the size of the corner changes?

In this project, you will attempt to answer these questions.

# Getting Started

Follow these steps to carry out your project.

- Design wastebaskets on paper and make models of your designs. Find the volumes and surface areas of your models. Keep accurate records of your results.

- Study the models looking for relationships among corner-square sizes, volumes, and surface areas. Discuss your preliminary findings with your group.

- Work out a plan for investigating the three design questions posed on the previous page. Among the tools you may decide to use are graphs, formulas, diagrams, tables, calculators, graphing calculators, and computers.

- Answer the questions posed on page A7.

- Actually build the preferred wastebasket using available materials like cardboard, plywood, or metal.

# Extensions

1. As you learned in Chapter 8, the graphs of quadratic functions are parabolas. The vertex of a parabola is the maximum or minimum point on the parabola. Suppose that the volume, $V$, of a wastebasket is given by the expression $V = Ax^2 + Bx + C$, where $x$ is the length of a side of the corner square. Find a formula for the value of $x$ that will produce the maximum volume.

2. Draw a pattern for a large piece of material that would be cut into smaller pieces to form wastebaskets that meet the criteria. Be sure to account for overlap when the wastebaskets are assembled. Also, try to minimize the amount that cannot be used.

3. Contact an engineer who designs packages for a manufacturer. Find out the kinds of problems the engineer is asked to solve. How does the engineer minimize manufacturing costs without sacrificing the attractiveness of the package?

4. Choose a consumer product sold in a package familiar to most people. Redesign the package in a way you think would attract more customers. Explain why you think your package is better than the one currently in use.

5. Investigate other shapes that are commonly used for wastebaskets. For example, consider wastebaskets shaped like cylinders and wastebaskets shaped like upside down pyramids with the vertex removed. How do the volumes and surface areas of these wastebaskets compare?

# Culminating Activities

Show what you have learned in this project by completing one of the following activities.

1. Imagine that you are a design engineer working for the wastebasket manufacturer. Write a report for your employer in which you answer the three questions posed under the problem "Design a Wastebasket."

2. Give an oral presentation to your class in which you outline the relationships between surface area and volume that you discovered in this project.

3. A wastebasket manufacturer plans to make open-topped wastebaskets from rectangular patterns measuring 32-by-40 inches. Write a report for the manufacturer recommending the best design to use and outlining the reasons behind your recommendation.

# The Million-Dollar Challenge

In 1856, the country of British Guiana issued a one-cent black-on-magenta postage stamp. On April 4, 1856, a letter bearing one of those stamps was posted. Today the stamp on that letter is the only surviving specimen of the 1856 British Guiana issue. Per pound, it may be the most valuable item on Earth. In 1980, an anonymous buyer purchased the stamp for $850,000. Using the formula $V_n = P(1 + r)^n$, we can determine that the stamp appreciated at an annual rate of about 16% during the 124 years that elapsed between 1856 and 1980. At $850,000 per $\frac{1}{20,000}$-lb weight for a typical stamp, the one-cent black-on-magenta stamp is worth about $17 billion per pound today.

The investor who bought the British Guiana stamp in 1873 for 84¢ made a very good investment. But not all stamps are good investments. Today's postage stamps, issued by the millions, will never become rare enough to appreciate significantly. Most will probably decrease in value year after year.

People with money to spend are always on the lookout for investments that will appreciate steadily and reliably over the years. Real estate is almost always a dependable investment. But what about purchases like stamps, coins, gold, paintings, antiques, baseball cards, cars—things you can keep in your home to enjoy while they appreciate? If you had a million dollars to invest in such items, which ones would you buy?

# Choose an Investment

An investor has turned over a million dollars to you. Your job is to choose three areas of investments and then to purchase items from those areas that can be kept in the investor's home. The investor is not interested in stocks, bonds, real estate, or other "paper" investments. Purchases must be items like those listed above—stamps, gold, and so on—that the investor can enjoy at home while they appreciate. In return for your recommendation, the investor will pay you one-half the amount the $1 million investment appreciates each year. There is an added incentive for you to choose the three areas wisely: should any of the items lose value, you will be required to pay back one-half the amount of yearly depreciation to the investor.

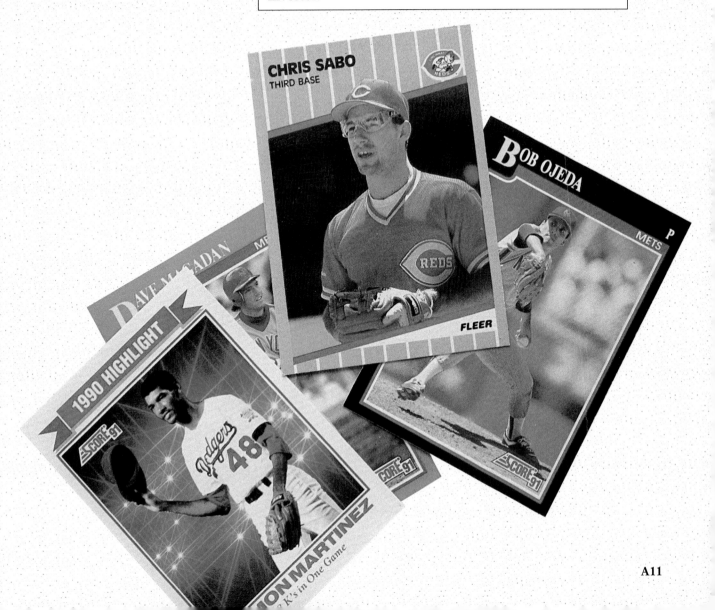

# Getting Started

Follow these steps to carry out your project.

- Brainstorm with your group. Choose at least six areas where you think good investments might be found.

- Collect data from recent years on prices of items in the areas you have chosen.

- Analyze the data. Use logarithms, formulas, graphs, calculators, or any other methods you choose to answer questions like the following.
  1. At what annual rate has the item appreciated in the past year? 2 years? 5 years? 10 years?
  2. Has the rate of appreciation been constant? If not, how can you predict future appreciation rates?
  3. What other factors besides past performance might affect future prices of the item?

- Choose the three areas where you will invest the million dollars.

- Create graphs, displays, tables, or any other materials that you can use to convince the investor that you have done your homework well.

- Decide exactly how you will spend the money.

## Extensions

1. The British Guiana stamp has appreciated at about 16% annually. Suppose a Roman coin that once belonged to the emperor Julius Caesar and was worth a penny at the time it was minted has appreciated at a modest 5% per year. First, guess which of the following choices is closest to the current value of the coin. Then calculate the value.
   a. about $20   b. about $2,000   c. about $2,000,000
   d. about $200,000,000,000,000,000,000,000,000,000,000,000,000,000.

2. Research and report on appreciation in the value of paintings by the Dutch artist Vincent van Gogh.

3. Research supply and demand. Write a paper explaining how the price of a commodity is affected by supply and demand.

## Culminating Activities

Show what you have learned in this project by completing one of the following activities.

1. Prepare a report for the investor. Outline and justify your recommendations on how the million dollars should be spent. Include any supplementary material such as graphs or charts that might support your conclusions. Develop a 10-year forecast on the growth potential for the investments you have recommended.

2. Write an article entitled "Investing in Collectibles" for a financial newspaper or magazine.

# Creating Order From Chaos

The inclination to sort and classify is probably as old as human nature itself. Archeologists have discovered fuel logs and animal bones neatly stacked by prehistoric peoples according to size. Today we arrange compact discs alphabetically and classify mountains by height. Behind the tendency of prehistoric and contemporary people alike is the human need to make sense of and gain some control over an apparently disorderly world.

Classifications can be harmful, as when people categorize or stereotype each other by race or religion. Other classifications of individuals can be sources of great insight. Some psychologists use a 4-dimensional system to classify people according to whether they are extroverted or introverted (E or I), sensing or intuitive (S or N), feeling or thinking (F or T), and judging or perceiving (J or P). Sixteen classifications result. Research suggests that about 13% of the population are type ESTJ and about 1% are type INFP. Knowledge of one's type can be a source of self-understanding and an incentive for personal growth.

Every branch of science has numerous classification systems. Astronomers classify stars by their surface temperatures. Geologists classify rocks by their crystal form, hardness, specific gravity, and several other categories. The mineral galena, for example, is cubical, 2.5 in hardness, and 7.5 in specific gravity. In a 3-dimensional [crystal form, hardness, specific gravity] classification system, galena could be classified as a [Cu, H2.5, SG7.5] mineral. The three dimensions in the classification distinguish galena from nearly every other mineral.

## 3-Dimensional Classification System

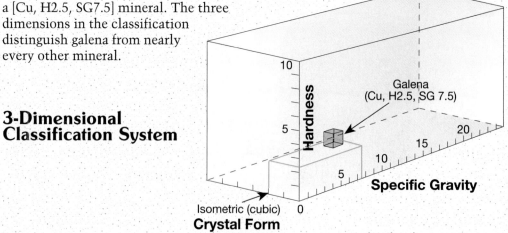

Galena (Cu, H2.5, SG 7.5)

Hardness

Specific Gravity

Isometric (cubic)
**Crystal Form**

# Create a Classification System

In this project, you will collect and analyze data in an area of interest to you. Then you will create a classification system that could be of use to people doing research in the area you have chosen. You will analyze your data to determine the frequency of each of the categories in your system and any other statistics you feel would be of value to people using your system.

## Getting Started

Follow these steps to carry out your project.

- Brainstorm with your group to choose an area of mutual interest for which you will create your classification system. The area can be from science, social studies, language arts, art, sports, or another topic you prefer. Be sure that a classification system is not already in use in the area you choose.

- Choose a number of important and distinguishing categories that you can use to classify items in your area. Examples of categories are specific gravity (rocks) and surface temperature (stars). You should choose an ample number of categories, since after collecting data you may find that one or more categories are not as distinguishing as you had anticipated. These can always be dropped.

- Collect data on your categories.

- Define your classification system. With the perfect system, an item's classification should place it together with other items that are similar to it in important and useful ways. Your system should have at least three distinguishing dimensions or categories.

- Calculate the portion of the entire population that is represented by each classification in your system. For example, in the psychology classification system discussed above, 13% of the population is classified ESTJ. Compile any other statistics that might help you demonstrate the importance and usefulness of your system.

## Extensions

1. Research and report on the taxonomic system for classifying plants and animals.

2. An $N$-dimensional classification system has $a$ classifications in the first dimension, $b$ in the second, $c$ in the third, and so on, through $z$ in the $Nth$. How many distinct $N$-dimensional classifications can be delineated using the system?

3. The Richter scale is a 1-dimensional classification system that classifies earthquakes according to their energy. Choose a familiar 1-dimensional system and propose a second dimension that could be used to classify objects of study more precisely.

## Culminating Activities

Show what you have learned in this project by completing one of the following activities.

1. Write a report describing your work on this project. Explain the importance of your classification system. Describe the system and tell how you collected your data and compiled your statistics. Give examples of how your system can be used.

2. Make an oral report to the class describing your classification system. Bring several examples of items in your area of study and demonstrate how each can be classified using your system.